Building-Construction Design – From Principle to Detail

José Luis Moro

Building-Construction Design – From Principle to Detail

Volume 2 – Conception

Springer Vieweg

José Luis Moro
Stuttgart, Baden-Württemberg
Germany

ISBN 978-3-662-71005-0 ISBN 978-3-662-71006-7 (eBook)
https://doi.org/10.1007/978-3-662-71006-7

This Springer Vieweg imprint is published by the registered company Springer-Verlag GmbH, DE, part of Springer Nature.
The registered company address is: Heidelberger Platz 3, 14197 Berlin, Germany

Contents

Contents

Generation of Surfaces

J. L. Moro, *Building-Construction Design – From Principle to Detail*,
https://doi.org/10.1007/978-3-662-71006-7_1

Contents

1.1 Formation of Continuous Planar Layer Surfaces from Individual Components

Envelope components, as well as individual layers thereof, must, as a rule, be **superficial** and **continuous** in order to fulfil their tasks. This presupposes the creation of large-area two-dimensional, mostly planar layers of material, for which various manufacturing, geometric, and constructional variants can be used, which are to be treated in the following. The first step in this chapter is to combine individual parts within the component surface, i.e. in the two dimensions, **length** and **width** (◘ Fig. 1.1). Individual leaves or layers of a multi-layered envelope structure will be investigated separately with regard to their specific constructional subfunctions (as on ◘ Fig. 1.2). The composition of the surface from identical parts is considered in each case. This corresponds to standard building practice. Constructional transitions between different elements within a particular functional leaf or layer (such as on ◘ Figs. 1.3 and 1.4) must be carefully planned and executed. The seam or joint formation itself is deepened in ***Volume** 3, Section 1* by incorporating geometric, mechanical, and sealing techniques. In *Section 2* the constructional composition of the assembled component is considered in detail with regard to the layering in the third dimension, i.e. the **thickness**.

***Vol. 3**, Sect. 1 Surface Seams*

Sect. 2 Build-Up of Envelopes

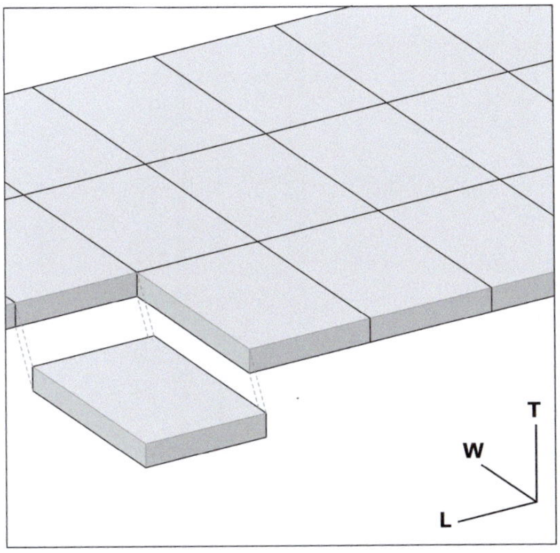

◘ **Fig. 1.1** Composing an envelope component in the component surface, defined by length **L** and width **W**

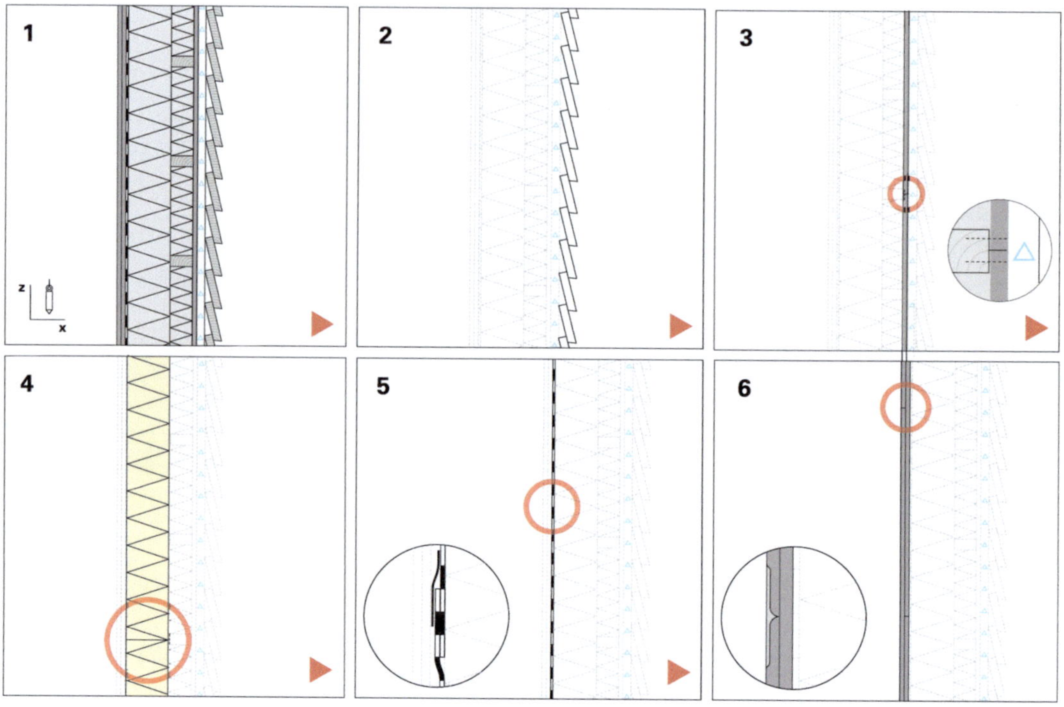

Fig. 1.2 Production of continuous functional leaves or layers in the composite envelope structure of a wooden panel wall (detail **1**) by composing and joining individual sections. The following constructional partial functions are considered: **2** sealing against moisture (first stage); **3** sealing against moisture (second stage); **4** thermal insulation; **5** blocking or retarding vapour; **6** inner space enclosure/force conduction. Each functional layer is interconnected by means of adequate joints to form a functional continuous surface

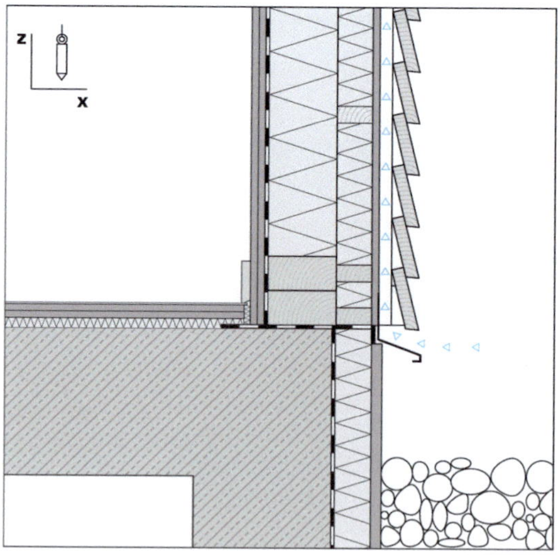

Fig. 1.3 Footing of a wooden panel wall

The nature of the composition of individual elements into a continuous surface is initially dependent on fundamental decisions regarding:

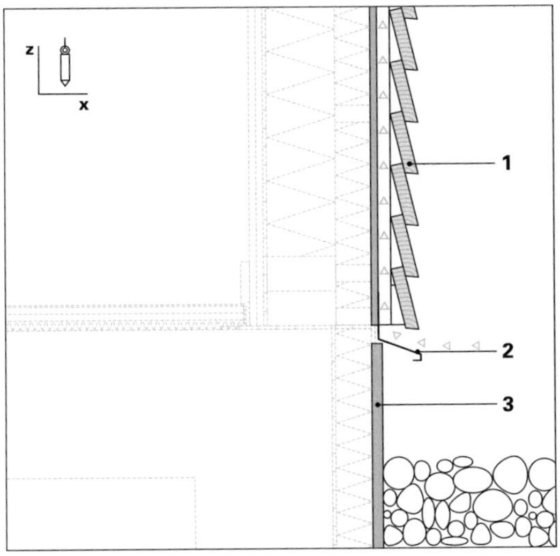

Fig. 1.4 Production of continuity of the functional layer 'weather skin' at the constructional transition between regular exterior wall and building base in the following stages: **1** bevel siding (weather skin of exterior wall, first stage)/sheathing (weather skin, second stage); **2** drip flashing; **3** fibre-cement plate in base area

- the geometrical position of the parts relative to one another, i.e. the **laying geometry**;
- the geometrical position of the **part edges** relative to each other, for example, whether these are butt-jointed or overlapping;
- the **surface geometry**, that is, the question of whether the surface is planar or curved, or in the latter case what kind of curvature the surface has;
- the nature of the **individual components** to be aggregated, which may result from the material nature or from the manufacturing process.

For the constructional design of the **component joint**, these definitions are also of major importance. These are treated, as mentioned above, in ***Volume** 3*, *Section 1*.

***Vol. 3**, Sect. 1 Surface Seams*

1.1.1 Dimensional Pre-Settings of Starting Elements

Depending on the type of material used according to the aforementioned reflections, either **shapable** materials are used, usually mixed from different raw materials, such as concrete, or otherwise **semi-finished solid** products, usually industrial products, such as steel profiles and timber parts, can be used as starting elements. Hence, the following elements can be used alternatively, whereby, in each case, the order of magnitude of the envelope component (wall, floor construction, etc.) is applied as a benchmark:

- **Mouldable** or **castable** materials which are cast in a **negative mould** and subsequently solidify to a fixed, flat component of a larger, theoretically even unlimited, extent (▪ Fig. 1.5, option **1**). The negative casting mould is usually either removed or sometimes left in place, if the circumstances suggest (so-called permanent formwork). Examples are reinforced-concrete walls, floors, screeds, cast floorings.
  There are also, in the case of some particular materials, casting processes which are not suitable for the production of envelope components of any size because the dimensions of the castings are limited for reasons of material properties or manufacturing—as in the case, for example, of cast steel. As a result, in spite of applying casting methods, it may nevertheless be mandatory for actual constructional use to operate with components of the other categories discussed below.
- **Band** or **strip-like** elements, in which a dimension (length) can be arbitrarily large, a further (width) is limited to a given average dimension for various reasons such as production, transportation, or handleability, and the third (height or thickness) is comparatively small—such as is the case with rolled flat steel or solid-wood plates manufactured in great lengths. Depending on the characteristics of the material and the thickness of the strip, the element can sometimes also be trimmed to larger lengths and rolled up subsequently—such as strip steel, textiles, plastic foils (▪ Fig. 1.5, option **2**).
- **Plate-shaped** elements the thickness of which is substantially smaller than the other two dimensions, length and width, which are also limited for certain reasons, such as the production process (float-glass or wood-based panels) (▪ Fig. 1.5, options **3** and **4**).
- **Rod** or **bar-shaped** elements which, for example, have a dimension (length) which is markedly larger than the other two (width, height) (▪ Fig. 1.5, options **5** and **6**), as a result of the material characteristic, as in the case of wood, or as a result of the manufacturing process. In some cases, the length can be discretional, provided the element is produced in an endless process. Individual parts are then cut to length, as required, by severing.
- **Block-like** elements, the three dimensions of which have similar magnitudes (▪ Fig. 1.5, option **7**), and which are then, obviously, limited by the smallest dimension of the envelope component to be composed, that is to say, by its thickness. Block-like elements can also emerge from a casting process. Various reasons can result in an even greater reduction in the dimensions of the building block, for example, restrictions from the production process, e. g. in the case of bricks.

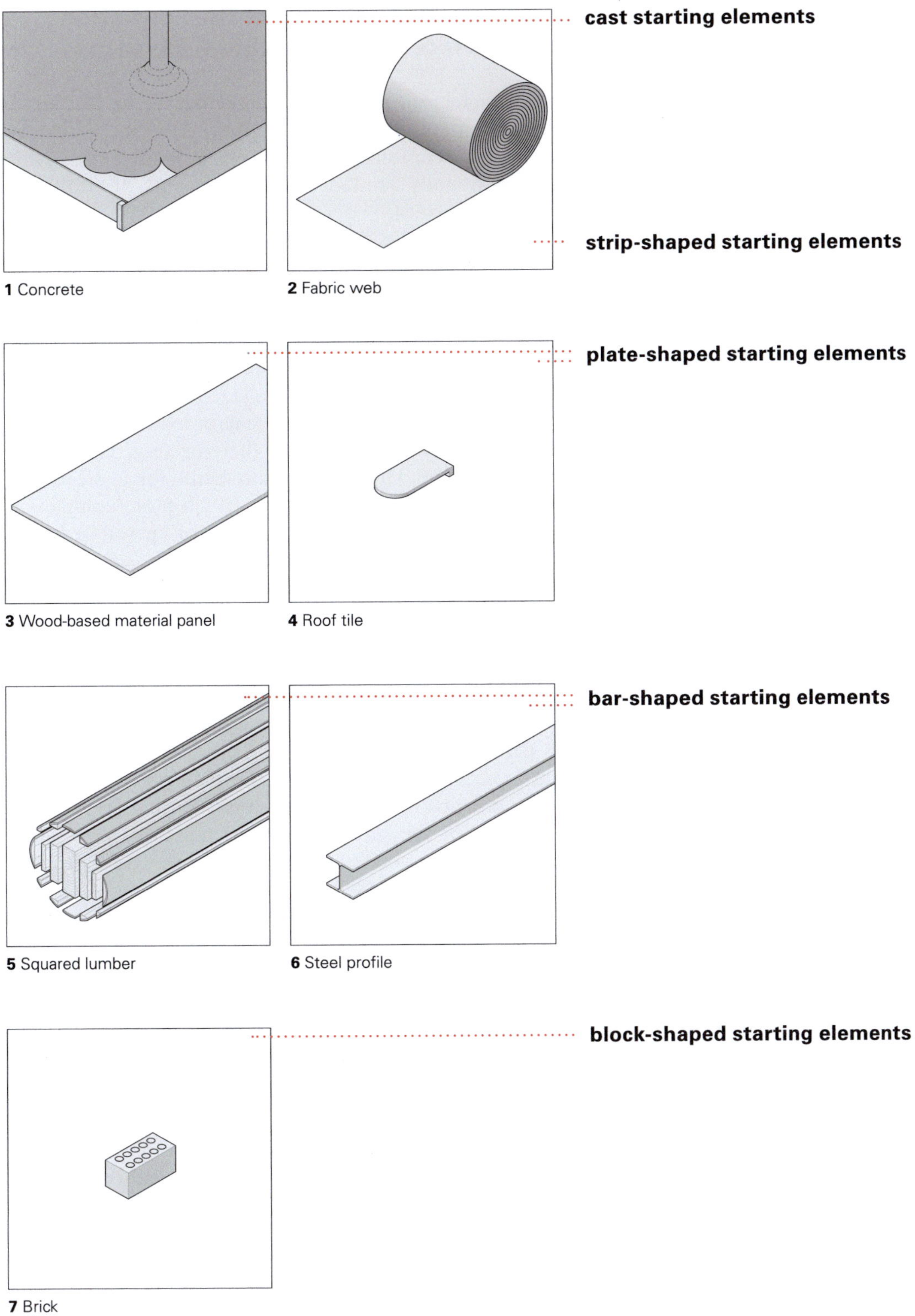

Fig. 1.5 Types of starting elements for generating a continuous functional surface

1.1.2 Geometric Principles of the Formation of Surfaces from Individual Elements

In order to create a certain component layer or leaf in the form of a continuous, gapless—initially assumedly planar—surface with discretional, theoretically unlimited dimensions from dimensionally limited starting elements, the following criteria are important:

- As a rule, it is expedient to produce a single-ply surface of **non-overlapping**, merely abutting elements. This allows covering a given surface with the least possible material expenditure, as well as doing so flush with the surface, without offsets, projections, etc. A prerequisite for this is that the—initially always identical—starting elements can be **parquetted** or **tesselated** in the general mathematical sense, i. e. have a geometry suitable therefor (e. g. ◘ Figs. 1.9, 1.10, and 1.11). The basic requirements for tesselation of the surface are discussed below. Much more complex are the conditions for the gapless closure of curved or faceted surfaces. Basic aspects of this are also discussed below; the complete filling of the space with spatial individual modules can likewise play a constructional role in some cases: for example, in the design of spatial frameworks.
- If the initial elements should **overlap**, one can also deviate from this requirement. Overlaps can be desired in construction, for example, for purposes of sealing or absorbing deformations. However, a limited degree of overlapping is generally sufficient for these purposes and desirable for reasons of material economy, so that certain basic geometries of the starting elements must still be observed for a continuous surface formation (◘ Fig. 1.6).In terms of the constructional implementation, the edges of the initial elements can be understood both as butt joints of flat components (◘ Fig. 1.7, left) and as bars of a latticework (◘ Fig. 1.7, right).

► *Sect. 1.2 Shape Definition of Continuous Curved Layer Surfaces*

► *Sect. 9.3.5 Spatial Trusses*

***Vol. 3**, Sect. 1 Surface Seams*

1.1.2.1 Tesselation of the Surface

A tesselation or **tiling** of a surface is defined as follows: [1]

- in mathematics, the overlap-free, complete covering of the plane with congruent regular polygons.

Certain forms of the basic element are excluded from the outset, such as the circle (◘ Fig. 1.8). Elementary variations of the area filling can be realised with polygons (◘ Figs. 1.9, 1.10, 1.11, 1.12, 1.13, 1.14, 1.15, 1.16, 1.17, 1.18, 1.19, and 1.20). The narrower definition of tesselation postulates a single regular polygon as the basic module. Since the corner an-

1 *Brockhaus Enzyklopädie*, 19th ed. (1991), Vol. 16, p. 550.

Fig. 1.6 Overlapping lay-out of individual elements for sealing against water. Despite the overlap dimension, the elements have a (roughly) tesselatable form

Fig. 1.7 (Left) Plate cladding: The geometric tesselation principle determines the positions of the joints. (Right) Area-filling latticework based on a modular tesselation principle (here, quadrangular meshes), which in this case (also) regulates the positions of the bars

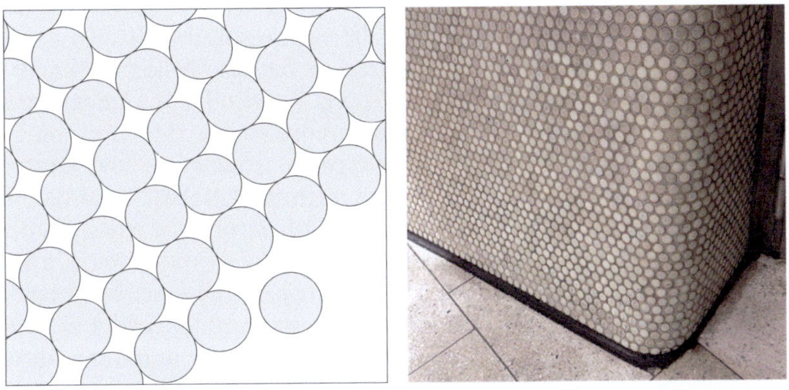

Fig. 1.8 (Left) A starting element with circular geometry is not suitable for tesselation. (Right) Knob ceramics: This technique involves laying round mosaic tiles and filling the spaces between them with mortar. The joint proportion is high in terms of surface, which limits the watertightness. Since there is no form-fitting interlocking of the individual elements, however, the mosaic units can be easily adapted to doubly-curved shapes

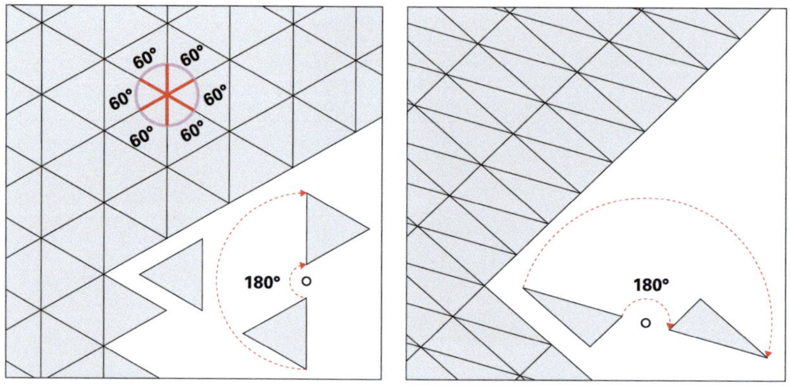

Fig. 1.9 (Left) Tesselation by means of **equilateral triangles**. This is one of the three tesselations with regular congruent polygons. The six abutting angles of 60° add up to 360°. The basic module is rotated by 180° to obtain the complementary position. (Right) Tesselation by any triangle. A triangular module is converted to the complementary position by a 180° rotation. In this case too, a single element is therefore sufficient for complete tesselation

gles at which the polygons meet must always add up to 360° and since all corner angles must be the same according to this narrower definition, there are only three **regular tesselations** of the surface:

- with **equilateral triangles** (Fig. 1.9 left): 6 abutting corner angles of 60° each give 360°. This variant is relatively rare for pavements and claddings in building practice. The proportion of joints on the surface is relatively large; the individual elements have sharp corners that are sensitive to breakage.
- with **squares** (Fig. 1.10): 4 abutting corner angles of 90° each give 360°. This variant has the greatest importance for building practice, especially in its variation with rectangles.
- with **regular hexagons** (Fig. 1.11): 3 abutting corner angles of 120° each result in 360°. This variant is of particular importance because it has the smallest circumference/area ratio, i.e. it includes the largest area with the smallest circumference, or in this case joint length. Similarly, with this type of tessellation, the area can be gridded for latticework with the smallest total bar length (for a given mesh size). With lattices, it also has the advantage that only three bars meet at one node. On the other hand, the lack of geometric stiffness of the hexagon—as opposed to the triangle—has a negative effect in structural terms. This variant occurs frequently in nature, for example, in foams, honeycombs, or cell tissue.[2]

More general definitions of tesselation from a single basic element include the following variants:

2 Thompson D W (1992) *On Growth and Form*, p. 499 ff.

Fig. 1.10 (Left) Tesselation by means of **squares**. This is one of the three tesselations with regular congruent polygons. The four abutting angles of 90° add up to 360°. (Right) Flooring with square ashlars

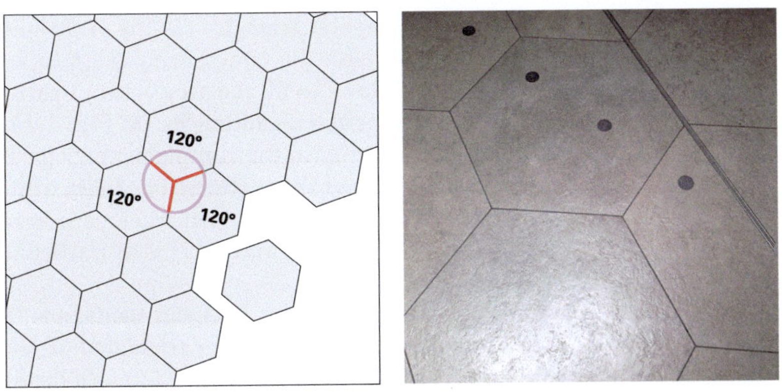

Fig. 1.11 (Left) Tesselation by means of **regular hexagons**. This is the last of the three tesselations with regular congruent polygons. The three abutting angles of 120° add up to 360°. (Right) Flooring made from hexagonal ceramic tiles

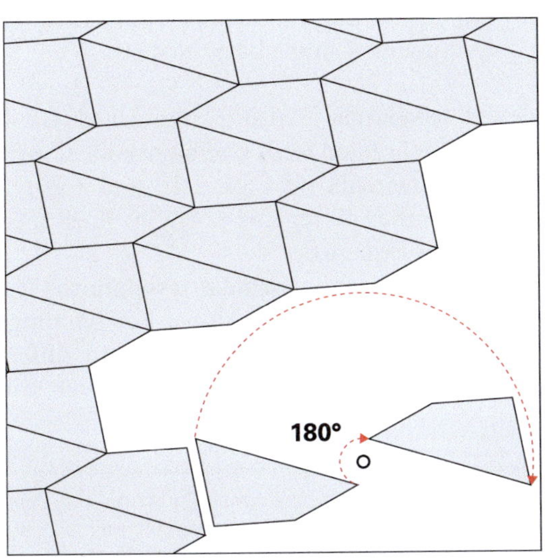

Fig. 1.12 Tesselation by means of any **irregular quadrangle**. The basic module must be rotated by 180° to obtain the complementary position. A single basic module is therefore sufficient for tesselation

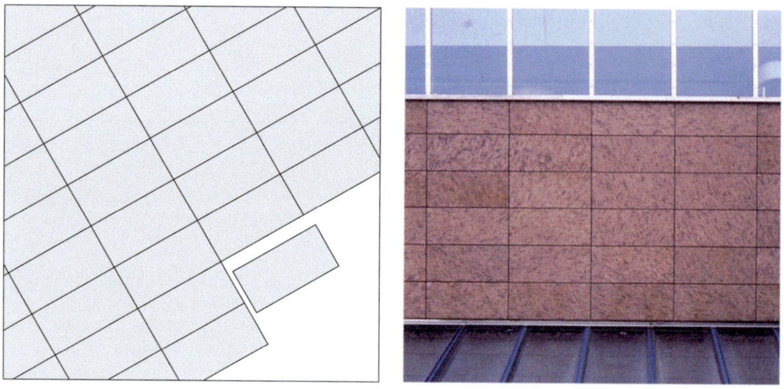

Fig. 1.13 (Left) Tiling of rectangular source elements by lateral butting. (Right) Façade cladding from rectangular panels in cross-joint grid

- Arbitrary **irregular triangles** (Fig. 1.9 right). Any triangle, together with the complementary position by rotating it by 180 °, allows to fill the area without gaps.
- Arbitrary **irregular quadrangles** (Fig. 1.12). Any quadrangle, together with the complementary position by rotating it 180°, allows the surface to be filled without gaps. The special case of rectangles with various side ratios (Figs. 1.13, 1.14, 1.15, 1.16, and 1.17) is of particular importance for building practice.
- Fifteen variations from **irregular pentagons** (Fig. 1.18).[3] Regular pentagons cannot be tesselated, as their corner angles (108°) can never give 360° in multiplication. It may be necessary to mirror the basic element, so that for practical purposes not one but two modular basic elements are required to fill the area (Fig. 1.18).
- Three variations from **irregular hexagons**.[4] Certain edge lengths and corner angles must be adhered to for an area filling consisting of a single basic element.

A **semi-regular tesselation** is also distinguished, i.e. where the surface is completely filled with combinations of two or more congruent basic elements (Figs. 1.19 and 1.20).[5] In addition, the same number of regular polygons must meet at each corner in the same sequence.

There are also numerous **irregular tesselations** (Figs. 1.21 and 1.22) with partly great importance for building practice. These can consist of various combinations of different modular basic elements, mostly polygons, regular or non-regular.

3 Until recently, 14 variants were known (Pottmann and Bentley (2007) *Architectural Geometry*, p. 152). A new variant was recently discovered (*Süddeutsche Zeitung* 12/08/2015 Das magische Pentagon). It cannot be ruled out that others will follow.

4 Pottmann and Bentley (2007), p. 152.

5 Ibid. p. 154.

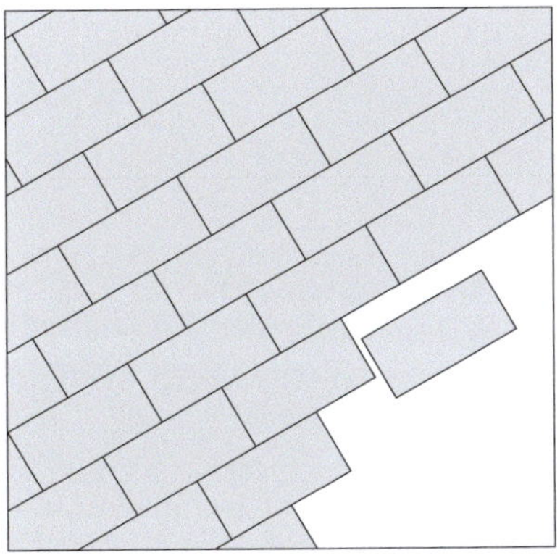

Fig. 1.14 (Left) Tiling of rectangular source elements by lateral butting and row-wise offsetting by half the element's length. (Right) Brick bond

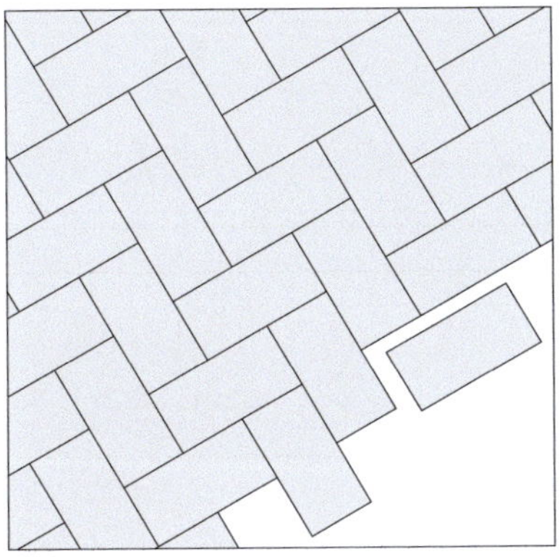

Fig. 1.15 (Left) Tiling of rectangular source elements in herringbone pattern. Rotation as on Fig. 1.19. (Right) Wooden herringbone parquet flooring

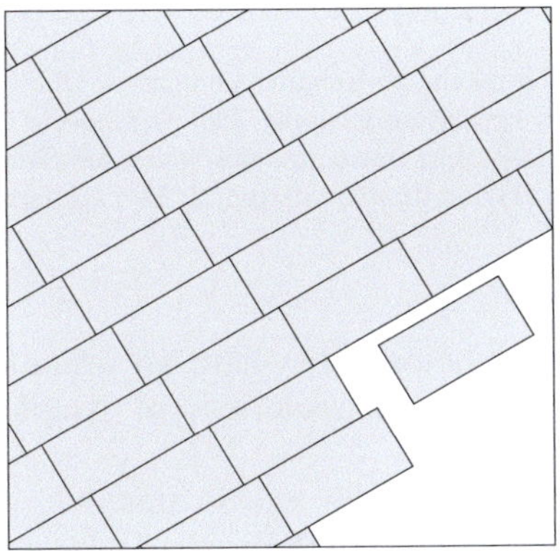

Fig. 1.16 (Left) Tiling of rectangular source elements by lateral butting and row-wise ofsestting by a differential length. (Right) Façade cladding from rectangular fibre-cement panels slightly offset from each other

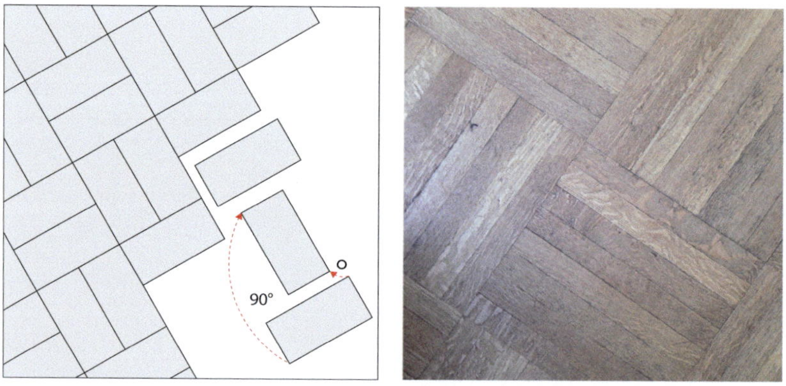

Fig. 1.17 (Left) Tiling of rectangular source elements with a side ratio of 1:2 by lateral butting and rotating by 90°. (Right) Parquet laid in checkerboard pattern

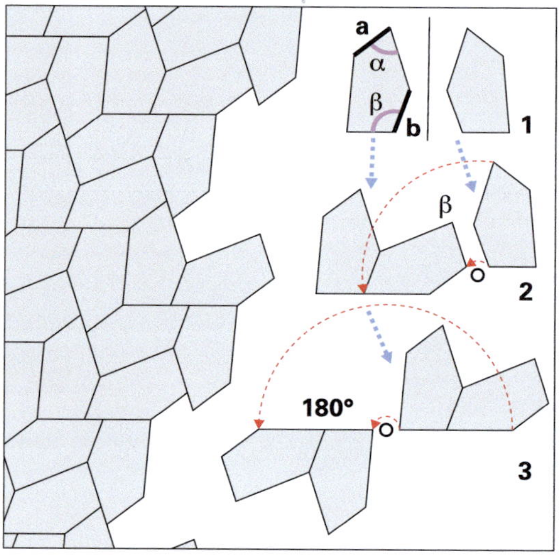

Fig. 1.18 Tiling by means of irregular pentagons. One of the 15 known possibilities is shown here as an example. The prerequisite is that **a** = **b** and the sum of $\alpha + \beta = 180°$. The basic modules which tesselate the surface are created by mirroring (**1**) and double rotation (**2**, **3**). Two mirrored basic modules are required

These types of tesselation often have an ornamental character and are particularly characteristic of Arabic architecture (Figs. 1.19 and 1.23).

Non-polygonal tesselation is also possible, provided that certain requirements are met (Fig. 1.22). Good examples of this variation are the figures of M. C. Escher (Fig. 1.24).

The presented tesselation patterns can also be transferred to non-planar surfaces with some restrictions (Figs. 1.25 and 1.26 right), but in most cases the requirement to use only identical and, in addition, geometrically regular elements need to be abandoned in that case. Area fillings from **irregular quadrilateral meshes** play a major role in the definition

► *Sect. 1.2.5 Meshes*

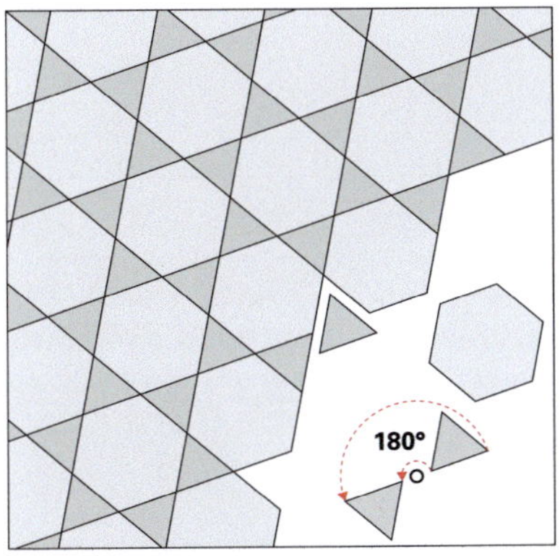

Fig. 1.19 (Left) **Semi-regular tesselation** with regular hexagon and equilateral triangle. The triangle is rotated by 180° to obtain the complementary position. Two different basic modules are required. (Right) Variation of the tesselation on the left from hexagons and six-pointed stars. The latter are created by combining a hexagon with six surrounding equilateral triangles

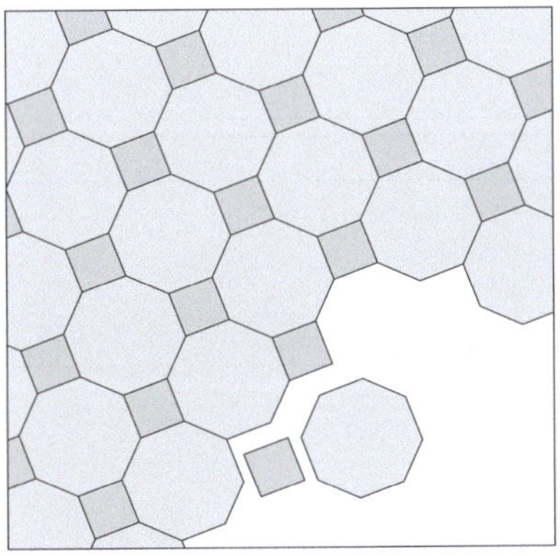

Fig. 1.20 (Left) **Semi-regular tesselation** with regular hexagon and square. Two different basic modules are required. (Right) Wooden decorative ceiling with surface-filling pattern according to the pattern on the left

of curved free-form surfaces, both as control polyhedra and as meshes. Irregular triangulations, i.e. partially modular or completely non-modular tesselations made of irregular triangles, are of great importance, especially as so-called **meshes** for curved free-form surfaces, as any surface can be facetted from them using planar elements. Furthermore, triangulations are always shear-resistant, which is an important aspect for structural latticework. Further considerations are made below.

▶ *Sect. 1.2.5*Meshes

1.1.3 Surface Formation by Combining Individual Elements

1.1.3.1 Cast Starting Elements

Due to their manufacturing principle, cast-construction methods allow, at least theoretically, producing seamless surfaces in unlimited dimensions. Modular and geometric questions

Fig. 1.21 Ornamental surface tiling in Arabic architecture. The patterns are based on simple modular, repetitive geometric structures, but vary the filling elements to generate complexity

Fig. 1.22 Surface filling on a façade with non-polygonal motifs by M. C. Escher

of area layout or questions of joint design between the composing elements are therefore largely non-existent in that case. The **integral** construction principle then prevails.

► *Sect.* 11.5.6 *Concreting Joints*

***Vol. 1**, Sect. 22.6.4 Cast Steel*

Nevertheless, in practice, most casting techniques encounter limits which make it necessary to subdivide the casting process of a larger area into individual sections or stages or to limit the dimensions of the component to be cast from the beginning. This, in turn, leads to joint formation between individual solid components and thus to the application of the **quasi-integral** or **differential** construction principle. The same questions are then to be answered with regard to element laying and joint formation as with non-casting techniques. However, there are specific conditions characteristic of casting that deserve closer examination.

Fig. 1.23 (Left) Tiling with two non-regular basic modules. (Right) Tiling of a door surface with non-regular but polygonal basic modules according to a repetitive pattern typical of Arabic architecture

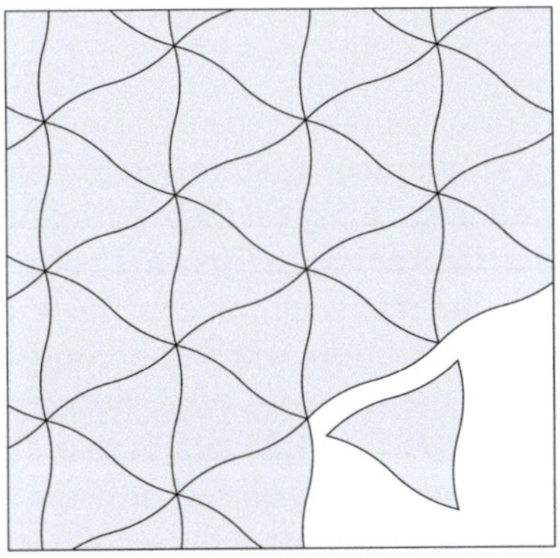
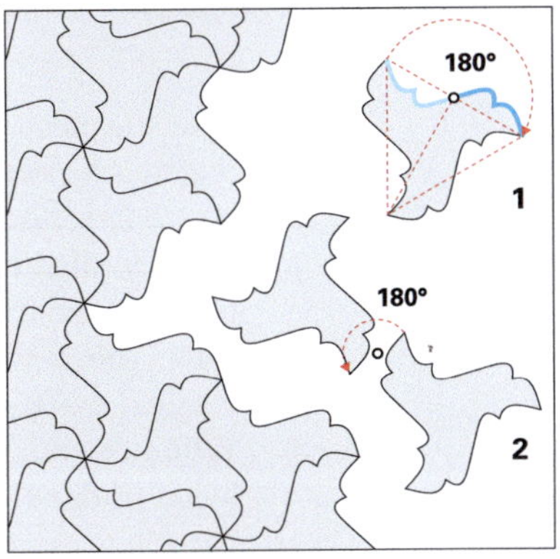

Fig. 1.24 (Left) Tiling with a non-regular basic module. In this case, its geometry is based on an equilateral triangle. The edge lines are curved, but are congruent with each other and are generated through rotation by 120°. The resulting basic module is then rotated by 180° to obtain the complementary position. A single basic module is therefore sufficient. (Right) Tiling with a non-regular basic module based on an equilateral triangle. A half of the curved edge line is rotated by 180° around the centre of the triangle side; the whole edge line is then rotated twice by 120° (**1**). The resulting basic module is then rotated by 180° to obtain the complementary position (**2**). A single basic module is therefore sufficient

Fig. 1.25 (Left) Curved surface (helical surface). The basic rules of area filling also apply (with modifications) to this non-planar surface. Here for example, the principle shown in Fig. 1.14, left, or Fig. 1.16, left, is realised. (Right) Polyhedral surface consisting of individual planar partial surfaces. Here too, basic principles of tesselation apply (here for example, the principle according to Fig. 1.13 left)

Fig. 1.26 (Left) Area-filling floor pavement made of irregular, practically unprocessed flagstones without straight edges. Here, there certainly is an area-filling, but neither complete nor semi-regular tesselation. There is an approximate fit of the formats as found. The likewise irregular joints are filled with grouting material (mortar). (Right) Surface-filling covering from broken tiles (trencadis). The principle is similar to that on Fig. 1.8

The simplest and most suitable type of joint formation is directly adjoining a subsequent casting section to a previously hardened one. The preceding casting section is either previously shuttered at the envisaged joint (Fig. 1.27) for the purpose of casting the next one, or can freely harden to receive the next casting section when the component is in a vertical position and the abutting edge, hence, is horizontal. When the next section (Fig. 1.28 left) is cast, the plastic material under these circumstances closely adheres to the already hardened edge surface, so that after the second section has set, a capillary or hair joint usually develops between both sections, or even a material quasi-continuum (Fig. 1.28 right; Fig. 1.29). Compressive forces can be transmitted via direct contact; tensile and shear forces either through embedded reinforcement or—provided the strength of the casting material is sufficient—by means of tooth- or dovetail-shaped edge formation.

If solid, already pre-cast individual components are to be joined, the joining techniques of the differential construction principle are available—for example, bolted connections. Alternatively, the same casting technique can also be used by forming a **grouting joint** or **pocket** between the abutting components, i.e. a joint space which is open at the top and is closed towards the other sides and which is poured out after the parts are laid (Fig. 1.30). The joint space and the joint flanks are designed in such a way that a uniform distribution of the potting compound is ensured. Standing joint spaces are, of course, substantially more difficult to fill than joint spaces lying open in their length (as in the example on Fig. 1.30). As a rule, a joint space cross-section with as obtuse angles as possible (>90°) is selected in order to facilitate the distribution of the filler. In addition, a minimum size of

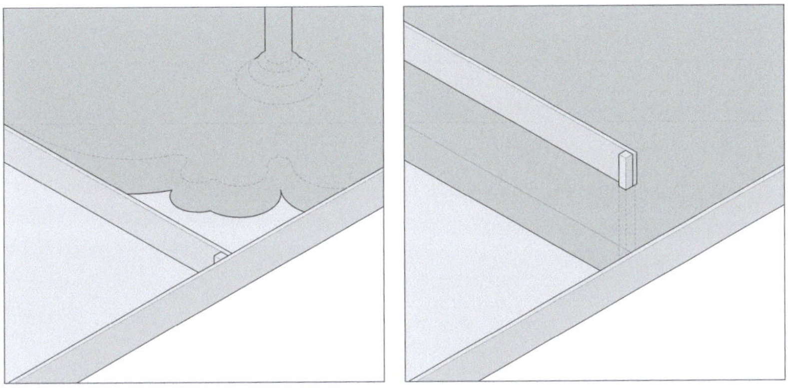

Fig. 1.27 (Left) Casting the first section. (Right) Removing the intermediate shuttering after setting

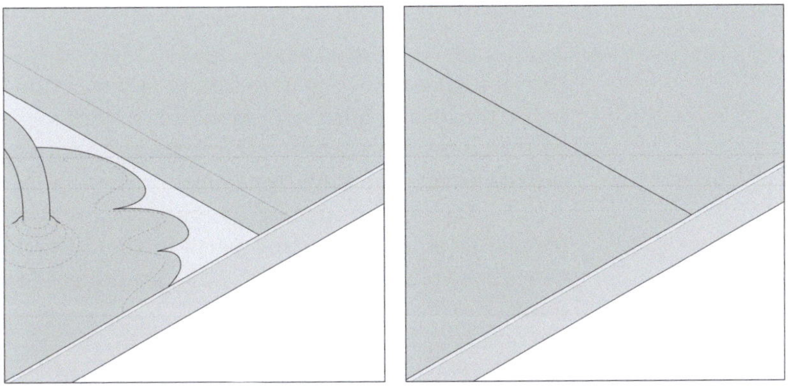

Fig. 1.28 (Left) Casting the second section. (Right) Contact joint between the two casing sections

Fig. 1.29 (Left) Bonding reinforcement in preparation for the pouring of another batch of concrete (flight and landing of stairs). (Right) Pouring joint: On the left is the irregular pouring joint with the second (upper) concreting section added directly; on the right, a triangular lath was inserted into the formwork at the height of the joint to achieve a clean appearance

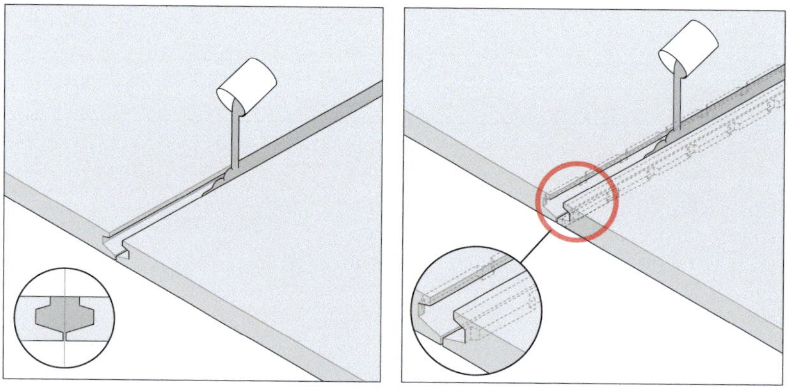

Fig. 1.30 (Left) Grouting joint between abutting elements. The joint cross-section permits shear forces to be transmitted perpendicular to the component surface if using a shear-resistant filler. (Right) As on the left, additionally with shear toothing by pocket formation in the flanks of the grouting joint space (effective in longitudinal direction of the joint)

the joint space must be maintained as a function of the viscosity of the plastic potting compound. In terms of the capability of the connection between the parts, the above applies regarding the direct adjoining by casting.

Overlaps at the edge of the parts, as are occasionally used for the joining of abutting parts in the differential construction principle, do not offer any advantages with a grouted joint.

1.1.3.2 Band-Shaped Starting Elements

Seam Execution

Larger coherent layers of strip material can be formed by simply placing widths together, as long as no further requirements are imposed on the joint, such as tensile strength and tightness (Fig. 1.31). Otherwise, the abutting edge has to be configured accordingly in order to absorb these stresses. Since, in particular, rolling-up strip material is generally comparatively thin, a force connection between narrow part edges is thus difficult to manufacture; in such cases an **overlapping joint** is preferable (Figs. 1.32, 1.33, and 1.34). The abutting strips remain in a plane; the overlapping strip, which is usually associated with one of the two lanes, is placed on the adjacent lane by cranking. The overlapping dimension determines the area which can be used for joining or sealing purposes, which in any case is greater than in the case of simple butting. The disadvantage of the thickening of the layer resulting from the overlaps is mostly negligible because of the small thickness of the layer.

This is the case, for example, with insulation panels in thermal-insulation layers

Similar conditions prevail when **covering** the joint with strips of the same or a related material (Figs. 1.35, 1.36, and 1.37). If the cover strip is mentally widened up to the width of the web, the solution is created with staggered, mul-

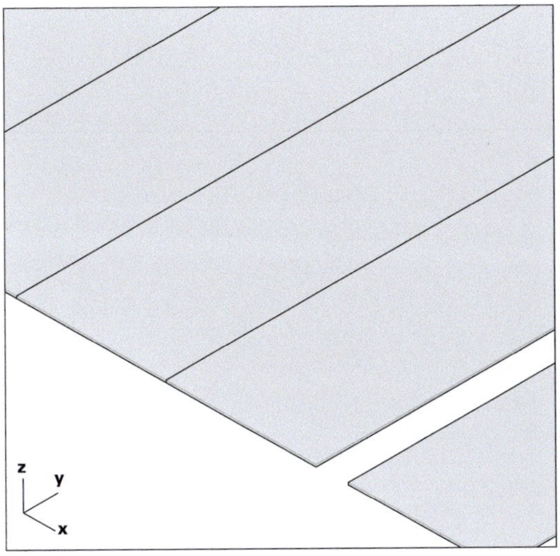

Fig. 1.31 Abutting webs

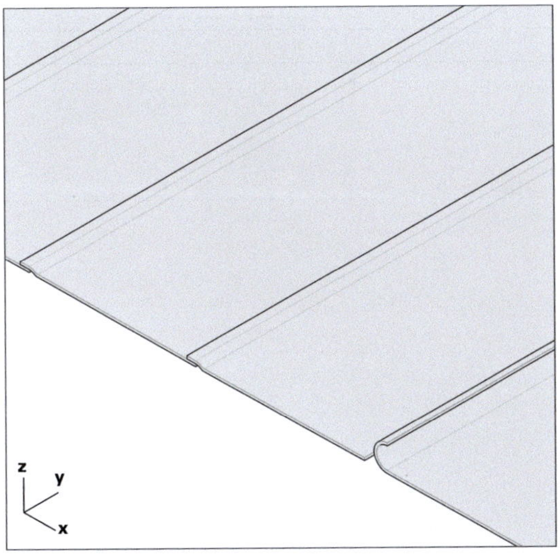

Fig. 1.32 Overlapping webs

tilayer webs (Figs. 1.38 and 1.39). The overlying membrane ensures the appropriate connection and, if necessary, sealing of the underlying joint, provided that—at least in the area of the joints—a full-surface connection (e.g. adhesive bonding) of the overlying membranes is made.

A joining area enlarged compared to the simple abutment can also be achieved by **bending up** the band edges (Figs. 1.40 and 1.41). The strips perpendicular to the layer surface allow a two-dimensional connection between the webs. Due to the projecting ribs or upstands, it is not possible to incorporate them within a layer sequence. The layer can at most be located on the outer side of a layer package. As a weathered

Fig. 1.33 Overlap-welded membrane strips, as in Fig. 1.32

Fig. 1.34 Overlap welding of roof waterproofing membranes, as shown in Fig. 1.32

horizontal or inclined surface, this solution offers the additional advantage that the exposed joint is raised from the water-bearing level. In most cases, the projecting strips additionally contribute to stiffening the otherwise thin and usually flexible band. As an alternative, the bent-up connection strip can also be bent to the side in the case of a flexible material (Fig. 1.42), whereby conditions are again similar to those of the simple overlap.

***Vol. 3**, Sect. 42.3 Conceptual-Design Measures*

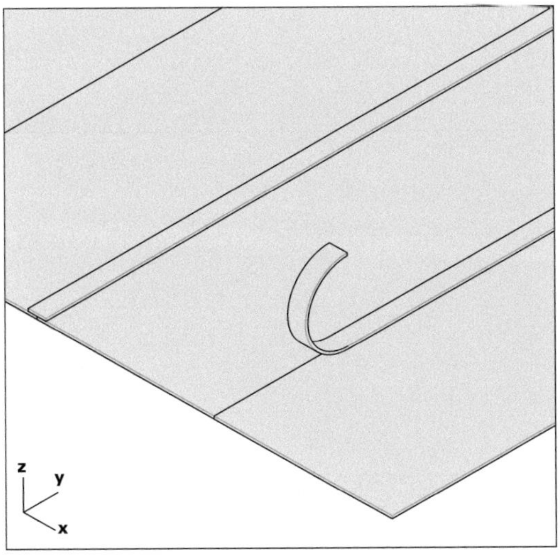

Fig. 1.35 Abutting webs with cover strips

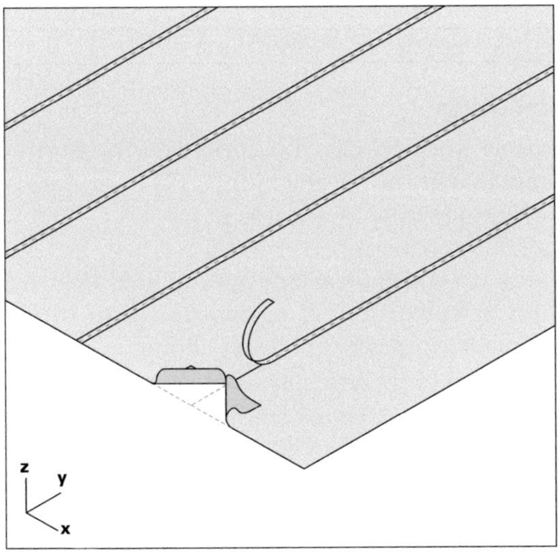

Fig. 1.36 Abutting panels with cover strips. Example: textile panels with sewn border

Layout Geometry

Due to the comparatively large band length and the fact that the mostly flexible bands are laid on full-surface support elements—not on grid-like substructures—the laying geometry of the webs does not play the same role as in the case of plate-shaped or rod-shaped starting elements. Otherwise, conditions are generally the same as those of the former.

Fig. 1.37 ETFE membrane webbing welded with seam tape extended prior to assembly, as shown in Figs. 1.35 and 1.36

1.1.3.3 Plate-Shaped Starting Elements

Seam Execution

A continuous surface can be produced by simply placing plates together with blunt edge joints (Figs. 1.43 and 1.44), but this solution is only applicable if:

- the plates are laid **full-surface** upon a load-bearing subplate. On a linear substructure, in contrast, it is to be expected that point or asymmetrical loads acting on a plate cause the edges to move perpendicular to the layer surface and to create an offset in the layer or leaf surface (Fig. 1.45).
- and an uplifting of an edge by unwanted deformation of a panel is prevented, imposing suitable **weighting** (e.g. by superimposed layers within a package) (Fig. 1.46).If the plate is resting on a **substructure** of rod-shaped elements, the butt joint is preferably placed on one of these rods (Fig. 1.47), in a way that:

- at this point, the load can be transferred to the substructure, and at the same time,
- by fastening the panel edges to this substructure, their mutual displacement or uplifting is prevented.

► *Sect. 1.1.3.3.2 Layout Geometry*

If the plate spans one-way between parallel supports, the other two edges oriented in the perpendicular direction remain unsecured, so that an out-of-plane **shear connection**, for example in form of a strip or profile attached at the bottom or rear, is required to avoid mutual displacement of the panel edges (in → **z**, as in Fig. 1.45) (Figs. 1.48, 1.49, and 1.50). This can be the case both with an unevenly distributed load

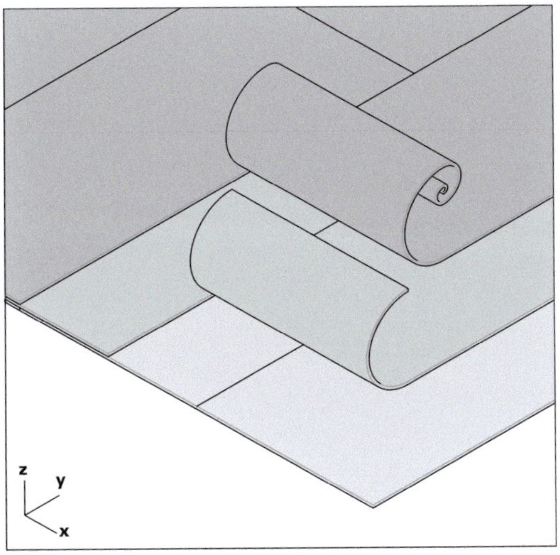

Fig. 1.38 Multi-layered, mutually offset webs

Fig. 1.39 Multilayer waterproofing membranes laid in layers offset from each other (see joint offset at the leading edge), as shown in Fig. 1.38

acting transverse to the surface (→ **z**) and with a shear load acting on the same surface (i.e. in the plane **xy**). In the latter case, a rigid **shear-resistant sector** is created on the surface.

Larger plate formats, which are often found on the market, mean that the length/width of the plate is significantly greater than the span which it can bridge, which is limited from the beginning due to its comparatively small thickness. This can be used to continue the plate over several supporting fields. At the same time, this makes it possible to utilise a structural continuity effect across the rods of the substructure and, consequently, to significantly reduce the bending moments on the plate (cf., for example, Fig. 1.45).

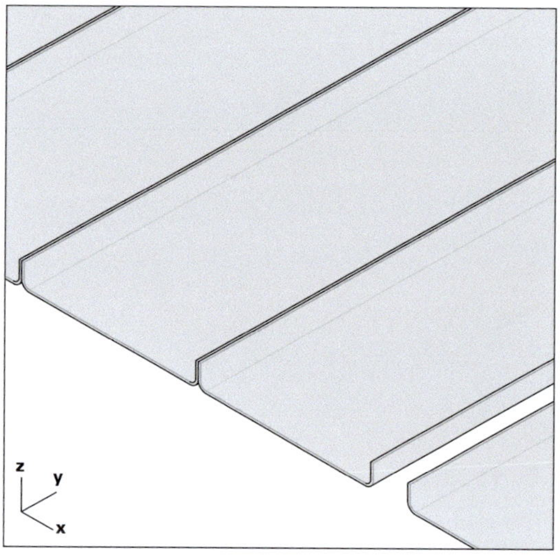

■ **Fig. 1.40** Folded-up web edges

■ **Fig. 1.41** Sheet-metal strips with raised seams, as in ■ Fig. 1.40

Vol. 3, Sect. 42.5.1 Prolonging the Joint Trajectory—Geometric Measures

Fastening the plate to a substructure also results in a **backing** of the joint, as well as in an **extension** and **twisting** of the joint trajectory, which is an advantage for sealing purposes. An overlap of the plate edges (■ Fig. 1.51) results in a comparable effect, but the geometrical conditions as well as the stiffness of the plate have to be taken into account. This can be inferred from contemplating ■ Fig. 1.52: The overlapping dimension **o** determines the size of the contact surface available for sealing purposes between abutting elements. At the overlapping surface, the parts are always in full contact whenever the support of the overlapping parts is planar. The overlapping dimension gives the free-spanning dimension **f** by

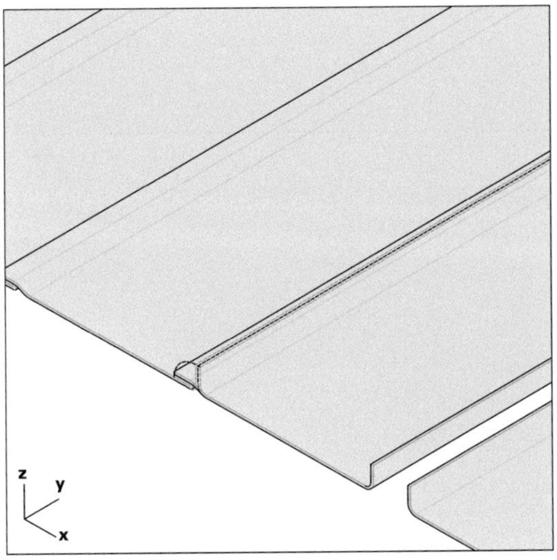

Fig. 1.42 Folded-up and flattened web edges

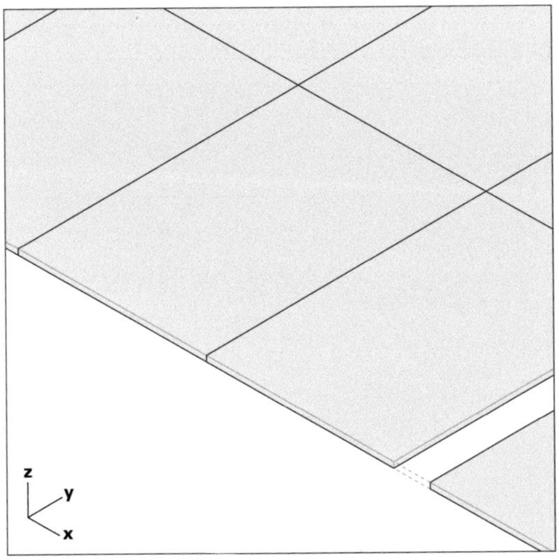

Fig. 1.43 Butting plates

deduction from the element length **l**. It is obvious that, with decreasing inclination angle α, the overlapping dimension **o** also becomes smaller until the limiting case, in which the elements contact only at the outermost edges (**l** = **f**; **o** = 0), a case which, however, is of no practical importance. For pragmatic reasons, the angle α is usually chosen such that the dimension **o** is sufficient for the intended **sealing** or **fastening** purposes. Everything that goes beyond it entails unnecessary material consumption.

Fig. 1.44 Area filling in the case of a wall's stone slab cladding by laying one slab on top of the other, as in Fig. 1.43

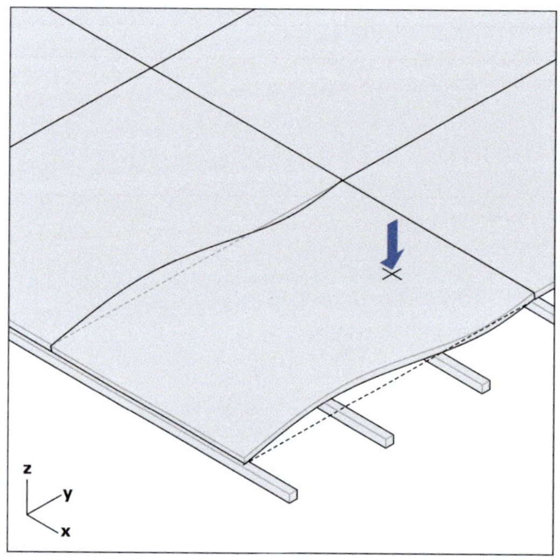

Fig. 1.45 Deformation of adjacent plates on a substructure due to non-uniform loading shown on the example of a three-span plate. Mutual offsetting of the plate edges

Applying these considerations to comparatively thin plates (large length/width, small thickness) (Fig. 1.53)—as measured against the proportions of the cross-section of a rod—it becomes clear that the free-spanning plate section **f** is very large in comparison to the thickness **d**—which governs the stiffness—so that, with heavy loading (e. g. live loads), moderate plate stiffness, or even larger plate dimensions, this usually leads to an undesirable deformation of the plate under bending (cf. Fig. 1.52 centre) and to a **gaping** of the lap joint by

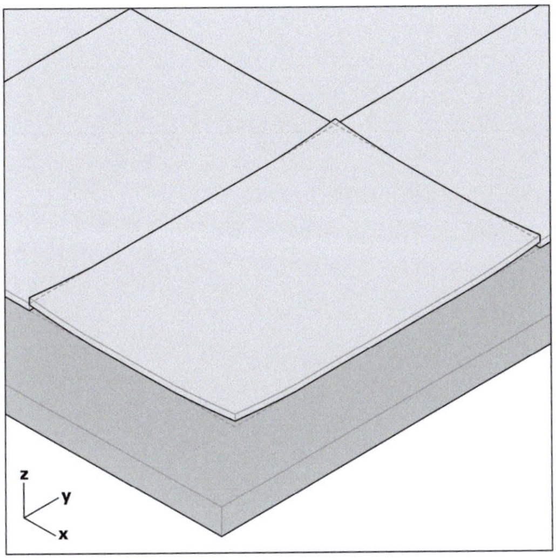

Fig. 1.46 Uplifting of abutted plate edges as a result of deformation of the plates

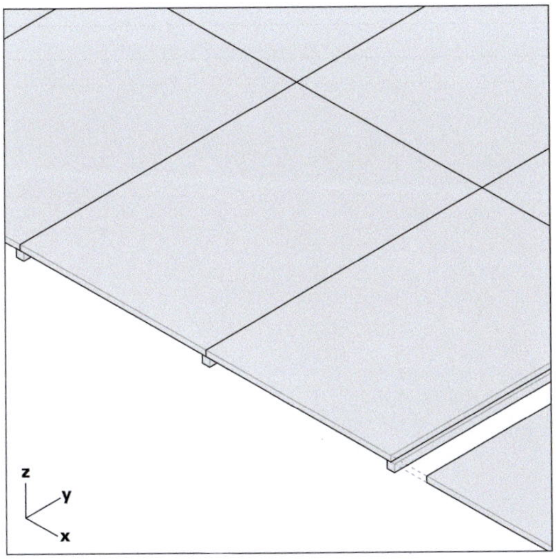

Fig. 1.47 Plates attached to substructure

an angle ϕ, something which is contrary to the purpose of the overlap (sealing, fastening).

However, this is not the case if the boundary conditions just mentioned do not apply. Insofar as elements are small and thin and are subjected to only small loads, for example, façade elements such as shingles, or also roof tiles, this type of overlapping or scaled surface formation with plates is possible. It should be borne in mind that a scaling on a curved base also results in gaping joints (see Fig. 1.53 below).

► *Sect. 1.1.3.3.2 Layout Geometry*

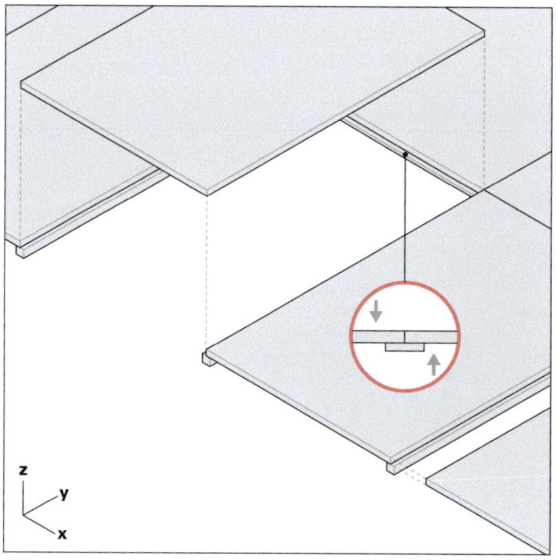

Fig. 1.48 Shear-resistant connection of the edges of plates that are not supported by the substructure, e.g. by means of added board strips or other backing strips

Fig. 1.49 Board cladding forming a surface between or under a supporting structure made of wooden ribs, without shear restraint (as in Fig. 1.48) at their butt joints (horizontal in the picture). The ratio of panel thickness to span width may allow for this solution according to Fig. 1.47

► *Sect. 1.1.3.2 Band-Shaped Starting Elements*

Plate **edges folded up** (Figs. 1.54 and 1.55) also increase the joint area which is available for fastening and sealing between abutting panels as compared to laid-out plates with butting edges. In addition, similar advantages apply as with bands. As a rule, this solution is particularly advisable in the case of plates which are produced in a manufacturing process which allows direct shaping of the upstands; for example, with cast parts such as prefabricated concrete parts or U-glasses. In

Fig. 1.50 Panel cladding forming a surface on a substructure of wooden ribs, as in Fig. 1.48. The cross bars (horizontal in the picture, above and below) provide a fixation for the transverse joints of the panels and prevent them from sliding against each other

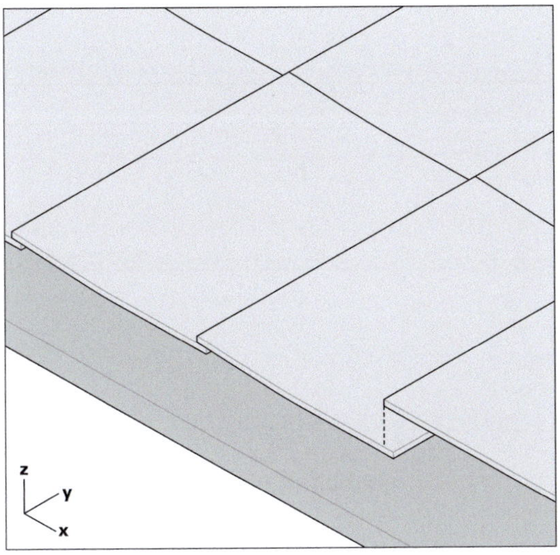

Fig. 1.51 A plate overlap is not useful with the usual proportions between width/length and thickness of the plates

production processes which essentially produce **flat sheet material** (for example, wood-based panels, float-glass), the retrofitting of this type of edge ribs usually offers no particular advantages.

A good bond between abutting plate edges is provided by **rabbets** (Figs. 1.56 and 1.57) as well as **tongue-and-groove joints** (Fig. 1.58). Both versions require a certain minimum thickness of the plate, as otherwise the delicate profiled edges

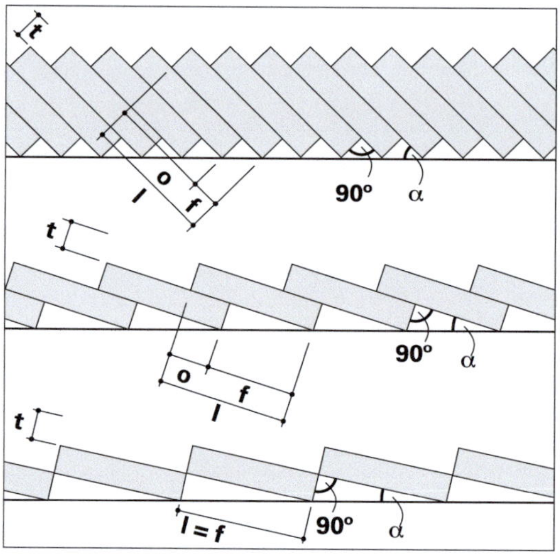

Fig. 1.52 Geometrical conditions when overlapping adjacent parts as a function of the laying angle α; **d** construction thickness, **o** overlap dimension, **f** free-spanning length, **l** overall length of the component

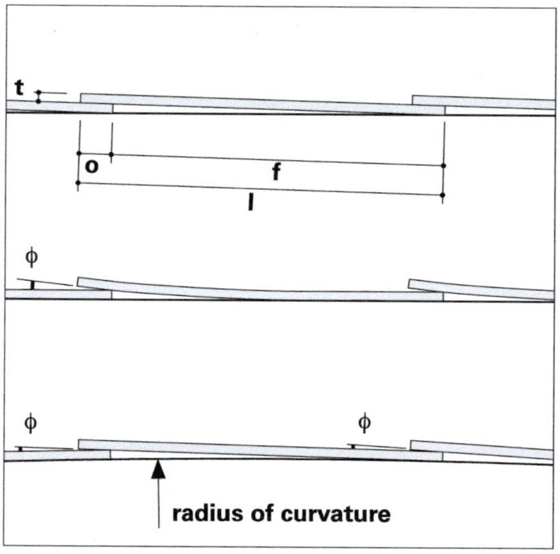

Fig. 1.53 Overlapping arrangement of thin plates

can be damaged during assembly or installation. Simple rabbets allow simple assembly, as the following element is always placed on top of the already laid (Fig. 1.56) from the top or from the outside. Rabbets on both edges (length and width) (Fig. 1.57) can also be realised. Connections with tongue and groove (Fig. 1.58), on the other hand, allow only one mounting direction within the plate's plane, which makes the assembly process more difficult. However, they create a continuous, flush surface. Furthermore, it must be ensured that no significant shifts occur between the edges to be joined dur-

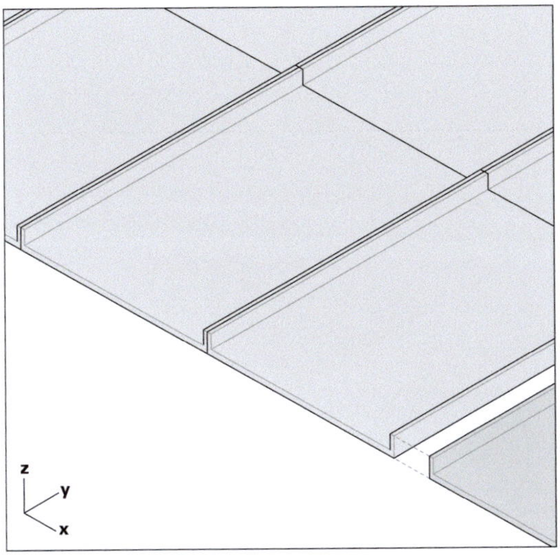

Fig. 1.54 Folded-up plate edges

Fig. 1.55 Profiled-glass façade with U-shaped cross-section of panes, as shown in the diagram in Fig. 1.54

ing assembly, for example, because of different deflections of the adjacent plates, so that the tongue can be inserted into the groove. In most cases, this results in this solution being chosen only with a certain minimum thickness of the plates or whenever parts lie flat on a supporting plate (for example, with flooring plates).

Due to the different edge geometries on opposite sides of a plate, larger **trimming scrap** than for simple butt joints is to be expected for both rabbet joints and tongue-and-groove joints, since residual pieces can often not be used due to improper edge profiling.

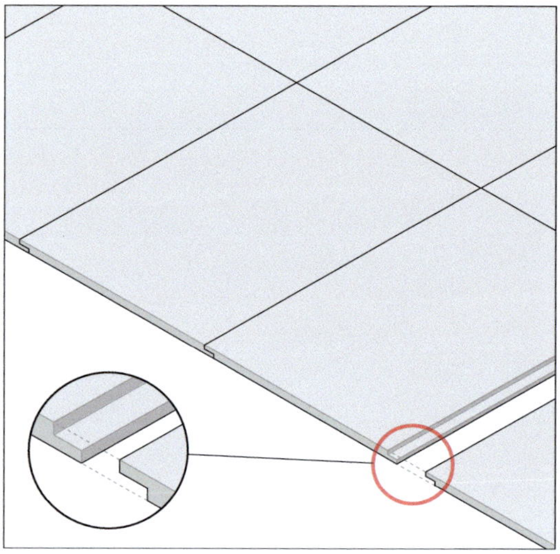

Fig. 1.56 Rebated plate edges

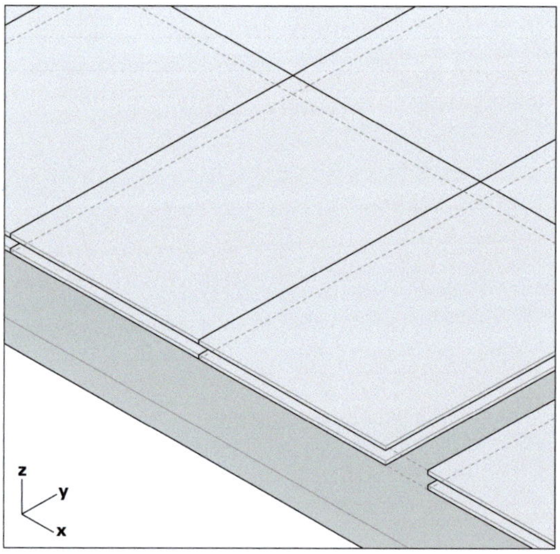

Fig. 1.57 Plate edges rebated all around

Layout Geometry

If the elements are located on a **full-surface supporting slab**, the laying geometries can initially be freely chosen. In the case of multi-layered laying, it is advisable to **offset** the two superimposed joint grids by a difference (Fig. 1.59), since the covering of the joint can offer advantages depending on the application, for example, in order to reliably prevent the edges of the lower layer from uplifting or to geometrically disentangle the connections within both layers.

If the plates rest on a substructure of rod-shaped elements, it is advisable in each case to lay the butt of two plates longitudinally on a bearing bar (Fig. 1.60). In the case of a **one-**

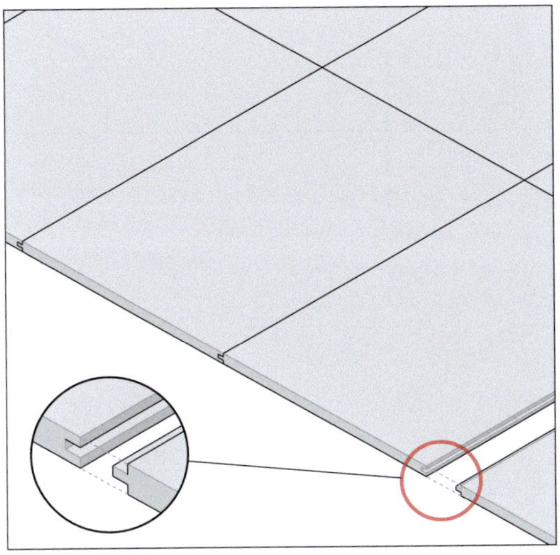

Fig. 1.58 Tongue-and-grooved plate edges

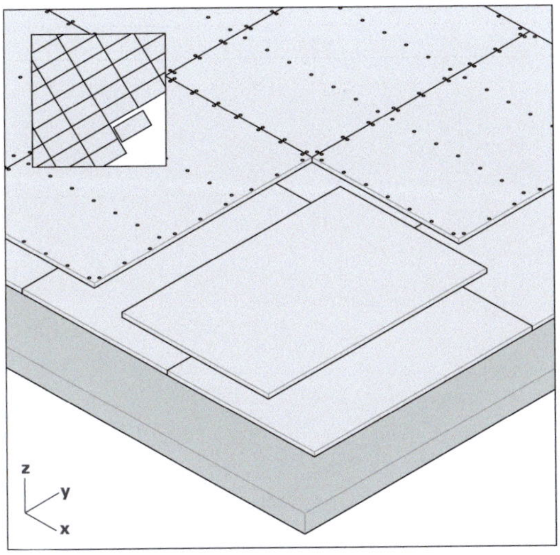

Fig. 1.59 Laying geometry with cross joints in two mutually offset positions on supporting plate

way substructure (in → **X**), the other edge direction (→ **Y**), which has no support, must be connected by coupling profiles **p** in order to prevent mutual out-of-plane edge displacement (see also Fig. 1.48). The plate spans one-way in this case (in → **Y**). The spacing **a** between the bearing rods depends on the load-bearing capacity of the plate, which depends in turn mainly on its thickness.

If one chooses to **offset the edges** (Fig. 1.61), it is nevertheless reasonable to adhere to the condition that joints parallel to the substructure (in →**X**) always lie on a bearing bar.

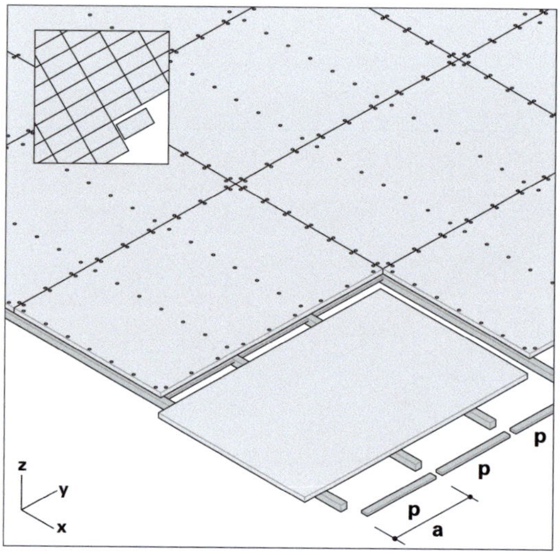

Fig. 1.60 Laying geometry with cross joints on a directional substructure. Coupling profiles **p** at the free edges

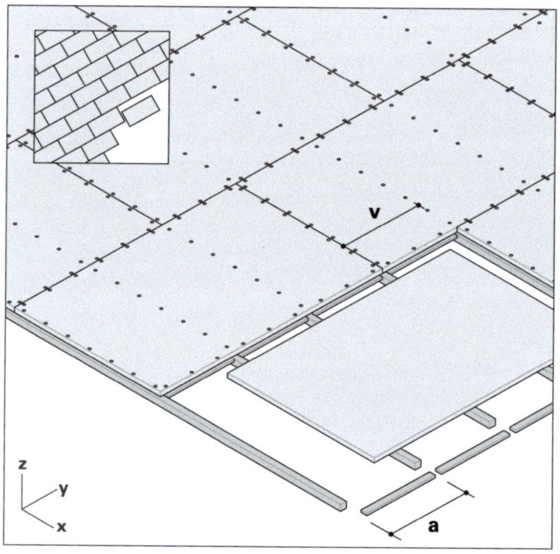

Fig. 1.61 Layout geometry with offset on unidirectional substructure. The joint offset **o** is identical to, or a multiple of, the spacing **a** of the substructure members

The joint offset **o** is therefore always a multiple of the distance **a** between the bearing rods.

In the case of two layers of panels, it is advantageous to disaggregate the layer joints by double offset, i.e. moving them in both main directions in each case (in →**X** and in →**Y**) (Fig. 1.62). The offset **o** transverse to the alignment of the substructure (i.e. in → **x**) must be based on the spacing **a** of its members, i.e. be equal to or a multiple of it. This ensures that all joints in this direction are supported by a bar. Due to the

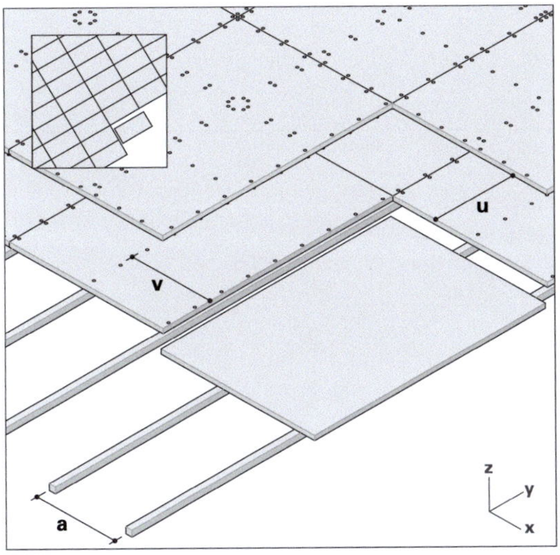

Fig. 1.62 Support of the panel on a unidirectional substructure. Offset between panel layers in both main directions. The offset **v** transverse to the alignment of the substructure must be equal to the distance between members **a** or a multiple thereof. Thus, the longitudinal joints always rest on a bar. The offset **u** is arbitrary

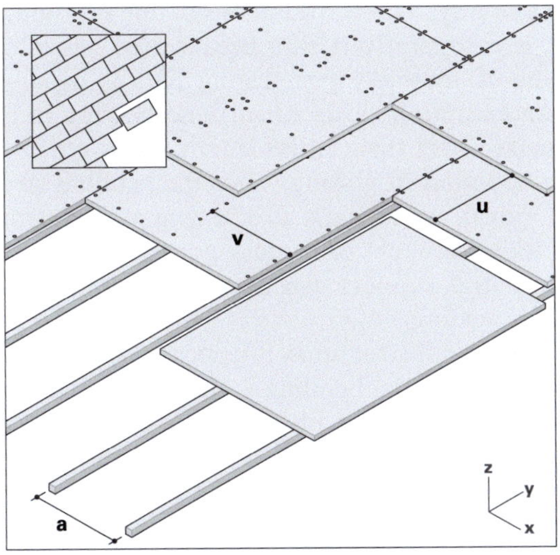

Fig. 1.63 Support as in Fig. 1.62. In this case, however, the panels of both layers are laid with an offset, i.e. with the end edges offset. The offset **v** transverse to the alignment of the substructure is again equal to **a** or a multiple thereof, the offset **u** arbitrary

double layer, the transverse joints (in → **x**) are always secured against uplift or displacement perpendicular to the plane without the need for further measures (like in Fig. 1.48) as they are always covered or supported by the respective other layer. Even within the panel layer itself, it is possible to lay the panel with an offset (Fig. 1.63).

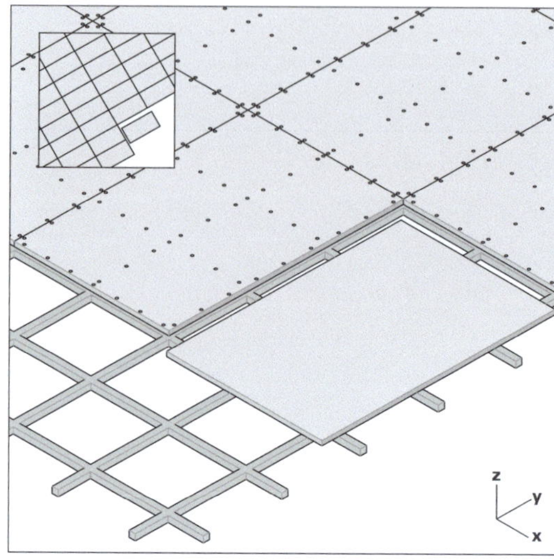

Fig. 1.64 Support of the plate on a bi-directional substructure

Sect. 3.2.1 *One-Way and Two-Way Load Transfer*

Example: wooden sleepers on a floor

Although **two-way spanning substructures** of intersecting sets of bearing bars (Fig. 1.64) are possible and also offer the advantage of a plate being linearly supported on all sides (i. e. spanning **two-way**, in → **x** and in → **y**), this solution, notwithstanding, is comparatively rare because the hassle associated with skeins of mutually penetrating bars is usually circumvented. An exception exists when these bearing rods lie on a carrier surface over their entire length. This type of support is also to be found in glazing, since the bending stress of the panes is critical, particularly in the case of glass surfaces, so that four-sided support of a glass pane may be necessary. A linear perimetral support can also be advantageous for reasons of joint sealing.

***Vol. 1**, Sect. 18 Glass*

For two-way substructures, it is recommended to make the fields **square**, as biaxial bending is the best way to make good use of the stiffness of the plate. Since glass panes are always butted over a linear bearing, his consideration, in glazing, leads to approximately square pane formats, which respectively cover a field between opposed bearing rods.

Cf. *Sect. 1.1.3.3.1 Seam Execution* above

***Vol. 3**, Sect. 42.6.5 Overlapping Joint*

Overlapping scaled arrangements (Fig. 1.65) are also possible under certain conditions. In particular, the overlapping is used for sealing purposes by laying it, in the case of standing or inclined weathered surfaces, according to the water-flow direction (Figs. 1.66, 1.67, and 1.68).

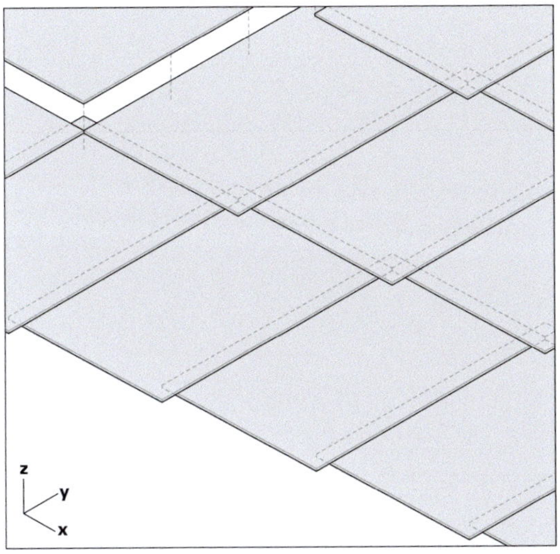

Fig. 1.65 Scaled arrangement of thin, small-format plates with low top-load. The plate edges overlap here in both edge directions

Fig. 1.66 Scaled laying of flat stone slabs on a roof with overlapping in the two main directions as in Fig. 1.65

1.1.3.4 Rod-Shaped Starting Elements

Seam Execution

The essential characteristic of a rod-shaped element is the comparable magnitude of two of its dimensions, width and height, compared to its substantially larger length. In the case of plates, on the other hand, width and length are of a comparable magnitude, while the thickness is substantially smaller. As a result, rods have a higher stiffness—obviously, as related to the width **b** of a single element—in the direction of

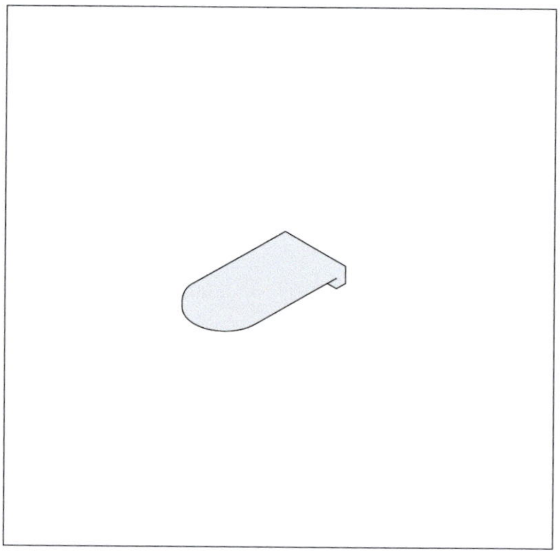

Fig. 1.67 Plain tile

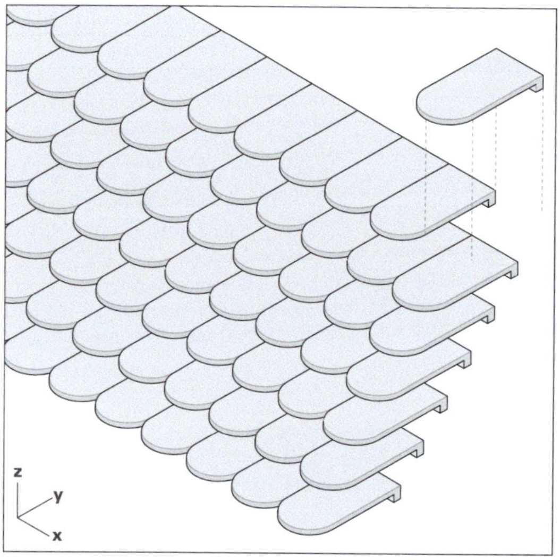

Fig. 1.68 (Left) Roofing from plain tiles. (Right) Roofing from plain tiles during construction

their width than plates, since their cross-sections always have more compact proportions (Fig. 1.69).

As with plates, in the case of rod-shaped starting elements, simply **adding** them **side-by-side** yields continuous surfaces (Fig. 1.70). It is also possible to leave gaps (Fig. 1.71), if, for example, deformations of the bars have to be absorbed (as in the case of wood). This does not make the surface truly waterproof, but such solutions can be used, for example, for external façade cladding where only rain resistance is required rather than total watertightness.

Cf. ***Vol. 3***, *Sect. 42.4.3 Closed Joint*
Cf. ***Vol. 3***, *Sect. 42.4.1 Open Joint*

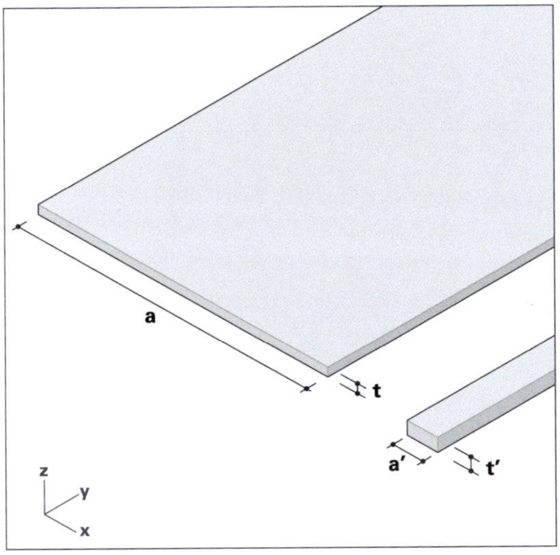

Fig. 1.69 In relation to the respective widths **a**/**a′** of plate and bar, the comparatively greater thickness **t′** of the bar leads to a higher stiffness in the direction of width, i.e. in → **x**

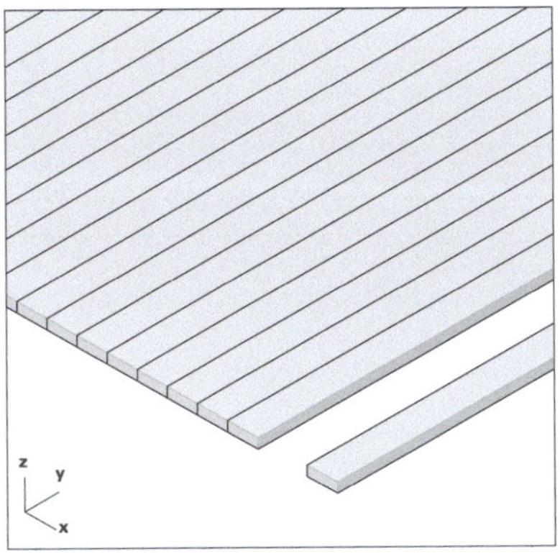

Fig. 1.70 (Left) Bars butted against each other. (Right) Façade cladding consisting of bars placed side-by-side

With regard to the action of a force perpendicular to the plane (→ **z**), the same restrictions apply as in the case of plates: Where there is no shear connection between abutting edges, these can be displaced out-of-plane against one another under different influences such as non-uniform loading or inherent deformation. **Rabbets** or **rebates** as well as **tongue-and-groove joints** (Fig. 1.72) provide a remedy and unite the surface to form a plate which is bending-stiff in the direction of the bars (→ **y**) having good **transverse load distribution**. In this way, point or uneven loads oriented perpendicular to the surface are always distributed among the neighbouring bars, so that there is cooperation and the edges do not move mutually perpendicular to the surface of the component. Introduc-

Cf. ***Vol. 3***, *Sect. 42.6.6 Rebated Joint* and *Sect. 42.6.7 Tongue-and-Groove Joint*

For the concept of lateral load distribution, cf. ▶ ▶ Sect. 3.2.1.1, particularly ▶ Fig. 3.41, left

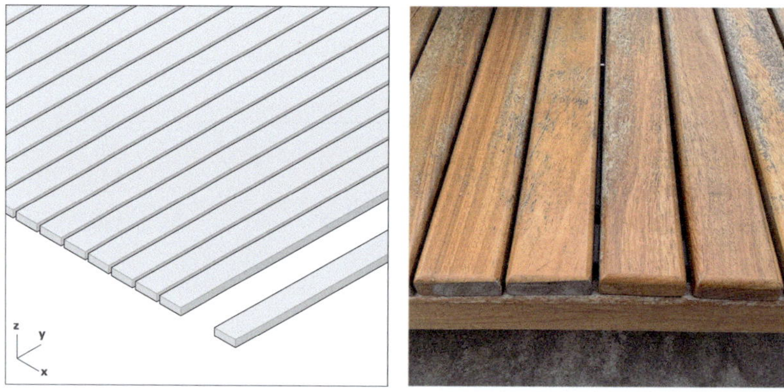

Fig. 1.71 (Left) Bars laid to gap. (Right) Open-jointed boarding according to the scheme on the left. The gaps can have different purposes, e.g. as here, surface drainage

Fig. 1.72 (Left) Bars laid with skewed overlapping. (Right) Skew-overlap siding

ing this kind of profiled edges, in the case of bars, proves to be problem-free in most cases as compared to larger plate formats, since bar-shaped elements are usually easier to handle.

Cf. ***Vol. 3***, *Sect. 42.6.3 Covered Joint*, *Sect. 42.6.4 Backed Joint*, and *Sect. 42.6.5 Overlapping Joint*

Overlapping, **scaled**, or **cuffed** arrangements (Fig. 1.73) are also possible with rods. The only slight tendency of the narrow rod to deformation in the width direction allows this type of laying. Overlapping arrangements made of rods alternately placed on top or below (Fig. 1.74) are also suitable for surface formation. Like the gaped installation (Fig. 1.71), these solutions allow for lateral deformation of the bars (in → **x**) and are therefore particularly suitable for timber construction as they neutralise the effects of the swelling and shrinkage characteristic of wood. If the overlapping joint is additionally secured against gaping, as is generally required, an effective **shear-resistant bond** is created between adjoining elements in both perpendicular directions (in → **z** and in → **−z**) so that **transverse load distribution** can also take place at right angles to the bar's axis within the plate's plane (i.e. in → **x**).

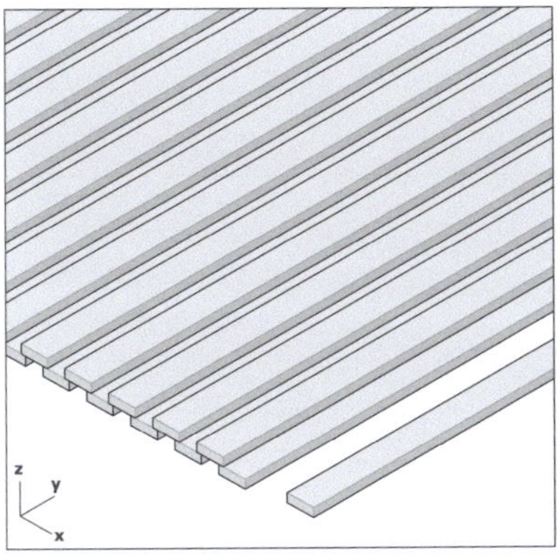

Fig. 1.73 (Left) Bars placed alternately over and under each other. (Right) The boarding consists of boards placed alternately at the front and at the back

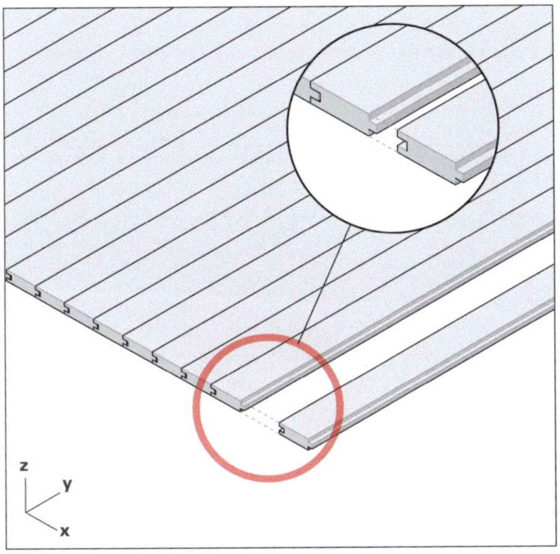

Fig. 1.74 Bars with tongue-and-groove joint

Layout Geometry

The laying of bar skeins according to a **cross-joint grid** (Figs. 1.75, 1.76, 1.77, 1.78, and 1.79) is the simplest possible laying geometry. If the leaf formed by the bars is fastened on a substructure of bars, it is obviously convenient to place the end-face seam lines of the bars axially upon the supporting profiles of the substructure, which then consists purposefully of a **unidirectional system** of mutually parallel bearing bars which are at right angles to the covering bar (→ **x**). This solution allows the attachment of the bar ends directly on the substructure and thus prevents the undesirable lifting of the bar's end edge, as would be the case if the bar were to overhang freely over the bearing element. The maximum distance **a** between the bearing bars is, as always in this type of system, determined by the load-bearing capacity of the bar. However, in the case of this solution, the rod can span over several fields of the substruc-

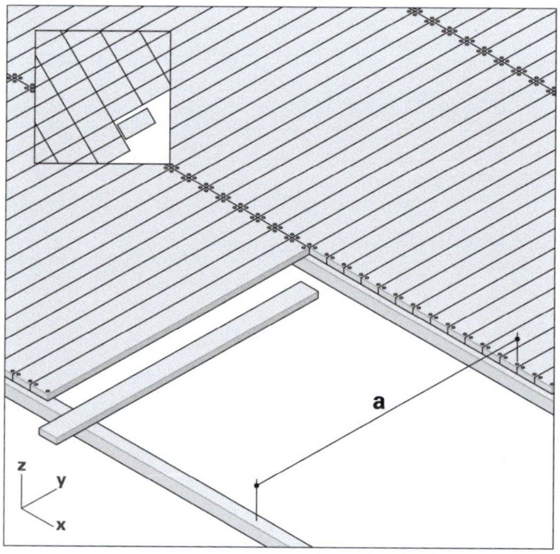

■ **Fig. 1.75** Bars laid against each other in cross-joint geometry

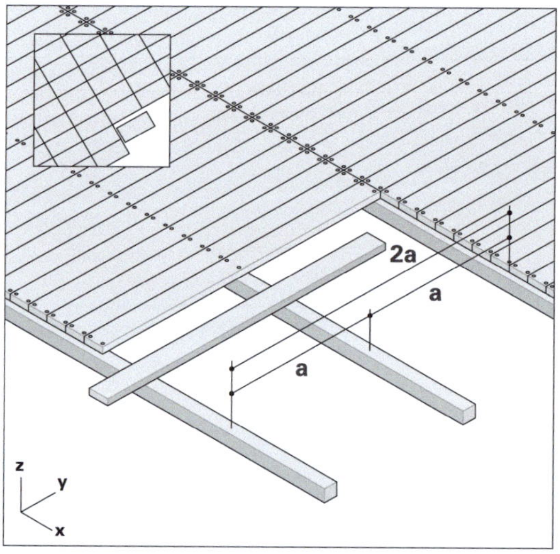

■ **Fig. 1.76** Bars laid against each other in cross-joint geometry. Continuity effect of the bar covering several spans (here two)

ture (■ Figs. 1.76 and 1.78), which activates a structural **continuity effect** and reduces the bending on the cover bar. This results in the possible lengths of a cover bar being a multiple of the distance **a** between bearing bars, in the example shown, 2**a** (■ Fig. 1.76). Varying the distance between bearing bars (■ Fig. 1.77) results, in general and all other conditions being equal, in a less efficient use of material with the cover bar, since its dimension is predetermined by the largest field to be spanned (i.e. **b**) and, hence, is over-dimensioned over smaller fields (i.e. **a**), provided the bar thickness is kept uniform.

Offsetting the end edges of adjacent bars (■ Fig. 1.80), which in each case is done in multiples of the spacing between

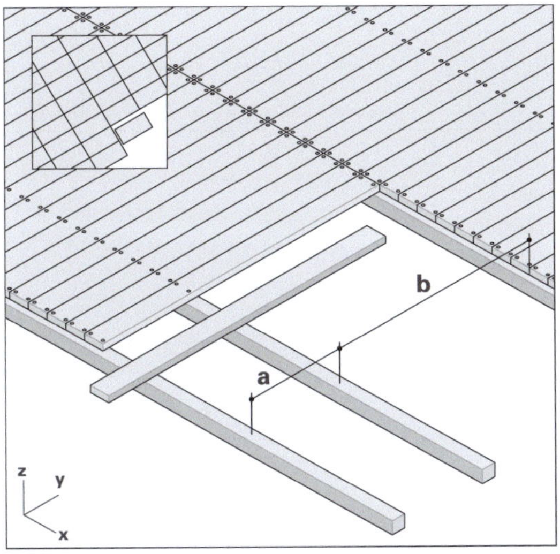

■ **Fig. 1.77** A bar spans two different fields **a** and **b**. The bending stiffness of the bar is not well utilised

■ **Fig. 1.78** Façade siding with gridded joints, presumably according to the diagram on ■ Fig. 1.75, or rather according to ■ Fig. 1.76

bearing rods, so that the front edges can always be fixed on a support, offers no significant design advantages compared with the previous solution, aside from the structural continuity of the bars covering more than one field. However, a visually less pronounced joint pattern results, which may be desired in some cases. In addition, the same staggered bar layout results (unlike the cross-joint solution of ■ Fig. 1.76) in a shear-resistant joint in its plane transverse to the bar axis (i.e. in → **x**), which can be advantageous in some applications (such as some floorings where there is no attachment to a substructure).

■ **Fig. 1.79** Gridded façade siding. Each board covers a total of five battens of the substructure frame, as can be seen in the nailing. Corresponds to the scheme of ■ Fig. 1.76, but with three intermediate bars

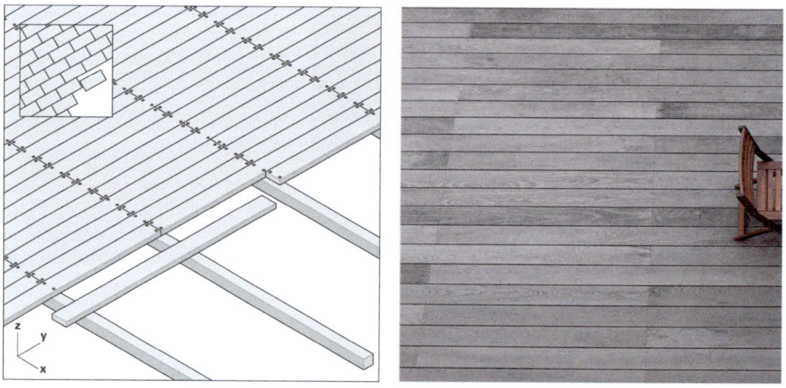

■ **Fig. 1.80** (Left) Bar layout, offset by one span length. (Right) Planking laid in accordance with the scheme on the left

The **fishbone**-shaped laying of bars on a substructure (■ Fig. 1.81) leads to conditions more unfavourable in constructional terms insofar as the cover bar, without recognisable benefit, needs to cover a larger span, namely the distance between bearing bars divided by $\cos 45°$, which results in 41% more. A structural continuity effect cannot be produced—at least over the butting angle of the cover bars. The width of the covering bar should at least equal the width of the bearing bar in a diagonal position in order to be able to secure it in a tilt-proof manner. In the case of covering bars coupled on the long side, e. g. in the case of tongue-and-groove joints, this condition is dispensed with since tipping is not possible. Herringbone geometries are particularly used as a visually interesting pattern when subjected to moderate stress. These limitations do not

Fig. 1.81 (Left) Laying the bars according to a herringbone pattern. (Right) Parquet laid in herringbone pattern on battens

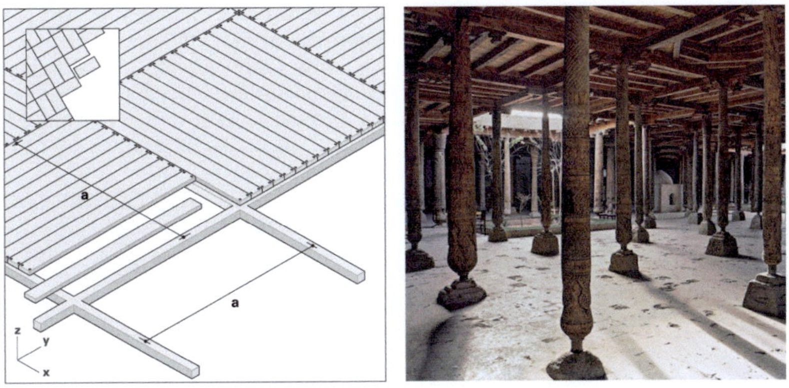

Fig. 1.82 (Left) Laying the bars on non-directional substructures changing the spanning direction from field to field. (Right) Checkerboard arrangement of the bars on a two-way support structure according to the scheme on the left to take advantage of a more favourable load transfer

apply, of course, when the bars rest with their entire surface on a load-bearing base, as in some parquet floors.

Checkerboard-like layout geometries on substructures with laying direction of the bars (Fig. 1.82) alternating from field to field can only be executed on a two-way system of bearing bars with two intersecting skeins of bars in both main directions. With regard to the load transfer to the substructure, the advantage, as compared to a one-way system made up from only one skein of bearing bars, is that each supporting bar has to bear the load of only a half load-infeed field **a** instead of the one of an entire field as in the case with one-way solutions. This is particularly important when the bearing bars do not lie on a full-face support, but are spanning freely as a bending beam. The Checkerboard arrangement of the bars thus allows an overall two-way load transfer to the substructure, despite the use of one-way load-bearing elements (covering bars). This solution is sometimes used for beam gratings (Fig. 1.82 right). In this case, it is obviously not possible to use a continuity effect of the covering bars.

Cf. ***Vol. 3***, *Sect. 55.4.2 Roofs and Floors from Beam Grillages*

Fig. 1.83 Almost flat vault made of voussoirs. The load-bearing effect results from their mutual engagement (the ring-shaped contact surfaces are conical, not cylindrical). It is a rare example of a self-supporting horizontal surface made of building blocks. The execution requires the most profound knowledge of stonemasonry (Monasterio de San Lorenzo del Escorial)

1.1.3.5 Block-Shaped Starting Elements

Seam Execution

Block-shaped elements are suitable only in exceptional cases for the construction of free-spanning, load-bearing components (Fig. 1.83). Their three dimensions, length, width, and height, all of similar magnitude, have a size that is governed by the thickness of the surface component, which is limited from the outset. In contrast to plate-shaped and rod-shaped starting elements, in which at least one dimension—that is to say the length and width of the plate, or the length in the case of the rod—can be used to generate planar surfaces, being a measure liable to be subjected to bending and, hence, suitable for covering spaces, block-like elements must usually be laid on a full-surface bearing support to generate continuous surfaces. Load-bearing substructures made of bars, such as are used with plate-shaped and rod-shaped starting elements, are not economically reasonable in this case because of the short distances between butt joints inherent in the system, which would result in an unreasonable density and complication of the substructure. This type of use of block-shaped elements placed on a supporting surface is found, for example, in horizontal paving (Fig. 1.84 right; Fig. 1.85 right; Fig. 1.86 right). Apart from compressive forces perpendicular to the layer plane, no appreciable stresses are to be expected here, so that no special requirements are placed on the joining between adjacent stones.

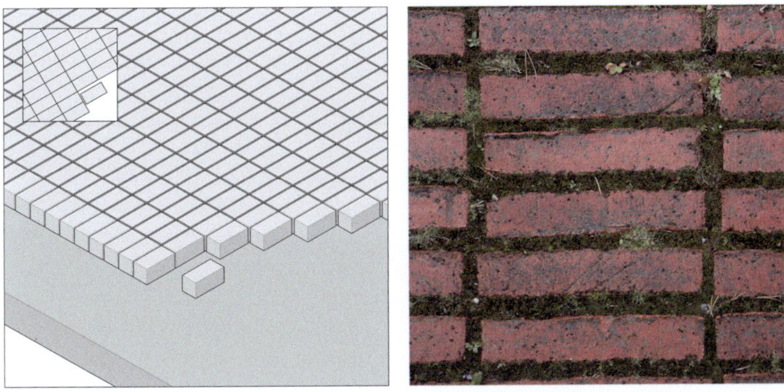

Fig. 1.84 (Left) Laying block-shaped elements on a flat base in cross-joint pattern. (Right) Brick paving, laid in a grid pattern, as the scheme on the left

Fig. 1.85 (Left) Laying block-shaped elements on a flat surface with offset joints (in a bond). (Right) Cobblestone paving with offset joints, as in the scheme on the left

Layout Geometry

In this type of use of block-shaped elements, the laying geometry plays no role with respect to the transfer of forces. In the case of floor coverings (Figs. 1.84, 1.85, and 1.86), for example, it has only a visual, decorative meaning or can be selected under certain conditions for reasons of minimising waste or also if an in-plane shear connection is desired resulting from a staggered offset (Fig. 1.85).

A single—albeit significant—exception is **masonry bonds** (Figs. 1.87 and 1.88 left). These yield essentially planar surface components made of block-like starting elements which are erected in a standing or, at most, slightly inclined position. Self-supporting horizontally lying planar surfaces made of block-shaped elements can be produced only with great effort, or not at all, for the above-mentioned reasons (Fig. 1.83). The contact joint between adjacent elements must be designed, for this purpose, to be tension and shear-resistant, which usually entails an unacceptable effort. Such a solution can also

An exception are, for example, segmental construction modes in prestressed concrete construction; also extremely rare historic straight, flat vaults of wedged voussoirs (cf. Fig. 1.83), in which, however, in principle no shear and tensile strength of the joint is required. Cf. on this issue ***Vol. 1**, Sect. 26.9.3.2 Bond—Lap Acting by Compression*, in particular Fig. 26.127

Fig. 1.86 (Left) Laying block-shaped elements on a flat base with laying direction changing in pairs. (Right) Blocks on a load-bearing base can be laid in any pattern, often chosen with decorative intent

Fig. 1.87 (Left) Brick. (Right) Bricking with offset head joints (in a bond) in the form of a standing wall

be ruled out from the outset whenever the material itself lacks tensile strength (as is the case with mineral materials).

On the other hand, in a standing position, a proper gearing—a so-called bond—of the stones elicits a load-bearing behaviour which makes the surface capable of bearing not only compressive but also shear and tensile forces acting within the plane and even, to a limited extent, bending stress acting at right angles to it. If the load within the plane of the component, that is to say, the stress which leads to compression in the surface, is large enough, tensile and shearing stresses within it, which would lead to the joint between stones gaping or slipping, can be neutralised or over-compressed. A combination of **vertical load** and **friction** within the horizontal joint or **bed joint** between blocks comes into effect. A prerequisite for the effectiveness is furthermore the bond, which entails offsetting vertical joints, or **head joints**, between adjacent blocks from course to course. In this way, load-bearing surface components can be produced.

Vol. 1, *Sect. 13.5 Mechanical Properties* as well as ***Vol. 3***, *Sect. 44.3.3.2 Tangential Force Closure (Frictional Closure)*; cf. also ► *Chap. 7 Masonry Construction*

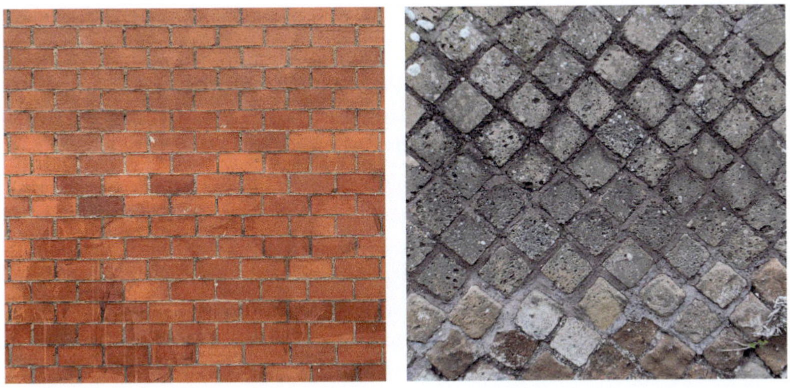

Fig. 1.88 (Left) Masonry—typical brickwork with staggered head joints. (Right) Roman *opus reticulatum*; stones arranged in a grid pattern: non-load-bearing lost formwork for a concrete wall core

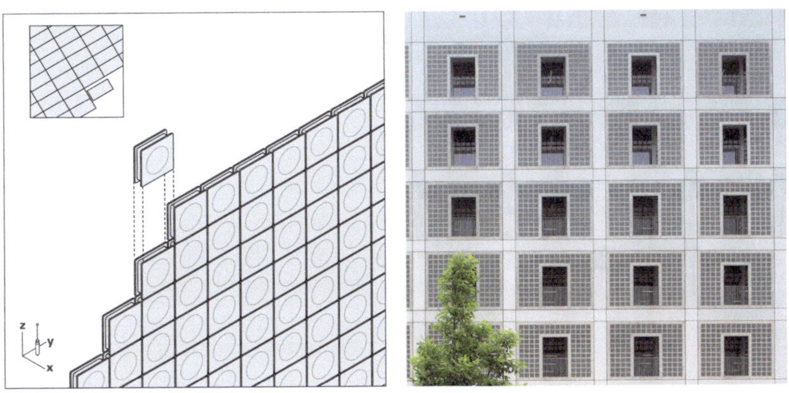

Fig. 1.89 (Left) Laying geometry with cross-joints in a non-load-bearing glass-brick wall. (Right) Glass blocks arranged in a grid pattern on a façade

Cross-joint geometries can also be realised in exceptional cases (Fig. 1.88 right; Fig. 1.89), provided they are non-load-bearing or only self-supporting surfaces (for example, glass-brick walls), with which, apart from the dead load, no appreciable additional stresses occur.

1.2 Shape Definition of Continuous Curved Layer Surfaces

If curved surfaces are to be constructed using commercially available basic elements in **strip**, **plate**, **rod**, and **block** shapes, special restrictions must be taken into account, which are to be examined in the following by means of some examples. It should be recalled that **cast components** are also bound to these restrictions if they are made using formwork or other negative moulds, which in turn consist of a solid base material.

E.g. wood boardings for concrete

1.2.1 Special Features of Curved Surfaces

Whereas the above-mentioned conditions were applicable extensively to planar surfaces, special conditions apply to curved surfaces, which are to be examined in more detail below. Although curved surfaces do not have the same relevance as planar ones in the practice of construction, they feature specific strengths as compared to them: In load-bearing applications, they can gain in stiffness and load-bearing capacity simply by virtue of their shape. For certain constructions such as shells, vaults, or domes, curvature is even an indispensable prerequisite for their load-bearing capacity. They also have a great potential for aesthetic expression. Sculptural building forms made of curved surfaces have been created in recent years in numerous, very spectacular and medially effective examples.

The variety of curved surfaces is immense. In addition to the **regular variants**, that is to say, those which form the basic repertoire of curved surfaces used in building construction, there are countless **free-form surfaces** which cannot be defined with simple means and which until now have been almost unrealisable because:

- they could be transposed from the drawing or the model to the scale 1:1 only with the greatest effort;
- because their **load-bearing** behaviour was hardly computable in advance;
- because these complex surfaces could only be manufactured with the greatest effort and expense from available, largely **planar** or rectilinear **rod-shaped** base elements. This also applies to the shuttering of moulded elements.

***Vol. 1**, Sect. 3.4.2 Application of New Digital Technologies of Planning and Manufacturing to the Building Sector*

Digital planning tools have resulted in a fundamental change in the basic conditions. Today, powerful design tools exist that allow to capture both the geometry and the load-bearing behaviour of complex free-form surfaces. Furthermore, modern CNC-controlled manufacturing facilities also offer the possibility to implement complex designs in terms of manufacturing technology. Although one has always to expect an additional expenditure as compared to planar components, it can still be claimed that curved surfaces are much easier and more cost-effective to manufacture today than just a short time ago.

Conversely, this does not mean that, in the face of modern construction and manufacturing technology, it has nowadays become indifferent how a curved surface is defined and generated geometrically. As with all other areas of design and construction, the availability of advanced tools does not open the doors to utter arbitrariness. It is of fundamental importance in this sense that the planner makes both the process of geometrical and technical development as well as the structural realisation as simple as possible within the coordinates of the

***Vol. 4**, Sect. 5.9 The New Freedom of Forms—An Appeal*

intended design. Applied to curved surfaces, this means that a regular surface must always be given preference over a free-form surface as long as the former implements the desired design form satisfactorily.

In order to understand the geometrical conditions of the most important curved surfaces of regular nature, some theoretical considerations are to be made in the following sections.

1.2.2 Geometric Conditions

The geometric definition of rectilinear and planar elements is trivial and requires only a few spatial coordinates—actually merely two points in space in the case of the former, three in the case of the latter. The situation is different with curves or curved surfaces, in particular free-form curves or surfaces, which all of them differ from the former, the planes and straight lines, in their **curvature**. In order to understand their geometry, it is therefore first necessary to clarify some basic concepts related to curvature.

1.2.2.1 Tangent Plane, Normal Vector

Through every point **P** of an arbitrary uniform, planar or continuously curved, surface **S** (Fig. 1.90), run **surface curves** ($\mathbf{s}_u$, $\mathbf{s}_v$) which emerge when the surface intersects with planes ($\prod_u$, $\prod_v$) containing this point. These planar surface curves can be assigned tangents $\mathbf{t}_u$, $\mathbf{t}_v$ at the point **P**, which define or deploy a characteristic **tangential plane** $\prod_T$. The tangential

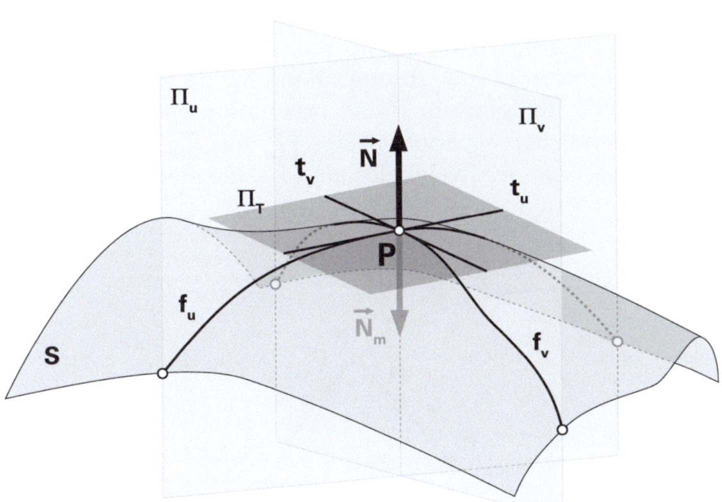

Fig. 1.90 Curved surface with tangential plane $\prod_T$ and normal vector **N**. The sectional planes $\prod_u$ and $\prod_v$ shown here are perpendicular to the surface and are therefore normal sectional planes. Both contain the normal vector **N** or the principal normal vector $\mathbf{N}_p$

plane contains the tangents of all conceivable surface curves through the point **P**. Exceptions to this rule are non-regular points, such as a cone tip. Curved surfaces are distinguished from planar ones by the fact that not every point on the surface has the same tangential plane.

The unit vector which is perpendicular to the tangential plane on **P** is also perpendicular to the surface in **P** and is called the **normal vector N**. Two oppositely oriented vectors fulfil this condition: on the one hand, the vector oriented to the convex side of the surface; on the other hand, the vector oriented to the concave side. In geometry, the latter, i. e. the one pointing to the centre of the associated osculating circle (see below), is chosen as the characteristic one and called the **principal normal vector** $\mathbf{N_p}$. If the selected planes Π_u and Π_v are perpendicular to the surface at the point **P**, both contain the normal vector **N** or $\mathbf{N_p}$ (like in Fig. 1.90). They are then referred to as **normal sectional planes**, the surface curves **u** and **v** lying on them as **normal surface curves**.

1.2.2.2 Curvature

The curvature of a curve **a** at a point **P** can be described as the deviation or change in direction of the curve from the tangent **t** in the vicinity of either side of the point (Fig. 1.91). It is equal to the curvature of the **osculating circle o** to the curve **a** at this point **P** and can therefore be calculated as $\mathbf{c} = 1/\mathbf{r}$, where **r** is the radius of the osculating circle.

The curvature of a surface at a point **P** can be described, analogous to the curve, as the change of direction of the surface with respect to the tangential plane in the vicinity of **P**. The greater the curvature, the greater the deviation. It depends on the respectively considered orientation of the associated surface curves $\mathbf{s_u}$, $\mathbf{s_v}$ through **P** in relation to a coordinate system **u/v** with centre in **P** and orientation along the mutually perpendicular sectional planes Π_u and Π_v (Fig. 1.92). These directions are freely selectable, i. e. the plane pair Π_u/Π_v

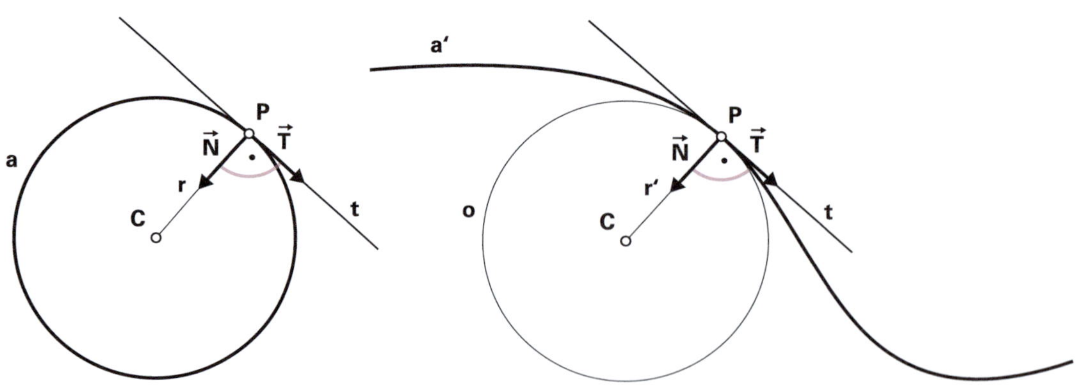

Fig. 1.91 **Curvature** of a curve **a** at a point **P**; left, circle: The curvature is constant and is defined by $\mathbf{c} = 1/\mathbf{r}$; in the generic curve **a'** on the right side, the curvature at **P** is defined as the curvature of the **osculating circle o** at **P**; thus $\mathbf{c'} = 1/\mathbf{r'}$. **N** principal normal vector, **T** tangential vector

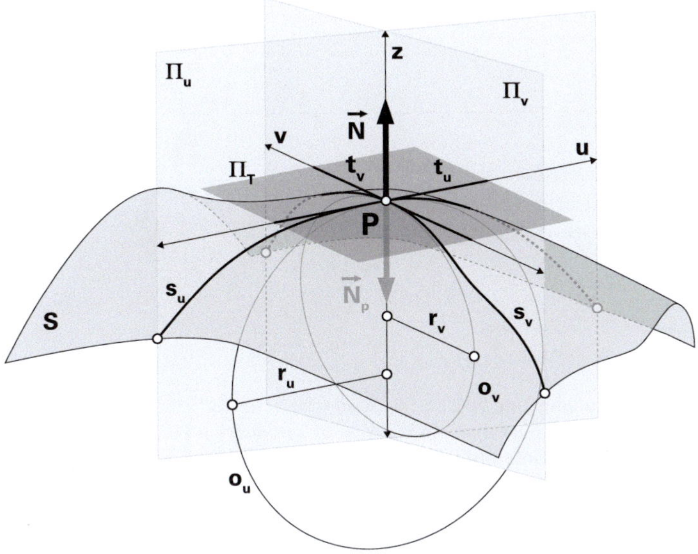

Fig. 1.92 **Curvature** of a surface **S** at a point **P** in alignment of a coordinate system **u/v**; the curvature of the surface at **P** is equal to the curvature of the associated osculating circles $\mathbf{o_u}$ and $\mathbf{o_v}$, i.e. respectively $\mathbf{c_u} = 1/\mathbf{r_u}$ and $\mathbf{c_v} = 1/\mathbf{r_v}$

initially rotates freely about the **z** axis. The curvature of the surface in a given direction **u** and **v** is again defined by the curvature of the associated osculating circle $\mathbf{o_u}$ or $\mathbf{o_v}$ of the respective surface curve. This is calculated in each case as:

$$\mathbf{c_u} = 1/\mathbf{r_u}$$

$$\mathbf{c_v} = 1/\mathbf{r_v}$$

where $\mathbf{r_u}$ and $\mathbf{r_v}$ are the radii of the two osculating circles $\mathbf{o_u}$ and $\mathbf{o_v}$ in the directions **u** and **v**.

Among the various selectable orientations **u** and **v**, there is an orientation in which the curvatures $\mathbf{c_u}$ and $\mathbf{c_v}$ assume maximum and minimum values, respectively. These are called **principal curvatures** $\mathbf{c_1}$ and $\mathbf{c_2}$ and are always perpendicular to each other (Fig. 1.93).

Gaussian Curvature

The Gaussian curvature **C** is defined as the product of the two principal curvatures $\mathbf{c_1}$ and $\mathbf{c_2}$:

$$\mathbf{C} = \mathbf{c_1} \cdot \mathbf{c_2} = \mathbf{1/r_1} \cdot \mathbf{1/r_2}$$

The concept of Gaussian curvature can also be derived from the **Gaussian image** of a surface, or a sector thereof (Fig. 1.94). Here, the unit normal vectors **N** of each point on the surface **S** are transferred to the centre of a sphere with radius 1, the so-called **Gaussian sphere**. They define or sweep a spherical cap **S*** of the sphere's surface. The size of this cap is a measure for the curvature of the surface **S**: the greater **S***, the

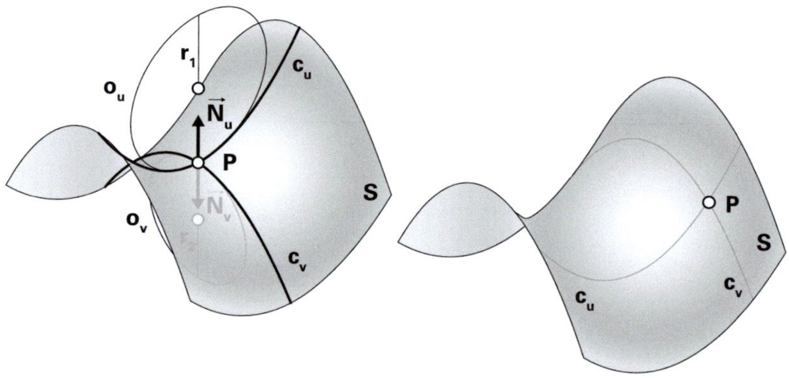

Fig. 1.93 **Main curvatures**: at each point **P**, the alignments **u/v** can be found where the curvatures $\mathbf{c_u}$ and $\mathbf{c_v}$ take a maximum and a minimum value respectively. Left, main curvatures in the axes of symmetry a **P** of a saddle surface; right, main curvatures at an arbitrary point **P**. For the saddle surface, the normal vectors $\mathbf{N_u}$ and $\mathbf{N_v}$ of the curvatures point in opposite directions, i.e. the point **P** is a hyperbolic point

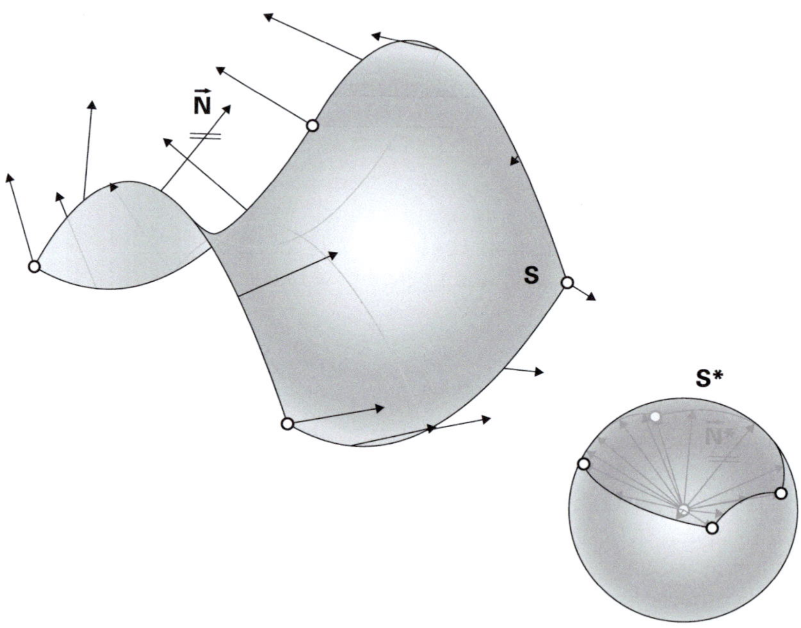

Fig. 1.94 Transfer of a surface **S** to a **Gaussian sphere**. The unit normal vectors **N** assigned to each surface point (here, only selected edge points are shown) are transferred to the centre of a sphere with radius 1. The resulting spherical segment is the **Gaussian image S*** of **S**. The greater **S***, the greater the deviation of the normal vectors from one another, i.e. the greater the curvature

greater the curvature (Fig. 1.95). The ratio between the area of the Gaussian image **S*** and the area of the original surface **S** is equal to the Gaussian curvature **C**:[6]

$$\mathbf{C = S^*/S = c_1 \cdot c_2 = 1/r_1 \cdot 1/r_2}$$

6 Pottmann and Bentley (2007), p. 496.

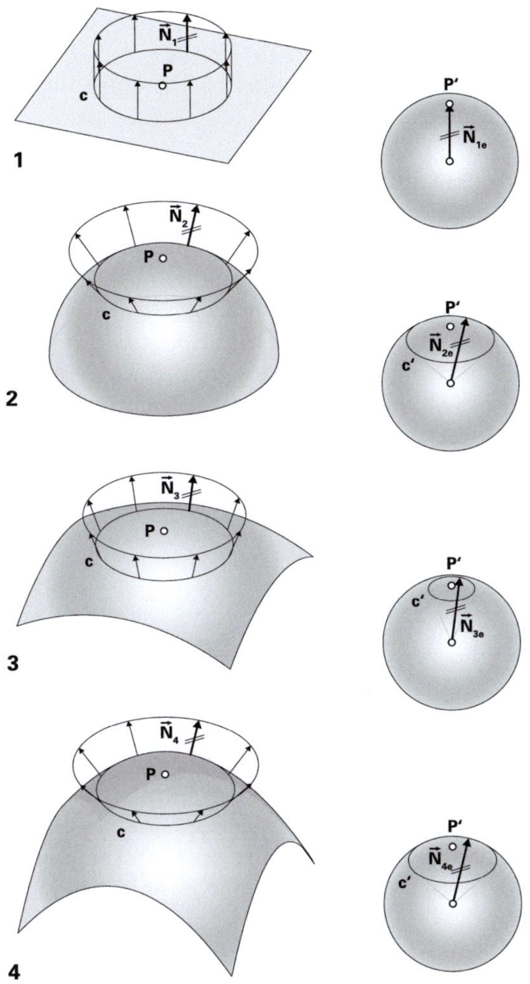

Fig. 1.95 Transfer of a surface **S** to a **Gaussian sphere**. **1 plane**: all normal vectors $\mathbf{N_1}$ have the same orientation. The Gaussian image **S*** is a point. **2 spherical cap**. **3 circular sector** of a slightly curved surface: small Gaussian image **S*** on the sphere. **4 circular sector** of a stronger curved surface: larger Gaussian image **S*** on the sphere

Special cases of Gaussian curvature are:

- **Planes** have Gaussian curvature $\mathbf{C} = 0$ (Fig. 1.95, case **1**), because both principal curvatures $\mathbf{c_1}$ and $\mathbf{c_2}$ are zero.
- **Uniaxially curved surfaces** have Gaussian curvature $\mathbf{C} = 0$, because one of the two principal curvatures $\mathbf{c_1}$ and $\mathbf{c_2}$ is necessarily zero. However, this can also be the case with singular points of biaxially curved surfaces, for example, at vertices of a torus surface. Points with a principal curvature equal to zero are referred to as **parabolic points** or **flat points** with respect to the curvature of the surface within them.

E.g. the points $\mathbf{P_2}$ and $\mathbf{P_4}$ on the torus surface on Fig. 1.106

- **Spherical surfaces** have equal principal curvatures c_1 and c_2, which are each equal to 1/**r** (▪ Fig. 1.95, case **2**); the Gaussian curvature of a sphere is therefore always:

$$c_1 \cdot c_2 = c^2 = 1/r \cdot 1/r = 1/r^2$$

- If both principal curvatures c_1 and c_2 point in the same direction in space, both values have the same sign; the Gaussian curvature is therefore positive. These are **synclastically** curved surfaces. The points on this type of surface are called **elliptical points** with respect to the curvature of the surface within them (▪ Fig. 1.95, cases **2–4**).
- If both principal curvatures c_1 and c_2 have opposite directions, both values have different signs; the Gaussian curvature is therefore negative. These are **anticlastically** curved surfaces or saddle surfaces. The points on this type of surface are called **hyperbolic points** with respect to the curvature of the surface within them (▪ Figs. 1.93 and 1.94).

Median Curvature

The mean or median curvature C_m is the arithmetic mean value of the two principal curvatures c_1 and c_2 (▪ Figs. 1.96 and 1.97).

$$C_m = (c_1 + c_2)/2$$

As a result, surfaces with equal, oppositely oriented principal curvatures have a mean curvature equal to zero. These are so-called **minimal surfaces**, which occur in the physical model of soap skins and are of particular importance in membrane construction.

► *Sect.* 4.3.3.2 *Membrane and Rope Structure, Mechanically Tensioned*, particularly ► ▪ Fig. 4.232, left. Cf. also ► *Sect.* 3.4.4.3 *Form-Finding of Surface Structures under Membrane Forces*

1.2.3 Regular Surface Types

Regular curved surfaces can be classified, for the present purposes, in a simplified way according to:

- the **nature** of its curvature;
- the possibility of transforming them into a plane, or **developing** them;
- its **law of generation**.

The subdivision into these large groups results from different perspectives. This means, for example, that surfaces which result from the same law of formation may have different types of curvature. Or surfaces with the same kind of curvature can be the result of different laws of generation simultaneously. Some surfaces can therefore be classified into different categories—hyperbolic paraboloids, for example, are at the same time ruled surfaces and translational surfaces.

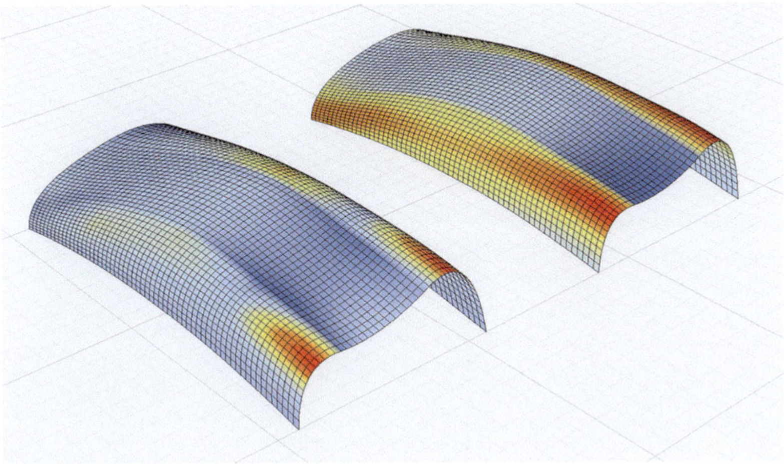

Fig. 1.96 Coloured representation of the Gaussian (left) and median (right) curvature of a free-form surface using a CAD program

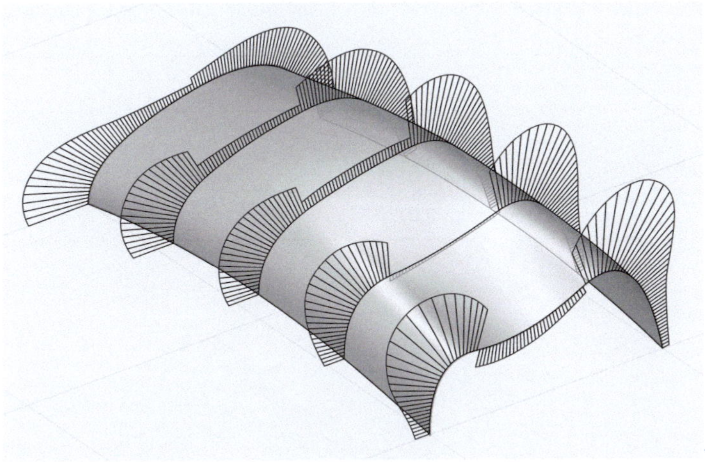

Fig. 1.97 Graphic representation of the median curvature of a free-form surface on five surface curves using a CAD program

1.2.3.1 By Type of Curvature

A difference is made between **uniaxially** or **one-directionally** (singly) or **biaxially** or **two-directionally** (doubly) curved surfaces. In the case of uniaxial curvature, the tangential plane defined at each point is simultaneously tangential to the surface along a straight line contained in the surface (Fig. 1.98). In other words, it is always possible to define at each point a straight line in the surface, all points of which always have the same tangential plane. This is not the case with biaxially curved surfaces: The tangential plane, in that case, touches the surface only at this precise point, and there is no

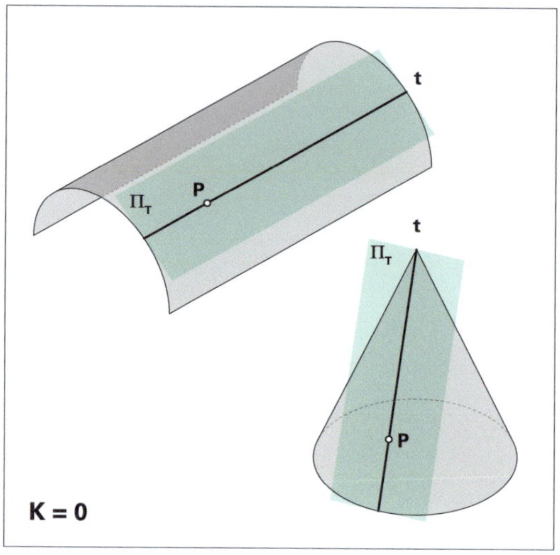

Fig. 1.98 In the case of **uniaxially curved** surfaces, for each tangential plane $\prod_T$ at any arbitrary point **P**, a tangent **t** can be found running through **P** at which the plane touches the surface tangentially. Its Gaussian curvature **C** is zero

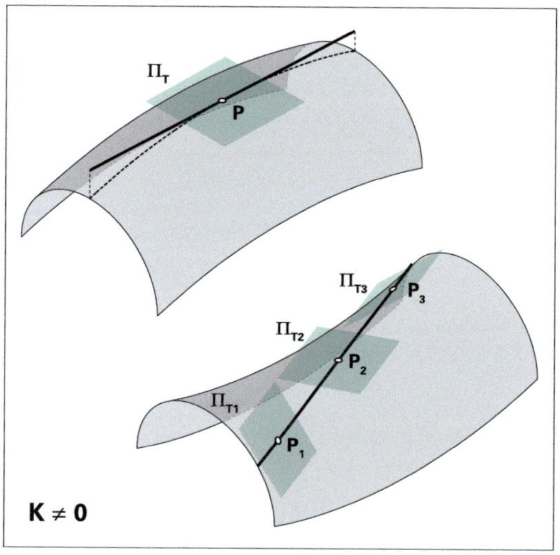

Fig. 1.99 With biaxially curved surfaces, either no straight line **t** through **P** which is contained in the surface (above) can be found from the outset; or otherwise, the points $\mathbf{P_i}$ of such a line always have different tangent planes ($\prod_{T1}$, $\prod_{T2}$, $\prod_{T3}$, ...) (bottom). Their Gaussian curvature **C** is not equal to zero (except for singular parabolic points, in some cases)

straight line contained in the surface in the first place (Fig. 1.99 top); or the points of a straight line possibly contained in the surface each have different tangential planes (Fig. 1.99 bottom).

See above ▶ *Sect. 1.2.2.2.1 Gaussian Curvature*

As discussed above, uniaxially curved surfaces always have a Gaussian curvature equal to zero.

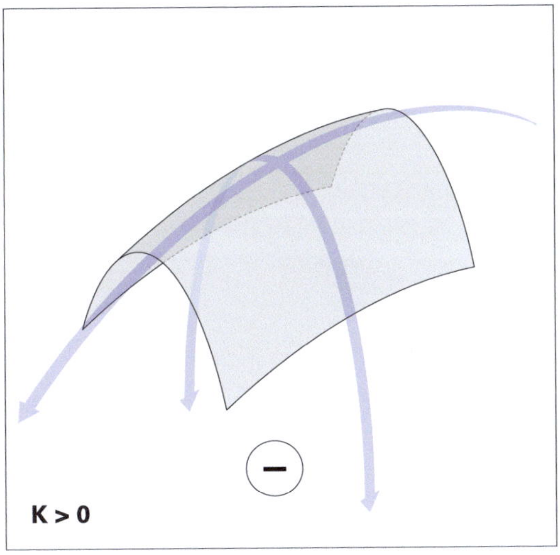

Fig. 1.100 Uniformly or synclastically curved surface. Its Gaussian curvature **C** is greater than zero

From this definition, it follows that all surfaces which are uniaxially curved are **ruled surfaces**. However, not all ruled surfaces have uniaxial curvature. Biaxially curved surfaces are also distinguished in:

See definition below in ▶ *Sect. 1.2.3.3 By Generation Law*

- **uniformly** double-curved (**synclastic**) surfaces (Fig. 1.100). The curvatures each point into the same half-space. These surfaces are approximately **dome-shaped**; their Gaussian curvature is greater than zero.
- **inversely** double-curved (**anticlastic**) surfaces (Fig. 1.101). The curvatures are each oriented towards different half-spaces. These surfaces have an approximate saddle shape. Their Gaussian curvature is smaller than zero.

Surfaces common in construction featuring a uniaxial or simple curvature are:

- **cylindrical** surfaces;
- **conical** surfaces.

Those with biaxial or double curvature:

- **spherical** surfaces or their cut-outs in the form of caps, such as **dome forms**;
- **saddle** surfaces such as **hyperbolic paraboloids**;
- **hose** or **toral** surfaces, etc.

In principle, uniaxially curved surfaces can be built with significantly less effort than biaxially curved ones. This can be attributed to the fact that they can always be produced from straight base elements, in approximation, even from planar pieces composed to form a polyhedron, in special cases, even

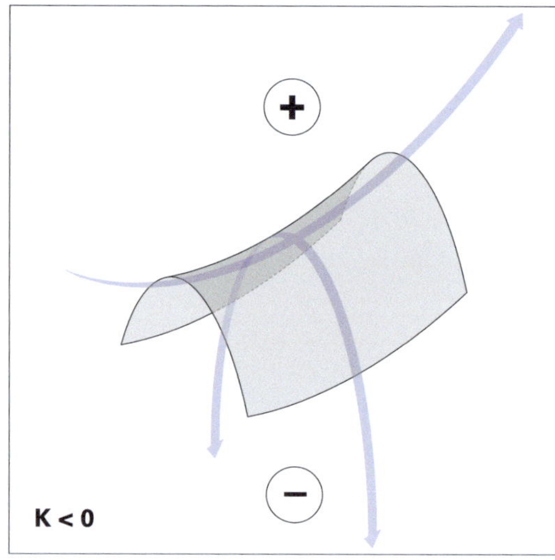

Fig. 1.101 Inversely or anticlastically curved surface. Its Gaussian curvature **C** is smaller than zero

from plates bent, in turn, to an uniaxially curved shape. Biaxially curved surfaces, on the other hand, can be produced from these elements only in exceptional cases, and then often only as rough approximations.

► *Sect. 1.3.2 Composing Biaxially Curved Surfaces*

1.2.3.2 By Developability on the Plane

In order for a curved surface to be unwound or developed onto a plane, it must have a **uniaxial curvature**. In principle, double-curved surfaces **cannot** be developed onto the plane. As already stated above, developability significantly facilitates the constructional implementation of a surface. Developable surfaces are always **torses**.

► *Sect. 1.2.3.1 By Type of Curvature*

Definition in ► *Sect. 1.2.3.3.2 Ruled Surfaces*

1.2.3.3 By Generation Law

Regarding the way in which surfaces are generated following a geometric rule, a distinction is made between the following variants.

Surfaces of Revolution

Surfaces of revolution are produced by rotation of a **generatrix**—a straight line, a planar, or a non-planar curve—around a **rotation axis**. Each plane perpendicular to the axis of revolution always cuts the surface of revolution in a circle, a so-called **circle of latitude** or **parallel**. Each plane containing the rotation axis always intersects the surface in an identical curve, the **meridian**.

Special surfaces of revolution are:

- the **circular cylinder**: If a straight line rotates around a straight line parallel to it, a circular cylinder (Fig. 1.102) is created. Each circle of latitude is identical in this case.

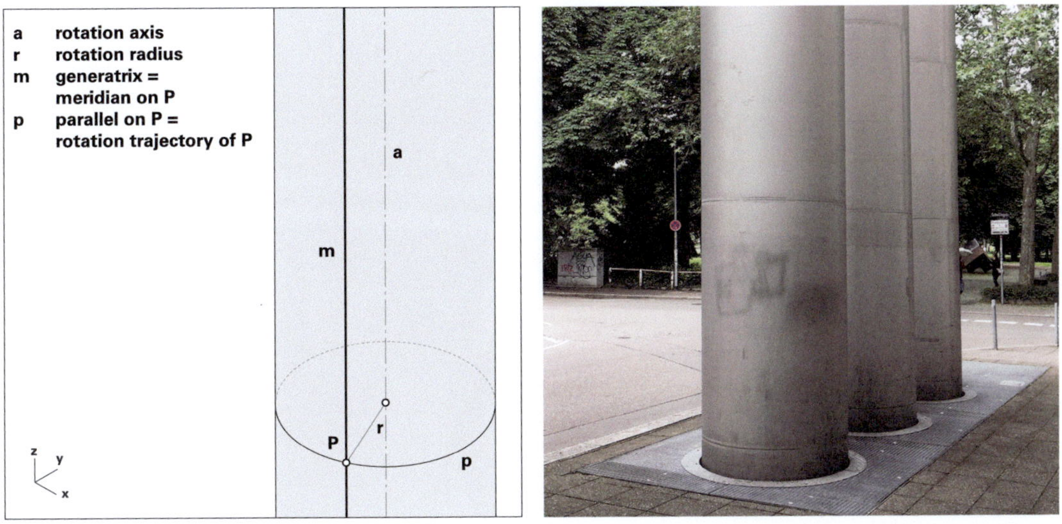

Fig. 1.102 (Left) Surface of revolution: circular cylinder. (Right) Circular cylinder made of sheet metal. The latitudinal circles are reflected in the weld seams

The meridian is a straight line (**generatrix line**). All meridian or sheath lines are parallel to one another in space. Being a special case of the generic cylindrical surface, the circular cylinder can also be classified as a **ruled surface**.

- the **circular cone**: If a line rotates around another one intersecting it, a circular cone emerges (Fig. 1.103). The meridians are straight lines (generatix lines) which intersect the axis in a single point, the cone vertex, so that they are not parallel to one another. The circular cone can also be regarded as a **ruled surface**.
- the **sphere**: If a circle rotates about a straight line contained in its plane and passing through its centre, a sphere (Figs. 1.104 and 1.105) is created. Meridians, like the parallels, are circles with a centre point on the rotation axis. Any plane that intersects the sphere creates a circle as a sectional line. Spheres occur in building construction in particular, as partial surfaces, usually as spherical caps.
- the **tubular** or **toral** surface: If a circle rotates around a straight line contained in its plane which does not pass through its centre, a tube or torus is created (Fig. 1.106). Meridians are thus either complete circles (open torus, Fig. 1.106 right; Fig. 1.107) or a circular section (closed torus, Fig. 1.106). Circles of latitude are either simple (closed torus) or double and concentric (open torus).

Surfaces of revolution which are generated when regular, mathematically easy-to-define curves such as conic sections rotate about an axis, also have some structural significance. They are a:

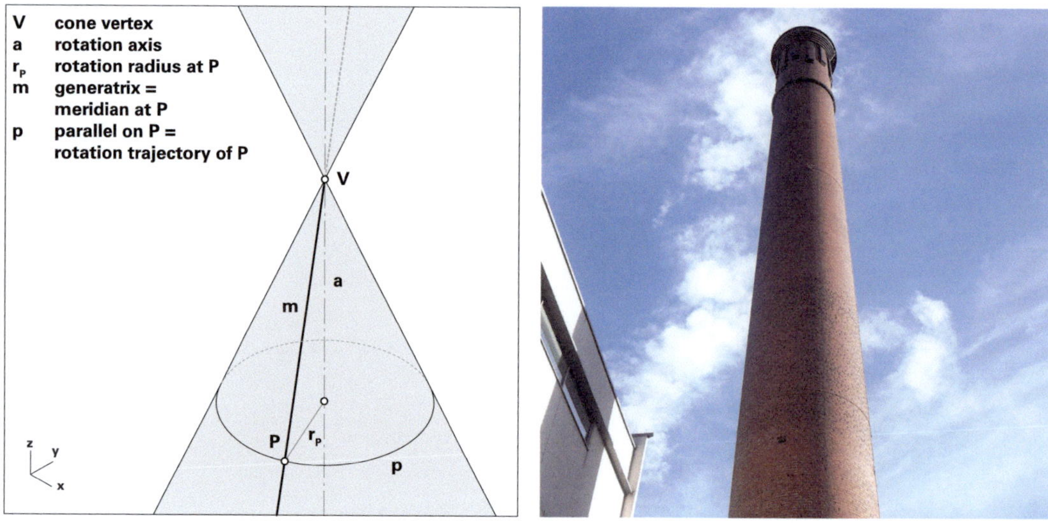

Fig. 1.103 (Left) Surface of revolution: circular cone. (Right) Circular conical brick chimney (truncated cone; cut-off tip)

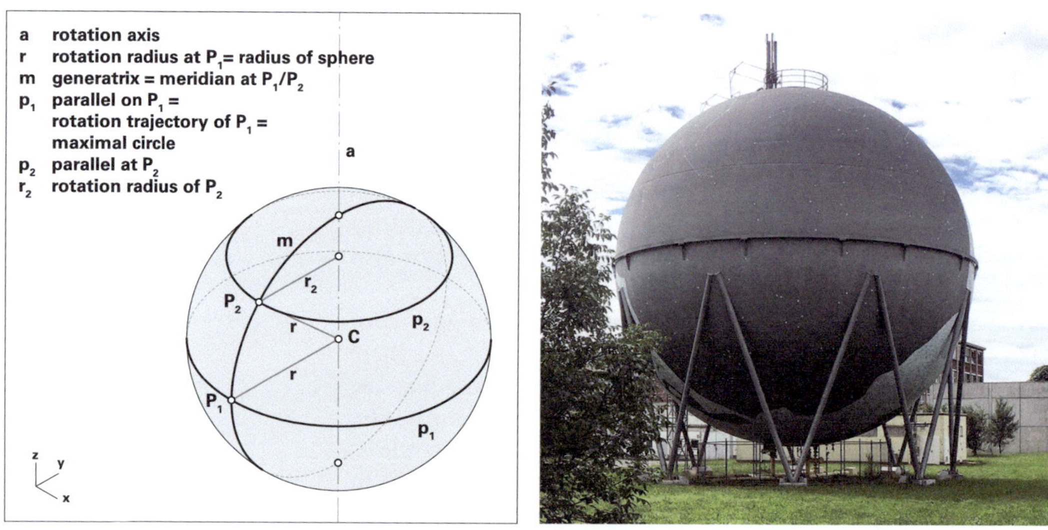

Fig. 1.104 (Left) Surface of revolution: sphere. (Right) Spherical reservoir

- **paraboloid of revolution**: If a parabola rotates about its axis, a paraboloid of revolution (Fig. 1.108) is created.
- **single-body hyperboloid of revolution**: If two hyperbola branches turn around their common axis of symmetry, a single-body hyperboloid of revolution (Fig. 1.109) is created. Meridians are therefore always the same hyperbola. A special feature of this surface is the fact that it can also be defined as a **ruled surface**: If a straight line that is not intersecting with the axis and runs oblique to it rotates, the same type of surface is produced as when rotating a hyperbola (Fig. 1.119).

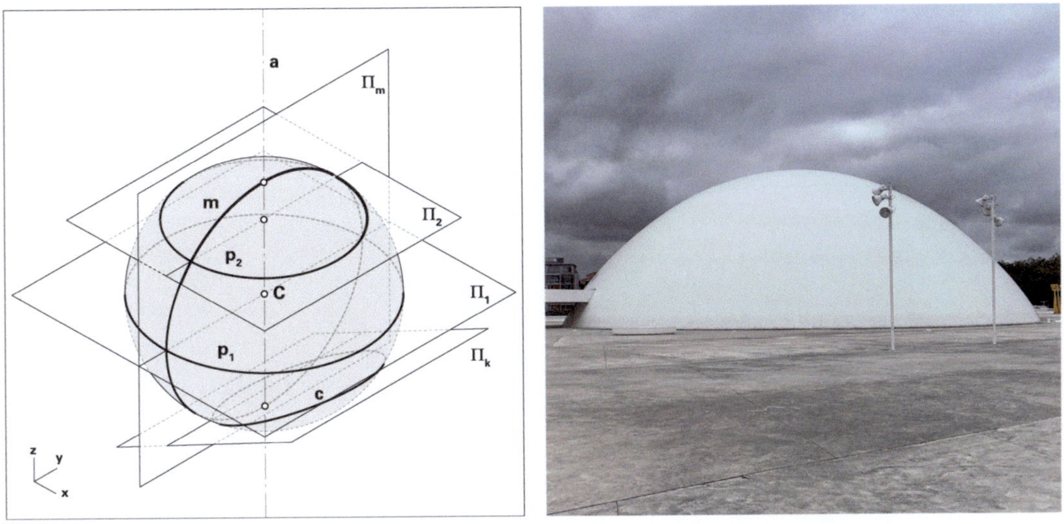

Fig. 1.105 (Left) Plane sections through a sphere create circles: maximal circles when the plane (Π_m, Π_1) passes through the centre point **C**. (Right) In construction, the sphere often appears as a cap, as in this dome. The flat section at ground level, i.e. the footprint, results in a circle

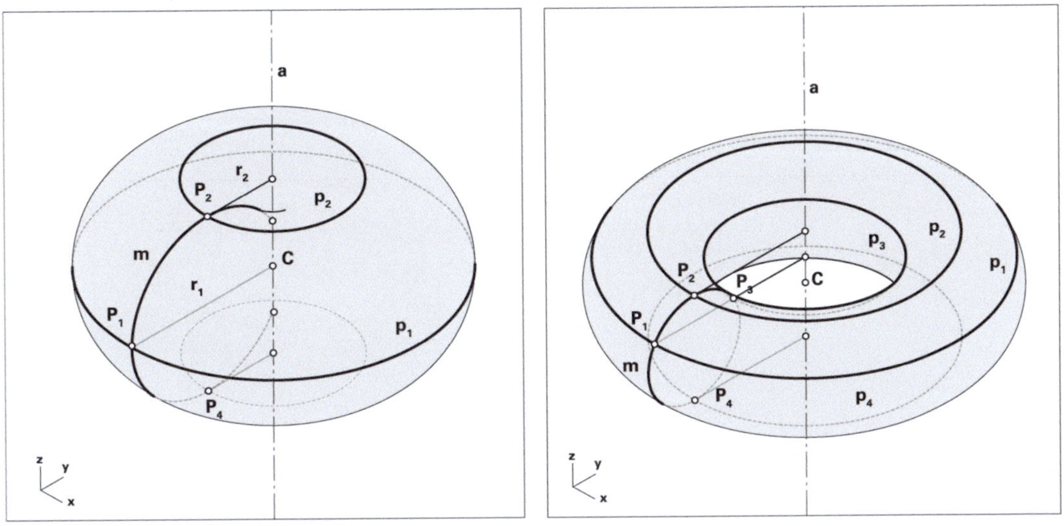

Fig. 1.106 (Left) Surface of revolution: closed torus. The generatrix is the circular section **m**. (Right) Surface of revolution: open torus. The generatrix is a complete circle **m**

- **double-body hyperboloid of revolution**: If both hyperbolas rotate around the common axis, a double-body hyperboloid of revolution (Fig. 1.110) is created.
- **ellipsoid of revolution**: If an ellipse rotates around one of its main axes, an ellipsoid of revolution (Figs. 1.111 and 1.112) is created.

Ruled Surfaces

Ruled surfaces are created by the movement of a generating line, or **generatrix**, in space, for example, by sliding along a guiding curve or **directrix** (Fig. 1.113).

Fig. 1.107 (Left) Open toroidal surface: ring vault (Palace of Charles V in the Alhambra of Granada). (Right) Open toroidal surface: column base

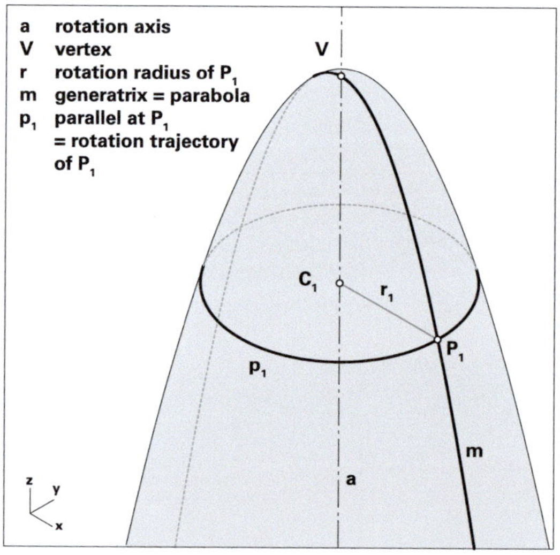

Fig. 1.108 Paraboloid of revolution

Special ruled surfaces are those in which the tangential plane remains unchanged along a straight generating line or straight generatrix. These surfaces are called **torses**—example: cylinders, cones, or surfaces tangential to a spatial curve.[7] Neighbouring generatrices lie either in one plane, i. e. they intersect at a point, as in the cone, or are spatially parallel to

7 Another surface that can be developed is the oloid, a ruled surface that leans against two equal guide circles (► ► https://de.wikipedia.org/wiki/Oloid).

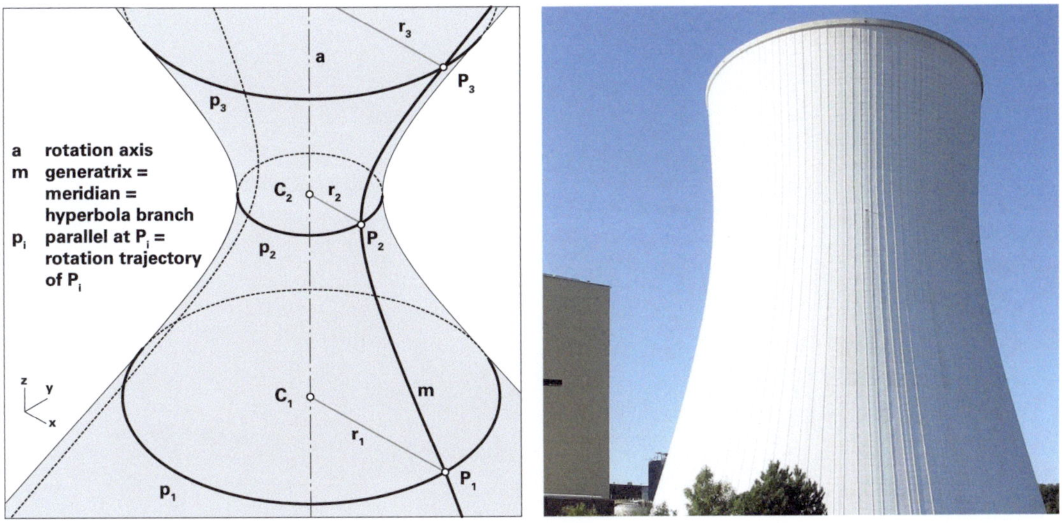

Fig. 1.109 (Left) Single-body hyperboloid of revolution from the rotation of a hyperbola. (Right) Cooling tower: single-body hyperboloid of revolution

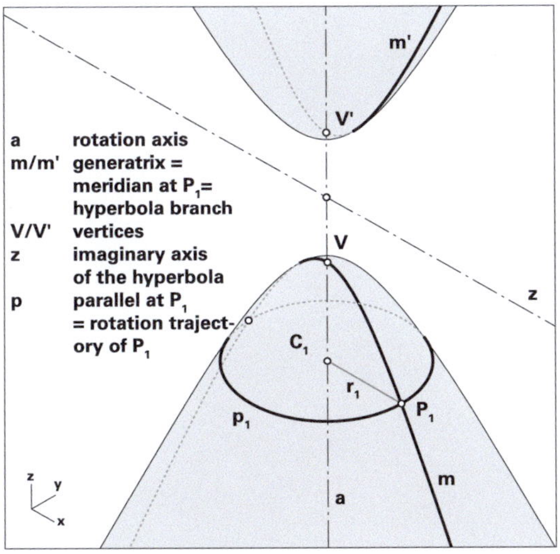

Fig. 1.110 Double-body hyperboloid of revolution

each other, i.e. they intersect at a point at infinity, as in the cylinder, for example, or they lie tangential to a spatial curve (Fig. 1.114). Torses have **uniaxial curvature** and can be **developed** into the plane.

In the case of the remaining ruled surfaces, the tangential plane of a point **P** changes when it shifts along a generating straight line; the tangential plane thus turns continuously around this mantle line. In other words, neighbouring generating lines are **skewed** against each other. This distinction between torses and generic ruled surfaces, i.e. the question of whether a surface can be **developed**, has fundamental significance for constructional applications.

► *Sect. 1.3 Constructional Implementation of Continuous Curved Layer Surfaces*

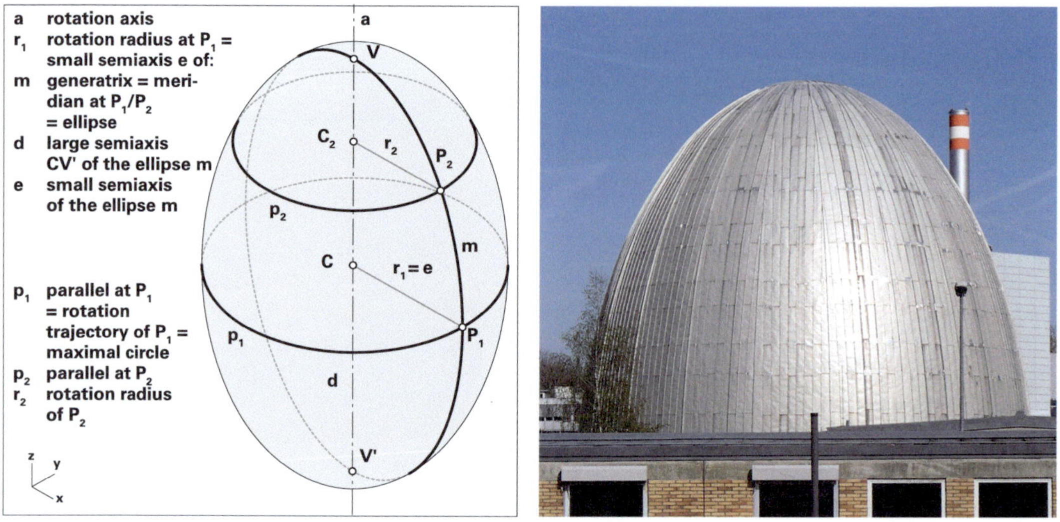

Fig. 1.111 (Left) Ellipsoid of revolution. (Right) Reactor encasing in the form of an ellipsoid of revolution

Fig. 1.112 Baroque dome shaped as an ellipsoid of revolution bisected along the main axis

The most common ruled surfaces for constructional use are:

- **generic cylindrical surfaces**: A straight line slides with a constant direction along a directrix (Figs. 1.115 and 1.116).
- **generic conical surfaces**: Straight lines always passing through a fixed point slide along a directrix (Fig. 1.117).

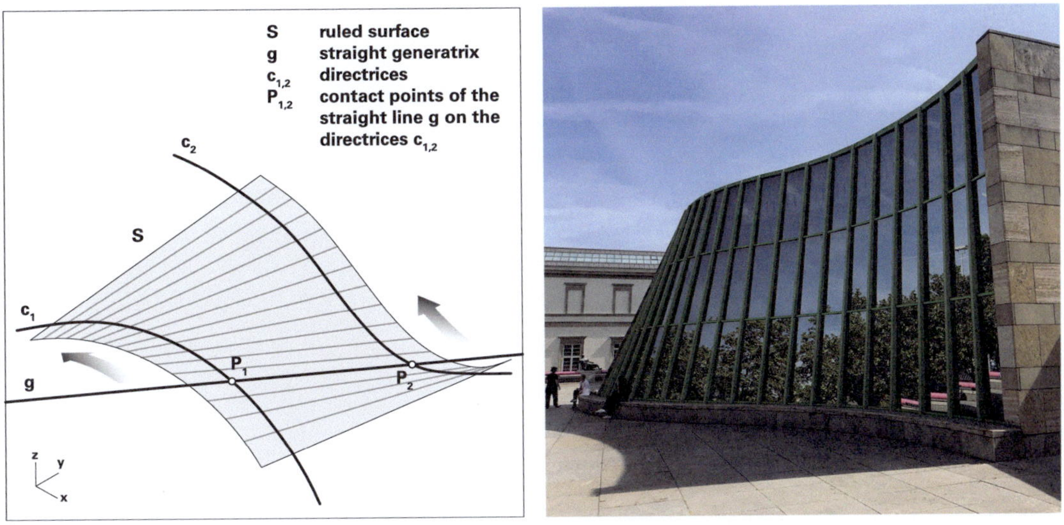

Fig. 1.113 (Left) Generic ruled surface. (Right) Glass façade with the shape of a generic ruled surface. The rectilinear mullions follow the straight generatrix lines. They are not parallel to each other. The resulting glass facets are therefore warped (not planar). For this reason, offsets are introduced at the transom joints, which can be seen in the shadows and reflections. (Staatsgalerie Stuttgart; arch.: J Stirling)

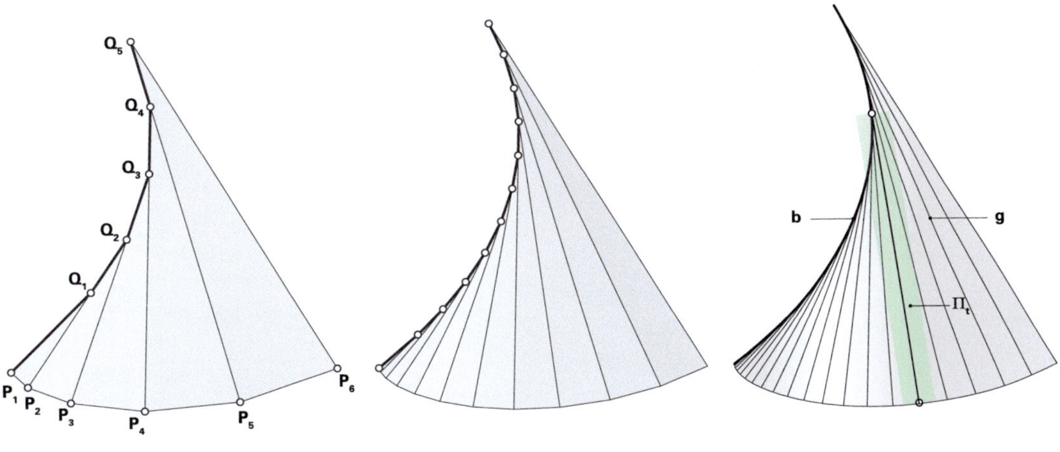

Fig. 1.114 A **surface tangential to a spatial curve** is uniaxially curved: The polyhedron on the left consists of flat triangular partial surfaces, since their sides intersect at one point each; for example, $\mathbf{P_1Q_1}$ intersects with $\mathbf{P_2Q_2}$ at point $\mathbf{Q_1}$. By progressive subdivision of the polyhedron surface (centre) one arrives at a continuously curved ruled surface with generatrices **g**, which lie tangential to a spatial boundary curve **b**. A tangential plane $\prod_t$ to one of the generatrices lies tangential to the surface at all points of this line and thus fulfils the condition of developability

- **conoidal surfaces**: A straight line slides along a curved directrix and a straight one (Fig. 1.118).
- **single-body hyperboloid**: A straight line slides along two ellipses or, in a special case (**a** = **b**), along two circles (Fig. 1.119). Or also: hyperbolas slide along a directrix, namely an ellipse or a circle, situated at the vertex of the hyperbola (Fig. 1.109).

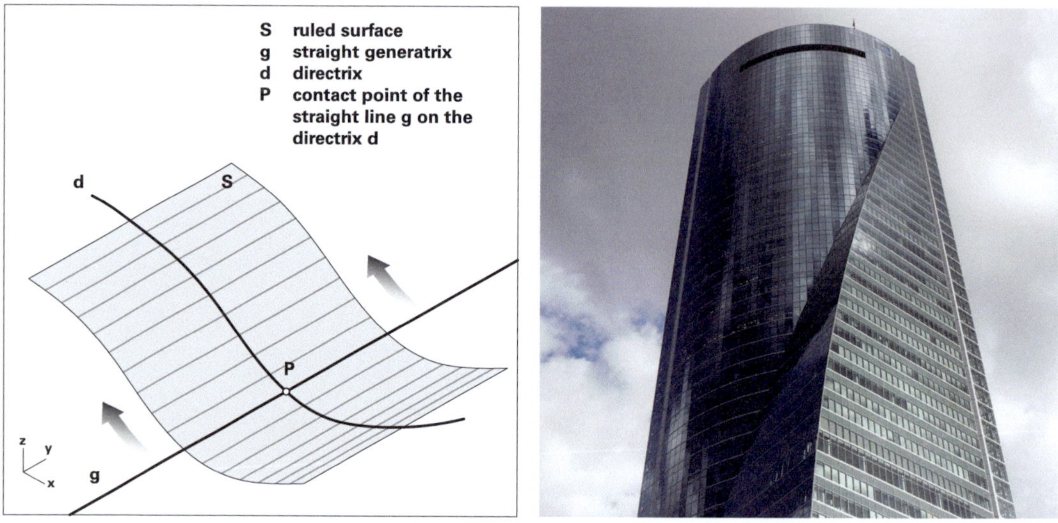

Fig. 1.115 (Left) Generic cylindrical surface. (Right) The tower's main façade and the intersecting curved side façade are both cylindrical surfaces. The vertical mullions of the main façade follow the straight generatrix lines, which run parallel to each other. On the intersecting curved side façade, it is the floor-slab edges that follow the generatrices. The surfaces, which have only uniaxial curvature in their entirety, can easily be faceted to generate planar partial surfaces, which are filled with flat glass

Fig. 1.116 Building consisting of two coupled cylindrical surfaces. The straight generatrix lines are parallel to the straight edge (Curitiba Museum, arch.: O. Niemeyer)

- (special) **helicoidal surface**: A straight line slides along a helix line (Fig. 1.137).
- **hyperbolic paraboloid**—also referred to as HP or hypar surface: A straight line that slides along two guiding lines skewed against each other in space (Figs. 1.120, 1.121, 1.122, and 1.123).

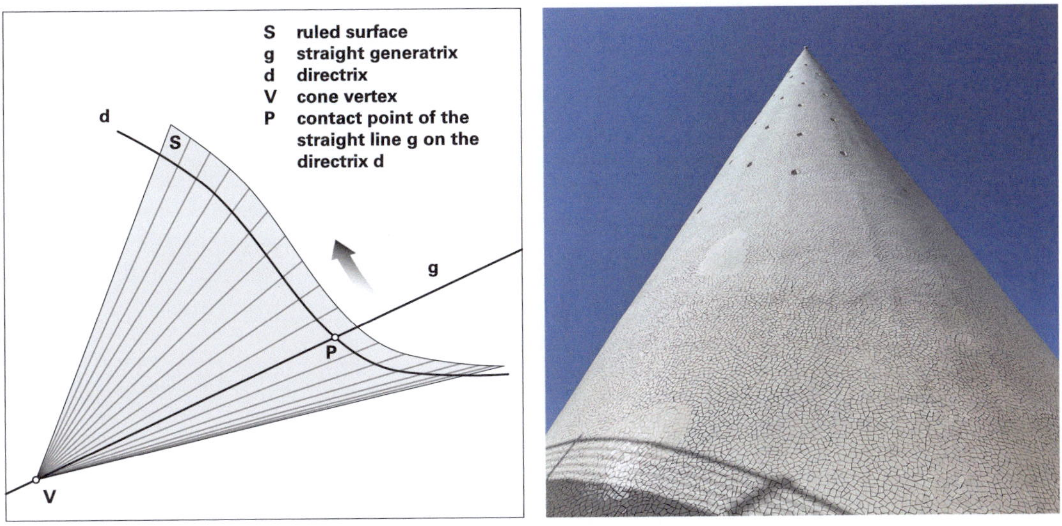

Fig. 1.117 (Left) Generic conical surface. (Right) Conical roof

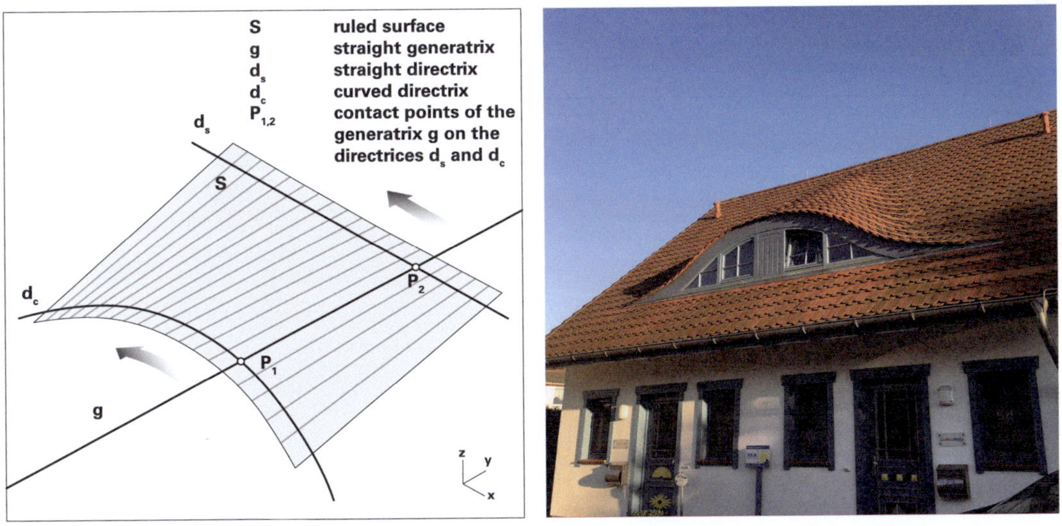

Fig. 1.118 (Left) Conoidal surface. (Right) Approximately conoidal-shaped dormer window

Translational Surfaces

Translational or sliding surfaces are generated by the parallel displacement of a **generatrix** curve on a **directrix** curve (Fig. 1.124). The generatrix and the directrix always intersect at one point. In the course of the displacement, the generatrix can remain unchanged or can also be continuously modified according to a defined law—by scaling or stretching.

In a special case, the generatrix or the directrix can be a straight line. In this case, a **ruled surface** is produced, namely a **cylindrical** surface if the generatrix remains unchanged, or a **conical** surface if the generatrix is continuously scaled. If both

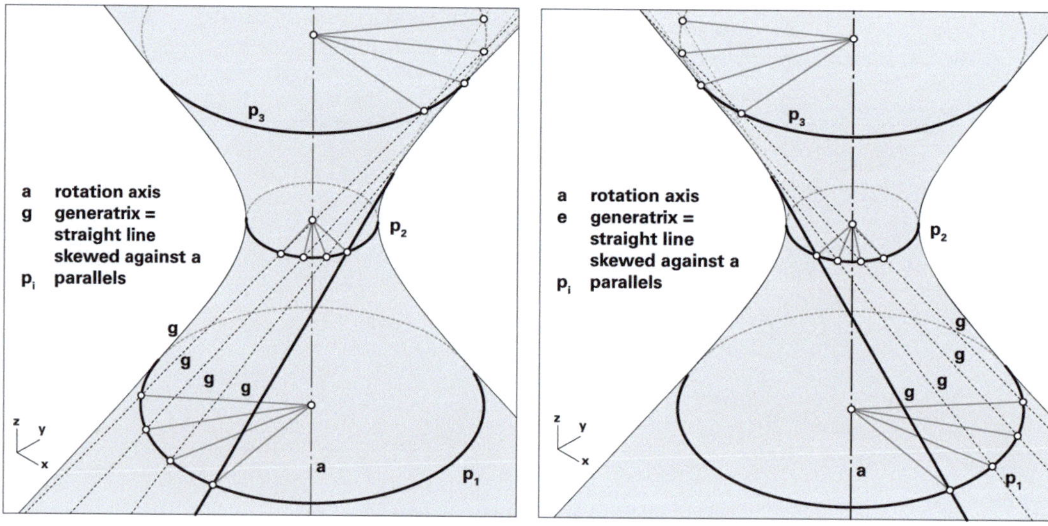

Fig. 1.119 Single-body hyperboloid of revolution from the rotation of a skewed straight line (cf. ruled surfaces), in each case in one of two opposing directions

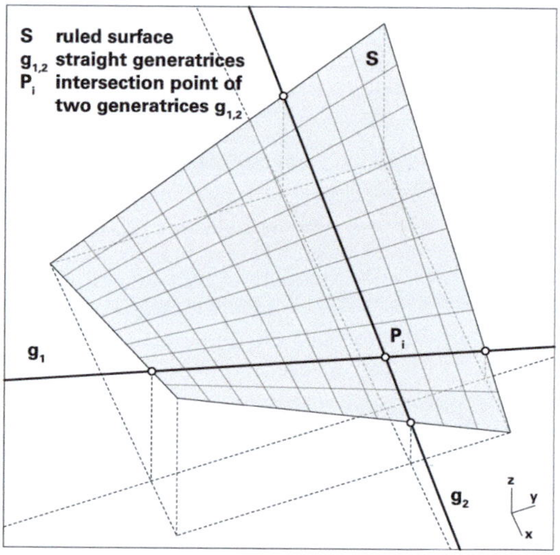

Fig. 1.120 Hyperbolic paraboloid as a product of two intersecting sets of straight lines

the generatrix and the directrix are each a straight line, a **plane** is created.

It is clear that by combining the simplest curves as generatrices and directrices, a very large range of translational surfaces is created. Some among them usual in building construction are the following:

- **cylindrical surfaces**: parallel shifting of a constant generatrix on a directrix (Fig. 1.115);
- **conical surfaces**: parallel displacement of a continuously scaled generatrix on a straight line (Fig. 1.117);

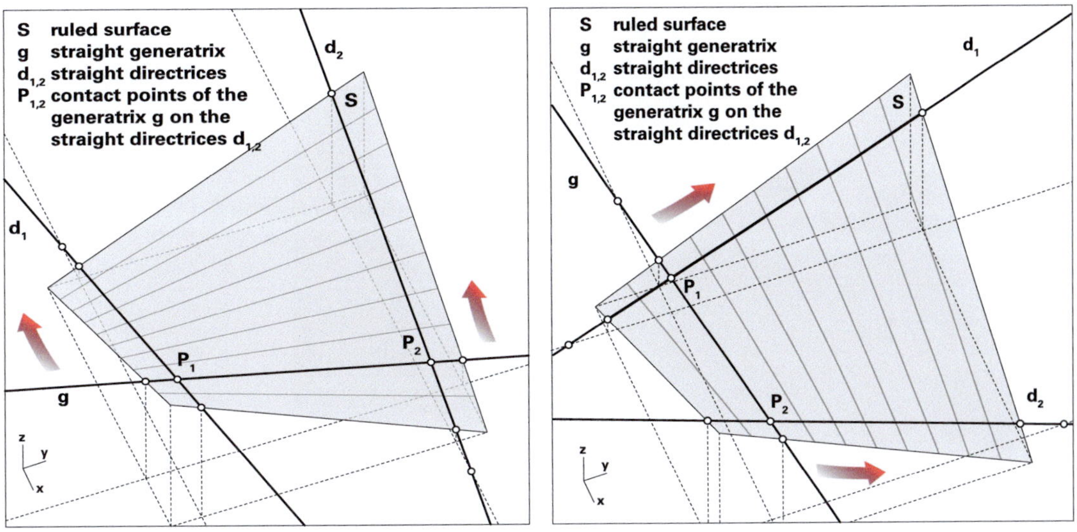

Fig. 1.121 (Left) Hyperbolic paraboloid as a ruled surface. A generatrix **g** glides along two skewed directrices $\mathbf{d}_{1,2}$. (Right) Direction reciprocal to the one of the illustration on the left

Fig. 1.122 Concrete shells shaped as a hyperbolic paraboloid. The ruled-surface characteristic is recognisable by the clearly visible traces of the rectilinear formwork boards

- **hyperbolic paraboloid**: is generated moving a parabola along another parabola (Figs. 1.125 and 1.126). Planar sections of the surface parallel to the coordinate planes **xz** yield **congruent parabolas** (Fig. 1.127 left), and likewise do those parallel to **yz** (Fig. 1.127 right). Planar sections parallel to **xy** yield **hyperbolas**—hence the term *hyperbolic* paraboloid—in the special case of the plane **xy** itself, two intersecting straight lines (Fig. 1.128).

Translational surfaces are of great importance for constructional applications (Figs. 1.129 and 1.130). Although they

1

■ **Fig. 1.123** Hyperbolic paraboloid cut in a straight line along generatrix lines (section as in ■ Figs. 1.120 and 1.121)

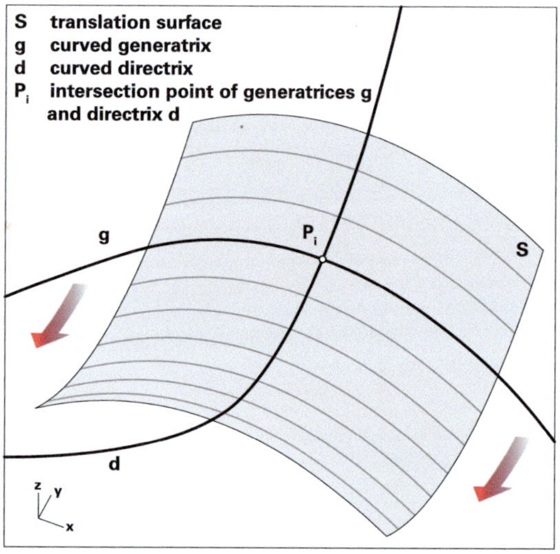

■ **Fig. 1.124** Generic translational surface. A generatrix **g** glides along a directrix **d**

feature a double curvature and are not developable, i.e. are not torses, they are nevertheless suitable for being converted into a **polyhedral surface** made from planar partial surfaces. Although this is only an approximation to the continuously curved translational surface, the difference, with small curvatures, sufficiently small-scale faceting, and the large scales customary in building, often can hardly be recognised with the naked eye.

In detail: both the generatrix and the directrix are transformed into **polygons** or **funiculars** (■ Fig. 1.131). Two adjacent corners of the generating polygon **A** and **B** are imaged by a parallel displacement or translation into the two points **A′** and **B′** (■ Fig. 1.132). It is true that the distances **AB** and

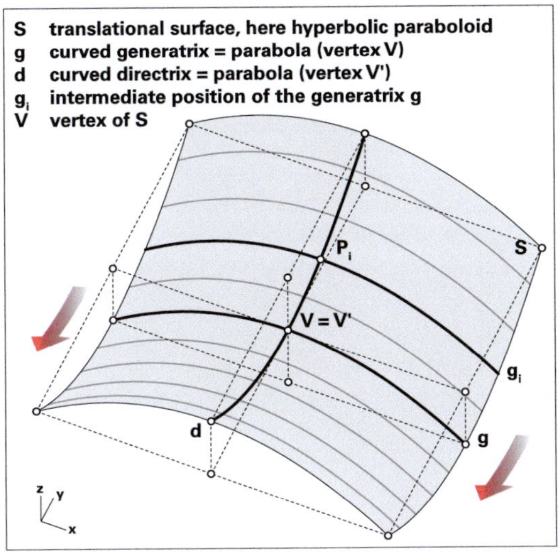

Fig. 1.125 Example of a translational surface generated from two parabolas. The generatrix **g** traces along **d** on its vertex **V**. Surface cut-out centred on the vertex **V** of the surface (= vertex **V** of the generatrix). It is a hyperbolic paraboloid

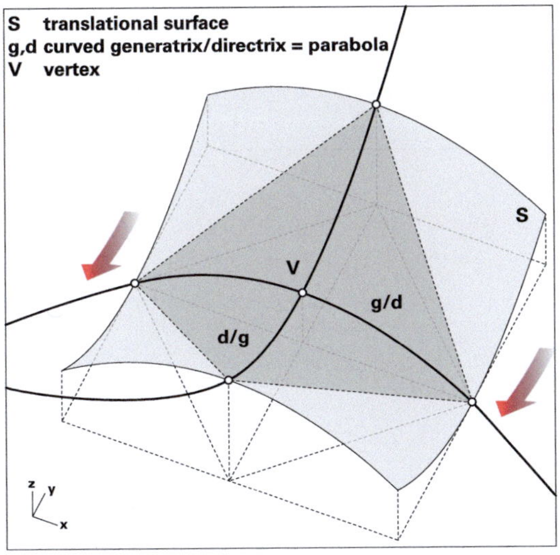

Fig. 1.126 Hyperbolic paraboloid as a translation surface with a parabola **g** as generatrix and a parabola **d** as a directrix. Each can be alternately considered as generatrix or directrix, respectively

A′B′ are spatially parallel to one another since they are components of mutually congruent, only (parallelly) shifted polygons. Thus, the condition for the area **ABA′B′** being planar is already satisfied. Since the generating polygon remains unchanged during displacement, **ABA′B′** is, moreover, a parallelogram, since the line segments **AA′** and **BB′** are also parallel to one another (Fig. 1.133).

1

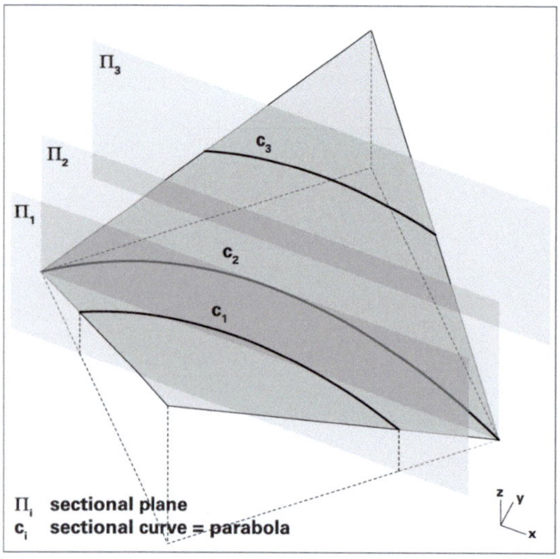

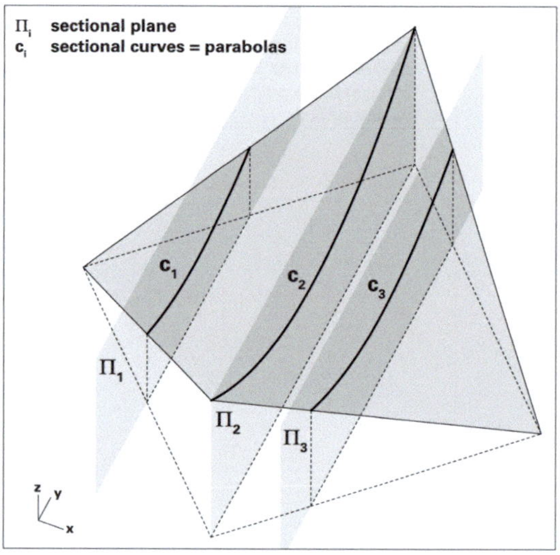

Fig. 1.127 (Left) Hyperbolic paraboloid: vertical planar sections yield parabolas. (Right) Direction reciprocal to the one on the left

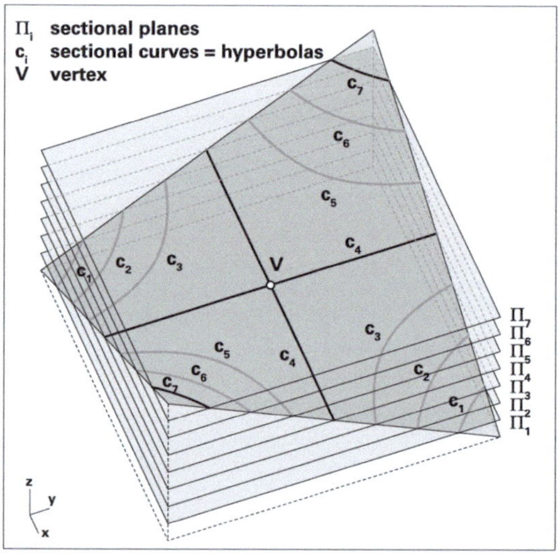

Fig. 1.128 Horizontal planar sections yield hyperbolas $\mathbf{c}_i$. The horizontal section through the vertex **V** yields two intersecting straight lines $\mathbf{c}_4$

Scaled Translational Surfaces

A translational surface can also be created by **shifting** and simultaneously **scaling** the generatrix (Fig. 1.134). It is also possible to only partially scale the generatrix, that is, a curve section thereof. A translational surface of this type can also be converted into a polyhedron. If the generating polygon is continuously scaled (Fig. 1.135), only **AB** and **A′B′** are parallel to each other, **AA′** and **BB′** lie on straight lines which intersect and thus lie on one plane forming an angle α with respect to one another, which is dependent on the scaling factor. The resulting partial surface **AA′B′B** is a (planar) **trapezoid**.

Fig. 1.129 Non-rectilinear section of a hyperbolic paraboloid (see diagram in Fig. 1.130) (Las Lomas de Cuernavaca Chapel, Cuernavaca, Mexico; arch: F Candela)

As a result, a very large variance of translational surfaces can be generated which can always be satisfactorily approximated in their constructional implementation by the use of **planar plate material** (Fig. 1.136).

Helicoidal Surfaces

Helicoidal surfaces are produced by the sliding-over of a generating straight line or curve on a **helix line**, which serves as a directrix (Figs. 1.137 and 1.138). A helix line is produced by the simultaneous **rotation** and **displacement** of a point around an axis or along an axis. A complete rotation of the point is assigned to the displacement by a fixed distance—the so-called **phase**. If the generatrix is a straight line, the special helicoidal surface can also be considered a **ruled surface**. Helicoidal surfaces, however, are not torses, so they **cannot** be **developed**.

1.2.4 Digital Methods of Surface Definition

The above-mentioned surfaces can be defined and described by means of their **law of generation**. They are often created by spatial displacement and, in singular cases, by simultaneous affine, i.e. similar, alteration of a starting line. In this way, certain singular sections of this surface can be described mathematically or geometrically from the beginning. In other words, dimensions suitable for factory execution can be **calculated** or, at least, **geometrically constructed** with sufficient accuracy. If characteristic sections of a surface are, for example, always a second-order parabola, any number of intermediate points can be determined by its function:

$$\mathbf{f(x) = a_1x^2 + a_2x + a_3}$$

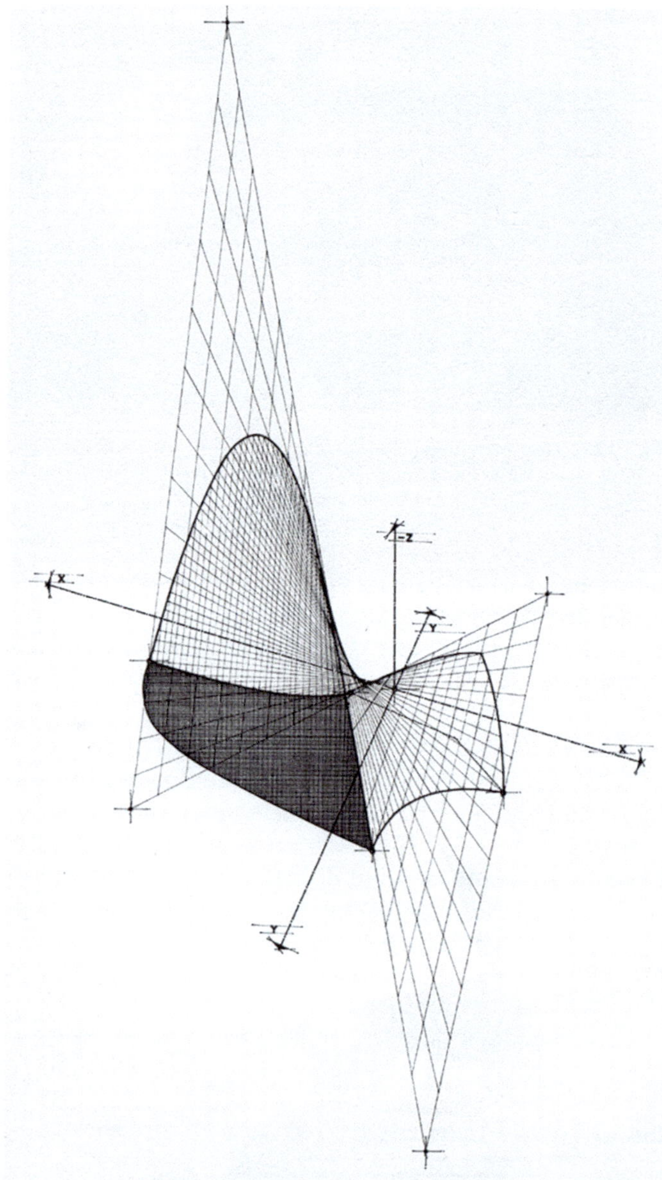

Fig. 1.130 Geometric derivation of the shell in Fig. 1.129. It results as a particular non-straight section, chosen according to design criteria, of the basic paraboloid shown here as a line drawing, itself cut in a straight line along the generatrix

after having defined the parameters $\mathbf{a}_1$ to $\mathbf{a}_3$. Even more simply, a surface can be produced by means of **artisanal means**, for example, by the gradual shifting of a reusable gauge—i. e. thereby creating a translational surface—on the construction site.

It is clear that a familiar law of origin which has been established in advance is a major facilitation for transferring designed forms into built structures. Furthermore, an exact knowledge of the geometry can also be a prerequisite for the

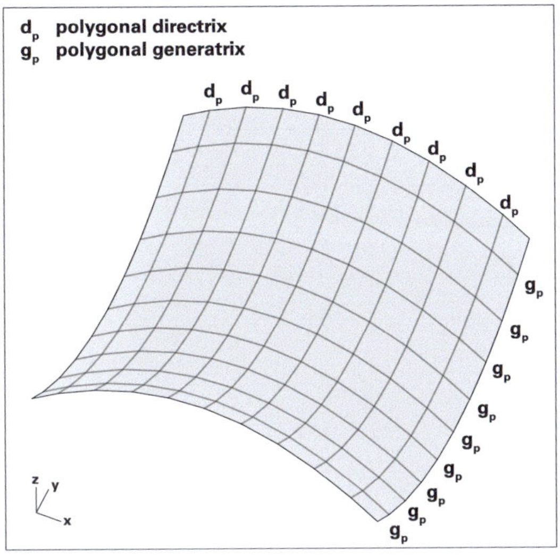

Fig. 1.131 (Left) Approximation to a translational surface by dissolving it into a polyhedron from single parallelograms, applied to the example on Figs. 1.124, 1.125, and 1.126. (Right) Grid shell: geometrically it is a translational surface. The glass facets are planar

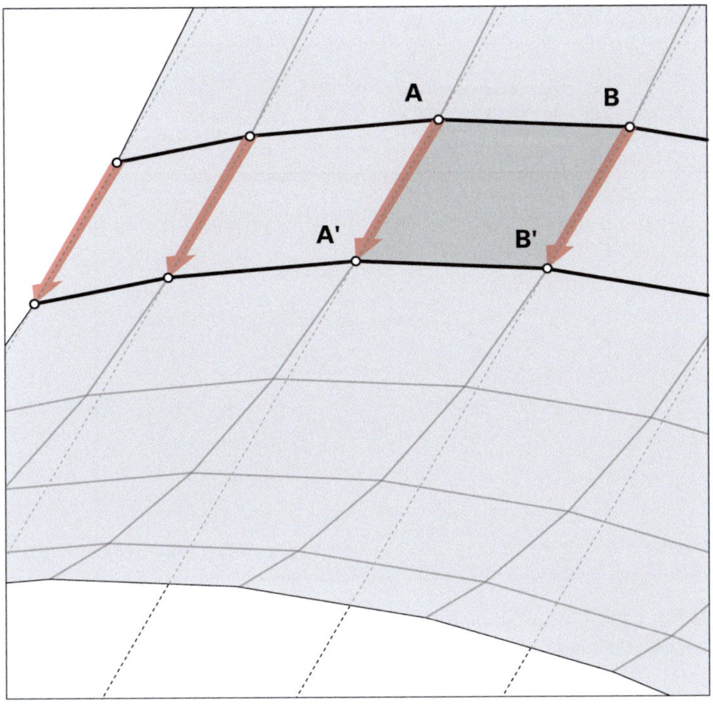

Fig. 1.132 Detail of the polyhedron on Fig. 1.131 with graphical representation of the translation process

structural modelling and pre-calculation of the **load-bearing behaviour** of such a shape.

This is the reason why surfaces of this type, within the group of curved forms, have been dominant in the construction industry for a long time and are so essentially still to-

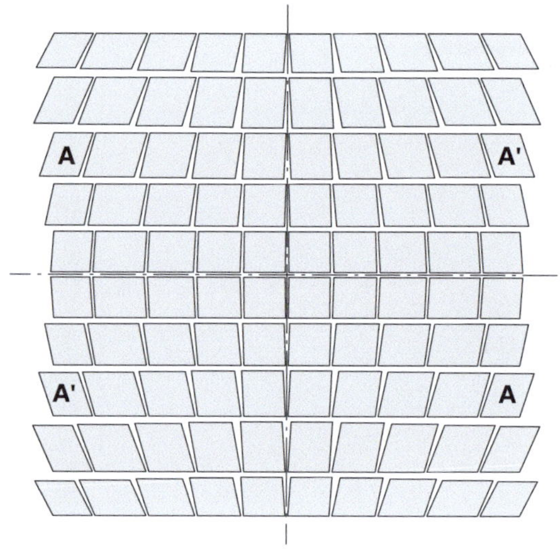

Fig. 1.133 The planar parallelograms of the polyhedron depicted on Fig. 1.131 (boundary fields are trimmed shorter) are developed here onto the plane. Since the vertex of the surface lies centrally in the section and both parabolas **g** and **d** are symmetrical, any element **A** occurs twice in identical and twice in mirrored form (**A′**). In a single quadrant, all elements are geometrically different from one another

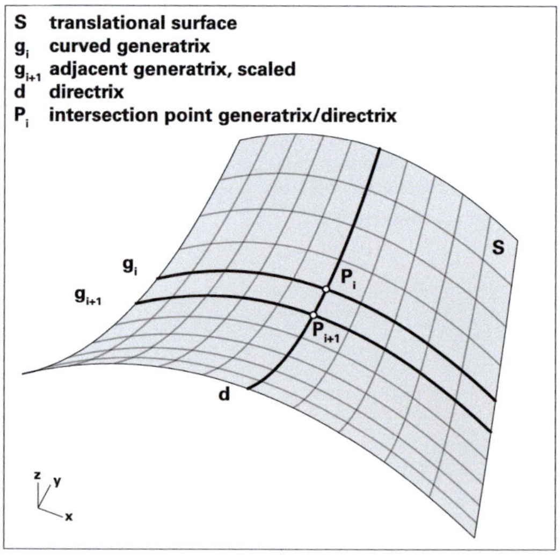

Fig. 1.134 Translational surface generated by the displacement and simultaneous scaling of a generatrix **g** along a directrix **d**

day. Although free forms were also often built in the past, two conditions had to be met for that purpose: Firstly, the highest level of manual craftsmanship was required, as has largely been lost today. A good example of traditional, purely handcrafted methods of building complex curved shapes is historical shipbuilding. In this craft, one procedure involved first defining rough guidelines by means of drawings, on the basis of which the surfaces were then manufactured by visual judgement (e. g. applying the hull's boarding along the formers);

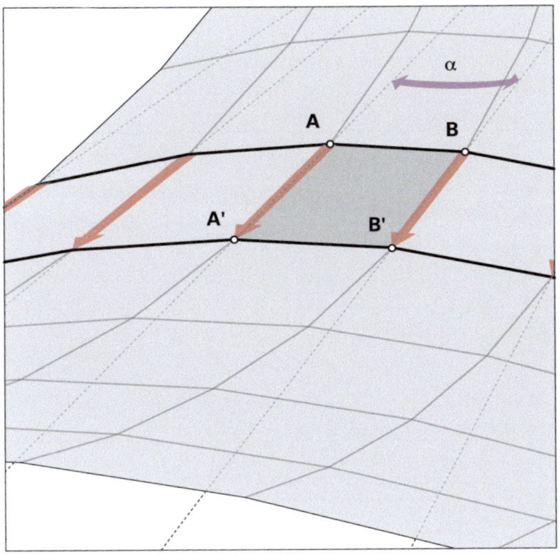

Fig. 1.135 Stretched translational surface converted into a polyhedron as on Fig. 1.134 (detail). **AA′B′B** is a trapezoid

Fig. 1.136 (Left) Translational surface enlarged by stretching along the curved guideline (see ridge line): lattice shell with tapering cross-section. Here, too, the glass facets are always planar (Berlin Central Station). (Right) General view of the lattice shell of the picture on the left

secondly, the other option was to define the free-form by means of a **scale-reduced model** and then record it for the production as precisely as possible by measuring it.

A fundamental change in these boundary conditions has resulted from introducing digital design aids such as **CAD** software for shape definition, computer-assisted **finite-element** calculation methods for determining the load-bearing behaviour, and software for automated production and assembly. Modern CAD programs allow the designer not only to virtually model conventional regular surfaces, as discussed in

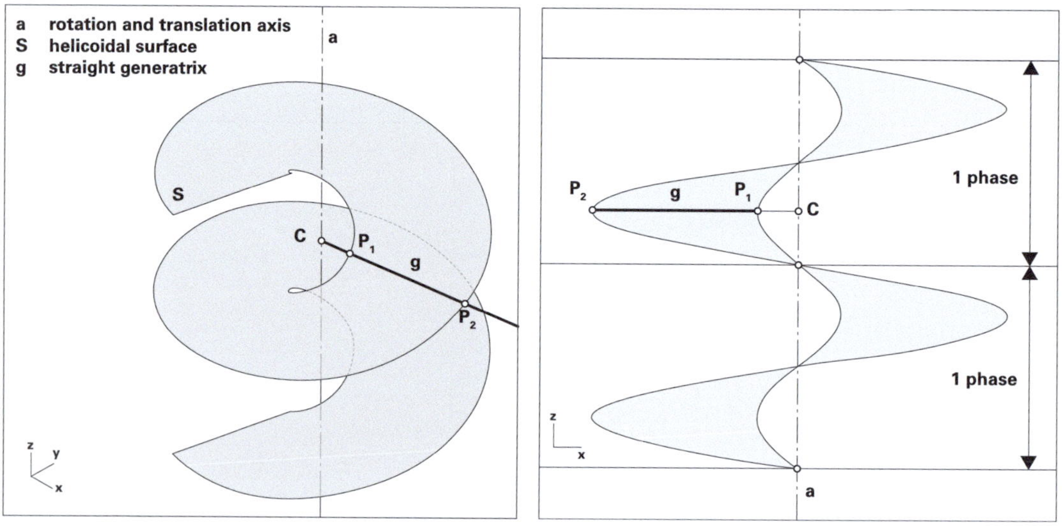

Fig. 1.137 (Left) Helicoid from the rotation and simultaneous displacement of a straight generatrix **g** along an axis **a**. (Right) Perpendicular elevation of the helicoid. The distance travelled in a complete 360° rotation is the so-called phase

Fig. 1.138 (Left) The soffit of this spiral ramp is a helical surface. The straight generatrix lines can be recognised by the formwork impressions. They are always aligned with the central axis of the supporting cylinder and intersect it at different heights (Niemeyer Centre, Avilés; arch: O Niemeyer). (Right) The flight of steps of this spiral staircase is a discretised approximation of a helical surface. The generatrix lines are also recognisable in the stepped edges of the soffit (Deutsches Historisches Museum, Berlin; arch.: I M Pei)

previous sections, in three-dimensional space, but also to generate discretionary surfaces by freely defining boundary conditions or even manually or interactively distorting a starting element. Even if, under these conditions, a mathematical definition of the surface is theoretically possible by defining a coordinate system and a characteristic mathematical function, this step is neither particularly effective nor compulsory for the designer because:

- the choice, a priori, of a mathematical function for generating a certain surface requires broad mathematical knowledge, which is usually not available to a sufficient extent to architects in the building practice, or can only be made usable in close cooperation with mathematicians;
- defining dimensions for the manufacturer or other involved specialist planners using alternative, simpler, and more illustrative methods of geometric definition can be queried at any time in any number and almost any exactness from the programme;
- in particular cases, the digital file can even be passed on to specialist consultants and to the manufacturer, who can then use the information regarding the form contained therein for their own purposes, for example, for structural calculations or for use in a **fully automated**, CNC-controlled or even fully robotised production facility.

This opens up an enormous new repertoire of forms for the building industry, at least geometrically in the sense of an exact definition of forms.

▶ *Sect.* 3.4 *Form-Related Issues of Axially Stressed Structures*

An important advantage of the digital CAD methods of shape definition is therefore the possibility of generating shapes visually and intuitively via manual input (mouse or tablet) in addition to a purely numerical definition, while maintaining precise mathematical data output. Specially developed shape elements allow the definition and gradual, tactile manipulation of curves and surfaces in order to optimise the shape according to the wishes of the designer, i.e. by the **visual assessment** by the designer. This process is essentially an analogue, non-digital process, which has great similarities with manual drawing. In this process of digital definition of form, the direct transformation of a design idea or form conception into a precise shape, a procedure familiar from analogue form definition—e.g. by means of a hand drawing or template— is thus at least partially preserved.

Despite the rapid increase in the computing power of computers in recent years, the generation of complex curved surfaces is still a computationally intensive and sometimes very time-consuming process. It is therefore important to keep the complexity of the digital model as low as possible while maintaining sufficient accuracy in order to limit the computing power required for form definition. This is particularly important for digital modelling, for instance, in animated films, where each frame of the film must be modelled and rendered separately. For this purpose, for example, a careful distinction is made between objects in the distance, which are modelled with correspondingly less precision and thus more data-efficiency, and those in the foreground, which are modelled with correspondingly greater precision and complexity.

Comparable restrictions exist in architectural design. Here, too, it is essential to keep the amount of data to be processed for a form definition within acceptable limits. Particularly computationally intensive methods, such as the generation of surfaces based on continuous curves, are used for particularly exposed or particularly complex curvatures, especially for those where seamless, continuous curvature is important for visual reasons; simpler, less computationally intensive methods, such as the approximation of curved surfaces by meshes, are used for less critical areas.

Unlike in the film industry, extremely high accuracies are generally not necessary in the construction industry, since highly finely graded curved surfaces hardly ever occur in construction anyway. This has to do with the geometrical constraints associated with the common basic elements of construction, namely rectilinear bars or planar panels (◘ Fig. 1.139). They impose by themselves a certain degree of **discretisation** on the surface model, i. e. a resolution of curved lines or surfaces into **broken polygon lines** or **faceted polyhedra** (◘ Fig. 1.140). This also coincides with the objectives of digital pro-

◘ **Fig. 1.139** Faceting or **discretisation** of a biaxially curved surface into individual planar glass sheets

◘ **Fig. 1.140** Approximate modelling of a dolphin by **discretisation** using a triangular mesh. This can be further refined, if necessary, for a more realistic representation of the curves

cessing, because these elementary components can be defined geometrically with the simplest means: a straight line with two points in space, a plane with three. This considerably reduces the necessary computing power. A continuous refinement of spatial polyhedra is then possible by different graphical methods, as far as the desired visual continuity of the curvature is still satisfactorily ensured, but on the other hand the facet size is still large enough to meet the requirements of manufacturing, statics, or assembly. Discretised surfaces composed of individual polygons (or three-dimensional building structures composed of polyhedra such as spatial frameworks) are called **meshes** or **polygon meshes** in computer-aided modelling (◘ Figs. 1.141, 1.142, and 1.143).

Note, however, that mesh polygons do not necessarily have to be planar. Naturally, this applies to polygons with more than three nodes, especially to the quadrilateral, which occurs most frequently in practical use (such as on ◘ Fig. 1.141). Triangular-mesh nets, in contrast, always have planar sections. Nevertheless, discretised polygon meshes can be generated on an optional basis in such a way that they consist exclusively of planar polygon surfaces. For this purpose, planarity can be set as a default in the mesh-generating algorithm.

As a rule, polygons or polygon meshes are initially defined in the design process as a rough approximation to the desired shape (◘ Fig. 1.144, step **1**, step **1′**) and then visually smoothed or refined by various methods (◘ Fig. 1.144, step **2**, step **2′**). The final result of this process can alternatively be finer polygons (as on ◘ Fig. 1.144, step **2**, step **2′**) or, as a limiting case after a sufficient number of refinement steps, non-discrete, continuously curved curves (e. g. Bézier curves, B-spline curves, NURBS curves) (◘ Fig. 1.144, step **3**) or three-dimensional surfaces generated therefrom (◘ Fig. 1.144, step **3′**).

Even if, strictly mathematically considered, the curves (◘ Fig. 1.144, step **3**) or continuously curved surfaces (◘ Fig. 1.144, step **3′**) can be regarded as a limit case of the refinement of polygons after theoretically infinitely many steps (◘ Fig. 1.144, step **2**, step **2′**), this process does not take place in design practice in this form because it is impracticable. One

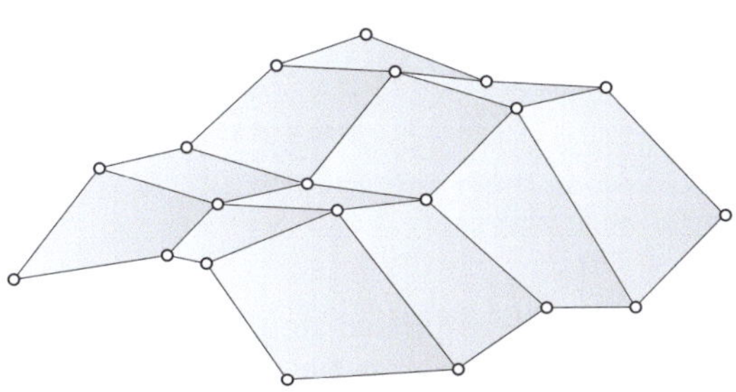

◘ **Fig. 1.141** Polygon mesh for approximate modelling of continuously curved surfaces

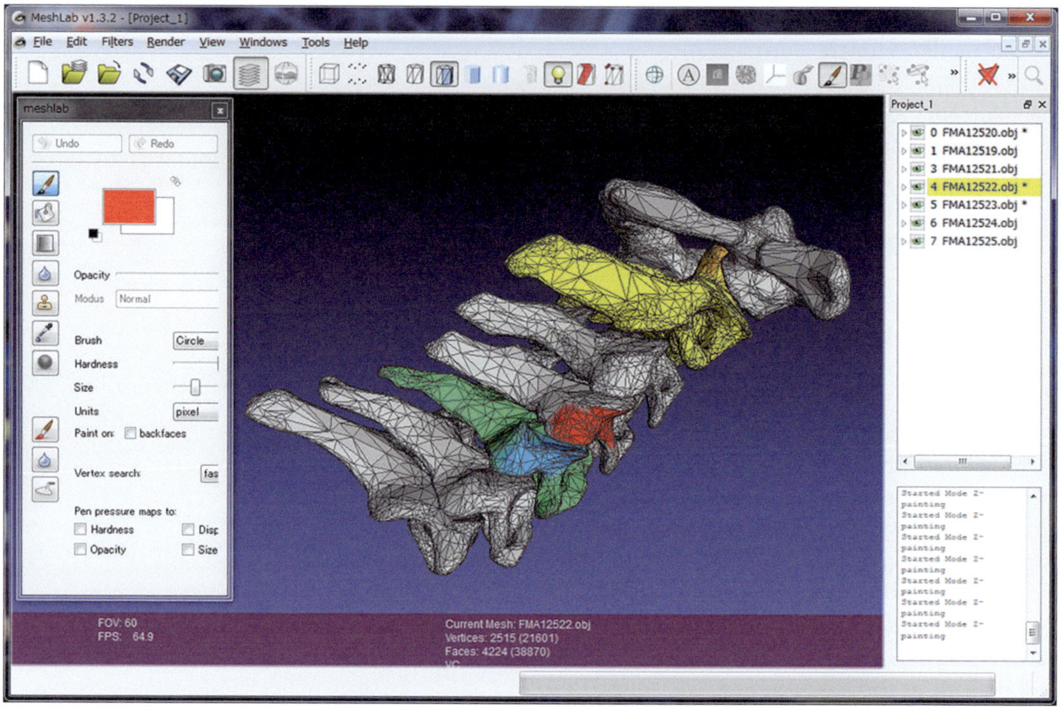

Fig. 1.142 With meshes, it is possible to model the most complex irregular shapes: here, vertebrae modelled in an approximation by meshes

Fig. 1.143 Free-form lattice shell with triangular faces, based on a digitally generated polygon mesh (Arch.: Foster Ass.)

will never refine a polygon mesh step by step until the discretisation disappears, but only as far as is necessary to build it from individual planar parts. This indeed corresponds to working with meshes. This practice saves computing time, since these networks are nothing more than a matrix with spa-

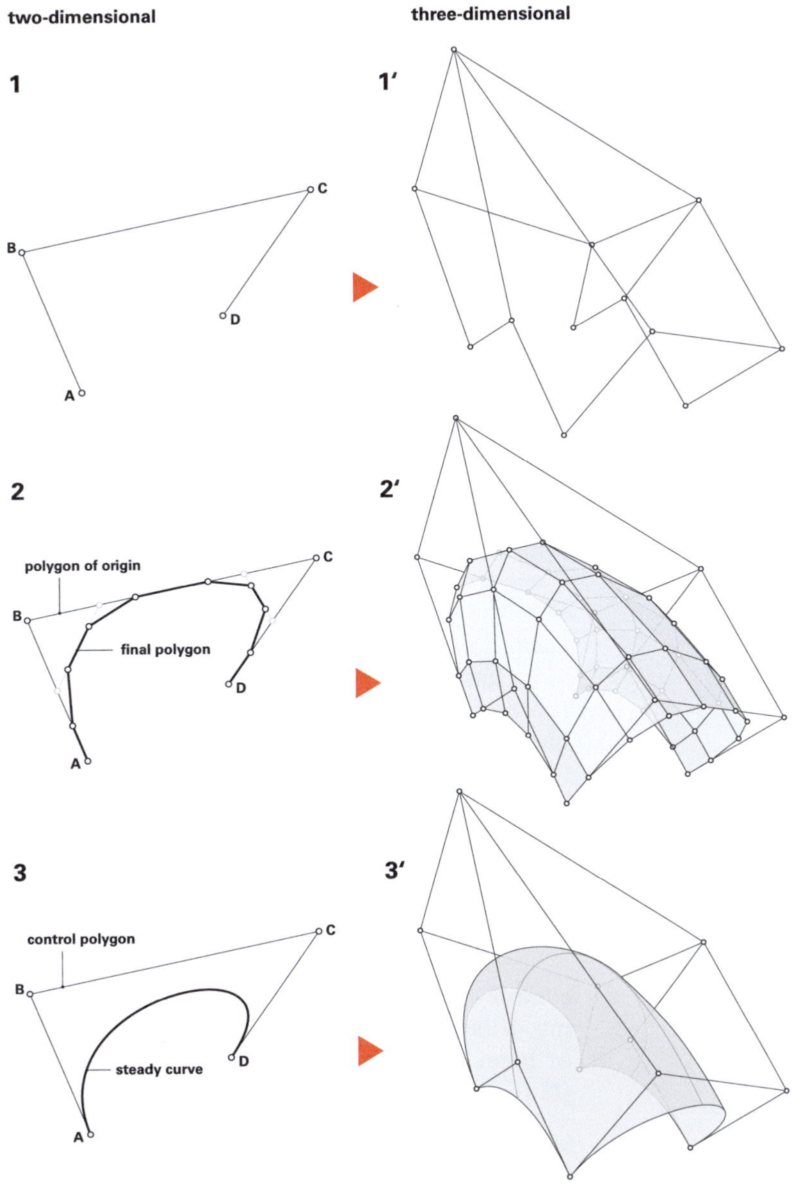

Fig. 1.144 Step **1**: Three-sided **polygon ABCD**: rough approximation to a targeted shape element. Step **1′**: Spatial **polygon mesh**: rough approximation of a target surface. Step **2**: Smoothing or refinement of the three-sided original polygon **ABCD** into a discretised, curve-like polygon via a defined mathematical procedure of **subdividing** the polygon sides. Step **2′**: Smoothing or refinement of the spatial origin polygon mesh into a discretised mesh using a defined mathematical procedure of **subdividing** the polygon sides. Step **3**: Conversion of the three-sided **control polygon ABCD** into a continuously curved, here not discretised curve via a defined mathematical procedure. Step **3′**: Transformation of the mesh of **control polygons** into a curved surface using a defined mathematical procedure

tial coordinates of the points involved, linkage data between them, and possibly other supplementary data.

Only in special cases where steady, non-discretised curvature is essential will curves be used (as on Fig. 1.144, step **3**, step **3′**; Fig. 1.145). These curves are derived directly from the associated so-called control polygons (see further below)

Fig. 1.145 Continuously curved, non-discretised surface with highest demands on precision and continuity (Cloud Gate, Chicago)

using an algorithm, without intermediate steps as in the previous case of polygon subdivision. The amount of data to be processed is much larger than in the previous case, as is the required computing power.

Since these curves are the results of the mathematical modification of the original polygons, they can be manipulated as desired (Fig. 1.146) by modifying these original polygons without having to intervene in the data set of the already refined final shape, which would possibly cause disturbances in the continuity of the surface.

For this reason, such polygons of origin are also called **control polygons**. They guarantee an overall steady, flowing, coherently consistent shape without local discontinuities or breakages, as are usually aimed for in architectural design (Fig. 1.146).

Nevertheless, there are also methods to manipulate meshes locally, by hand or by eye, so to speak. In this way, for example, localised bulges or depressions of any shape can be worked into a surface without reference to, or implications for, the overall shape. This process can be compared to the impression of a finger in plastic clay. This is in contrast to the bending deformation of a thin strip of wood when pressed with the finger, where the entire element deforms continuously, not just locally. The latter is a process that corresponds more to the deformation of a surface by control polygons. Local, so to speak, plastic deformation of free-form surfaces, however, occurs primarily in the modelling of living forms, such as in film animations, but less frequently in architectural design (Fig. 1.142).

In addition to the planar partial surfaces that are often used in the meshes discussed above for the purpose of easier construction, **developable surfaces** are of particular impor-

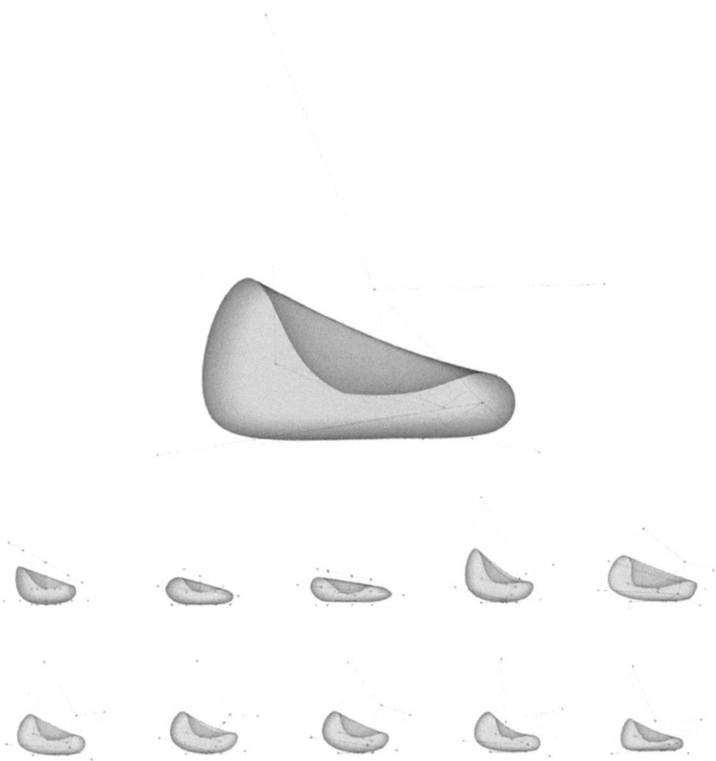

Fig. 1.146 Manipulation of the shape by changing the **control polygons** compared to the initial shape above. Polygon control guarantees an overall consistent shape without disruptive breaks

tance for the design and building practice. They are only uniaxially curved and can be assembled or bent from flat starting elements such as plates, bands, or strips, provided that the material is sufficiently flexible. Since developable surfaces are always ruled surfaces, they can also be built relatively easily from straight bar-shaped material. More complex biaxially curved surfaces can therefore also be composed, as an alternative to the discrete meshes of planar polygons discussed above, of strip-shaped developable surfaces that are connected to each other at their edges. This type of, so to speak, partially facetted surfaces is called **semi-discrete**.

It goes without saying that continuously biaxially curved, non-discrete surfaces can also be implemented in construction, for example, by casting techniques or by thermal deformation of flat starting material such as glass. Already, non-planar quadrilateral faces of a mesh are biaxially curved surfaces that cannot be developed and require corresponding manufacturing processes. However, these forms entail more constructional effort. In these cases, the increased manufacturing costs must therefore always be taken into consideration, which often rule out such constructional solutions from the outset.

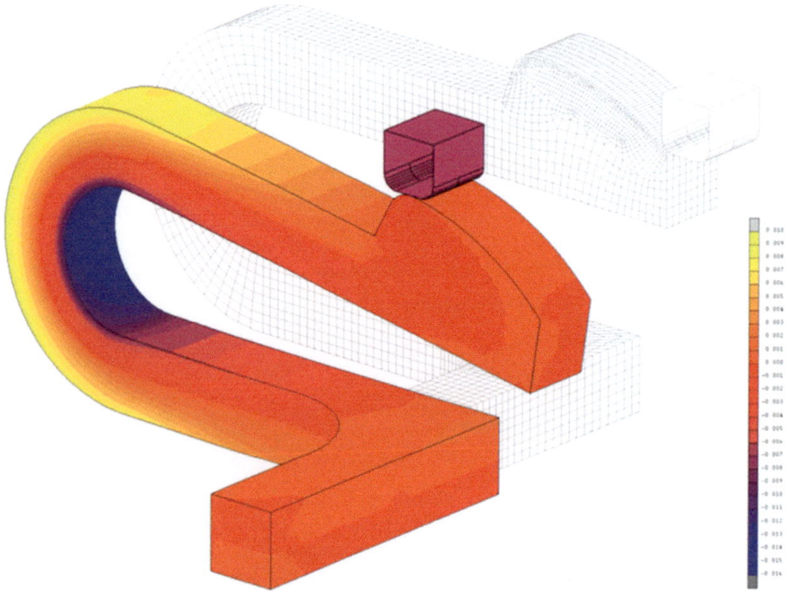

Fig. 1.147 **Finite-element** model (FE model) of a bent component under load with representation of the local stress levels (see colour-hue scale). In the background, the underlying polygon mesh

Despite very great freedom in the design of constructional shapes using digital methods, however, the static behaviour of the surfaces must always be analysed as independently load-bearing surface structures or latticework. Only through statically compatible shaping or subdivision can the load-bearing reserves of a curved surface or lattice-type supporting structure be fully exploited. Here too, modern software exists for the purpose of determining a statically optimised design.[8] This represents a replacement for experimental methods, which until recently were exclusively used for the purpose of statically optimised form-finding using physical models.[9] Finite-element calculation methods, which are also based on polygon formation, and which normally use the digital network models developed for form-finding for static calculation, now allow much more extensive structural analyses than was previously possible using physical models (Fig. 1.147).

1.2.5 Meshes

Meshes are the simplest, most efficient, and flexible method, in terms of computing effort, for digitally modelling complex free-form surfaces with sufficient accuracy for construction purposes. They are the elementary tools of computer-aided

8 Cf. the seminal theoretical work by Prof. K. Linkwitz.

9 Cf. for instance *IL 18 Seifenblasen*, *IL 34 Das Modell*, publications of the Institute für Light Surface Structures, University of Stuttgart.

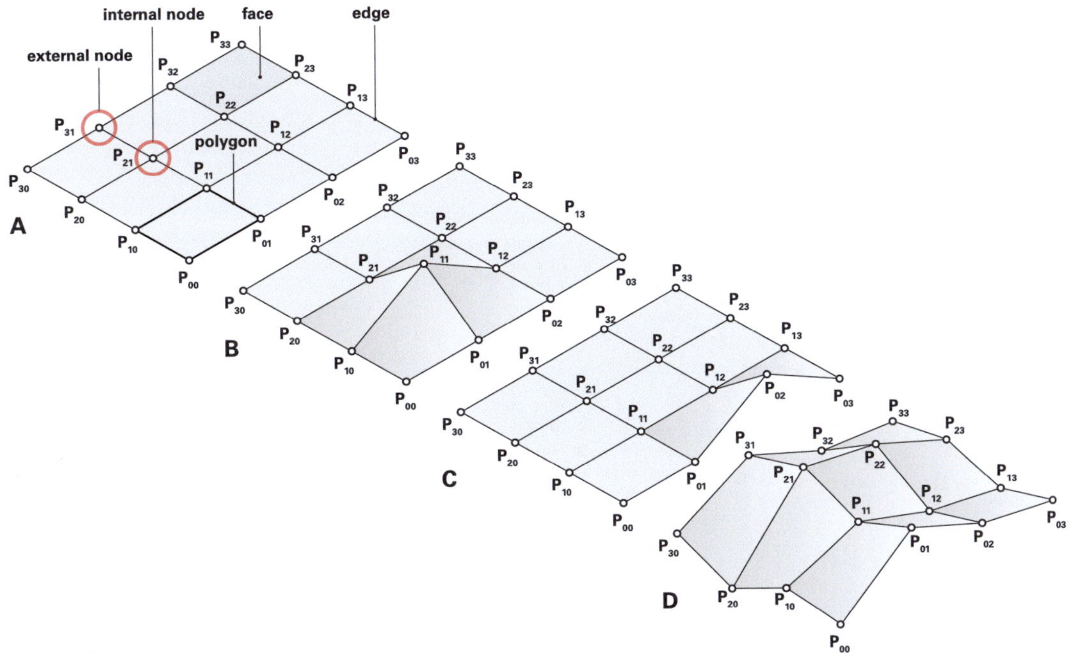

Fig. 1.148 Polygon mesh made of **quadrangular** faces: **A** plane; **B**, **C** single points shifted out of the plane (polygons like $\mathbf{P_{00}P_{10}P_{11}P_{01}}$ are not planar); **D** freeform

design and construction, and can be refined as far as necessary by various procedures. These methods are explained in more detail in the following sections.

► *Sect. 1.2.6 Refinement of Polygon Meshes—Approximation to a Curvature*

Meshes are made up of **vertices**, **edges**, and enclosed **faces**. Nodes are connected by edge lines in a certain type of **connectivity**, which is characteristic of the respective mesh. A group of edges surrounding a face forms a **polygon**. Faces are joined together at edges in such a way that a continuous faceted surface is produced (Figs. 1.148 and 1.149).

In the practical constructional execution, faces can be designed as self-supporting plates or panels, with edges corresponding to the butt joints (Fig. 1.150); meshes can also be implemented as latticework, where edges represent a bar, mesh nodes a constructional node and faces in turn a covering flat component (Fig. 1.151).

Meshes can have different degrees of regularity (Fig. 1.152). In architecture, completely unregulated meshes without a recognisable linking rule between the nodes are normally not useful for generating free-form surfaces (Fig. 1.152, **1**). The most common are structured meshes, where there is a clearly identifiable rule of connection between the nodes, e.g. the number of edges converging at a node (Fig. 1.152, **2**). Regular meshes with additional meshing rules (e.g. the angles of the edges converging at a node) (Fig. 1.152, **3**, **4**) occur with special surfaces (e.g. uniaxially curved ones).

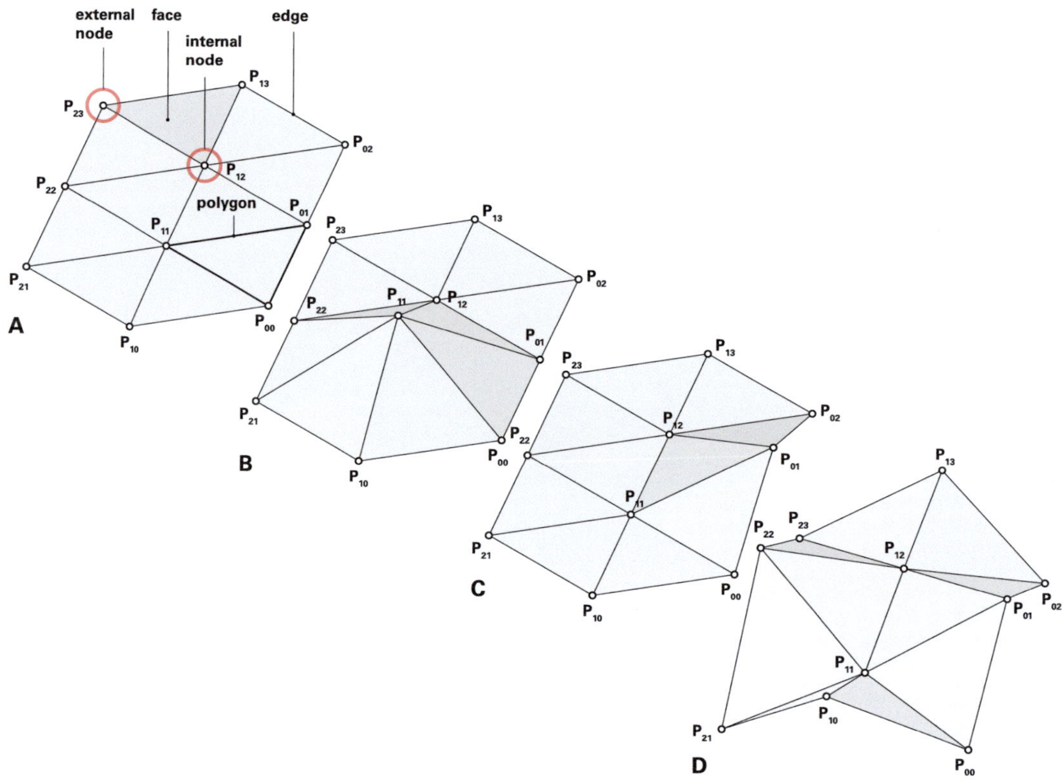

Fig. 1.149 Polygon mesh of **triangular** faces: **A** plane; **B**, **C** single points shifted out of the plane (all polygons remain planar); **D** free-form

Fig. 1.150 Realisation of a mesh by plates cut to polygons in a self-supporting shell structure

The most common polygon meshes in architectural design are **triangular** and **quadrangular** or **quadrilateral meshes** (Figs. 1.148 and 1.149). Hexagonal meshes are only used in exceptional cases. The following characteristics of each subdivision of the mesh network are of interest:

- **Triangular meshes** (Fig. 1.149): Already with the tessellation of the plane it became clear that a surface, in the

Fig. 1.151 Realisation of a mesh in the form of a latticework with covering glass plates supported on it

simplest form, can be composed of triangles. Any mesh of quadrilateral faces may be converted into a mesh of triangular faces by inserting a diagonal bar into a face. Six edges converge at each regular inner node, i.e. the nodes are hexavalent. Exceptions to this rule are called irregular nodes.

► *Sect. 1.1.2.1 Tesselation of the Surface*

Triangular faces of meshes are always planar, which is of great importance for their constructional implementation. Furthermore, bar triangles are always shear-resistant, which is statically advantageous particularly in the case of lattice shells. A disadvantage, on the other hand, is the relatively large number of bars (six) that meet in the node of a latticework, which increases the constructional complexity.

- **Quadrilateral meshes** (or short **quad meshes**) (Fig. 1.148): Squares or rectangles also tesselate the plane in an elementary way. Four quadrangles each meet at a regular inner node, i.e. always four edges meet there (nodes are quadrivalent). Exceptions to this rule are called irregular nodes. This is the case if (mostly for reasons of geometric topology) locally different mesh faces (non-quadrangles) must be introduced. As noted, quadrilateral meshes are not necessarily planar (Fig. 1.148 **B–D**).

► *Sect. 1.1.2.1 Tesselation of the Surface*

Quadrilateral meshes are the mesh geometry most frequently used in architectural design. It results in panel formats that are easy to manufacture and handle, and produces a relatively low joint density or, in the case of bar grids, a relatively small number of bars. It is possible to de-

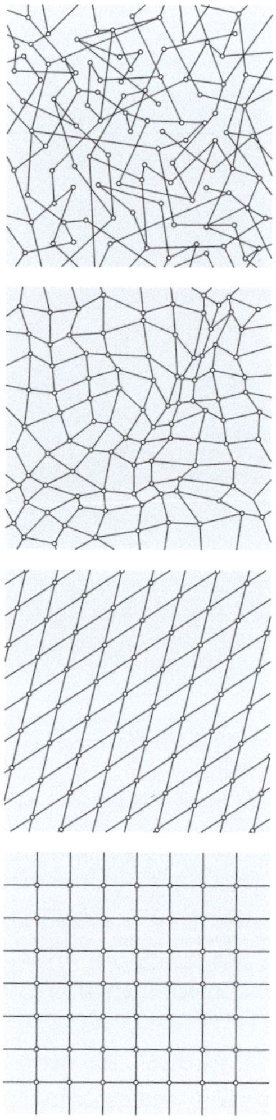

Fig. 1.152 Different degrees of regularity of nets: **1** without any recognisable order; **2** structured; **3** structured and regular; **4** structured, regular, and orthogonal

velop mesh geometries in such a way that all quadrangular faces are planar, which greatly simplifies production. In lattice structures, quadrilateral meshes, in contrast to triangular meshes, are not shear-resistant. This is usually a disadvantage from a structural point of view, but, under certain circumstances, can also be advantageous from a constructional one.

Cf. ▶ *Sect.* 4.3.2.5 *Dome from Bars*

1.2.6 Refinement of Polygon Meshes: Approximation to a Curvature

Polygon meshes can be **refined** using various methods to better adapt them to a desired free-form and achieve a more pleasing visual appearance. There are various methods of refinement, for example a **subdivision** (◘ Figs. 1.153 and 1.154). This involves introducing additional nodes according to a specific rule. This opens the possibility to relocate these nodes in space for better adaptation and thus to refine the mesh network. This process can be repeated indefinitely until one theoretically obtains a continuously curved surface; but this is impractical and is not practiced in this form.

Cf. also ▶ *Sects. 1.2.7.1 Subdivision Curves and* ▶ *1.2.8.1 Subdivision Surfaces*

There are also algorithms that generate **continuously curved** curves based on a polygon (control polygon). On this basis, three-dimensional, also continuously curved, non-discrete surfaces can then be generated.

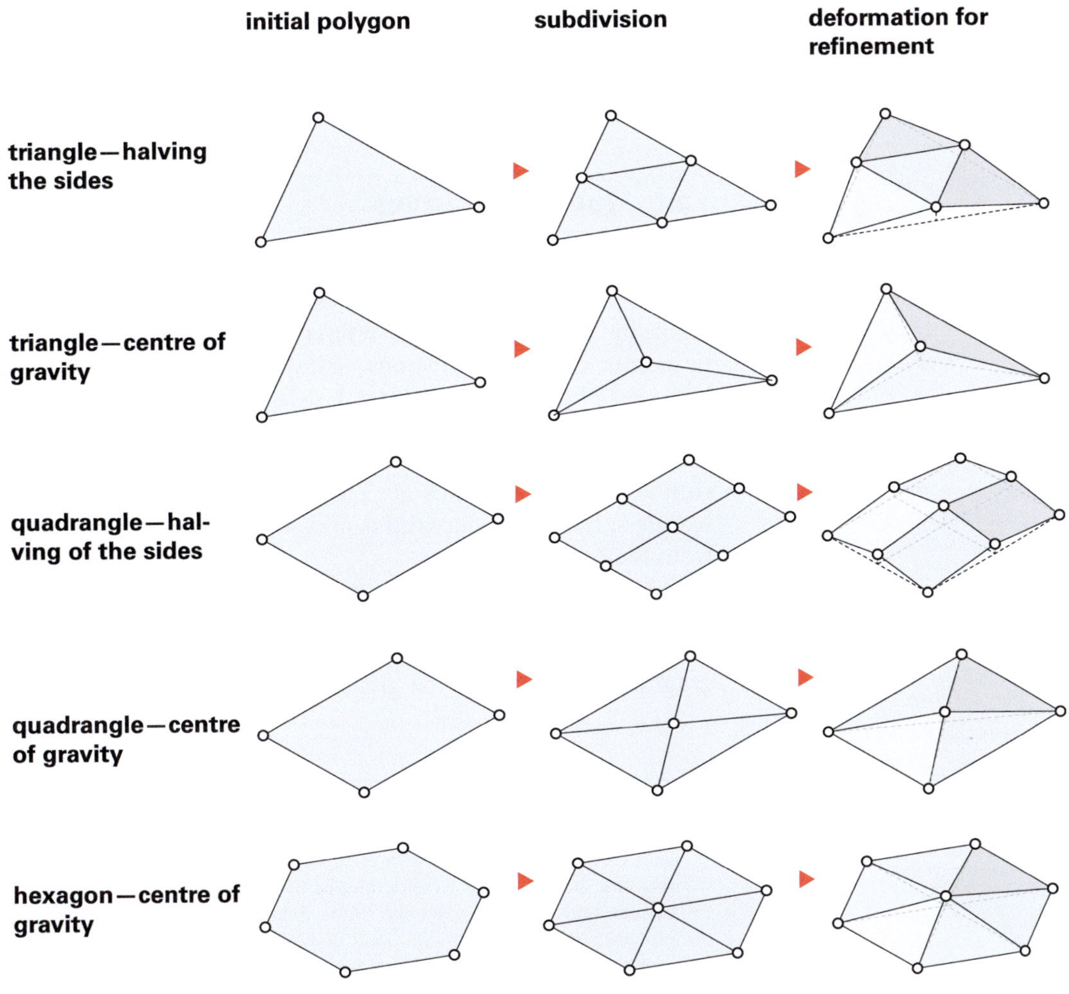

◘ **Fig. 1.153** Possibilities of **refining** a mesh face by subdividing it to better approximate curved geometries. Planar initial faces are shown here, but the methods are also applicable to non-planar faces

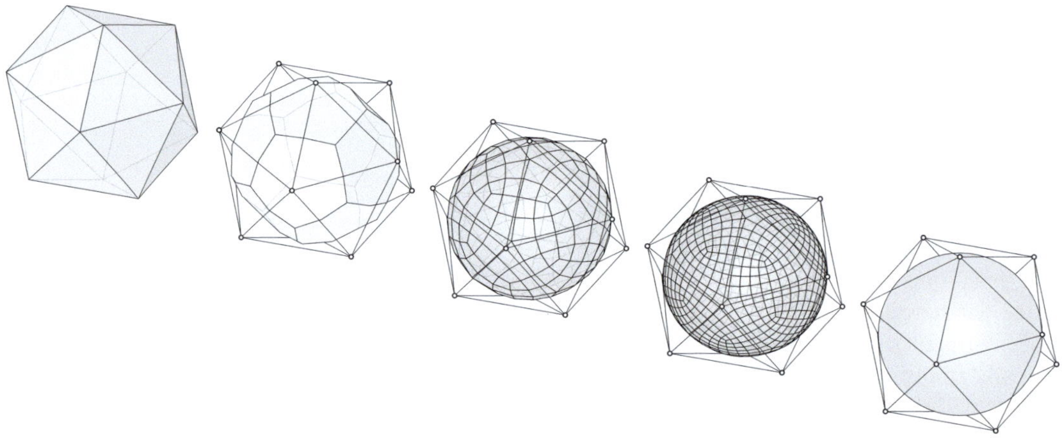

Fig. 1.154 Transformation of an icosahedron (20-side surface) into a sphere in a refinement process by subdivision (Catmull-Clark algorithm)

Starting from two-dimensional free forms, various refinement methods will be investigated in the following which produce both discretised surfaces (using subdivision curves) and non-discretised surfaces (using free-form curves).

1.2.7 Free-Form Curves

Free-form curves provide the basis for the generation of continuously curved surfaces. They can be easily manipulated by hand by changing certain reference elements—characteristic points, enclosing polygons, or tangents. At the same time, however, the full mathematical definability of the form is guaranteed with the aid of data sets of any desired precision, since these curves are based on mathematically defined algorithms. These algorithms are programmed in such a way that seamless, harmonic curvatures and transitions are guaranteed from the outset.

These shape elements are derived from the **subdivision curves** and **Bézier curves**, which were developed in the mid-twentieth century for automobile design and later for computer graphics.[10] Because of the great importance they have gained

10 The method of subdivision curves was developed in 1947 by G. de Rahm on the basis of the division ratio 1/3 to 2/3 and newly deduced later (1974) by G. Chaikin with the division ratio 1/4 to 3/4 for computer graphics. Bézier curves were developed by the French car manufacturers Citroën and Renault in the late 1950s. Even before that, comparable approaches existed in the USA. Paul de Casteljau (Citroën) developed the underlying algorithm in 1959. However, this was not published for reasons of secrecy. Pierre Bézier (Renault) developed it independently from Casteljau and published it in 1962. Therefore, the curves are called Bézier curves (cf. Pottmann and Bentley (2007), p. 259, 280).

in the meantime thanks to the spread of CAD software for the building industry and because they are the basis for numerous free-form surfaces, their essential characteristics, as well as the surfaces that can be derived from them, will be discussed in the following.

The following curve types have been developed for designing with digital tools or are used in common CAD programmes:

- **subdivision curves**: smoothing a polygon to a curve by linear interpolation of the polygon sides;
- **Bézier curves**: generated by graphical interpolation based on the Casteljau algorithm;
- **B-spline curves**: more versatile manipulability with local shape control;
- **NURBS curves** (Nonuniform Rational B-Spline): for the most complex curve definition; more wide-ranging manipulation of the curve by applying additional parameters.

For the Casteljau algorithm, cf. ► *Sect. 1.2.7.2 Bézier Curves*, in particular ◘ Fig. 1.160

These curve types will be examined in more detail below.

1.2.7.1 Subdivision Curves

Subdivision curves are created in a simple smoothing process by progressively cutting off polygon corners. The polygon sides are divided in a given ratio so that the newly found points create a new, more strongly smoothed polygon. Different algorithms are distinguished:[11]

- **Chaikin algorithm** (◘ Figs. 1.155 and 1.156): The polygon sides are divided at both ends in the ratio 1/4 to 3/4. The intermediate points thus created are connected to each other and produce a more strongly smoothed polygon. After successive steps, a continuous curve is created which touches the sides of the original polygon. The result is a quadratic B-spline curve (like on ◘ Fig. 1.163**B**).
- **Lane-Riesenfeld algorithm** (◘ Fig. 1.157): assumes a bisection of the sides of the starting polygon and a double interpolation of the polygon corners. The final result approaches a cubic B-spline curve. The generalisation is that **n** interpolations result in a B-spline curve of order **n** + 1 (like on ◘ Fig. 1.163**C**).

See further below ► *Sect. 1.2.7.3 B-Spline Curves*

See further below ► *Sect. 1.2.7.3 B-Spline Curves*

1.2.7.2 Bézier Curves

A Bézier curve is defined by a control polygon, the modification of which indirectly modifies the curve (◘ Fig. 1.158 **B**, **C**). Depending on how many sides or vertices the polygon has, one distinguishes Bézier curves of the first (◘ Fig. 1.159), second (◘ Fig. 1.158 **A**), third (◘ Fig. 1.158 **B**), etc. order. By linear interpolation as a function of a parameter **t**, the points

11 Pottmann and Bentley (2007), p. 279ff.

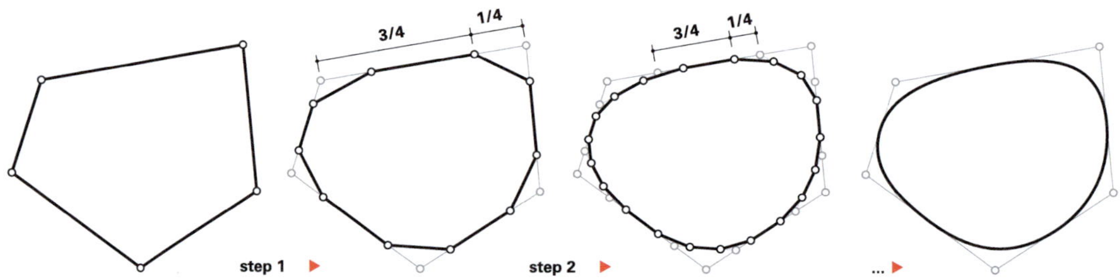

Fig. 1.155 **Chaikin algorithm**: progressive refinement and smoothing of a closed polygon (left) by cutting off the corners according to a given method. In this case, a linear interpolation is applied where the polygon sides are shortened in the ratio 1/4 to 3/4. After a sufficient number of subdivision steps, the continuous **subdivision curve** is formed (right)

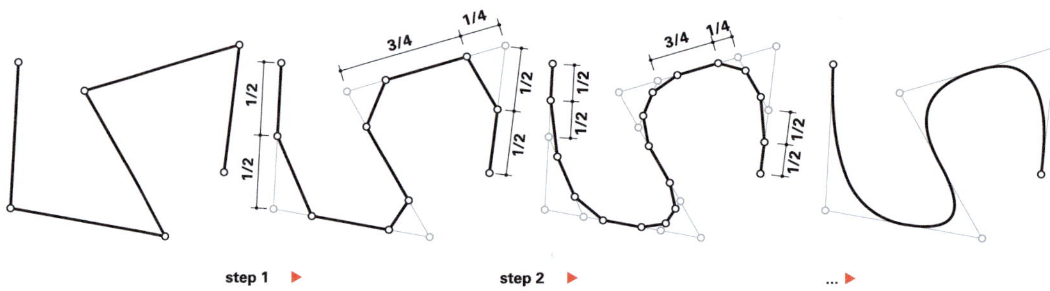

Fig. 1.156 **Chaikin algorithm**: Smoothing an open polygon (left) by cutting off the corners using the chaikin algorithm as in Fig. 1.155. With open curves, the first or last polygon side at both ends is not divided according to the specified ratio, but halved

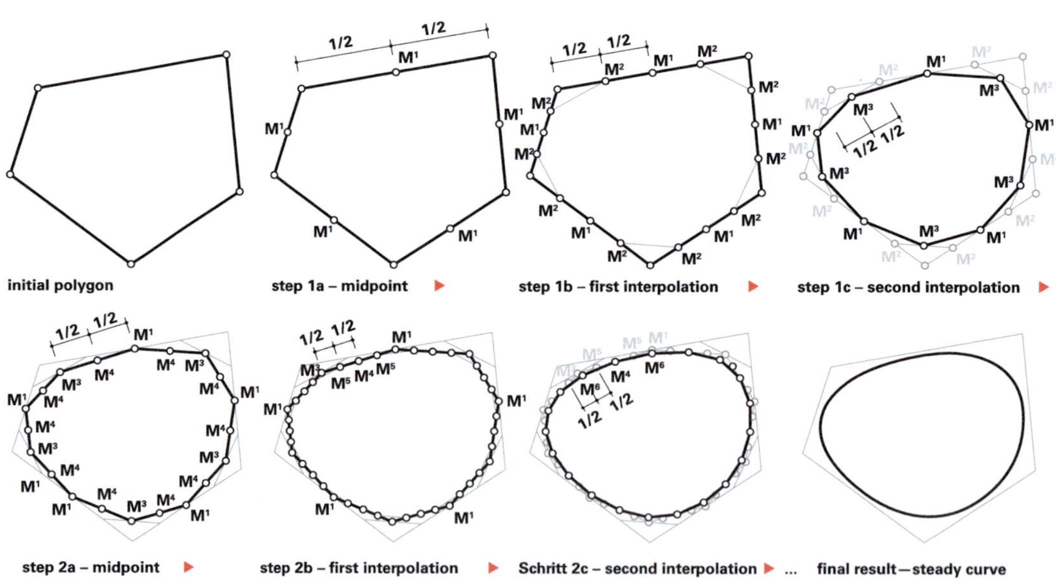

Fig. 1.157 **Lane-Riesenfeld algorithm**: Smoothing a polygon (left) by halving the polygon sides (sub-step **a**) and interpolating twice (sub-steps **b** and **c**). The new polygon is constructed by connecting the first determined point (**M¹** in step **1**) with the last determined point (**M³** in step **1**)

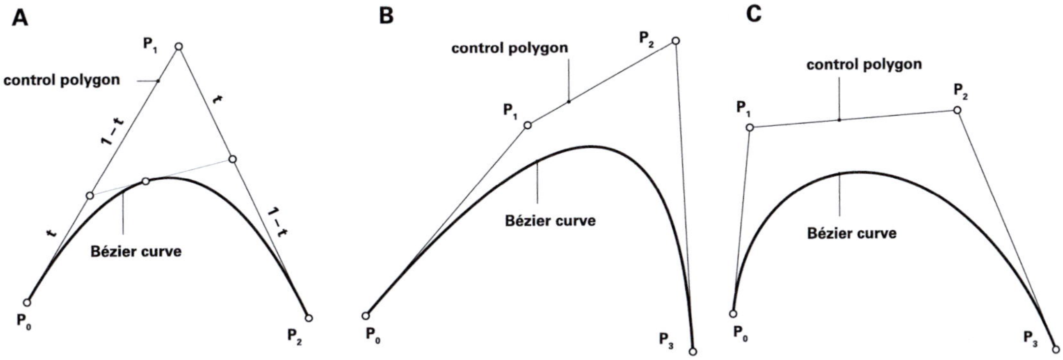

Fig. 1.158 A: Bézier curve of **second order**: Parabola branch. Using the Casteljau algorithm as on Fig. 1.160 results in the conventional thread construction of the parabola. **B**: Bézier curve of **third order**. The control polygon is manipulated manually by the designer and defines the course of the curve. **C**: If the control polygon of the curve is changed as compared to Fig. 1.160 **B**, the curve shape changes accordingly. Already when changing a control point, **all** points of the Bézier curve change

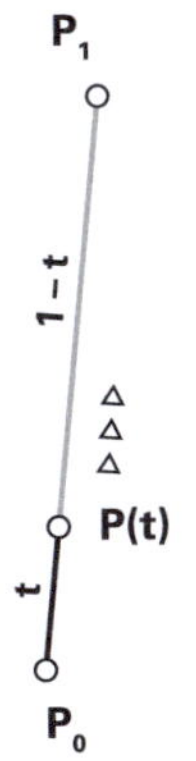

Fig. 1.159 Bézier curve of **first order**: straight line segment. The generation is done by applying the Casteljau algorithm as in Fig. 1.160, but in a single step. The curve is identical with the control polygon (here control-line segment)

of the Bézier curve are obtained by successive application of the Casteljau algorithm (Figs. 1.160 and 1.161). Bézier curves also appear spatially (Fig. 1.162).

1.2.7.3 B-spline Curves

A B-spline curve[12] is composed of several Bézier curve sections. In common CAD software, the sections are evenly distributed (uniform B-Spline). As a result, unlike a continuous Bézier curve, where changing even one control point changes all points of the curve (cf. Fig. 1.158**C**), the Bézier seg-

12 B-Spline = *Basis-Spline*. In manual drawing, the English term spline refers to a flexible template that is brought into the desired shape by several fixed points. Such stencils were used in architecture and shipbuilding until recently (cf. Pottmann and Bentley (2007) p. 256, 269).

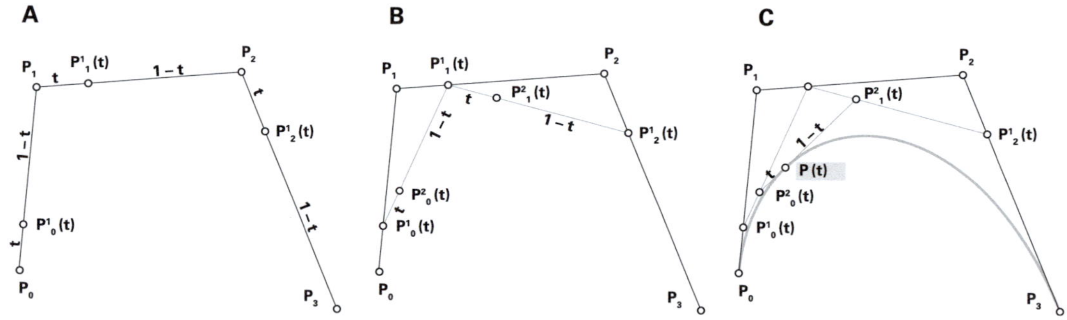

Fig. 1.160 A: Generation of the Bézier curve on Fig. 1.158**C** by **graphical interpolation** (Casteljau algorithm). **Step 1**: Starting from the control polygon $\mathbf{P_0}$–$\mathbf{P_3}$, definition of the three intermediate points $\mathbf{P^1{}_0(t)}$, $\mathbf{P^1{}_1(t)}$, and $\mathbf{P^1{}_2(t)}$ by dividing the polygon sides in the ratio **t**: (1 – **t**), where $0 < \mathbf{t} < 1$. **B**: **Step 2**: Starting from the polygon $\mathbf{P^1{}_0(t)}$–$\mathbf{P^1{}_2(t)}$, definition of the two intermediate points $\mathbf{P^2{}_0(t)}$, $\mathbf{P^2{}_1(t)}$ by dividing the polygon sides in the same ratio **t**: (1 – **t**). **C**: **Step 3**: Subdivision of the section $\mathbf{P^2{}_0(t)}$–$\mathbf{P^2{}_1(t)}$ in the same ratio **t**: (1 – **t**) by defining the intermediate point **P(t)**. **P(t)** is a point on the Bézier curve for the parameter **t**. $\mathbf{P^2{}_0(t)}$–$\mathbf{P^2{}_1(t)}$ is the tangent at curve point **P(t)**, $\mathbf{P_0P_1}$ at point $\mathbf{P_0}$ and $\mathbf{P_2P_3}$ at point $\mathbf{P_3}$. Since this case is a third order curve, there are also three intermediate point hierarchies, i. e. $\mathbf{P^1{}_i(t)}$, $\mathbf{P^2{}_i}$ **(t)**, and **P(t)**

ments can be modified locally, thereby affecting only one section or selected sections, if desired (Fig. 1.164**A**). This significantly increases the flexibility of the design and eliminates the tedious manual assembly of the Bézier curve sections. Furthermore, the B-spline algorithm guarantees that the Bézier curve sections always merge harmoniously, i. e. without abrupt changes in curvature.

A B-spline curve is defined by: [13]

- the number **m** of sides of the control polygon;
- the number of control points **m** + **1**;
- the order **n** of the Bézier curve sections and the B-spline itself (always identical);
- the node vector.

Unlike Bézier curves, the order of the B-spline curve is not determined by the number of sides of the polygon, but can be freely selected (Fig. 1.163). However, the maximum order **n** of a B-spline curve is equal to **m**. A B-spline curve can be open (Fig. 1.163**C**, Fig. 1.164**B**) or closed (Fig. 1.164**C** and 1.165).

1.2.7.4 NURBS Curves

NURBS curves are characterised by an additional shape parameter, the **weighting g** of single control points of the control polygon. They are **n**-dimensional central projections of an **n + 1**-dimensional B-spline curve; e. g. on Fig. 1.166 the two-dimensional projection of a three-dimensional B-spline curve. (Accordingly, three-dimensional NURBS curves are projections of four-dimensional B-spline curves.) The individual definition of the weights of the individual control

13 Pottmann et al. (2010), p. 262.

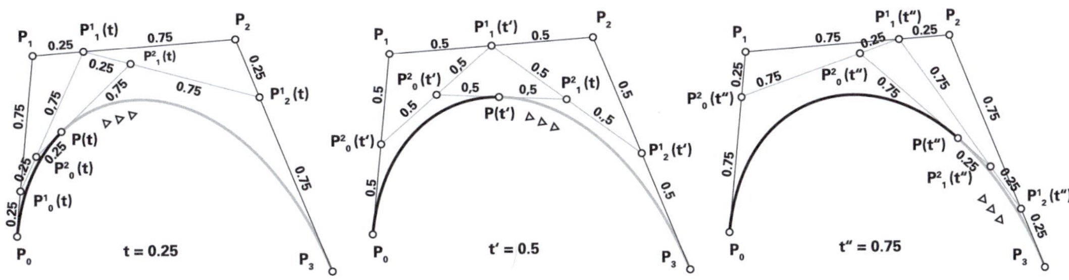

Fig. 1.161 Generation of the Bézier curve in Fig. 1.158**C** by successively applying the Casteljau algorithm on Fig. 1.160, each for different values of the parameter **t** from 0 to 1. Three exemplary intermediate steps for **t** = 0.25, **t** = 0.5 and **t** = 0.75 are shown here

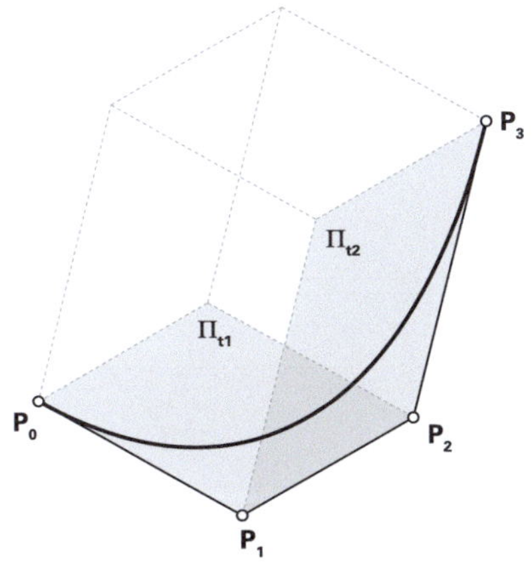

Fig. 1.162 **Spatial** Bézier curve. The control polygon is $\mathbf{P_0}$–$\mathbf{P_3}$; the planes $\prod_{t1}$ and $\prod_{t2}$ are each tangential planes to the curve at points $\mathbf{P_0}$ and $\mathbf{P_3}$

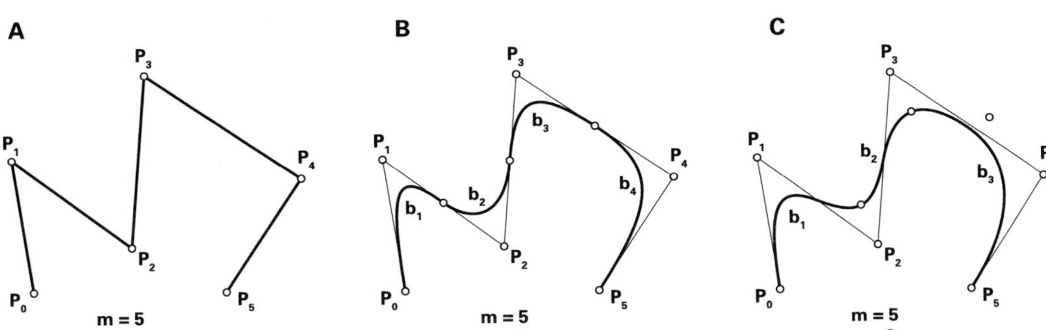

Fig. 1.163 A: **B-spline curve** with five-sided control polygon (**m** = 5), **first-order** curve (**n** = 1): Curve is identical with the control polygon (see also Fig. 1.159). **B**: **B-spline curve** with five-sided control polygon (**m** = 5), **second-order** curve (**n** = 2): The B-spline curve is composed of four Bézier curve sections $\mathbf{b_1}$ to $\mathbf{b_4}$. C: **B-spline curve** with five-sided control polygon (**m** = 5), **third-order** curve (**n** = 3): The B-spline curve is composed of three Bézier curve sections $\mathbf{b_1}$ to $\mathbf{b_3}$. Note that the curve increasingly detaches itself from the control polygon. This effect increases with increasing order of the B-spline curve

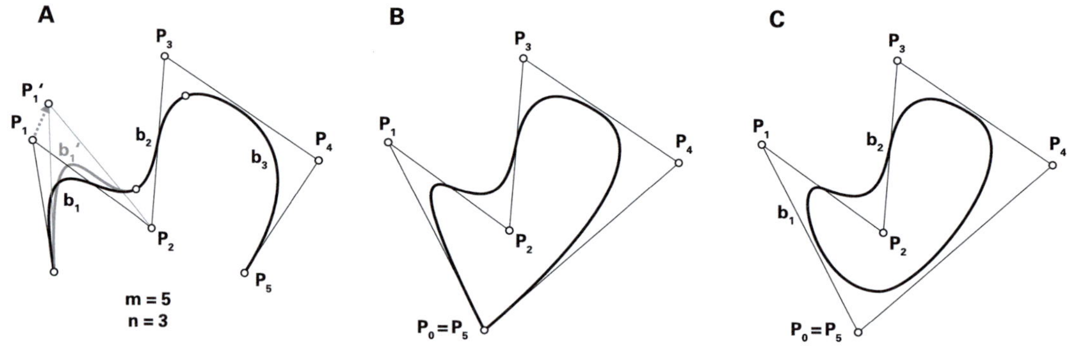

■ **Fig. 1.164** A: When changing a control point $\mathbf{P_1}$ of a B-spline curve, e.g. the one in also ■ Fig. 1.163**C**, only the corresponding curve segment $\mathbf{b_1}$ is changed. The rest of the curve ($\mathbf{b_2}, \mathbf{b_3}$) remains unchanged. **B**: **Open** B-spline curve with identical start and end points $\mathbf{P_0}$ and $\mathbf{P_5}$; at these, there is no steady curvature; a corner is formed. **C**: **Closed** B-spline curve with identical start and end points $\mathbf{P_0}$ and $\mathbf{P_5}$; the curve has constant curvature at all points

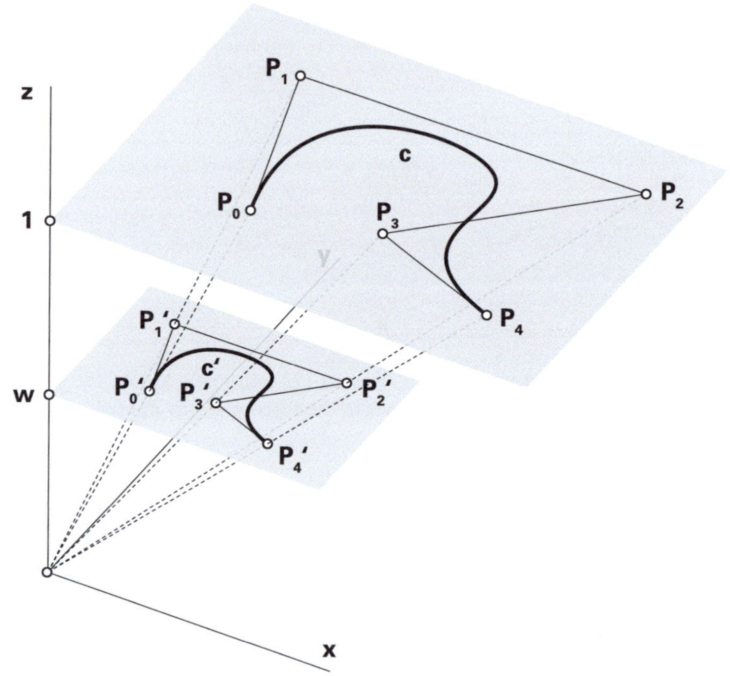

■ **Fig. 1.165** The planar NURBS curve **c** is a central projection of the spatial B-spline curve **c′** onto the plane with **z**-coordinate 1. Each point of **c′**, i.e. all $\mathbf{P'_i}$, is assigned a weighting $\mathbf{w_i}$. These $\mathbf{w_i}$ are each identical with the **z**-coordinate of the points $\mathbf{P'_i}$. In this particular case, the weights of all points $\mathbf{P'_i}$ are equal, namely **w**, so that the B-spline curve **c′** is located on the plane $\mathbf{z = w}$. In this case, **c** is an affine representation of **c′**. A B-spline curve is therefore a special NURBS curve with equal weightings (Pottmann and Bentley (2007), p. 405)

points of the associated B-spline curve allows a differentiated manipulation of the NURBS curve for design purposes (■ Fig. 1.166). ■ Figure 1.167 shows the effect of changing a single weighting on the shape of a flat NURBS curve. In CAD programs, weightings are set numerically or graphically (e.g. using sliders) in dialogue windows.

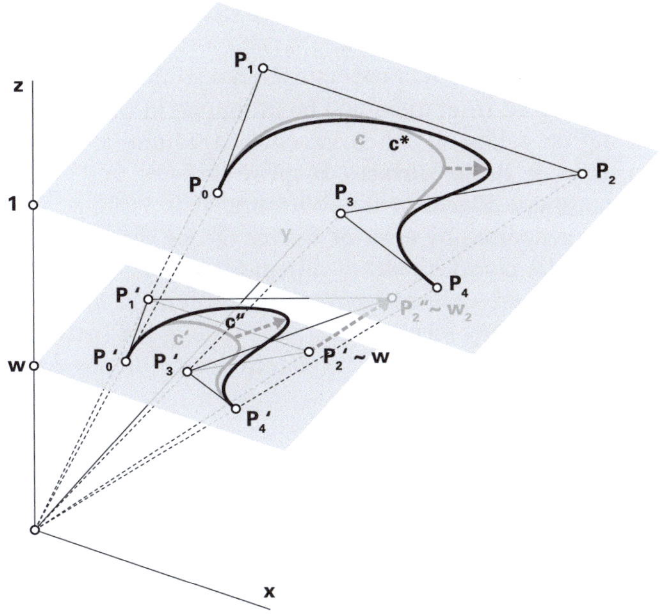

Fig. 1.166 Normal case of a NURBS curve: The weightings **w** of the points $\mathbf{P'}_i$ are set individually according to the designer's wishes: here, for example, $\mathbf{P'}_2$ is moved upwards, i.e. assigned a new weighting $\mathbf{w}_2$. The B-spline curve **c′** moves out of the plane **z** = **w** and becomes spatial; the NURBS curve changes accordingly. The change of the curve **c** occurs in the direction of the point $\mathbf{P}_2$. By individually setting the values $\mathbf{w}_i$, the NURBS curve can be deliberately manipulated (Pottmann and Bentley (2007), p. 405)

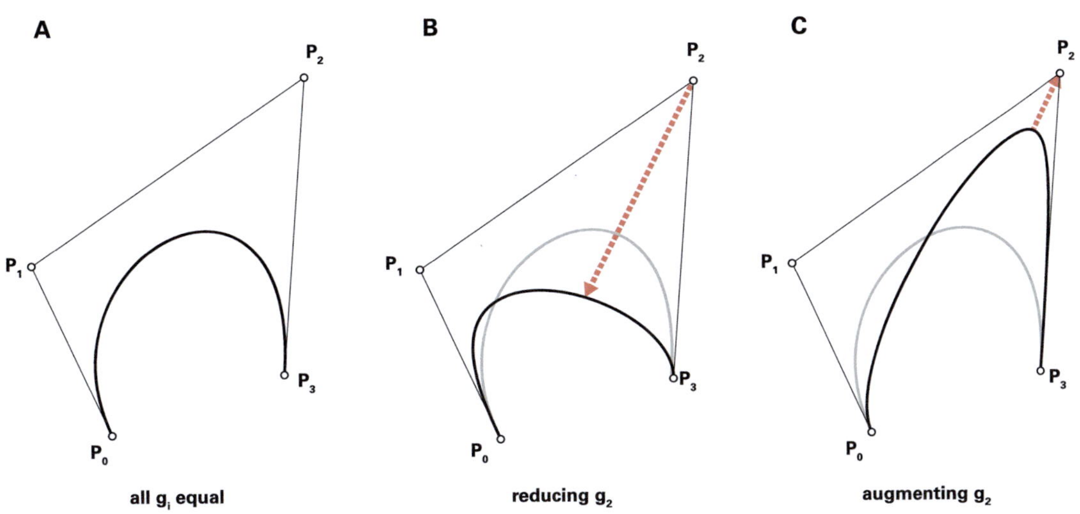

Fig. 1.167 A: **NURBS curve** with four-sided control polygon. All weightings $\mathbf{w}_1$ to $\mathbf{w}_4$ of points $\mathbf{P'}_1$ to $\mathbf{P'}_4$ of the (imaginary) associated B-spline curve are equal here. **B**: Reduction of the weighting $\mathbf{w}_2$ of point $\mathbf{P'}_2$. The curve is deformed away from point $\mathbf{P}_2$. **C**: Increase of the weighting $\mathbf{w}_2$ of point $\mathbf{P'}_2$. The curve is deformed towards point $\mathbf{P}_2$

1.2.8 Free-Form Surfaces from Curves

Based on the free-form curves considered so far, **surfaces** that can be produced from them will be examined in the following. Adopting the categories of curves considered until now, **subdivision surfaces**, **Bézier surfaces**, **B-spline surfaces**, and **NURBS surfaces** are considered in the following. It is assumed that a surface is generated by a set of curves in one direction (**u**) as well as in the complementary direction (**v**). As with curves, these surfaces are defined by control polygons in both directions. Together they form a mesh surrounding the surface, the corner points of which can be manipulated for design purposes.

1.2.8.1 Subdivision Surfaces

► *Sect. 1.2.7.1 Subdivision Curves*

Similar to subdivision curves resulting from the smoothing of an original polygon by systematically cutting off the corners according to a fixed algorithm, surfaces defined by polygons, applied to the third dimension, can also be smoothed by gradual subdivision into finer-meshed surfaces, thus visually more closely approaching continuously curved surfaces.

As with subdivision curves, many different algorithms can be applied. In ◘ Fig. 1.168, the application of the Chaikin algorithm is shown as an example, which has already been applied to curves above (◘ Figs. 1.155 and 1.156). Other algorithms are in use, such as the Doo-Sabin algorithm, the Catmull-Clark algorithm, the loop algorithm, triangle-based subdivision, and multi-resolution modelling. [14]

The algorithms can be configured so that subdivisions always produce planar faces. This is important for architectural applications because, under such conditions, flat material can be used, which makes the manufacturing process much easier. [15]

The subdivision is carried out step by step until a satisfactory result is found or face sizes suitable for constructional purposes are created (◘ Figs. 1.169 and 1.170).

1.2.8.2 Bézier Surfaces

The general Bézier surface is defined by the network of surrounding control polygons (◘ Figs. 1.171, 1.172, and 1.173). A distinction is made between **row polygons** in the direction **u** and **column polygons** in the direction **v**. These are not necessarily congruent with each other, nor must they form planar quadrangular meshes. The surface curves can be derived by applying the Casteljau algorithm to the individual control polygons (cf. ◘ Fig. 1.171). (In CAD programs, however,

14 Pottmann and Bentley (2007), p. 397 ff.

15 Ibid., p. 405.

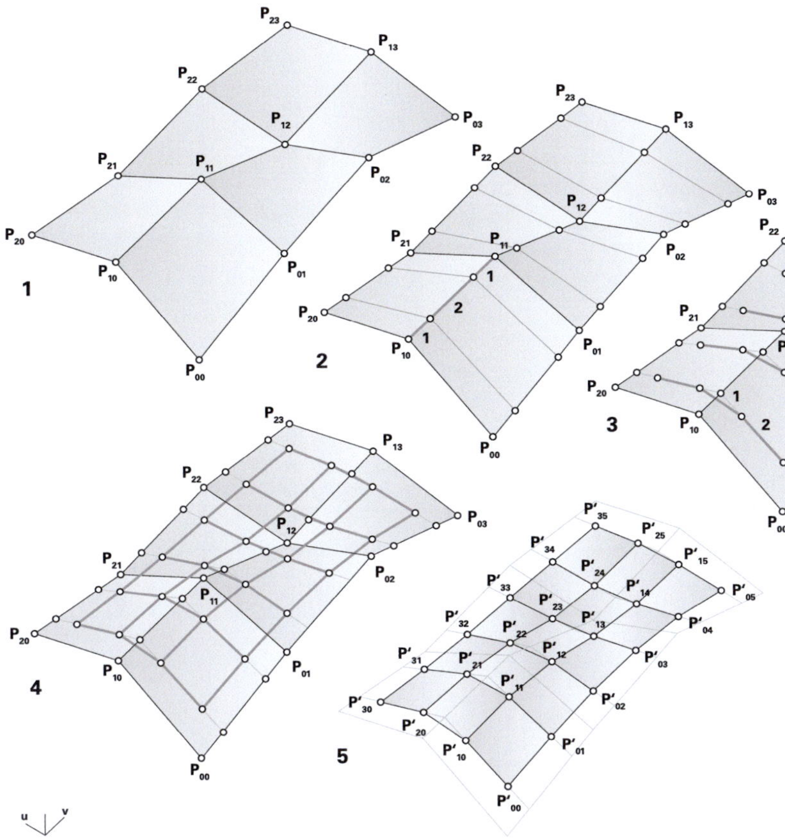

Fig. 1.168 Creation of a **subdivision surface**. Starting from a surface (**1**) defined by polygons (non-planar quadrilateral meshes), the refined, more smoothed subdivision surface (**5**) is generated. This is done in this case by applying the Chaikin algorithm to the polygon sides. The intermediate points ($\mathbf{P'}_{ij}$) determined in this first interpolation stage in both directions **u** and **v** are then connected with straight lines and form the refined quadrilateral mesh. Since the Chaikin algorithm is used here, after a sufficient number of subdivision steps, a second-order B-spline curve is generated from an original polygon or a second-order B-spline surface from the original surface. During this process, the original polygons are transformed, in their original geometry, into the control polygons of the resulting B-spline curves (Pottmann and Bentley 2007). (1) Initial surface; the quadrilateral meshes are hyperbolic paraboloids (ruled surfaces). (2) The polygon sides in the v-direction are divided in the ratio 1:2:1 (Chaikin algorithm) and connected transversely (in the u-direction) with straight lines; the straight lines are generators of the ruled area and therefore lie on it. (3) The determined straight lines are divided again in the same ratio and the corners are cut off. (4) The newly determined polygons in u-direction are completed in v-direction to form a mesh. (5) The subdivision surface found has a larger number of smaller meshes

this is automatically implemented.) Only the Bézier curves of the four-edge polygons are included in the area. The notation of the Bézier surface results from the order of the polygons, each in **u**- and **v**-direction, e.g. **3,2** (like on Figs. 1.172 and 1.173).

Translational Bézier Surfaces

A special case of a Bézier surface is a translational Bézier surface. It results from the sliding of a Bézier curve along another (Fig. 1.174). A distinction is made between two translation directions **u** and **v** and the corresponding control poly-

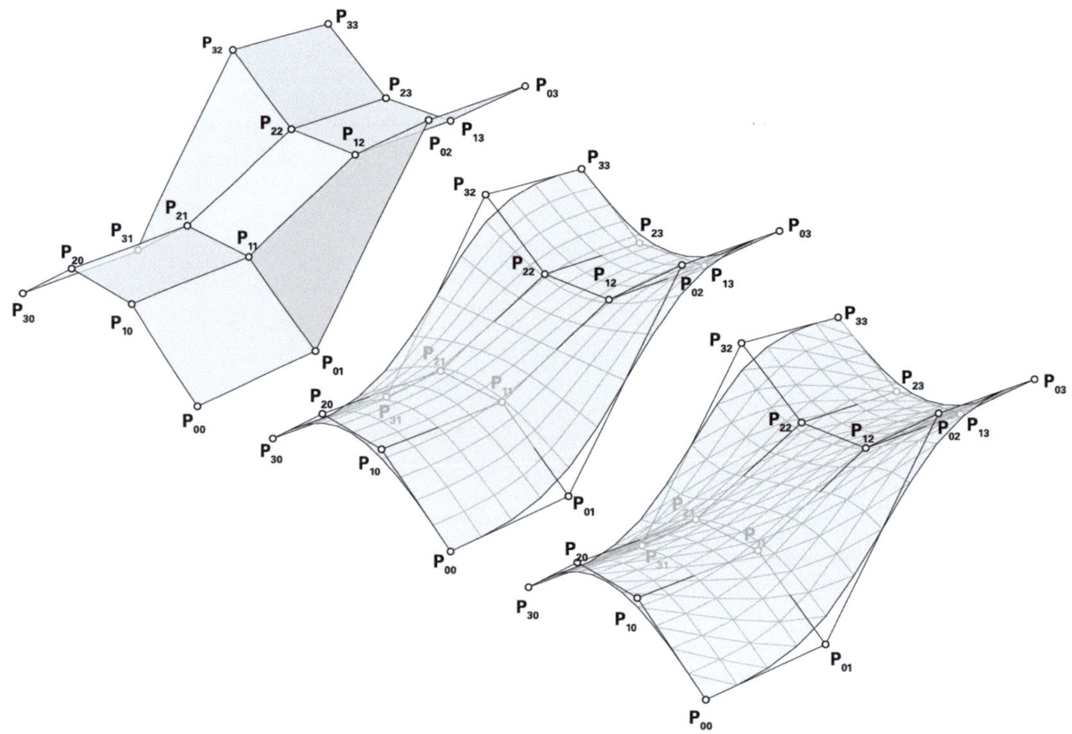

Fig. 1.169 Creation of a **subdivision surface**. Left, the original polygon mesh; centre, smoothing using Catmull-Clark algorithm; right, smoothing by loop algorithm

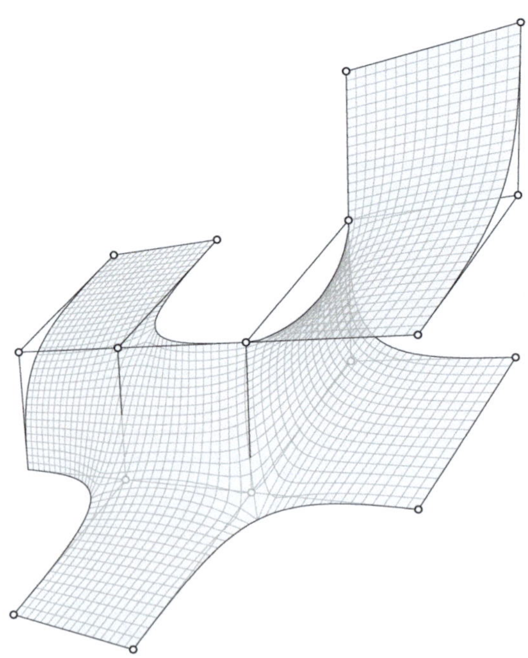

Fig. 1.170 **Subdivision surface** based on the Catmull-Clark algorithm

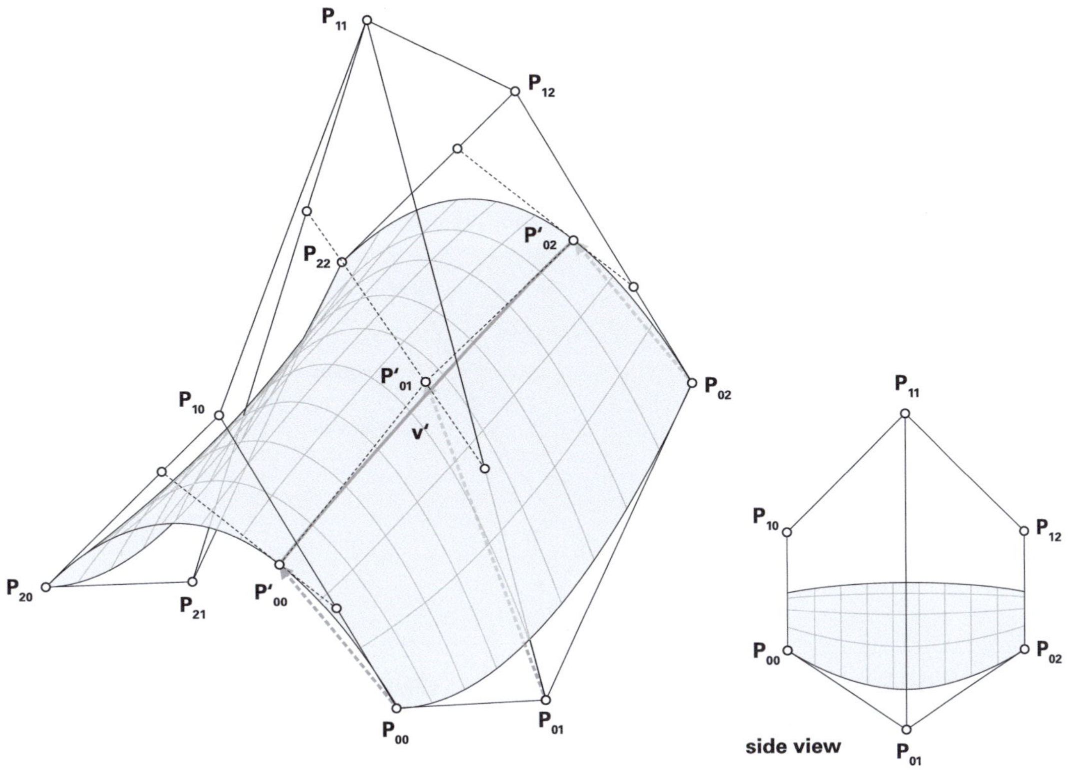

Fig. 1.171 **Bézier free-form surface** (order **2,2**). The surface is defined by the surrounding mesh of control polygons with the vertices $\mathbf{P}_{ij}$. The control polygons are not necessarily congruent; the quadrangles of the mesh not necessarily planar. The area contains mesh curves in direction **u** (belonging to the row polygons) and those in direction **v** (belonging to the column polygons). Applying the Casteljau algorithm to the line polygons creates intermediate points $\mathbf{P'}_{0j}$. These create a control polygon (here $\mathbf{P'}_{00}\ \mathbf{P'}_{01}\ \mathbf{P'}_{02}$), which defines the surface curve **v'** (again by applying the Casteljau algorithm). The same process can be performed in complementary direction **v** using the column polygons

gons. Alternatively, the Casteljau algorithm can be applied to the row polygons in direction **u** and the column polygons in direction **v** (Fig. 1.175).

Translational Bézier surfaces are a special case of the Bézier free-form surfaces. All control polygons in one direction (**u**, **v**) are congruent with each other; the same applies to the generating curves $\mathbf{u}_i$ and $\mathbf{v}_i$. Quadrangular faces of the control polygon mesh are always planar. As with all translational surfaces, this type of surface can be converted relatively easily into a faceting of planar partial surfaces. This significantly simplifies their constructional execution.

► *Sect. 1.2.3.3.3 Translational Surfaces*

Ruled Bézier Surfaces

Again, a special case of Bézier surfaces is ruled Bézier surfaces. They occur when the Bézier curves in the **u** and/or **v**-direction are first order. These are then identical with straight lines. If the Bézier curve is only in one direction of order 1 and no further specifications are made to the control polygons

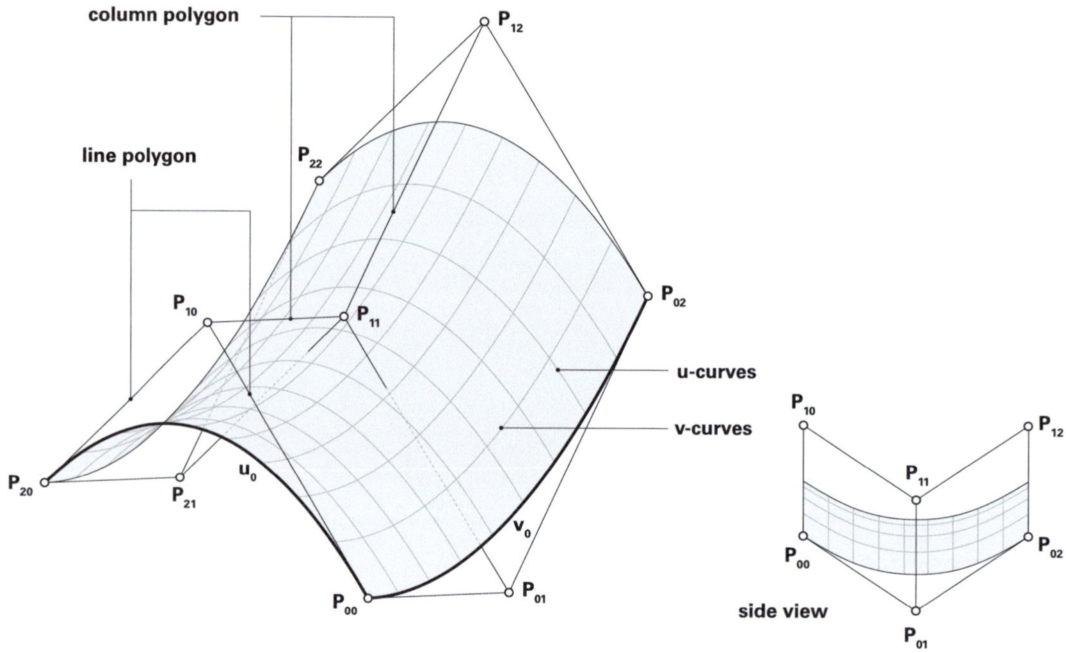

Fig. 1.172 Generic Bézier free-form surface (order **3,2**). Through largely free definition of the control polygons, a variety of free-form surfaces can be generated for design purposes

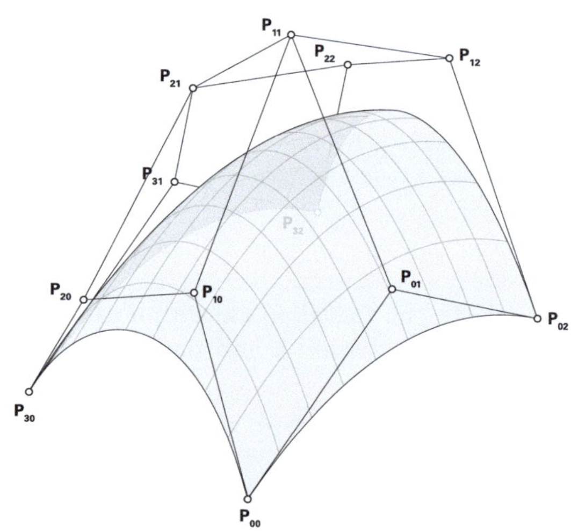

Fig. 1.173 Generic Bézier free-form surface (order **3,2**). Variant to the surface on Fig. 1.172

in the opposite direction, **generic ruled surfaces** are created (Fig. 1.176). If, on the other hand, more far-reaching specifications are made for the control polygons in the opposite direction, special cases arise:

- with congruent and parallel control polygons in the opposite direction, the result is a **generic cylindrical surface** (Fig. 1.177);

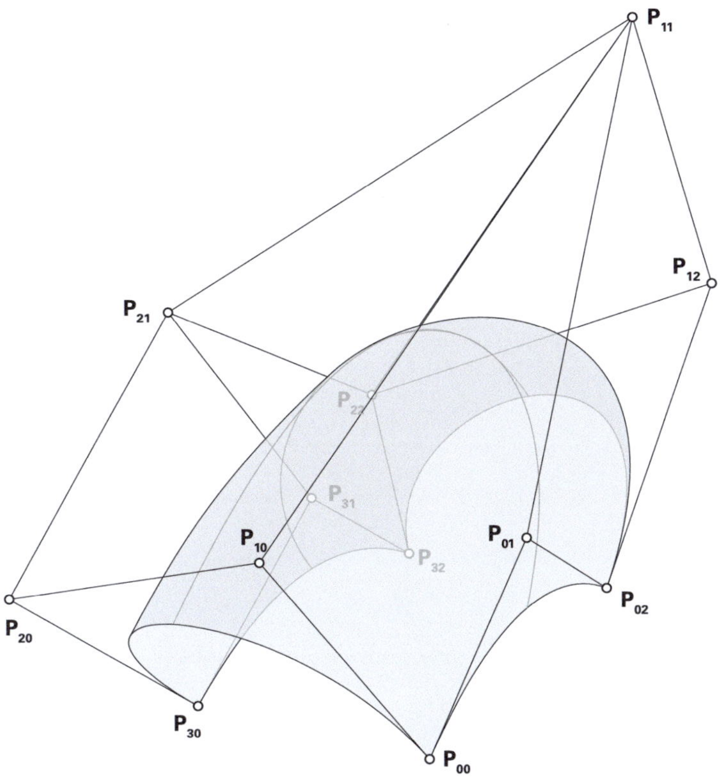

Fig. 1.174 Generation of a **Bézier translational surface**: Two Bézier curves of second order $\mathbf{u}_0$ and $\mathbf{v}_0$ define the surface (order **2,2**). It is created both by $\mathbf{u}_0$ sliding along $\mathbf{v}_0$ and vice versa by $\mathbf{v}_0$ sliding along $\mathbf{u}_0$. The control polygons in direction **u**: $\mathbf{P}_{i0}\ \mathbf{P}_{i1}\ \mathbf{P}_{i2}$ are each congruent (row polygons). The same applies to the control polygons in the direction **v**: $\mathbf{P}_{0i}\ \mathbf{P}_{1i}\ \mathbf{P}_{2i}$ (column polygons). It follows from the translational movement that all **u**-grid curves are congruent among themselves as well as all **v**-curves. Only the Bézier curves of the edge polygons $\mathbf{P}_{00}\ \mathbf{P}_{10}\ \mathbf{P}_{20}$ and $\mathbf{P}_{02}\ \mathbf{P}_{12}\ \mathbf{P}_{22}$ are included in the surface. The same applies in complementary direction **v**

- with affine and parallel control polygons in the opposite direction, a **generic conical surface** is created (Fig. 1.178).

If the curves in **u**- and **v**-direction are of order 1, a hyperbolic paraboloid is formed (Fig. 1.179).

1.2.8.3 B-Spline Surfaces

The characteristic properties of B-spline curves can also be transferred to B-spline surfaces: They are more flexible than Bézier surfaces because they can be better modified locally; the order of the curves in both directions **u** and **v** are freely selectable—independently of the side number of the polygon **m**, but not greater than this (Fig. 1.180).

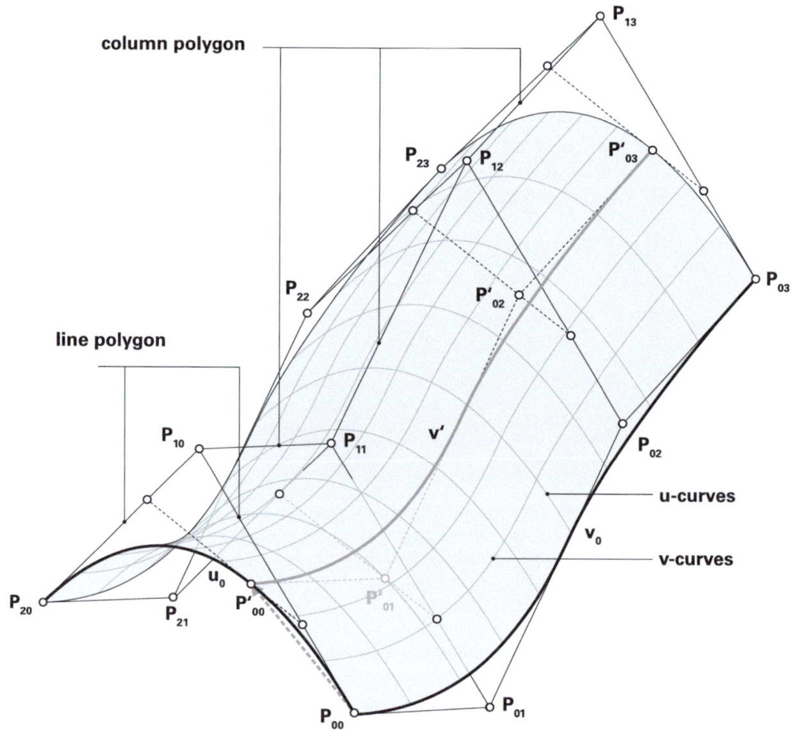

Fig. 1.175 Generation of a **Bézier translational surface** from a second order $\mathbf{u}_0$ Bézier curve and a third order $\mathbf{v}_0$ Bézier curve (order **2**,**3**). In addition to the translation process as described on Fig. 1.174, which can be transferred analogously to this case, the surface can also be created by applying the Casteljau algorithm to the control polygons: By dividing the segment in the line polygon $\mathbf{P}_{00}\,\mathbf{P}_{10}\,\mathbf{P}_{20}$, this gives the intermediate point $\mathbf{P'}_{00}$, to which a **v**-curve is moved. The interpolated intermediate points $\mathbf{P'}_{00}$ to $\mathbf{P'}_{03}$ yield the control polygon of the applied curve **v'**. The surface is created by successive interpolation of points. The same process can be applied to the **v**-direction by applying the Casteljau algorithm in a similar way to the column polygons. Only the Bézier curves of the edge polygons $\mathbf{P}_{00}\,\mathbf{P}_{10}\,\mathbf{P}_{20}$ and $\mathbf{P}_{03}\,\mathbf{P}_{13}\,\mathbf{P}_{23}$ are included in the surface. The same applies in complementary direction **v**

1.2.8.4 NURBS Surfaces

Similar to the NURBS curves, the weighting **w** of the individual nodes can be freely adjusted. In addition to the choice of the order of the curves, this opens up more diverse possibilities of shape manipulation for design purposes (Fig. 1.181).

1.2.9 Free-Form Surfaces from Developable Partial Surfaces

Many of the surfaces, discretised to polygon meshes, considered so far pursue the purpose of creating an acceptable approximation to a curved surface on the one hand, but on the other hand, for reasons of better and simpler constructional implementation, to be realisable from planar partial surfaces if possible. Minor deviations from planarity can sometimes be accepted, depending on the elasticity or plasticity of the material, because minor distortions of the partial surfaces can be

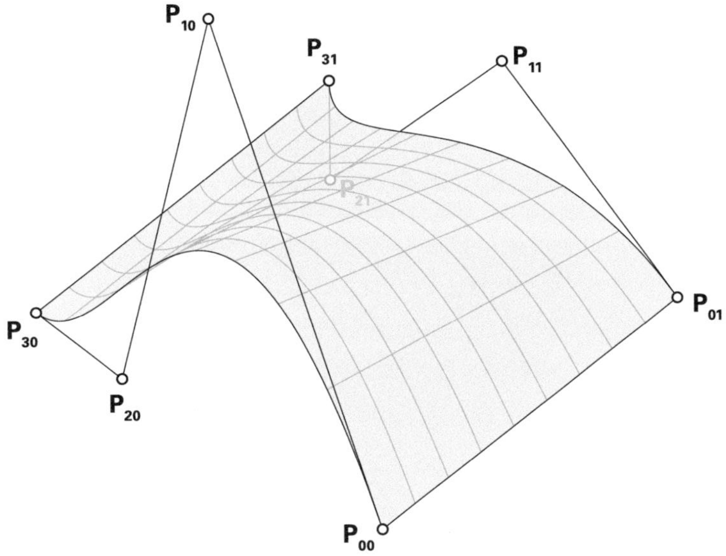

Fig. 1.176 The **Bézier free-form surface** of order **3,1** is a ruled surface. Since in this case both line polygons are different, this is a **(generic) ruled surface**, i.e. the generating lines in the **v**-direction are neither parallel nor do they meet at a point

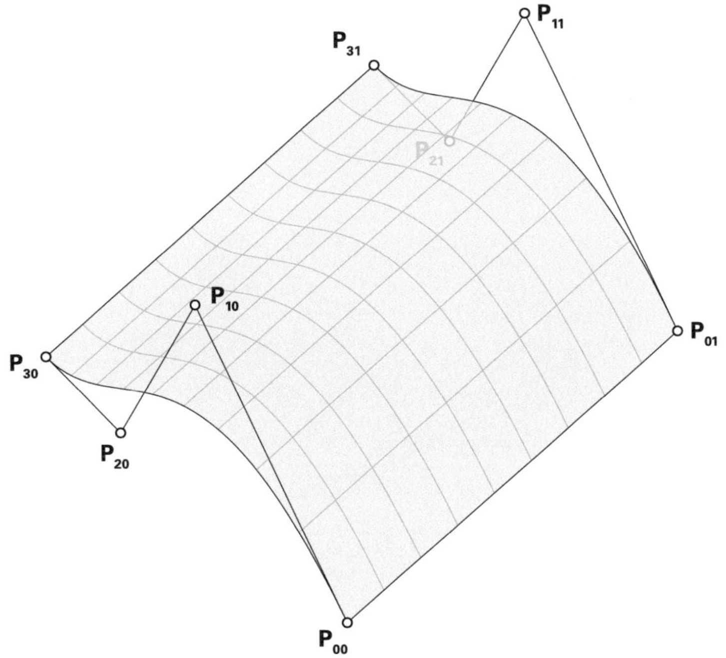

Fig. 1.177 This **Bézier free-form surface** of order **i,1** is also a ruled surface. Since in this case both line polygons are congruent in the **u**-direction, this is a **(generic) cylindrical surface** (order **3,1**), i.e. the generating lines in **v**-direction are parallel to each other

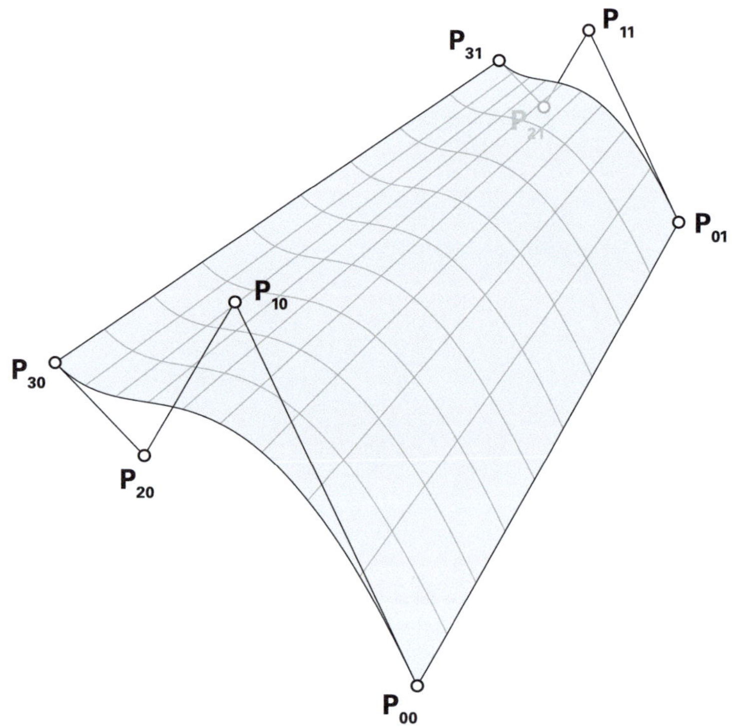

■ **Fig. 1.178** **Bézier free-form surface** of order **3,1**. Since in this case both line polygons are affine, there is a **(generic) conical surface**, i.e. the generating lines in the **v**-direction meet at one point

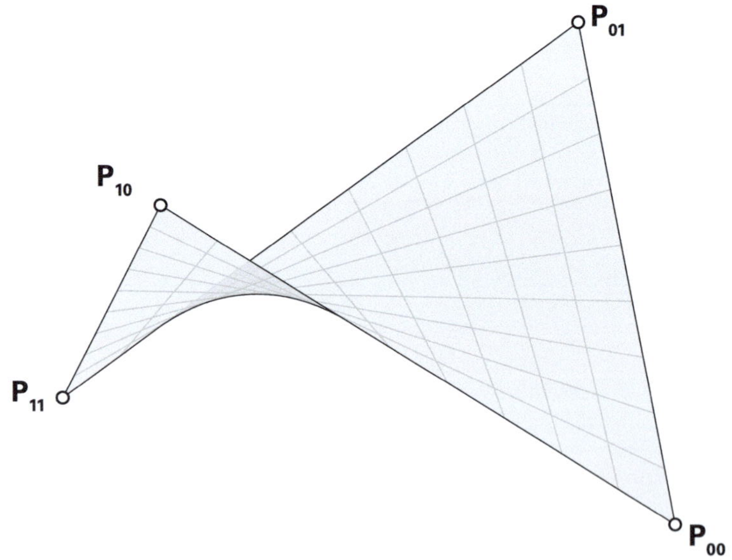

■ **Fig. 1.179** A **Bézier free-form surface** of order **1,1** is a **hyperbolic paraboloid**, which is created by this method as a control surface by straight lines as **u**- and **v**-curves

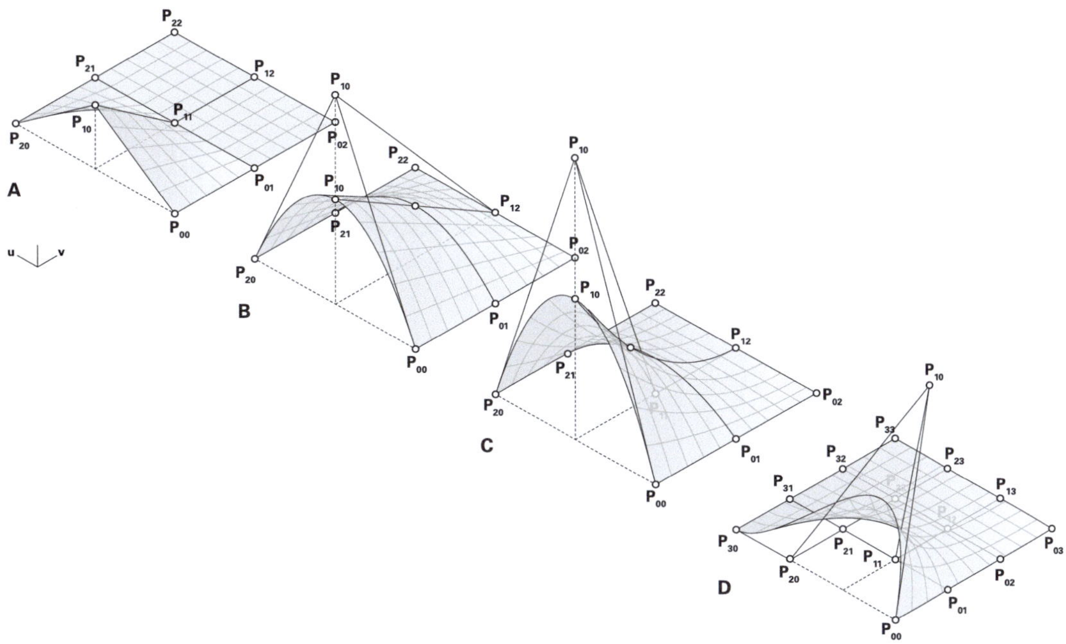

Fig. 1.180 **B-Spline surfaces**: **A** order **u 1**, **v 1**; the displacement of point $\mathbf{P_{10}}$ gives two planar quadrangles and two hyperbolic paraboloids $\mathbf{P_{10}}$. **B** order **u 2**, **v 1**; parabolas in **u**-direction and straight lines in **v**-direction: a ruled surface, a conoid. **C** order **u 2**, **v 2**; Second order curves in **u**- and **v**-direction. **D** order **u 3**, **v 3**; second or third order curves in **u**- and **v**-direction

forcibly induced by wilful deformation, so to speak, by pressing them into the desired shape. This is the case, within certain limits, with laminated glass, for example.

However, flat starting materials can sometimes be bent freely, so that they are able to generate spatially curved surfaces. This applies, for example, to thin wood-based panels, textiles, foils, or thin sheets. In addition to the characteristic elasticity of the material, the ratio of the thickness to the dimension in the bending direction is also decisive. The element must be thin enough to be deformed to smaller radii of curvature. In the physical model, paper strips, for example, can be used for the purpose of approximate shape determination in early design phases (Fig. 1.182).

More detailed considerations on the creation of curved surfaces from individual elements can be found in the following ► *Sect. 1.3 Constructional Implementation of Continuous Curved Layer Surfaces*

As discussed above, only **uniaxially curved surfaces** can be produced by this method, i.e. **ruled surfaces**, and within this group only the **torses**. This is, so to speak, the reversal of the development process, i.e. the conversion of a planar surface element into a curved surface (Fig. 1.182). Since the developable curved surfaces can be cut out in the plane at their side edges along any desired line, it is still possible to generate non-developable, biaxially curved surfaces, including free-form surfaces, by discretised assembly of flexible, or per-se uniaxially curved, surfaces (Fig. 1.183). In this process, they are joined together at their side edges, resulting in a surface that is continuously curved in one direction and discretised or faceted in the complementary one. These are called **semi-discrete surfaces**.

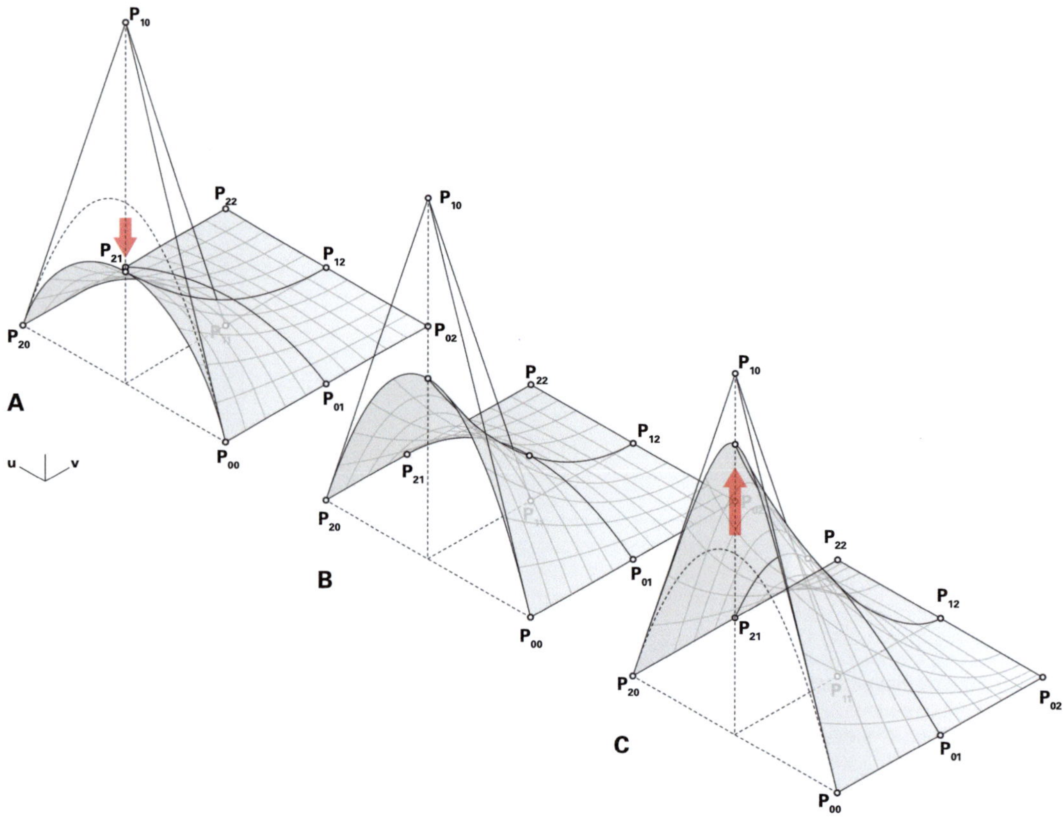

Fig. 1.181 **NURBS surfaces: A** Weighting of node $\mathbf{P_{10}}$: $\mathbf{w_A < 1}$. **B** Weighting of node $\mathbf{P_{10}}$: $\mathbf{w_B = 1}$. **C** Weighting of node $\mathbf{P_{10}}$: $\mathbf{w_C > 1}$

Similarly, a double-curved surface can be covered with flexible, planar strips without creases (Fig. 1.184). The stripes follow **geodesic lines**, i.e. the shortest connections between two points on the surface. According to this method, such surfaces can be covered without gaps in the constructional implementation with developable strips placed next to each other, e.g. thin sheets or thin wood-based panels (Figs. 1.185, 1.186, and 1.187). However, the side edges must be cut accordingly so that no gaps occur. For practical implementation, it is best to choose the greatest curvature of the surface for the alignment of the strips, as they are most flexible in their longitudinal axis. Cross-wise, in the direction of the smaller curvature, the edges from the discretisation appear less visible.

1.3 Constructional Implementation of Continuous Curved Layer Surfaces

Starting from the theoretical considerations regarding the geometry of curved surfaces as were the subject of the last sections, we shall now examine how the most important regular,

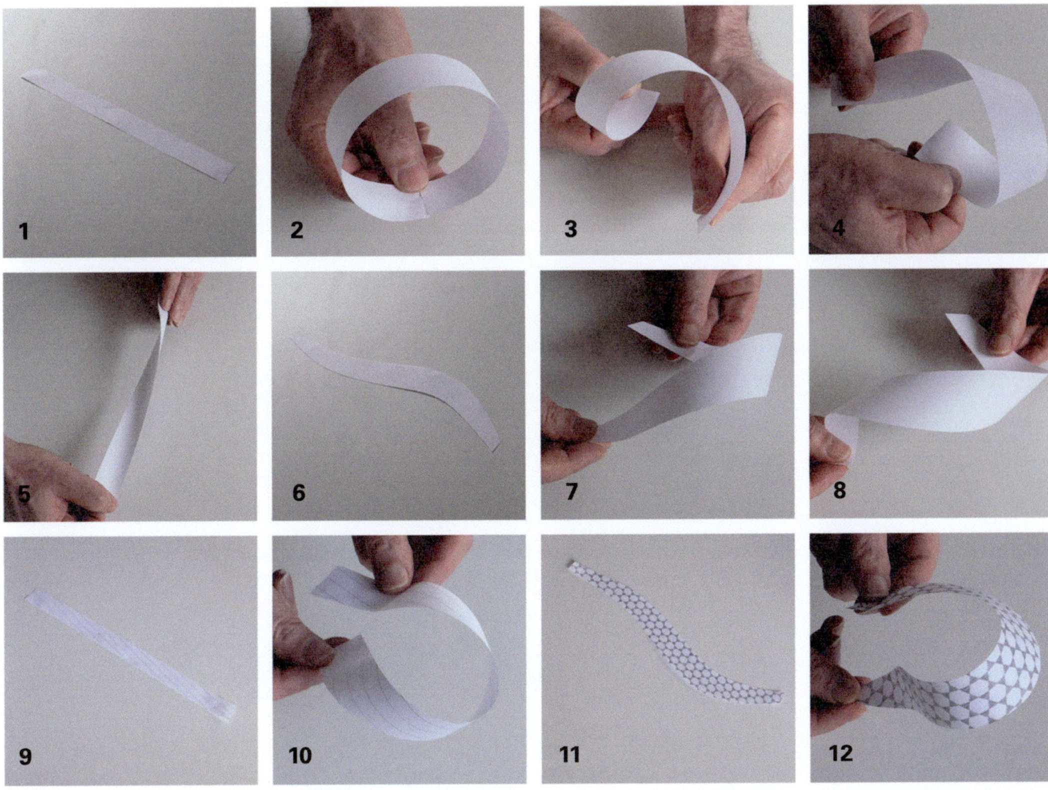

Fig. 1.182 Reversal of the development process: Unrolled planar elements, which are thin and elastic enough to be bent (here paper strips), can assume a variety of curved shapes. The result is always uniaxially curved surfaces: cylinders, cones or tangential surfaces to a curve. **1–5** Strip with parallel straight side edges. **6–8** Strips with arbitrary, non-parallel side edges. Patterns applied on the developed surface remain distortion-free on the curved surface. Modular or repetitive patterns allow the surface to be filled with repeating, planar elements when executed in building practice. **9**, **10** Strips with parallel lines with any angle to the side edges (**9**). In this way (**10**) the curved surface can be covered with strips of always the same width without gaps. As the straight lines on the development (**9**) are the shortest connection between two points lying on them, they are also the shortest connection on the curved surface (**10**): i.e. they all are geodesic lines. **11**, **12** Any area-filling patterns on the development (**11**) remain undistorted on the curved surface (**12**). This can be covered with the same tiles, for example

uniaxially and biaxially curved geometric shapes can be converted into built form using customary structural elements. In accordance with the classification already introduced, it is to be distinguished between band, plate, rod, and block-shaped elements.

1.3.1 Composing Uniaxially Curved Surfaces

1.3.1.1 Band-Shaped Starting Elements

In the case of **cylindrical surfaces**, strip-shaped material can be laid in the direction of the greatest curvature, that is to say along the **generatrix curve**, lying full-surface, without wrinkling, in butting lanes running parallel to each other (Fig. 1.188 left). In the case of a circular cylinder, this layout direc-

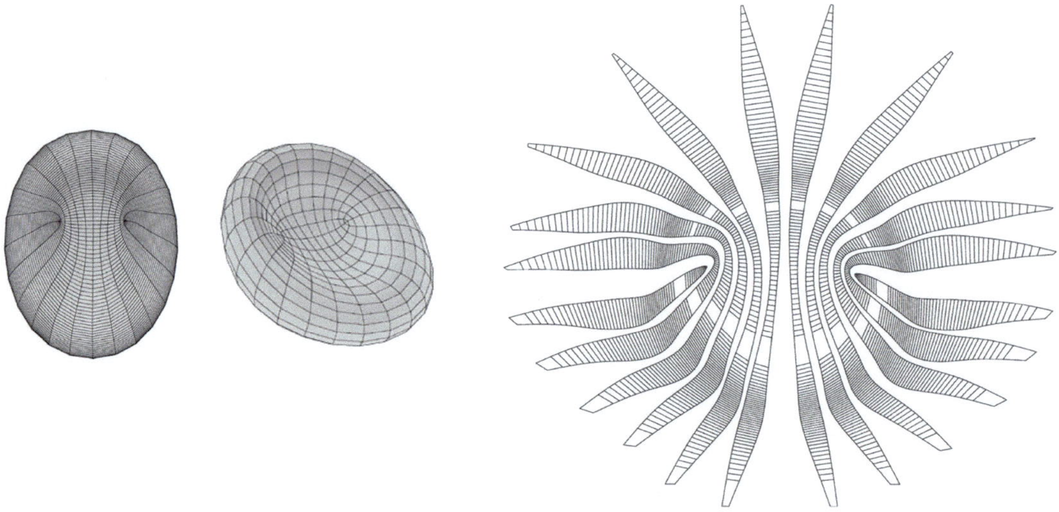

Fig. 1.183 Left, double-curved object with semi-discretised surface of developable strips. On the right, exemplary cutting pattern of the planar strips from which such objects can be assembled

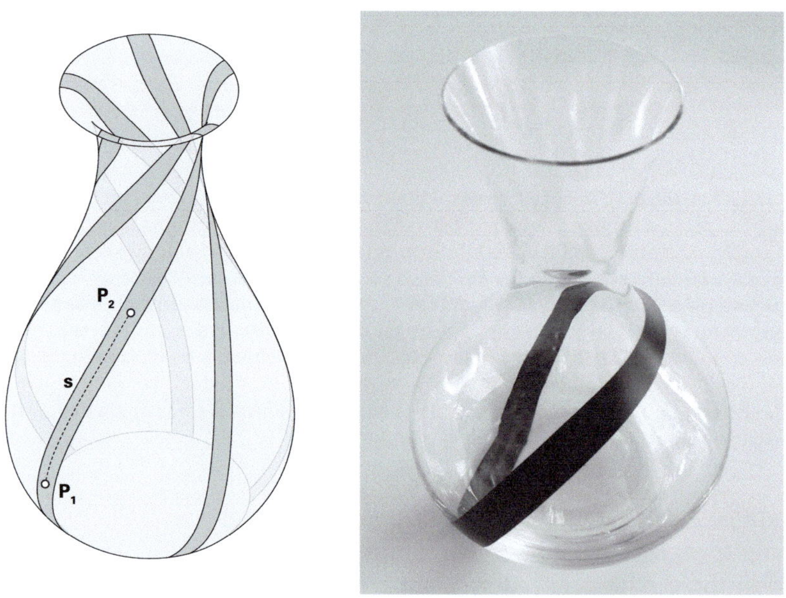

Fig. 1.184 (Left) Developable strips can be laid out wrinkle-free along a double-curved surface, provided they follow a **geodesic line**, i. e. the shortest connection distance **d** between two points $\mathbf{P_1}$ and $\mathbf{P_2}$ on the surface. (Right) Line routing alternative to the one on the left. The adhesive strip was placed freely along its axis, without pulling sideways, on the double-curved surface of the vase. The starting point (bottom) and direction were initially chosen arbitrarily, then the direction resulted automatically. The end meets the beginning eventually. The strip automatically follows a **geodesic line**

tion is the **circle of latitude** (Fig. 1.188 left). Even along the **straight generatrix line**, that is, in the direction of zero curvature, strip material can conform to the curvature of the cylindrical surface (Fig. 1.188 right) because of its flexibility in rolled-out condition. In this way as well, covering the surface

Fig. 1.185 Sculpture made of bent developable strips (ICD/ITKE, University of Stuttgart)

Fig. 1.186 Pavilion of spatially interwoven developable strips of wood-based material (ICD/ITKE, University of Stuttgart)

with strip material in parallel, mutually adjacent lanes is possible. In addition, strip material can also be rolled out on cylindrical surfaces along **geodesic lines** running obliquely to the straight generatrix line (Fig. 1.189 left). Since the cylinder surface can be developed onto the plane, a band can be flat-

► *Sect. 1.2.9 Free-Form Surfaces from Developable Partial Surfaces*

Fig. 1.187 Surfaces covered with developable metal-sheet strips (Walt Disney Concert Hall, Los Angeles; arch: F. Gehry)

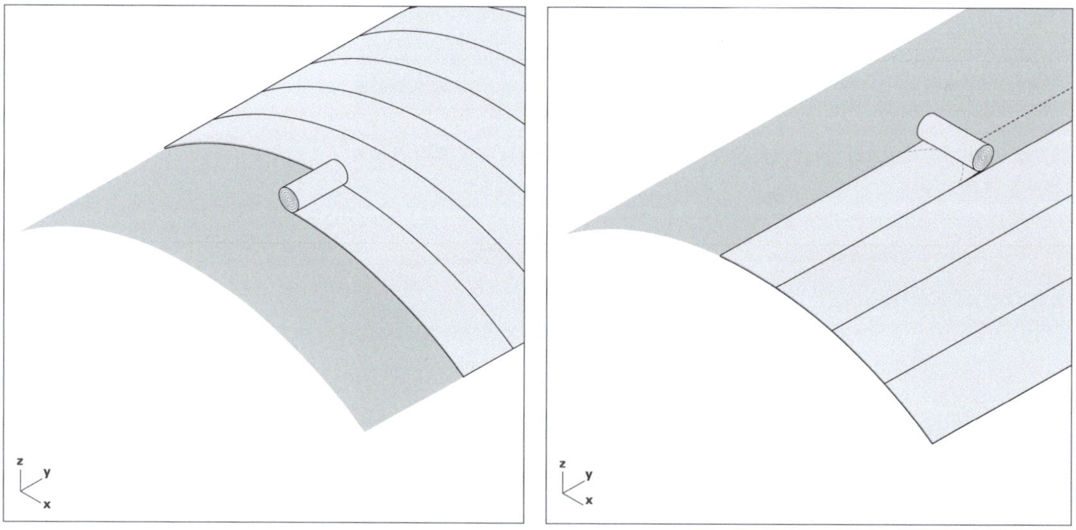

Fig. 1.188 (Left) Strip material can be unrolled without creasing along the directrix of a cylinder surface (here, parallels of latitude). (Right) Strip material can also be laid without wrinkling along the straight generatrix line of the cylinder surface

tened in any oblique position $\mathbf{P_1P_2P_3P_4}$. During the conversion into a cylinder surface, the course of the strip results in a spatial helix line ($\mathbf{P'_1P'_4}$, $\mathbf{P'_2P'_3}$), on which the belt wraps the cylinder in a wrinkle-free manner. Given the comparatively flat cut-outs of cylindrical surfaces, as are often found in construction, the trajectory of the strips appears to be almost linear in the plan (Fig. 1.189 right; see also Figs. 1.190, 1.191, and 1.192).

In terms of construction, this means that cylindrical surfaces can be covered without a crease even at any arbitrary

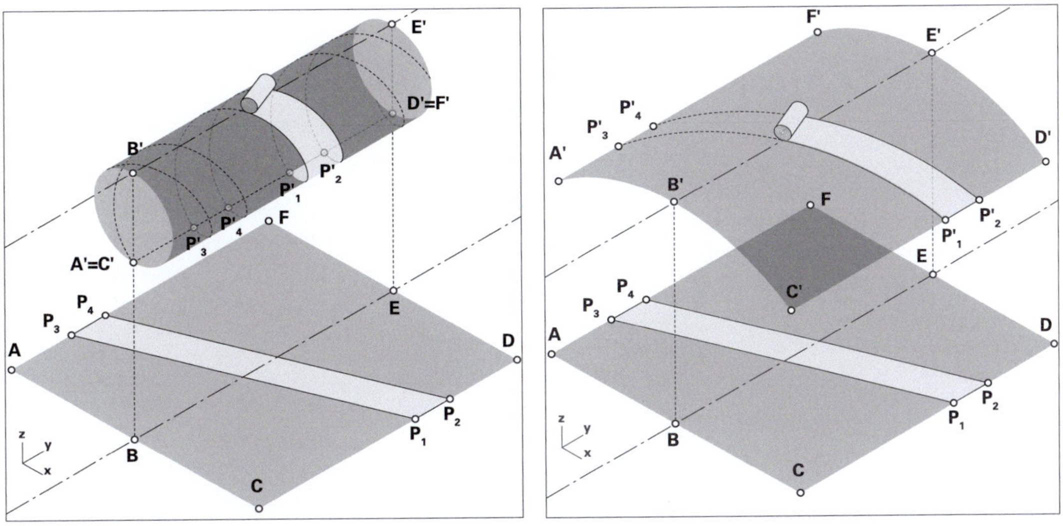

Fig. 1.189 (Left) Crease-free wrapping of a cylinder surface with a strip $\mathbf{P_1P_2P_3P_4}$ obliquely to the cylinder axis along a spatial helix line $\mathbf{P'_1P'_2P'_3P'_4}$. Below, in unwound state. It is a geodesic line. At the bottom, in unrolled condition. (Right) In the case of shallow cut-outs of cylindrical surfaces, the trajectory of the strip along a geodesic line approaches a straight line in the plan. Below, cylinder surface in unwound state

Fig. 1.190 Cylindrical surface along the parallels of latitude laid out with developable strips. Example for variant on Fig. 1.188, left

Fig. 1.191 Cylinder made of originally flat metal-sheet strips wound helically along geodesic lines

angle, provided that the strip follows the helix-shape geodesic line resulting from the respective roll-out angle.

The use of band material on **conical surfaces** is somewhat more restricted. On cones, it can no longer be laid free of wrinkles along the generating curve, in the case of regular circular cones, along the **circles of latitude**, since it is not a geodesic line (Fig. 1.193 left). Each of the two lateral edges of the strip runs along different circles of latitude $\mathbf{p_1}$ and $\mathbf{p_2}$, each of which has different radii. The band edge which follows the smaller circle of latitude $\mathbf{p_1}$ necessarily throws folds. In developed state, the circle describes a circular arc in the plane, which the band cannot follow once rolled out flatly. On the other hand, fold-free rolling-out along a generatrix line $\mathbf{l_g}$ is possible (Fig. 1.193 right). In the case of a belt with a constant width, the edges then no longer follow the straight generatrix line. For this to happen, the strip must be cut in the shape of a wedge along two generatrix lines $\mathbf{l_1}$ and $\mathbf{l_2}$, resulting in an offcut surface **O**. In this case, butting wedge-shaped belts can be laid

Strips as shown on Fig. 1.195, left, however, lie flat when developed and do not wrinkle when transferred to the conical surface

Fig. 1.192 Conical roof surface covered with flexible developable strips along parallels of latitude (ESC Armadillo, Glasgow; arch.: Foster Ass.)

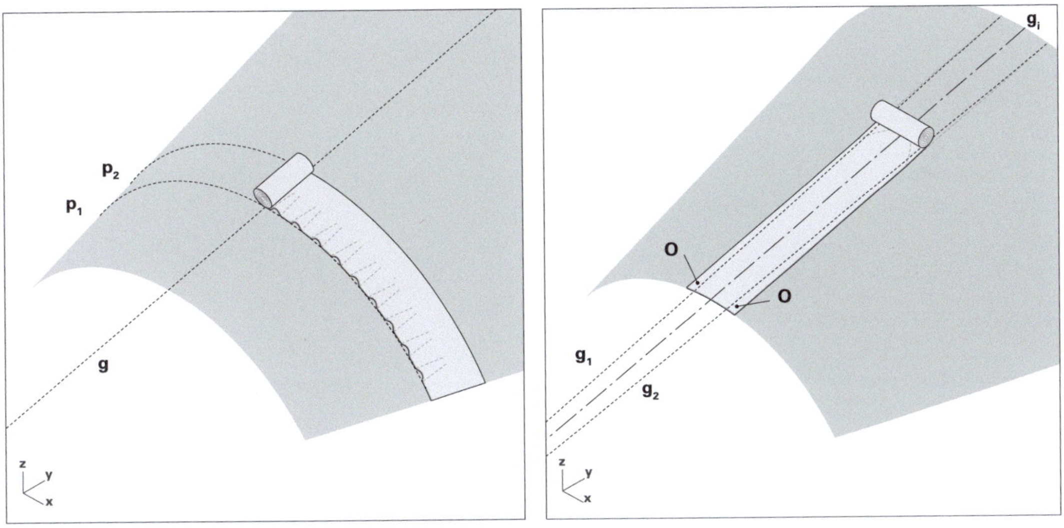

Fig. 1.193 (Left) A strip laid along a curved directrix (here, parallels of latitude $\mathbf{p}_i$) creases on one side (not a geodesic line). (Right) Strips laid along a generatrix line $\mathbf{g}_i$ are fold-free. The edges, however, do not follow the generatrix lines $\mathbf{g}_1$ and $\mathbf{g}_2$. This results in an offcut surface **O**

over the entire surface (Fig. 1.193 left), with the edges running along straight generatrix lines $\mathbf{l}_{gi}$. However, the strip can also be laid completely without cut-off edges in parallel adjoining lanes. All surface-covering laying patterns in the planar

1

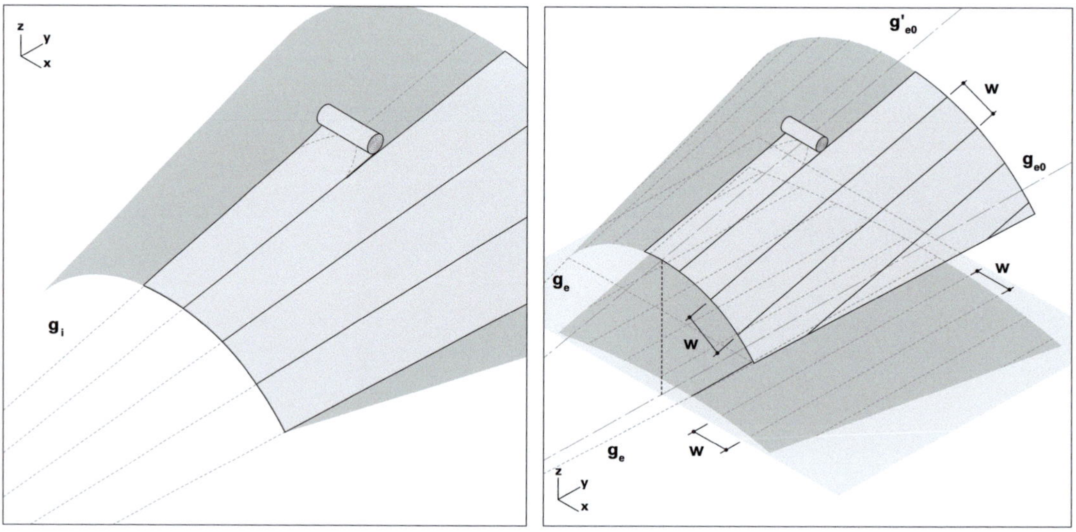

Fig. 1.194 Strips cut in wedge shape can be laid along straight generatrix lines $\mathbf{g_i}$ without creasing. (Right) Strips laid parallel to a selected generatrix line $\mathbf{g_0}$. The continuous flat laying pattern of the unrolled strips can be transferred to the conical surface, where it covers the latter without creases (along geodesic lines)

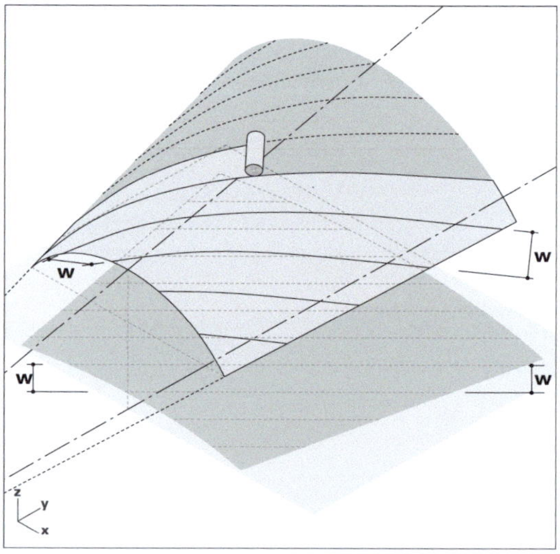

Fig. 1.195 Parallel strips laid at an arbitrary angle on the flat development cover the associated conical surface without creases (along geodesic lines)

development can be accordingly transferred to the curved conical surface without any crease. In this context, it is indifferent whether the mutually parallel lanes follow a selected straight mantle line (Fig. 1.193 right), or whether they adopt an arbitrary angle (geodesic lines) (Figs. 1.194 and 1.195). The covering of the conical surface is always free of creases. Consequently, depending on the position of a belt linearly rolled-out on the development, a directrix can always be found on the as-

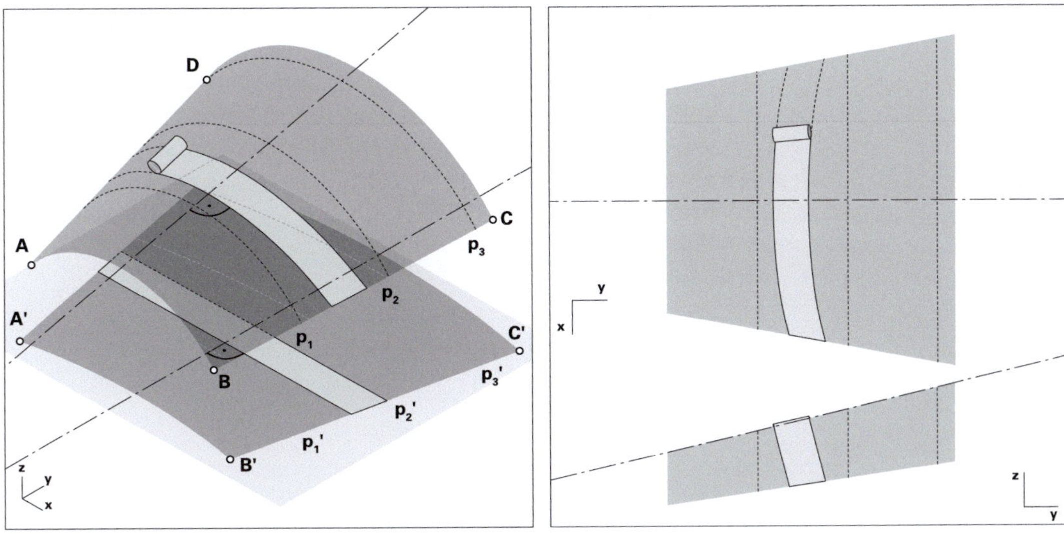

Fig. 1.196 (Left) Any desired position of a strip straightened-out in the flat projection can be transferred to the conical surface, so that the strip can be developed (along geodesic lines). (Right) Two-panel projection of the conical surface shown on the left with rolled-up strip

sociated conical surface which allows the belt to be rolled out completely crease-free on the conical surface (Fig. 1.196).

Transferring this to a complete conical surface, this means that—similarly to the cylindrical surface on Fig. 1.189 left—a conical surface can be continuously covered with a band free of wrinkles following a helix line—in this case with a constantly changing helix radius.

1.3.1.2 Plate-Shaped Starting Elements

Plate-shaped material can be placed on a **cylindrical surface** in such a way that, in each case, a single plate touches the surface on a straight generatrix line (Fig. 1.197). In the depicted arrangement, a gap is created under the butt joint with a constant distance δ from the cylinder surface. Each individual plate section is a **rectangle**.

If the joint distances are chosen identical, the individual plate sections are all identical likewise. If, in addition, the joint distances in both directions—the straight generatrix line and the generating curve—are chosen to be the same, the plates become **squares**. It is obvious that the denser the faceting is, that is, the narrower the joint distances are, the closer the approximation of the resulting polyhedron to the cylinder surface results. Here, the characteristic of the cylinder as a **translational surface** is taken advantage of ► *Sect. 1.2.3.3.3*.

On a **conical surface**, plate-shaped material can be placed on top, in analogy to the cylinder surface (Fig. 1.198, right). The plate sections are tangent to the conical surface on a **straight generatrix line** $\mathbf{l_g}$. Again, a distance δ_i emerges at the plate seams, which in this case, however, is variable: It becomes larger the further away from the cone vertex it is. The individual plate elements become trapezoidal in their cutting pattern (Fig. 1.198, left). This follows from the property of the conical surface as a (scaled) **translational surface**. A row along a **directrix curve c_d** can consist of identical elements, provided the angles between the joint lines in the direction of the generatrices are the same.

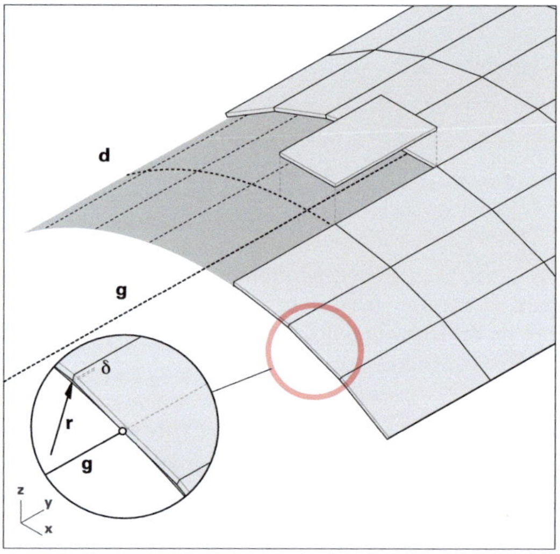

Fig. 1.197 Formation of a polyhedral surface from rectangular plate sections as an approximation to a **cylindrical** surface

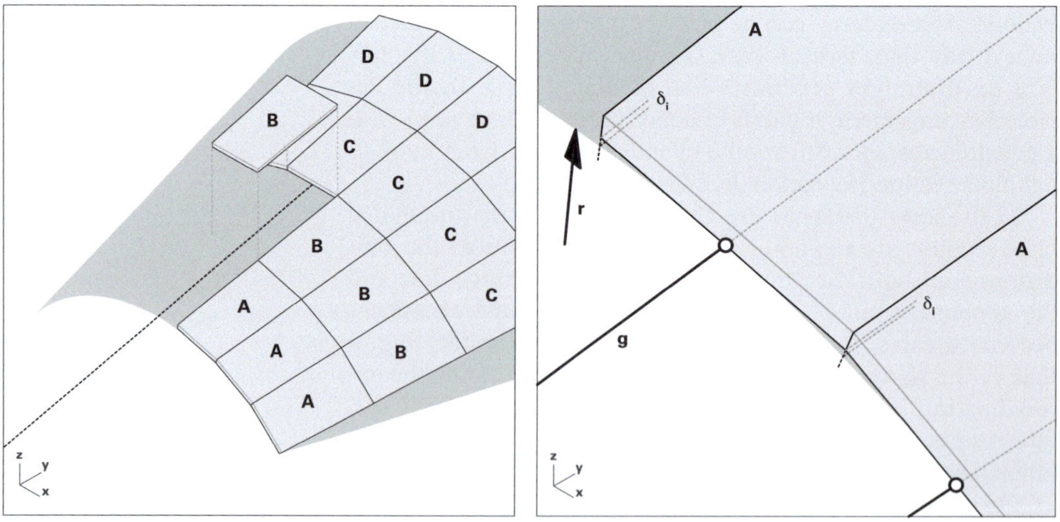

Fig. 1.198 (Left) Polyhedral surface from trapezoidal plate elements as an approximation to a **conical** surface. (Right) Detail of the polyhedral surface. Tangential contact on a straight generatrix line **g**, variable distance δ_i to the cylinder surface at the plate seams

1.3.1.3 Rod-Shaped Starting Elements

Rod-shaped material allows good adaptation to a cylindrical surface when the rods run in the direction of the straight generatrix line $\mathbf{l_g}$ (◘ Fig. 1.199 left). Precisely, the small width of the rod material allows a narrow faceting of the polyhedron and, consequently, a good fit to the curved surface of the cylinder. Analogously to plate material, the individual rod also rests on a straight generatrix line $\mathbf{l_g}$ on the cylinder surface and touches it there. At the butt joint, spacings δ to the cylinder surface are also formed, which, however, are only minimal because of the fine faceting. In the case of flexible material, they can possibly also be pressed close as a result of the fastening.

Furthermore, a cylindrical surface can basically also be covered in the direction of the **directrix curve** $\mathbf{c_d}$ with rod-shaped material placed parallel to one another (◘ Fig. 1.199 right). This may, for example, be necessary if the water-flow direction along the greatest curvature, i.e. the directrix, is to be taken into account, and as few joints as possible should lie transversely thereto. The precondition for this is, of course, that the rod material can cope with the curvature of the surface in the direction of the directrix curve $\mathbf{c_d}$. This requires either thin rods, flexible material, or very small curvatures. The covering then lies upon the entire surface of the cylinder. If the curvature is too large for such a covering pattern, rod material can also be laid obliquely to the cylinder axis (◘ Fig. 1.200). The curvature to which the rods are subjected can be gradually controlled by selecting the laying angle. However, it must be taken into account that the rods running diagonally experience a **twisting** along their axis, which the material must be able to withstand. The geometrical conditions can be compared with those of a belt on a cylindrical surface (cf. ◘ Fig. 1.189).

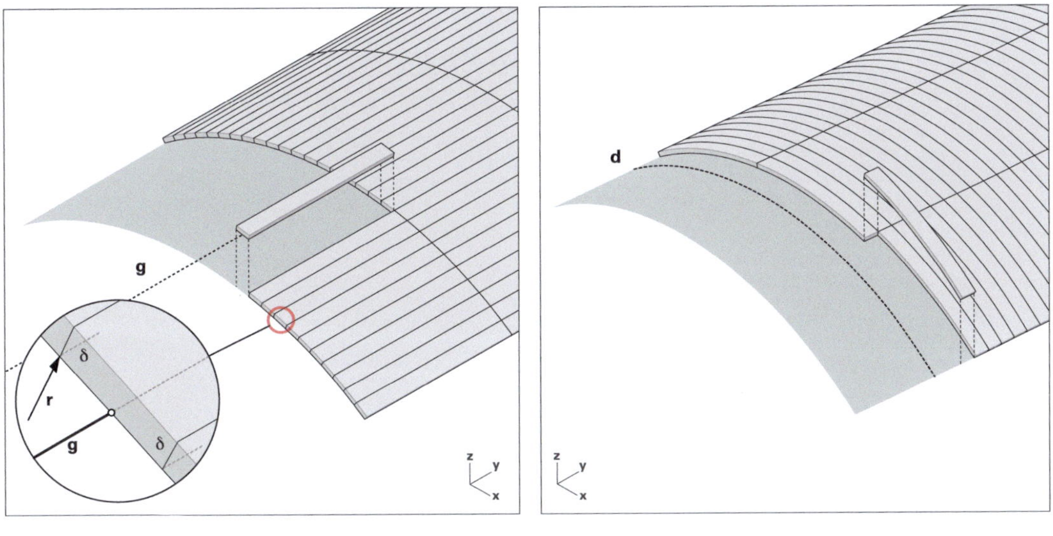

◘ **Fig. 1.199** (Left) Laying rod-shaped material on a cylindrical surface along the straight generatrix line **g**. (Right) Laying rod-shaped material on a cylindrical surface along the directrix **d**, that is, along the greatest curvature

On a **conical surface**, the simplest approximation using bar-shaped material is to lay out the rods along the straight generatrices (■ Fig. 1.201 left). In analogy to plate material, the bars must also be trimmed in trapezoidal form in order to avoid gaping joints. Since rod material is usually bound to a **maximum width w**, the rods can also be laid with **offset joints** (■ Fig. 1.201 right) in order to avoid excessive widths and too

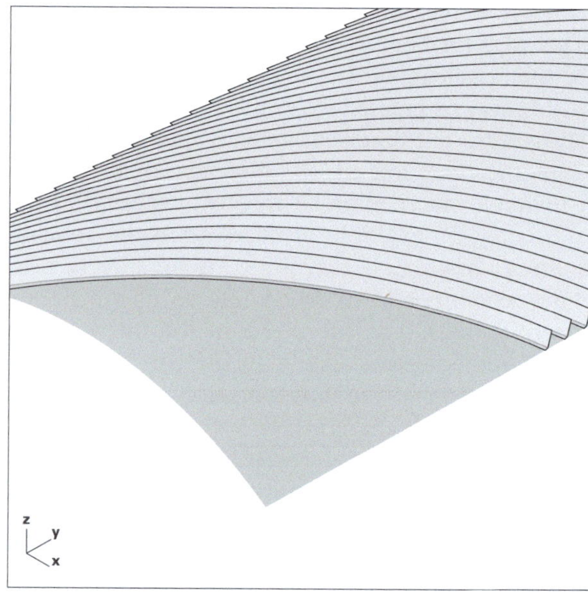

■ **Fig. 1.200** Laying rod-shaped material on a cylindrical surface obliquely to the cylinder axis. The rods are twisted

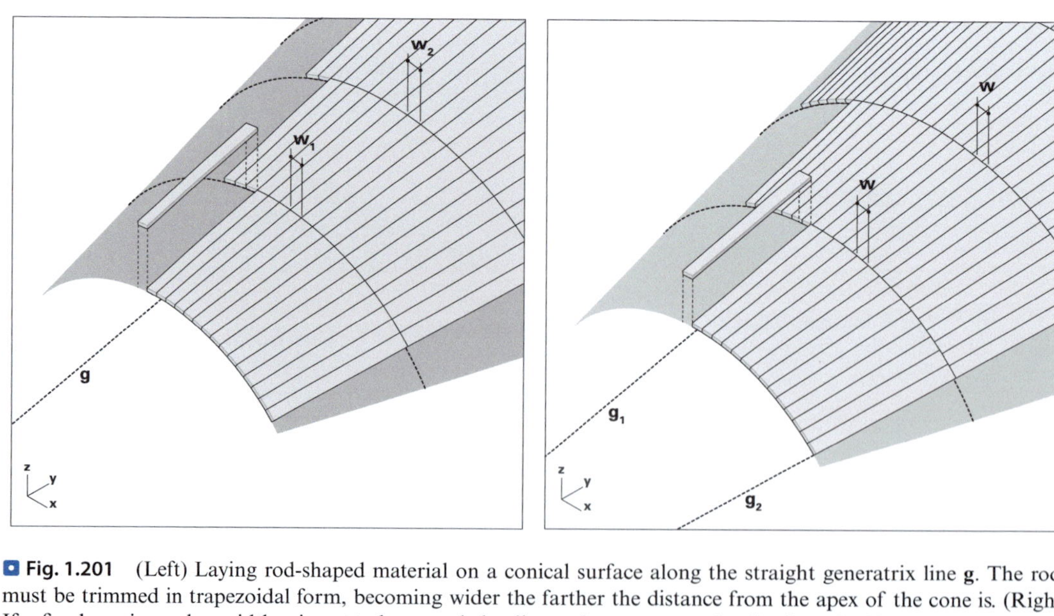

■ **Fig. 1.201** (Left) Laying rod-shaped material on a conical surface along the straight generatrix line **g**. The rods must be trimmed in trapezoidal form, becoming wider the farther the distance from the apex of the cone is. (Right) If a fixed maximum bar width **w** is not to be exceeded, adjacent bar rows must be distributed with **offset joints**, for example, starting out of a common generatrix line (like $\mathbf{g}_2$)

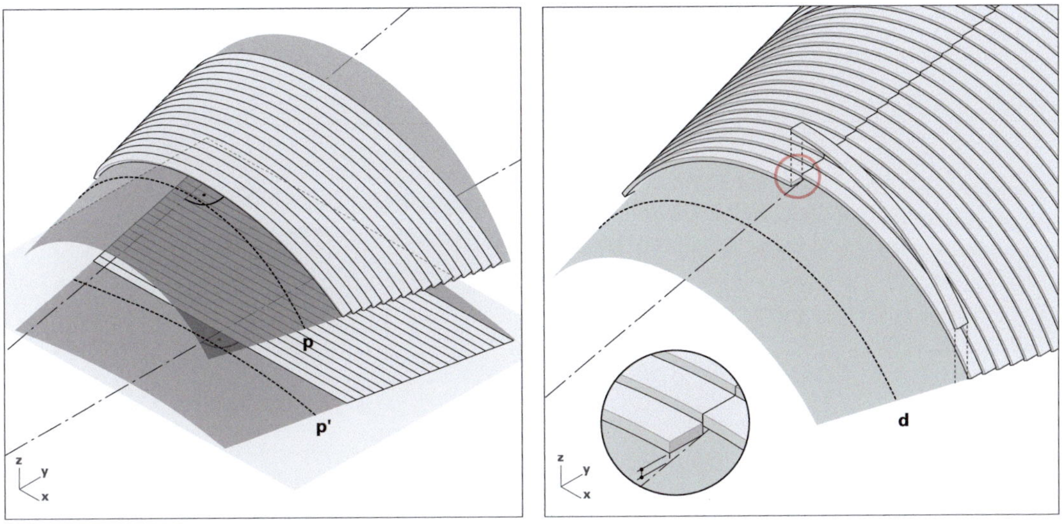

Fig. 1.202 (Left) Laying rod-shaped material on a conical surface flat along the curvature. The planar rectilinear laying pattern on the development changes into a curved one on the curved conical surface (along geodesic lines). The laying direction can also be realised with any other angle. (Right) Laying rod-shaped material on a conical surface along the greatest curvature, i.e. along a curved **directrix d** (here, a parallel of latitude). This type of laying results in a staggered arrangement of the bars

many different formats. Starting out from a common **reference line** (straight generatrix line $\mathbf{l}_{g0}$), the bars—then always identical—can be laid offset in continuous rows.

If the rods cannot be laid along the generatrix line $\mathbf{l}_g$ for special reasons—for example because of the water-flow direction—laying patterns can also be implemented in the same way as covering the conical surfaces with strip material in which the rods are laid out flat **following the curvature** (Fig. 1.202, left). They then follow a line which results from transferring the rectilinear trajectory on the developed surface into the curved conical surface (geodesic line). As with band material, the laying angle can be freely selected.

The laying of the rods along the **directrix curve** $\mathbf{c}_d$, i.e. at right angles to the generatrices (Fig. 1.202, right), impinges on boundaries in the cone, since the difference in length between the directrices under both side edges of a rod cannot generally be absorbed by the latter. If a rod is located with a side edge on a directrix of the cone, a **gap** to the cone surface is created on the opposite side. The consequence of this is a staggered abutment of the adjoining rods (Fig. 1.202, right, detail). The basic prerequisite for this type of layout is that the bar material can cope with the curvature.

1.3.1.4 Block-Shaped Starting Elements

If uniaxially curved surfaces, such as a cylinder or a cone, are to be covered or formed by means of block-shaped elements, the single module can assume different positions in relation to

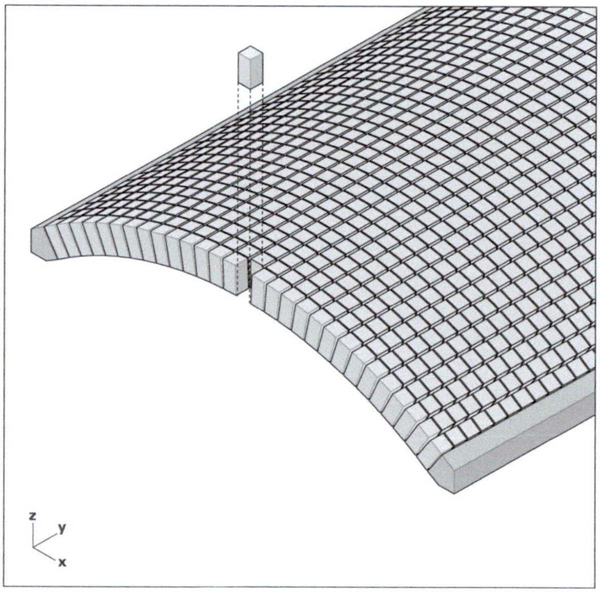

Fig. 1.203 Vault from upright blocks laid in cross-joint geometry. For the purpose of transferring forces between adjacent blocks, these must either be made wedge-shaped or the joints must be filled with filling material

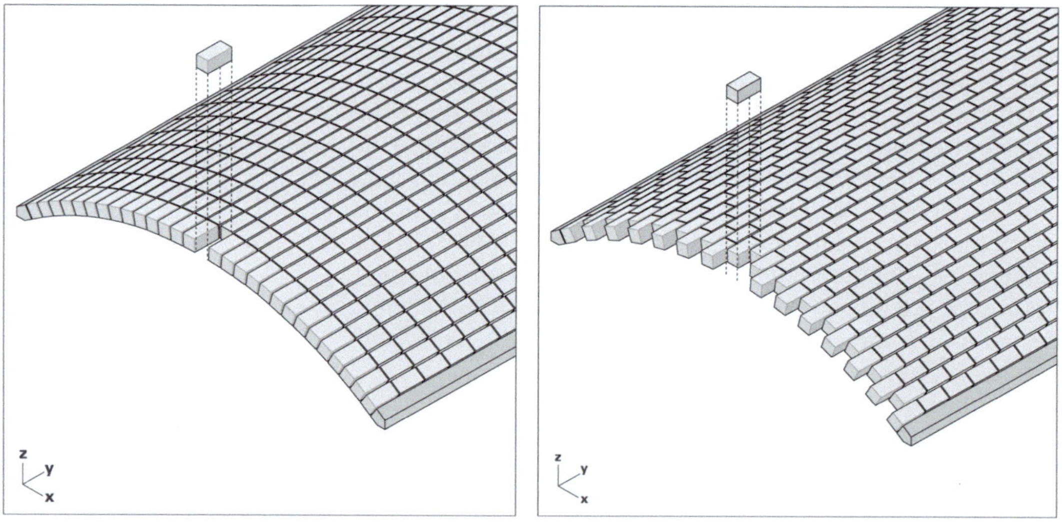

Fig. 1.204 (Left) Vault of horizontally laid stones in **cross-joint geometry**. It can be regarded as the addition of individual arches due to the continuous transverse joints. (Right) Vault made of horizontally laid blocks in a **staggered bond**. The interlocking of adjacent blocks in the longitudinal direction of the arch activates a load-bearing effect of the vault as a uniform shell and not just as the addition of parallel arches

the surface: Since its dimensions length/width/height—unlike those of bands, plates, or rods—are not substantially different from each other, the building block can occupy any desired position (Figs. 1.203 and 1.204 left) in relation to the surface (in perpendicular direction: length, width, or height, regardless). Because of the freedom when selecting the position of the module, the **thickness** of the covering or the self-sup-

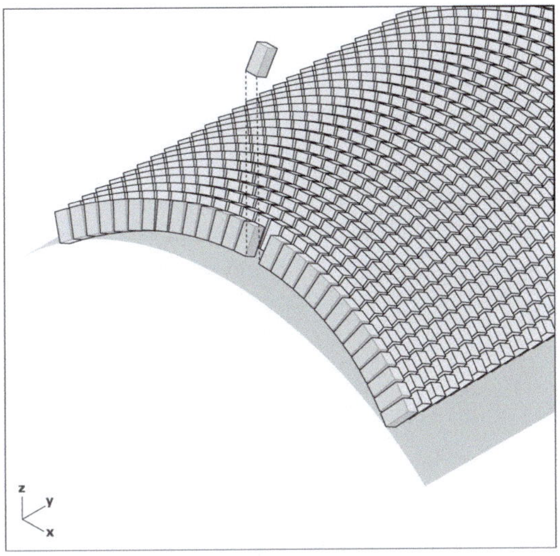

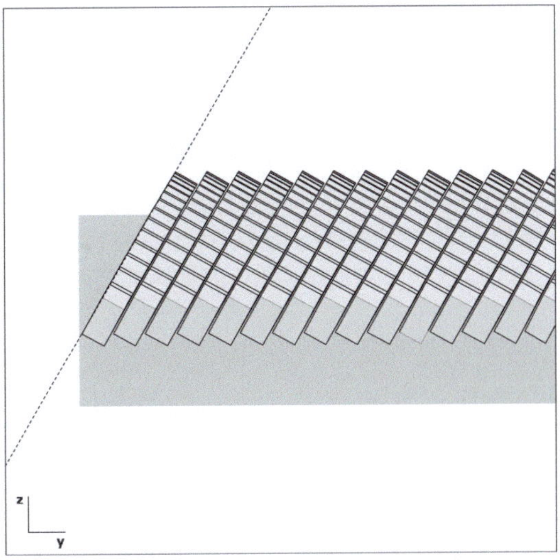

Fig. 1.205 (Left) Ring-shaped building-block layout in parallel planes which are inclined to the cylinder axis. In a regular circular cylinder, the rings follow an ellipse. (Right) Perpendicular elevation of the arrangement on the left

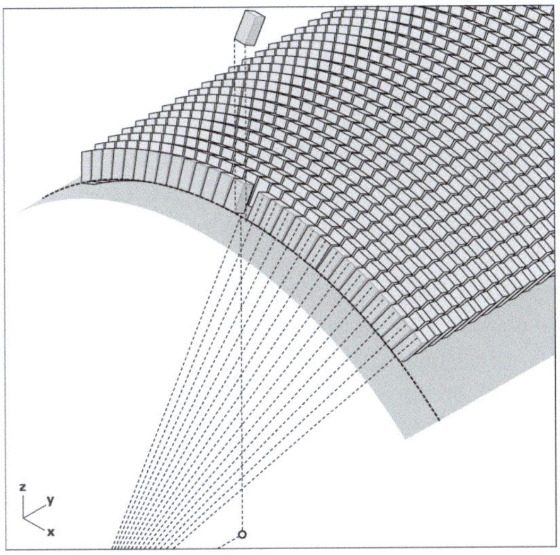

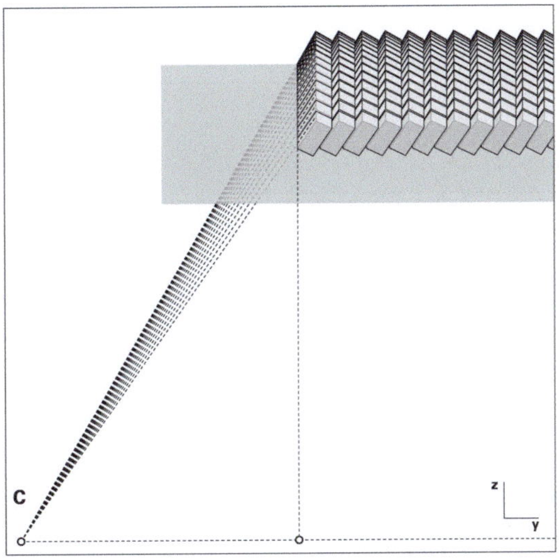

Fig. 1.206 (Left) Ring-shaped building-block layout along generatrices in tilted position. The individual building blocks of a ring are oriented towards a common centre **C**. Their side faces, overall, form a conical surface. (Right) Perpendicular elevation of the arrangement on the left

porting shell of individual elements can be determined very simply, i.e. alternatively equal to the length, width, or height of the building block. The small dimension of the starting element also allows good adaptation to curvatures, whatever the position of the module to the surface. Arrangements can be implemented along the straight generatrices or the curved directrices (Figs. 1.203 and 1.204) as well as diagonal orientations (Fig. 1.207 left).

In principle, the application of block-shaped components allows for a great adaptability to a wide range of surface ge-

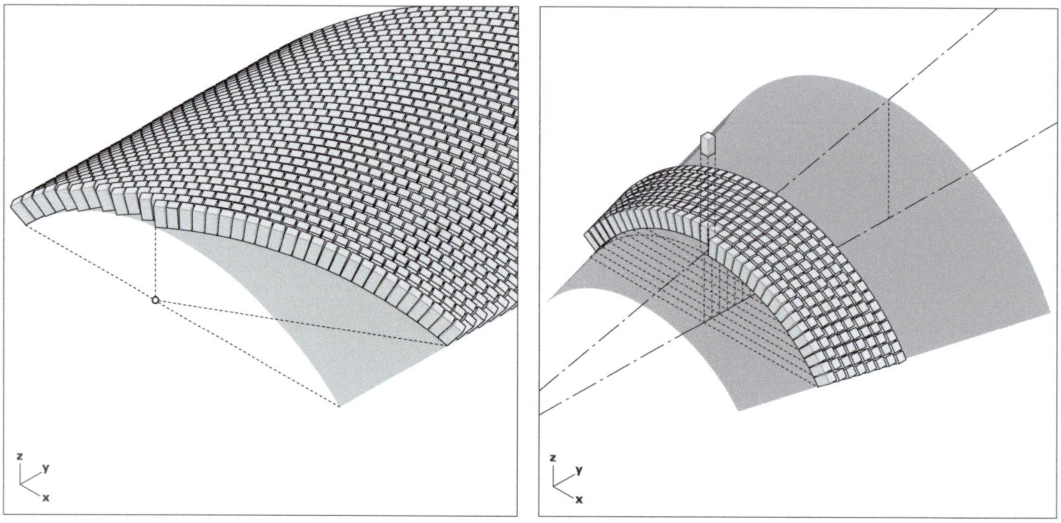

Fig. 1.207 (Left) Ring-shaped arrangement of building blocks in vertical planes running obliquely to the cylinder axis. With a regular circular cylinder, rings acquire an elliptical shape. (Right) Building-block laying on a conical surface in rings along directrices (here, parallels of latitude). In the case of a circular cone, these are each circular arcs with varying radius

ometries. The curvature can be accommodated both by the suitable **cutting pattern** of the module, which in certain cases can also require individual trimming and correspondingly high constructional hassle, as well as by **wedge-shaped joints** when using one recurring block format. The latter option is used, in particular, with prefabricated elements such as bricks; the former, for instance, with wedge-shaped stone voussoirs.

The variety of conceivable building-block arrangements on an uniaxially curved surface is very large, which is why only a few possibilities will be shown here. Non-right-angled positions of the block as to the surface, as well as diagonally lying block courses can be also realised (Figs. 1.205 and 1.206), something which can sometimes be advantageous for special constructional considerations. In these cases, it may be necessary to consider that, with circular cylinders or circular cones, blocks must be laid along lines which are no longer circular and which consequently also entail different block formats (with wedge cut) or varying joint widths (with wedge joints).

An example are Nubian vaults with inclined stone courses

The formation of curved surfaces, both with uniaxial and biaxial curvature, by means of block-shaped elements is of great importance, especially in the case of load-bearing applications in **masonry construction**. In this case, spans are bridged in the direction of the greatest curvature, that is to say the directrix, with freely spanning arch or vault-shaped constructions. Laying the building blocks along a continuous **ring** in this direction, i. e. in the form of an arch, is fundamental to the load-bearing capacity of these structures. The laying geometries, one would speak of **bonds** in masonry construction, are also decisive for the carrying behaviour of arch, vault, or

Fig. 1.208 Cylindrical vault with cross-joint pattern of bricks according to Fig. 1.204, left

dome constructions. Different block arrangements (Figs. 1.203, 1.204, 1.205, 1.206, and 1.207) are primarily explained by the static behaviour and the production of this type of construction (Figs. 1.208, 1.209, and 1.210).

► *Chap. 7 Masonry Construction* or ► *Sect. 4.2.2 Systems under Compression—Pitched Roofs and Vaults*

1.3.2 Composing Biaxially Curved Surfaces

Biaxially curved surfaces present particular difficulties when covering them with components in band, plate, or rod shape with maximal geometric approximation. The vast abundance of conceivable doubly curved surfaces forbids this subject from being treated even halfway comprehensively at this point. Instead, two representative doubly curved surfaces, such as the **hyperbolic paraboloid** and the **sphere**, which have previously been comparatively frequently used in construction, will be investigated in detail for their constructional realisability. Spherical surfaces were extremely important in historical vault construction; hyperbolic paraboloids were often realised in modern concrete shell construction.

As already noted, the hyperbolic paraboloid can simultaneously be regarded as a **translational surface** and a **ruled surface**. Many of the aspects discussed can be transferred by analogy to other surfaces from these categories. Similarly, the sphere can be regarded as an example of a **surface of revolution**, so that many statements pertaining to the sphere can also be extended to other surfaces of revolution.

► *Sect. 1.2.2 Geometric Conditions*

1.3.2.1 Hyperbolic Paraboloid

Considered as a **translational surface**, the hyperbolic paraboloid results from the sliding of a generating parabola $\mathbf{c}_g$ along a guiding parabola $\mathbf{c}_d$ (Fig. 1.211 left), whereby each of the

Fig. 1.209 Cylindrical vault with staggered bond according to Fig. 1.204, right

two parabolas can correspondingly assume the opposite function (Fig. 1.211 right). Where the vertices of both parabolas meet, the **vertex V** of the surface is located.

Alternatively, this surface can also be regarded as a **ruled surface** which is generated from two sets of straight lines **g′** (Fig. 1.212). Unlike uniaxially curved surfaces such as cylinders and cones, two adjacent straight lines are **skewed** against each other. This fact is of great importance to the constructional implementation of the surface.

Band-Shaped Starting Elements

When covering a hyperbolic paraboloid with band-shaped material, restrictions are even narrower than with a conical surface: Tapes rolled out along the generatrices $\mathbf{c_g}$ as well as the directrices $\mathbf{c_d}$ throw folds (Fig. 1.213 right). Only such bands which extend along parabolas near the vertex **V** can accommodate the distortions free of wrinkles, assuming a

Fig. 1.210 Conical vaults with stone rows along the parallels of latitude, corresponding to Fig. 1.207, right (Trulli in Apulia)

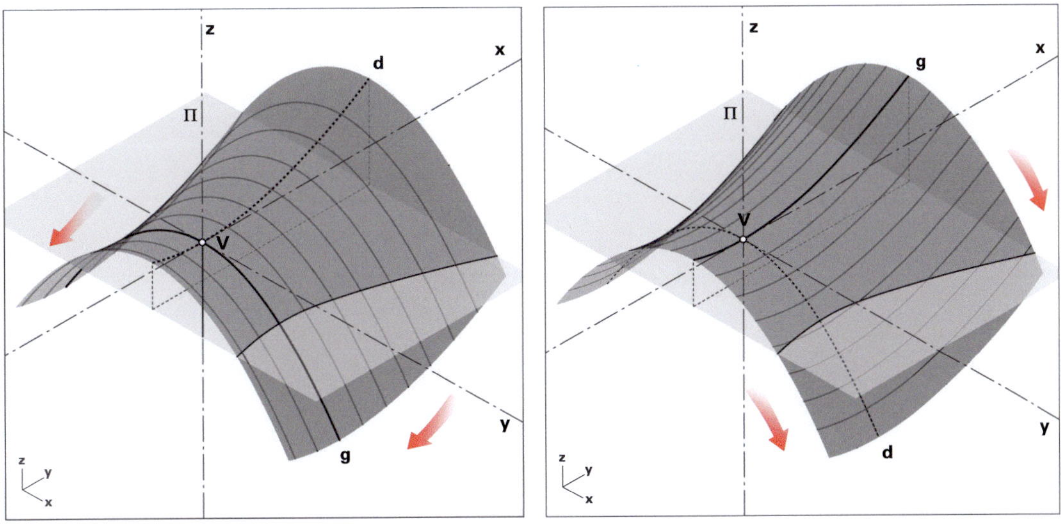

Fig. 1.211 (Left) The hyperbolic paraboloid as a translational surface: A parabolic generatrix **g** glides along a parabolic directrix **d**. (Right) It is also possible for the respective other parabola to assume the function of the generatrix **g** or directrix **d**

certain elastic or plastic characteristic of the material (Fig. 1.213 left). Laying bands along the straight generatrix $\mathbf{l}_g$ (Fig. 1.214) is simpler, though. However, it should be noted that even if the centre axis of the band follows the straight line $\mathbf{l}_g$, the strip nevertheless distinctly **twists** around this axis. In that case, it depends on the elastic or plastic deformability of the material of the band surface itself, to what extent this can happen without wrinkles. Furthermore, it must be taken into account that the bands cannot be cut with a con-

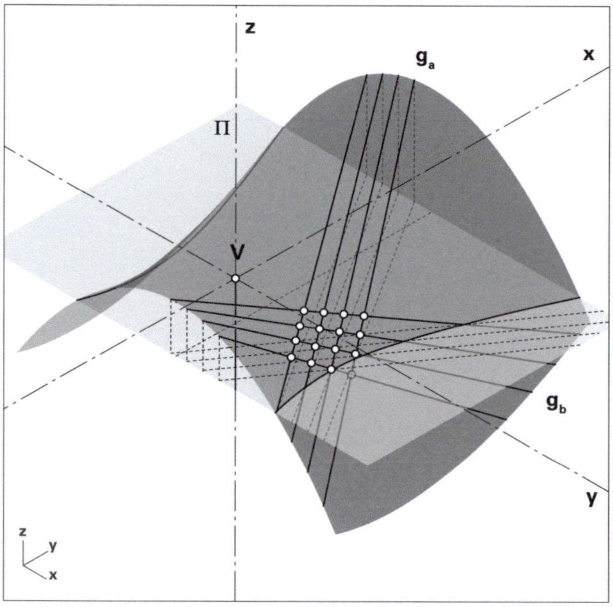

Fig. 1.212 The hyperbolic paraboloid as a ruled surface: two sets of intersecting straight generatrices $\mathbf{g_a}'$ and $\mathbf{g_b}'$

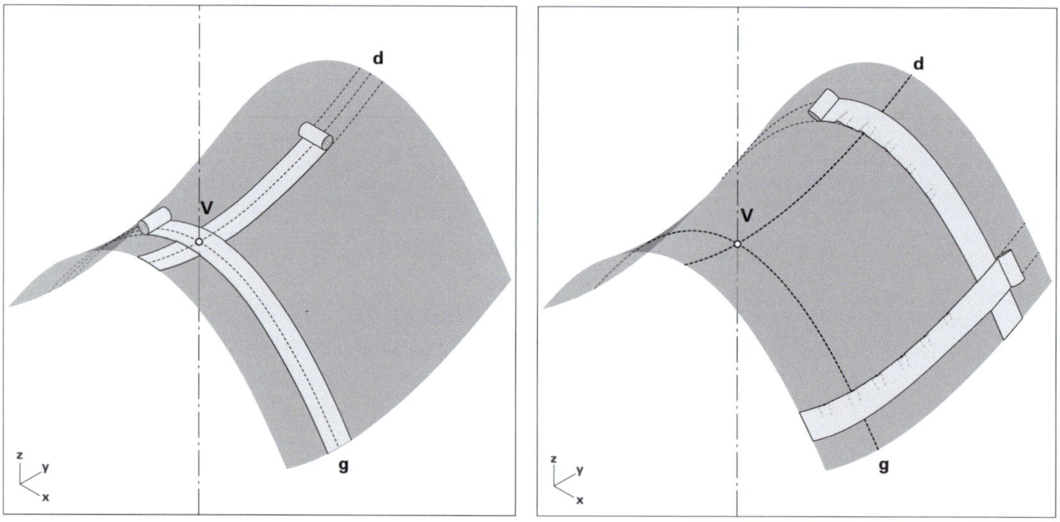

Fig. 1.213 (Left) Strips laid along the parabolas adjacent to the vertex can be unrolled almost wrinkle-free. (Right) If laid-out along parabolas far-off from the vertex **V**, the strip wrinkles

stant width as a result of the skewed position of neighbouring straight generatrices, which is the cause of the mentioned twist, but rather have to be cut with variable widths $\mathbf{w_1}$ to $\mathbf{w_3}$.

Cf. also considerations on coverings from rod-shaped elements in ▶ *Sect. 1.3.1.3 Rod-Shaped Starting Elements*, particularly Fig. 1.201, left

In principle, unlike with **developable surfaces** such as cylinders or cones, there is no way of unwinding a strip along a certain, even curved, line on the hyperbolic paraboloid totally without creases, since it is a non-developable surface; unless the deformability of the belt still allows it along a geodesic line (cf. Fig. 1.184).

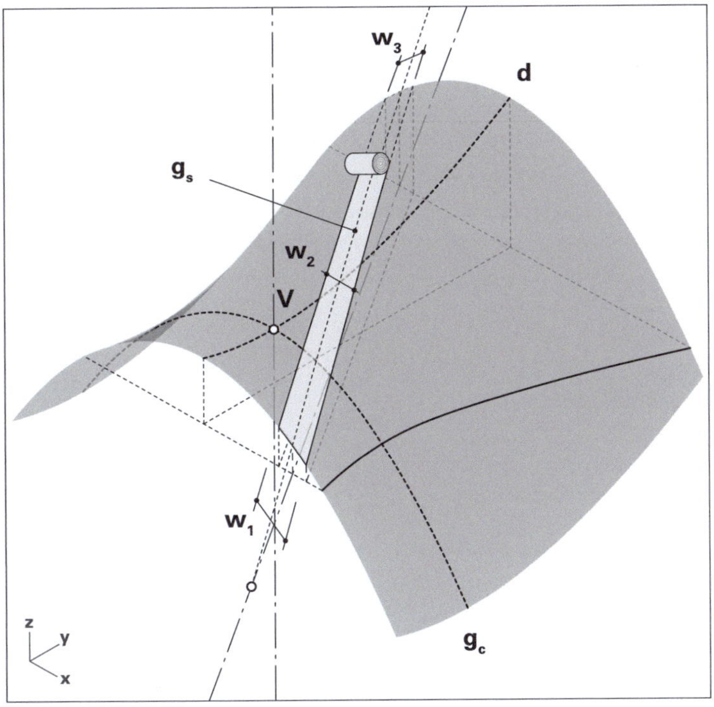

Fig. 1.214 If laid along a straight generatrix $\mathbf{g_s}$, the strip is twisted and has to be cut with variable widths $\mathbf{w_1}$ to $\mathbf{w_3}$

Plate-Shaped Starting Elements

If plate-shaped material is to be used, the simplest approximation to the hyperbolic paraboloid is the **polyhedron** resulting from the dissolution of the generating parabolas and directing parabolas into **funicular polygons** (Fig. 1.215). Two pairs of adjacent intersecting funiculars each define a trapezoidal, planar single element. As already noted (Fig. 1.133), for the sake of symmetry, only two equal and two mirror-inverted elements can be found. Otherwise, all the partial surfaces have different cutting geometries.

► *Sect. 1.2.3.3.3 Translational Surfaces*

Rod-Shaped Starting Elements

The conditions mentioned in the case of band-like material are more pronounced in the case of rod-shaped material. Laying it along the **generating parabola** $\mathbf{c_g}$ or the **directing parabola** $\mathbf{c_d}$ can only be achieved with flat, thin rods and with flexible material. The problems are similar to those of the cone, when laying out along the directrix curve (see Fig. 1.202 left): Gaping joints arise between neighbouring rods.

Laying bars along one of the two sets of **straight generatrices** is more convenient. In this case, the straightness of the covering bars can be preserved, but other geometrical problems arise: If the bars are laid as on Fig. 1.216, i.e. for instance, tangentially to a **main parabola** $\mathbf{c_{g0}}$ through the vertex **V**, with a side edge $\mathbf{P_iP'_i}$ along a straight line $\mathbf{l_g}$ and always

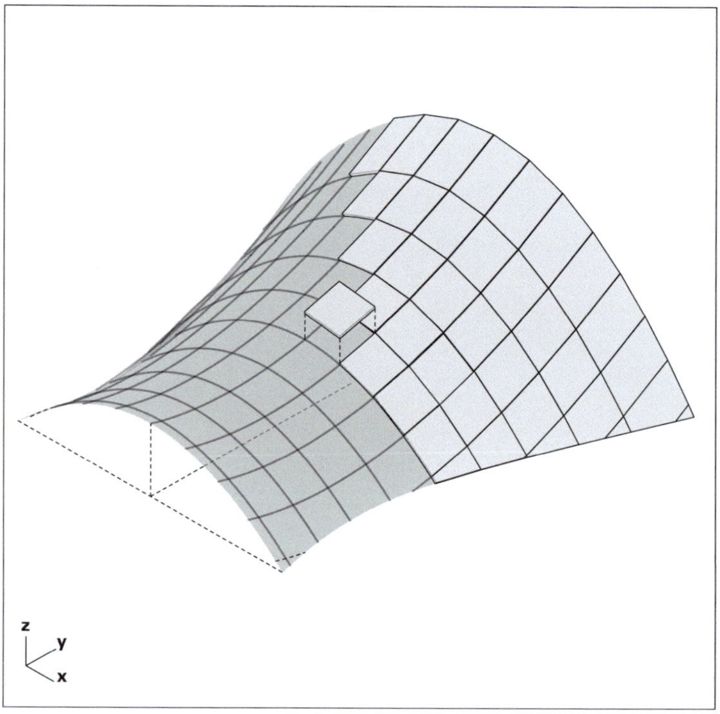

Fig. 1.215 Covering the hyperbolic paraboloid with plates, using its property as a **translational surface**

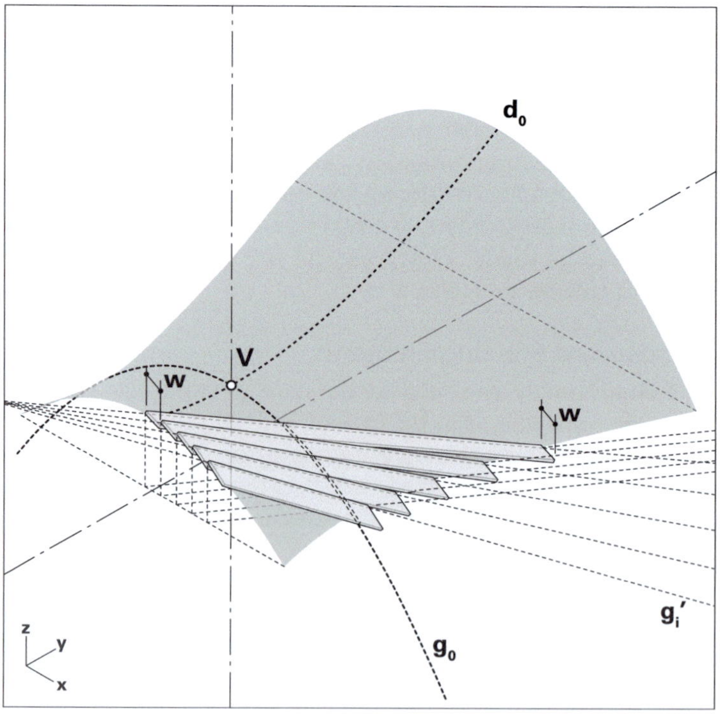

Fig. 1.216 Covering a hyperbolic paraboloid with rods tangential to a main parabola $\mathbf{g}_0$

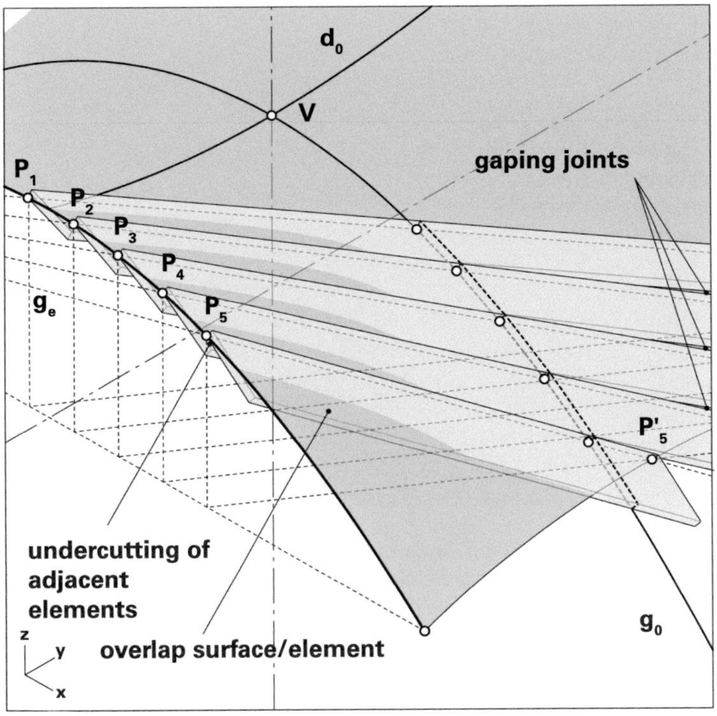

Fig. 1.217 Detail. Gaping joints and undercuts are the result of the double curvature of the hyperbolic paraboloid

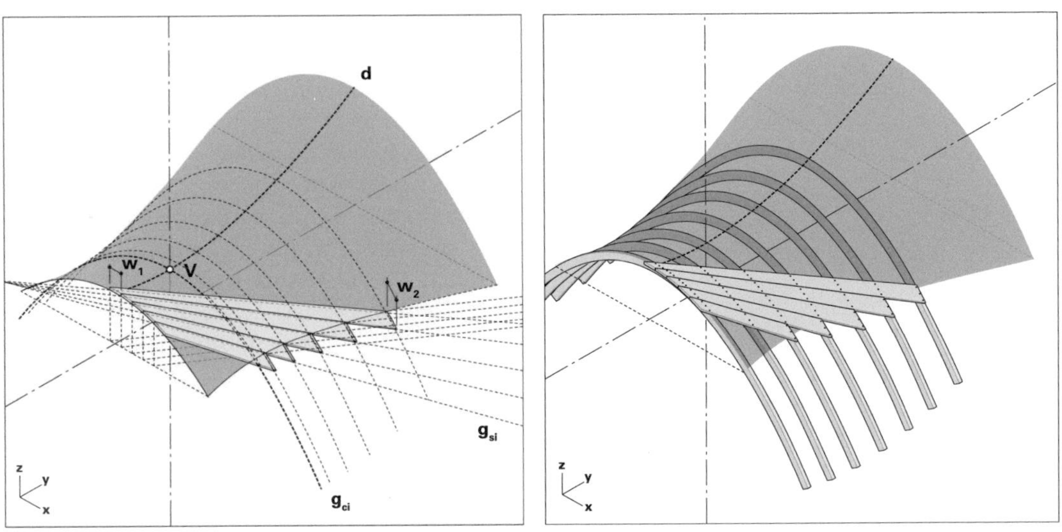

Fig. 1.218 (Left) The rods can be twisted and cut into a wedge shape to form a closed surface. (Right) Possible constructional implementation using a substructure running along the parabolic generatrices

with recurring rod width **w**, the following can be observed (■ Fig. 1.217): On one side of the main parabola, the rods separate from each other; gaps arise which become wider the wider the distance from the main parabola. On the other side, the rods penetrate the surface of the hyperbolic paraboloid and underlaps between adjacent rods emerge.

Fig. 1.219 Four-module hyperbolic paraboloid under construction: The formwork boards visible at the edges show the directions of the generating straight lines. It can be seen how some wedge-shaped boards are inserted between groups of parallel boards (e.g. bottom right) to practically solve the problem of the generatrices not being spatially parallel (Bolsa de México, arch: F Candela)

However, a continuous surface of bars may possibly be formed if the rods can be **twisted**. By their twisting, the open joints can be closed. By twisting, the rods are, so to say, forced into the inclination of adjacent generating parabolas $\mathbf{c}_{gi}$ (Fig. 1.218 left). A prerequisite for this is, however, sufficient flexibility or plasticity of the material which, depending on the case, is sometimes given within certain limits, as well as the **wedge-shaped trimming** of the bars, since these also have different widths $\mathbf{w}_1$ and $\mathbf{w}_2$ at different ends. This can be done, for example, by laying flat boards on a substructure consisting of mutually parallel identical parabolas (Fig. 1.218 right). The fastening itself forces the flexible boards into their desired position. They are to be cut into wedge shape, or at least alternately mounted with wedge-shaped intermediate pieces (Fig. 1.219). As with facets made of plate material, a better approach to the surface is possible when using slimmer rods (Fig. 1.220 left). In principle, both directions of straight line sets are suitable for this type of covering (Fig. 1.220 right).

Cf. Herzog T, Moro J L (1992) *arcus 18: Interview with Félix Candela on May 10, 1991, in Madrid*, Cologne

Block-Shaped Starting Elements

The small size of block-shaped starting elements and the resulting narrow-meshed faceting of a surface composed of them allows for a relatively simple adaptation to any doubly curved surface, including the hyperbolic paraboloid, as discussed in this section. However, this variant has only a marginal constructional significance since practical applications of doubly curved surfaces from building blocks are extremely rare. The only construction that is based on the use of bricks,

The only example known to the author are the shells of Eladio Dieste, which are made of bricks. The joints between the bricks are reinforced to absorb the unavoidable tensile forces.

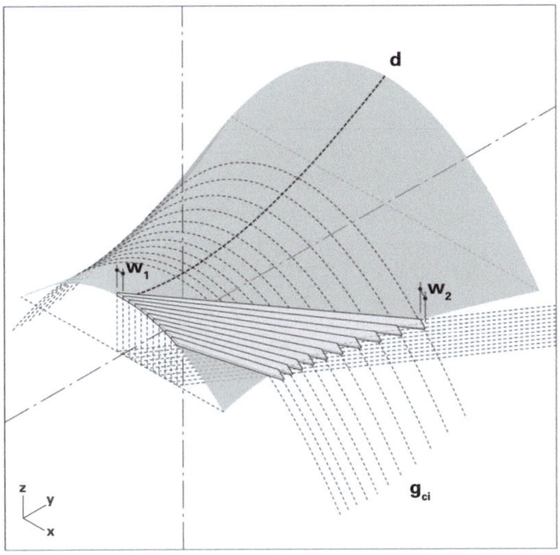

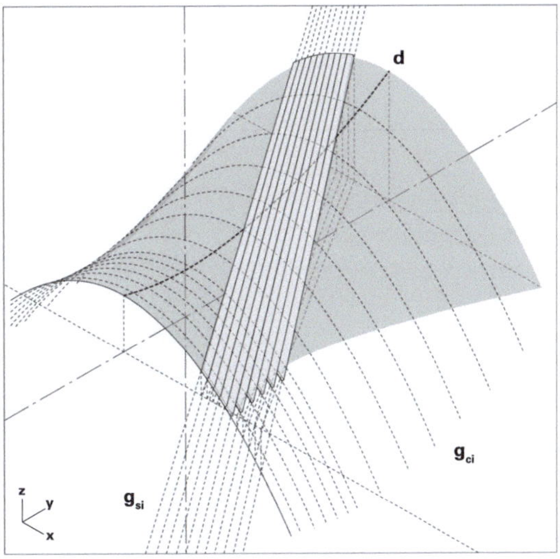

Fig. 1.220 (Left) More dense subdivisions with narrower bars allow for better adaptations to the paraboloid shape. (Right) The direction of the opposed set of straight lines $\mathbf{g}_{si}'$ can also be used for a covering with bars

namely classic brickwork with masonry bonds, is unsuitable for creating free-spanning, anticlastically curved surfaces since it cannot absorb the inevitable tensile forces.

The situation, however, is different in the case of synclastic surfaces made of building blocks, for example, spherical surfaces, which were very important in historical dome construction. They will be discussed in the following.

1.3.2.2 Sphere

Like the hyperbolic paraboloid, the spherical surface, too, is a **non-developable**, **doubly curved** surface, but a **synclastically** curved one. For constructional implementation, it is necessary to subdivide this spherical surface according to a geometrical pattern which is as favourable as possible. Subsequently, the subdivision can serve as a grid for a covering with planar elements such as plates, or alternatively as a reference for the creation of a supporting lattice of bars (Figs. 1.221 and 1.222).

► *Sect.* 4.3.2.5 *Dome from Bars*

Since the sphere is a **surface of revolution**, the most obvious way to divide the sphere is the radial pattern of **meridian circles $\mathbf{m}_i$** and **circles of latitude p** (Fig. 1.223 left). The resulting fields diminish significantly in size towards the **pole** S of the sphere, which amounts to a certain disadvantage, in particular if, in the case of a grid shell, bars converge there in the meridian direction. These fields can then alternatively be merged to multiples (Fig. 1.223 right). Another way of conveniently dividing a sphere is by intersecting with parallel planes (Fig. 1.224 left). In each case, **circular arcs $\mathbf{a}_i$** result which, in the case of regular division, appear in **symmetrical pairs $\mathbf{a}_i$** and **$\mathbf{a'}_i$**. Their radii are naturally different and decrease with increasing distance from the median circle segment $\mathbf{a}_0$. If the spherical surface is additionally cut with parallel planes in

Fig. 1.221 Spherical dome made of bars along meridian- and latitude-circles, corresponding to the variant on Fig. 1.223, left

Fig. 1.222 Spherical dome made of bars in triangular array, corresponding to the variant on Fig. 1.225

the orthogonal direction, cut-out panels with similar edge dimensions are formed (Fig. 1.224 right).

In principle, two variants are conceivable:

- The distances between the intersecting planes are always the same. As a result, the lateral edges of the fields become steadily longer with increasing inclination of the surface, i. e. increasing distance from the point **V**.
- The distances between the intersecting planes decrease continuously with increasing distance from the point **V**, so that the side edges of the fields, that is to say circle segments, have *approximately* the same lengths. Exactly iden-

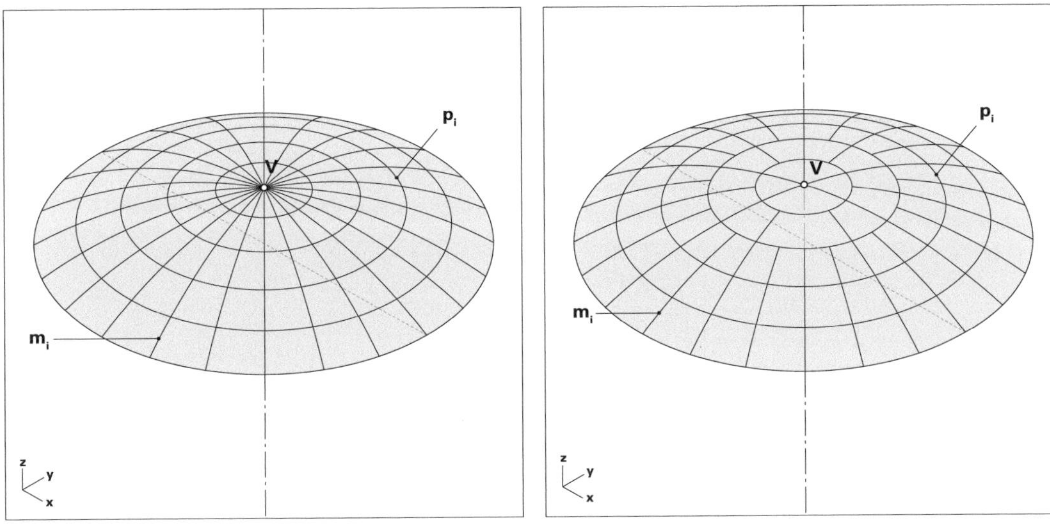

Fig. 1.223 (Left) Radial division of the spherical surface with meridian circles $\mathbf{m}_i$ and circles of latitude or parallels $\mathbf{p}_i$. (Right) The small fields occurring in the area of the vertex **V** can be grouped into larger multiples

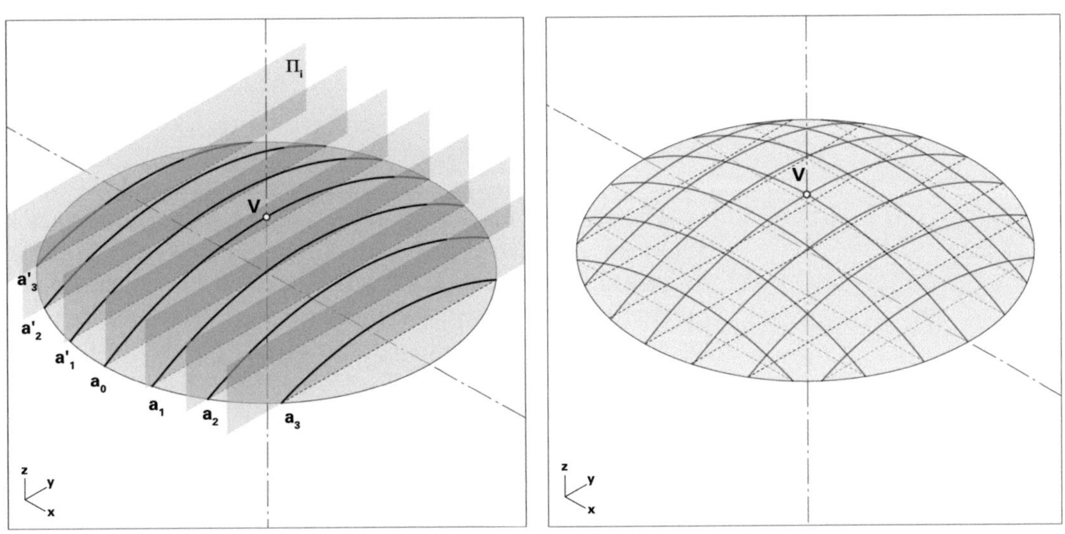

Fig. 1.224 (Left) Division of the spherical surface by intersecting with parallel planes $\prod_i$, which produce circular arcs $\mathbf{a}_i$. (Right) Division of the spherical surface by two sets of circular arcs which are perpendicular to each other in plan

tical segment lengths cannot be achieved with the spherical surface using this subdivision. The **translational surface** provides an approximate approach which can be used for constructional purposes with always the same side lengths.

► *Sect. 1.2.3.3.3 Translational Surfaces*

Furthermore, it is important to note that the corners of the spherical segments, following this segmentation, do **not lie** in **one plane**. In the case of load-bearing grids of bars which follow this subdivision geometry, there may also be disadvantages with respect to load-bearing behaviour, since square arrays of rods are not rigid without additional measures.

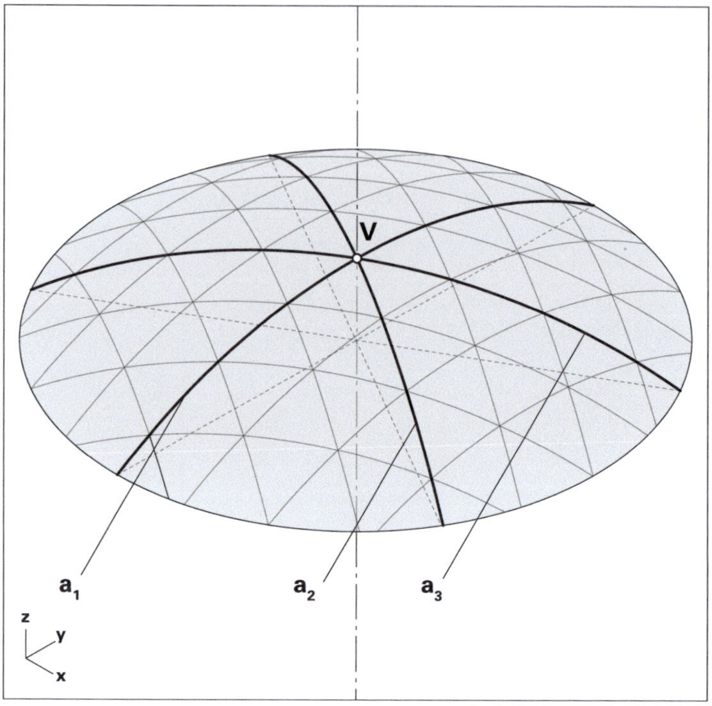

Fig. 1.225 Division by three sets of circular arcs $\mathbf{a}_{1-3}$, each in a 120° triangle pattern in the plan projection

Therefore, it may be useful to cut the spherical surface at angles of **120°** with three sets of planes respectively parallel to each other (Fig. 1.225). This creates a **triangulated**, i.e. inherently shear-resistant latticework. The resulting circular arcs, i.e. three sets $\mathbf{a}_{1i}$, $\mathbf{a}_{2i}$, and $\mathbf{a}_{3i}$, are parallel to the three main segments $\mathbf{a}_1$, $\mathbf{a}_2$ and $\mathbf{a}_3$, respectively, which run through the vertex **V**. Fields with an approximately triangular shape are formed, each of which is delimited by three circular segments. If, for example, a grid shell of bars is produced according to this pattern, the fields are therefore triangular and consequently rigid without additional measures.

As a substitute for the sphere, a **translational surface** can also be used under certain conditions, which results from the sliding of a generating circular segment $\mathbf{c}_g$ on a directing circle segment $\mathbf{c}_d$. Both circular segments initially have the same radius. The resulting translational surface closely approaches the spherical surface with the same radius, as long as a partial section in the form of a flat spherical cap is contemplated (Fig. 1.226). Only diagonally to the cross from $\mathbf{c}_g/\mathbf{c}_d$, this differs somewhat more from the spherical surface. The flatter the cap, the smaller the deviation. In the case of a hemisphere (Fig. 1.227), however, the differences are more pronounced. Like all other translational surfaces, this offers significant advantages in constructional implementation with the aid of planar components.

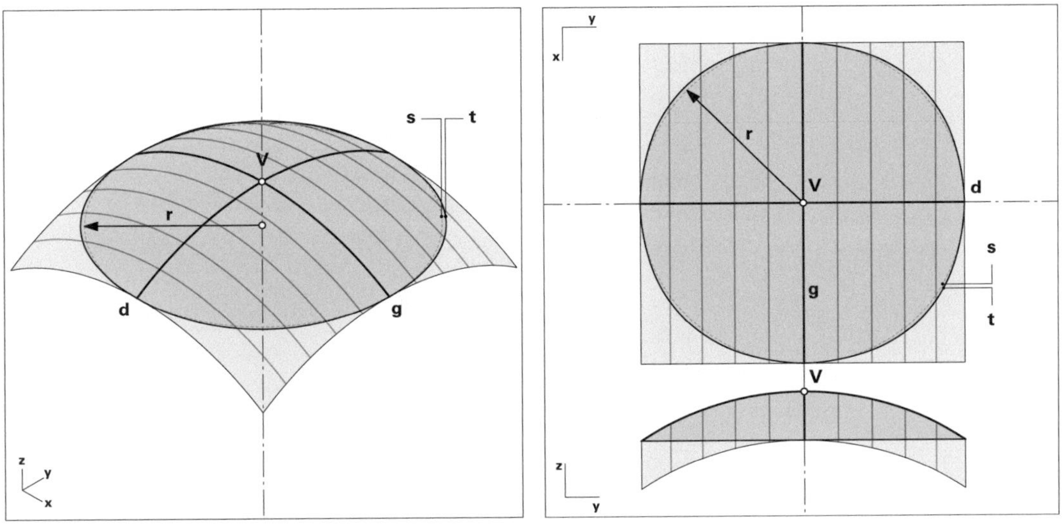

Fig. 1.226 (Left) Replacement of the spherical surface by a translational surface. The intersection of the translational surface with a horizontal plane (**t**) has only a small deviation from the intersection of the sphere therewith (**s**). (Right) Two-panel projection of the superimposed spherical and translational surface as on the left

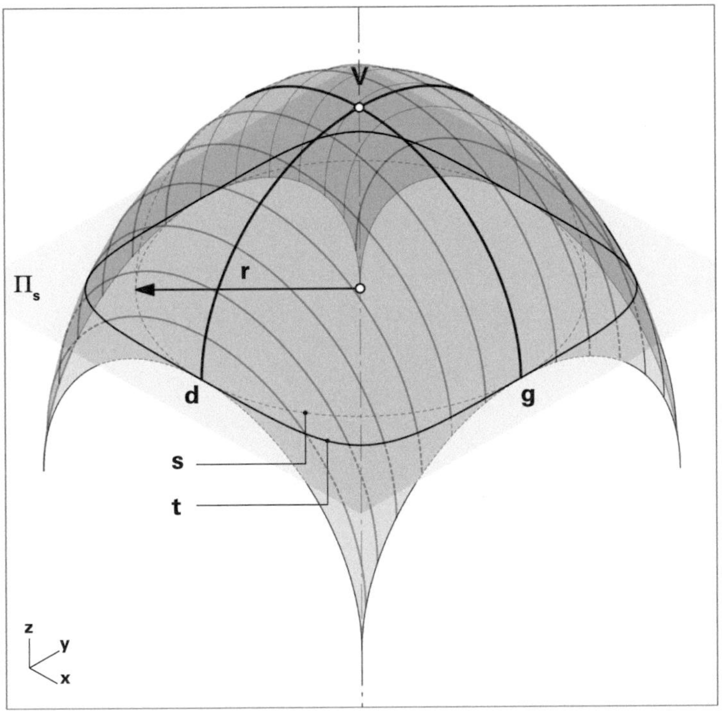

Fig. 1.227 The deviation between the translational surface and the sphere is markedly greater in the case of less flat cut-outs, as shown in the deviation between the intersection of the translational surface (**t**) and the intersection of the sphere (**s**)

Fig. 1.228 Spherical dome with band covering along meridian circles, corresponding to Fig. 1.230

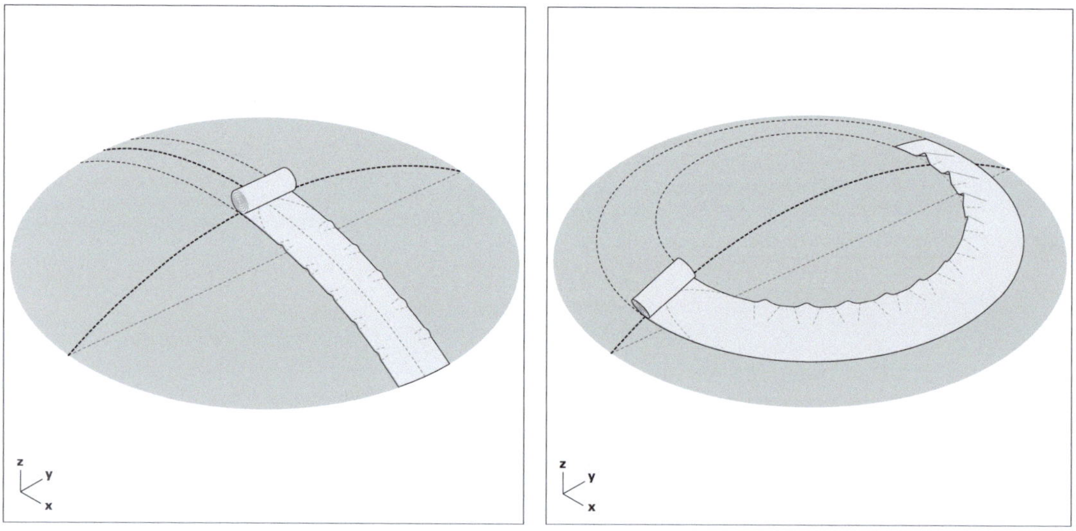

Fig. 1.229 (Left) Rolling out a strip on the spherical surface along a meridian circle. (Right) Rolling out a strip on the spherical surface along a parallel circle

Band-Shaped Starting Elements

If the surface of the sphere is to be covered by means of band-like material (Fig. 1.228), the band can be rolled out along its median axis along a meridian (Fig. 1.229 left), but due to the double curvature, the edges of the strip strike creases. This can be neutralised with certain materials by a corresponding deformability of the material and makes itself less noticeable with small band widths or large radii of

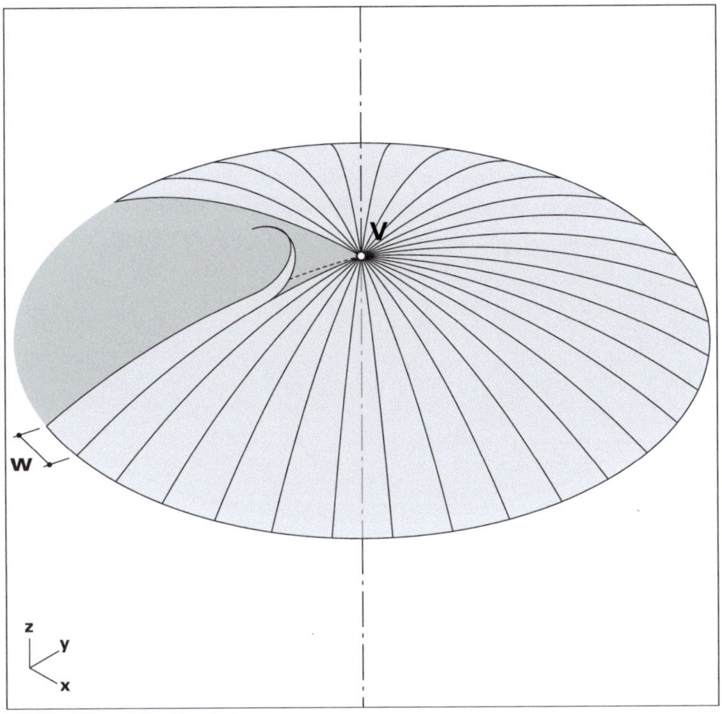

Fig. 1.230 Covering a spherical surface along the meridian circles with wedge-shaped strips

curvature. If it is laid along a circle of latitude (Fig. 1.229 right), the band will also crease at the side which is closest to the apex, since the circle of latitude has a smaller radius there. Since the spherical surface **cannot be developed**, it is also not possible in this case to roll out a strip completely without folds.[16] A radial solution is often chosen for practical purposes, in which the bands follow the meridians (Fig. 1.230) and the bands are cut triangularly. Often, this laying geometry is also justified by the flow direction of the rainwater from the vertex **V** downwards along the meridians. The double curvature must then be accommodated by the elastic or plastic deformability of the bands. The smaller the strip width **w**, the easier the double curvature can be neutralised.

Like a dome covered with standing-seam sheet metal.

Plate-Shaped Starting Elements

If coverings of plate-like material are designed according to the radial pattern, the individual faces are each covered with a trapezoid-shaped planar element (Figs. 1.231 and 1.232). The plate may lie in such a way that it lies on the sphere tangentially at a point $\mathbf{P_T}$ (Fig. 1.233 left), or it may rest with the corner points on the intersections of meridians $\mathbf{m_i}$ and

16 Except along geodesic lines, for small band widths and laterally cut band strips, where at the end, depending on the orientation of the bands and the steepness of the cap, a singular point may be unavoidable.

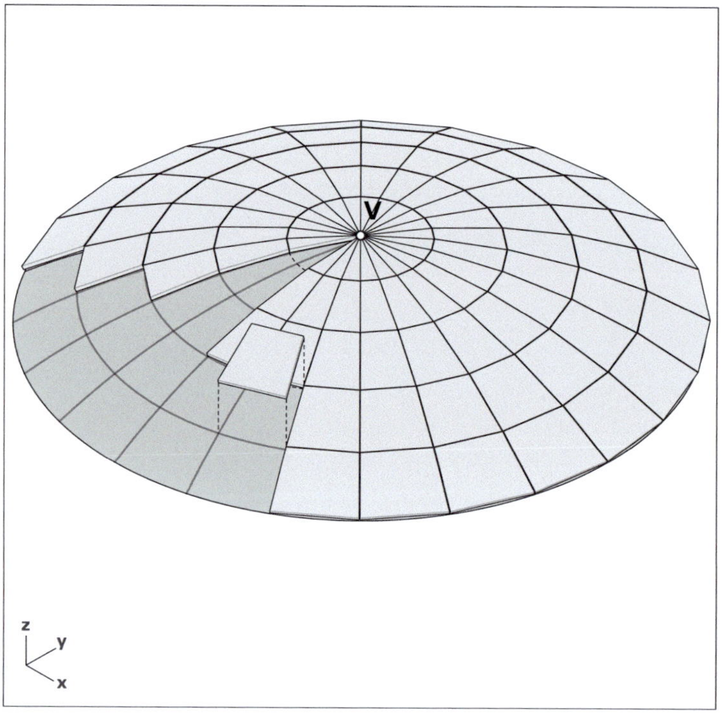

Fig. 1.231 Covering a spherical surface in a radial pattern by means of plate-shaped material

Fig. 1.232 Approximation to a spherical surface by means of plate-shaped material, here glass panes, according to the variant on Fig. 1.231 (Dome of the Reichstag, arch.: Foster Ass.)

parallel circles $\mathbf{p_i}$ (Fig. 1.233 right). The side edges on the circles of latitude are parallel to one another, the edges on the meridians necessarily intersect on the middle axis through **V** (cf. Fig. 1.231), so that all edges of the plates lie on one plane.

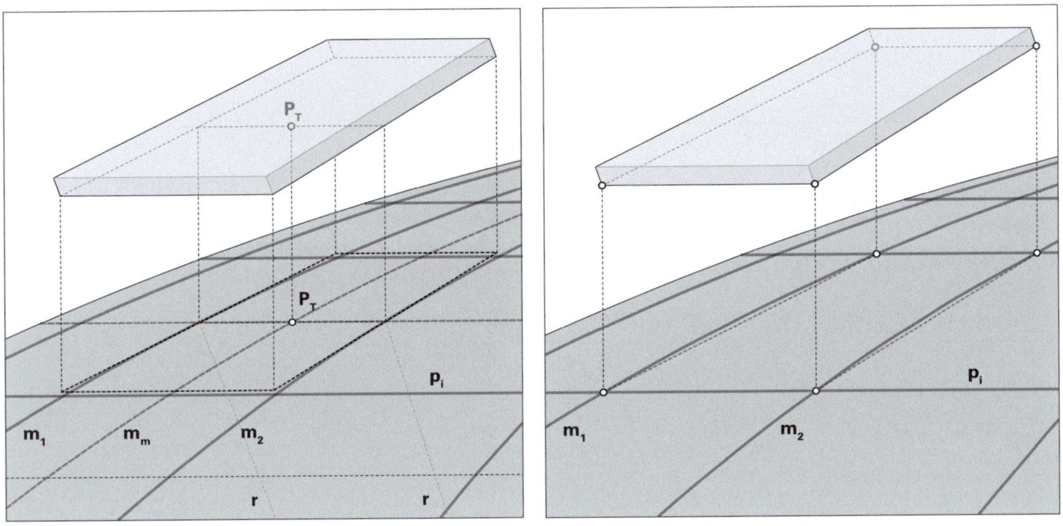

Fig. 1.233 (Left) Covering the spherical surface with a plate which touches it at a point $\mathbf{P_T}$ (and only at this point). (Right) Alternatively, the intersections of the meridian circles $\mathbf{m_i}$ and the parallel circles $\mathbf{p_i}$ can serve as a reference

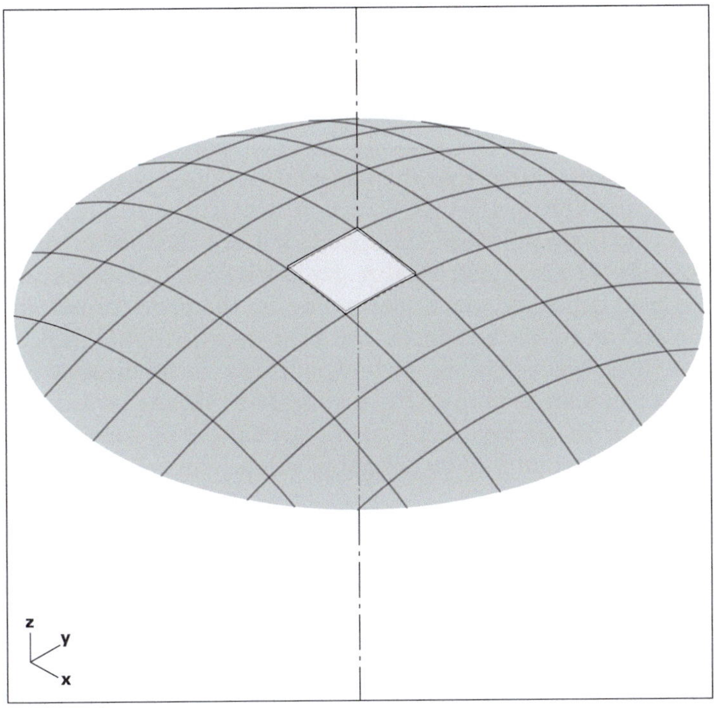

Fig. 1.234 Covering the sphere with plate elements as divided by perpendicularly intersecting circular arcs

If the spherical surface is alternatively subdivided along parallel circular segments in a square pattern (Fig. 1.234), faces are formed which are similar to one another but not identical. The distances of the intersecting planes Π_i can be chosen so

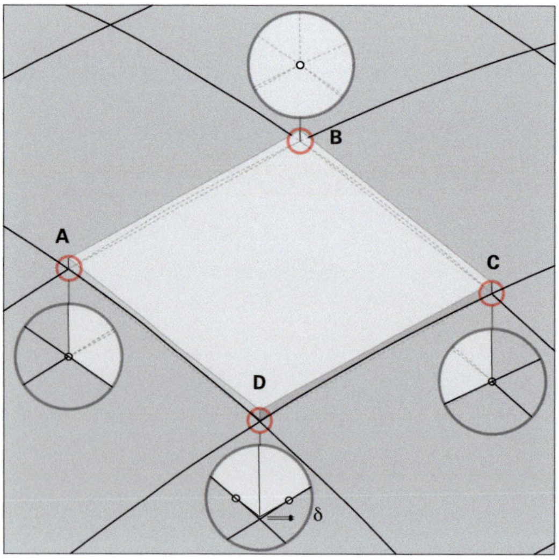

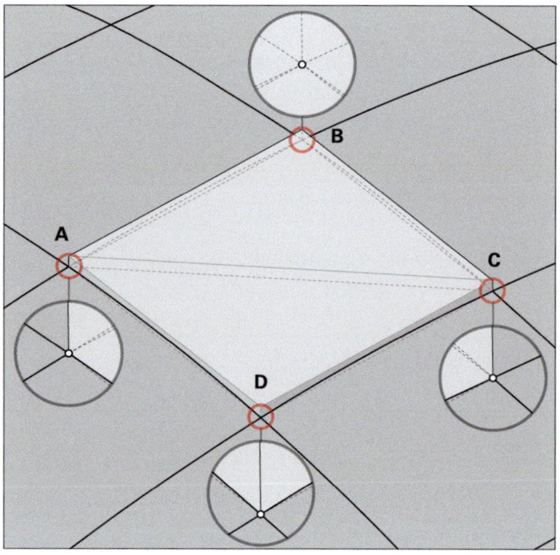

Fig. 1.235 (Left) The intersections of the circular arcs are not on a plane in the case of the sphere. At the point **D**, a deviation δ is produced. (Right) By dividing the plate into two triangular sections, the point **D** can be made to coincide with the intersection of the circular arcs

that the circular segments which bound the fields are always of the same length, but the angles at which they intersect at a point will always be different. Each individual resulting field can be occupied with a planar plate analogously to the radial pattern. However, this can only be done with a certain deviation, which results from the fact that the three corner points of the plate element (**A**, **B**, **C**) can be brought into coincidence with the three points of the sphere, but not the fourth corner point **D** (Fig. 1.235 left). A gap **d** between the sphere and the plate emerges, which, depending on the plate format and the curvature of the sphere, can pose constructional difficulties. To prevent this, the plate would have to be divided into two triangular sections **ABC** and **ACD** by means of a diagonal joint **AC** (Fig. 1.235 right), so that the corner point **D** could be brought into its desired position. This creates a triangular pattern.

Alternatively, a regular triangular grid according to Fig. 1.225 can be selected from the outset, in which the corner points of the planar triangles can always lie on the spherical surface (Fig. 1.236 left). This type of subdivision can provide significant advantages in structural terms, in particular, with grid shells made from individual bars (Figs. 1.237 and 1.238).

Vol. 1, *Sect. 26 Force Transmission*

Plate-shaped material can be laid more simply on a **translational surface** from circular segments (according to the principle on Fig. 1.227). Contrary to the sphere, the rectangular fields can be covered with planar elements with approximately the same formats, without dimensional deviations of the corner points from the surface, as shown in Fig. 1.235, left (Fig. 1.236 right).

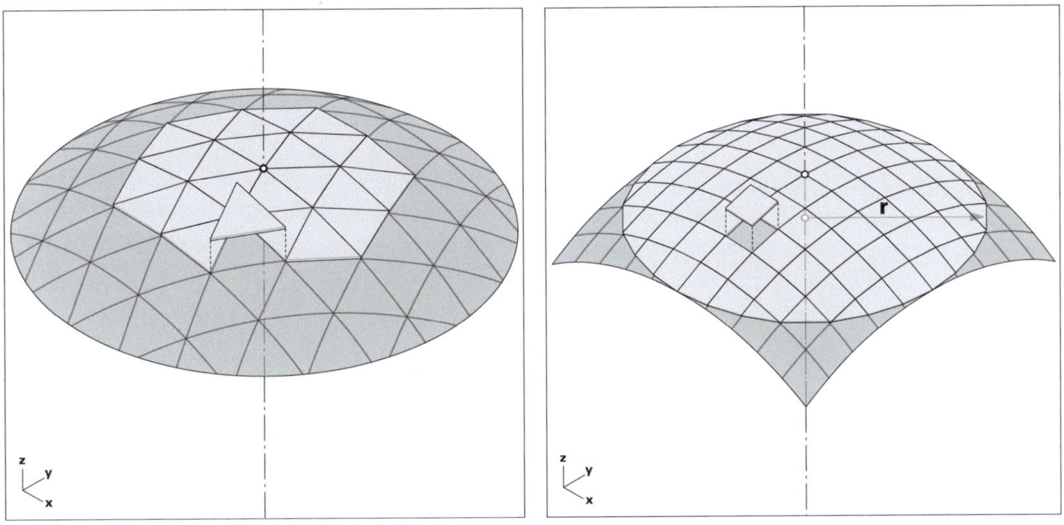

Fig. 1.236 (Left) Covering a spherical surface with triangular plate elements. (Right) Alternatively: covering a translational surface with approximately square plate elements

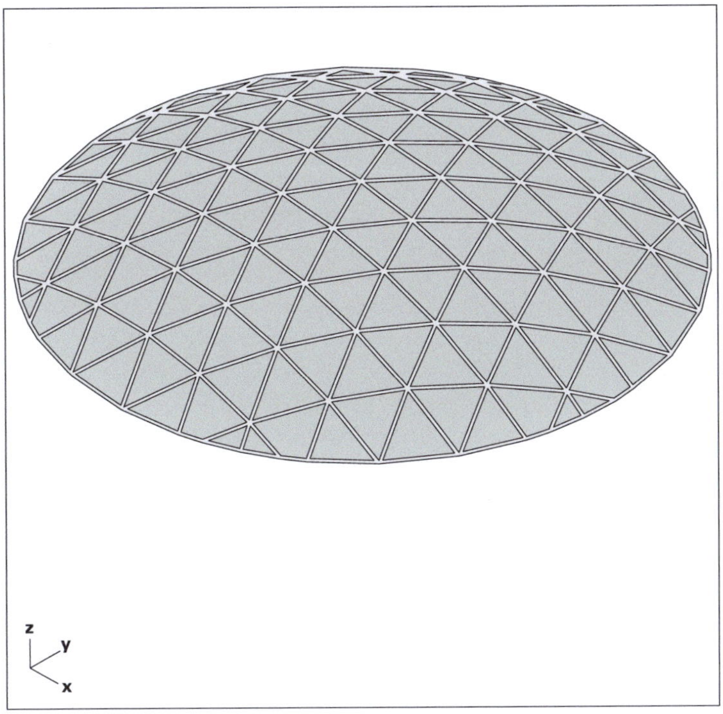

Fig. 1.237 Grid shell made of bars according to a triangular grid

Rod-Shaped Starting Elements

If a continuous covering of a spherical surface is to be produced from rod-shaped elements, first, a layout according to **meridian circles** (Fig. 1.239 left), i.e. those with an identical centre point as the sphere itself, is suitable. The prerequisite is that the rod be flexible or plastic enough to conform to the curvature of the spherical surface by means of uniax-

Fig. 1.238 Division of the sphere according to a triangular grid, analogous to Fig. 1.236, left. In this case, the special division geometry of the geodesic dome was utilised (Parc de la Villette, Paris; arch.: B. Tschumi)

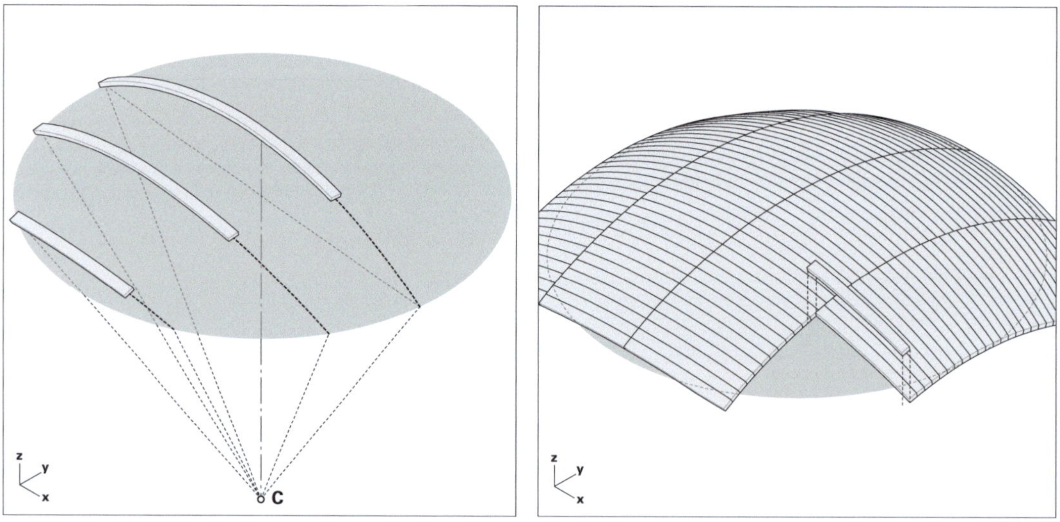

Fig. 1.239 (Left) Rods on a spherical surface. Uniaxial bending along the meridian circles. (Right) Full coverage of the spherical surface with rods along meridian circles

ial bending along its central axis. The rod touches the spherical surface along the corresponding meridian circle. The radii of the meridian circles, and hence the radius of curvature of the rod, are, of course, always the same, namely equal to the radius of the sphere. A full-surface covering of the sphere with rods follows the radial principle, similar to band-shaped material (Fig. 1.230), whereby the pole of a sectional spherical cap does not necessarily have to lie on it (as on Figs.

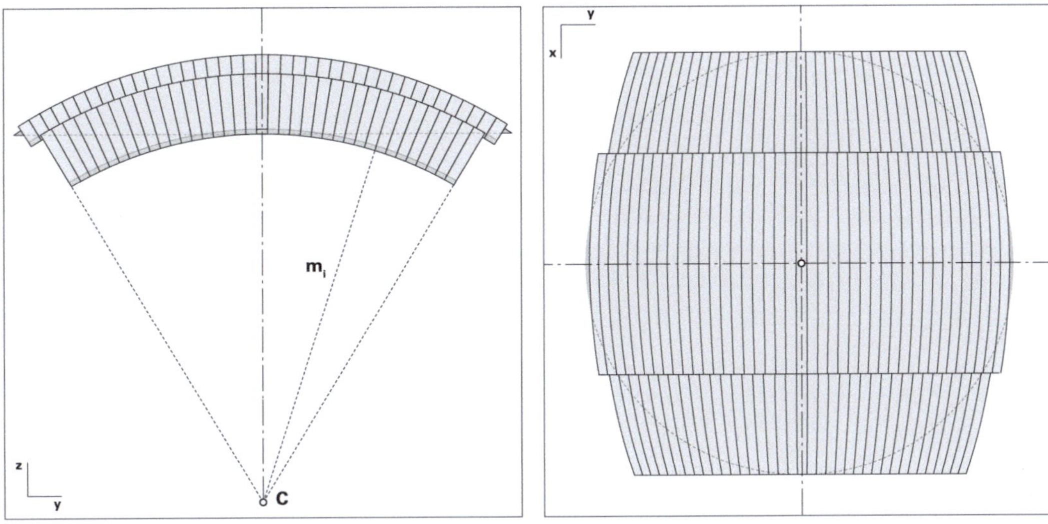

▪ **Fig. 1.240** (Left) Perpendicular elevation of the surface on ▪ Fig. 1.239, right. Meridian circles $\mathbf{m}_i$. The rods are wedge-shaped. (Right) Plan projection of the surface depicted on ▪ Fig. 1.239, right

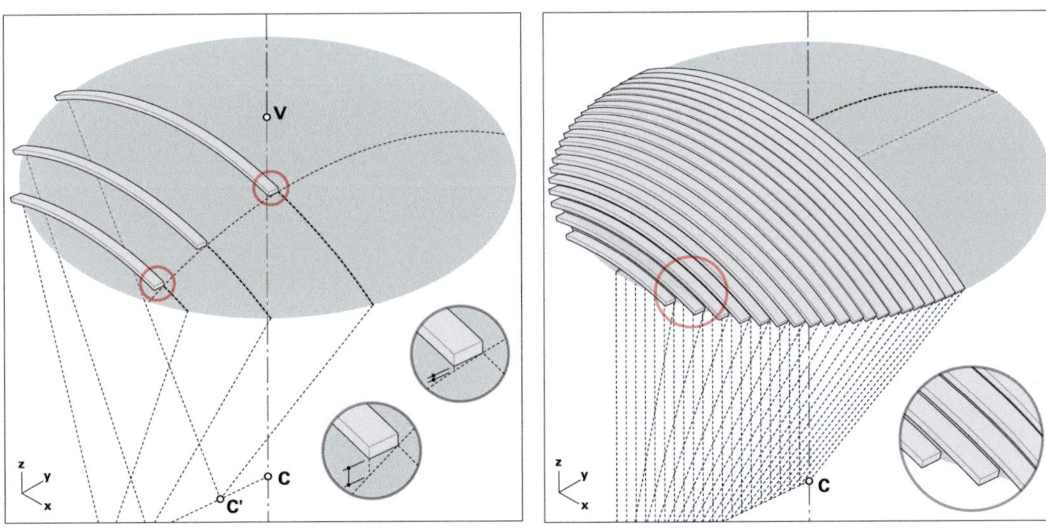

▪ **Fig. 1.241** (Left) Laying of rods on a spherical surface along parallel circles without twisting. (Right) Due to the double curvature of the sphere, gaping joints are created in the outer areas

1.239 and 1.240). The bars must be cut **wedge-like** and, in order to prevent excessive bar widths, may be laid in successive rows with **joint offset**.

Alternatively, a spherical surface can also be covered with rod-shaped elements along the **circles of latitude**. These lie on mutually parallel planes and have varying radii. Correspondingly, each position of a rod also has a different radius of curvature. If the rod is bent only uniaxially on one side in the direction of the circle of latitude, it can be laid so that it touches the spherical surface at a side edge (▪ Fig. 1.241, left). Due to the double curvature of the sphere, the other edge cannot be

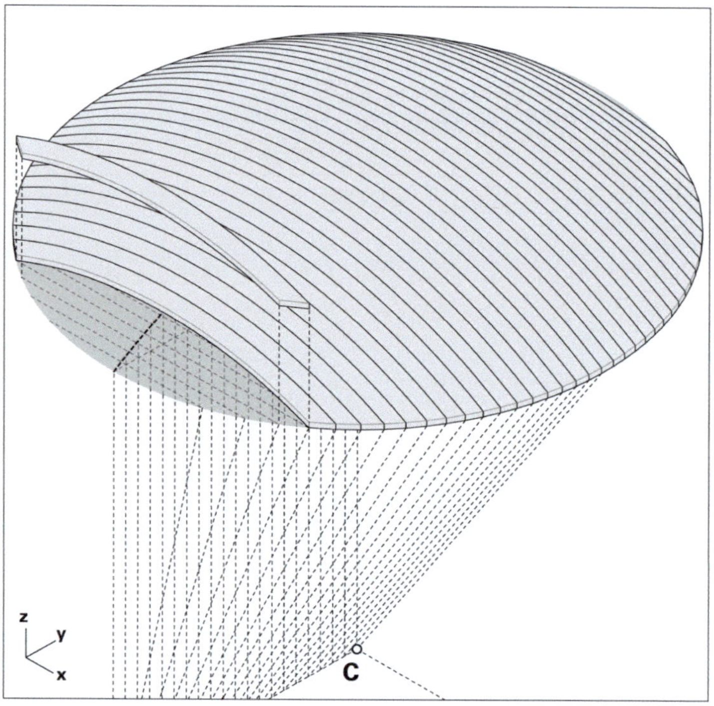

Fig. 1.242 For a full-surface covering with closed joints, the rods must be twisted

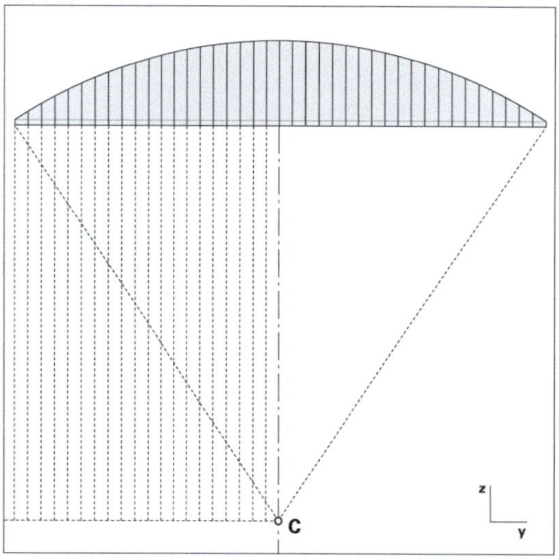

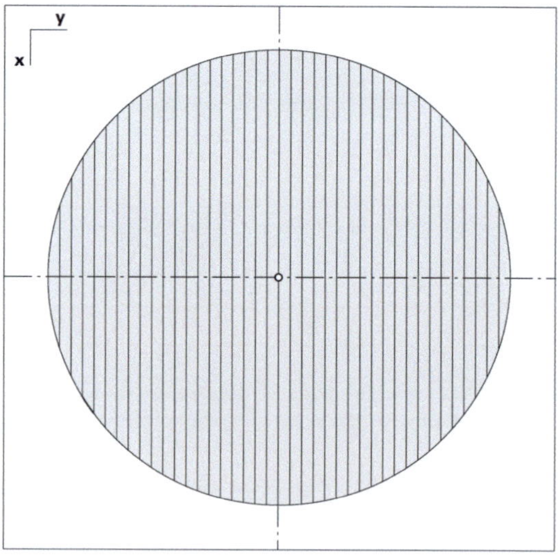

Fig. 1.243 (Left) Perpendicular elevation of the surface depicted on Fig. 1.242 with the location of the parallel circles. (Right) Plan projection of the surface depicted on Fig. 1.242

made to touch it, so that a gap between the rod and the sphere emerges. The gap width increases continuously with increasing distance to the apex, until the joint between adjacent bars—depending on the thickness of the bar—gaps (Fig. 1.241 left, detail; Fig. 1.241 right).

Fig. 1.244 Dome vault constructed in ring courses, according to variant on Fig. 1.245

If a jointless covering is to be achieved according to this laying principle, the rod must be forced onto the spherical surface in such a way that these joints close. This assumes a biaxial warping of the rod, i.e. both a **bending** and a **twisting** along the rod axis (Figs. 1.242 and 1.243). The more the circles of latitude approach the pole, the more this effect becomes manifest. In the case of most customary materials, this deformation of the rod is only possible within narrow limits, which is why only spherical sections with small curvatures can be covered according to this geometry.

Block-Shaped Starting Elements

Generating a spherical surface with block-shaped elements is, in analogy to the examples with uniaxial curvature as discussed above, geometrically comparatively simple because the relatively small initial element permits very **small-scale facets**. If questions of load-bearing are excluded, as with non-bearing floorings on a supporting spherical underlay, there are many conceivable full-surface covering patterns. They will not be discussed at this point because of their scarce constructional significance. The laying patterns of building blocks, however, have a completely different meaning in the case of self-supporting shell-like spherical surfaces, for example, in the case of **bricked dome constructions** (Fig. 1.244). Although these have rather historical than contemporary interest in terms of building construction, some examples are given below nonetheless. In the case of masonry dome constructions, not only does the load-bearing function play a part in determining the laying pattern, but also the construction process, a major goal of which is usually to reduce the scaffolding work.

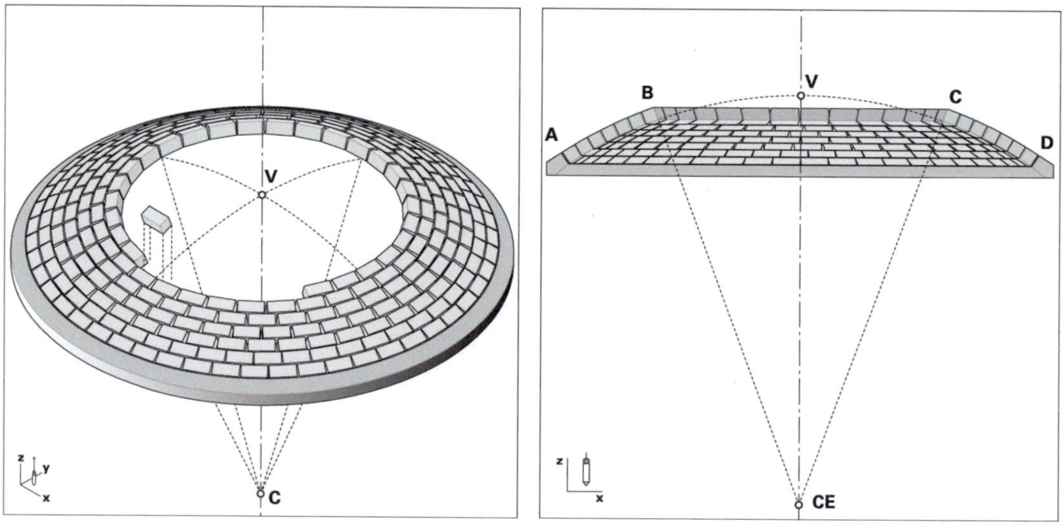

Fig. 1.245 (Left) Laying of building blocks along latitude circles in the form of annular layers, which act as stable compression rings during the construction process as soon as they are closed, preventing the inclined blocks from slipping down on the fresh mortar. (Right) Section along a meridian circle. The stones are radially fanned from the centre point **CE**

The simplest geometric organisation of the building blocks follows the horizontal **circles of latitude** (Fig. 1.245, left). Each course of stones follows a circle of latitude, has a different radius, and consequently also usually a different number of building blocks as well as different joint widths. For reasons of load-bearing capacity, it is necessary that a minimum overlay length, that is, a sufficient **interlocking** of the blocks of superimposed courses, is maintained.

As soon as a so-called **ring** or **annular course** is closed, it acts as a compression ring and prevents the collapse of the unfinished arch strip **AB** or **CD**, which acts in the finished state (**ASD**) along a meridian line (Fig. 1.245, right). To ensure that this curved line in the final state carries the load as a structural arch, the building blocks in the meridian section must fan **radially**. Thus, the bricks of the last ring course illustrated in the figure are to be laid tangentially to an imaginary cone with generatrices **MB** and **MC**. As long as the directrices **MB** and **MC** are flat enough, blocks can be bricked free of centring, i. e. freehand, under certain conditions, before the aforementioned ring closure occurs in the ring course and the blocks can no longer slide off for geometrical reasons because they wedge each other. Only after a certain critical inclination—before reaching the ring closure in each ring course—are auxiliary measures such as scaffoldings or centrings necessary to prevent the blocks from slipping.[17]

17 The innovative masonry techniques Brunelleschi developed for this purpose during the construction of the Florentine dome are interesting: He had vertical, radially extending intermediate webs bricked up at small distances in each ring course. The lying bricks of the ring course were laid in

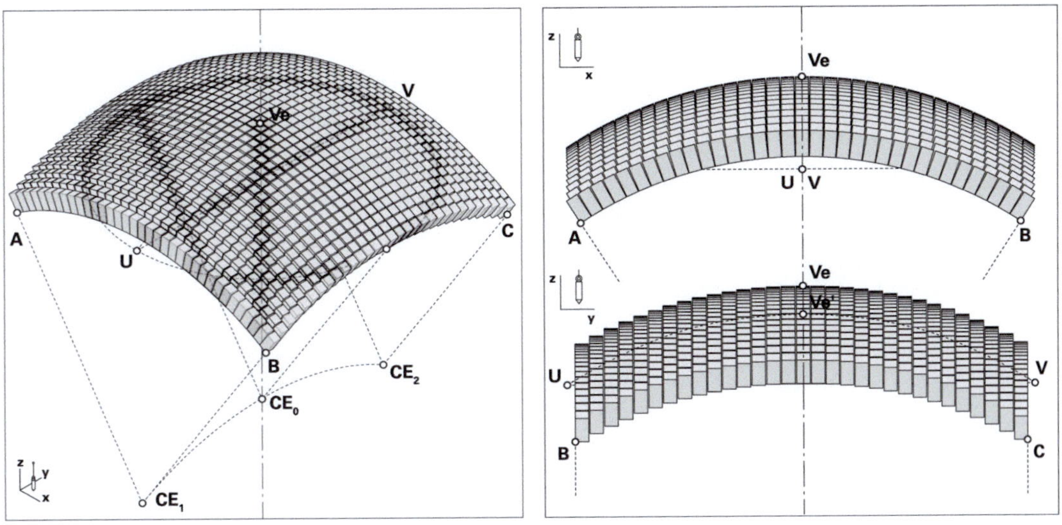

Fig. 1.246 (Left) Individual courses **AB** are displaced parallel to one another along a circular guideline **UVeV**. A translational surface is formed. (Right) Different side views of the geometric configuration on the left from two directions → **x** and → **y**. The block course **BC** is not fanned and therefore does not act as a genuine arch

Numerous historic domes have been erected according to this principle, in which geometrical conditions are closely interwoven with the requirements of load transfer and erection. In some cases, other geometrical organisational principles were also used. Some alternatives, for which there are examples from the highly developed Byzantine vaulting technique,[18] are discussed below.

Solutions based on translational surfaces as on Fig. 1.246, left, are produced by successive bricking of vertically arranged arch strips, such as **AB**, along a circular-segment-shaped directrix **US′V**. In terms of construction, this can be accomplished for an individual arch with simple means such as a narrow reusable gauge.

Geometrically, this variant solves from the beginning the conflict between the spherical shape and the supporting wall cube, which historically has produced diverse makeshift solutions. On the other hand, the disadvantage is that the construction is unambiguously **uniaxial**, since the individual arches have a load-bearing effect only in one direction—their own direction of span—and consequently, no biaxial shell effect is activated. In a structural regard, the missing interlocking between parallel individual arches is disadvantageous. This arrangement pattern also contradicts the point-symmetric ba-

Like squinches and spandrels: cf. ▶ *Sect.* 4.3.2.6 *Full-Leaf Shell, Synclastically Curved, Punctually Supported*

the spaces between them. The wedge shape of these spaces prevented the bricks from slipping on the fresh mortar. The wedging effect of the ring layer closed at the end was achieved in this way in an imaginative way even after smaller bricking sections. Supporting scaffolding work for slippery masonry blocks within these smaller areas (not for the complete ring course) was accordingly simple and economical. Cf. Mislin (1997).

18 Read more in Choisy A (1883) *L'Art de Batir Chez les Byzantins*, Paris.

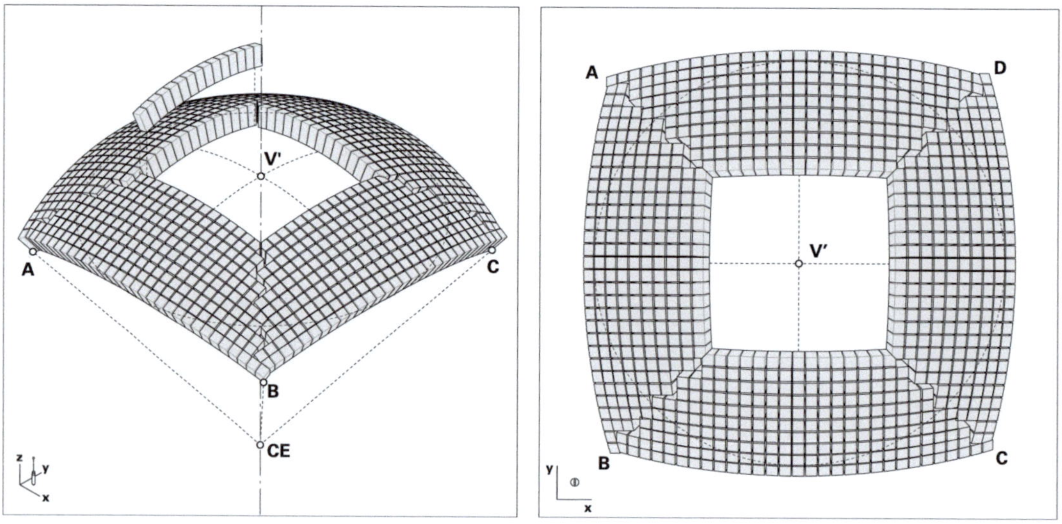

Fig. 1.247 (Left) Arrangement of single arches such as **AB** and **BC** in radially fanned planes with pivot point **CE**. (Right) Top view of the dome shell on the left in unfinished state

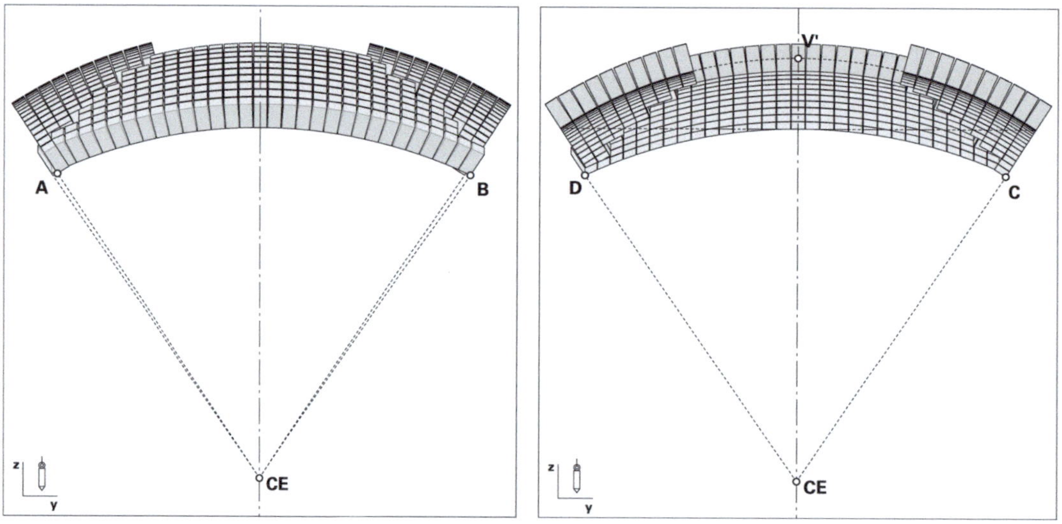

Fig. 1.248 (Left) Side view of the dome shell on Fig. 1.247, left. (Right) Section through the dome shell on Fig. 1.247, left

sic geometry of the spherical surface. Correspondingly, the faceting of the surface is also different in both orthogonal directions (Fig. 1.246 right), which may also appear visually.

An answer to these drawbacks is dissolving the dome surface into four independent, mutually identical caps which meet at the diagonals and can be interlocked there, albeit not entirely without geometrical conflicts. Figures 1.247 and 1.248 show an example with arches in planes **MAB** or **MBC** that fan radially from **M**. They therefore follow the respective meridian circles of the sphere. Since the blocks of the individual arches, in turn, are also fanned, in the final state, real

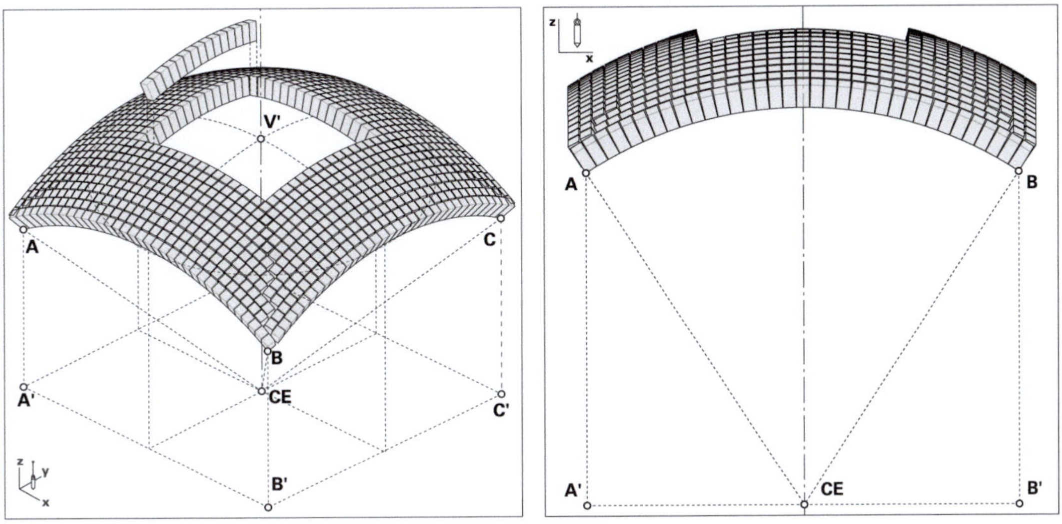

■ **Fig. 1.249** (Left) Single arches in vertical planes such as **AA′BB′** or **BB′CC′**. Blocks tilted, tangential to the conical surface **MAB** or **MBC**. (Right) Side view of the dome shell on the left

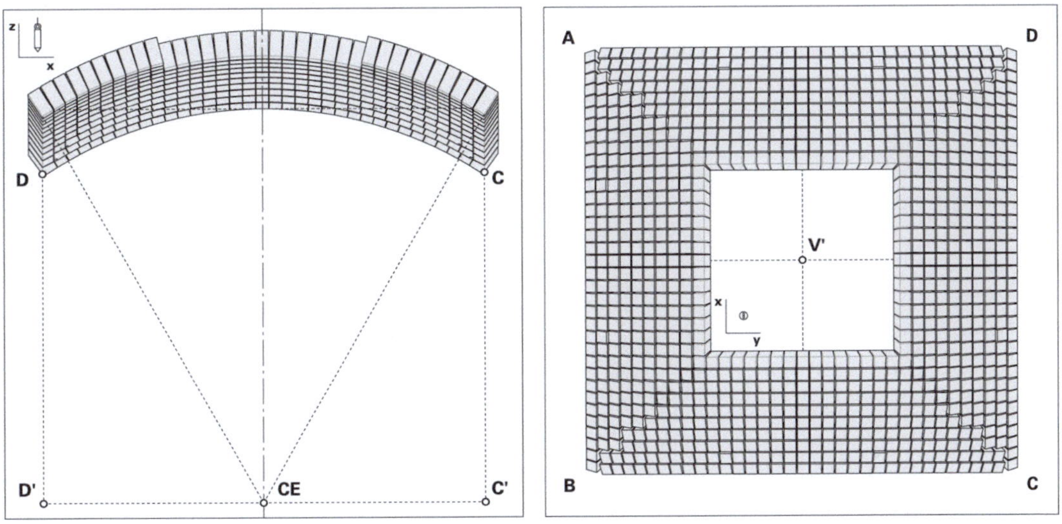

■ **Fig. 1.250** (Left) Section through the dome shell on ■ Fig. 1.249, left. (Right) Top view on the dome shell on ■ Fig. 1.249, left

arches emerge in both main directions of the shell, that is, parallel to **AB** and **BC**. However, the changing joint width between the fanning individual arches is a disadvantage here. It becomes continuously smaller towards the ends. Likewise unfavourable is the curved trajectory of the edge curves **AB**, **BC**, **CD**, and **DA** in the footprint, which also entails bearing walls being curved in the ground plan.

In ■ Figs. 1.249 and 1.250, a variant is shown in which single arches from fanned building blocks, such as **AB** and **BC**, are also used. However, these are arranged in vertical planes such as **AA′BB′**, or **BB′CC′**, unlike in the previous example. They again follow a circular directrix, so that a **translational surface** is formed. Furthermore, the dome shell is divided into four identical caps which diagonally abut one another. The disadvantages of the first variant in the form of a translational surface (Fig. 1.246) are circumvented by the fact that the blocks are not laid vertically, but tilted, analogous to the vault in Fig. 1.206. Despite the vertical position of the arch planes, a **biaxial fanning** of the blocks results at each point of the surface. The tilted blocks are easier to wall up freehandedly than vertical ones; the marginal arches can rest on straight walls.[19]

Radial organisational patterns of building blocks along the meridian circles are rather rare, especially when the pole lies on the dome surface, i. e. on its apex. Due to the acute-angled intersections between adjacent meridian circles, geometrical conflicts arise which are difficult to solve with building blocks.

Bibliography

Adriaenssens S et al (2016) *Advances in Architectural Geometry 2016*

arcus XVIII (1999) – *Zum Werk von Felix Candela, Die Kunst der leichten Schalen.* Müller Rudolf

Becker K, Pfau J, Tichelmann K (2004) *Trockenbau Atlas 1. Grundlagen, Einsatzbereiche, Konstruktionen, Details.* 3., revised and expanded ed. Müller, Cologne

Becker K, Pfau J, Tichelmann K (2005) *Trockenbau Atlas 2. Einsatzbereiche, Sonderkonstruktionen, Gestaltung, Gebäude. Grundlagen, Einsatzbereiche, Konstruktionen, Details.* Müller, Cologne

Bläsi W (2008) *Bauphysik*. 7th ed.- Haan-Gruiten: Verl. Europa-Lehrmittel Nourney, Vollmer

Block P (2015) *Advances in Architectural Geometry 2014*, Cham, Springer

Ceccato C (2010) *Advances in Architectural Geometry 2010*, Vienna, Springer

Choisy A (1883) *L'Art de Bâtir Chez les Byzantins*, Paris

Häupl P, Willems W (ed) (2013) *Lehrbuch der Bauphysik: Schall - Wärme - Feuchte - Licht - Brand - Klima.* 7th fully revised and updated ed. Springer Vieweg, Wiesbaden

Hesselgren L (2013) *Advances in Architectural Geometry 2012*, Vienna, Springer

Klix W D, Nickel H (1990) *Darstellende Geometrie*, Verlag Harri Deutsch; Thun, Frankfurt/M

Mislin M (1997) *Geschichte der Baukonstruktion und Bautechnik*, Vol. 1, *Antike bis Renaissance*, 2ndedition, Werner, Düsseldorf

Pottmann H, Bentley D (2007) *Architectural Geometry*, Exton, PA, Bentley Institute Press.

Pottmann H (ed) Asperl A, Hofer M, Kilian A, (2010) *Architekturgeometrie.* Ambra und Springer, Vienna, New York

Pottmann H, Wallner J (2010) *Computational Line Geometry.* Springer, Vienna, New York

Schüle K, Gösele W (1985) *Schall, Wärme, Feuchte*, Bauverlag. Wiesbaden/Berlin

Wilson E (2004) *Islamic Designs*, British Museum Press, London

19 Choisy A (1883), p. 99 ff.

Build-Up of Envelopes

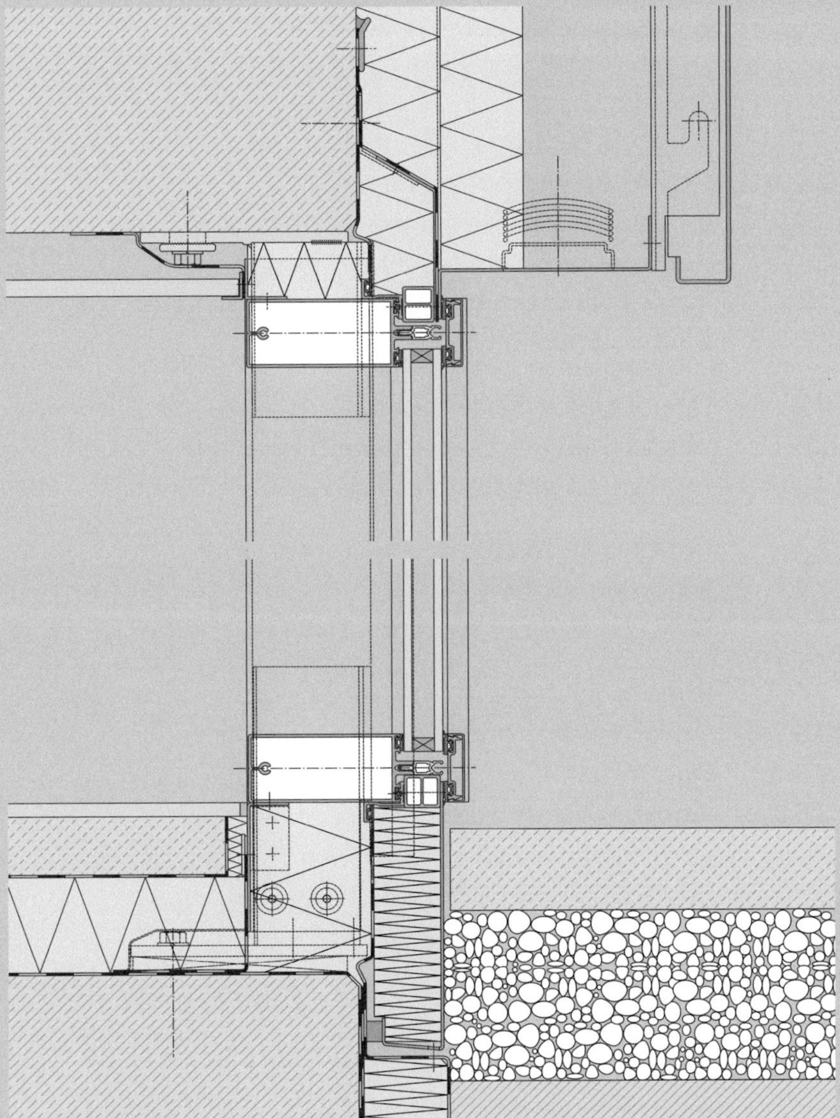

J. L. Moro, *Building-Construction Design – From Principle to Detail*,
https://doi.org/10.1007/978-3-662-71006-7_2

Contents

2.1 Constructional Implementation of the Constructional Partial Functions: Basic Solution Principles

The constructional implementation of the basic load-bearing and protective functions as described in ***Volume 1*** as well as associated constructional and building-physics-related questions will now be discussed in more detail. For this purpose, various alternative **constructional build-ups** of envelopes are analysed. The functions are fulfilled by these in the interaction of the individual components of the envelope construction.

Vol. 1, *Chapt. VI Functions*

As already noted, it cannot be assumed here that a partial function can always be assigned unequivocally to a separate material component of the building construction. Rather, it is often the case that a component, or a component's layer, is supposed to fulfil several partial functions at the same time. A solid concrete leaf can, for example, exert functions of dissipating forces, noise protection, fire protection, etc. Conversely, it can neither be assumed that a single partial function is always satisfied by a single component or a single layer. Often, a **layered sequence** consisting of several layers arranged behind or one above the other is responsible for ensuring a specific partial function. An example of this are multi-stage seals.

Vol. 1, ► *Sect. 2.2.2.2 According to Constructional Partial Functions*, as well as ► *Sect. 25.3 Allocating Partial Functions to Building Components*

In most cases, the constructional designer must count on the fact that different constructional elements with extremely different functional assignments sometimes converge in minuscule spaces. These must be coordinated with one another **spatially** and **geometrically**, but also **functionally**. Conflicts between partial functions occur frequently enough: Elements with load-bearing function (e. g. façade anchors) jeopardise, for example, the thermal-insulation function (by thermal bridges) and the protective function against moisture forming within the construction (condensation water is formed); or thermal-insulation panels deteriorate the noise-protection function of a wall-leaf due to their excessive bending stiffness.

In order to elucidate the not infrequently complex relation between the material component to be executed physically and the function, or rather functions, allocated to it, the essential **constructional implementation principles** of the partial functions mentioned will be discussed in more detail below using an exemplary planar leaf component, namely contemplating the **interplay** of the individual functions within the overall construction. The way in which the component is made up of individual constructional elements, i. e. its **basic constructional constitution** or **build-up**, which is to be investigated in various variants in the following, plays an essential role.

2.1.1 Definitions of Terms

The above-mentioned partial functions of envelope elements are fulfilled in most cases by a **constructional package** consisting of individual **layers**. Each individual layer has specific physical properties which enable it to take over one, or even more than one, task. Layers with a greater bending stiffness are referred to as **leaves**.[1] Unlike layers lacking leaf properties, they can absorb loads and pass them on to their bearings. This definition essentially corresponds to a *constructional* definition of the leaf concept, on which the main classification in ► Sect. 2.1.2 was based. There is also an acoustic definition of the leaf. Leaves are, in this particular sense, all layers which are capable of acting as a vibrating mass in a springy system (mass-spring-system). Consequently, layers, which in the constructional sense are not considered as leaves, must nevertheless sometimes be regarded as acoustically effective leaves. One example of this is a render layer on a thermal-insulation composite system.

Vol. 1, ► *Sect. 28.3.3.3.2 Double-Leaf Components*

While the constructional build of a component describes the way it is constituted from individual parts, a surface component can also be provided with a **lining** or **additional build-up**—sometimes referred to simply as a **build-up**—i.e. a layer sequence that is placed on or attached to the component. In the context of the present classification, we will speak of an (additional) build-up only if a thermal-insulation layer forms part thereof.

2.1.2 Basic Structures of Envelopes

After discussing the questions which arise from the seamless composition of a component layer *within* its *area*, defined by the dimensions of length and width, in *Section 1*, the entire *constitution* or *composition* of an envelope component, including all relevant constructional elements, will be discussed below with consideration of the three dimensions of length, width, and thickness (◘ Fig. 2.1). Since the component's **thickness** is that dimension which separates outer from inner space in the case of outer envelope components, or two inner spaces in the case of inner ones, questions of construction physics play an important role in addition to questions of geometry, load transfer, and manufacturing. For this rea-

1 As in other cases, the concept of leaf, which in this context, from the point of view of the structural composition of an envelope component, is a layer surface with a sufficient stiffness, must be distinguished from the structural concept of the shell as described in ► *Sects. 3.4.4* and *3.4.5*. In addition, also the acoustic concept of the leaf exists—as mentioned in the text.

son, consideration is given to relevant aspects of hygrothermal, noise, and fire-protection partial functions.

Cf. **Vol. 1**, *Sections 25 through 30*

In the current perspective of this section, the convergence of the various constructional partial functions such as load-transfer, sound insulation, or heat protection, which have been examined separately from each other in ***Volume 1***, *Chapter VI*, within one single component will be at the focus of our attention. In addition to the partial functions defined and treated there, also elementary geometric issues of assembly play a fundamental role, as is always the case with construction design. This point of view is also at the focus of the following considerations, although the geometrical compatibility of the structural elements was not explicitly defined as a separate requirement or function.[2]

Cf. **Volume 1**, *Chapter VI Functions*

In this context, we distinguish the following **basic structures** of envelope components, partly adopting the practice of the classification already introduced in *Sect. 26*, which follows the principles of load transfer[3]:

Vol. 1, ▶ *Sect. 26.9 Realising the Force-Transmission Function Within the Element—Structural Principle of the Component*

- **single-leaf systems**, that is to say, those consisting essentially of a single (full-walled) leaf, optionally supplemented with a layering;
- **double-leaf systems**, i.e. those consisting of two leaves, which are either coupled or separated from one another, either with or without intermediate layers;
- **multilayer composite systems**, i.e. those in which the leaf characteristics are reproduced by a mostly material-saving, planar surrogate construction;

2 Geometric questions of constructional build-up are often not properly weighted in their fundamental importance for engineering design work. The definition of a separate subfunction termed *geometric compatibility* would be quite justified.

3 The purpose of the classification of basic constructions of envelope components used here is to facilitate understanding the logic of fundamentally different approaches. Like all other classifications, it is always to be applied carefully to the individual case. Thus, it happens often that different constructional principles are realised in one identical envelope construction. As a rule, different principles are applied in such cases at different hierarchical levels of construction. An example will illustrate this: A mullion-and-transom façade, which is fastened to floor slabs of a primary structure, is considered first as a ribbed system according to the present classification. We would, however, designate any existing sandwich elements spanning between mullions or transoms as multilayer composite systems. The latter are, however, not fastened to the primary structure (floor slab), but to parts of the secondary structure and are thus located in a different hierarchical level of the constructional design. Furthermore, individual elements of the constructional system can appear in a modified form. Ribs, as have been introduced in this context, may also appear as latticed trusses, as cable girders, or comparable elements. Conceptually, however, nothing changes in its function within the overall system.

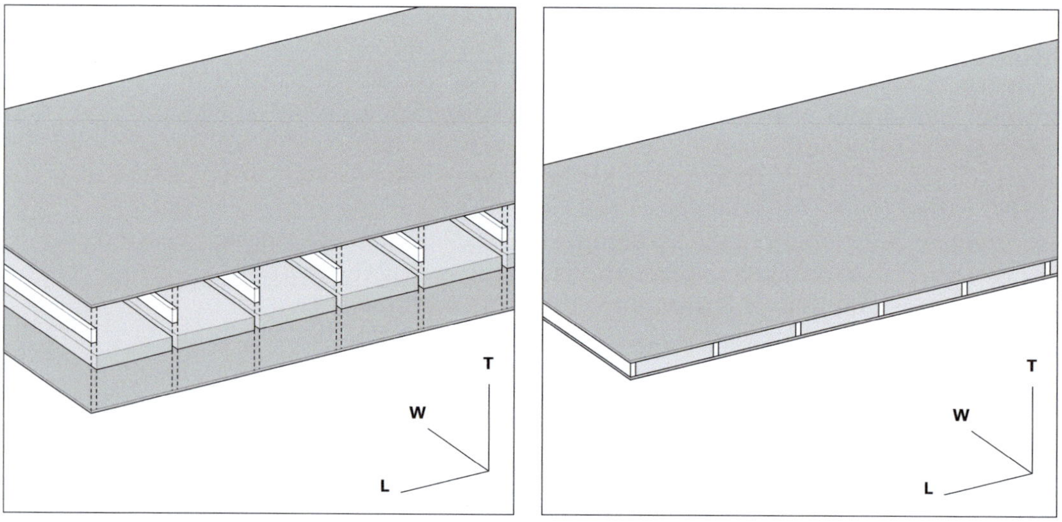

Fig. 2.1 Building up an envelope component, regarded in its thickness dimension **T**

- **ribbed systems**, that is, all systems in which the surface is made up of one or more skeins of mostly force-conducting rod-shaped elements and a separate surface-forming sheathing structure;
- **membrane systems**, i.e. those which consist of one or more membrane layers. They are either mechanically or pneumatically prestressed and are thereby capacitated to absorb forces.

2.2 Single-Leaf Systems

2.2.1 Single Leaf without Lining

Vol. 3, ▶ Sect. 12 Full-Leaf Systems

The task of encapsulating or enclosing a space can first be accomplished in the simplest form using **single-leaf envelopes** made of only one material (Fig. 2.2). The space-enclosing, load-bearing function and other, for example building-physics functions, are fulfilled by the same, structurally not further differentiated component, i.e. the main leaf. This also reveals a significant advantage of this variant, namely, its **constructional simplicity** and its **insensitivity** to constructional problems or building-physics-related problems resulting from the interplay of various constructional constituents.

A substantial advantage of the single-leaf design with regard to the load-bearing function lies in the **homogeneous structure** of the component, in which the forces or the resulting stresses can unfold freely without dangerous concentrations. In particular, questions of the bracing of load-bearing structures, which often involve separate measures in the case

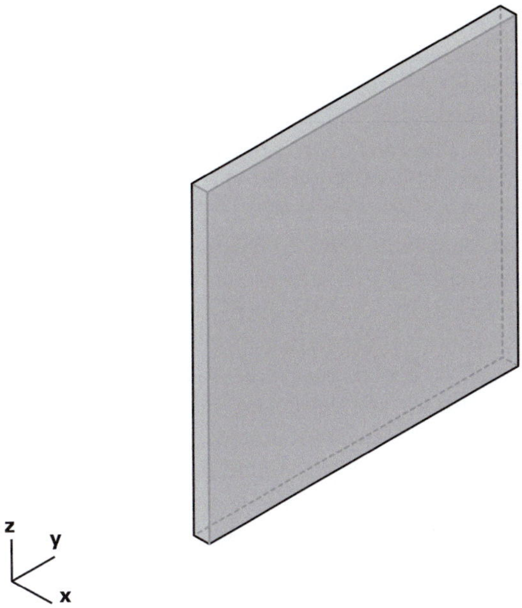

Fig. 2.2 Single leaf without addition

of bar frameworks or ribbed systems, can be solved with simple means taking advantage of the racking stiffness of a full-walled component. This also applies to the following examples, in which the main load-bearing function is also assigned to a leaf.

If the single-leaf component consists of a single layer, in this case all the partial tasks which are posed on the leaf component are to be met by the material of this layer, which in this case is identical to the leaf. In the case of outer envelope components, this is only possible in rare cases without additional protective layers such as barrier sheets, overlays, and renderings. Individual requirements, which, in the single-leaf and single-layered component, must be met by a single material, are often in conflict with one another: e. g. thermal protection, which benefits from materials as light and porous as possible, and moisture protection as well as load-bearing, two partial functions the fulfilment of which is fundamentally favoured by dense material structure. In particular, the extremely high requirements posed today on envelope components can scarcely be met under such adverse conditions. For this reason, it can be said that single-leaf and single-layer outer envelope components presently play only a very limited constructional role. On the other hand, certain inner leaf components, in exceptional cases, can be designed according to this pattern, for example, un-rendered brick separation walls. Particularly where only mass or material density is required for a specific partial task such as, for example, sound or fire protection, a single-layer design can prove to be advantageous.

Cf. ▶ *Sects. 2.2.2 Single Leaf with One-Sided Addition Without Substructure* and ▶ *2.2.3 Single Leaf with One-Sided Addition with Substructure*

2.2.1.1 Outer Envelope Components

A variant, relevant to building practice, of a single-leaf, but multi-layered component is the exterior wall made from brickwork, rendered on one side, but mostly on both sides (◘ Fig. 2.3). The partial functions of load-bearing and thermal protection are effected by the main leaf. The materials used for the leaf, such as, for example, aerated brick material or aerated concrete, offer a tolerable compromise between the partial tasks which are contradictory in the aforementioned sense. Both load-bearing capacity and thermal protection are therefore limited, which is why this design variant is constricted to specific applications (low multi-storey housing construction). The rendering layers essentially take over the partial tasks of moisture and wind protection.

Vol. 1, ▶ *Sect. 28.3.3.3.1 Single-Leaf Components*

From a point of view of **sound protection**, this component also is a single-leaf system whose air-borne and impact-noise protection is solely dependent on the area-related mass and the dynamic stiffness of the main leaf. As we have seen, the mass, which has a favourable effect on sound transmission, is limited for reasons of thermal protection. The bending stiffness of the component, on the other hand, is comparatively high for reasons of the required load-bearing capacity. These conditions must be assessed as rather unfavourable from an acoustic perspective, although such exterior walls offer better sound insulation than lightweight-construction walls in ribbed construction.

Vol. 1, ▶ *Sect. 27.3.7 Single-Leaf Exterior Wall from Aerated Brickwork*

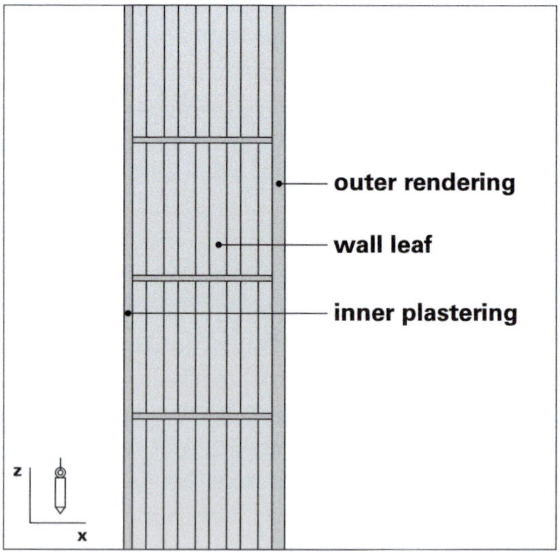

◘ **Fig. 2.3** Single-leaf exterior wall made of lightweight brickwork with plaster on both sides, an example of a single-leaf wall with protective layers but without additional build-up

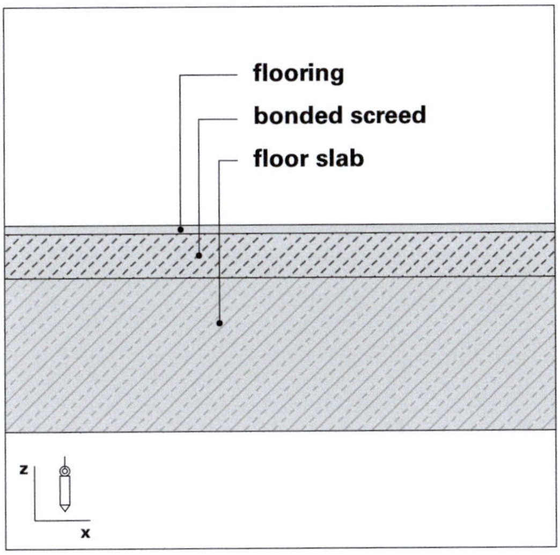

Fig. 2.4 Solid floor slab with bonded screed. The example represents a transitional form between the variants of the single-leaf element without and with addition

2.2.1.2 Inner Envelope Components

Whenever questions of thermal protection are not important, as in the case of partitions between uniformly tempered interior spaces (Fig. 2.4), the mass and density of the component can be increased as much as required for high load-carrying capacity—e. g. for load-bearing inner walls. In this case, the area-related mass also increases in the case of the single-leaf surface component, so that an adequate noise protection can also be achieved.

The fire protection of the surface component is ensured primarily by the fire resistance of the main leaf. Since this is chiefly a component made of mineral materials (masonry, concrete), sufficient fire protection is provided by the component thickness already required for static reasons. With modern solid-wood walls, the burn-off thickness must be taken into account, which must be added to the statically required cross-section.

2.2.2 Single Leaf with One-Sided Lining Lacking a Substructure

For the reasons just mentioned, it is generally necessary to differentiate **functional layers** within the envelope component in order to meet increased requirements. For this purpose, the full-walled leaf is usually assigned primarily the task of load-bearing due to its bending and shear stiffness, which is a given from the beginning. This leaf is then supplemented with an **addition** in the form of a **lining**, i.e. with additional

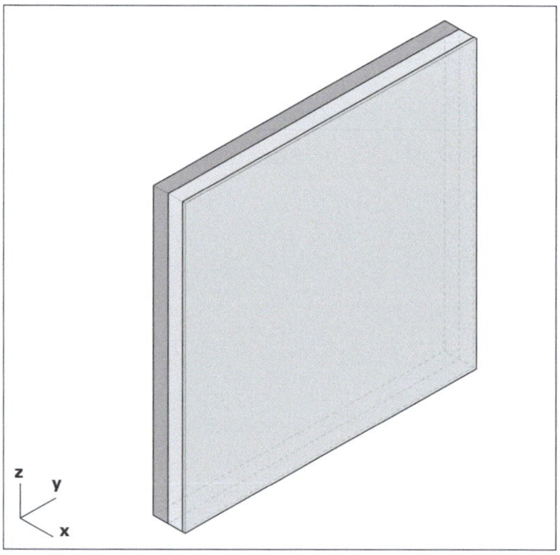

Fig. 2.5 Single leaf with external addition without substructure

functional layers (Fig. 2.5). This is almost without exception in turn composed of several layers. It may alternatively be applied to either of the two sides of the leaf, or otherwise on both sides, but the latter occurs in the rarest cases; for it is mostly possible, and much easier in constructional terms, to fulfil the tasks assigned to the added package when it is only one-sided. (Partition walls made of solid wood are a certain exception, namely those where sufficient fire protection must be ensured on both sides without burning, e.g. by fire-protection boards). This applies in the same way to all the variants discussed below featuring a lining. In the case of outer envelope components, the lining is almost without exception to be found on the outer side for reasons of building physics.

2.2.2.1 Outer Envelope Components

An essential functional layer of the lining, for outer leaf components, is the **thermal-insulation layer**. Within an envelope construction, it is determinant in the same way as a load-bearing leaf, particularly considering the thickness of the insulation which is customary today. This is not the case for most other functional layers. In addition, the most serious conflicts between individual requirements, in particular between thermal insulation and load-bearing as well as moisture-protection, arise at the thermal-insulation layer. In addition to the latter, which is indispensable in temperate geographic latitudes, at least one further layer in the form of a **weather skin** or **weather screen** is necessary for moisture protection from the outside. This is essential, if only because otherwise the porous insulation layer will soak up water and thereby lose its functionality. Other layers, such as the leaf situated behind, must also be protected from moisture.

► *Sect. 2.1 Constructional Implementation of the Constructional Partial Functions—Basic Solution Principles*

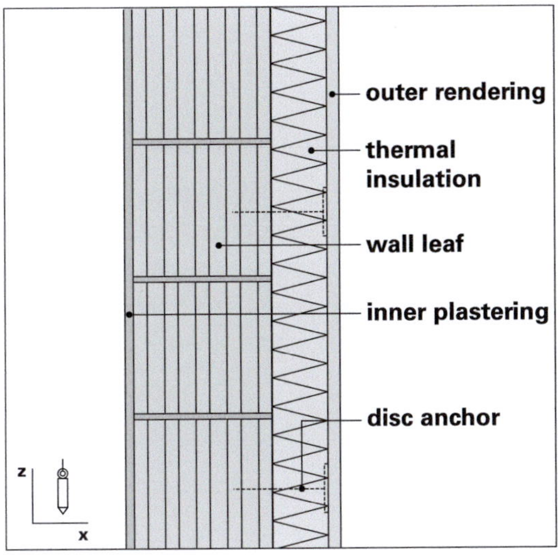

Fig. 2.6 Thermal-insulation composite system on an outer wall; an example of a single-leaf outer wall with an external addition

Ideally, in addition to the necessary waterproofness, the external weather skin also has (at least a certain) water-vapour-diffusion capacity. This allows moisture from the inside (water vapour) and any moisture that has penetrated the structure from the outside (e.g. through cracks or gaps) to diffuse out through the weather skin. Renderings from external thermal-insulation composite systems, for example, are capable of this. Rear ventilation as an alternative option for removing moisture is not possible with this construction variant due to the design, as the weather skin must adhere to the entire surface of the thermal insulation in this case, as there is no substructure.

On the other hand, it is not possible to realise a diffusion capability of the weather skin to the outside in unventilated flat roofs, which represent the effective standard today (Fig. 2.7). (The weather skin is the waterproofing in this case). Appropriate precautionary measures are therefore taken. For example, the additional installation of a vapour retarder or even a vapour barrier on the room side is essential in these cases. Otherwise, there is a risk of a vapour trap, because penetrated moisture cannot diffuse out and forms blisters on the inside of the weather skin or destroys it in the long term.

Vol. 3, ► *Sect. 12.3.3 Flat Roofs on Load-Bearing Leaves*

Vol.1, ► *Sect. 27.1.1.1 Single-Step Moisture Protection*

In the simplest type of constructional execution, the lining consists of a stratified layer package consisting of thermal-insulation layer and weather skin (Figs. 2.6 and 2.7). With regard to load-transfer, this means that the forces impinging externally on the envelope component must be transmitted across the layer package of the lining to the primarily force-conducting inner leaf. In this case, this does not take place by means of individual webs, ribs, or anchors, but by

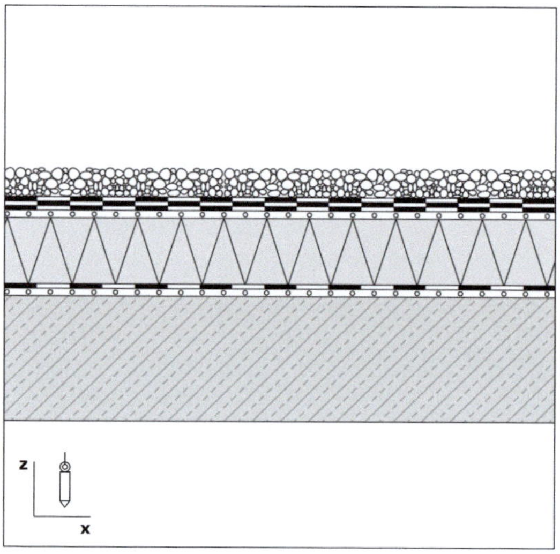

Fig. 2.7 Flat-roof construction, an example of a single-leaf envelope component with an external addition

the strength of the lining itself, i. e. first of all, of the thermal-insulation layer, which has to be regarded as an essential component of the addition. Thus, surface loads which impinge on the component are transmitted continuously across the thermal-insulation layer to the supporting leaf. In the case of linear and punctual loads, a certain force distribution takes place within the structure, depending on the overall limited stiffness of the layer packet. Both effects are advantageous. However, a structure whose core layer, for reasons of thermal protection, necessarily consists of a porous, comparatively soft material cannot absorb high force concentrations as a rule, since otherwise strong deformations, for example, compression, would occur. This represents a disadvantage which must not be underestimated for an outer component, in particular when the effect of man loads or impact loads on leaf components is taken into account. This applies, for example, to thermal-insulation composite systems (Fig. 2.6), which are extremely sensitive to mechanical damage.

Vol. 3, *▶ Sect. 12.3.1.1 Exterior Walls with External Thermal-Insulation Composite System (ETICS)*

The following factors are crucial for the ability of the lining to reliably transfer concentrated loads to the supporting leaf:

- The **strength** of the **core layer** of the lining, i. e. the thermal-insulation layer. Both the compressive and the tensile strength are important here since both stresses are not transferred to fastening elements.
- The **anchoring** of the lining to the supporting leaf. While pure compressive forces are transmitted by simple contact between leaf and lining, the latter must be secured against tensile stress and shear at the interface between the leaf and the lining. Depending on the load to be expected, this

can be achieved by simple friction, i.e. where no tensile stress, but possibly limited shear stress, is to be expected (a). If more stress is applied, adhesion may be required (b). Even a ballast can neutralise uplifting forces, as long as components are horizontal (c) (◘ Fig. 2.7). Alternatively, individual wire anchors can also be used, which, for example, in the form of disk anchors, press the thermal-insulation layer against the supporting leaf (d) (◘ Fig. 2.6). Although this fastening principle initially contradicts the reasonable concept of completely dispensing with fastening elements that penetrate the insulating layer, the local, only limited thermal bridges due to the anchors are nevertheless accepted.

(a) As with a floating screed, for example

(b) As with a flat-roof build-up (see Fig. 2.7), in which the layers are bonded together to prevent lifting)

(c) As with a gravel layer on a flat roof

(d) As with a thermal-insulation composite system

- The **stiffness** as well as the **compressive** and **tensile strength** of the **outer skin**. For conceptual reasons, this layer ought, as a matter of principle, not to be particularly heavy with inclined or standing components, since it must be able to be carried by the thermal-insulation layer without further fastening elements. Otherwise, a strong shear stress on the thermal-insulation layer is elicited, since the outer layer then slides off. The thicker the insulation is, the stronger this effect will be, since the offset moment between the outer layer and the supporting leaf increases.

► *Sect. 2.3 Double-Leaf Systems*

An exception is **horizontal components**, provided that it can be assumed that, apart from compressive forces, no appreciable additional loads act on the outer layer. This is the case when the outer layer is designed as a heavy leaf with a dead weight which counterweighs any uplifting forces, for example, in the case of concrete screeds on insulation layers. In such a case, the outer layer can be designed as a leaf with sufficient strength and bending stiffness for good load distribution. This is then in a position to distribute larger compressive forces, as well as punctual loads, on the soft support. Conceptually, this corresponds rather to the case of a double-leaf build-up to be discussed below and will only be considered there.

► *Sect. 2.3 Double-Leaf Systems*

With regard to **noise protection**, depending on the mass of the outer layer, it may be a **double-leaf system** of vibrating masses in an acoustic sense. Heavier coatings, such as render layers on thermal-insulation composite systems or gypsum-plasterboard leaves (◘ Figs. 2.8 and 2.9) fulfil this requirement. Factors determining the sound insulation of the overall structure are:

Vol. 1, ► *Sect. 28.3.3.3.2 Double-Leaf Components*

- the **mass distribution** on both leaves. Here, the main leaf will be much heavier for conceptual reasons.
- the **dynamic stiffness** of the thermal insulation. Soft, elastic layers have a favourable sound effect, but they can only poorly fulfil a major task of the structure, namely, to transmit forces to the supporting main leaf: a further ex-

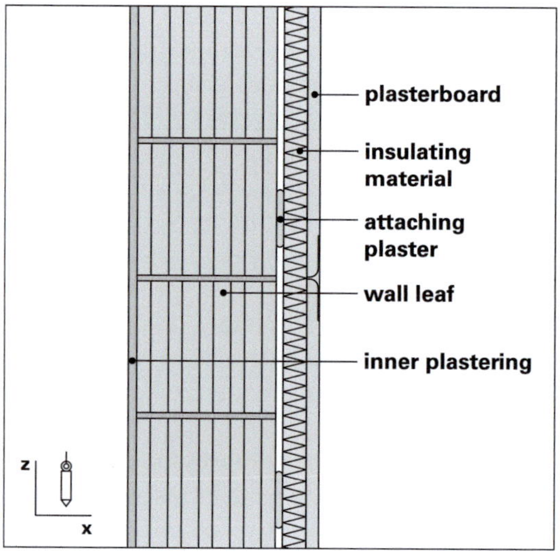

Fig. 2.8 A supplementary structure made of insulating material and gypsum boards added to a wall on the inside: an example of an internal envelope component with a one-sided lining

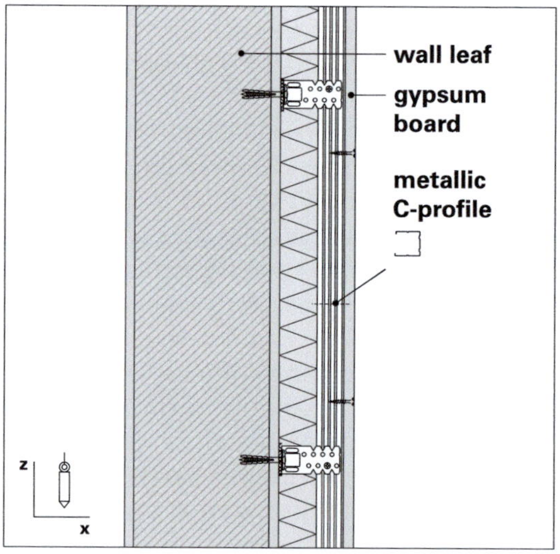

Fig. 2.9 Added leaf made of a flexible plate (gypsum board) and a rib system made of a soft-elastic metal substructure to improve the sound protection of a wall: an example of a single-leaf component with ribbed lining

ample of a constructional target conflict. When insulation layers are too stiff, such constructions added on leaves therefore sometimes rather have a negative effect on noise protection.

- the **distance** between the main leaf and the outer coating. This, at the same time, equals the insulation-layer thickness. Larger distances (and thicknesses of insulation layers) have a favourable effect on both thermal and sound protection.

The aim of the fine-tuning of the mentioned parameters is to shift the resonant frequency of the oscillating *mass-spring system* so far into the range of higher or lower frequencies that it is prevented from coinciding with the relevant audible noise frequencies.

These vary from case to case: in the case of traffic noise, for example, the medium frequencies are prevalent

In summary, the described constructional system can be regarded as a particularly advantageous solution, especially in terms of **thermal protection**, when used in outer envelope components since:

- a multilayer component can be realised so that a **high-performance thermal-insulation layer** with comparatively large layer thickness can be integrated into the construction. In some cases, in particular in the case of horizontal components, it is even possible to realise insulation layers of *any* desired thickness.
- this thermal-insulation layer is not affected in any way by any fastening elements which penetrate through it. Consequently, there is also no risk of condensation at thermal bridges.

However, the thickness and the weight as well as the type of execution of the external weather skin are limited. As a rule, these are thin coatings or films which lie on, or adhere to, the entire surface of the thermal insulation. These are, as a consequence, mostly sensitive to mechanical damage and represent a certain **weak point** of this constructional principle.

Cf. external thermal-insulation composite systems or waterproofing membranes on flat roofs, which have to be protected at rather great expense.

2.2.2.2 Inner Envelope Components

Single-leaf components with added construction are rare as **inner wall components**. The main functions in this application, namely sound and, if necessary, fire protection, are usually already fulfilled by the characteristics of the main leaf. Such constructions can be found only in the case of retrofitting of envelopes which do not reach the required values, in particular as an additional soundproofing measure (◘ Fig. 2.9). Partition walls with a thermally insulating lining can also be used between differently tempered rooms. Sometimes, as mentioned above, partition walls made of solid wood are fitted with facing leaves on both sides in order to improve either sound insulation or fire protection, which the unclad leaf alone cannot guarantee.

The parameters to be observed were referred to above in connection with outer building components (► *Sect. 2.2.2.1*)

On the other hand, **floor constructions**, which, executed as solid slabs, consist of a supporting main leaf, often have to be provided with an additional layering in order to ensure sufficient impact-sound protection. Floating screeds can be understood as bending-stiff facing leaves which are simply placed on an insulating layer without fastening elements. They represent, as already indicated above, a border case to double-leaf systems and are to be discussed elsewhere.

► *Sect. 2.3.1 Two Leaves with Intermediate Layer*

2.2.3 Single Leaf with One-Sided Lining Containing a Substructure

2.2.3.1 Single Rib Layer

In this case, the structure which is attached or installed in front or on top of the supporting leaf essentially consists of a system of **ribs**, an outer thin, at least a partly bending-stiff, **leaf** attached thereto, as well as an **insulation layer** (◘ Figs. 2.10 and 2.11) incorporated into the cavities between the ribs. This construction differs from double-leaf systems as described below in that the outer leaf does not have sufficient bending stiffness in order to get along without or with minimal anchors, but must be attached to a linear, rib-like substructure. This has a lasting effect on the constructional circumstances.

▶ *Sect. 2.3 Double-Leaf Systems*

In this case, a further **differentiation** of functions in the added package therefore takes place in comparison with the case discussed above. The discharge of external loads from the outer leaf or weather skin, which takes them up in the first place, onto the supporting leaf no longer occurs via the insulation as with the rib-less construction, but via the considerably stiffer rib system. As these ribs are made of a rigid, dense material for reasons of force conduction, they must be regarded as a **thermal bridge** within the insulation layer, or also as a **sound bridge**, depending on the requirements and the application. Unlike in the case of punctual anchoring of double-leaf systems, these are *linear* thermal bridges, which must be assessed as much more critical than punctual ones.

The distance **a** between the ribs (◘ Fig. 2.10) results from the bending stiffness of the outer leaf. Insofar as plate ma-

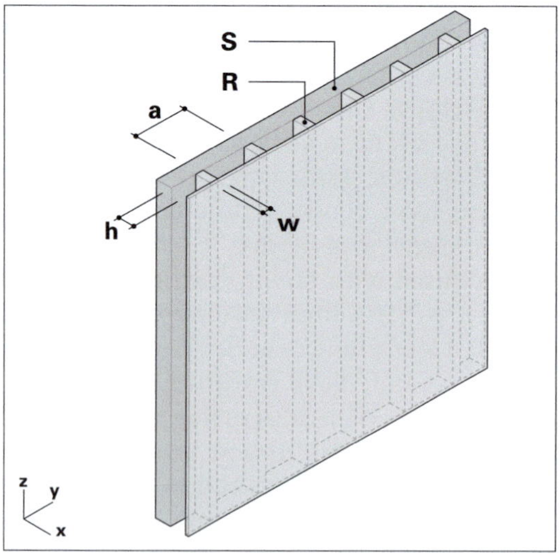

◘ **Fig. 2.10** Non-load-bearing addition from a single-ply rib system **R** on a supporting leaf **S**; **a** rib spacing; **w** rib's cross-section width; **h** rib's cross-section height

terial with thicknesses between 2 and 3 cm is involved, this measure is frequently in the range of 40 to 80 cm.

The ribs can rest on the supporting leaf linearly or can be attached to it as often as desired so that they only experience **compression/tension** (at right angles to the component plane, along coordinate axis → **x**) or shear stress (along the rib axis → **z**), but no significant bending. Their cross-section height (◘ Fig. 2.10) is thus not determined by the bending stress, but, in the context of the added construction design, depending on the application:

- by the required **thickness t** of the **thermal-insulation layer** laid between the ribs. A very large insulation-layer thickness results in large rib dimensions, which may not be reasonable since:
  - as the insulation thickness increases, the building-physics-related problem of the **thermal bridge** at the rib is fundamentally exacerbated;
  - with increasing cross-section width **b** of the rib—this must be in an appropriate ratio to the height **h**—the effect of the **thermal bridge** over this rib again increases;
  - in the case of inclined or standing components, a comparatively heavy outer skin produces strong **shear forces** in the ribs due to the offset moment between the skin and the supporting leaf. This also requires dimensioning the ribs correspondingly, which strengthens their effect as a thermal bridge and, overall, unnecessarily increases the weight and the material consumption of the constructional lining.
- or by a minimum distance **s** between the leaves, which may be necessary for reasons of **noise protection**.

***Vol. 1**, Sect. 28.3.3.3.2 Double-Leaf Components*, as well as *Sect. 28.3.3.4. Construction Variants of Double-Leaf Envelope Components* > see: one bending-stiff, one bending-soft leaf

In principle, a tolerable compromise is to be found in determining the **rib spacing (a)**, since the greater the distances become, the fewer thermal bridges occur on the one hand, but the heavier the outer leaf will be, which, in turn, entails heavier ribs with a larger cros,s-section width **w** and, consequently, a more severe thermal-bridging effect.

With this type of construction, any moisture that may have penetrated into the building—whether precipitation water that has entered from the outside through cracks or gaps or water vapour that has penetrated from the inside—does not have to be dissipated to the outside (only) through the diffusion capacity of the weather screen (as in the previous variant without ribs in the added lining), but can be removed more reliably with the help of **rear** or **bottom ventilation**. For this purpose, it is sufficient not to completely fill the depth of the rib (i.e. the height **h** of the rib's cross-section (in → **x**) with thermal-insulation material, but to leave a cavity between it and

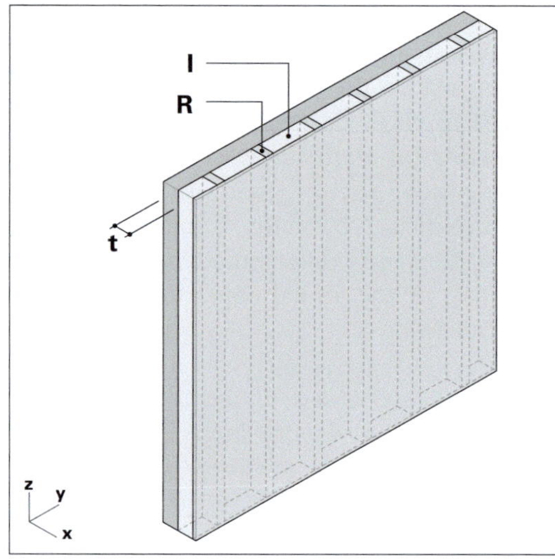

Fig. 2.11 The cavities between the ribs **R** are filled with insulation material **I** (thickness **t**), either completely or leaving an air space

the weather screen (Fig. 2.12). In the case of **exterior enclosures**, this requires the ribs to be vertical. In the case of **ventilated roofs**, where this variant is also used, it must be ensured that the ventilated cross-section is sufficiently dimensioned for allowing air movement on a slope or—in the case of flat roofs—even horizontally. If necessary, water-vapour penetration from the interior must be additionally controlled by means of suitable vapour retardants or vapour barriers applied on the inside (Fig. 2.13).

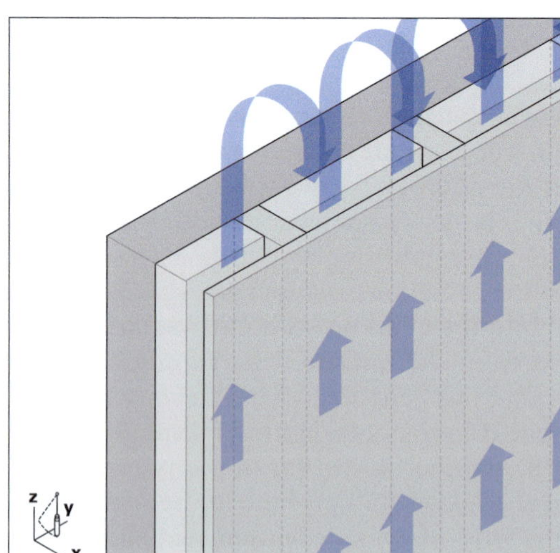

Fig. 2.12 Back or under-ventilation of the outer leaf in the case of a single-ply rib system added to the main leaf by only partial filling of the cavities with insulation material

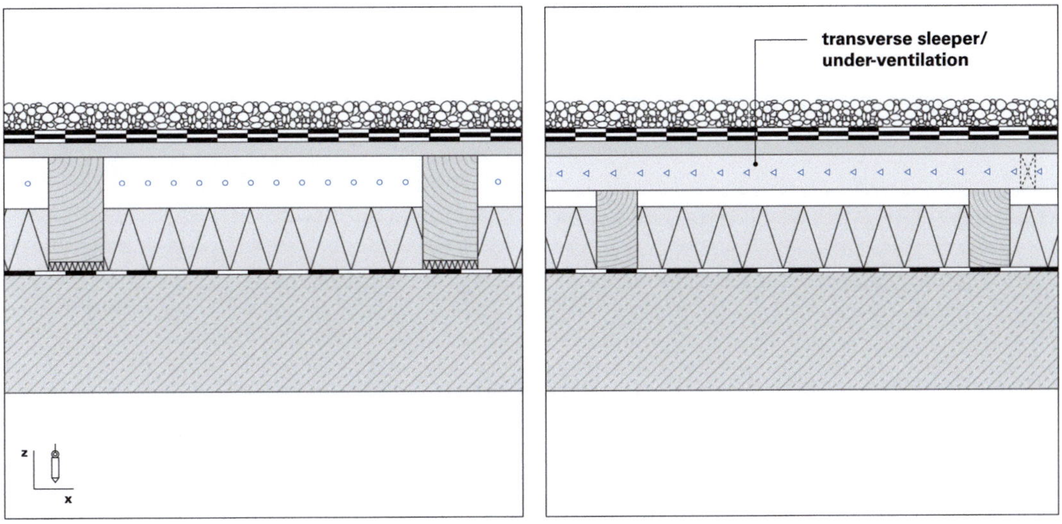

Fig. 2.13 (Left) Ventilated flat roof construction on a concrete slab with a set of wooden sleepers. The ventilation is parallel to these within the cavity between the insulation and the upper leaf. (Right) Ventilated flat-roof construction on a concrete slab with two stacked sets of wooden sleepers. The ventilation is parallel to the upper ribbed layer

From an **acoustic** point of view, this is a double-leaf vibrating mass-spring system. The type of connection of the rib system with the supporting main leaf plays a decisive role. The more flexible the coupling is, the better the noise protection will be. Point connections of the rib at large distances have a favourable effect, whereas linear connections are disadvantageous since they act as sound bridges. As with the variant in ► Sect. 2.2.1, such a structure, being too rigid, can even lead to a deterioration of the noise protection. If the cavities are filled with a sufficiently flexible insulation material—which, in this case, unlike in ► Sect. 2.2.1, is possible because of the force conduction via the ribs, not via the insulation layer—cavity resonance is not to be expected.

With respect to the use of the constructional lining discussed for **interior wall components**, the same as in ► Sect. 2.2.1.2 applies.

2.2.3.2 Double Rib Layer

In the case of very thick insulation layers, which are standard in today's highly insulated envelopes, the single-layered ribbed system can be replaced by a **double-layered** ribbed system (Fig. 2.14 left) in order to avoid the above-mentioned difficulties with constant enlargement of the insulation layer's thickness. Thereby, a cross-oriented skein is attached to the ribs on the outside. This construction has the advantage, thermally, that the heat bridges due to the ribbing are reduced to the crossing surfaces of the superposed rib layers, as long as the interstices are filled with insulating material in both rib layers (Figs. 2.15 and 2.16).

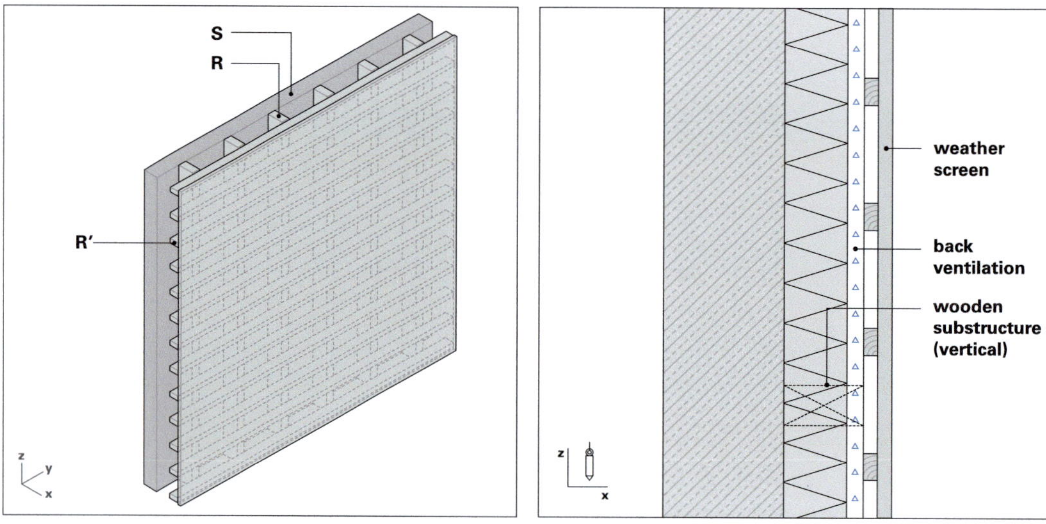

Fig. 2.14 (Left) Construction of a two-ply ribbed system with cross-wise placed rib layers **R**, **R'** on supporting leaf **S**. (Right) Exterior wall with solid backup wall and light wall addition from two ribbed layers. Back ventilation allowed for by the protrusion of the rear set of ribs (vertical) beyond the insulation layer

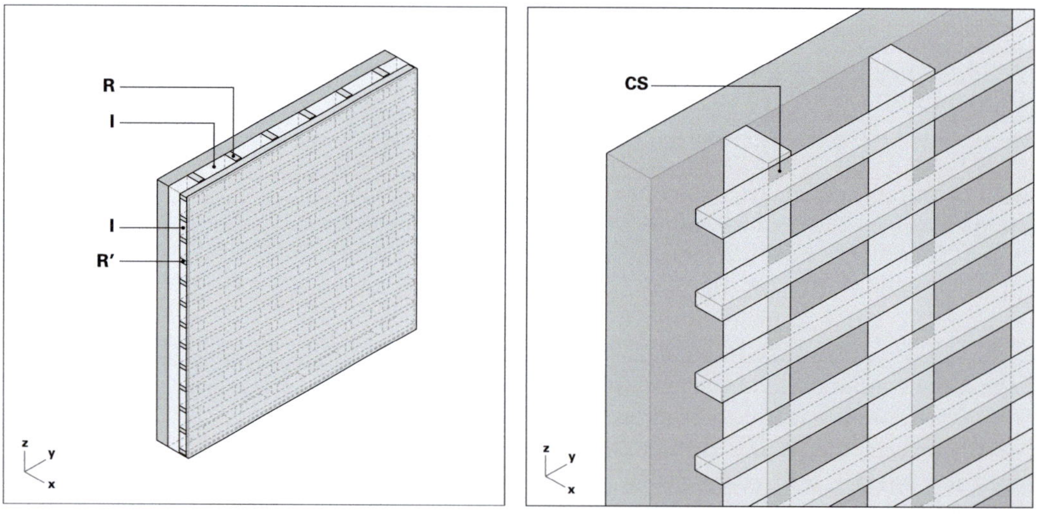

Fig. 2.15 (Left) Element as on Fig. 2.14, left; The interstices of the two ribbed layers **R**, **R'** are filled with insulating material **I**. (Right) In the case of a two-ply rib structure, the thermal bridge is reduced to the contact surface **CS** between two intersecting ribs

Although in the case of single-ply rib systems it is also possible to form an **air layer** by making the thickness of the insulation layer smaller than that of the available cavity (Fig. 2.12), care must be taken in such a case that the insulation panels are secured in position and do not possibly clog the air space by peeling off and falling down, which is possible, in particular, with standing components. In this case, the two-ply ribbed system, in which the insulation layer and the

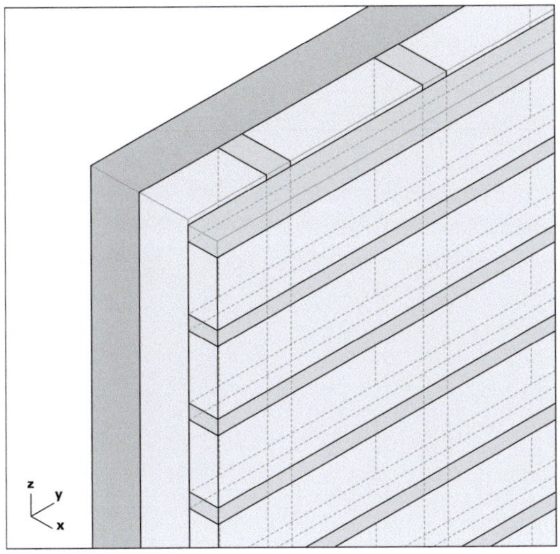

■ **Fig. 2.16** Detailed view of the structure depicted on ■ Fig. 2.15, right

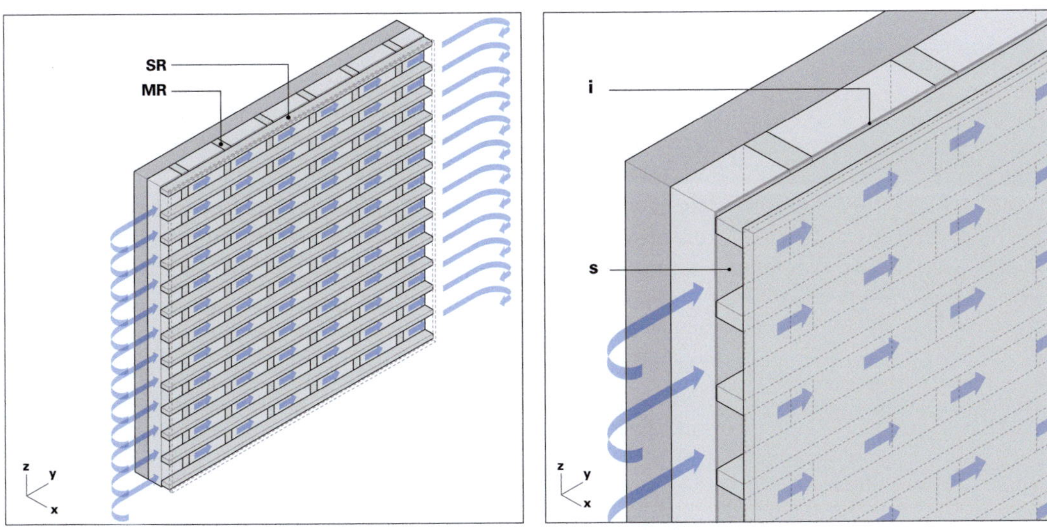

■ **Fig. 2.17** (Left) Clear constructional separation of the insulation layer in the space between the main ribs **MR** and the air layer in the cavities between the secondary ribs **SR**. See detail on the right. (Right) A diffusive, water-repellent thin sheet or film **s** at the interface between the two rib layers (**i**) protects the insulation from moisture from the outside. At the same time, it allows the possibly moistened insulation layer to dry out by vapour diffusion into the air layer

air layer are separated in constructional terms (■ Fig. 2.17), provides a higher degree of safety.

The **orientation** of the rib layers, which is discretional with respect to the load-bearing direction perpendicular to the component plane (in → **x**), since we assume that the main ribs lie linearly on a supporting leaf or can be attached thereto, can nevertheless be predetermined by the direction of flow of the air due to thermal buoyancy whenever an air

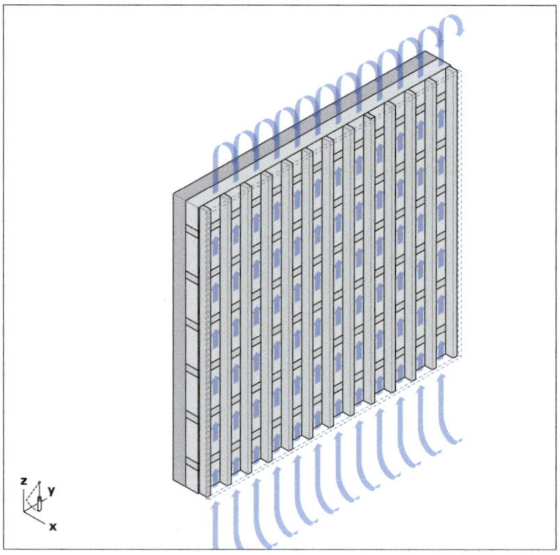

Fig. 2.18 The orientation of the rib layers may be predetermined by the position of the component. For inclined and upright components, the flow direction of the air due to thermal buoyancy must be taken into account

layer with contact to the outer space is involved. This applies to **inclined** components as well as, in particular, to **vertical** ones (Fig. 2.18). With horizontal components, different factors can lead to giving preference to a special rib orientation; with regard to air movement within the cavity, the orientation ought to be chosen in which the pressure differences between opposite ventilation openings are largest.

Vol. 3, ▶ *Sect. 12.3.3.6 Ventilation*

From a **building-acoustical** point of view, a double-ribbed layer which is locally connected at the intersection points is tantamount to a double-leaf oscillating mass-spring system with a larger spring action than with a simple rib layer. In principle, it therefore provides an improved noise protection.

2.3 Double-Leaf Systems

2.3.1 Two Leaves with Intermediate Layer

Vol. 3, ▶ *Sect. 12 Full-Leaf Systems*

In this variant, the envelope component consists of two **bending-stiff**, **planar** leaves which are arranged parallel to each other (Figs. 2.19 and 2.20) with a certain clearance. In contrast to the variants discussed above, each of the two leaves has sufficient stiffness in order to deflect the resulting forces without requiring a more or less complicated substructure, such as, for example, a rib system. Both leaves nevertheless can, but need not, be (at least partially) force-fittingly coupled to one another (Fig. 2.21).

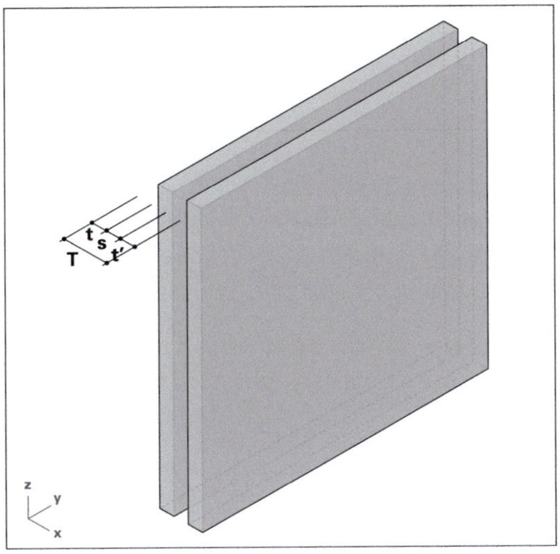

Fig. 2.19 Two leaves without filling of the intermediate cavity. **T** Overall thickness of the construction; **t**, **t'** thicknesses of the individual leaves; **s** spacing of the leaves, equal to the thickness of the inter-leaf space

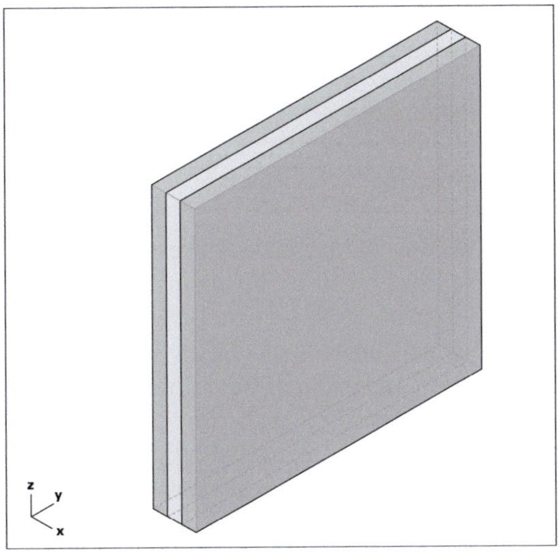

Fig. 2.20 Two leaves with filled cavity

From the viewpoint of load-bearing, in the basic construction principle of this variant, therefore, no webs are provided which could connect both leaves to form a **structurally co-operating system** (according to the I-principle) consisting of web and two chords, i. e. the two leaves themselves. Each of the two leaves thus acts on its own, whereby their respective bending stiffnesses against force attack at right angles to the envelope plane (→ **x**) can certainly be incremental in certain cases—as long as compressive and/or tensile forces can be passed over the gap—but the theoretically available **structural depth** equal to the overall thickness **T** (Fig. 2.19) is not fully

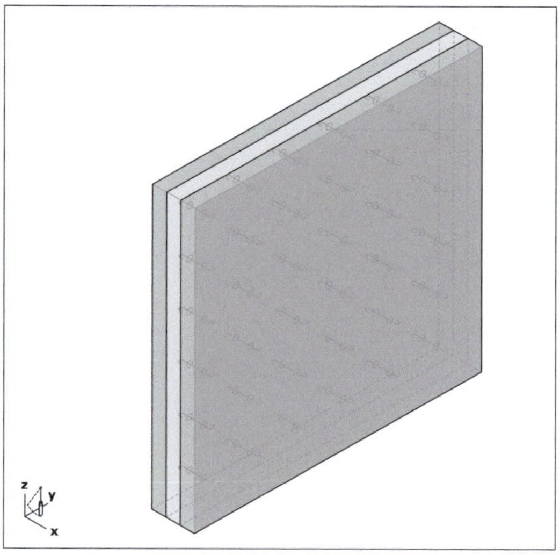

Fig. 2.21 A re-anchoring of the outer leaf to the inner—here shown exemplarily as two-leaf masonry—necessarily penetrates the intermediate layer

utilised when the component is subject to **bending stress**. With respect to load-bearing, the double-leaf solution, hence, does not offer any particular advantage as measured against its entailed constructional complexity.
Depending on the position of the envelope component, characteristic difficulties may also arise:

- **Horizontal** component: As we have seen, the bearing capacity of the horizontal component is not a result of the moment of inertia of two structurally interacting leaves. The bending stiffnesses of both separate, structurally non-cooperating leaves must be computed separately, or, if they are placed on top of each other, in their simple arithmetic sum. In the case of a horizontal component position, for the sake of simplicity, usually only **one single leaf** is designed to be load-bearing. Almost without exception, it is the *lower* leaf which is designed as the only load-bearing element, so that all further layers or leaves can simply be placed or built up on this supporting leaf. The upper leaf then bears on the lower layer, whereby it usually rests on one or more intermediate layers—for exterior components, at least one thermal-insulation layer—(for example, a screed on an insulating layer on a solid slab). Although their bending stiffness does not influence the overall load-carrying capacity of the component, it is noticeable by its **load-distributing effect**, which has a favourable influence on the overall structural behaviour under large point loads.
- **Inclined** or **upright** component: In this case, as a result of the dead load of the leaves, it must be taken care of how

the load components parallel to the envelope surface can be borne. Since it is mostly the inner leaf which is attached directly to the primary structure, or is already part of it, it is first of all necessary to clarify how the outer leaf is held. The sum of the external loads can be converted into two force components (Fig. 2.22 right; Fig. 2.23): one at right angles to the envelope plane (**F**) and one parallel to it

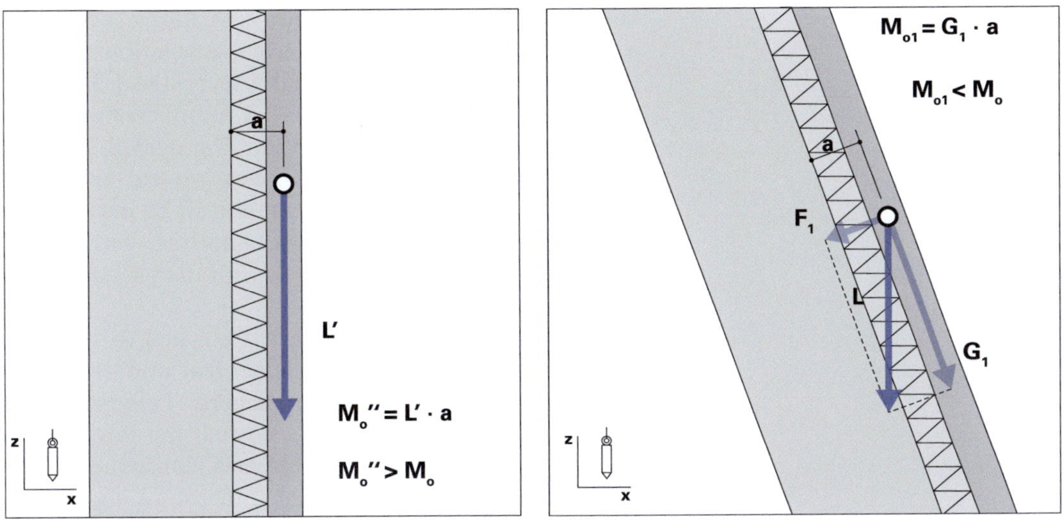

Fig. 2.22 (Left) Also with increasing load **L″**, the offset moment $\mathbf{M_o}$ increases, the lever arm **a** remaining the same. (Right) On the inclined leaf, the load **L** is split into two components $\mathbf{F_1}$ and $\mathbf{G_1}$. The force component $\mathbf{G_1} < \mathbf{L}$ produces a reduced offset moment $\mathbf{M_{o1}}$

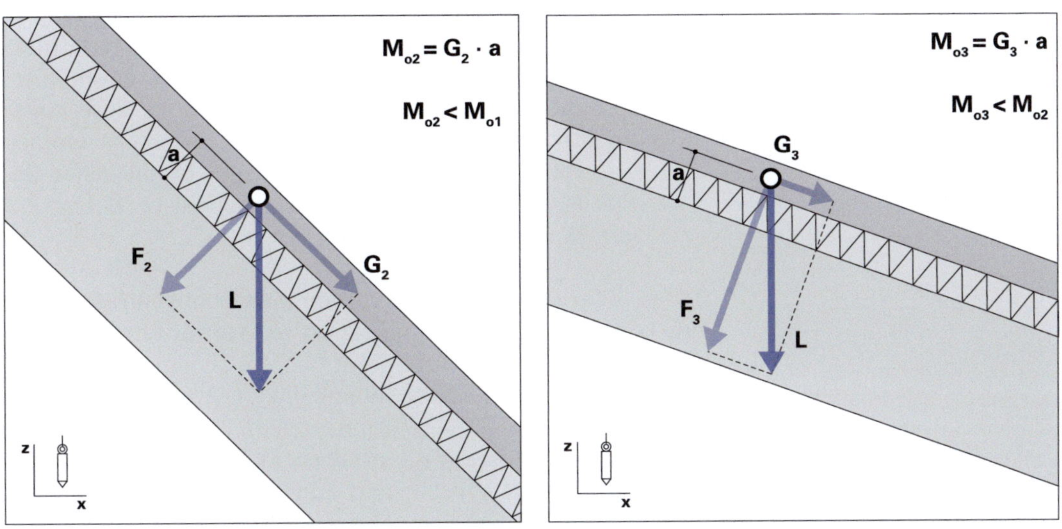

Fig. 2.23 (Left) With decreasing inclination, the force component **G** becomes smaller ($\mathbf{G_2}$) and thus also the offset moment $\mathbf{M_{o2}}$. (Right) Even smaller inclinations lead to continuously decreasing force components **G** and offset moments $\mathbf{M_o}$

(**G**). What was said for the lying components also applies to component **F**. The component **G** can be accommodated in two ways:

- By means of a **separate bearing** for the outer leaf, which, located at the outer side of the insulation layer, carries the load parallel to the leaf's plane and down into the primary support or the foundation—similarly to the rear-ventilated outer wall on ◘ Fig. 2.27. This solution has the advantage of not requiring a major force-fitting connection between the leaves, something which has a **favourable effect** on the thermal insulation (because there are none, or only minimal heat bridges). However, this solution presupposes that a supplementary supporting structure or a foundation is provided, which is largely independent of the load-bearing leaf and which is stable on its own. Since the outer leaf, by design, does not have a force-transmitting connection with the inner one—and thus in most cases with the primary structure—it cannot be held by it.
- The force component **G**, which is parallel to the envelope surface, is directed to the inner leaf and thus to the primary support via suitable **anchors**. This is conveniently done in limited spacings so that the force can be distributed as much as possible and, consequently, the anchors can be limited in their dimension, and in particular in the size of their cross-section, since it is always to be considered that these connections represent **thermal bridges** through the insulation layer.
  The lever arm between the force component **G** and the inner leaf, to which this force is to be deflected, produces an **offset moment**, which is the larger, the greater the distance between the leaves is (◘ Fig. 2.24). As a result, very thick insulation layers, such as are required in today's highly insulated envelopes, also entail correspondingly large anchoring cross-sections and, consequently, thermal bridges. Naturally, the offset moment also increases with increasing weight of the outer leaf, that is, with increasing force component **G** (◘ Fig. 2.22 left; ◘ Fig. 2.25).

These observations reveal a certain **immanent contradiction** of this construction in its application to exterior envelope components: from a structural point of view, the material of both leaves cannot be utilised economically, since either:

- the **inner leaf** is always attached to the primary structure, or belongs to it itself, and consequently performs a **primary support function**. The outer leaf is then, as a matter of fact, a *dead mass* and must be carried by the inner one. Except for horizontal components, heat bridges are inevitably created.

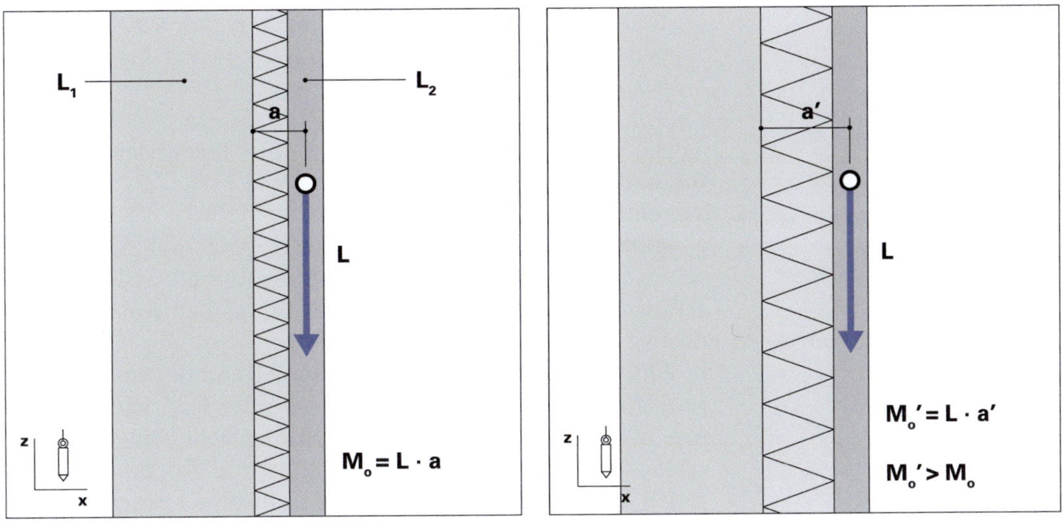

Fig. 2.24 (Left) The dead load **L** of the outer leaf $\mathbf{L_2}$, which is anchored to the load-bearing backup leaf $\mathbf{L_1}$, produces an offset moment $\mathbf{M_0}$ as a result of offset **a**. (Right) As the distance **a′** increases, the lever arm increases and thus also the offset moment $\mathbf{M'_0}$

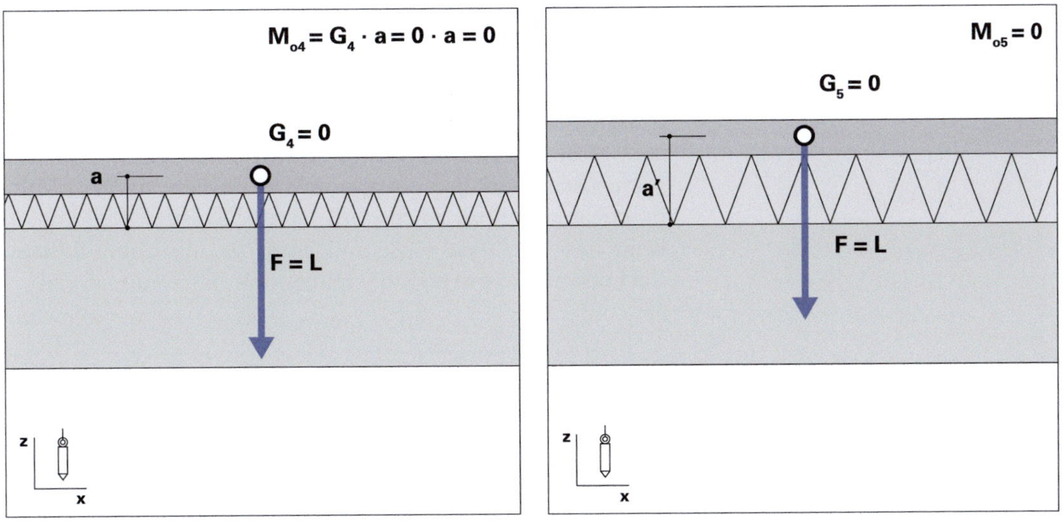

Fig. 2.25 (Left) In a horizontal position, the force component parallel to the envelope surface is $\mathbf{G} = 0$ and thus also the offset moment $\mathbf{M_0}$. (Right) In a horizontal position, the distance **a′** between leaves, in this case equal to the insulation thickness, has no influence on the force transmission so that the insulation can be dimensioned according to building-physics requirements

- or the outer leaf is made capable of standing on its own or otherwise must be carried by means of a largely separate **supplementary support structure**, which entails corresponding structural complications: as it were, a doubling of the load-bearing leaf structure. It is often simply impossible to design the outer leaf to be stable without utilising the stiffness of the primary supporting structure by anchoring it back, as it would not itself be load-bearing due to its extreme slenderness, which, as it were, is inherent to the very constructional concept, especially at greater heights in the vertical position.

This construction can often be found in **double-leaf solid outer walls**, for example, in masonry (◘ Figs. 2.27 and 2.28) or reinforced concrete. The outer leaf always acts solely as a **weather skin**, a function which, in principle, also can be achieved with a light cladding or sheathing. Specific properties of the brick version result in further constructional problems which are difficult to solve.

Leaf with lining, see ▶ *Sects. 2.2.2 and 2.2.3*
***Vol. 3**, Chap. III Exterior Envelopes*

Only in the case of the **horizontal** component does this problem solve itself, so to say: The outer leaf, which lies on top, being a bending-stiff cover without heat bridges, can exert a beneficial load-distributing function for large concentrated loads—such as for use on trafficable terraced roofs—which justifies its thickness and heavy mass (◘ Fig. 2.26).

In an **acoustic** sense—as also in structural terms—this version is a double-leaf system: a vibrating mass-spring system. Problematic cavity resonances in the interspace are not to be expected as a rule, since the intermediate layer is almost always filled with insulating material. Once again, factors decisive to good sound insulation are:

- the **mass distribution** of the leaves. Here, different area-related masses of the two leaves have a favourable effect on sound protection. This also corresponds essentially to the normal practical situation, since both leaves also perform different tasks in a *structural* sense and, as a rule, also have different thicknesses and are possibly made from different materials.

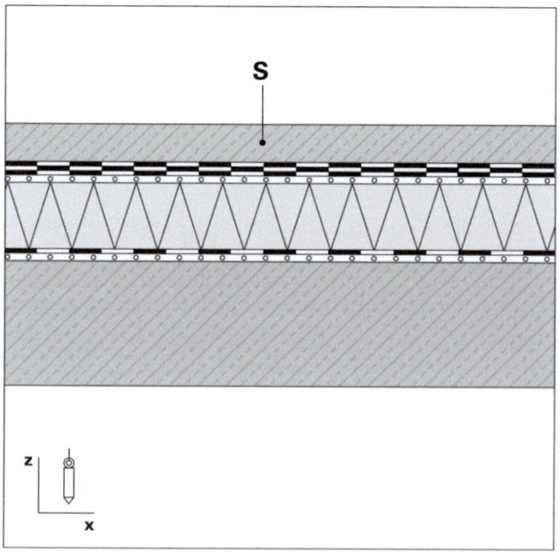

◘ **Fig. 2.26** Layering on a flat terrace roof: load-spreading screed leaf **S** made of reinforced concrete, lying over its entire surface on the insulation layer: an example of a horizontal covering leaf in double-leaf construction

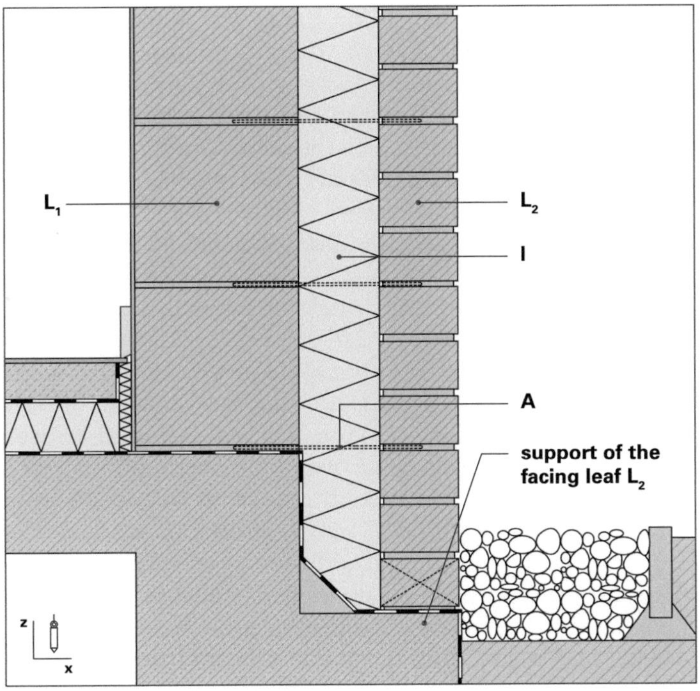

Fig. 2.27 Double-leaf masonry with core insulation **I**, without back ventilation: support of the outer leaf $\mathbf{L_2}$ on a foundation (or basement wall) for ablation of the vertical force component **G** in separation from the back wall $\mathbf{L_1}$. The wire anchors **A** transmit the horizontal force component **F**. Thermal insulation made of water-repellent material

- the **dynamic stiffness** of the thermal insulation. In the case of vertical components, soft insulation materials can be used accordingly. Only in the lying position, it may be necessary to choose stiffer insulation layers in order to transfer the load of the overlying leaf to the lower one. In this case, the noise protection must be ensured by influencing other parameters.
- the **distance** between leaves. As discussed above, associated questions of load-carrying have to be clarified in the case of larger leaf distances, which, as a matter of principle, have a favourable impact on building acoustics.

A **force-fitting coupling** between the leaves affects noise protection adversely, since this results in sound bridges. The smaller the cross-sections of connections and the more elastic they are, the more favourable the conditions. A good example of the building-acoustic effectiveness of the oscillating mass-spring system is the **floating screed** on a supporting slab, whereby the screed leaf can be considered as a boundary case between a real leaf and a light additional structure due to its small thickness and mass.

From a **hygrothermal** point of view, two variants are conceivable for the double-leaf element in temperate regions:

- There is a vapour barrier on the inside of the thermal-insulation layer. This prevents the penetration of water vapour into the insulation package and thus also dew-water accrual. This solution can be found, for example, in flat roofs (Fig. 2.26), in which the incorporation of a vapour barrier is not a real additional hassle since it can be simply laid out horizontally.
- A vapour-pressure equalisation is achieved by diffusion over the entire layer package. This solution can be found in double-leaf exterior walls made of mineral materials (masonry, concrete) (Figs. 2.27 and 2.28). In this case, the double-leaf element is to be regarded as a unit in a hygrothermal sense: since there is no back ventilation in the layer structure under consideration, vapour diffusion between interior and exterior space can only occur through both leaves. As expected, condensation can occur on the inside of the—cold—outer leaf under adverse conditions. The choice of suitable vapour-diffusion resistance of the layers must ensure that no permanent moisture is produced within the component. Conflicts of goals may occur on the outer leaf which, from the point of view of weather protection, should have a very dense fabric structure. This is usually difficult to reconcile with high vapour-diffusion perme-

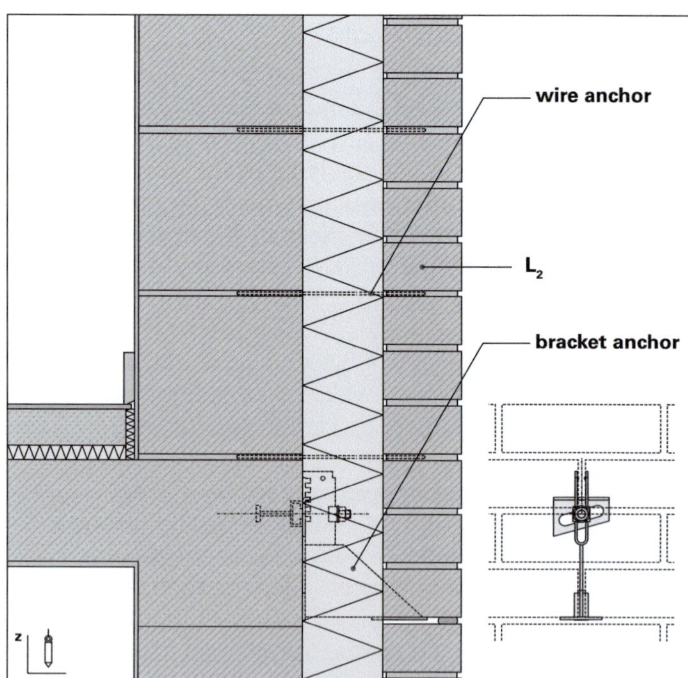

Fig. 2.28 Anchoring the veneer leaf $\mathbf{L}_2$ by means of bracket anchors (vertical force component **G**) and wire anchors (horizontal force component **F**). In this version (support on 2/3 of the brick width), the veneer leaf is supported at each two floors according to *EN 1996-1-1*, 8.4.3

ability. Higher vapour-diffusion resistance of the insulation layer itself also has a favourable effect in this respect.

2.3.2 Two Leaves with Intermediate Layer and Air Layer

If the cavity between the leaves is not completely filled up with thermal-insulation material (◘ Fig. 2.29 left), it is possible—as in the case of the variants with ribs in the lining—to implement a rear or bottom ventilation of the construction in order to remove any moisture that may have penetrated from the inside, or outside, to the outside. This prevents condensation from forming on the inside of the outer leaf, as was potentially the case with the previous variant.

The introduction of a **back ventilation** between the filling from insulation material and the outer leaf (◘ Fig. 2.29 right) entails—with a given insulation thickness—an **increase** in the **distance** between the leaves. This additional dimension must be so large that an effective air movement can be expected. The increase in the distance between the leaves aggravates the aforementioned problems of load-carrying in the double-leaf envelope component, whereas rear ventilation generally results in an improvement of the element's behaviour in terms of building physics.

The transmission of compression acting upon the outer leaf to the inner leaf and the primary structure via the insulation layer (in → **x**) is not possible directly and over the entire surface, as is the case with the solution discussed above, because there is no contact between the insulation layer and

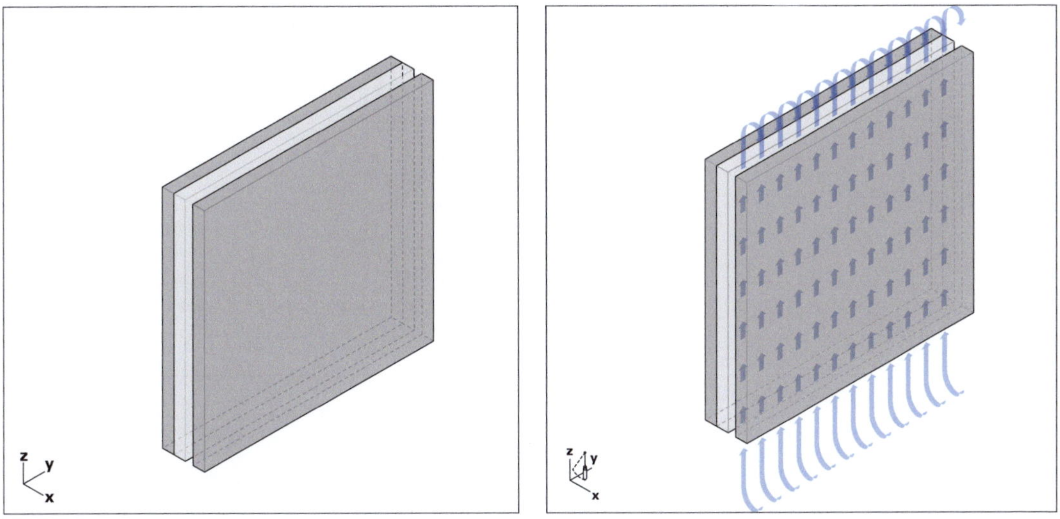

◘ **Fig. 2.29** (Left) Two leaves with intermediate layer and air layer. (Right) Rear or under-ventilation in the interspace. The position of the component with respect to the vertical must be taken into account. It affects the flow velocity of the air decisively

the outer leaf. This must be effected by connecting elements subjected to compression, which must be dimensioned more strongly than tensile elements (such as wire anchors) as a result of the danger of buckling, and have a negative influence as thermal bridges (◘ Fig. 2.21). For this reason, this solution is frequently used with **vertical** envelope components (◘ Figs. 2.30, 2.31, and 2.32), more rarely with horizontal ones (◘ Fig. 2.33) where the entire dead weight of the outer leaf must be transferred in addition to the external loads.

In the case of **horizontal** components, this solution is therefore hardly applicable because, in the case of a horizontal position, the problem of transmitting forces from the upper to the lower leaf through the core layer is already exacerbated for geometrical reasons because of a larger force component **F** at right angles to the component surface (◘ Fig. 2.25 left). Since, in this case, there is no surface contact between the upper leaf and the core layer because of the interposed air layer, force transmission must happen by means of suitable local supports, which—at least with heavy loads—inevitably become heat bridges as they have to penetrate the insulation layer as far as to the lower leaf. The fundamental advantage of this solution, namely, the possibility of an under-ventilation of the envelope construction, in the case of a horizontal position, can unfold its effect reliably only in the case of a larger air-layer thickness

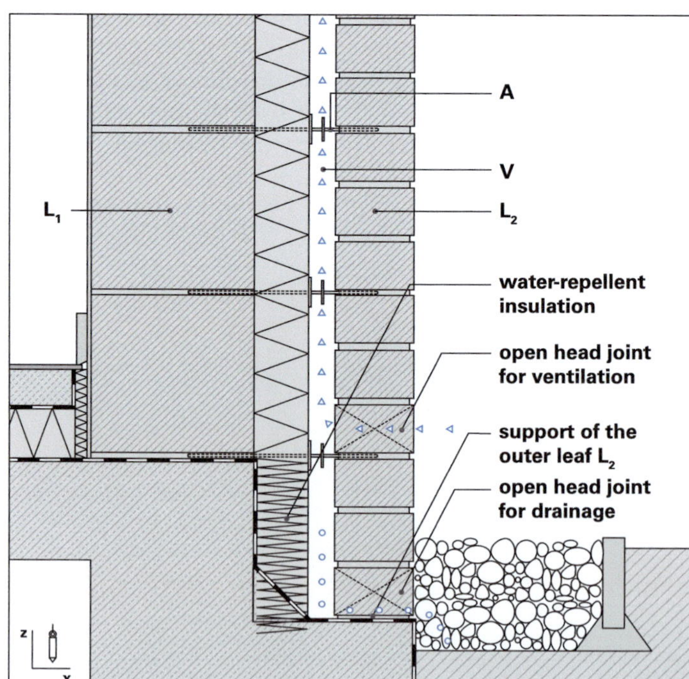

◘ **Fig. 2.30** Double-leaf masonry with back ventilation **V**: support of the outer leaf $\mathbf{L_2}$ on a foundation (or basement wall) for removal of the vertical force component **G** in separation from the back wall $\mathbf{L_1}$. The wire anchors **A** conduct the horizontal force component **F**

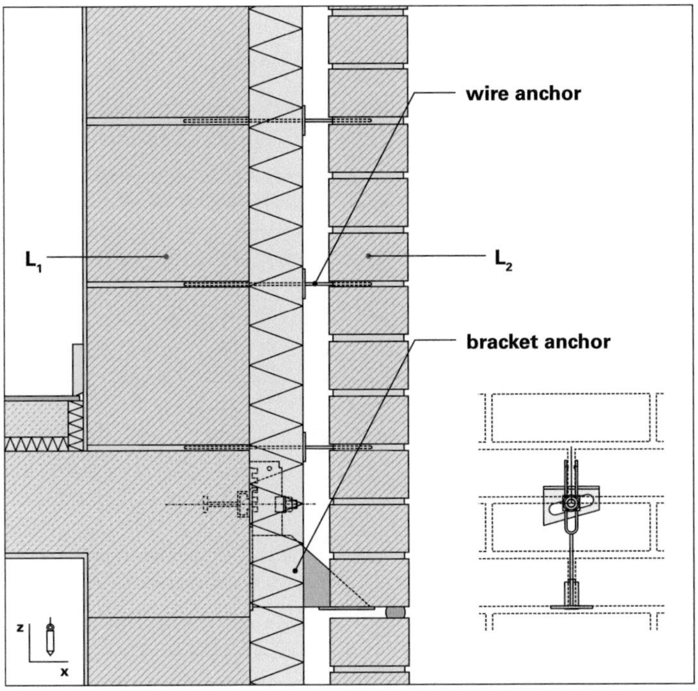

Fig. 2.31 Anchoring of the veneer leaf $\mathbf{L_2}$ by means of bracket anchors (force component **G**) and wire anchors (force component **F**). In this version (2/3 bearing), the veneer leaf is supported by bracket anchors at each two floors, according to *DIN 1053-1*, 8.4.3

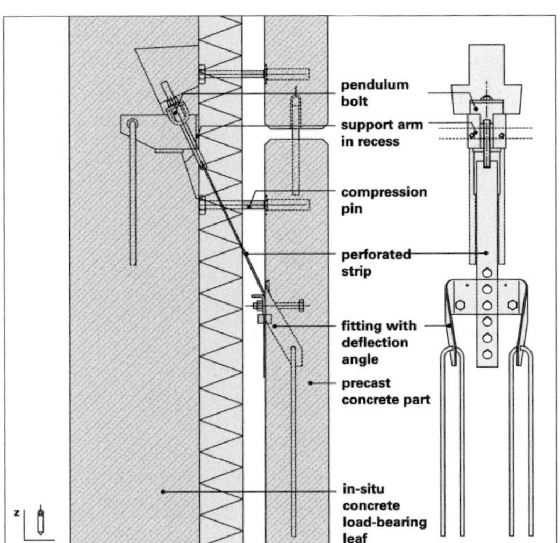

Fig. 2.32 Anchoring of a concrete veneer slab to a supporting concrete leaf by means of a commercially available anchoring system (*Halfen*)

and thus larger leaf spacing, owing to a lack of thermal buoyancy in horizontal position.

As with all other versions with back-up or under-ventilation, the expected **direction of flow** of the air must always be taken

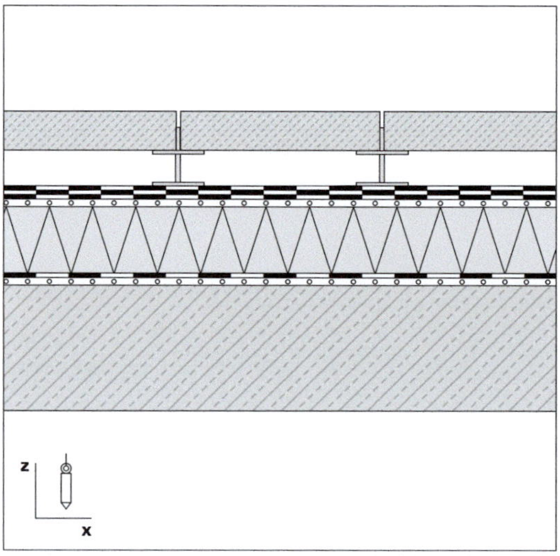

Fig. 2.33 Terrace surfacing mounted on a roof standing on props. Loads are introduced into the insulation layer via the supports. Wide disk-like foot plates ensure sufficient force distribution

into account (Fig. 2.29 right). It must not be significantly hindered by anchoring elements placed within the air layer.

From the point of view of **building acoustics**, there is no significant change compared to the variant discussed in ▶ Sect. 2.3.1. Cavity resonances are prevented by a soft-elastic insulation layer. However, a disturbing transmission effect can emerge via the air layer (telephony effect), which in the case of vertical components is possibly to be prevented by floor-wise segmenting of the air layer. This has corresponding effects on the back ventilation of the outer leaf, so that, as the case may be, adequate ventilation openings are to be provided on each floor.

2.4 Multilayer Composite Systems

Load-carrying leaves, as required for the above-described systems for supporting and fixing all further components of the envelope, are characterised in most cases by high weight or increased material expenditure. Particularly in the case of concrete leaves, one of the few materials which are suitable at all for forming solid homogeneous leaves, high dead weight must be expected. This need not always be a disadvantage. Like in the case of floor constructions, for example, this helps ensuring good soundproofing between storeys, or achieving good thermal storage capacity for internal components. In many cases, however, large material consumption and high dead weight are rather undesirable in constructional applications.

Vol. 1, ▶ *Sect. 28.3.3.3 Airborne-Sound Behaviour of Components*, or *Sect. 28.3.4.2 Impact-Sound Behaviour of Floor Constructions*

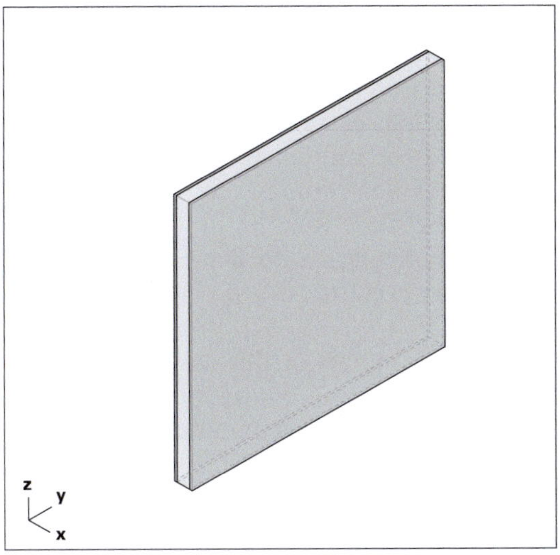

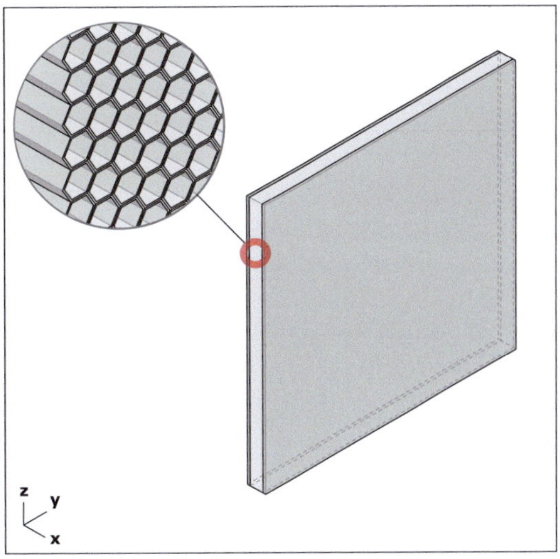

Fig. 2.34 (Left) Sandwich element. (Right) Honeycomb-core element

As an alternative to genuine homogeneous leaves, multilayer composite systems are therefore sometimes advisable, which essentially combine significant advantages of the single leaf and, at the same time, feature reduced dead weights. These may be, for example:

Vol. 3, ▶ *Sect. 13 Multilayer Composite Systems*

- **sandwich** systems (Fig. 2.34 left);
- **honeycomb-core** systems (Fig. 2.34 right).

Both are discussed in more detail below.

2.4.1 Sandwich Systems

Sandwich systems (Fig. 2.34 left) are described elsewhere with regard to their functional principles of load-bearing and building physics. In this type of envelope, the mass is, as it were, *thinned out* or replaced by a less structurally efficient, lighter, but better heat-insulating material where it does not offer any advantages either structurally or physically, i. e. in the **leaf core**. The outer thin leaves or layers have a denser material structure, have load-bearing capacity and, as a rule, are also tight against moisture, both in liquid and gaseous state. From a static point of view, they are analogous to a **I-system**, a system made of a thermally insulated core and a one-stage moisture and vapour barrier on both sides. This results in an extremely efficient, material-saving envelope element devoid of interfering thermal bridges, since the individual layers are coupled to one another by adhesion or adhesive bonding and no stiffening webs or ribs are required within the insulation layer.

Vol. 1, ▶ *Sect. 26.9.7 Multilayer Composite Element*

Vol. 1, ▶ *Sect. 26.9.7 Multilayer Composite Element*

Vol. 1, ▶ *Sect. 27.2.2 Basic Combination Patterns of Functional Layers Relevant to Moisture*

In contrast to the double-leaf systems discussed in ► Sect. 2.3, there is **no hierarchy** between the leaves in the sense of force conduction, for example, in the form of a supporting leaf and a supported leaf. As a rule, both must absorb (alternately) compressive and tensile forces within their plane. In contrast to double-leaf components, the core layer is assigned a load-bearing function, in particular bearing shear forces. All three elements—both leaves and the core layer—work together from a static perspective in a co-operating system.

Vol. 1, ► *Sect. 26.9.7 Multilayer Composite Element*

Thanks to the extreme simplicity of this design variant, the sandwich element can be installed in a horizontal and vertical position, without altering its constructional design (◘ Figs. 2.35, 2.36, and 2.37).

From a **building-acoustics** point of view, a sandwich, despite its differentiated multi-leaf-like construction, can nevertheless be regarded as a **single-leaf oscillating system**, and consequently behaves essentially as a single leaf, since the layers are connected to one another in a force-fitting, full-surface manner. The static mode of operation of the sandwich requires a core layer with high dynamic stiffness, which is not conducive to the noise protection of the element. Because of the extreme structural efficiency of the sandwich—a significant advantage of this construction—there is also no significant mass which could dissipate the sound energy. For these reasons, it can be said that, with multilayer composite systems, circumstances are very unfavourable in terms of building acoustics. Envelopes in sandwich construction are therefore only applicable where poor noise protection is tolerable, for example, in industrial construction.

2.4.2 Honeycomb-Core Systems

Alternatively, the core of the envelope element can also be made of a light honeycomb structure (◘ Fig. 2.34 right). This produces a large number of small-sized air chambers which can achieve a good thermal-insulation value due to a greatly reduced convection. If there is a shear-resistant connection between the honeycomb core and the cover leaves, a similar I-characteristic can be realised as with the sandwich element. However, a thermal bridging effect of the honeycomb webs producing a direct heat-conducting bond between the exposed cover leaves can prove a disadvantage.

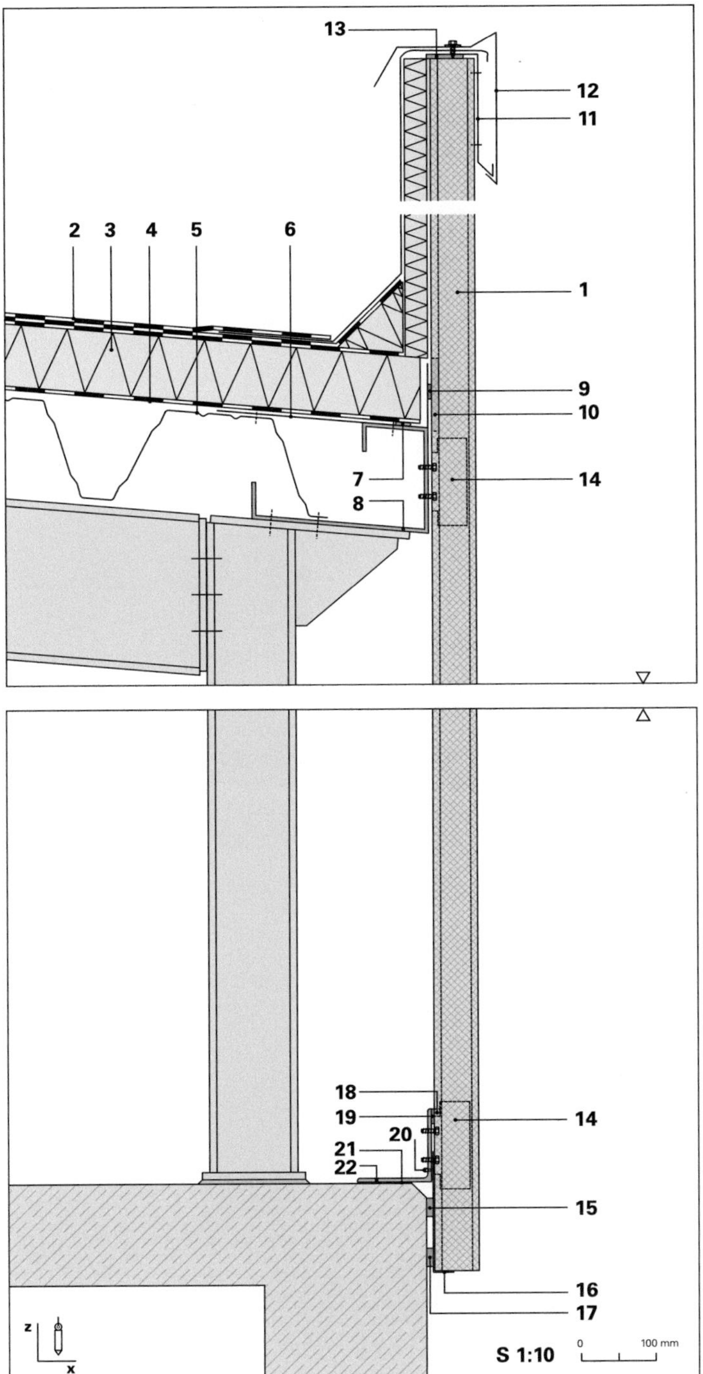

Fig. 2.35 (Top) Sandwich-type outer wall, vertical laying direction, roof connection detail. (Bottom) Sandwich-type outer wall, vertical laying direction, wall-base detail. (**1**) sandwich panel; (**2**) waterproofing; (**3**) thermal insulation; (**4**) vapour barrier; (**5**) load-bearing leaf, trapezoidal sheet; (**6**) edge sheet; (**7**) sealing tape as vapour barrier; (**8**) edge angle; (**9**) sealing tape, continuous; (**10**) sealing tape, only in the joint cavity; (**11**) supporting fascia bracket; (**12**) fascia profile; (**13**) sealing tape; (**14**) integral-connection bracket for the panel; (**15**) sealing tape; (**16**) if needed, mounting support; (**17**) plinth profile; (**18**) permanently elastic caulking in the joint cavity; (**19**) sealing tape, continuous; (**20**) mounting aid; (**21**) sealing tape; (**22**) angle bracket

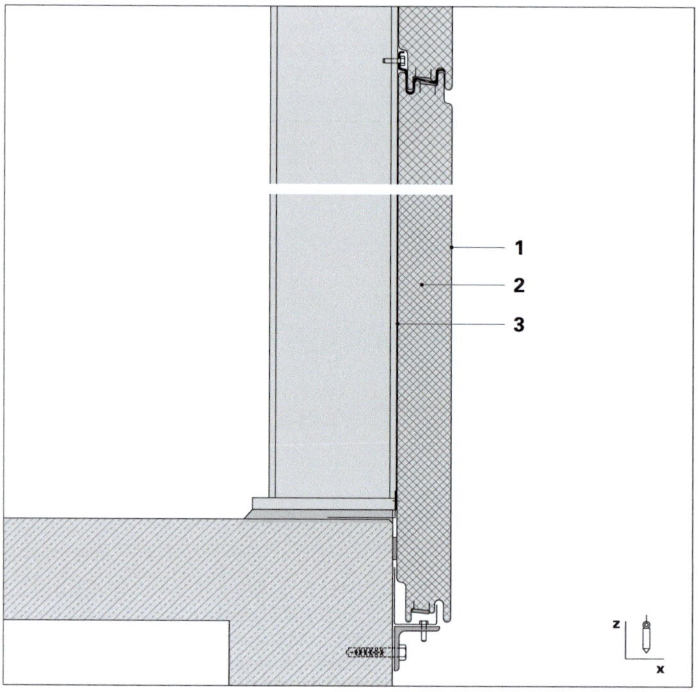

Fig. 2.36 Base detail of a sandwich-type outer wall in horizontal laying direction (system *ThyssenKrupp Hoesch*). (**1**) band-coated outer cover plate; (**2**) PUR insulation core; (**3**) band-coated inner cover plate

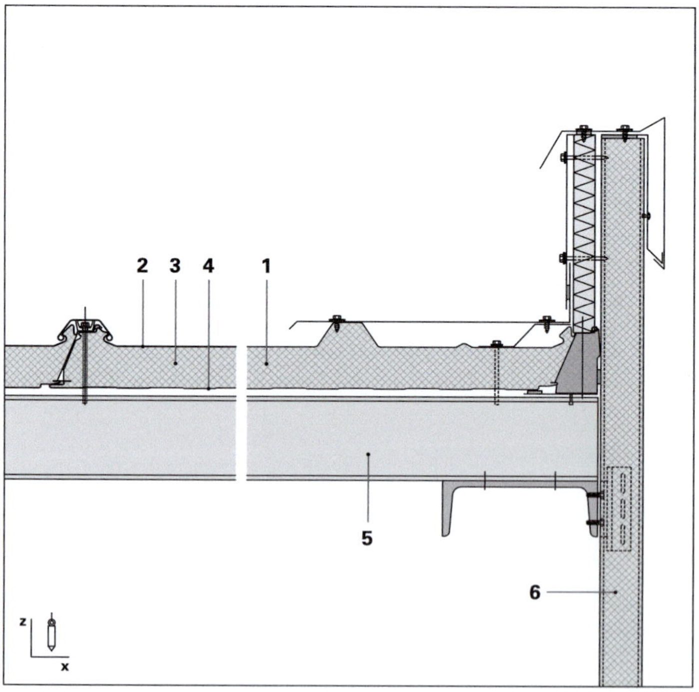

Fig. 2.37 Sandwich-type roof element (section crosswise to the gradient) showing the fascia connection (*ThyssenKrupp Hoesch* system). (**1**) sandwich roof element; (**2**) band-coated outer cover plate; (**3**) PUR insulation core; (**4**) band-coated inner cover plate; (**5**) primary structure; (**6**) sandwich wall panel

2.5 Ribbed Systems

2.5.1 Constructional Principle

In a manner similar to the sandwich and honeycomb systems discussed above, but with other means, ribbed systems (◘ Fig. 2.38) provide a substitute for full-walled leaves which, although they fulfil the fundamental constructional task of being areal and therefore envelope-forming already on account of their specific morphology, are often characterised by high material consumption, high weight, and sometimes only moderate thermal-insulation capacity.

► *Sect. 2.2.1 Single Leaf without Addition*

In the ribbed system—in contrast to full leaves as well as to the multilayer composite systems discussed above—a **functional separation** takes place between **surface formation** and **force conduction**. In ribbed elements, the surface is formed by one or two thin plates, which, as measured by the entire surface which they cover, cannot (and need not) achieve the stiffness necessary for the overall element on their own—unlike the simple and double-leaf systems discussed above. In this property, ribbed systems are quite similar to multilayer composite systems. However, in the case of ribbed systems, the force transmission between the surface-forming cover leaves is not effected by a simultaneously surface-forming element, such as the sandwich core, but by non-areal, i.e. linear, elements which are specialised for this purpose: by one or more **skeins of bars** which are attached to these plates and provide them the lacking stiffness. Hence, the usual description of these rods as **ribs**, analogously to the anatomical concept.

Vol. 3, *► Sect. 14 Ribbed Systems*

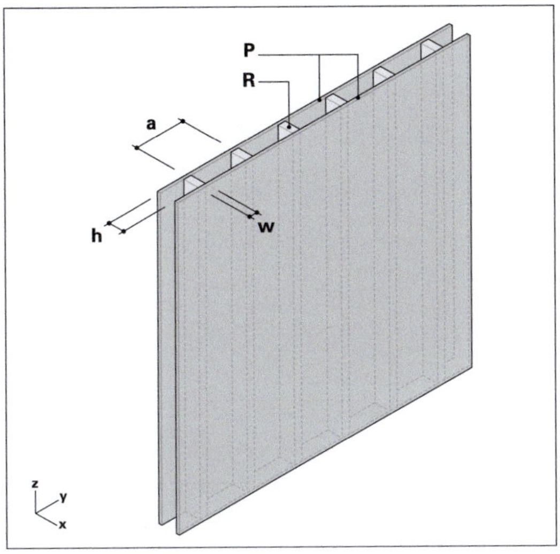

◘ **Fig. 2.38** Ribbed element from a rib layer **R** and a two-sided surface-forming plate **P**. **a** rib spacing; **w** rib's cross-section width; **h** rib's cross-section height

The thin, surface-forming cladding or sheathing thus only stretches from rib to rib and, in this way, even with a small thickness, also provides the stiffness required for force conduction in conjunction with the rib structure. Nevertheless, it can, if necessary, also be designed to be sufficiently flexible to ensure a good noise protection by virtue of realising a mass-spring system. Further aspects of the comparatively complex interaction of both elements are discussed below.

In principle, this results in a very lightweight envelope component with the potential for good thermal insulation and good soundproofing, which, however, partly consists of a more or less complex framework and is often associated with increased constructional complication and potential coordination problems between individual components.

2.5.2 One-Way and Two-Way Ribbed Systems

Since the sheathing thin plates can be sufficiently supported in a one-way manner, that is to say by uniaxially spanning ribs, the rib structure of these elements usually consists of a single rib skein. Two or more intersecting ribs in the same element and in the same plane (◘ Fig. 2.39) generally improve the carrying behaviour of the structure. Due to the resulting two-way load transfer by the thin sheathing, they also allow for somewhat larger spans—i.e. larger rib spacings while keeping the same plate thickness—or, alternatively, somewhat thinner plate material. The load-bearing capacity of the entire element can also be improved by the two-way design of the rib substructure, both for forces within the element plane (**yz**) and at right angles to it (→ **x**). An exam-

See the detailed discussion of biaxial ribbed elements from the point of view of force transmission in ***Vol. 1***, ► *Sect. 26.9.5 Element Composed of Ribs Oriented Two or Multiple-Way*

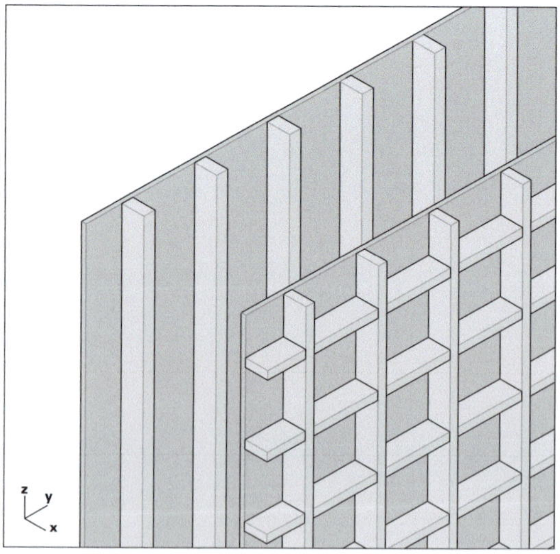

◘ **Fig. 2.39** Ribbed elements, with rib sets respectively spanning one (back) and two-way (front)

ple of this is provided by floors from beam grillages. This, however, comes at the expense of a considerable increase in constructional hassle elicited by the numerous penetration points of the ribs, a factor which, in most cases, prohibits such a multiaxial rib structure. Nevertheless, there are isolated cases in which, for other reasons, such intersecting ribs nevertheless are executed. The emphasis is always on utilising the two-way load transfer under bending stress, such as in the case of beam grillages.

These cases are dealt with in more detail in: beam grillages: ▶ *Sects. 4.3.1.3* and *4.3.1.4*; see also ***Vol. 3***, *Sect. 14.4.1 Lattice or Frame Walls*

On the other hand, inter-crossing, thus biaxially spanning ribs, is always relatively easy to construct in a relatively simple manner if the rods cross in different planes, that is, do not penetrate each other. The rods can then be passed continuously without seams. This is advantageous, in particular, in those cases where no bending-stiff, but rather an articulated connection at the bar node is required. A bending-stiff connection at the node would be much more difficult to execute with a rod arrangement in different planes, as is the case with continuous rods. Mutually crossing, non-intersecting ribs are found with some grid shells (▶ *Sect.* 4.3.2.5 *Dome from Bars*) as well as, in principle, with rope nets. In the latter, ropes act as ribs. In contrast to bars, these are not bending-stiff, but otherwise perform the same task within the envelope element. Textiles, which in constructional use are, for example, coated membrane fabrics, are a special modification of the rib principle with crossing, but not penetrating, again not bending-stiff, rib skeins, namely, warp and weft.

▶ *Sect. 4.3.3 Tensile-Stressed Systems*

2.5.3 Ribbed Systems with Integrated Envelope Construction

Similar to ▶ Sect. 2.2.3, this design variant consists of a system of parallel ribs which, in this case, are lined or clad with a thin leaf on both sides (◘ Fig. 2.38). In contrast to the light non-load-bearing design which is used in the variant discussed in ▶ Sect. 2.2.2 (see ◘ Fig. 2.10), a **load-bearing secondary construction** is present in the combination of the rib and the thin plate which is capable of deflecting external loads on its own over a certain span to a bearing or to the primary structure. Provided there is a shear-resistant bonding at the connection between the rib and the thin leaf—for example, when gluing in wooden-panel construction—both elements also work together composite-like under bending stress on the rib: Ribs and leaves connect to a **co-operating system** in which the entire element thickness **T**, as well as the material of the components, are statically fully utilised. A certain **co-operating leaf width** on either side of the rib, in this case, behaves in a manner analogous to a compressive or tensile chord. The available structural depth is thus increased by the thickness of both leaves.

Sect. ▶ *2.2.3 Single Leaf with One-Sided Addition with Substructure*

Vol. 1, ▶ *Sect. 26.9.4 Element Composed of Ribs Oriented One-Way*, Fig. 26.180 as well as Fig. 26.206 to Fig. 26.208

The variant investigated in this section is based on a (at least partial) integration of the rib skein and the enveloping construction—in particular the most relevant layer, the insulation—within the same plane. In the case of outer envelope components, the insulation layer primarily fulfils a thermal-insulation function, and possibly also a sound-insulation function. But it also appears in the case of inner envelope elements, where it is generally used for building-acoustical purposes as cavity damping.

The ribs act as **bending beams**, so that their cross-sectional height **h** is decisive for the bending stiffness of the envelope component against loads at right angles to the envelope plane. The larger the rib's cross-sectional height is, the more space there is for thermal insulation within the rib cavity in the case of outer envelope elements. Since the load-bearing function, which was entrusted solely to the inner supporting leaf in the variant in ▶ Sect. 2.2.3, is now mainly assigned to the ribs, the thusly increased structural depth of the load-carrying rib can be used beneficially for increased thermal insulation within the cavity (◘ Fig. 2.40). This is a clear advantage of this solution for outer envelopes.

The distance **a** between the ribs (◘ Fig. 2.38) is, in turn, predetermined by the bending stiffness of the sheathing. Since this generally consists of comparatively thin plates, this dimension is usually in the range below 1 m.

In addition to the concurrence of the cladding as beam chords of the rib, both leaves also contribute to securing the rib, which is relatively slender in its width **w**, against **tipping** or **buckling** within the envelope's plane (**yz**, i.e. in → **y**). The force-fitting connection of the rib with the shear-plate-

Tipping: ***Vol. 1***, ▶ *Sect. 26.9.4 Element Composed of Ribs Oriented One-Way*, Fig. 26.212, Fig. 26.213

Buckling: ***Vol. 1***, ▶ *Sect. 26.9.4 Element Composed of Ribs Oriented One-Way*, Fig. 26.166

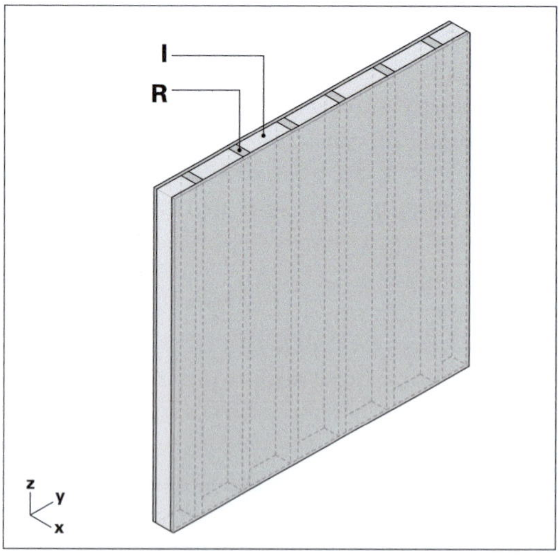

◘ **Fig. 2.40** Ribbed element as in ◘ Fig. 2.38. The interstices of the ribs **R** are filled with insulation **I**. The load-bearing and enveloping functions are integrally combined in a single plane

like component of the sheathing prevents a lateral deflection (in → **y**) of the slender profile. In particular, buckling is to be feared (in → **z**) if strong axial compressive forces are introduced into the rib, for example, with load-bearing wall elements.

Likewise, the sheathing has a **bracing** effect on the overall ribbed system, which is not secured from the outset against shear stress within the leaf's plane (**yz**). The racking stiffness of the two leaves transforms the envelope component into an element rigid against shear. Compressive stresses, which build up in the thin leaves as a result of this shear stress and in extreme cases can lead to **buckling** of this sheathing, are again neutralised by the bonding of the thin plate with the rib. The plank is held at right angles to the envelope plane (→ **x**) by the rib, which in this direction activates its great stiffness due to its cross-sectional height **h**. Out-of-plane buckling of the thin plate or sheathing is not possible.

Vol. 1, ► *Sect. 26.9.4 Element Composed of Ribs Oriented One-Way*, Fig. 26.186

It becomes clear how the components of this design interact very advantageously in an **overall structural system**, both for forces at right angles to the element's plane (→ **x**), causing bending, as well as for those within the plane itself (**yz**), causing shear within the rigid cladding. In addition, this is a case in which structural necessities (the rib's cross-sectional height **h**) at the same time also result in conditions which are reasonable in terms of thermal protection, i. e. to a large insulation thickness.

It is obvious that the rib represents a **thermal bridge** from the point of view of thermal protection, which is why this constructional system can only be realised, in its simplest expression, in timber construction, where the rib has only a moderate thermal conductivity. If other materials, such as metals, are used, or if higher thermal requirements are set, a thermal separation, i.e. a doubling of the rib system, a disconnection of internal and external elements or a superposed lining is unavoidable. Variants of this type are discussed in the following ► Sect. 2.5.3.1.

► *Sect. 8.3.4 Timber Frame Construction*, as well as ***Vol. 3***, *Sect. 14.2.1.1 Wooden Frame Walls*

As in the case of the single-leaf variant with cladding without substructure, it is advantageous to have **rear** or **bottom ventilation** between the thermal-insulation layer and the back of the outer panel (◘ Fig. 2.41) and/or sufficient water-vapour-diffusion capacity of the outer panel to remove any moisture that may have penetrated into the construction. This is achieved, for example, by a sufficiently diffusion-permeable outer cladding (◘ Fig. 2.41 right). As a general rule, water-vapour penetration from the interior should be controlled by suitable vapour retardants or vapour barriers applied to the interior layers.

► *Sect. 2.2.2 Single Leaf with One-Sided Addition without Substructure*

In **acoustic** terms, in spite of the obvious double-leaf construction, a *single-leaf* oscillating system, which is not unlike a simple leaf, is present in the first place. The linear bond between rib and cladding, which is necessary from the point of view of load transfer, makes the element an altogether stiff

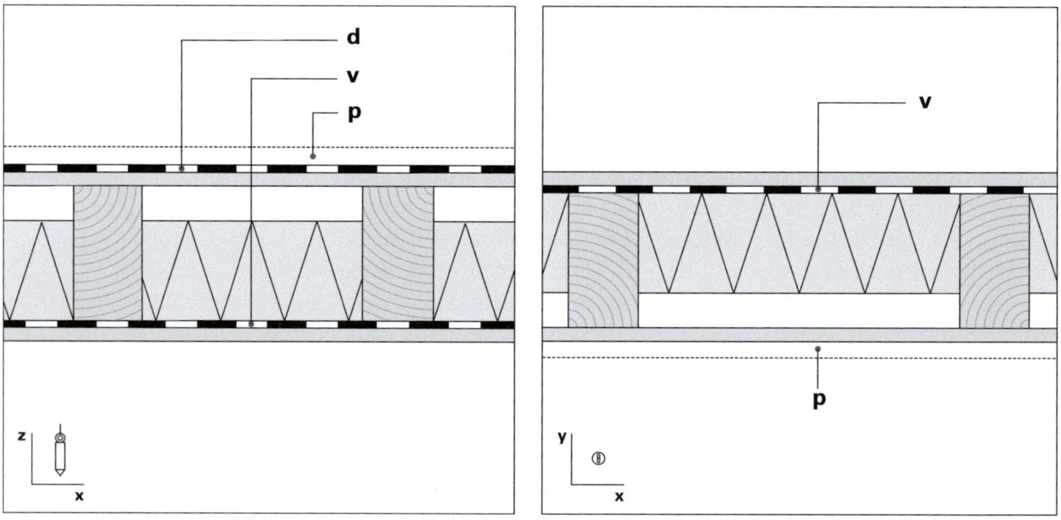

Fig. 2.41 (Left) Pitched roof with the simplest construction. Additional protective layer **p** (roofing) made, for example, from bitumen shingles. Vapour retardant or barrier **v**. Diffusive layer **d** for improvemente of tightness. (Right) Wooden ribbed wall with the simplest structure. Additional protective layer **p** (weather skin), for example executed as a rendering. Vapour retardant or barrier **v**

system. Moreover, paradoxically because of its static efficiency, it has no significant damping mass. Worse still, the small mass present is furthermore distributed in an inhomogeneous manner, namely, on ribs and sheathing, which is why it cannot unfold its acoustic effect. As a result, the noise protection is only moderate to low. In order to avoid additional deterioration due to cavity resonance, a damping of the cavities between the ribs by means of a soft-resilient insulating material is required in any case.

2.5.3.1 Ribbed System with One-Sided Lining Lacking a Substructure

Analogous to the variant in ▶ *Sect. 2.2.2 Single Leaf with One-Sided Lining without Substructure*

In this variant, the structure of the lining consists of an insulation layer as well as an outermost layer or thin leaf, which adheres to it, or rests on it when the element is horizontal. There is no substructure as well as *no fasteners* penetrating the insulation layer. For this reason, outer layers of upright components can only be light, comparatively thin skins; the leaf can also be designed to be heavier with horizontal components, since no sliding is to be feared in that case, and its load can be transferred to the load-bearing rib element over the entire surface across the insulation layer. Basically, beam floors with floating screeds (Fig. 2.42) can be counted in this category. In that case, the insulating layer assumes the acoustic decoupling of the screed from the load-bearing element, that is, the main floor structure; sound bridges are excluded in the absence of an attachment of the upper leaf.

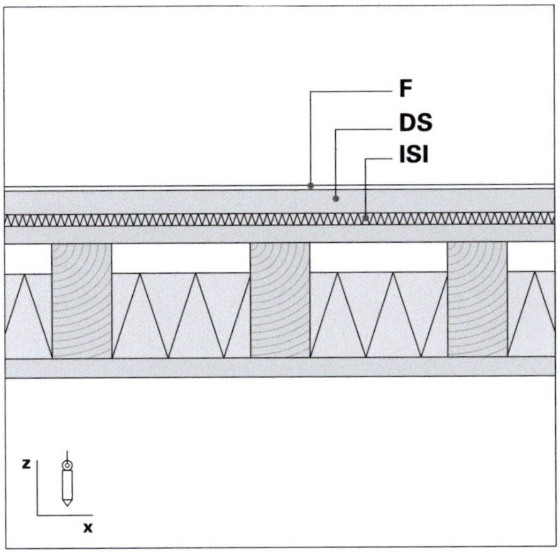

■ **Fig. 2.42** Floor structure in wooden-panel construction. The floor construction on top of the load-bearing ribbed element consists of impact-sound insulation (**ISI**), dry screed (**DS**) and flooring (**F**)

The continuous insulation layer of the structure (■ Fig. 2.43) neutralises the thermal bridge of the ribs in outer leaf components and, overall, increases the thickness of the insulation package beyond the dimension of the rib's structurally necessary cross-sectional height **h** (■ Fig. 2.44 left). As in the case of a full leaf with a lining lacking a substructure, care must also be taken in this case to ensure that any mois-

▶ *Sect. 2.2.2 Single Leaf with One-Sided Lining without Substructure*

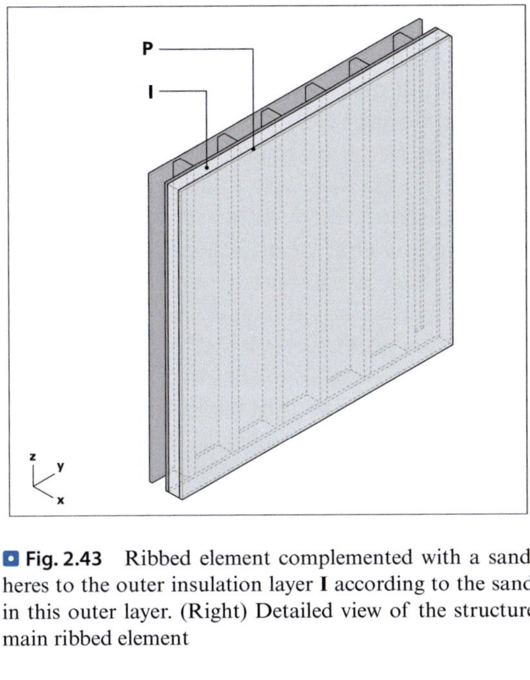

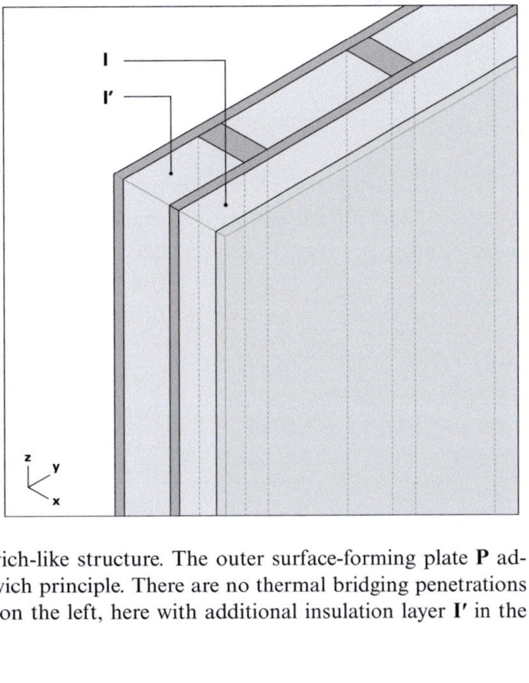

■ **Fig. 2.43** Ribbed element complemented with a sandwich-like structure. The outer surface-forming plate **P** adheres to the outer insulation layer **I** according to the sandwich principle. There are no thermal bridging penetrations in this outer layer. (Right) Detailed view of the structure on the left, here with additional insulation layer **I′** in the main ribbed element

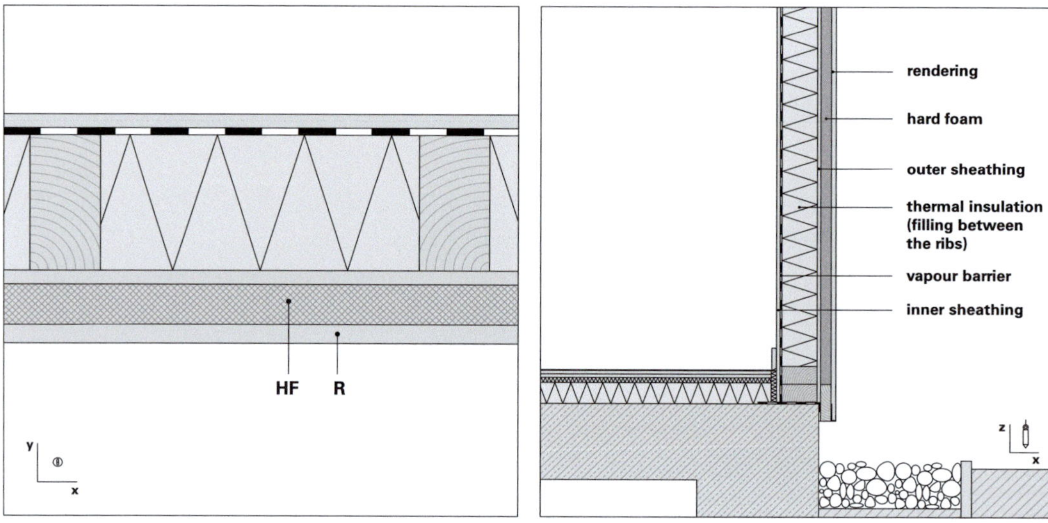

Fig. 2.44 (Left) Outer wall in wooden ribbed structure. External structure of hard-foam board (**HF**) and rendering (**R**). (Right) Outer wall in wooden ribbed construction as on the left, showing a possible execution of the base area

ture that has penetrated the structure can be evacuated to the outside. This in turn requires sufficient water-vapour-diffusion capacity of the outer layers, i.e. in this case the outer cladding, the thermal-insulation layer, and the weather skin. In timber construction, such a solution can be found in the form of ribbed walls with a thermal-insulation composite system (Fig. 2.44). In an inclined position, this is, for example, a ribbed roof with over-rafter insulation; in a horizontal position, a flat beam roof with an unventilated structure. In the latter case, water-vapour diffusion through the weather skin (in this case, waterproofing) is not feasible from a constructional point of view (as is also the case with a flat roof on a leaf). Additional measures such as a room-side vapour barrier are also essential here.

Vol. 3, ▶ *Sect. 14.3.2 Pitched Roofs*

Vol. 3, ▶ *Sect. 14.3.3 Flat Roofs*

Overall, conditions in terms of building physics are comparable to those of the simple leaf with a lining, whereas the mass of the main leaf significantly decreases in the present case of the ribbed element. Due to its heat-storage capacity, the mass of a main leaf can guarantee a desired thermal inertia and also contribute significantly to a good noise protection.

▶ *Sect. 2.2.2 Single Leaf with One-Sided Addition without Substructure*

From an **acoustic** viewpoint, a two-leaf vibrating mass-spring system is acting, one leaf representing the ribbed element, the other the added layer or thin leaf of the lining. As a result of the risk of slipping, the mass of this outer layer of the lining, for conceptual reasons, is only small for upright elements, which is why their noise-protection effect is correspondingly moderate. This is somewhat different in the case of horizontal components. There, even heavier cover leaves—such as floating screeds—can be used.

2.5.3.2 Ribbed System with One-Sided Lining Containing a Substructure

In this case, the ribs of the structure have the primary task of creating an additional cavity above or in front of the load-bearing ribbed element (◘ Fig. 2.45 left), which can either be left unused (◘ Fig. 2.46) or otherwise be used either for **back ventilation** (◘ Fig. 2.47 right) or **thermal-insulation** purposes (◘ Fig. 2.45 right). The ribs either span transversely to the main rib system (→ **y**) (◘ Fig. 2.45 left) or parallel thereto (→**z**) (◘ Fig. 2.47), in which case the secondary rib comes to rest on the main rib longitudinally for direct force transmission.

Analogously to the solution in ▶ Sect. 2.2.3.2 (◘ Fig. 2.15), the thermal bridge is reduced to the contact surface of two superposed ribs when the rib system is doubled up and the newly created cavity is filled with thermal insulation, which is why, in this sense, the transverse orientation of the rib skein is preferable to the longitudinal laying (see also ◘ Figs. 2.15 and 2.16). The total insulation thickness can be considerably increased by this design (◘ Fig. 2.45 right).

▶ Sect. 2.2.3.2 Double Rib Layer

From an **acoustic** perspective, the system results in a mass-spring system, consisting of the ribbed element and the added rib skein, which cannot provide any significant soundproofing mass. This is why it is poor in terms of building acoustics. The coupling between the rib element and the added skein should be as elastic as possible for a slightly better noise protection.

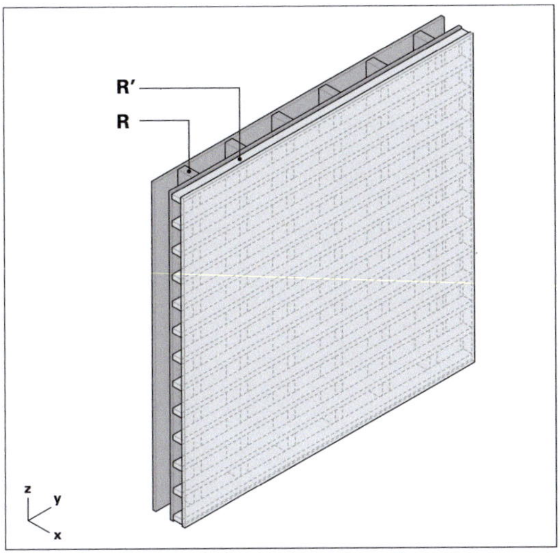

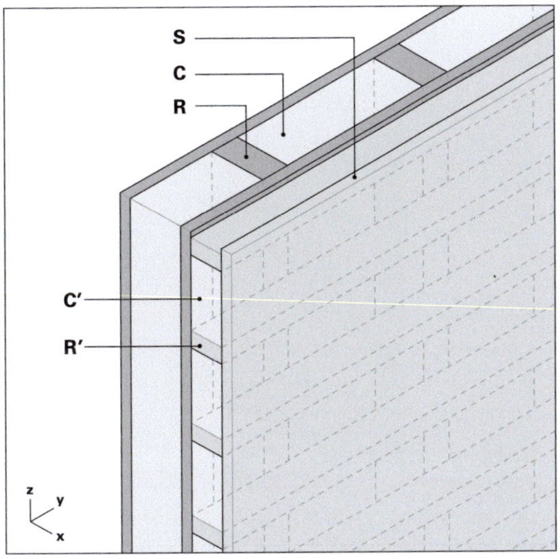

◘ **Fig. 2.45** (Left) Ribbed element with added external structure in ribbed construction; **R** main rib set, **R'** secondary rib set. (Right) Detailed view (compartments **C** and **C'** filled with thermal insulation). Unlike in the case of ◘ Fig. 2.43, the outer sheathing **S** (usually the weather skin) is anchored to the inner rib set by the outer rib layer **R'**. Thermal bridges are, however, reduced to the contact surfaces between both rib layers **R** and **R'** (analogous to the case on ◘ Fig. 2.15, right)

2

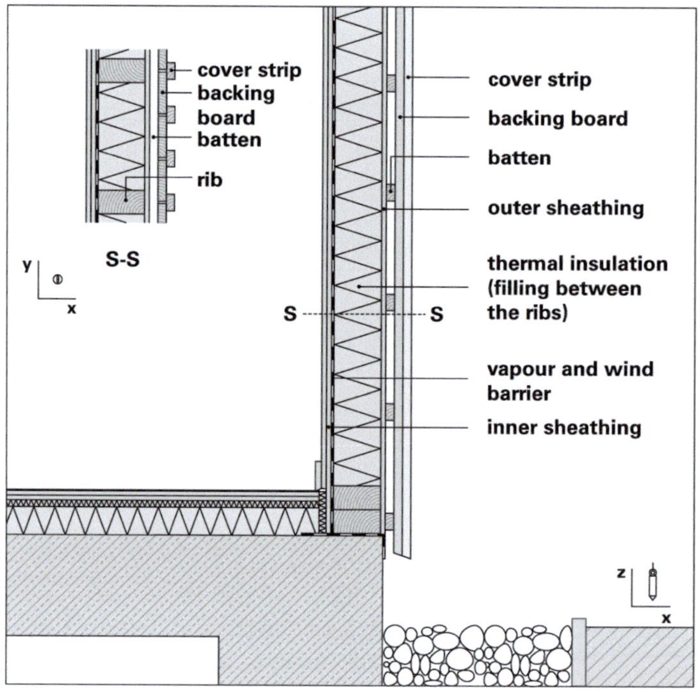

Fig. 2.46 Outer wall in wooden ribbed construction with representation of the base area. Transverse lathing on the rib element for fastening the external siding (in this case, a covering-strip boarding). No back ventilation of the weather skin is provided, which presupposes only a moderate driving-rain exposure—for example, with sufficient protection by a roof overhang

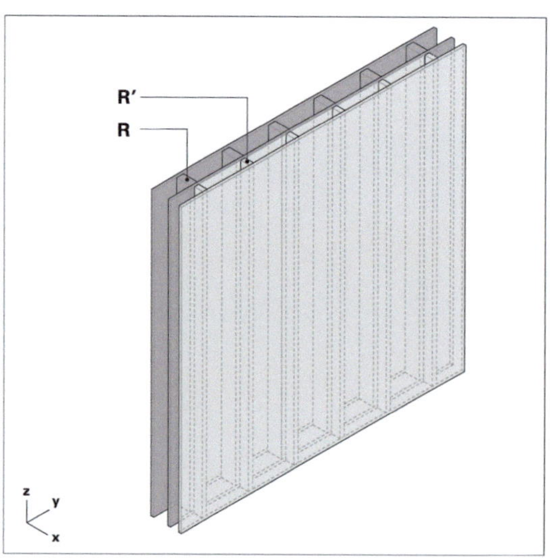

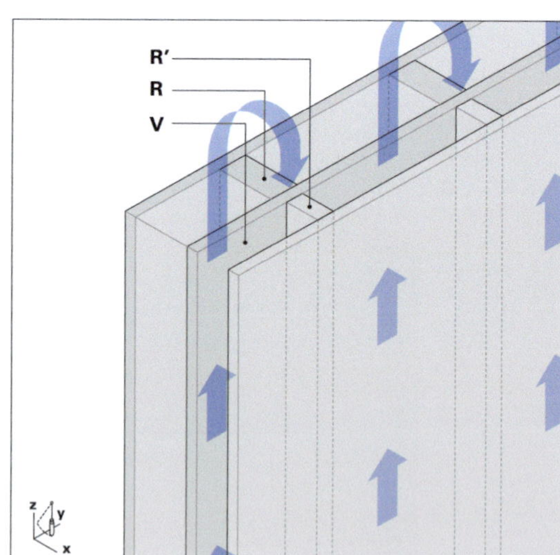

Fig. 2.47 (Left) Outer rib **R'** layer mounted on the inner one **R**, not across but lengthwise. The former has no longer a load-bearing function, but serves merely as a lining. (Right) Detailed view. Under certain circumstances, this orientation of the secondary ribs **R'** (the same as the one of the main ribs **R**) may prove to be advantageous, for example, to enable back ventilation **V** when cross-aligned ribs (as in Fig. 2.45, left) would be in the way

2.5.3.3 Ribbed System with Single-Sided Lining Containing a Double Ribbing

An additional structure in the form of a further rib layer which is placed perpendicular to the first secondary rib skein (◘ Fig. 2.48 left) can usually no longer be justified by the necessity to realise a larger insulation thickness or to minimise thermal bridges. However, it may sometimes be useful to create a cavity in order to back-ventilate or under-ventilate the construction (◘ Fig. 2.48 right). As mentioned, it is then necessary to ensure that the alignment of the cavities, which at their ends are in contact with external air, supports or facilitates the air movement: that is, for example, in the case of inclined or upright envelope elements, following the direction of thermal buoyancy (◘ Figs. 2.49, 2.50, and 2.51).

▶ *Sect. 2.2.3.2 Double Rib Layer*

The double-ply lining clearly complicates the mass-spring system in terms of building acoustics, since, as it were, there is a **double resonance system**, which is extremely difficult to control in terms of design. In comparison to the simple rib structure, no appreciable improvement in the soundproofing behaviour can be achieved.

2.5.4 Ribbed Systems with Separation of Envelope Construction and Ribbing

The function of load transfer is assigned to a system of ribs which, in this case, is formed separately from the actual surface-forming envelope construction (◘ Fig. 2.52 left). For

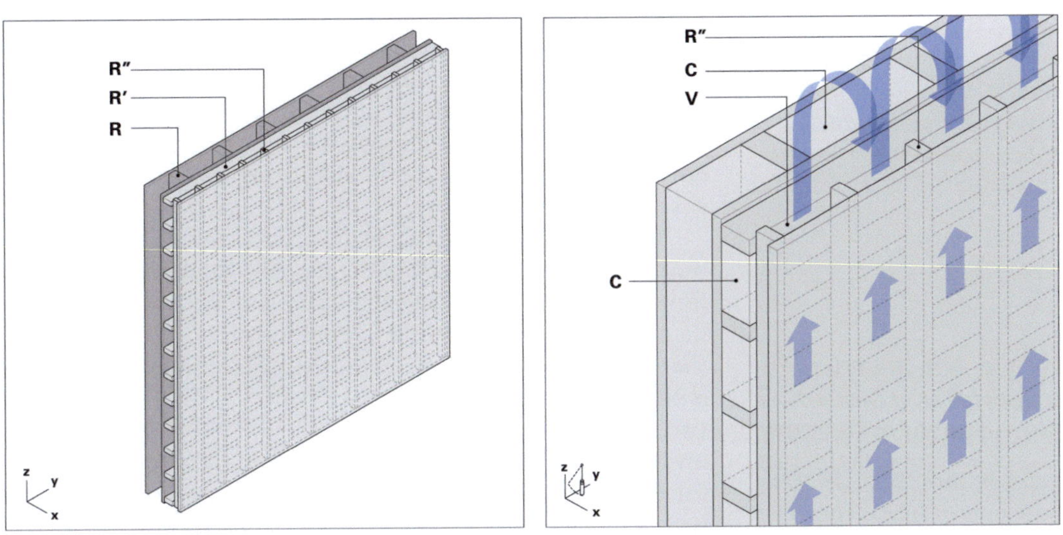

◘ **Fig. 2.48** (Left) Ribbed element with added external double structure in ribbed construction. Main rib set **R**; two secondary rib sets **R′** and **R″**. (Right) Detailed view (compartments **C** filled with thermal insulation). The outermost rib layer **R″** can be used for the under- or back-ventilation **V** of the construction. The position of the component with respect to the vertical is again to be considered

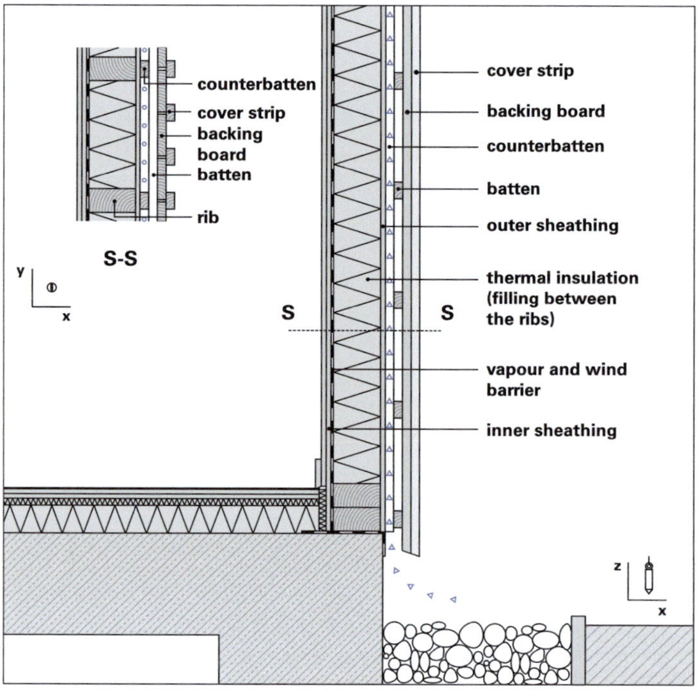

Fig. 2.49 Outer wall similar to the solution in Fig. 2.46, but with additional counter-battening and introduction of vertical continuous rear ventilation. The sequence of layers of battens and counter-battens is reversed compared to the illustration on Fig. 2.47, right, but there is no significant difference in construction

the present consideration, it is at first irrelevant whether the former is made of one or more successive rib hierarchies (as on Fig. 2.52 right): The structural construction principle discussed remains unchanged. On the ribs, a **space-enclosing leaf** is fastened on one side, which has a just sufficient thickness and bending stiffness in order to bridge the spacing **a** between the ribs. For thermal reasons, in the case of outer envelope components, the load-bearing rib layer is in most cases in the interior.

Like for instance with slatted glazing, see ***Vol. 3***, ▶ *Sect. 14.3.1.2.4 Thermal Separation*

The orientation of the main group of ribs is dependent on their most important task, namely, **load transfer**. When force is applied within the element plane (**yz**), the ribs use to be aligned along this force (i.e. either in →**y** or in →**z**). If the force acts at right angles to the element plane (→ **x**) and is therefore a **one-way spanning system** of bending beams on two supports, the orientation of this ribbed layer depends on its bearing: The ribs then span from support to support; for example, with a wooden beam floor, from wall to wall or, with a mullion-and-transom façade, from floor to floor.

Vol. 1, ▶ *Sect. 26.9.4 Element Composed of Ribs Oriented One-Way*, Fig. 26.201 to Fig. 26.214

Vol. 1, ▶ *Sect. 26.9.4 Element Composed of Ribs Oriented One-Way*, Fig. 26.163 and following

▶ *Sect. 2.5.3 Ribbed Systems with Integrated Envelope Construction*

The cross-sectional height of the rib **h**, as in the case of integrated rib systems, is given firstly by questions of force conduction, in particular the necessary **bending stiffness** of the

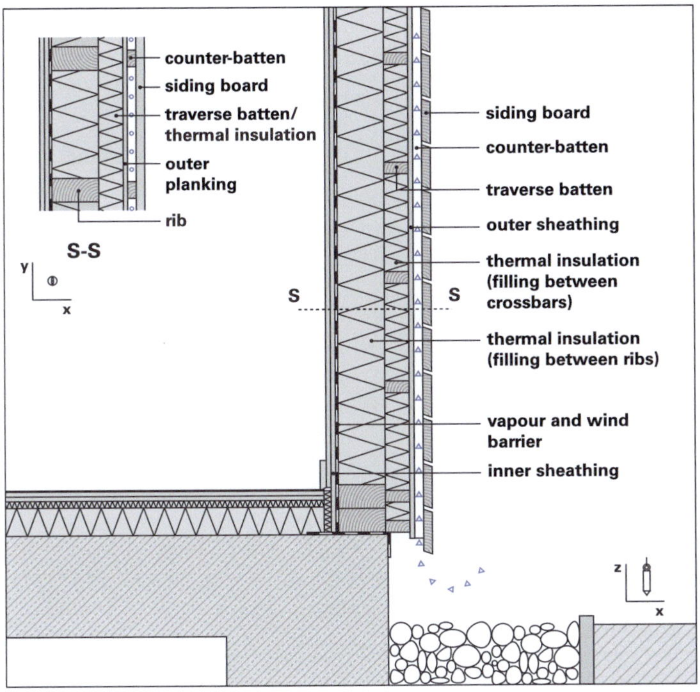

Fig. 2.50 Exterior wall in wooden rib construction according to structure on Fig. 2.47, right. The plane of the transverse lathing is used for a doubling of the thermal insulation, and that of the counter-lathing for a back ventilation of the weather screen, the boards of which accordingly must go horizontally

rib, provided it is subject to bending stress. This is, of course, especially true for lying components, such as floors (Fig. 2.53 left) or roofs (Fig. 2.53 right), since there the force component oriented at right angles to the element surface (→ **x**), such as, for example, dead load, usually dominates. Since, in this design variant, the load-bearing elements are clearly separated from the space-enclosing envelope construction, there are no conflicts of goals between issues of load transfer and other partial functions covered by the continuous layer construction (such as heat and sound insulation).

The height **h** of the rib's cross-section, as well as—if the rib is not additionally held laterally—its width **w**, must also be a necessary minimum dimension in the case of an **axial load** which may be imposed on the ribs (→ **z**), for example compression, in order to prevent buckling (in → **y**).

Vol. 1, ▶ *Sect. 26.9.4 Element Composed of Ribs Oriented One-Way*, Fig. 26.166 as well as Fig. 26.168 to Fig. 26.178

The parallel rib system is not inherently stiff against **shear stress** in the component plane (**yz**). Shear can, in this case, be absorbed by non-displaceable bearings. Alternatively, the **racking stiffness** of the ribbing, braced by the sheathing, can also stiffen the system—as with a timber beam floor with a shear-resistant panelling.

As with a mullion façade attached to a storey floor

The distance **a** between the ribs, for conceptual reasons, is predetermined by the bending stiffness of the outer leaf.

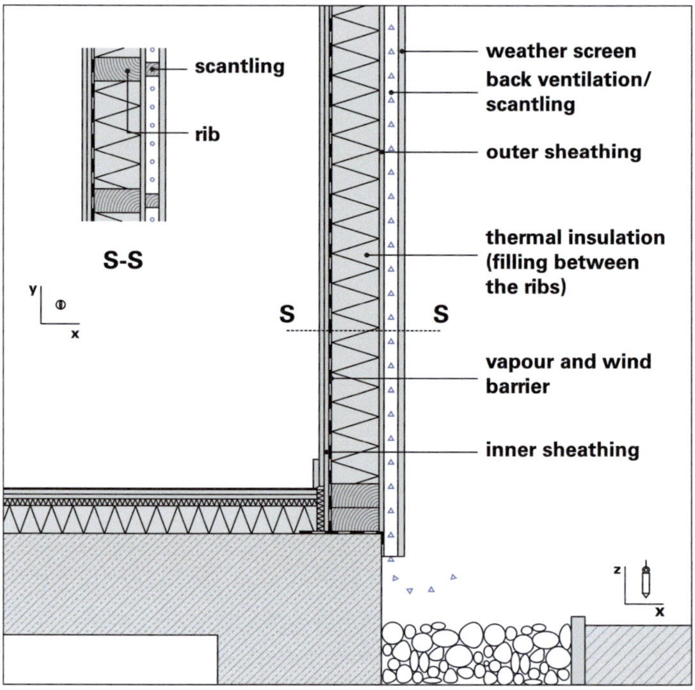

Fig. 2.51 Outer wall in wooden ribbed construction: A scantling is attached lengthwise in front of each rib, creating a cavity for the back ventilation. For this component position, it must necessarily be oriented vertically, therefore parallel to the rib set

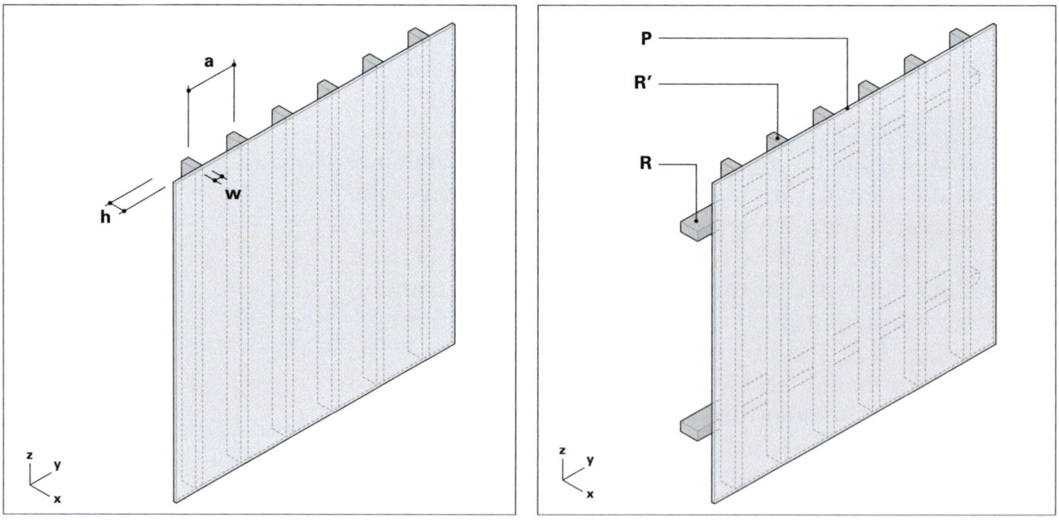

Fig. 2.52 Ribbed element with separation of load-bearing rib set and space-closing leaf. **a** rib spacing; **w** rib cross-section width; **h** rib cross-section height. (Right) Alternative with two rib hierarchies: two sets extending in right angles to each other: main rib **R**, secondary rib **R′** as well as surface-forming plate **P**

Vol. 1, ▶ *Sect. 26.9.4 Element Composed of Ribs Oriented One-Way* as well as ▶ *Sect. 4.2.1.2 Flat Covering from Bar Skeins*

At right angles to the element plane (i.e. → **x**), external loads are applied to the rib skein by bending as a one-way-spanning plate, i.e. one spanning between two adjacent ribs.

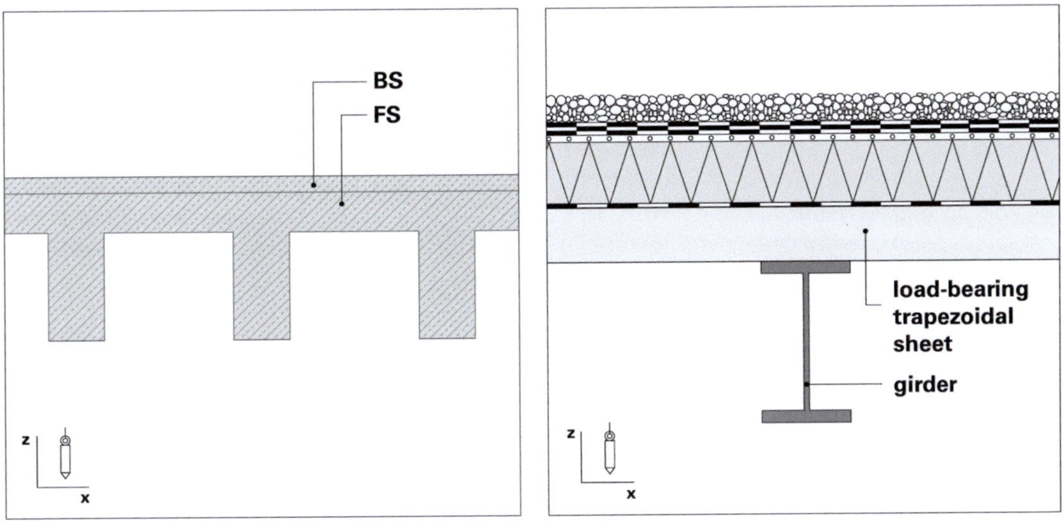

Fig. 2.53 (left) Beam-and-slab floor with equalising bonded screed (**BS**) on the raw floor slab (**FS**) as an example of a ribbed element without an additional light structure—for example for a floor structure. (Right) Roof with load-bearing steel construction and light roof construction as an example of a ribbed element with a light build-up package

If the comparatively slender ribs are additionally held (Fig. 2.54 left) laterally by **cross-bars**, their load-bearing capacity is improved in two respects:

- In the event that the ribs, additionally or alternatively to the bending stress emerging as a result of, for example, wind force, are subjected also to a **compressive load** at right angles to the envelope surface → **x**, i.e. along the rod axis parallel to the envelope plane → **z**, a buckling in the di-

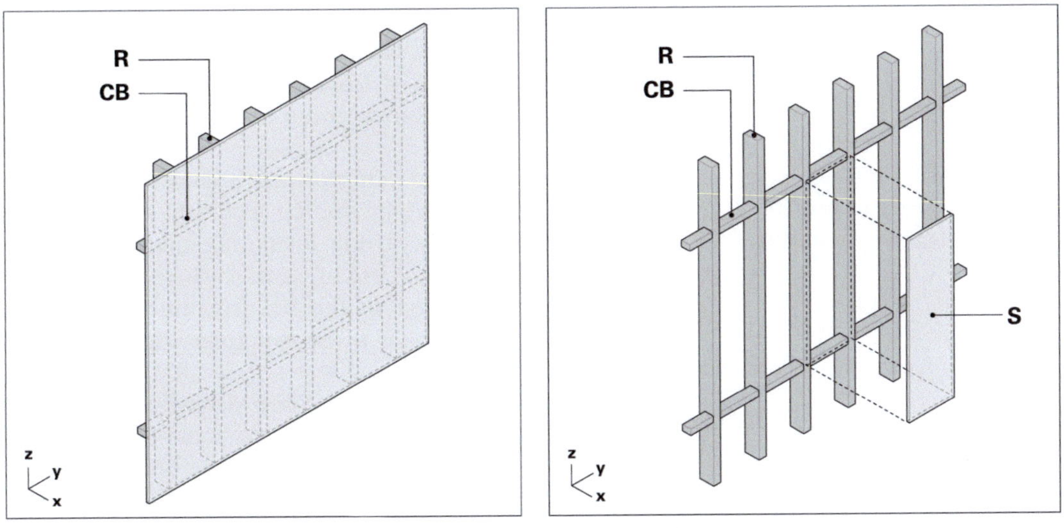

Fig. 2.54 (Left) Ribbed element as on Fig. 2.52, left, but with crossbars **CB**. (Right) All-round bordering of a sheathing element **S** in the presence of ribs in both main directions, i.e. of main ribs **R** and crossbars **CB**

rection of their weak cross-sectional dimension, i.e. their width **w** in direction of → **x**, is made more difficult.

- Individual loads oriented transversely to the bar axis (→ **x**) which impinge on a single rib are distributed by the cross-bars to adjacent ribs, since there is a shear-resistant connection between the rib and the cross-bars or, otherwise, the cross-bar continues above the rib. This results in relief of the affected rib. This effects a **cross-distribution** of point loads.

Vol. 1, ▶ *Sect. 26.9.4 Element Composed of Ribs Oriented One-Way* as well as ▶ *Sect. 4.2.1.2.5 Cross-Distribution of Loads in Bar Systems*

In addition, ribs and cross-bars arranged externally flush in both main directions offer the possibility of **linearly** framing planar elements (◘ Fig. 2.54 right). This can be of significance in the case of mutual abutting of surface-forming coverings, for example, in the case of mullion-and-transom façades to ensure the necessary tightness of the envelope.

Cf. glazing in ***Vol. 3***, ▶ *Sect. 14.3.1.2 Mullion-and-Transom Façade*

From the perspective of **thermal protection**, it is required that the necessary thermal insulation be provided by the single *full-surface* component, namely the thin outer leaf. For the same reason, **sound insulation** can also only be provided by this element. For this purpose, it is practically unavoidable to equip the outer skin with a **lining**, or to form it in a **multi-leaf** manner. Alternative types of execution of this sort will be examined in the following.

2.5.4.1 Ribbed System with Leaf and Lining Lacking a Substructure

▶ *Sect. 2.2.2 Single Leaf with One-Sided Lining without Substructure*

The same conditions as described above (◘ Fig. 2.5) apply analogously (◘ Fig. 2.55 left). This solution is advantageous, in particular, in the case of horizontal components (◘ Fig. 2.56), since a slipping of the sensitive lining is not to be feared in this particular case. Besides, this solution can also be used with inclined envelope components (◘ Fig. 2.57). However, the securing of the set-up package may have to be carried out with the help of fasteners penetrating the insulation layer—something contravening the basic concept (as in ◘ Fig. 2.57).

Cf. Wooden-beam floor with floor substructure executed as floating screed

In the case of vertical components, it is also necessary to prevent the outer layer from slipping. For thin skins, the adhesion to an adequately strong insulating material is sufficient (◘ Fig. 2.58). In the case of heavier outer leaves, re-anchorage is again required to the bearing inner leaf or rib skein, which in turn implies a substructure. These cases will be investigated in the following ▶ Sects. 2.5.4.2 and 2.5.4.3.

Such as the render layer of a thermal-insulation composite system

From the point of view of **building acoustics**, in the case of this variant, a double-leaf vibrating mass-spring system emerges consisting of the ribbed leaf on the one hand and a layer or thin leaf of the lining on the other hand, provided that there is a resilient coupling between both. The layer or thin leaf of the superadded structure must have a certain min-

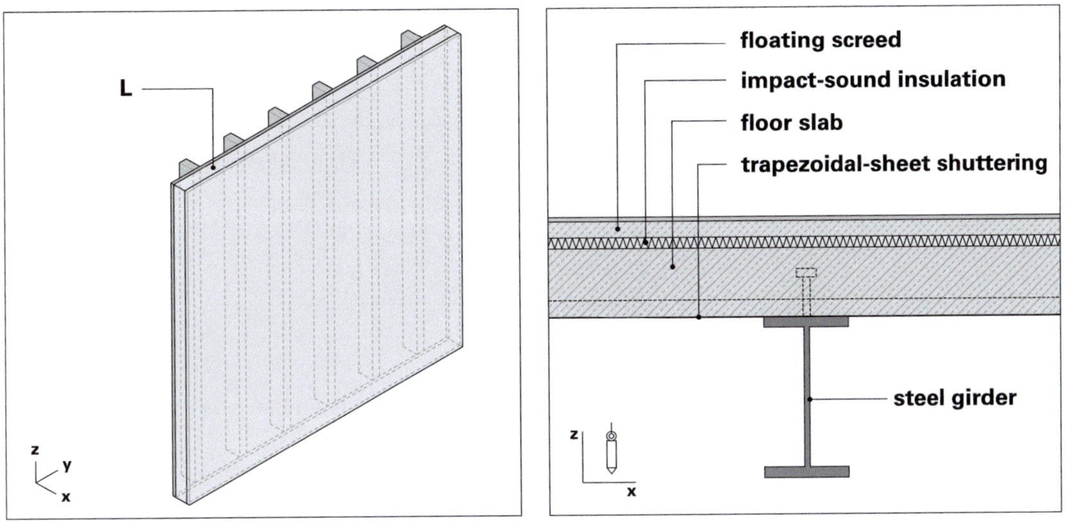

Fig. 2.55 (Left) Ribbed element with light lining **L** lacking substructure. (Right) Steel-concrete composite floor construction with floating screed as an example of a ribbed element with additional layering without substructure

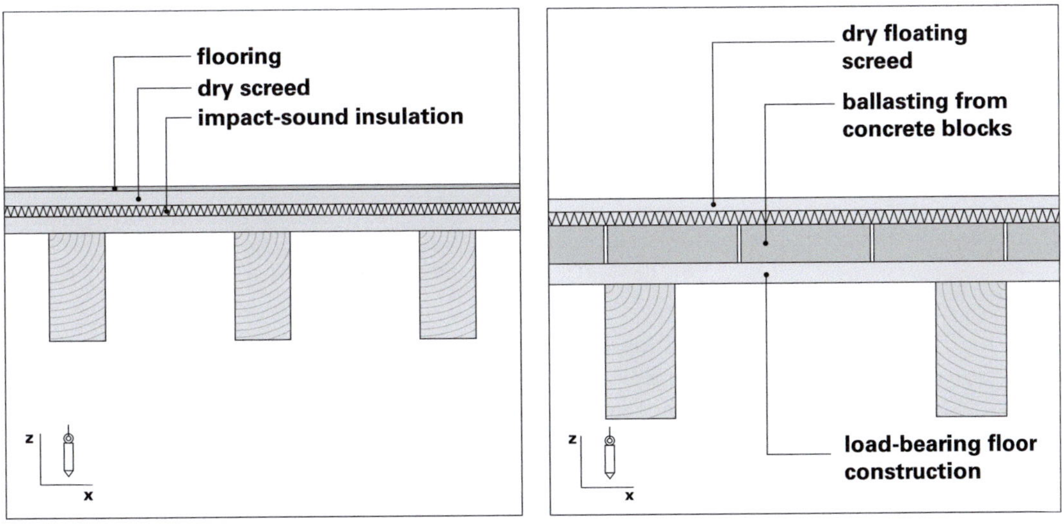

Fig. 2.56 (Left) Wooden beam floor with floating dry screed as an example of a ribbed element with added layering without substructure. (Right) Floor construction as on the left, but with an acoustically effective ballasting, as another example of a ribbed element with an added layering without substructure

imum mass. As a rule, it is formed by the outer two-dimensional finish of the lining. Usually, with walls, it is the weather skin; with floors, a screed plate.

The basic circumstances are less favourable than for the comparable system with a full-leaf because the mass distribution within the ribbed leaf is inhomogeneous and therefore is not advantageous for noise-protection purposes. Also, as a rule, the mass of the ribbed leaf is smaller than in the case of the full-walled leaf since the former is more efficient

► *Sect. 2.2.2 Single Leaf with One-Sided Addition without Substructure*

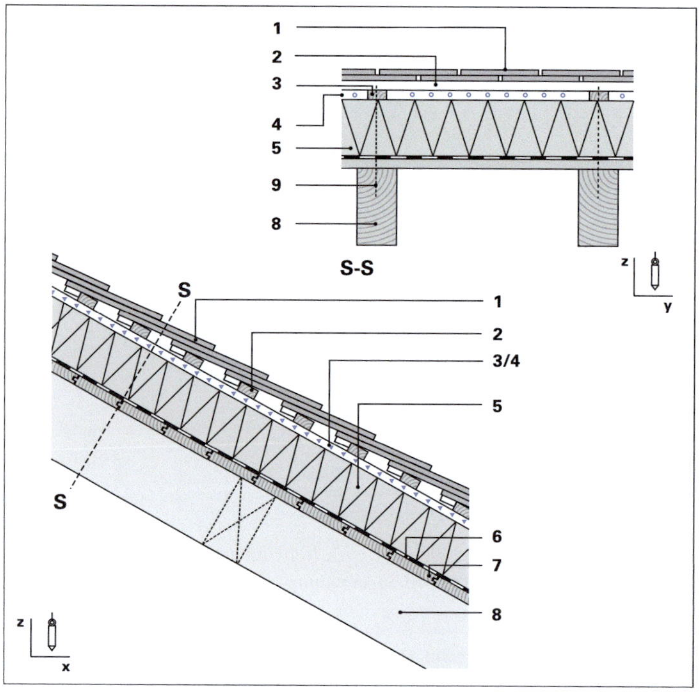

Fig. 2.57 Pitched roof construction with attached, continuous thermal insulation as an example of a ribbed system. The anchoring of the battens and roof covering to the load-bearing construction is carried out by means of continuous rafter nails. Almost complete elimination of thermal bridges. In this variant, the waterproofing is made below the closed-cell hard-foam insulation. (**1**) tile roofing; (**2**) batten; (**3**) counter-batten; (**4**) moving air layer; (**5**) thermal insulation (closed-cell foam all over); (**6**) waterproofing/vapour and wind barrier; (**7**) boarding; (**8**) rafter; (**9**) nailing

and material-saving in structural terms. As a consequence, it can be said that the variants mentioned in this section are rather unfavourable in terms of noise protection.

2.5.4.2 Ribbed System with Leaf and Lining Containing a Simple Transverse Ribbing

▶ Sect. 2.2.3 Single Leaf with One-Sided Lining Containing a Substructure

Here again (Fig. 2.59 left) the same conditions apply as in ▶ Sect. 2.2.3 (Fig. 2.10). The transverse set of ribs provides a certain lateral distribution of point or non-uniform loads acting on adjacent main ribs.

2.5.4.3 Ribbed System with Leaf and Lining Containing a Double Ribbing

▶ Sect. 2.2.3.2 Double Rib Layer

See ▶ Sect. 2.2.3.2 (Fig. 2.14 left). Here, too, this structure permits a clear separation between an insulation layer (in the intermediate spaces of the rear rib layer) and a possible air layer (in the cavities of the front rib layer) (Fig. 2.59 right; see also Fig. 2.60).

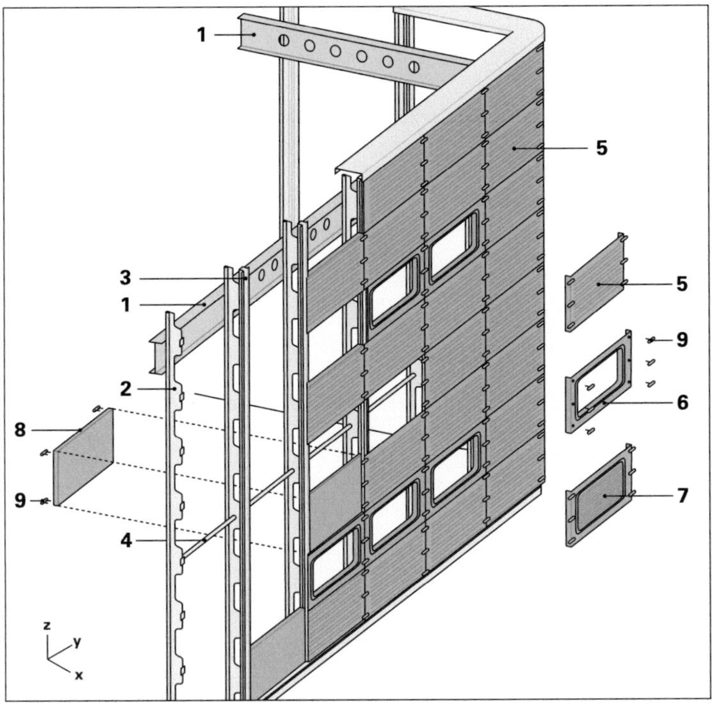

■ **Fig. 2.58** Façade in ribbed construction (arch. N. Grimshaw). (**1**) primary structure; (**2**) façade mullion; (**3**) mounting rail; (**4**) transversal stiffening of the mullions to prevent buckling; (**5**) external panel, closed; (**6**) external panel, with window; (**7**) external panel with ventilation grille; (**8**) internal panel; (**9**) fixing clip

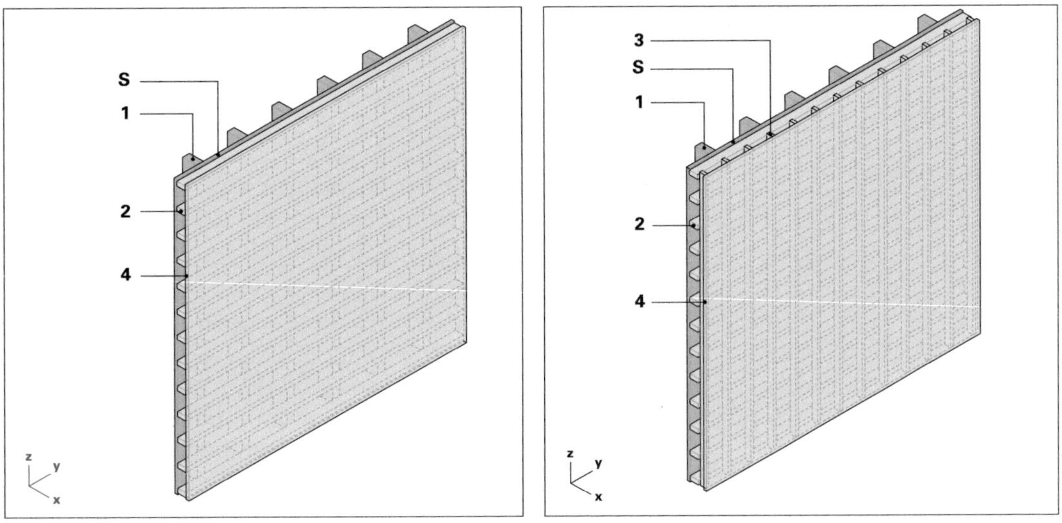

■ **Fig. 2.59** (Left) Ribbed element with added structure of single-layered cross-ribbing **2**. The sheathing **S** between main rib **1** and secondary rib **2** is optional. (Right) Ribbed element with an added two-ply, alternately transversal ribbed structure **2** and **3**. The sheathing **S** between the main rib **1** and the secondary rib **2** is optional

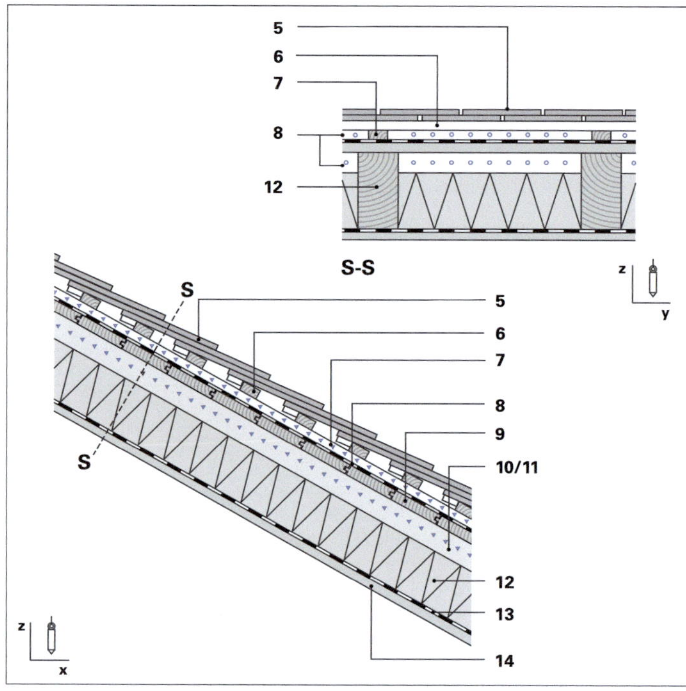

Fig. 2.60 Inclined ventilated roof, construction according to Fig. 2.59, right: the inner sheathing, as contained in this construction, does not normally assume a load-bearing function, which is why it was dismissed when depicting the corresponding basic structure on Fig. 2.59, right. The layer sequence of lattice and counter-battens is reversed against the pattern on Fig. 2.59, right. This, however, does not alter the basic constructional principle. Key: (**1**) main ribbing; (**2**) secondary ribbing of the first order; (**3**) secondary ribbing of the second order; (**4**) surface-forming sheathing; (**5**) tile roofing; (**6**) batten; (**7**) counter-batten/moving air layer; (**8**) waterproofing; (**9**) boarding; (**10**) moving air layer; (**11**) rafter; (**12**) thermal insulation (filling between the rafters); (**13**) vapour and wind barrier; (**14**) inner sheathing; (**15**) mullion; (**16**) elastic joint profile; (**17**) spacer; (**18**) pressing strip; (**19**) double-pane insulation glass

2.5.4.4 Ribbed System with Leaf and Lining Containing a Longitudinal Ribbing

► *Sect. 2.5.4 Ribbed Systems with Separation of Envelope Construction and Ribbing*

In this variant, the ribs of the lining are located on the main ribs and aligned **along** them (Fig. 2.61 left). From a constructional point of view, this variant equals the first (Fig. 2.59 left), since these superadded secondary ribs do not increase the load-bearing capacity of the main rib. They also do not provide a lateral distribution of point loads such as does the solution with transverse ribs located in the lining (Fig. 2.59 left).

Ribbed System with Surface Element

This solution is used in a modified form when the space-enclosing leaf is divided into individual elements butting over the main rib. The secondary rib of the superstructure is then split into two abutting individual rods separated by a joint

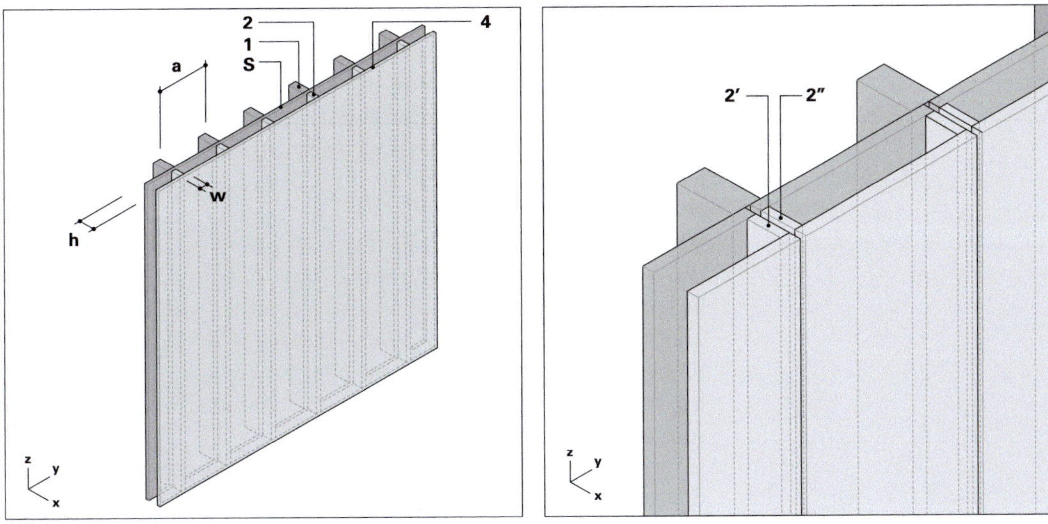

■ **Fig. 2.61** (Left) Ribbed element with added lengthwise aligned rib set **2**. The sheathing **S** between main rib **1** and secondary rib **2** is optional. **a** distance between main ribs; **b** width of main rib; **h** height of the added structure. (Right) Detailed view of the constructional principle depicted on the left. Here, a variant unitised field-by-field is shown in which the longitudinal rib **2** is divided into two half-leaves **2′** and **2″**. The basic principle corresponds to that of conventional mullion-glazing as depicted in ■ Fig. 2.62, left

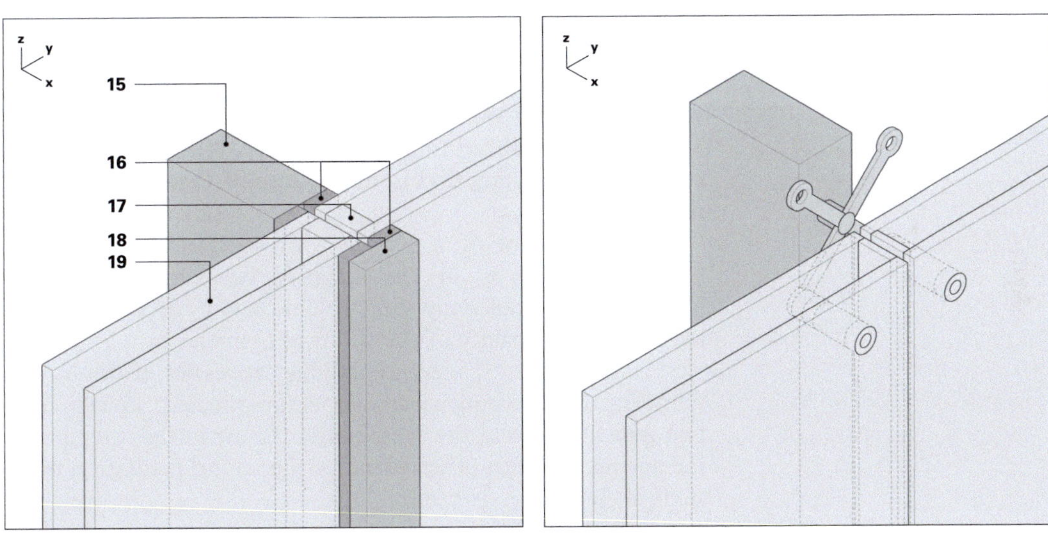

■ **Fig. 2.62** (Left) Further specification of the construction principle shown on ■ Fig. 2.61, right, in form of a pressed-edge glazing. (Right) The glass fixture with punctiform connectors instead of a continuous rib (as on the left) modifies the basic construction principle only insignificantly. (for key, cf. ■ Fig. 2.60).

(■ Fig. 2.62 left), each of which is assigned to the respective abutting element. Via this edge rib, the **compressive forces** necessary for the attachment of the leaf can be transferred to the main rib.

A mullion façade, or mullion-and-transom façade, with **slatted glazing**, or alternatively with **sandwich elements** (■ Fig. 2.62 left), is a special case of this constructional solution and, at the same time, a very illustrative example. In the mullion façade, transversal ribs (i.e. the transoms) are not existent in glazed areas between the main ribs (here, the mullions), not being envisaged in the constructional concept, since the leaf must appear transparent and without disturbing visible bars. This means that the spacing between the mullions is, of course, limited by the **load-bearing capacity** of the **glass pane**, which acts here as a plate on two opposed linear supports. In the example of the mullion-and-transom façade, a **four-sided bearing** of the glass pane is also possible, which—depending on the side ratio of the pane—improves its load-bearing capacity. This solution then also brings about a transverse bracing of the ribs (i.e. of the mullions), as described above in this section.

Vol. 1. ▶ *Sect. 23.4.1 Insulating Glass*

From the point of view of **thermal protection**, in the case of the two-pane glazing, the **inter-pane cavity**, i. e. the **inert-gas layer** contained therein, assumes the task of thermal insulation, which is why this cavity, from the present perspective, is regarded as an insulating layer. This is only possible because convection can be largely prevented by limiting the interspacing distance to between 10 and 20 mm. The partial constructional disadvantage of not being able to couple the two glass panes force-fittingly in the field between the mullions by means of webs or the like for the purpose of maintaining transparency proves, however, to be an advantage with respect to thermal protection, since no thermal bridges are produced there. Only at the joint above the main rib do these emerge, that is, in the case of the insulating glass pane, in the spacer.

In the variant of a façade with sandwich elements, the mullion spacing **a**, or the distance between any supporting transoms that may be present, again depends on the load-bearing capacity of the sandwich panel.

From a point of view of **building acoustics**, a double-leaf vibrating mass-spring system is again present, in which the two panes represent the masses and the air in the interspace is the spring. As with other variants, the sound insulation of the surface element is determined by:

- the **mass distribution** between the panes. Here, different surface-related masses—in this case, therefore, different pane thicknesses—have an advantageous effect.

Vol. 1, ▶ *Sect. 28.3.5 Particularities of Sound Protection of Windows*

- the **dynamic stiffness** of the resilient layer. In the normal case, with insulating glass panes, for reasons of thermal protection, it is made up of dynamically inert gases, which at the same time have a favourable impact on building acoustics. Here, the designer has little room for manoeuvre.

Vol. 1, ▶ *Sect. 28.3.5 Particularities of Sound Protection of Windows*

- the **distance** between both leaves (i.e. the glass panes). This can only be varied within narrow limits.

At the pane joint, a sound bridge results which can hardly be avoided, since in this case the functions of thermal protection and load-bearing have priority over noise protection.

A further variant of this constructional principle is the **point-mounting** of the insulating glass panes, which in this case appear as surface-forming elements (see ◘ Fig. 2.62 right). Here, the sealing element (spacer of the glazing) and the fastening element or force-transmitting part are separated. The sealing function is carried out by a caulking which is not subjected to compression. The mounting of the glazing is no longer linear, so, from a structural point of view, it is in this case a point-mounted plate or pane. Since the mullion is also relieved of the sealing function (as is still necessary in the case of ◘ Fig. 2.62 left), it can assume freer constructional forms. In this variant, a further **differentiation** of the construction takes place: In addition to separating the load-carrying and space-delimiting functions, as is typical for rib systems, in particular for the last one discussed, a separation of force conduction and sealing functions also takes place within the connection detail.

Vol. 3, *► Sect. 15 Point-Supported Glass Envelopes*

Sealant joint according to the principle of filling with flank adhesion; cf. ***Vol. 3***, *► Sect. 1.4.3.3 Joint with Flexible Filling and Flank Adhesion*

► Sect. 2.5.4 Ribbed Systems with Separation of Envelope Construction and Ribbing

In terms of **building acoustics**, the glass panes, which are held at points, have a slightly more favourable effect than the ones held linearly, since the panes can oscillate more freely and the sound-transmitting cross-section is generally smaller.

2.6 Complementing Functional Elements or Leaves

All the basic constructions of envelope elements considered so far can also appear equipped with supplementary elements (◘ Fig. 2.63) which perform a specific function or form an **air cavity** with a specific purpose between them and the primary leaf construction. In the case of double-leaf systems with an intermediate layer and air layer as described in ► Sect. 2.3.2, the outer leaf already constitutes a similar functional element.

Vol. 3, *► Sect. 16 Added Functional Elements*

Complementary surface elements or leaves can fulfil the following tasks:

- create an **additional weather skin** with back ventilation;
- provide **sun protection** (◘ Fig. 2.64);
- provide **anti-glare protection**;
- provide **visual communication**;
- provide improved **noise protection**;
- provide improved **thermal protection**;
- create a specific **visual appearance** of the envelope;
- create an **enclosed cavity** for various purposes such as:

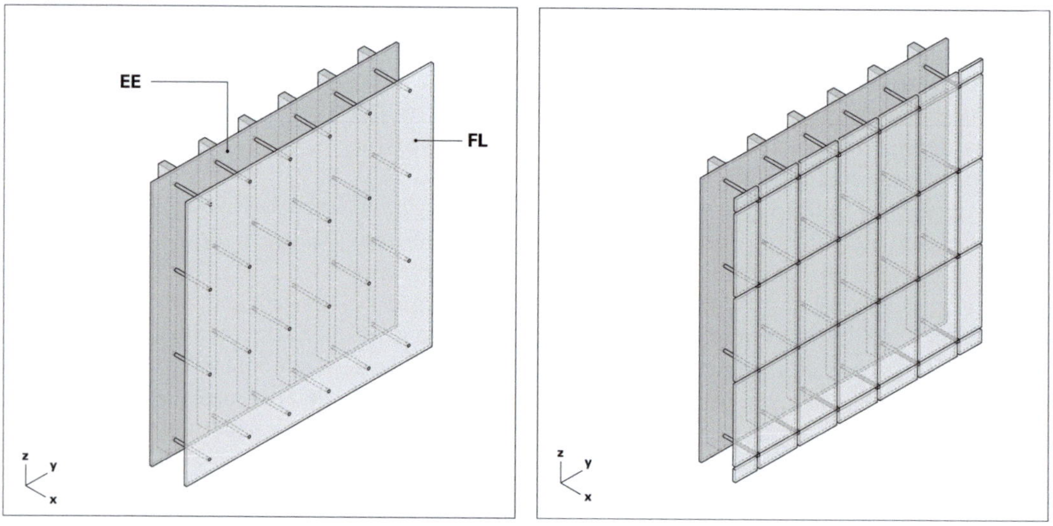

Fig. 2.63 (left) Enveloping element **EE** (of arbitrary constructional design) with functional leaf **FL** added on the outside. (Right) Unitised variant of the solution on the left, as used, for example, in façade construction

- routing of **ducts** (Fig. 2.65 left);
- **ventilation**, as is the case, for example, with **double-glazed façades** (Fig. 2.65 right). There, the added leaf usually also takes on additional functions among the listed ones such as sun protection, glare protection, sound insulation, and thermal protection.

Such added structures may be substantially independent from the construction mode or from the type of execution of the main envelope. This in turn means that added constructions or leaves can be combined with virtually any of the above-described envelope variants.

The back-anchoring or, depending on the position of the envelope element in space, the support of the added leaf usually is tantamount to an additional **thermal** or, as the case may be, also **acoustic bridge**. Under certain circumstances, specific layers of the main envelope can also be impaired in their functionality by these force-conducting elements. An example of this is a raised terracing covering on a flat roof, in which care must be taken that the roof seal is not damaged by punctiform plate bearings.

Added leaves can have an influence on the **functioning** in terms of **building-physics** and, thus, on the **constructional design** of the main envelope element under given circumstances. Added closed outer leaves, such as occur in the case of double-glazed façades, protect the main envelope from direct weathering and therefore make it possible to considerably simplify their constructional design.

The combined envelope structure consisting of the main envelope and the added leaf must always be regarded as a ho-

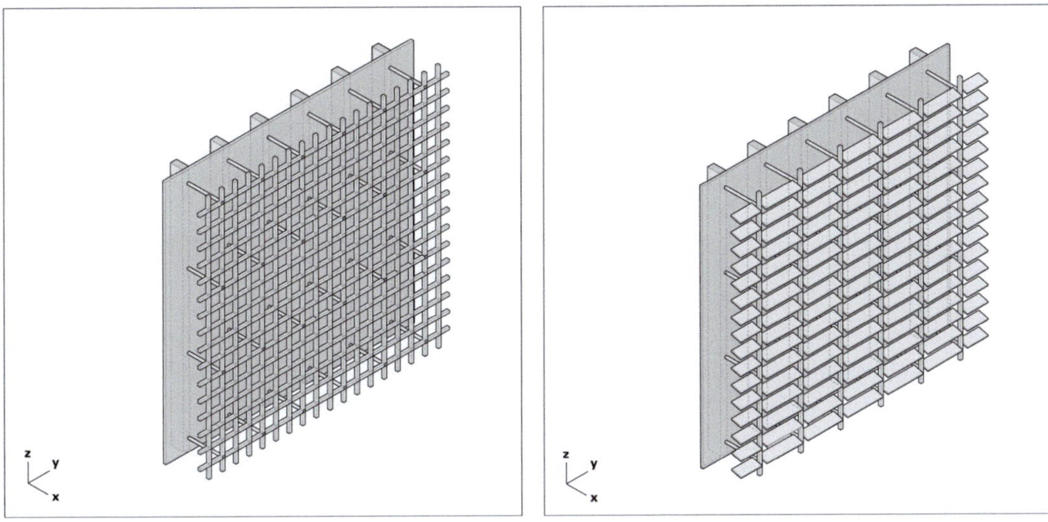

Fig. 2.64 (Left) Added grid-like framework for various functions such as driving-rain protection, partial visual protection, or sun protection. (Right) Added fixed or adjustable lamellae system

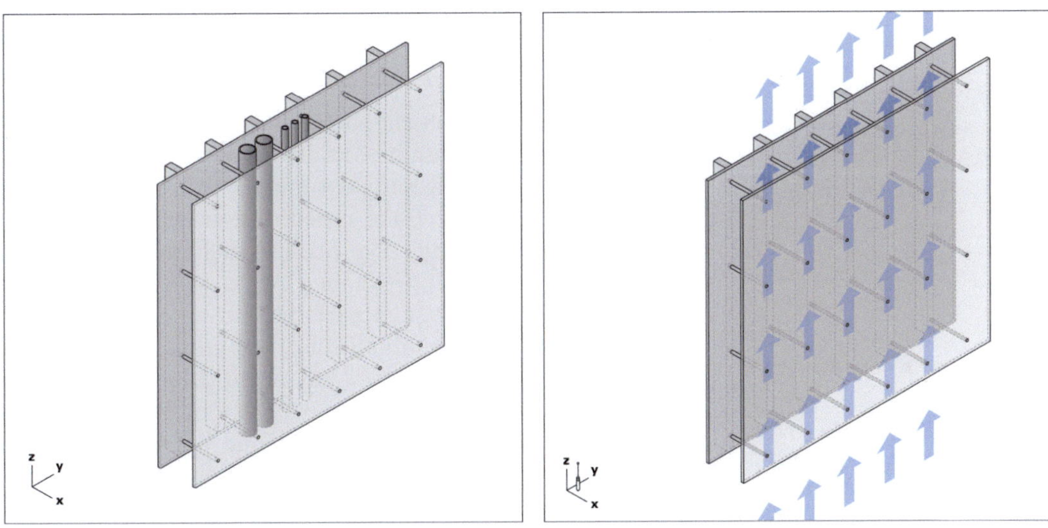

Fig. 2.65 (Left) Added leaf with duct routing in the interspace. (Right) Added leaf with air conduction in the cavity. This constructional and functional principle is used in double-glazed façades

listically functioning system from an operational and physical viewpoint. Sometimes, the addition of a leaf causes certain undesirable effects such as the conduction of fire in a façade cavity, which acts like a shaft in the event of a fire, or so-called acoustic *telephony effects* between floors. In principle, such combined systems are difficult to analyse in terms of building acoustics, and their behaviour can only be inadequately anticipated since they are always series circuits of resonance systems with complex acoustic interactions.

2.7 Membrane Systems

When used as envelope elements, membranes can be employed either:

- as **single-layer sheets** without insulation function for temporary structures or simple canopies, or:
- as insulating **multi-ply sheets** with air interspace or filling from suitable insulation materials.

The partial functions of force conduction as well as protection against wind and rain are assigned to the membrane itself. If a thermal-insulation function is present, it is to be provided by an air buffer between membrane layers, which should not exceed a maximum width in order to suppress as far as possible the convective phenomena which reduce the insulation value of the envelope construction. Alternatively, suitable fibre-insulation materials or granules can also be used as fillings.

2.7.1 Mechanically Prestressed Membrane Systems

Vol. 1, ▶ *Sect. 26.4.2 Movable Systems*, as well as *Sect. 26.9.8* and *26.9.9*

Membrane systems, which adopt their form due to a mechanical pretension, require anticlastic geometries with double curvature. With the simplest means, this can be effected by punctual supports of the membrane in form of suspensions or mast propping. Typical for such membrane constructions are the scallop-shaped edges, which form freely on their own as a consequence of the cutting pattern and the pretension and, as a result of the membrane curvature, always produce spatially curved, i.e. non-planar, curves. Edge connections to the ground or to any flat enclosing surfaces, for example, façades, are therefore one of the most difficult and elaborate—as well as also most aesthetically problematic—details of such membrane constructions. However, they are indispensable if the membrane is to enclose a defined space continuously in concurrence with other enveloping elements, at least with the floor. These edge connections are, of course, to be thermally separated themselves.

▶ Sect. *1.2.3.1 By Type of Curvature*
▶ *Sects. 3.4.5.2 Membranes and Rope Nets* and ▶ *4.3.3 Tensile-Stressed Systems*

In the case of multi-layered membranes with a heat-insulating function, the individual layers are to be guided as parallel to one another as possible, which means that they are fixed to one another at the correct distance at the attachment points where they are mutually coupled by means of spacers. The physically favourable air interspace of around 10 to 15 mm is very difficult to achieve with membranes, which are usually wide-spanning, so that high insulation values can hardly be realised activating a thermal insulation by air due to the almost unavoidable convection occurring within the air interspace, which tends to be too wide. Improvements can be achieved by granulate (such as silica-aerogel) or fibre-insulation filling.

Favourable energy balances can be realised with passive solar-energy gains for membranes made of transparent plastics (for example, ETFE). The energy yield as well as the heat-transfer resistance can also be increased by low-e coatings of the plastic membranes.

In terms of **building acoustics**, membranes are associated with a fundamental drawback due to the minimal mass of the construction, which is inherent to the structural type, but which is partly compensated for by the low dynamic stiffness and the material damping of conventional membrane materials. Depending on the surface weight, simple membranes achieve sound-insulation values of 3–12 dB, two-layered ones about 17–18 dB with a basis weight of between 0.8 kg/m^2 and 1.5 kg/m^2.[4]

In principle, membranes offer very favourable conditions for creating climatically effective envelopes because of their elementary constructional simplicity as well as the nearly absent thermal-bridging and, apart from seams or adhesive edges, jointless envelope surface. Only their moderate insulation values and their, depending on the material choice, sometimes comparatively short life under weathering influence limit their suitability for building purposes. Their limited sound-insulation capacity is also worth pointing out in this context.

2.7.2 Pneumatically Prestressed Membrane Systems

Constructional conditions are similar in the case of pneumatically prestressed membranes as in the case of mechanically stressed ones. Straight edges for the purpose of simpler connections can, however, be realised in this variant with comparatively simple means, since pneumatic cushions are often executed in any case with a fixed edge mount, which in most cases is implemented linearly for the sake of simplicity. Multi-layered cushions designed for achieving good thermal-insulation values by air buffering are technically unproblematic, but cannot be carried out at constant, and not at all minimal, layer distances for geometrical reasons because of the pneumatically induced mutually opposing arching of the layers. This, in turn, somewhat reduces the achievable insulation value.

In terms of building acoustics, pneumatic cushions offer comparatively favourable values. Like mechanically pre-

4 Figures provided by the Department of Building Physics, University of Stuttgart, Prof. S. R. Mehra.

2

stressed membranes, they have low dynamic stiffness and high material damping. Pneumatic membranes also act as sound absorbers which already swallow part of the sound energy on their surface. The enclosed airspace also has a favourable sound-protection effect, which can be measured, but not yet explained in detail. Cushions can achieve sound-insulation values in the range of 25 dB at surface weights of the individual membranes of around 1 kg/m^2.[5]

■ Standards and Guidelines

DIN 4095: 1990-06 Planning, design and installation of drainage systems protecting structures against water in the ground

DIN 4102: Fire behaviour of building materials and building components

Part 1: 1998-05 Building materials; concepts, requirements and tests

DIN 4108 Thermal protection and energy economy in buildings

Part 2: 2013-02 Minimum requirements to thermal insulation

Part 3: 2018-10 Protection against moisture subject to climate conditions—Requirements, calculation methods and directions for planning and construction

Part 4: 2020-11 Hygrothermal design values

Part 10: 2021-11 Application-related requirements for thermal insulation materials

Supplement 2: 2019-06 Thermal bridges—Examples for planning and performance

DIN 4109: Sound insulation in buildings

Part 1: 2018-01 Minimum requirements

DIN 7863: Elastomer glazing and panel gaskets for windows and claddings—Material requirements

Part 1: 2022-02 Non cellular elastomer glazing and panel gaskets

Part 2: 2019-12 Cellular elastomer glazing and panel gaskets

DIN 18195: 2017-07 Waterproofing of buildings—Vocabulary

DIN 18533: Waterproofing of elements in contact with soil

Part 1: 2017-07 Requirements and principles for design and execution

DIN 18534: Waterproofing for indoor applications

Part 1: 2017-07 Requirements and principles for design and execution

DIN/TS 18599: Energy efficiency of buildings—Calculation of the net, final and primary energy demand for heating, cooling, ventilation, domestic hot water and lighting

5 Figures provided by the Department of Building Physics, University of Stuttgart, Prof. S. R. Mehra.

Part 12: 2021-04 Tabulation method for residential buildings

EN 1996: Eurocode 6—Design of masonry structures

Part 1-1: 2019-09 General rules for reinforced and unreinforced masonry structures, prEN

Part 1-2: 2022-10 General rules—Structural fire design, prEN

Part 3: 2021-10 Simplified calculation methods for unreinforced masonry structure, prEN

EN 13022: Glass in building—Structural sealant glazing

Part 1: 2014-08 Glass products for structural sealant glazing systems for supported and unsupported monolithic and multiple glazing

EN 13967: 2017-08 Flexible sheets for waterproofing—Plastic and rubber damp proof sheets including plastic and rubber basement tanking sheet - Definitions and characteristics

EN 13970/A1: 2007-02 Flexible sheets for waterproofing—Bitumen water vapour control layers – Definitions and characteristics

EN 14509: 2013-12 Self-supporting double skin metal faced insulating panels—Factory made products—Specifications

EN 15651: Sealants for non-structural use in joints in buildings and pedestrian walkways

Part 1: 2017-07 Sealants for facade elements

EN 15812: 2011-06 Polymer modified bituminous thick coatings for waterproofing—Determination of crack bridging ability

EN 15814: 2015-03 Polymer modified bituminous thick coatings for waterproofing—Definitions and requirements

EN ISO 6781: Performance of buildings—Detection of heat, air and moisture irregularities in buildings by infrared methods

Part 1: 2023-11 General procedures

VDI 6203: 2017-05 Façade planning—Criteria, degree of difficulty, assessment

IVD-Merkblatt Nr. 22: 2014-11 Anschlussfugen im Stahl- und Aluminium-Fassadenbau sowie konstruktivem Glasbau – Einsatzmöglichkeiten von spritzbaren Dichtstoffen

RAL-GZ 996: 2014-11 Fassadenbefestigungstechnik—Gütesicherung

Atlas Gebäudeöffnungen: 2015-06 Atlas Gebäudeöffnungen—Fenster, Lüftungselemente, Außentüren

Bibliography

Bögle A, Schmal PC, Flagge I (2003) *leicht weit – Light Structures. Jörg Schlaich, Rudolf Bergermann*. Prestel, Munich, Berlin, London

Bollinger K et al (2011) *Atlas Moderner Stahlbau: Material, Tragwerksentwurf, Nachhaltigkeit*. Institut für Internationale Architektur-Dokumentation, Munich

Cremers J, Binder M, Bonfig P, Hartwig J, Klos H, Leuschner I, Sohn E, Stark T (2015) *Atlas Gebäudeöffnungen: Fenster, Lüftungselemente, Außentüren*. Detail, Institut für Internationale Architektur-Dokumentation, Munich

Herzog T, Krippner R, Lang W (2016) *Fassaden Atlas: Zweite überarbeitete und erweiterte Auflage – Grundlagen, Konzepte, Realisierungen*. Detail, Institut für Internationale Architektur-Dokumentation, Munich

Hestermann U, Rongen L (2015) *Frick/Knöll Baukonstruktionslehre 1*. 36th ed. Springer Vieweg, Wiesbaden

Hestermann U, Rongen L (2018) *Frick/Knöll Baukonstruktionslehre 2*. 35th ed. Springer Vieweg, Wiesbaden

Hugues T, Steiger L, Weber J (2012) *Holzbau: Details, Produkte, Beispiele*. Detail, Institut für Internationale Architektur-Dokumentation, Munich

Kaufmann H, Krötsch S, Winter S (2017) *Atlas mehrgeschossiger Holzbau*. Detail Business Information GmbH, Munich

Kind-Barkauskas F, Kauhsen B, Polónyi S, Brandt J (2009) *Stahlbeton Atlas: Entwerfen mit Stahlbeton im Hochbau*. Institut für Internationale Architektur-Dokumentation, Munich

Knippers J, Cremers J, Gabler M, Lienhard J (2010) *Atlas Kunststoffe. Membranen, Werkstoffe und Halbzeuge, Formfindung und Konstruktion*. Institut für Internationale Architektur-Dokumentation, Munich

Küttinger G, Fritzen K (2014) *Holzrahmenbau: bewährtes Hausbau-System*. Holzbau Deutschland, Bund Deutscher Zimmermeister im Zentralverband des Deutschen Baugewerbes. Bruderverlag, Cologne

Kummer N (2017) *Masonry construction*. Birkhäuser, Basel

Mittag M (2012) *Baukonstruktionslehre – Ein Nachschlagewerk für den Bauschaffenden über Konstruktionssysteme, Bauteile und Bauarten*. 18th rev. ed. Springer Vieweg, Wiesbaden

Natterer J, Herzog T, Schweitzer R, Volz M, Winter W (2003) *Holzbau Atlas*. 4th ed., newly edited. Birkhäuser, Basel

Pech A, Gangoly H, Holzer P, Maydl P (2015) *Ziegel im Hochbau: Theorie und Praxis*. Birkhäuser, Basel

Pottmann H, (ed), Asperl A, Hofer M., Kilian A, (2009) *Architekturgeometrie*. Ambra and Springer, Vienna, New York

Raso I (2010) *GlasDoppelFassaden: Am Beispiel von fünf verschiedenen Gebäuden*. VDM Verlag Dr. Müller, Saarbrücken

Russ C et al (2008) *Sonnenschutz: Schutz vor Überwärmung und Blendung*. Freiburg Fraunhofer Solar Building Innovation Center SOBIC, Stuttgart

Saxe K, Stronghörner N, Uhlemann J (ed) (2016) *Essener Membranbau Symposium 2016 (Berichte aus dem Bauwesen)*. Shaker, Herzogenrath

Saxe K, Stronghörner N (ed) (2018) *Essener Membranbau Symposium 2018 (Berichte aus dem Bauwesen)*. Shaker, Herzogenrath

Fundamentals with Dr. Matthias Weißbach

J. L. Moro, *Building-Construction Design – From Principle to Detail*,
https://doi.org/10.1007/978-3-662-71006-7_3

Contents

3.1 Preconditions

3.1.1 Load-Bearing Structure and Overall Design

Among the numerous constructional tasks which a building structure must fulfil to secure durable usability, special attention must be given to the basic task of **load-bearing** in the primary structure during the conception and subsequent design of a building. In contrast to other partial tasks, which are often fulfilled mainly at the component level and rarely have an effect on the overall design, the design of the load-bearing structure has a dominant influence on the overall building concept and the architectural appearance on the hierarchical level of the complete building. For this reason, the considerations set out in ***Volume 1*** on the force-conducting behaviour of an individual component will be pursued in this chapter at a higher hierarchical level in the context of the **primary load-bearing structure**, or some of its relevant basic modules. If individual elements as described in *Sect. 26*, this is done in this case—unlike in ***Volume 1***—specifically contemplating their use as components of a primary load-bearing structure of a building and taking into account their position in relation to the plumbline—as, for example, as load-bearing walls or floors.

Vol. 1, *Sect. 25.1.3 Main Functions of Buildings*, as well as *Sects. 27, 28, 29, and 30*

The basis for understanding this chapter is the content of ***Vol. 1***, *Sect. 26 Force Transmission*

3.1.2 Functional Interrelationships

In the same way that components can be designed as mono- or multifunctional also in the case of primary load-bearing structures, apart from considering their main function, namely transmitting forces, sometimes, on the level of the overall building, also other functions must be taken into consideration which can appear in a closely interconnected manner. This applies, for example, to wall construction modes, where two-dimensional components have to simultaneously fulfil various building-physics functions alongside load-bearing. Likewise, the structures themselves are also determined by the needs of building use—such as appropriate daylighting and ventilation of interior spaces. Such dependencies must always be carefully incorporated into the reflections of structural design.

Vol. 1, *Sect. 2.2.2.2 According to Constructional Partial Functions* as well as *Sect. 25.3 Allocating Partial Functions to Building Components*

3.1.3 Space Formation

The basic necessity of space formation, and hence the delimitation of space, for the proper usage of a building is the fundamental reason for generating **two-dimensional components** characteristic of building construction, which are a basic prerequisite for the definition of space (◘ Fig. 3.5). In most relevant cases these have a fully closed air- and watertight surface; only in rare cases are they pervious (◘ Fig. 3.1).

Vol. 1*, Sect. 25.1.2 Basic Functions of Buildings*

Therefore, two elementary tasks of buildings, from which a variety of partial tasks relevant to the construction and the load-bearing structure can be derived, and which, for this reason, deserves particular attention in this context is:

- providing a walkable **floor** and a protective **covering**, as well as mostly
- all-round lateral **enclosing** of rooms.[1]

► *Sect. 3.1.6 The Elements of the Structural Cell*

From elementary constructional considerations such as will be discussed below, usually a **differentiation** between the coverage and the lateral enclosure of space results, which also manifests itself conceptually and linguistically in the distinction between **roof**, or **ceiling**, and **wall** (◘ Fig. 3.2). It is certainly often the case that structures have only a coverage, either without lateral containment while the cover is curved and thus encloses space (◘ Fig. 3.3), or loads from the coverage are transmitted into the ground by punctual supports, such as columns or pillars, but otherwise have no other encasement (◘ Fig. 3.4). However, enclosures without coverage are rarely to be found (or one would not even consider such an enclosure to be a building in the first place.), if any as a supplement to a covered structure, as in the case of a separate courtyard (◘ Fig. 3.5), since the protection from rainfall is an almost indispensable task of building construction (◘ Fig. 3.6).

From the point of view of the load-bearing structure, the following restrictions apply: A room enclosure need not necessarily also be part of the load-bearing structure. The separation of load-bearing structure and space-creating envelope is a feature of skeletal construction (◘ Fig. 3.7). Nevertheless, structural axes very often provide predestined places for room separations, since even singular supports or beams, etc. usually already have a clear effect as spatial caesuras and moreover are often predestined places for force-conducting connections of non-load-bearing envelope components like separation walls.

► *Sect. 3.1.6 The Elements of the Structural Cell*

In addition, the influence of the constructional design of the **coverage**, as compared with that of the enclosure, on the overall form of the building will become clear in the course of deepening engagement with the subject matter. It is not for nothing that the types of load-bearing structures are usually referred to precisely according to the coverage (arch structure, tensile structure, etc.).

1 It is obvious that the distinction between *enclosing* and *covering* is derived from the conventional distinction between vertical components (walls) and horizontal components (floors, roofs). Although factually this is not forcible (there are curved envelope constructions in which there is no differentiation between vertical and horizontal), it nevertheless does justice to the vast majority of buildings that are constructed in practice. Cf. ◘ Fig. 3.4 and ***Vol. 4****, Sect. 5.3 The Dominance of the Cuboid.*

■ **Fig. 3.1** Coverings for space enclosure or shadowing. They are not intended as weather protection

■ **Fig. 3.2** Clear differentiation between wall and roof surfaces in traditional masonry construction (house at the hillside, arch: Schmitthenner)

■ **Fig. 3.3** Continuous enclosure of space without differentiation between wall and roof (*Centro Niemeyer*, Avilés, Spain; arch: O Niemeyer)

Fig. 3.4 Coverages without side enclosure

Fig. 3.5 Creating surfaces is an elementary task and goal of building construction

Fig. 3.6 Courtyard: lateral enclosure without coverage

Fig. 3.7 The concentrated load transfer by means of rod-shaped columns in skeletal construction exonerates the flat lateral enclosure of load-bearing tasks and allows freedom in defining the relationship between interior and exterior, something which is lacking in wall construction

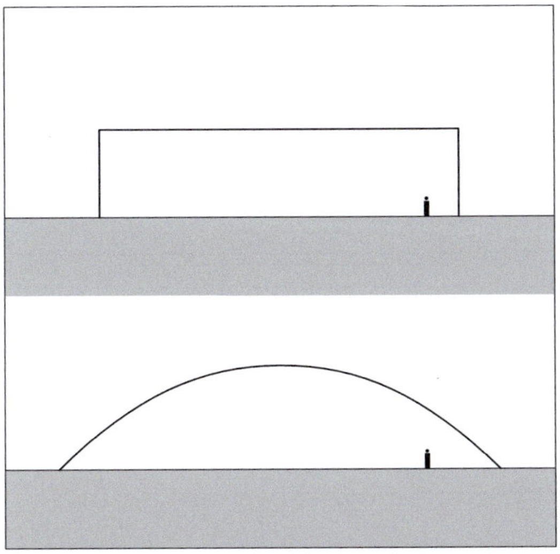

Fig. 3.8 Elementary structures are identical to the basic building cell that constitutes them

3.1.4 Elementary and Composite Structures

In the following, we will call a spatial unit which is delimited by a coverage with a defined span, and possibly also by a lateral enclosure, a **basic structural cell** or, simplifying, a **structural cell**. In this sense, single-room buildings are often identical to the basic structural cell (Figs. 3.8 and 3.9). However, many buildings consist of an **addition** of individual structural cells (Figs. 3.10 and 3.11), the corresponding space units of which can, however, be coupled to each other spatially through suitable openings. This interconnection of space units can go so far that a partition between

Fig. 3.9 Two simple buildings, each consisting, respectively, of a cuboidal and a curved basic cell

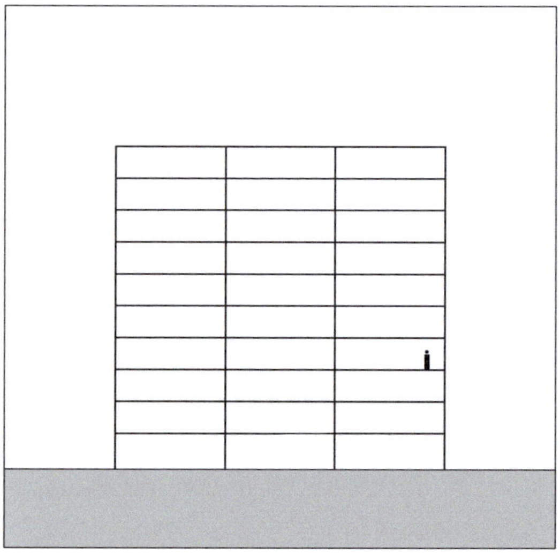

Fig. 3.10 Modular composite structures consist of the addition of structural basic cells

Fig. 3.11 Multicell building

adjacent rooms almost completely vanishes so that only punctual supports remain, i.e. columns. It may then happen that single-space buildings are also composed of a combination of several individual structural cells.

An interesting historical example is the interior space of the Hagia Sophia, cf. Figs. 4.258, 4.259, and 4.260 in ▶ *Sect. 4.3.2.6 Full-Leaf Shell, Synclastically Curved, Punctually Supported*

The following reflections are each based on a space-encompassing basic structural cell, which is investigated in various forms with regard to its construction and its load-bearing behaviour. As noted, this basic cell can, in the case of elementary structures, be identical with the primary structure itself (◘ Fig. 3.8) or, alternately, the modular basis for a **composite structure** resulting from the addition of the cell in one or more spatial dimensions (◘ Figs. 3.10 and 3.11). The consideration of the basic cell has didactic character and does not claim to fully reflect the load-bearing behaviour of a primary structure in its entirety, especially with composite structures, because:

- the load-bearing behaviour of a composite structure is also determined *by the manner* in which the underlying elementary cells are assembled;
- many structures are composed, but they are **not modular**, they do not consist of a constantly recurring, always identical basic module;
- structures cannot be statically analysed independently from their **scale.**

***Vol. 1**, Sect. 2.1.3 Order According to Constructional Aspects* as well as ***Vol. 4**, Sect. 1 Scale*

In spite of all necessary qualifications, the study of such a cell allows grasping the basic load-bearing behaviour of elementary structures as well as realising a basic categorisation of composite primary structures as are frequent in building construction, and achieving a fundamental understanding of load transfer within its constructional framework.

3.1.5 Design Principles of Adding Structural Cells

In the following, a distinction is made between the **horizontal** addition—in one or two directions in the plane—and the **vertical**. A horizontal addition allows the creation of theoretically unlimited space sequences while preserving a largely accessible floor level. Although the vertical addition of cells satisfies the need for a walkable total area overall, it seldom creates a coherent space, since the storey heights must always be overcome by staircases, ramps, etc., and space incisions are almost invariably created. Nevertheless, there are some exceptions in the form of singular spatial and structural concepts.

Cf. the 'Raumplan' (Space Plan) of Adolf Loos

3.1.5.1 Horizontal Addition

The **spans** of spatial coverages are always restricted by the limits of material, technical, and economic feasibility. This means that the largest realisable dimension of a single space is specified in at least one of the two spatial directions, i.e. in the direction of span of the space covering (this is the case with one-way spanning coverings; right-angled to this, as will be seen below, the dimensions of the spatial cell are, as a matter of principle, unrestricted). This is usually the smaller room dimension, as the smaller span is always selected for reasons of structural rationality; or to put the other way round: Once a ceiling span has been defined, it is usually advantageous and appropriate to dimension the room in the other main direction—now for reasons of use-related rationality—larger than the ceiling span, up to the room proportions that still appear appropriate for a particular use. These can be rather compact in normal spatial cells, such as common rooms in flats, but they can also be extreme, for example in commercial buildings such as industrial halls.

In the presence of two spanning directions (i.e. with two-way spanning room coverings), even both room dimensions are limited by a maximum span for this reason—in construction practice, this is the case with two-way spanning solid floor slabs, for example. Since certain not too extreme room proportions are usually desired for utilisation-related reasons, in the first case (the one-way spanning covering) the actually freely selectable room dimension (that transverse to the spanning direction of the covering) is nonetheless limited likewise in most construction-relevant cases, as noted. The resulting, always limited maximum size of basic structural cells cannot always satisfy the need for usable space. This is the reason why the principle of **horizontal addition** of a basic structural cell in one or two directions is of fundamental importance to the development of building types and forms (◘ Figs. 3.12, 3.13, and 3.14). The grouping of individual buildings with independent use within a dense village

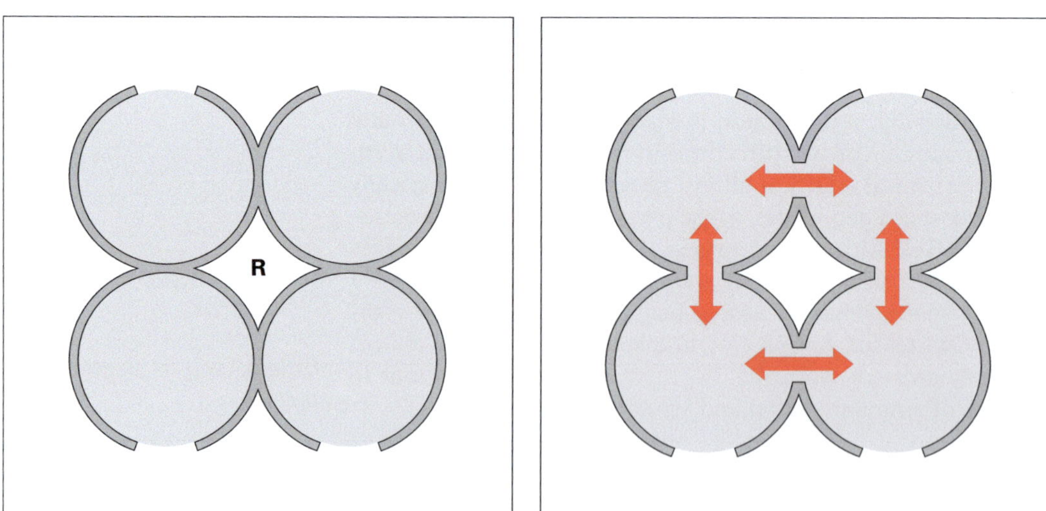

◘ **Fig. 3.12** (Left) Addition in ground plan of round basic cells. Hard-to-use residual spaces (**R**) are unavoidable. (Right) Addition of round basic cells. Possible connection of the rooms at the contact points via narrow door openings

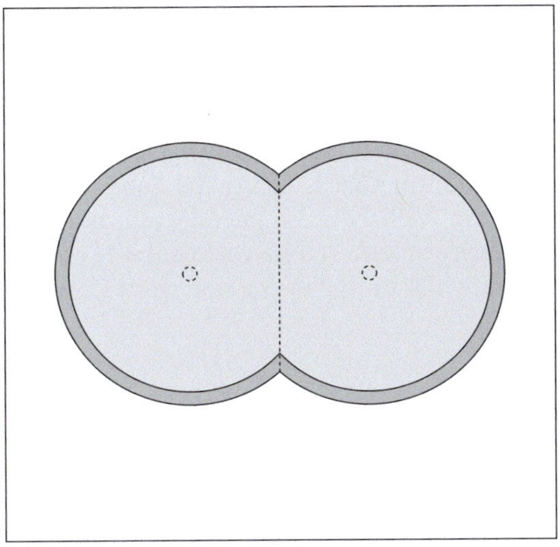

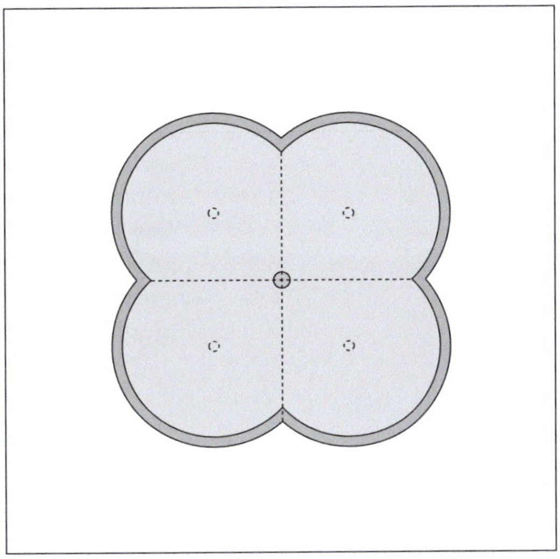

Fig. 3.13 (Left) Addition of two basic cells with round footprint by mutual intersection and creation of a spatial interconnection. (Right) Addition of four round basic cells. Rooms connected at the intersections. Load discharge in the centre by column

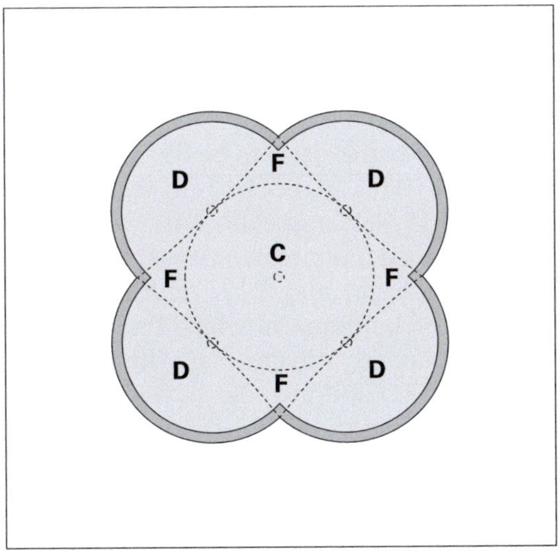

Fig. 3.14 (Left) Addition of four round basic cells with dome shape (**D**) supplemented by central raised dome (**C**) and four filling caps (**F**). (Right) Cottage with a uniformly curved envelope

or town structure usually also requires building in immediate vicinity to an adjacent building. The basic structural cell is governed by the technically and economically reasonably realisable dimension and can be coupled to form a larger interconnected space by agglomeration to a connected building cluster or through corresponding openings in the separating components (Fig. 3.12, left). Geometrical conditions are analogous to those of **tesselating** continuous surfaces from individual elements.

► *Sect. 1.1.2 Geometric Principles of the Formation of Surfaces from Individual Elements*

3

3.1.5.2 Vertical Addition

Moreover, rooms can also be added **vertically**, i.e. stacked. The resulting multiple use of land is of great economic importance. Similar questions as when adding a structural base cell in the surface also arise when stacking it (◘ Fig. 3.10). In contrast, the interfaces between the storeys, that is to say the **floor constructions**, unlike the partition walls between laterally abutting cells, are limited by the necessity of safe and easy accessibility of floor surfaces which are totally flat on the upper side, or at least only minimally inclined.

3.1.6 The Elements of the Structural Cell

Whereas the design of **lateral space delimitations** is derived primarily from functional, spatial, and aesthetic considerations, and the boundary conditions of their load-bearing behaviour rarely are determinant for form and design, the design of the **coverage** is often dominated by structural issues which derive from the mere necessity to bridge spans safely and permanently with self-supporting building material (◘ Fig. 3.15). This dominance is all the clearer the greater the span to be bridged.

An exception is the bracing of a building, which must be included in the spatial concept at an early design stage. An example of this is the mutual transverse bracing of planar wall panels

From a structural point of view, the essential constructional difficulty of the coverage derives from the inevitable necessity of **deflecting loads** over the space to be covered to dissipate them into the ground.[2] This applies to both vertical loads due to gravity (◘ Figs. 3.16, left and 3.17, left), which need to be deflected away from the plumbline, as well as to horizontal loads (◘ Figs. 3.16, right and 3.17, right). Although there are some examples of buildings in which the main part of the loads is in fact rectilinearly directed, i.e. without deflection, through the construction and into the ground, they represent—from the current perspective—functionally restricted buildings, which can only meet very limited usage requirements (◘ Fig. 3.18).

For this reason, in the following sections, in particular, coverages in numerous variants will be investigated with regard to their load-bearing behaviour. The lateral **enclosure** of the basic structural cell is usually assumed to be a closed wall (◘ Fig. 3.19) in the discussed examples, but is, as far as not otherwise noted, fundamentally interchangeable by different constructions as discussed in *Chap. 26* (◘ Figs. 3.20, 3.21, and 3.22).

***Vol. 1**, Sects. 26.9.2, 26.9.3, 26.9.4, and 26.9.5*

Coverage and lateral **enclosure** are closely linked and are mutually interdependent in many respects. The design of a coverage determines:

2 See the magnitude BIC in: Institute for Lightweight Structures, University of Stuttgart (ed) (1971) *IL 21*, p. 43 ff.

Fig. 3.15 Building free-spanning coverings is usually more difficult than erecting upright enclosures

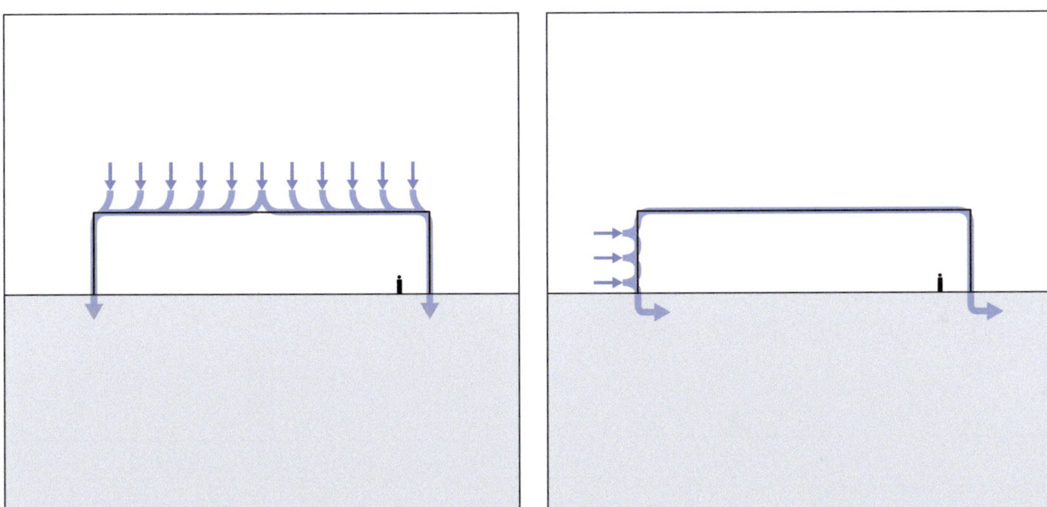

Fig. 3.16 (Left) Deflecting vertical loads in a cuboid-shaped space-forming building structure. (Right) Deflecting horizontal loads in a cuboid-shaped space-forming building structure

- the **main dimensions** of the basic structural cell, because the maximum coverable span is fixed at the same time;
- likewise, mostly its **geometrical orientation** or the symmetry conditions in question within the ground plane, because the support of the load-bearing structure is thereby defined—as a result, directional or non-directional structures emerge.

Conversely, the pre-determined definition of a certain ground-plan geometry of the **enclosure** of a structural cell may some-

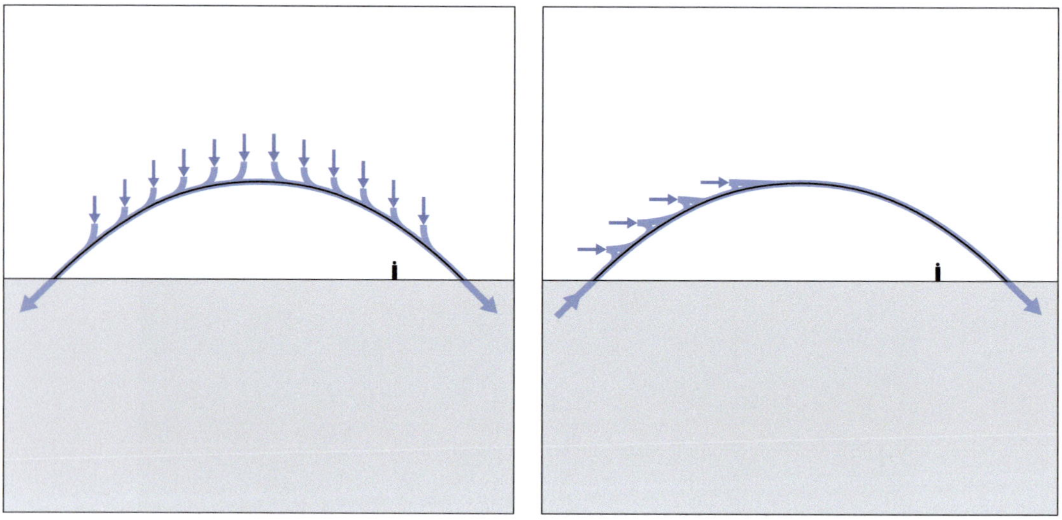

Fig. 3.17 (Left) Deflecting vertical loads in a curved space-forming building structure. (Right) Deflecting horizontal loads in a curved space-forming building structure

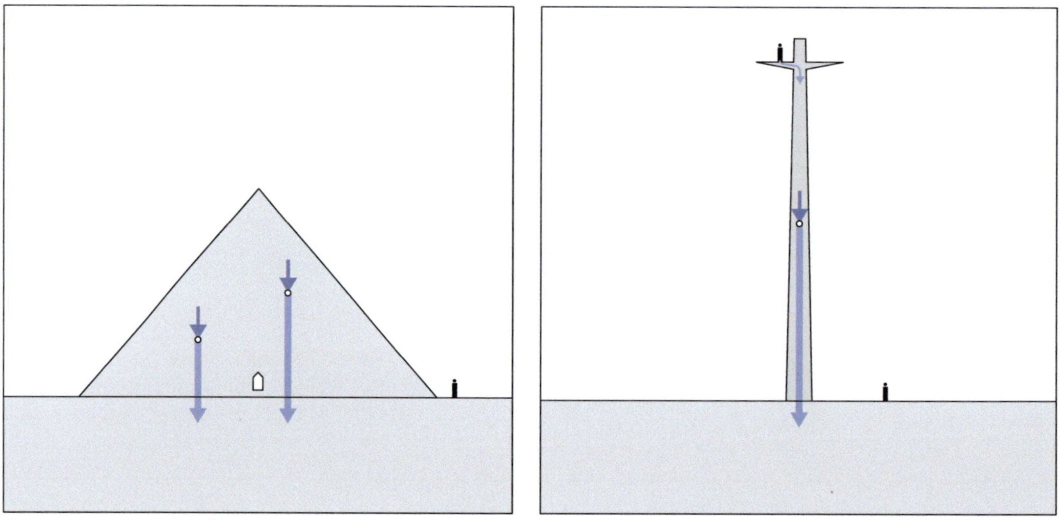

Fig. 3.18 (Left) Direct, i.e. deflection-free transmission of loads into the ground entails constraints in space formation (example: pyramid). (Right) In this example (tower), which transmits the vertical (though not the horizontal!) main loads mostly directly in rectilinear form, the upper platform already assumes a force deflection

times also set clear boundary conditions for the coverage. Pre-settings in this regard may result in a limited selection of coverages, and in any case fix the spans to be covered. Thus, for example, the geometrical proportioning of the side edges of a rectangular cell provides the respectively most favourable—because shorter—span.

Interestingly, such cases are more frequent in the historical development of building designs than might be assumed in the case of which the spatial enclosure is chosen accord-

Fig. 3.19 Gradual transition from the cubic shape of the building at the bottom to the spherical one at the top (mosque in Pocitelj, Bosnia-Herzegovina)

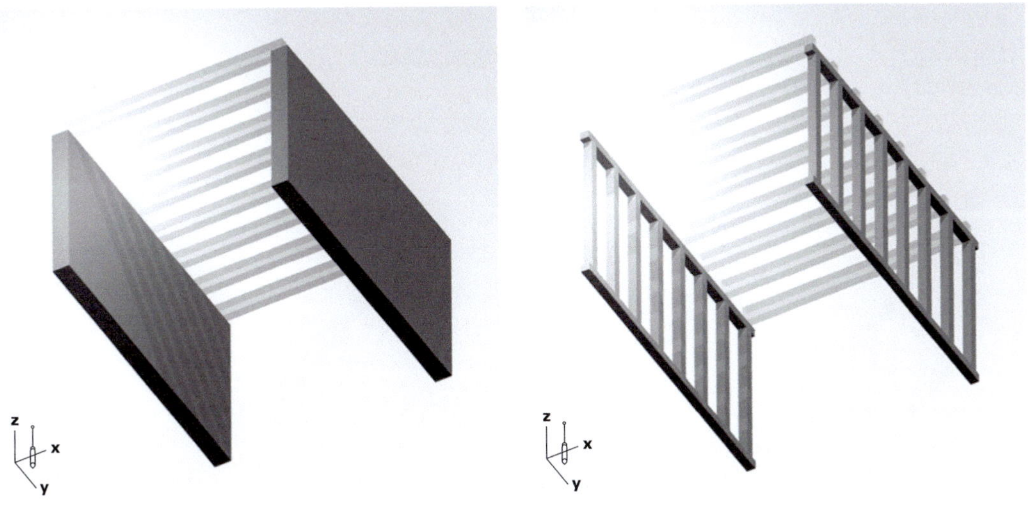

Fig. 3.20 (Left) An example of a structural basic cell made up from enclosure and covering. (Right) Substitution of a ribbed wall for the solid walls of the cell on the left

ing to criteria of *additivity*—as we will see, with rectangular geometry—the coverage placed on top, on the other hand, being based on considerations of *structure* and *materials*, is designed according to a totally contrary geometry. An excellent example of this is the combination of an easily addable cubic substructure and a wide-spanning dome-shaped coverage, which was comparatively easy to manufacture—sometimes completely executed without centring—as well as a predominantly compressed load-bearing element which offered decisive advantages when using brittle mineral materials (Fig. 3.19). The study of the geometric and technical solutions developed by the historic master builders, some of which were highly imaginative, are an interesting and informative lesson in conceptual and constructional design.

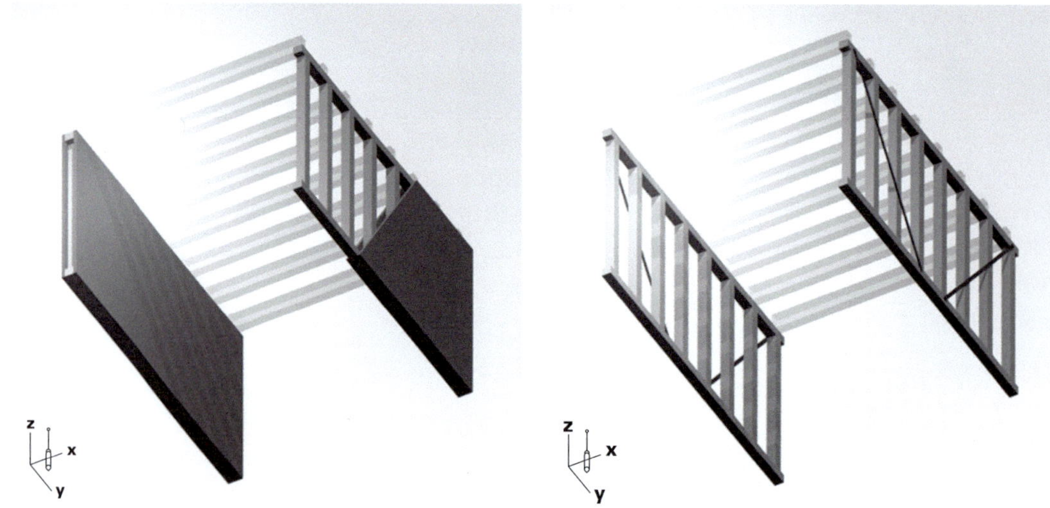

Fig. 3.21 (Left) Shear-plate formation by sheathing both sides of the ribbed wall of Fig. 3.20, right. (Right) Bracing of the ribbed wall by diagonals

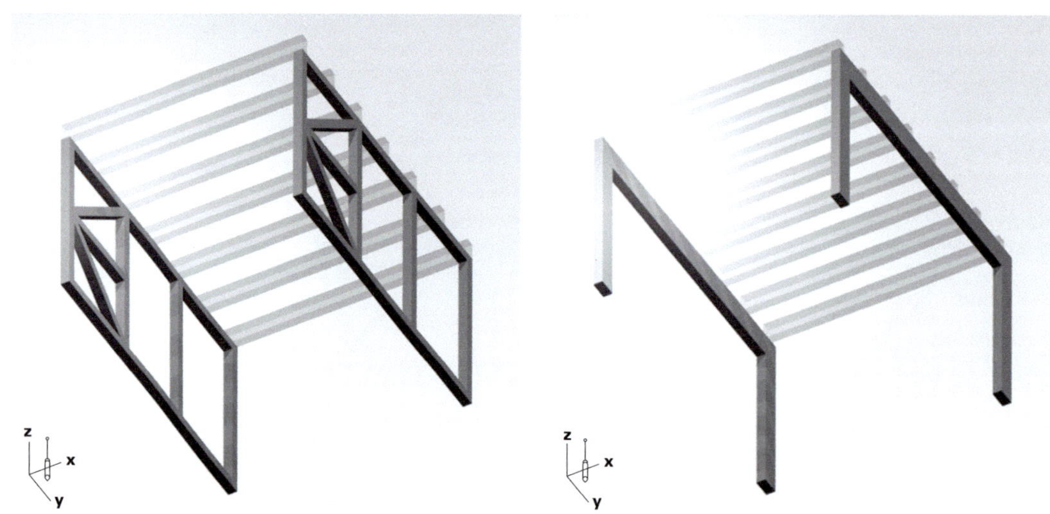

Fig. 3.22 (Left) Substitution of a braced stud wall for the solid walls of Fig. 3.20, left. (Right) Substitution of a portal frame for the solid walls of Fig. 3.20, left

Read more in Schlaich, Heinle (1996); cf. also ▶ *Sect. 4.3.2.6 Full-Leaf Shell, Synclastically Curved, Punctually Supported*

3.1.6.1 The Vertical, Planar Space-Enclosing Element

In the simplest form, the task of enclosing a space can be fulfilled by a **dome-shaped** structure which limits this space in a continuous half-spherical shell shape. Such elementary designs are very old and were executed in stone as well as in wood. They are found in examples as in Fig. 3.14, right,

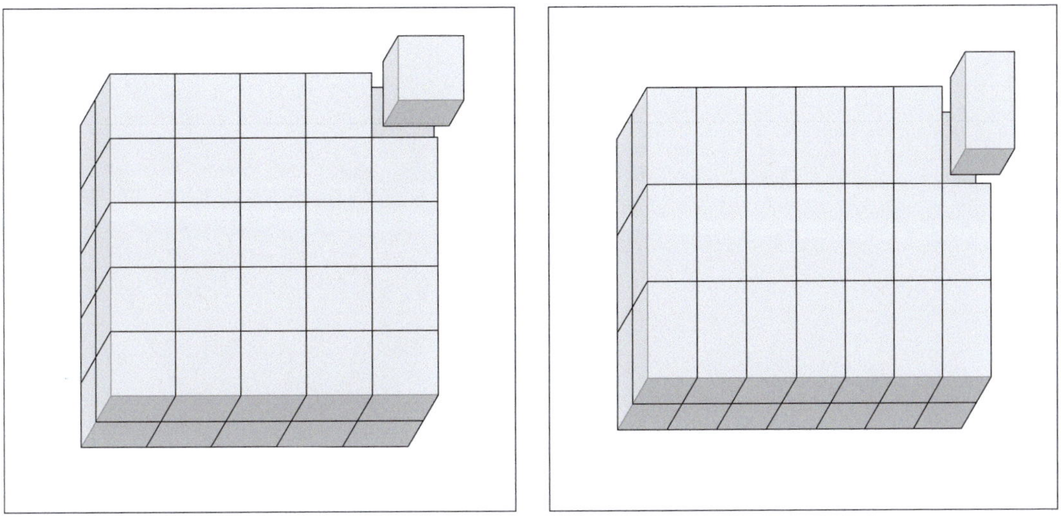

Fig. 3.23 (Left) Gapless modular packaging of space made of regular polyhedra: cubes. (Right) Gapless modular spatial packaging from polyhedra: parallelepiped or cuboid with rectangular footprint

are characterised by great constructional simplicity, and have very favourable relations between the encapsulated space volume and the surface of the enclosing components. On the other hand, they are afflicted by the disadvantage that they cannot be expanded in any direction by simple continuation of the constructional principle. Moreover, they cannot be added to a larger arrangement without residual spaces (Fig. 3.12), neither in the surface nor in the height, and allow only a limited grouping by suitable intersections and penetrations (Figs. 3.13 and 3.14, left). Even **vaulted buildings** as in Fig. 3.14, right, can be continued linearly in direction of the vault axis conforming to the system, but laterally, they cannot be added to a form coherent larger space, just as the dome.

The so-called A/V-ratio between the enveloping surface (A) and the enclosed building volume (V) is an important measure of both the economy and the energy efficiency of a building; cf. also ***Vol. 4**, Sect. 1.5.1 The Influence of the Scale on Physical Conditions*

As in ▶ *Sect. 4*, ▶ Fig. 4.95

The geometric inference from applying the addition principle in the surface is using **vertical planar enclosing walls** instead of inclined and/or vaulted ones as in the case of the above-mentioned dome-like or arch-like structures, as well as **orthogonal ground-plan geometries** (Figs. 3.23, 3.24, and 3.25). Enclosure walls between abutting individual cells are transformed, by the addition, into partition walls between two adjacent modules (Figs. 3.26 left, 3.27, and 3.28; see also Figs. 3.29, 3.30, 3.31, 3.32, 3.33, 3.34, 3.35, and 3.36). Planar rectangular walls have important advantages over curved walls:

- Neighbouring rooms can adjoin **without wasted residual areas** and can be mutually connected by means of large openings (doors or other kinds of penetration) (Figs. 3.26, right, 3.27, and 3.28).

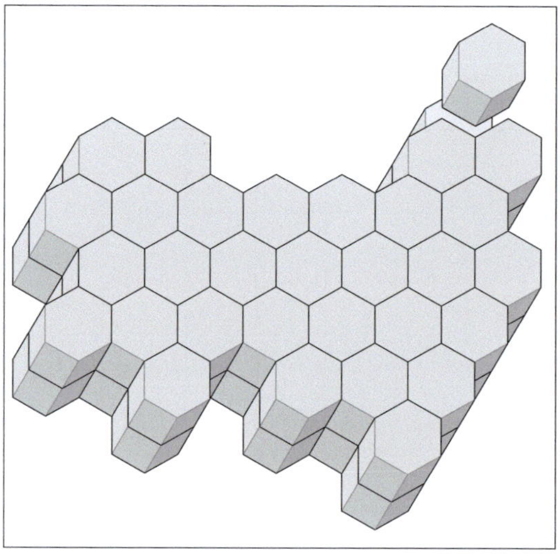

Fig. 3.24 Gapless modular spatial packaging from polyhedra: prism with hexagonal base

Fig. 3.25 Gapless modular packaging made of regular polyhedra: so-called truncated octahedra as they occur in nature with foaming (acc. to Stevens)

- The wall geometry, which is rectilinear in the ground plan, allows two **similar**, **neutrally shaped spaces** to be created on both sides of the wall (Figs. 3.26, 3.27, and 3.28, left).
- In contrast to inclined or curved room boundaries, the **floor surfaces** close to the wall can be used well, since people can stand upright there (Fig. 3.37). This is not possible with vaults starting at ground level, for example.
- The walls, in this process, are usually arranged in **orthogonal patterns**, since these lead to identical—and therefore spatially equally well usable—**right angles** on either side of a wall node (Fig. 3.38).

Fig. 3.26 (Left) Addition in ground plan of rectangular basic cells. No residual rooms. (Right) Spatial connection in one direction via narrow door openings

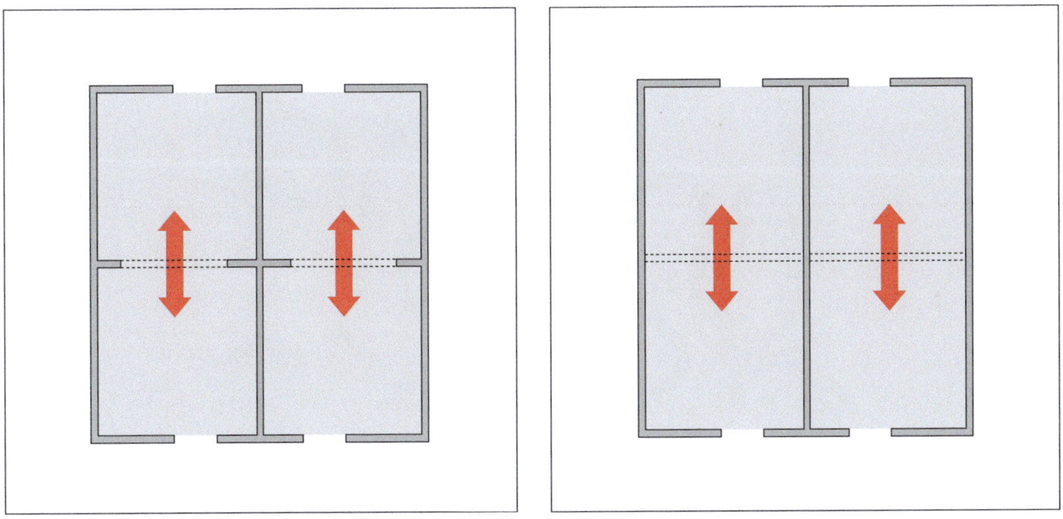

Fig. 3.27 (Left) Larger openings allow a broad spatial connection. (Right) Complete removal of a partition. Undisturbed spatial flow between individual cells

Surface tesselations made of hexagons as often found in nature (Fig. 3.24), the interfaces of which intersect at equal angles of 120° (this is related, among other things, to the force equilibrium of thin membranes) and enclose a maximum space with a minimum surface, are however remarkably rare in building construction. Modularly composed space packages made of regular polyhedra, as can be observed in some natural forms (Fig. 3.25), can be produced exclusively from **prisms** in building construction, since, for functional reasons, a flat floor surface which extends continuously between modules is always required for unrestricted

For example, cf. Stevens (1974) *Patterns in Nature*

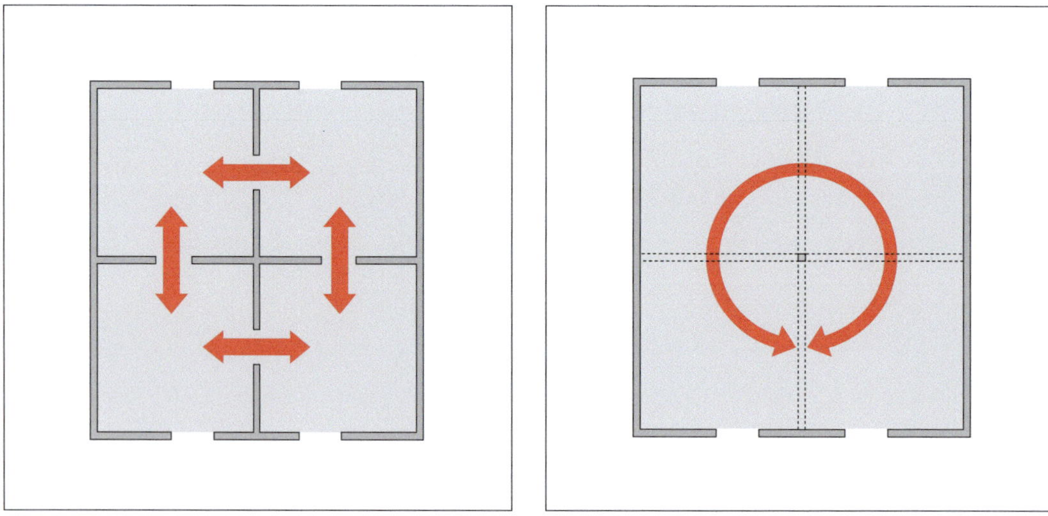

Fig. 3.28 (Left) Spatial connection of adjacent cells in two directions via door openings. (Right) Spatial interconnection of four adjacent cells with a central column. The floor or roof construction must be carried by means of at least one beam

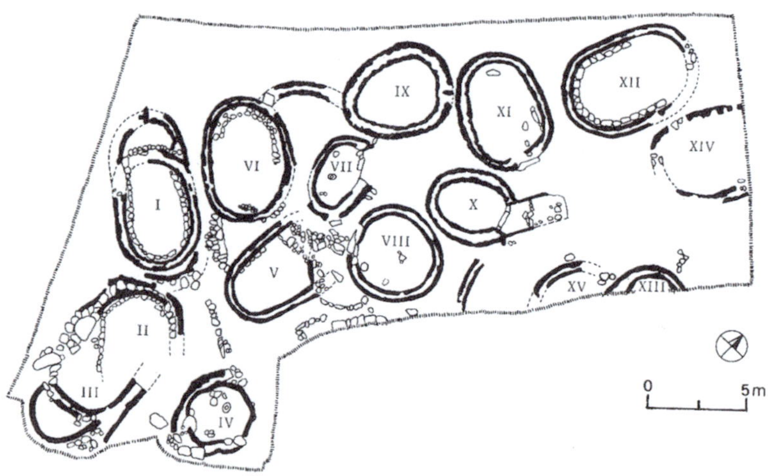

Fig. 3.29 Loose addition of individual buildings with a circular or oval shape in a stone-age settlement

usability of the interconnected room (Figs. 3.23 and 3.24). This is obviously not realisable with the truncated octahedron depicted in Fig. 3.25, for example. The footprint of the prism must, of course, be liable to be tesselated. For the sake of simplicity and due to the practical advantages of the right angle (see above), almost only **cuboids**, i.e. prisms with a rectangular or square base area, are used in multistorey buildings. Other conceivable space-filling polyhedral groupings are useless for purposes of building construction, in particular because of geometric and functional restrictions to ensure easy accessibility.

According to Stevens (1974), there are only 22 conceivable gapless space packages of regular and semi-regular polyhedral. They occur in nature for example in foams and crystal lattices

Fig. 3.30 Grouping of trulli in Apulia

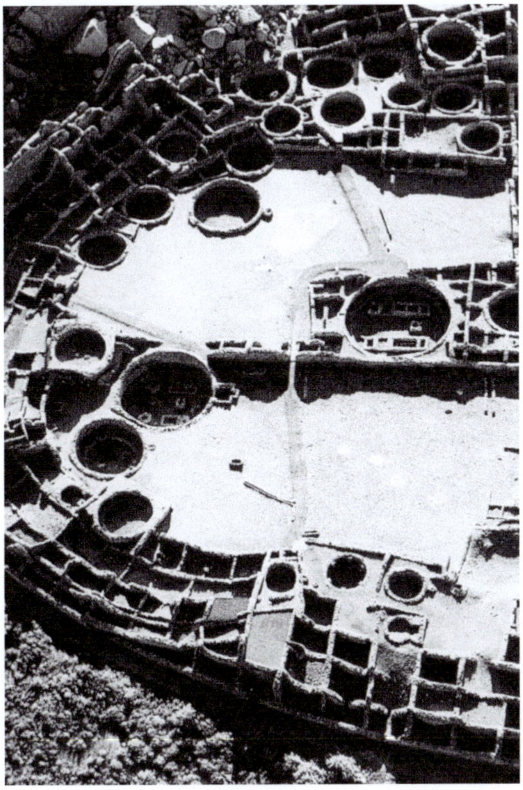

Fig. 3.31 Combination of orthogonal and circular basic forms in a pre-Columbian settlement in America

A consequence of creating a basic structural cell from a configuration of planar surfaces is the emergence of a **corner** or **edge** between two abutting planes. There may also be an edge between the surrounding element and the coverage—a fascia, eaves, or a roof verge. In mechanical respect (impact sensitivity), as well as in terms of building physics and construction,

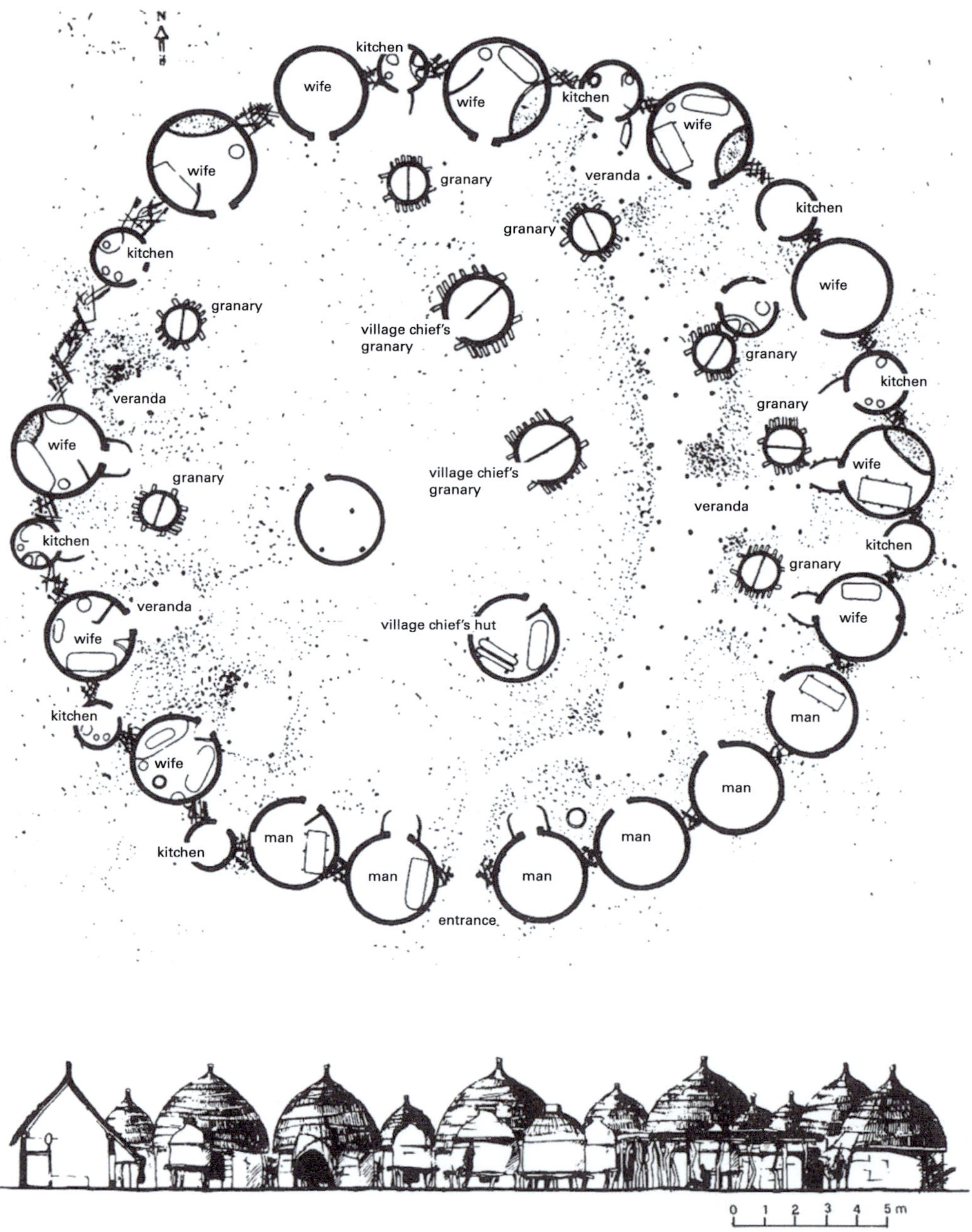

Fig. 3.32 A circular settlement structure from round individual cottages near a village in Cameroon. Most of the buildings are grouped linearly and thereby form a kind of wall. In the interior, some common storage buildings and the huts of the tribal chief are arranged freely

special conditions arise at the corner, which deserve special attention in conceptual and constructional design.

***Vol. 1**, Sect. 26.9. Realising the Force-Transmission Function Within the Element—Structural Principle of the Component*

In the simplest embodiment, the surrounding element consists of a supporting wall, in which the carrying and enveloping function is fulfilled by the same element. In principle, though, all constructional variants of planar surface components are

Fig. 3.33 Grouping of adobe houses in Egypt. Simple cuboidal designs that can be attached to one another without gaps

1 street
2 courtyard
3 salon
4 kitchen
5 latrine
6 alcove
7 stable
8 storage room
9 cistern
10 vestibule

Fig. 3.34 Nested, densely tesselated floor plan of rectangular basic cells (Cieza, Spain)

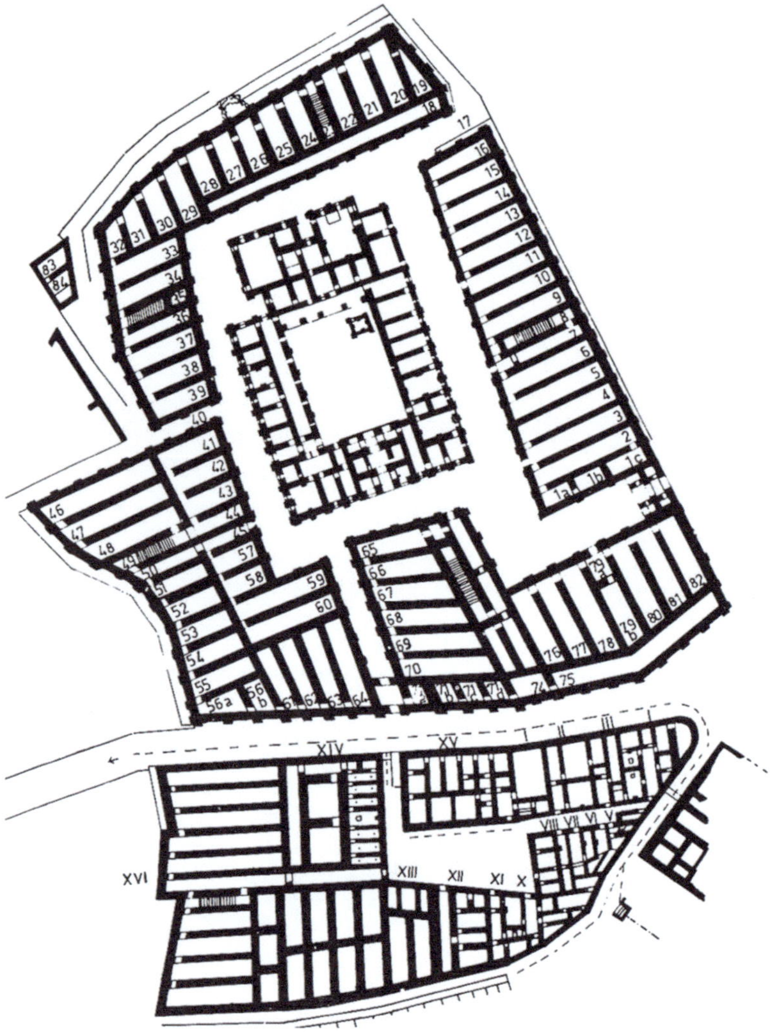

Fig. 3.35 Linear grouping of narrow rectangular basic cells into string-like sequences. The length of the narrow sides is dictated by the limited bridgeable span (main temple in Hattusa, Hittite empire)

conceivable. However, it is also possible to use a **framework** or **truss** made of columns or posts and beams or girders which are located within a plane (Figs. 3.20, 3.21, and 3.22). The actual two-dimensional space enclosure, then, is a **non-load-bearing wall** which, as mentioned, does not necessarily have to be at the same location as the load-bearing component.

3.1.6.2 The Covering

The upper space termination is formed by a separate element in the basic cell made from vertical enclosures: a coverage. With the single-storey basic cell, only a few boundary conditions for the shape of the coverage can be derived from the use of the interior space. At a height which is removed from the direct range of the user, a wide variety of designs are conceivable: plane horizontal or inclined, but also, for example, curved vault or

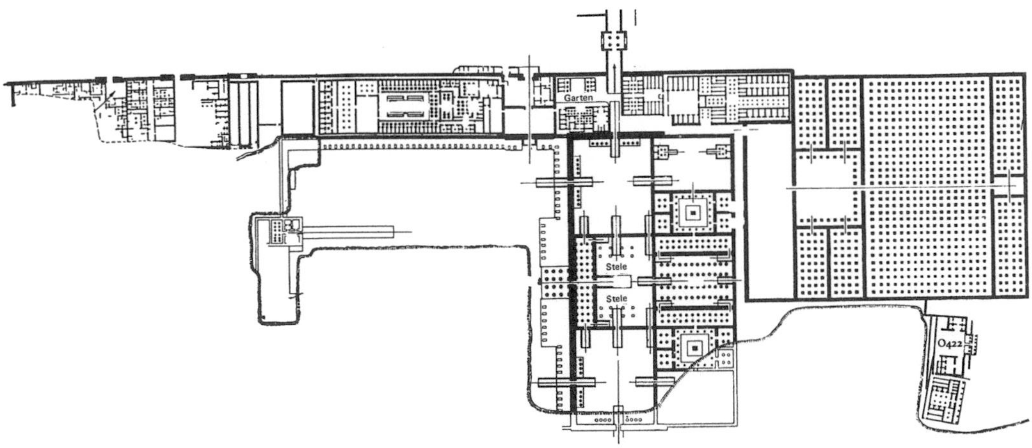

■ **Fig. 3.36** Early example of creating large contiguous spaces by adding a square or rectangular basic module (here, intercolumnar space) in two directions (palace complex in Tell el Amarna, Egypt)

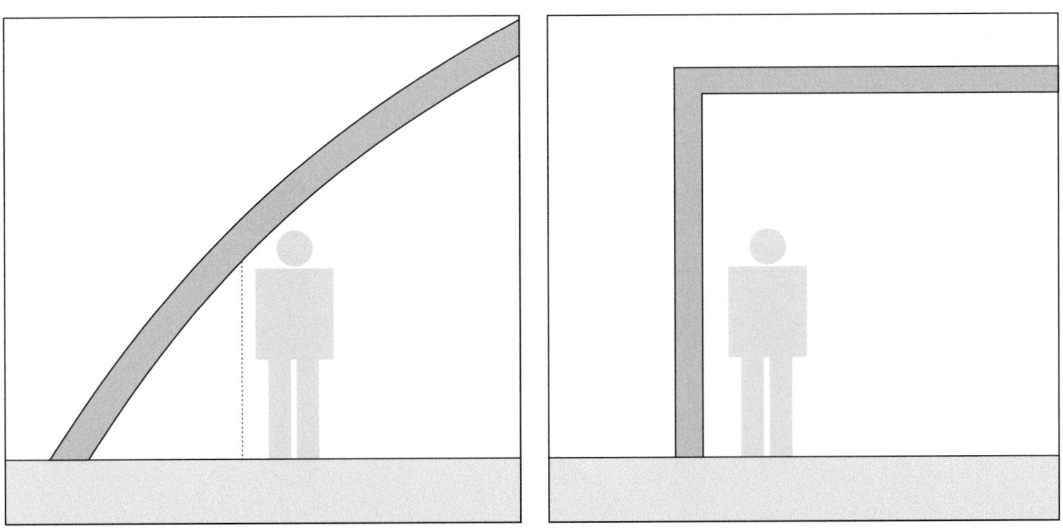

■ **Fig. 3.37** (Left) Restricted use of the areas adjacent to the springing of inclined or arched enclosing walls. (Right) Better use of space with vertical enclosures

dome-shaped. Formal aspects, questions of load transfer, such as the creation of a large structural depth on inclined roofs, or issues of weather protection, such as the ability to quickly evacuate precipitation water, play a decisive role; likewise, questions of overall design, room design, as well as daylighting.

The fact that, unlike vertical enclosing walls, the dominant loads acting within the coverage are **perpendicular** to the component plane, which leads to a material-inefficient **bending stress**, particularly in the case of flat floor structures, is important for the structural design of the coverage. Among other factors, this circumstance is responsible for the fact that other materials or constructions are frequently used in coverages than with enclosing walls, and therefore a clear,

Cf. the comparison in ***Vol. 1**, Sect. 26.3 Comparison Between Bending Moments/Shear Forces and Axial Stresses, or Accordingly Membrane Stresses*

3

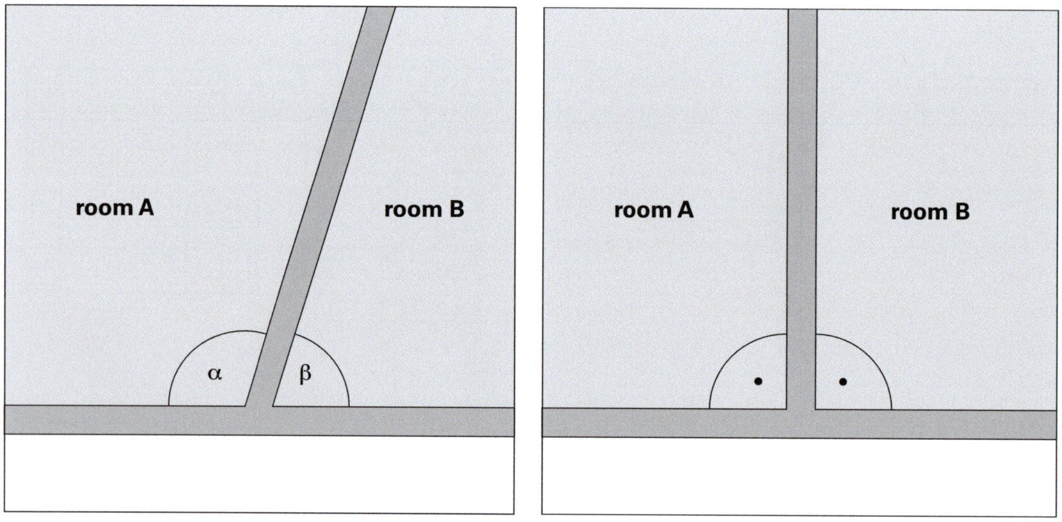

Fig. 3.38 (Left) Oblique partition connection to a straight wall leads to various, not equally usable space layouts. (Right) Perpendicular partition wall connection allows the same space layouts on both sides

Fig. 3.39 (Left) Stackability with a flat ceiling without any residual space. (Right) Stackability with vaulted ceiling not without residual spaces **R**

even formal distinction between the two elements of the base cell, i.e. between the **wall** and the **floor**, or the **wall** and the **roof**, took place. A good example of this is traditional masonry construction, where the roofs and floors conventionally consisted almost always of wooden beams (Fig. 3.2).

In addition to their coupling in the plane, it may also be necessary to stack basic structural cells vertically, as already mentioned, for economic reasons. The coverage of the lower space serves as the floor of the adjacent one (Fig. 3.39, left). This requirement to generate a **level**, **accessible**, i.e. **horizontal**, surface for the upper storey can, in principle, be fulfilled by a vault with a horizontally executed filling or mounting on the upper side (◘ Fig. 3.39, right). The easiest way to do this, however, is to use a **flat floor element** from the outset.

Cf. also the example in *Sect. 4*, ► ◘ Fig. 4.109

In addition to flat coverages, **inclined roofs** must also be taken into account in the context of areal structural and enveloping elements, in which an inclination is not only unobjectionable—the accessibility of a roof is only of secondary importance—but is indeed desired, because:

- **statically usable depth** is obtained, which can be used for load-carrying purposes. In certain roof structures, such as rafter roofs, the slope can be used to convert a part of the statically unfavourable **bending stress** on the roof beams (rafters) into a statically more efficient **longitudinal force**.
- the technical impossibility of executing a roof surface which is really **tight against rainwater** has traditionally favoured a roof pitch as steep as possible.
- in simple buildings, an inclined roof was often used to extract the **smoke** from the domestic stove. An inclination also promotes the **ventilation** of a roof structure, as long as suitable continuous air spaces permit this.
- occasionally, pitched roofs are used for the **daylighting** of interior spaces (example: shed roofs).

3.1.6.3 The Basic Cell Composed of Enclosure and Covering

The familiar configuration of vertical and horizontal or more or less flat inclined, essentially planar components is not only a consequence of the fundamental elementary planning considerations mentioned above, but also of the principles of **manufacturing technology** as well as of the practice of **traditional construction methods**, in which straight rectilinear solid walls and rod-shaped straight steel or wood components are processed.

Chap. IV Construction Modes

Since these planar basic elements still represent the essential components of a building, even in modern building modes, they have been examined in detail in their individual load-bearing behaviour in ***Volume 1***, *Sect. 26*. Together with some selected curved structures, they will now be grouped together in their essential variants and combinations to form elementary structures, i.e. basic structural cells, and classified into morphological groups.

Vol. 1, *Sect. 26 Force Transmission*

► *Sect. 4.1 Overview of Elementary Structures*

Planar surface elements, i.e. both wall and floor components, can in principle assume all constructional designs as described in ***Volume 1***, *Sect. 26*. The structural variants examined there can, in principle, also be transferred to curved surface components, whereas a load-carrying behaviour differing from the planar variant can usually be expected. The conceivable combination possibilities of the different variants of encasement and coverage in the elementary structural cell are very numerous and can—and will—not be exhaustively investigated in this context. The respective load-bearing behaviour of planar components with specific

Vol. 1, *Sect. 26.9 Realising the Force-Transmission Function Within the Element—Structural Principle of the Component*; cf. also the examples of the variants in ◘ Figs. 3.26, 3.27, and 3.28

constructional compositions is dealt with in the mentioned section, so that the reader can transfer the considerations of the following sections analogously to similar cases.

3.1.7 The Structural Tasks

For the breakdown of loads into load components, cf. ***Vol. 1***, *Sect. 26.2.2 External Load*, Figs. 26.12, 26.13, 26.14, 26.15, 26.16, and 26.17

The essential static task of a load-bearing structure is to transfer loads into the ground. These can be oriented differently, so that they are divided into two basic components, namely in:

- **vertical** loads and
- **horizontal** loads.

In this context, the structural components of the enclosure and the coverage act conjointly. They transfer the loads within a basic structural cell and pass them on either to the foundation, and thus to the ground, or to adjacent basic cells, which in turn are stable and capable of absorbing these loads.

For a safe fulfilment of this static task, sufficient:

- **load-carrying capacity** and
- **stiffness**

of the components and their connections need to be ensured.

3.1.7.1 Stability

► *Sect. 5.1.2 Requirements*

The load-bearing capacity guarantees the structural stability of the basic constructional cell and, thus, of the entire structure under the action of:

- **usage loads** such as dead weight, live, wind, and snow loads, which must be increased with a **safety factor** γ_f, as well as under:
- **exceptional loads** such as from impact, earthquake, fire.

The associated requirements are defined by the safety level stipulated in the code.

3.1.7.2 Serviceability

The **stiffness** of the components and of their connections influences the type and size of the deformations of the structures and thus their serviceability. It must allow bearing usage loads with **limited deformations** so that the user's activities can unfold fully within the basic cell and thus throughout the building. Furthermore, the stiffness of the load-bearing structure ensures the functionality and integrity of the non-load-bearing envelope construction as well as the mechanical equipment of the building.

3.2 Load Transfer

The load transfer in the direction of the bearings, where loads must ultimately end up, is always associated with a **force deflection**, as is typical for building construction and has been discussed above. The way in which this load deflection takes place within the load-bearing structure, as well as, in particular, the directions or load paths along which the force is transmitted to the bearings, profoundly affects not only the construction, but also the entire configuration and geometry of the building. The governing static laws and the constructional and geometrical consequences of the load transfer will now be discussed in more detail.

▶ *Sect. 3.1.6 The Elements of the Structural Cell*

3.2.1 One-Way and Two-Way Load Transfer

There is a basic difference between a **one-way** and a **two-way load transfer**. One-way or uniaxial load transfer is characteristic of so-called **directional** structural systems; two-way or biaxial load transfer of **non-directional.** This basic classification of load-bearing structures based on their particular type of load transfer will also underlie the following section.

The respective peculiarities of the load-bearing effect in both cases will be explained below using the example of longitudinal and transverse bending of an edge-supported generic flat component under load perpendicular to its surface. The statements on this bending-stressed system can, however, also be transferred to axially compressive- and tensile-stressed systems.

▶ *Sect. 4 Types*

3.2.1.1 Load-Carrying Behaviour

An imaginary, on two sides linearly supported layer consisting of individual parallel, mutually unbonded strips (◘ Figs. 3.40, left, 3.41, 3.42, and 3.43) illustrates a purely **one-way** transfer of perpendicular loads to the opposing linear supports: i.e. the already mentioned **redirection of the force** of the external load, which initially acts at any arbitrary location, to the two spatially localised supports, where the force is brought into equilibrium by corresponding reactions.

key

- $\mathbf{B_m}$: bending in the middle
- $\mathbf{B_e}$: bending at the edges
- **TD**: transverse distribution of loads
- **TB**: transverse bending
- $\mathbf{f_m}$: deformation in the middle
- $\mathbf{f_e}$: deformation at the edges

If the area is subjected to a point load, only the strip directly affected by the point load deforms according to the principle of one-way load transfer. If, however, the strips are coupled to one another on the long side, for example by means

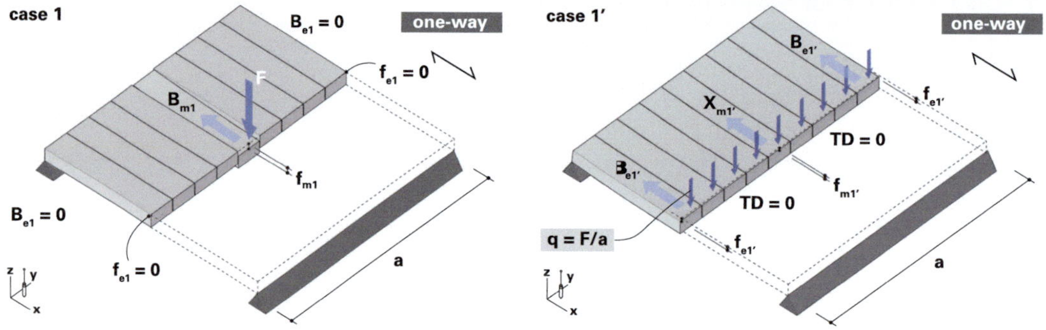

Fig. 3.40 (Left) Sectioned functional model of a directional system (half span) composed of independent strips. One-way transfer of the single load **F** without cross-distribution of loads in the case of non-cooperating parallel strips. The bending deformation $\mathbf{f_{m1}}$ in the middle is assumed to be 1. Since the edge strips are not affected by the single load because of the lack of cooperation, the edge deformation $\mathbf{f_{e1}}$ is equal to zero ($\mathbf{f_{m1} = 1; f_{e1} = 0}$). (Right) The same system as on the left, but subjected to a linear load **q** which in total is equal in magnitude to **F** for comparison purposes. As a result of the even distribution of the load along the length **a**, the central deformation $\mathbf{f_{m1}}'$ and the edge deformation $\mathbf{f_{e1}}'$ are the same. They are smaller in their magnitude than in case **1** because of the load distribution on many individual strips ($\mathbf{f_{m1}' = 1/9; f_{e1}' = 1/9}$)

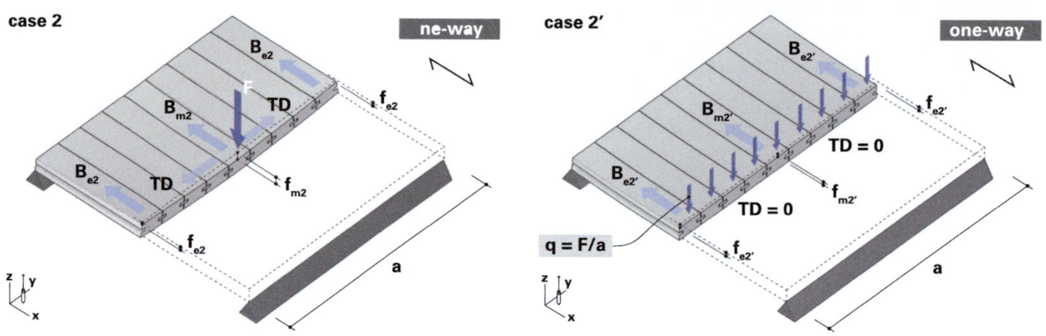

Fig. 3.41 (Left) One-way load transfer with transverse distribution (**TD**) of the acting single-load **F** on adjacent strips by shear connection. The coupling between strips is assumed to be hinged, i.e. no bending moments are transmitted there (only shear forces), so that only transverse distribution occurs, but no transverse bending. The deformation in the centre $\mathbf{f_{m2}}$ is greater than the edge deformation $\mathbf{f_{e2}}$ since the effect of the central force **F** decreases towards the edge ($\mathbf{f_{m2} = 1/3; f_{e2} = 1/30}$). (Right) Same system as left, but subjected to a linear load **q**. The strip family, assumed here to be interconnected with a shear-resistant joint as in case **2** on the left, behaves as lacking a shear-resistant bond, i.e, as in case **1′** above ($\mathbf{f_{m2}' = 1/9; f_{e2}' = 1/9}$)

of an (hinged) tongue-and-groove connection (Fig. 3.41, left), the adjacent elements are also forced to deform, and, hence, to bear load, by the stressed strip. The latter is thereby relieved; the point load is thus distributed sideways (in → **y**). Although this is still referred to as a one-way load transfer, in this case the load is also distributed transversely, i.e. there is a **load cross-distribution**. The effect can be described as a forced participation of adjacent components not directly subjected to load. This allows for a more efficient utilisation of the (always) available overall cross-section. The further away the strips are from the applied point load, the less load they take on and the less they deform.

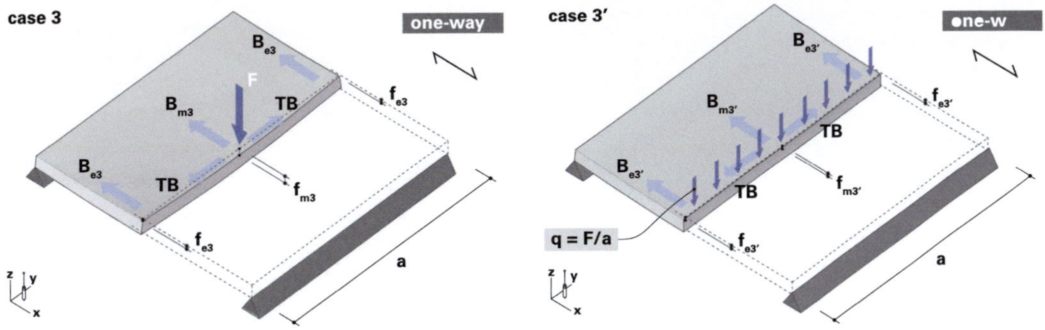

Fig. 3.42 (Left) One-way load transfer with transverse bending (**TB**) with homogeneous cross-section. The deformation $\mathbf{f_{m3}}$ in the centre decreases in comparison to the previous cases as a result of the bending stiffness of the cross-section in the transverse direction (i.e. → **y**), which opposes the transverse bending stress. Since the cross-section is inherently more rigid than in case **2**, the edge deformation $\mathbf{f_{e3}}$ is, however, greater than that there ($\mathbf{f_{m3} = 1/7}$; $\mathbf{f_{e3} = 1/10}$). (Right) Bending $\mathbf{B_{m/e3}}'$ in the spanning direction → **x** leads to a compressive stress on the upper cross-sectional zone, which consequently also expands transversely thereto (in → **y**). As a result, a slight negative bending of the cross-section results, despite the uniformly distributed load **q**, which manifests itself in a smaller centre deformation $\mathbf{f_{m3}}'$ than the edge deformation $\mathbf{f_{e3}}'$ ($\mathbf{f_{m3}' < 1/9}$; $\mathbf{f_{e3}' > 1/8}$)

The same effect of load cross-distribution occurs for all loads that are **not evenly distributed** across the strip alignment (in → y).

If a **homogeneous plate cross-section** (Fig. 3.42, left) is created instead of the shear-distributing, but otherwise hinged connections between separate strips, the material is flexed in this direction due to the bending stiffness thus activated in → **y**. This results in a recognisable continuous **bending deformation** in → **y**, **transversely** to the original spanning direction → **x**, an indication of the presence of a **transverse bending**. Similar to the hinged strips (case **2**), the deformation will be the smaller, the farther the contemplated plate region is located away from the loaded point, the smallest, naturally, being at the edge. The amount of deformation, both in the middle as well as on the edge, decreases compared to cases **1** and **2**.

__Vol. 1__, Sect. 26.9.5 Element Composed of Ribs Oriented Two- or Multiple-Way > Bending

If, in contrast to the cases discussed above, in which the component is placed on two opposing supports, a further two supports are introduced following, and additionally supporting the lateral, previously unsupported, edges (Fig. 3.43, left), a **two-way load transfer** with **two equivalent spanning directions** results. Since a deformation at the edges ($\mathbf{f_e}$) is now prevented by the bearing, the deformation at the centre ($\mathbf{f_c}$) is further reduced.

__Vol. 1__, Sect. 26.9.5.1 Element Ribbed Two-Way on Linear Supports and *Sect. 26.9.5.2 Element Ribbed Two-Way on Punctual Supports*

In the two cases without a transverse bending (cases **1′** and **2′**, Figs. 3.40, right and 3.41, right), a distributed linear load results in a continuous deflection with the same centre and edge deformation. If the cross-section is homogeneous, it bends slightly convexly upwards in → **y** (Fig. 3.42, right). A distributed load **q**, in the case of a two-way load transfer with otherwise unaltered force magnitude, yields the effect of reduced centre deformation (Fig. 3.43, right).

Cf. the explanatory statement in the caption to Fig. 3.42, right

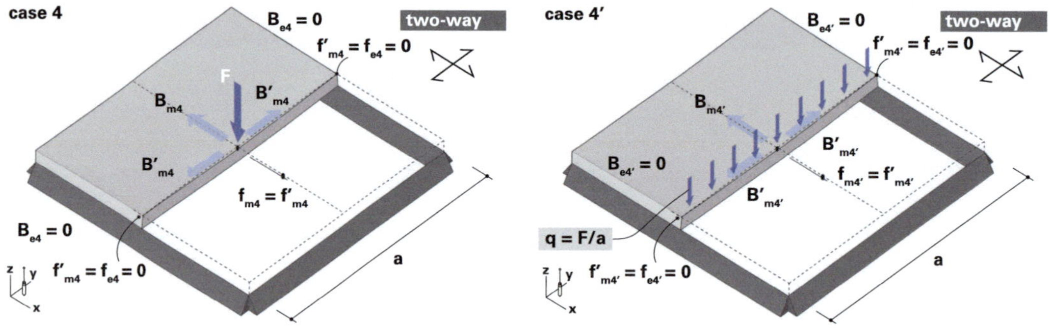

Fig. 3.43 (Left) Two-way load transfer. The deformation in the centre $\mathbf{f_{m4}}$ is the smallest in comparison with the other cases shown. The edge deformations $\mathbf{f_{e4}}$ are equal to zero because of the linear edge support ($\mathbf{f_{m4} = 1/15}$; $\mathbf{f_{e4} = 0}$). (Right) Same system as left, but under a linear load **q**. The centre deformation $\mathbf{f_{m4}'}$ is smaller than in case **4** on the left. If the load is uniformly distributed over the entire surface = **a · a**, in terms of magnitude = $\mathbf{F/a^2}$ (not shown here), the centre deformation $\mathbf{f_{m4}'}$ further decreases to 1/42 ($\mathbf{f_{m4}' = 1/27}$; $\mathbf{f_{e4}' = 0}$)

Cf. next section

Under certain conditions, longitudinal and transverse bending have the same order of magnitude in the case of two-way load transfer: This is referred to as a **non-directional** structural system.

It is possible to see on the diagrams that the strips **twist** (case **2**), or that the imaginary plate strips do so (cases **3**, **4**, and **4′**): This means two parallel sections are twisted relative to one another about the axis of the strip under consideration. Due to the torsional stiffness of the material, the load thus meets additional resistance.

For the cooperation of plate strips of slabs spanning two-way, cf. ***Vol. 1**, Sect. 26.9.1.1 Slab on Four Linear Supports*

The described effects of **cross-distribution**, **transverse bending**, or **two-way load transfer** result in a more favourable load-bearing behaviour compared to the first case, which makes effective use of the material particularly in the case of non-uniform loads.

3.2.2 Influences of Load Transfer on the Geometry of The Structural Cell

Issues of load transfer affect the footprint design of the basic cell made up from enclosure and coverage as well as the geometric organisation of skeins of bars of a coverage. This will be discussed in the next sections. The following considerations each assume a **direct load transfer** to the supports, in which the loads are transferred directly, i.e. vertically, from the support to the foundation (or the next lower primary support structure) (see Fig. 3.99, left). Although they refer to bending-stressed flat coverages—slabs, sets of beams—they can also be analogously applied to one-way spanning, axially stressed coverages such as arches or ropes.

With indirect load transfer, the load is first transferred from the support via another beam in the transverse direction (e.g. via a carrier beam); cf. Fig. 3.99, right

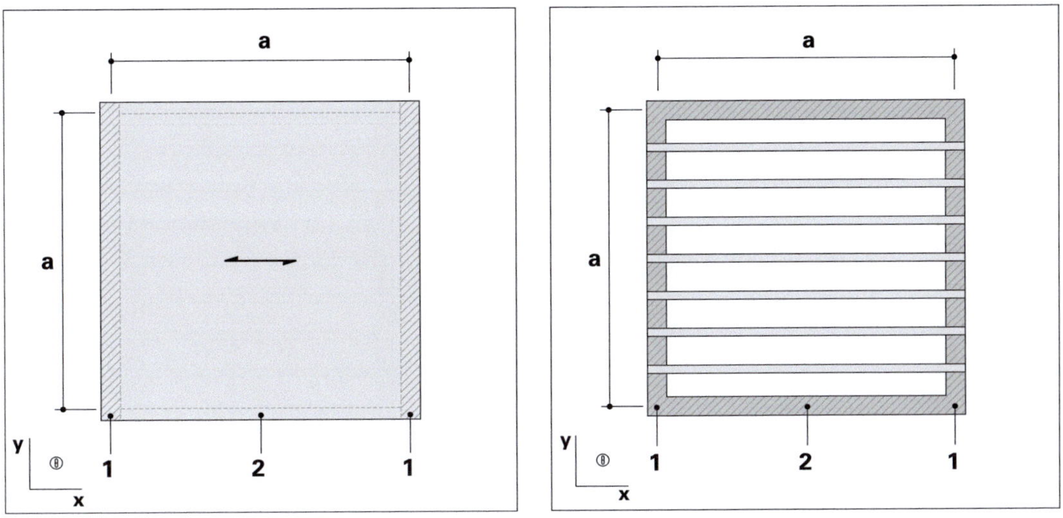

Fig. 3.44 (Left) Room cell covered with a slab, sides **a** equal. (Right) Room cell covered with bars, side lengths **a** equal. (**1**) load-bearing wall, support, (**2**) non-load-bearing wall, no support; if necessary, bracing function for load-bearing wall **1**

3.2.2.1 Choice of the Spanning Direction for One-Way Load Transfer

For reasons of expediency, the **smallest** possible span is first chosen for all structures of coverages, unless no other considerations advise against. If two optional spanning directions are available as in the case of a walled rectangle with two possible opposed pairs of supports, it is, of course, advisable to choose the shorter one from an economic point of view.

For this reason, when designing systems with one-way load transfer, square footprint proportions and the resulting equal spans in two directions are fundamentally avoided from a structural point of view—unless there are other reasons for such a solution, as noted. Instead of square, the quadrangles are conventionally designed as rectangles with a **short spanning direction** (Figs. 3.44, 3.45, and 3.46). It is then—at least—possible to create rooms equally well usable as those with square footprint, but with a significantly reduced span.[3] This principle applies to both one-way spanning slabs as well as to skeins of bars. A practical example for spatial cells with coverage from bar skeins is conventional masonry construction with wooden beam floors (Figs. 3.47 and 3.48).

3 The direct comparison between the shown cases 1, 2, and 3 results in beam depths of 100%, 64%, and 156%, assuming the same base area of the cell and approximately the represented room proportions, provided the weight of the beams is neglected. With increasing scale, the conditions in cases 1 and 3 are even more unfavourable, taking into account the dead weight.

3

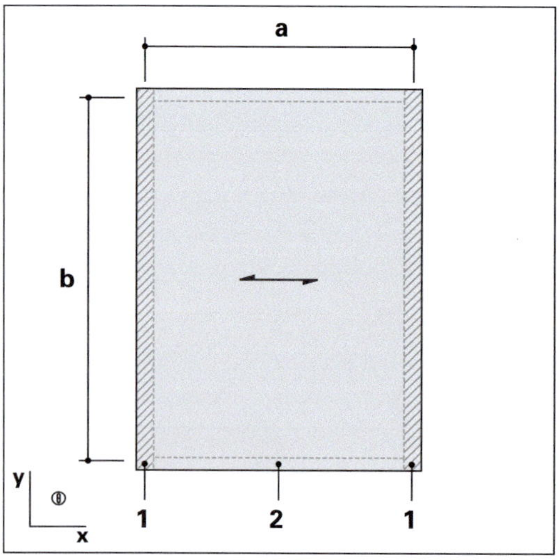

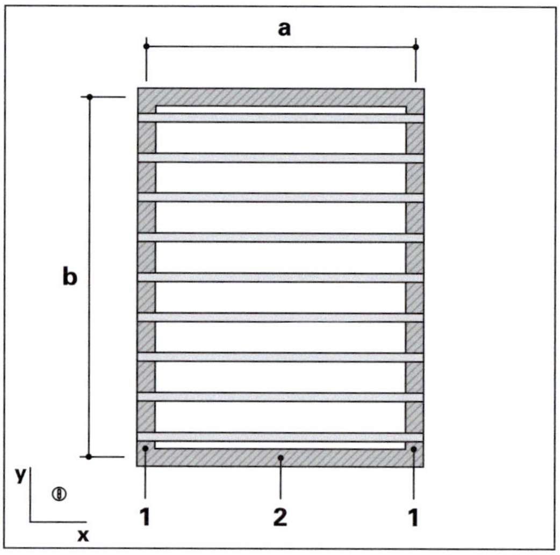

Fig. 3.45 (Left) Rectangular space cell with a short span **a**, which is covered with a plate; **a** is selected here as the spanning direction of the covering. (Right) Rectangular space cell with a short span **a**, which is covered with a bar skein; **a** is chosen here as the spanning direction of the covering

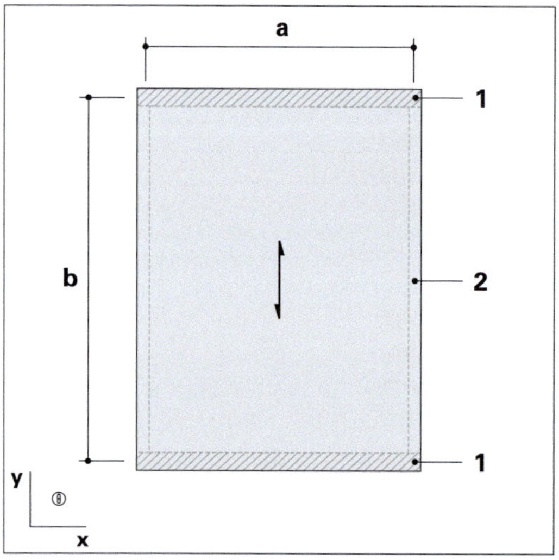

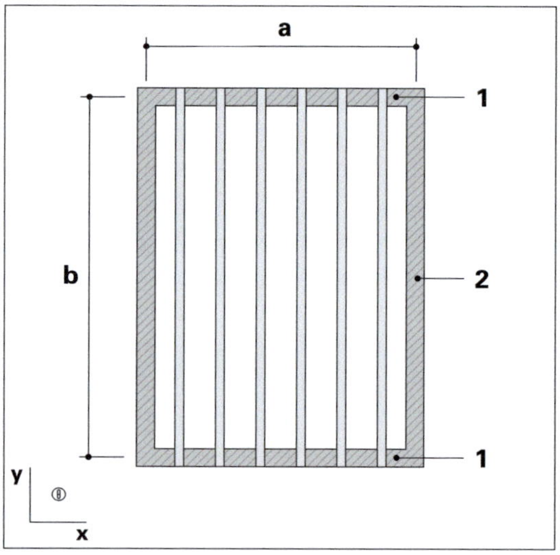

Fig. 3.46 (Left) Rectangular space cell covered with a plate; here, the long side **b** is selected as the spanning direction of the covering. This arrangement contradicts the principle of structural efficiency and can only be justified by other requirements. (Right) Rectangular space cell spanned with a bar skein; here, the long side **b** is selected as the spanning direction of the covering, as on the left

In order to minimise spans, it is also recommended to arrange beam positions **perpendicular** to the parallel supports (Fig. 3.50 left). Angled bar skeins have to bridge larger spans and, consequently, must be dimensioned stronger without resulting benefit (Fig. 3.50, right). In the case of plates, however, the load transfer self-acting results from the shortest possible span (Figs. 3.49 and 3.50).

Fig. 3.47 Conventional masonry construction made up of several added room cells, which are covered with bar skeins—i.e. wooden-beam floors

3.2.2.2 Interaction Between Span, Structural Depth, and Ground-Plan Geometry

The mutual dependence between the **span** of a coverage spanning one-way, i.e. the distance between the two supports on the enclosure, and its **structural depth**, raises further elementary, system-immanent questions as soon as the ideal-type principle of equally distant parallel straight bearings is abandoned (Fig. 3.51, case **1**). In this latter case, a maximum of structural efficiency can be achieved with the same structural depth and construction height. Other solutions are, however, also conceivable in principle, as explained below.

Adjacent Bays with Different Spans

If, on the other hand, a continuous plate or beam skein is stretched over multiple adjacent fields with *different* spans (Fig. 3.51, cases **2** to **4**), it appears initially obvious to adapt the structural depths of the plates or beams to the actual spans from case to case. The coverage can be arranged flush at the bottom (case **2**) or flush on top (case **3**). The latter variant is particularly suitable for forming continuous topmost roof or floor surfaces. If an invariable design depth is desired, the largest span is of course determinant, so that the beams are inevitably oversized over the shorter spans (case **4**).

► *Sect. 4.2.1.1.3 Extendibility* as well as *Sect. 4.2.1.2.1Extendibility*

3

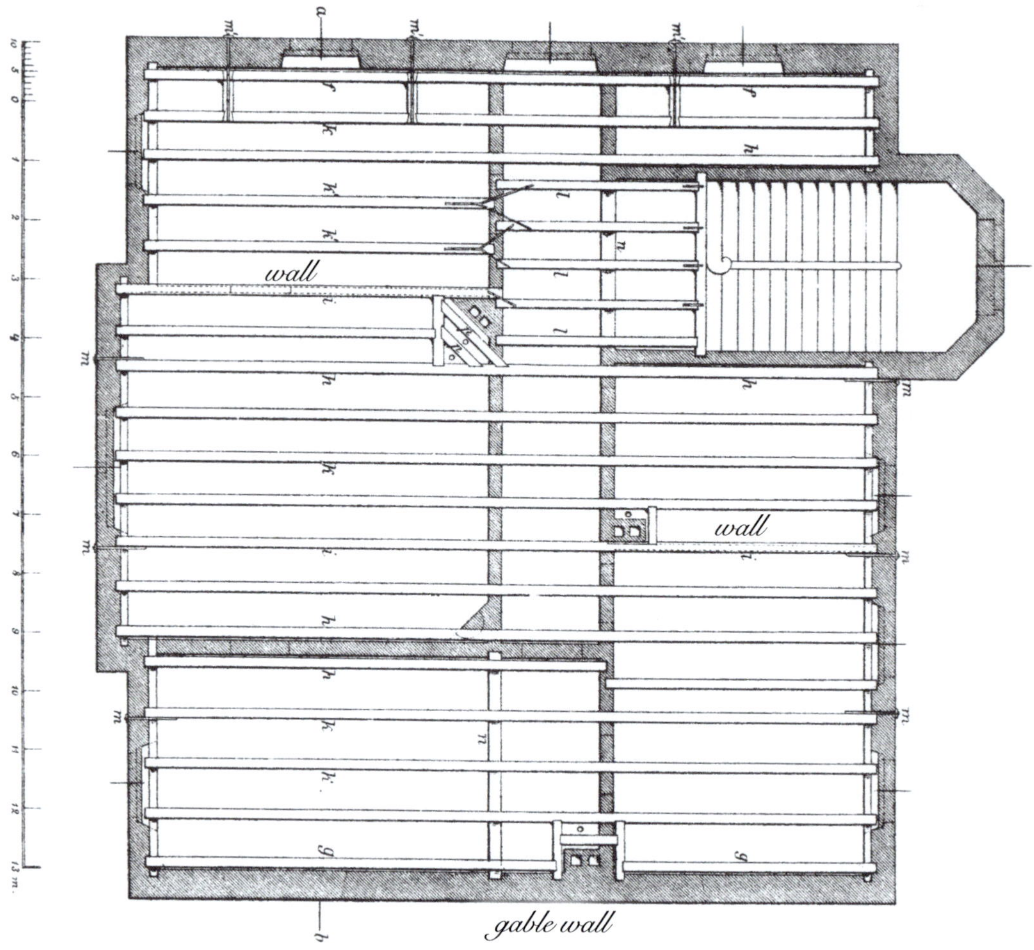

Fig. 3.48 Beam arrangement in a traditional masonry building. It can be seen that beams are usually—but not consistently—oriented over the shorter span. In this case non-structural considerations, for example from a streamlined structural organisation, can be the decisive factor

Non-Parallel Supports

Non-parallel supports necessarily result in changing spans of the coverage (Fig. 3.52), so that it must be decided whether the structural depths are correspondingly staggered or kept constant. In the case of slabs, the latter would mean either to continuously vary their thickness or to tier it step by step. If the thickness of the slab remains the same, it is also possible in concrete construction to react to the different stresses resulting from varying spans by means of adapted reinforcement. In the case of bar skeins, it is necessary to decide whether the beam depths are staggered, or all beams are given the same depth and, as a result, are oversized except for the one covering the largest span.

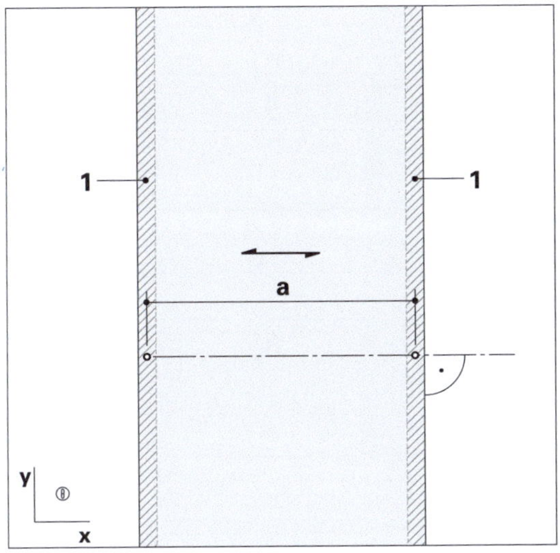

Fig. 3.49 Plate spanning one-way: load transfer perpendicular with respect to the supports: The load is always discharged over the shortest possible span **a**

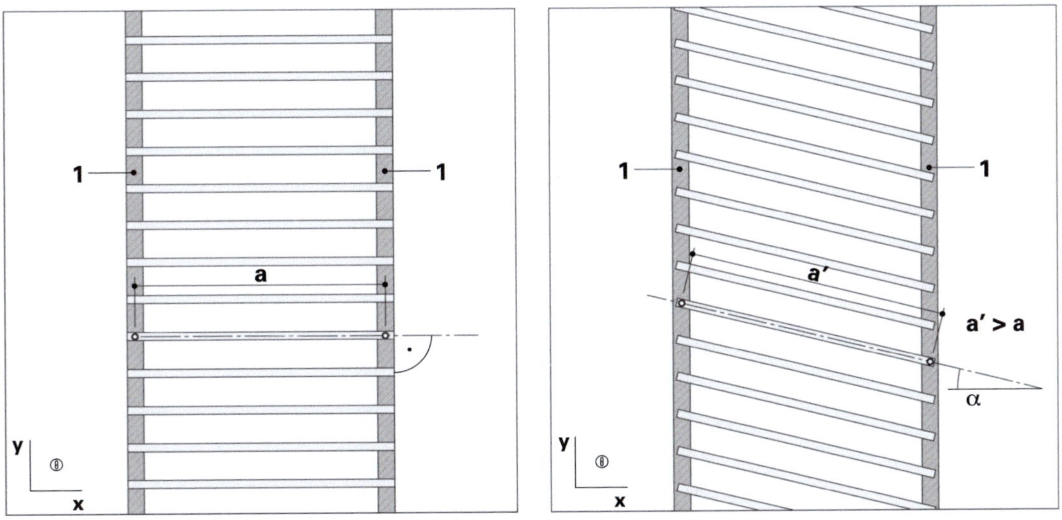

Fig. 3.50 (Left) Perpendicular orientation of a beam layer with respect to the supports: shortest possible span **a**. (Right) Oblique orientation of the beam layer with respect to the support: increased span **a'** with identical space width as measured between walls

Concentric Curved Supports

If beams are fanned keeping a constant span, i.e. are arranged radially over concentrically circular or comparably curved supports (Figs. 3.53 and 3.54, left), the load (linear load **q**) changes over the length of the beam due to the approximately trapezoidal load-infeed range (Figs. 3.53, right and 3.54, left). The effect increases with increasing curvature (Fig. 3.54, left). There is the choice between continuous structural depth of the beam (with ensuing over-dimensioning) or

▶ *Sect. 3.3.6.1 Bar Grillages*

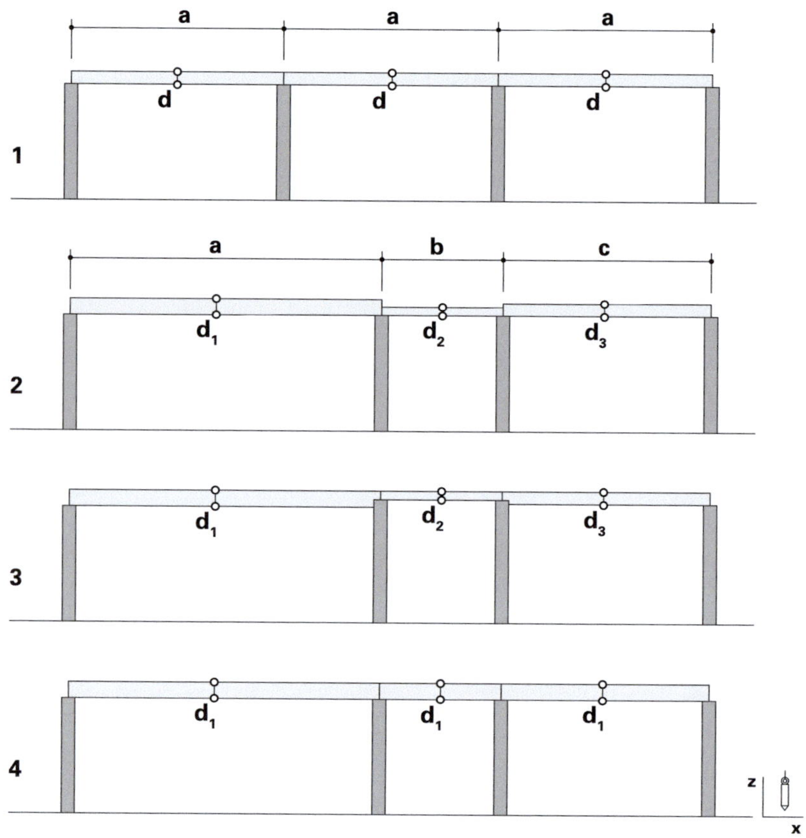

Fig. 3.51 Possible design variants for the construction depth, with three adjacent bays of different spans: **case 1**—three equal bays **a** are covered with a construction with a **constant** structural beam depth **d** determined according to static conditions. This allows for maximum structural efficiency of the covering under the given conditions. **case 2**—**three** different bays **a**, **b**, and **c** are assigned to three beam depth $\mathbf{d_1}$, $\mathbf{d_2}$, and $\mathbf{d_3}$, each of which is dimensioned according to statics. Beams are flush at support height. **case 3**—**statically** calculated beam depths as above; beams flush on the top to create a stepless roof or floor level. **case 4**—uniform beam **heights** = $\mathbf{h_1}$, dimensioned according to the largest span **a**; beams over **b** and **c** are over-dimensioned

load-conformal wedge-shaped adjustment of its structural depth. The same applies to a plate (Fig. 3.53, left).

Parallel Bar Skein on Curved Supports

Figure 3.54, right, shows the geometric conditions prevalent in this case, which inevitably result in changing spans with parallel bar positions. Similar considerations as in the cases discussed above are also applicable to this configuration. Slabs, instead, always carry loads self-acting in the direction of the shortest span, i.e. always radially (as on Fig. 3.53, left).

With issues of this kind, a large number of non-construction-related design factors always play a role. These sometimes can also result in cases where the mentioned structural disadvantages are deliberately accepted. All border conditions must be carefully considered by the designer.

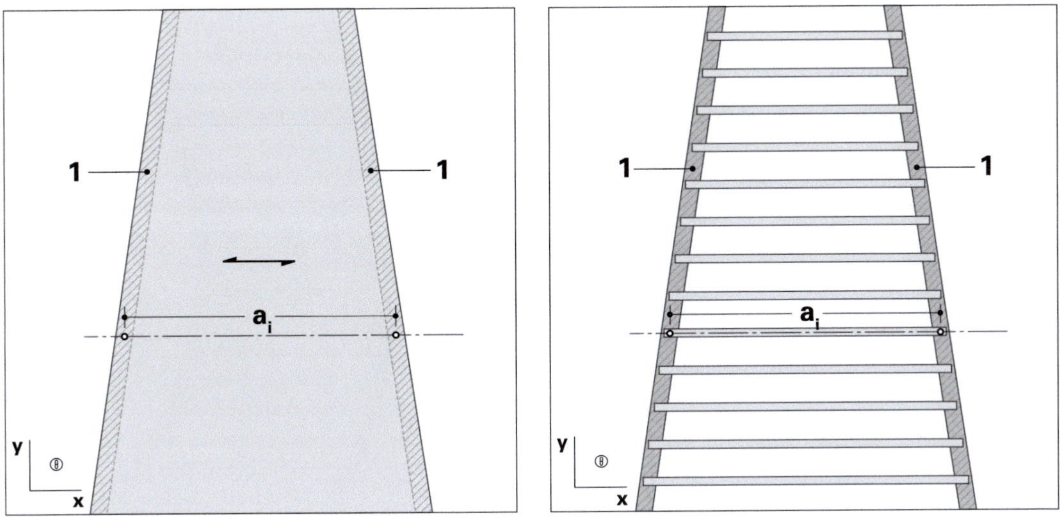

Fig. 3.52 (Left) Permanently changing spans $\mathbf{a}_i$ as a consequence of non-parallel course of the supports; cover by plate. (Right) Permanently changing spans $\mathbf{a}_i$ as a consequence of non-parallel course of the supports; cover by beam skein. (**1**) load-bearing wall, support

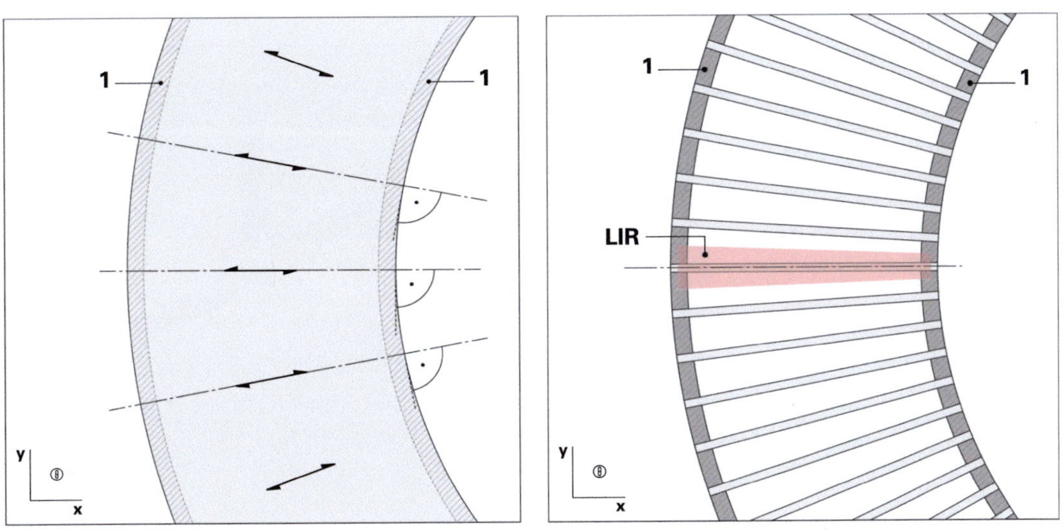

Fig. 3.53 (Left) Plate spanning one-way resting on curved supports. The perpendicular load transfer in the shortest possible way is established by itself in radial orientation. With a constant surface load, the plate experiences a lower loading on the side with the smaller radius. (Right) Radial or fan-shaped alignment of the beam layer. Even with parallel or concentric progression of the supports, the load distribution on a beam is not constant over its length (**LIR** = load infeed range). **1** load-bearing wall, support

3.2.3 Ratio of Spans in Two-Way Load Transfer

If both load directions in a system with two-way load transfer have the *same* spans and degrees of stiffness, the load is distributed **equally** to both orientations (→ **x** and → **y**). Two intersecting identical beams—as exemplarily illustrated in

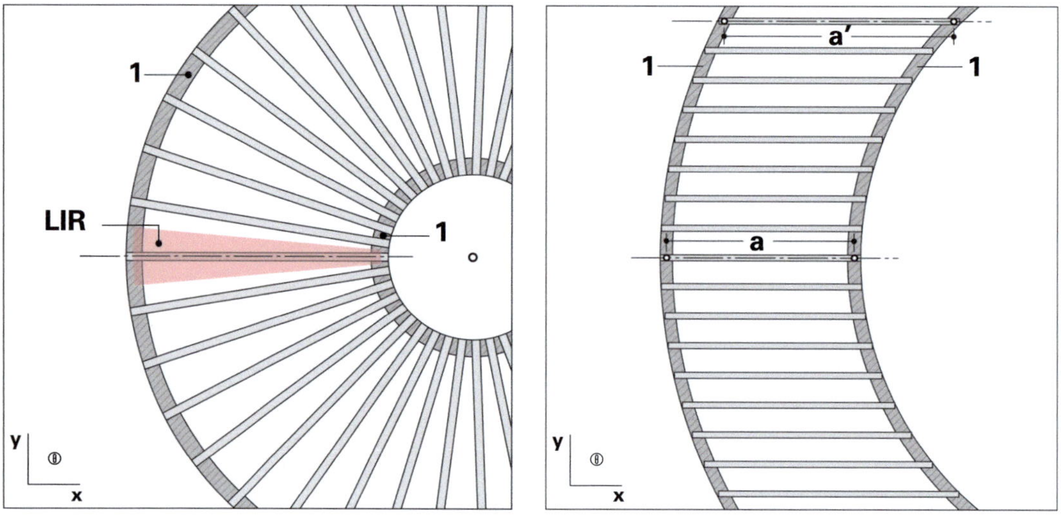

Fig. 3.54 (Left) The effect shown on Fig. 3.53, right, increases with greater curvature. In an extreme case, the space for the beam support inside can become scarce. (Right) Parallel bar skein on an annular support. Minimum span **a** and continuously increasing spans (as **a′**); (**1**) load-bearing wall, support

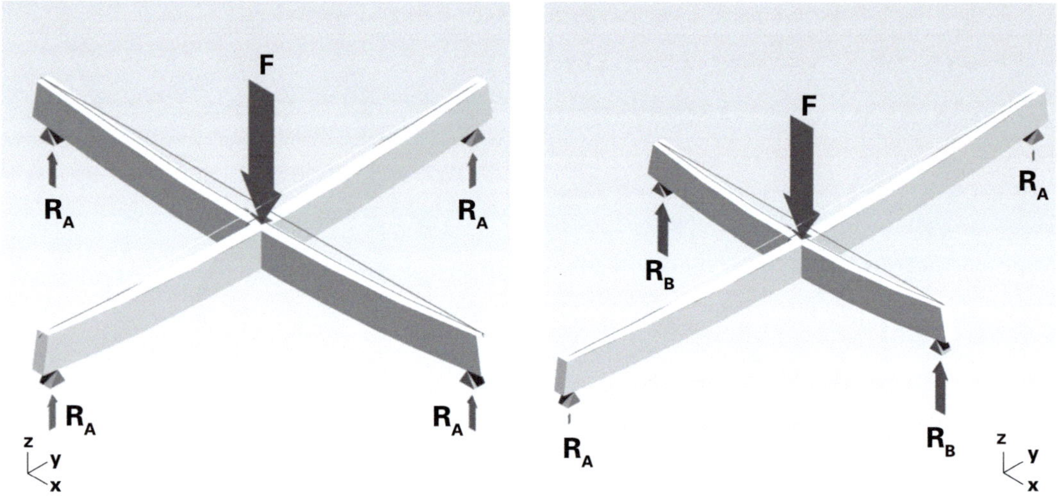

Fig. 3.55 (Left) Genuine two-way load transfer with two equal spans. The reaction forces $\mathbf{R}_A$ acting at the four supports are identical. (Right) Limited two-way load transfer with two distinctly different spans. The shorter, i.e. stiffer, beam section also carries the largest loads = $\mathbf{R}_B$. The other two supports, however, are hardly loaded with the smaller reactions $\mathbf{R}_A$. The material utilisation is poorer than on the left

Fig. 3.55, left—transfer the point load in **equal proportions** to the supports.

However, if there are *different* spans or degrees of stiffness in each direction, the force will take the path of greater stiffness or the smaller span (Fig. 3.55, right) to a substantially greater extent. The system is **statically indeterminate** or **over-braced**.

Beam grillages as discussed in ▶ *Sect.* 4 are genuine non-directional structures. They have the same beam skeins in both spanning directions, which penetrate each other through the nodes and which are connected to each other in a bending-stiff manner. In order to optimally utilise the advantages of the two-way load transfer, grillages have **square proportions** in the ground plan, i.e. both spans are of equal dimension. The load distribution, which is more efficient as compared to a directional system, allows for significantly smaller structural depths. The elaborate node formation, in which at least one rod must always be interrupted and subsequently connected in a bending-stiff manner, often proves to be disadvantageous in the case of grillages.

▶ *Sects. 4.3.1.3* and *4.3.1.4*

3.2.4 Load Transfer and Usage

Loads typical for load-bearing building structures can be divided into two broad groups:

- **Permanent** loads. These are, in particular, dead loads which, according to the definition, are always distributed **continuously** over the entire supporting structure.
- **Changing** loads such as traffic, snow, wind loads, etc. These can appear continuous in individual cases—as in the case of a uniform snow layer on a roof, for example—but are usually either punctually or locally limited or asymmetrically distributed. It can be said that this type of load is **discontinuous**. Changing loads include, in particular, **usage loads**, the safe and usage-compliant transfer of which constitutes a substantial part of the purpose of a load-bearing structure.

Considering cases **1** to **4** and **1′** to **4′** on ◘ Figs. 3.40, 3.41, 3.42, and 3.43 from this point of view, it is clear that the latter, i.e. cases **1′** to **4′**, those which originate from a *uniform* load distribution, are a special case, while cases **1** to **4**—with *discontinuous* loads—are rather the normal case in building. Pure one-way spanning systems without lateral load distribution as in case **1** are not suitable for building construction, since local deformations occur which would hinder usage. A minimum degree of cross-distribution is required to ensure a continuous, stepless component surface.

The conditions shown in the cases **1′** to **4′** can also be transferred to a continuous surface load (= $\mathbf{F}/\mathbf{a}^2$) over the entire component surface

Furthermore, pure one-way systems operate **without** the **involvement** of adjacent support members, which in principle leads to poor material utilisation and large—albeit only local—deformations. For reasons of efficiency of load transfer and more favourable deformation, a transverse load distribution, or transverse bending as in case **3**, or even a fully fledged two-way load transfer as in case **4**, is therefore also desirable.

The capability of cross-distribution of loads is already inherent to the component structure in the case of *homogeneous* cross-sections, as in cases **3** and **4**, and does not require any further constructional measures.

The situation is different in the case of components made of **bars** (cases **1** and **2**), in which the cross-distribution of loads must be activated by suitable constructional means. A conceivable measure—a shear-resistant connection by a tongue-and-groove joint—is shown in example **2**. However, this has only limited structural significance. Further measures are imaginable which can force a group of separate, one-way spanning bars which are not necessarily laid without gaps, to co-operate. These are, for example:

▶ *Sect. 4.2.1.2.5 Cross-Distribution of Loads in Bar Systems*

- an added shear-resistant plate (variant **2** in the following section);
- transversely laid-out rods which penetrate the longitudinal bars and are connected to them in a bending-stiff manner (variant **3** in the following section);
- transversely laid-out rods which are connected to the longitudinal bars only shear-proof but otherwise hinged (variant **4** in the following section).

Numerous structures in building construction consist of a framework of bars, or non-rigid linear elements such as ropes, which fulfil the main load-bearing functions and are flat-lined with a surface element which is of a limited thickness, a thin plate, a glass pane, etc. They must therefore be enabled by one of the above-mentioned structural measures for **load cross-distribution**.

The choice of the constructional principle used for this purpose determines the design of the structure as well as its load-bearing behaviour, its geometrical organisation, and, last but not least, its formal appearance. For this reason, the discussed variants will be investigated in more detail and characterised in their respective peculiarities in the following sections.

3.3 The Constructional Composition of The Space-Enclosing Surface Element

Superficial surface components with different **structural designs** are considered with regard to load-bearing in ***Volume 1***. This takes place there in an abstract and isolated form, detached from superordinate influences from the interplay of components within an overall load-bearing structure. In the following, however, surface elements will be considered as components of a load-bearing structure and their practical constructional design will be analysed in more detail, taking into account the direction of gravity.

Vol. 1, *Sect. 26.9 Realising the Force-Transmission Function Within the Element—Structural Principle of the Component*

Cf. the introductory remark in ▶ *Sect. 3.1.1 Load-Bearing Structure and Overall Design*

The classification of the variants selected in ***Volume 1*** for the sake of the greatest possible generality and applicability also underlies, in a slightly modified form, the following considerations, with particular reference to the conditions prevalent in primary load-bearing structures. Particular attention is paid to **full-leaf** and **ribbed systems** because of their great constructional importance. Variants investigated in ***Volume 1*** which are of little or no relevance for primary structures are dismissed; others, in contrast, are accordingly further differentiated.

In contrast to ***Volume 1***, this section also discusses structures made of **curved components**—at least in their essential characteristics. Although the design variants **1** to **4** (◘ Fig. 3.68), which are examined in the following, are discussed mainly in their planar, predominantly bending-stressed design, the considerations are, however, also analogously applicable to curved, predominantly axially compressive- or tensile-stressed components as depicted in the overview on ◘ Fig. 3.69. The tables on ▶ ◘ Figs. 4.4, 4.5, 4.6, 4.7, and 4.8 in ▶ *Chap.* 4 provide a comprehensive overview of the conceivable elementary structural alternatives (see also various examples of full-leaf and ribbed systems from different technical disciplines on ◘ Figs. 3.56, 3.57, 3.58, 3.59, 3.60, 3.61, 3.62, 3.63, 3.64, 3.65, 3.66, 3.67, and 3.68, 3.69).

3.3.1 Full-Leaf Surface Element (Variant 1)

Space-enclosing surface elements, which are always an integral part of a building, consist in their simplest form of a continuous single-leaf, solid-walled planar **slab** or **shear-plate**, or of a comparable curved element, i.e. a **shell** (◘ Fig. 3.70). As they consist of a continuous flat or curved solid leaf, they are referred to collectively as **single-leaf** or **full-leaf systems**. The advantage of its constructional simplicity is counteracted by the disadvantage of its comparatively **large dead weight**. This is discussed elsewhere. The larger the scale of the component under consideration, the greater is the proportion of the dead load under otherwise unchanged boundary conditions, so that, starting from a certain point, its load-bearing reserves are consumed to a significant extent merely to bear their own weight. This physical law is exemplified, for example, by the solid floor slab

***Vol. 1**, Sect. 26.9.1 Solid-Leaf Element*

▶ *Sect. 2.2 Single-Leaf Systems*

***Vol. 1**, Sect. 26.9.1 Solid-Leaf Element*

▶ *Sect. 4.2.1.1 Slab, Spanning One-Way*

***Vol. 1**, Sect. 28.3.3.3 Airborne-Sound Behaviour of Components*

***Vol. 1**, Sect. 29.9 Factors Influencing Fire Resistance*

The comparatively large mass of solid-walled components can, on the other hand, have a favourable effect with respect to different types of functions. This is the case under specific conditions, for example, for noise protection and fire protection.

Furthermore, in some cases, it may be relevant that certain fundamental physical effects of the solid-walled component can possibly not be achieved with the required efficiency by it alone. This includes, for example, thermal protection, which is

Fig. 3.56 Full-leaf system

Fig. 3.57 Full-leaf system

often realised only to a limited extent in the case of full-walled elements lacking an additional thermal-insulation layer, or can only be achieved by using specific—for example, aerated—materials within the solid-walled surface element itself. As a result, the load-carrying capacity of the component is often simultaneously impaired.

Furthermore, the good load distribution within the homogeneous structure of the slab or plate is a significant advantage. This is evident, for example, in the good **transverse distribution** of point loads or asymmetrical loads in the case of plates, but in particular in the **shear-plate** or **shell effect**, which is available without further structural bracing measures. This property, which is significant from a structural point of view, can be lost when the full-walled surface ele-

Fig. 3.58 Bar system

Fig. 3.59 Bar system

ment is converted into a rod structure, as will be discussed below. It may then be necessary to restore the shear-plate or shell effect by means of additional elements, such as, for example, diagonal braces or shear-resistant sheathing. These stiffening measures are dealt with further below.

▶ Sects. 3.2.2, 3.2.3, and 3.2.4

▶ Sect. 3.3.5 Bracing of Bar Systems Within Their Surface

Fig. 3.60 Bar system

Fig. 3.61 Bar system

3.3.2 Element From Bar Skein and Plate (Variant 2)

A significant reduction in the dead weight at a comparable load-bearing capacity can be achieved by converting the solid slab into an element from a skein of bars with a surface-forming flat plate which is much lighter than the full plate (Fig. 3.71). This process of breaking up the homogeneous leaf into a bar framework produces a load-bearing structure that is collectively referred to as a **ribbed system**. Its load-bearing behaviour differs fundamentally from that of the full-leaf system. The following variants **3** and **4** are also considered bar frameworks and therefore ribbed systems.

▶ *Sect. 2.5 Ribbed Systems*

Fig. 3.62 Cable system

Fig. 3.63 Cable system

In the variant considered here, the skein oriented along the spanning direction assumes the main load-bearing action. The plate can be made substantially thinner than the rib-less full slab, and in the area between the bars it takes over the **load transfer** to the latter, as well as the **transverse distribution** of load on several adjacent bars.

Fig. 3.64 Bar system

Fig. 3.65 Bar system

This has already been discussed in ***Vol. 1***, *Sect. 26.9.4 Element Composed of Ribs Oriented One-Way*

If a material continuum or a force-fitting connection exists between the rod and the plate, a **structural interaction** of both elements can be activated. For example, a limited plate strip can be used over a bar as a compression chord.

Fig. 3.66 Bar system

Fig. 3.67 Textile fibre system

There is a mutual dependence between the bar distances and the thickness of the plate. It is clear that the smaller the distances, the thinner the plate can be (Fig. 3.91). However, its thickness must be adequately dimensioned so that both main functions of the plate, i.e. longitudinal load transmission and load cross-distribution, are fulfilled. It is in accordance with the logic of this type of load-bearing structure that the bar spacings which define the span of the surface-forming plate are substantially smaller than the span of the overall structure. Further considerations are made below.

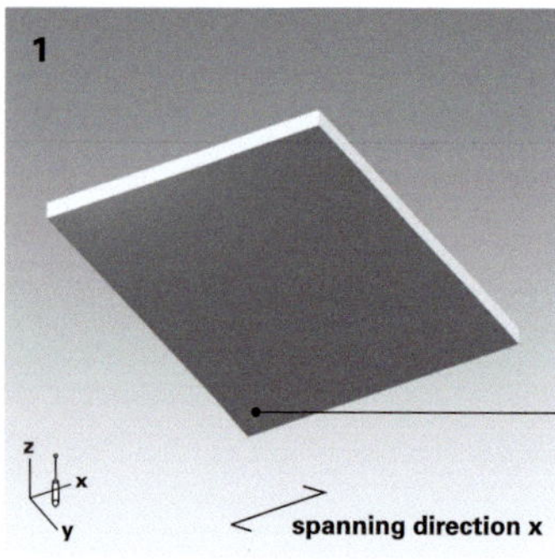

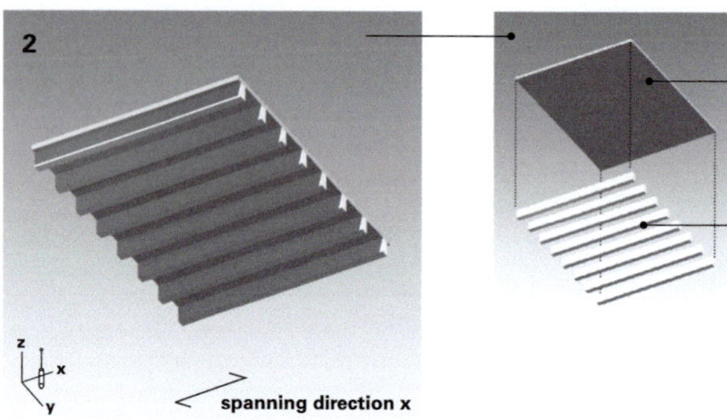

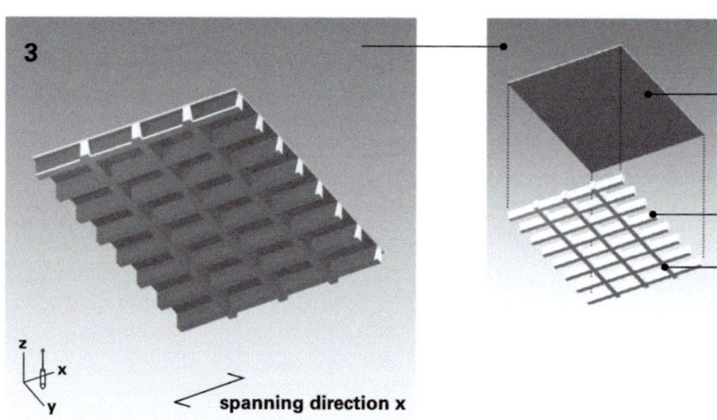

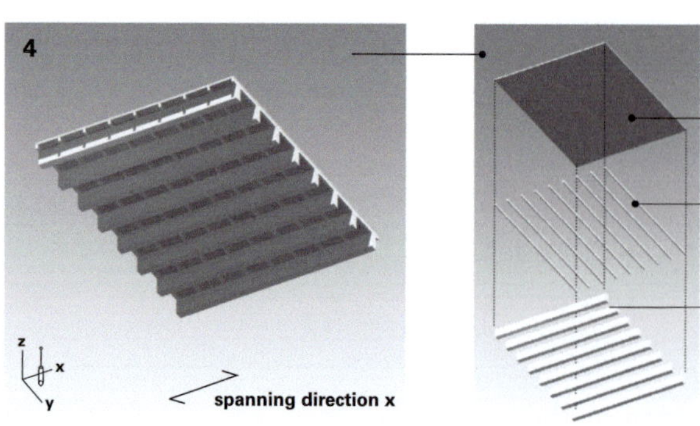

	surface formation	main load-bearing action	load cross-distribut-ion
1 solid slab			
slab	●	●	●
2 bar skein and plate			
plate	●		●
bar skein		●	
3 bar grillage and plate			
plate	●		
longitudinal bar skein		●	
transversal bar skein			●
4 bar skein, transversal skein on top, and plate			
plate	●		
transversal bar skein, placed on top (2nd hierarchy level)			●
longitudinal bar skein (1st hierarchy level)		●	

Fig. 3.68 Four structural variants of a planar element with constructional build-up and load-carrying functions of the components

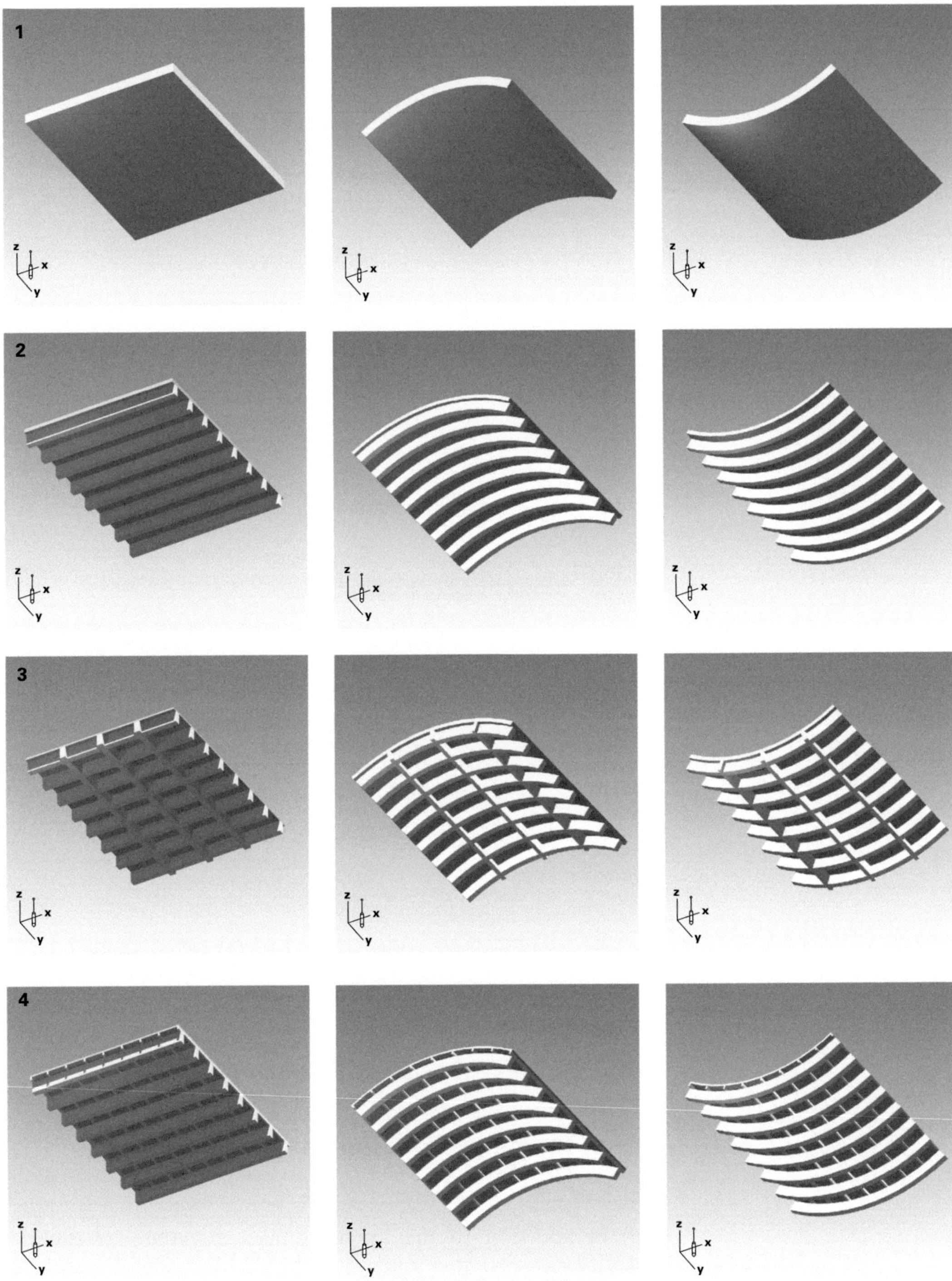

Fig. 3.69 Planar and concave or convex-curved forms of appearance of the considered structural variants of surface elements

3

Fig. 3.70 Full-leaf surface element

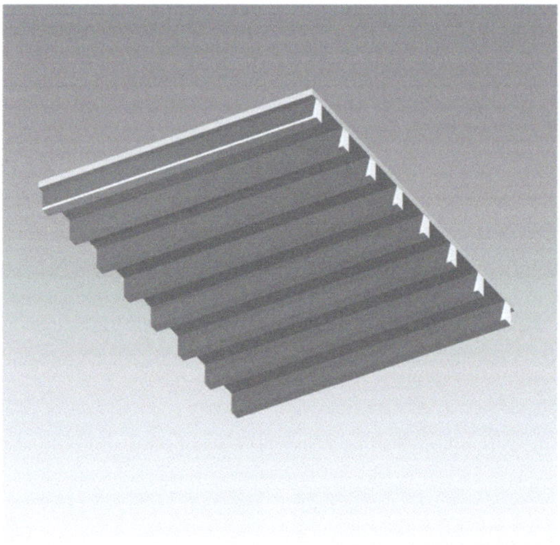

Fig. 3.71 Element from bar skein and plate

▶ Sect. 3.3.7 Some Basic Design Considerations about Bar Skeins

3.3.3 Element From Bar Grillage and Plate (Variant 3)

If, in addition to the longitudinally directed bar skein, a skein cross-oriented against the main spanning direction of the element is introduced, a **grid** of interpenetrating rods is formed, topped by a surface-forming plate (Fig. 3.72). The cross-skein provides the cross-distribution of the load on the longitudinally oriented skein. The plate closes the surface and transfers loads in the area of the interstices be-

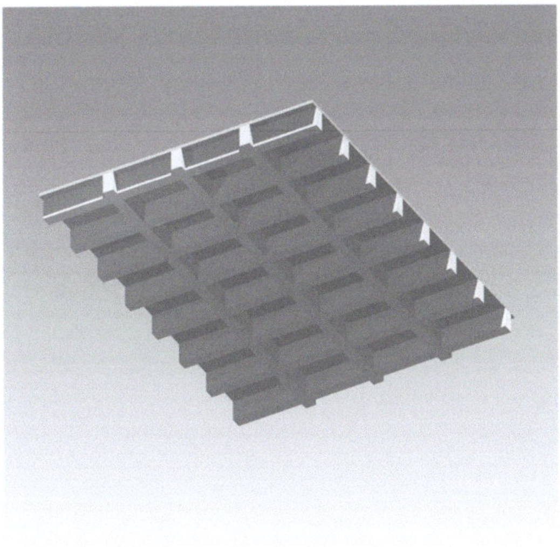

Fig. 3.72 Element from bar grillage and plate

tween rods. This happens reasonably, although not necessarily, both to the longitudinal as well as to the transverse bars, which are flush on the plate side, and mostly also on the opposite side. As a result, the plate can be linearly supported by a four-sided linear frame of bars, which means that it can span two-way, which is why it can generally be thinner than in variant **2**.

Cf. also ► *Sect. 3.3.6.1 Bar Grillages*

3.3.4 Element From Bar Skein, Subordinate Cross Skein, and Plate (Variant 4)

In contrast to variant **3**, the external load from the surface-forming plate is not transferred directly to the longitudinal bar skein, but firstly to the cross skein, which in turn delivers the load onto the longitudinal skein (Fig. 3.73). According to this principle, **additional bars** are either placed on the beams, suspended between them or also below them (Fig. 3.74, see also Figs. 3.75 and 3.76), namely *transverse* (perpendicular) to their spanning direction to use the shortest available span. One can therefore distinguish different **component hierarchies**. The load-collecting longitudinal bars are located at a higher hierarchy level than the downstream transverse rods. The surface-forming sheathing is in turn associated with an even further downstream level than the transverse rods. We then speak of a **hierarchical grading** of different **positions** of bars or, at the end, **surface elements**. According to the same construction principle, elements can also be formed with more than two bar hierarchies.

Cf. Fig. 3.50, left

► *Sect. 3.3.7 Some Basic Design Considerations about Bar Skeins*

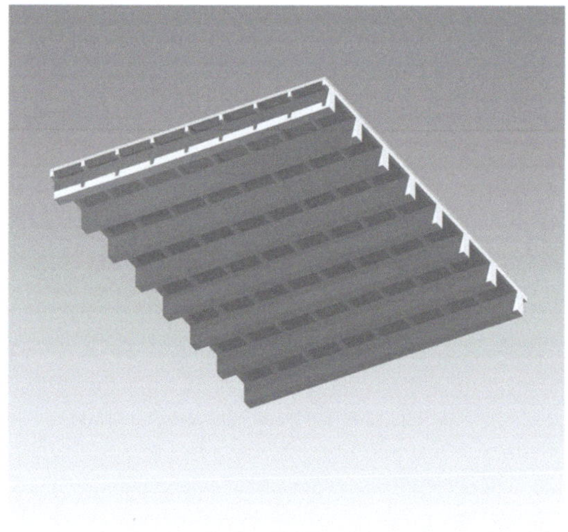

Fig. 3.73 Element from bar skein, subordinate cross skein, and plate

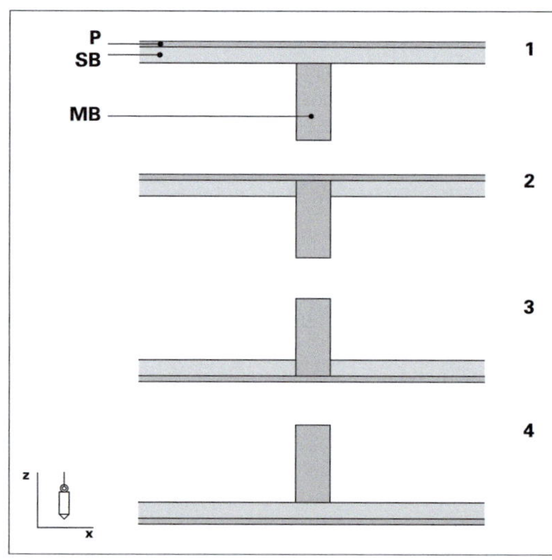

Fig. 3.74 Variations of structural variant **4** (**P** plate, **MB** main bar, **SB** secondary bar): (**1**) secondary bars placed on top, (**2**) secondary bars are hung between main rods and placed flush with the top. The final sheathing has a four-sided support in each field provided by main and secondary bars. (**3**) secondary bars are flush at their bottom with the main bars, so that similar (although inversed) conditions prevail as in **2.** (**4**) secondary bars hung below the main bars. The final panelling in each field, in this case, as in case **1**, has always a two-sided support on secondary bars

3.3.5 Bracing of Bar Systems Within Their Surface

If a full-walled surface element as described in ▶ *Sect. 3.2.1* is converted into a sheathed bar system as described in ▶ *Sects. 3.2.2*, 3.2.3, and *3.2.4*, the shear stiffness within its surface—

Fig. 3.75 (Left) Connection of two secondary beams to a continuous main beam. Even if they are placed flush at their tops, this connection is not able to create a continuous action of the abutting beams, which is a prerequisite for the two beams' position being of equal rank and thus for a non-directional structure. (Right) Connection of two secondary beams to a continuous main beam (variant **2** on Fig. 3.74). In this case, the secondary beams are designed to be flush with the main beam so that no continuity effect can be achieved

Fig. 3.76 An example of a suspended subcarrier layer (variant **4** on Fig. 3.74). The shear-plate of the roof is executed in the subcarriers plane, so that the main girders are in the open

which is given in the full leaf of variant **1** (Fig. 3.70) from the start—may have to be secured by **additional bracing measures**. There are basically the following possibilities:

3.3.5.1 Triangular Meshes

If the bar system itself is structured according to a **triangular grid**, a (triangulated) framework of geometrically stiff triangular meshes is provided (Fig. 3.77). No further elements are required for in-plane shear stiffening.

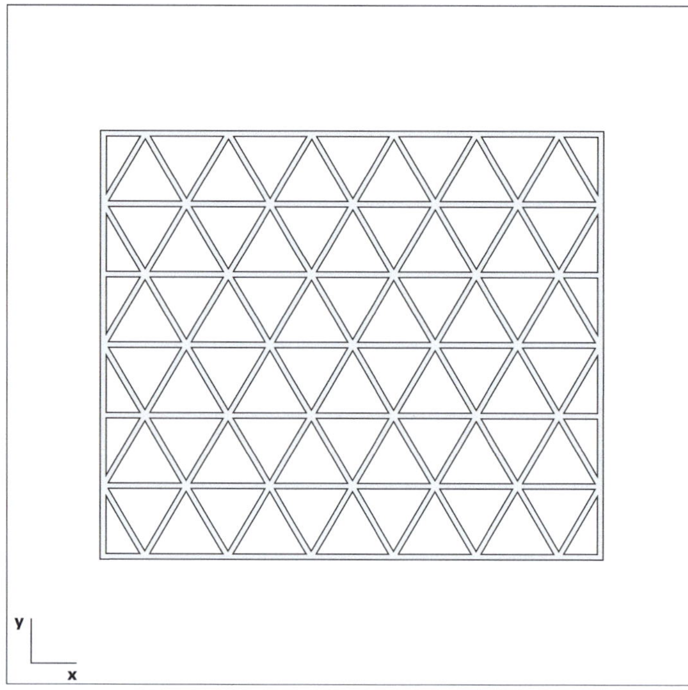

■ **Fig. 3.77** Shear-resistant bar framework based from a triangular grid (truss)

3.3.5.2 Diagonal Bracing

Stiff triangles can also be formed by introducing diagonal braces into quadrangular meshes. In this case, the non-stiffened meshes formed by the hinged coupled longitudinal and transverse bar layers are triangulated by means of diagonal members and consequently stiffened.

Vol. 1, *Sect. 26.9.4 Element Composed of Ribs Oriented One-Way*, Fig. 26.185

The diagonal bracing of a rectangular field can be effected by one diagonal rod or by two counter-current ones (■ Fig. 3.78). With larger dimensions and correspondingly larger cross-sectional forces acting within the diagonals, several coupled diagonal fields can also be formed in order to distribute the stress (■ Figs. 3.79, 3.80, 3.81, and 3.82). Alternatively, a circumferential, frame-like strip of diagonally braced fields can also achieve the necessary stiffening (■ Fig. 3.83).

Vol. 1, *Sect. 26.9.4 Element Composed of Ribs Oriented One-Way*, Figs. 26.187, 26.188, 26.189, 26.190, and 26.191

Diagonal bracings develop their effect with hinged connections between longitudinal and transverse sets of bars. For this reason, they are particularly important for surface elements made from hierarchically stepped bar layers according to variant **4**, in which there is basically no bending-stiff connection between crossing bars. Beam grillages subjected to bending according to variant **3**, however, act differently, because their nodes between bar skeins are bending-stiff perpendicular to the component plane, as a systemic given. In addition, this node can also be made bending-stiff within the component plane if specific reasons suggest so, which would lead to an inherently stiffened frame element. On the other

► *Sect. 3.3.4 Element from Bar Skein, Subordinate Cross Skein, and Plate*
► *Sect. 3.3.3 Element from Bar Grillage and Plate*

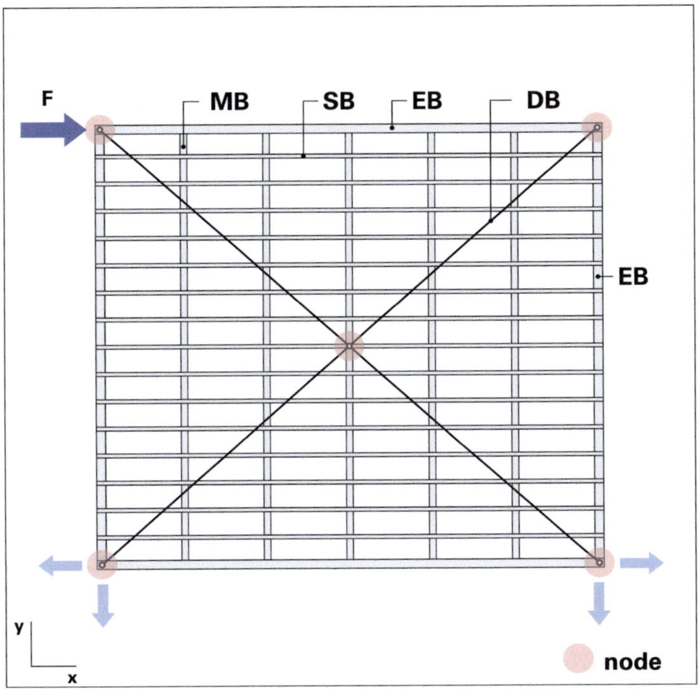

Fig. 3.78 Bracing of a bar framework from square meshes using edge bars (**EB**) and a set of diagonal bars (**DB**). In this case, the main (**MB**) and the secondary bars (**SB**) are not involved in the bracing. The diagonal force components are introduced only in the nodes at the four corners and in the field centre. The resulting axial forces in the edge bars require a larger cross-section than in the case of the equally oriented main and secondary bars

hand, diagonal bracings are again used with shell-like curved lattice structures with hinged bar nodes.

Corresponds also to the structural variant **3**

The **node formation** from a longitudinal, a transverse, and a diagonal rod, or even, sometimes, two of them, is sometimes complicated in constructional terms. On the one hand, it is desirable to have the rod axes converging in one point as far as possible in order to avoid offset moments. On the other hand, disentangling the node into different planes is often unavoidable, since otherwise difficult penetrations must be realised (Figs. 3.84 and 3.85). In such cases, the diagonal members are preferably designed as tension bars, since these are not endangered by buckling, their cross-sections are correspondingly slender and may be designed freely in their geometry, for example as flat steels or ropes.

► *Sect. 4.3.2.5 Dome from Bars*

***Vol. 1**, Sect. 26.9.4 Element Composed of Ribs Oriented One-Way*, cf. Fig. 26.191

3.3.5.3 Shear-Resistant Sheathing

Since the rod element is almost always finished off with a surface-forming sheathing, this can basically be used for the shear-stiffening of the bar framework. The prerequisite for this is the necessary **shear** or **racking stiffness** of the sheathing, something which, for example, generally can hardly be realised with a glazing. On the other hand, in those cases in

Because the forces on the bearings are difficult to transmit. This is further complicated by the requirement to be able to replace glass panes when necessary

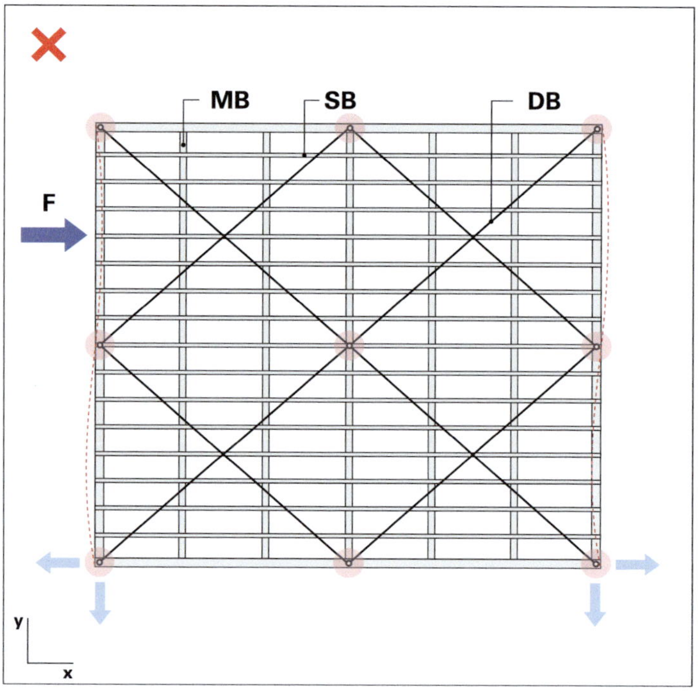

Fig. 3.79 A bar framework with several diagonal braces. The deformation pattern is the same as in the element on Fig. 3.81 despite additional diagonal rods. As is the case with the variant depicted on Figs. 3.78 and 3.81, this element is neither braced to overall form a shear plate. An applied force **F** as indicated results in the large deformations shown. Only the forces in the individual diagonal rods are reduced

which rods merely constitute a bending-stiffening of an otherwise comparatively thick, continuous plate (example: beam-and-slab floors), the racking stiffness is already ensured by the existence of the plate (Fig. 3.86), which in this case usually assumes the primary support function anyway.

The division of tasks between bars and light sheathing is as follows: A thin sheathing provides the shear stiffening of the bar framework within the element surface; the bars in turn stiffen the thin plate against buckling. This cooperation between the two parts is described in more detail elsewhere.

Vol. 1, *Sect. 26.9.4 Element Composed of Ribs Oriented One-Way*, as well as Fig. 26.186

Attention must be paid to the seams of the elements of the sheathing, since these must always be designed to be resistant to shear (Fig. 3.87). A simple solution to this task is to always place the joints on a rod and fix the elements of the sheathing thereto (Fig. 3.87).

3.3.5.4 Frame Formation

The bracing of a superficial element by forming a frame can be achieved by means of a bending-stiff connection of preferably equal-ranking rod skeins, i.e. those with similar lengths and bending stiffnesses. It is particularly suitable for bar or beam grids according to the principle of variant **3**. Compared to diagonal braces, it is associated with a

According to Fig. 3.68 in ▶ *Sect. 3.3 The Constructional Composition of the Space-Enclosing Surface Element*

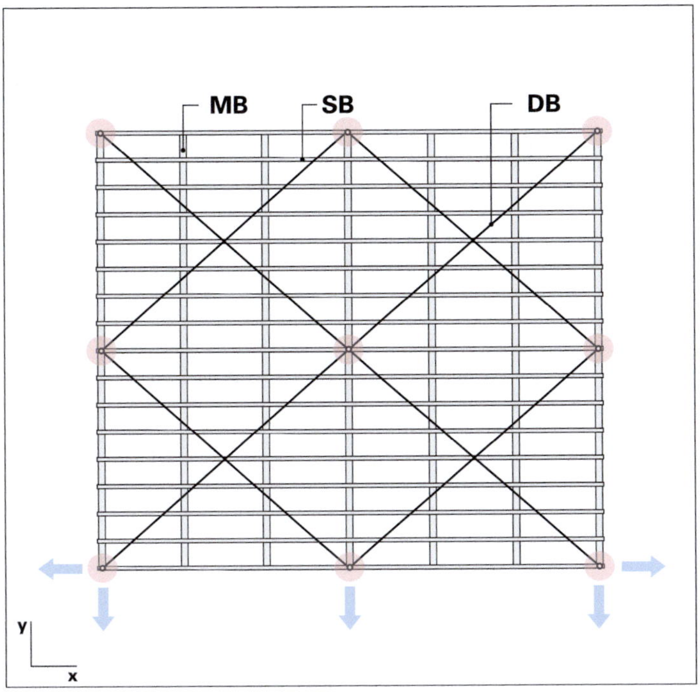

Fig. 3.80 Bracing of the bar frame with several diagonal braces as on Fig. 3.79, but with an additional support in the middle axis. The diagonal forces, and thus the elongations of the diagonal members, can thereby be markedly reduced. This makes it clear that a direct connection of the diagonal rods to the supports is always useful

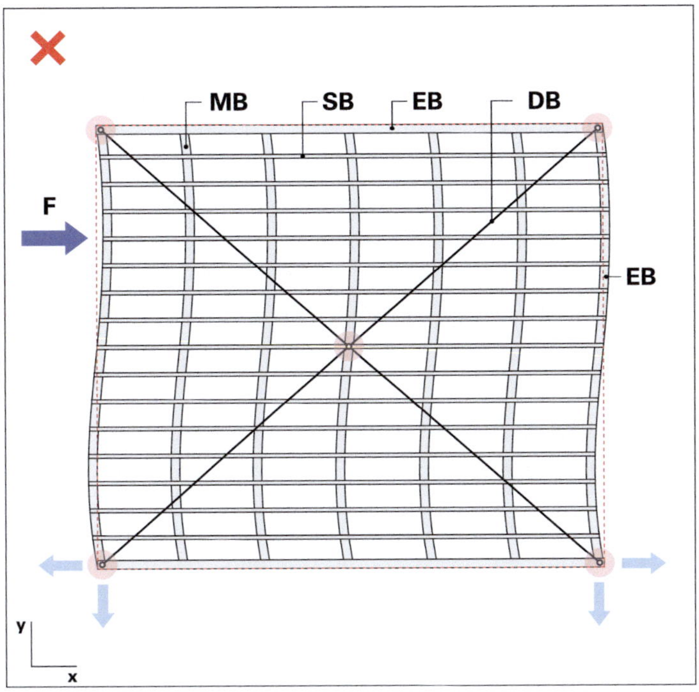

Fig. 3.81 Element as on Fig. 3.78 on the left under a point load **F** attacking at any location. The load must be introduced by bending the main bars into the nodes of the diagonal rods (only in the corners and in the middle). Large deformations occur. The element is not braced to form a shear-plate

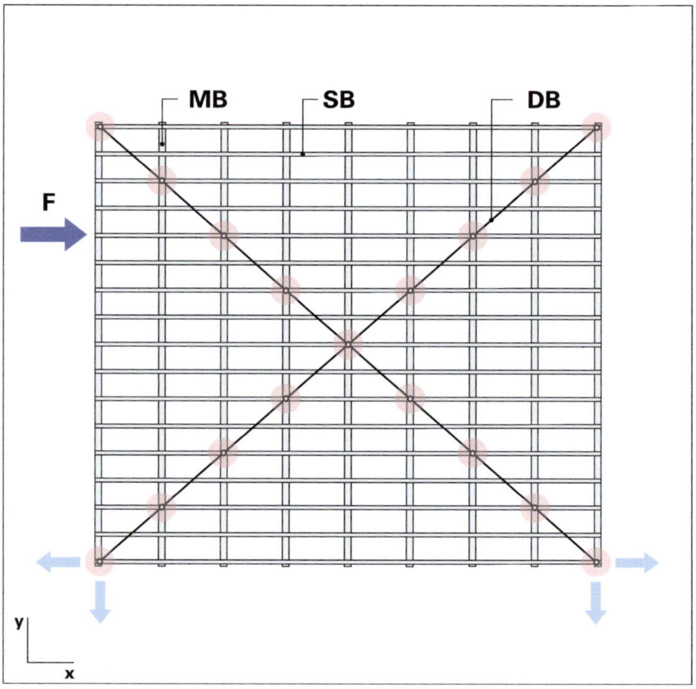

■ **Fig. 3.82** Bracing of the bar framework by connecting the main (**MB**) and the secondary (**SB**) to the diagonal bars (**DB**) at their crossing points (nodes). By the multiple attachment of the secondary bars to the diagonals, the load can be removed with considerably smaller deformations than in the cases on ■ Figs. 3.78, 3.79, and 3.81. The element acts as a rigid shear-plate. It becomes clear how efficient it is to repeatedly connect the diagonals to the nodes of the basic element. The basic prerequisite is the correct design of the lattice geometry: By an appropriate selection of the gridding, intersections between longitudinal, transverse, and diagonal bars must be created. In the variants on ■ Figs. 3.78 and 3.79, this is the case only in the corners and in the centre, or respectively in the edge-bar centre

high expenditure of materials and is suitable for surface elements of low and medium spans in which the formation of frames with bending-stiff rod nodes is given from the beginning, without additional expenditure, by being designed as a grillage.

3.3.6 Completing Bar Systems to Surfaces Using Sheathing

While the surface of solid-walled elements according to variant **1** is produced, so to say, by the load-bearing component itself and loads are evenly distributed on the support by the homogeneous material structure, the bar skeins of elements according to variants **2** to **4**—which for reasons of material economy almost without exception are laid at intervals—in order to fulfil both tasks of surface formation and load transfer, need to be completed to a closed surface with a mostly

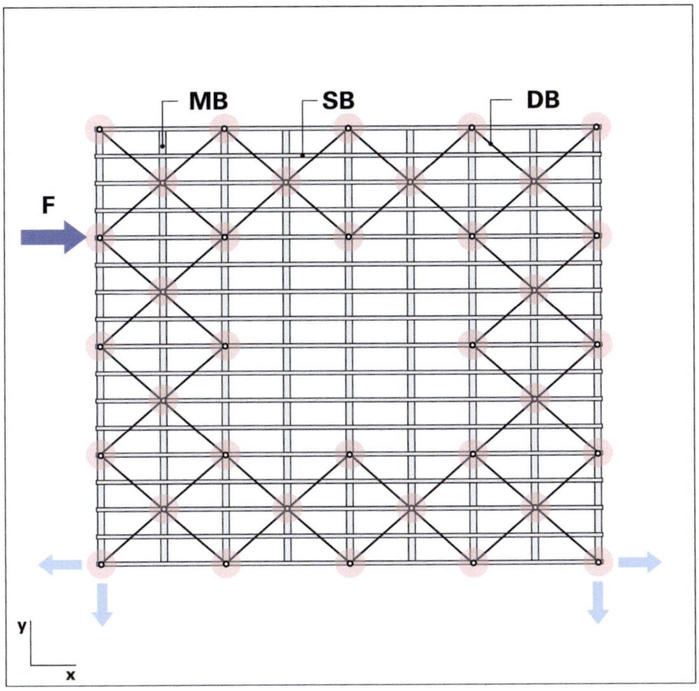

Fig. 3.83 Bracing by a circumferential, frame-like strip of diagonal braces. The element is stiffer than the variants on Figs. 3.79 and 3.81, but cannot be viewed as a rigid shear-plate. Notwithstanding, this type of circumferential diagonal bracing can be used well for stiffening very large-sized coverings, or for bracing those which are to remain unobstructed in the middle region, for example, for daylighting purposes

thin skin. Some important constructional questions need to be clarified in this context and will be discussed below.

3.3.6.1 Bar Grillages

Bar grids as in variant **3** consist of a longitudinal and a cross-bar skein. If the transverse bars are individual, load-distributing bars placed at intervals (Figs. 3.88 and 3.89), a final covering then spans essentially one-way between the bars of the longitudinal skein (Fig. 3.89, left) or with a larger bi-axial share (such as Fig. 3.89 right), depending on the field proportions (respectively in → **x**, → **y**). If, on the other hand, the two skeins—the longitudinal and transverse bar skeins—are equivalent in terms of their static order of rank, as in the case of a grillage, and if they are of equal depth and are flush, they offer a four-sided linear bearing (Fig. 3.90) in each field to be covered. The sheathing consequently spans two-way and, as a rule, can be made thinner and lighter than with one-way load transfer (as shown on Fig. 3.88).

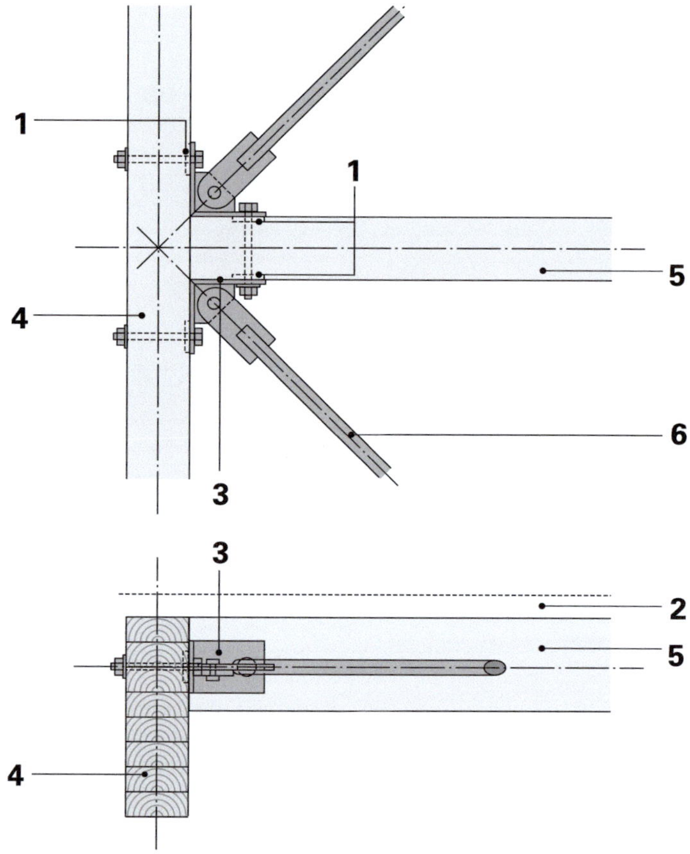

Fig. 3.84 Connection of main to secondary beam with diagonal bracing. (**1**) Single-sided connector with threaded bolts; (**2**) further addition; (**3**) connection bracket from steel; (**4**) main beam, (**5**) secondary beam; (**6**) round bar from steel

3.3.6.2 Orthogonally Stacked Hierarchical Bar Skeins

Elements as described as variant **2** or **4** in ► *Sect. 3.2.4* consist of a bar layer, or of a plurality of stepped hierarchies of bar positions respectively oriented perpendicular to each other, and a concluding sheathing (Fig. 3.91). The essential constructional goal of this arrangement is to gradually reduce the spans of the elements to such an extent that the surface-forming element—the sheathing—only covers a small dimension and can be made correspondingly thin and light. With this means, the weight of the surface element, which is necessarily the greatest in the case of the full-walled component according to variant **1**, is intended to be reduced as far as possible. Basically, it applies in this respect:

The bar spacing within a set of bars of a particular hierarchy level is equal to the span of the elements of the subordinate hierarchical level (Fig. 3.92).

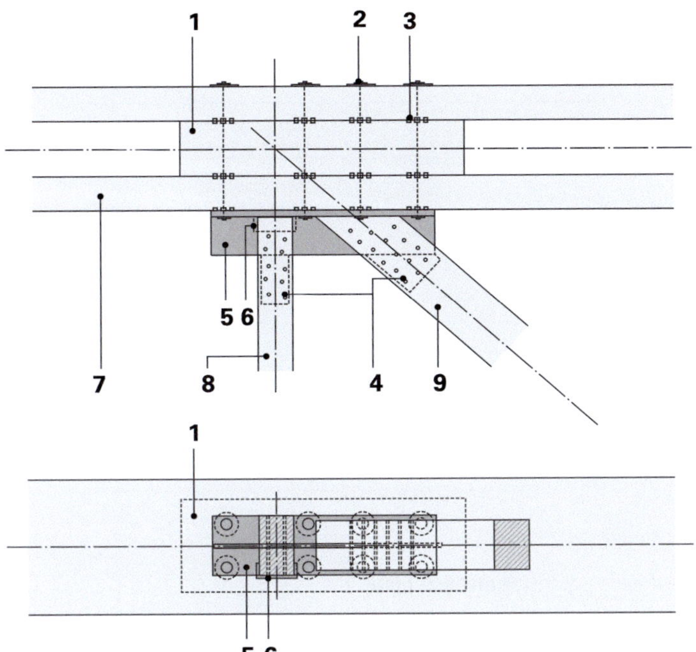

Fig. 3.85 Connection between a pincer-like twin main beam, secondary beam, and diagonal rod (acc. to Halász/Scheer). **(1)** Filler block; **(2)** double-sided connector with bolt safety; **(3)** threaded bolt; **(4)** rod dowel; **(5)** halved IPB profile; **(6)** L-profile, welded; **(7)** main beam; **(8)** secondary beam; **(9)** diagonal bar

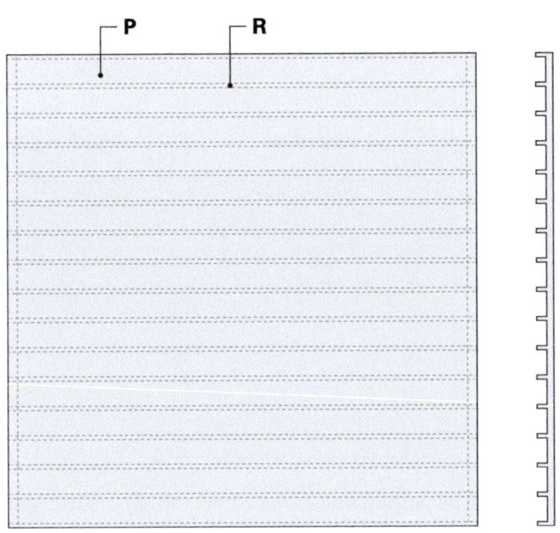

Fig. 3.86 A shear-resistant surface element made of a monolithic plate (**P**) and ribs (**R**) according to variant **2** (Fig. 3.71), for example a concrete beam-and-slab floor

As a rule, the final sheathing spans between the bars of the hierarchically directly superordinated bar position—mostly placed immediately below—that is, **one-way**.

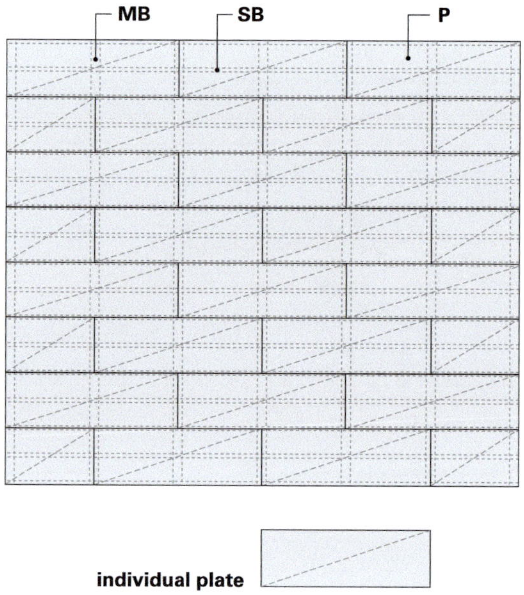

Fig. 3.87 Shear-resistant sheathing (**S**) of a bar element consisting of main bars (**MB**) and secondary bars (**SB**). Offset of plates from field to field for better shearing strength

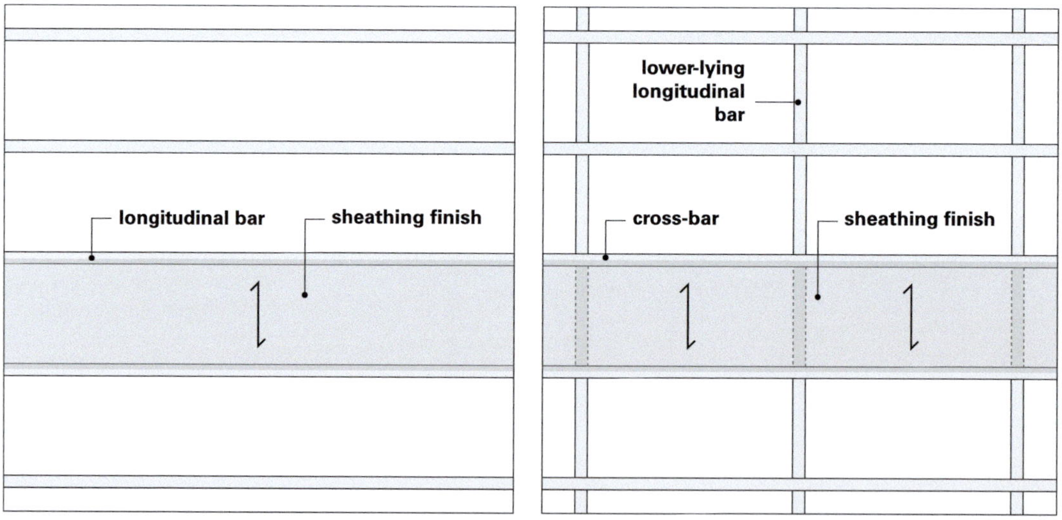

Fig. 3.88 (Left) Bar skein without cross-bar skein with sheathing finish placed on top. The latter spans one-way between the parallel bars (i.e. a double-sided support). (Right) Double-sided support of the final sheathing as above, here on the cross-bar skein. There is a set of longitudinal bars of higher hierarchy, which is arranged further below and is not used for the support of the sheathing

For various reasons, such as, for example, depth limitation of the construction, it may be necessary to execute rods of two hierarchies in a flush manner, that is to say, for example, by inserting secondary beams between main beams. A limited **four-sided support** of the final thin plate is then also possible. If the proportions of the fields delimited by main and second-

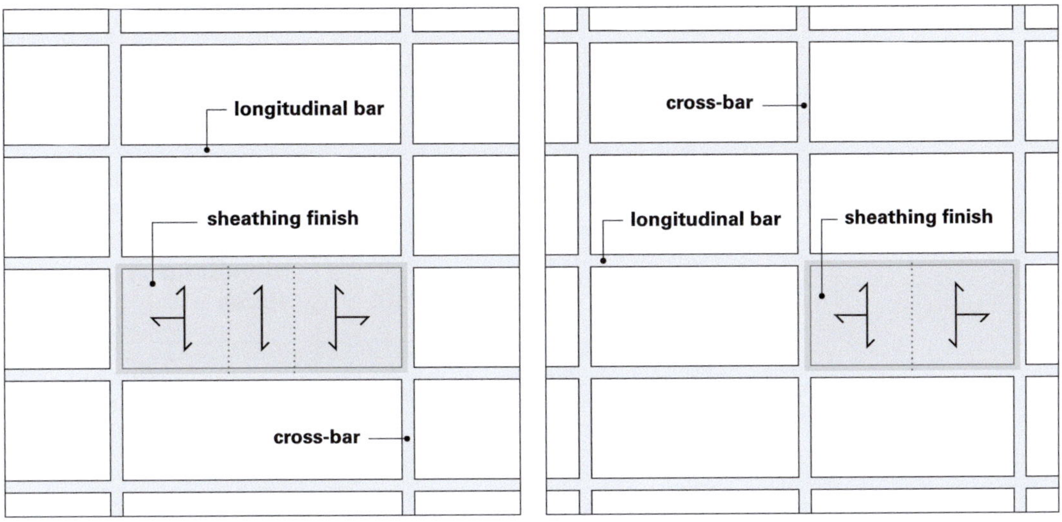

Fig. 3.89 (Left) Predominantly one-way load transfer of the final sheathing, which is placed on the load-bearing bar grillage made of longitudinal bar skein and cross-bar skein, the latter being arranged at relatively large distances and flush with the longitudinal bars—despite four-sided support. This is due to the markedly rectangular proportion of the field. There is a limited two-way load transfer only in both side areas, whereas the centre strip acts one-way. (Right) Conditions as above, but more compact field proportions. Two field areas with limited two-way load transfer are formed

ary rods are approximately square, a true two-way load transfer of the final plate can also be taken advantage of in this directional design.

3.3.7 Some Basic Design Considerations About Bar Skeins

As mentioned, there is a close interrelationship between the **bar spacings** and the **dimensioning** of the respective element placed on top, i.e. subordinate bar or plate. Although the following considerations are illustrated by means of bars under bending stress, they also apply, in principle, even for compressive- or tensile-stressed members, although with substantial differences in the magnitude of their respective stress.

The distances between main bars are initially **freely selectable**. They determine the span to be bridged by the subcarriers. In other words:

- **Closely spaced main bars** require only **slender secondary bars**, since the latter's span **a** is small (Fig. 3.92, left).
- **Further distant main bars** require **more powerful secondary bars**, since they have to bridge a large span **A** (Fig. 3.92, right).

It is obvious that main bars placed further apart have a **larger load-infeed area**—as computed between the centres of the adjoining inter-spacings (Fig. 3.93)—and conse-

3

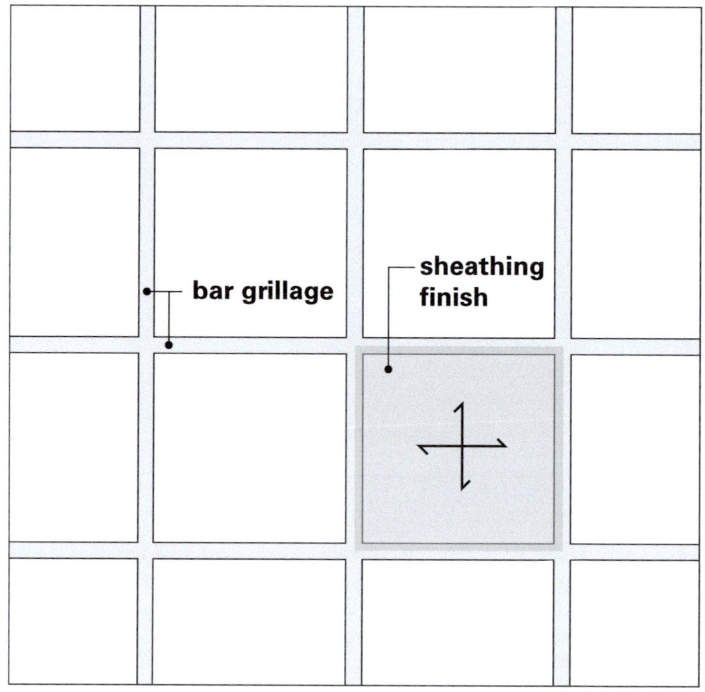

Fig. 3.90 Bar grillage with equal-ranking longitudinal and cross-bar skeins. Since the fields are square, a genuine two-way load transfer is activated in the final sheathing

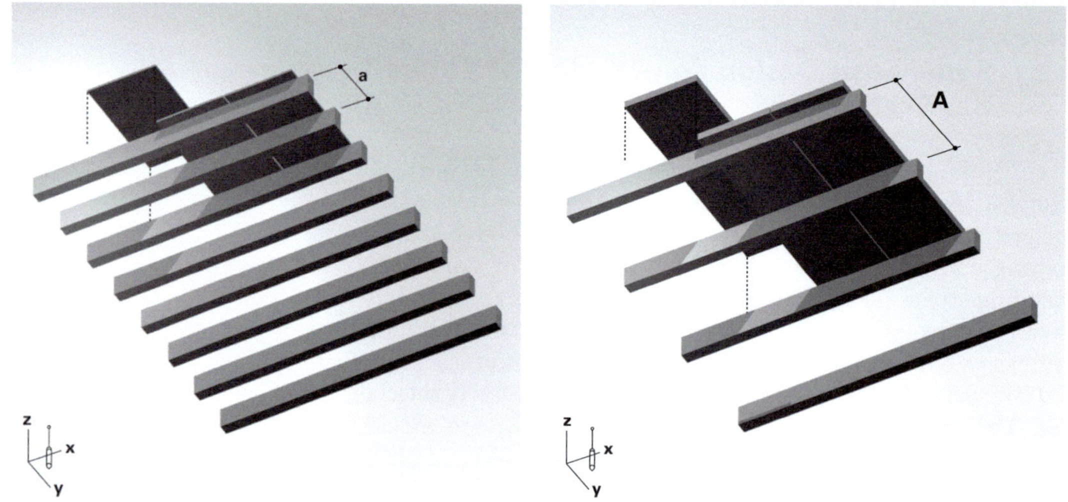

Fig. 3.91 (Left) Covering the interspaces between the beams with thin plates. The beam spacing **a** must be matched to the load-carrying capacity of the plate. (Right) Thicker and thus more load-bearing plates permit larger beam spacing **A**. Or vice versa: larger beam spacings result in thicker plate material

▶ *Sect. 3.2.2.2.3 Concentric Curved Supports*

quently are dimensioned stronger than if they were closer together, despite bridging the same span (in → **x**) as in the other variant—i.e. the distance between the main supports.

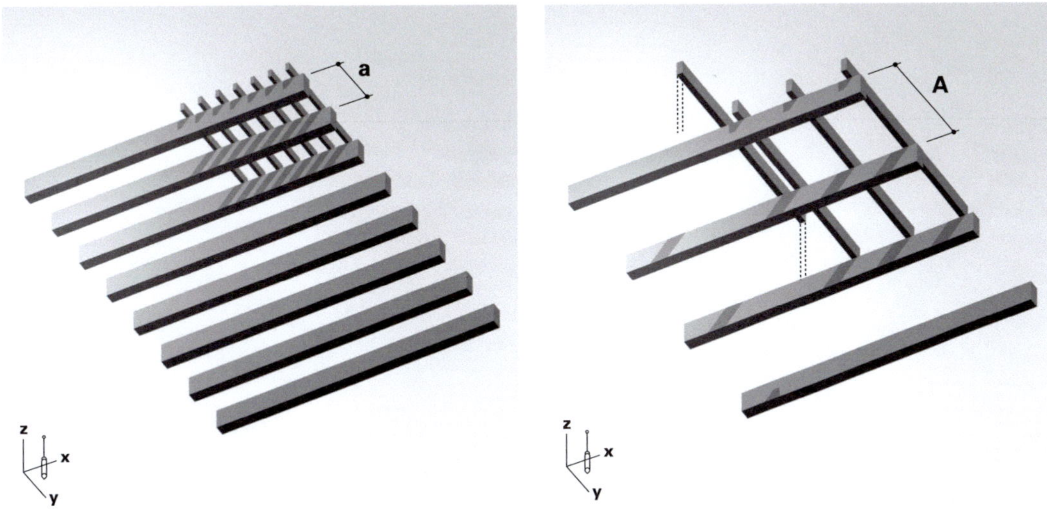

Fig. 3.92 (Left) Main beams with close spacing require only small secondary beams, since these are bridging only a small span (**a**). (Right) Large main beam spacings require thicker secondary beams since they are laid over a larger span (**A**)

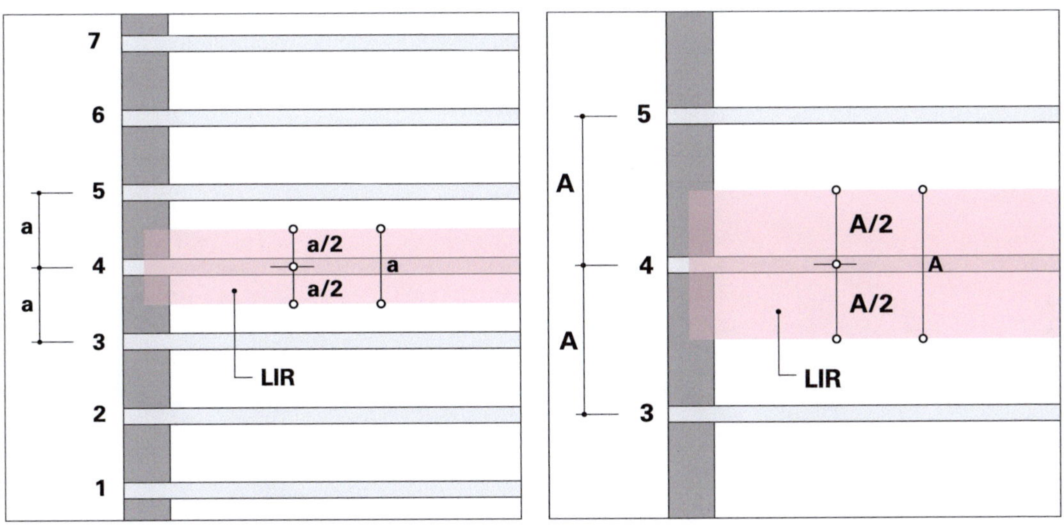

Fig. 3.93 (Left) Load-infeed range (**LIR**) of the beam **4** at beam distance **a**. (Right) Load-infeed range (**LIR**) of the beam **4** at beam distance **A**

Again, it results in principle from the constructional logic of this system that the respective subordinate rod layer, or plate, bridges a **smaller span** than the superordinate one. If this were not the case (as hypothetically depicted on Figs. 3.94, left, and 3.95, left), it would be sensible to prefer for the subordinate skein the shorter span in the direction of the superordinate one (as on Figs. 3.94, right, and 3.95, right). This, however, would reduce this principle to absurdity, since the system, as it were, would thereby be transformed into variant **2**. There is nevertheless an exception to this rule in the case of **indirect load transfer**, as can be inferred from the case on Fig. 3.96, left.

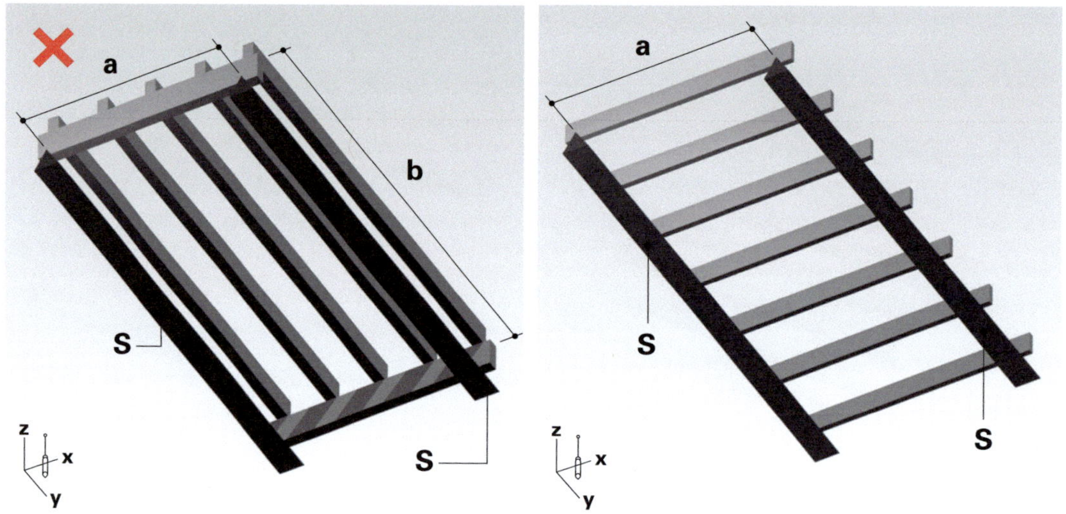

Fig. 3.94 Field of a floor construction according to design variant **3** (final panel not shown) with secondary beams over the larger span **b** and with main beams over the smaller span **a**. If a support option such as the illustrated linear supports **S** exists (direct load transfer is then possible), this bar arrangement is nonsensical from a structural point of view. Instead, it is recommended to use the smaller span **a** as shown on the right (see also the visualisation of the load-transfer path on Fig. 3.95, left). (Right) Bar arrangement with only one bar position over the short span **a** as a useful alternative to the bar arrangement on the left (see also the visualisation of the load-transfer path on Fig. 3.95, right)

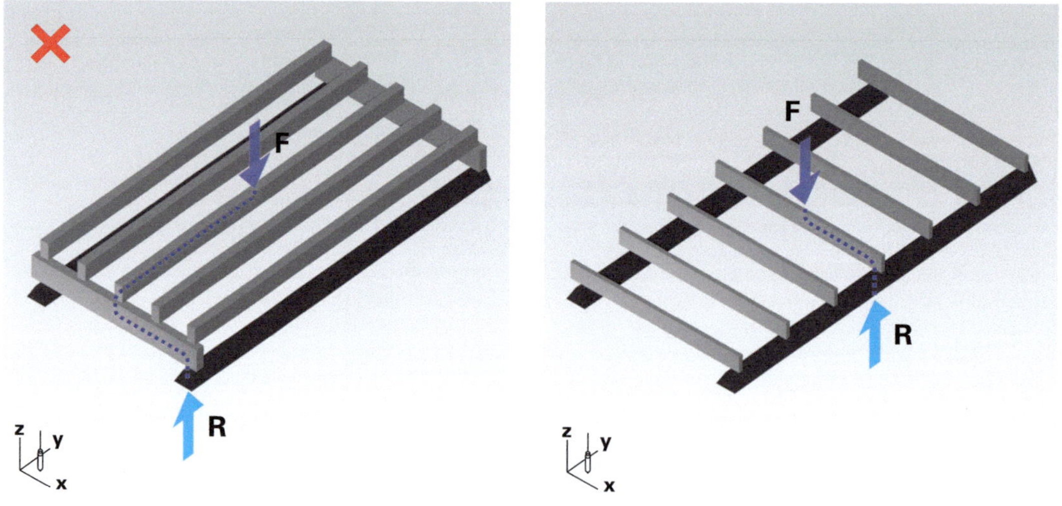

Fig. 3.95 (Left) Load transfer within the structure on Fig. 3.94, left. Long path from load application **F** to the bearing reaction **R**. (Right) Load transfer within the structure on Fig. 3.94, right. Significantly shorter force path from the attacking load **F** to the bearing reaction **R**

The choice of a certain combination of spans and bar distances can be made out of different considerations:

- **Structural depths**: In principle, the rule is that excessive bar depths can be avoided if a **main bar**, which has a larger

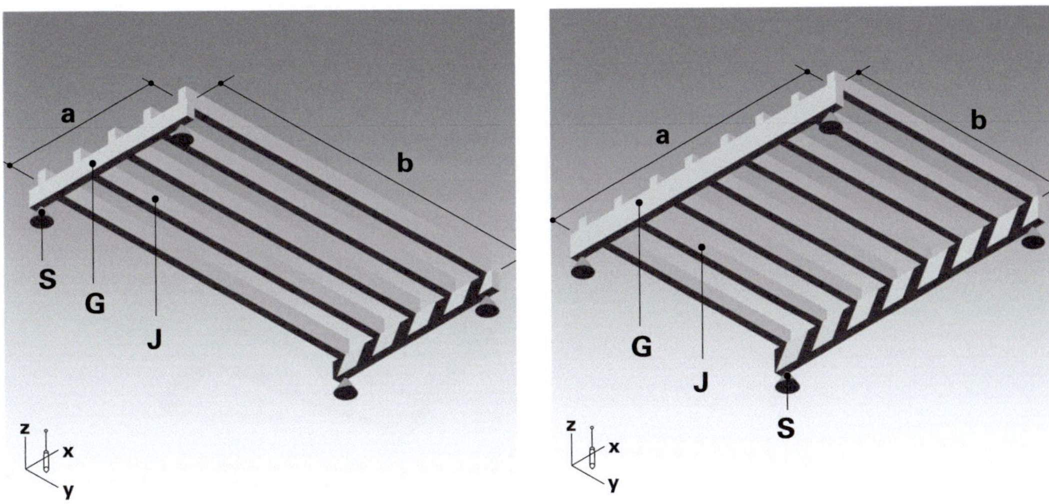

Fig. 3.96 (Left) In the case of a punctual support, it may be useful to place the main carrier over the longer span (**b**), the secondary carrier over the shorter (**a**). This is an **indirect load transfer**. The span of the girders **G** is significantly smaller than that of the joists **J**. As a result, the girders are subjected to considerably less stress than on the right and can be better assimilated to the joists in their dimensions. The structural fields are then oblong rectangular. **S** point support. (Right) The girders are comparatively heavily loaded since they must bear the added load of the joists. The spans of the girder and the joists are here of a comparable magnitude, the structural field is therefore square to oblong rectangular. The girders **G** must be dimensioned much stronger than the joists **J**

load-infeed area than the secondary bars for systemic reasons, and is thus more heavily loaded from the beginning, is placed over the **shorter span** (Fig. 3.96, left, → **x**). If it spans over the longer one, it assumes disproportionate dimensions because its required structural depth increases in the second power while the span increases only linearly (Fig. 3.96, right, → **x**). These conditions, which are causally related to the punctual support of the system with indirect load transfer, contradict, as already mentioned above, in some way the otherwise valid principle which postulates the reduction of the spans with a decreasing hierarchical level.

Cf. the comparison on Fig. 3.96

- Due to its structural depth and material nature, the top-end **areal component**—for example, a boarding for a floor or a brick tiling on a roof—predetermines a reasonable span—and thus a **bar spacing** for the next lower bar position, here for the secondary bars. It is reasonable to choose a much wider span for them. In this way, specific pre-settings for the design can be derived.
- **Bottom view** of the ceiling or the roof: From aesthetic and spatial considerations, it may prove to be useful to form stronger caesura in the construction—by means of powerful, further distant main bars—or, conversely, to make the set of bars more uniform and continuous—with narrow rod distances.
- Under certain conditions, the integration of **supply** and **disposal ducts** within the design can set certain bar depths

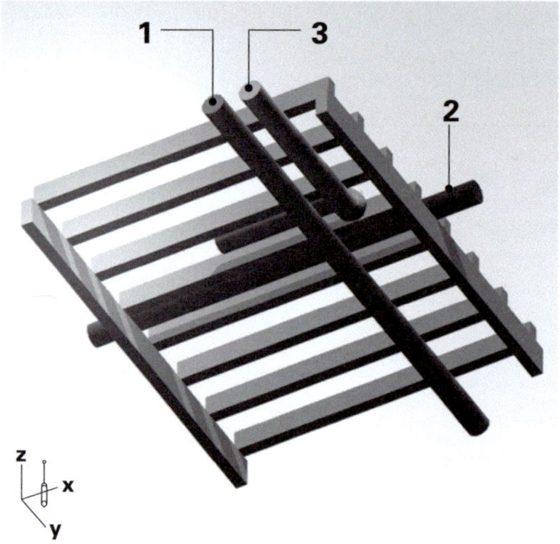

Fig. 3.97 Stacked carrier layers, as often occur in directional systems, permit simple routing of ducts in the deck construction. Both positions enable installing in two perpendicular directions: Pipes can either be routed in the lower rod position along the length (**1**), in the upper across (**2**), or by cranking downwards and upwards in the two perpendicular directions (**3**)

to provide free space for the management of the conduits (Fig. 3.97). From these structural depths, useful bar spans and spacings can then be derived.

The hierarchical grading of stacked bar positions can be carried out analogously to the principle according to variant **4**, in which the following sequence occurs (cf. the example on Fig. 3.98, left):

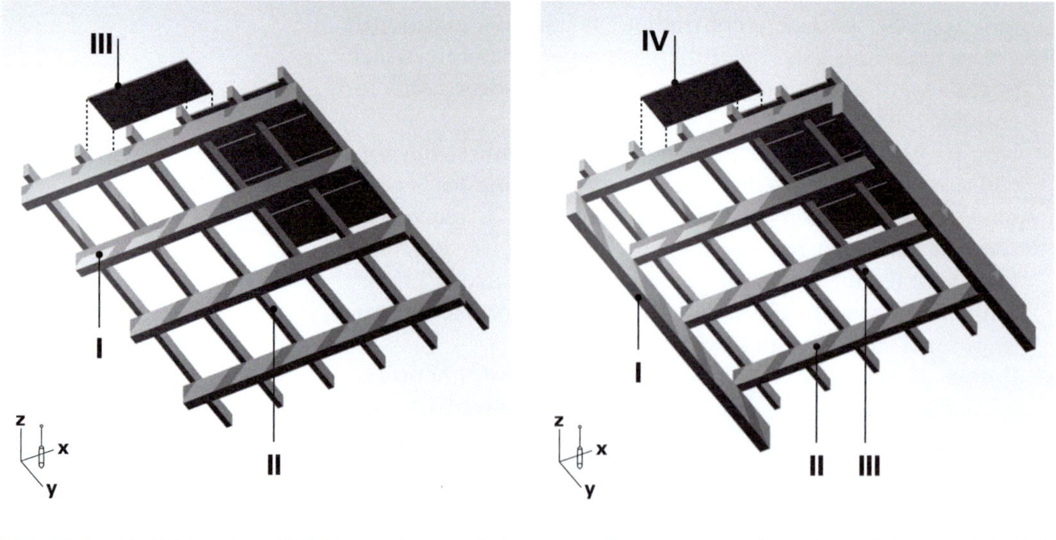

Fig. 3.98 (Left) Covering with 3 hierarchies: a main beam layer (**I**), a secondary beam layer (**II**), and a plate (**III**). (Right) Covering with 4 hierarchies: a main beam layer (**I**), a first secondary beam layer (**II**), a second secondary (or tertiary) beam layer (**III**) and a plate (**IV**)

- **I main** beam;
- **II secondary** beam;
- **III** final **plate**.

It can also be executed with more than three hierarchy levels. In the example shown on ◘ Fig. 3.98, right, a further bar position was introduced so that **four** hierarchical levels can be identified there:

- **I girder**;
- **II main** beam of first order;
- **III subcarrier** of second order and, additionally,
- **IV** finishing **plate**.

The principle of essentially **one-way load transfer** in the systems of hierarchically stepped bars, which are also referred to as **directional systems**, has already been described above in ▶ *Sect. 3.2.1*. When transferred to graded bar systems, it comes into effect as follows (◘ Fig. 3.99):

***Vol. 1**, Sect. 26.9.4 Element Composed of Ribs Oriented One-Way*, cf. Figs. 26.209, 26.210, and 26.211

- The topmost flat element, the final plate, is subjected to a **point load**. This is redirected by means of bending along the spanning direction of the plate, i.e. **one-way**, onto the two adjacent bars belonging to the next lower bar layer, on which the plate rests. In the case of load cross-distribution, further adjacent bars are also loaded.

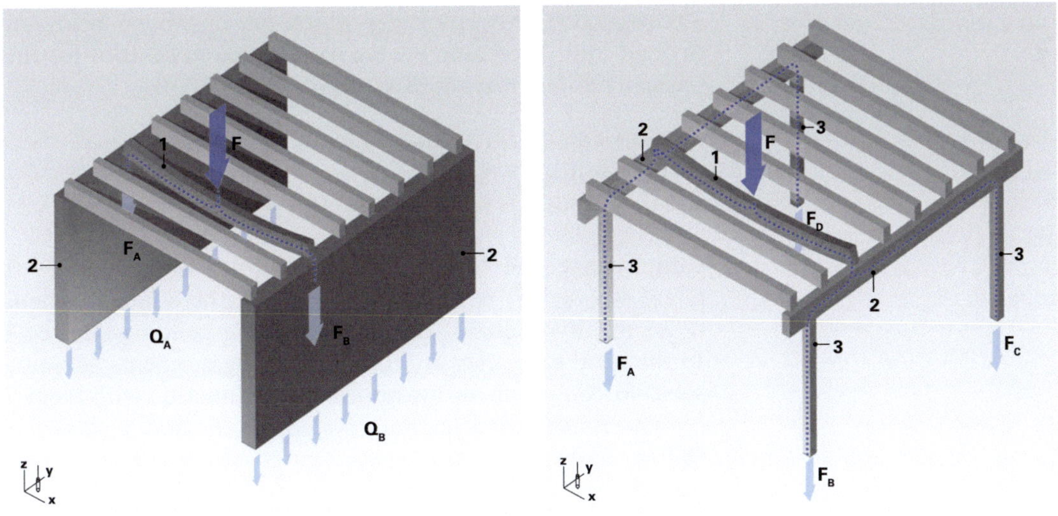

◘ **Fig. 3.99** (Left) **One-way** load transfer is characteristic of directional systems: It happens longitudinally along the beam to the load-bearing wall panels (**direct** load transfer). **1** joist, **2** wall panel. (Right) Hierarchically tiered load-bearing system: The load transfer is successively carried out in three stages: from the joist (**1**) to the girder (**2**) and from there into the column (**3**) (**indirect** load transfer)

- These two bars, possibly including further collaborating ones, in turn transfer their load share via bending along their own rod axis—that is, **rotated** by **90°** but again **one-way**—to the hierarchically superior rod position placed underneath, etc.

The advantages of **stacked systems** are:

- The possibility to mutually connect rods of different hierarchies at right angles by simply **placing** them on top of each other. No complicated connections are required.
- The possibility of designing supported beams as **multi-span carriers**, whereby the **bending moments** can be significantly reduced. Prerequisites are bars that can bear moments with alternating signs, such as wooden square cross-sections.
- The possibility of routing **ducts** within the bar spacings (◘ Fig. 3.97). Since, usually, at least two bar hierarchies are staggered in height, one can change the direction by 90° by simply routing the line into the next higher bar position.
- The possibility to **conduct air** within the bar spacings for the purpose of ventilating the construction. As in the case of ducts, air movement can be carried out in two directions with at least two bar positions.

Vol. 3*, Sect. 14.2.3 Flat Roofs*

3.4 Form-Related Issues of Axially Stressed Structures

The structures for coverages which are examined below are divided into three broad categories, which are linked to three basically different types of **stress** (◘ Fig. 3.100):

- predominantly **compressive**-stressed structures (**c**);
- predominantly **bending**-stressed structures (**b**);
- predominantly **tensile**-stressed structures (**t**).

Compressive and tensile-stressed load-bearing structures (**c**, **t**) are stressed by **normal** forces and are therefore considered to be **axially** stressed—in contrast to the bending-stressed **b**. In the ideal case, they are free of bending moments as well as shear forces, which always appear in conjunction with them.

The schematised support systems **c**, **b**, and **t** shown on ◘ Fig. 3.100 are already special cases of compressive, bending, and tensile-stressed structures, namely due to their specific combination of shape and support (arch—beam—rope). They will, however, be henceforth used as substitutional recognition symbols for the three groups because of their importance for the covering of rooms in building construction.

Whereas the group of bending-bearing structures (**b**) transmits loads primarily via the **internal** lever arm, i.e. by the **structural depth** of the component, in the case of axially

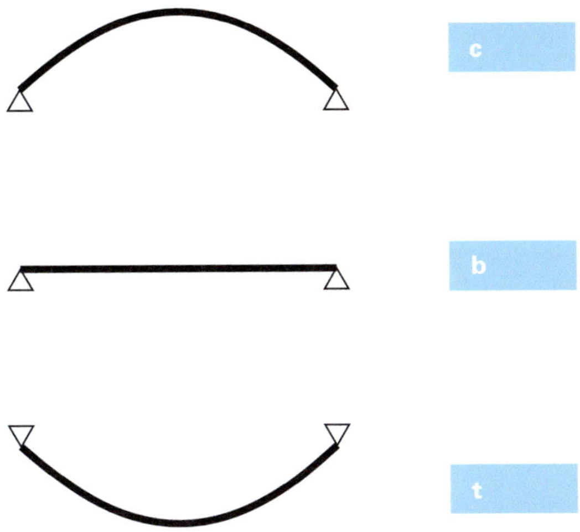

■ **Fig. 3.100** Structures subjected to three types of stress: compression **c**, bending **b**, and tension **t**

stressed ones (**c**, **t**) it is the **external** lever arm (or **camber f**), as measured between the normal force within the cross-section and the reaction at the bearing, which play a decisive role during load transfer (■ Figs. 3.101 and 3.102). Consequently, in the case of the latter, the **thickness** of the component is not decisive, but rather its **shape**. For this reason, the importance of *shaping* of axially stressed structures will be particularly emphasised in the following sections.

This becomes clear from the observation in the two charts on ■ Figs. 3.103 and 3.104; cf. also the diagram on ■ Fig. 3.100

3.4.1 Interrelations Between Support, Form, and Stress

In ***Volume 1***, basic considerations were already made as to the influence of the **shape** of load-bearing structures on the **support reactions** and **stresses** on the component's cross-section in the case of exemplary compressive- bending- and tensile-stressed load-bearing systems. The following charts on ■ Figs. 3.103 and 3.104 take up these reflections. The graph in ■ Fig. 3.103 again reproduces the case examined in ***Volume 1***. The conditions described there for these structures, which usually are referred to in building practice as strutted frames (compression-stressed), beams (bending-stressed), and suspension structures (tensile-stressed), assume an external loading in form of a **punctual load F**, respectively centred. Strutted frames and suspension structures are each shaped in such a way that they do not experience bending stress under the given load conditions.

Vol. 1, *Sect. 26*; cf. Fig. 26.43

■ Figure 3.104 shows the analogous case for a modified external loading, namely for a **uniformly distributed** load **q**. The supporting systems considered are the arch (compressive-stressed), again the beam (bending-stressed), as well as

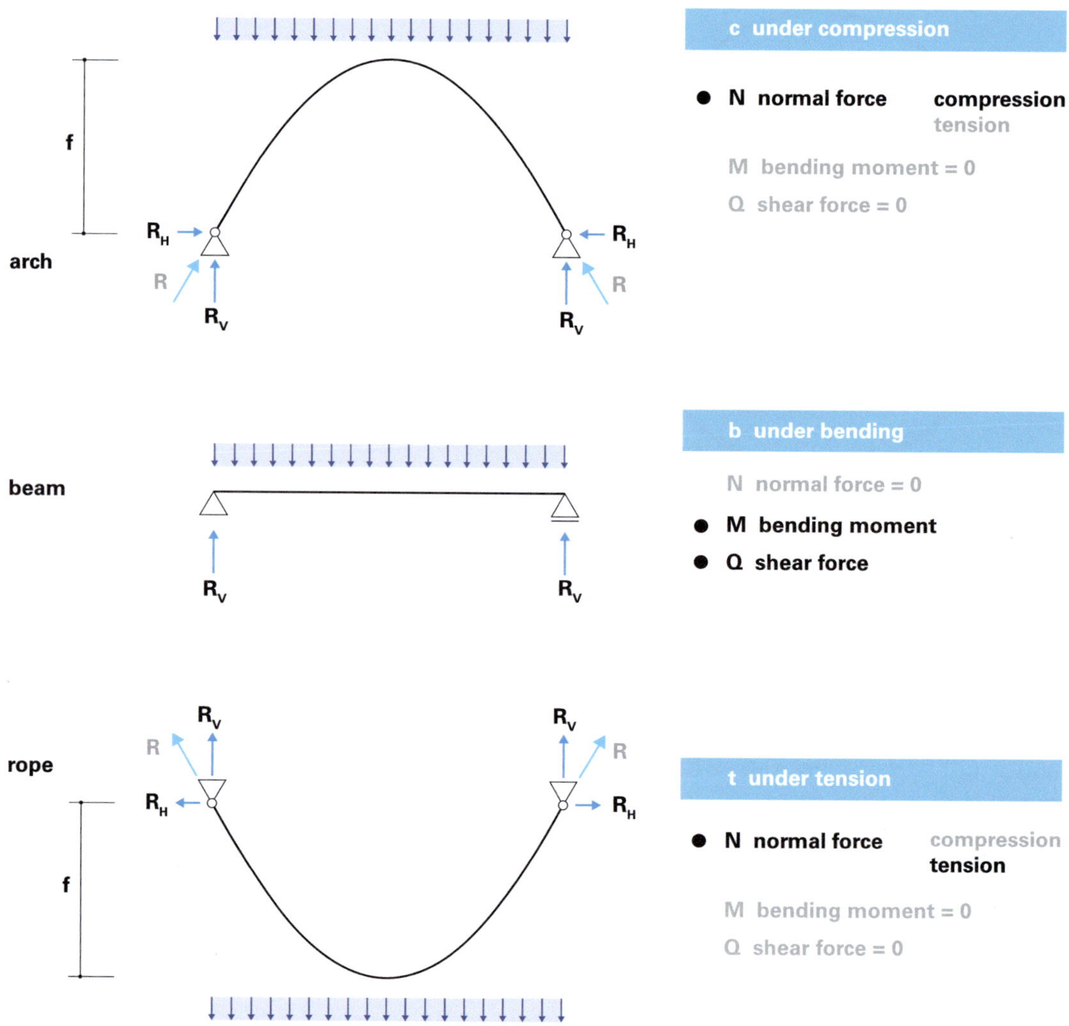

Fig. 3.101 Basic load-bearing behaviour of the three main categories of coverings as underlie the following considerations. The **dominant stresses** (**N**, **Q**, **M**), which are decisive for the understanding of the load transfer, are presented respectively. In the case of the three load-bearing structures, a **double-sided** support is always assumed (in the case of two-way load bearing: four-sided), as is the case with the overwhelming majority of coverings in building construction. Gradual intermediate steps with different camber measures **f** are examined on Fig. 3.104

► *Sect. 3.4.2 The Concepts of Funicular Line and Thrust Line*

the rope (tensile-stressed), or the tensile member shaped like a hanging rope. These are also shaped in such a way that no bending moments occur.

The statements in ***Volume 1*** addressing these case studies are valid here as well. Because of their extraordinary structural significance, they will be recalled again at this point:

- **Bending** strains the cross sections, and thus the material, very unevenly compared to **axial** stress, i.e. pure compression or tension. Material efficiency, as is particularly necessary in lightweight construction, requires avoiding bending as much as possible. In this respect, the beam (**b**, see diagram below left) performs very badly.

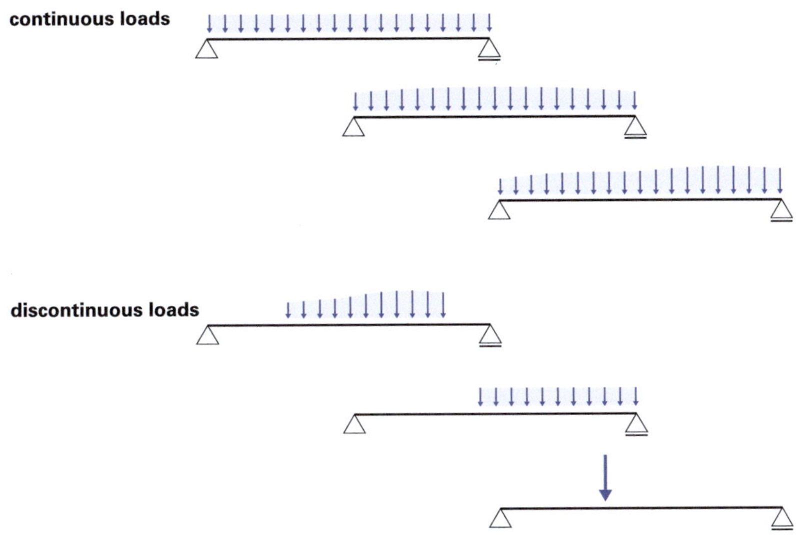

Fig. 3.102 Examples of continuous or uniform and discontinuous or non-uniform loads

- However, the examples show that this advantage of axially stressed systems comes at the expense of increased **structural height**. The compressive and tensile-stressed structural systems shown presuppose a static height or **camber f**, which is, under comparable boundary conditions, in any case larger than the structural depth **h** of the bending-stressed beam. Under certain conditions of use of the load-bearing structure, this may lead to preference being given to bending-stressed systems despite their poor material utilisation.

  Cf. Figs. 3.103 and 3.104

  ▶ *Sect. 3.1.6.2 The Covering*
- A further disadvantage of axially stressed systems (**c**, **t**) is the existence of **horizontal** forces **H** at the supports, which in the case of compressive- and tensile-stressed variants are respectively oriented oppositely (Fig. 3.105). They become larger with decreasing height, i.e. camber **f**;
- Coverages are always connected with **force deflection**. The larger the amount of force deflection, i.e. the span, the larger the stress acting on the material. Direct force transfer as in cases **1** and **9** on Figs. 3.103 and 3.104 is economical in material terms, but is not suitable for coverage and can only be used at component level (as in the case of lattice-girder bars).

  ▶ *Sect. 3.1.6 The Elements of the Structural Cell*
- Conditions with compressive (**c**) and tensile-stressed (**t**) structures are **analogously symmetrical** with respect to the horizontal line, namely in part with reversed signs: Compressive and tensile stresses are quantitatively comparable. The same applies to the horizontal support reactions **H**, which respectively have opposite signs. On the other hand, the vertical support reactions **R** have the same sign.

Fig. 3.103 Illustration of the stress on a planar bar structure when the geometry changes step by step. On this illustration, the transition is shown from the axially loaded bar, through a strutted frame or suspended system made up of straight bars, to a beam loaded transversely to its axis. Analogously, the transition from the straight bar to the curved arch or tensile member or rope is shown on Fig. 3.104. The essential differences of stress acting on the component cross-section in the respective stages are discussed in detail in ***Volume 1***, *Chap. 26 Force Transmission*. The diagram shows the efficiency of bending-free load-bearing systems compared to those subjected to flexural stress. The cases working under tensile stress (in each case, the lower halves of the diagrams) are particularly efficient since they are not subject to buckling, unlike the compressed ones (upper halves). The stresses presented naturally apply to the defined boundary conditions such as loading and support. Here, idealised conditions are assumed for concentrated load (**F**); on Fig. 3.104, for distributed load (**q**). Distributed or continuous loads **q** in construction are, for example, dead loads. Concentrated or discontinuous loads (**F**) are often live loads. In both diagrams, the components follow the respective thrust lines of the load cases, i.e. with the exception of the beam, they can be regarded as bending-free. Deviations of the component geometry from these thrust lines necessarily lead to bending and thus to increased stresses in the cross-section (cf., in this respect, Fig. 3.111), to an increasing extent the greater the deviation. Analogously, when the geometry of the component is unchanged, changes in the load lead to comparable additional bending moments (cf. Fig. 3.112)

- There is **no continuous transition** between compressive-stressed and tensile-stressed load-bearing structures (**c**, **t**) on the one hand and bending-stressed load-bearing structures (**b**) on the other. Rather, a *qualitative* change, or a kind of **snapping-through** into the state of the beam, occurs.[4] This term is derived from the arch statics, and is actually applied to extremely flat arches, which run the risk of failing by reverting into the non-intended beam form—in this case, not virtually but effectively in mechanical terms.

Under a given load, as already mentioned, not only the **form** of the load-bearing system determines the stresses, but the combination of **form** and **support**. This is clearly shown by the two examples on Fig. 3.106, each having the same shape, but different bearings. They then act in each case as curved bending beams with a sliding bearing (cases **A** and **B**) and as an arch or a rope with two hinged (non-displaceable) bearings (cases **A′** and **B′**). The differences in stress are obvious.

3.4.2 The Concepts of Funicular Line and Thrust Line

3.4.2.1 Funicular Line

The tensile-stressed structural systems (**t**) shown on Figs. 3.103 and 3.104 in the lower halves of the diagrams are each formed according to the **funicular lines** associated to the acting external loads, i.e. **F** for strutted frame/suspension structure and **q** for arch/rope. The funicular line reproduces the geometry of an imaginary hanging rope under a specific load and given boundary conditions—like the suspension points

4 The term 'snapping-through' is understood here in the figurative, not in the statical theoretical sense. It is merely intended to denote a virtual intermediate state in our sequential observation.

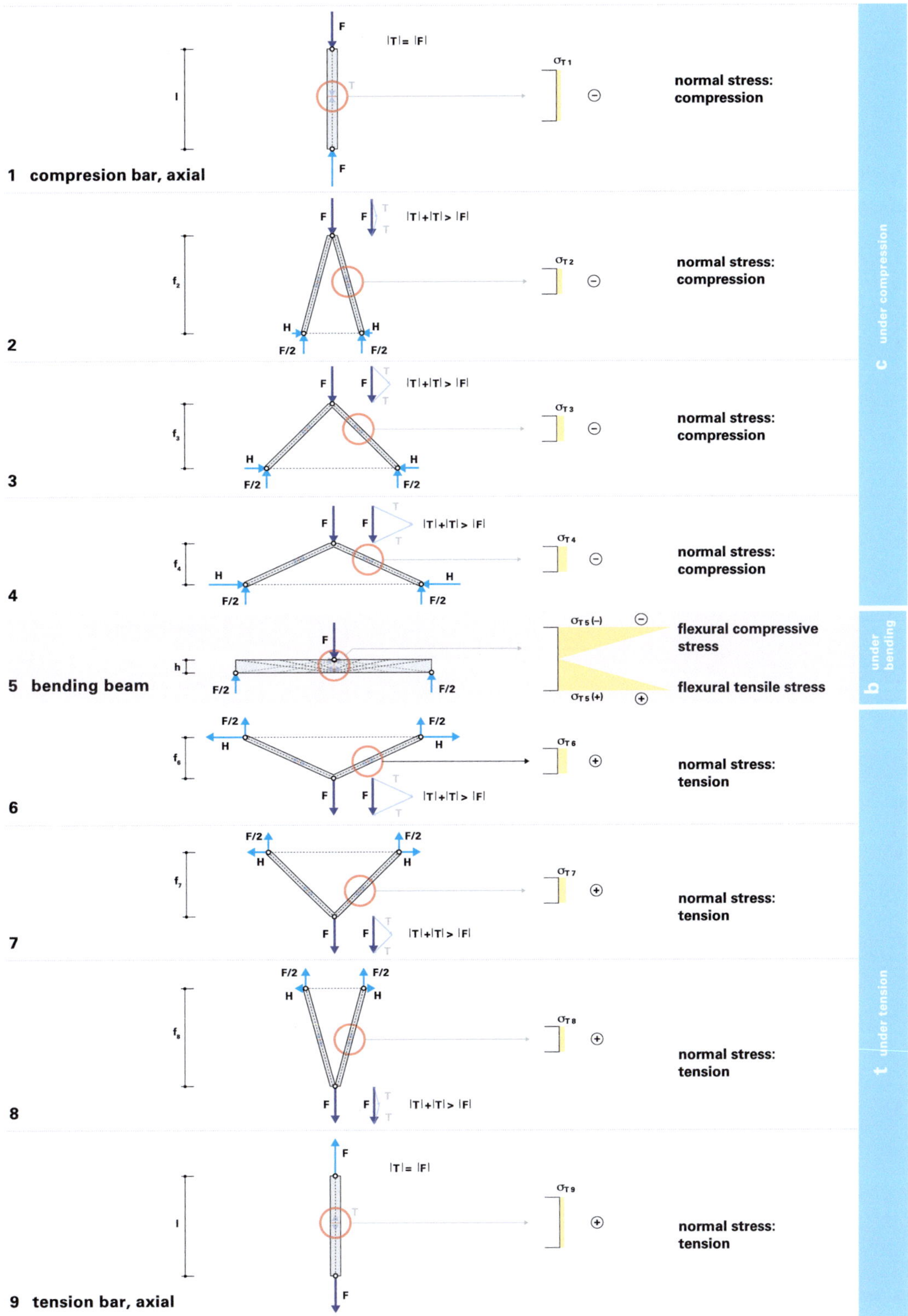
1 compresion bar, axial
|T| = |F|
normal stress: compression
2
|T|+|T| > |F|
normal stress: compression
3
|T|+|T| > |F|
normal stress: compression
4
|T|+|T| > |F|
normal stress: compression
5 bending beam
flexural compressive stress
flexural tensile stress
6
|T|+|T| > |F|
normal stress: tension
7
|T|+|T| > |F|
normal stress: tension
8
|T|+|T| > |F|
normal stress: tension
9 tension bar, axial
|T| = |F|
normal stress: tension
c under compression
b under bending
t under tension

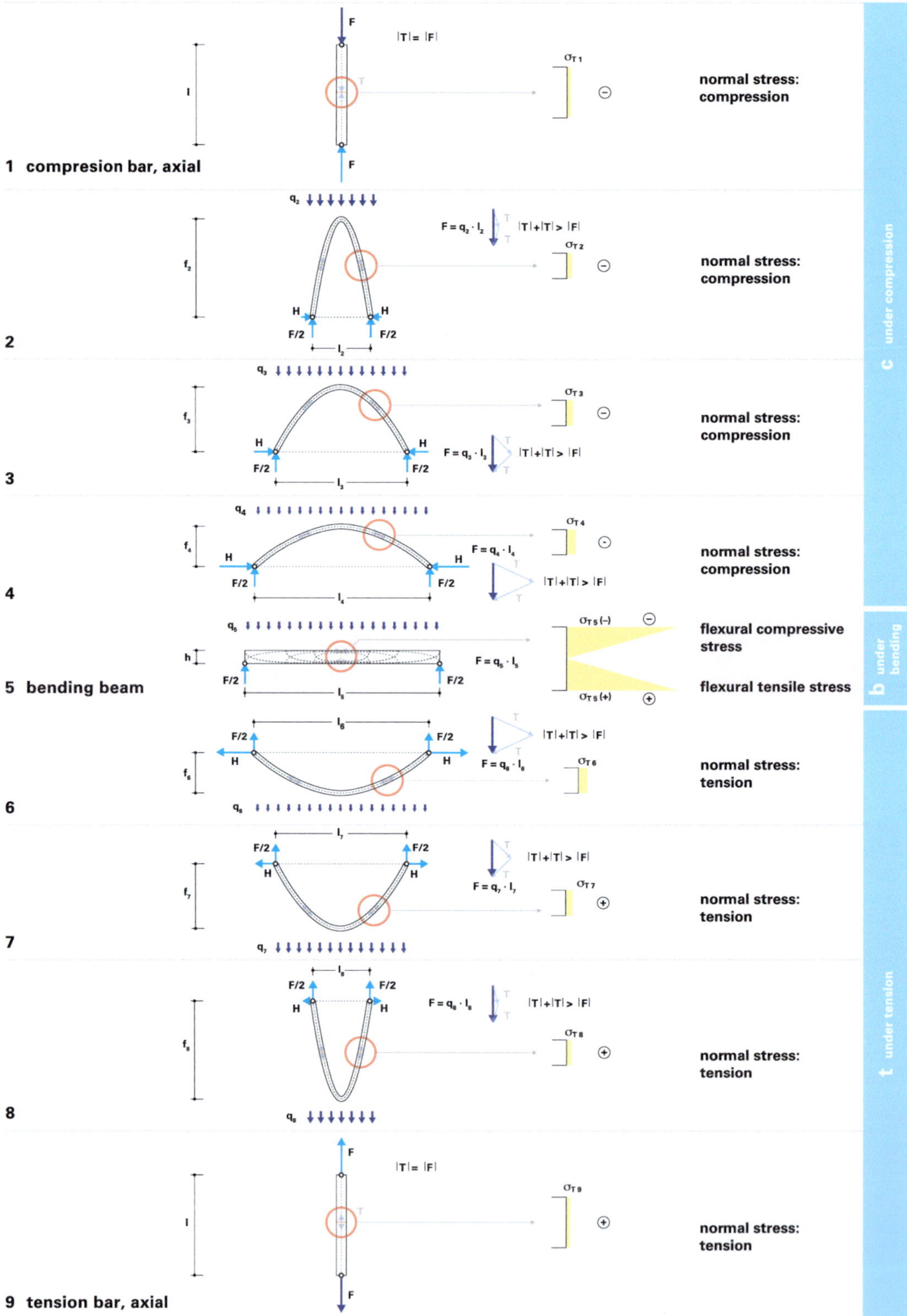

Fig. 3.104 See caption of Fig. 3.103

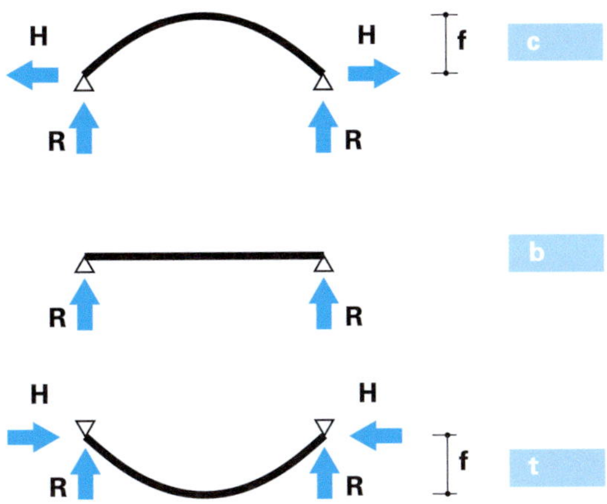

Fig. 3.105 Reaction forces at the supports of the three main types of structures

and the rope length. A rope is by definition **not bending-stiff**, so it cannot absorb bending moments. The suspension form of the funicular line is thus **free of bending**. Even if a (rigid) tensile member is loaded instead of a rope, it remains almost free of bending, as long as it follows the geometry of the funicular line.

Consequently, the tensile stressed systems (**t**) of the two diagrams on Figs. 3.103 and 3.104 are valid for a non-bending-stiff rope as well as for a bending-stiff tensile member. The conditions are identical

If the loading on the rope structure changes, it will adopt a new form under the new load pattern. Due to its lack of bending stiffness, it adapts to the respective load conditions by **strain-free deformation** and, in turn, adopts the form of the funicular line—then modified—appropriate to the new loading conditions (Fig. 3.107). A rope cannot be different from the bending-free funicular line, since it lacks the bending stiffness required to counteract the force effect by means of a resistance (resistance moment).

On Fig. 3.107, different funicular lines for changing load patterns are shown

For this reason, rope structures belong to the movable load-bearing structures; cf. ***Vol. 1***, *Sect. 26.4.2 Movable Systems*

A **rigid tensile member** (Fig. 3.108) behaves differently from a rope. If the load pattern changes, according to whose corresponding funicular line the structure is shaped, it *cannot* adapt to the new conditions, i.e. to the new funicular line, because of its bending stiffness. It essentially retains its shape, undergoes **bending stress** and consequently only (minor) deformation. In addition, **shear forces S** emerge within the cross-section. The larger the deviation of the new funicular line from the form, i.e. the system axis of the load-bearing structure, the larger the bending moments and shear forces will be—as the formula in the marginal column makes clear. This deviation can be detected locally at each point of the system axis and is referred to as **eccentricity e**. The effective bending moment is then calculated from the product of **normal force** N and **eccentricity e**.

$\mathbf{M} = \mathbf{N} \cdot \mathbf{e}$, where **M** is the bending moment, **N** the normal force, and **e** the eccentricity

3.4.2.2 Thrust Line

In an analogy often surprising for non-specialists, the conditions described for tensile-stressed systems can be transferred

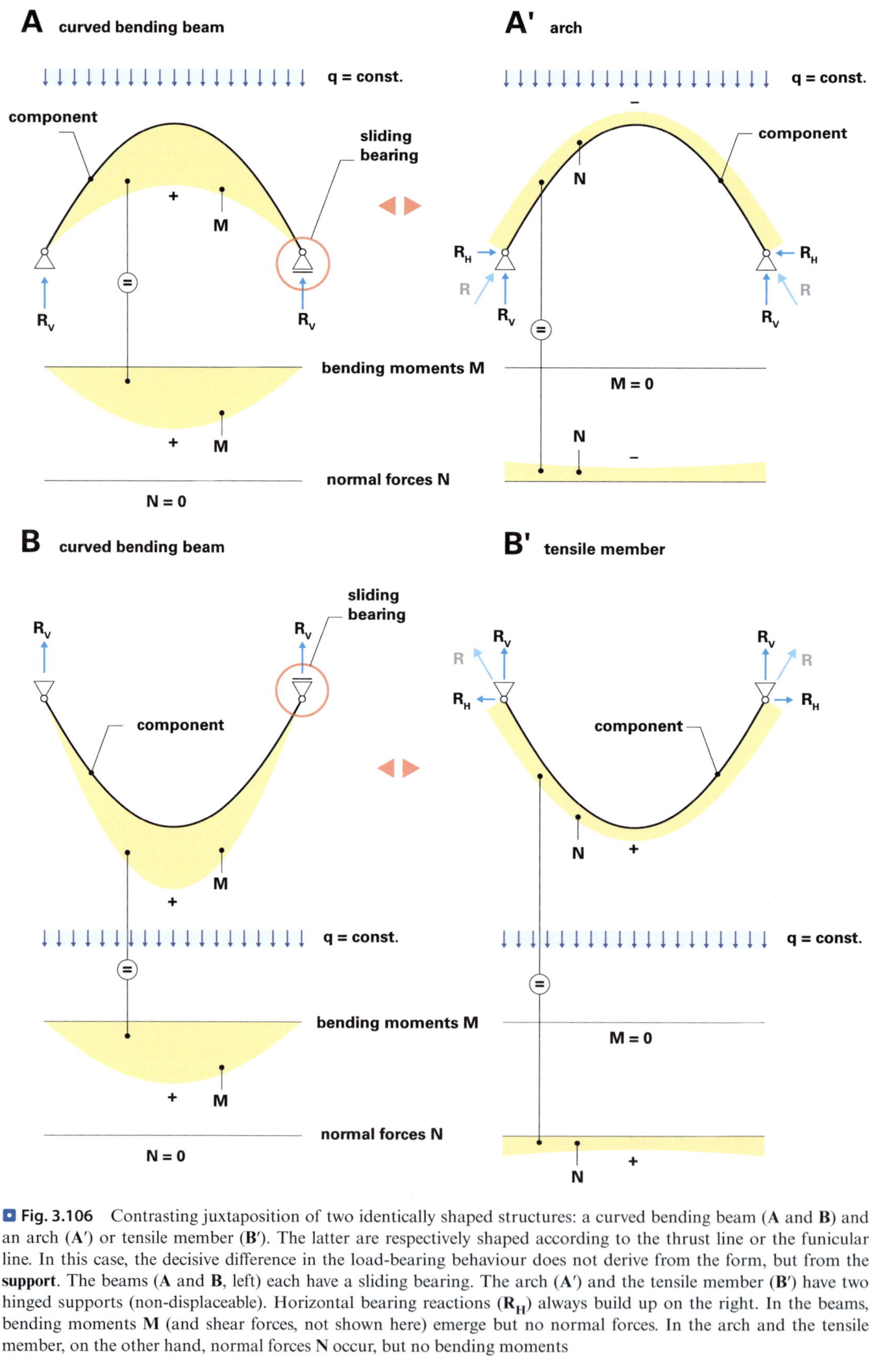

■ **Fig. 3.106** Contrasting juxtaposition of two identically shaped structures: a curved bending beam (**A** and **B**) and an arch (**A′**) or tensile member (**B′**). The latter are respectively shaped according to the thrust line or the funicular line. In this case, the decisive difference in the load-bearing behaviour does not derive from the form, but from the **support**. The beams (**A** and **B**, left) each have a sliding bearing. The arch (**A′**) and the tensile member (**B′**) have two hinged supports (non-displaceable). Horizontal bearing reactions ($\mathbf{R_H}$) always build up on the right. In the beams, bending moments **M** (and shear forces, not shown here) emerge but no normal forces. In the arch and the tensile member, on the other hand, normal forces **N** occur, but no bending moments

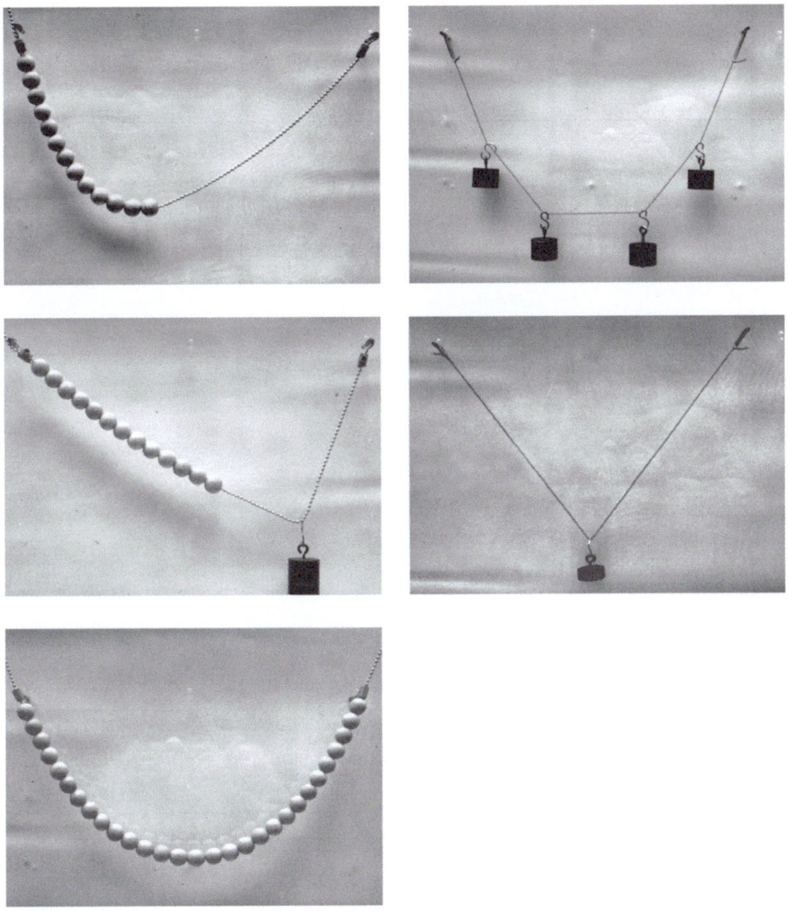

■ **Fig. 3.107** Different funicular lines resulting from different load patterns

to compressive-stressed (**d**) by **reversing** (■ Fig. 3.110). If a compressive-stressed load-bearing system is formed under a specific load and given boundary conditions (particularly support, hinges, etc., see, e.g. ■ Fig. 3.106) according to the **inverted shape** of the funicular line, i.e. its mirror image on the horizontal line, which a rope would assume under the same boundary conditions and identical loading, the structure is free of bending. There are also no shear forces. The reversal of the funicular line is called a **thrust line**.

■ Figure 3.110 shows various related funicular and thrust lines for changing load patterns

If the load pattern on the load-bearing structure changes again, adapting to the new thrust line, i.e. the reversal of a new imaginary funicular line, is naturally no longer possible by changing the shape. A rope cannot be subjected to compression because it would elude it due to lack of bending stiffness. Compression-bearing structures are therefore always resistant to bending. The shape of the load-bearing structure is retained and **bending moments M** and **shear forces S** arise within the cross-section, similar to the bending-stiff tensile member. These do not give rise to deformations without strain

3

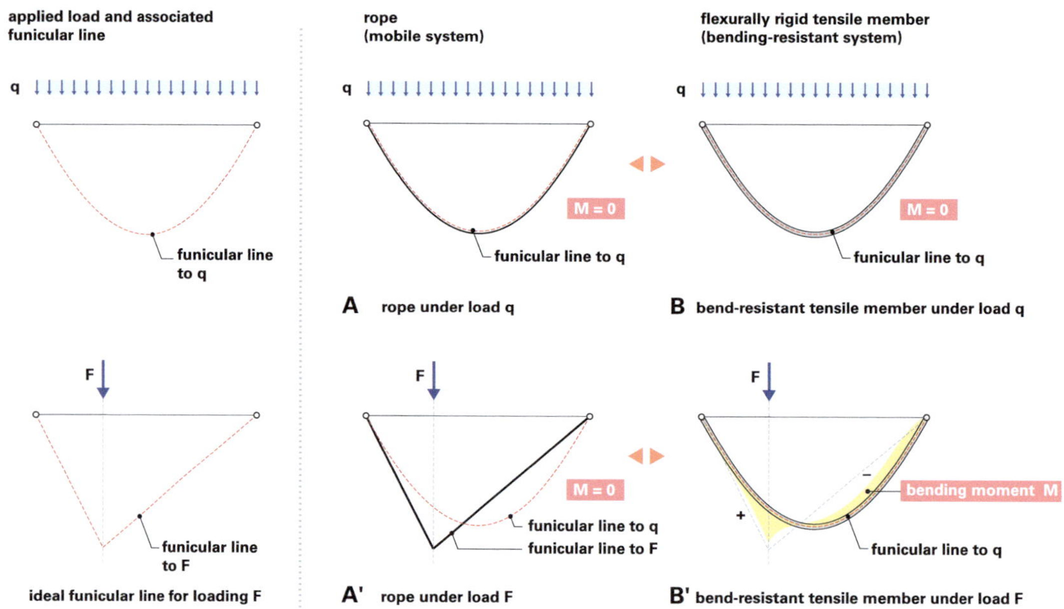

Fig. 3.108 Contrasting juxtaposition of the load-bearing behaviour of a (flexible) **rope** (mobile system) and a **rigid tensile member** of fixed geometry under varying loads **q** and **F**. Due to its lack of bending stiffness, the rope adapts itself to the respective funicular line and therefore always remains free from bending stress (**M = 0**). It is for this reason that it is considered a *mobile* structure. The bending-stiff tensile member (shaped according to the funicular line associated to **q**) reacts to the load **F**, and the thus altered funicular line, with bending moments (**B'**). The rope is therefore always minimally loaded in its cross-section (because of pure axial tension) and can therefore be regarded as extremely efficient. The **shape changes** associated with this favourable load-bearing behaviour, on the other hand, are often an obstacle for the convenience of the load-bearing structure. Rope structures must therefore often be stabilised additionally, as is the case in the structure shown on Fig. 3.109

Fig. 3.109 Tensile-stressed rope construction with combined stiffening by ballast (concrete slabs) and rope subtension (eng.: Schlaich, Bergermann & P)

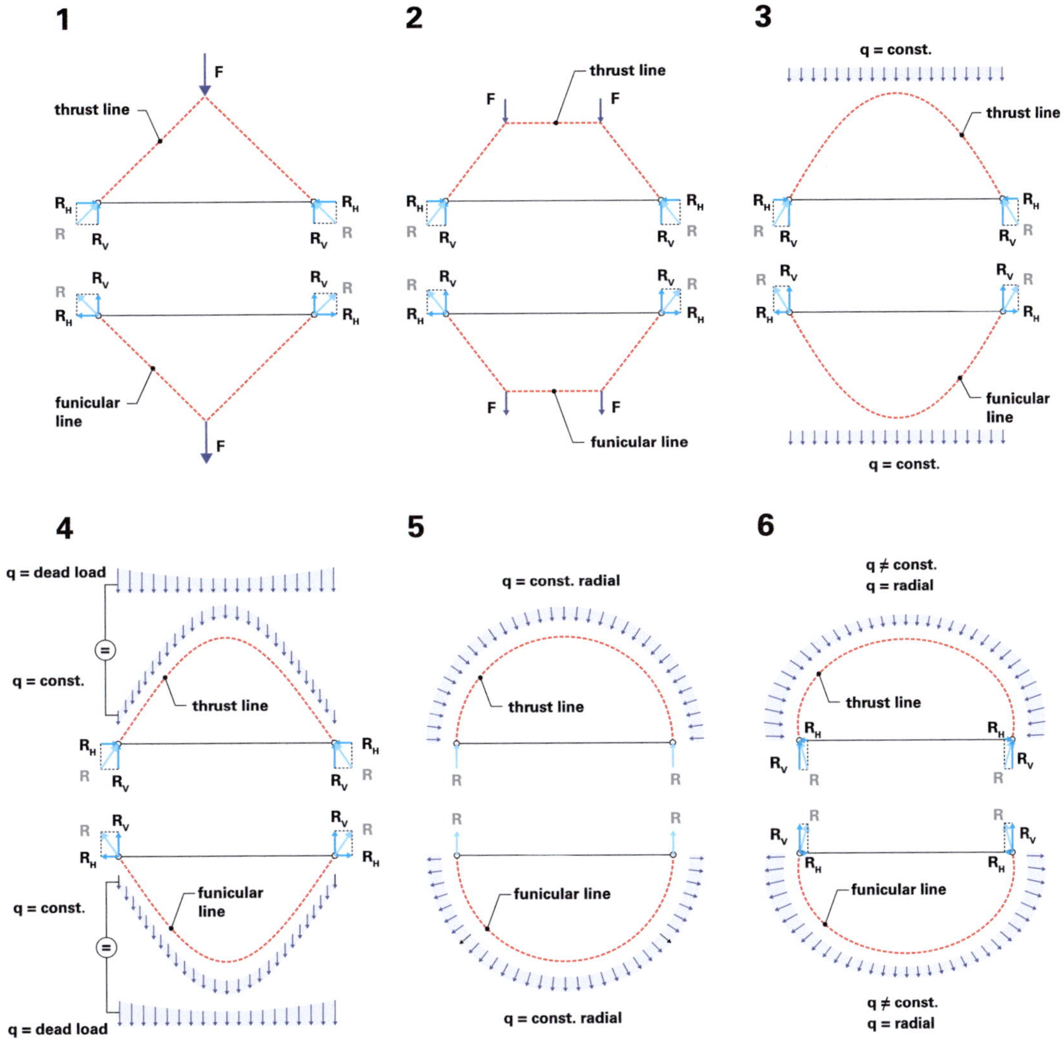

Fig. 3.110 Forms of **thrust lines** (above) and correlative **funicular lines** (below) under different loads with similar rod/rope lengths. The six cases: **(1)** single load, centred: straight trajectory (truss); **(2)**two single loads: polygonal funicular line; **(3)** constant horizontal linear load: second degree parabola; **(4)** own load: catenary (hyperbola); **(5)** constant radial linear load: circular arc; **(6)** non-constant radial linear load: ellipse

like in case of the rope. Again, the greater the deviation or **eccentricity e** between the thrust line and the system axis of the structure, the larger the bending moments and shear forces.

3.4.3 Deviations from the Thrust Line

Tensile-stressed structures are often **rope constructions** in building practice, which take advantage of the extreme material efficiency of this type of load transfer. The ideal state of a perfect correspondence between the funicular line and the system axis of the component is always given. For reasons of *usability*, not of *load-bearing capacity*, only the undesired ef-

Fig. 3.111 Deviations of the component geometry from the thrust line cause bending moments which are proportional to the eccentricity. The examples shown assume a constant horizontal linear load **q**; The thrust line is thus a quadratic parabola. Depending on which side the deviation from the thrust line takes place, positive or negative moments are produced. At the intersections between system and thrust line (**I**), the points of contraflexure are located. In case **0**—coincidence between the thrust and system lines of the component—there are **no moments** nor **shear forces**. Only **normal forces** then act in the component's cross-section. The material utilisation is optimal under these conditions

fects of the change in shape must be compensated for; or they must be kept within tolerable limits by suitable measures.

This includes, for example, a **bracing** or **ballasting**. With the latter, it can be ensured that a constant load pattern—namely the ballast—dominates over any changing loads. The deviations between the resulting funicular lines can be minimised in this way, and consequently also the associated shape changes.

We have seen that compression-bearing structures are always rigid and cannot adapt themselves to modified thrust lines. Deviations of the system axis from the thrust line are therefore possibly critical in this type of construction and must be carefully considered by the designer (Figs. 3.111 and 3.112). An *exact* correspondence between system and thrust line is difficult to achieve in construction practice for various reasons:

The favourable influence of the permanent load when there is a load **F** which does not conform to the thrust line Fig. 3.113 (e.g. an alternating load) can be seen by comparing examples 5 and 6 on Fig. 3.112

- As a rule, there is **no invariable** thrust line from the outset which could be taken into account when shaping the component, since load patterns in building practice almost invariably change over time. This is due to the fact that there are not only permanent loads such as dead load, but also live loads or other changing loads such as wind or snow. The significant factor is the *ratio* between permanent and changing loads—the so-called **load ratio**. The larger the proportion of permanent loads, the smaller the deviations from the thrust line.
- **Deformations** of an arch under load, i.e. elongations, will inevitably lead to a deviation from the thrust line when the component has been shaped according to it. The eccentricity between the system and the thrust line increases, in particular, if the load remains the same in its *distribution*—the thrust line thus remaining unchanged—but increases in *amount*—and the deformation as a result. Here, too, the disturbing effect of varying loads manifests itself, even if the load patterns are *affine* to one another. Displacements or settlements at the supports cause the same effect.
- Historic masonry arches, almost without exception, have a **circular** shape, and thus differ from the actual, mostly parabolic, thrust line. The reason for the preference of the circular geometry lies undoubtedly in the difficulty of producing non-circular, e.g. parabolic, arches, which entails custom-cut voussoirs. Circular arches, on the other hand, can be built from the same voussoir. Moreover, the

The conditions prevalent with the circular masonry arch are given on Figs. 3.113, 3.114, 3.115, and 3.116

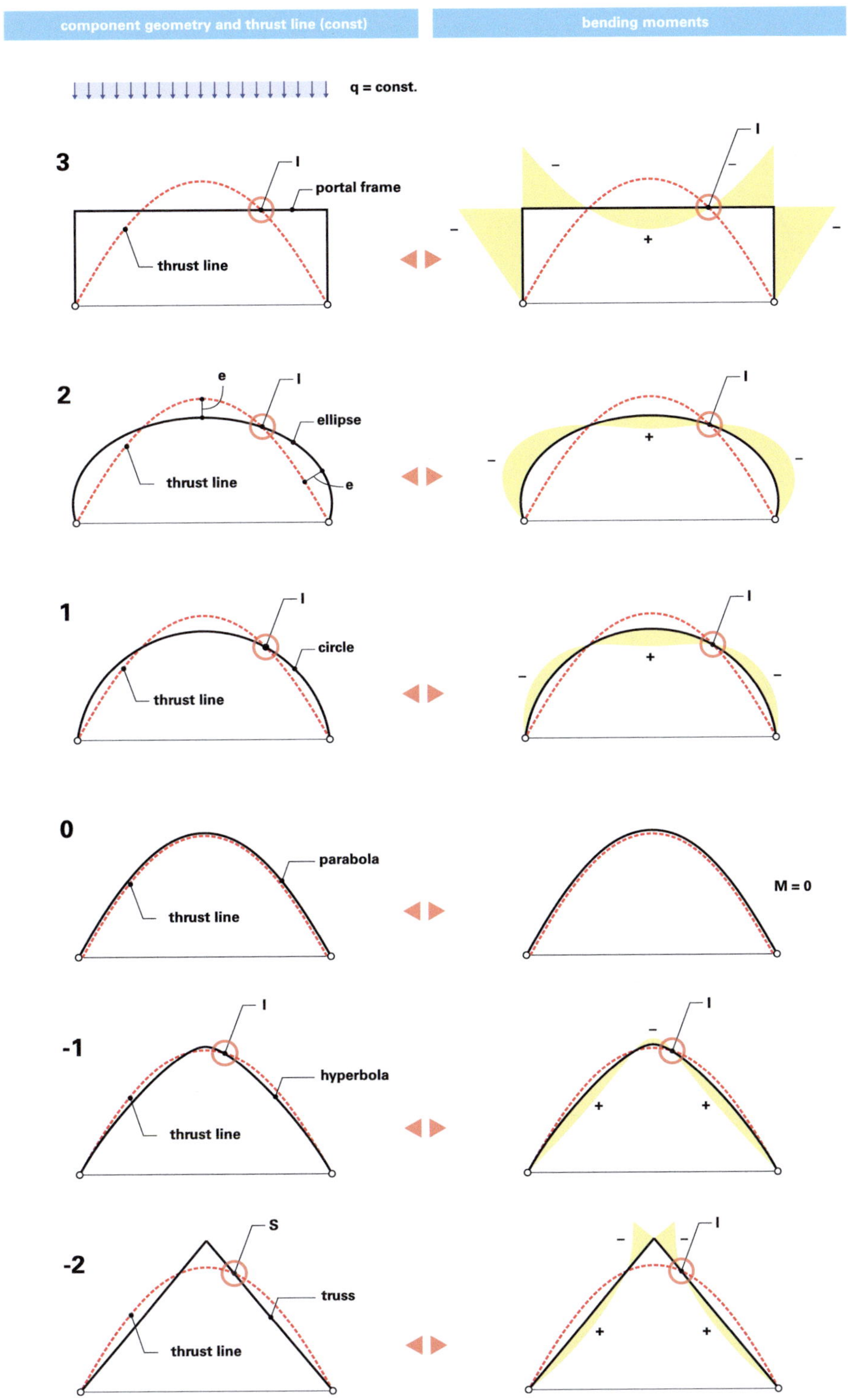
component geometry and thrust line (const)
bending moments
q = const.
3
I
portal frame
thrust line
2
e
I
ellipse
thrust line
e
1
I
circle
thrust line
0
parabola
thrust line
M = 0
-1
I
hyperbola
thrust line
-2
S
truss
thrust line

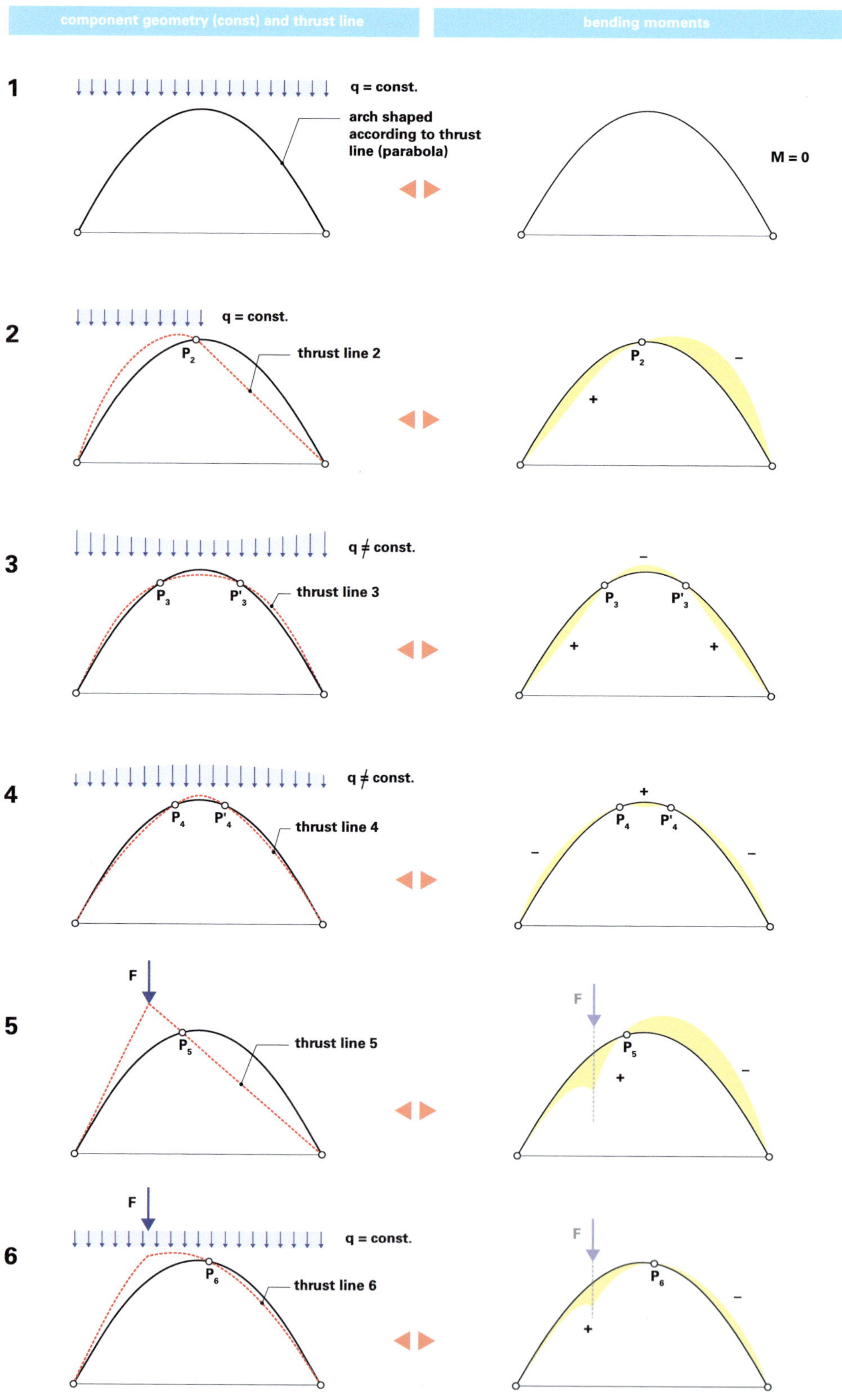
component geometry (const) and thrust line
bending moments
1
q = const.
arch shaped according to thrust line (parabola)
M = 0
2
q = const.
P2
thrust line 2
P2
+
–
3
q ≠ const.
P3
P'3
thrust line 3
–
P3
P'3
+
+
4
q ≠ const.
P4
P'4
thrust line 4
+
P4
P'4
–
–
5
F
P5
thrust line 5
F
P5
+
–
6
F
q = const.
P6
thrust line 6
F
P6
+
–

Fig. 3.112 Deviations of the component geometry from the thrust line. In this case, the component geometry is kept invariable. It corresponds to a quadratic parabola according to the load in case **1** (constant linear load). In cases **2** to **6**, the **load** is varied in each case so that deviations occur between the fixed system line of the component and the changing respective thrust line, which lead to bending moments. Alternating loadings, as usually occur in load-bearing structures, lead to this kind of bending stress, and in most cases require a minimum bending stiffness of the arch. In the case of combined loads (as shown in case **6**) of a constant linear load (**q**) and other loads (e.g. the single load **F**), the bending stress decreases increasingly the smaller the disturbing load (here **F**) is relative to the linear load (**q**)

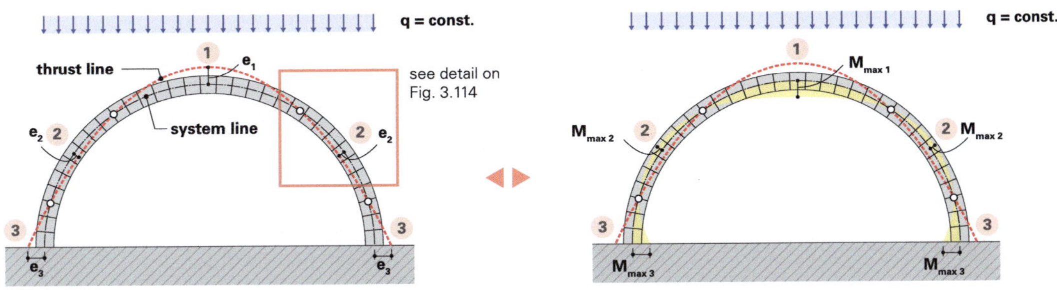

Fig. 3.113 Load-carrying behaviour of a circular arch from radial voussoirs with restrained supports (conventional masonry arch) under a constant linear load **q**. The deviations **e** of the circular system line from the parabolic thrust line result in bending moments in three areas (**1** apex, **2** arch extrados, and **3** impost). See detail on Fig. 3.114

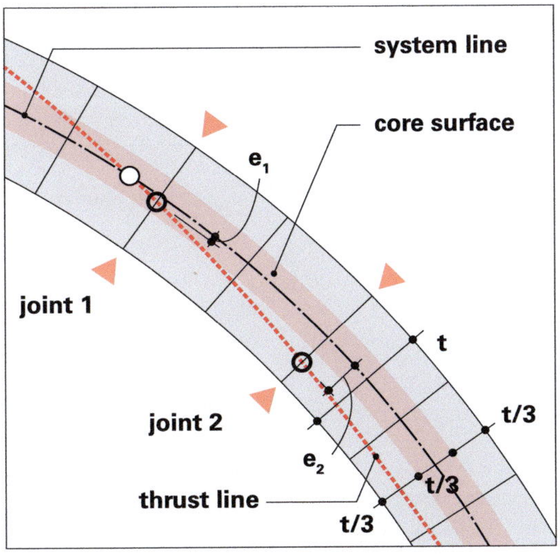

Fig. 3.114 Detail of the masonry arch shown on Fig. 3.113 with the thrust-line trajectory and its deviation **e** from the system line. The effect of the eccentricities on the cross-sectional stress of joints **1** and **2** is shown on Fig. 3.115

ideal shape of the arch which conforms to the thrust line seems to have been recognised by master builders only in rare cases. However, the fundamentals of its load-bearing behaviour were well understood as well as the methods for correct handling of the weaknesses of arch construction. The negative effects of the deviation between the thrust and the system lines were compensated for by means such as, for example, large structural depth or sta-

3

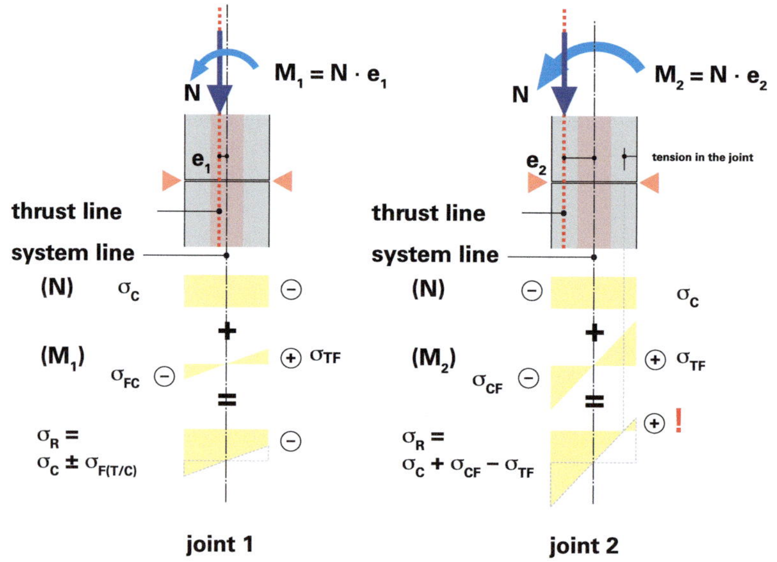

Fig. 3.115 Overlay of the stresses in the joint cross-sections **1** and **2** (see Fig. 3.114): compressive stress (σ_C) as a consequence of the normal force **N** as well as flexural compressive stresses (σ_{FC}) and flexural tensile stresses (σ_{FT}) from the bending moment **M = N · e**. Due to the large eccentricity $\mathbf{e_2}$ (force axis outside the core surface), tensile stresses occur in the cross-section of the joint **2** (Cf. also ***Vol. 1***, *Sect. 26.9.3.2 Lap–Bond Acting under Compression,* Fig. 26.140

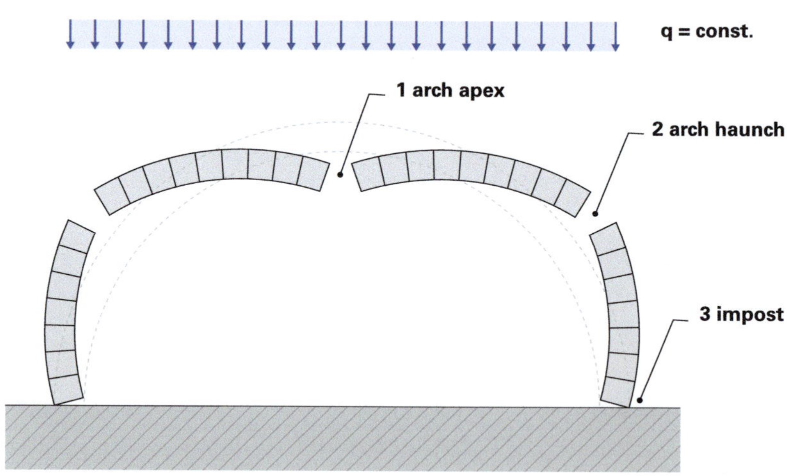

Fig. 3.116 Failure of a circular masonry arch by breaking open of the stone joints in the areas which are subjected to the greatest bending stress (see Fig. 3.113, right)

bilising arch spandrels. Also, fillings between adjacent arches in the impost area which applied additional external compression to the arches had a similarly favourable effect by, as it were, forcing the thrust line into the system line via influencing the load pattern.

Stone arches made of single stones were, for centuries, the only structurally feasible predominantly compressive-stressed, nearly bending-stress-free structures which could exploit the

Fig. 3.117 Heavy conventional circular masonry arch

Fig. 3.118 Slender, wide-spanning ashlar arches of the Pont du Gard with stiffening spandrels

Fig. 3.119 The infill between the arch haunches, the arch spandrel, applies a stabilising lateral counterpressure on the slender circular arches, so that the thrust line is, as it were, *forced* into the circular shape of the arches by the load pattern altered thereby. At the same time, the spandrel allows the routing of the path across the arch interspace

Fig. 3.120 The arch spandrels are dissolved in this bridge into a relatively dense pillar row which carries the roadway. On the arch—here made from concrete—the load is punctiform but is nevertheless fairly well distributed and approximates a uniform load. As a result, the thrust line is polygonal, but still lies within the arche's cross-section (Sandö bridge, Sweden)

mentioned advantages of axially stressed systems in a material-compliant manner—with brittle minerals, thus subjecting them to compression. For this reason, larger spans, for centuries, could only be bridged with this kind of structure (Figs. 3.117, 3.118, 3.119, 3.120, and 3.121).

Fig. 3.121 Polygonal arch, braced by the roadway deck. The guideline of the arch is polygonal because of the punctual load introduction by the columns. It therefore coincides essentially with the thrust line of the loads (Menn: Viamala bridge)

Fig. 3.122 Reinforced concrete arch: a triple-hinged arch with strong arch panels in the area of the symmetrical arch sections which are subjected to bending as a result of changing traffic loads. They gradually taper towards the three hinges (two bearing and one apex hinge) (Maillart: Arve Bridge near Vessy, 1936)

The availability of tough materials, especially of steel and reinforced concrete—wood being hardly suitable for arches due to its rectilinear growth—opened up completely novel possibilities, not only in tensile but also in compressed constructions. Arches made of tough, ductile material are capable of absorbing compressive as well as tensile stresses. These combined stress conditions are typical of bending stress, so that modern steel or reinforced-concrete arches are considerably less sensitive to eccentricities than historical masonry arches. Consequently, arches with non-thrust-line-conforming

Fig. 3.123 Modern trussed arch in steel: triple-hinged arch with strong discrepancy from the thrust line. Bending moments from these deviations as well as from changing loads can be absorbed by the large bending stiffness of the trussed half-arch sections (Grimshaw: Waterloo Station)

geometries can also be realised, which in some cases feature large eccentricities and are sometimes not dissimilar to portal-frame shapes. In this way, functional requirements from clearance gauges can be better fulfilled, for example by avoiding areas difficult to use near flat arch springs (Figs. 3.122 and 3.123).

3.4.4 Surface Structures Under Membrane Forces

► *Sect. 3.4.2 The Concepts of Funicular Line and Thrust Line*

The interrelationship between form, support, and stress as was explained for bar frameworks and ropes can be analogously applied to **surface structures**, i.e. to **shells** and **membranes**. For the structural efficiency of the load-bearing structure, the aim of the design process is, again, to carry off the dominant loads exclusively by means of **compression** and **tension**, not by bending nor shear. If this is the case, this type of load transfer within the shell or membrane surface is referred to as a **membrane state**.[5] It can be compared in analogy with the axial stress of bars. Under such conditions, which will be discussed in more detail below, curved surface structures achieve their maximum structural efficiency, for example, as measured by the ratio between structural

5 A distinction must therefore be made between the *abstract* concept of the membrane structure, i.e. the surface structure that is subject to pure axial stress, regardless of compression or tension, and the *physical* membrane structure consisting of a technical membrane skin.

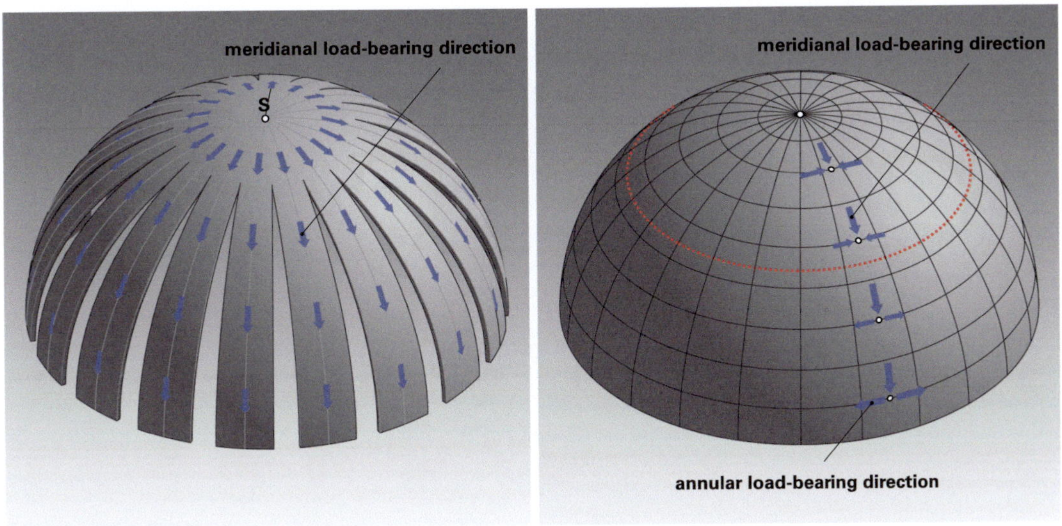

Fig. 3.124 (Left) Load-bearing structure consisting of radially arranged individual arches without connection to each other. The structural mode of action corresponds to that of a two-dimensional arched supporting structure. The corresponding rules for shaping the arches in accordance with their thrust line apply. The load-bearing action is always uniaxial in the direction of the meridian. (Right) Homogeneous shell structure: In addition to the meridianal direction of load-bearing, the homogeneity of the surface also results in a **ring** or **annular load-bearing direction** along the circles of latitude. This can be understood as a stiffening force that prevents the (imaginary) individual arches from deviating from the given shape. The shell supporting effect is therefore two-way (in meridianal *and* annular direction). In this case, a spherical surface is shown, where ring compression forces occur in the upper shell area and ring tensile forces in the lower shell area (cf. ► *Sect.* 4.3.2.4 *Full-Leaf Dome*)

depth and span, an efficiency which is not anywhere nearly achieved by any other type of structure.

In the case of two-dimensional, i.e. planar, shapes of equilibrium which have hitherto been considered, i.e. arch and rope, the force was transferred axially along the linear component. With the transformation of the structural system into a three-dimensional one, one can first imagine that a shell shape, for instance, consists of individual arch sections placed next to each other (Fig. 3.124, left). Under these conditions, the load transfer would still always take place over these individual arch sections in the meridian direction. This would be nothing other than the additive expansion of the two-dimensional arch structure into the third dimension. Accordingly, pure axial forces would occur only if the individual arches were respectively shaped according to the thrust line of the acting load.

Cf. also the derivation of the load-bearing behaviour of a dome shell in ► *Sect. 4.3.2.4 Full-Leaf Dome*

The conditions for a real shell, however, are different. This is because the imaginary arch sections do not act separately from one another in the case of a truly areal, homogeneous shell structure. Instead, there is a link between them (Fig. 3.124, right). In addition to the load-bearing effect in the meridian direction, a second bearing direction in the **annular direction**, or the one along the circles of latitude, can be expected. Illustratively, the effect of this additional bearing direction can be equalled to a blocking force effect which

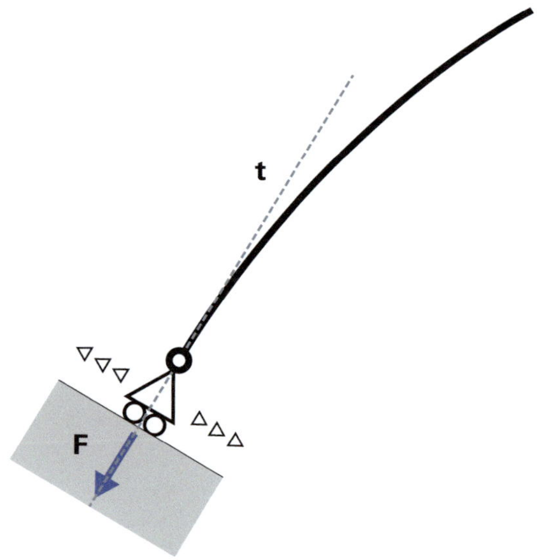

Fig. 3.125 Idealised sectional view of the support area of a shell: **membrane-compliant** support by orientation of the support surface perpendicular to the tangent **t** of the shell and allowance for free movement in this direction

prevents the (imagined) individual arch sections from shirking under excessive compression or bending stress as can occur with the two-dimensional system of the singular arch whenever the system and the thrust line do not coincide. This effect increases the load-bearing capacity of the shell significantly compared to an additive system of single arches.

Although the terms 'meridian' and 'parallel of latitude' (ring) are derived from the geometry of surfaces of revolution, which represent only a special case of curved surface structures, they can nevertheless be applied as orthogonal main directions to any surface structure. The meridian direction denotes the vertical orientation, the latitudinal-circle direction the horizontal one

The two-way load transfer via meridian and annular normal forces combined with sufficient shear bearing capacity—the internal static indeterminacy—thus allows the shell to transfer not only a single fixed load pattern for which it was formed in the membrane state but also different **surface loads** deviating therefrom, provided they are **distributed continuously**. The direct coupling of the load-bearing structure's *shape* to a particular *load*, the principle of the thrust line of the arch or the funicular line of the rope, is therefore cancelled for curved surface structures (!). In other words, a curved surface structure can also carry loads in the membrane state even if these deviate from the shape-determining load. It should be noted that curved *rod-shaped* components such as arches or ropes are not capable of this.

However, **discontinuous loads**, such as partial surface loads or single loads, also in the case of curved surface structures, lead to disturbance of the membrane state. The essential boundary condition for the emergence of the membrane state within the surface is a **membrane-compliant support** of the load-bearing structure. This means a discharge of the normal forces tangentially to the membrane surface, without deformations at right angles thereto, which otherwise leads to edge-bending defects of the surface structure (Fig. 3.125).

3.4.4.1 Membrane State

The membrane state, which sets in under the stated conditions, is characterised by tensile and compressive forces in the shell's centre surface. These lead to normal stresses evenly distributed over the thickness of the shell, the so-called **membrane stresses**. In the case of areal components such as are found in spatial structures, this means that the forces are **tangential** to the surface. Non-membrane forces are, in turn, bending and shear forces perpendicular to the surface. These create a so-called **bending state** in the shell. They stress materials much more. Some exemplary force actions on a double-curved component are shown on ◘ Fig. 3.126, characterised by respectively replicating a membrane or bending state.

The centre surface of the shell is that which always passes through the centre of the cross-section

If these are described, as usual, in the shell's coordinate system, membrane forces are referred to when following **meridian** and **ring** directions ($\mathbf{n}_\varphi$, $\mathbf{n}_\theta$) and likewise the corresponding shear forces ($\mathbf{n}_{\varphi\vartheta} = \mathbf{n}_{\vartheta\varphi}$).

Cf. ▶ ◘ Fig. 4.179, right, in ▶ *Sect. 4.3.2.4 Full-Leaf Dome*

3.4.4.2 Bending State

A disturbance of the membrane state, either by a non-membrane-compliant support or a discontinuous loading, leads to **bending stress** of the curved surface structure. This is associated with large deformations of the shell, which, however, subside with increasing distance from the disturbance. They are therefore also referred to as **subsiding bending disturbances** of shells which, however, are decisive to the local and global stability of the shell and the cross-sectional dimensioning.

3.4.4.3 Form-Finding of Surface Structures Under Membrane Forces

The experimental form-finding using a physical **model** serves to study the linking of form and load. This is permitted, as shown by the analogy between the arch and the rope (◘ Figs. 3.127 and 3.128), since the shape-forming stress distribution within a structure is a function of its geometry, support, and stress, but not of the scale. The scale decisively influences the ratio of dead weight to live load, and must therefore not be dismissed in a static analysis *from the outset*.

▶ *Sect. 3.4.2 The Concepts of Funicular Line and Thrust Line*

In experimental form-finding, a distinction is made between different methods. These include the **flow** method, **pneumatic** method, and **inversion** method. The latter has already been explained for the thrust-line determination of the arch. Its transfer to curved surface structures requires a flexible and shear-soft surface, for example a net or cloth, which finds its equilibrium under a defined load and defined support conditions in a suspended form. This is the only possible form that the model can adopt under the chosen boundary conditions under gravity. As already noted, its not only tensile but also compressive-resistant reversal is able to absorb continuous stresses in the membrane state, even if they

***Vol. 4**, Sect. 1 Scale*

▶ *Sect. 3.4.2.2 Thrust Line*

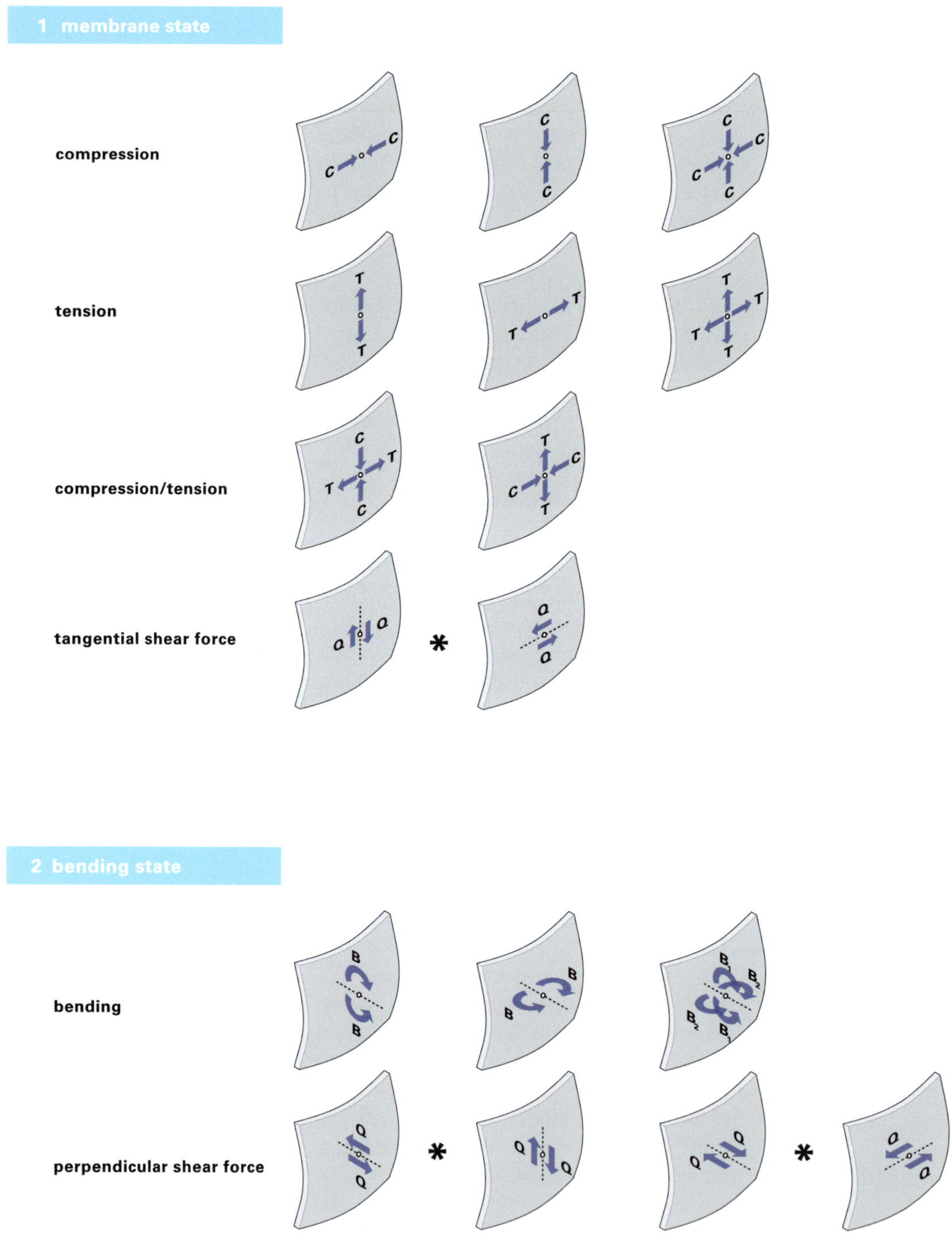

Fig. 3.126 Representation of exemplary **membrane stresses** (compressive, tensile, shear force) and **non-membrane stresses** (bending, shear force) oriented transversely to the component, using the example of a doubly curved shell-like component. The shear-force pairs marked with * always occur together

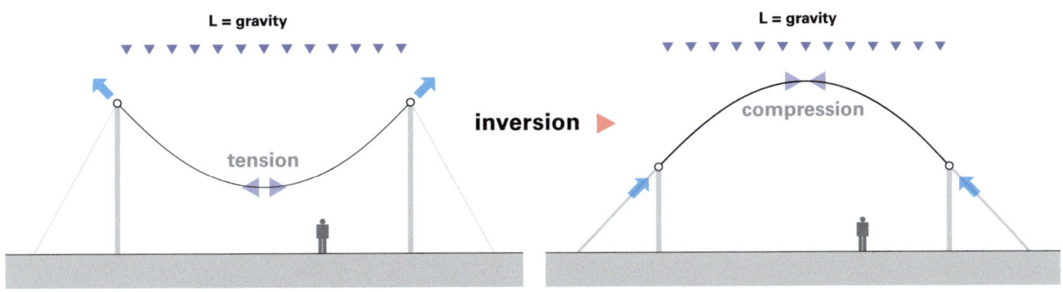

Fig. 3.127 (Left) Three-dimensional **tensile-stressed suspended form** in sectional view under vertical load. (Right) Associated three-dimensional **compressive-loaded reversed form** in sectional representation under a vertical load, which is now inverted against the component

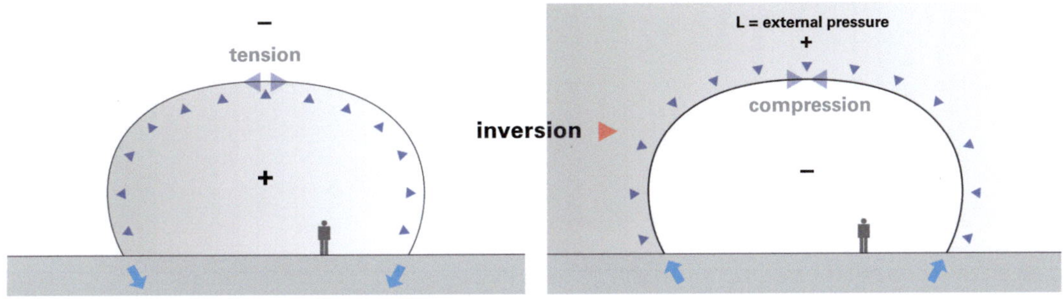

Fig. 3.128 (Left) Three-dimensional **tensile-stressed pneumatic membrane form** in sectional view under internal pressure. (Right) Associated three-dimensional **compressive-stressed reversed form** in sectional view under external compression

deviate from the form-finding. This means that the form-found shell is subjected to **pure compression** under the shaping load when neglecting axial deformations ($\mathbf{E_A} \rightarrow \infty$), whereas any other membrane-compliant load generates **tensile** and **compressive** forces. Only in this respect is it possible to speak of **thrust surfaces** or **funicular surfaces** in the case of freely shaped, form-found surface structures. However, this is generally not customary.

Form-found: This technical term denotes the shell shape corresponding to the hanging form

A particular type of form-found surface structures are **isotensoids**, i.e. membrane surfaces which have the **same membrane stress** at all points of their surface and in all directions. The simplest example of this is a weightless spherical shell subjected to hydrostatic over or under-pressure acting on all sides.

Minimising the internal potential, while applying the variance calculation, results in surface structures characterised by a minimum area ($\iint dA = \min$) and at each point of the area by a mean curvature $\mathbf{c_m} = 0$. These surfaces are called **minimal surfaces**. All non-planar minimal surfaces are thus characterised at every point by two curvatures which are **equal** in **magnitude** but **opposed** in **orientation**. For better understanding, a minimal surface can be exemplified by open, quasi-weightless soap skins with an arbitrary border line.

On the concept of minimal surfaces: cf. ▶ *Sect. 1.2.2.2.2 Median Curvature*, as well as ▶ *Sect. 4.3.3.2 Membrane and Rope Structure, Mechanically Tensioned, Linearly Supported*, particularly ▶ Fig. 4.232, left

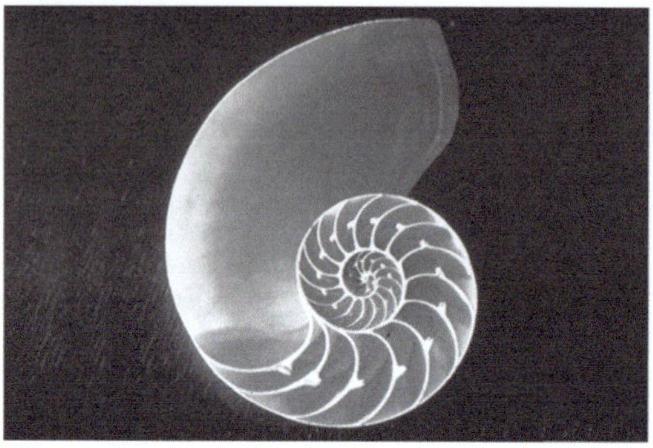

Fig. 3.129 (Left) Natural shell: nautilus. (Right) Technical shell: restaurant in Xochimilco, Mexico (arch: F. Candela)

3.4.5 Constructional Variants of Surface Structures Under Membrane Forces

3.4.5.1 Shells

The concept of the shell also exists in ordinary language, and denotes thin, mostly curved building envelopes with a certain, though limited, stiffness. Familiar everyday examples are eggs or nut shells (Fig. 3.129, left). Shells, in building practice, are structures with comparable properties: They are characterised by curvature and have extremely favourable ratios between spans and shell thickness as compared to conventional structures made of planar components (Fig. 3.129, right).

Shells are, according to their constructional definition, curved surface structures the load-bearing capacity of which is determined by their **compressive** and **tensile strength** *within* the shell surface. Their dimensional and structural stability are ensured by the bending stiffness of the shell.

According to their shape, shells are distinguished which are bent in the **same** direction and in **opposite** directions:

- **uniformly** or **synclastically** curved shells, which are known as vaulted and domed structures, chiefly feature **compressive** membrane stresses (Figs. 3.130, 3.131, and 3.132). However, continuous surface loads non-compliant with thrust surfaces, in addition to compressive membrane

Fig. 3.130 Classic dome shell in solid construction (Frauenkirche Dresden)

Fig. 3.131 Dome of the Pantheon in Rome

stress, also produce **tensile membrane stress**, which primarily occur as **tensile ring forces** in domes and vaults (Fig. 3.124 right). The construction of shells is very diverse and has markedly influenced the development of architecture. The first vaulted structures made of masonry were already erected in pre-antiquity. They were executed as exclusively compressed solid vaults. Its characteristic nature of transferring loads under almost pure compressive stress benefits, in addition to concrete, also brittle mineral stone materials, which for centuries were the most powerful and most durable available building materials. In particular, spaces

Fig. 3.132 Co-curved shell (Atocha station, Madrid; Arch: R. Moneo)

Fig. 3.133 Grid shell (Multihalle Mannheim; Arch: K. Mutschler, F. Otto)

with larger spans were virtually impossible to cover with any other structural system. It is certainly no coincidence that the widest-spanning historical building, the Pantheon in Rome, is actually a shell construction (Fig. 3.131).

The development of continuous shell structures achieved an apex with concrete and reinforced-concrete construction. This material, which can be cast in any shape into deliberately formed shuttering and is characterised by its inherent compressive strength and a tensile strength which can be added by reinforcement, resulted in a large variety of new shell forms.

▶ *Sect. 3.3 The Constructional Composition of the Space-Enclosing Surface Element*

Grid shells are created by transforming the shell into net and grid structures made of steel and wood (Fig. 3.133). In the already defined sense, they again correspond to the construction principle of variant **3**. They are finally covered with a thin plate, which merely stretches over a single field of the grating.

Fig. 3.134 Extremely light and delicate lattice shell of steel and glass (Bosch-Areal, Stuttgart; eng.: Schlaich, Bergermann & P)

Grid shells opened up new possibilities for shell construction in the recent past and replaced concrete shell construction in the era of shutterless, industrial construction methods. Modern grid shells are characterised by their high transparency and extraordinary lightness in their appearance (Fig. 3.134). In these constructions, the formation of triangular meshes is a prerequisite for securing tangential or in-surface shear stiffness. Quadrangular meshes, as occur with translational shells, therefore require a bracing from pretensioned ropes or a stiffening effected by glass or other coverings;

► *Sect. 1.2.3.3.3 Translational Surfaces*

- **Conversely** or **anticlastically** curved shells have only been used since the middle of the twentieth century and have so far mainly been designed as hyperbolic-paraboloid (*hypar*) shells (Fig. 3.129, right). Recently, various free-forms or mathematically not easily definable forms of equilibrium have also been implemented. They carry off their loads in the membrane state equally by means of tensile and compressive forces to edge chords, which collect the vertical loads and guide them into the foundation. The hypar surface can be generated as a ruled surface by the translation of straight generatrices. This is a decisive advantage especially for the production of reinforced-concrete shells.

► *Sect. 1.2.3.3.3 Translational Surfaces*

Fig. 3.135 Tent membrane

In contrast to basic structural cells from planar components, i.e. plate-shaped vertical enclosures and slab-shaped horizontal coverages, in which the advantages of axial stresses can only be used in the surrounding wall panels, but never in the covering slabs, shells are able to fully exploit the efficiency of membrane forces and to combine this with the ability to encircle spaces on all sides thanks to their curvature. Thus, they combine the ability to efficiently *carry* loads and, at the same time, to *cover* and to *envelop* spaces.[6]

3.4.5.2 Membranes and Rope Nets

Vol. 1, *Sect. 26.4.2 Movable Systems*

Curved surface structures of flexible, exclusively tension-resistant structural elements (Fig. 3.135) are **movable** structural systems according to the already discussed definition, which are always stable in terms of load-bearing capacity, but are prone to great deformations. These are often in contradiction with the building's usage requirements and the durability of the materials employed.

► *Sect. 2.7 Membrane Systems*

Vol. 1, *Sect. 26.9.9 Mechanically Pre-Stressed Membrane*

The dimensional stability of exclusively tensile-stressed constructions can be ensured by a **mechanical** or **pneumatic prestressing**. It additionally enables the curved surface structure to absorb compressive forces by depleting the prestress. Mechanical prestressing produces an opposed double curvature of the rope net or the membrane. Typical examples

6 This fitting wording is adopted from the book Heinle, Schlaich (1996) *Kuppeln aller Zeiten – aller Kulturen*, p. 207.

Fig. 3.136 The saddle surface typical of mechanically prestressed membranes

are hypar surfaces. For the stabilisation of the membrane or of the cable network, a **two-way** or **biaxial pull** is thus required which can only be applied to the membrane or the net if both pulling directions are mutually *opposite*. The typical mechanically pretensioned membrane shape is therefore the double **conversely** or **anticlastically** curved surface, which is also referred to as a **saddle** surface (Fig. 3.136). In each case, each pulling direction requires its own set of tensile members—in the case of fabrics, respectively warp and woof; in the case of nets, two intersecting rope skeins. The two orientations are distinguished according to the **bearing** direction (hanging) and **prestressing** direction (standing).

Pneumatic prestressing usually leads to a **uniform** or **synclastic** double curvature of the surface structure, as is generally the case with air-domes. In the interior of the hall, overpressure of a few centimetres of water head, which is secured by air sluices at the entrances, suffices.

Vol. 1, *Sect. 26.9.8 Pneumatically Pre-Stressed Membrane*

The pneumatic pre-tensioning is used for surface structures made from **technical membranes**. These include polyester or glass-fibre fabrics which are coated with PVC or PTFE. A PVC-coated polyester membrane is suitable for convertible, pneumatically pre-stressed constructions because of its low kink sensitivity. It is therefore foldable.

PTFE-coated glass-fibre membranes are characterised by their favourable fire behaviour (non-flammable) and their durability. Its kink sensitivity requires a particularly careful design of the connection details, as well as a skilled and fold-free execution. They are used exclusively for mechanically pre-stressed membranes.

Cf. the diagrams on ◘ Figs. 3.103 and 3.104, lower half (tension): a cable tensioned straight (camber **f** = 0, pre-tensioning force theoretically infinite, cannot build up a reaction force transverse to its axis

As with ropes, which must have a **minimum camber f** in order to be able to bear forces transversely to their axes, the membrane or the net requires a minimum camber **f** or—which is synonymous—a **minimum curvature** in order to unfold its load-bearing capacity. As is the case with the rope, where tensile forces become larger the smaller the camber **f**, it likewise applies to membranes and nets that the smaller the curvature, the larger the tensile forces within the element will be. This also is true for the sensitivity to deformation.

The same applies to **pneumatically stressed** membranes. The curvature has the same importance for the stability of the load-bearing structure as for mechanically stressed membranes (◘ Fig. 3.137).

Orthotropic = orthogonally anisotropic

Rope nets as well as membranes consist of intersecting ropes or roving skeins. Textile membranes are woven or stitched fabrics. They are characterised by pronounced **orthotropic** properties, i.e. they behave mechanically differently in warp and weft directions. Thus, the membrane cutting pattern and the alignment of the fabric (warp and weft directions) play an important role within the surface structure.

The rope harnesses of the nets can, however, be designed as equal-weight bearing directions. Two or more rope coils are conceivable (◘ Fig. 3.138).

Rope nets can span very large areas. However, they need a **coating** or **cladding** to generate a closed surface (◘ Fig. 3.139). This can be applied as a non-bearing wood boarding, a sheet shingling, or a film. The substitution of the con-

◘ **Fig. 3.137** Pneumatically tensioned membrane, here used as secondary supporting structure of the façade (SSE Hydro, Glasgow; arch.: Foster & P)

Fig. 3.138 Rope-net construction (Olympic Stadium Munich; arch.: G. Behnisch, F. Otto and others)

Fig. 3.139 The rope skeins crossing each other in the rope net are covered with a flat material on top (Olympic Stadium Munich)

tinuous surface component of the self-supporting membrane by a load-bearing net as a primary structure, to which a non-supporting flat film or a comparable secondary element is then applied for the purpose of surface formation, is analogous to the already discussed substitution of solid-walled plates or leaves by corresponding rod systems. According to the principle, a rope net which is covered two-dimensionally and consists of mutually transversely oriented, equal-ranking load-bearing rope skeins can be counted among element constructions according to variant **3**.

▶ *Sect. 3.3 The Constructional Composition of the Space-Enclosing Surface Element*

For textile membranes, free spans are rather limited. For this reason, they are frequently arranged as a secondary structure within a primary support of, e.g. masts or rope skeins. By a modular design of the surface support, it is possible to limit the size of the individual fields of the membrane and thus also to limit the tensile stresses to be absorbed within it.

Bibliography

Choisy A (1899) *Histoire de l'Architecture*, reprint 1987 by *Slatkine Reprints*. Geneva, Paris

Leicher G W (2002) *Tragwerkslehre in Beispielen und Zeichnungen*. Düsseldorf, Werner Verlag

Stevens P S (1974) *Patterns in Nature*. Boston, Toronto; Little, Brown & Co

Thompson D W (author), Bonner JT (ed) (2007) *On Growth and Form*. Cambridge, Cambridge University Press

Weischede D, Stumpf M (2018) *Krümmung trägt – Ein Handbuch zur Tragwerksentwicklung mit Stabwerksmodellen*

Types
with Dr. Matthias Weißbach

J. L. Moro, *Building-Construction Design – From Principle to Detail*,
https://doi.org/10.1007/978-3-662-71006-7_4

Contents

4.1 Overview of Elementary Structures

The matrices on ◘ Figs. 4.4, 4.5, 4.6, and 4.7 represent the attempt to illustrate elementary structural forms by arranging them according to a few elementary distinction criteria. As a representative of the overall structure, the **covering** is examined, being the most significant component for performing demanding structural task, namely supporting loads perpendicular to its axis. It is also for this reason that the different variants studied are classified according to the design of the covering. The vertical enclosure elements are involved in the analysis to the extent that they participate in the structural behaviour of the basic structural cell and therefore interact in structural terms with the covering. The analysis is differentiated according to the latter's **shape** and the way in which it is **supported**.

This procedure does not claim to capture the boundless variety of conceivable structural forms. Neither is it intended to curb the imagination of the designer from the outset. The classification rather aspires to merely provide a simple aid for a better understanding of the mode of operation of the most elementary structural systems. The more important variants for today's building practice are deepened subsequently and are graphically highlighted in the matrices. Some of the historically important structures are also discussed because of their special interest, but only contemplating their basic traits. Other variants can indeed be derived from this systematic, but because of their limited constructional significance, they are dismissed within this context. The variants relevant for the present reflection, which are treated in more detail in the following sections, are summarised in the overview on ◘ Figs. 4.8 and 4.9.

As already discussed previously, the analysed load-bearing variants are divided into three broad categories:

► *Sect. 3.4 Form-Related Issues of Axially Stressed Structures*

- predominantly **compressive**-stressed structures (**c**) (◘ Fig. 4.1);
- predominantly **bending**-stressed structures (**b**) (◘ Fig. 4.2);
- predominantly **tensile**-stressed structures (**t**) (◘ Fig. 4.3).

As already indicated, the type of stress is fundamentally dependent on three parameters: the **external load**, the **shape**, and the **support** (◘ Figs. 4.4, 4.5, 4.6, 4.7, 4.8 and 4.9).

***Vol. 1**, Sect. 26.2. Basic Concepts*

For the categorisation mentioned above, external loads are assumed as being predefined, namely in form of **distributed vertical loads** typical of building practice. As a rule, this load component usually dominates the overall load pattern, especially in heavy constructions. It goes without saying that other load components must also be included in a detailed individual analysis. In this chapter, especially **horizontal load compo-**

Fig. 4.1 Compressive-stressed structure

Fig. 4.2 Bending-stressed structure

nents are additionally considered because of their relevance. Each of the structural variants contemplated in this chapter is therefore examined in terms of its load-bearing behaviour under both **vertical** and **horizontal** loads.

The other two parameters, **form** and **support**, are varied in the matrices on Figs. 4.4, 4.5, 4.6, and 4.7, in order to systematically derive a selection as representative as possible of elementary, practically relevant structural variants.

Fig. 4.3 Tensile-stressed structure

4.2 Directional Systems

Directional structural systems are characterised by **one-way load transfer**. They are extremely important in building practice. Compared to non-directional ones, they are generally distinguished by constructional simplicity, inexpensive manufacture, and often also by internal static determinacy. On the other hand, the force paths, already for geometric reasons of the—merely—one-way load transfer, are longer than in the case of non-directional systems, something which from the outset reduces the efficiency of the load-bearing structure. The interaction of individual structural members within a static overall system is also less pronounced in one-way structures, so that components with a larger dimension are often required than with two-way load bearing.

4.2.1 Bending-Stressed Systems

Flat coverings, which are mainly subjected to bending, are extremely important in building construction for the reasons already mentioned, and will therefore be dealt with in detail in the following in various variants.

► *Sect. 3.1.6.2 The Covering*

4.2.1.1 Slab, Spanning One-Way

Slabs, spanning one-way (Fig. 4.10) or, as discussed in more detail below, two-way, are the most frequently used coverings in building construction. Although they can also be realised in other materials—in recent years also in wood—the slab or

► *Sect. 4.3.1.1 Slab Spanning Two-Way, Linearly Supported*

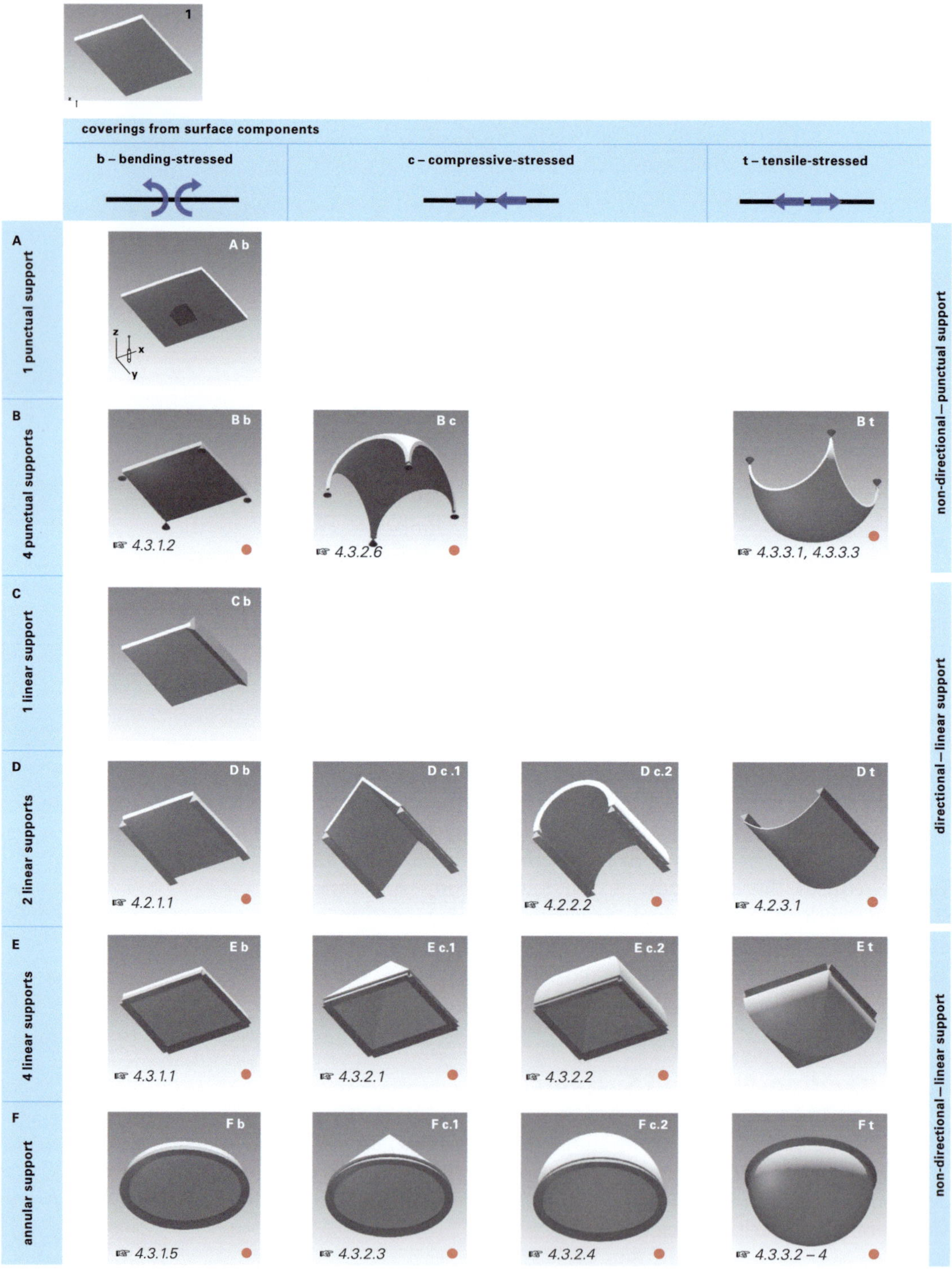

Fig. 4.4 Overview of elementary load-bearing structures made from solid-walled surface components. The variants marked with dots are of particular interest and are examined in more detail in the following sections

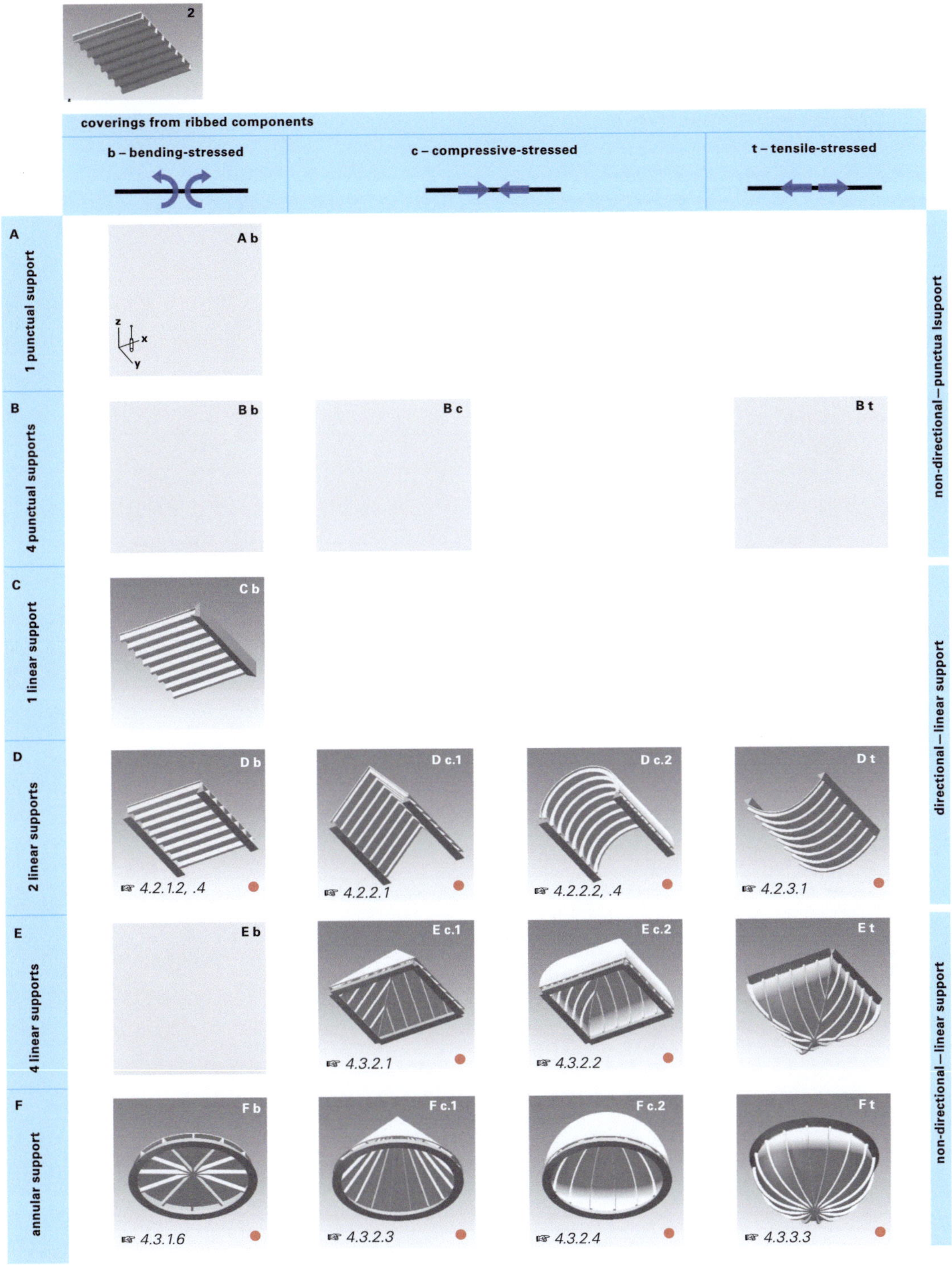

Fig. 4.5 Overview of elementary structural forms from one-way ribbed components. (variants with grey fields are not meaningful)

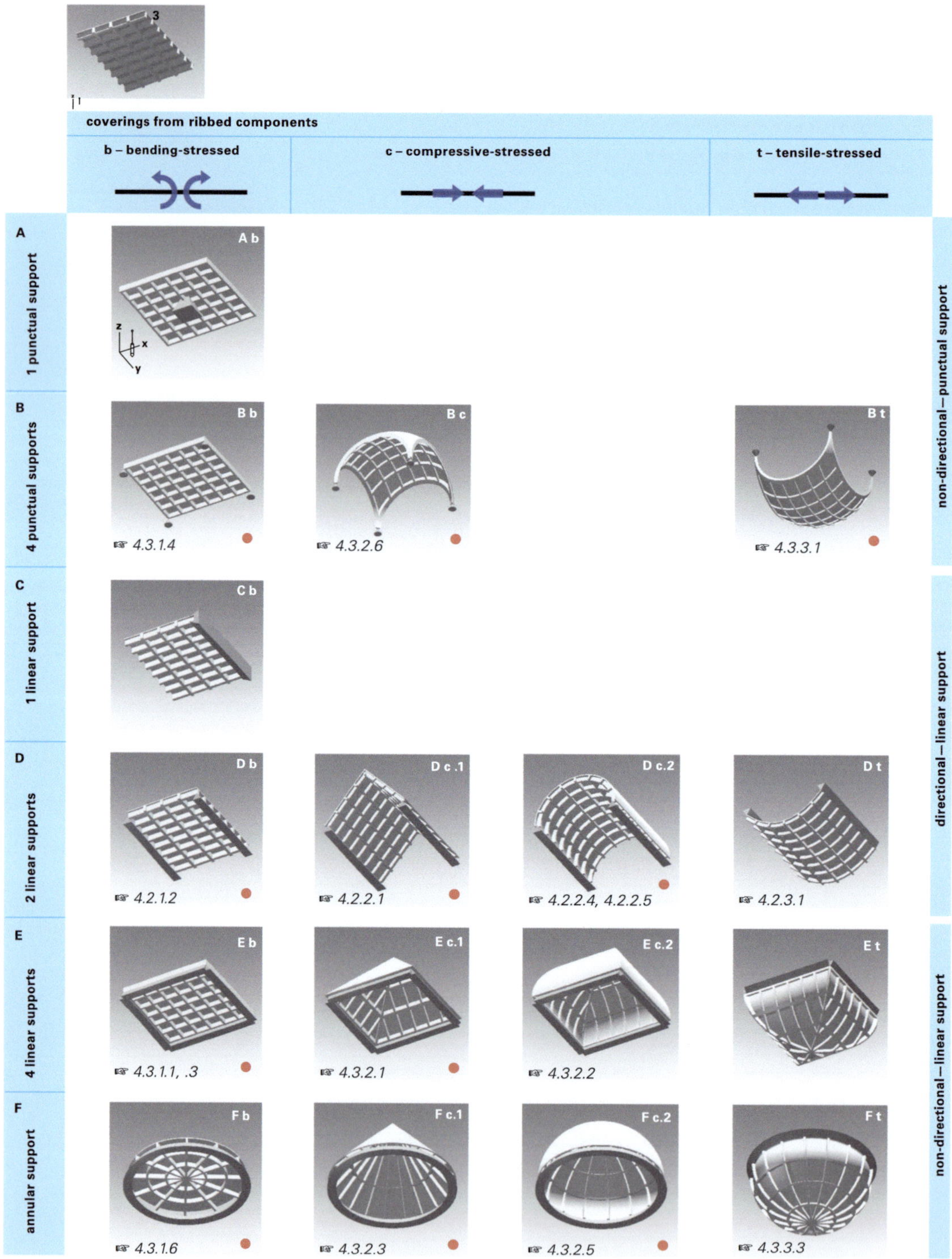

Fig. 4.6 Overview of elementary structural forms made from two-way ribbed components

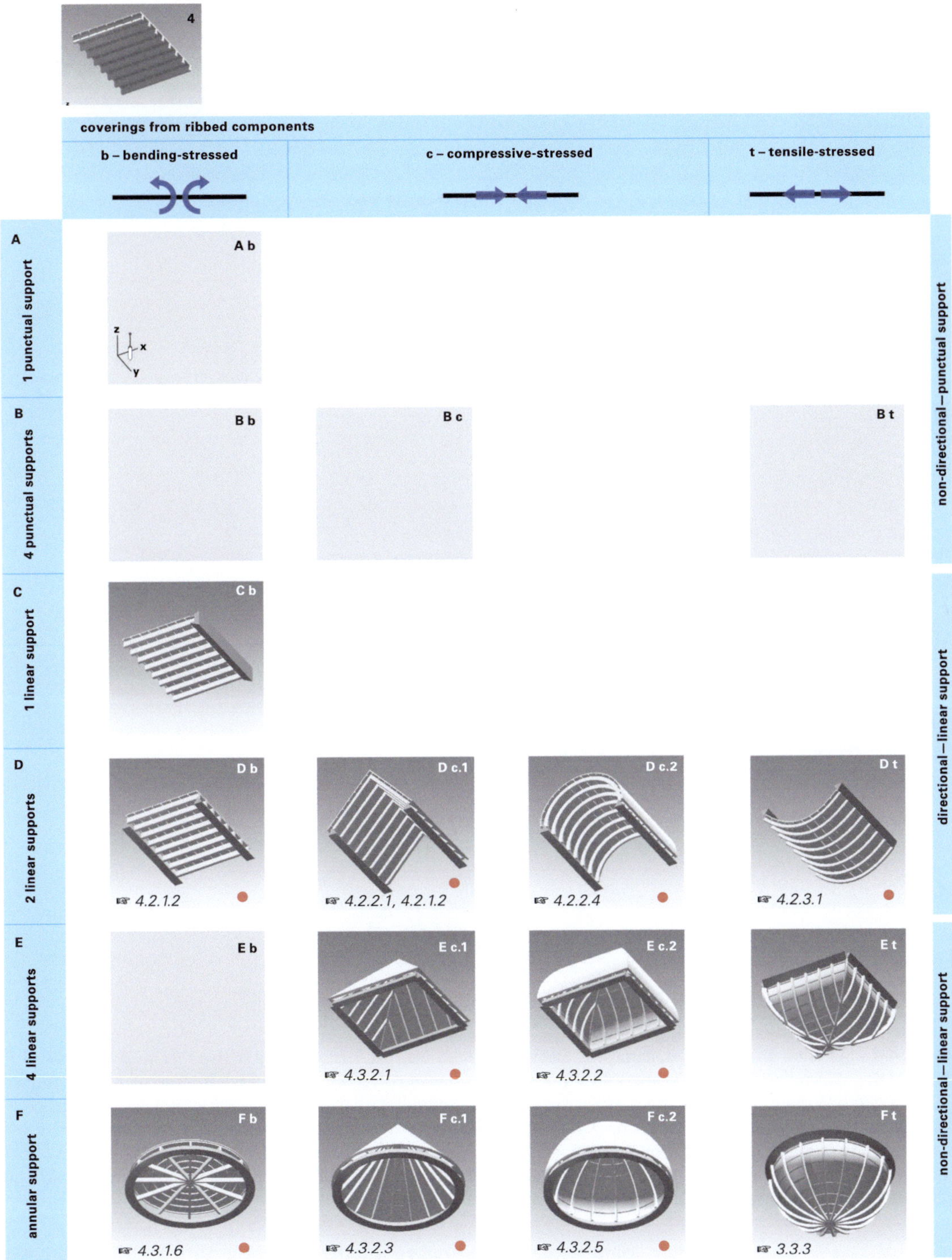

Fig. 4.7 Overview of elementary structural forms from one-way, hierarchically tiered ribbed components (variants with grey fields are not meaningful)

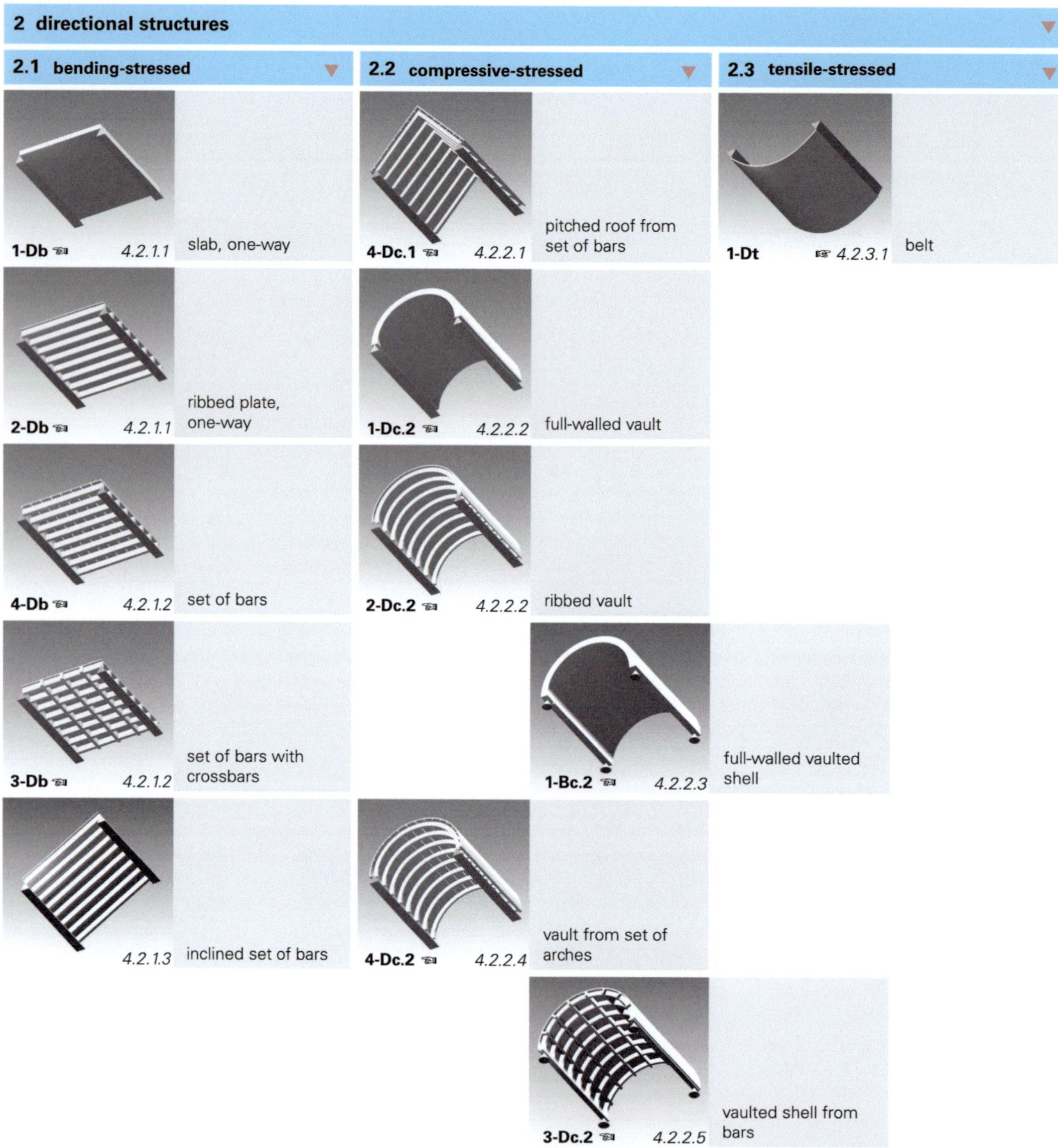

Fig. 4.8 Selection from the systematics of elementary load-bearing structures as derived from Figs. 4.4, 4.5, 4.6, and 4.7, depending on stress, shape, and support (directional structures). These types of structures are examined in more detail in the following sections. The corresponding section numbers and section titles are listed in this compilation on the right of the respective icon. On the left, the order codes for the selected structural variants are given as assigned in the overviews on Figs. 4.4, 4.5, 4.6, and 4.7. In each case, many alternative structural designs of the surface-forming components (variants **1** to **4** according to ► *Sect.* 3.3) are conceivable for each load-bearing variant, although only a single type of execution is shown here. See also Fig. 4.9

plate generally appears most frequently as a **solid component** from **reinforced concrete**. Today, this type of slab is used especially in form of a semi-prefabricated unitised slab.

► *Sect. 11.6.5.1 Unitised Floors*

From a developmental viewpoint, slabs are recent designs. They have only been realisable in constructional terms since the development of modern reinforced concrete. Only the combined ability of reinforced concrete to form a continuous,

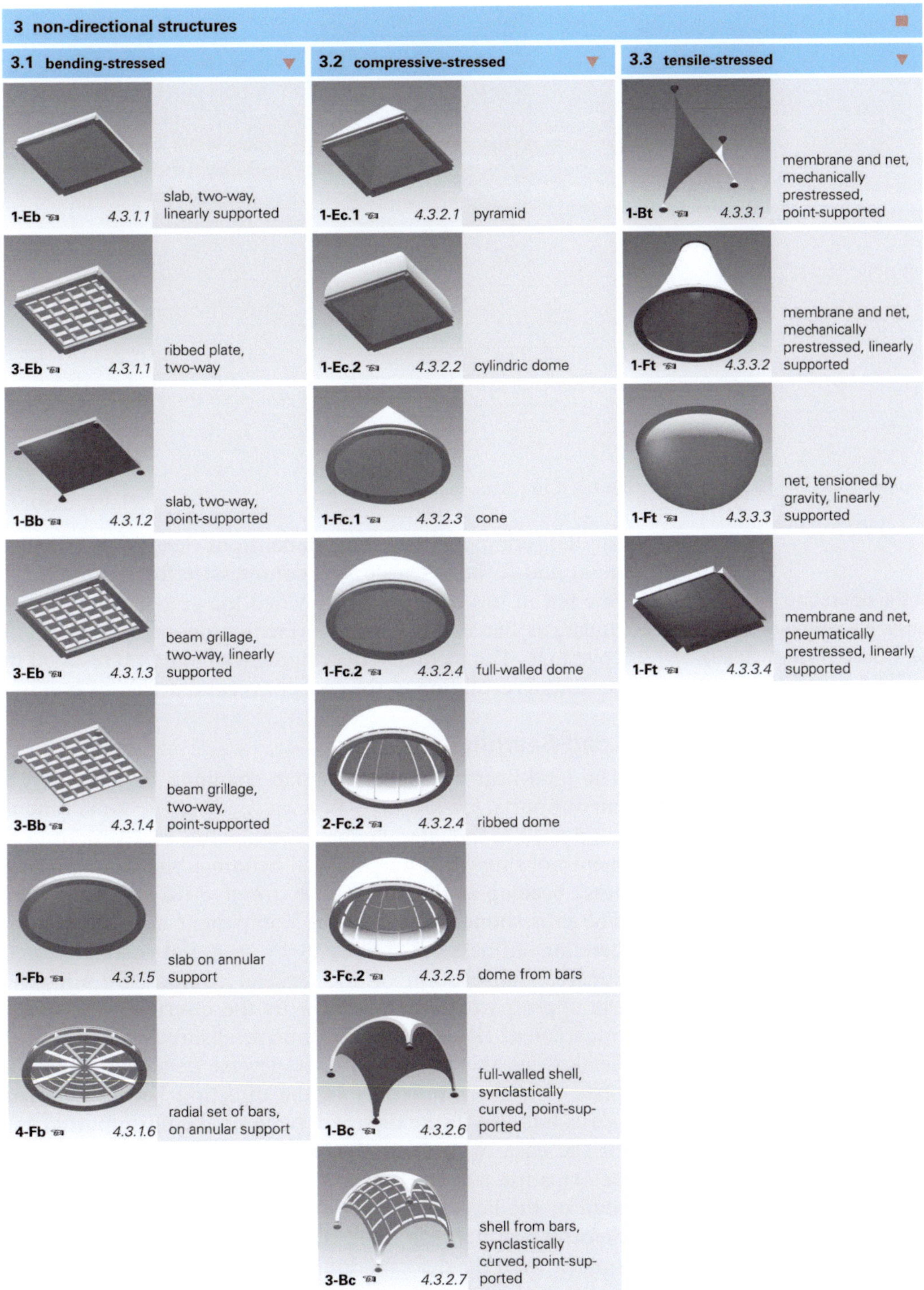

Fig. 4.9 Selection as on Fig. 4.8 (non-directional structures)

Fig. 4.10 Slab, spanning one-way

jointless component in three dimensions (length, width, thickness) and to bear tensile and compressive forces, not only in one but in *two* directions, provided the precondition for producing a flat bending-stiff surface component ideal for flat coverings. The technical development of the slab resulted in a profound revolution in the conception and design of buildings.

For the operating principle of transverse bending, see also ***Vol. 1***, *Sect. 26*, particularly *Sect. 26.9.1.1 Slab on Four Linear Supports*

Load-Bearing Behaviour

The load-bearing effect of a slab spanning between two linear supports, for example, two walls, i.e. at the same time unsupported at the lateral edges, corresponds to a structural element not only with longitudinal bending, but also with **transverse bending** as discussed in ▶ *Chap.* 3 (Fig. 4.11, right). The cross-bonding within the component and the resulting bending stiffness are created by the **material continuity** of the concrete (absorption of the flexural compressive stresses in the upper part of the plate) or by the inserted transverse reinforcement (R-mesh reinforcement, absorption of the lower flexural tensile stresses). The transverse deflection causes the plate to deform not only in the direction between the supports, but also in the direction between the unsupported ends. In the case of a **discontinuous** load (Fig. 4.12, left), this deformation **a** at the free end (**A**), which is closer to the single load or the larger force component, will be stronger than the deformation **b** at the more distant one (**B**), but in both cases smaller than the deformation **c** at the point of force attack, or at the area of the largest load concentration. If, on the other hand, the force is centred (Fig. 4.11 right), the deformation distributes uniformly to both free ends **A** and **B**.

▶ *Sect. 3.2.1 One-Way and Two-Way Load Transfer*

This is definitely a directional load-bearing system which, however, has a significant **transverse load distribution** as well as

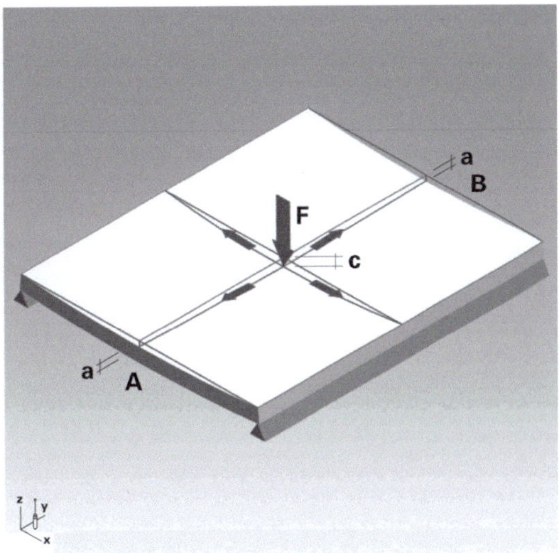

Fig. 4.11 (Left) Theoretical model of a slab spanning one-way sliced into individual strips according to ► *Sect. 3*, ► Fig. 3.40, left. The slab effect is lost, the system acts as a densely packed group of individually load-bearing bars. Pure one-way load transfer. (Right) If the strips are fused into a homogeneous slab, a two-way load transfer occurs, which is typical of the slab effect. Consequently, deflections (**a**) which are smaller than those at the point of force application occur at the non-supported edges (**A**, **B**)

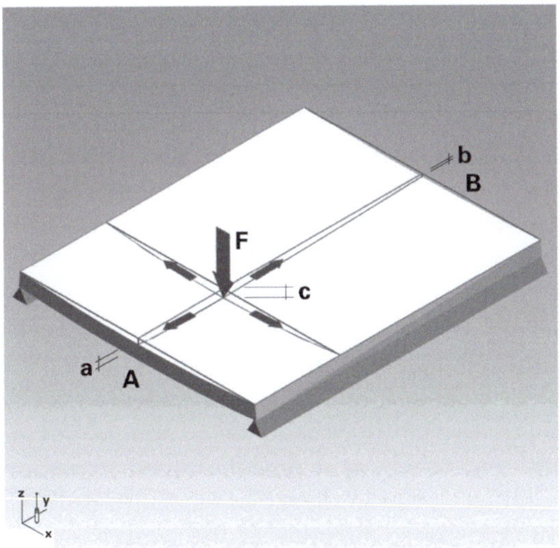

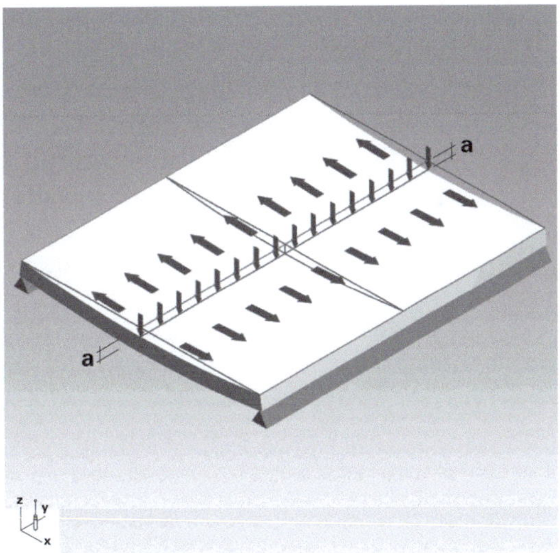

Fig. 4.12 (Left) An asymmetric load application results in different edge deflections **a** and **b** as compared to the symmetrical arrangement on Fig. 4.11, right. This means that even the farthest edge (**B**) is still involved in the load transfer. (Right) A load application strictly uniform relative to the supports, like the linear load shown here, does not benefit from the ability of the slab to cross-distribute loads. In this particular case, the load-bearing effect is identical to that on Fig. 4.11, left, and the load transfer is strictly one-way

transverse bending, which has a favourable effect when point loads or discontinuous loads are present.

However, the transverse deflection does not offer any decisive advantages in the case of a load that is evenly distributed parallel with the supports (Fig. 4.12, right).

Cf. also ► *Sect. 3*, ► Fig. 3.42, right

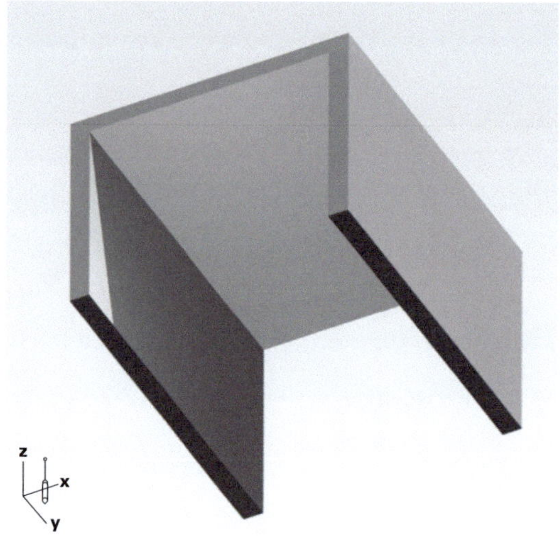

■ **Fig. 4.13** A slab spanning one-way (in →**x**) is fundamentally similar to a directional bar system according to ■ Fig. 4.29 what the type of load removal is concerned. However, it has a distinct ability to cross-distribute non-uniform loads (in →**y**)

The Basic Structural Cell

Like other directional systems, the slab spanning one-way (■ Fig. 4.13) can be combined within a structural cell with a perimeter of supporting solid walls or, alternatively, be supported on skeletal systems where the slab rests on the linear support of a beam (■ Fig. 4.23, left). Girder and slab can be made **monolithic** in solid construction, that is, jointly cast, so that the effective structural depth of the beam is increased by the thickness of the slab (■ Fig. 4.24, left).

Extendibility

The basic structural cell, like all other directional systems, can be extended only by **adding** individual bays **laterally** (■ Fig. 4.14) (in →**x**) at a given slab thickness, which should not exceed certain limits, and, thus, at a predetermined maximum span range. By contrast, it can in theory be **expanded endlessly** (■ Fig. 4.15) in the transverse direction of the span (in →**y**).

► *Sect. 4.2.1.2.1 Extendibility*

In the case of extensions in height, similar conditions prevail as in the case of bar skeins. The stacking of cells creates multi-storey buildings (■ Fig. 4.16). The discharge of the vertical loads over the walls is to be considered taking into account the floor slabs bound into the wall plane. The incorporation of the floor structure into the cross-section of the vertical wall is unproblematic in the case of solid slabs, since a solid slab can either be monolithically connected to a concrete wall (■ Fig. 4.17, left) or integrated into a brick wall as a concrete strip (■ Fig. 4.17, centre). The **transverse compression** of a continuous solid slab (as on ■ Fig. 4.17, right) in-

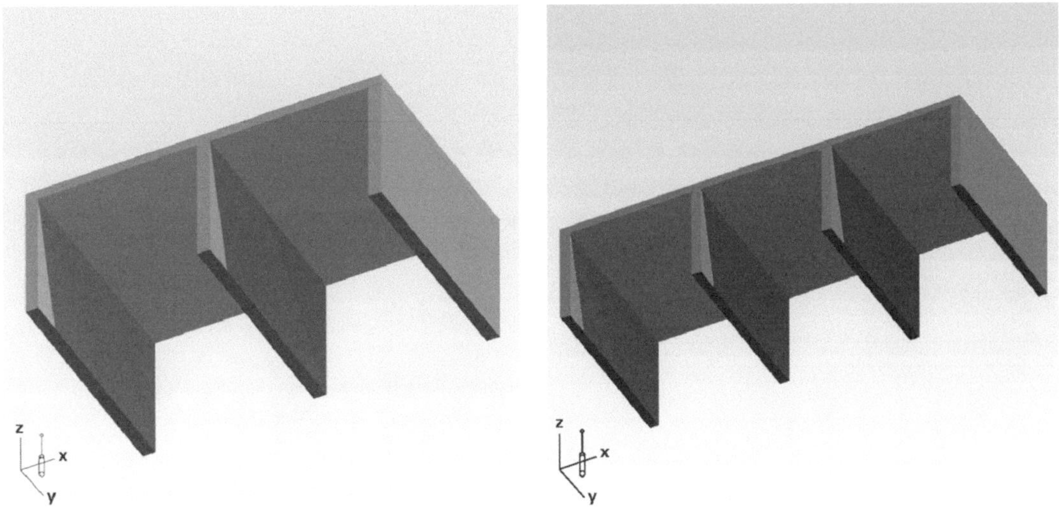

Fig. 4.14 (Left) The slab spanning one-way must also be extended in the spanning direction (→**x**) by adding bays. (Right) Discretional extensibility in the spanning direction (→**x**) by successive addition of bays

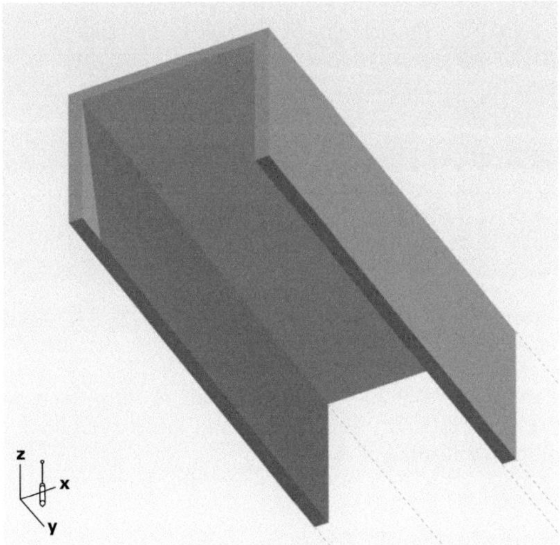

Fig. 4.15 Unlimited extendibility in the longitudinal direction (→**y**), i.e. transversely to the spanning direction, by simply prolonging the system

corporated within the wall support is uncritical and can even be advantageous. It improves the end-anchoring of the tensile reinforcement and increases the shear-force-carrying capacity of the slab. Furthermore, a certain restraint at the end support is caused by the superimposed load of the wall, and the rigidity of the slab against twisting is activated. On the other hand, this type of transverse compression is critical in the case of solid-wood plates which are integrated into a load-bearing wall.

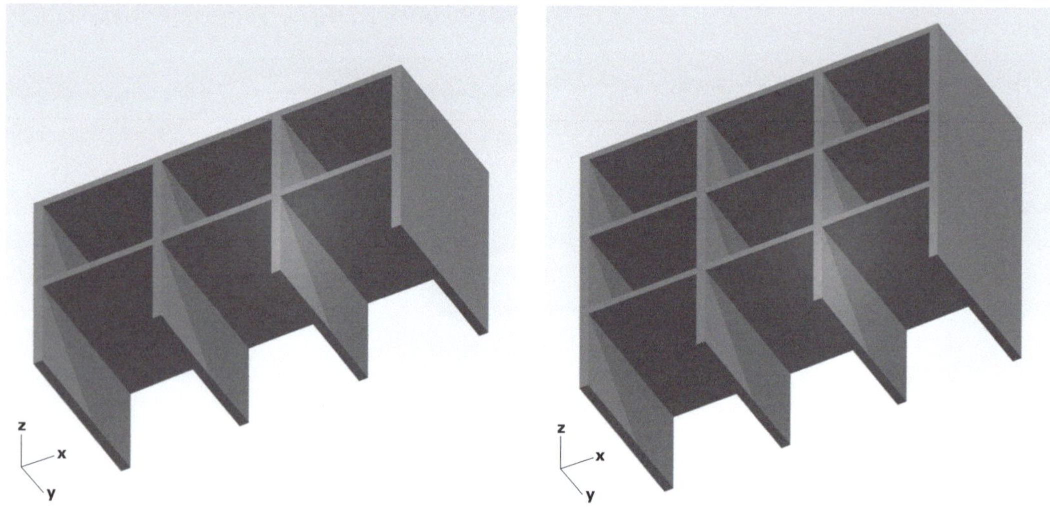

Fig. 4.16 (Left and right) Extension by stacking cells. Multi-storey structures emerge

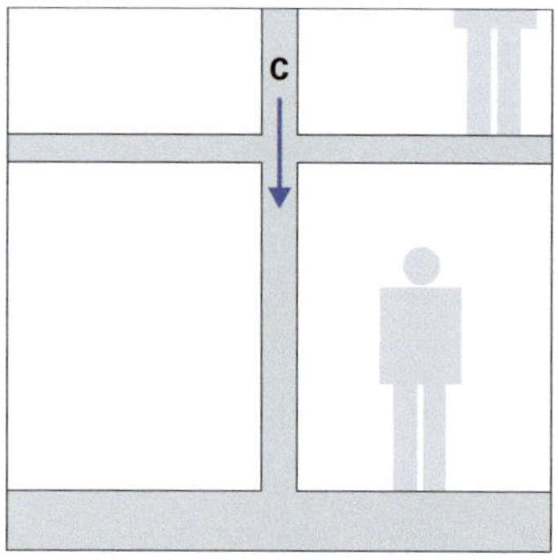

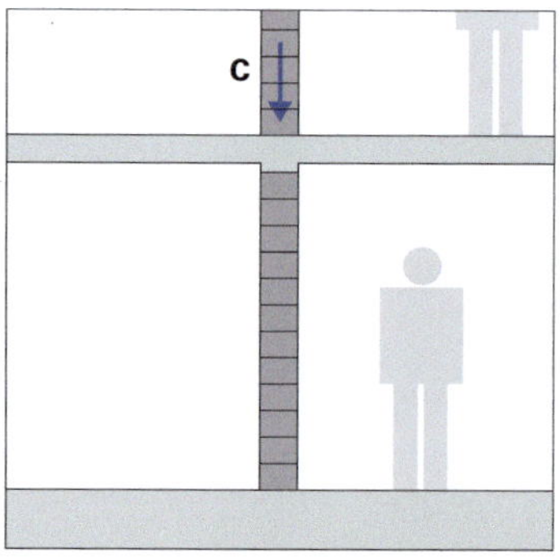

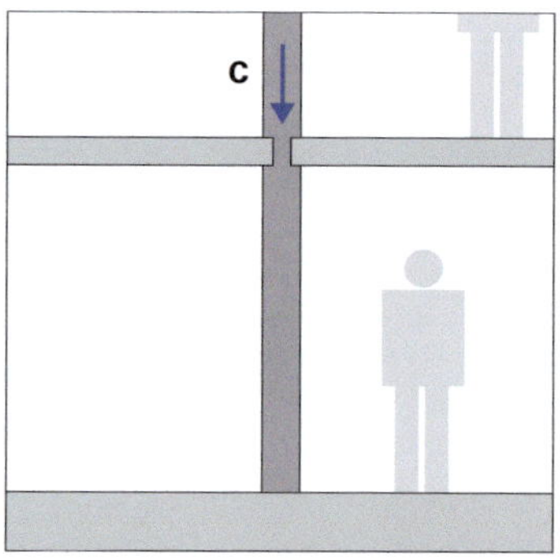

Fig. 4.17 (Left) Monolithic incorporation of a solid floor slab into a concrete wall. (Centre) Incorporation of a solid floor slab into a brick wall as an integrated ring anchor. (Right) Incorporation of a non-solid floor, for example made of wood, into a load-bearing wall. Cross-sectional weakening of the wall

► *Sect. 4.2.1.2.1 Extendibility*

With regard to wall offsets (in →**x**), the same applies as for bar skeins.

Ribbed Slab, Spanning One-Way

Slabs have favourable ratios between thickness and span (about 1: 30). However, because of their relatively **large areal weight**, the thickness of solid slabs meets relatively narrow limits. Beyond a certain span, which is reached at around 8 m and involving a thickness of about 30 cm, a good part of the load-bearing capacity of the solid slab is merely used to bear its dead load, so that it becomes uneconomical.

Cf. ***Vol. 4****, Sect. 1.6.1 The Influence of the Blueprint*

Various **constructional variants** of the slab provide bypasses to this limitation. In these, the massive slab is converted into a hollowed-out slab, which represents a first step towards a true ribbed element. The ribs remaining between the cavities and the surface-forming plate are thereby monolithically connected and act together statically, so that a plate strip along the rib acts as a compression chord thereof, the rib, or rather its steel reinforcement, in turn acting as a tension chord. Surface-forming slabs can be arranged on one side on top of the ribs (as in beam-and-slab floors) or also on both sides of the ribs (as in hollow-core slabs).

Ribbed slabs enable a significant weight reduction and an increase in spans, but require larger structural depth. Some of the most common types of execution are discussed below.

Execution Variants

Solid slabs and ribbed slabs may appear in building practice in the following variants:

- Solid slabs:
  - **Solid slab** cast on site. Conventional type of execution.
  - Semi-prefabricated **unitised floor slab**: most widely used today. For design and manufacturing reasons, this design requires slightly larger plate thickness and more reinforcing steel than the in-situ concrete slab. On the other hand, it is very economical because of the rapid and simple erection (no dedicated formwork, only lost formwork) (▣ Fig. 4.18).

    Vol. 3, Sect. 20.5.1.1 In-Situ-Concrete Slabs
  - **Hollow-core slab** (▣ Figs. 4.19, 4.20, 4.22, and 4.23): They are produced as prefabricated parts and have cavities in the middle of the plate's cross-section, where flexural compressive and tensile stresses are low.
  - **Beam-and-slab** floors/**ribbed slabs** (▣ Figs. 4.18 and 4.21): Single ribs give the slab the required stiffness to bridge larger spans. The slab itself only stretches between the ribs, so that its thickness—and thus the overall weight of the entire slab—can be greatly reduced. An effective width of flange also acts as a compression chord of the rib.

    ► *Chap. 10*, ► ▣ Figs. 10.52, 10.53, and 10.54, as well as ***Vol. 3**, Sect. 20.6.3 Floors in Ribbed Reinforced-Concrete Construction*

    Ribbed slabs are well suited for prefabrication, whereby they are produced especially in the **stressing-bed** process.

    Hollow-core slabs are already a transitional form to a **ribbed element** (structural variant **2**). Although ribbed slabs can already be regarded morphologically as ribbed elements, they are characterised by a monolithic bond between the rib and the covering plate and thus retain certain typical features of the full-walled leaf (structural variant **1**) (cf. also ▣ Figs. 4.22, 4.23, 4.24, 4.25, and 4.26).
- **Steel-Concrete Composite Slabs**: A cooperation between a steel profile and the surface-forming concrete slab is acti-

  ***Vol. 3**, Sect. 20.6.2.2 Steel-Concrete Composite Floors*

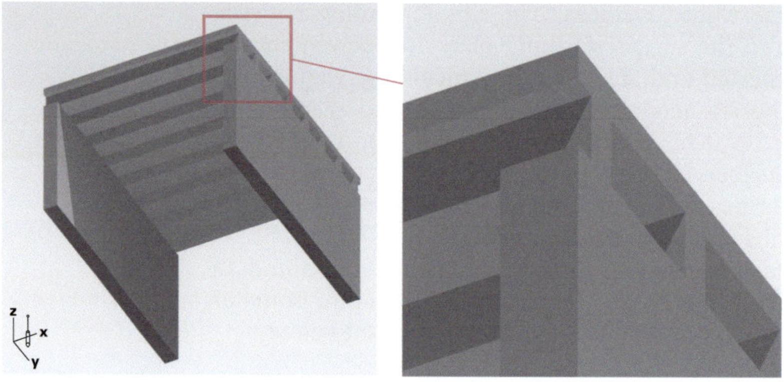

Fig. 4.18 (Left) Reduction of the dead weight of the slab by forming ribs. The slab and the rib are fused together and act conjointly in load-transfer (beam-and-slab construction or ribbed plate in reinforced concrete). (Right) Detailed view

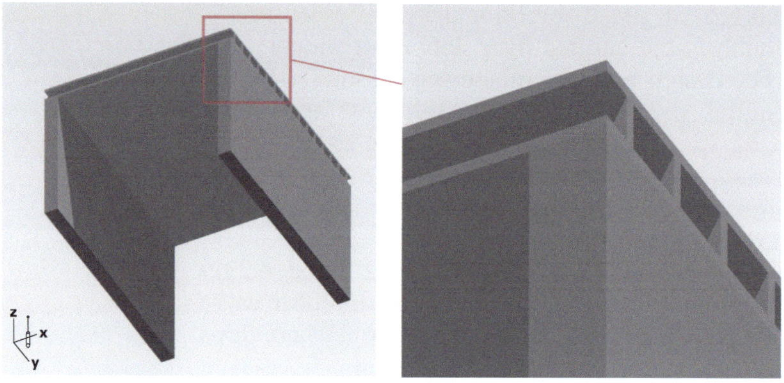

Fig. 4.19 (Left) A hollowing out of the plate component results in a similar load-bearing effect as in Fig. 4.18 with a significant weight reduction (hollow-core slab in reinforced concrete construction). (Right) Detailed view

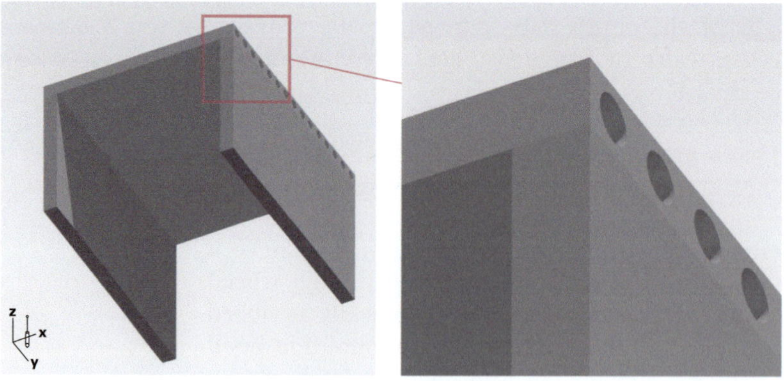

Fig. 4.20 (Left) Hollow-core slab covering similar to Fig. 4.19, but with cylindrical cavities. (Right) Detailed view

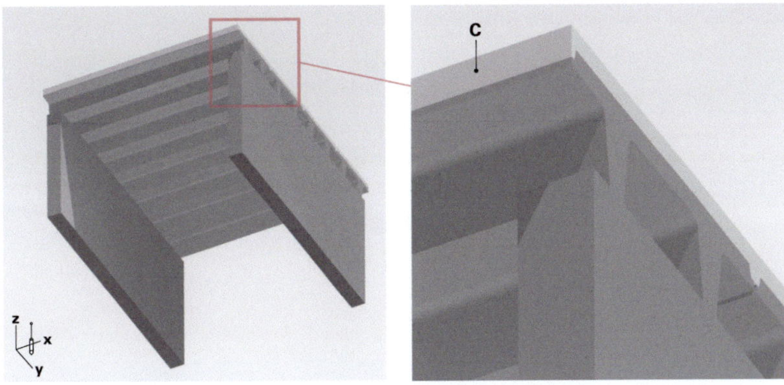

Fig. 4.21 (Left) Unitised beam-and-slab construction of double-rib elements. A combined plate and shear-plate action is achieved by applying a top concrete layer **C**. (Right) Detailed view

Fig. 4.22 Mounting of a slab element spanning one-way (hollow-core slab)

vated. The steel profile essentially takes over tensile forces and the concrete slab compressive forces.

- **Wood-Concrete Composite Slabs**: The interaction between wood and concrete is analogous to that between steel and concrete, whereby wood—like steel—essentially absorbs the tensile forces.

***Vol. 3**, Sect. 20.6.1.6 Wood-Concrete Composite Floors*

- **Solid-Wood Plate**: This is the relatively rare case of a slab that is not made of concrete. In this case, the floor slab is made of modern wood-based materials, be it stacked-plank timber, micro-laminated timber or cross-laminated timber. While the first two designs act as one-way plates due to the uniaxial orientation of their fibres, the latter variant can be used to create a quasi-bidirectional plate, i.e. with discharge in two directions, due to the cross-con-

***Vol. 3**, Sect. 20.5.1.5 Solid-Wood Floors*

Fig. 4.23 (Left) Hollow-core slab, cantilevering over joist. (Right) Hollow-core slab. Grouting joints visible

Fig. 4.24 (Left) Roof element spanning one-way. (Right) Girders (parallel to the viewing plane) and roof elements spanning one-way (transverse to the viewing plane) flush at the **same height**. Despite the lack of stacking, the load transfer is nevertheless hierarchical. The roof elements manufactured in the stressing bed could not have been made to be continuous even with stacking (factory hall in Como, Italy, arch.: A. Mangiarotti)

nected board layers. When the board layers are glued together, the plate is also able to act as a shear-resistant plate.

- **Timber-Frame Floors** and floors made of **wooden construction elements**: These are other comparatively rare cases of non-concrete slabs. Timber-frame floors are made from beams sheathed on both sides forming a panel-like element with flat boundary surfaces. Thus, it shows some, albeit rather superficial, resemblance with hollow-core concrete

Vol. 3, *Sect. 20.6.1.4 Wood-Panel Floors* and ibid. *Sect. 20.6.1.5 Floors from Wooden Construction Elements*

Fig. 4.25 Structure of a prefabricated concrete building during construction: one-way spanning prefabricated floor slabs on girders, prefabricated columns continuous over three-storeys with cleat supports

slabs, which is why it is at least mentioned at this point. However, there being a constructional separation between rib, i.e. beam, and surface-closing sheathing, it differs as a matter of principle from a monolithic element as dealt with in this section. Therefore, it could just as well be considered a ribbed element as discussed in the next section.

For wooden construction elements, both solid-wood panels as described above as well as ribbed elements are used. It also applies to the latter that they can be regarded as a transitional form to a ribbed element, depending on the type of execution.

Bracing

Thanks to their homogeneity and isotropy, solid slabs are (unlike bar systems as discussed in the following section) shear-resistant within their plane (**xy**) and can be used without additional measures as **shear-plates** for purposes of bracing of buildings. In the context of a basic structural cell, solid slabs are therefore well suited for the force-fitting connection of the cell to external fixed points such as, for example, building cores.

■ **Fig. 4.26** Side view of the structure shown on ■ Fig. 4.25. The ribs of the floor elements are perforated for the passthrough of installations

Basic cells stacked storey-wise (■ Fig. 4.16) require bracing against horizontal forces in the floor's span direction (in → **x**). This can be achieved by connecting them to bracing cores, for which the solid slabs, which are anyway acting as shear-plates, are well suited. Bracing by means of a portal-frame effect via the transverse bending stiffness of the walls/floors and via the bending-stiff design of the wall-floor connection is not possible as a matter of principle. For this purpose, stiffer components, such as columns and girders, are necessary, which can be combined to form load-bearing portal frames. Here, reinforced-concrete or prefabricated composite portal frames are often used.

► *Sect. 3.1.6.1 The Vertical, Planar Space-Enclosing Element*, in particular ► ■ Figs. 3.20, 3.21, and 3.22

As already mentioned in ► *Sect.* 3, the case must also be taken into account that the plate does not rest on wall plates but on bar-shaped girders. This is the case with skeletal structures. There is then an **indirect load transfer**.

► *Sect. 4.2.1.2.4 Bracing*, as well as ► *Sect. 4.2.1.3 Bracing of Skeletal Structures*

The conditions for bracing basic structural cells with slabs spanning one-way are analogous to those with rod skeins on panel and skeletal systems and are discussed further below.

Usage in Building—Design Aspects

The **solid slab** has a very great importance in modern building construction because it combines advantages that other floor constructions have only partially. For this reason, these will be discussed in more detail below. They also apply in the same way—in some cases even more so—for **two-way** solid slabs as discussed below.

▶ Sect. 4.3.1.1 Slab Spanning Two-Way, Linearly Supported

Solid slabs have advantages, for example, with regard to **load transfer**:

- **Racking Stiffness** in the horizontal plane. This is usually used for the bracing of the building. It results by itself without any additional measures thanks to the monolithic, isotropic component produced in the cast construction method and a two-way oriented reinforcement.
- Very favourable ratio of structural **thickness** to **span** in the usual span range of less than 8 m. In particular, where building height and enclosed space are to be kept minimal—as in the case of multi-storey housing construction—this aspect is extremely important. Two-way spanning plates can even be made slimmer than those spanning one-way.
- The good **cross-distribution** of loads allows for a great freedom of design regarding the support of the slab and the arrangement of slab perforations. For example, non-regular arrangements of supports can be implemented. Also in this regard, plates spanning two-way show additional advantages compared to those spanning one-way.

▶ Sect. 4.3.1.1 Slab Spanning Two-Way, Linearly Supported

Or in regard to **manufacture**:

- **Geometrical conditions** for shuttering and casting are relatively **simple**, since the slab is a flat, unprofiled component. In particular, the *lying* casting position, in comparison to the *standing* one like in the case of walls, is a significant advantage of the solid slab, since one complete formwork surface becomes superfluous thereby, namely the upper side. The proper compaction of the unset concrete in the bottom part of the formwork, which with standing formworks sometimes is critical because difficult to control, is also unproblematic with the slab cast horizontally because of the very small concreting depth (identical to the thickness of the slab).
- Thanks to the use of **system formwork** and semi-prefabricated slabs, the cost-intensive formwork labour has become much simpler. Solid slabs are also competitive in terms of cost.
- Singular elements such as perforations, recesses for the fit-out (e.g. for luminaires) or ducts (conduit lines for electrical installation, pipe sections) can be flexibly integrated and implemented relatively easily in production.

▶ Sect. 11.6 Formwork Technology

▶ Sect. 11.6.5 Semi-Prefabricated Parts

Last but not least, there are significant **usage-related** benefits to be mentioned:

Vol. 1, *Sect. 28.3.3.3.1 Single-Leaf Components*

- Good **airborne-sound protection** thanks to the large surface weight of the solid slab.
- Sufficient **impact-sound protection** when combined with bonded screed and soft-resilient flooring. This is a significant point of view because, in administrative buildings, only this floor construction ensures the necessary **free displacement** of partition walls while maintaining sufficient impact-sound protection. Floor constructions with floating screeds are not suitable for this purpose. Under these conditions, the total **mass** of the floor plays a decisive role.

Vol. 1, *Sect. 28.3.4.2 Impact-Sound Behaviour of Floor Constructions* and *Sect. 28.3.4.3 Improvement of Impact-Sound Protection by Floorings*

- Good **fire protection**. Already the statically required slab thickness results in a **fire-resistant** component.
- Large **thermal-storage mass**. Especially in the summer, this property proves to be very important in order to temporarily store excess heat and to cool the structure. However, the thermal inertia of a solid slab can also be required during transitional periods, particularly in the case of large-area glazing and with the high insulation standard applied today, in order to prevent temporary overheating in interior spaces.

In its variant as a one-way-spanning, linearly supported plate, the solid slab is predestined especially for those cases where, for reasons of usage, wall panels are used as an encasement, in particular in the case of **cellular constructions** as are customary in residential building. The useful spans in the range between 5 and 8 m, which are reasonable with this type of floor, can be well reconciled design-wise with the following exemplary usage variants:

- the width of two single rooms;
- the width of **three small individual cells** as are realised, for example, in combination office concepts;
- the width of **three garage parking spaces** (3 × 2.5 m = 7.5 m). In the case of underground garages, these must usually be integrated into the same load-bearing grid as the usages in the floors above.

▶ *Sect. 3.2.2 Influences of Load Transfer on the Geometry of the Structural Cell*

Some fundamental considerations concerning the **design definition** of **spans** in directional coverings are given in ▶ *Chap.* 3.

4.2.1.2 Flat Covering from Bar Skeins

A much older way, from a developmental and historical perspective, of covering a space flatly than a solid slab is the **beam floor** of one or more stacked bar skeins (◘ Fig. 4.27). It is very old, especially in the variant of the wooden beam floor (◘ Fig. 4.28). In its simplest form, a layer or skein of mutually parallel beams is stretched between two walls (◘ Fig. 4.29), which extend parallel to one another. The floor construction thus created can be regarded as a **directional ribbed system** as defined in *Chap. 26*, whereby the dominant external

Vol. 1, *Sect. 26.9.4 Element Composed of Ribs Oriented One-Way*

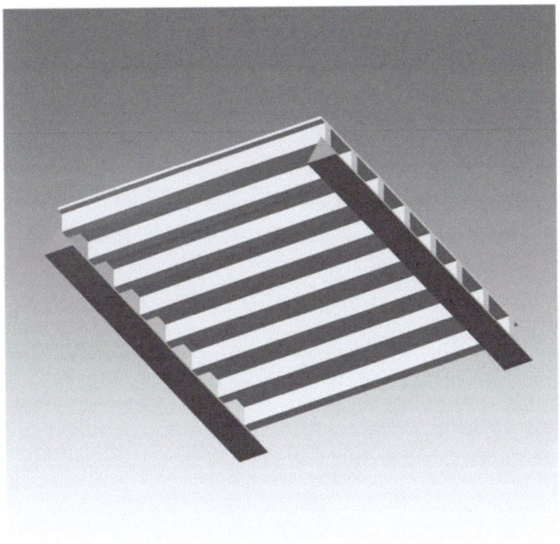

Fig. 4.27 Flat covering from bar skeins

Fig. 4.28 Flat covering made of a set of bars: wooden beam ceiling in the Regensburg townhouse

loads, i.e. dead and live loads, are in this case directed *perpendicular* to the element surface. In the sense of the present consideration of the construction of a planar element, the beam floor, depending on its type of execution, belongs to version **2** or **4**.

▶ *Sect. 3.3 The Constructional Composition of the Space-Enclosing Surface Element*

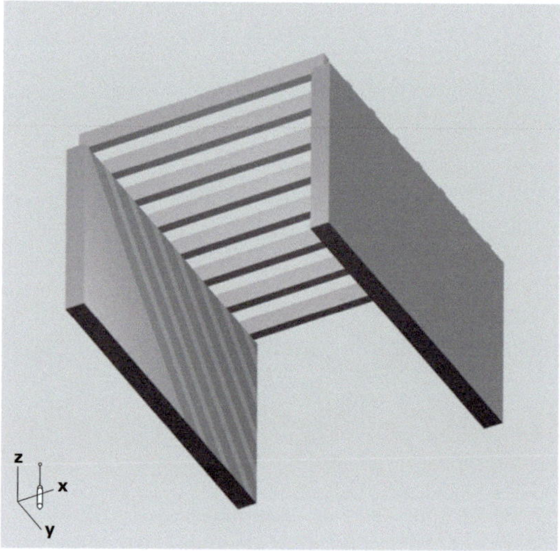

Fig. 4.29 Covering from a flat bar skein between supporting wall panels

Extendibility

▶ Sect. 4.2.1.1.3 Extendibility

This basic cell can only be expanded by **adding** another cell **laterally** in the spanning direction of the beams (→**x**) as a result of their limited span, analogously to the one-way-spanning plate (Fig. 4.30, left). Consequently, parallel *bays* (Fig. 4.30, right) are formed.

In the direction of the wall (→**y**), however, this basic cell is **indefinitely extendible** in principle, since only the walls have to be lengthened and the set of joists to be supplemented (Fig. 4.31).

Vol. 1, *Sect. 15.4 Mechanical Properties*

In the height (→**z**), these types of cells can be expanded by simple **stacking** (Fig. 4.32). Again, as with slabs, it is necessary to clarify the incorporation of the bar skein into the load-bearing wall and the transfer of vertical loads across the connection between wall and floor. As a rule, vertical loads are prevented from being transmitted across the beams by compressing them (Fig. 4.33). This applies in particular to wooden beams, which do not tolerate any significant transverse compression, since the wood lacks the required stiffness and strength perpendicular to the grain direction. Instead, a wall cross-section sufficient for force transmission should be left within the wall core between the beams abutting from both sides (in that case, no continuous effect of the carriers is possible Fig. 4.33 centre, right) or alternatively to transmit the load in the wall sections between adjacent carriers (Fig. 4.33, left).

In the case of end supports of beam floors, a hinged connection by means of a centring strip is required to accommodate deformation-induced angles of deflexion of the beams at the support. In addition to the avoidance of transverse compression in wooden beam floors (see above), this also necessi-

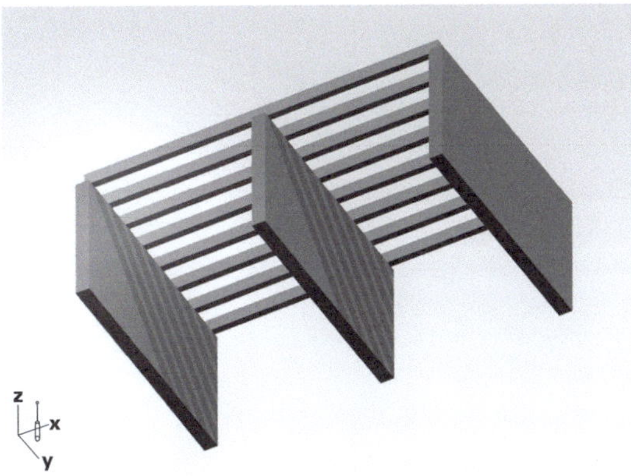

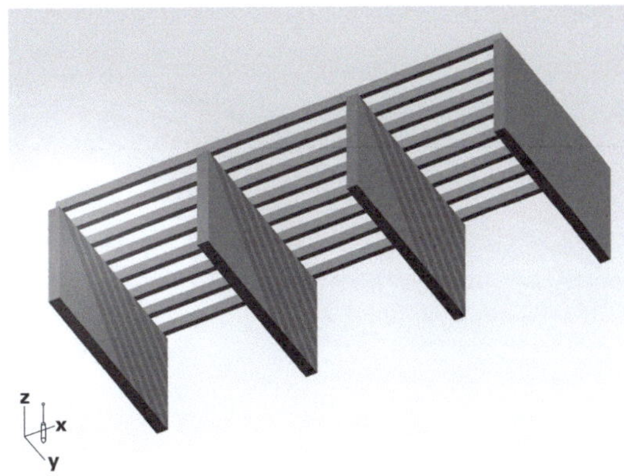

Fig. 4.30 (Left) Extension of the basic cell on Fig. 4.29 in the spanning direction of the skein by adding a parallel connected bay. (Right) Discretional extension of the basic cell in the spanning direction of the skein, but putting up with a segmentation of the enclosed spaces

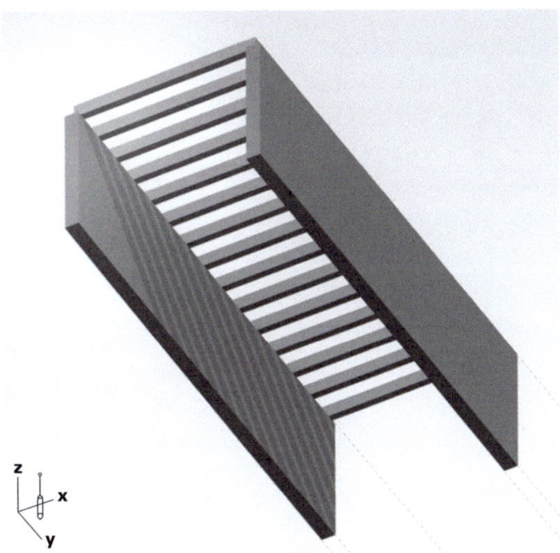

Fig. 4.31 Unlimited extension of the basic cell on Fig. 4.29 in the direction of the wall, i.e. transversely to the direction of the bars, simply by prolonging the construction principle of the bay

tates a decoupling of the floor support from the vertical load transfer of the wall. A common example of this is floor end supports in masonry walls on projecting ledges or corbels (Figs. 4.34 and 4.35). Furthermore, due to the strict structural separation of wall and beam in wooden beam floors, the wood is protected from rot due to building moisture contained in the masonry.

Offsets between the enclosing walls of superimposed storeys, so that offset walls come to rest upon floors, significantly increase bending moments and shear forces acting on the bar skein and annuls the bracing shear-plate action of the wall (Fig. 4.36). There are hardly any contexts where this might be desirable, which is why such offsets should be avoided as

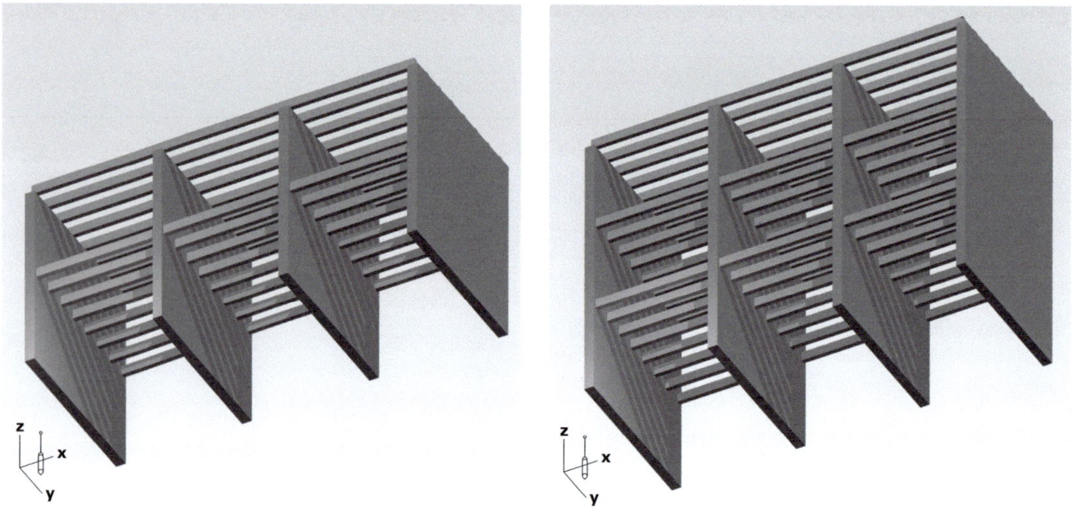

Fig. 4.32 Extension of the basic cell on Fig. 4.29 in height (→z)

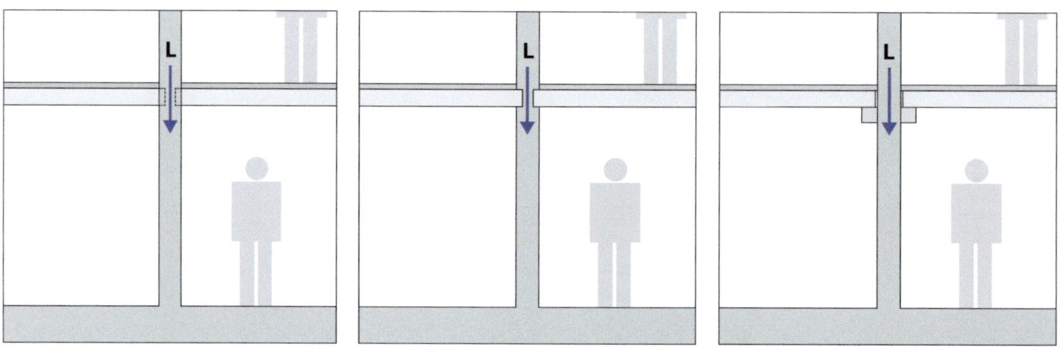

Fig. 4.33 (Left) Discharge of the load **L** across the wall cross-section in the remaining wall section between beam axes. Beams inserted in the wall within individual pockets. (Centre) Discharge of the load **L** over a continuous wall cross-section remaining between beam ends. Beam position supported in a continuous linear recess: weakening of the wall cross-section. (Right) Discharge of the load **L** via the undisturbed wall cross-section. Support of the beams on a separate element (corbel, trimmer beam)

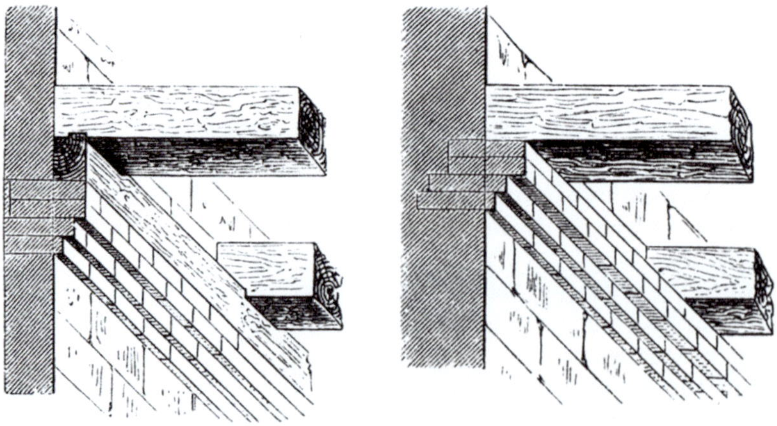

Fig. 4.34 Support of wooden carriers of a wooden-beam floor upon a wall: The support is carried out on a continuous ledge made of projecting ashlars. The wooden beam is well-ventilated all around and does not run the risk of rotting within the wall core due to the effect of building moisture (see also variant on Fig. 4.33, right)

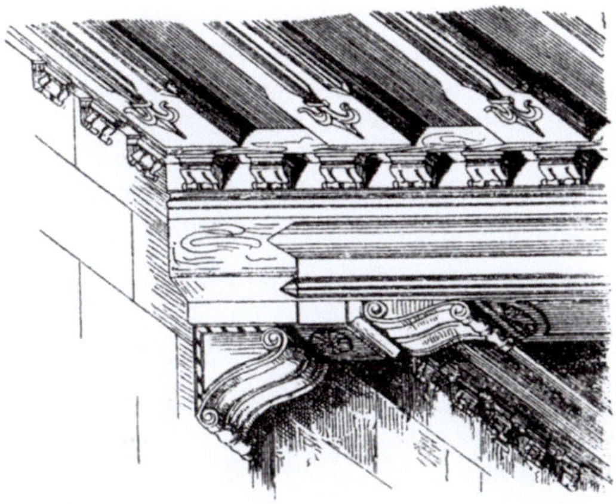

Fig. 4.35 An historical example of a support of a girder on a cantilevered corbel stone which is inserted in the wall. The conditions are similar to those on Fig. 4.34

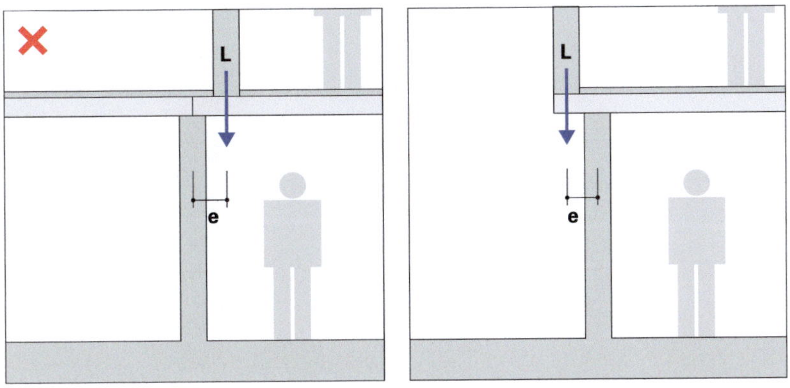

Fig. 4.36 (Left) Offset between overlying walls. Additional bending moment **L · e** acting on the beam (or the floor) on the right side of the wall, as well as shear stress, which is otherwise not present. The wall can no longer unfold a bracing shear-plate effect over several storeys. This solution must always be avoided. (Right) Offset between overlying walls. Discharging moment **L · e** acting on the beam with cantilever. Shear-plate effect of the wall not possible, as on the left

a matter of principle. In rare cases, as with girders with cantilevers, wall offsets (see half-timbered structures: projecting façades) can relieve the stresses on the beams (Fig. 4.37).

Spatial Merging of Adjacent Cells

In order to connect adjacent bays, for example for purposes of utilisation, sufficiently large **openings** within the separating wall panels suffice. If, however, these bays are to be fused into a spatially connected unit, a separating wall can also be replaced by a **truss** of columns and beams (**T–T**) (Fig. 4.38). In this way, large interconnected space units can be created (Fig. 4.39). The directional bay structure remains recognisable in the column positions. The flat ceiling design—in contrast to, for example, a vaulting—is conducive to the creation of a continuous space.

▶ *Sect. 3*, ▶ Figs. 3.26, 3.27, and 3.28, left

Fig. 4.37 Projection of the wall frames typical of half-timbered construction, each of which rests on the cantilevering floor-beam ends

The **bracing effect** of the wall panels replaced by trusses is consequently lost: Suitable replacement measures are necessary in the wall or beam direction (→ **y**). These inner trusses can also be connected force-fittingly to external shear-resistant wall panels by means of the **racking stiffness** of the floor. The shear-plate action of a wooden-beam floor must be secured by a circumferentially arranged **annular beam**. (This is described in more detail below.) It binds the individual beams together and distributes incoming horizontal loads onto the floor joists. The beams are appropriately formed over the supports as a multi-field carrier so that the field moments, and thus the statically required carrier depth, can be reduced.

Cf. *'Bracing'*; also Fig. 4.50, right

As with all buildings with large depths of space, the question arises of how to provide the interior spaces with appropriate **daylighting** and **ventilation**, which can no longer be ensured by windows in the perimetral walls alone. This question will be pursued in the following section.

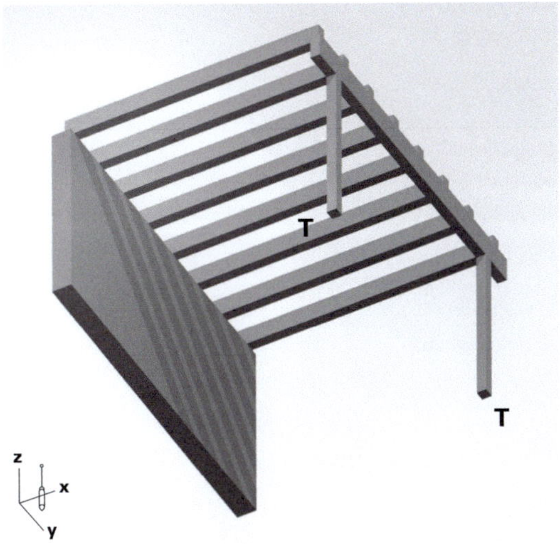

Fig. 4.38 Conversion of a load-bearing and space-enclosing wall panel into a truss **T-T** (right side) composed of two columns and a beam

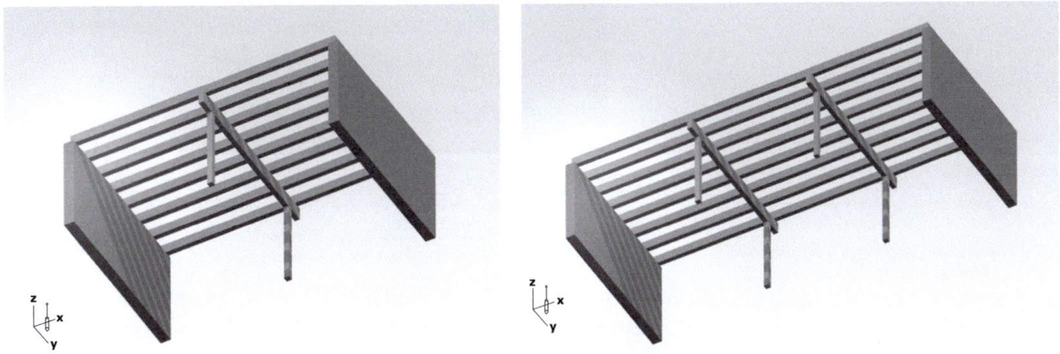

Fig. 4.39 (Left) By this solution, the otherwise segmented bays of the directional system can be connected to form a continuous space volume. (Right) The bays are then discernible only by the column rows and, if applicable, visible ceiling incisions generated by the girders

The Problem of Daylighting and Ventilating Added Cells

As already mentioned, the peripheral areas adjacent to the outer walls can be daylit and ventilated using window openings. On the other hand, the interior areas are cut off from this possibility. The more bays are added laterally, the severer the problem poses itself.

The basilican cross-section was developed in response to this necessity. This designation is based on the building type of the **basilica**, which originated in antiquity and is characterised by the elevated centre bay (Figs. 4.40 and 4.41) or nave, or by a staggered bay sequence ascending from the sides to the centre. On the side walls of the prominent central bay, window openings can be installed which supply the central

Fig. 4.40 Example of a building concept as shown on Fig. 4.41, right (San Miniato al Monte, Florence)

area with light and air. Ventilation is supported by the natural thermal buoyancy: Warmer indoor air rises in the upper middle bay and escapes through the upper windows, while the cooler fresh air flows in through the lower windows of the lateral bays.

Even though this particular building configuration for lighting and ventilating the interior areas is not absolutely necessary today, as, unlike in the past, it is currently possible to construct larger openings for lighting and ventilation purposes in a roof area, it is also sometimes used in modern buildings. Although the term *basilican* suggests that it is an exclusively sacral construction, this building cross-section has often been realised also for many other uses, including in industrial buildings.

In some of the exemplarily depicted load-bearing structures (Fig. 4.41, right), the beam ceiling was additionally equipped with a sloping roof surface for improved weather protection, both in the middle bay and over the side bays.

Bracing

In the direction of the walls (→ **y**), the system can be regarded as braced against horizontal forces as long as they act as **she-**

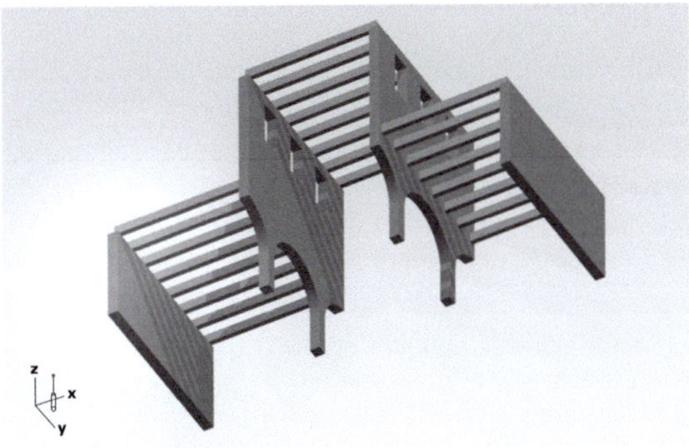

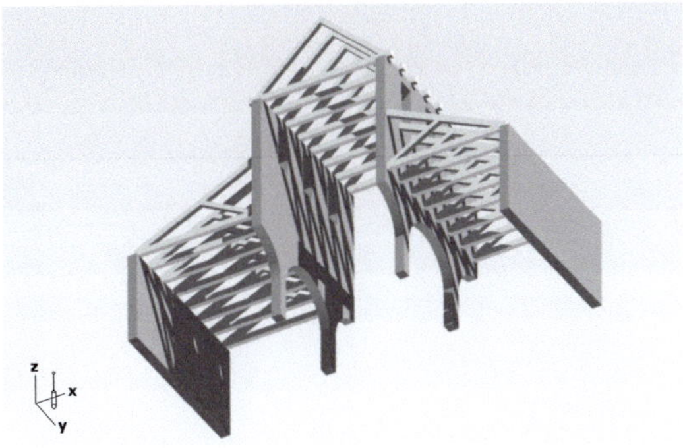

Fig. 4.41 (Left) Superelevation of an inner bay for the daylighting and ventilation of the multi-bay interior. (Right) Covering of the structural bays with sloped roof structures. Three-bay or three-aisle system as in the case of historical basilican constructions

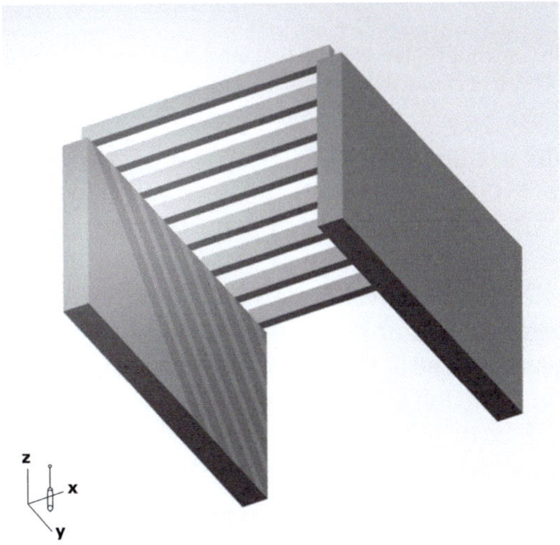

Fig. 4.42 Bracing of the basic cell against horizontal forces in the direction of bars (→ **x**) by means of heavyweight walls

ar-plates within their plane. A prerequisite for this is the necessary tensile and compressive strength of the wall panels, or sufficient **superimposed load**, in order to remove the horizontal loads into the ground completely without tensile stress—as is particularly required in masonry work. Transversely to the wall plane (→ **x**), however, the system is not braced. The following measures are conceivable for this purpose:

- The walls can be carried out as **heavyweight walls** (Fig. 4.42) and thus stabilise themselves. The horizontal forces are, as it were, *over-compressed* by a combination of large dead load and a large wall thickness. This solution is associated with a great consumption of material and plays no role in modern building practice.

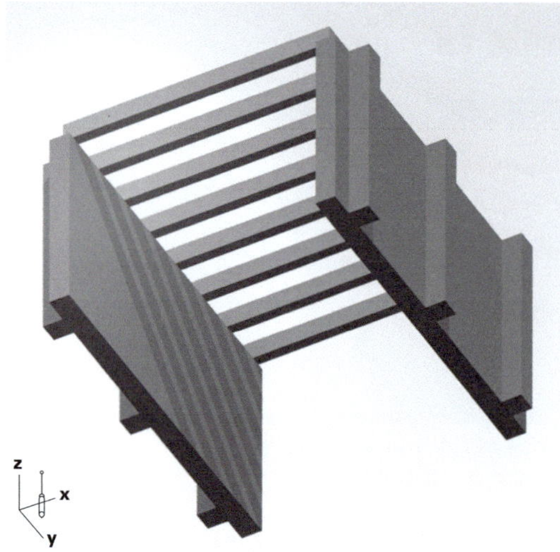

Fig. 4.43 Bracing of the basic cell against horizontal forces in the direction of bars (→ **x**) by means of wall piers (buttresses)

- If walls are concrete panels capable of absorbing flexural tensile forces, a **restraint**—against bending stress in the plane **xz**—at the foundation is possible up to a certain degree of slenderness.
- **Wall piers**, alternatively external or internal, can stiffen the wall against horizontal forces (Figs. 4.43 and 4.44).
- A **transverse wall** stabilises the two longitudinal walls (Fig. 4.47, left). It is necessary to have a bonding at the seam of both panels (Fig. 4.45). At the same time, several transverse walls can serve to divide the space into sections, i.e., spatial cells (Fig. 4.47, right). The same bracing effect is also obtained by transverse wall panels located on the outside of the spatial cell (Fig. 4.48). The laterally unsupported wall sections beyond the bracing transverse wall (**free ends**), as well as the wall sections between neighbouring cross-walls, are nevertheless subjected to bending and are therefore dimensionally limited in their span. This applies in particular to walls made of masonry.
- Spaces enclosed all-around can be formed by the parallel **sequential arrangement** of **transverse walls** (Fig. 4.47, right). The drawback of the previous variant, namely, that the *free ends* of the longitudinal wall panels are not held laterally, and that they would be jeopardised (for example) in masonry construction, is compensated by the fact that, in this case, the transverse and longitudinal walls stabilise each other. Free ends are consistently bypassed by the formation of butt, cross, or corner nodes (Fig. 4.45). A **box** or **cellular construction** is created. From this basic cell, it becomes clear that there are two categories of walls in this directional system (Fig. 4.46):

***Vol. 1**, Sect. 26.9.3.2 Bond—Lap Acting by Compression*

Fig. 4.44 Wall piers

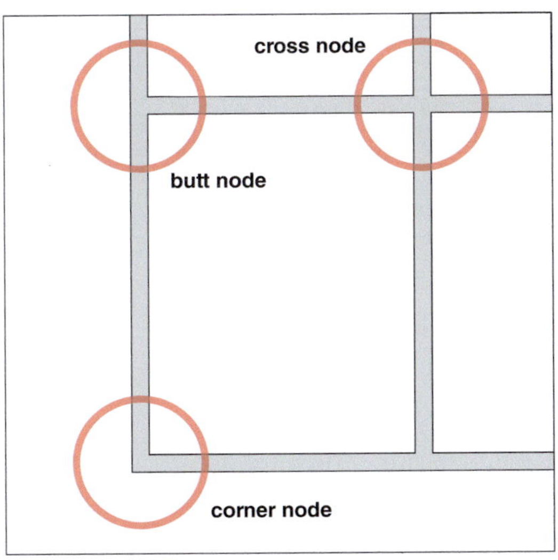

Fig. 4.45 In brickwork-type **box construction**, free, i.e. non-retained, walls are avoided by forming butt, cross, and corner nodes

- **load-bearing walls (L)** on which the beams rest, and:
- **bracing walls (B)** that do not support beams and only laterally support and stabilise the aforementioned load-bearing walls. These too, however, require sufficient vertical load in order to be able to develop their bracing effect. If necessary, at least a part of the resulting floor loads can be introduced into these walls for this purpose by changing the beams' spanning direction in subsequent floors.

▶ *Sect. 7 Masonry Construction*

However, it should be noted that load-bearing walls can also perform a bracing function, if required within

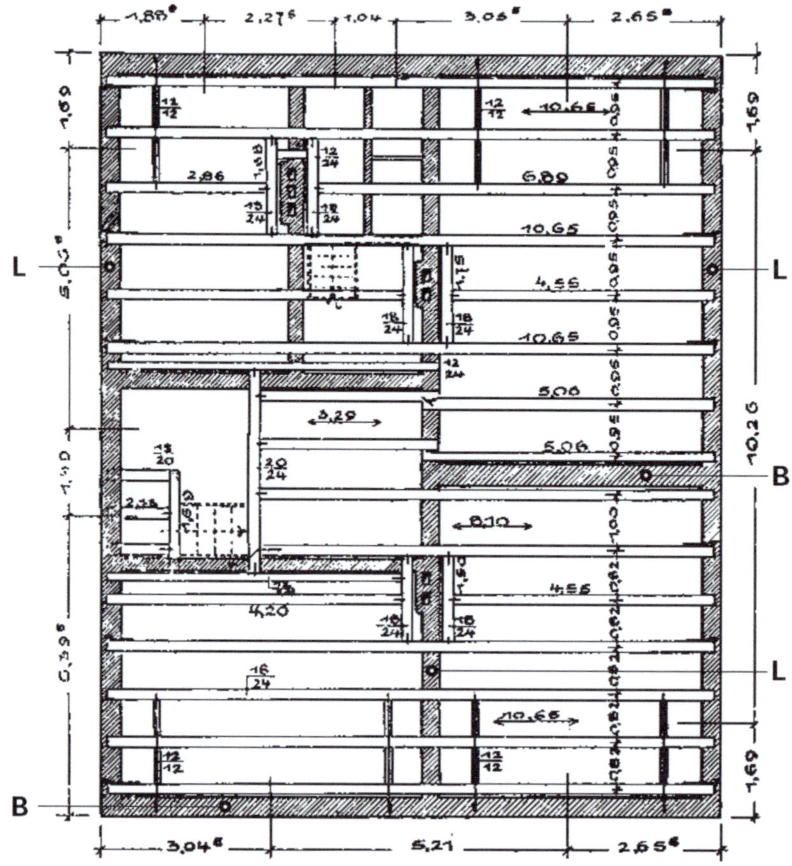

Fig. 4.46 Differentiation between **load-bearing** walls (**L**) and **bracing** walls (**B**) as a function of the support conditions of the sets of beams in the illustrated storey. However, beam spanning directions are often carried out changing from storey to storey, e.g. with the purpose of introducing loads into the bracing walls, which are otherwise not loaded

the load-bearing structure, due to their inherent racking stiffness. In this respect, speaking of bracing walls means they are *merely* bracing, so to speak.

In the case of the transverse bracing of wall panels (Fig. 4.47, left), it is important to take into account that a wall panel which is loaded transversely to its plane, for example by a wind force, and is braced by a plurality of transverse supports, by cross-walls or buttresses, for example (Figs. 4.47, right and 4.48), can only bridge a limited span between supports—as already noted. This is because the wall has to distribute the horizontal loads transversely upon the supporting components by means of its bending stiffness. A masonry wall can do this on its own only to a very limited extent. The beam layer of the floor is not capable of bracing the wall by itself without additional measures, not being shear-resistant, in contrast to a floor slab. To stiffen the wall also at its upper edge in the absence of a

Cf. the failure pattern in ***Vol. 1**, Sect. 26.9.3.2 Bond—Lap Acting by Compression*, Fig. 26.145

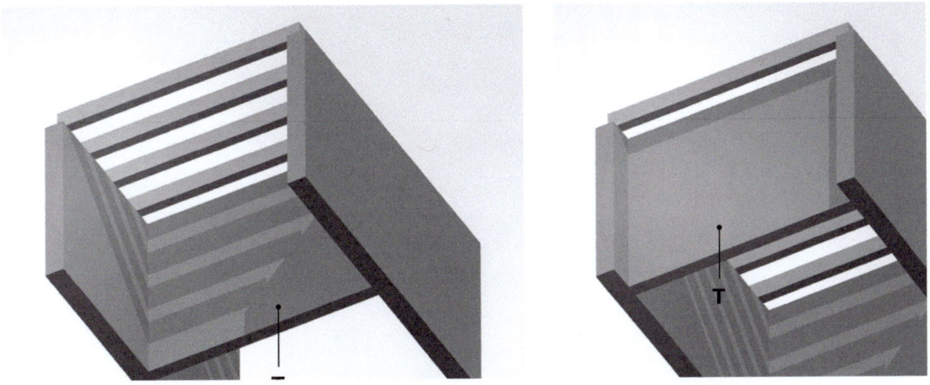

Fig. 4.47 (Left) Bracing of the basic cell against horizontal forces in the direction of the bars (→ x) by means of a **transverse wall (T)**. When executed in masonry, however, the horizontal forces can only be transmitted to the bracing component, i.e. the transverse wall, if an annular beam (see Fig. 4.50, right) discharges these forces via bending, or alternatively a floor shear plate as on Fig. 4.49, right, and Fig. 4.50, left. The same applies for the examples on Fig. 4.47, right, and Fig. 4.48. (Right) Several bracing transverse wall panels (**T**) produce enclosed space cells. The result is a closed, rigid box

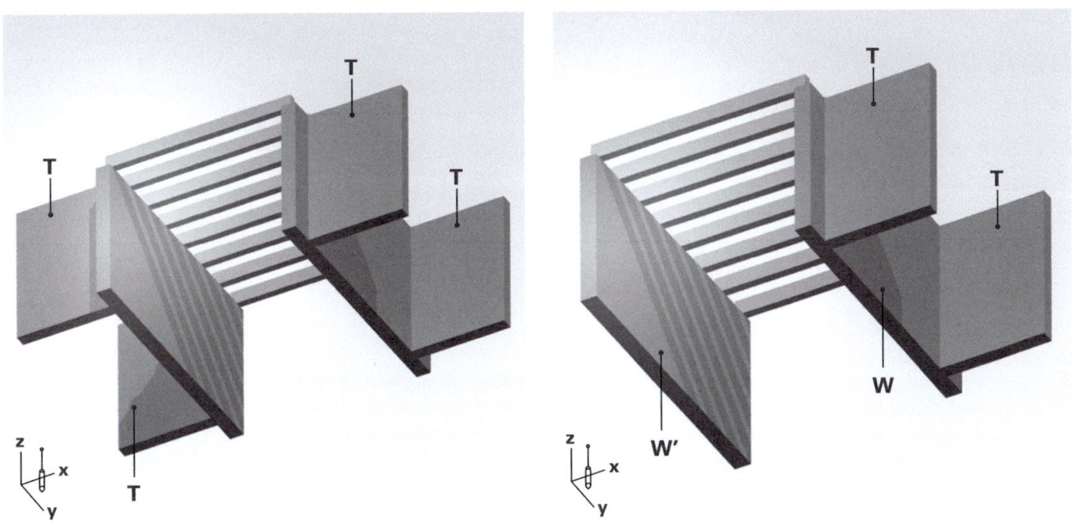

Fig. 4.48 (Left) Bracing transverse walls (**T**) can be arranged externally. (Right) Tensile and compressive forces acting on the longitudinal, laterally braced wall (**W**) to the right are discharged to the transverse panels (**T**) via the wall connection. Horizontal forces in beam direction (→ **x**) acting on the unbraced wall panel on the left (**W'**) are transferred axially through the set of beams to the bracing transverse walls (**T**) to the right in the asymmetrical example shown

rigid slab, a brick wall can be provided with a **ring beam** or **annular beam** (Fig. 4.50, right) which can deliver the load to the fixed points, such as, for example, transverse wall panels, via bending. A perimetral **ring anchor** or **annular anchor**, i.e. a perimetral tensile member without bending stiffness, also has a certain effect. The annular beam binds the joist floor together forming a shear-plate. For this load-distributing bending behaviour, an over-compressing superimposed load is also useful on masonry or reinforced-concrete walls.

► *Sect. 7.4.2 Stiffening and Stabilisation of Walls in Masonry Construction*

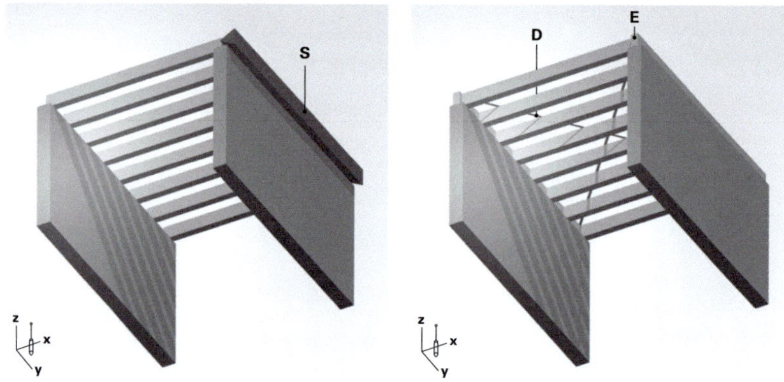

Fig. 4.49 (Left) Bracing of the masonry panels in the transverse direction (→ **x**) by **external support** S (here symbolically shown as a bearing) in the roof or floor plane, for example by a floor shear-plate and core. (Right) Formation of a floor shear-plate by **diagonal bracing (D)** and **edge bars (E)**. In order to transfer the horizontal forces to the shear-plate, the beams must be connected to the diagonals at the crossing points

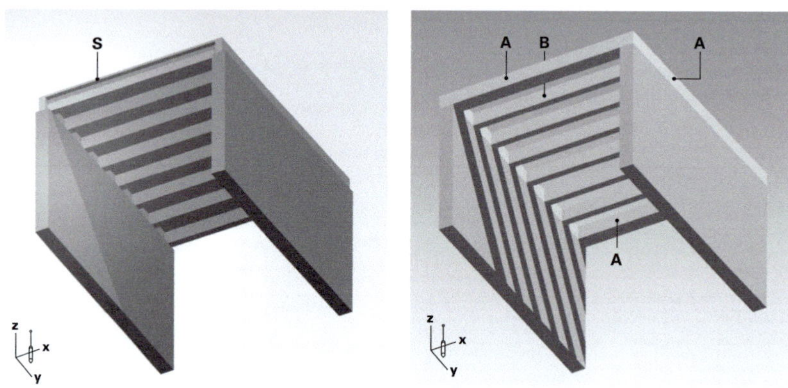

Fig. 4.50 (Left) Shear-plate formation by means of **sheathing (S)** applied on the top of the set of beams. In addition, a circumferential annular anchor is required because of the usually not sufficiently tensile-resistant sheathing materials. (Right) Shear-plate formation by arranging a circumferential **annular beam (A)** which deflects horizontal forces by bending in the plane of the covering (**xy**) upon the wall panels respectively extending in the direction of forces. (→ **x** or → **y**). **B** roof or floor beam

- A **shear-plate floor** which is connected to one or several **fixed points**—for example to a bracing core or to wall panels arranged adequately outside the structural cell—can stabilise the overall construction of the structural cell, including the enclosing walls (Figs. 4.49, 4.50, and 4.51). Constructional measures for transforming a non-shear-resistant bar system into a shear-resistant plate have been discussed above.

▶ *Sect. 3.3.5 Bracing of Bar Systems Within Their Surface*

Cross-Distribution of Loads in Bar Systems

For an understanding of the mode of operation of this system, it is crucial that, in the idealised case of a **statically determinate system**, i.e., a combination of single-span beams, only a single beam is affected by a single load and this beam alone transfers the load onto the elements placed beneath. This means that there is no concurrence of adjacent beams of the beam layer concerned. There is therefore **no cross-distribution** of loads as shown in the schematic model in ▶ *Sect.* 3. In the next-lower beam position, in principle only the two beams are also responsible for the load transfer on which the loaded beam of the next lower beam hierarchy, i.e. the topmost beam layer, rests. Their direct neighbours will not receive a load.

▶ *Sect. 3.2.1 One-Way and Two-Way Load Transfer*, ▶ Figs. 3.40, 3.41, 3.42, and 3.43

In this type of load transfer, the beam layers need not be stacked. Single-span beams mounted flush also transfer loads in this way. The type of **support** of the beams is decisive.

▶ *Sect. 3.3.7 Some Basic Design Considerations about Bar Skeins*, in particular ▶ Fig. 3.99

It is true that this principle allows to easily predict the trajectory of the load, because the competences, as it were, are clearly regulated; But it is evident that the absence of a co-operation between neighbouring members is a **poor economic principle**. For this reason, these systems are never used in pure form in practice, but, one way or another, it is ensured that loads can be removed in several ways, which is the way how **statically indeterminate systems** operate.

▶ *Sect. 3.3.4 Element from Bar Skein, Subordinate Cross Skein, and Plate* (variant 4), ▶ Fig. 3.74; cf. also ▶ Figs. 3.75 and 3.76

In the following, certain steps will be investigated with an **increasing load cross-distribution**:

Cf. also the reflections in ▶ *Sect. 3.2.4 Load Transfer and Usage*

Shear-Resistant Plate

If the topmost surface-forming sheathing has sufficient bending and shear stiffness against shear perpendicular to its plane, it can take over the task of—at least a partial—cross-distribution of loads between adjacent beams. This is the case, in particular, with intermediate stages between a plate and a bar system, such as, for example, in the case of a ribbed slab from reinforced concrete. Too heavy a finishing sheathing rather contradicts the principle of the rod systems; its load-bearing behaviour rather resembles that of a simple plate.

Skeins of Continuous Beams

The first and most obvious measure in the design of floor systems is executing a beam layer as a set of continuous **multiple-bay beams** (Figs. 4.52, 4.53, and 4.54). This takes advantage of the fact that the beams are stacked—or suspended—and pass uninterrupted above, or beneath, the support provided by the main beam. The bending and shear stiffness of this multiple-bay beam forces neighbouring superordinate beams—i.e. main beams—to cooperate, provided that an adequate coupling exists between the two beam layers. Variants as **hinged girders (Gerber beams)** also yield the same result and are widely used in timber construction (Fig. 4.53, right).

This type of support is **statically indeterminate** and leads, in principle, to the distribution of an imaginary point load which is applied to a continuous beam over *a number* of neighbouring next-lower beams on which it rests.

4

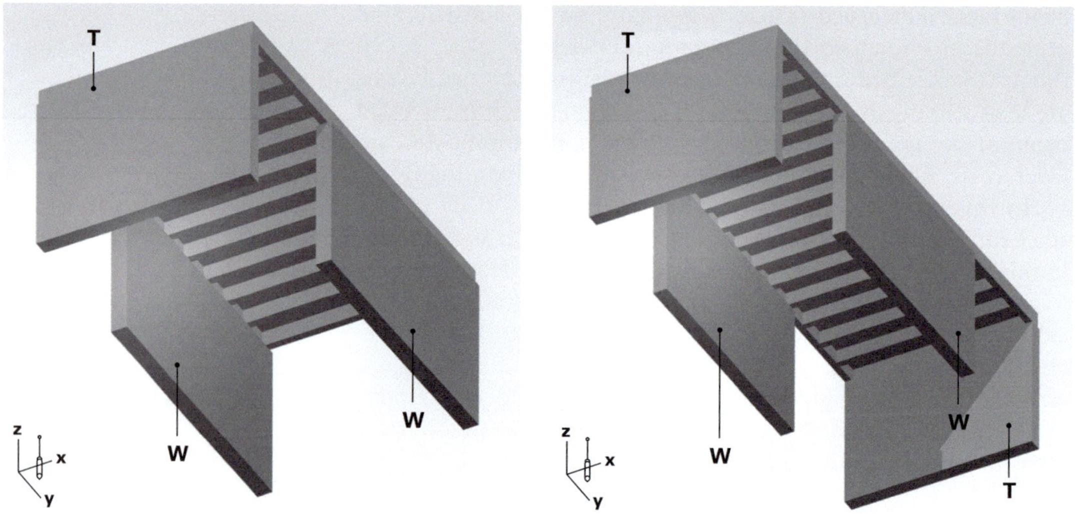

Fig. 4.51 (Left) A transverse wall panel (**T**) can exert a bracing effect on a longitudinal wall panel (**W**) when a roof or floor shear-plate (in this case, a sheathed beam position according to Fig. 4.50, left) connects both force-fittingly. (Right) Twists, which can occur with unfavourable proportions in the system on the left, are excluded in the illustrated symmetrical system by two transverse wall panels **T**

Fig. 4.52 (Left) Continuous subcarrier position, placed on main carriers. (Right) Tilt stabilisation of beam positions with slender upright cross-section by means of a diagonal bracing (see ***Volume 1***, *Chap. 26*, Fig. 26.200)

Transverse Ribs

A transverse rib or cross-rib (Figs. 4.55 and 4.56) which is interposed between the joists of a beam layer forces the adjacent beams of a particular member to participate in its deformation and thus to take over a part of the load acting on it (Note, however, that the transverse beam is not supported at its ends and therefore does not participate itself in the load transfer, only indirectly by redirecting and distributing the load onto a number of neighbouring beams). As with the continuous beam, the prerequisite is again the bending and shear stiffness of the cross-rib. Here, a clear case already appears of a **co-operation** of neighbouring beams. This effect will weaken the more strongly the further away the neighbours are from the loaded beam. A certain transverse distribution of a single load takes place which is not present in the pure one-way load transfer.

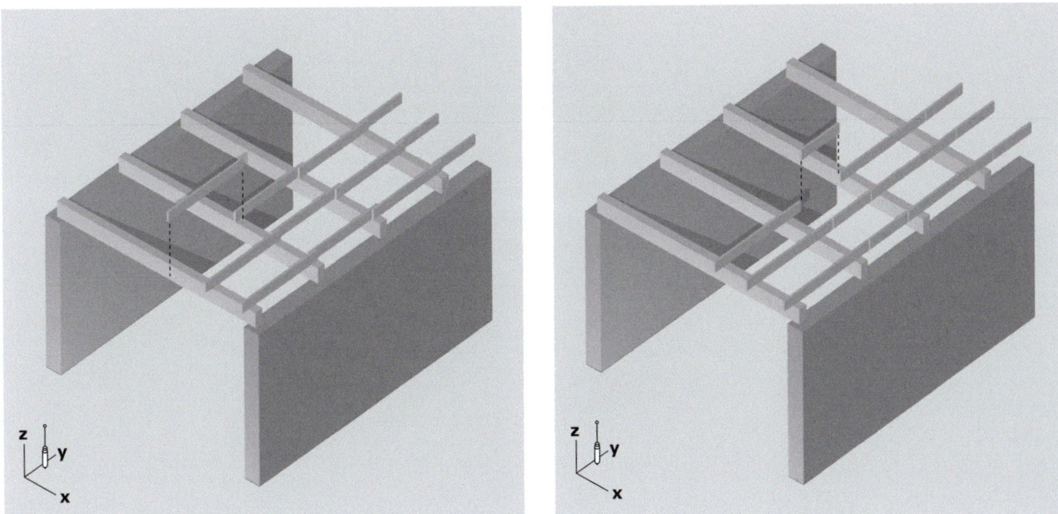

Fig. 4.53 (Left) Stacked carrier layers allow utilising a continuity effect and a significantly reducing the bending moments in the topmost set of carriers. The continuity effect is achieved in wood construction, for example, by overlapping and lateral nailing—as in the case of so-called tie purlins. (Right) Solution analogous to the one on the left. The continuity effect is achieved by hinged coupling of the carriers at the points of contrafexure (articulated beam)

Fig. 4.54 Example of a load-bearing structure with stacked beam skeins in timber construction

For the sake of completeness, it should be pointed out once more that a cross-distribution is only of importance and useful when loads are **punctual** or at least **asymmetrical**. A linear load distributed evenly across a beam position, or a uniform surface load, produces the same load-bearing behaviour with and without a cross-distribution.

Cf. the analogous case of the slab, Fig. 4.12, right

If the individual transverse rib is supplemented by further ones, a gradual transition takes place to surface elements with two perpendicularly oriented, equal-ranking beam skeins, as discussed elsewhere under variant **3**. At the end of the logical sequence towards an ever increasing cross-distribution of loads in bar systems are the **beam grillages** as described below.

▶ *Sect. 3.3.3 Element from Bar Grillage and Plate* (variant **3**)

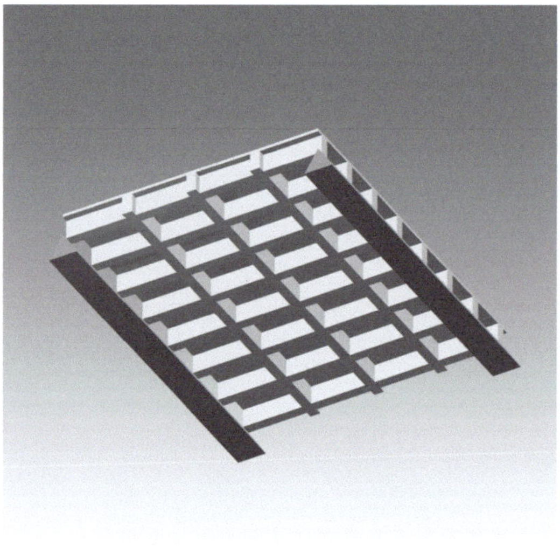

■ **Fig. 4.55** Beam covering with transverse ribs

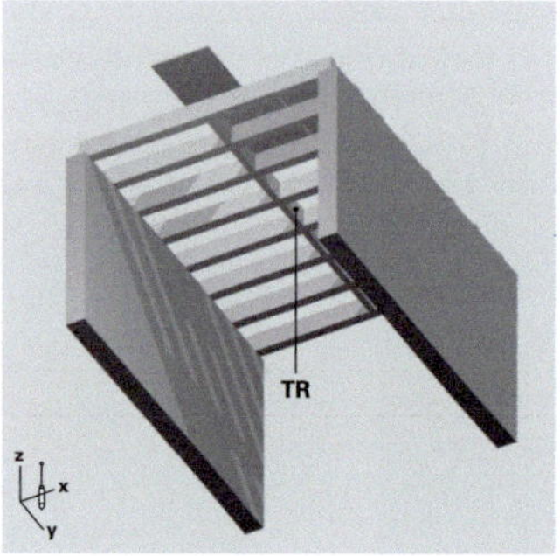

■ **Fig. 4.56** (Left) Directional load-bearing system with a load-distributing **transverse rib TR**. (Right) Example of a load-bearing system in concrete with a transverse rib. The load distribution of large mobile point loads (vehicles!) naturally plays an important role in a garage

▶ *Sect. 4.3.1.3 Beam Grillage Spanning Two-Way, Linearly Supported* as well as ▶ *Sect. 4.3.1.4 Beam Grillage Spanning Two-Way, Punctually Supported*

Directional Bar Systems in Skeletal Structures

There is no significant change in the load-bearing behaviour of one-way floor systems if they are not mounted on a masonry wall, as previously assumed, but are so linearly on a bar—a beam or girder—which in turn rests on **punctual** column supports (■ Fig. 4.57, left). The wall is thus replaced by a truss made from columns and beams, so a further carrier hierarchy is added, as it were. However, it must be taken into account that, unlike with the supporting wall panel, this is an **indirect load transfer**.

▶ *Sect. 3.3.7 Some Basic Design Considerations about Bar Skeins*, in particular ▶ ■ Fig. 3.96, left

Typical of directional floor systems in skeletal structures are **rectangular** support fields, i.e. those with two distinctly different spans (■ Figs. 4.57, right and 4.58). In this case, the main beams are laid over the *small* span and the secondary beams over the *long* one, based on considerations of the flow of force within rod systems. In this way, their respective beam depths are equalised. Thus, the:

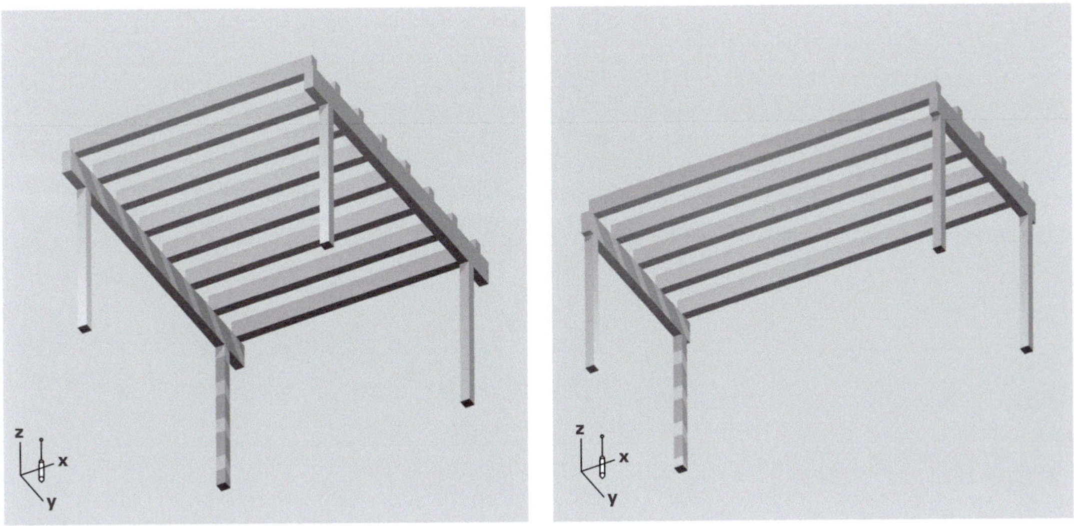

Fig. 4.57 (Left) Basic cell made of a bar skein on girders instead of on masonry walls. (Right) Rectangular structural field with clearly different spans typical of directional bar systems

Fig. 4.58 Example of a directional structure from hierarchically stepped beam skeins

- **subcarriers** span over a larger field, while the:
- **main carriers**, conversely, have to bear considerably more load because their load-infeed area is larger.

However, the decision how the individual beam hierarchies are oriented can also result from entirely different considerations, such as for example the use of the building or the visual appearance of the soffit.

Checkerboard arrangements of beam skeins (Fig. 4.59), which are used particularly in timber construction, are a hybrid of one-way and two-way systems. The spanning direction of the joist skeins changes Checkerboard-like between adjacent square support fields. In this way, a main carrier receives the load portion of only one joist skein. In the case of equally oriented joists, the load of two skeins would be charged on it instead. It is thus possible to assimilate the edge carriers in their structural depth to the joists.

► *Sect. 8.4.1.5 Four Beams Butting with End Sides on Column*

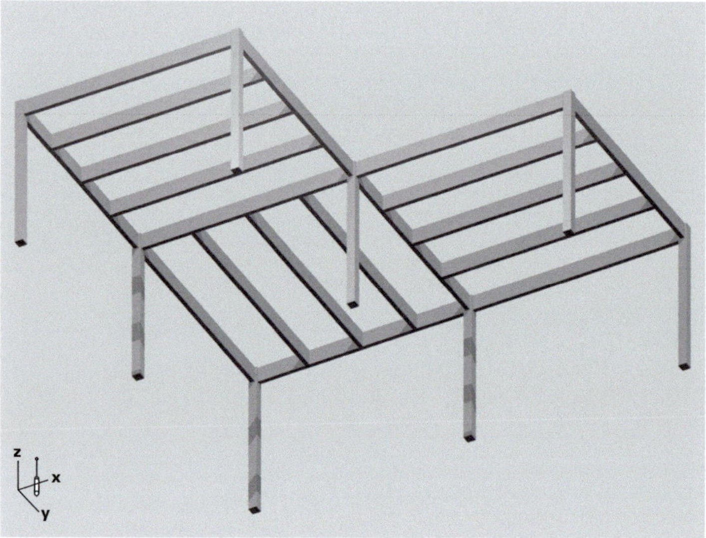

Fig. 4.59 Example of a partially directional and non-directional system. In this system of timber construction, directional sets of bars are laid alternatingly in the shape of a checkerboard, so that, overall, a sort of (at least geometrically) non-directional system emerges. The girders are thereby relieved, since they always only receive the load of one carrier load-infeed field (instead of two fields)

4.2.1.3 Bracing of Skeletal Structures

Issues of bracing have a special significance in skeletal structures, since the bracing effect of wall panels, which were so far presupposed as a systemic component oriented in both contemplated spatial directions (→ **x**, → **y**) belonging to an interlaced wall structure, is dispensed with from the outset. The racking stiffness against horizontal loads must therefore, as a result, be reconstituted, as it were, by means of additional measures, i.e. by an additional **bracing construction** (Figs. 4.60 and 4.61, left). In this context, the type of load transfer within the covering, that is to say the question whether it is one-way or two-way, is irrelevant to this requirement. The following considerations therefore apply to skeletal structures with one-way as well as two-way load transfer, that is to say both for directional and non-directional skeletal systems.

***Vol. 4**, Sect. 8.5.2 Horizontal Bracing*

***Vol. 4**, Sect. 8.5.2 Horizontal Bracing*

The most important bracing measures are:

- Attachment of the skeleton to **fixed points** by the bracing effect of shear-resistant **floor plates**. Racking stiffness is already inherent to solid floor slabs and similar shear-resistant surface elements. For bar systems as described in this section, a shear-resistant panel can be produced by means of measures as described above. Fixed points can be either:
  - Single **wall panels** integrated into the skeletal construction (Fig. 4.61, right). There must be panels in both main directions → **x** and → **y** of sufficient length to receive the horizontal loads. Their axes must not intersect

► *Sect. 4.2.1.1.6 Bracing*

► *Sect. 3.3.5 Bracing of Bar Systems Within Their Surface*

Fig. 4.60 Diagonal bracing in the façade plane

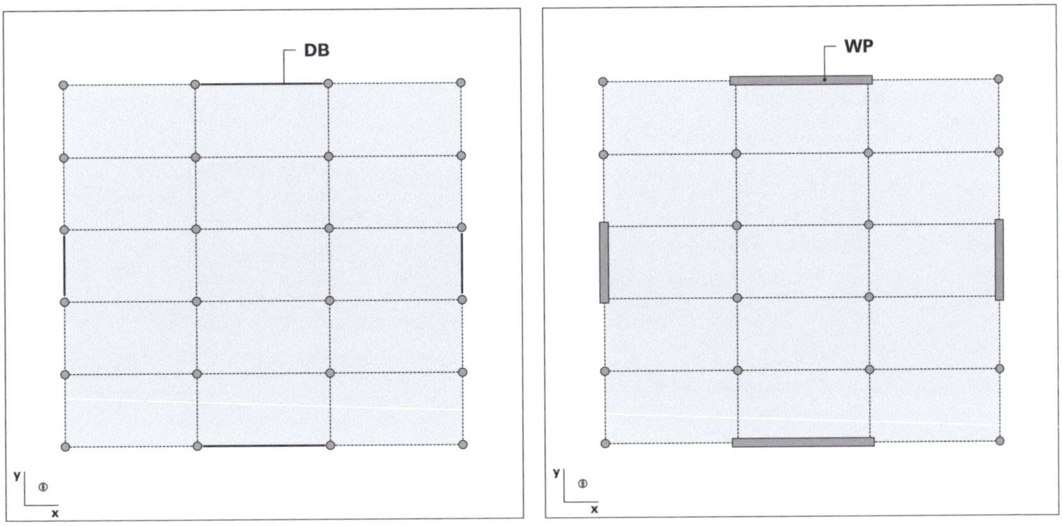

Fig. 4.61 (Left) Bracing of a skeletal structure by means of diagonal bracings (**DB**). A prerequisite for this case, as well as for the one on the right, is a shear-resistant floor panel. (Right) Bracing of a skeletal structure using wall panels (**WP**)

at one point, since otherwise twists cannot be locked. Their arrangement must be designed in such a way that horizontal forces are absorbed in both main directions of the building (→ **x**, → **y**) and moments around the vertical axis (→ **z**) of the structure are precluded.

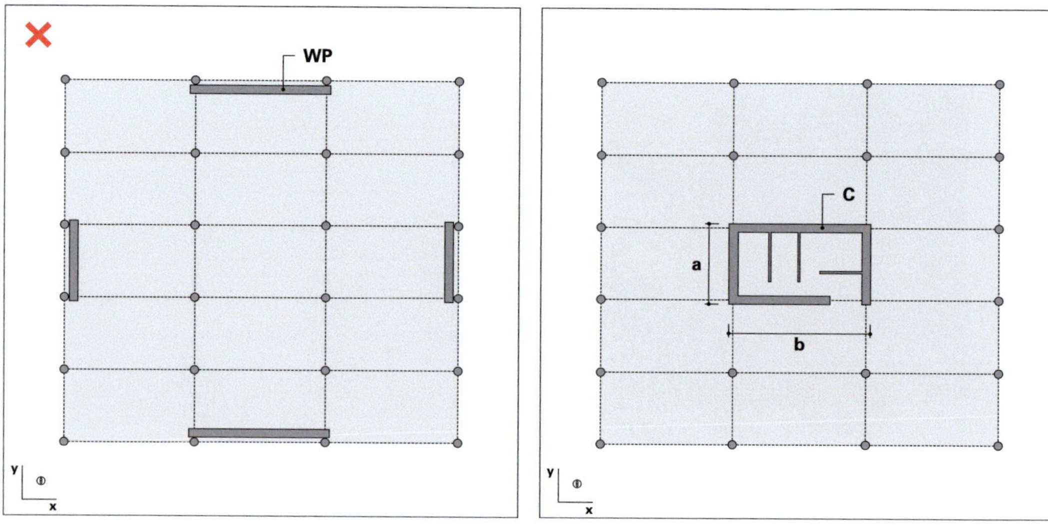

Fig. 4.62 (Left) Unlike in Fig. 4.61, right, the wall panels are offset from the structural axis. The loads are mainly discharged by means of the columns arranged close to the wall panels. The panel does not receive the floor load required for the bracing function. (Right) Bracing by a **building core** (**C**). The footprint proportions of the core (**a** : **b**) are unfavourable, since one side is considerably shorter than the other. As a result, there is less shear-plate length in the direction of → **y** than in the direction of → **x**. The façade surfaces, and therefore the wind-attack surfaces in both directions, however, are almost the same

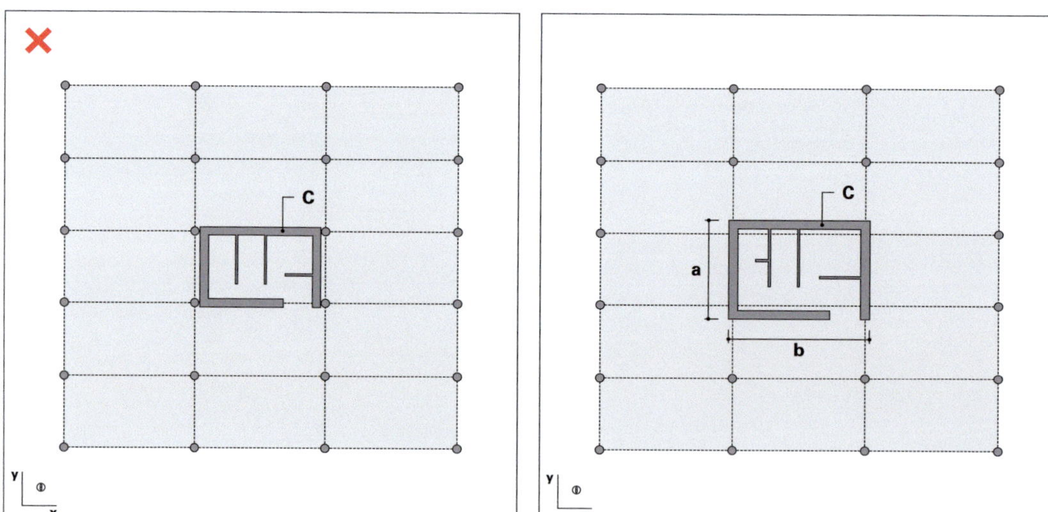

Fig. 4.63 (Left) Similarly to Fig. 4.62, left, too little floor load is introduced into the core in this core design since this is intercepted by the columns arranged close to the core. (Right) Core design with more favourable footprint proportions **a** : **b** in comparison to the solution on Fig. 4.62, right

- Or also **building cores** (Fig. 4.62, right). These are boxes stiffened in themselves by means of transversely arranged wall panels. They must have sufficient moment of inertia in both bracing directions → **x** and → **y** to oppose the acting horizontal loads as well as rotational movements (Fig. 4.63, right). Cores located centrally in the building allow for even, non-restraining horizontal

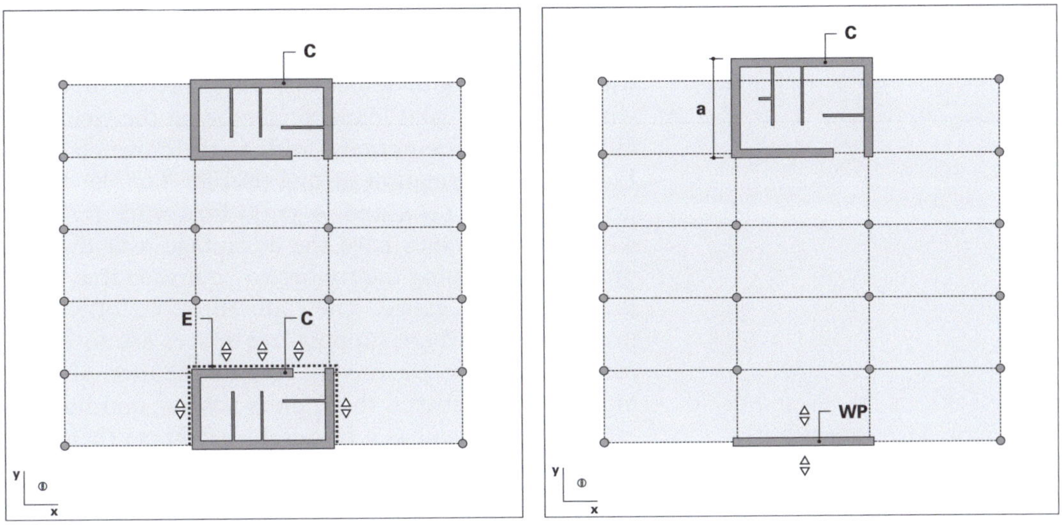

Fig. 4.64 (Left) With this core arrangement, attention must be paid to the possible constraints between the two rigid fixed points. Possibly, expansion joints (**E**) are to be provided. (Right) Bracing alternative to the solution on the left with wall panel **WP** instead of one of the cores. This wall can absorb the deformations which are perpendicular to its plane without causing constraints. The resulting reduced stiffness of the system in → **y** may be compensated for by an enlarged core dimension **a**

deformations of the construction, which can unfold freely radially from the core. Multiple cores situated at the building's periphery can be critical, especially at the front ends of elongated buildings. This can result in restraints and, as a consequence, in cracks. If such a core arrangement, out of non-static considerations, is unavoidable, expansion joints must possibly be inserted in the vicinity of the core, so that the connection of the floor slab to the core is rescinded at least in one direction (Fig. 4.64, left).

▶ *Sect. 5 Deformations*

As already noted, both bracing wall panels as well as building cores require sufficient **superimposed load** to fulfil their bracing task. It is therefore appropriate to integrate these as fully fledged load-bearing main elements into the load-bearing grid as a substitute for column supports. If columns are arranged in their vicinity, the required **floor load** on the wall panel or core (Figs. 4.62, left and 4.63, left) is missing in that case.

▶ *Sect. 4.2.1.2.4 Bracing*

- Stiffening of individual trusses made of columns and beams by means of **diagonal bracings** (Fig. 4.61, left). Since this is certainly a *replacement* for shear-plates, the same requirements apply to the arrangement of diagonal bracings as for wall panels (see above). Appropriate locations for diagonal bracings are to be defined; design criteria often play an important role either from aesthetic considerations or functional ones, because openings like, e.g. for windows or doors in fields containing diagonal bars are only possible to a limited extent.

From a structural viewpoint, it is advisable to place diagonal bracings as close as possible to the location where the horizontal forces to be absorbed occur, that is, with respect to wind loads, preferably in the vicinity of façades which extend *transversely* to their own plane.

- **Portal-frame** formation in one (◘ Fig. 4.65, left) or in both directions (→ **x** and → **y**) (◘ Figs. 4.65, right and 4.66). Portal frames have the advantage over diagonal bracings of keeping the trusses to be braced free within their clearance gauge. They are therefore often used in larger halls where support-free spaces are to be created. In this case, the mostly even distribution of bending moments between the frame's vertical and horizontal members also has a favourable effect, so that excessive carrier depths can be avoided in the case of larger spans. Transversely to the frame plane, diagonal bracings as described above are often inserted (◘ Fig. 4.67) in order to prevent them from tipping sideways.

  In the case of portal frames, it is rather disadvantageous that, at least with larger components, bending-stiff erection joints must usually be carried out. If reasonable transport dimensions are adhered to, it is sometimes also possible to pre-fabricate in the factory complete frame segments—in particular three-hinged frame halves—so that no bending-stiff erection joints are necessary. It is also preferable to avoid joints where the greatest bending moments occur, in particular at the rigid frame corner.

Vol. 1, Sect. 26.7.2 Composite Linear Components

Vertical members of planar portal frames—unlike hinged columns—require a greater stiffness within the frame plane than transversely to it. As a result, narrow oblong cross-sec-

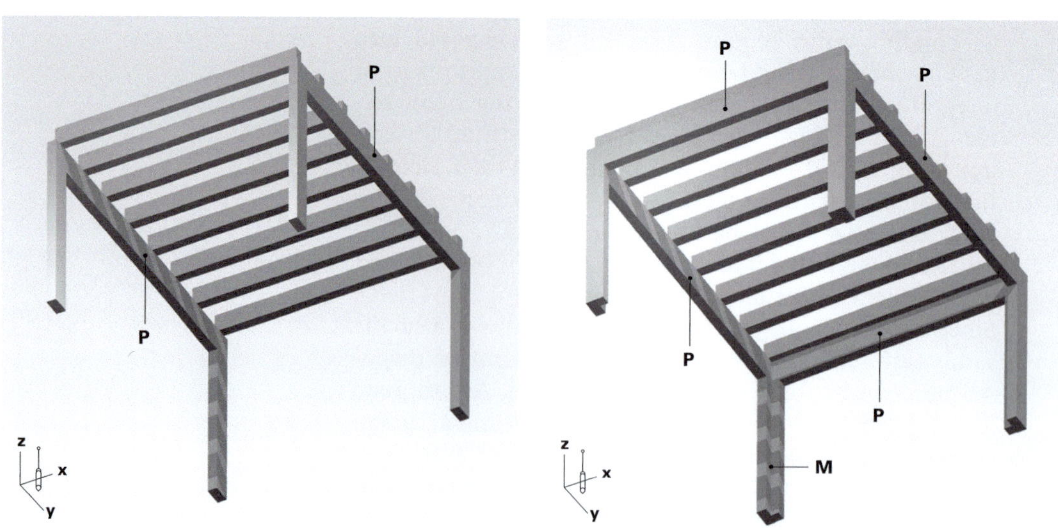

◘ **Fig. 4.65** (Left) Portal-frame (**P**) formation in → **y**; direction → **x** not braced. (Right) Portal-frame (**P**) formation in → **x** and → **y**, vertical frame members **M** have to be bending-stiff in both directions

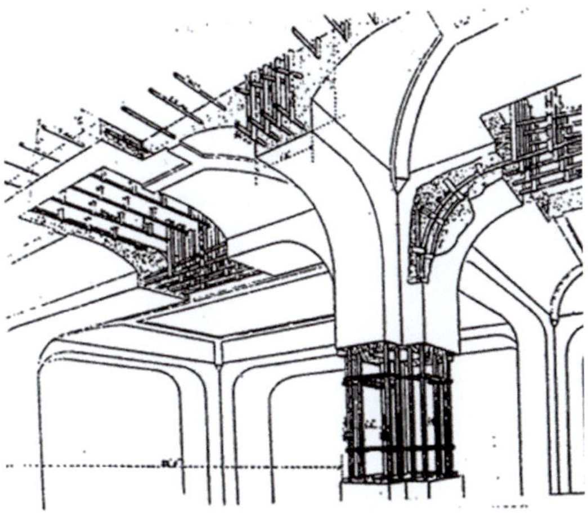

Fig. 4.66 Portal-frame formation in → **x** and → **y** exemplified by an in-situ concrete construction (system Hennebique)

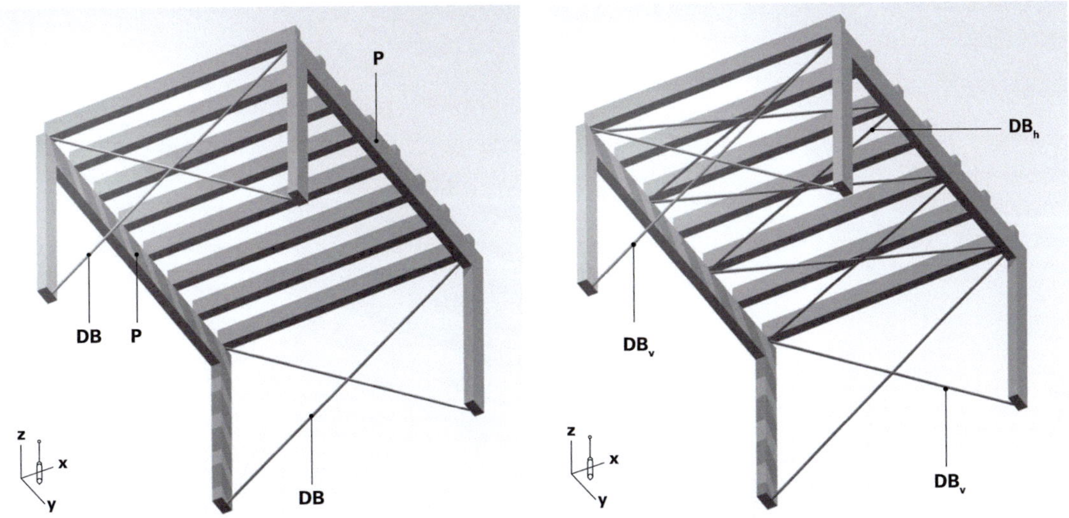

Fig. 4.67 (Left) Portal frame formation in → **y**; diagonal braces in → **x**. (Right) Bracing with vertical diagonal bracings (**DB**$_v$) as on the left, additionally horizontal diagonal bracings (**DB**$_h$) in roof or floor plane (**xy**)

tions are often used (Fig. 4.65, left). Portal-frame systems can in principle also be executed in two directions (→ **x** and → **y**) (Fig. 4.65, right) and can be added vertically to form multiple stacked frames in multi-storey buildings (see also examples on Figs. 4.68, 4.69, 4.70, 4.71, 4.72, 4.73, and 4.74.

4.2.1.4 Inclined Planar Coverings from Bar Skeins

Purlin Roof

If, in the case of a flat, directional beam floor covering two bays with a central transverse-oriented main carrier, the latter is lifted in its position so that the beams resting on it end up inclined towards the edges, the structural system which un-

Fig. 4.68 Portal-frame corner in a multi-storey building with erection joint **E**

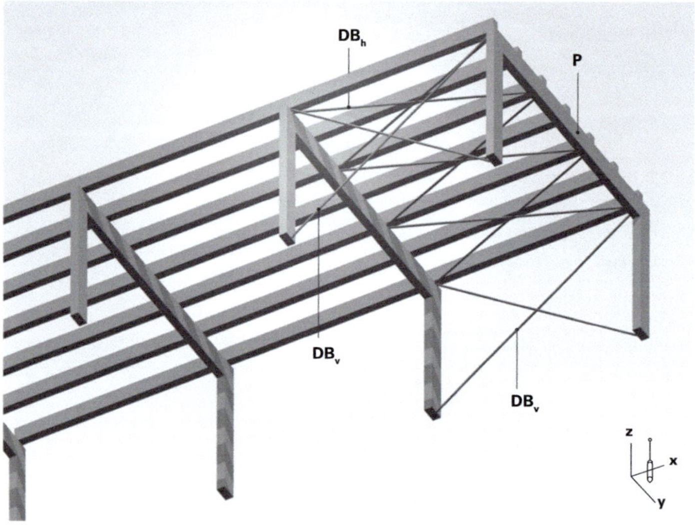

Fig. 4.69 Sequence of portal frames (**P**): One particular portal-frame bay is stiffened with vertical and horizontal diagonal bracings (**DB_v**, **DB_h**). The non-diagonalised bays, however, are connected to the stabilised bay by the carrier layer and thus, in turn, are braced against forces acting in → **x**

derlies the conventional purlin roof (Figs. 4.75 and 4.76, left) results. This constructional variant will be contemplated in more detail in the present section using the example of conventional wooden roof structures.

The term '**purlin**' denotes the carrier which is oriented transversely to the beams' spanning direction, i.e., in the ridge direction (→ **y**), which in the illustrated case (Fig. 4.76, left) is referred to as a **ridge purlin**. The sloping beams, which are called **rafters** in conventional roof trusses, are placed on top

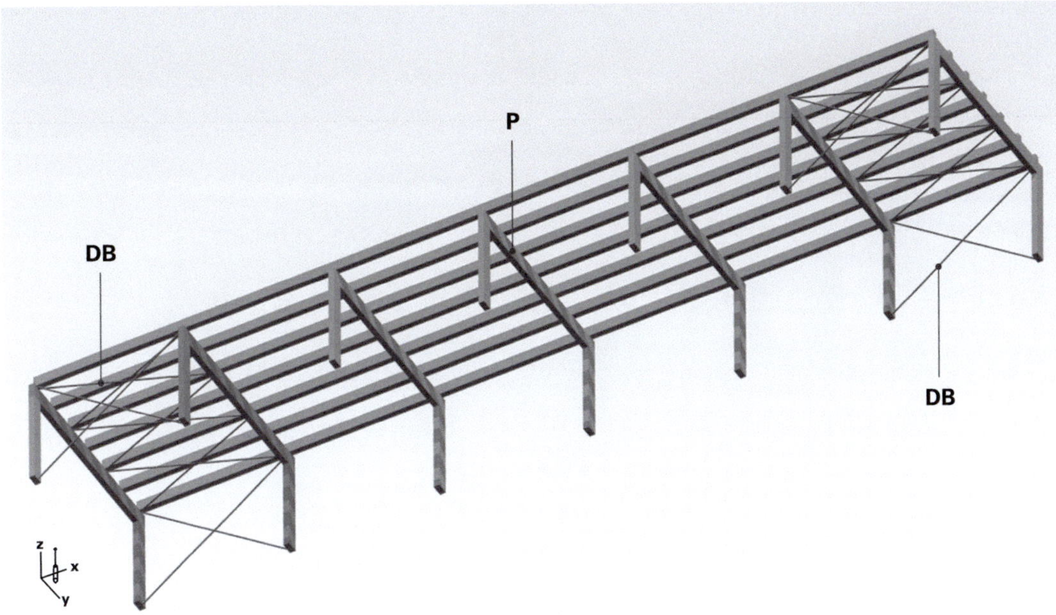

■ **Fig. 4.70** Design of a complete structural hall unit from portal frames according to the principle depicted on ■ Fig. 4.69. The end bays are each braced with diagonal bracings (**DB**) in the wall and roof plane, and so are, by the tying effect of the carrier position in → **x**, also all other bays

■ **Fig. 4.71** Portal-frame hall (ice-skating rink in Munich, arch.: K. Ackermann)

of the wall crest also on wooden parts (**base plates**). These then do not have a load-carrying function, since they rest in their entirety flat on the wall, but rather an equalising function between the diverging tolerance ranges of the wall and the timber structure.

The type of stress of this roof construction is not essentially different from that of a flat roof. Here, too, the central truss

Fig. 4.72 Portal-frame hall. The diagonal bracings in the frontal wall plane are clearly recognisable

Fig. 4.73 Portal-frame hall. The frames are retained transversely to their plane by a restraint. The hierarchically subordinate carrier transverse to the frame plane is suspended (Crown Hall, IIT Chicago, arch.: Mies van der Rohe)

composed of posts and purlin, or the so-called **purlin sequence** or **strand** consisting of more than one such trusses, assumes the function of an imaginary central wall, that is, to bear vertical loads (and in most cases also the wall's associated bracing function against horizontal ones; cf. the next item to be discussed). The beams or rafters, also here, act—at least predominantly—as bending beams and transfer their load to the truss. In this system, their inclined position does not fundamentally change this. Typically, however, there are normal forces in the rafters (Fig. 4.79), which, however, do not constitute the main design load.

Fig. 4.74 Portal-frame hall. In historic timber architecture of East Asia, the load-bearing structures are not braced by means of diagonal bracings, as is the case mostly in Europe, but by portal-frame formation from pillars and frame beams (Forbidden City, Beijing)

Fig. 4.75 Purlin-roof slope

As with the roof structures discussed further below, the structure is closed on the top by further constructional layers: For this purpose, the rafter skein, in conventional roofs, is executed so tightly spaced that the rafter distances can be bridged by a light, transversely oriented position of **laths** (Fig. 4.77). The brick tiles, or other covering elements, are then hooked or otherwise engaged into these.

The illustrated example (Fig. 4.76, left) represents the **simple-supported** purlin roof, Fig. 4.76, right, shows the **double-supported**, in which two purlin strands support the rafters,

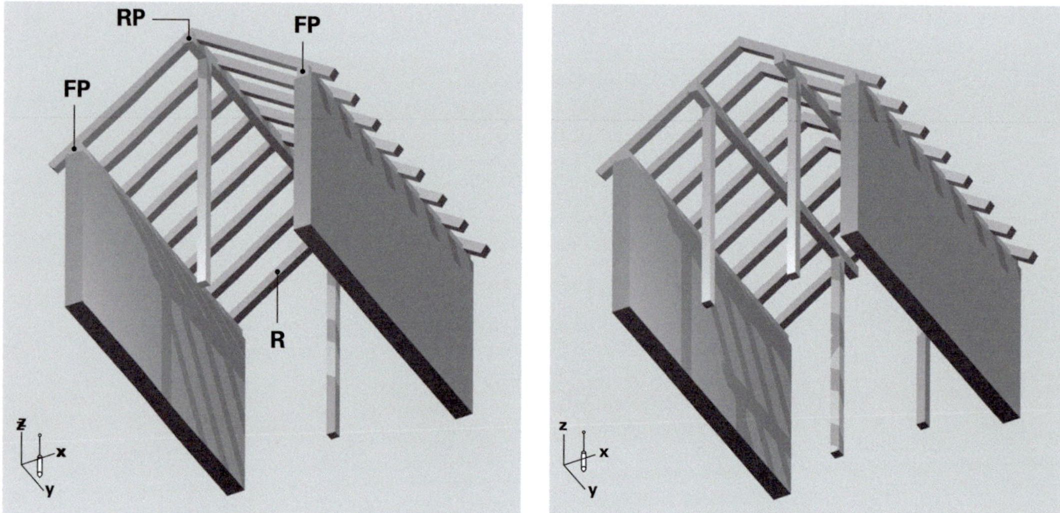

■ **Fig. 4.76** (Left) Schematic diagram of a purlin roof with a ridge purlin (**RP**), foot purlins (**FP**), and rafters (**R**). (Right) Schematic diagram of a double-string purlin roof with overhanging rafters

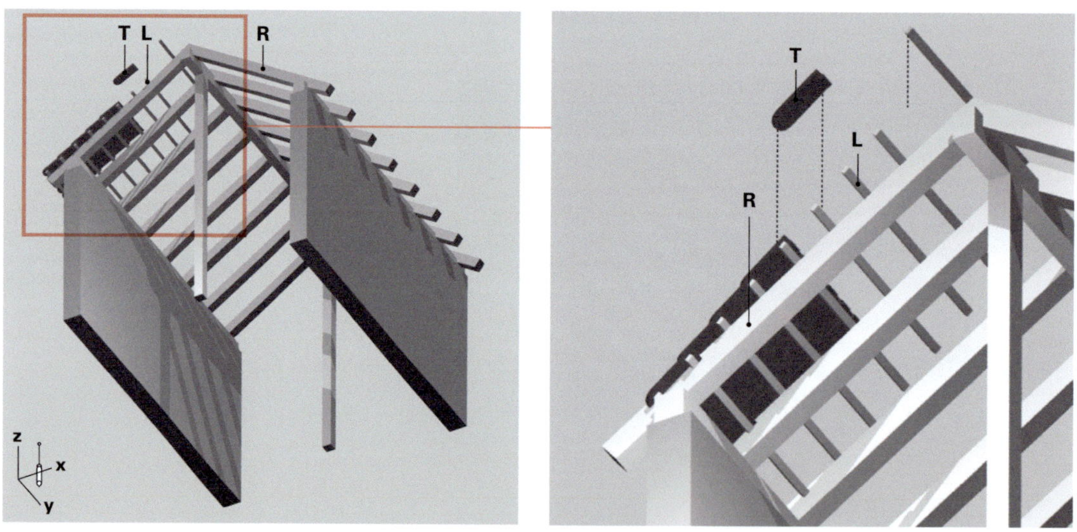

■ **Fig. 4.77** (Left) The spacing of the rafters (**R**) is adjusted to the load-bearing capacity of the lath skeins (**L**). These, in turn, are laid at intervals which are governed by the length of the roof tile (**T**) and its overlap. (Right) Detailed view

which in turn cantilever freely towards the ridge and are coupled at their ends (see also ■ Figs. 4.78 and 4.79).

Bracing

As mentioned with flat coverings featuring inner trusses instead of inner wall panels, these lose the bracing effect of the wall panels transversely to the beams' spanning direction and therefore require additional bracing measures. This applies in the same way to trusses from purlins and posts in purlin roofs. The traditional solution to this problem in timber construction is adding **angle braces** to the post heads, which transform

Fig. 4.78 Characteristic flat-sloping south-European purlin roof

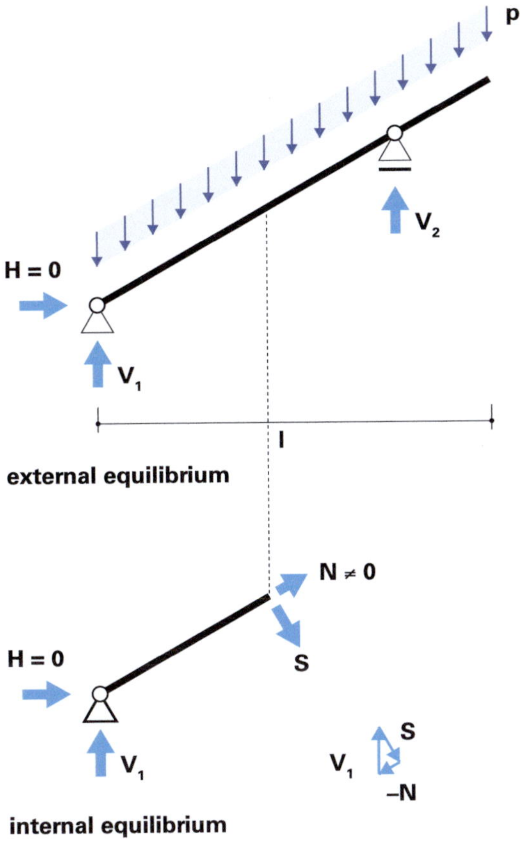

Fig. 4.79 Representation of the structural system of a purlin roof (top). In the case of a vertically acting load **p**, the horizontal reaction **H** at the support is equal to zero. If the sectional forces **S** and **N** are considered (below), a **normal force N** unequal to zero emerges in addition to the shear force **S** and the associated bending moment, as a consequence of the inclination of the bar to the horizontal line. Compared to normal forces in rafters of rafter roofs (▶ *Sect. 4.2.2.1*), however, this force is subordinated

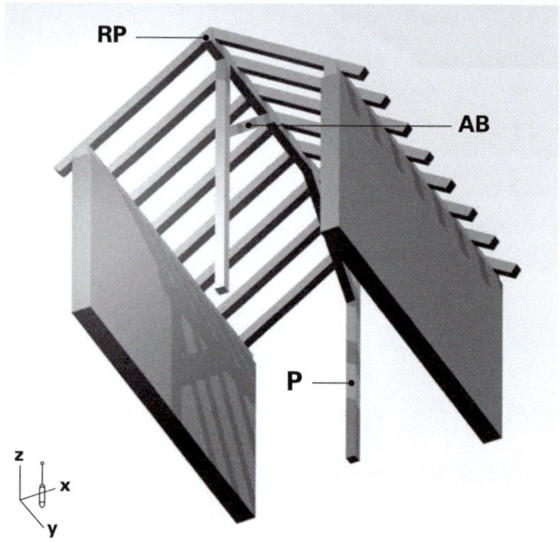

■ **Fig. 4.80** Bracing of the roof truss in ridge direction (→ **y**) by means of a purlin string from ridge purlin **RP**, post (**P**) and angle braces (**AB**)

■ **Fig. 4.81** Purlin string with posts and angle braces

the purlin string into a rigid frame (■ Figs. 4.80, 4.81, and 4.82). By virtue of the corners having been made bending-stiff thereby, horizontal forces can be absorbed in the direction of the purlin (→ **y**).

In the opposite direction (→ **x**), the roof frame constitutes, by itself, already a rigid triangular frame composed of the sloping rafters and the horizontal floor, but the horizontal load is instead introduced into a triangular, and therefore inherently stiff, reinforcing **roof truss** consisting of a post, a lower rafter section, and floor beams in order to avoid directing excessive loads on the rafters (■ Fig. 4.83 top right picture). To further alleviate the load on the rafter, instead of the rafter, a strut

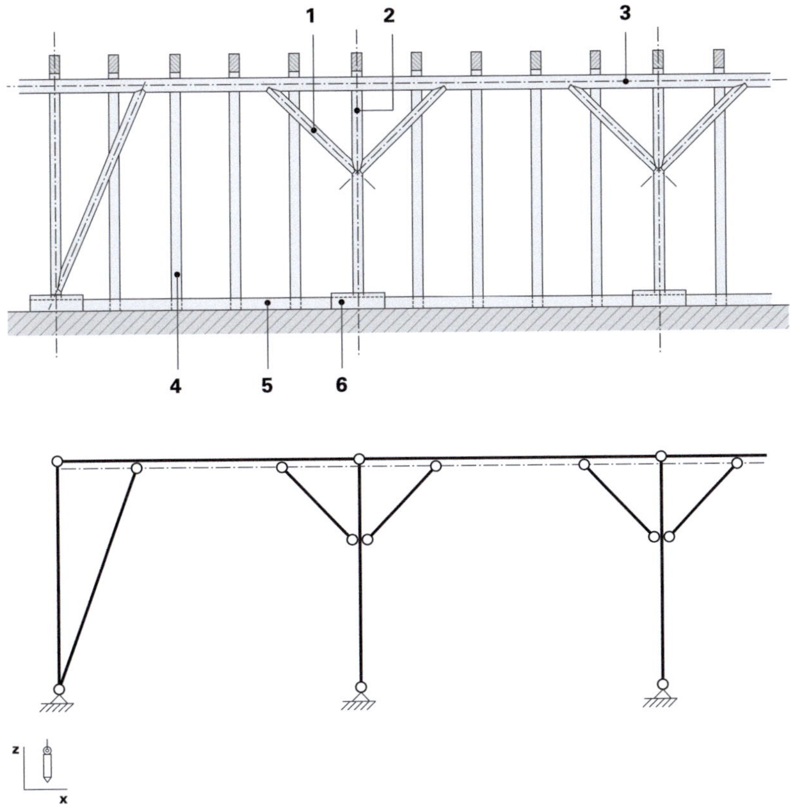

■ **Fig. 4.82** **Purlin string** with posts and angle braces. Below: structural system. (**1**) Angle brace; (**2**) post; (**3**) middle purlin; (**4**) rafter; (**5**) base purlin; (**6**) sill plate

placed besides it acting as a diagonal bar can also be inserted (■ Fig. 4.84). Roof trusses are also executed totally without columns or posts so that the roof space remains free of supports. The loads from the purlin node are then transferred into the support at the base plates by means of a frame-like truss made of a horizontal **tie-bar**, oblique **struts** and angle-brace-like stiffening struts (■ Fig. 4.83 bottom row of pictures).

Truss Roof

From here, it is only a small step to the lattice-trussed roof, in which the elements of the rafters, tie-beam, tie-bar, and struts are joined together to form a **truss** acting as a triangulated latticed frame structure (■ Figs. 4.85 and 4.86). These very stiff—because very high—lattice girders are put up at larger distances, so that an additional beam (purlin) in ridge direction (→ **y**) is required.

This kind of roof also appears in modified form, e.g. with brick arches instead of wooden trusses (■ Figs. 4.87 and 4.88). The horizontal thrusts of the arches, in the case of the shown historical examples, are alternatively accommodated with internal or external masonry piers or abutments.

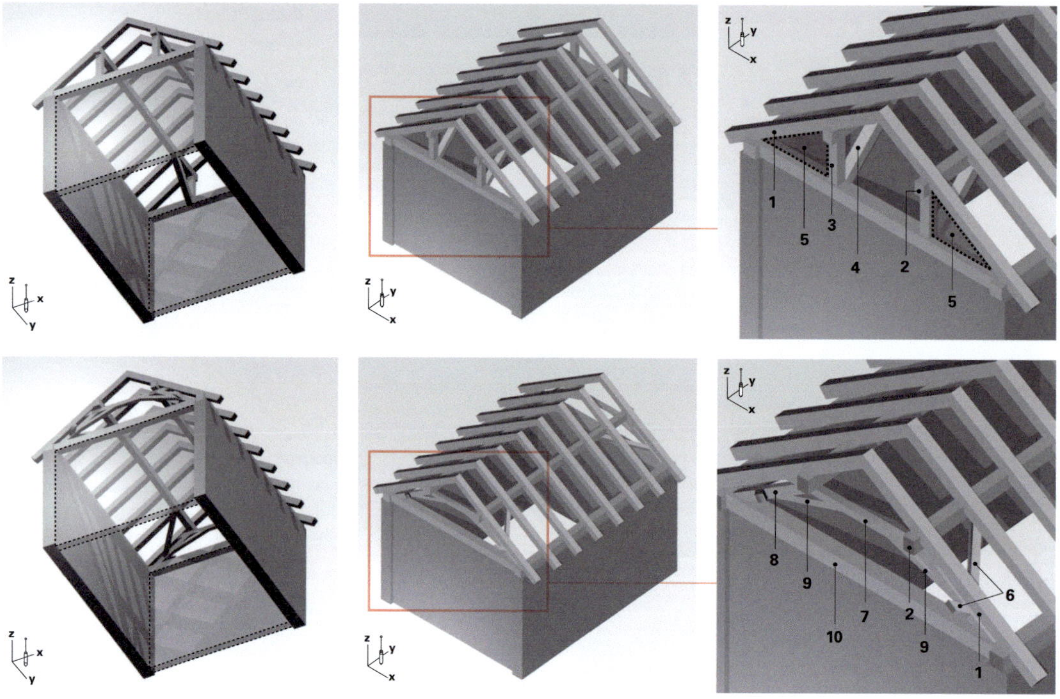

Fig. 4.83 Schematic diagram of a double-stringed purlin roof with rafters cantilevering on the ridge side. Both of the purlins are mounted on posts in the upper example; in the lower, on a frame-like truss consisting of a strut, a tie-bar, and an angle brace, so that a column-free roof space is created. The support thrust generated by this frame must be short-circuited by means of a tie-beam, analogously to a rafter roof. **(1)** Rafter; **(2)** purlin; **(3)** post; **(4)** angle brace **(yz)**; **(5)** rigid triangular truss; **(6)** angle brace (roof plane); **(7)**tie-bar (→ **x**); **(8)** strut **(xz)**; **(9)** angle brace **(xz)**; **(10)** tie-beam

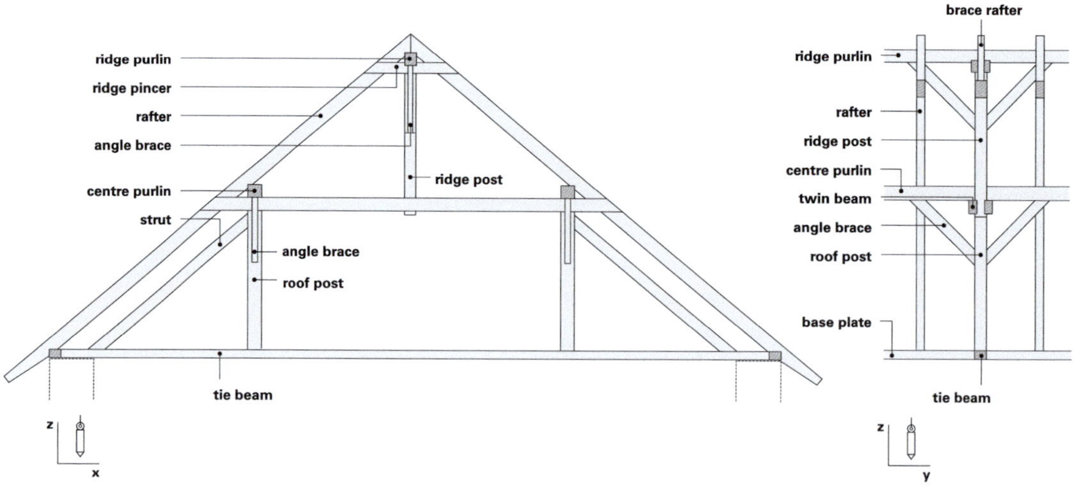

Fig. 4.84 Example for the execution of a purlin roof with wind truss

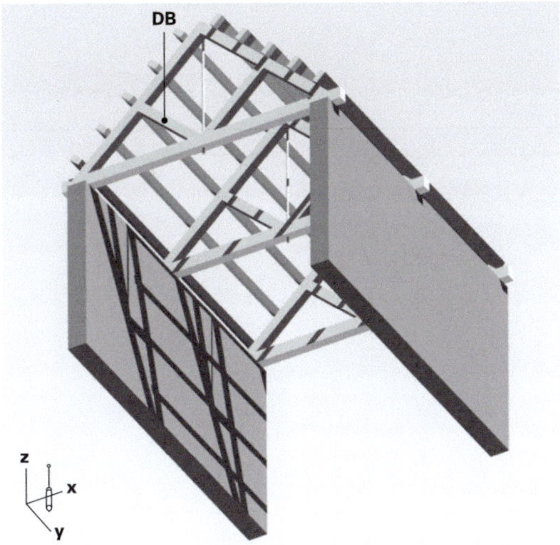

Fig. 4.85 Design of a truss-like **roof girder** generated by introducing diagonal braces (DB). The latter prevent the slender rafters from buckling under compression and at the same time reduce their deflection

Fig. 4.86 (Left) Example of a roof girder in timber construction. (Right) Example of a girder roof in steel construction with cables acting as tensile chords (Polonceau truss)

4.2.2 Systems Under Compression—Pitched Roofs and Vaults

4.2.2.1 Pitched Roof from Bar Skeins

Rafter Roof

Rafter or couple roofs operate without purlin strands, but are still part of the category of **directional bar structures** (Figs. 4.89 and 4.90, left). Their constructional principle can be derived in two ways:

- The beams of a simple basic structural cell with a flat roof are replaced by **triangular trusses** composed of three bars: two inclined rafters and a horizontal beam tying their bases

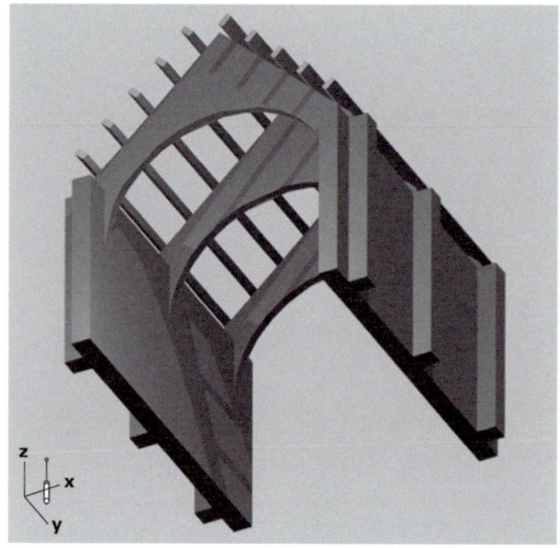

Fig. 4.87 Masonry arches acting as a replacement for the main girders as on Fig. 4.85. The longitudinal purlins are placed on these

Fig. 4.88 Construction as on Fig. 4.87 (Abbey Church San Miguel de Cuxá, Spain)

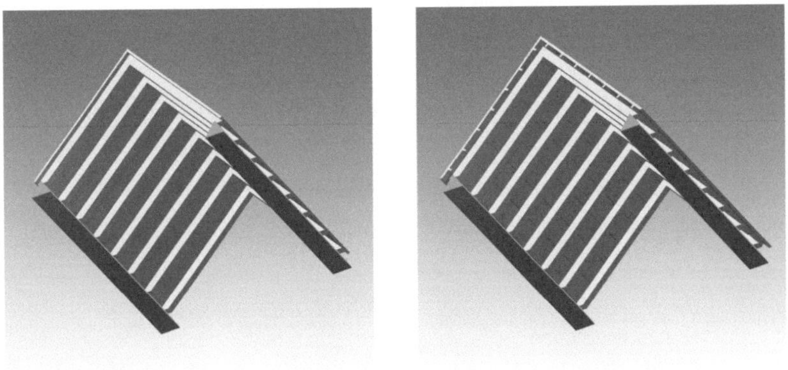

Fig. 4.89 Rafter roof. Left: without lathing; right: with lathing

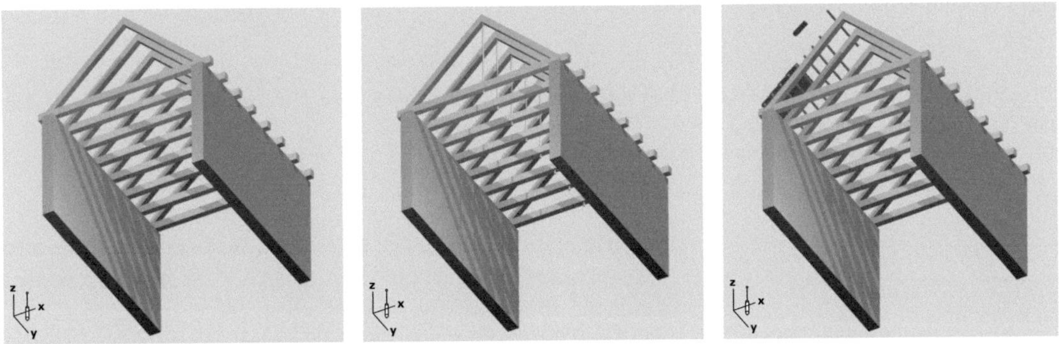

Fig. 4.90 (Left) Schematic diagram of a **rafter roof**. (Centre) Suspension of the **tie-beam** from both rafters to prevent sagging. (Right) Rafter roof with lathing and roofing

together. External loads are no longer transmitted by pure bending, as is the case with a horizontally spanning beam (e.g. of a flat floor), but, due to the inclination and type of support of the rafters, they are converted (at least partially) into longitudinal **normal force** within them, a stress which is more materially economical than bending. The steeper the rafters are inclined, the smaller are the longitudinal forces acting within them. The triangle thus essentially has the load-bearing behaviour of a **triangulated truss**, in which normal forces are predominant over bending stress. Thus, the inherently large structural depth of the roof is advantageously activated for load-bearing purposes.

__Vol. 1__, Sect. 26.3 Comparison Between Bending Moments/ Shear Forces and Axial Stresses, or Accordingly Membrane Stresses

- In order to partially convert the strong **bending** of the beams in horizontal position into **normal force** and to make them slenderer, the beams are arranged inclined so that they mutually support each other at the ridge. The normal force thereby produced within the rafter, which is significantly smaller in the purlin roof, tends to force the wall crest, which is extremely sensitive to horizontal thrust, outward. To prevent this, a further beam is arranged which connects the two rafter bases horizontally to each other and prevents them from shifting outwards. Understandably, this beam,

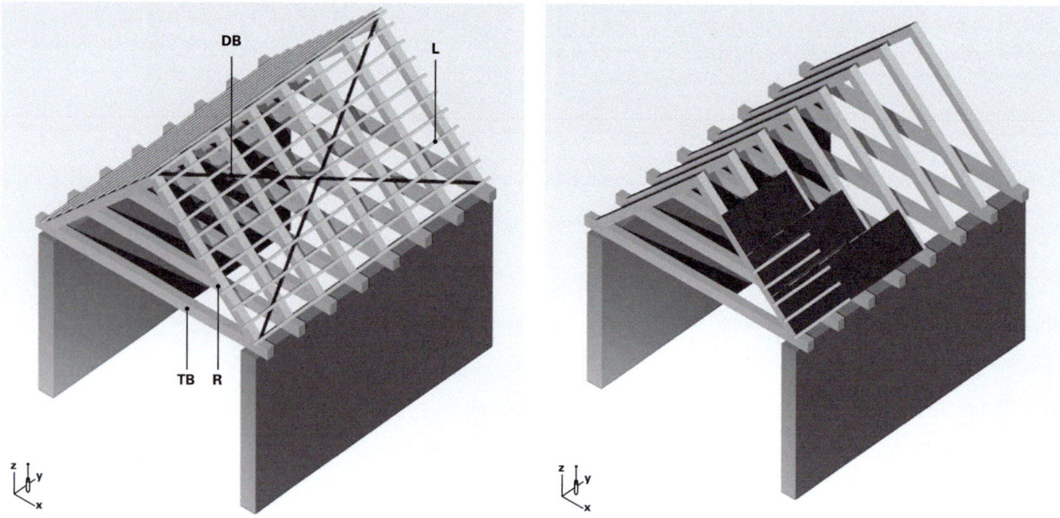

Fig. 4.91 (Left) **Bracing** of a rafter roof in ridge direction (→ **y**) by creating a shear-resistant truss made of diagonal bracings (**DB**), rafters (**R**) and laths (**L**). The triangular truss of two rafters (**R**) and tie-beam (**TB**) is inherently stiff in planes parallel to the gable (**xz**). (Right) Bracing of a rafter roof in ridge direction (→ **y**) by forming a plate by means of a sheathing

in carpentry construction, is called a **tie-beam**: its main task is to absorb tensile forces acting as a tie-bar. Only vertical loads are then transferred to the wall crest.

An inclination is also necessary for evacuating precipitation water from the roof surface as quickly as possible. From the point of view of evolutionary history, pitched roof structures can also be regarded as constructional responses to the limited sealing properties of scaled roof coverings.

***Vol. 3**, Sect. 14.2.2.5 Roofing*

In order to prevent an unsightly sag of the wide-spanning tie-beam, the latter is sometimes suspended from the ridge node by means of a **tie-rod** (Fig. 4.90, centre).

In a further step, the rafters are supported in their midst by oblique struts (similar to Fig. 4.85) fanning from the centre-point of the tie-beam. In this way, they act as double-span beams and, moreover, their buckling length under longitudinal loading is reduced. Also, in the case of conventional **collar-beam roofs**, an intermediate strut, the **collar beam**, is inserted in a similar manner in order to couple the rafters together and to introduce a buckling safeguard against the normal force in the rafter. At the same time, the collar-beam skein allows an intermediate floor plane to be created in the usually comparatively high-clearance space delimited by steeply inclined rafters.

Bracing

Rafter roofs are usually braced in ridge direction (→ **y**) by generating a shear-resistant truss within the roof plane. This is usually done by means of diagonal belts called **cocking pieces**. These are tensile belts which are nailed cross-like diagonally onto the rafters and are connected in particular with the gable trusses (Fig. 4.91, left). Together with the rafters, acting here as posts, and the laths, acting here as cross-bars, or a ridge beam or base plate, they form a **shear-resistant truss** with great racking stiffness which braces the rafter roof in the ridge direction and thus allows for bearing wind loads impinging on the gable side. The same effect is achieved by a flat panel sheathing covering the roof (Fig. 4.91, right). In the plane of the triangular rafter trusses (**xz**), the construction is inherently stiff.

► *Sect. 3.3.5.2 Diagonal Bracings*

► *Sect. 3.3.5.3 Shear-Resistant Sheathings*

4.2.2.2 Full-Leaf Vault

In analogy to solutions with flat floors or roofs, directional systems can also be designed as **vaults** (Figs. 4.92 and 4.93). Here, too—with linear supports at the imposts—there

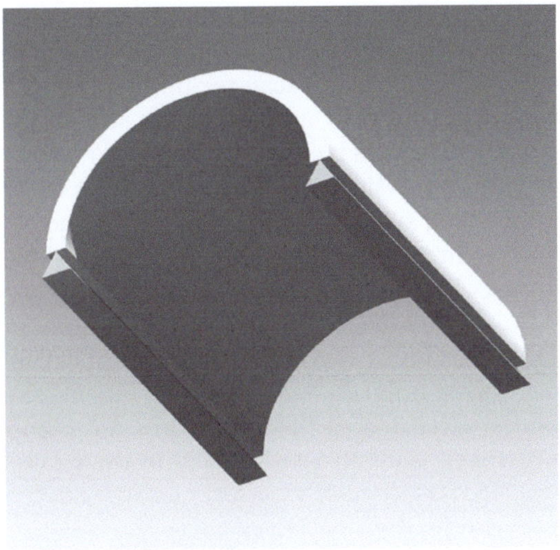

Fig. 4.92 Full-leaf vault

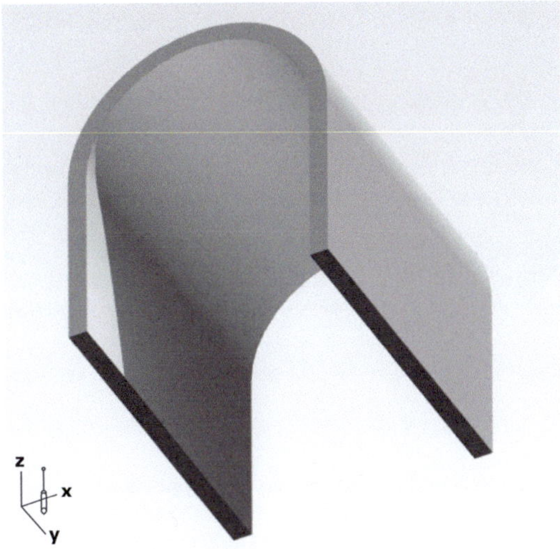

Fig. 4.93 Covering from a **vault** which spans between two walls

is a clear spanning direction. As with beam floors, vault span widths are limited by the possibilities of the material and the particular type of construction. However, the conditions of load transfer are not the same as those of flexed flat coverings as described in ▶ *Sect. 4.2.1*. Vaults belong to the category of predominantly axially stressed structures, in which the **form** plays a decisive role—analogous to the structural depth in the case of flexed flat coverings. Shape-related issues of axially stressed load-bearing systems are discussed elsewhere.

▶ *Sect. 3.4 Form-Related Issues of Axially Stressed Structures*

Load-Bearing Behaviour

In this context, the term *full-leaf vault* or *solid-walled vault* should be understood to mean those which consist of a continuous curved surface component or one profiled, typically at its intrados, by means of protruding vaulting ribs. Depending on the particular design, these structures act, in terms of load transfer, either as a sequence of individual arches, or as a continuous **shell** in a structural sense. Crucial for this is both the ability to distribute loads transversely and the tensile strength of the surface component. This will be discussed later in ▶ *Sect. 4.2.2.3*.

▶ *Sect. 3.4.5.1 Shells*

Vaults without a shell action unfold their characteristic load-bearing effect, such as single arches, by virtue of a **non-sliding support** or firm **skewback**, that is to say, by the blocking of **vault thrusts**, i.e., of the horizontal component of the normal force flowing within the component and impinging at the impost. Only an almost deformation-free support in horizontal direction at the impost (→ **x**) allows an efficient arch action with primarily axial forces flowing within the component. A yielding skewback leads to bending within the vault, which, at

▶ *Sect. 3.4.1 Interrelations Between Support, Form, and Stress*, in particular ▶ Fig. 3.101

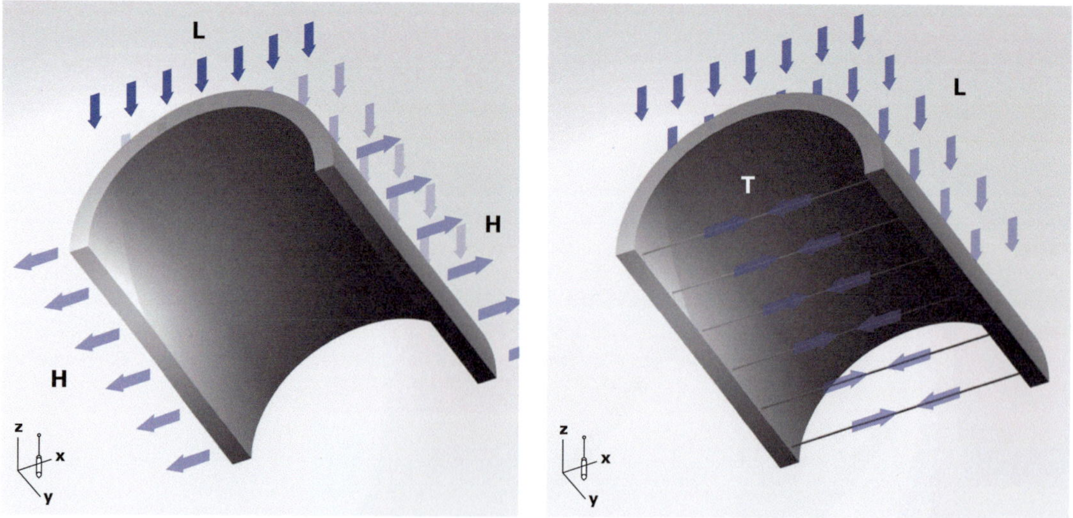

Fig. 4.94 (Left) A vault shell under load **(L)** produces a **vault thrust (H)** on the supports which must be accommodated at the support in addition to the vertical load component (not shown here). (Right) **Tie members (T)** in the plane of the bearings can absorb the vault thrust. When there is no vertical enclosure, they can be integrated into the floor and are therefore invisible

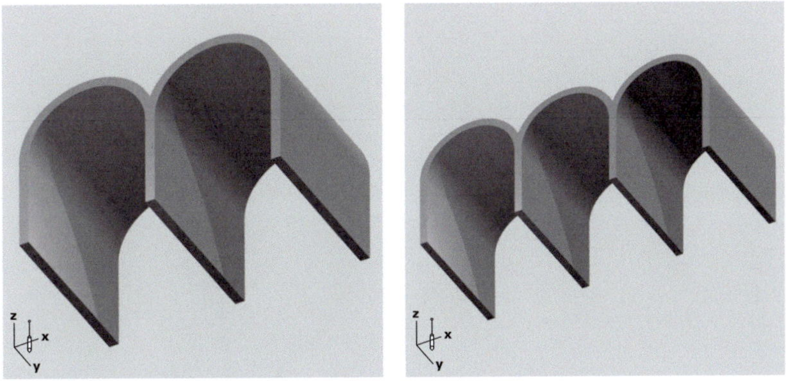

Fig. 4.95 (Left) Similarly to the directional system of a flat floor composed of bars, the vault can only be extended in the spanning direction by lateral addition of bays. (Right) Characteristic segmentation of the spaces when expanding a directional vault system in the spanning direction

least in the case of masonry arches, is undesirable and can even be critical for stability. The bending within a vault can be reduced to a minimum by adequate shaping in the best possible adaptation to the thrust line of the acting loads.

In the case of imposts situated at ground level, absorption of vault thrusts can be effected alternatively by direct **introduction** into the **ground** or by reconnection by means of a **tie-bar** (Fig. 4.94). In the case of elevated imposts, for example when vaults are placed on vertical enclosures, the bearing of thrusts is usually a question which is delicate both in terms of conceptual and constructional design, something which did cause many headaches, especially to historic master builders. The simple reconnection of the thrusts by means of tie-members visible in the interior (Figs. 4.105, right and 4.121) is indeed feasible in principle, but was never seriously considered as a standard feature, for example, for historic church naves—presumably for spatial and aesthetic reasons. An *external* removal of vault thrusts can be carried out either by **adjacent vaults**, by mutually cancelling the thrusts of adjacent barrel vaults so that only a **vertical load component** remains, or by other additional measures. This is discussed below.

▶ *Sect. 3.2.2.2.4 Bracing* and ▶ *Sect. 4.2.2.2.5 Bearing Vault Thrusts*

Vaults without a genuine shell effect are, in particular, **brick vaults** as occur in historical buildings. They act as a sequence of single arches (Fig. 4.95, left), particularly whenever they have been bricked in this way, i.e., in individual segments using several times slim centrings, a very economical construction method for barrel vaults. As a rule, arches and vaults show that they are more strongly influenced by the **building process** than other types of load-bearing structures, since a large amount of preparatory work such as centring is usually required until the construction achieves its final load-bearing capacity. With regard to the shaping and load-bearing behaviour of this type of vault made of masonry, what has been already said about bricked arches applies in essence.

▶ *Sect. 1.3.1.4 Block-Shaped Starting Elements*, ▶ Figs. 1.202 and 1.203, left

▶ *Sect. 3.4.3 Deviations from The Thrust Line*, in particular ▶ Figs. 3.113, 3.114, 3.115, and 3.116

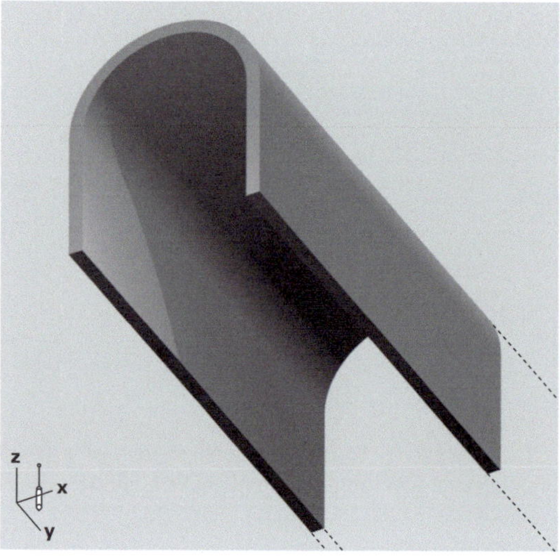

Fig. 4.96 Discretionary extension of a vault in longitudinal direction, analogous to the flat bar system

Extendibility

These directional load-bearing systems made from vaults can be extended analogously to those with flat covers, namely:

- laterally (in → **x**) by **adding** further cells or bays (Fig. 4.95); in churches, these are called *naves*;
- or transversely to the spanning direction (i.e. in → **y**) by simply **continuing** the construction (Fig. 4.96).

In contrast to flat coverings such as slabs or bar skeins, in which the magnitude of the span has only an effect on the **structural depth** of a plate or a beam, in the case of vaults, there is a close interplay for purely geometric reasons between the span and the structural form, and thus also the **vault rise** or **vault camber**. In the case of a circular system line, particularly close interrelations apply: In semi-circular barrels, the vault rise is always equal to half the span. Variations of this rule are only possible reducing the cut-off angle of the vault shell, such as happens with a segmental arch. More freedom of shaping is available with non-semi-circular system lines, like for example underlie the Gothic pointed arch or ogive, an important invention of arch construction which has several advantages: erection in two circular arch segments made of always the same voussoir; extensive independence from the span when choosing the arch rise (Fig. 4.97); and better adaptation to the thrust line of the acting loads than with semi-circular arches.

Spatial Merging of Adjacent Cells

▶ *Sect. 4.2.1.1 Slab, Spanning One-Way* and ▶ *Sect. 4.2.1.2 Flat Covering from Bar Skeins*

In analogy to plate or beam-covered basic structural cells, as described above, adjacent vaulted bays can also be spatially interconnected. The constructional solution suitable for stone material for supporting the vault's spring or impost is, in

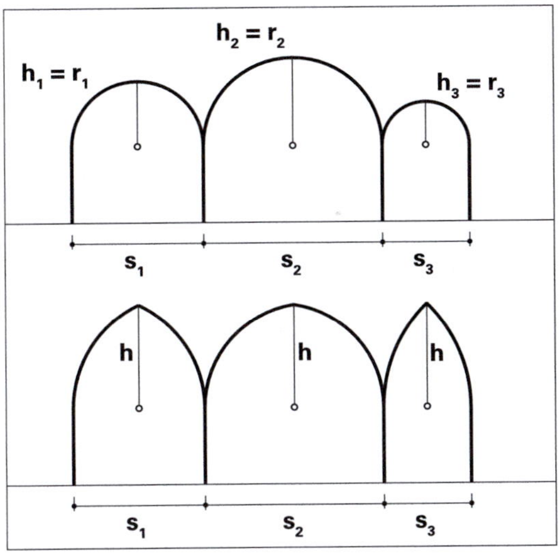

Fig. 4.97 Arch height and radius are necessarily the same in semi-circular arches. Pointed arches provide greater design freedom: In the illustrated example, the same height **h** can be maintained despite different spans $\mathbf{s_1}$, $\mathbf{s_2}$, and $\mathbf{s_3}$

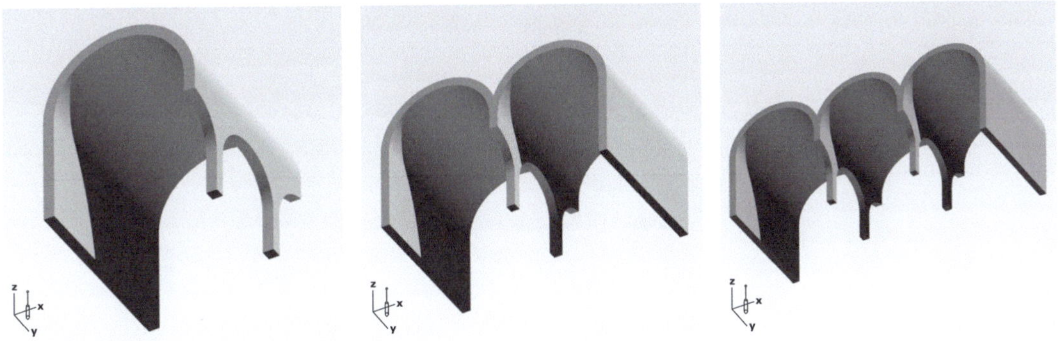

Fig. 4.98 (Left) Similar principle as on Fig. 4.38, applied to a vault. A wall panel is opened up here in form of an arch. (Centre) Spatial interconnection of two bays through arch openings. (Right) Discretionary continuation of the principle in the centre

turn, the **arch**, which can be placed on both pillars or columns (Fig. 4.98). It must be taken into account that this arch, like the vault resting on it, also produces horizontal thrusts at the (own) imposts, but in this case parallel to the vault axis (→ **y**). These neutralise themselves mutually by the addition of arches, but must be counteracted at the ends of the arch sequence by means of suitable abutments (Fig. 4.99).

Due to the particular geometry of the vault, the **spatial differentiation** of the individual bays is stronger than with flat ceilings. Otherwise, the same principles apply as for added systems with beam ceilings. Vaulted spaces can also be executed in the form of a **basilican cross-section** (Fig. 4.100). As with flat ceilings, the central nave is designed so high that window openings for **daylighting** and **ventilation** of the central area can be accom-

Fig. 4.99 Bracing the walls against vault and arch thrusts by means of **buttresses**: both on the side walls—support of the barrel vaults of the lateral naves, effect in axial direction → **x**, (see Fig. 4.99, right)—as well as on the front walls—support of the arched arcades between the central and lateral naves, effect in axial direction → **y**) (Basilica of San Salvador, Northern Spain)

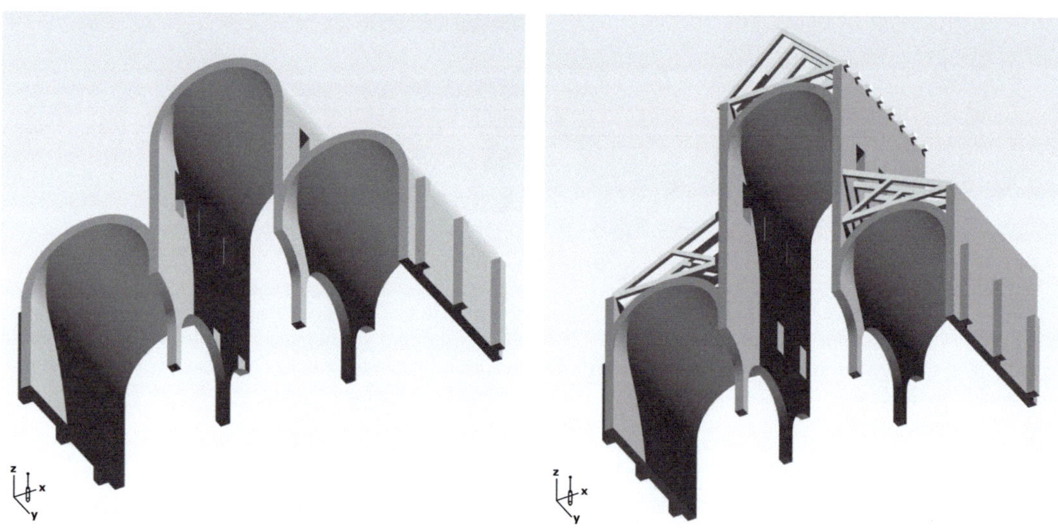

Fig. 4.100 (Left) Three-aisled system with lateral discharge of the vault thrusts by means of buttresses. The vault thrusts between central and lateral naves do not neutralise each other here because of the difference in height between them. Appropriate structural measures are necessary for this purpose, such as, for example, minimising the thrusts by means of steep vaults or short-circuiting them by tensile members. (Right) Three-aisled system with over-vaulting and additional roofing by means of a roof truss. Frequent solution in historical church construction

modated in the side walls. This forces the impost of the central vault to be arranged above the upper edge of the window openings, which results in comparatively large building heights.
Furthermore, this solution results in the fact that the thrusts of the central vault do not compensate themselves anymore with those of the laterally adjoining ones (Fig. 4.100, left), since both act at different heights. The result is considerable

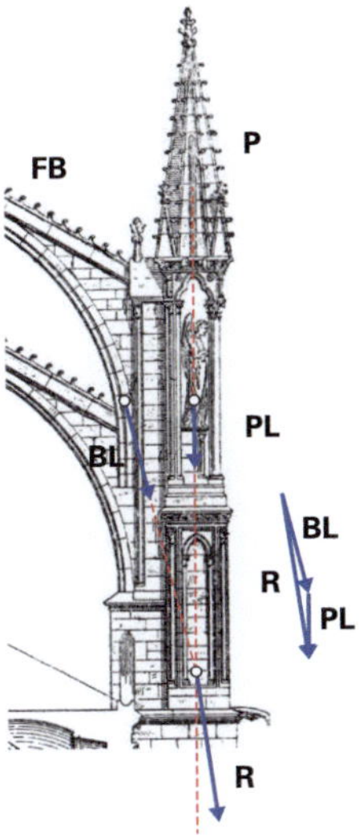

■ **Fig. 4.101** (Left) **Over-compressing** of the horizontal thrusts in the case of church vaults by additional loads superimposed upon flying buttresses in Gothic church architecture, here by added pinnacle **P**. The resultant **R** of the buttress-arch load **BL** and the perpendicular pinnacle load (**PL**) is deflected more strongly into the vertical and thus more strongly into the abutment cross-section at the bottom. (Right) Load-generating pier toppings of St Bénigne Cathedral in Dijon. The completely unornamented design of the topping clearly shows its intended function as pure ballast

horizontal thrusts both outwards at the height of the central vault's impost and inwards at the height of those of the lateral vaults. The latter tend to push the inner bearing walls inwards, into the space of the central nave. These horizontal forces are counteracted by the increased load of the larger central nave, which, as it were, deflects the resulting force further downwards, thus diminishing the lateral thrusts. In some cases, the dead weight of the walls of the central nave was deliberately increased for this purpose (in analogy to ■ Fig. 4.101). Transverse stiffening compression bars across the central nave, which would have short-circuited these inwards oriented forces, would not be desirable under the given spatial conditions, although this solution is indeed sometimes to be found.[1] Instead, as will be addressed below, external shoring would later be introduced (■ Fig. 4.102).

Bracing

The vault with regularly arranged diaphragms (■ Fig. 4.103) is rigid and requires no additional bracing transversely to its

1 For example, in the Cathedral of San Sabino in Bari (Italy), where transversal arches are placed in the middle of the nave. For a detailed and very vivid description of the essential structural questions pertaining the development of medieval church architecture, see Choisy, Auguste (1899) *Histoire de l'Architecture*, reprinted 1987 by Slatkine Reprints, Geneva, Paris. Cf. also ***Vol. 4****, Sect. 5.7.1.4 The Mediaeval Basilica.*

Fig. 4.102 Cathedral of Amiens. Discharge of the vault thrusts of the naves by a system of **flying buttresses** as on Fig. 4.106, right

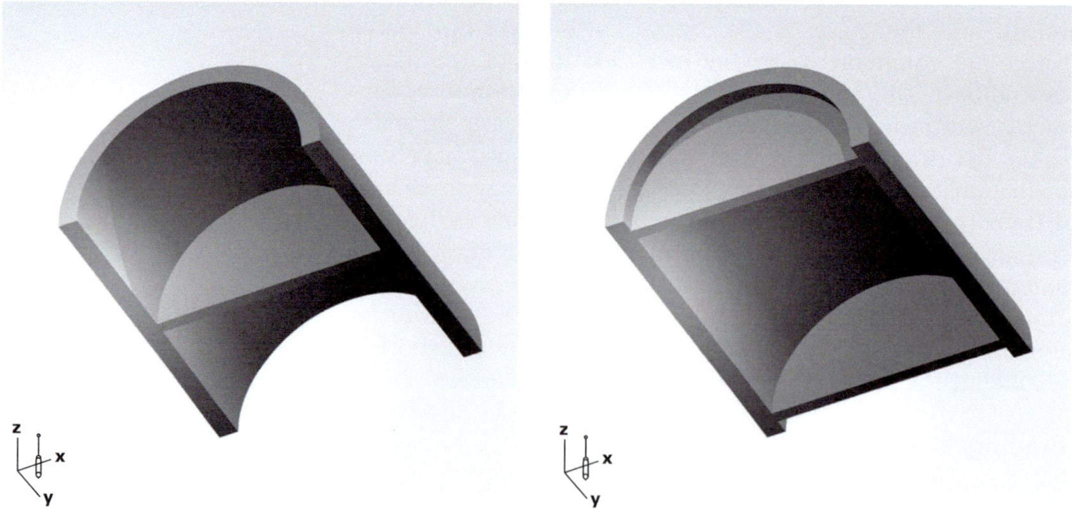

Fig. 4.103 (Left) Stiffening of the vault shell by a **transversal diaphragm**. (Right) Stiffening cross-diaphragms can be used functionally for generating spaces

axis (in → **x**). In the longitudinal direction (→ **y**), the same applies if sufficient shear indentation exists between individual vault strips or segments. This applies in the case of masonry vaults, especially in the case of **staggered bonds**. Shell vaults created thereby are stiff in the direction of the main axis.

Staggered bond: ▶ *Sect. 1.3.1.4 Block-Shaped Starting Elements*, in particular ▶ ◘ Figs. 1.203, right and and 1.208, as well as further below, ◘ Fig. 4.110, right

Conditions change, however, when the vault rests on vertical enclosing walls. In the direction of the main axis (→ **y**), the stability of the construction is usually ensured by the racking stiffness of the longitudinal walls. In this case, the total construction must be braced transversely to the vault axis (→ **x**) by means of supplementary measures. Since the problem of the transverse bracing of the basic structural cell cumulates with that of diverting vault thrusts, both questions will be discussed conjointly in the next section.

Bearing Vault Thrusts

In addition to the external horizontal forces against which the structural cell has to be braced, systemic **horizontal thrusts** of the vault must be expected. In the case of vaults with imposts at ground level, they are introduced directly into the foundations or are recoupled by means of tie-members below ground level. In the case of raised vaults on vertical enclosing walls, they must in principle be diverted by added structural elements (◘ Fig. 4.104, left). In classical vault construction, both force components are diverted *conjointly* over the same bracing elements. Thrusts occur at the vault base or impost area, i.e. at right angles to the wall plane at the level of the wall crest (→ **x**), where the enclosure is particularly sensitive to horizontal loads.

When individual bays of the same size are added, the oppositely directed thrusts of adjacent arches **neutralise** each other, so that in this case only **vertical loads** occur at the bearings (◘ Fig. 4.104, right). The thrusts of the end vault must, however, be absorbed by the wall element or an additional construction.

We have seen that the structural cell can be regarded as braced in vault axis (→ **y**) due to the racking stiffness of the walls. If the wall panel is transformed into a sequence of arched openings, the arch thrusts at the front ends must be removed by means of struts or similar elements.

Transversely directed horizontal forces (in → **x**) and vault thrusts can be taken for individual vaults or flanks of vault sequences by the following measures:

- **Wall piers**, in vault construction so-called **abutments** or **buttresses**: mostly external, since they then essentially absorb compressive forces (◘ Fig. 4.105, left). This coincides with the load-bearing behaviour of the masonry constructions customary in vaults, in which tensile stresses should be avoided as far as possible. Moreover, interior piers would get into a geometric conflict with the interior vault surface at the impost area. For these reasons, they are rarely to be found under these conditions. The abutments' depth can be

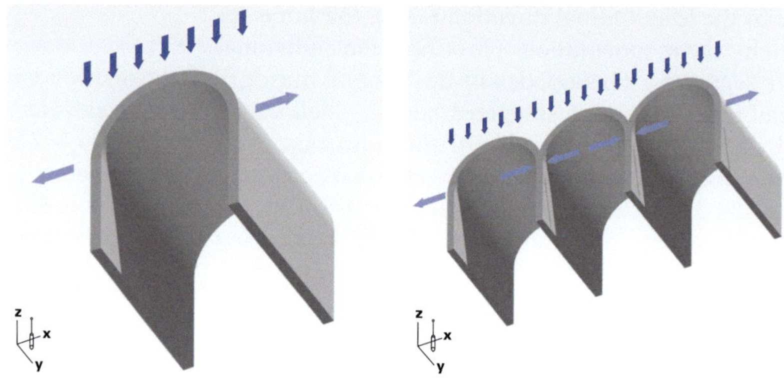

Fig. 4.104 (Left) In contrast to planar coverings composed of bars, the vault generates a **horizontal thrust** at the supports or imposts (only forces in the foremost plane are shown here). (Right) The arch thrusts neutralise each other in the lateral addition of aisles at the contact points of the vaults. Only the thrusts at the two ends of the system must be discharged by other means

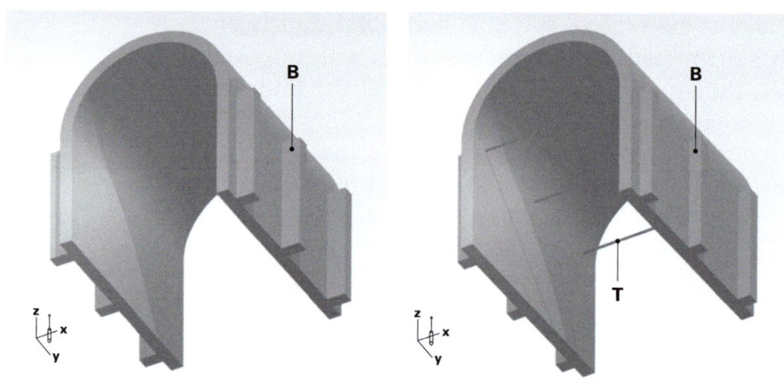

Fig. 4.105 (Left) Horizontal bracing of the vaulted basic cell by means of **wall piers** or **buttresses (B)** siding the wall panels. In contrast to flat coverage systems, these have to accommodate both the systemic horizontal vault thrusts as well as the external horizontal loads. They are therefore to be made deeper in the vault's spanning direction than in that case. (Right) Neutralisation of the vault thrusts by means of **tie-bars (T)** and horizontal bracing by **wall buttresses (B)**, which are relieved from load as compared to the vault on the left and can thus be dimensioned smaller. From a developmental point of view, this variant is a rare exception because of the undesired spatial effect of the visible tension members

graded according to the tipping moments. They can also be aided by internal tie-bars, which significantly reduce the thrusts at the imposts, leaving mostly horizontal forces from wind to be borne by the piers (Fig. 4.105, right).

- Supporting **flanking vault shells**, mostly in the form of quadrant-shaped half-shells (Fig. 4.106, left), deflect the thrusts diagonally in the direction of the ground and can then, in turn, be braced by piers or be supported by further semi-barrels. *Cascading* vault sequences are then formed, each lateral vault spanning a separate bay or nave. This solution is to be found in five- or multi-nave mediaeval church construction.
- A **sloping strutting system** of arches (Fig. 4.106, right), so-called **flying buttresses**, shows an effect similar to vault

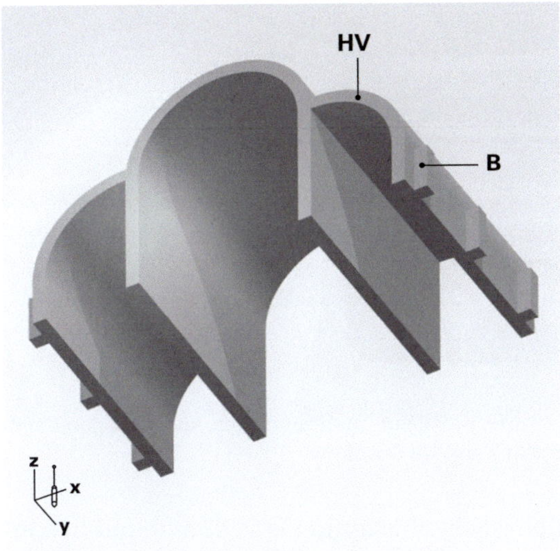

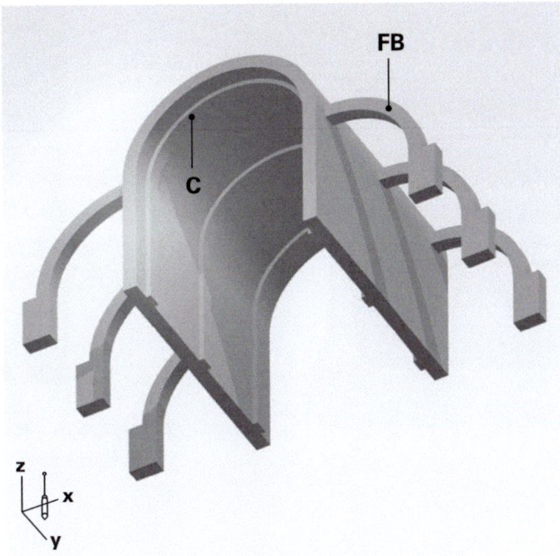

Fig. 4.106 (Left) The thrust of the central vault is discharged by lateral **half-vaults (HV)** and **buttresses (B)**. Formation of three parallel naves. (Right) The thrust of the central vault is absorbed by lateral **flying buttresses (FB)**. Stiffening of the vault shell by **chords (C)** for load concentration on the flying buttresses. A lateral nave can be accommodated under the arches

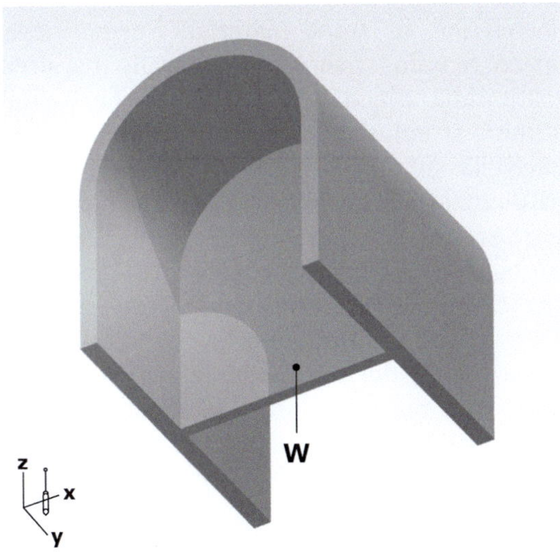

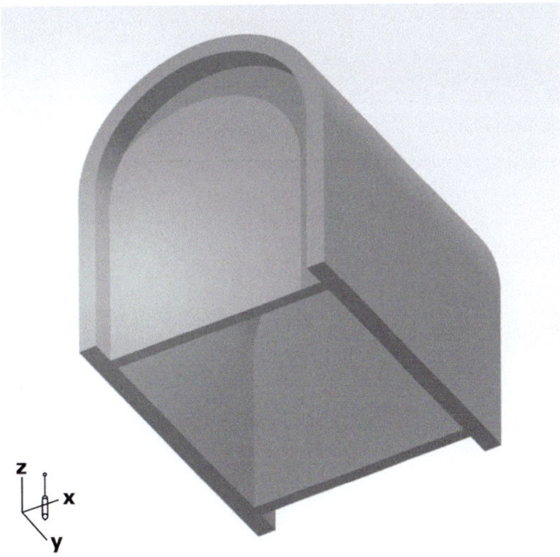

Fig. 4.107 (Left) Horizontal Bracing of the vaulted basic cell by a **transverse wall (W)** which at the same time reinforces the vault shell against buckling (see also Fig. 4.103, left). (Right) Bracing of the basic cell shown on Fig. 4.93 by several space-forming transverse walls (see also Fig. 4.47, right). In particular, the front walls were traditionally used for this purpose, since the vaulted rooms were often designed continuous as in the case of church naves, and no separating walls were desired

shells, but the forces are bundled in each main axis of a strut. This solution is therefore particularly suitable for ribbed vaults or for the groined vaults of Gothic church architecture.

- **Internal transverse walls** are also conceivable as bracing and thrust-absorbing elements (Fig. 4.107). However, thrusts again create **horizontal** tensile stresses in this trans-

Fig. 4.108 Room covering made of vaulted concrete shells on point supports (weekend house near Paris, Le Corbusier)

verse wall, which is why this option is seldom found in conventional masonry construction without other elements for transverse bracing—unlike flat floor systems; also because the transverse walls segment the space, which is often undesirable, as was the case in mediaeval churches.

- **External tranverse walls**. This position results in **compressive stresses** on the bracing wall panels, which favours the load-bearing behaviour of stone material. Furthermore, the vaulted space remains free of structurally required transverse partitioning, which would hamper the spatial design. The support structure made of wall panels can be used to create cell-like side rooms. As with all bracing panels made of brittle material, **superimposed load** is beneficial to fulfilling their task.

In summary, it can be said that vault systems are of great importance for the developmental history of numerous historic designs. Over a long period of time, they were the only way of bridging large spans with durable material, that is, at that time, stone masonry. In a biaxial version, so to say, they alternatively appeared in the related design of the dome, as discussed further below. Their structural function as compressed constructions matches the properties of mineral materials. However, with the advent of modern materials such as steel or reinforced concrete, these designs quickly lost their importance. Today they play a subordinate role in construction practice (see some examples on Figs. 4.108 and 4.109).

Ribbed Vault

► *Sect. 1.3.1.4 Block-Shaped Components*, ► Fig. 1.202, right

Already the interlocking of the masonry bond in the direction of the vault axis (→ **y**), a **staggered bond** (Fig. 4.110, right), represents a first measure for the activation of an—at least partial—shell-like load-bearing effect. Thereby, an improved shear resistance within the shell surface is achieved. An enhanced shell load-bearing effect can be attained by reinforcing the surface component by means of **vaulting ribbing**, i.e. localised strip-like thickening of the shell (Fig. 4.111).

■ **Fig. 4.109** (Left) Multi-storey building with vaulted floors from brick shells (see right). The stiffening wall piers are recognisable at the end wall (apartment-house in Madrid, arch.: Luis Moya). (Right) The vaulted brick floor of the building on the left

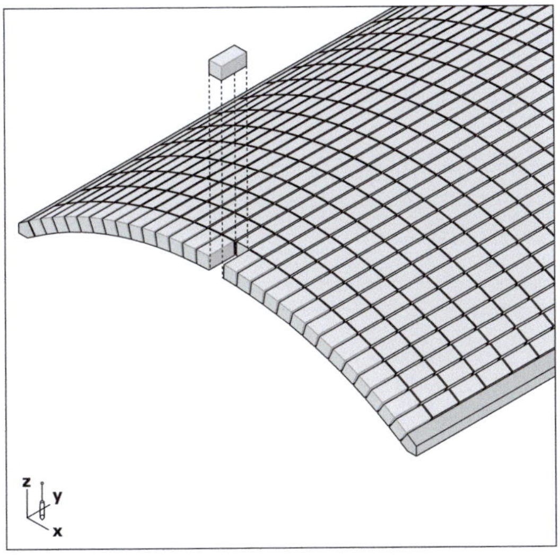

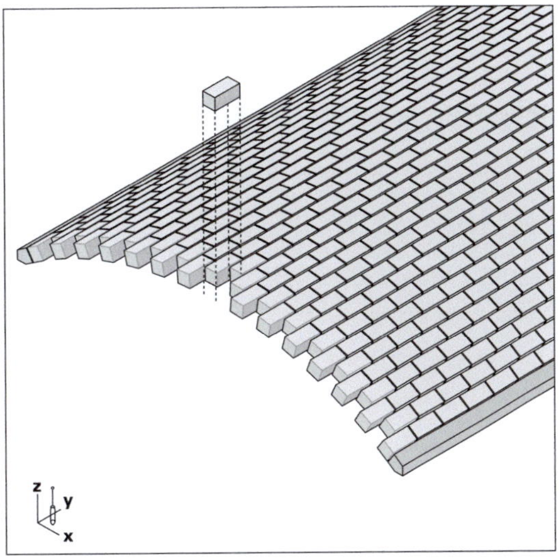

■ **Fig. 4.110** (Left) Masonry vaults from separate arched strips. (Right) Masonry vault with **interlocking staggered bond**

Closed **transverse diaphragm walls** (■ Fig. 4.103, left) produce the same effect, actually in even greater measure due to their very large racking stiffness. They are, of course, space-forming and are particularly suitable as end-faces (■ Fig. 4.112, left).

4.2.2.3 Full-Leaf Vaulted Shell

If the barrel vault is constructed in reinforced concrete, the structure is not only an isotropic, continuous surface component with a good ability to distribute loads transversely, but is also capable of absorbing tensile forces. It therefore acts as a genuine **cylindrical shell**, i.e. as a uniaxially curved shell surface (■ Fig. 4.112). Its main spanning direction is not necessarily the arched system line (→ **x**), as in the case of the conventional vault, but two main directions emerge within the shell, which

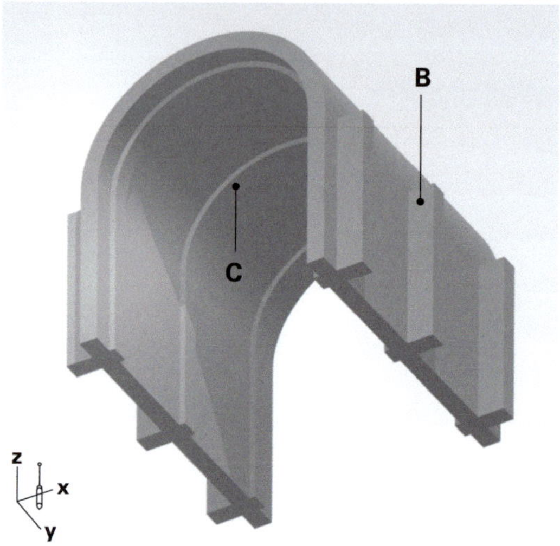

■ **Fig. 4.111** Stiffening of the vault shell by **chords** (**C**) which concentrate loads on the stiffening buttresses (**B**)

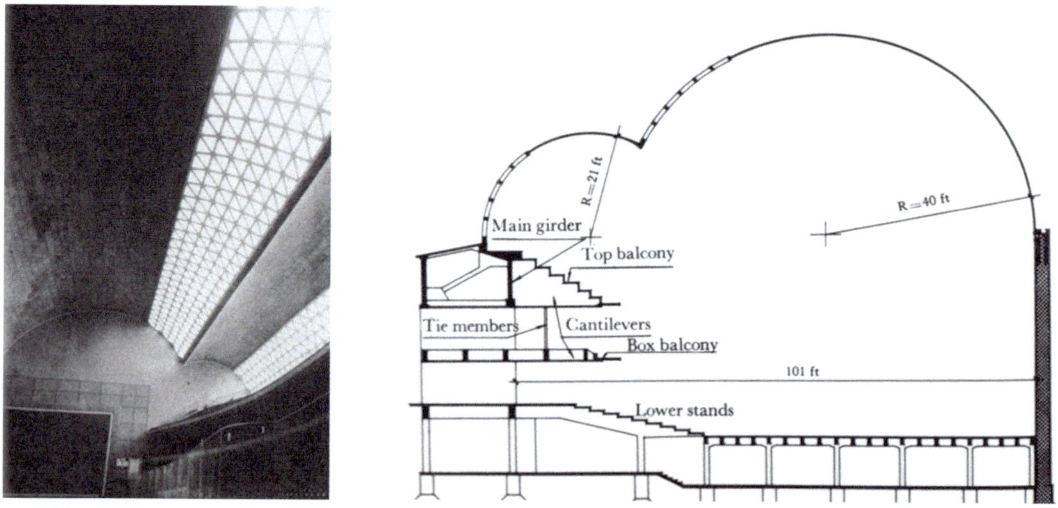

■ **Fig. 4.112** Cylindrical concrete shell of the Frontón Recoletos in Madrid; interior (left) and cross-section (right) (eng.: E. Torroja)

change their orientation over the shell surface as a function of the loading and the bracing diaphragms (■ Fig. 4.113).

The two-way load-bearing capacity of genuine vault shells—in contrast to non-shell-like vaults—allows for a **point support** of the load-bearing structure. The load transfer takes place similar to synclastically curved, point-supported shells as discussed below. In order to activate the shell effect of the merely *uniaxially* curved surface structure, the vault has to be stiffened at regular intervals by means of transverse diaphragms or similar measures, such as radial cable bracings (■ Fig. 4.114), since it lacks **curvature**, and hence stiffness, in direction of the rectilinear generatrix (→ **y**).

► *Sect. 4.3.2.6 Full-Leaf Shell, Synclastically Curved, Punctually Supported*

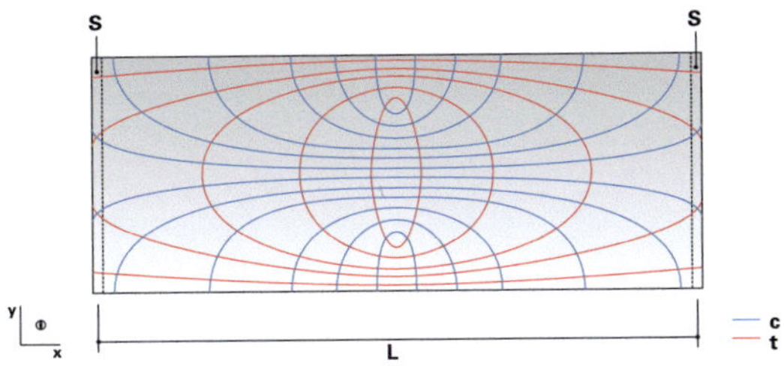

Fig. 4.113 Flow of forces in a vault shell, spanning freely over a distance **L** (main stress trajectories), top view. **S**: support; **c** compression; **t** tension

Fig. 4.114 Cross-diaphragm stiffening of a vault shell: here in form of a radial cable bundle under prestress (Bosch-Areal, Stuttgart)

4.2.2.4 Vault from Arch Skeins

While solid-walled vaults, as discussed above, are characteristic of surface-forming constructions such as masonry or concrete, vaults made of parallel arrays of individual arches (Figs. 4.115, 4.116, 4.117, 4.118, and 4.119) are typical of construction methods with compressive- and tensile-resistant materials such as steel or wood, sometimes also concrete, which are based on the use of mainly linear, rod-shaped components. It is obvious that the mode of operation of this kind of structure can be traced back to individual arches, which, in their idealised sequence forming a continuous barrel vault, initially work as a one-way spanning system without cross-distribution of loads. As in the case of flat covers made of bar skeins, the latter must be provided by suitable additional elements which span crosswise to the arches. These can be top-mounted straight rods (purlins) running in the direction of the vault axis (→ **y**) (variant **4** as discussed elsewhere), or also cross-bars (variant **3** in that very place) connected to the arches in a shear and bending-stiff way. The spans, which are

▶ *Sect. 3.3 The Constructional Composition of the Space-Enclosing Surface Element*

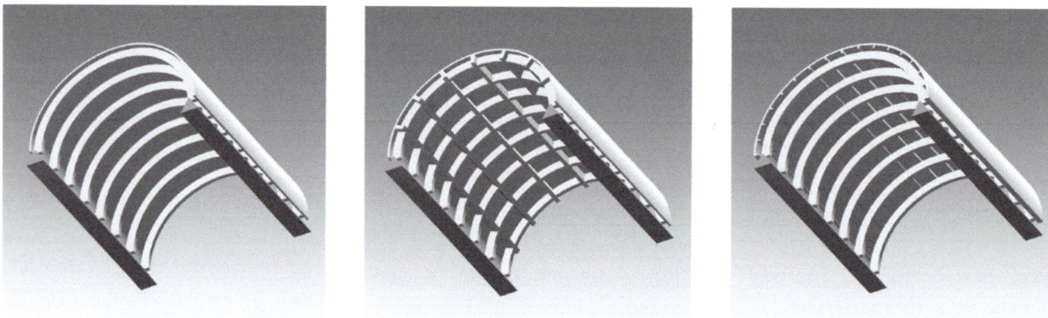

■ **Fig. 4.115** Vault from arch skeins

■ **Fig. 4.116** Barrel vault of a set of parallel arches in steel (Design Centre Linz, arch.: Th. Herzog). Coupling of the individual arches by means of longitudinal beams (purlins)

■ **Fig. 4.117** Two bays braced by diagonals, centre and right (Design Centre Linz, arch.: Th. Herzog). The stiffening principle is shown on ■ Fig. 4.120

reduced by the framework of arches and transverse bars, can finally be covered by a thin sheathing. Glass coverings are often used for this load-bearing variant.

Bracing

Vaults composed of individual arches (■ Fig. 4.121) are stable in the direction of their transverse axis (→ **x**, see ■ Fig. 4.120), as are all single arches within their own plane. Along

■ **Fig. 4.118** Glazed barrel vault from parallel arches (Gallery Vittorio Emmanuele in Milano)

■ **Fig. 4.119** Arched structure from reinforced concrete (Municipal Bath Heslach, Stuttgart)

the vault's longitudinal axis (→ **y**), on the other hand, the design has to be braced due to the lack of tipping stability of the individual arches. The position of transversely oriented bars, which is already necessary for the transfer and cross-distribution of loads, together with the arches, causes that:

- rectangular fields emerge which can be stiffened by **diagonal bracing** as shown on ■ Fig. 4.120. With appropriate distribution of triangulated fields, the arch surface can be stiffened along its longitudinal axis. As a rule, the diagonal bracing is preferentially located only in singular trans-

verse bays composed of an associated pair of arches in such a way that, there, horizontal forces can be introduced into the foundation. From the viewpoint of load-bearing, these braced bays are preferably arranged as close as possible to where the wind loads to be absorbed occur, typically directly behind the terminating end façades, with the aim of keeping the force paths within the structure as short as possible. The remaining arches are then stabilised as described below.
- The individual arches are **coupled** to one another transversely to their plane by rods running parallel to the vault axis (→ **y**) (therefore referred to as **coupling rods**), so that it is sufficient to connect such a tied-together arch array to fixed points, such as, for example, the diagonalised pairs of arches described above, in order to stabilise the overall structure. The coupling bars are subjected to purely axial compressive or tensile stress, do not experience any bending due to vertical loads, and do not even have to be in contact with the actual vault shell.
- The same coupling function can also be fulfilled by **purlins** that run at right angles to the arches and are top-mounted or suspended from them (◘ Figs. 4.116 and 4.117). In addition to the axial compressive and tensile forces to which they are subjected as coupling rods, they are also subjected to bending due to vertical loads. They must also be connected horizontally to fixed points, such as braced pairs of arches. They are referred to as **coupling purlins**.
- Under certain conditions, a covering surface component placed on top of the arches can also couple together the array of arches. This is achieved, for example, with a trapezoidal sheet-metal shell. The thin shell must be able to absorb the horizontal forces without the risk of buckling, which generally limits the permissible force level. This shell must also be connected to a fixed point.

All the transverse elements mentioned—linear bars or areal shells—can also provide important lateral buckling stabilisation for the individual arches, at the same time as exerting their coupling function. Under heavy loads, they prevent the compressed rod, which the arch ultimately amounts to, from swerving sideways. This lateral support is particularly important for slim, high arch cross-sections (◘ Figs. 4.116, 4.117, and 4.118).

For elevated vaults from arch arrays placed on vertical enclosing walls, the same considerations apply as for solid-walled vaults in respect to bracing and taking shear stresses (◘ Figs. 4.120 and 4.121).

4.2.2.5 Vaulted Shell from Bars

► *Sect. 3.3 The Constructional Composition of the Space-Enclosing Surface Element*

An interesting modification of a vault from longitudinal and transversal bars (analogous to variant **3** elsewhere) is a cylindrical curved lattice of mutually connected hinged individual bars which acts as a **shell** in the structural sense. This is called

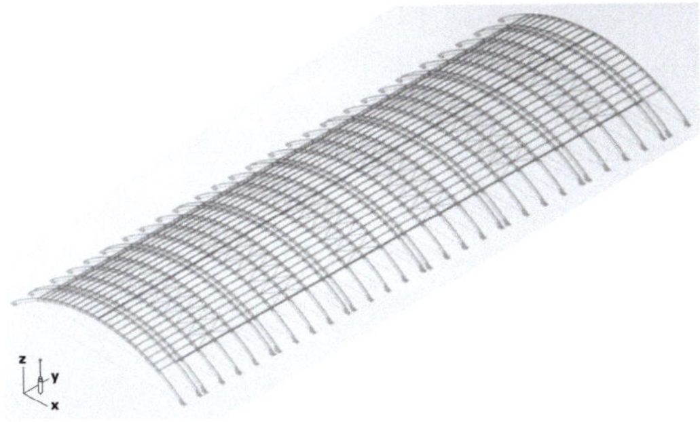

Fig. 4.120 Design Centre Linz. Bracing of the barrel vault from individual arches by means of single diagonally braced bays (arch. Th. Herzog)

Fig. 4.121 Arched structure: short-circuiting of the arch thrusts by means of tie-bars

a **grid** or **lattice shell**. It can also be regarded as the transformation of a full-walled shell vault, as discussed in ► *Sect. 4.2.2.3* above, into a latticework. It is essential in this context that the main load-bearing effect does not emerge in the direction of transversely oriented main arches (→ **x**), but rather **two-way** in both directions of the grid rods. The required stiffness of the rod structure is achieved either by using a **triangular grid** (Fig. 4.124) or by **diagonal bracing** of the individual fields in the case of quadrangular grids (Fig. 4.123). In order to ensure a full-surface two-way load transfer, *all* fields are diagonally stiffened as a rule. In order to avoid a plurality of complex node connections, ropes are preferably tensioned continuously penetrating the nodes in two diagonal directions.

► *Sect. 3.3.5.1 Triangular Meshes*

In this way, a self-supporting lattice shell is formed which is stressed in its surface only by **membrane forces**. For the lattice bars, this means that they are only subjected to **axial forces** (compression, tension). As a result, they are exceptionally slim, especially when compared to vaults from arch arrays. Such grid shells are therefore often used in combination with full glazing.

Such grid shells must, like solid-walled vault shells, also be stabilised at regular intervals by **diaphragms** in planes perpendicular to the vault axis (**xz**), as shown on Fig. 4.103, left, to maintain their shape stability. In the case of lattice shells,

Fig. 4.122 Grid-shell-like barrel vault with square meshes and diagonal rope bracing. Shell structure with equidistantly arranged arch ribs for stiffening, which can be recognised here as stronger truss ribs. These assume a similar stiffening function for the grid shell as cross-diaphragms or radial cable stiffeners. The shell can bear any uniform load via normal forces (Petrówskij-Passage, today a department store, Moscow; eng.: Suchow)

Fig. 4.123 Glazed grid shell. The square grid is stiffened by the effect of a diagonal rope bracing (Museum of History of Hamburg, eng.: Schlaich, Bergermann & P.)

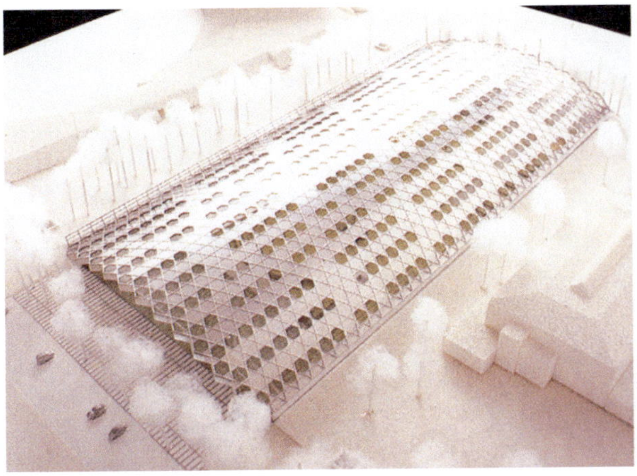

Fig. 4.124 (Left) Vaulted grid shell with triangular grid. The shear stiffness is thereby ensured without additional measures (project Waldstadion Frankfurt/Main, arch.: N. Foster). (Right) Structure of the grid shell on the left

closed diaphragms which would impair the fine-cut appearance of the construction are not normally used, but rather fan-shaped cable stiffeners which compress the shell and achieve the same effect (Fig. 4.123) or, alternatively, local bending-stiff arch ribs (Figs. 4.122, 4.123, and 4.124).

4.2.3 Tensile-Stressed Systems

4.2.3.1 Belt

Belts are uniaxially curved load-bearing structures suspended between two parallel linear supports (Fig. 4.125). In some respects, they represent the reversal of a compressive-stressed barrel vault as discussed above. Executed as a non-bending-resistant structure (rope or membrane structure), it automatically adopts the funicular line of the prevailing loads and thus operates under pure tensile stress. If it is made of bending-stiff structural members, it can be formed according to, at least, the main loads, and in turn experiences mostly pure tensile stresses (Fig. 4.125).

▶ *Sect. 4.2.2.2 Full-Leaf Vault* and ▶ *Sect. 4.2.2.4 Vault from Arch Skeins*

Belts without bending stiffness are movable structures and, hence, respond to changes of load pattern, like are unavoidable in construction, with elongation-free **changes of shape**. As already mentioned, these shape changes must remain within tolerable limits, which is why, as with all other movable systems, **stiffening measures** are essential, which have a decisive influence on the design of the load-bearing structure. Belts can be stiffened by:

▶ *Sect. 3.4.5.2 Membranes and Rope Nets* as well as ***Vol. 1**, Sect. 26.9.9 Mechanically Pre-Stressed Membrane*

- a prestressed **rigging** from below (Fig. 4.126). This is possible with belts only *along* the spanning direction (→ **x**, see Fig. 4.130), since *transversely* there is no curvature (→ **y**). This measure is equivalent to a **prestressing**, that

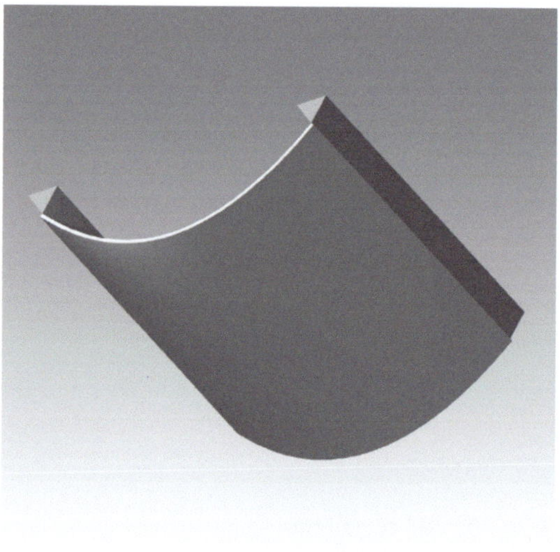

Fig. 4.125 Belt

is to say, to introducing an additional load which also affects the load pattern and thus, accordingly, the shape of the funicular line. If the prestressing forces are introduced at points, as is the regular case, the load-bearing form changes towards a polygonal line. The load-bearing rope and under-tensioning rigging result in a kind of rope truss which is capable of absorbing alternating loads.

- increased dead weight, i.e., **ballasting**. This is called a **self-weight-stiffened** membrane structure. Similar to compression-bearing structures, in which the unfavourable effect of alternating loads, which leads there to a deviation of the system line from the thrust line and, consequently, to bending moments, is limited by increasing the dead weight and thus improving the **load ratio**, also in the case of belts, the increased constant load share from ballasting reduces the effects—shape changes—of the changing load shares.
- sufficient **bending stiffness** of the structural members. However, a great advantage of these structures, namely the extremely high material efficiency due to the prevalence of pure tensile stresses, is thereby given away in some respects because then the bending stress comes back into play—as it were, through the back door—and makes the structure again material-consuming and cumbersome. Sometimes, the most expedient way is a combination of these measures (as in the examples on Figs. 4.126, 4.127, 4.128, and 4.129).

▶ *Sect. 3.4.3 Deviations from The Thrust Line*

As we have seen, the missing curvature *transversely* to the spanning direction is characteristic of belts. In this direction (→ **y**, see Fig. 4.130), belts are rectilinear. A transversely oriented prestressing of the suspension structure as in the case of classical mechanically pretensioned, anticlastically curved membranes is therefore not possible and hence not envisaged

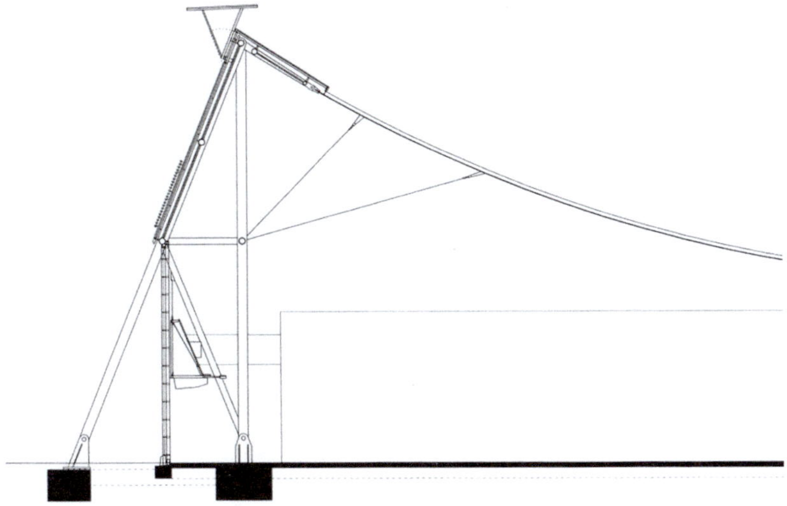

Fig. 4.126 Additional guying of a ballasted belt. Trade-Fair Hall Hannover (arch.: Th. Herzog and Partner)

Fig. 4.127 New trade-fair halls in Stuttgart: a one-way spanning suspended construction (arch.: T. Wulf und Partner). Lattice trusses can be seen in a tangential position to the roof at its edges, which provide support to the belt and support the thrusts caused by the suspended structure

at all. In this direction → **y**, the question of **cross-distribution** of loads arises—as with all surface structures. This can be ensured transversely to the spanning direction only by sufficient bending stiffness of a secondary element which is transverse to the main support elements (e.g. ropes). In general, this is only appropriate with an adequate roof construction, for example with double-leaf wooden cassettes. This results in a structural composition according to the construction principle of variant **2** (see elsewhere), or also with a straight cross-bar array (variant **4**, at the very same place). Otherwise, ballasting is more appropriate, with no cross-distribution of loads required.

▶ *Sect. 3.3 The Constructional Composition of the Space-Enclosing Surface Element*

4

Fig. 4.128 New trade-fair halls in Stuttgart, side view of the transverse façade

Fig. 4.129 *Lowara* branch in Montecchio, Italy, from Renzo Piano Building Workshop. Example of the use of belt roofs in a single-storey office building

Extendibility

Regarding extendibility, the same applies as to other directional systems: In the direction of span (→ **x**, see Fig. 4.130), only an addition of single bays is possible; transversely to the direction of span (→ **y**), in theory, an indefinite continuation of the construction is possible (Fig. 4.130).

Bracing

The construction of a belt is not braced in either direction (→ **x**, → **y**, see Fig. 4.130). The basic structural cell from the actual belt and two elevated linear supports, including the bracing structure, must be considered. In the spanning direction (→ **x**), the bracing structures must in any case be de-

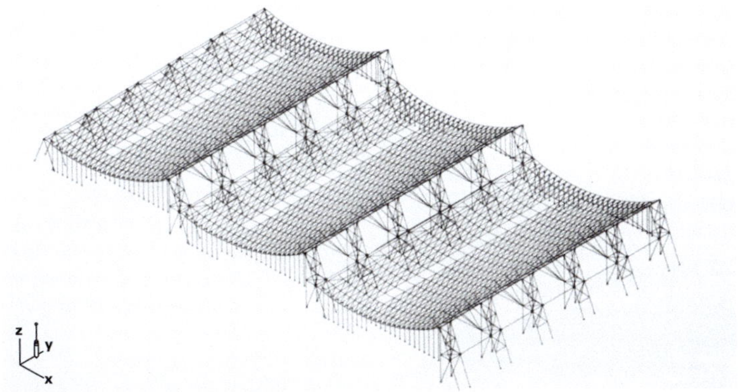

Fig. 4.130 Extension principle of a belt. Like all directional systems, individual modules are added in the spanning direction (→ **x**). Transversely (→ **y**), an uninterrupted continuation of the construction is possible

Fig. 4.131 Suspended construction from curved bending-stiff wood profiles (factory hall für *Wilkhahn*, arch.: F. Otto)

signed as stable frameworks, as they must absorb the horizontal force component from the belt structure (Fig. 4.130). This results in an immediate stabilisation of the overall construction. These supports can in turn be stabilised transversely to the spanning direction (→ **y**) by means of appropriate measures. The belt itself must be able to transmit horizontal loads to the bearings by means of sufficient **shear stiffness** within its surface.

Usage in Buildings—Design Aspects

Belts allow covering large spans with extremely reduced construction depth. They consequently are often used for roofing larger hall spaces (Figs. 4.131, 4.132, and 4.133). In a sense, the ease and constructional simplicity of the belt itself is attained at the expense of increased structural complexity

Fig. 4.132 Olympic buildings 1964 in Tokyo. The tensile members oriented transversely to the centre axis are not ropes, but curved steel beams (arch.: K. Tange)

Fig. 4.133 Olympic buildings 1964 in Tokyo, interior

for generating elevated linear supports and, in particular, for counteracting the systemic horizontal forces occurring there (see Figs. 4.127 and 4.128).

4.3 Non-Directional Systems

Whereas previously, directional —therefore one-way acting— structures were at the centre of the consideration, we shall examine in the following those which transfer loads **two-way** in the sense of our definition, namely at first under the assumption of a four-sided support, both in the variant of a linear

▶ *Sect. 3.2.1 One-Way and Two-Way Load Transfer*

and a punctual one, although supports with more than four sides or more than four punctual supports are also possible. Although with four-sided support or four-point support essentially two spanning directions exist, the designation of **non-directional structures** has become nonetheless generally established. Structures with more than two spanning directions and ring-like supports are also considered to be two-way, since the flow of forces can already be fully described computing the two main force components within the surface.

In general, non-directional structures are more complex in terms of manufacture than directional ones, especially when executed as a bar structure. They are, as a matter of principle, more efficient and less material-consuming in terms of load-bearing behaviour, since a stronger cooperation of the components among each other is activated. They are therefore usually characterised by internal static indeterminacy. Non-directional structures, or basic structural cells made from composite non-directional structures, are already recognisable as such by their—at least approximately—point-symmetrical geometry.

4.3.1 Bending-Stressed Systems—Flat Coverings

First, flat coverings, which are extremely important for building construction, will be discussed (◘ Fig. 4.134). There are, of course, far-reaching coincidences with the already mentioned flat coverings spanning one-way.

► *Sect. 4.2.1.1 Slab, Spanning One-Way* and ► *Sect. 4.2.1.2 Flat Covering from Bar Skeins*

4.3.1.1 Slab Spanning Two-Way, Linearly Supported

The unsupported side edges of the one-way spanning version of the slab deflect under load (◘ Fig. 4.11, right). This is pre-

► *Sect. 3 Fundamentals*, ► ◘ Fig. 3.42; in this section, cf. ◘ Fig. 4.11, right, and ◘ Fig. 4.12

◘ **Fig. 4.134** Slab spanning two-way

4

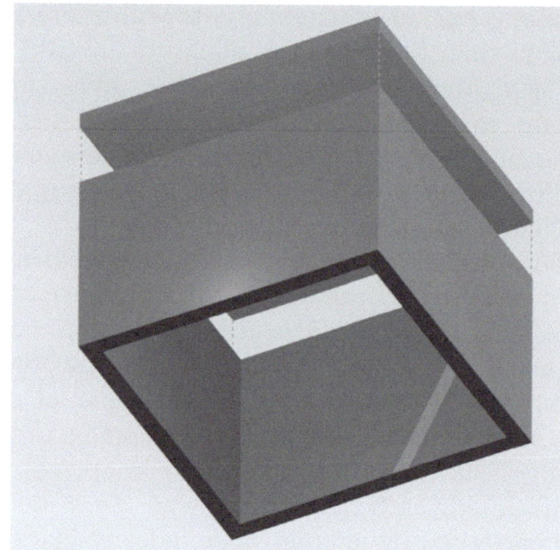

Fig. 4.135 **Slab** linearly supported on four sides with two-way load transfer. Similar conditions apply as for beam grillages (see ▶ *Sect. 4.3.1.3*): the best material utilisation is achieved with equal spans in both directions, i.e. with square structural fields

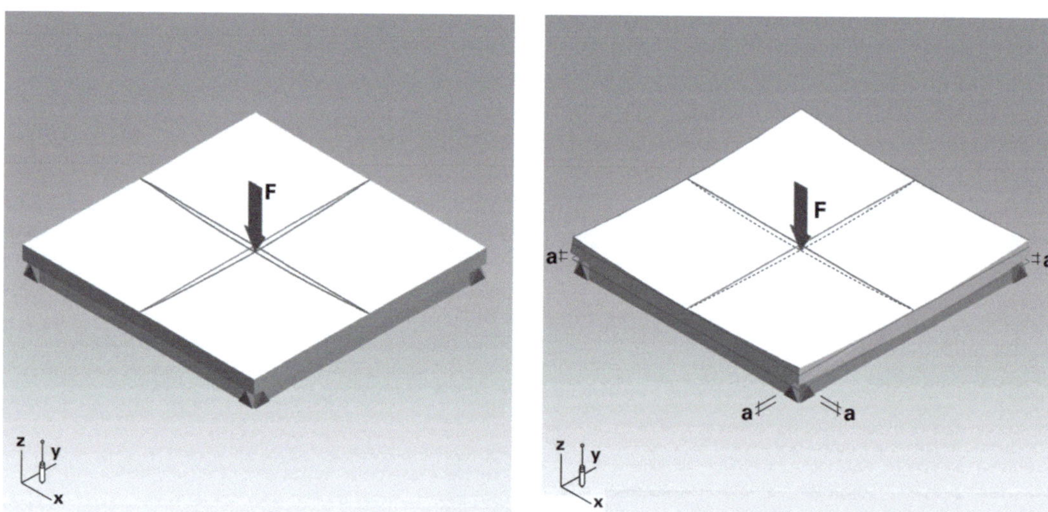

Fig. 4.136 Two-way load transfer of a centric single load **F** in a square-shaped slab with four linear supports. With equal spans, the deflections are smaller than those on Fig. 4.11, right. (Right) Lifting of the corners in a slab not rigid against twisting

▶ *Sect. 3.2.1 One-Way and Two-Way Load Transfer*

vented by the support in the case of the slab supported on all sides (Figs. 4.135 and 4.136, left) so that a transverse bending can be fully activated, not only from the load cross-distribution (case **3** and **3' elsewhere**), but in *direct* load transfer at the support (cases **4** and **4'**, at the very same place). In the case of square plate proportions, a true **two-way load transfer** occurs. Concrete can be utilised more efficiently in this way since its compressive strength comes into effect in two directions. The transverse tensile stresses are absorbed by a trans-

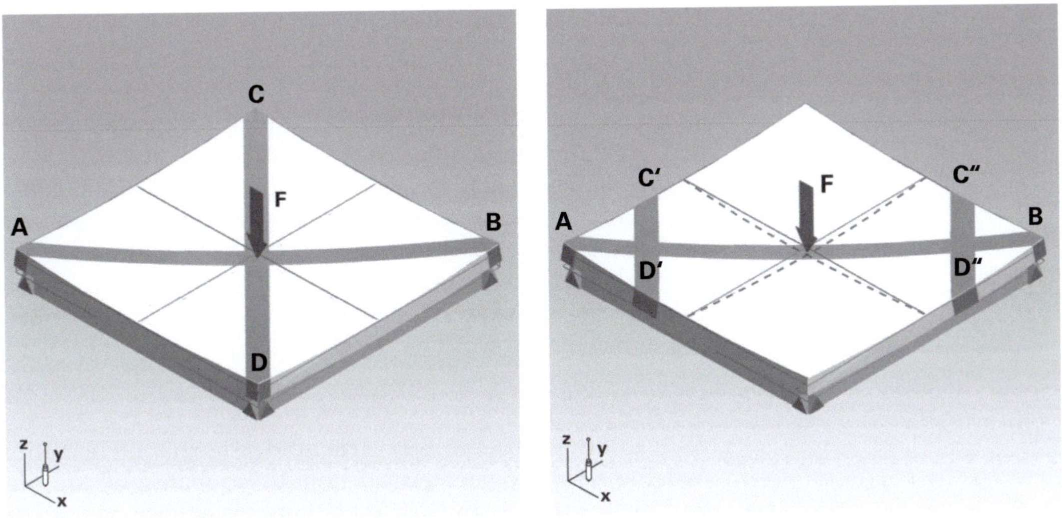

■ **Fig. 4.137** (Left) Contemplation of the diagonal strips (**AB** and **CD**) of the slab. (Right) The diagonal strips (**AB** shown here) are supported by the much stiffer (because shorter) transverse strips (**C'D'** and **C"D"**), so that a kind of cantilever action is established: The ends of the diagonal strips (**AB** here) lift off

verse reinforcement (integrated in the mesh reinforcement), which only comes with a moderate additional expense.

Load-Bearing Behaviour

A characteristic deformational behaviour of the slab spanning two-way causes the **corners** of the plate to uplift as a consequence of the load acting on the central field (■ Fig. 4.136, right). This can be explained as follows: Considering a diagonal plate strip (**AB**, ■ Fig. 4.137, left), it is supported by the imaginary transverse plate strips (**CD**), and thus partially relieved, according to the rules of two-way load transfer. This leads to a reduction in its deflection in the field; in the corner area, however, the supporting transverse strips (**C'D'** and **C"D"**, ■ Fig. 4.137, right) are very short, and consequently also very stiff. They then act as a rigid support, which causes the diagonal plate strip **AB** to behave as a beam with two cantilever arms: Its cantilevered end portions lift off the support.

__Vol. 1__, Sect. 26.7.3.8 Element with Hinged Linear Support on Four Sides (Bending Plate) Subjected to Perpendicular Surface Load

If this deformation is hindered by suitable measures, for example by superimposed load, the ends of the diagonal strip **AB** are restrained, which results in an additional improvement in the load-bearing capacity of the slab.

Cf. the diagram in ***Vol. 1***, *Sect. 26.7.3.8*, Fig. 26.85; cf. also the torsional restraint of the edge area of a beam grillage

As already described, a slab spanning two-way, with direct load transfer, acts only as such when the proportion of its edges is approximately **square** (■ Fig. 4.135). Rectangular ground-plan formats lead to the load being mainly transferred in the **short** spanning direction, namely in the **stiffer** one. The two-way utilisation of the structural depth of the plate is then not given. As a rule of thumb, this is already the case with ratios of **1: 1.5**.

The Basic Structural Cell

Linear supports of the slab can, in principle, consist of vertical enclosing walls, which create a cell-like space enclosed all-around (Fig. 4.135). Since the spans of the slab are limited, an extension of this type of construction results in a sequence of individual closed, dimensionally restricted spatial cells. This is typical of housing construction. As an alternative, the slab's support can also be carried out on a skeletal system consisting of columns and girders. The latter then provide the linear support for the slab. Systemically, the girders run in two directions.

Extendibility

As a result of the two-way structural principle of operation and the resulting geometrical non-directionality of the slab and the bearing, both main directions are equivalent with regard to the extendibility of the construction: Because of the spans being limited, it can only be extended in both directions by **adding** additional structural bays. In the case of skeletal systems made of columns and beams, this results in column positions with always square structural fields.

In the case of the most common type of execution of the two-way spanning slab, the solid reinforced-concrete slab, **structural continuity** can be implemented in both directions without any restrictions.

Bracing

Structural cells made of surrounding wall-like enclosures as discussed above are braced in both directions (→**x**, →**y**), since the covering—at least when executed as a solid slab—also has a stabilising racking stiffness. Combined with a skeletal system, the basic cell must be empowered to absorb horizontal loads by means of suitable bracing measures, i.e. through diagonal bracing, by connecting it to fixed points by means of the racking stiffness of the covering, etc. As an alternative, a framework effect of both elements column and beam which are then combined into a portal frame is also possible, especially in the embodiment as a monolithic reinforced-concrete construction. With this two-way spanning covering, a **two-way frame effect** can consequentially be created. The double-oriented, rigid frame corners can be monolithically cast in one piece. This construction was a decisive developmental step in the beginnings of reinforced-concrete skeletal construction, but today it has no practical significance because of the associated high formwork and reinforcement expenses. In some respects, it can be said that this solution was replaced by modern point-supported flat floors.

Cf. Fig. 4.66

► *Sect. 4.3.1.2 Slab Spanning Two-Way, Punctually Supported*

Usage in Building—Design Aspects

Essentially, the same statements apply as to the one-way spanning slab. The two-way slab provides an even more far-reach-

► *Sect. 4.2.1.1 Slab, Spanning One-Way*

Fig. 4.138 Ribbed slab, spanning two-way

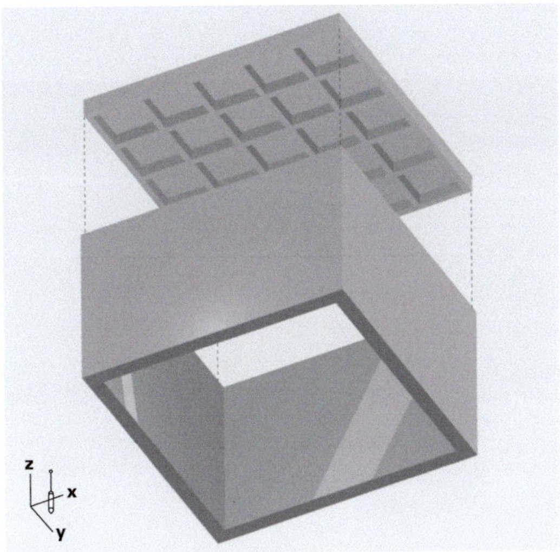

Fig. 4.139 The slab linearly supported on four sides with two-way load transfer can also appear in ribbed design according to the pattern on Fig. 4.18. In reinforced concrete construction, these components are referred to as **coffered floors**

ing freedom in the design of the supports than in the case of the one-way version. The pronounced ability of the slab to distribute loads permits, within certain limits, a free arrangement of linearly supported and free-spanning slab sectors. Irregular support configurations which are not bound to geometric grids can also be realised.

Cf. the considerations above in ► *Sect. 4.3.1.1.4 Bracing*

Ribbed Slab, Biaxially Spanning (Coffered Floor)

As with the one-way spanning slab, the two-way spanning one becomes, beyond a certain span, so heavy that it consumes its

Fig. 4.140 Individual formwork elements made of steel, laid out on a supporting formwork, before concreting a coffered floor

Fig. 4.141 Coffered floor from in-situ concrete (arch.: Louis Kahn)

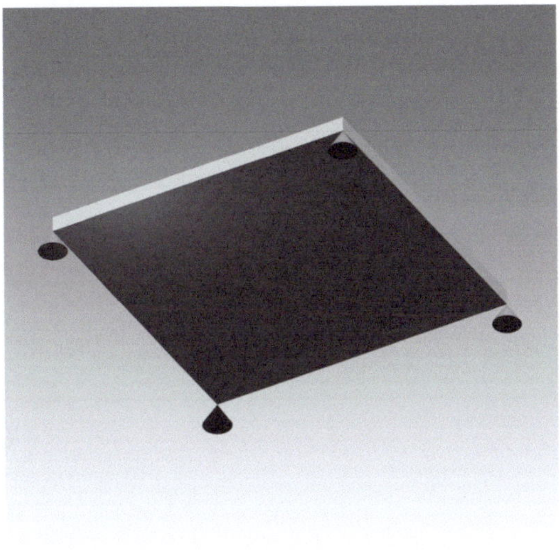

■ **Fig. 4.142** Slab spanning two-way on punctual supports

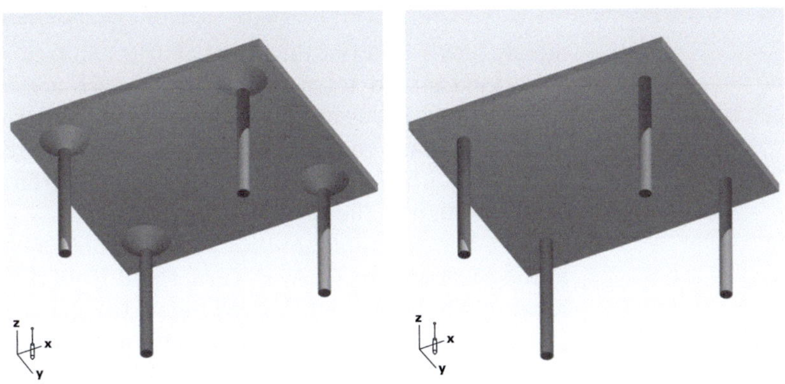

■ **Fig. 4.143** (Left) **Mushroom slab**. (Right) **Point-supported flat slab**, often also referred to as mushroom slab. The load-bearing effect is the same as for the mushroom slab on the left, but the mushroom heads are missing. The strong shear forces at the columns are absorbed by suitable reinforcement (■ Fig. 4.144)

load-bearing reserves merely to bear itself. It is then also possible to realise **ribbed slabs**, two sets of ribs forming a **coffered slab** (■ Figs. 4.138 and 4.139).

Coffered slabs are usually poured on site. In this way, the numerous mutual penetrations of ribs can be carried out monolithically without further connection problems. Industrialised formwork systems exist for this purpose (cf. ■ Fig. 4.140, see also ■ Fig. 4.141).

4.3.1.2 Slab, Spanning Two-Way, Punctually Supported

The punctual support of the slab (■ Fig. 4.142) does not initially benefit its load-bearing behaviour since the loads have to be transferred in the support area from the slab to the column within a geometrically very restricted area. The compar-

__Vol. 1__, Sect. 26.7.3.10 Element with Hinged Punctual Support on Four Sides (Bending Plate) Subjected to Perpendicular Surface Load

Fig. 4.144 Inserting a prefabricated punching shear reinforcement made of double-headed studs and spacer bars into the formwork (prod.: *Halfen®*)

atively small material thickness of the slab (see above), which is conceptual, so to speak, and otherwise a great advantage of the slab floor, is a weak spot from this point of view since high **shear-force concentrations** emerge in the area around the support point. Hence, there is a risk of **perforation** or **punching-through** of the supports.

The constructional answer to this problem was developed, amongst others, by C. A. P. Turner and Robert Maillart in the form of the **mushroom slab** (Fig. 4.143, left). As a result of the mushroom-shaped widening of the column head, the so-called **critical circular section** is enlarged in the area near the support where the greatest punching shear forces occur. This mitigates the effects of the point support.

Since this type of floor is associated with a comparatively high formwork expense, today it is preferably carried out as a **point-supported flat slab** without mushroom heads (Fig. 4.143, right). The increased shear forces are absorbed by means of a **punching shear reinforcement** (Fig. 4.144). This can be integrated into the regular slab thickness so that a continuous slab cross-section and a totally flat soffit are maintained. The formwork costs are then comparable to those of a conventional solid slab.

Nevertheless, it is often referred to even today as mushroom floor

The structural depth of the plate is optimally utilised in two directions if the intercolumniation fields are approximately the same in both directions, i.e the fields are **square**. However, unlike in the case of the *linearly* supported two-way spanning slab with strongly diverging spans, a **two-way load-bearing effect** nevertheless is present also in case of rectangular column fields with different spans as a result of the indirect load transfer. From the point of view of reinforcement, an arbitrary single load is first transferred to the stiffer short edge strip over the longer slab strip, and then diverted

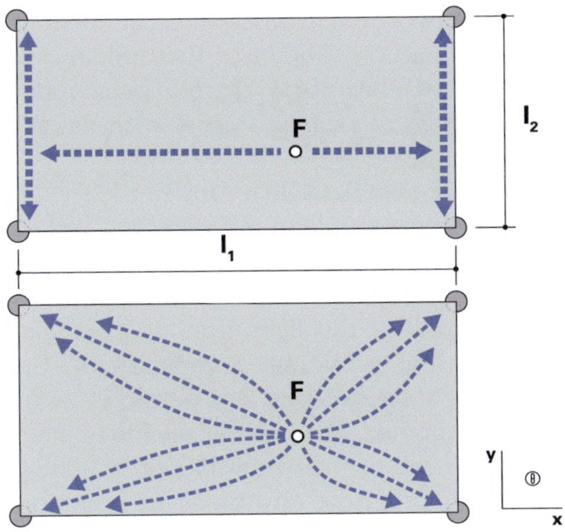

Fig. 4.145 Indirect load transfer in a rectangular slab: above, as viewed from the point of view of reinforcement; below, main stress trajectories

from the latter to the supports (Fig. 4.145). However, the material is not as efficiently utilised as with identical spans in both directions.

The Basic Structural Cell

The development of the point-supported flat slab allowed for economical use of the **girder-free solid slab** in modern skeletal construction, in its most powerful variant as a system spanning two-way. The basic structural cell, a combination of slab and four supports, is initially not braced: A frame effect between support and plate cannot be realised due to the very small thickness of the slab. Consequently, additional reinforcing elements such as **building cores** are required for such structures. Due to the horizontal racking stiffness of the slab, the skeletal structure can be connected force-fittingly to these fixed points.

Usage in Buildings—Design Aspects

Where the combination of a solid slab with a skeletal structure is convenient, for example, in **modern administrative buildings**, the point-supported flat slab is extensively used. It has some significant advantages:

- The usual requirements for a floor in building construction are met:
  - **airborne**-sound insulation;
  - **impact**-sound insulation;
  - **fire protection.**
- Furthermore, the **thermal-storage capacity** turns out to be an advantage in current administrative buildings. The high-heat-insulating façades and, in particular, the high internal heat loads (from lighting, office machines, etc.) can

lead to excessive room temperatures at medium to high outdoor temperatures. The large thermal-storage mass of the floor deploys a favourable thermal behaviour.

- In terms of **usage**, a skeletal system with flat floors offers the advantage of:
  - **major ground-plan flexibility**. Due to the few fixed points in the ground plan (columns, cores), a wide variety of floor-plan divisions can be realised without difficulty by freely distributing or relocating light partition walls.
  - great freedom in the definition of column positions. Thanks to the pronounced capacity of the slab of cross-distribution of loads, it is possible—within certain limits—to provide **free column positions** which do not necessarily have to follow an axial grid. The point-supported plate has this advantage in common with all other two-way load-bearing systems, such as, for example, the beam grillage. Boundaries are, of course, set by the limited span of the slab—up to just over 7 m.
  - **unimpeded installation** of ducts under the ceiling, since no girders hinder the duct routing.
  - **ensuring** good **impact** and **airborne-sound protection** combined with high variability of the positions of **partition walls**. For this purpose, a bonded screed or a screed on a separation layer is installed, which in combination with the slab's mass and a soft-resilient floor covering (e.g. carpet flooring) ensures the required impact-sound protection.
  - a large usable **thermal radiating area** in the form of undressed solid floor or ceiling surfaces. For this purpose, the slab or the bonded screed is equipped with a system of heating and cooling pipes. This is called a **thermal component activation**.

Point-supported flat slabs can now also be implemented using semi-prefabricated **unitised slabs**.

4.3.1.3 Beam Grillage Spanning Biaxially, Linearly Supported

Load-Bearing Behaviour

***Vol. 1**, Sect. 26.9.5.1 Element Ribbed Two-Way on Linear Supports* as well as ibid., Fig. 26.225

The basic load-bearing behaviour of linearly supported beam grillages is discussed in detail elsewhere. The advantages of the two-way load transfer can best be exploited, analogously to the homogeneous slab, in the presence of a *direct* load transfer of this structural system, by **square** grillage formats with equal spans in both main directions (◘ Fig. 4.146). Beam grillages with two different spans as in ◘ Fig. 4.148, left, result in a preference for a particular spanning direction and, thus, to an insufficient utilisation of individual structural members, or overloading of others. The advantages of the two-way load transfer cannot be fully exhausted in this way (◘ Figs. 4.147 and 4.148).

The phenomenon of the **uplifting corners** is the same as with the linearly supported plate (▫ Fig. 4.136, right).

▫ **Fig. 4.146** Beam grillage spanning two-way

▫ **Fig. 4.147** **Beam grillage** with slender ribs made of wood-based panels. The rib at the column is dimensioned stronger to receive the concentrated shear forces. Note the exceptionally slender rib cross-sections, which are made possible because the numerous longitudinal and transverse ribs stiffen each other at short intervals. (Right) Beam grillage with triangular grid geometry

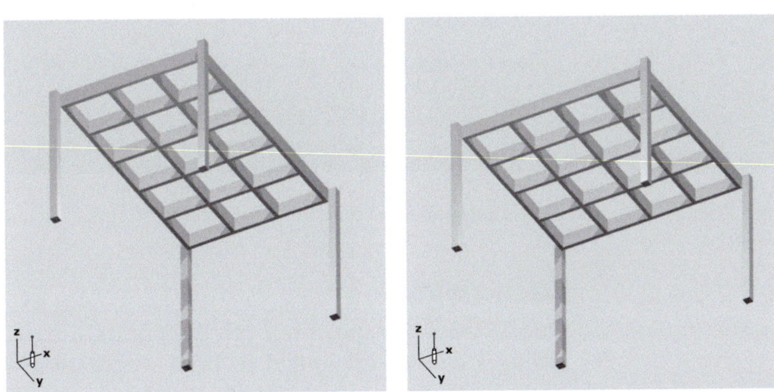

▫ **Fig. 4.148** (Left) From the considerations in ► *Sect.* 3, ► ▫ Fig. 3.55, it follows that in the case of non-directional support systems, different spans in both directions have an unfavourable effect on the load-bearing behaviour. (Right) The advantages of the two-way load transfer are best used in **square structural fields** with equal spans in both spanning directions

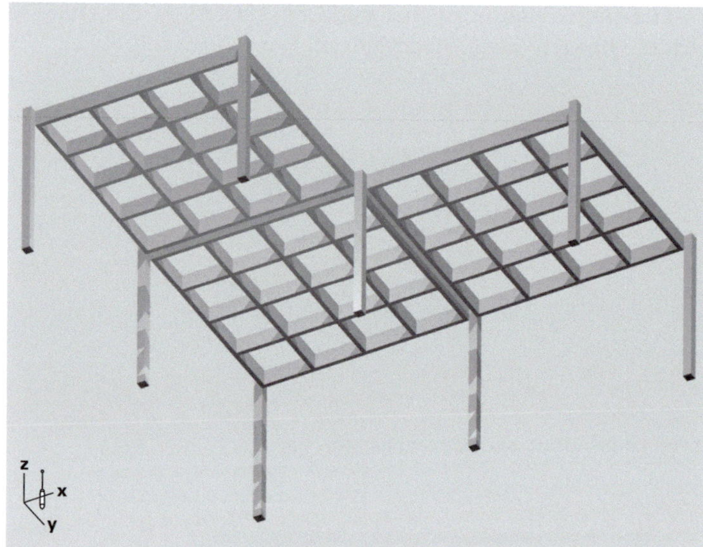

Fig. 4.149 A system of beam grillages can be expanded by addition in two equivalent directions

Fig. 4.150 Beam grillage composed of multiple added modules made of steel truss girders (school building Solothurn, arch.: F. Haller)

Extendibility

Due to the same spans in both directions, a structure of beam grillages can be expanded in both orientations under identical conditions by simply adding an element (Figs. 4.149 and 4.150). A **continuity effect** over several supporting fields can be easily realised, as in the case of solid slabs.

Execution Variants

The **node design** is always a delicate constructional detail in the case of beam grillages, which is always associated with a

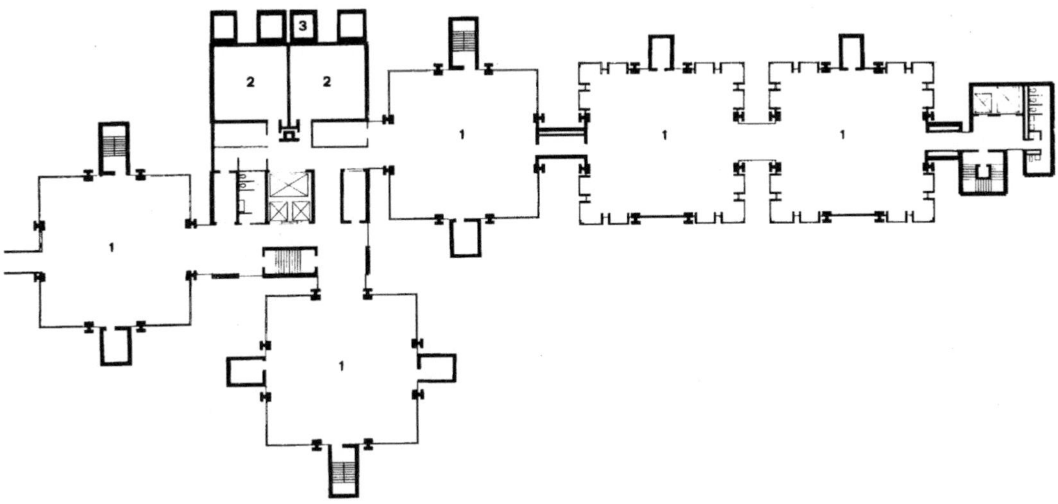

Fig. 4.151 *Medical Research Centre*, floor plan. The individual grillage units can be recognised on the floor plan of the individual buildings as square floor areas (arch.: Louis Kahn)

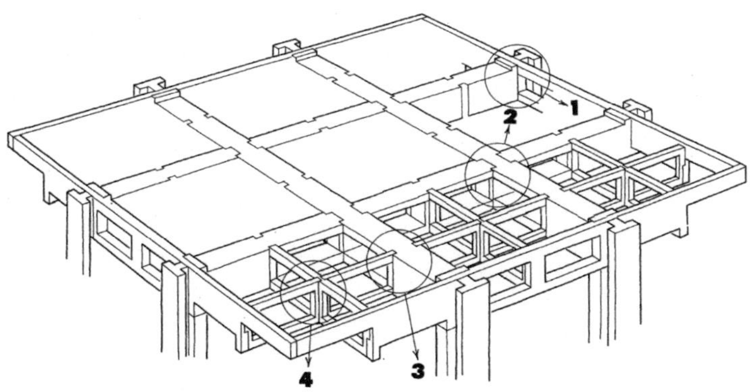

Fig. 4.152 *Medical Research Center*, spatial representation of the beam grillage in prestressed concrete

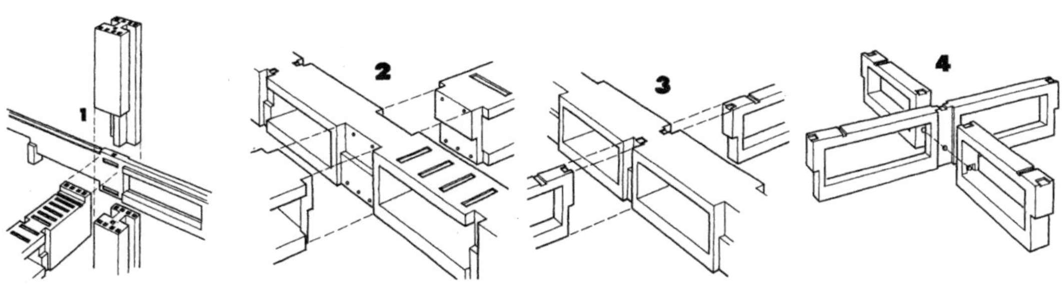

Fig. 4.153 Node details with respectively continuous and transversely butted rib

comparatively large manufacturing expense because of its frequency, which is so to say system-conditioned in the element. Since grillages in the rarest cases have transport dimensions, these connections must also be designed as **erection joints**, which usually precludes certain joining techniques from the

Fig. 4.154 Pictures of the erection process

Fig. 4.155 A load-bearing structure made of wood-based panels with curved contours, but with a basic structure and load-bearing behaviour similar to that of a beam grillage (Roof at Plaza de la Encarnación, Sevilla; Arch.: J. Meyer-H)

outset—for example in steel construction, in most cases welding, or in wood construction gluing. It is also aggravating that, in turn, the joints must always be **bending-stiff**, which is always associated with increased technical complexity.

One way to reduce the manufacturing effort required for the nodes is to make one of both ribbed arrays continuous in the nodes and only interrupt the transversely running set of ribs at the nodes. In this way, the continuous rib skein mounted in the first step acts as a working scaffolding for the transversely oriented rib sections to be subsequently fitted in between (Figs. 4.152 and 4.153). An interesting example in prestressed concrete construction, in which the problem of nodes was solved by post hoc prestressing, is shown on Figs. 4.151, 4.152, 4.153, and 4.154.

Cf. also the representation of this beam grillage in ▶ *Sect. 10 Prefabricated-Concrete Construction*, ▶ Figs. 10.56, 10.57, 10.58, and 10.59

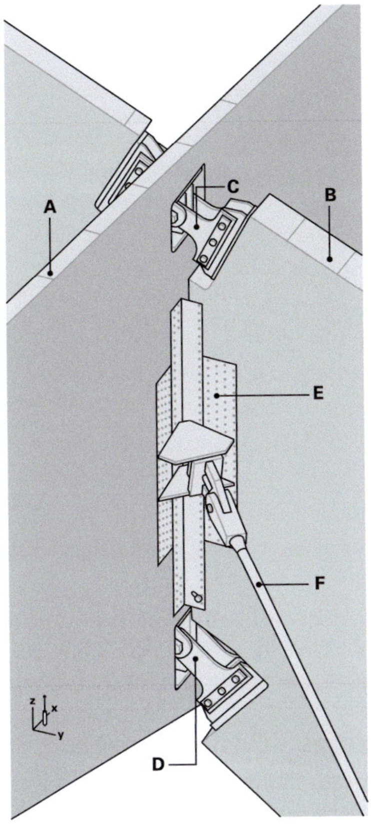

Fig. 4.156 Node of the grillage on Fig. 4.155. One rib (**A**) passes through while the other (**B**) is butted. The bending-induced tensile and compressive force is transmitted via steel parts through an opening in the continuous rib (**C**, **D**). Shear forces are absorbed with sheet-metal connectors (**E**). The bar of a cross-bracing connects diagonally (**F**)

In timber construction, it is difficult to transfer the forces acting at the node transversely to the continuous rib in this constructional solution, i.e. to transfer transverse flexural compression and tension from the transverse rib section, across the aforementioned continuous rib's cross-section, since wood does not tolerate such stresses transverse to the grain. Instead, the forces have to be passed through by way of steel connections, for example by slotted gussets (Figs. 4.155 and 4.156). In timber construction, realising beam grillages is made more difficult from the outset because, in that particular construction mode, generating bending-resistant connections is always especially complicated.

Vol. 1, *Sect. 15.4 Mechanical Properties*

The simplest way to execute the nodes of beam grillages monolithically is by pouring in reinforced concrete. The step of monolithically connecting the covering plate to the rib array is then obvious, so that eventually a **coffered slab** is produced as discussed above.

► *Sect. 4.3.1.1.6 Ribbed Slab Spanning Two-Way (Coffered Floor)*

4

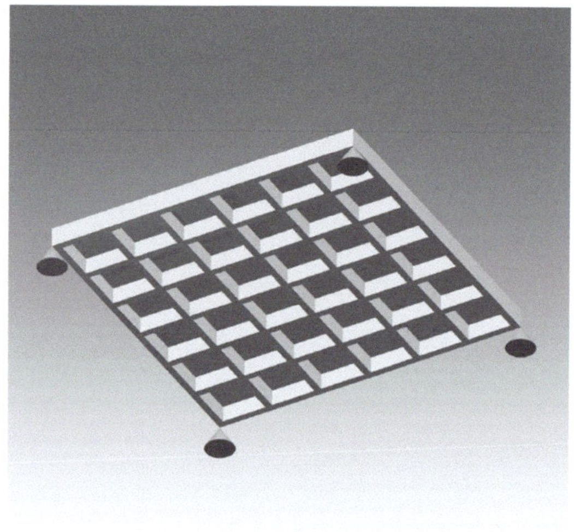

Fig. 4.157 Beam grillage spanning two-way, on punctual supports

The mentioned issues of the constructional design of beam grillages are, of course, also applicable to **point-supported beam grillages** as discussed in the following section.

Usage in Buildings—Design Aspects

Linearly supported beam grillages are, in the sense of the present disquisition, those which are supported all around on load-bearing wall panels. A support on a skeletal system consisting of columns and beams which provide a linear support to the grillage is conceptually nonsensical, since the grate, by its double orientation, has itself the capability of indirect load transfer and hence can itself assume the function of a hypothetical supporting beam. Linearly supported beam grillages are found wherever they can deploy their abilities, i.e., when covering large rooms with rather compact spatial proportions in footprint. Square ground-plan geometries can be realised, as well as triangular and hexagonal ones if triangular grids are used. Due to its good material efficiency, the two-way load transfer allows the rib depth to be significantly smaller than for one-way spanning beam systems.

Cf. the point-supported beam grillages in the next ▶ *Sect. 4.3.1.4*

Their ability, similar to the slabs', to distribute forces within their structure and to allow for different alternative force paths, offers a comparatively great freedom in the design of the supports, so that linear bearings with larger, largely freely arranged openings or largely free column positions can be realised.

4.3.1.4 Beam Grillage Spanning Two-Way, Punctually Supported

***Vol. 1**, Sect. 26.9.5.2 Element Ribbed Two-Way on Punctual Supports*

The basic load-bearing behaviour of point-supported beam grillages is discussed in detail elsewhere. In contrast to the linearly supported beam grillage as discussed in the previous section, there is an **indirect load transfer** in this case (Fig.

4.157). The ground-plan proportions of the intercolumniation field of a point-supported beam grillage, i.e. the ratio between the two span lengths, do not play the same role as with linear support. Analogously to the point-supported slab, a single point load acting at an arbitrary location initially takes the path of the **longer rib**, in order then to be transferred in the transverse direction by the stiffer **short edge rib** onto the supports (Fig. 4.158). Although the other ribs are also partly involved in this load transfer, this is nevertheless the main load path within the grillage. The larger the divergence between the spans, the more the short rib—assuming identical rib depths—will be oversized. Nevertheless, in all cases—regardless of the field proportions—there is basically a **two-way load transfer**.

► *Sect. 3.2.2 Influences of Load Transfer on the Geometry of the Structural Cell*; in particular ► Fig. 3.99, right

Extendibility

Conditions are comparable to the linearly supported beam grillages in ► *Sect. 4.3.1.3*.

Execution Variants

The same applies as with the linearly supported beam grillages in ► *Sect. 4.3.1.3*.

Usage in Buildings—Design Aspects

Point-supported beam grillages (Figs. 4.159, 4.160, 4.161 and 4.162) are suitable for use in skeletal structures to cover

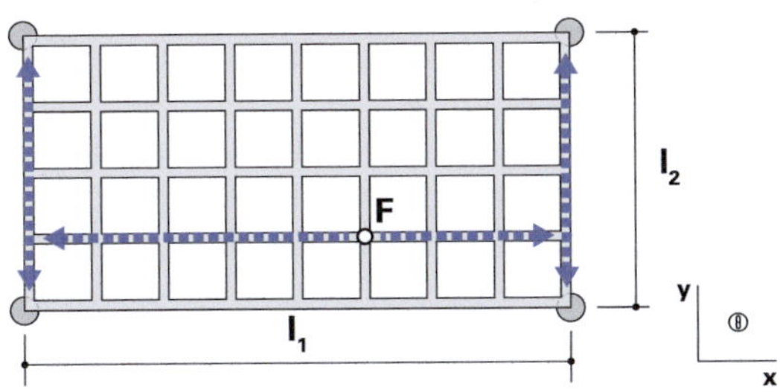

Fig. 4.158 Indirect load transfer in a rectangular beam grillage (assumed to be not rigid against twisting)

Fig. 4.159 Beam grillage from truss girder elements in steel (construction system Maxi, arch.: F. Haller)

Fig. 4.160 Square beam grillage in steel (New National Gallery Berlin, arch.: M. v. d. Rohe)

larger contiguous spaces by adding individual grillages in two directions. In particular in the case of usages, or building concepts, for which identical intercolumniations in two orientations are advantageous, beam grillages can play their strengths.

If the individual fields are coupled to one another in a bending-stiff manner over the column axes, the bending moments in the span can be reduced thanks to the structural continuity effect. Even in individual fields, the bending stress—and thus the overall depth—of the beam grillage can be reduced by cantilever action if choosing a suitable column position (Fig. 4.161).

Such as in the *Neue Nationalgalerie Berlin*

In the case of beam grillages which are point-supported, there is comparatively great freedom in defining the supports, since, as in the case of the slab, it is also a load-bearing structure with a pronounced ability to distribute loads. Supports are preferably arranged at a rib node.

Fig. 4.161 1(Left) Because of the easier execution of the welded erection joints for the various nodes, the beam grillage was pre-assembled on the ground and raised in one piece together with the columns (New National Gallery Berlin, arch.: M. v. d. Rohe). (Right) Reduction of bending moments acting on the grillage by shifting the columns inwards and creating a cantilever (National Gallery Berlin, arch.: M. v. d. Rohe)

■ **Fig. 4.162** (Left) Beam grillage in reinforced concrete. (Right) Beam grillage in timber

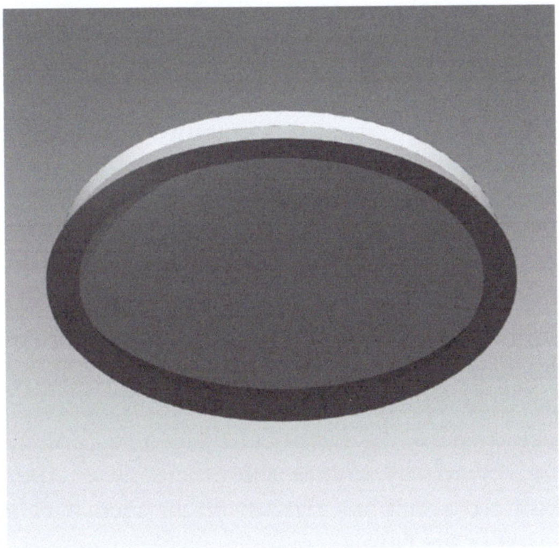

■ **Fig. 4.163** Slab on annular support

4.3.1.5 Slab on Annular Support

While in the case of the load-bearing structures considered so far in ▶ *Sect. 4.3* with four-sided linear bearings two spanning directions were present, many spanning directions can be identified with a radial bearing. Structures of this kind are nevertheless considered to be **two-way** in statics because, for the static analysis of the load transfer, it is sufficient to split it into two orthogonal load directions. The basic considerations regarding two-way load-bearing structures apply analogously.

▶ *Sect. 3.2.1 One-Way and Two-Way Load Transfer*

Flat slabs on annular bearings (■ Fig. 4.163) are rather rare in building construction. The usage-related problems generally associated with round building shapes, which are presumably causally related to this fact, have already been discussed above. Circular plates, however, have a more favourable bearing behaviour than rectangular ones and even than square plates: There are no corners that could lift off. There is

▶ *Sect. 3.1.6.1 The Vertical, Planar Space-Enclosing Element* no difference between diagonal and orthogonal spanning direction. It can be argued that plates on annular supports are an ideal case with regard to shape and bearing.

Furthermore, when placed on vertical walls, the systemically pre-set cylindrical drum geometry is inherently stable against horizontal forces and, consequently, the basic structural cell does not require any further bracing measures.

4.3.1.6 Radial Skein of Bars, Annularly Supported

▶ *Sect. 3.2.2.2.4 Parallel Bar Skein on Curved Supports*; ▶ Fig. 3.54, right

Circularly supported coverings cannot be executed in a reasonable manner with parallel bar skeins, which is why **radial bar patterns** are usually employed (Fig. 4.165). This geometrical form of organisation elicits a central penetration point at which all the ribs meet. Since, of course, at most only a single rib can be continuous here, all rods are usually interrupted at this point and are force-fittingly connected to each other. In the case of a large number of bars, i.e. a small angle of the circular sectors defined by the bars, there may be slanting cut-outs between bars at this juncture which are sometimes difficult to cope with in constructional terms. More favourable geometric conditions can be achieved by introducing a **centric connecting ring** or **cylinder**. It must be able to absorb the bending moments at the rod connections, which reach their maximum value exactly at this centre. In the upper area, the cylinder is compressed radially, and pulled in the lower part. This is called a **flipping moment** (Figs. 4.164 and 4.164).

▶ *Sect. 3.2.2.2.3 Concentric Curved Supports*; ▶ Figs. 3.53, right, and 3.54, left

As in the case of all fanned bar arrangements, changing load-infeed areas of the beams are, so to say, inherent in the system and invariably reduce the efficiency of the structure if the beam cross-sections remain unchanged over its length, which is usually the case.

Fig. 4.164 Radial skein of bars on annular support

Fig. 4.165 (Left) Example of a **radial beam position** on a tambour-like substructure from masonry. The beam in the centre of the image is the only one that is continuous. (Right) Radial set of ribs on the hairpin curve of the famous car ramp in the Fiat factory building in Lingotto (arch.: M. Trucco). In this case, the ribbed skein is supported by a centric pillar

Fig. 4.166 Spoke wheel

In an interesting variation, this load-bearing system can be designed as a **ring-rope system** or **spoke wheel** subject only to tensile stress (Fig. 4.166). This will be examined in more detail elsewhere.

A spoked wheel can also be understood as a mechanically prestressed rope structure as discussed in ▶ *Sect. 4.3.3.2 Membrane and Rope Structure, Mechanically Tensioned, Linearly Supported*; cf. also ▶ *Sect. 9.3.7.4.2 Radial Ring Rope Structures—Spoked Wheels*

4.3.2 Compressive-Stressed Systems

In this category, coverings from spherical **domes** and **cones**, in particular, will be considered in the following. They can each be regarded as the point-symmetrical form of revolution of

Fig. 4.167 Pyramid on four linear supports

▶ Sect. 4.2.2 Systems under Compression—Pitched Roofs and Vaults

the already mentioned inclined roofs and vaults. However, they display a qualitatively different load-bearing behaviour compared to their linear axis-symmetrical counterparts.

4.3.2.1 Pyramid

▶ Sect. 4.2.2.1.1 Rafter Roof

Pyramid-shaped coverings (Fig. 4.167) are similar in their basic load-bearing behaviour to inclined planar roofs as discussed above (Fig. 4.168). Deviating from pitched roofs as described there, pyramids have a load-bearing structure which is shear-resistant in both main directions (→ **x** and → **y**). Additional bracing measures are therefore superfluous. The individual planar boundaries of the pyramid unfold a combined shear-plate and bending-plate effect. They are coupled to each other via the ridges and thus stiffen each other. The mutual rigid support, together with the bottommost bearing element, for example a supporting wall, corresponds in total to a completely circumferential linear support. Because of their triangular shape, they do not need any further stiffening elements in their plane.

As is the case with rafter roofs and as a result of the oblique force attack, a horizontal force component, which is to be borne at the support, also arises here in addition to the vertical component. These thrusts can, for example, be introduced into a shear-plate-like floor construction at support height—with *annular anchors* in the edge region—or can also be accommodated all round by an *annular beam*. Horizontal thrusts can also be short-circuited within the system, in the absence of a horizontal bearing, for example by a peripheral annular tensile member in the roof plane.

4.3.2.2 Cylindrical Dome

Four-sided linearly supported domes of mutually intersecting cylindrical surfaces (Fig. 4.169) are designated as **co-**

Fig. 4.168 (Left) Tower top with pyramidal geometry on octagonal plan. (Right) Pyramidal roof on octagonal tambour (Baptistery of the Duomo of Florence, see below)

Fig. 4.169 Cylindrical dome

ved vaults in historical vault building (Fig. 4.170). In terms of their development, they can be interpreted as a compromise between two basic geometries, similar to some types of domes (inner, outer-circle dome): the curved shell, which is approximately axially stressed in the direction of the curvature and the quadrangular bearing, for instance upon a wall square, which offers some advantages in terms of usage and addibility.

Cf. next ▶ *Sect. 4.3.2.6 Full-Leaf Shell, Synclastically Curved, Punctually Supported*

▶ *Sect. 3.1.5 Design Principles of Adding Structural Cells*

A disadvantage of four-sided coved vaults is that the cylindrical caps are only uniaxially curved and are particularly vulnerable in the lower region, where the span between the ridges is greatest. The only slightly curved caps act almost like a pla-

Fig. 4.170 Dome consisting of cylindrical caps on octagonal plan (Santa Maria del Fiore, Florence, arch.: F. Brunelleschi)

nar plate there. Their stiffness can be significantly improved if they also have a slight elevation in the horizontal direction, and thus a curvature. They then act as a double-curved shell. Alternatively, the quadrangular floor plan can be converted into an octagonal one, to reduce the span widths of the partial shell caps, a solution which was used in the famous Florentine dome of Brunelleschi (Fig. 4.170).

The ridges have a reinforcing effect, as in the case of folded plates, for example the pyramidal roof. A ring tension in the lower region does not occur here—unlike with the spherical dome. Instead, an in-plane plate compression prevails from the mutual transverse support of the vault caps.

4.3.2.3 Cone

Conical shells (Figs. 4.171 and 4.172) can, analogously to the dome, be virtually cut into individual meridian strips which, unlike the doubly curved dome surface, are rectilinear in the case of the cone (Fig. 4.173), to illustrate their basic load-bearing behaviour. Under vertical load, such as from dead weight or snow, these strips are subjected to axial stress, which results in a **compressive force** in the **meridian** direction. On the other hand, there is a **ring compression**—in contrast to that of the spherical dome, over the *entire height* of the cone—owing to the load portion directed transversely to the

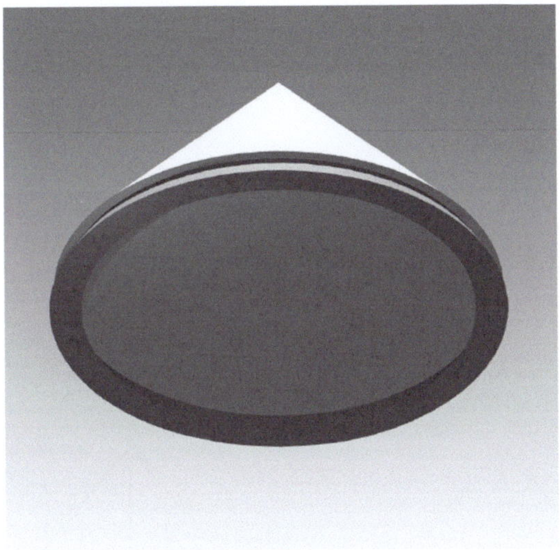

Fig. 4.171 Cone on annular support

Fig. 4.172 Trulli are old conically shaped buildings from stepped corbelled stone rings (corbelled domes) (Apulia, Italy)

strip, which deforms the strip from its straight axis downward. Ring tensile forces are non-existent with a distributed vertical load. Forces acting along the axis of the imaginary strip, e.g. the load of a mounted lantern, excite pure axial compression in the meridian direction; ring forces are then null.[2] In this type of loading, it can be said that the conical surface is then shaped according to the *ideal thrust surface*.

With regard to manufacturing, it is important that the cone—as opposed to the sphere—is geometrically a (developable) **ruled surface** with only **uniaxial curvature**—in the direction of the circles of latitude. The advantages of this type of construction when produced of rod- or plate-shaped material—this also applies to concrete formwork—are discussed elsewhere.

▶ *Sect. 1.2.3 Regular Surface Types*

▶ *Sect. 1.3.1 Composing Uniaxially Curved Surfaces*

2 Heinle and Schlaich (1996) *Kuppeln aller Zeiten – aller Kulturen*, p. 211.

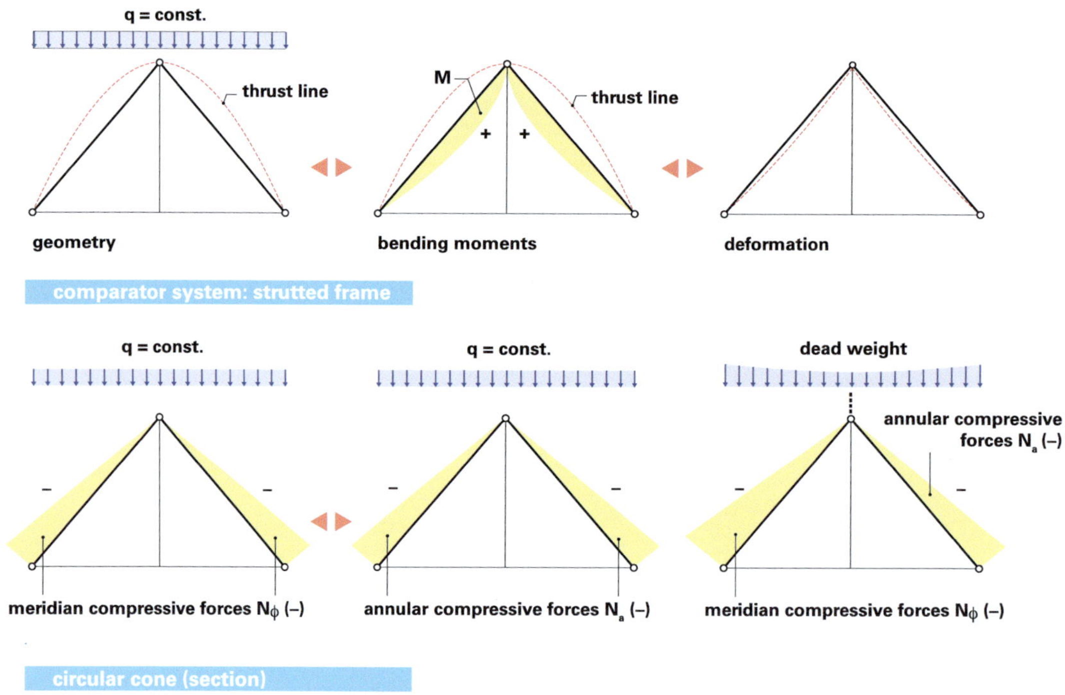

Fig. 4.173 Illustration of the basic load-bearing behaviour of a circular-cone-shaped dome compared to the strutted frame of the same sectional geometry. Exclusively meridian compressive forces N_φ and annular compressive forces N_a occur (see also the load-bearing behaviour of the spherical dome on Fig. 4.179, righ)

4.3.2.4 Full-Leaf Dome

The geometrical issues related to spherical domes are discussed in ▶ *Sect.* 1.3.2.2 *Sphere*

Example of an exception: The Nubian vaults from inclined ring courses, cf. the example on ▶ Fig. 1.204

Domes (Fig. 4.174) can be considered **shell-like structures**. They have a synclastic curvature. Their surface can generally not be generated directly from straight lines—especially in the case of the classical spherical dome. Approaches in the form of polygonal grids are, however, possible and, in constructional terms, result in easily realisable grid shells. Spherical and related curved component geometries can be easily manufactured from smaller building blocks. There are a number of geometrically related alternatives to the pure spherical dome.

Domes have a paramount architectural significance. As predominantly compression-bearing structures, they are, like vaults, a form of load-bearing structure predestined for brittle stone or concrete material. The dome has even a few manufacturing advantages over the vault as it can be built mostly without centring when being successively erected from compression-resistant ring courses, as when bricking or forming layer-wise. Such scaffold-less vaulting is only possible to a limited extent with barrel-shaped vault shells. For centuries, in the Western cultural area, the largest and most demanding structures were covered with domes.[3]

3 This is true for the European, African, and Central Asian cultural spheres, but not for East Asia or pre-Columbian America.

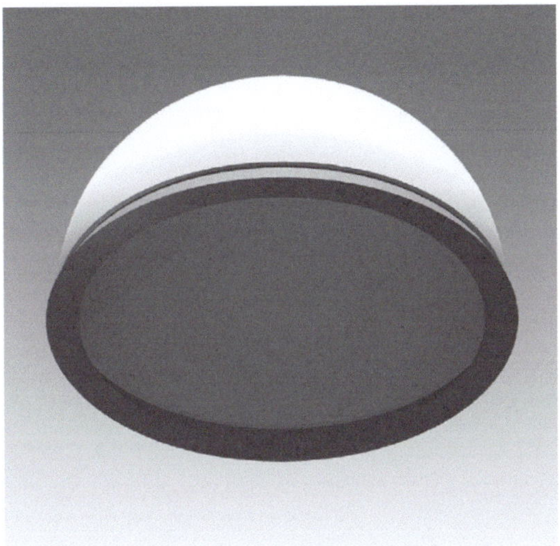

Fig. 4.174 Spherical dome

Load-Bearing Behaviour

The load-bearing behaviour of a dome can be illustrated by a comparison with the arch. A dome can, in a first step, be understood as a radial sequence of individual arches (Fig. 4.180, left), analogous to a vault. If the dome is thus conceived, each arch carries the load individually by pure compressive stress (**meridian compression N**φ), provided that it is shaped on the thrust line of the acting load pattern as viewed in its respective plane. In the case of the usual distributed vertical loads, roughly parabola-shaped guidelines and the expected horizontal forces (thrusts) at the supports appear. Assuming these ideal conditions, as well as a suitable bearing which takes up the horizontal thrust, the dome shell simulating a thrust surface (Fig. 4.175) resulting from the revolution of the thrust line carries the load under pure compressive stress along the meridian lines (Fig. 4.176). Under these cir-

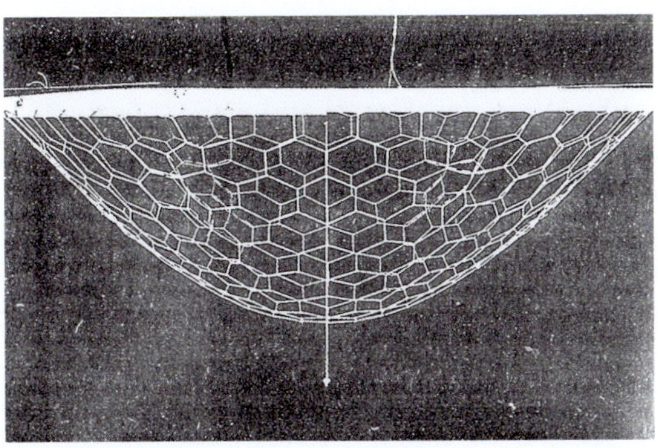

Fig. 4.175 Determination of the **membrane surface** or **thrust surface** of a dome using a hanging model. The approximately catenary-shaped profile is clearly recognisable (F. Otto)

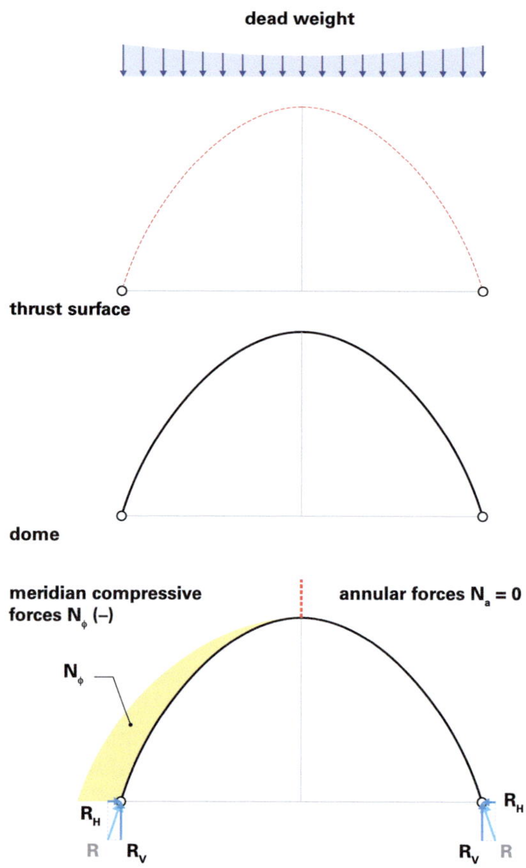

Fig. 4.176 Dome shaped according to the thrust surface of the acting forces. A catenary-like profile emerges under the assumed load. Under these idealised conditions, only **meridian compressive forces** $\mathbf{N}_\varphi$ occur which give a *non-vertical*, though only slightly inclined, reaction force **R** at the support, and consequently a horizontal force component (dome thrust) $\mathbf{R}_H$. There are no annular forces $\mathbf{N}_a$

▶ *Sect. 3.4.4.3 Form-Finding of Surface Structures under Membrane Forces*

cumstances, there is *no* stress, neither tension nor compression, transversely to this direction—i.e. along the circles of latitude.

This ideal condition is, however, rarely achieved in practice for various reasons which will be discussed subsequently. Deviations from these theoretical conditions, which were first recognised by Wren and Gaudí and were constructed in an approximate form (Figs. 4.177 and 4.179) and which are also found in some autochthonous constructions (Fig. 4.178, right), produce transversely oriented forces along the circles of latitude, which in the special—but significant—case of the spherical dome split into (Figs. 4.179 and 4.180, right):

Ring forces are often also referred to—in analogy to the meridian forces $\mathbf{N}_\Phi$—with the notation $\mathbf{N}_\Theta$

- **compressive forces** (annular compression $\mathbf{N}_a$–) in the upper area, above the so-called **fracture joint**;
- **tensile forces** (annular tension $\mathbf{N}_a$+) in the lower area, below the fracture joint.

Fig. 4.177 Dome of St. Paul's Cathedral in London by Christopher Wren. The overall construction is triple-shell; the actual load-bearing dome shell is the middle one and is shaped according to a **thrust surface** (thrust line graphically superimposed). It was the first scientifically substantiated application of the thrust-line theory, which had already been developed by Hooke and Newton, to dome construction

Fig. 4.178 (Left) Continuous shell formed according to a membrane surface (by means of a hanging model) on a square support consisting of four columns. It can thus be added to form a larger interconnected hall (eng.: H. Isler). (Right) African clay dome shaped similar to a real membrane surface (Chad)

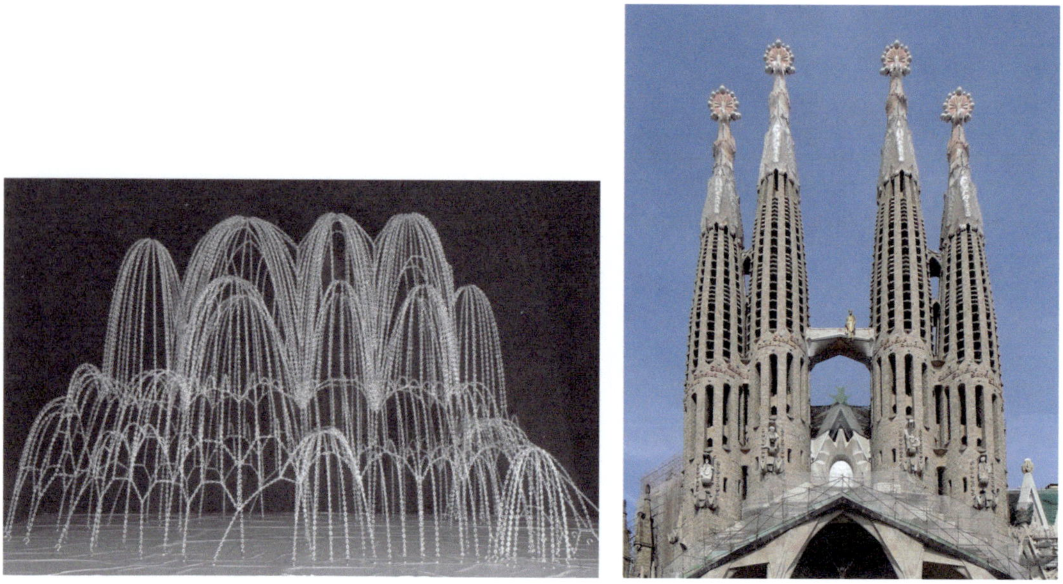

Fig. 4.179 (Left) Determination of the thrust surfaces of the Crypta Güell domes using a hanging model (replica of the original model by A. Gaudí, shown in reverse). (Right) Sagrada Familia in Barcelona (A. Gaudí)

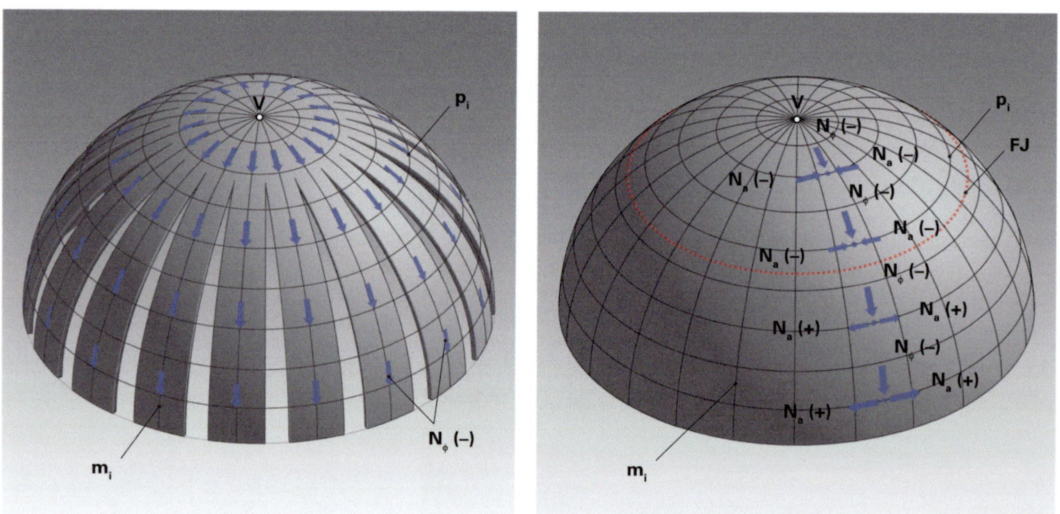

Fig. 4.180 (Left) Dome subdivided into individual (imaginary) arch strips. Compressive forces (N_{φ}), which are indicative of arches, act in the meridianal direction ($\mathbf{m}_i$); $\mathbf{p}_i$ parallels. (Right) Force distribution in the spherical dome. In meridianal direction ($\mathbf{m}_i$), compressive forces $N\varphi$ act; in the parallel direction ($\mathbf{p}_i$) above the fracture joint **FJ**, compressive forces N_a (–) emerge; and below the fracture joint **FJ**, tensile forces N_a (+). The order of magnitude of the force amount is given on Fig. 4.180

This can be clearly seen from the image of a hemisphere compressed from above, which tears in the lower region—as a result of pulling—and bulges at the top as a result of compression. The concept of the fracture joint is naturally derived from masonry technique. The existence of a fracture joint in the spherical dome can be derived as follows:

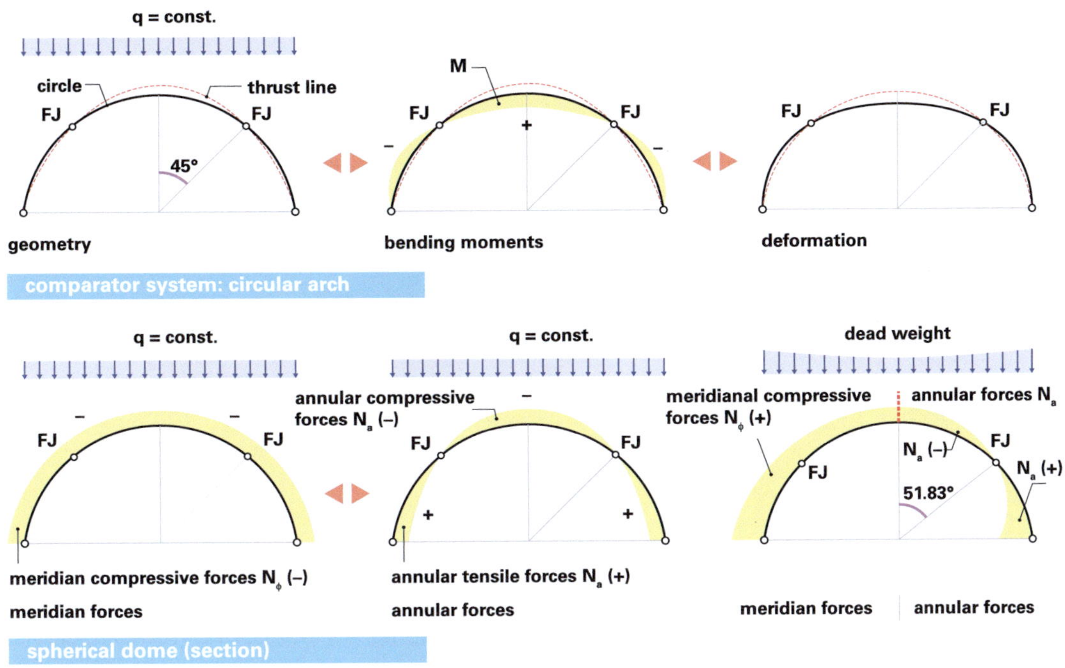

Fig. 4.181 Illustration of the basic load-bearing behaviour of a (three-dimensional) spherical dome compared to a (two-dimensional) arch of the same radius. The fracture joint **FJ** is located at the point of inflection between the bulging and the depressed arch section (deformation pattern top right). Assuming a uniformly distributed load **q**, the fracture joint **FJ** emerges at an angle of 45°. With respect to the own load (bottom right), it is 51.83° and the meridian forces $\mathbf{N}_\varphi$ and annular forces $\mathbf{N}_a$ vary correspondingly (according to Heinle, Schlaich (1996) p. 210.)

Circular arches, which can be regarded as the basic element of a spherical dome imaginarily cut open radially, differ from the ideal thrust line of a customary load as shown on Fig. 4.180. Under severe load, the arch is deformed in such a way that the deviation from the thrust line becomes even greater. The lower portion of the arch tends to yield outwards, so that, in the direction of circles of latitude, **tension**, a so-called **annular tension**, is produced at this height. The upper portion of the arch, on the other hand, tends to deflect downwards from the thrust line. In the annular direction, the individual arch strips, consequently, press against each other, resulting in an **annular compression**.

Cf. also ▶ *Sect. 3 Fundamentals*, ▶ Fig. 3.112

In the spherical dome, the following force distribution can be observed:

- **Above** the fracture joint, **compression** acts on any section of the dome shell in both the meridian as well as in the annular or ring direction;
- **Below the** fracture joint, on any part of the dome shell, the following forces act:
  - in meridian direction: **compression** (meridian compression) ($\mathbf{N}_\varphi$–);
  - in annular direction: **tension** (annular tension) ($\mathbf{N}_a$+).

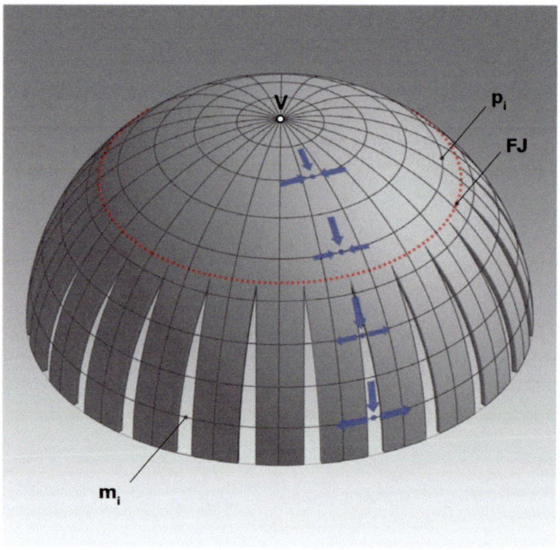

■ **Fig. 4.182** For many spherical domes made of brittle material such as masonry, a typical crack pattern (shown here magnified) emerges which results from the lack of tensile strength below the fracture joint (**FJ**). The mode of action in this region then equals the one of a radial sequence of individual arches as shown on ■ Fig. 4.180, left, in an abstracted manner

These conditions can, of course, be reversed with a modified guide geometry of the dome. Under the idealised condition of a constant, uniformly distributed load, biaxially acting **membrane forces**, which are tangentially oriented with respect to the dome surface, emerge.

It is interesting to note that many historic dome structures actually do have radial cracks below the fracture joint because the stone material could not absorb the accruing tensile forces (■ Fig. 4.182).[4] The domes then *effectively* act in the lower region as a radial arrangement of individual arched strips separated from each other. Their remaining stable was due to the fact that the thrust line remained within the arches' cross-section.

The problem of the annular tensile forces was certainly sometimes recognised, but domes that were actually reinforced against ring tension could only be realised in the eighteenth century. Domes made of compression- and tension-resistant material, which are technically feasible today, can not only absorb the resulting tensile forces, but, in contrast to the arch, also can transfer any continuous load via pure membrane forces, i.e. forces acting axially within the cross-section.

Ribbed Dome

A differentiation of the continuous dome shell into localised stiffer ribs and thinner shell webs spanning between them (■ Fig. 4.183) can already be observed on historical examples of brick domes. The most important aim of this structural de-

4 For example, the dome of the Pantheon in Rome.

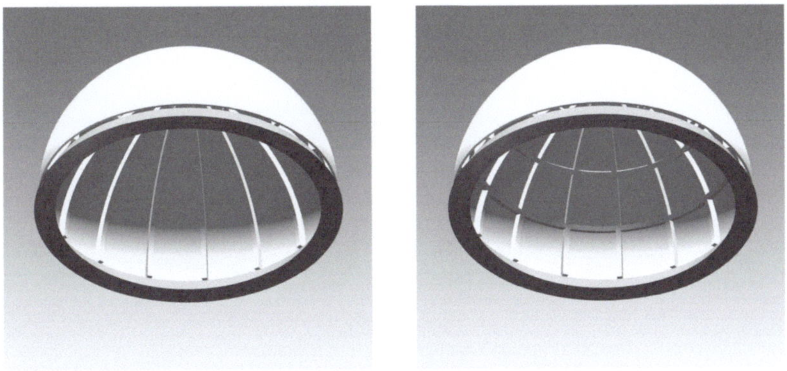

Fig. 4.183 Ribbed dome

Fig. 4.184 Dome of the Hagia Sophia featuring meridianal ribs. The meridianal forces are deflected and concentrated on the ribs just above the rib support over the dome caps and window arches, creating a window ring which gives the dome an almost hovering character

sign is to reduce the weight of the dome while maintaining sufficient stiffness, which is ensured primarily by the stiffer ribs. In this respect, this structural design is not fundamentally different from other surface elements made of bar arrays. There are examples of domes with ribs only in the meridianal direction (Fig. 4.184) and those with a grid of ribs in meridional direction and rings in the parallels' direction (Figs. 4.185 and 4.186, right). Reducing weight was a factor, in par-

Cf. variants **2**, **3** and, with qualifications, also **4** in ▶ *Sect. 3.3 The Constructional Composition of the Space-Enclosing Surface Element*

E.g. variant 2, examples in each case on Figs. 4.184 and 4.186, right

Fig. 4.185 Dome of the Pantheon in Rome with a coffer-like subdivision. In the narrower sense, these are not ribs, since the concrete dome acts like a continuous shell. The cavities provide a certain weight reduction

ticular, in the case of domes with large spans, a factor that decisively influenced the size of the thrusts occurring in the impost area and ultimately decided on the constructional viability of the dome.

The different stresses occurring within these two rib groups are significant for domes: **meridian ribs** are subjected to compressive stress, **ring ribs** above the fracture joint likewise, but below the fracture joint the latter experience tension, and this is the truly critical point with domes from brittle building material. Ring ribs could therefore not be used for the transmission of annular tensile forces before the availability of truly useful tensile members.[5] As compression members,

5 Cf. Heinle and Schlaich (1996): Some attempts to incorporate tensile elements within the dome base were relatively early (e.g. Brunelleschi in Florence), but they were hardly effective. The first truly effectual tensile members, according to Heinle/Schlaich, were installed during the renovation of the dome of St. Peter in Rome in 1748 by Poleni.

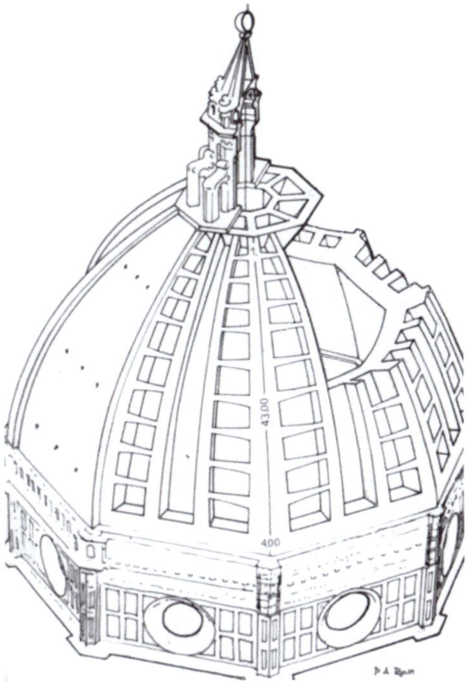

Fig. 4.186 (left) Dome of St. Peter's Basilica with ribbing in meridian direction. In this building, effective tensile members were installed in the tensile annular zone of the dome for the first time. (Right) Dome of the Florentine cathedral with ribs in meridian and parallel directions (see also Fig. 4.170). The polygonal domical vault on an octagon unfolds a dome-like support behaviour (for the polygonal groined vault, see also ▶ *Sect. 4.3.2.1 Pyramids* and ▶ *Sect. 4.3.2.2 Cylindrical Domes*)

however, they can help to laterally stiffen extremely slim and heavily stressed meridian ribs against buckling. However, the ring tension, which is dangerous for bricked domes, does not *necessarily* occur: The relation between the meridian and the thrust line is the decisive factor. The dome of the Florentine cathedral, for example, has only negligible ring tension thanks to the steep domed form and the weighting at the apex produced by the lantern (Fig. 4.186, right), two factors that contributed to the assimilation of the profile of the dome to the thrust line of the acting loads. This does not apply to St. Peter's Basilica, which for this reason suffered damage early on and had therefore to be renovated (Fig. 4.186, left).

Subdividing the dome shell into ribs and thinner shell webs results in forces concentrating in the stiffer elements, the ribs, and, at least partially, annuls the areal shell effect of a continuous surface element. Advantages can be obtained with domes made of brittle material, as mentioned above, a material which anyway cannot absorb the biaxial combined compressive and tensile force typical of shells for lack of tensile strength. The situation is different in the case of reinforced-concrete domes, which are predestined shells not only because of their ability to absorb compression *and* tension simultaneously, but also because of their continuous seamless

■ **Fig. 4.187** (Left) In this modern example of the Palazzetto dello Sport by Nervi, a shell effect is also present. The ribs are primarily derived from the erection process (permanent-formwork shell elements from ferro-cement). (Right) Framed dome from reinforced concrete: Century Hall in Breslau (arch.: Max Berg)

material structure (■ Fig. 4.187, right). Nevertheless, and despite the increased formwork costs for ribbed domes made of reinforced concrete, numerous examples can be found of this solution. Particularly noteworthy are the domes of Nervi, which were cast on ferro-concrete moulds (■ Fig. 4.187, left).

4.3.2.5 Dome from Bars

This type of dome is considered to be a **lattice shell** with a main load-bearing structure composed of bars manufactured from a compressive and tensile-resistant material such as wood or steel. They are covered by a thin surface element—for example, wood cladding, glass, or a membrane—which is not usually involved in the main load-bearing function.

Similar to the ribbed dome, the radial distribution basically originates a division of tasks between the rods in such a way that the meridian ribs transmit compressive forces while the annular ribs respectively bear compressive forces above the fracture joint or tensile forces below them. With this type of structure, quadrangular individual fields which are not inherently stiff result between bars (■ Fig. 4.188, left). Shear stress on the grid shell, e.g. from wind or discontinuous load, can under these circumstances only be absorbed by the bending stiffness of the rods and the rod nodes. The outcome is frames and, accordingly, a **framed dome** overall.[6]

A diagonal bracing of the fields (**Schwedler dome**, ■ Fig. 4.188, right) has a considerably more efficient effect on the shear bracing of the lattice structure, especially when it consists of pure tensile members, which visually are almost imperceptible. This is a substantial issue particularly in the case of glass-covered domes—an important domain of latticed domes.

Cf. bracing variants of grids in ▶ *Sect.* 3.3.5.2 *Diagonal Bracing*

Hitherto, a radial division of the dome surface has been assumed, as suggested by the distinction between meridians

6 Heinle and Schlaich (1996) p. 154.

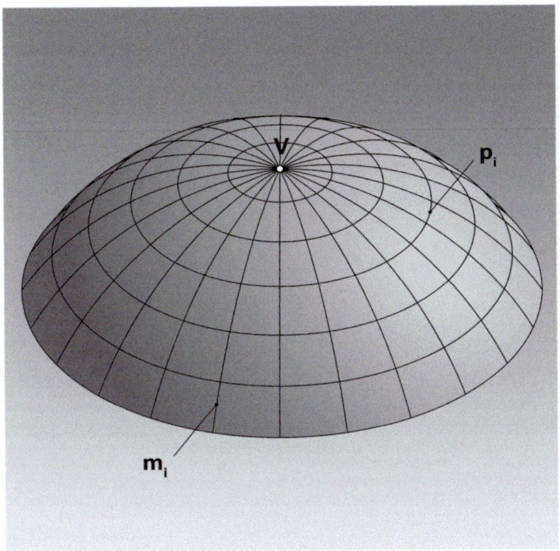

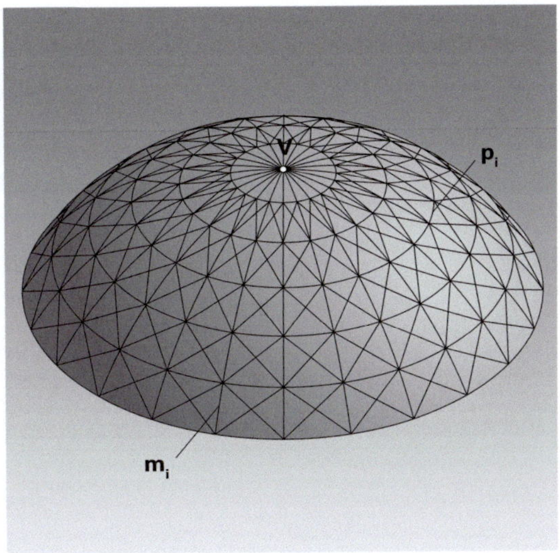

Fig. 4.188 (Left) Radial subdivision of a spherical surface by means of meridianal circles ($\mathbf{m}_i$) and parallel circles ($\mathbf{p}_i$). Quadrangular meshes are created, which can be made strong enough to form a frame (framed dome) capable of absorbing shear forces within the shell surface. (Right) Stiffening of the dome of square meshes (as on the left) by diagonal braces (Schwedler dome)

and parallel circles. The geometrical difficulties of such a surface division into roughly trapezoidal individual facets have already been addressed elsewhere. ► *Sect. 1.3.2.2 Sphere* With regard to the flow of forces, it should be noted that the radial bar organisation results in a large bar density at the apex where the loads to be removed are, however, the smallest and where it is difficult to make these rods converge at one point. In the impost area, where the loads, conversely, are the largest and a larger number of bars would be helpful at that point, the bar density, however, is lowest. Alternative divisions of the dome surface that might elude or mitigate this effect are examined elsewhere. ► *Sect. 1.3.2.2 Sphere* An interesting option is to divide the dome surface by intersecting with mutually parallel planes, which results in a division into approximately quadrangular fields (Fig. 4.189, left) or triangular fields (Fig. 4.189, right). Thereby, rod concentrations are avoided. Triangular fields are even shear-resistant without further measures and support the shell action of the lattice.

Nevertheless, the lack of shear stiffness of quadrangular grid meshes can, under certain circumstances, be an advantage in dome construction. Square lattices with hinged nodes allow the dome surface to be brought into any shape by lozenge formation (Fig. 4.191). For doing so, it is sufficient to flatten the grid on the ground (Fig. 4.190, left), so to say developing it onto the plane, in order to assemble it, and then to lift it into the desired shape (Fig. 4.190, right). The prerequisite for this is that the rods are so slender and elastic that they can generate the curvature of the shell surface through elastic deformation, i.e. bending. This solves from the outset

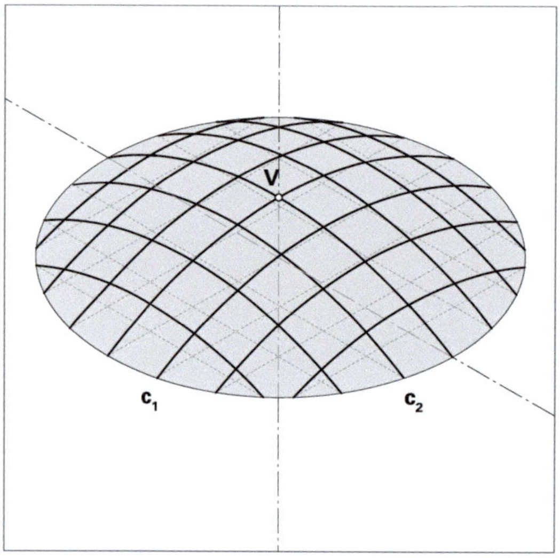

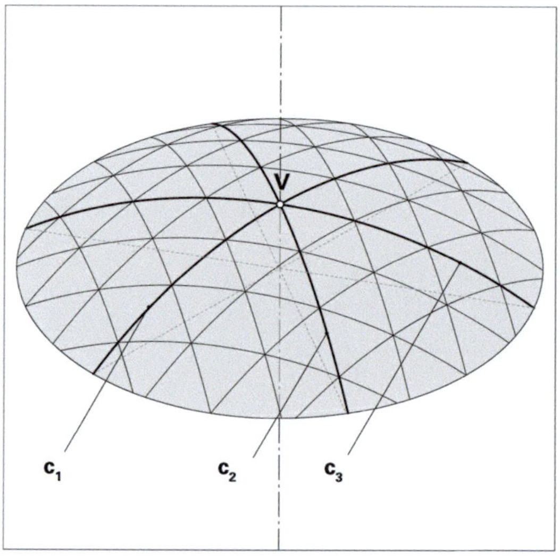

Fig. 4.189 (Left) Subdivision of a spherical surface by means of two sets of planes, respectively parallel to each other. Again, non-rigid quadrangular meshes are formed, but the individual rods are distributed uniformly over the dome surface, unlike in the case of radial subdivisions, and have approximately the same length. For geometrical reasons, this division is particularly suitable for flat domes. (Right) Subdivision of a shallow spherical dome like on the left, but with three sets of planes, so that rigid triangular meshes emerge (lamellar dome)

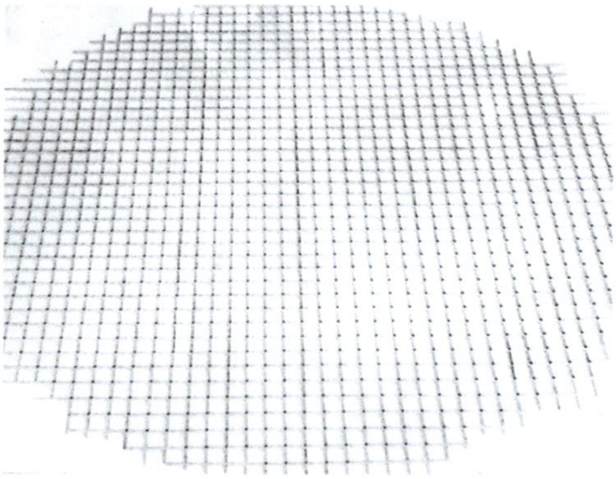

Fig. 4.190 Grid shell from a square mesh grid as shown on the left extended flat. Due to the rotational nodes, it can be brought into any desired shape during erection by jacking it up (right)

the problem of the double curvature that is always difficult to generate with other shell designs. The slender bars are placed under bending prestress in advance by the bending, thus already using up part of their bending stiffness, but this capability is only of secondary importance for the load transfer in the shell anyway. This principle of curving the surface also has the advantage that there are no kinks at the nodes, as would be the case if the bars were straight and the shell surface faceted. The nodes are therefore always tangential to the shell surface. The slender bars are not butted at the junctions, but are guided past each other in different layers (Fig. 4.191). This eliminates the need to produce countless joints at the nodes, something which significantly simplifies the construction of the shell. Thanks to the free rotation of the articu-

Fig. 4.191 Articulated node formation of a lattice shell made of wooden rods crossing each other with the help of stud bolts. Bar penetrations are consistently avoided by this design principle. During assembly, the bolt connections are freely rotatable. Afterwards, they are tightened and the shell is diagonalised. Depending on the stresses, two (top) or four rods (left) are laid (Multi-Hall Mannheim)

Fig. 4.192 Multi-Hall Mannheim (arch.: K. Mutschler and F. Otto)

lated nodes, the meshes automatically adopt a locally different rhombic shape (Fig. 4.191) which is necessary for the three-dimensional shell geometry to be generated. Then the nodes are fixed, the grid is braced with diagonal ropes against shear and the shell is subsequently capable of load-bearing (Fig. 4.192). It is possible to realise genuine membrane sur-

Fig. 4.193 (Left) Grid shell with triangular mesh. (Right) Vertex node of a framed dome in timber construction

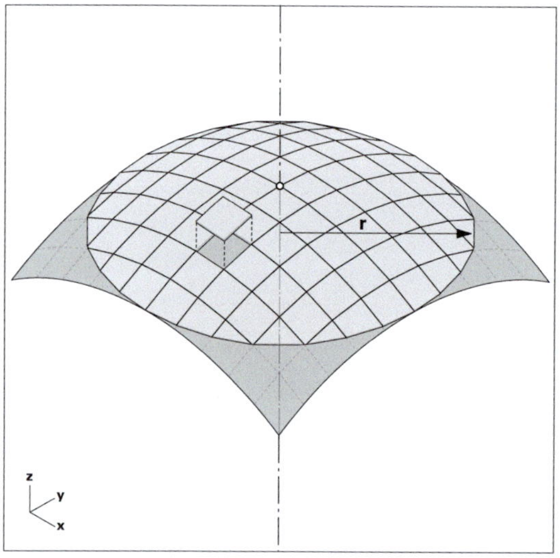

Fig. 4.194 (Left) A substitutional form of the spherical dome is the **translational surface** generated by a circular guideline and a circular directrix. The square meshes are flat when the circles are substituted by polygons. In the case of shallow domes, the difference in shape to a pure spherical surface is negligible. (Right) Example of a lattice dome with diagonally cable-braced rectangular meshes. Their geometry corresponds to a translation surface (AQUAtoll, Neckarsulm, eng.: Schlaich Bergermann & P)

faces, which are difficult to define mathematically in advance, by transferring the geometry of a hanging model to the grid shell. According to this construction principle, the most delicate grid shells have been erected (Fig. 4.195, right).

Dome surfaces of approximately equal quadrilateral or triangular meshes, so-called **lamellar domes** (Fig. 4.193), are also associated with geometric difficulties. In particular, it proves difficult to execute them from planar plate material and to equalise the bar lengths and node angles. This type of geometric division is only suitable for flat spherical caps.

▶ *Sect. 1.3.2.2 Sphere*, in particular ▶ Figs. 1.223, right, and ▶ 1.224

Fig. 4.195 Grid shell with square meshes and rope bracing (roofing Bosch area Stuttgart, eng: Schlaich Bergermann & P.)

Alternatively, a **translational surface** is also suitable for such flat dome shells, which largely solves the problems mentioned: It can be produced from planar plates, allows a considerable reduction in the variance of rod lengths, and is visually scarcely distinguishable from a pure spherical surface (Figs. 4.194 and 4.195, left), provided the dome is rather flat (Fig. 4.195).

► *Sect. 1.3.2.2 Sphere*, in particular ► Figs. 1.225 and 1.226

Last but not least, the **geodesic division**[7] of the dome should be mentioned in this context, which makes it possible to work with a minimum of different rods and nodal angles. It was developed primarily to deal with the manufacturing difficulties of countless different parts. This problem has, however, been alleviated in the last few years by digitised manufacturing methods (CNC). Geodesic domes are generated by projecting an icosahedron (regular solid made from twenty equal polygons) on the spherical surface from the centre of the sphere. The 20 triangular caps thus formed are divided

***Vol. 1**, Sect. 3.4.2 Application of New Digital Technologies of Planning and Manufacturing to the Building Sector*

7 Although the geometric definition of the geodesic dome (1954) is often attributed to Buckminster Fuller, it was already developed in 1922 by Walter Bauersfeld for the planetarium in Jena.

4

◘ **Fig. 4.196** (Left) Olympic Palace in Mexico City, Distrito Federal (arch.: E. Castañeda, A. Peiri, eng.: F. Candela). (Right) Geodesic division of the sphere into a bar grid (USA Pavilion at the Expo in Montreal, B. Fuller)

into hexagons, and, at their respective corners, pentagons result (◘ Fig. 4.196, right).[8]

4.3.2.6 Full-Leaf Shell, Synclastically Curved, Punctually Supported

► *Sect. 3.4.5.1 Shells*

► *Sect. 4.2.2.2.1 Load-Bearing Behaviour*

On four points on the diagram on ◘ Fig. 4.197

► *Sect. 1.2.3.3.2 Ruled Surfaces* and ► *Sect.* 1.2.3.3.3 *Translational Surfaces*

► *Sect. 4.3.2.4 Full-Leaf Dome* as well as ► *Sect. 4.3.2.5 Dome from Bars*

Fundamental issues of the static mode of action of **shells** are discussed elsewhere. Shells allow a variety of shapes and cannot be included in a single category within the chosen classification. Shell structures can adopt the form of barrel vaults (cylindrical shells), but they are then able to rest on point supports, in contrast to conventional barrel vaults. They can also be shaped like spherical domes, either linearly supported all around—every dome with annular tensile strength acts as a shell—or even with cut-outs in such a way that the loads, in whole or in part, are delivered to individual points. Furthermore, shells can also assume anticlastic curvatures (e.g. as hyperbolic paraboloids).

Because of their constructional significance as compressive-stressed concrete shells, this section will treat **synclastically curved shell structures** in their *point-supported* variant (◘ Fig. 4.197). Above, the *linearly supported* variant has already been investigated under the category of domes.

Load-Bearing Behaviour

From an historic and developmental perspective, these types of shell structures are relatively old. They resulted from the necessity to reconcile the structural performance of rotation-

8 Further literature on geodesic domes: Wester Ture (1985) *Structural Order in Space—The Plate-Lattice Dualism*, Kopenhagen; Heinle and Schlaich (1996).

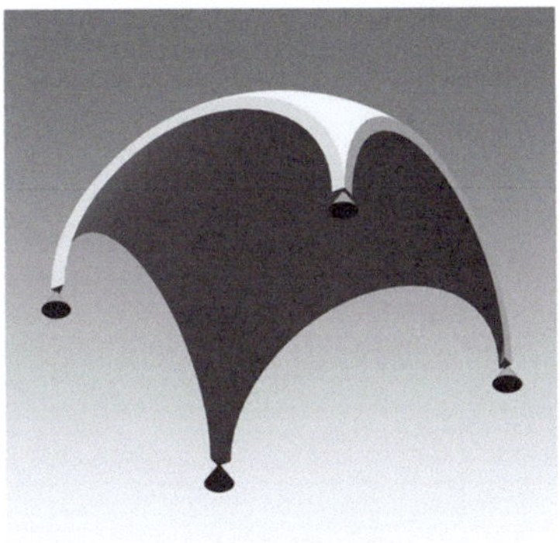

Fig. 4.197 Full-Leaf dome on punctual supports

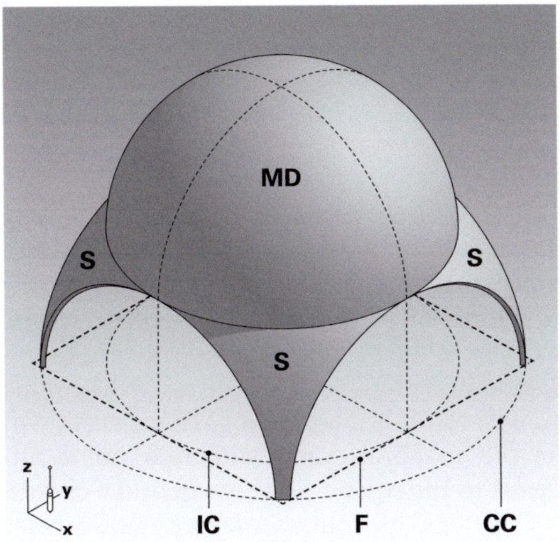

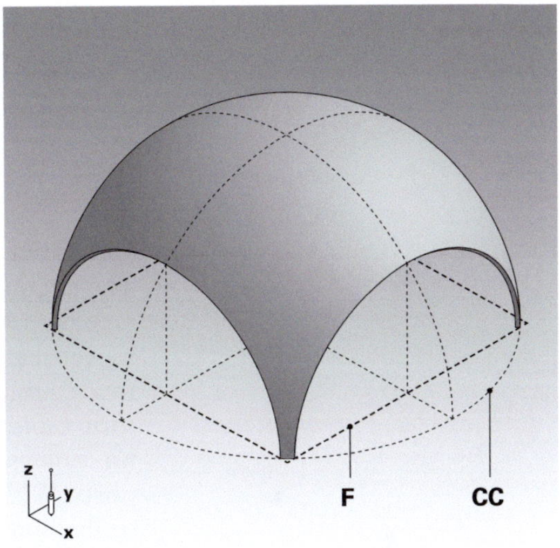

Fig. 4.198 (Left) **Spandrel dome** (**CC**: circumscribed circle, **IC** inscribed circle, **F** footprint, **S** spandrel, **MD** main dome = inscribed-circle dome). (Right) **Circumscribed-circle dome** (**CC**: circumscribed circle, **F** footprint)

ally symmetrical, compression-loaded shells with the superior addability and functional suitability of orthogonal building geometries. With this objective, two variants of domes were developed during the historical evolution of this type of load-bearing structure, which in the broader sense belong to the category of this section:[9]

► *Sect. 3.1.6.1 The Vertical, Planar Space-Enclosing Element*

- The dome resting on **spandrel** or **pendentive** shells (Fig. 4.198, left). It results from the continuous linear bearing of a dome, an *inscribed-circle* dome in relation to the underlying square defined by the four corner supports, on four spherical caps—the *spandrels*—which direct the loads to the four support points. The four spandrels are cut-outs

9 A good description of the load-bearing behaviour of spandrel and circumscribed-circle domes can be found in Heinle and Schlaich (1996) p. 219f.

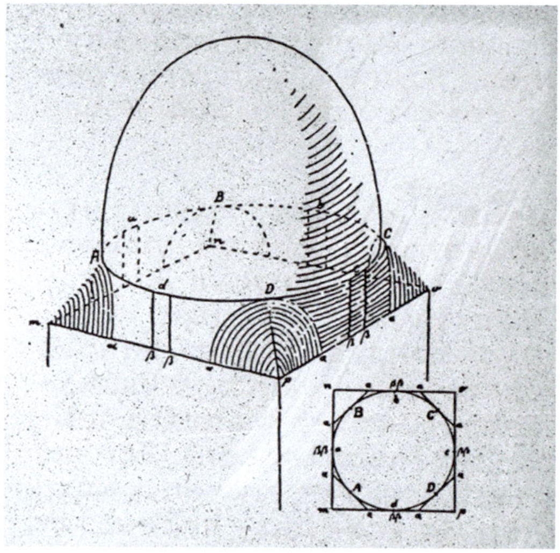

Fig. 4.199 Transition between dome and square substructure by means of squinches

from a *circumscribed-circle* dome as described below. A variation of the spandrels are the *squinches* (Fig. 4.199).

- The **circumscribed-circle** dome (Fig. 4.198, right): Its diameter is the diagonal of the inscribed square. It is cut off by vertical planes over the sides of the square, resulting in four circular lateral openings—called **buttress arches** in historic buildings. The result is a continuous dome shell, with no valleys like the spandrel dome, with four point supports. With regard to the square to be covered, the circumscribed-circle dome has, of course, a larger radius, hence being much flatter than the spandrel dome.

▶ Sect. 4.3.2.4.1 Load-Bearing Behaviour

The basic load-bearing behaviour of these cut-out, or composed, spherical domes derives from that of the pure spherical dome. A dome on spandrels behaves like a circumferentially supported dome, as long as the spandrel ring is evenly stiff: As in the case of the simple spherical dome, forces are delivered to the support continuously and uniformly over the cross-section. In historic masonry constructions, conditions were, however, more complicated because of the differences in stiffness between the more rigid supports of the buttress arches (comparable to arch chords) and the more compliant spandrel shells, load relocations had to take place within the dome shell. In essence, the load-bearing behaviour of such composite shells was decisively determined by the respective degrees of stiffness of the elements involved in the dome, of the spandrels, of the valley between both, and of the buttress arches.[10] In this context, the spandrel shells could have a supporting

10 Heinle and Schlaich (1996), p. 219f.

effect against the dome's thrusts—in the case of cracked masonry domes or in the case of flat dome caps—and so did any lateral apsidal domes (Figs. 4.200, 4.201, and 4.202).

Cf. Hagia Sophia on Fig. 4.201

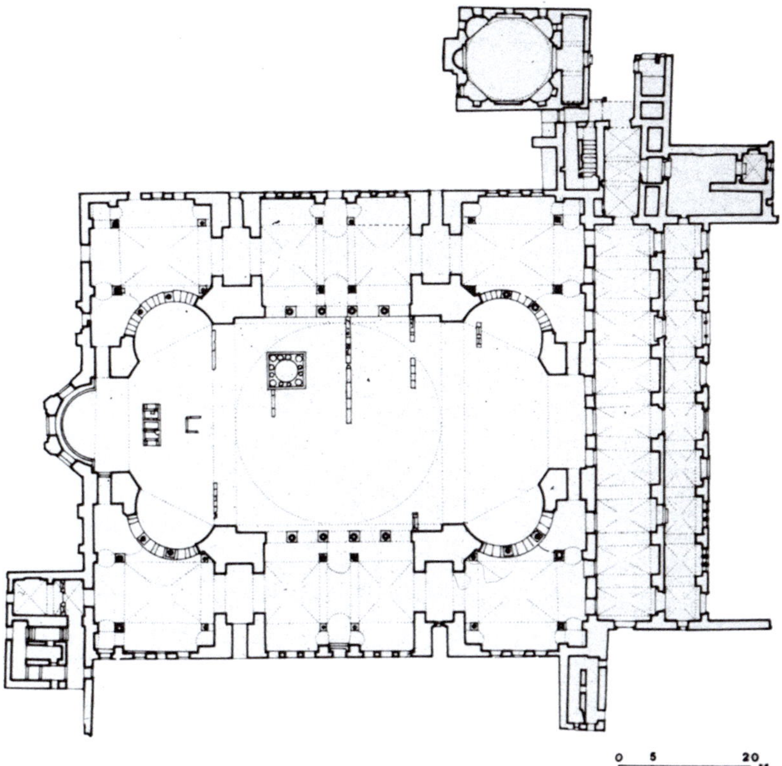

Fig. 4.200 Hagia Sophia, ground plan

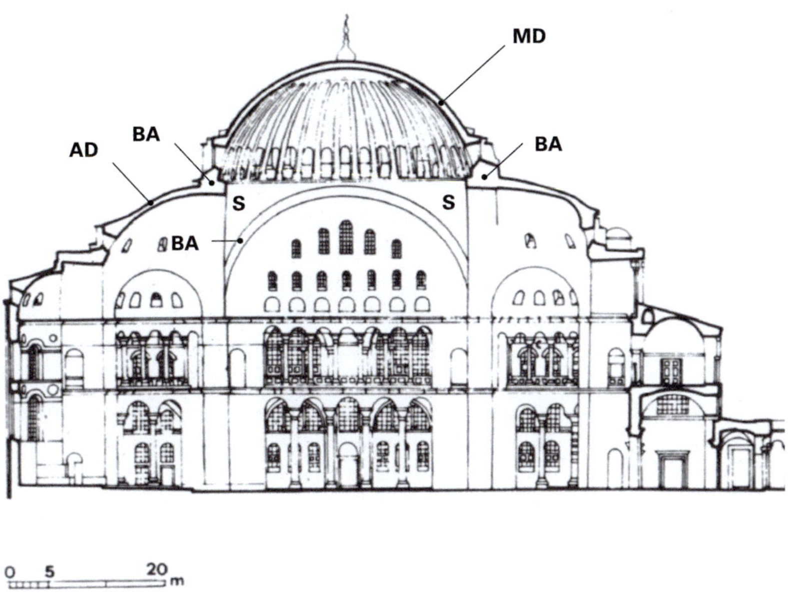

Fig. 4.201 Hagia Sophia, longitudinal section (**MD** main dome, **AD** apsidal dome, **S** spandrel, **BA** buttress arch)

Obviously, according to a source[11], the first dome of Hagia Sophia, which collapsed, was a circumscribed-circle dome, of which the spandrels are still preserved. The main dome still existing today is therefore a steeper inscribed-circle dome resting on these spandrels

Fig. 4.202 Hagia Sophia, interior space

In the case of circumscribed-circle domes, the inclined meridian forces arriving at the edge of the circular side cut-outs must be absorbed both in their perpendicular component via the arch compression and in their horizontal direction. The circular sections, in historic arched structures, were built to reinforce the shell as buttress arches. However, an arch can only limitedly accommodate the horizontal load component oriented transverse to its plane, so that, for this purpose, either adjoining barrel-vault shells or supporting apsidal shells were required. Overall, the flatter pitch of the dome had a rather detrimental structural effect.

This is different from the behaviour of a shell formed by a genuine **membrane surface** which results from the inversion of a hanging form. These are, under the given load conditions of vertical loads, no longer spherical surfaces, but freely shaped surface structures the geometry of which is mathematically no longer easy to describe and must be determined via a physical or digital model and, if necessary, measured. Uniform force distributions result in the shell, so that the force flows without undesirable concentrations as well as without tension and, in particular, without bending, from the apex as directly as possible to the four supports. As a result of the forces which then always attack obliquely at the point bearings, a **horizontal**

Cf. also the statements on the linearly supported dome in ▶ *Sect. 4.3.2.4.1 Load-Bearing Behaviour*

11 Heinle and Schlaich (1996), p. 30ff.

force component results, i.e., a **thrust**. In the case of membrane shells under vertically acting external loads, exclusively vertically oriented support reactions such as on the hemispherical shell (with ring tensile strength) are not possible.

Extendibility

It can be argued that the two historical dome variants of the spandrel dome and circumscribed-circle dome were specifically developed with the aim of simplified extendibility. Both variants can be expanded indefinitely in both directions (◻ Figs. 4.203, right and 4.204). In the modern variant of the membrane dome, added shapes can also be produced from a basic module. Directional groupings of modules on a rectangular ground plan can also be realised (◻ Fig. 4.205).

Usage in Buildings—Design Aspects

Today, modern, full-walled membrane shells are almost exclusively made of reinforced concrete (◻ Figs. 4.205 and 4.206). Very high slenderness can be achieved in the range of up to a shell thickness equal to **l/500** (**l** = span). Such shells also show the extremely efficient load-bearing behaviour of axially—in this case compression—loaded membrane structures. With synclastic curvature, they can be carried out being subjected

For comparison: An eggshell has a slenderness of about 1/100

◻ **Fig. 4.203** (Left) Spandrel of the Mihrimah-Mosque in Istanbul (Sinan, 1555). (Right) Circumscribed-circle dome (Bibliothèque St. Geneviève, Paris)

Fig. 4.204 The addition of circumscribed-circle domes—here in an approximation as a groined polygonal vault—generates a large hall (Yerebatan cistern in Istanbul, 532 CE)

Fig. 4.205 (Left) Market hall in Algeciras (eng: E. Torroja). (Right) Synclastically curved concrete shells with three-point support (eng.: H. Isler)

to almost pure compressive stress. Although reinforced concrete is also able to absorb tension, this compression-bearing effect nevertheless has the advantage that the concrete is under uniform compression and therefore does not crack. In this way, waterproof shells can be produced which do not require any further sealing.

On the other hand, the insufficient thermal protection of a naked concrete shell is a hindrance to the use of shells in building construction. At least one side of the shell has therefore to be covered with an insulation layer for this purpose, which impairs the formal lightness and elegance of these highly efficient structures.

In analogy to the point-supported, tensioned membrane structures, also in the case of compression-loaded shells, the lateral space enclosure within the openings is technically feasible, but difficult to solve from an aesthetic viewpoint. Shells which

Fig. 4.206 Modular addition of point-supported shells along an axis to form a continuous hall (eng. H. Isler)

are enclosed by planar side walls often appear dull and clumsy. Recessed walls with wide, projecting edges of the shell instead tend to preserve the free-floating character of these structures.

The advantage of easy manufacture from a viewpoint of survey and shuttering, as in the case of geometrically easily definable spherical shapes, is lost when using suspension forms, i.e. non-regular, empirically found membrane surfaces. Nevertheless, this task has been considerably simplified in the last few years by the possibilities of computer-assisted manufacturing. There are also sophisticated methods for the pneumatic production of formwork surfaces which significantly reduce the formwork costs.

***Vol. 1**, Sect. 3.4.2 Application of New Digital Technologies of Planning and Manufacturing to the Building Sector*

4.3.2.7 Shell from Bars, Synclastically Curved, Punctually Supported

In addition to the design in concrete, point-supported shells can also be designed as **grid** or **lattice shells** (Fig. 4.207).

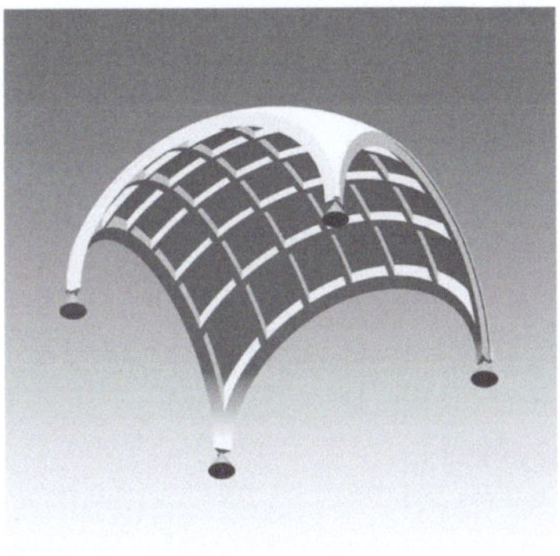

Fig. 4.207 Shell from bars, synclastically curved, punctually supported

Rods made of tension and compression-resistant materials such as steel or wood are used. Morphologically, these surface structures belong to the structural variant **3** as introduced in ► *Chap.* 3:

► *Sect. 3.3 The Constructional Composition of the Space-Enclosing Surface Element*

For such point-supported shells, the procedure described above for linearly supported lattice shells is also applicable: Two equal-ranking, intersecting skeins of bars are laid in a two-way spanning grid. The meshes are subsequently covered with a thin sheet or skin, usually glass or membranes. In this context, the lack of shear stiffness of these square grids, which are initially not triangulated, is a great advantage during erection since such a grid with rotatable nodes can be brought into any conceivable continuously curved shape by being jacked, and hence also into an ideal **thrust-surface shape** which has been computationally determined beforehand. The lattice meshes abandon their square shape and transition into lozenge shapes with variable angles, each adapted to the ensuing surface curvature. After creating this membrane form, the nodes are fastened and the meshes are braced diagonally with ropes. The footings are secured against horizontal thrust. Subsequently, the desired shell effect is established.

► *Sect. 4.3.2.5 Dome from Bars*

4.3.3 Tensile-Stressed Systems

After having studied tensile-stressed systems spanning one-way in the form of belts, two-way spanning, i.e. non-directional or biaxial, variants will now be investigated. **Biaxial tension** appearing in the construction is therefore characteristic of all structures to be discussed hereafter.

As a structural variant for the surface-forming component in the sense of the present definition, variant **3** appears in this category of load-bearing structures, i.e., an element from equivalent, mesh-like mutually crossing—in this case non-intersecting—linear elements: In this variant, these are not bending-stiff rods but, instead, flexible fibres or ropes as occur in **woven membranes** and **rope nets**. In order to create a dense surface in fabrics, or a closed surface in nets, the textile braid is coated and the meshes of the netting are covered with a suitable film for this purpose, which then only provides weather protection and does not have any structural effect.

► *Sect. 3.3 The Constructional Composition of the Space-Enclosing Surface Element*

The structural variant **1** also appears sometimes, that is to say, a completely homogeneous flat element under biaxial tension, such as, for example, **films**.

4.3.3.1 Membrane and Rope Structure, Mechanically Tensioned, Punctually Supported

According to the chosen classification shown on ◘ Figs. 4.4, 4.5, 4.6, and 4.7, linearly supported membrane structures are discussed separately in ► *Sect. 4.3.3.2*

In this category, tensile-stressed structures with predominantly punctual support will be discussed (◘ Fig. 4.208). The most important variant of this group are the **mechanically pre-**

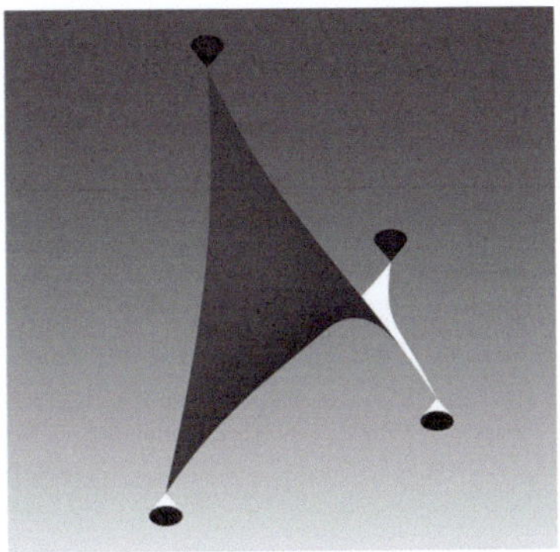

■ **Fig. 4.208** Membrane and rope structure, mechanically tensioned, punctually supported

stressed and stiffened **membranes** and **rope structures** (■ Fig. 4.209). Their basic load-bearing behaviour is discussed in detail in ***Volume 1***. Some further aspects are important in this context and will be addressed in the following:

Vol. 1, *Sect. 26.9.9 Mechanically Pre-Stressed Membrane* as well as ► *Sects. 4.3.3.1* and *4.3.3.2*

- In order to produce the sufficient anticlastic curvature essential to the load-bearing capacity of these types of structures, it is necessary to create a series of low and elevated supports or bearings. The height differences of these bearings and their proper arrangement in space tension the membrane into the required **saddle form** (■ Fig. 4.210, left). It goes without saying that these points must not lie within a single plane, not even approximately.

■ **Fig. 4.209** Solar protection in the form of mechanically prestressed membranes (Expo 92, Seville)

Fig. 4.210 (Left) Rope net with two mutually curved cable skeins (load-bearing and prestressing direction). The more prominent square fields at the top are created by the seams of the plate covering (canopy of the Olympic Stadium Munich). (Right) Suspension of the membrane (here, rope net) from externally arranged, guyed masts, which are mounted like hinged columns (Olympic roofing Munich)

High points must be created by suitable building elements, such as **masts** (Fig. 4.210, right). Membranes can also be supported by linear elements: for example, compressive-stressed ones as **arches** or tensile-stressed ones as **support ropes**. Depending on the attack of the membrane tensile forces at a mast point, supporting masts can be stabilised either by the membrane itself (Fig. 4.228, left), or additionally by separate **stay rigging** (Figs. 4.210, right and 4.211, left). The latter is necessary, in particular, with masts arranged outside the membrane. In doing so, the angle of inclination of the mast and stay cables must be observed in order to keep the forces as small as possible. It is often necessary to create an appropriate clear space in the vicinity of the building for the necessary anchorages, which must be taken into account during planning. In this type of stabilisation, the masts are only subjected to axial compression, thus acting as **hinged rods**.

Cf. further below ▶ *Sect. 4.3.3.1.2 Extendibility*

In principle, masts can also be executed restrained at their footing in order to absorb force components acting at the mast tip transversely to their axis without staying. However, this solution is found only for smaller constructions because of the comparatively large bending stress elicited thereby. Heavily bent, and therefore sturdy, these masts tend to counteract the pronounced light-weight character of these constructions.

- Free membrane edges, as they occur mainly at the edge scallops, must be reinforced with edge elements such as sewn-in ropes. An **edge rope** serves for continuous load introduction from the areal membrane (Fig. 4.212). A rope hem or doubling of the membrane protects the sensitive cutting edge.
- Membranes are extremely sensitive to **large point loads**, which are almost inevitable in the case of point support for obvious reasons. In addition to suitable reinforcements or

Cf. remarks in ▶ *Sect.* 3.4.5.2 *Membranes and Rope Nets*

Fig. 4.211 (Left) Close-up of the masts on **Fig. 4.208**. (Right) Node design of a cable net with two continuous, intersecting cables (eng.: Schlaich, Bergermann & Partner)

Fig. 4.212 Point support of a membrane and an edge rope contained in a hem to introduce the tensile stresses of the membrane into the support

Fig. 4.213 A star-shaped cutting pattern of a membrane at its suspension with reinforcing effect. The linear membrane connection to a ring is also recognisable. View from below:

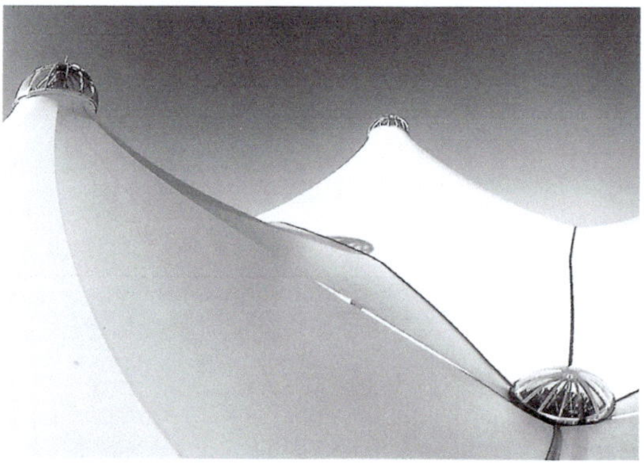

Fig. 4.214 Membrane surface with high and low points. Support ropes are clearly visible at the ridges and valleys. At the low point in the lower right side, the rainwater is collected and drained

duplications, such as can be fabricated for smaller stresses on the membrane itself (Fig. 4.213), the following solutions can also be implemented:

- Radial orientation of the membrane's stronger warp direction in main force direction by means of a radially fanned seam pattern (Figs. 4.212 and 4.227, right).
- Separated meridian ropes or straps incorporated in membrane hems (Fig. 4.214).
- **Rosettes**: In this case, the membrane is tied at several points and connected (Fig. 4.215) with a cable ring to the bearing point (for example a mast tip).
- **Rope loops** which are connected to the mast tip and introduce the tensile force into the thin membrane over the entire circumference of the loop. Since the rope loop

Fig. 4.215 Suspension of a membrane in rosette form by six rope loops. View from below

Fig. 4.216 Behind the projecting tip, the rope-loop fixation of the rope net to the mast can be seen

is not bending-resistant, it adapts tangentially to the membrane surface (Fig. 4.216).
- The force can also be introduced analogously via bending-stiff **rings** (Fig. 4.213).
- **Saddles** or **spherical caps**. In this case, forces are introduced superficially over evenly curved support elements.
- And, last but not least, the deliberate avoidance of a punctual membrane attachment already beforehand.

Fig. 4.217 Determination of a membrane surface using a soap-skin model (IL, University of Stuttgart, Frei Otto)

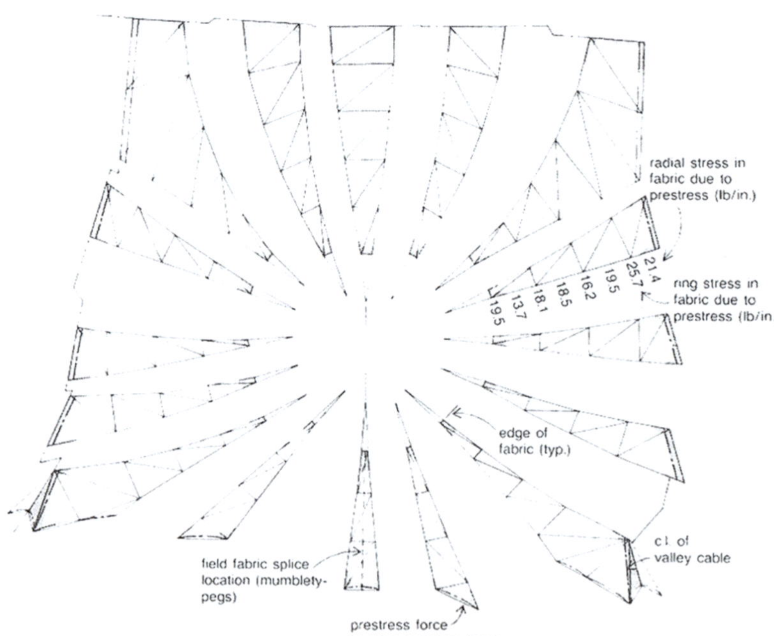

Fig. 4.218 Example of a membrane cutting pattern in strips

- Under defined boundary conditions such as the type of support, the nature of the membrane, and the applied pretension, membranes adopt a specific—and *only* this—**form**. It can be determined beforehand by means of mate-

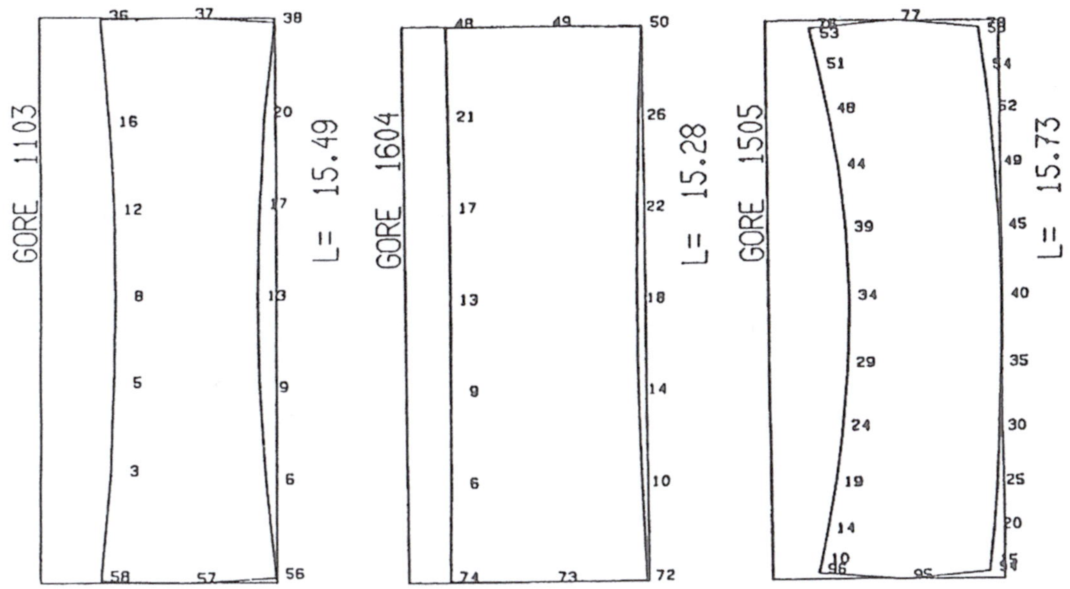

Fig. 4.219 Blanks of fabric webs. The rectangles represent the uncut web

rial or digital models (Fig. 4.217). In order for this form to be adopted in the final constructed state and the calculated stresses not to be exceeded within the membrane, a **cutting pattern** of the membrane surface is to be derived from individual webs as a function of the final shape since tissue or film is only available in limited widths for manufacturing reasons and sewing, gluing, or welding of individual strips is hence necessary for fabricating membranes. The cutting geometry must be adjusted to the maximum available web width. Faults in the cutting pattern can lead to wrinkling and, ultimately, tearing of the membrane (Figs. 4.218 and 4.219).

- The free edges of a membrane assume the characteristic concave curve shape of a **scallop**. It arises because of the tensile force acting in the membrane at right angles to the edge rope, which always pulls the thin skin inwards away from the edge. The tensile forces acting within the membrane tangentially along the edge rope can, as a matter of principle, not be introduced into the rope via friction. For this purpose, an additional **edge belt** is required which is attached at both ends to the punctual bearings. Neither rectilinear nor convexly curved membrane edges projecting outwards can be produced without linearly connecting the membrane to appropriately shaped **bending-stiff edge members** (Fig. 4.220). Examples of this solution are given below.
- The prestressing force, which is essential to the load-bearing capacity of the membrane, has to be transferred into the ground via the foundation. The arising, sometimes not

Cf. also ***Vol. 3**, Sect. 17 Membrane Systems*, Fig. 17.23, left, and Fig. 17.24

► *Sect. 4.3.3.2 Membrane and Rope Structure, Mechanically Tensioned, Linearly Supported*

Fig. 4.220 Straight-line mounting of a membrane on a bending-stiff edge element (in this case, a concrete wall)

insignificant, forces often lead to elaborate, often heavy foundations which stand in sharp contrast to the fineness and elegance of the visible membrane construction.

The Basic Structural Cell

A basic form of membrane construction is the **four-point membrane**. It occurs when a membrane is stretched between four points not lying within a plane (Figs. 4.221 and 4.222). It features the **saddle surface** characteristic of membranes. For obvious geometrical reasons, four-point membranes cannot enclose a space for themselves on all sides. For this purpose, additional elements such as, for example, light space delimitations from bending-stiff elements, or else from further membrane surfaces, are suitable. The connections to the main

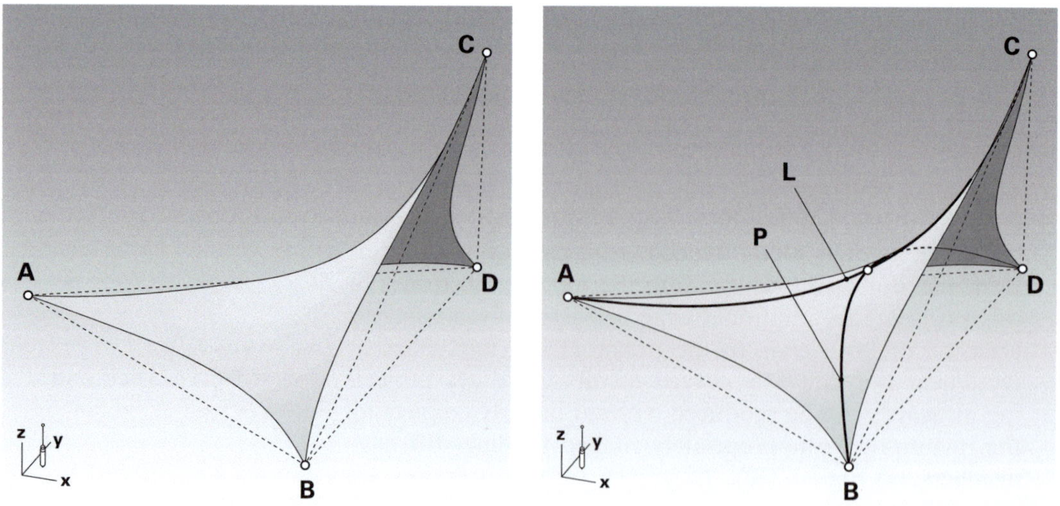

Fig. 4.221 (Left) **Four-point membrane**. The vertices **A** to **D** do not lie in one plane, so that the required double anticlastic curvature is produced. (Right) **Four-point membrane.** Load-bearing direction **L** and prestressing direction **P** of the biaxial tensile force in the membrane

■ **Fig. 4.222** Four-point membrane

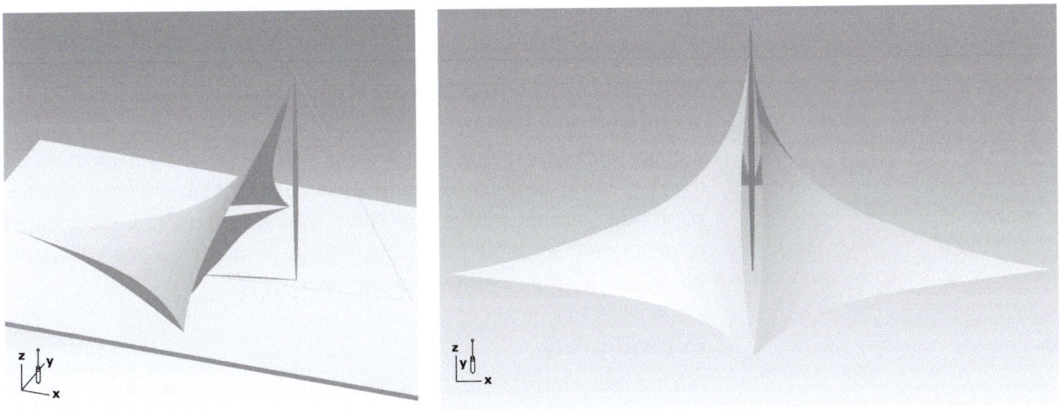

■ **Fig. 4.223** (Left) Four-point membrane. The high point is held by a guyed mast (hinged column). (Right) Four-point membranes are a basic module of membrane construction and can be modularly combined into a wide range of combinations. Here: two four-point membranes attached to a common mast

membrane can be carried out by means of sewn or welded membrane flaps or also pneumatic cushions. However, combinations of several four-point membranes can definitely be formed into enclosing envelopes. For this purpose, edge scallops can be made so low that residual openings can be easily closed (as in the example on ■ Fig. 4.223, left). Furthermore, a linear support of membranes is also possible (■ Fig. 4.220), whereby the interior can be encapsulated without additional enclosing surfaces (■ Fig. 4.224, right).

Cf. also ► *Sect. 4.3.3.2 Membrane and Rope Structure, Mechanically Tensioned, Linearly Supported*

Although laterally enclosing surfaces which are added to membranes are technically feasible, they nevertheless strongly impair the distinctive and original appearance of these structures. This applies in particular to planar enclosures which are in visual conflict with the formally very dominating membrane

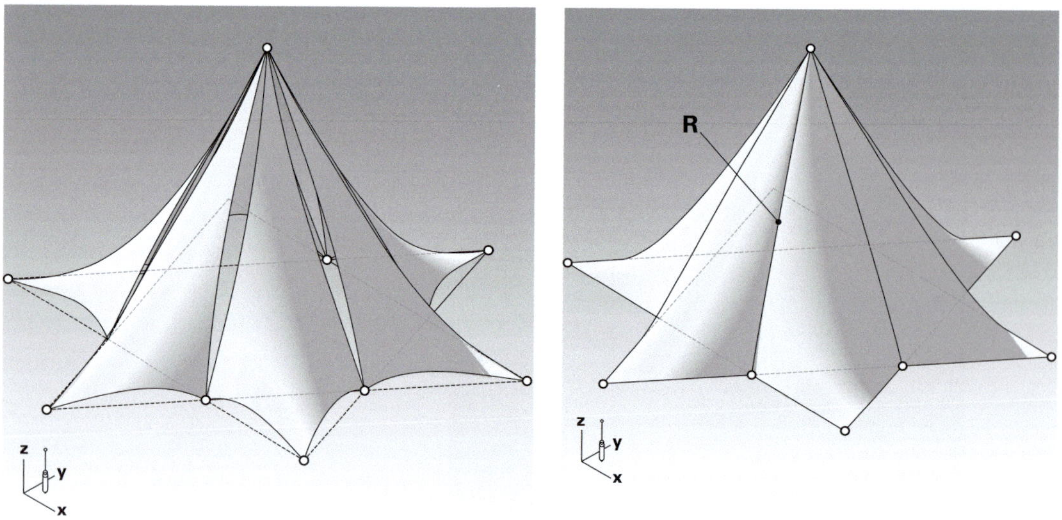

Fig. 4.224 (Left) Six four-point membranes around a hexagon. The free lateral edges between the modules form double scallops. (Right) Six four-point membranes as on the left. The two side edges are connected to a rope **R** here and form a closed membrane surface. The lower edges are also held in linear shape

curvature. In their constructional design, it must be taken into account that, although *rigid* membrane edges certainly can be connected immovably to bending-stiff envelope components, it is not permissible to restrict *movable* membrane edges, such as rope edges or free edges, in their mobility. In the latter case, it is necessary to incorporate movable connecting constructions between the membrane and the enclosure.

Extendibility

Although membrane structures, at first glance, give the impression that they could be formed in any shape and unrestrictedly combined into discretional arrangements, some restrictions still apply:

- **Four-point membranes** as defined above are the basis for numerous membrane structures. These arise from their simple addition (Fig. 4.224). In this way, a simple, easy-to-grasp base element can be added like a basic structural cell to more complex forms also in membrane construction—analogously to other types of structure discussed here. By maintaining the saddle surface, the necessary anticlastic curvature is guaranteed;
- The limited strength of membranes also limits their free span between supports. For large spans, therefore, a strengthening by segmentation with **belt ropes** is required at which the thin skin is held linearly and by means of which the forces can be concentrated onto the punctual supports without exceeding the permissible stresses on the membrane. This results in **ridges** and **grooves** or **valleys** on

Fig. 4.225 (Left) Modular sequence of membrane surfaces tensioned between support cables: ridge cables **R** and valley cables **V** tension the membranes and create the necessary curvature. (Right) Lateral enclosure of a space under a tent membrane

the membrane surface (Figs. 4.215 and 4.225, left). In this way, a membrane surface is segmented into different **modules**, which visually show clearly (Fig. 4.225, left).

- It is certainly true that membranes can bridge large spans with a minimum of material, but it must always be borne in mind that, because of the curvature which is indispensable for fulfilling their load-bearing function, a sufficient **camber** is necessary for this purpose, not only for the **bearing direction** but also for the conversely curved **prestressing direction**. In particular, the curvature line of the prestressing direction is relevant to the spatial profile and is therefore also determined by requirements of the use of the building. These necessities result in a comparatively large overall building height, which is usually the sum of the cambers of both the pre-tensioning and the load-bearing curvature (Fig. 4.225, right). Only when the prestressing curvature is closed in ring form—as in the case with radially symmetrical membrane forms, the tents—the load-bearing curvature alone is decisive for the building

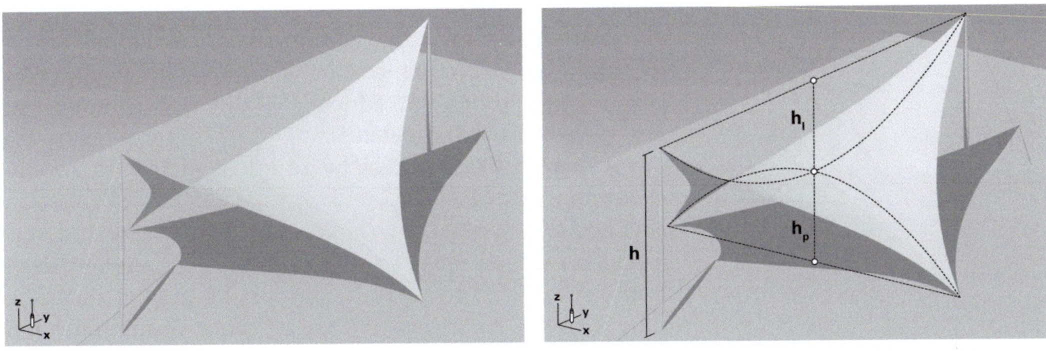

Fig. 4.226 (Left) Four-point membrane with two high points attached to guyed masts. (Right) The total height of the construction **h** here is the sum of the camber of the prestressing curvature $\mathbf{h_p}$ and that of the load-bearing curvature $\mathbf{h_l}$

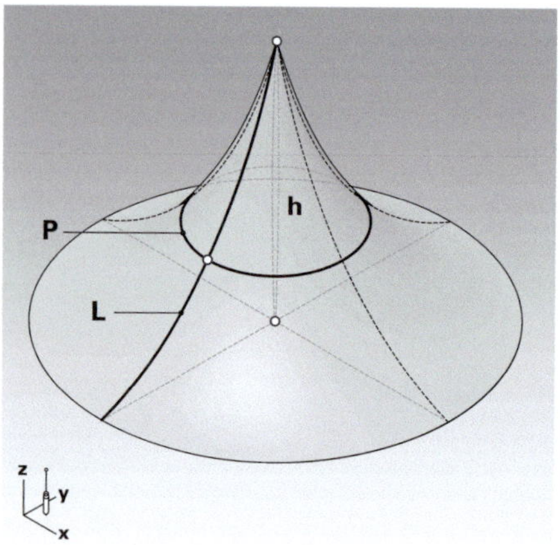

Fig. 4.227 (Left) **Tent membrane** with annular closed prestress force **P**. The building height **h** results here solely from the load-bearing direction **L**. (Right) Tents

Fig. 4.228 (Left) Tent of the Institute for Lightweight Structures of the University of Stuttgart (arch.: F. Otto). (Right) Membrane roof as sun protection (‚Palenque', arch.: J. M. de Prada Poole, Expo 92, Seville)

height (Figs. 4.227, left and 4.228). Herein, membranes are similar to other axially stressed structures, which also derive their load-bearing capacity from the form—like vaults and domes.

An enlargement of the curvature, i.e. a reduction in the radius of curvature, can be achieved with a predefined span—i.e. without introducing additional ground supports—by **suspended columns** on rope supports. For this purpose, membranes must be supplemented with suitable rope support structures.

- The above-described interaction between height and span leads, in turn, to the necessity, in the case of **shallow** membrane structures, to provide support at smaller distances, that is to say to reduce the spans, since a minimum curvature must not be fallen short of. At alternating high and low points, compressive and tensile forces have to be thereby introduced alternately into the ground. At low points, the membrane surface must also be drained from rainwater.

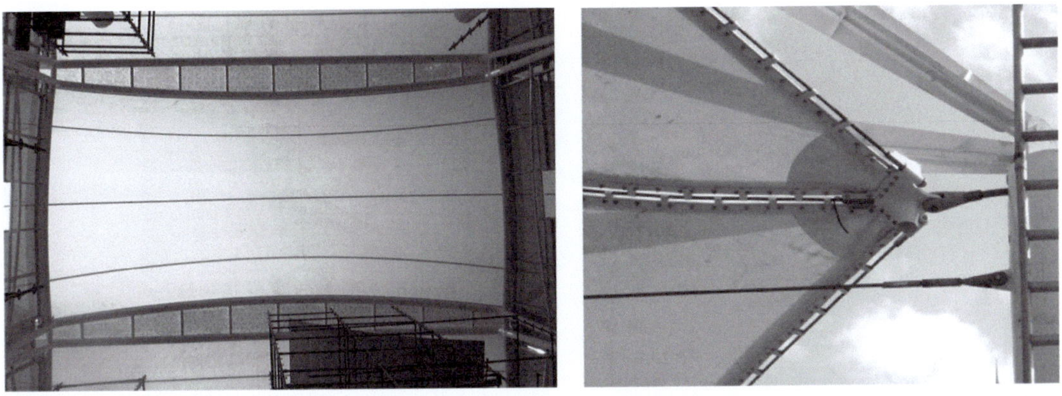

Fig. 4.229 (Left) Closure of a scallop opening between two abutting membranes by means of a transparent filling element. (Right) Edge connection of two membranes

- It must be taken into account that **biaxial tensile forces** always act at the connection areas between membrane modules—as is the case everywhere else in the membrane—which pull the edges towards the centre of the membrane. The edges can be coupled in a linear force-locking manner—by stranding or linking—so that the tensile forces are transferred between adjacent membrane panels (Figs. 4.224, right and 4.229, right). It is also possible to create oval openings from two mirrored scallops, which can be closed with a filling piece—membrane or transparent plates. This is a solution for the purpose of easy exchangeability of individual membrane fields should it be necessary (Figs. 4.224, left and 4.228, left).

Usage in Buildings—Design Aspects

Membrane constructions allow to cover spaces with a minimum of material consumption. No other construction can compete with them in this respect. Membranes are easy to transport when packaged and allow quick on-site assembly. Owing to the detachable connections which are usually employed in this type of construction, membrane structures can also be dismantled without difficulty. For these reasons, load-bearing membrane structures are predestined for mobile, temporary structures (Fig. 4.230). This character is additionally supported by certain disadvantages, namely their sensitivity to mechanical damage and the **limited durability** of thin membrane skins, at least as compared to the usual life-cycles of conventional structures. In this respect, however, no really sound statements can be made, since the life expectancy of modern technical membranes such as, for example, PTFE or silicon-coated glass-fibre fabrics, is difficult to assess since they are subject to very short innovation cycles.

Regular uses, such as uses of permanent structures, are also difficult to achieve because membranes are subject to restrictions in terms of **building physics**:

Further remarks can be found in ***Vol. 3***, *Sect. 17.4 Issues of Building Physics*

Fig. 4.230 Membrane roof (Expo 92, Seville)

- They provide only **minimal noise protection**—especially in single-layer versions without additional measures—and are also problematic in terms of **room acoustics**.
- They can only be implemented with **limited thermal protection**. Multi-layered membranes encapsulate a more or less static air layer, provided the distances between the skins are not too large. This prevents strong convection and achieves a moderate level of insulation. Multi-layered mechanically stressed membranes, for this purpose, are mounted with suitable spacers. Pneumatically tensioned membranes can be made more than two-ply by inserting intermediate layers in the interior of the cushions. Insulation fillings are also possible which offer improved thermal protection but often diminish, or even completely eliminate, the translucence of the membrane, depending on the type of insulation material used.

In addition to their extreme material efficiency in terms of load-bearing capacity, there are other advantages:

- Most materials used for membranes are **translucent** or **transparent**, so that a good diffuse natural lighting of interiors is possible without additional measures.
- Membrane films and fabrics are **waterproof** and **airtight**. Consequently, diverging partial functions such as force conduction and sealing are covered with the same component.

- The strong curvature, the large building heights which are systemically prescribed, and the smooth surface of membranes favour air movements in the interior caused by **thermal buoyancy**. By suitable arrangement of air-intake openings in the lower area—for example, at the scallop openings—and exhaust air openings at high points, efficient natural ventilation is possible.

4.3.3.2 Membrane and Rope Structure, Mechanically Tensioned, Linearly Supported

In addition to analysing *point-supported*, mechanically stressed membranes and rope structures, these must also be examined in their *linearly supported* version (Fig. 4.231). The same statements apply analogously to their load-carrying behaviour as well as to other aspects. The following particular diverging points must, however, be taken into account:

▶ *Sect. 4.3.3.1 Membrane and Rope Structure, Mechanically Tensioned, Punctually Supported*

Although, strictly speaking, intermediate supports or supporting ropes can also be regarded as bearings for the membrane, they should rather be considered part of the membrane structure itself and are therefore not contemplated in this section (cf. remarks in ▶ *Sect. 4.3.3.1 Membrane and Rope Structure, Mechanically Tensioned, Punctually Supported*)

- The linear bearing or mount of a membrane is fundamentally conducive to its load-bearing behaviour, since the forces acting within the thin skin are thus better distributed and local force concentrations, as in the case with point supports, can be avoided. If the membrane skin is anchored to the ground, it can have a stabilising function under tilting hazard for possible linear supports propping them up (e.g., arches) as well as masts.
- Linear connections of membrane edges solve the problem of **space enclosure**—as mentioned above—by themselves. This allows membranes to be connected without gaps to bending-stiff enclosing walls or even to solid foundations.
- It is obvious that linear connections require according **structural elements** such as linear foundations or bending-stiff edge elements which are associated with an increased structural complexity. In the simplest and most elementary membrane form, the point-supported one,

Fig. 4.231 Membrane and rope structure, mechanically tensioned, linearly supported

these elements are reduced to a minimum (point foundations, individual masts). However, the associated additional expenditure can sometimes be justified, in particular, from considerations of the use of the building (e.g. space enclosure, see above).

- In the case of ring-shaped supports, which act as closed **compression rings** as a result of the tensile force applied by the membrane, which is usually oriented towards the centre of the ring, the force component can be **short-circuited** within the annular plane, so that only forces acting *perpendicular* to the ring plane—usually uplifting vertical loads—are applied (◘ Figs. 4.232 and 4.233, right). Depending on the effect of the membrane force, the ring can adopt different geometries: circular, oval, etc. Also, **non-planar**, only approximately annular rim bearings, such as two intersecting arches, achieve a similar effect (◘ Fig. 4.234). They

◘ **Fig. 4.232** Waste-treatment plant in Vienna (arch.: Lang, eng. J. Natterer)

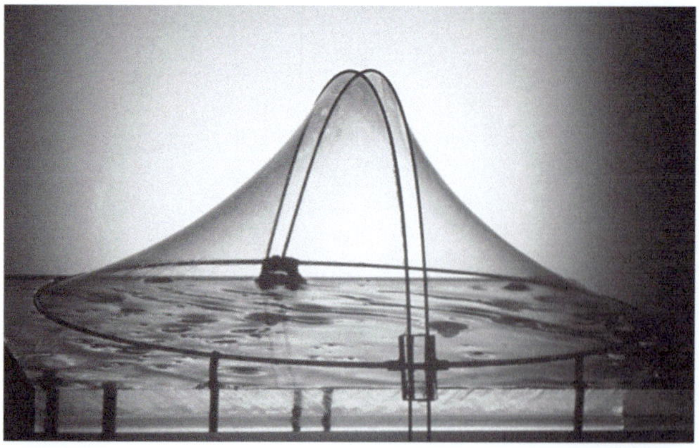

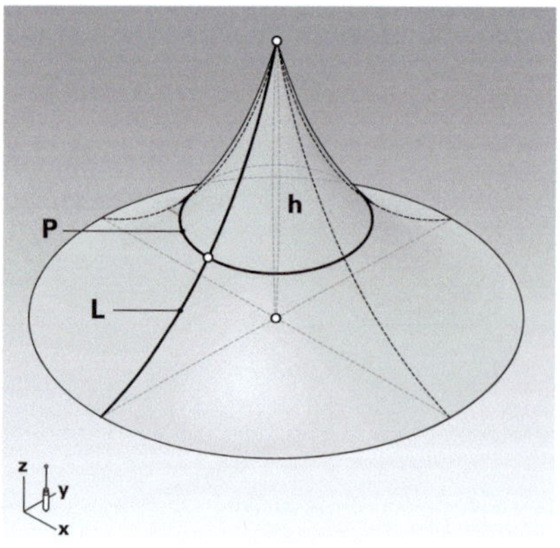

◘ **Fig. 4.233** (Left) Experimental form-finding of a **minimal surface** using a soap skin Exemplified by a linearly supported membrane (IL, University of Stuttgart, Frei Otto)

Fig. 4.234 Stadium in Raleigh: saddle-shaped cable net tensioned between tilted arches (arch.: H. Stubbins)

can furthermore also ensure an anticlastic curvature of the membrane at the same time and make additional support elements such as arches or masts superfluous.

The group of mechanically prestressed, linearly supported cable structures also includes the **ring-cable** systems or **spoke-wheel** systems, which are comparable in their structural geometry with radial bar structures (Figs. 4.235, 4.236, and 4.237). In comparison with the rigid bar framework, each bar is transformed into two tension members, i.e. ropes, that are spread out against each other at one end with a compression bar. The result is a construction with an approximately triangular cross-section. To enable the rope system to absorb forces transverse to its axis, a minimum angle of inclination of the spread as to the horizontal is required. The comparable bar system subjected to bending stress is thereby converted into a **load-bearing rope structure**. In order to prevent a group of ropes from sagging under unfavourable load conditions, they are **pretensioned**. The prevailing load conditions then only result in a partial reduction of the prestressing of one of the two rope groups, not in their sagging.

► *Sect. 4.3.1.6 Radial Skein of Bars, on Annular Support*; for ring cable systems, cf. also ► *Sect. 9.3.7.4.2 Radial Ring Rope Structures—Spoked Wheels*

The three corner points of the triangle are occupied by three ring-shaped elements essential for the load-bearing system, which run perpendicular to the triangle's plane. They short-circuit the tensile stress on the cables and therefore require no external anchoring. In terms of force, the load-bearing system is therefore self-contained. An inner hub (single or double) always acts as a **tension ring**; on the outside, a circumferential closed (single or double) **compression ring** is required. The three necessary rings can be combined in various ways:

- a double outer compression ring and a single inner hub or tension ring (Fig. 4.235 **A**);
- a double inner tension ring and a single outer compression ring (Fig. 4.235 **B**).

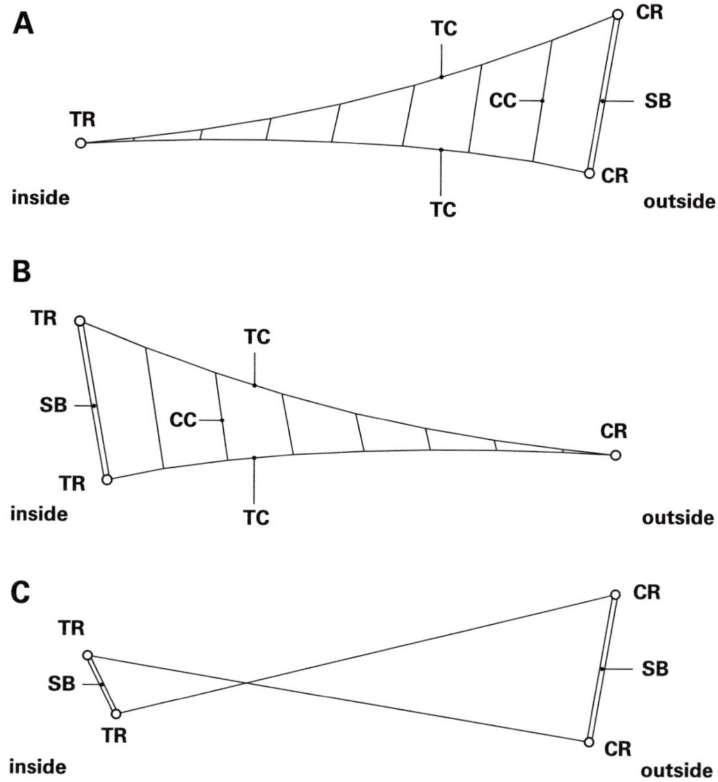

Fig. 4.235 Design variants of **ring cable systems.** Cross-section through one half of the load-bearing structure. **CR** compression ring; **TR** tension ring; **TC** tying cable; **SB** spreader bar; **CC** coupling cable (reduces the deformations and forms a cable truss together with **TC**)

Crossed systems with double tension ring and double compression ring are also possible (Fig 4.235 **C**).

4.3.3.3 Gravity-Prestressed Membrane or Rope Structure

Although the stiffening force is usually produced, as with most membranes, by mechanically applied pretension, which results in the characteristic anticlastic curvature, linearly fastened membranes can alternatively be stiffened (analogously to belts) also by **dead weight** (Fig. 4.238).[12] Ballasting is then again necessary, as in the case of belts, since the dead weight of the membranes or rope nets is not sufficient by far for this purpose. A **synclastic curvature** (Fig. 4.239), which is rather untypical for mechanically pretensioned membranes, forms under these conditions.

12 This applies in principle in the same way to point-supported membranes, but in that case the force concentrations at the suspensions are normally too high due to the heavy ballast. The author is not aware of any examples of this.

Fig. 4.236 Spoked-wheel-like structure of a stadium canopy under construction. The double compression ring can be seen outside; inside, the tension ring from several parallel ropes is visible (eng.: Schlaich, Bergermann & P.)

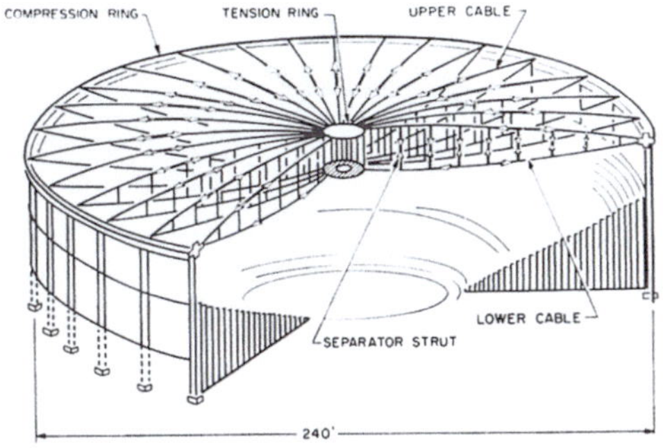

Fig. 4.237 (Left) Spoke-wheel-like roof structure; auditorium in Utica, USA (eng.: Lev Zetlin). (Right) Spoke wheel on the left: The radial ropes are formed into a cable girder by spreading them apart with struts. This is a reversal of the cable truss on Fig. 4.233 **A** and **B**, which instead is produced by coupling cables

Fig. 4.238 Gravity-stressed membrane or rope structure

4.3.3.4 Membrane and Rope Structure, Pneumatically Prestressed

► *Sect. 3.4.5.2 Membranes and Rope Nets* as well as ***Vol. 1**, Sect. 26.9.8 Pneumatically Pre-Stressed Membrane* and *Sect. 26.4.2 Movable Systems*

Pneumatically pretensioned membranes (◘ Fig. 4.240), as well as mechanically pretensioned ones, can in principle be supported both in point and in linear fashion. Their basic specific properties have been described elsewhere. The pneumatic pretension is caused by a pressure gradient between both membrane sides. Although this pressure gradient can also be produced with *open* pneumatically stressed membranes (for example, a ship's sail), this type of pneumatic structure has no significance in building. In contrast, it is *closed* cushions

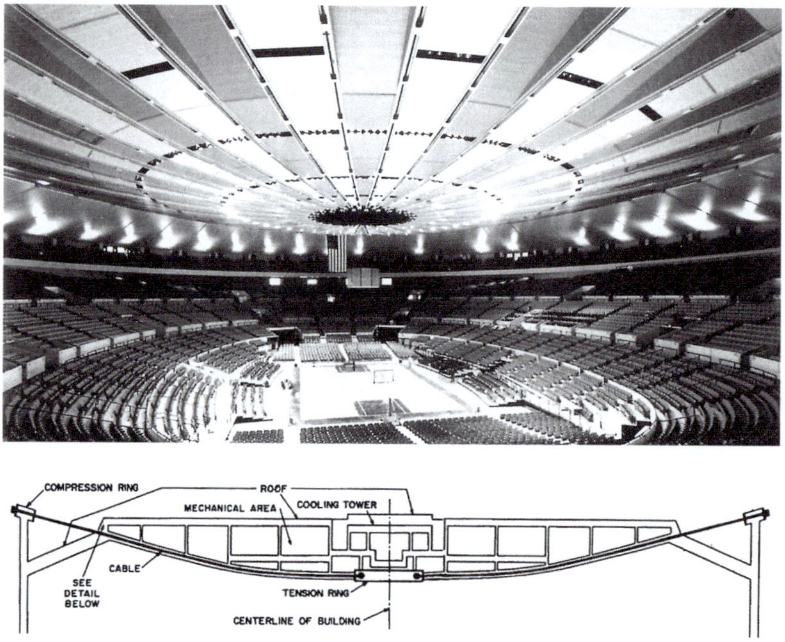

◘ **Fig. 4.239** Weight-braced ring-supported suspension structure. The necessary load is ensured by the superimposed technical floor (see cross-section below) (Madison Square Garden Arena, New York, eng.: Lev Zetlin)

◘ **Fig. 4.240** Membrane and rope structure, pneumatically tensioned

which are the rule in building construction. A difference is made between **single** and **double-membrane** systems. In both variants, a **closed space** is subjected either to over-pressure or under-pressure. Although a closed membrane envelope can be created without any additional elements, the rule, however, is that **linear boundary elements** such as strip-shaped foundations or bending-stiff edge rims are involved. For this reason, pneumatically prestressed membranes will be discussed under the category of **linearly supported** structures. Purely ring-shaped bearings are a special case in this context; as is the case with shells, different types of line bearings with a wide range of geometries are also conceivable.

***Vol. 1**, Sect. 26.4.2 Movable Systems*, Figs. 26.46 and 26.47

Load-Bearing Behaviour

Pneumatically prestressed membranes or cushions unfold their load-bearing capacity—as is the case with other axially stressed load-bearing structures—by virtue of their shape, i.e. thanks to their **curvature**. Contrary to mechanically prestressed membrane systems, where the curvature is generated by conversely oriented, mutually prestressing fibres or rope coils, in the case of cushions this is produced by a supporting medium: in construction practice, mostly air. The curvature is almost invariably **doubly curved** in the same direction or **synclastic** (■ Fig. 4.241), at least in over-pressure systems. The same **biaxial tensile stress** occurs in the membrane skin as in the case of mechanical stiffening.

Two basically different types of cushions are distinguished: **low** and **high-pressure** systems. In the case of **high-pressure cushions**, these are closed linear components with a strong curvature in one direction and nearly rectilinear in the other (hoses) (■ Fig. 4.241). They have a characteristic support and are generally loaded externally along or transversely to their axis. They represent a special case and will not be discussed further in this context. They can also be used as supporting members of **low-pressure cushions**. The latter are the normal case for pneumatic buildings.

According to a source[13], values around 10–100 mm water column correspond to low pressure; high pressure is in the range between 2000 and 70,000 mm

There is a direct relation between the radius of curvature of a membrane and the acting pressure (■ Fig. 4.242) according to the formula shown on ■ Fig. 4.243.[14] Accordingly, and given a predetermined pressure **p**, the larger the radius of curvature **r** is, the larger the tensile stress **T** within the membrane is, that is, the shallower the membrane is. Too large radii of curvature **r**, i.e. extremely flat cushions, are therefore avoided in most cases, that is, in order to rule out excessively high tensile stresses **T** in the membrane material at a constant pressure level **p**. As a means for this, punctual, but in particular linear, **intermediate supports** such as ropes, beams or arches

Cf. also the (inverse) conditions for the dome, in ▶ *Sect. 4.3.2.4*

13 Op. cit. p. 17.
14 Herzog (1976) *Pneumatic Structures – A Handbook of Inflatable Architecture*, p.8.

Fig. 4.241 (Left) Erection of the hose construction of the Fuji Pavilion at the Expo 70 in Osaka: a high-pressure system with a pressure between 1000 and 2500 mm water column (Herzog (1976) *Pneumatic Structures – A Handbook of Inflatable Architecture*, p.76) (eng.: M. Kawaguchi). (Right) Fuji-Pavilion

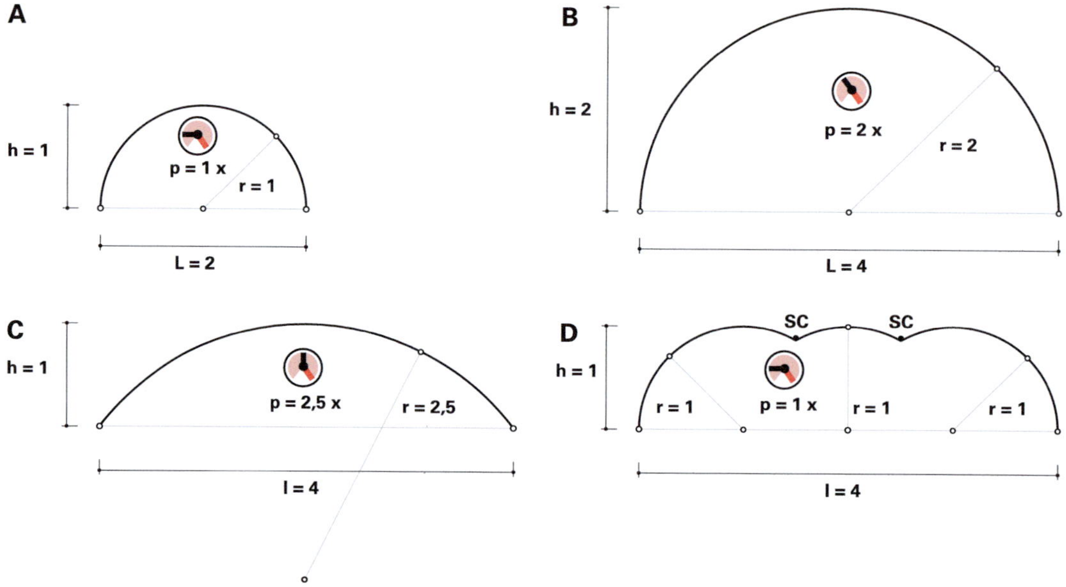

Fig. 4.242 Relationship between pressure, curvature, height, and span in a pneumatic structure. The internal pressure **p** increases proportionally with the radius of curvature **r** (thus with decreasing curvature) (Herzog (1976), p. 17). **A:** A small radius of curvature **r** bridges small span **l** at output pressure **p. B:** A double radius of curvature doubles the height **h** and the internal pressure **p**. The interior volume increases significantly (room conditioning!) and likewise the building height **h** (wind load!). As a consequence of the doubled radius of curvature, the internal pressure **p**, which must be maintained by increased energy consumption, also doubles. **C:** Reduction of the height to the original value (**h** = 1) with the same span **l**. The radius of curvature increases to the 2.5-fold value of the original one, and consequently also the internal pressure **p** (Op. Cit. p. 18). **D:** By introducing support cables **SC**, which constrict the membrane and reduce the radius of curvature to the original value (**r** = 1) with the same span **l** = 4, the internal pressure can be reduced to the value **p** = 1 ×. These examples show the need to increase the curvature of the membrane by constriction and segmentation to cover larger surfaces

are available. If, for example, a pneumatically stressed skin is constricted by ropes into several individual sections, the radius of curvature is obviously drastically reduced by the local bulges created between supporting ropes (Fig. 4.244). This means can also be used to avoid excessive building heights, as

1 For generic membrane surfaces

$$T = \frac{p}{(1/r_1 + 1/r_2)}$$

T = tensile stress
p = pressure
r_1 = largest radius of curvature of the surface
r_2 = smallest radius of curvature of the surface

2 For spherical membrane surfaces ($r_1 = r_2 = r$)

$$T = \frac{p \cdot r}{2}$$

3 For cylindrical membrane surfaces ($r_1 = \infty$ in direction of the straight generatrix, r_2 in annular direction)

$$T = p \cdot r_2$$

Fig. 4.243 Membrane stresses under pneumatic pressure

Fig. 4.244 Exhibition pavilion of the USA at Expo '70 in Osaka. A free-spanning single membrane with spans of 83 m and 142 m. The extremely flat curvature was only possible by introducing a supporting cable net (eng.: D. Geiger)

the considerations regarding mechanically prestressed membranes are valid here as well: Due to a minimum curvature being necessary, large spans also forcibly entail large overall construction heights in the first place. In some sense, cushions experience an additional mechanical pretension by means of supporting ropes. These ropes should in turn also have a certain minimum curvature, since in their case it likewise applies that the larger the radius of curvature, the greater the tensile force will be.

► *Sect. 4.3.3.1.2 Extendibility*

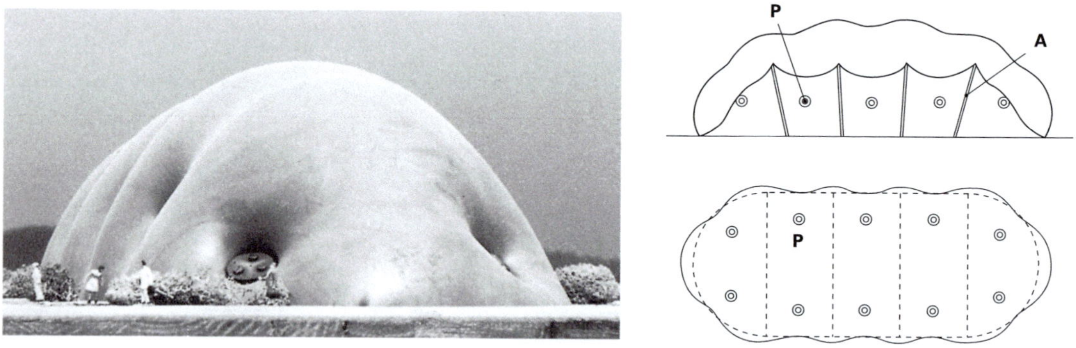

Fig. 4.245 Project for a pneumatic hall. Double membrane system with supporting arches **A** under the inner membrane and point connections **P** with the outer (design: A. Sklenar)

Fig. 4.246 Negative-pressure structure, double membrane: externally held by a bar frame (design: A. Sklenar)

▶ *Sect. 4.2.3.1.3 Usage in Buildings—Design Aspects*

Again, as in the case of mechanically stiffened membranes, secondary intermediate supports, either punctual or preferably linear, are virtually unavoidable (Fig. 4.245) in the case of cushions used for covering larger surfaces.

Although over-pressure systems tend to be the rule, a pneumatic membrane can also be stiffened by **under-pressure** (Fig. 4.246). Under-pressured cushions, together with tensile-stressed suspended structures, have the disadvantage of their **concave curvature** (as regarded from above) which, in order to cover spaces, requires extensive secondary support structures at the high points and makes the dewatering of roof surfaces more difficult, especially in the case of synclastic concave curvature.

Like mechanically stabilised membranes, cushions also assume only one characteristic membrane shape under given boundary conditions, under which pure tensile stresses, in the ideal case always in the same magnitude, prevail. The ideal membrane shape can be determined in a physical model from soap skins or rubber membranes and can be recorded by measurement technology, or alternatively it can also be created using a digital model. On this basis, appropriate **cutting patterns** are then determined.

Usage in Buildings—Design Aspects

The idea of entrusting the load-bearing capacity of a structure to a medium like air and, at the same time, bridging extremely large spans with ultra-thin skins has stimulated the imagination of building designers, especially in the 1960s and 1970s. During this period numerous, sometimes very spectacular, pneumatic structures or unrealised projects emerged. As a result of long-term operation, some weaknesses of this construction type have also become clear, eventually leading to a more sober assessment. In particular, the high maintenance costs, which are not least due to the need to continuously maintain an over-pressure or under-pressure by consuming energy, have become noticeable. Furthermore, the same functional restrictions apply as with mechanically stiffened membrane structures.

▶ *Sect. 4.3.3.1.3 Usage in Buildings—Design Aspects*

Nevertheless, recent spectacular examples show that pneumatic structures or components can have great advantages for specific applications (■ Fig. 4.247). The disadvantages of limited thermal protection can be mitigated by multi-layered cushions. Membrane bulkheads within cushions can, in addition to their thermal effect, also serve to increase the curvature of a cushion by constriction and consequently to reduce tensile stresses. The synclastic bend typical of cushions also favours the rapid dewatering of weathered surfaces.

More recent examples are not single but double membrane systems, namely executed as secondary structural elements

■ **Fig. 4.247** Pneumatic cushions, attached to a secondary steel structure (Allianz Arena, Munich; arch: Herzog & deMeuron)

in combined use with primary structures from other materials, especially steel. In these cases, the cushions are designed as surface-forming elements which are rimmed all around in a bending-stiff and watertight manner (◘ Fig. 4.247). New high-performance, very durable membrane materials such as, in particular, ETFE (ethylene-polytetrafluoroethylene) offer favourable conditions for a more extensive use of such structures in building construction.

Bibliography

Heinle E, Schlaich J (1996) *Kuppeln aller Zeiten – aller Kulturen.* Stuttgart, Dt. Verl.-Anst.

Herzog Th (1976) *Pneumatic Structures. New York, Oxford University Press*

Mislin M (1997) *Geschichte der Baukonstruktion und Bautechnik, Vol. 1. Antike bis Renaissance*, 2nd ed. Düsseldorf, Werner

Mislin M (1988) *Geschichte der Baukonstruktion und Bautechnik: von der Antike bis zur Neuzeit*; eine Einführung 1st ed. Düsseldorf, Werner

Deformations with Dr. Matthias Weißbach

J. L. Moro, *Building-Construction Design – From Principle to Detail*,
https://doi.org/10.1007/978-3-662-71006-7_5

Contents

5.1 Causes and Characteristics of Deformations

5.1.1 Causes

Structures and the components of which they are composed **deform** under external influences. These include:

Vol. 1, Sect. 11.11 Deformation

- **external loads**, for example:
  - **permanently acting loads** such as dead loads;
  - **variable loads** such as traffic loads, wind loads, snow loads, etc.;
  - **exceptional loads** such as earthquake loads.

  The deformations resulting from external loads are referred to as **load-dependent** deformations.
- or **environmental influences** such as:
  - changes in temperature;
  - changes in relative humidity;
- or by the effects of **material behaviour**.

Vol. 1, Sect. 11.11.1 Thermal Expansion as well as ibid. *Sect. 11.11.3.1 Load-Independent Plastic Deformation*

Deformations due to environmental influences and material behaviour are referred to as **load-independent** or also **impressed deformations**.

5.1.2 Requirements

Although **statics**, the theory of the forces acting on resting systems, pursues the aim of rendering buildings unmoveable, deformations of buildings are nevertheless an unavoidable phenomenon. They must be determined quantitatively and limited according to the requirements of:

- **stability**, as well as:
- suitability for use or **serviceability**,

because unhindered deformations of the load-bearing structure can increase the stresses acting on it and endanger its stability.

With regard to **serviceability**, deformations must be limited to a size compatible with the use of the structure. They must also be taken into account in the design of non-load bearing components such as façades or mechanical equipment such as elevators or supply lines. If necessary, constructional precautions must be taken to ensure that they can unfold free of constraints—that is, free of damage.

The requirements for safeguarding suitability for use are, of course, more far-reaching than those for ensuring stability, since they limit deformations during operation—i.e. stability being already assumed. This applies, for example, to the permissible deflection of a floor construction.

5.1.3 Definition

Deformations are **external state variables** and describe the displacement and rotation of the components within the structure as well as the overall structure itself.

Deformations result from **shape changes** inside the components, and these are regarded as **internal state variables**. They are therefore considered at a differential part of the component. Shape changes emerging during deformations as a result of the action of force are determined qualitatively by the nature of the **internal stress**, and are distinguished as follows:

- **elongation** under normal stress (◘ Fig. 5.1);
- **yielding** under shear stress (◘ Fig. 5.2);
- **deflection** due to bending stress (◘ Fig. 5.3);
- **twisting** under torsional shear stress (◘ Fig. 5.4).

The nature of the internal stress acting within the component is determined by the external effects as well as the support and geometry of the load-bearing structure. Thus, for example, an axial vertical load on a ceiling column generates normal stresses and thus a negative strain, i.e. a shortening. The transverse loading of the ceiling column, for example from wind loads acting on a connecting façade structure, produces bending stresses and thus a deflection of the column (◘ Fig. 5.5).

Cf. also the combination of normal stress and bending stress for an eccentrically loaded compression bar in ***Vol. 1****,Sect. 26*, Fig. 26.144, and an eccentrically loaded arch in ► *Sect.* 3, ► ◘ Figs. 3.114 and 3.115

A uniformly distributed load, for example from a supported floor construction, produces bending stresses in a bar-like girder and, consequently, a deflection, whereas within a parabolic arch it is transformed instead into normal compressive stresses, which result in a shortening. Consequently, the inner shape changes of these two components, despite being

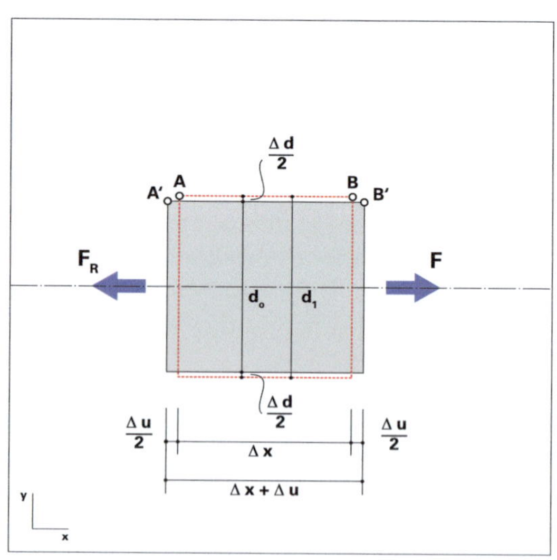

◘ **Fig. 5.1** **Elongation** due to normal stress from axial force action **F**. A **longitudinal expansion Δu** and a **transverse expansion Δd** with reversed sign occur

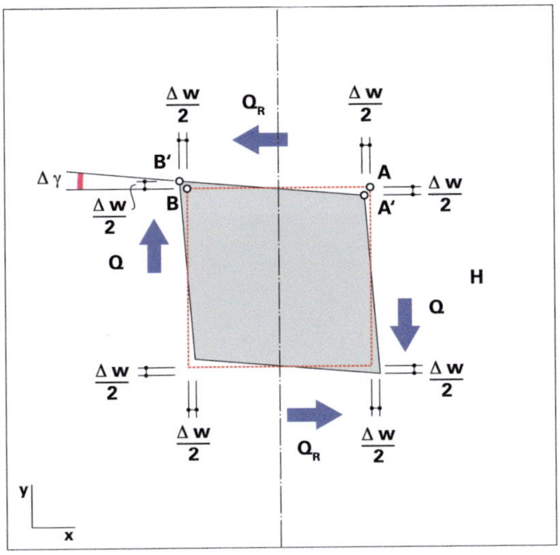

Fig. 5.2 **Shearing strain** as a result of shear stress from shear-force action **Q**. It is measured by the sliding dimension **Δw** or by means of the angular change Δγ

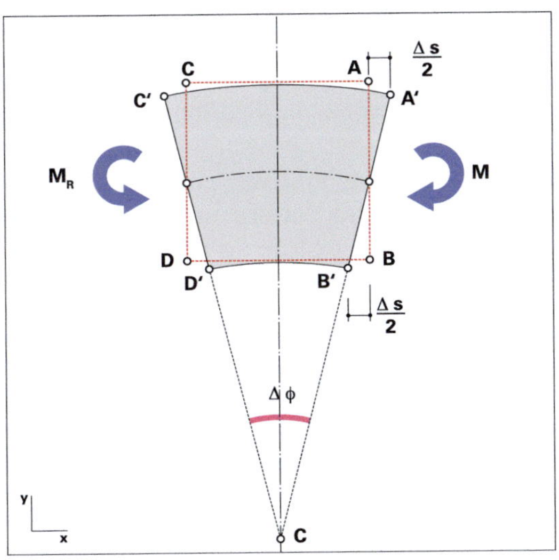

Fig. 5.3 **Bowing** due to bending stress from a bending moment **M**. Elongation and shortening of the cross-section fibres on both sides of the axis of gravity lead to an angular change Δφ

the consequence of the same external action, differ *qualitatively* from each other (Fig. 5.6).

A special case are **elongation-free deformations**, as occur with movable structures, i.e. rope nets or membranes. These systems are characterised by their lack of bending stiffness. For this reason, the term *elongation-free* deformation is justified, even though slightest strains occur nonetheless.

Cf. also the comparison of a curved bending beam and a parabolic arch in ▶ *Sect. 3*, ▶ Fig. 3.106

***Vol. 1**, Sect. 26.4.2 Movable Systems*

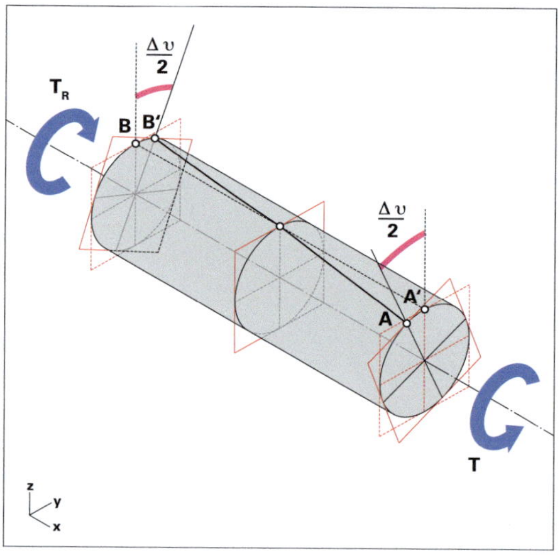

Fig. 5.4 **Twisting** due to torsional shear stress from a torsional moment **T**. It is measured on the basis of the angular change $\Delta\vartheta$

These deformations occur in principle whenever the shape of the load-bearing structure is **not affine** to the load. As a result of this discrepancy, bending-stiff and movable load-bearing structures react *fundamentally* differently:

- **Movable load-bearing systems** automatically adapt their shape to the new load pattern in the event of a changed load and, in this way, restore an affinity between form and load. There is then only pure tension in the system. This affinity is the reason for the extraordinary structural efficiency of movable systems, which, however, comes at the expense of comparatively large deformations—however, without strain. For certain building uses, the magnitude of these deformations is not tolerable.
- **Bending-stiff systems** cannot adapt with their shape to a changed load pattern precisely because of their **bending stiffness**. As a result of the non-affinity between form and load, which is rather the rule for bending-stiff systems, bending stress arises. This significantly reduces the efficiency of these structures. On the other hand, deformations remain relatively limited, which in principle is beneficial to usability. Furthermore, the shape of the load-bearing structure can be determined largely independently from anticipated load patterns, and thus dependently from *different* considerations—for example, of aesthetic or use-related nature. This freedom of design is the main reason why bending-stiff systems clearly dominate over movable ones in most building tasks.

Vol. 1, *Sect. 26*, Fig. 26.45

A distinction between elastic and plastic deformations is made in ***Volume 1***, *Chapter IV*.

Vol. 1, *Sect. 11.11.2 Elastic Deformation* and *Sect. 11.11.3 Plastic Deformation*

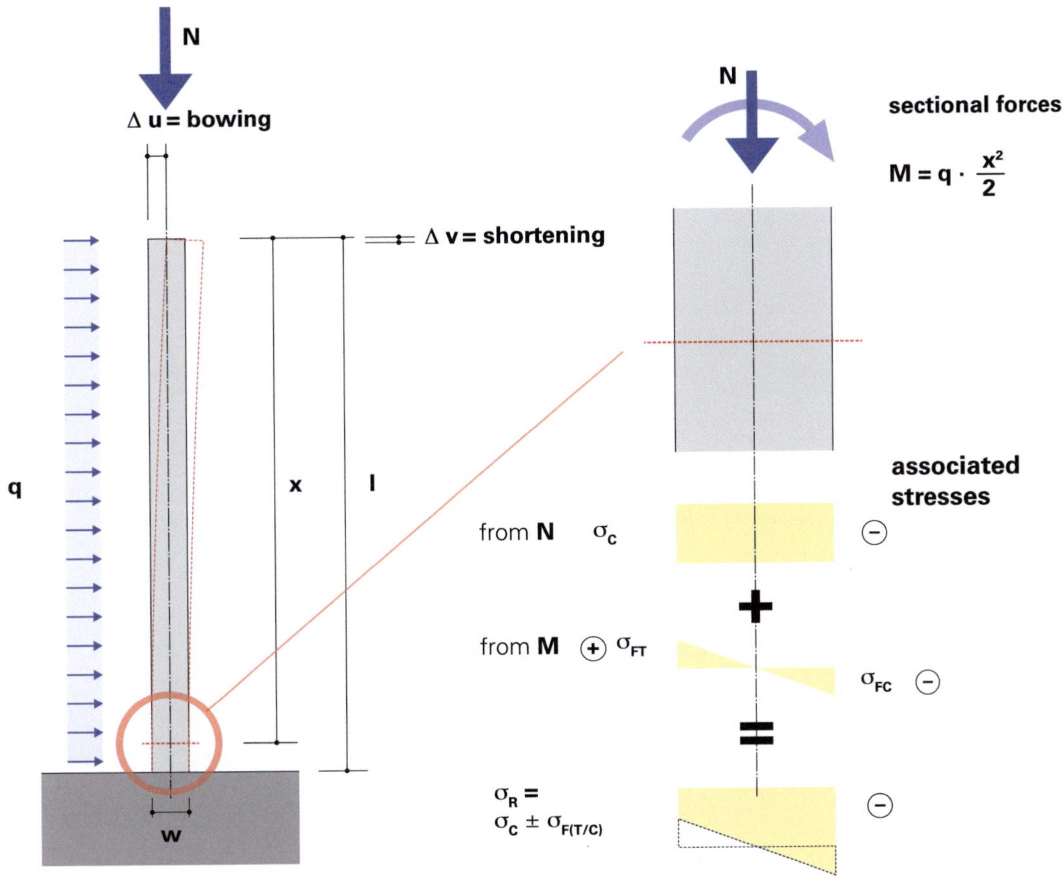

Fig. 5.5 Cantilevering column under normal force **N** and bending moment **M** from the effect of the transverse load **q**. In the cross-section, the compressive stresses σ_C from the normal force **N** are superimposed on the flexural compressive and tensile stresses σ_{FC} and σ_{FT}. The resulting stress σ_R is the summation of these components. There is a combination of bowing and shortening

5.1.4 Stress and Strain

The magnitude of the internal change of the shape—i.e. elongation, yield, deflection, or twist—is determined in deformations due to the action of force by:

***Vol. 1**, Sect. 11.11.2.1 Stress-Strain Diagram*

- the magnitude of the associated **internal stress**;
- the **stiffness** of the material.

This relationship, also known as **material law**, is characterised by Young's modulus, or modulus of elasticity, and the shear modulus of a material and is quantified for **elastic deformation** using **Hooke's law**:

$\sigma = \mathbf{E} \cdot \varepsilon$ for **compression** and **tension**

σ = normal stress; **E** = Young's modulus; ε = elongation

$\tau = \mathbf{G} \cdot \gamma$ for **shear**

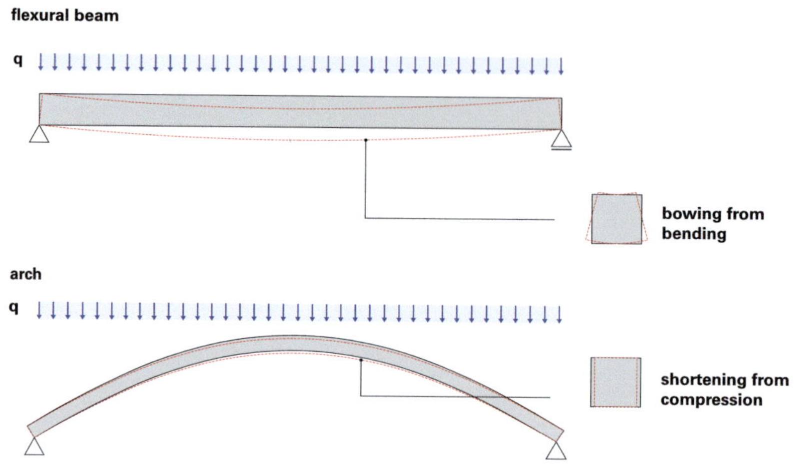

Fig. 5.6 Qualitatively different deformations of the flexural beam and of the arch under the action of the same load **q**. However, **shape** and **support** of both structural systems differ fundamentally from each other

Cf. an overview of elastic and shear moduli on Fig. 5.7

The large modulus of elasticity of steel compared to that of wood, for example, means that, in a steel component, significantly less internal shape changes can be expected under identical loading than with a wooden component.

The magnitude of the internal **stresses** depends, in turn, on the **effective cross-section** of the component over which they are distributed. This results in:

- **normal stresses** of an axially stressed column and, thus, its internal shape change from the acting normal force **N** distributed over the column's cross-section of the area **A**;

  $\sigma = \mathbf{N/A}$

- **bending-induced stresses** due to the transverse loading of the column from the acting force and from the **moment of inertia I** of the column's cross-section. Thus, for a rectangular column cross-section, for instance, bending around the strong axis results in significantly smaller bending stresses, and thus also deflection, i.e. internal shape changes, than in bending around the weak one (Fig. 5.5).

Starting from the **internal change** in **shape**, which refers to a component's cross-section, the **outer deformation** is obtained by the **summation** of the internal shape changes (in the present example, a shortening due to the normal force and a deflection due to bending) along the component axis, using kinematic relations (Fig. 5.5). The change in the length—displacement **u**—of an axially stressed column results from the summation of the elongations over the length of the column. In this process, the bearing conditions of the support must be taken into account (Fig. 5.8).

		material	E_{0cm}	E_{cm}	[8]	
concrete		C 12/15	25,800	21,800		
concrete		C 16/20	27,400	23,400		
concrete		C 20/25	28,800	24,900		
concrete		C 25/30	30,500	26,700		
concrete		C 30/37	31,900	28,300		
concrete		C 35/45	33,300	29,900		
concrete		C 40/50	34,500	31,400		
concrete		C 45/55	35,700	32,800		
concrete		C 50/60	36,800	34,300		
			$E_{\parallel}$	$E_{\perp}$	G	G_T
wood [1]	conifer [2]	S 7/MS 7	8,000	250	500	350
wood [1]	conifer [2]	S 10/MS 10	10,000 [3] [4]	300	500	330
wood [1]	conifer [2]	S 13	10,500 [3] [4]	350	500	330
wood [1]	conifer [2]	MS 13	11,500 [3]	350	550	360
wood [1]	conifer [2]	MS 17	12,500 [3]	400	600	360
wood [1]	conifer [2]	spruce (FI), pine (KI), fir (TA), larch (LA), Douglas fir (DG), southern pine (PIR), western hemlock (HEM), yellow cedar				
wood [1]	deciduous wood [5]	group A	12,500	600	1,000	400
wood [1]	deciduous wood [5]	group B	13,000	800	1,000	660
wood [1]	deciduous wood [5]	group C	17,000 [6]	1,200 [6]	1,000 [6]	660 [6]
wood [1]	deciduous wood [5]	group **A**: oak (EI), beech (BU), teak (TEK), keruing (YAN) group **B**: afzelia (AFZ), merbau (MEB), angelique (AGQ) group **C**: azobé (bongossi) (AZO), greenheart (GRE)				
wood [1]		The computation values are to be reduced by 1/6 for all-sided weathering, and 1/4 for continuous dampening				
steel		**rolled steel** and **cast steel**, all sorts	E		G	
steel			210,000		81,000	
glass		**soda-lime-silicate glass** (construction glass)	E		[8]	
glass			70,000 – 75,000			
			E [7]		[8]	
masonry		**brick**	$3{,}500 \cdot \sigma_0$			
masonry		**lime-sand brick**	$3{,}000 \cdot \sigma_0$			
masonry		**light concrete block**	$5{,}000 \cdot \sigma_0$			
masonry		**concrete block**	$7{,}500 \cdot \sigma_0$			
masonry		**AAC block**	$2{,}500 \cdot \sigma_0$			

[1] Solid wood as well as construction solid wood from conifer S 10, acc. to *DIN 1052-1* and *DIN 1052-1/A1*
[2] Sorting class acc. to *DIN 4074-1*, the quality classes III, II, and I acc. to *DIN 4074-2* are equivalent to the sorting classes S 7, S 10, and S 13
[3] For wood which is installed with a wood moisture content ≤ 15%, the values can be increased by 10% for deflection calculations.
[4] For roundwood: $E_{\parallel} = 12{,}000\ MN/m^2$
[5] Medium quality, minimum sorting class S 10 in the sense of *DIN 4074-1* or quality class II as defined in *DIN 4074-2*
[6] These values are independent from wood moisture
[7] Computational value, secant modulus from total elongation at about 1/3 of the masonry compressive strength
[8] Brittle materials are considered to be non-shear-resistant, so they are not assigned a shear modulus.

Fig. 5.7 Overview of **Young's moduli (E)** and **shear moduli (G)** of the most common building materials (acc. to Schneider (2004) *Bautabellen für Architekten*, 16th ed.), all values in N/mm^2

$$\varepsilon = \mathbf{du/dx};\ \text{for}\ \varepsilon = \text{const. and}\ \mathbf{dx} = \mathbf{L} \rightarrow \mathbf{u} = \varepsilon \cdot \mathbf{L}$$

The maximum displacement **u(L)** results at the column head assuming a rigid bearing (**u**(0) = 0) at the column footing.

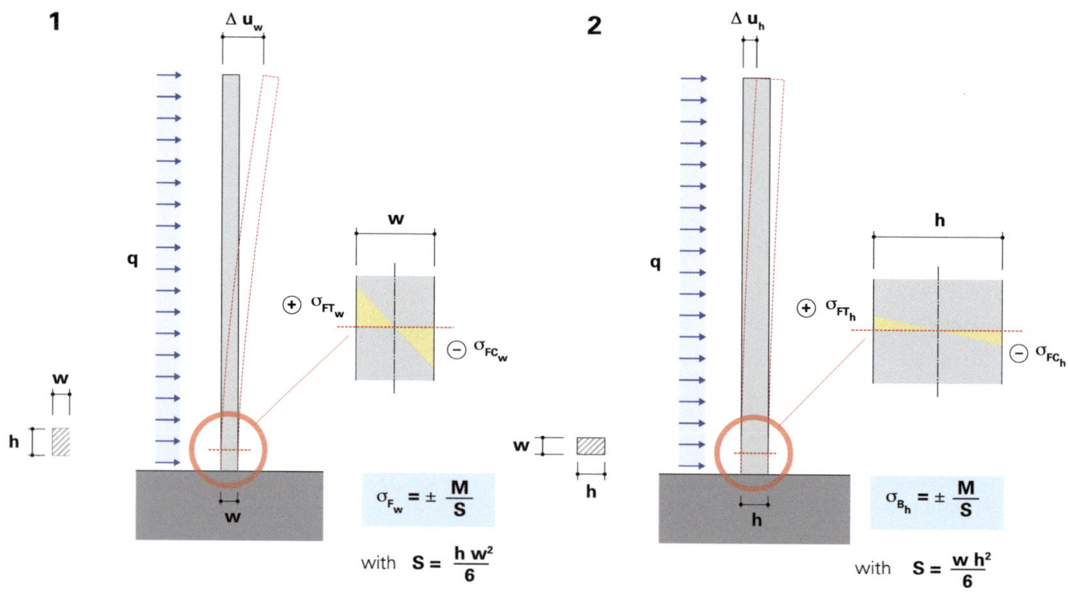

Fig. 5.8 Deformation **Δu** of a cantilevering column under horizontal linear load **q**, in one case in the direction of the cross-section's width **w** (case **1**), in another in the direction of the cross-section's height **h** (case **2**). The deformation Δu_h in direction of the stronger axis, i.e. the cross-sectional height **h**, is, of course, significantly smaller. The larger dimension of the cross-section's height **h** is included in the calculation of the section modulus **S** in the second power. The flexural tensile and compressive stresses σ_F are accordingly smaller in case **2**

5.1.5 Imprinted or Load-Independent Internal Shape Changes

Imprinted internal deformations of structures are caused by environmental influences such as temperature changes, which lead to changes in length and possibly warping of individual components or of the entire load-bearing structure. In this case, the geometry and the support conditions of the load-bearing structure play a decisive role—even more so than in the case of load-dependent deformations.

If the internal shape change can freely unfold within the structure, this is referred to as a **statically determinate system**. There are no internal stresses, independently of the degree of material stiffness and the cross-section's resistance.

Vol. 1, Sect. 26.1.2 Assigning Force-Transmitting Functions to Components as well as ibid. *Sect. 26.2.3 Support*

If, however, the internal deformation within the structure is hindered by its support, this is referred to as a **statically indeterminate system**. A **constraint** then builds up in the structure. This means that an elastic inner shape change is opposed to the imprinted inner shape change, so that the overall shape change becomes zero because it is prevented from unfolding. Depending on the magnitude of the material stiffness and the cross-sectional resistance, the elastic shape change results in internal stresses within the load-bearing structure (Fig. 5.9).

Internal shape changes can also gradually build up as a consequence of time-dependent material properties. The drying

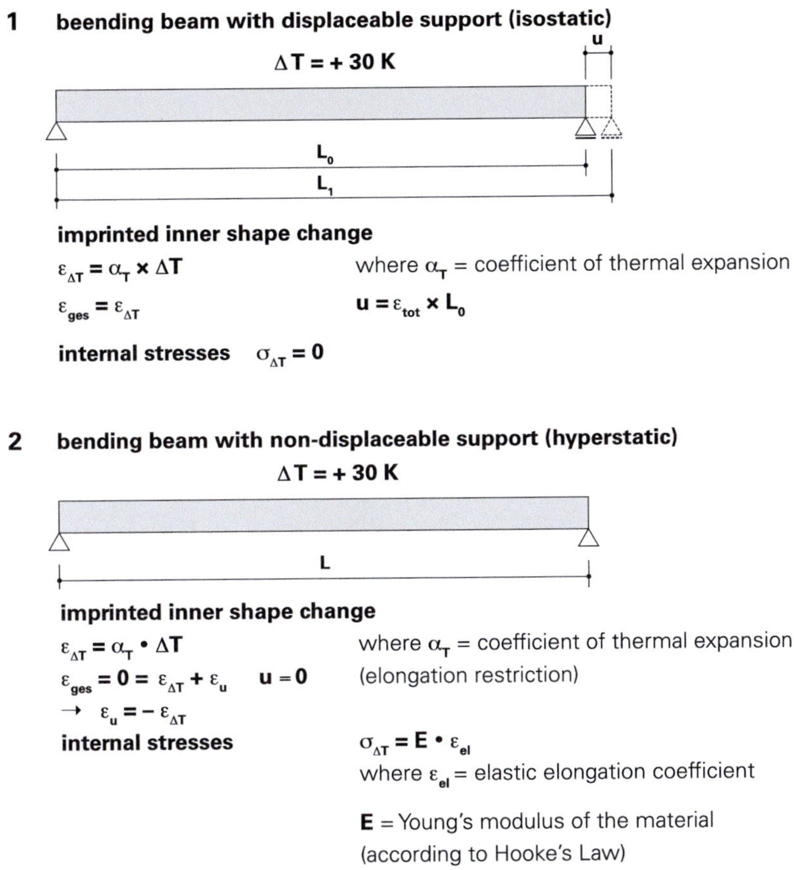

Fig. 5.9 Comparison between a displaceably and an immovably supported bending beam under temperature change of $\Delta T = 30$ K. The latter is subject to restraint stress due to the elongation being impeded by the bearing and experiences a stress $\sigma_{\Delta T}$

of wood or the curing of concrete yields a reduction in volume, which is expressed by a shortening of the components. This phenomenon is called **shrinkage** of building materials.

Vol. 1, Sect. 11.11.3.1 Load-Independent Plastic Deformation

Vol. 1, Sect. 11.11.3.2 Load-Dependent Plastic Deformation

Materials such as concrete or plastics tend to **creep** under permanent stress. At the same time, the elastic deformations which occur at the time of loading gradually increase with time. This is referred to as a **creep elongation**, if this can adjust freely—i.e. in the statically determinate system. The permanent stress which produces creeping remains constant in this process. If the deformation resulting from the creep elongation is hindered—in the statically indeterminate system—the result is a reduction of the permanent stress generating creeping. This is referred to as **relaxation** (Fig. 5.10).

Exceptional effects, such as non-uniform settlements of a building, impose an external imprinted deformation state upon the load-bearing structure. In this case, the external ef-

Sect. ▶ 6.2.4 Deformations of the Ground

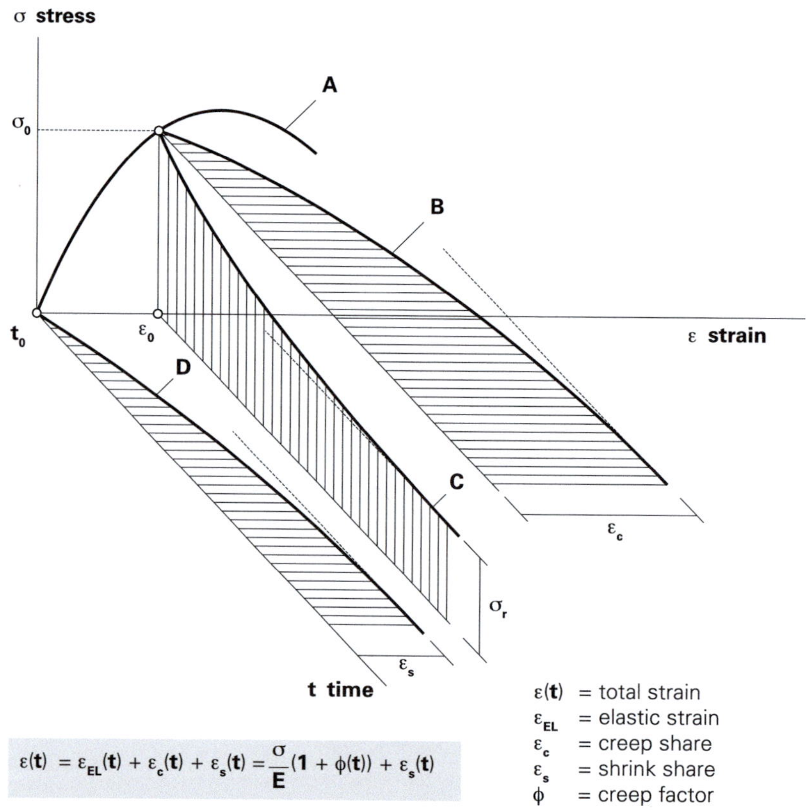

Fig. 5.10 Stress-strain behaviour as a function of time **t** (using the example of a mineral material such as concrete) in a spatial diagram with the coordinate axes σ, ε, and **t** showing the relaxation **C** (acc. to Heller H (1998) *Padia 1—Grundlagen Tragwerkslehre*, p. 344 bottom). **A** stress-strain-line, $\mathbf{t}_0$ = constant (time-independent); **B** creeping, σ_0 = constant; **C** relaxation, ε_0 = constant; **D** shrinking/swelling, $\sigma_0 = 0$

fects are not loads but deformations. The geometry and the bearing of the load-bearing structure also play a decisive role in this case (Fig. 5.11).

5.2 Impact of Deformations on Building Construction

The deformations of primary structures affect the load-bearing capacity as well as the functionality of buildings. They must be calculated applying the relevant load-case combinations and are to be viewed globally at the level of the load-bearing structure as well as locally for each component and its connections. In this process, also the transitions to secondary load-bearing structures, e.g. façades, and non-bearing structures of the fit-out must be considered.

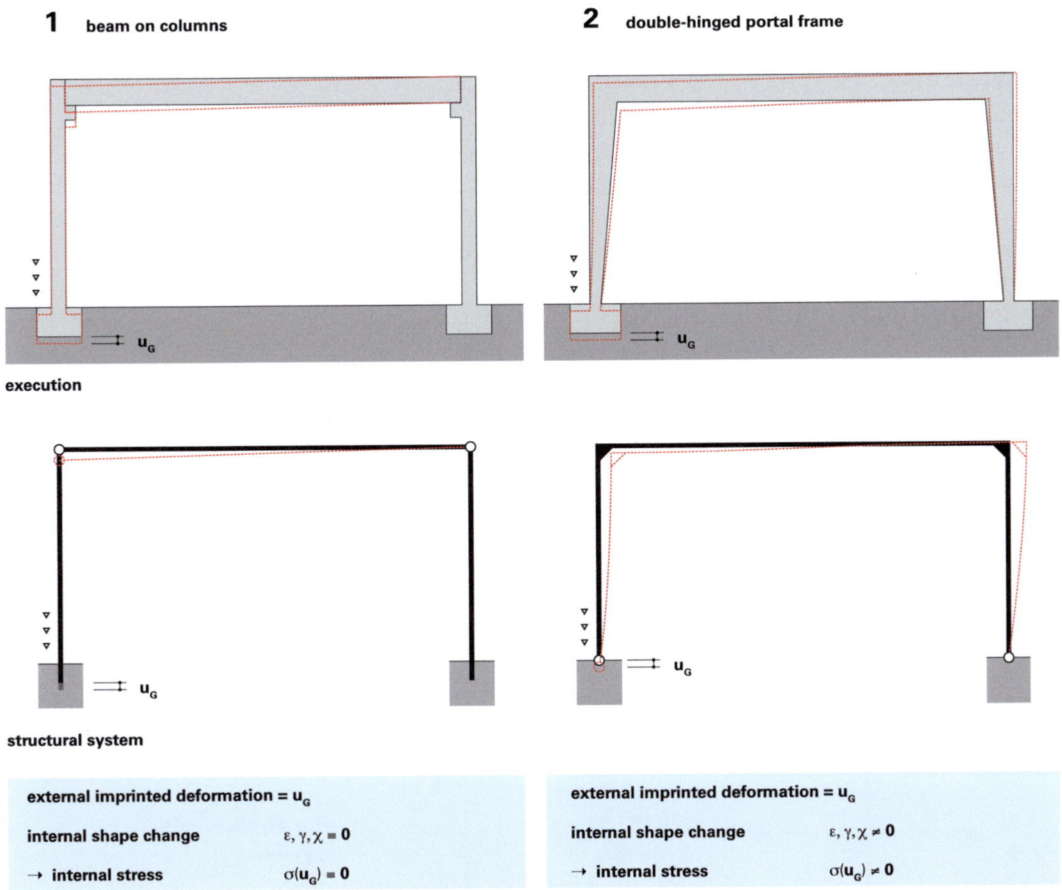

Fig. 5.11 Comparison between the deformational behaviour of a statically determined (case **1**) and a statically indeterminate system (case **2**) in the case of soil settlements

The conceptual and constructional design of connection details of load-bearing structures, building envelopes, and fit-out elements requires an accurate knowledge of the deformations of the load-bearing structure, taking into account manufacturing tolerances of the various trades.

***Vol. 1**, Sect. 4.4 Dimensional Allowances—Dimensional Coordination at Component Joints*

5.2.1 Impact on Load-Bearing Capacity

The generally freely adjusting deformations from external loads must be distinguished in dependence of:

- **global deformations** of the entire structure;
- and **local deformations** of individual components.

To account for the load-bearing capacity of a building, the global **horizontal deformations** must be investigated, which lead to skewing of vertical components such as columns and walls. They mainly result from wind loads on the building en-

5

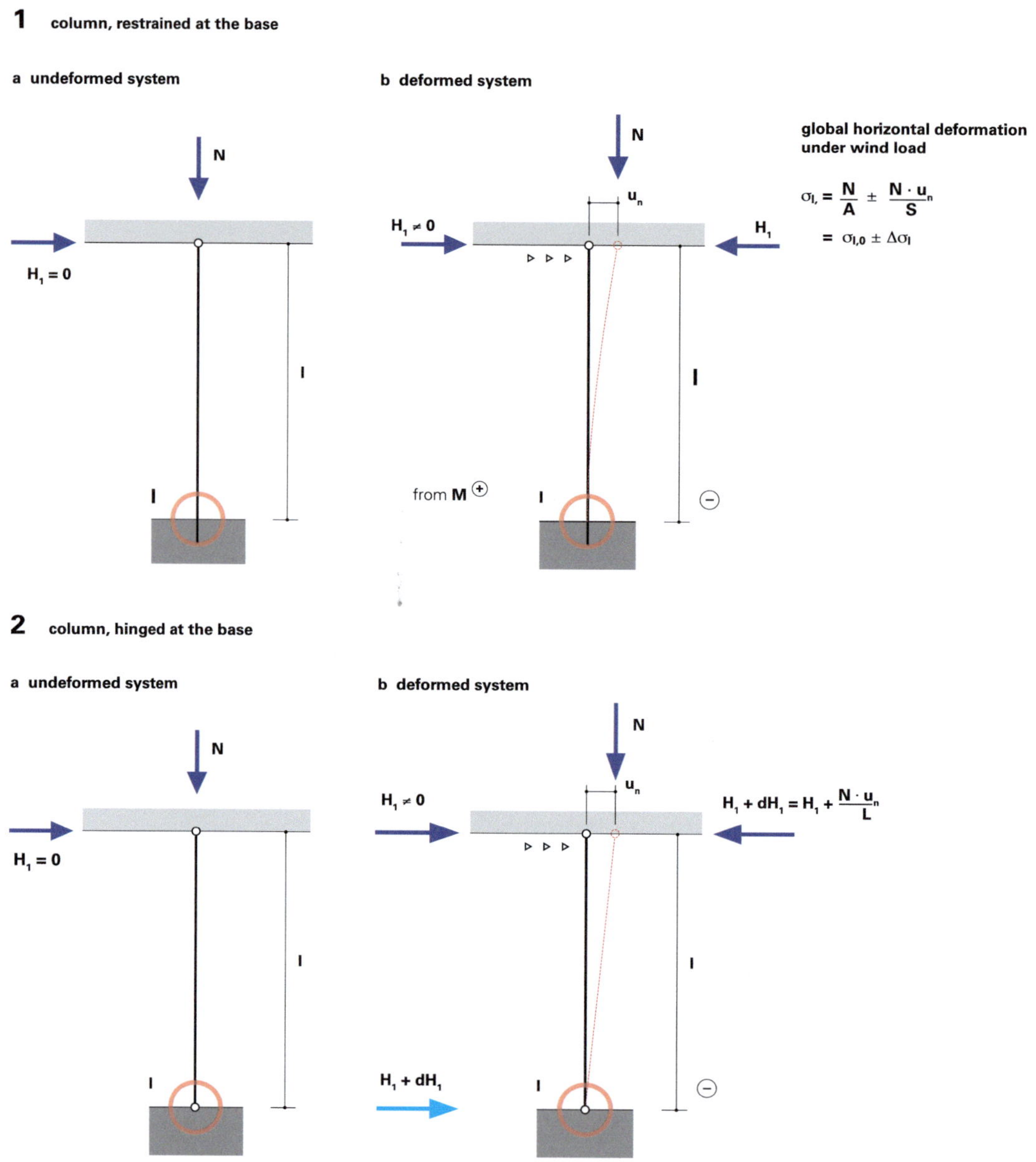

Fig. 5.12 Column under pure vertical load (on the left) and skewed position under combined vertical and horizontal load (on the right)

velope. The misalignment leads to an **additional stress** $\Delta\sigma$ on the columns and walls, or to **drifting forces** which must be transmitted into the shear-resistant floors (Fig. 5.12). By means of a suitable **bracing** of the primary structure, these additional loads can be largely avoided. For this purpose, exemplary bracing concepts are presented for different structures of building construction in ▶ *Sect.* 4.

For shear-plate construction modes: cf. ▶ *Sect. 4.2.1.2.4 Bracing*; for skeletal structures: ibid. ▶ *Sect. 4.2.1.3 Bracing of Skeletal Structures*

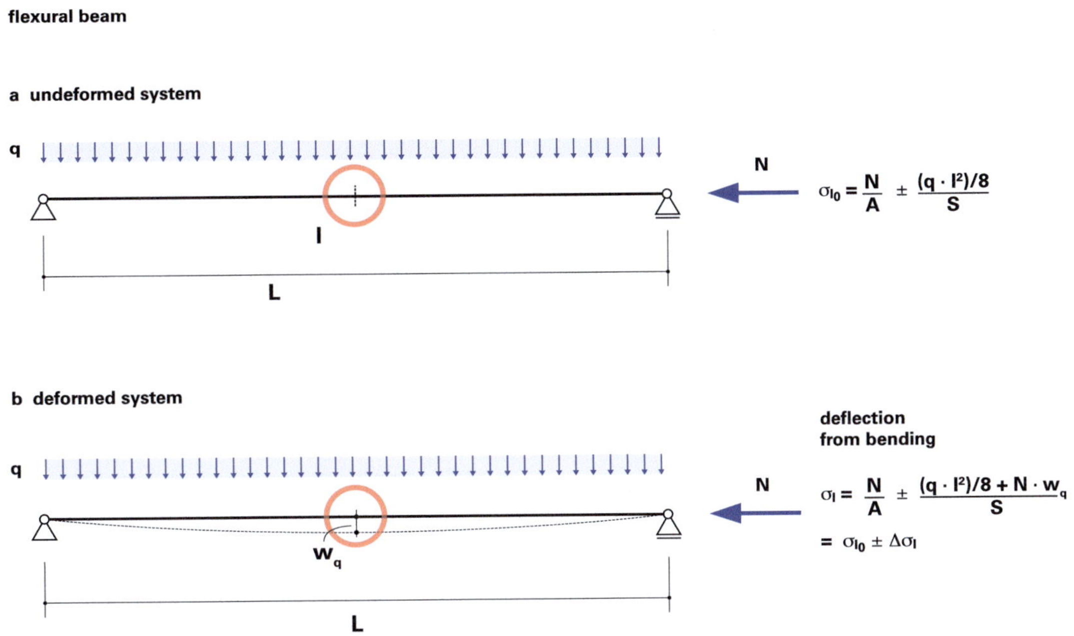

Fig. 5.13 Additional stress $\Delta\sigma_l$ on a flexural beam under bending and normal force when it deforms

The local deformations of individual components of the load-bearing structure have to be taken into account when checking the load-bearing capacity of compressive-stressed components, since they can lead to a stress increase in the cross-section or to a stability failure (Fig. 5.13).

Cyclic deformations resulting from temperature changes and time-dependent deformations from shrinkage and creep, can be treated, as far as they can unfold freely—i.e. unhindered—within the structure, in analogy to the load deformations what load-bearing capacity is concerned.

A partial or complete obstruction of these imprinted deformations results in considerable constraints within the load-bearing structure which, in their order of magnitude, can significantly exceed the actual loading locally or globally. Therefore, in the course of the design, structural concepts have to be developed which largely prevent the occurrence of constraints within the structure. The following is an example typical of building construction (Fig. 5.14). Further constructional measures are described in ▶ *Sect.* 6.

Sect. ▶ *6.3 Types of Foundation*

5.2.2 Impact on Functionality

The deformations of structures also have an effect on the **function** of the building and therefore on its use. For example, vertical deformations of a roof structure must be made compatible with the dewatering concept. The deflections of the floor construction must be adapted to the deformability of the

5

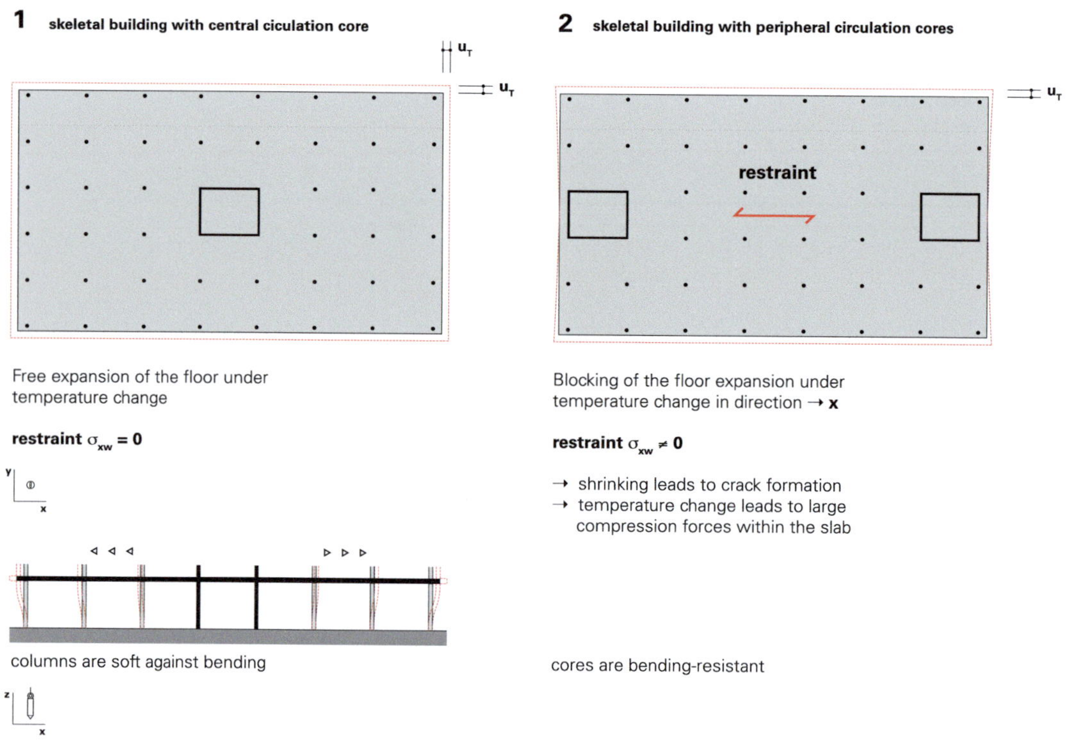

Fig. 5.14 Effects of alternating arrangements of bracing circulation cores in skeletal-building construction

subfloor as well as of the vertical partition walls and façade elements attached thereto. In this case, these envelope components must be connected to the primary structure in such a way that the deformation of the load-bearing structure can freely adjust. Otherwise, load relocations may occur within these non-load-bearing structures and thus to their impairment or even destruction (Fig. 5.21).

The horizontal deformations of primary structures result in a skewing and twisting of the connected enveloping elements (Fig. 5.15). In particular, deformation-sensitive façades made of insulating glazing can be decisive for the dimensioning of the main structure's bracing. Alternatively, long façades can be divided by transverse joints (Fig. 5.21, bottom picture).

5.3 Structural and Constructional Solutions of Building Construction

DIN 18197, DIN 18195-8, DIN 18540

The deformations determined within the framework of the structural calculations form the basis for the design and the connection details of the secondary structures of the envelope and the mechanical building equipment. Thus, the characteristics and the size of the deformations of the overall structure

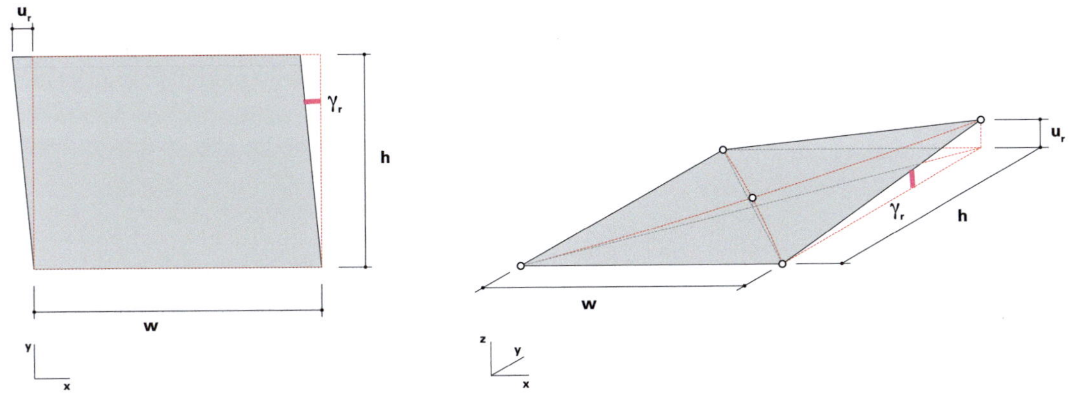

Fig. 5.15 **Skewing** (left) and **warping** (right) of façade elements as a consequence of horizontal deformations of the primary structure

as well as of individual components must be carefully determined. As explained in ▶ *Sect. 5.1*, the exact modelling of the external effects—the load models—and the realistic computation of the material behaviour—the material properties—play a fundamental role.

The essential foundations for a building being also economical in terms of deformation are already laid during the course of the design of the load-bearing structure. With respect to horizontal deformation, a distinction is made between **displaceable** and **non-displaceable** systems:

- **Displaceable** systems include **portal-frame** systems. The comparatively large deformations of the frames under horizontal loads can be absorbed by the compliant envelope constructions used. The interior fit-out of these buildings, which are generally designed as halls, is largely decoupled from the structure. The resulting deformations produce additional stresses on the portal frame (Fig. 5.11).
- **Non-displaceable** systems include **shear-plates** and **diagonal bracings**. They allow a low-deformation absorption of horizontal loads. This results in no additional stresses in the load-bearing structure (Fig. 5.20).

Examples of displaceable and non-displaceable structural concepts, as well as frequently occurring hybrid forms, are presented in ▶ *Sect.* 4.

▶ *Sect. 4.2.1.3 Bracing of Skeletal Structures*

Constructional measures for neutralising deformations in the load-bearing design include **joints**, **bearings**, and **hinges** (Figs. 5.16, 5.17, 5.18, and 5.19). As a result, obstructions of deformations—which cause constraints—are cancelled, so

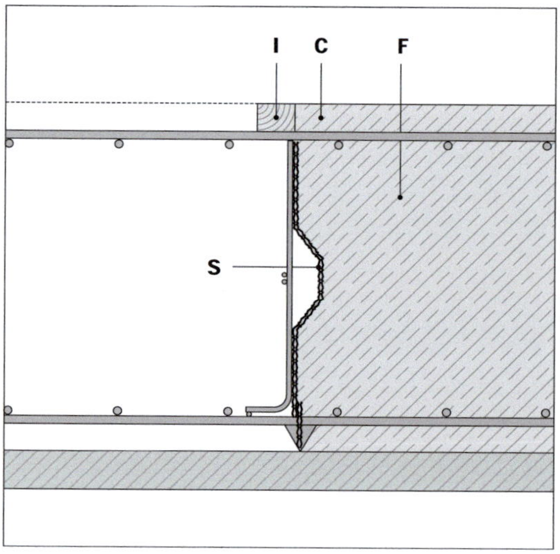

Fig. 5.16 Shuttering element **S** for an expansion joint in a concrete floor **F**, self-supporting with toothing. The concrete covering **C** is designed with an insert **I** to create a break point

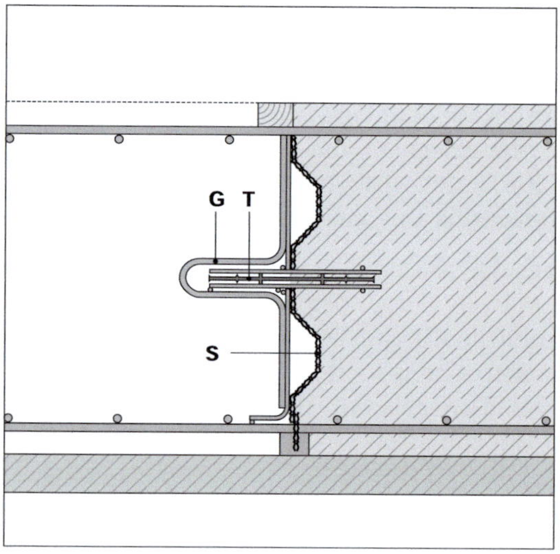

Fig. 5.17 Shuttering element **S** as on Fig. 5.16, with double-toothed joint and joint tape **T** in a guide **G**

that deformations from temperature and shrinkage can freely adjust (Fig. 5.20). The image in Fig. 5.21 shows a high-rise structure with a circumferential expansion joint around a core (case **2**) or with a transversal expansion joint along the middle row of columns with girder (case **3**).

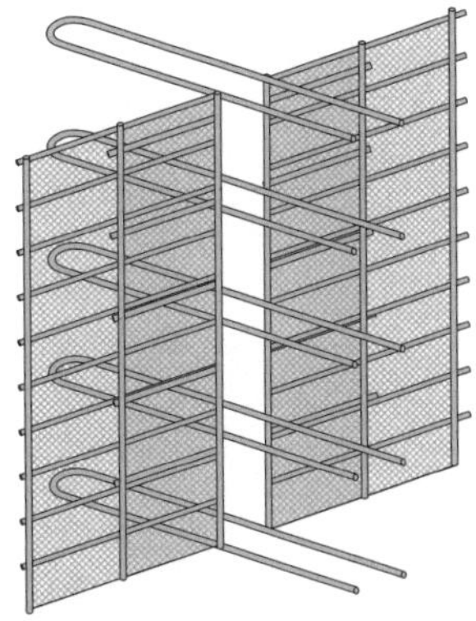

Fig. 5.18 (Left and top right) Shuttering elements with joint tapes for a wall. (Bottom right) Shuttering elements for a floor before concreting of a second section

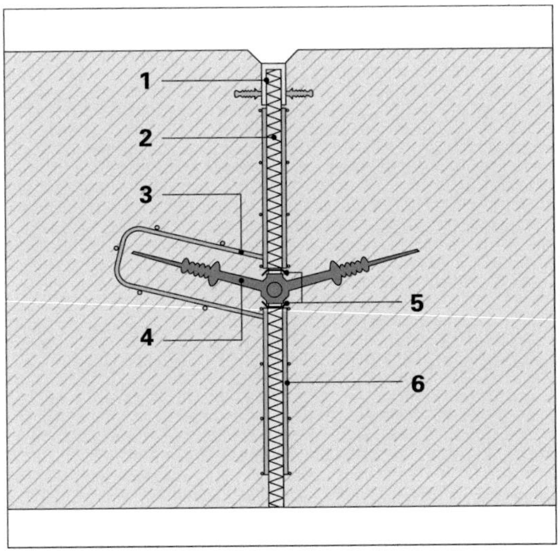

Fig. 5.19 Shuttering element for expansion joints with joint strips (manuf.: *Peca*®): (**1**) sheet metal folded for joint finishing strip; (**2**) separation joint filling as required; (**3**) joint tape cage for receiving the expansion joint band, tilted upwards in case of horizontal joints; (**4**) central tube of the joint tape; (**5**) DF-fixer for fixing the expansion joint tube below and above the band; (**6**) bearing mat on both sides of the joint insert as a retaining structure

5

■ **Fig. 5.20** Diagonal brace: an example of a system which is **non-displaceable** in terms of horizontal deformation

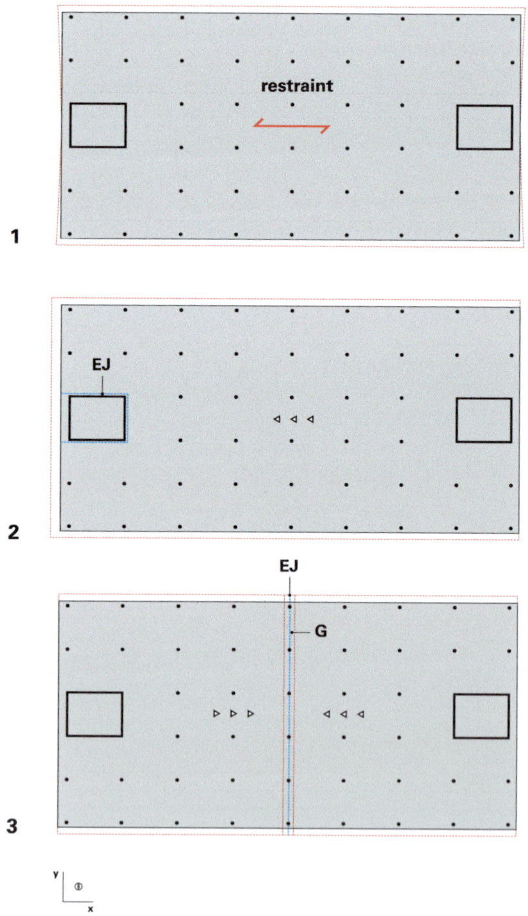

■ **Fig. 5.21** Constructional measures to absorb constraint in a skeletal building in case **1** (see ■ Fig. 5.14): (**2**) Circumferential separation of a core from the floor, deformations extending from the fixed point of the opposite core (**EJ** = expansion joint). (**3**) Transversal joint **EJ** over the central column row with girder **G**, deformations starting from both cores

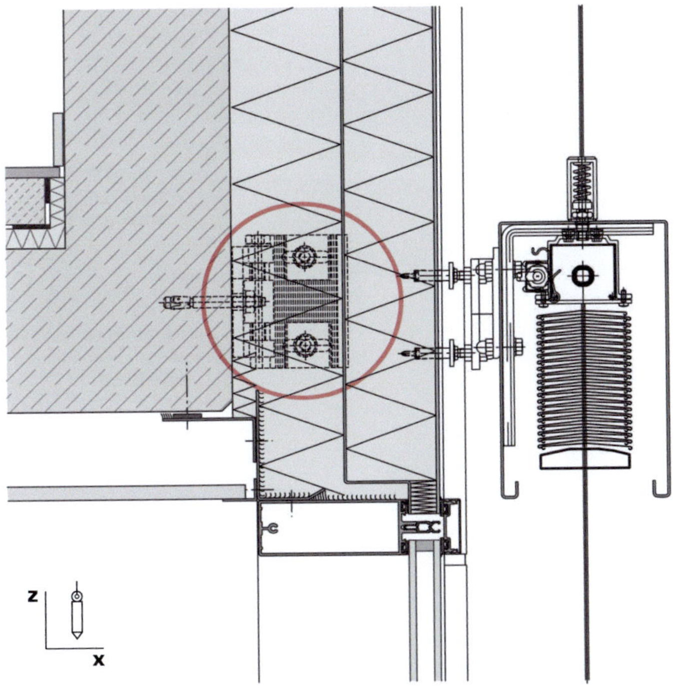

Fig. 5.22 **Deformational decoupling** between a floor slab (primary load-bearing structure) and a mullion-and-transom façade (secondary load-bearing structure) by means of a sliding connection (cf. details in ***Vol. 3***, *Sect. 14.3.1.2*)

Hinges and joints are often used for decoupling deformations between the primary and the secondary structure, where large differences in stiffness can also result in high constraint stresses building up (Fig. 5.22).

Joints also allow for the segmentation of long constructions in which imprinted inner shape changes, for example from temperature changes, can accumulate to large deformations over the whole length (Fig. 5.23).

Standards and Guidelines

DIN 18195: 2017-07 Waterproofing of buildings—Vocabulary

DIN 18197: 2018-01 Sealing of joints in concrete with waterstops

DIN 18540: 2014-09 Sealing of exterior wall joints in building using joint sealants

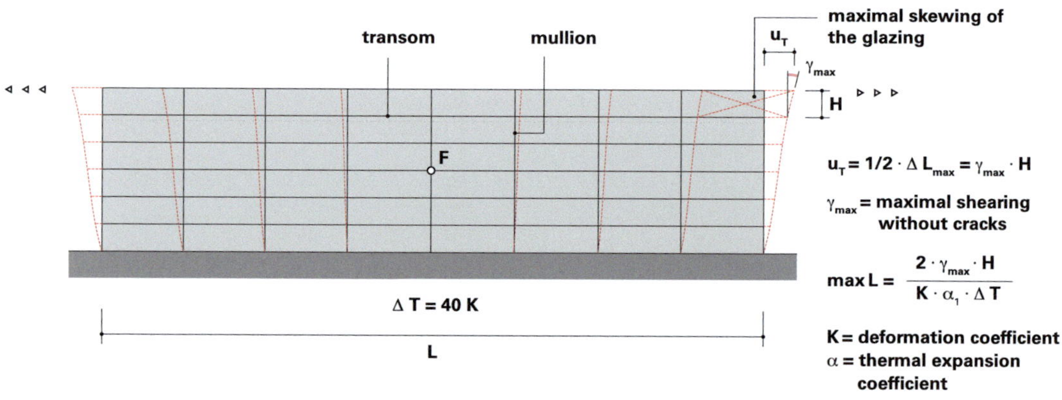

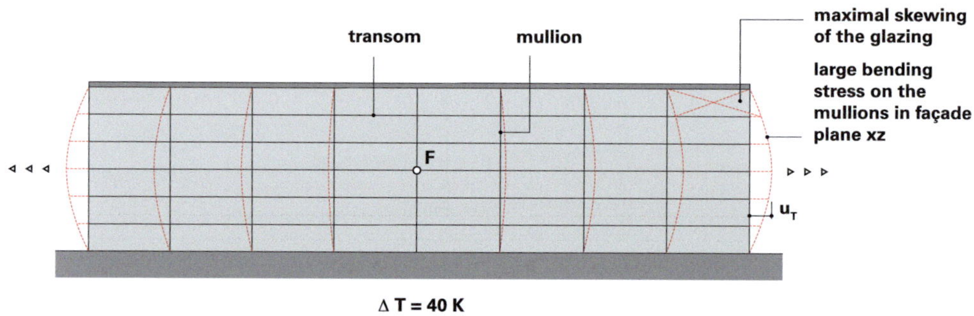

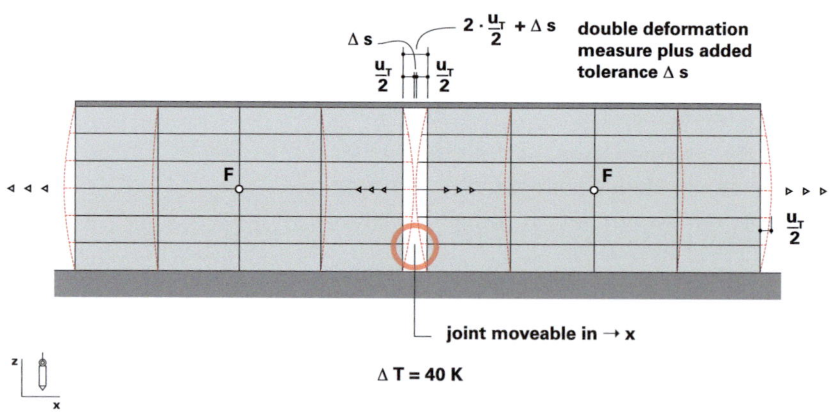

Fig. 5.23 Deformation of a hall façade not retained at its top (top diagram) and a hall façade linearly retained at its upper and lower ends (central and bottom diagram) under temperature change **ΔT**, in both cases extending from the fixed points **F**. Façade retained on both sides, on top shown without expansion joint, at the bottom with expansion joint (deformation in →**x** magnified)

Bibliography

Derler P, Koch J, Piertyas F (2018) *Erfolgreiche Bauwerksabdichtung: Neubau – Sanierung*. Kissing, WEKA

Leibinger-Kammüller, Nicola (2005) *Faszination Blech: ein Material mit grenzenlosen Möglichkeiten*. Würzburg, Vogel

Foundation with Dr. Matthias Weißbach

J. L. Moro, *Building-Construction Design – From Principle to Detail*,
https://doi.org/10.1007/978-3-662-71006-7_6

Contents

6.1 Generalities

The support of a primary structure significantly determines its stresses and deformations. At the same time, the abstracted static concept of the **support** of a load-bearing structure is to be understood, in final consequence, as a discharge of the loads into the ground.

In addition to the **own weight** of the building, **utility** and **snow loads** must be transferred into the ground by way of vertical soil compression. Furthermore, the loads from **horizontally** acting loads, such as wind, must also be safely accommodated in the ground.

A **foundation** must be constructed for safely transferring these building loads into the ground (Figs. 6.1, 6.2, and 6.3). It is part of the load-bearing structure and represents the structural connection between the building and the ground. The very different bearing and deformation behaviour of, respectively, the foundation and the construction of the building necessitates a suitable foundation structure which ensures the stability and serviceability of the building. *DIN 1054*

While the materiality and composition of the primary structures of the building construction as discussed in this chapter are essentially determined by the decisions of the design engineer, the foundation of a structure is determined by the encountered physical and structural constitution of the ground. The basic soil conditions of the site as well as the technically feasible foundation concepts derived therefrom are therefore to be taken into account during the design process from the beginning as an essential boundary condition for the conception of a building and its primary structure. *DIN 4020, DIN 4020* supplement 1

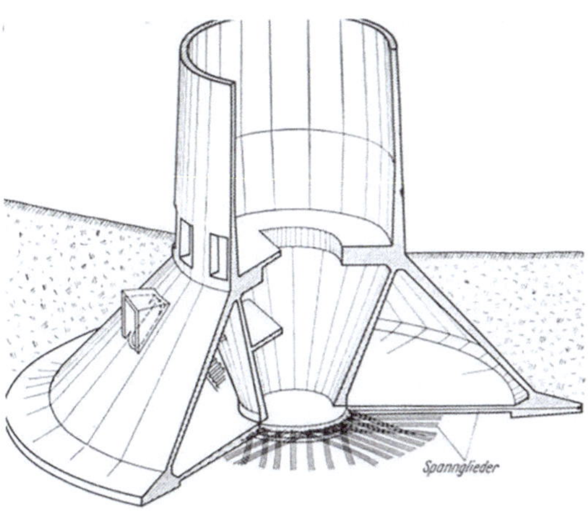

Fig. 6.1 Foundation cone of the television tower in Stuttgart (eng.: F. Leonhardt)

Fig. 6.2 Stripped strip foundation (left), shuttering of a strip foundation (top right)

Fig. 6.3 Individual foundations of a skeletal structure after stripping

However, the subsoil does not only carry loads and absorb deformations of the load-bearing structure, but, in turn, also represents a stress acting on the load-bearing structure. Thus, ground and water pressures, as well as settlements of the surrounding soil, can lead to relevant stresses on the load-bearing structure. Furthermore, being a continuous half-space, the subsoil conducts dynamic effects over long distances and transfers these to buildings. Thus, for computing the behaviour of primary load-bearing structures, it is of great importance to consider the **structure-ground interaction**. This will be discussed in more detail below.

6.2 Interaction Between Structure and Foundation

6.2.1 The Building Subsoil

DIN 18196, EN 1997-1, EN 1997-2

In geotechnology, a distinction is made between **soils** in the sense of loose rock, and **rock** in the sense of solid rock. This separation is based on the fundamentally different mechanical properties of these soil materials. The load-bearing behaviour of rock is scientifically formulated by means of the still rather young rock mechanics for mining, tunnelling, and dam building.

For the primary load-bearing structures of building construction considered in this chapter, the subsoil, however, generally consists of **loose-rock layers**, so-called **soils**. These are composed of three phases:

- The **solid substance**, also known as **grain composition**, is determined in its properties by the weathering process at the point of origin and the transport process to the deposition site.

  With regard to this, a distinction is made between:

  - **non-cohesive** soils (◘ Fig. 6.4 left)—for example, from gravel or sand, and
  - **cohesive** soils (◘ Fig. 6.4 centre)—for example, from clay or loam.

  In addition, soils may contain **organic constituents**, e. g. foul sludge or peat (◘ Fig. 6.4 right).

The grain size or specific surface area of the solid substance substantially influences the mechanical properties of the soils. For example, the comparatively large grain size of non-cohesive soils produces friction between grains within the aggregate. The very small grain size, and thus the large specific surface area of cohesive

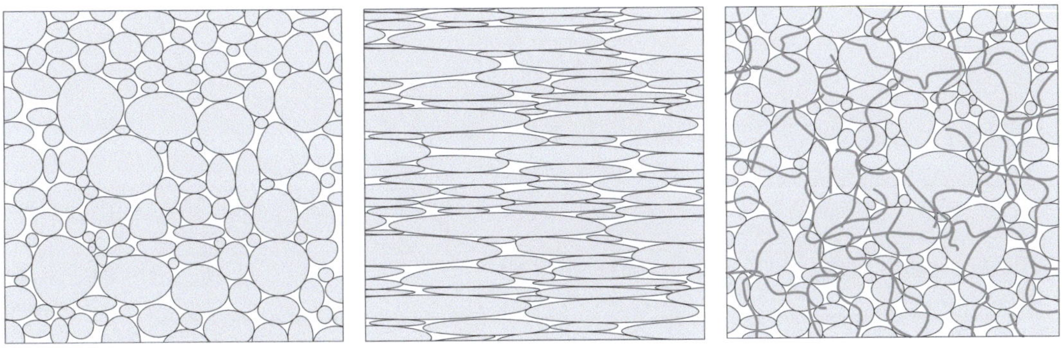

◘ **Fig. 6.4** (Left) **Non-cohesive soil**—granular soil structure. (Centre) **Cohesive soil**—platelet-like texture. (Right) **Organic soil**—interspersed with fibre mesh

Vol. 1, *Sect. 11.9.1.2.2 Loam Products* as well as ibid. *Sect. 11.11.3.2.2 Gliding > Water-Bound Granulates*

soils, results in a dominance of cohesive forces as a result of adsorption.

The grain-size distribution describes the composition of the soil from different grain sizes, represented by grain-size curves in weight percent (Fig. 6.5). Soils are further classified, for example, in coarse, mixed, or fine-grained soils or also in narrow, medium, or wide-graded soils.

The grain-size distribution is essential for the compressibility as well as the frictional capability, i. e. the load-bearing capacity, of the soil, its filter resistance, and frost hazard.

6

The pore space of the soil is enclosed within the grain structure as well as between the grains. The pore ratio of the soil—in percent by volume—as well as the pore type and pore size influence the properties of the soil.

In the pore space of the soil, **air** and **water** are found as other essential phases of the soil:

- **Air** is located in the closed as well as open pore space of the soils. In the closed pore space, the air influences the weights of the solid substance components and thus also the hydraulic stability of the grain structure.
- **Water** enters the soil in three different states. A distinction is made between:
  - **molecularly bound** water,
  - **surface** or **capillary** water and
  - **free groundwater**.

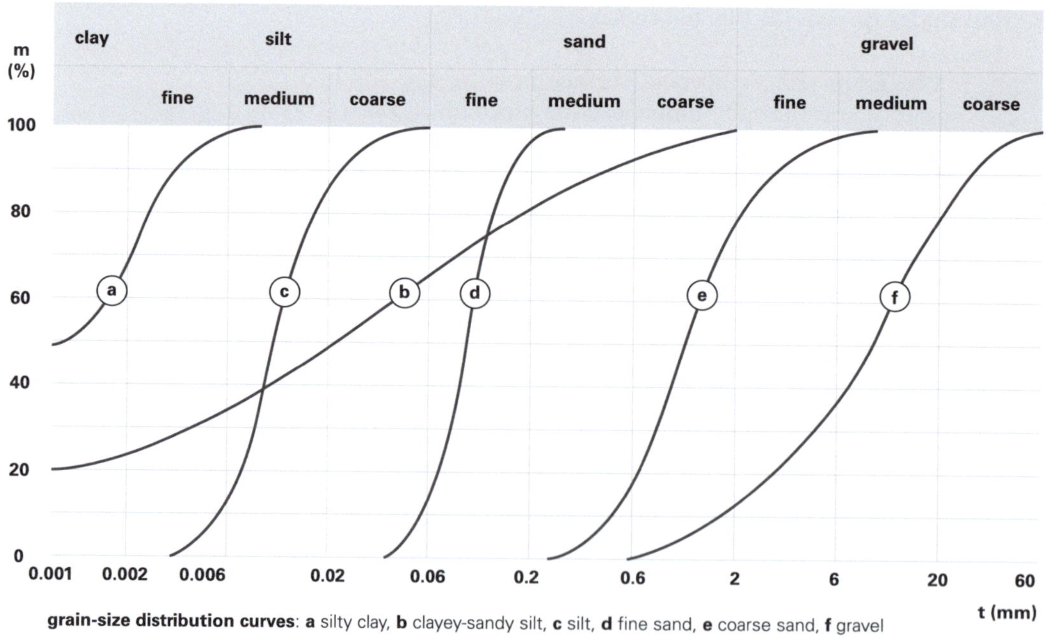

Fig. 6.5 Diagram of the **grain-size distribution curves** for different soil types

The proportion of water present in the subsoil and its state is influenced by the pore proportion as well as the pore type and pore size.
For example, narrow-graded, coarse-grained soils with a correspondingly large and coarse pore structure contain mostly free water. They have high water permeability and filter resistance and are generally frost-insensitive. Their load-bearing capacity is relatively slightly affected by the water content. It is only when the water saturation level—soil layers in ground water—is reached that, with a constant superimposed load, the effective stresses between the grains subside. This reduces the load-bearing capacity of the soil, which is generated by friction within the grain structure.
On the other hand, highly graded, fine soils with a consequently fine capillary pore structure are characterised by a high proportion of capillary and surface water, which can already lead to water saturation. This is associated with low water permeability and high frost sensitivity. The load-bearing capacity of these soils, which is based on cohesive forces between the soil particles, is influenced by the shape of the particles and their position relative to each other and thus by the water content. On the one hand, these soils do not require a superimposed load to develop a high load-bearing capacity. On the other hand, they are very sensitive to load changes and frost-thaw changes.

6.2.2 Load Transfer Between Structure and Foundation

The interaction of the structure and the foundation is determined by the ratio of the strengths and stiffnesses of both partners. Due to its relatively low load-bearing capacity in comparison with the load-bearing structure, the building ground requires **distributing** the forces which arrive concentrated in the columns and walls to an **enlarged area**. To create this enlarged surface is the main task of foundation elements.

From the deformations permissible for the structure and from the fracture conditions of the subsoil, permissible ground pressures are obtained which provide the base for calculating the required **footprint surface** of walls and columns.

The task of distributing the concentrated stresses upon this surface is thus assumed by the **foundation construction**. The utilisation of the permissible soil compression and the bending stiffness of the foundations influence the distribution of the ground pressures under the foundation footing and thus the utilisation of the bearing capacity of the ground with minimal deformations (◘ Fig. 6.6). For this reason, a soil breaking resistance of $\eta = \mathbf{2.0}$ is generally assumed for real foundation stiffness.

For ground break, cf. ◘ Fig. 6.17

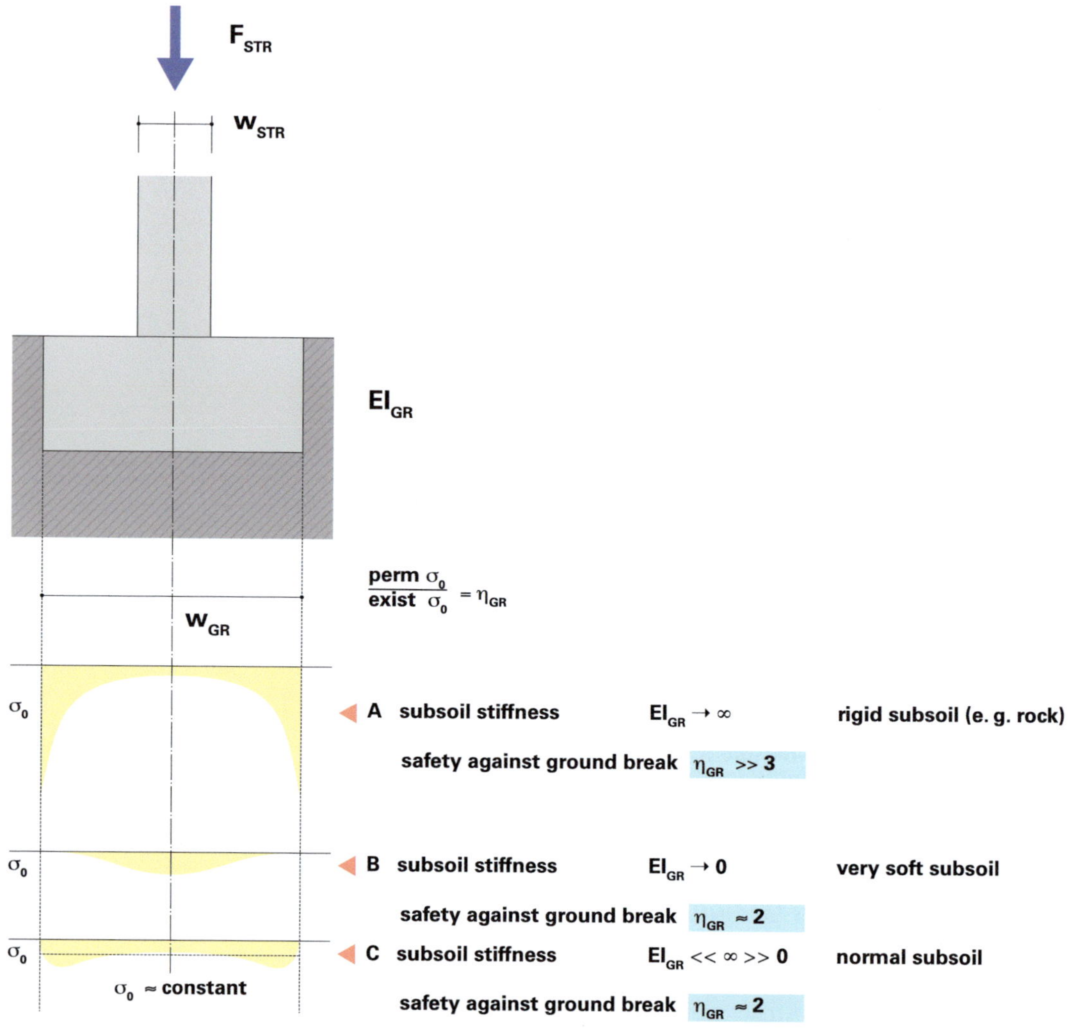

Fig. 6.6 Distribution of soil compression under a foundation in non-cohesive soil

Parts of the structure are integrated into the subsoil. As a result, they are exposed to **earth pressure** and **hydrostatic pressure** from ground water. These loads must be borne locally and globally by the primary structure of the building.

The **earth pressure** is the spatial stress condition in the continuous half-space of the subsoil, which must be sustained by the part of the building which penetrates into the subsoil. It results from the soil's own weight and from existing superimposed loads. The magnitude of the earth pressure increases with the depth below the top edge of the terrain. The horizontal component of the earth pressure burdening the perimeter of the structure depends on the strength properties of the ground and the deformations between building and ground. The latter leads to the distinction between:

- **active** earth pressures,
- **resting** earth pressures,
- **passive** earth pressures,

which differ significantly in the amount of the stresses on the primary support structure (Fig. 6.7).

The **resting** earth pressure equals the stress state in the undisturbed subsoil and therefore requires a quasi-rigid, non-displaceable structural body.

If under the effect of earth pressure there are significant deformations or displacements of the building, or of individual components thereof, which are generally directed away from the ground, the impact is clearly reduced. This is equal to the load of the earth wedge slipping behind the building body—this is then called the **active** earth pressure.

If the structural body or parts of it move against the ground under external loads (load, temperature, etc.), which implies pushing up a ground wedge, the resting earth pressure increases significantly. This is then termed **passive** earth pressure.

The stiffness and damping properties of the subsoil on the one hand and the dynamic coupling of the load-bearing structure to the subsoil—the foundation construction—on the other are of decisive importance for the transmission of dynamic loads of the subsoil, for example from earthquakes. The latter can be designed by means of suitable **damping** and **obliterating constructions** in such a way that the input of earthquake loads is reduced or the in-fed earthquake energy is consumed without structural damage and thus absorbed.

6.2.3 Load Propagation Within the Soil

The loads discharged to the foundation ground at the foundation bottom are passed on via the different soil layers. This results in a spatial spread of the load with increasing depth, which causes a reduction in the ground stresses. The angle at which the load area of the ground increases with increasing depth depends on the friction or cohesion within the individual soil layers (Fig. 6.8). Cohesive soils show significantly better load propagation. This is due to the load-bearing capacity of these soils, which is based on cohesion and not upon the superimposed load.

The spreading of the building loads in the soil is of great importance for the prognosis of the **settlements** to be expected (Fig. 6.9). For a settlement calculation, the existing pre-loads and the additional loads resulting from the building to be constructed must be determined in each soil layer. The very different stiffnesses and time-setting curves of these soil layers, with partially significantly varying layer thickness, require an accurate determination of the load propagation. For instance, a pre-loading of the building ground can have enormous advantages but also catastrophic consequences for a

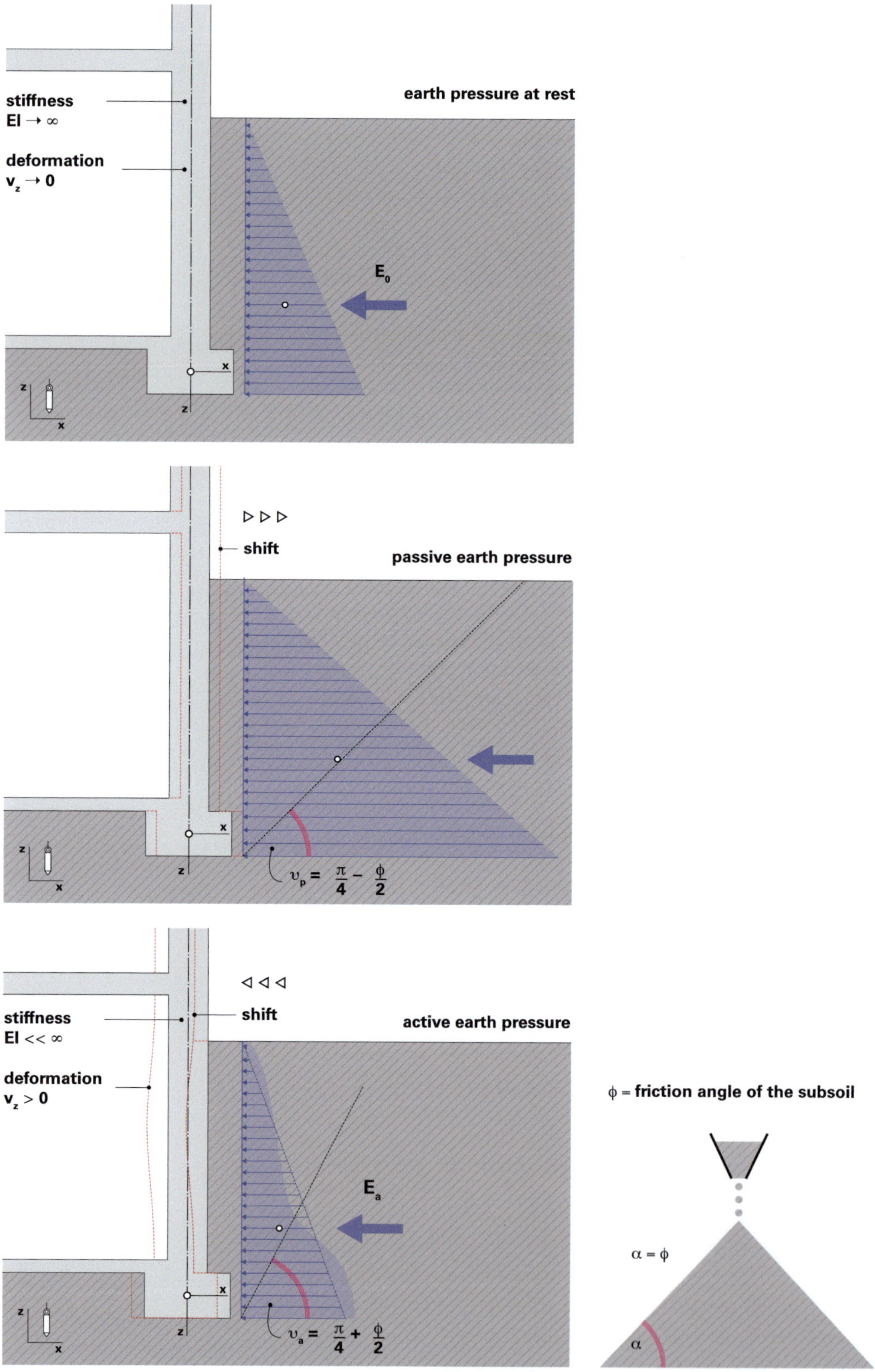

Fig. 6.7 Earth pressure at rest (top), passive (centre) and active (bottom) earth pressure

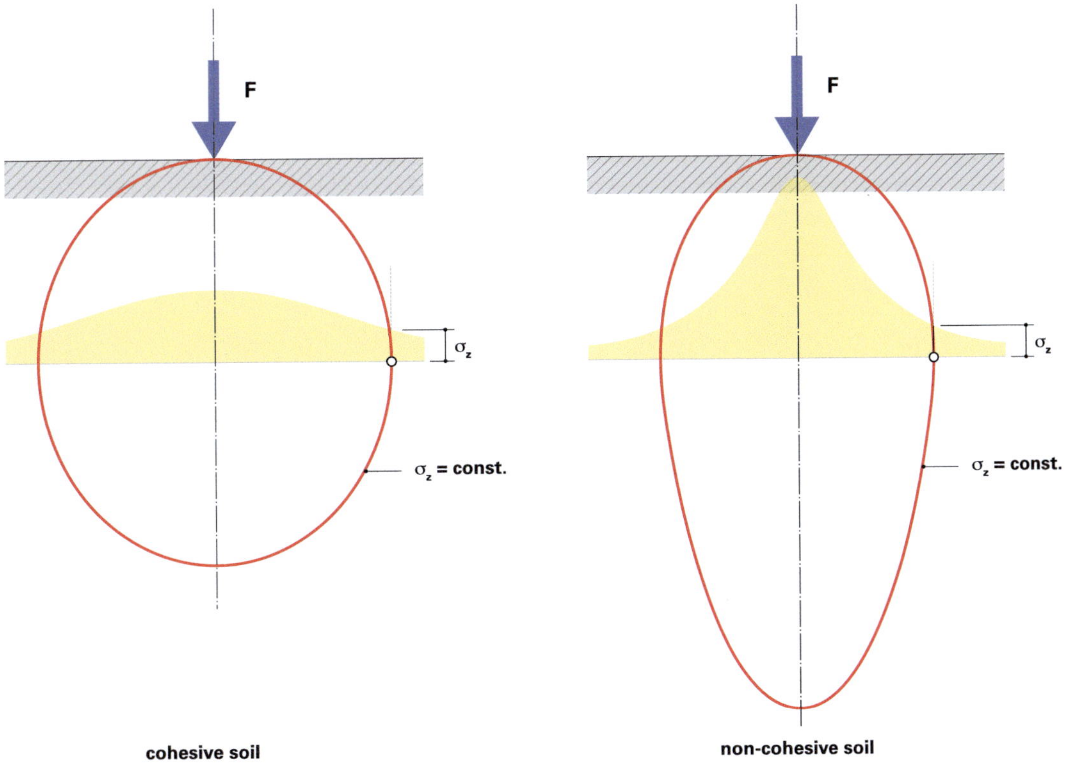

Fig. 6.8 Lines of identical σ_z-stresses, so-called σ_z-isobars

building to be erected (Fig. 6.10). For example, by mechanical compaction or over-moulding of surfaces to be built, the stiffness and load-bearing capacity of the subsoil can be deliberately improved before construction begins. On the other hand, uneven pre-loading of the subsoil from existing neighbouring buildings causes stiffness differences, which can result in differential settlements in the case of additional loads from a new building (Fig. 6.9).

For assessing the load-bearing capacity of the subsoil, tracking of the load propagation can also be important in some cases. In general, however, a sufficient proof of load-bearing capacity is provided by keeping permissible ground pressures at the location of the load introduction—the foundation footing.

If deep-lying subsoil strata which are not directly loaded and have a significantly reduced or variable load-bearing capacity exist in the foundation profile, a load-bearing capacity calculation must also be carried out on the basis of the load propagation. In case of ground terraces in the immediate vicinity of the building, a sloping down of the ground body along the critical shear joint is to be investigated. For this purpose, the ground stresses occurring as a result of the load propagation must also be determined.

▶ *Sect. 6.2.5 Failure of the Ground*

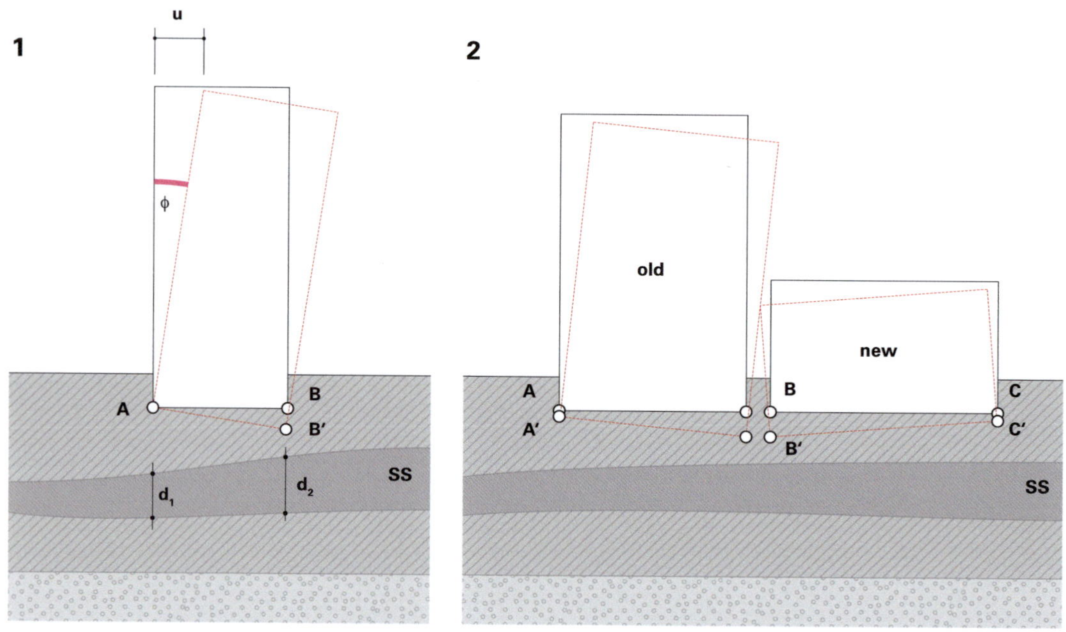

Fig. 6.9 Settling difference of first (**1**) and second (**2**) order: (**1**) Unequal layer thickness $\mathbf{t_1}$, $\mathbf{t_2}$ of the setting-sensitive soil layer **SL**. (**2**) Different pre-loading of the setting-sensitive soil layer **SL**

6

6.2.4 Deformations of the Ground

Deformations of the subsoil occur primarily as a result of load effects, but can also have non-load-related causes. The various causes of deformation are analysed in more detail below.

6.2.4.1 Load-Induced Deformations

Since soils are compressible substances, they react to any change in load through a change in shape. These deformations are referred to as **settlements**, as far as vertical movements due to compression of the grain structure are concerned. They occur preferably in cohesive soils and are not linear. They depend on the pre-loading of the soil layer considered and the loading time.

The development of the setting deformation depends on the type of soil. Whereas in the case of non-cohesive soils, the settlements appear immediately after load application as a result of the compaction of the grain structure, additional loads are absorbed in cohesive soils to a considerable extent as pore water over-pressure. This subsides very slowly and results in a relocation of loads on the grain structure. The resulting drainage of the loaded soil layer considerably increases the settling deformations. Because of the time dependency, in the case of cohesive soils, therefore, the term '**soil consolidation**' is used, which can lead to a lift-up of the ground after removal of the load (Figs. 6.11 and 6.12).

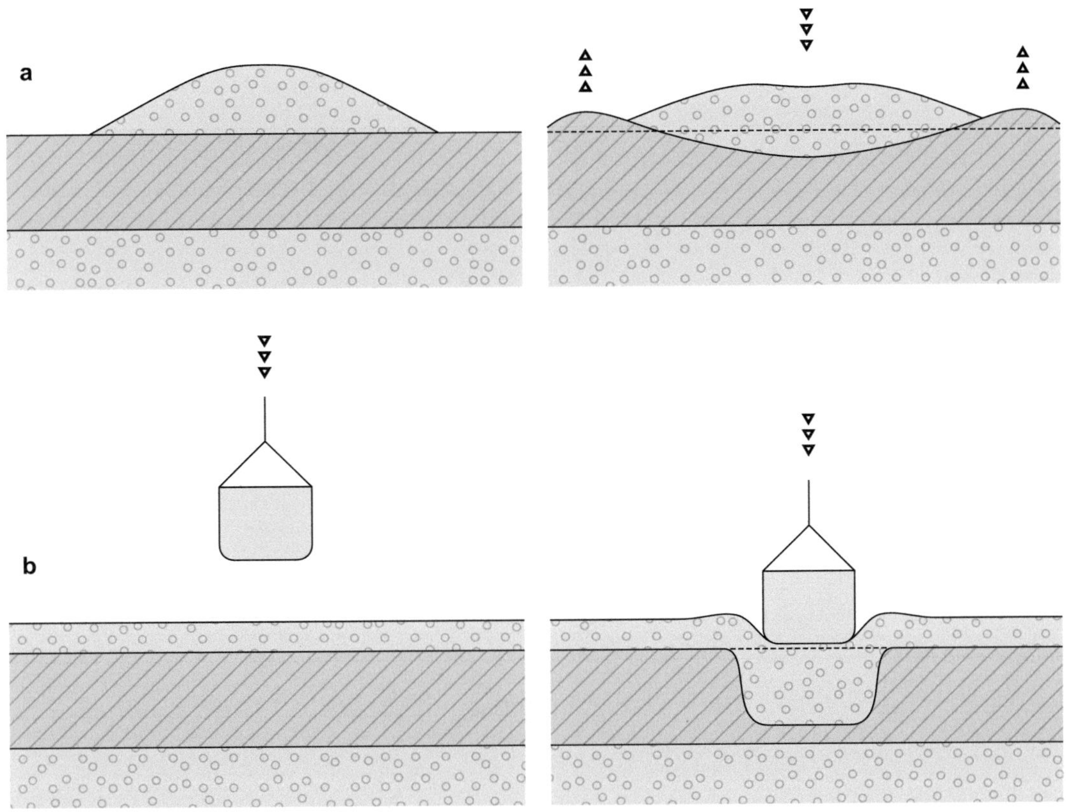

Fig. 6.10 Displacement of soft soil layers through the effect of ballast **a** (longer-term effect over a period of months) and impact **b**

This non-linear behaviour of the different soil layers of a subsoil requires a location-related, experimental determination of the settling properties in the foundation footing. For this purpose, undisturbed soil samples from the individual layers are removed by drilling in the course of a soil assessment and are tested in the laboratory using compression-setting devices.

The settlements determined on the basis of the building loads again represent a stress on the structure. In this respect, a distinction is made between:

- **uniform** settlements and
- **non-uniform** settlements.

Uniform settlements lead to a subsidence of the building, which generally does not result in any stress on the structure. This vertical movement of the building must be calculated and taken into account in terms of functional aspects, such as the connection of the building to its surroundings. Typically, these are vertical shifts of a few centimetres.

Unequal or **non-uniform** settlements, also referred to as **settling differences**, lead to deformations and stresses in the primary structure which must also be taken into account for

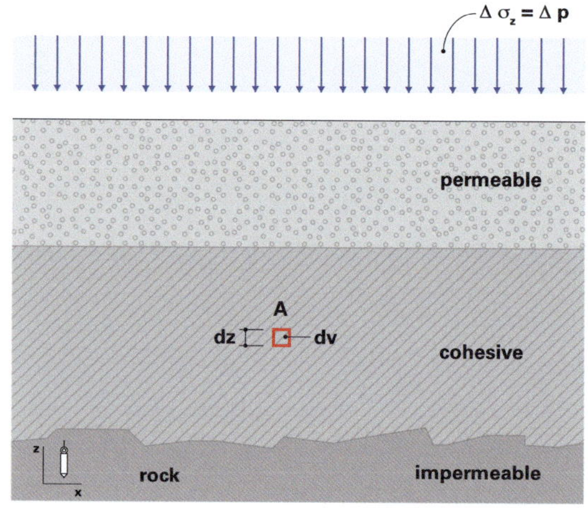

Fig. 6.11 Setting of the foundation bottom (in this case, point **A**) due to the constant top load **Δp** (t_0) (subsoil consolidation)

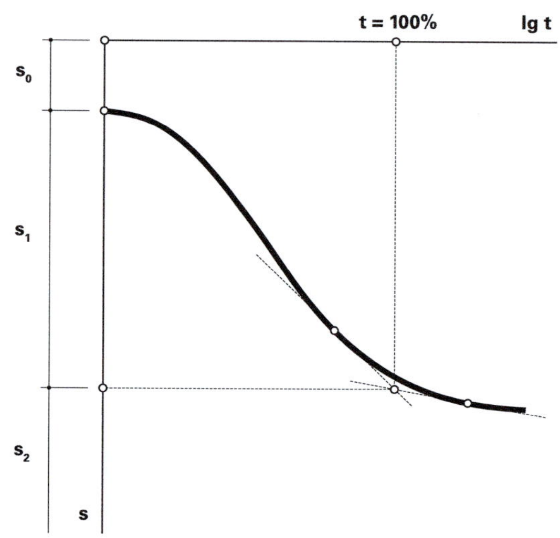

Fig. 6.12 A semi-logarithmic time-setting diagram related to the point **A** on Fig. 6.11. s_0 immediate setting (time-independent), s_1 primary setting (consolidation), s_2 secondary setting (creeping)

adjacent components of the building envelope and the fit-out (Figs. 6.13 and 6.14). These constraint stresses due to settlements acting on the primary structure result in a change in the load transfer and thus in the foundation loads which, in turn, change the ground settlements. This interaction requires a realistic modelling of the stiffness ratios of the load-bearing structure and its foundation construction for a soil that is sensitive to setting.

6.2.4.2 Frost-Induced Deformations

Deformations and stresses are also caused by the **uplifting** effect of **frost** in the subsoil. In temperate geographic latitudes,

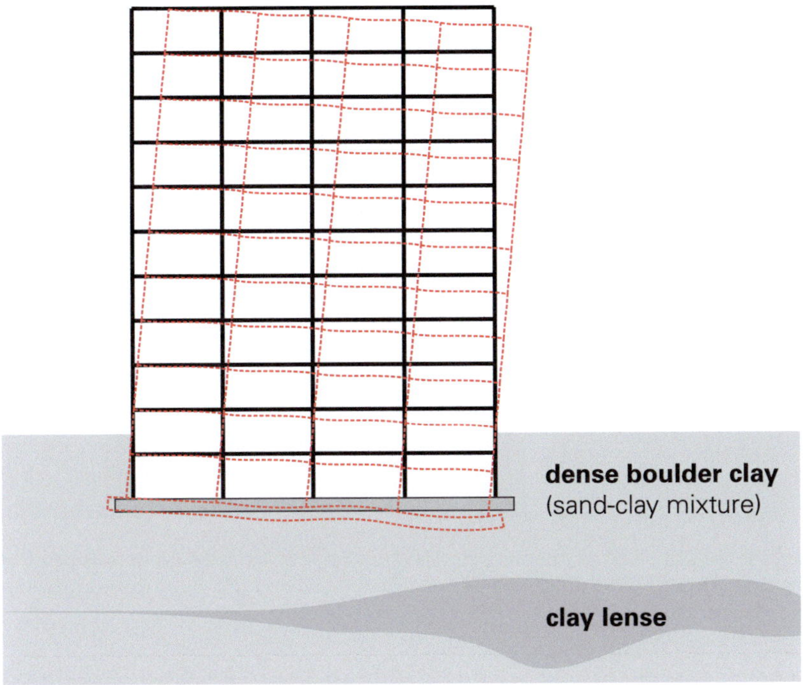

Fig. 6.13 Unilateral setting due to a clay lens leads to stresses and deformations in the frame structure of the building

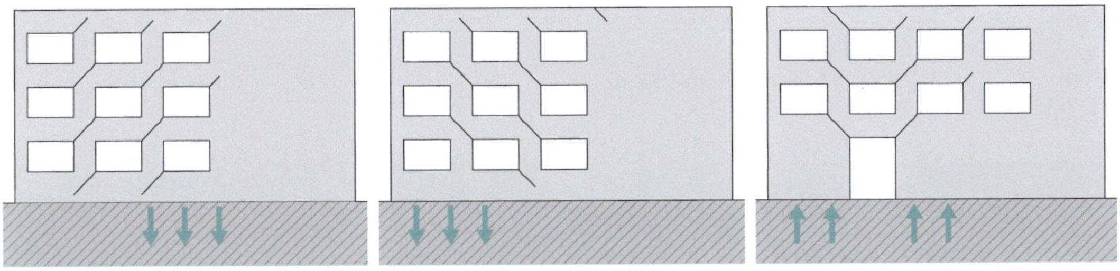

Fig. 6.14 Setting damage to building constructions depending on the type of setting using the example of a building in wall construction

it is to be expected that the ground freezes in winter to a depth of maximum 80 cm (**frost penetration**). This means that the pore water within the soil becomes ice and its volume therefore increases by up to 10%. This results in an expansion of particularly water-saturated soil areas and to a mostly uneven elevation of the superficial layers of the ground (local **ice lenses**) (Fig. 6.15 left). This effect is all the more pronounced the more fine-grained the soil texture is, and decreases significantly with increasing pore size. The effect of the frost is harmless whenever:

- the pore size offers sufficient space for the freezing soil water to expand, or whenever it can be reliably drained;

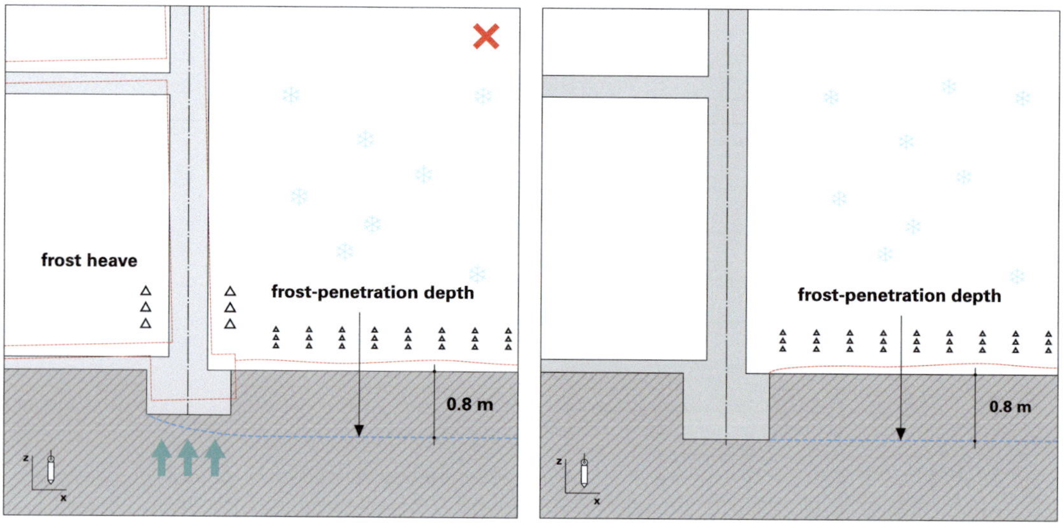

Fig. 6.15 (Left) Non-frost-proof foundation of an exterior wall: The foundation is not extended to frost-penetration depth and is pushed up by frost. (Right) Frost-proof foundation: The foundation is extended to frost-penetration depth. A frost heave is not to be expected

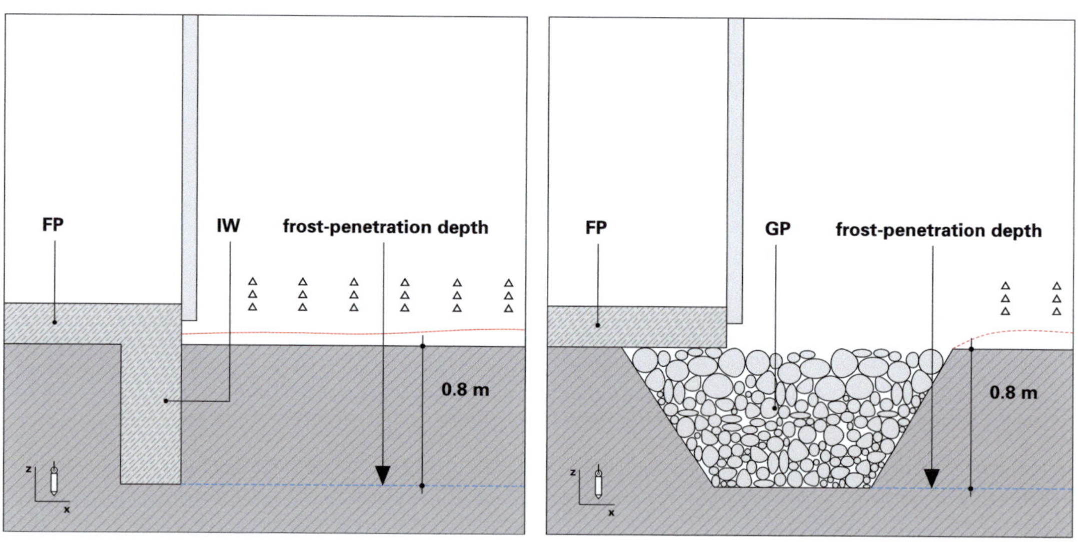

Fig. 6.16 (Left) Frost-proof execution of a floor plate **FP** by means of a concrete **ice wall IW** extended to frost-penetration depth. (Right) Frost-proof execution of a floor plate **FP** by means of a gravel filter packing **GP**

- the footing of the foundation is lower than the maximum frost penetration. Below this level, there is no longer any frost and, consequently, no expansion of the ground layer.

In the European code, the frost penetration is fixed at 80 cm, but in practice it is often set up to a minimum depth of 100 cm (Fig. 6.15 right).

The edge areas of floor slabs which are exposed to frost can alternatively be protected from freezing by means of an **ice wall** made of concrete (Fig. 6.16 left) or a **filter pack** made of gravel (Fig. 6.16 right).

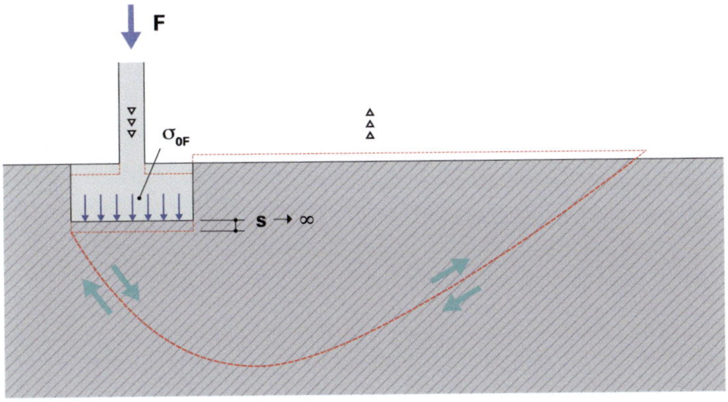

Fig. 6.17 **Ground break** as the result of an overloaded foundation

6.2.5 Failure of the Ground

As with any construction material used in building construction, the **strength** of the soil material is the decisive factor for the load-bearing capacity of the building ground. As a criterion of breakage, the **shear failure** of the soil is assumed. This corresponds equally to the cohesion- and friction-induced load-bearing behaviour of the ground and is in harmony with its pronounced non-linear load-deformation behaviour. *DIN 18122-1, DIN 18122-2*

This means that most of the failures of the building ground are based on an excess of tolerable **shear stresses** in the ground. Under the action of building loads, this results in the soil slipping along a predefined shear joint. In general, this is termed a **ground break** which can remove the designed support from the primary structure and thus provoke the failure of the building (Fig. 6.17).

The shear strength of each soil layer depends on its composition, its pre-loading history, and changing influencing variables such as water content or frost hazard.

It is the task of load-bearing-capacity calculations to determine the sliding joint across the different soil layers in the half-space of the subsoil under the effective additional loads of the structure which leads to the minimum shear strength at a maximum shear stress. If this shear stress exceeds the shear strength, uncontrolled sinking of the foundation into the ground will occur leading to a **ground failure**.

Further failures are the tilting and slipping of the building or a building part on the building site. In this case, the shear strength of the subsoil is once again the decisive resistance variable. These failures are of importance for primary structures which must bear large horizontal loads such as wind or one-sided earth pressure.

Danger of a **sliding failure** of the foundation (Fig. 6.18) is imminent if the horizontal force of the structure from **wind** $\mathbf{H_a}$ and **earth pressure** $\mathbf{E_a}$ to be delivered at the foundation footing is larger than the counter-acting **footing friction** of the founda-

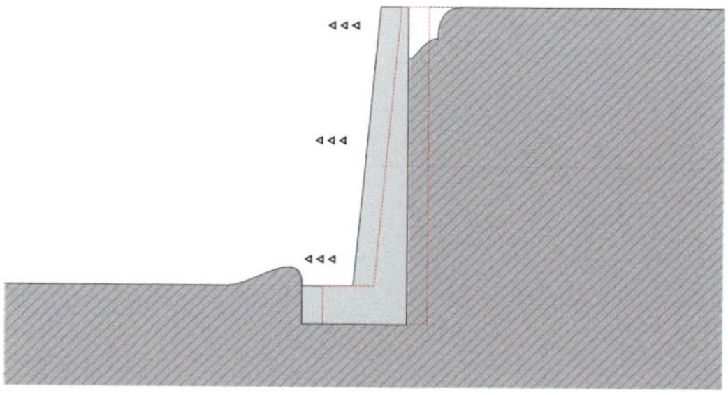

Fig. 6.18 **Sliding** of a retaining wall under the effect of earth pressure. This phenomenon is typical of retaining walls, but can also occur in buildings on slopes

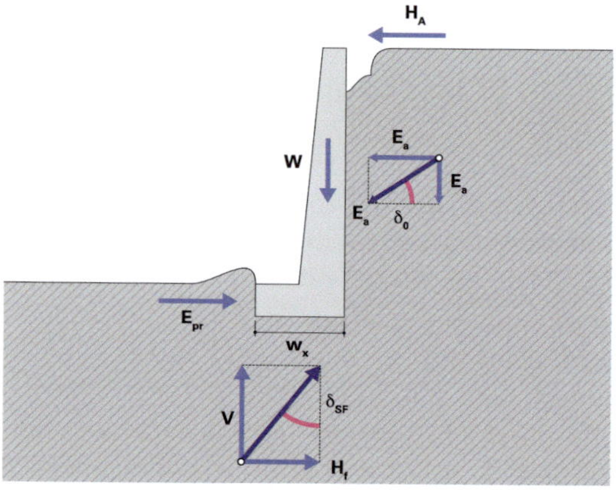

Fig. 6.19 Force conditions at the retaining wall as shown on Fig. 6.18 with finite displacement

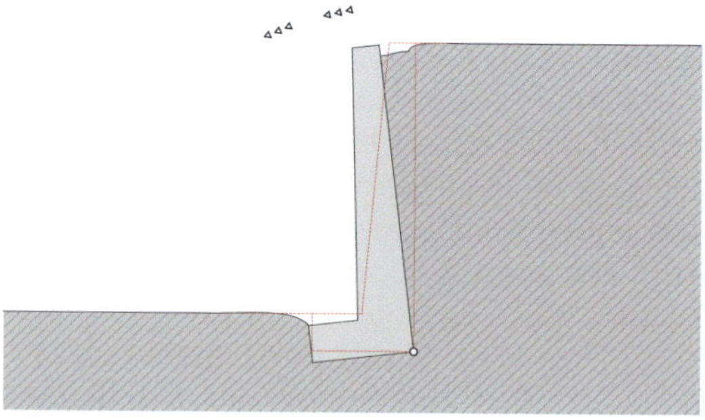

Fig. 6.20 **Tipping** of a retaining wall under the effect of earth pressure. In order to prevent tilting, the foundation must be arranged under the retaining wall in such a way that the resultant of the forces remains in the core area of the foundation bottom

tion $\mathbf{H_f}$ (Fig. 6.19). The horizontal **earth pressure** $\mathbf{E_{pr}}$, which counteracts the sliding, may be assumed as a stabilising factor when a horizontal displacement of the foundation is permitted.

The **tipping** of a building in the foundation results from the lever arm as measured from the horizontal loads acting on

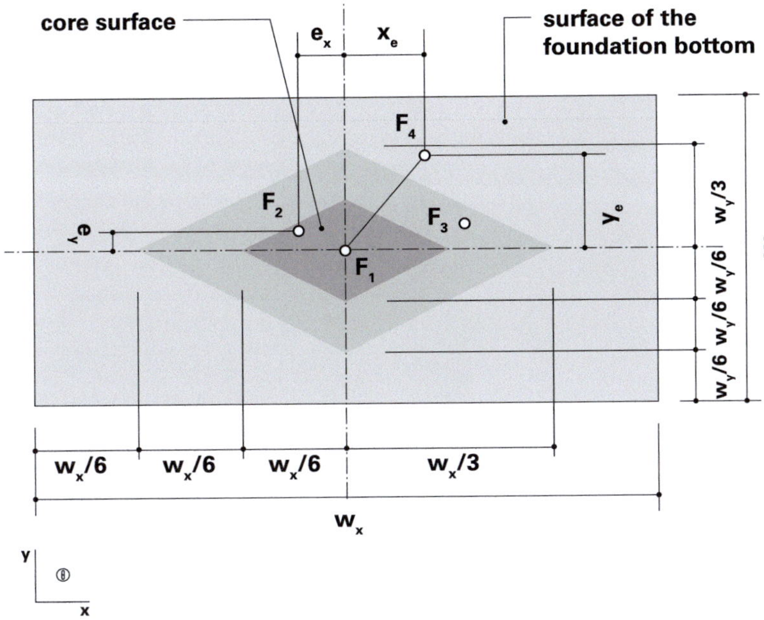

Fig. 6.21 Stress acting on a point foundation: If the force ($\mathbf{F}_1$) is applied at the centre of gravity of the foundation, a constant bottom-pressure distribution results. Eccentricities $\mathbf{e}_{x,y}$ of the force attack ($\mathbf{F}_2$) within the core surface (dark grey) lead to a modified bottom-pressure distribution, but the foundation remains over-compressed. In the case of a force application outside the core surface, the bottom stress on the tensile side becomes zero—a gaping joint is created. For the force application ($\mathbf{F}_3$) with eccentricity $\mathbf{e} \leq \mathbf{w}_{x,y}/3$, the foundation is still stable. In the case of a force application ($\mathbf{F}_4$) with eccentricity $\mathbf{e} > \mathbf{w}_{x,y}/3$, there is a risk of tipping over

the structure to the foundation bottom (Fig. 6.20). The resulting moment must be borne in the foundation by means of the footing's compressive stresses—uplifting tension cannot be taken up. In this process, the permissible ground pressures on the subsoil of the building must not be exceeded, and it must be avoided under constant loads that the foundation be **opened up**, thereby relieving parts of the foundation (Fig. 6.21).

A foundation model with a prognosis of the strength and stiffness properties of the individual soil layers is to be made for designing the foundation structure with sufficient certainty against these failure forms. This is generally done by means of geotechnical investigations within the framework of a soil assessment for the respective site of the building. In the case of buildings which are simple in terms of structural behaviour and foundation technology, empirical values can also be used.

Overall, the properties of the naturally grown, very inhomogeneous subsoil are subject to considerable fluctuations. These cannot be fully comprehended with a reasonable exploration effort even by geotechnical investigations. For this reason, a high **level of safety** is required for the design of foundation structures in comparison to the above-ground structures from steel or concrete.

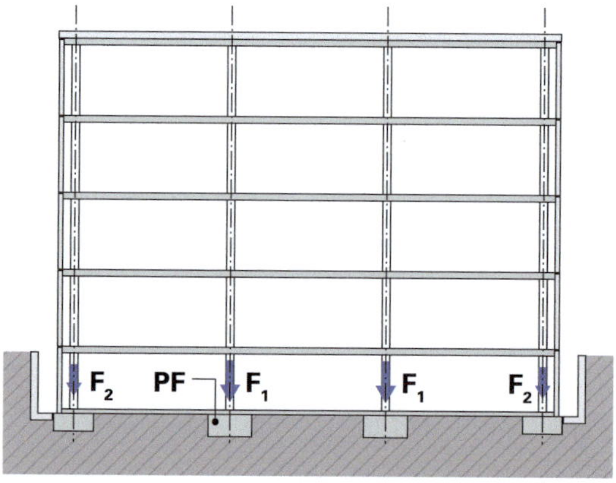

Fig. 6.22 **Shallow foundation** of a skeletal structure from point foundations **PF**

6

6.3 Types of Foundation

6.3.1 Foundation as a Link Between the Building and the Ground

The foundation of a building has the task of reconciling the structural requirements with the characteristics of the existing building subsoil. This includes safely dissipating building loads into the ground, limiting settlements to deformations compatible with the building use, and minimising or safely transferring loads from the subsoil to the building.

The subsoil conditions are the decisive basis for the development of foundation concepts and related constructions. They can even affect the entire load-bearing concept of the building.

If load-bearing soil strata are located near the top edge of the site or in the area of the building footing, executing a **shallow foundation** is reasonable (Fig. 6.22). This is understood as a broadening of a wall or a column-base in order to distribute the arriving loads to the softer subsoil. This form of foundation is then the most cost-effective, since it requires only a small additional excavation beyond the actual building volume and can be produced by simple beams or slabs of concrete or reinforced concrete. Therefore, shallow foundations are the standard type used in building construction.

If soil layers capable of load-bearing exist only at a great depth below the building footing, **deep foundations** can be used on piles, wells, or in form of caissons (Fig. 6.23). In this case, they prove to be economical because, in order to carry out a shallow foundation, measures for soil improvement or even a soil exchange for the upper layers would otherwise be necessary.

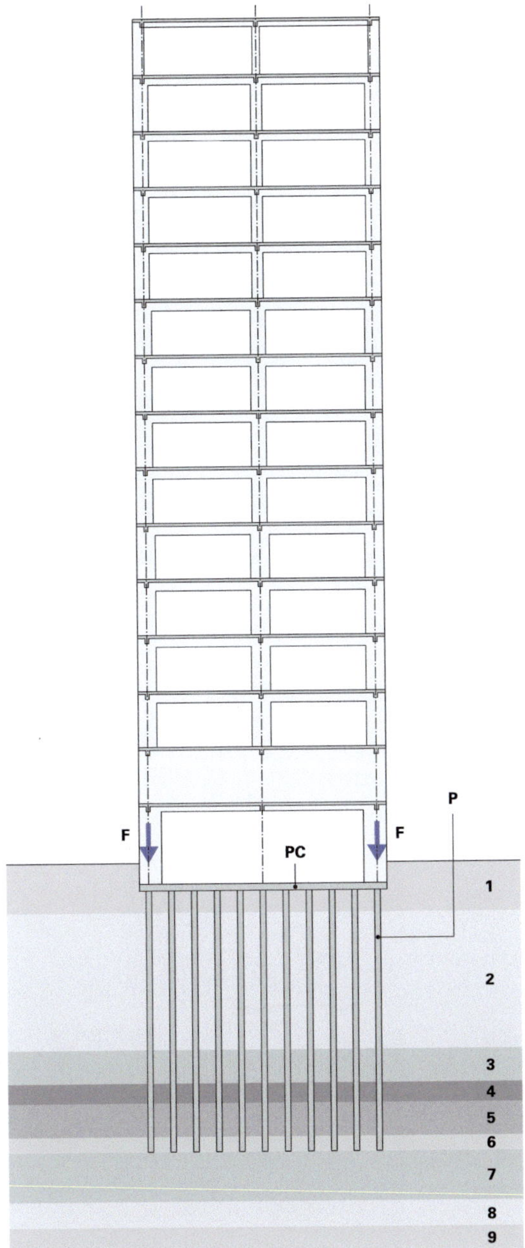

Fig. 6.23 **Deep foundation** of a high-rise building made of piles **P** and pile-cap **PC**, in soil layers **3–9**

6.3.2 Subsoil Improvement and Replacement

Measures to improve the building-ground properties are time-consuming and cost-intensive. They are therefore rarely used.

If, however, no soil able to bear loads is available at an economically justifiable depth, or if soil layers with strained

6

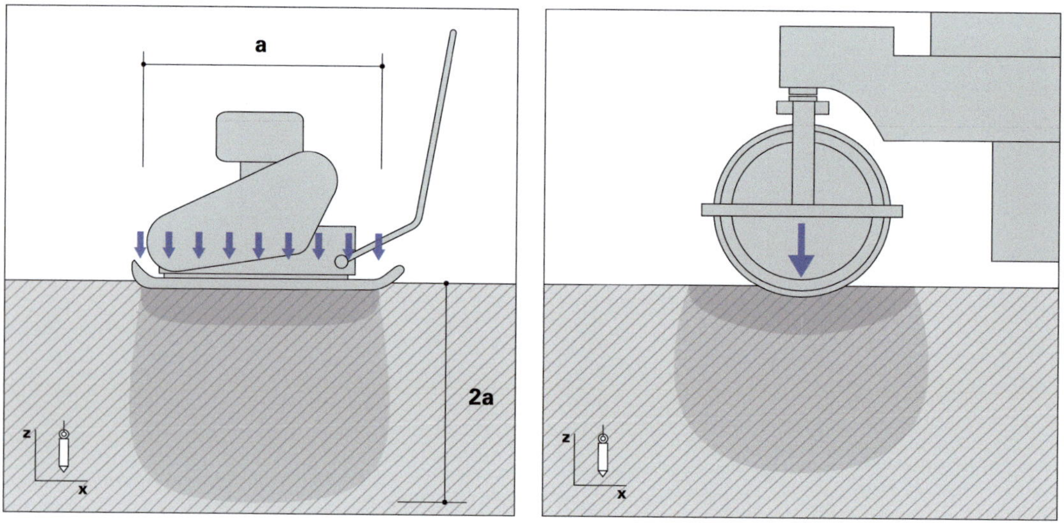

Fig. 6.24 (Left) **Compaction zones** under surface compaction pressure of a surface vibrator. (Right) **Compaction zones** under linear compaction pressure of a roller

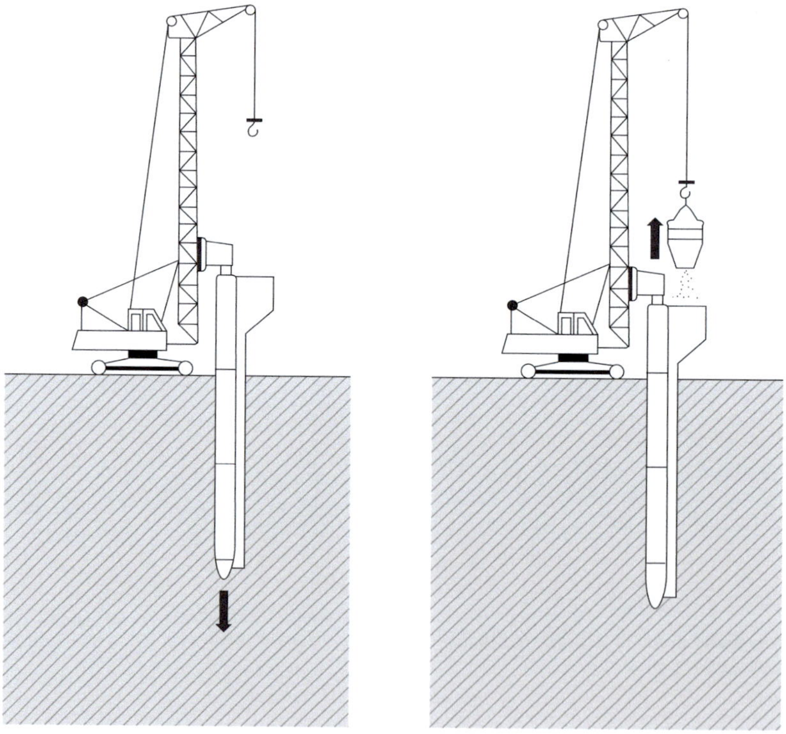

Fig. 6.25 **Spud vibrator**: left advancing, right pulling

groundwater or drinking-water-protection regulations preclude a deep foundation, measures must be taken for targeted soil improvement at the uppermost strata. In order to increase the shear strength and stiffness of the ground, different methods can be used depending on the involved ground layer.

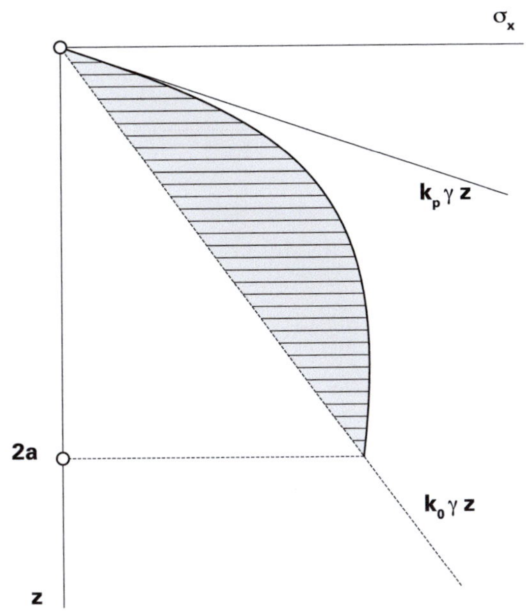

Fig. 6.26 **Compaction pressure**: additional stress by compaction, as a function of the depth of action **2a** as on Fig. 6.24 left

Loosely deposited, rather non-cohesive soils with high pore proportion can be transferred to a load-bearing site by **compaction** and **displacement**. For this purpose, vibrating plates (Fig. 6.24 left) or vibrating rollers are used (Fig. 6.24 right). In the case of very thick soil layers, a so-called **depth vibrator** can be used to improve the depth effect (Fig. 6.25). At a depth of up to 30 m, vibrational energy is introduced thereby into the soil, resulting in a densification of the grain structure (Fig. 6.26). The soil missing as a result of sinking is replenished when pulling on the vibrator head.

Cover-filling of the terrain before the beginning of construction causes a targeted pre-load of the ground. As a result, settlements are anticipated and the soil layers are transformed into a denser stratification. This results in increased strength and stiffness of the ground. This measure, which is also referred to as **pre-consolidation**, requires for non-cohesive soils only a pre-run of a few months before the beginning of construction works. Cohesive soils, on the other hand, take up the pre-load from the cover-filling as pore-water pressure, which only gradually subsides by draining water into surrounding areas, and thus loads the grain structure. In this case, an acceleration of settling and soil compaction can be achieved by means of ground dewatering with vertical drainage (Fig. 6.27).

The grain structure of the soil is compacted by **injecting** or pressing-in suspensions, emulsions, or pastes. The resulting soil-injection mixture can be deliberately influenced in terms of its properties. One of the most common applications

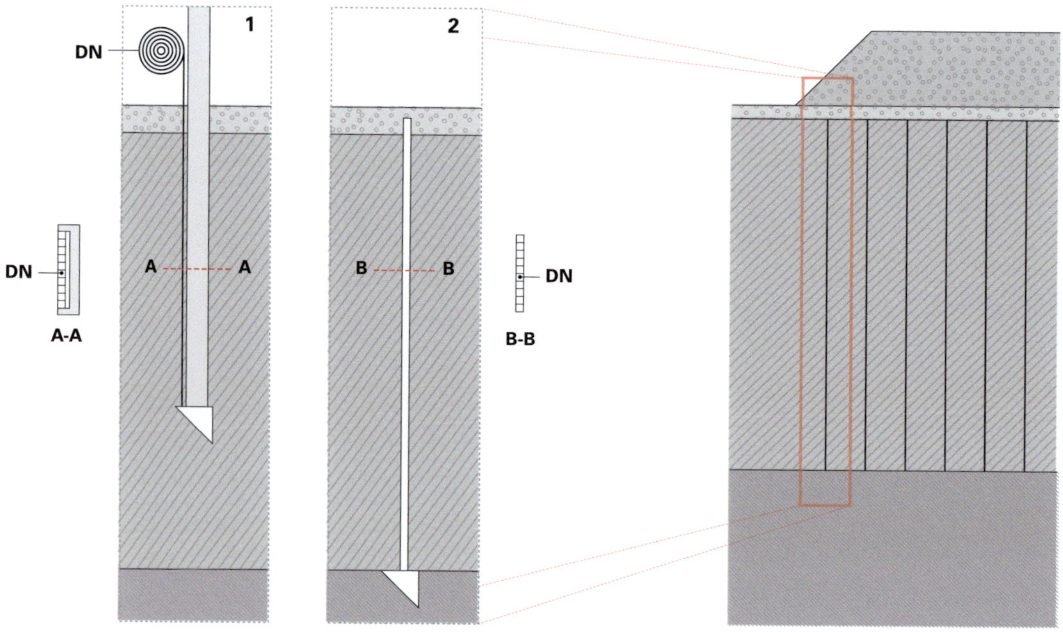

Fig. 6.27 Vertical drains. DN drainage non-woven. Sketch **1** shows the production, sketch **2** the final state of the vertical drain

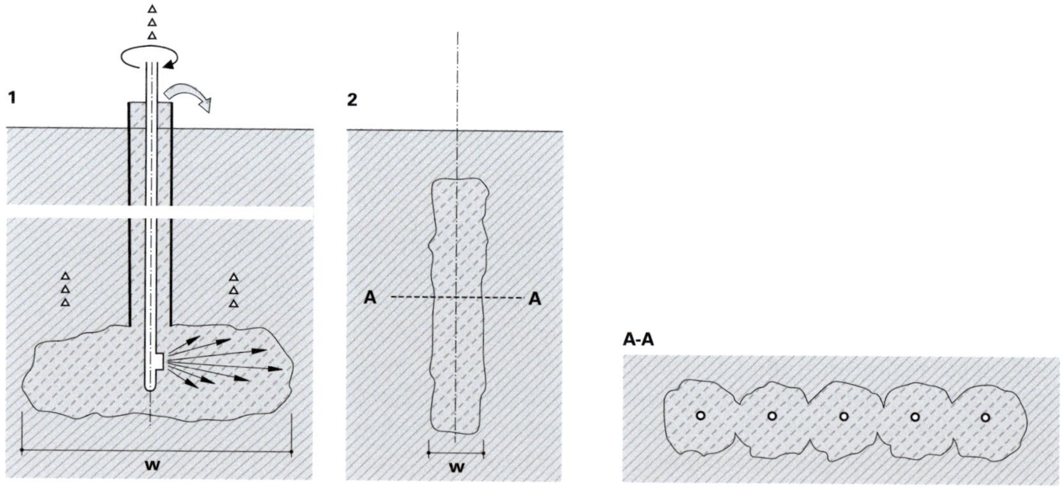

Fig. 6.28 Soil improvement by **high-pressure injection** of fine-cement suspension. Sketch **1**: manufacturing process by rotating injection from the bottom upwards. Sketch **2**: finished soil columns

of this process is the high-pressure injection of a cement suspension (*soilcrete*) (Fig. 6.28). A continuously rotating injection lance within a support tube injects with up to 500 bar a water-fine cement mixture into the soil layer to be improved. After hardening, lean-concrete columns form in the subsoil, which lead to a closed, very firm soil layer when the injection spacing is correspondingly narrow.

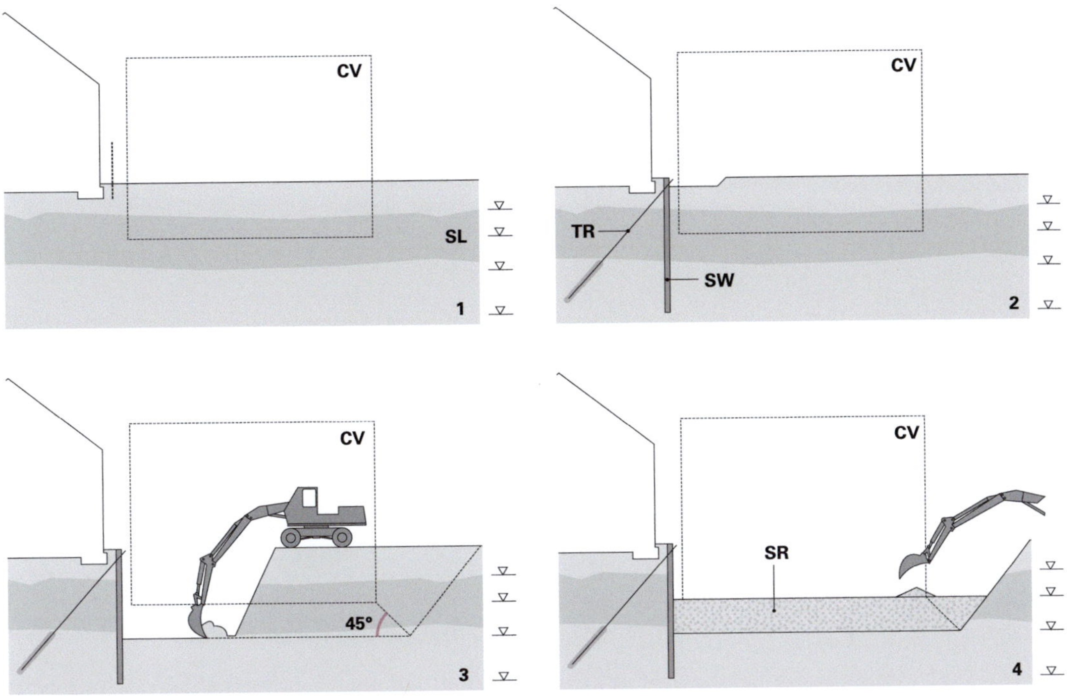

Fig. 6.29 Example of a **soil replacement**. (**1**) Initial state: planned construction activity in the vicinity of existing buildings, construction volume **CV**. Poorly load-bearing soft layer **SL** in the subsoil, which must be replaced. (**2**) Protection of the neighbouring building by sheet pile wall **SW**, secured by tie rod **TR**. (**3**) Excavation of the soil to be replaced until complete removal of the soft layer. Border of the excavation site defined as a function of the load-spreading angle (45°) of the new construction measure. (**4**) Filling and compacting the excavation pit with load-bearing soil replacement **SR** up to the height of the bottom of the planned building

When using appropriate injection substances, this method can also be used to change the permeability or tightness of the subsoil at defined depths.

In the case of ground layers or lenses close to the surface featuring very low strength and high setting sensitivity, **soil replacement** can also prove economical. In doing so, unsuitable soil material is removed and replaced by load-bearing soil, often gravel or sand (Fig. 6.29). This is then compacted by vibrating rollers or depth vibrators.

6.3.3 Shallow Foundations

Shallow foundations are surface foundations which are produced in a frost-free depth on a load-bearing subsoil in a water-free excavation pit. They take up the concentrated forces of vertical support elements of the primary structure and distribute these to the required contact surface in accordance with the prevalent soil conditions. Accordingly, shallow foundations must be sufficiently **bending-stiff** (Fig. 6.30).

6

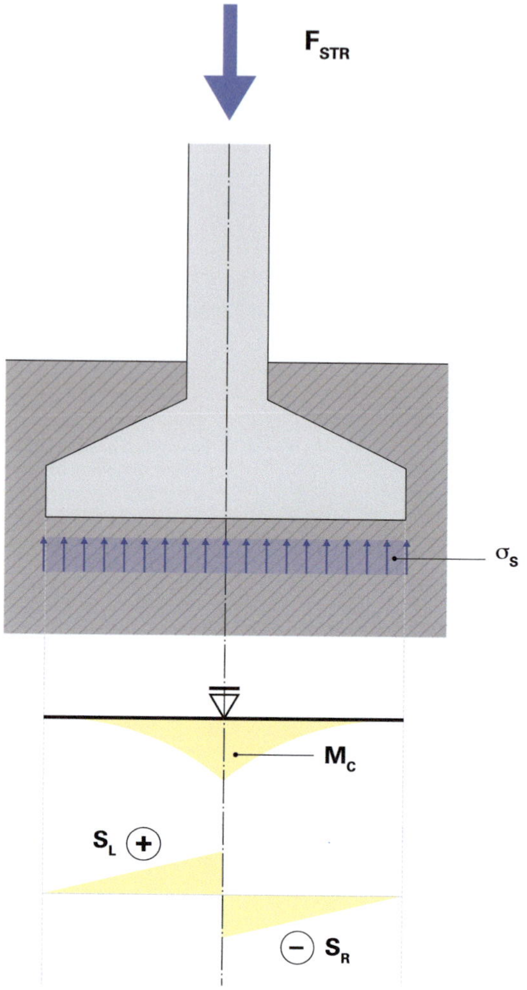

Fig. 6.30 Structural system of a foundation and its sectional forces: bending moment $\mathbf{M_C}$ and shear force $\mathbf{S_{L/R}}$

Point foundations transfer the loads of rod-shaped structural elements, such as pillars or columns, to the ground (Figs. 6.31, 6.32, 6.33, and 6.34). They consist mainly of unreinforced or reinforced concrete. They have a cuboid shape (Fig. 6.31 left), sometimes bevelled at the top (Fig. 6.31 right). They are designed according to the type and direction of the stress to be borne, so that the bottom of foundation is largely over-compressed, since a shallow foundation cannot transmit tensile forces into the ground. For this reason, a foundation under a constant bending stress projects on the compression side (Figs. 6.32 right and 6.33). A massive foundation pillar (Fig. 6.32 left) creates the necessary weight to absorb a variable horizontal force in the foundation floor by friction.

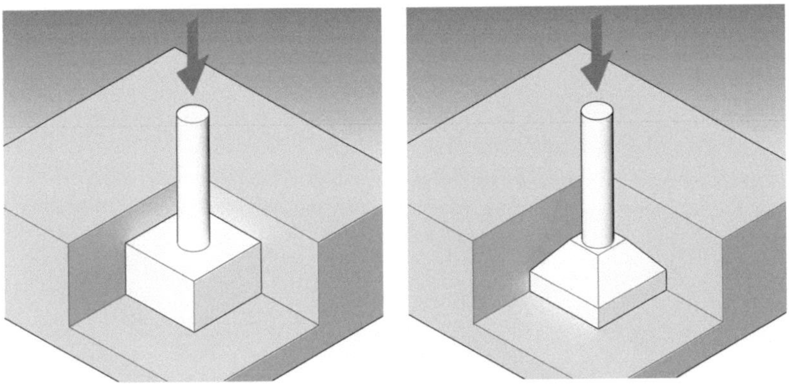

■ **Fig. 6.31** (Left) Point foundation—block foundation. (Right) Point foundation—bevelled

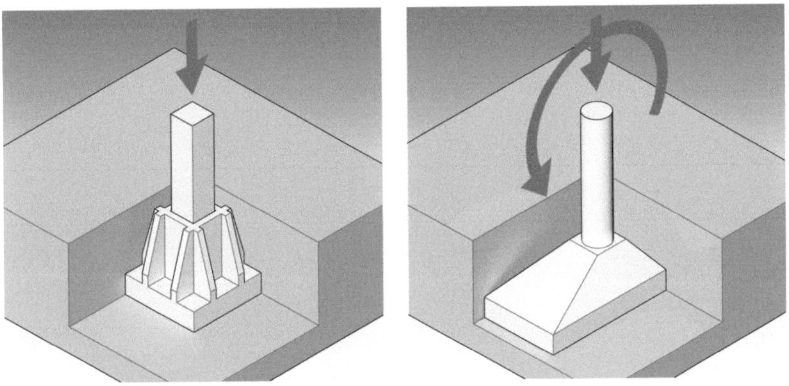

■ **Fig. 6.32** (Left) Point foundation—prefabricated, e. g. sleeve foundation. (Right) Point foundation in case of restraint against one-sided bending stress

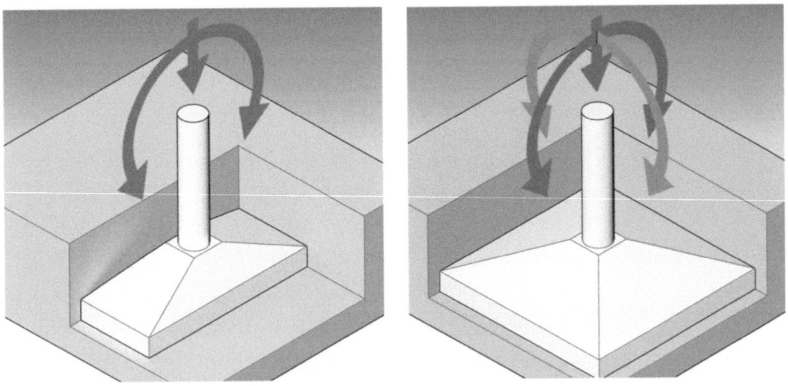

■ **Fig. 6.33** (Left) Point foundation in case of restraint against two-sided bending stress. (Right) Point foundation in case of restraint against all-round bending stress

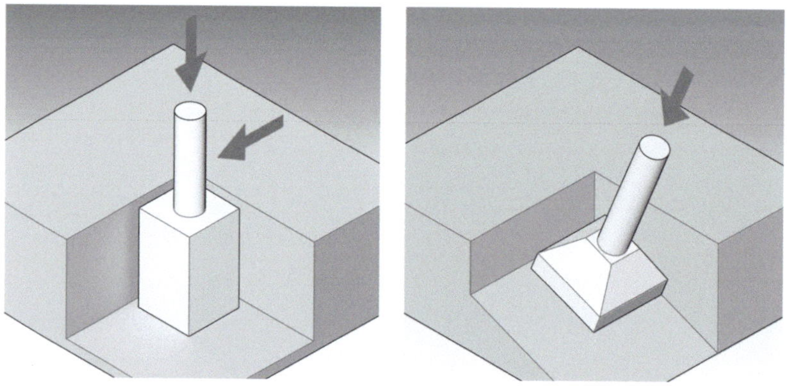

6

Fig. 6.34 (Left) Point foundation—pillars with variable horizontal load. (Right) Point foundation—inclined under oblique constant load application

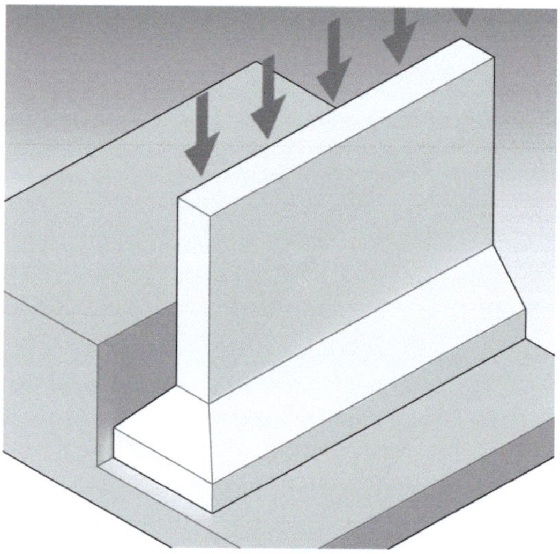

Fig. 6.35 Strip foundation for vertically loaded wall

Strip foundations are arranged underneath panel-shaped structural elements, for example walls, and are also predominantly made of unreinforced or reinforced concrete (Fig. 6.35). Under a constant moment stress or horizontal forces attacking perpendicular to the wall plane, strip foundations also project on the compression side.

Several individual and strip foundations form the **foundation construction** of a building. They can be combined into **foundation groups** to improve the load transfer into the foundation. In this way, a **tie member** between foundations serves for compensating horizontal forces, which would otherwise have to be short-circuited over the ground, thereby risking deformations (Fig. 6.36).

Foundation beams (Fig. 6.37 left) connect single foundations as well as strip foundations to a bending-stiff **foundation grate** (Fig. 6.37 right). In this context, the beams can

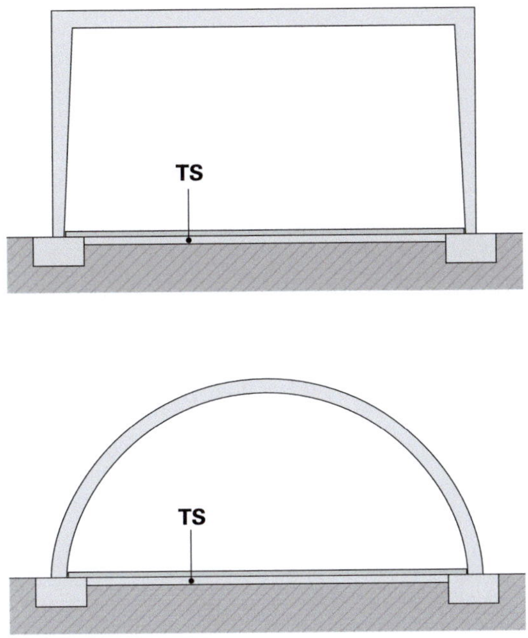

Fig. 6.36 **Tie straps (TS)**, e. g. for short-circuiting shear forces at foundation level

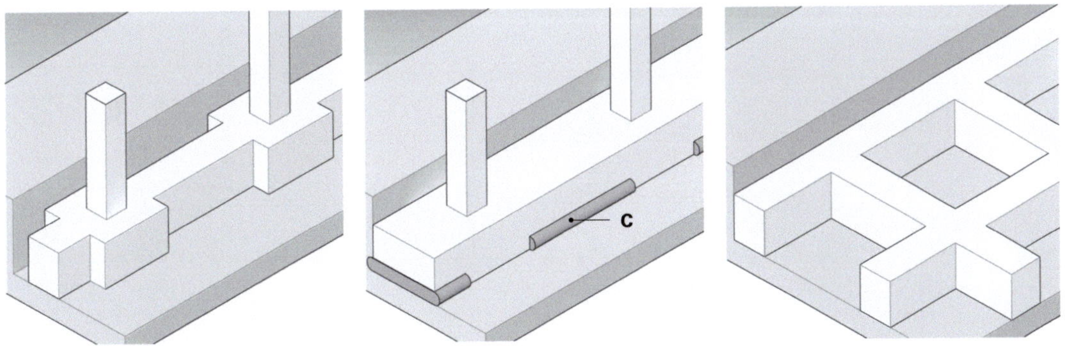

Fig. 6.37 (Left) Foundation beam. (Centre) Foundation beam with cushions **C**. (Right) Foundation grate

be involved in transferring loads into the ground in the same way as strip foundations, or can otherwise be deliberately separated from the ground by padding and thus limited to their function as flexural beams (Fig. 6.37 centre). They enable the transfer of large local loads to neighbouring foundations and thus also reduce uneven settlements of the structure.

If, however, the existing ground is not able to support a foundation group, a pad foundation in the form of a continuous **foundation slab** can be formed. This is particularly economical in connection with a necessary sealing against pressing water and with very irregular soil conditions in the foundation bottom.

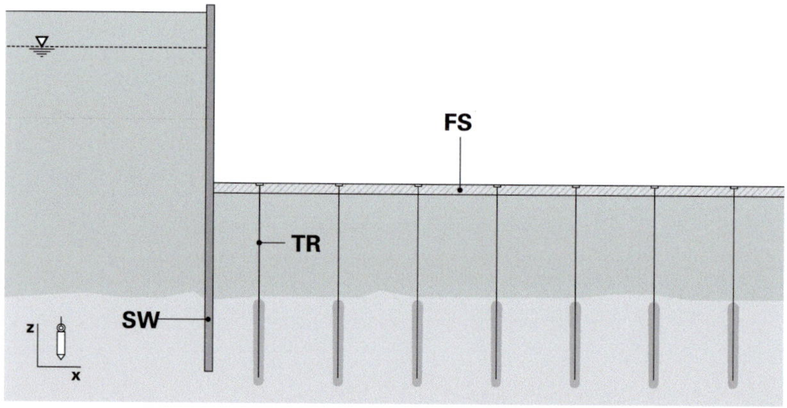

6

Fig. 6.38 Floating foundation with securing by tie rods. **FS** foundation slab, **SW** slurry wall, **TR** tie rods

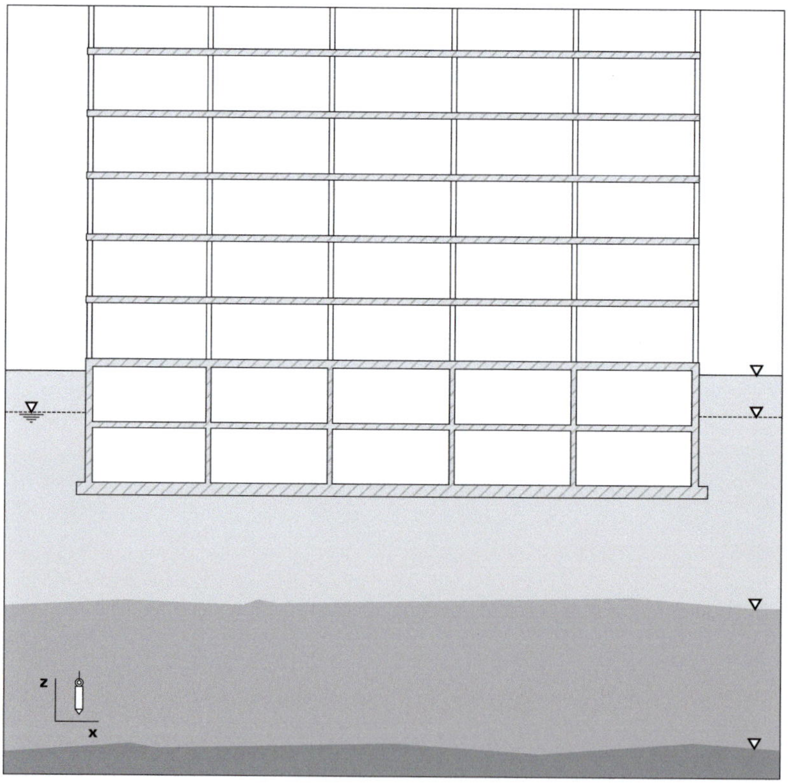

Fig. 6.39 **Foundation box** of a high-rise building

► *Sect. 6.3.4 Deep Foundations*

In the case of very soft, setting-sensitive soils, the foundation plate can be laid so deep that the load from the building acting on the foundation bottom is smaller than the excavation weight. This is then termed a **floating foundation**, which must be anchored by tensile piles in particular when exposed to buoyancy due to groundwater (Fig. 6.38) .

In addition to the loads to be transferred, the dimensioning of a shallow foundation depends primarily on the **stiffness** of the subsoil. The softer and more irregular the soil is, the more bending-stiff the flat foundation must be. In this context, the setting sensitivity of the primary load-bearing structure and of the adjoining components, for example the enveloping structure, must be taken into account.

Thus, forming a quasi-rigid **foundation box** consisting of the foundation plate, the wall panels and ceilings of the cellars, can be necessary (◘ Fig. 6.39). Such a basement box is necessary, in particular, in the case of high-rise buildings in order to distribute the large loads into the building ground with a primary structure consisting of a simultaneously slim and compliant skeleton.

6.3.4 Deep Foundations

If firm soil layers are only available at a very great depth, the design of a flat foundation is not economical and often technically impossible. This results in a vertical distance between the building bottom and the subsoil capable of bearing a foundation, over which the forces from the primary structure are to be transported. A deep foundation is then required. **Piles** are used for this purpose (◘ Fig. 6.23). In building construction, they consist predominantly of reinforced concrete and are produced as precast-concrete piles or cast-in-situ piles.

For the transport of loads from the building into the deep-seated supporting ground, the pile uses the **jacket friction** to the surrounding ground and the **bearing pressing** on the pile foot (◘ Fig. 6.40). Accordingly, piles which transport the predominant part of their load to the pile foot and deliver it there directly to the load-bearing ground, are called **tip-pressure piles** in foundation engineering.

If the firm ground strata cannot be reached by the pile, **floating piles** may be considered as a foundation. They gradually transfer their loads by jacket friction to the soft, poorly load-bearing ground, which subsequently directs the stresses to the deeper-seated soil layers.

This type of load transmission by friction between the pile and the surrounding ground allows forming **tensile piles** (◘ Fig. 6.41).

As a result, in contrast to a shallow foundation, also **uplifting forces** can be anchored in the ground by means of piles. This is used in buoyancy control of pad foundations below groundwater level and for introducing tensile forces from rope anchorages.

EN 1537

▶ *Sect. 6.3.3 Shallow Foundations*

Piles are also distinguished according to the way by which they are introduced into the soil. For the foundations of building construction, both **drilled** or **displacement piles** are mainly carried out:

6

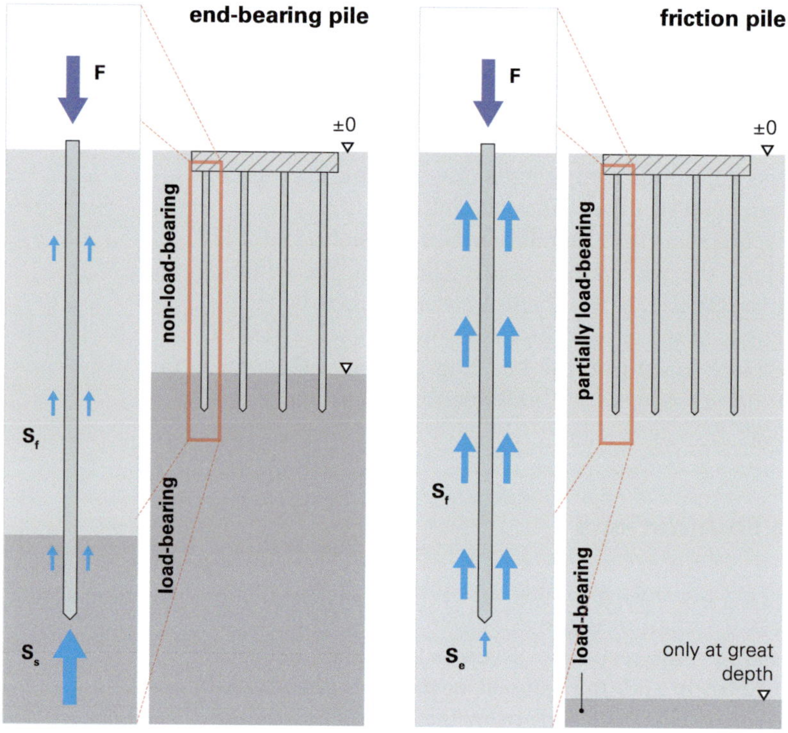

Fig. 6.40 **Load-bearing principles** of piles

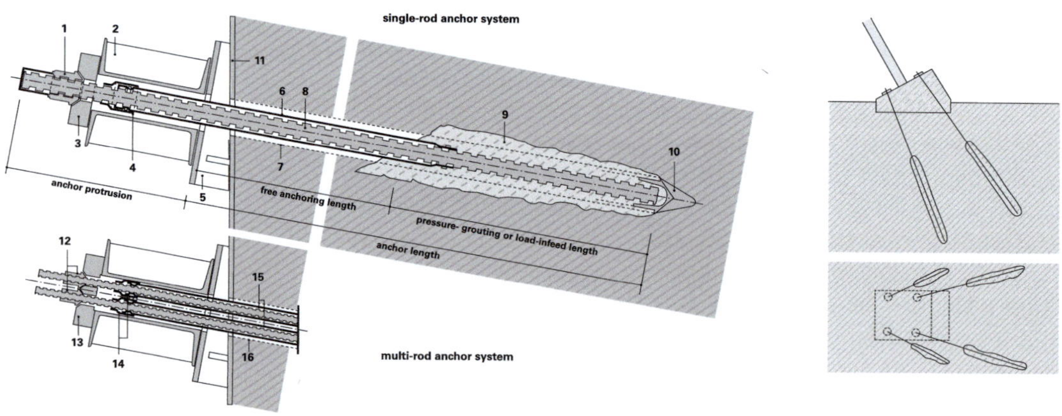

Fig. 6.41 **Tensile pile** (grouted anchor); bottom left, anchor group. (**1**) Anchoring nut with articulated cone, (**2**) waler with web bracing in the anchor area, (**3**) anchor-head plate, (**4**) sleeve for sealing the jacket tube, (**5**) wedge filling piece, (**6**) jacket tube, (**7**) bore-hole tubing, (**8**) threaded bar, (**9**) grouting volume (cement mortar), (**10**) drill bit, (**11**) sheet pile wall, (**12**) anchoring nut, (13) anchor head plate, (**14**) closure sleeve, (**15**) jacket tube, and (**16**) bore-hole tubing

EN 12699

- During **displacing** or **pressing-in**, the pile or propulsion tube displaces the ground (Fig. 6.42). The latter is thereby compacted and thus allows a higher jacket friction for load transfer. The energy to be introduced into the ground can lead to settling damage in neighbouring

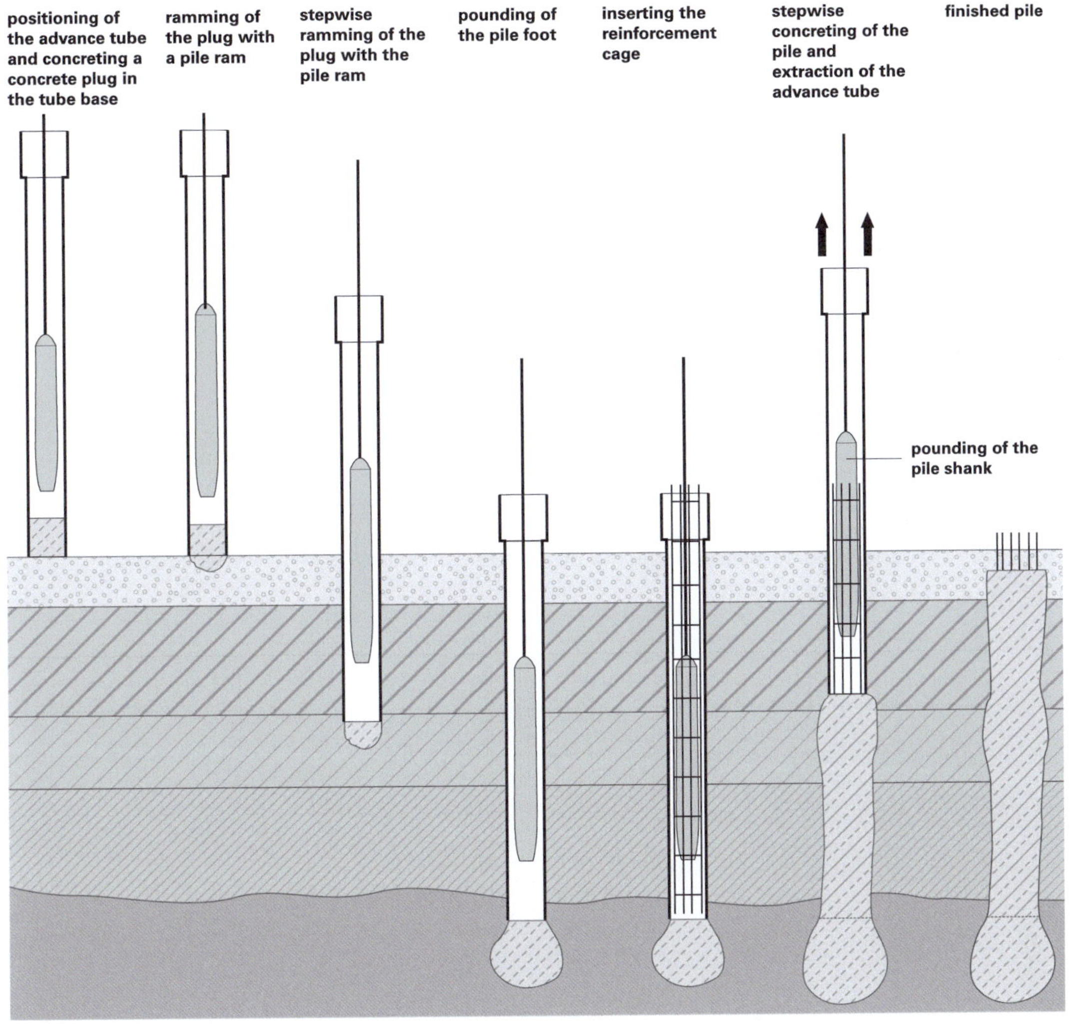

Fig. 6.42 Production of a **cast-in-situ driven concrete pile**

buildings. Displacement piles are limited in diameter and length. They are suitable in connection with a foundation plate, a so-called **pile cap**, especially for building loads distributed over a large area.

- **Drilled piles** do not displace soil because their volume has to be removed from the ground by drilling or rinsing before the pile is produced in the well (Fig. 6.43). Drilling and concreting are usually done with a piping of the borehole. Drilled piles can be produced with very large diameters (<2m) and into very large depths. They are particularly suitable for high concentrated building loads. *EN 1536*

The foundation of a building consists of a group of individual piles. These are connected to each other via a pile cap or a pile grate about 2m thick and connected to the primary structure of the building (Fig. 6.44). The pile cap, or the pile grate, dis-

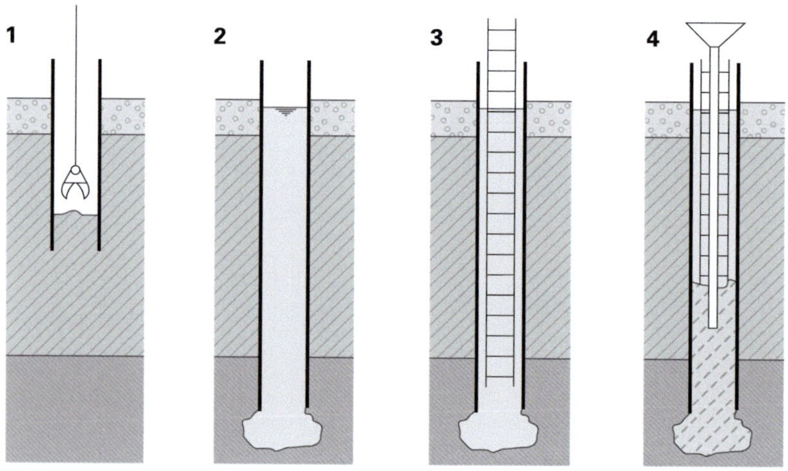

Fig. 6.43 Production of a **bored pile** with support tube (remaining in the ground) and water overpressure due to pressurised ground water. (**1**) Start of excavation within the support tube, (**2**) self-filling of the drilling space by ground water, (**3**) introducing the reinforcement, and (**4**) filling the concrete using a funnel and a downpipe

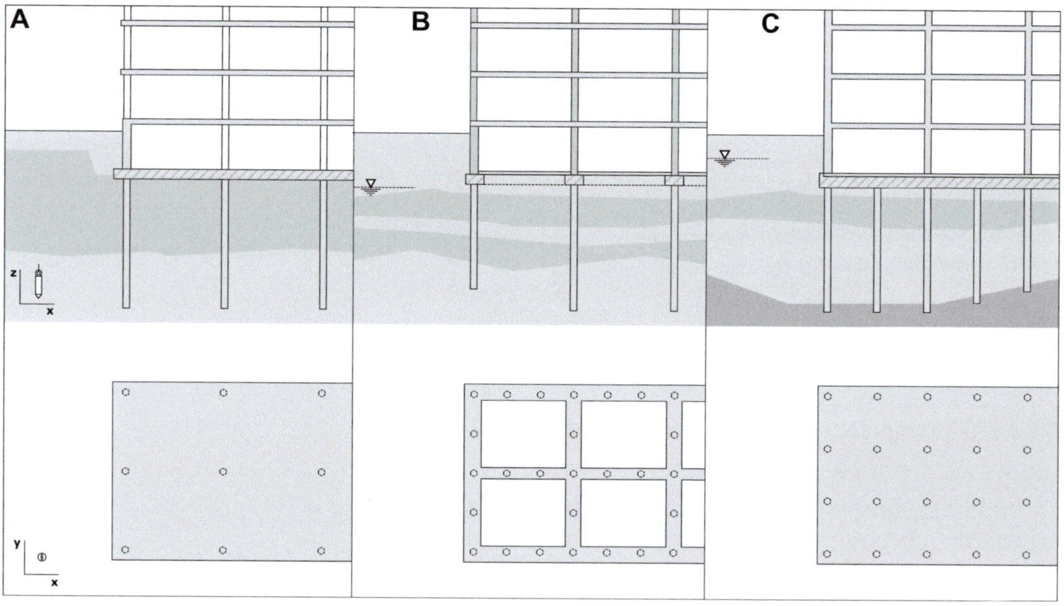

Fig. 6.44 **Perpendicular pile grates**. **A** Skeletal construction on bored piles, **B** masonry construction on driven piles, and **C** concrete box on bored piles

tributes the construction loads according to the individual pile stiffnesses and thus compensates for differences in settlement within the subsoil and load concentrations from the building.

The introduction of very large, concentrated forces into a subsoil with a very deep-seated supporting layer results in very large, inefficient pile diameters and a statically inefficient load distribution over a powerful pile cap in order to maintain the structural minimum distances between the piles. This

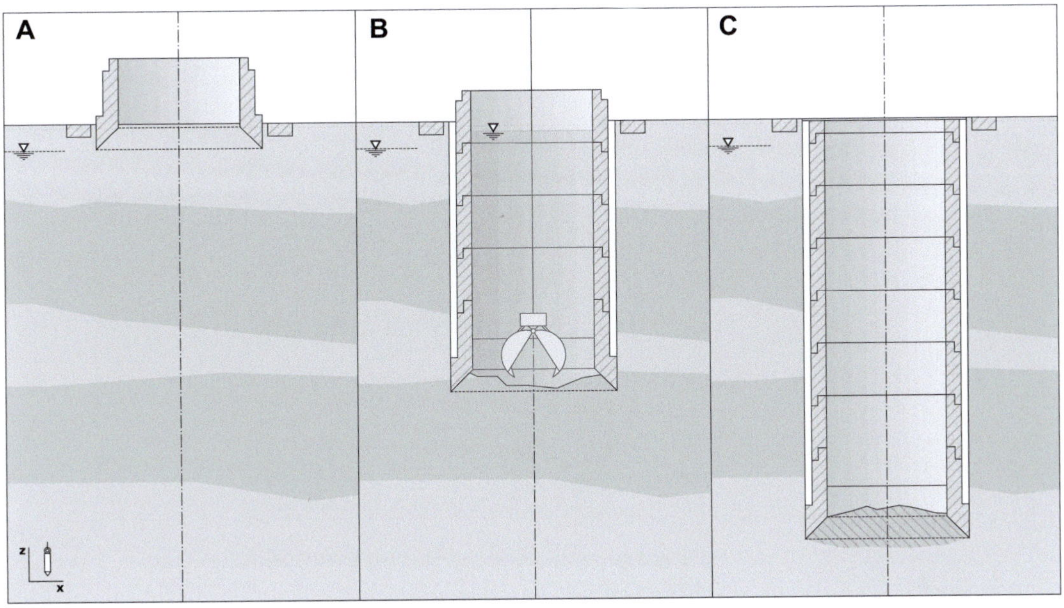

Fig. 6.45 Drawdown well. **A** Putting on the cutting part, **B** lowering with the sliding sleeve, and **C** closing the bottom

relates in particular to tower-like structures and is necessary in the case of a jacket friction insufficient for a floating pile foundation. Here, particularly pillar-like deep foundations, such as **well foundations** or **caissons**, are suitable:

- **Wells** are made in the sinking procedure (Fig. 6.45). Diameters and depths, in this process, are basically unlimited. Wells consist of prefabricated-concrete parts which are mounted in sections and are lowered into the ground by means of a **cutting bearing** under the action of their own weight. The process is advanced by means of a bentonite-supported sliding jacket, the internal excavation, and a controlled flushing under the cutting edges.
  When the foundation horizon is reached, the sliding mantle is injected with cement suspension and the well bottom is closed with concrete. Finally, the well is lined from inside or, if necessary, filled up to increase the superimposed load. This results in a foundation pillar with high load-bearing capacity by jacket friction and tip-pressure.
- **Caissons** are closed, reinforced-concrete structures quasi-rigid in themselves which are also brought to a deep-seated foundation bottom in the sinking procedure (Fig. 6.46). In contrast to the well, the excavation and rinsing process takes place within a working chamber, which is protected against the penetration of ground water by compressed air. Therefore, air locks are required for people, materials, and equipment. Beyond a pressure differential of 20 m water column, the stress on the personnel and the health-related air-lock times increase, so that only un-

Bentonite is a mixture of clay minerals. While it is stirred, it is liquid, but solidifies when at rest. It is used in underground construction as a supporting fluid.

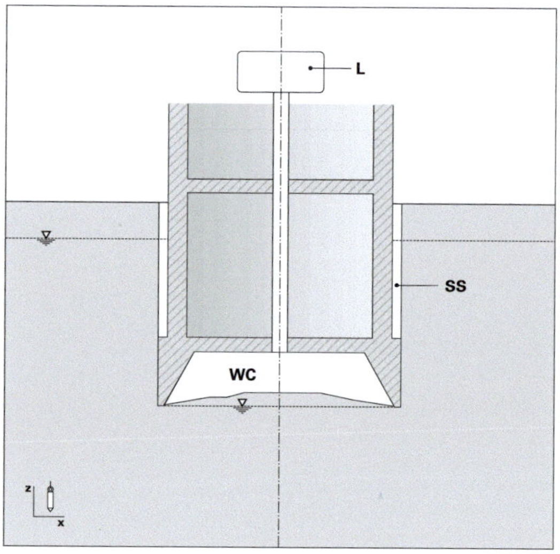

Fig. 6.46 Compressed-air lowering. **WC** working chamber, **L** lock, **SS** sliding sleeve

manned work in the compressed air chamber becomes possible.

With this method, in the case of unfavourable foundation conditions, the cellar box of a building consisting of several storeys can be constructed at site level in concrete cast on site and lowered to the deeper level of the foundation bottom. This means that caissons are conceptually similar to both a deep as well as a shallow foundation.

▪ Standards and Guidelines

DIN 1054: 2021-04 Subsoil—Verification of the safety of earthworks and foundations—Supplementary rules to *DIN EN 1997-1*

DIN 4020: 2010-12 Geotechnical investigations for civil engineering purposes—Supplementary rules to *DIN EN 1997-2*

Supplement 1: 2003-10 Aids to application, supplementary information

DIN 4124: 2012-01 Excavations and trenches—Slopes, planking and strutting breadths of working spaces

DIN 18122: Soil, investigation and testing—Consistency limits

Part 2: 2020-11 Determination of shrinkage limit

DIN 18196: 2011-05 Earthworks and foundations—Soil classification for civil engineering purposes

EN 1536: 2015-10 Execution of special geotechnical work—Bored piles

EN 1537: 2014-07 Execution of special geotechnical works—Ground anchors

EN 1538: 2015-10 Execution of special geotechnical work—Diaphragm walls

EN 1997-1: Eurocode 7—Geotechnical design
Part 1: 2022-10 General rules
Part 2: 2022-10 Ground properties

EN 12699: 2015-07 Execution of special geotechnical works—Displacement piles

EN 12794: 2007-08 Precast concrete products—Foundation piles
Corrigendum 1: 2009-04

EN 14199: 2015-07 Execution of special geotechnical work—Micropiles
Corrigendum 1: 2016-09

EN ISO 17892: Geotechnical investigation and testing—Laboratory testing of soil
Part 12: 2022-08 Determination of liquid and plastic limits

EN ISO 22477: Geotechnical investigation and testing—Testing of geotechnical structures
Part 5: 2016-10 Testing of grouted anchors

Bibliography

Decker H, Garska B (ed) (2018) *Ratgeber für den Tiefbau*. Cologne, Bundesanzeiger

Lang H J, Huder J, Amann P, Putrin AM (2011) *Bodenmechanik und Grundbau: Das Verhalten von Böden und Fels und die wichtigsten grundbaulichen Konzepte*. Berlin, Heidelberg; Springer

Mehlhorn G, (1995) *Der Ingenieurbau: Grundwissen Hydrotechnik, Geotechnik*. Berlin, Ernst und Sohn

Rübener R H (1985) *Grundbautechnik für Architekten*. Düsseldorf, Werner

Rübener R H, Stiegler W (1998) *Einführung in Theorie und Praxis der Grundbautechnik*. Werner Ingenieur-Texte, Vol. 49, 50, 67. Düsseldorf, Werner

Savidis S (1995) Vorlesungsskript *Grundbau und Bodenmechanik I und II*. Technical University of Berlin

Smoltczyk U, Witt K J (2017) *Grundbau-Taschenbuch. Geotechnische Grundlagen, Teil 1*. Berlin, Düsseldorf, Munich; Ernst & Sohn

Masonry Construction

J. L. Moro, *Building-Construction Design – From Principle to Detail*,
https://doi.org/10.1007/978-3-662-71006-7_7

Contents

7.1 Generalities

Vol. 1, *Sects. 1.3.1.1–1.3.1.5*

The concept of **construction mode** is defined and explained in *Chap. I*. In this sub-chapter and the following ones, some of the construction modes that are representative of today's building practice will be examined. The goal is, in particular, to describe the linkage and the interrelationship between:

- **structural**,
- **functional**, and
- **aesthetic**

aspects of design and construction within the framework of the characteristic rules of a particular construction mode.

7.2 Categorisation of Construction Modes

A fundamental categorisation of constructions, which is also applicable to the examples investigated in this chapter, distinguishes between two basic types of vertical **load transfer** and **space confinement**.

A distinction is made between:

Vol. 4, *Sect. 8.2 Wall and Skeletal Systems*

- **Wall construction**: The functions of vertical force transfer and space enclosure are provided by the same component, namely, the wall, a surface component. Loads are largely evenly **distributed** within the vertical surface component. Wall constructions are, in the static sense, **shear-plate structures** (◘ Fig. 7.1).

Vol. 1, *Sect. 2.2.2.1 According to Main Functions*

Vol. 1, *Sect. 26.1.1 Categories of Load-Bearing Structures*

- **Skeletal construction**: The functions of vertical force transfer and enclosure are assigned, respectively, to *separate* components or subsystems. Loads are transmitted within vertical columns, i.e. rod-shaped components (◘ Fig. 7.2). Spaces are encased by walls, that is to say surface components, which no longer exert a primary force-transfer function. As a result, the loads are transmitted—in contrast to wall constructions—in a **concentrated** manner within the cross-section of the column.

This means that primarily vertical components such as walls and columns are considered. This differentiation focusses, in particular, on the interior spatial conditions: In wall-construction methods, the interior space is segmented by walls into individual room cells; with skeletal constructions, clear room profiles are created in usable individual room widths in both main directions. The constructional and aesthetic design of exterior walls is also different for wall and skeletal structures. In this respect, the distinction is clearly one centred on utilisation, only secondarily construction-related. Roofs and storey floors are not taken into account in this classification as they are almost exclusively both areal and load-bearing in conventional building construction.

■ Fig. 7.1 **Masonry wall construction** from the nineteenth century in Stuttgart. Vertical and horizontal loads are generally transferred via wall panels acting as rigid shear-plates (outer and inner walls), which reinforce and stabilise each other in an orthogonal arrangement. Openings with limited dimensions are worked into the panes and allow the necessary lighting and ventilation of the interior. They are dimensioned just large enough that they do not excessively weaken the supporting wall structure

It goes without saying that such classification is basically related to orthogonal and cuboidal designs of building construction constituted by horizontal and vertical main components. The developmental significance of the transition from historical wall structures to modern skeletal structures in Europe and America is discussed in more detail in *Chap. IX*.

***Vol. 3**, Sect. 10.3 Historical Development of Building Enclosures*

▶ *Sect. 3.1.6 The Elements of the Structural Cell*

7.2.1 Wall Construction

The vertical structural elements of wall constructions—also termed shear-plate constructions—are two-dimensional. This allows for a linear support of the floors and a **uniform** and **distributed transfer** of vertical loads while avoiding high load concentrations. The latter are detrimental to the load-bearing behaviour of the wall panels, contradict the inherent logic of this particular construction method and are consequently not provided for in the structural system: A concentrated load on the wall (■ Fig. 7.3 left) is associated with an increased risk of

Fig. 7.2 Early **skeletal construction**—one of the first examples: Wainwright Building, 1890–91, St. Louis (USA) (architect: L. H. Sullivan). The outer walls are still made of masonry, but do not bear any floor loads, only themselves. This can be seen from the columns located close to the façade directly behind the wall, which are responsible for this task. At this height, however, the outer walls due to their extreme slenderness would not be able to withstand horizontal loads such as wind on their own, which is why they have to be connected to the braced skeletal load-bearing structure at the edges of the floors

buckling, which always threatens in the **weak direction** of the wall, i.e. **transverse to the wall plane** (→ **x**). To eliminate this risk without additional elements, the wall thickness, and thus its bending stiffness, would have to be increased at least at the point of application of the point load (Fig. 7.3 centre) or, alternatively, even over the entire length of the wall (Fig. 7.3 right). The former solution corresponds to a localised wall pillar that protrudes beyond the wall surface, already has a visual and spatial effect, and represents a kind of embryonic form of skeletal support; the latter would result in an unnecessary oversizing of the wall thickness over the entire wall surface beyond the point of application of the concentrated load, something which blatantly contravenes the general principle of material economy governing each and every construction method. Already this simple reasoning illustrates the absolute necessity of always distributing loads as widely as possible in wall-construction methods.

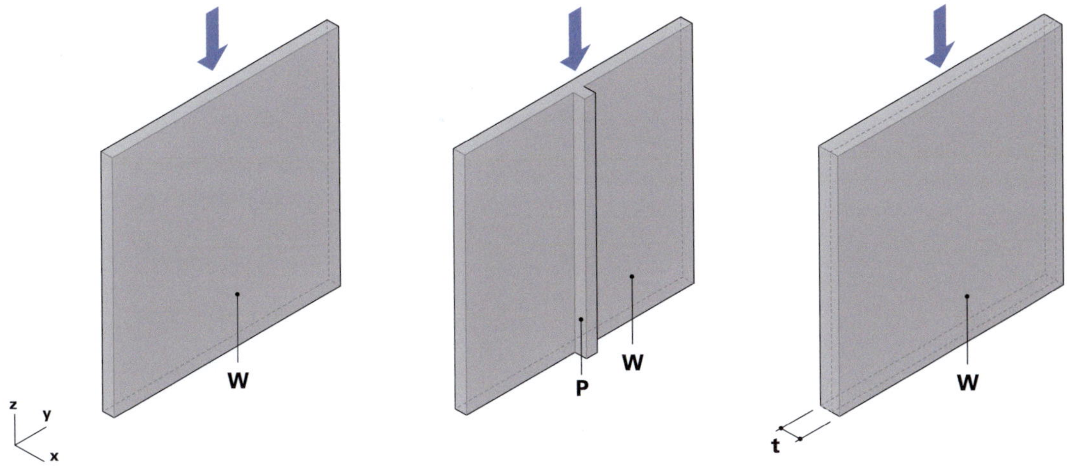

Fig. 7.3 (Left) Large single load on a wall **W**: There is a risk of buckling in the weak direction of the wall (→ **x**). (Centre) Reinforcement of the wall **W** by local thickening in the pillar **P** thus increasing the buckling stability in the weak direction (→ **x**). This solution corresponds to a **wall pillar** or **pilaster**. (Right) Thickening the wall **W** overall to thickness **t**, which also increases the buckling stability in the weak direction (→ **x**). However, this solution is material-intensive and contradicts the wall-construction principle

Because no high load concentrations occur, it is also possible with this construction mode to use materials of limited load-bearing capacity—such as, for example, light masonry blocks or also lying timber parts which are compressed transversely to the grain—which, however, have good thermal-insulation properties. Traditionally, this has always been conducive to the inherent double function of the wall, namely, being at the same time responsible for load transfer and for thermal insulation (Fig. 7.4). This aspect has been relativised today by the technical development of highly insulated multilayered and multi-leaf wall structures, since the insulation function is not provided anymore by the supporting leaf itself, but by an added specialised non-load-bearing layer, the thermal-insulation layer. Nevertheless, even today, the good load distribution of wall construction is a basic prerequisite for using high-insulation single-leaf walls made of masonry in low-rise housing construction.

***Vol. 1**, Sect. 2.2.2.2 According to Constructional Partial Functions*

Light vertically perforated bricks in ***Vol. 1**, Sect. 20.2.6 Shapes of Bricks* as well as ***Vol. 3**, Sect. 12.2.1.3 Single-Leaf Exterior Walls from Aerated Masonry*

The linear supports provided by the walls are well suited for supporting floors, in particular, solid floors as are almost exclusively used in wall construction today. Overall, a system of vertical and horizontal panels with great stiffness results from a structural viewpoint.

The **stabilisation** or **bracing** of the building against horizontal loads is achieved—at least partially—by using the **racking stiffness** of the flat wall components, i.e. the **shear stiffness *within* their plane**. The wall-construction method is not designed to absorb larger loads at right angles to the wall plane, since the thin wall elements, in this direction of force, cannot (and are not supposed to) provide sufficient stiffness.

Fig. 7.4 The hygrothermal functions of an exterior wall: protection, insulation, and thermal storage

Thus, in both possible horizontal main directions of application of the load, there must be at least one wall panel which transfers the loads via the racking stiffness of a shear-plate (not bending-plate stiffness, which the wall does not necessarily have to a large extent). Or in other words: Each wall panel, which has sufficient stiffness within its own plane to absorb forces in exactly this direction, must be, in contrast, supported transversely to its plane (where it is extremely sensitive to the application of forces) by a kind of *partner wall* (or at least through a wall pillar as on Fig. 7.3 centre) connected at right angles, which in turn absorbs the applied force in its own plane and dissipates it through its own racking stiffness.

The distances between the stiffening transverse walls (or, if applicable, between stiffening wall pillars as shown on Fig. 7.3 centre) must be small enough that the wall thickness already provided for the transfer of vertical loads is also sufficient to distribute the out-of-plane horizontal loads impinging in the sector between the transverse walls laterally to both of them via plate stiffness, i.e. bending stiffness (Fig. 7.5; cf. also Fig. 7.32). Only in this respect is a (limited) bending-plate stiffness of the wall required. If this maximum distance is exceeded, additional stiffening measures are required. There is already a horizontal support on the lower bearing of the wall, for example, at floor level. At the upper edge of the wall, however, which is not laterally stiffened (Fig. 7.32), further stiffening measures may be required. This can be a horizontally acting bending beam (Fig. 7.33) or a floor acting as a shear-plate (Fig. 7.37).

At the same time, the lateral stiffening of the thin wall panels by transversely positioned wall panels that are much stiffer in their own direction prevents lateral buckling at right angles to the wall plane due to the effects of the vertical load. This is the direction in which the wall panels are particularly

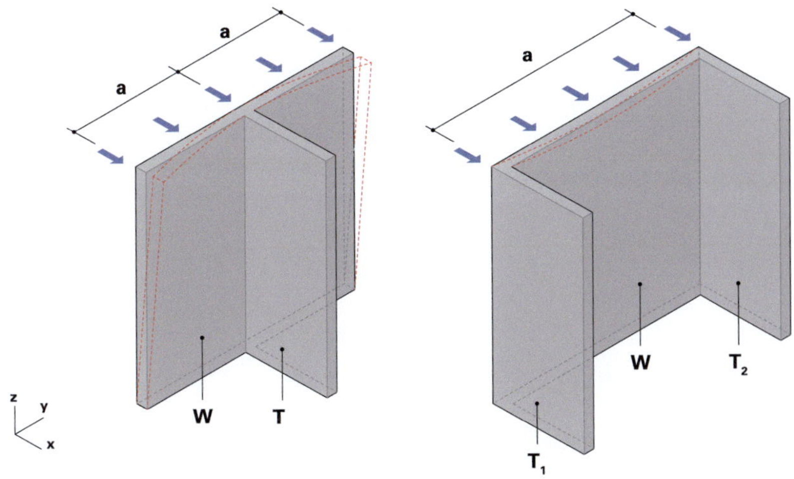

Fig. 7.5 (Left) Necessary plate stiffness of the wall **W** (i.e. in → **x**) so that the forces acting on the stiffening transverse wall **T** can be distributed laterally (i.e. in → **y**). The dimension **a** is limited (bottom linear support assumed). (Right) Conditions comparable to those on the left also apply to two stiffening transverse walls $\mathbf{T_1}$ and $\mathbf{T_2}$. Here, the distance **a** between the transverse walls is limited (also assuming lower linear support)

thin and consequently have the smallest stiffness against buckling. In addition to **bracing** against horizontal loads, the transverse wall panels hence also provide **buckling stabilisation** against vertical loads.

This results in a structure composed of thin wall panels arranged at right angles to each other, stiffening each other in a cooperative manner, so to speak, elements which are extremely rigid within their plane, but extremely sensitive at right angles to it. In the pure form of wall construction (box construction), there are sufficient wall panels located transversely to one another which shore, and thus stabilise, themselves mutually by node formation and force-locked coupling.

More details in ▶ *Sect. 7.5.1 Box Construction*

▶ *Sect. 4.2.1.2.4 Bracing*, in particular ▶ Figs. 4.47, 4.48, and 4.49 left

The mutual shoring can alternatively also take place, without directly connecting walls, by the cooperation of roof panels—as e.g. with detached-panel construction.

More details in ▶ *Sect. 7.5.3 Detached-Shear-Plate Construction*

In principle, the following construction modes are suitable for wall construction among the ones considered in this chapter:

- **masonry** construction (Figs. 7.1 and 7.6);
- **timber** construction in form of block, frame, prefabricated frame, and solid-wood construction;
- **reinforced-concrete** construction in form of in-situ concrete (Fig. 7.7) and prefabricated panel structures (Fig. 7.8).

Steel constructions belong almost exclusively to skeletal construction modes.

7.2.2 Skeletal Construction

In contrast to the extensive load distribution that characterises wall construction, the essential feature of skeletal con-

Fig. 7.6 Traditional mud-brick building in Saudi Arabia

Fig. 7.7 Gerrit Rietveld's Schroeder house in Utrecht, Netherlands, built 1924, is the prototype of the spatially open panel construction, which was novel in the 1920s

struction is the **concentration** of **vertical loads** in columns and the clear separation of the vertical supporting load-bearing structure (the columns) from the constructional elements intended for other purposes—e. g. from the envelope and the finishing work.

As a rule, we only speak of skeletal structures when the vertical load-bearing elements (in this case, the columns) are spaced in both main spatial directions at distances large enough for a standard ceiling span and thus for a regular ser-

Fig. 7.8 Semi-prefabricated element walls as basement-level wall panels just before erecting

viceable single-room width. A tightly packed stud arrangement in one direction (as with half-timbered or ribbed walls) is generally already regarded as a characteristic of a wall-construction method and therefore excludes being classified in terms of construction method as a skeletal construction, since—at least in this main direction—there is no free space profile available for unhindered utilisation. For this reason, **indirect load transfer** is generally present with ceiling constructions in skeletal structures (provided they are directional bar frameworks).

Floors and roofs, on the other hand, are designed—at least in conventional building construction—to be load-bearing over a full surface for obvious functional reasons, as is the case with wall-construction methods. In this respect, they are—as noted—usually excluded from the distinction between wall and skeletal construction.

Just as wall-construction methods are linked to certain materials that are particularly suitable for them (primarily mineral materials), skeletal structures are also closely associated with a certain group of materials, namely, the main group of materials that are both compressive and tensile-resistant, i.e. tough materials. In the oldest skeletal structures, this was mainly wood. The modern skeletal construction is primarily made of steel or reinforced concrete, but has recently been appearing more and more frequently also in timber.

The functional specialisation in, on the one hand, supporting rod-shaped column elements and, on the other, non-loadbearing two-dimensional enveloping elements represents an important qualitative step in the direction of the generally increasing functional and constructional differentiation typical of modern building structures. It allows, on the one hand, utilising modern high-performing materials such as steel or reinforced concrete for columns. This significantly reduces the load-bearing cross-section required for vertical loads—and thus the fixed points in the floor plan—in their total sum compared to wall-construction methods. At the same time, it provides a wide range of freedom in the design of the enveloping elements as they are relieved from the task of primary load-bearing. This includes, in particular, their simple omission for the purpose of creating larger interior spaces that are connected in the two main spatial directions, as well as the large-scale glazing of façades. Skeletal structures thus offer a high degree of flexibility of use due to the few vertical load-bearing elements (columns and possibly bracing building cores or a few shear walls or diagonal bracings), since the non-load-bearing room enclosures (then only required for *functional*, no longer purely for structural reasons) can be freely organised or even relocated in the floor plan. This does not apply to the same extent to wall structures. The strict requirement for wall-construction methods to place load-bearing and bracing walls exactly on top of each other floor by floor in order to enable continuous vertical load transfer does not exist in skeletal constructions—at least what walls are concerned, although it does apply to columns. Hence, with the latter construction mode, non-load-bearing partition walls can also be organised in varying positions floor by floor.

The transition from wall construction to skeletal construction is similar to that from the full-leaf system to the ribbed system. While wall construction and full-leaf element are both based on load distribution, skeletal construction and rib element are characterised by load concentration. However, wall/skeleton construction is a differentiation based on the overall building structure and—as noted above—is more use-related (see flexibility of use), whereas the full-leaf/ribbed system is a purely construction-related differentiation in terms of the build-up of an individual component. In this respect, the naming wall/skeleton construction is a pragmatic naming custom based on conventional building construction practice and traditional building terminology, which, however, does not take into account certain constructional features—such as the build-up of the floors or roofs. Due to its widespread use, it will nevertheless continue to be used in this context—with the appropriate reservations.

Of course, as mentioned above, this categorisation is limited to conventional cuboidal construction forms consisting of vertical and horizontal building elements, as they predominate in building construction, especially in multi-storey construction. Curved designs such as **shells** do not distinguish between walls (or columns) and ceilings and are therefore more precisely characterised as full-leaf systems or ribbed systems.

Since, contrary to the wall-construction mode, the constructional principle of skeletal construction does not provide any supporting shear-resistant wall plates, the skeleton—a pure framework—must be braced against horizontal loads, if required, by suitable measures. In addition to the skeletal structure, a **bracing construction** of components suitable for this purpose must be provided. An exception is portal-frame systems in which columns and beams are coupled to form a rigid load-bearing structure.

▶ *Sect. 4.2.1.3 Bracing of Skeletal Structures*

The following construction modes are suited as skeletal construction:

- **timber** construction or **wood-concrete composite** construction (◘ Fig. 7.9);
- **steel** construction or **steel-concrete composite** construction (◘ Fig. 7.10);
- **reinforced-concrete** construction (◘ Fig. 7.11).

▶ *Sect. 8.5 Wood-Concrete Composite Construction*; ***Vol. 3***, *Sect. 61.5.1.4 Wood-Concrete Composite Floors*

Although masonry construction, in principle, can also be carried out as a skeletal construction with masonry pillars, that type of construction is of no practical importance.

▶ *Sect. 9.3.1.1 Composite Construction* as well as ▶ *Sect. 9.3.3.1 Floors in Composite Construction*; cf. also ***Vol. 3***, *Sect. 20.6.2.2 Steel-Concrete Composite Floors*

7.3 Fundamentals of Masonry Wall Construction

Some essential features of wall construction—also referred to as shear-plate construction—will be further deepened in the following for the purpose of a better understanding of the constructional and form-related principles to which they are subject using the example of masonry structures. Reinforced-concrete wall constructions, however, are subject to dif-

▶ *Sect. 10 Prefabricated Concrete Construction* and ▶ *Sect. 11 In-Situ Concrete Construction*

◘ **Fig. 7.9** Timber skeleton

Fig. 7.10 Steel skeleton

Fig. 7.11 Reinforced-concrete skeleton

ferent rules because, unlike masonry, they are able to absorb tensile stress. They are discussed elsewhere.

▶ *Sect.* 11 *In-Situ Concrete Construction*

7.3.1 Interaction of Load-Bearing and Enveloping Functions of the Wall

The partially complementary, but sometimes conflicting, fundamental tasks of a wall, namely, **load-carrying** and **enveloping**, are fundamental in their mutual complex interplay for an understanding of the forms of wall construction in general and of classical masonry construction in particular.

The essential tasks of a wall component deriving from the **load-bearing** function are

- bearing **vertical** loads;
- bearing **horizontal** loads, either within the component plane or—with restrictions—at a right angle to it.

An essential factor for understanding the basic static and constructional principle of masonry construction is the great stiffness of the involved wall panels *within* their plane (i.e. their great racking stiffness) and the weakness of these components against forces *transverse* to their plane or with regard to the risk of buckling under vertical load in this direction (i.e. their only limited bending stiffness or buckling and tipping stability).

Cf. ***Vol. 1***, *Sect. 26.9.3.2 Bond—Lap Acting by Compression*, in particular Figs. 26.133, 26.134, 26.141, and 26.148

The **enveloping** function associated with load-bearing leads to a whole series of other additional tasks of the envelope:

- Provision of window or door openings. Since interior spaces must be adequately daylit and ventilated, it is always necessary to provide the outer wall with sufficient **openings**, so to say cut-out from the wall surface, except for rare exceptions (◘ Figs. 7.1 and 7.12). In conventional wall constructions, this is also referred to as a **perforated façade**. These necessary openings amount to a significant **weakening** of the load-bearing structure of the wall panel. This is the case because:
  - Only a usable **residual cross-section** of the wall after deduction of the widths of the openings, i.e. the sum of the cross-sections of the wall piers in between, remains for the removal of vertical loads (◘ Figs. 7.12 and 7.13). As a result, the compressive forces acting on the remaining wall cross-sections increase accordingly. In conventional masonry construction, therefore, pier widths between window openings must adhere to certain minimum dimensions.
  - For transferring horizontal loads, the closed wall panel is thereby transformed into a perforated panel or even into a largely diaphanous, frame-like element (◘ Figs. 7.14 and 7.15), depending on the window-to-

Vol. 3, *Sect. 18 Openings*

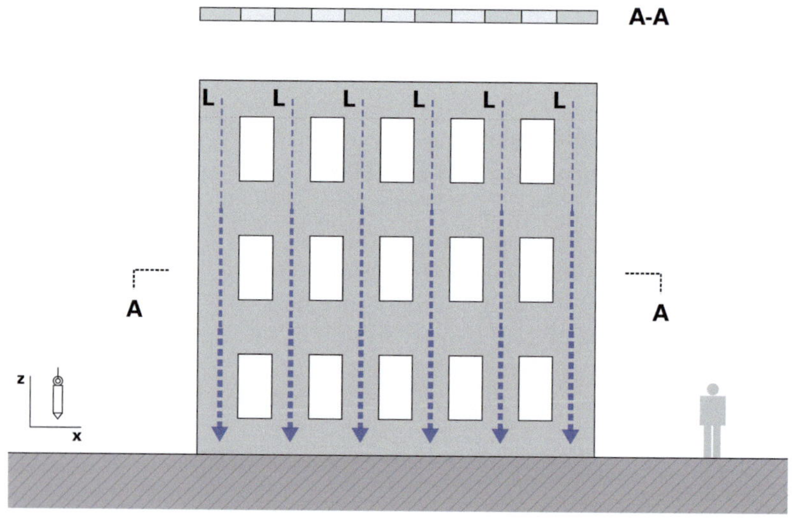

Fig. 7.12 Perforated façade. Narrow, high window formats allow wide intermediate wall pillars to be used to discharge the loads **L**. Load-bearing residual cross-section in section **A-A**

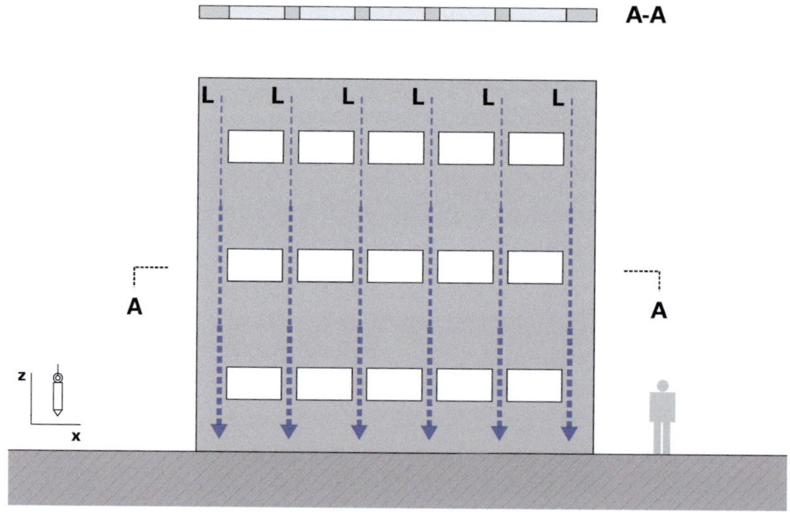

Fig. 7.13 Perforated façade. Wide, narrow-spaced window formats result in narrow wall pillars and high load concentrations

wall ratio. This can prove detrimental to the structural shear-plate effect of the wall component.

Localised **door openings**, which are indispensable for the operation of an interconnected unit of usage in the case of internal partition walls, also can, under certain conditions, result in an impairment of the structural functionality of a masonry wall. For transferring vertical loads, what has been said for perforated façades applies. For the bracing effect of a wall panel, a sequence of superimposed, particularly storey-high door openings—as may result

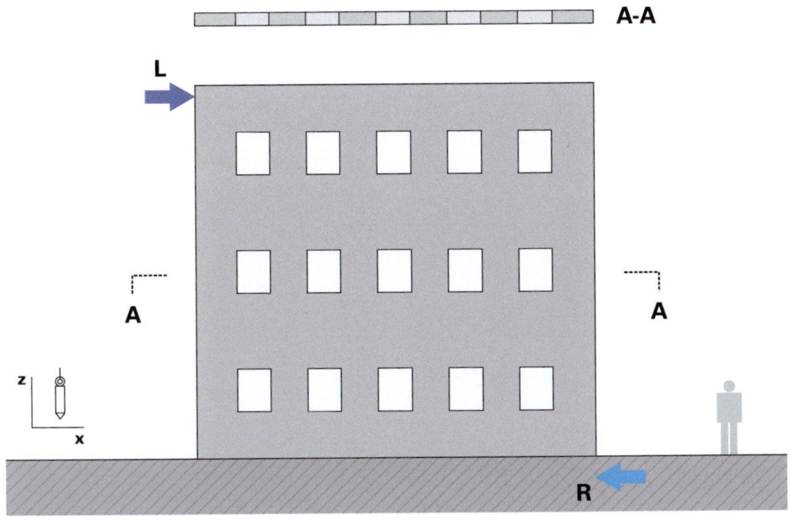

Fig. 7.14 Perforated façade. High wall-to-window ratio; the wall panel is mostly preserved

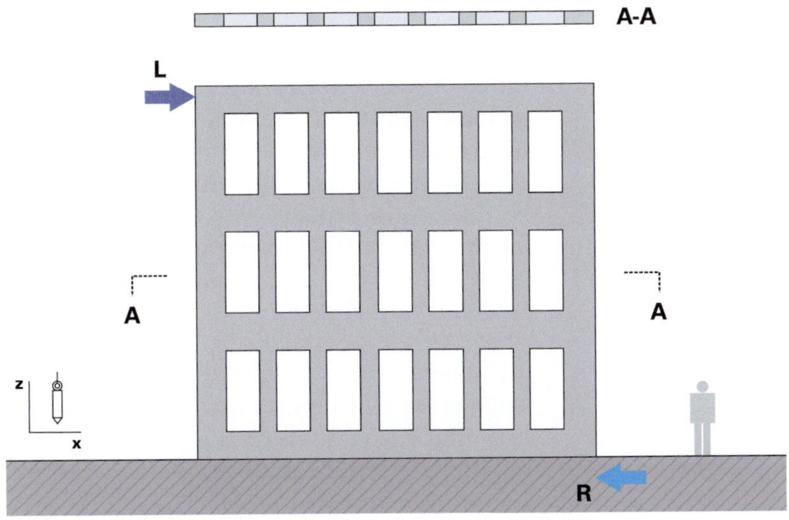

Fig. 7.15 Perforated façade. Small wall-to-window ratio; the wall panel is mostly converted into a frame-like element

from the stacking of identical floors like is customary in wall-building—can be critical, since the wall surface is thus slit open, so to say, into two narrower halves (Fig. 7.16), thereby also halving the available lever arm of the reaction moment opposing the tipping moment.

- Fulfilment of **building-physics** (Fig. 7.4) functions like:
  - **sound** protection;
  - **moisture** protection;
  - **thermal storage**;
  - as well as ensuring adequate **fire protection**.

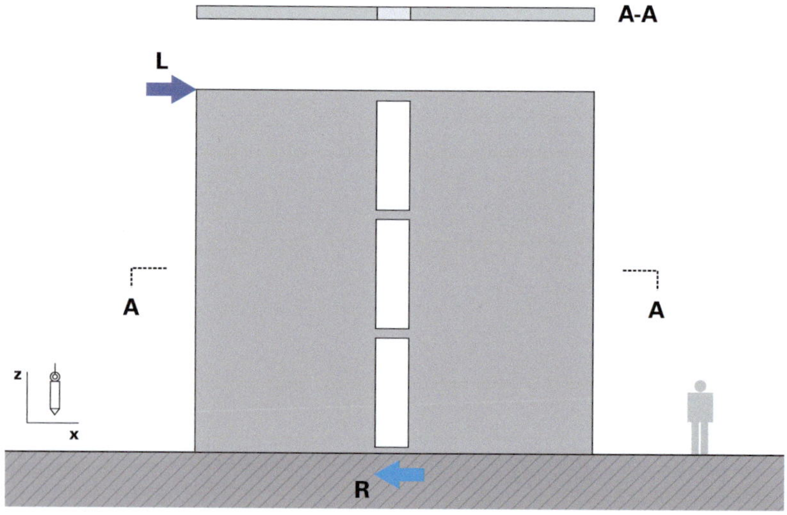

Fig. 7.16 Separation of the wall panel into two narrow halves by a sequence of high, overlying door or window openings

The objectives of these functions essentially coincide with those of the **load-bearing** function. Both **mass** and high **material density** of the wall are basically conducive to all of the above-mentioned tasks, although, of course, for some partial tasks, additional material properties have to be taken into account, for example, stiffness or strength for the load-bearing function. There are, at least in this respect, no fundamental conflicts of objectives between the load-carrying and the enveloping function. Among other things, this is a cause of the long-standing success of solid wall construction in diverse regions of the world under widely varying circumstances.

- Ensuring adequate **thermal protection**. Regarding this partial function of the envelope, conditions are quite different from those discussed above. Here, the requirements of the load-bearing function, which presuppose solid, dense material, are in **sharp conflict** with those of the thermal-insulation function, which instead requires a light, porous material. As long as the applicable thermal-protection standards were moderate—this applies to the long historical dissemination period of solid stone construction—the material properties of average brick material were sufficient for both purposes: for adequate load bearing and thermal insulation. This changed drastically with the introduction of stricter requirements of thermal protection: In the case of single-leaf brick walls, limitations of functionality, or additional measures to mitigate them, were unavoidable in view of the changed boundary conditions:

- either by limiting thermal protection and maintaining a better load-bearing capacity by using denser, heavier brick material—it was therefore unavoidable to relocate the thermal-insulation function from the load-bearing leave to an additional lining;
- or, alternatively, by securing good thermal protection and at the same time accepting a limited load-bearing capacity—this route is followed in low-rise, highly insulated housing construction using uninsulated single-leaf walls made of lightweight material.

In the case of the **single-leaf** variant of wall construction in building, the old advantages of conventional solid construction, namely, its structural simplicity and robustness, have at least partially been preserved under the aforementioned limitations. However, as noted, a result of this development was also **multilayered** and finally **multi-leaf** massive exterior wall structures. They all transfer the heat-protection function away from the load-bearing leaf to an additional thermal-insulation layer that is specialised for this purpose and hence much more efficient. As a result, constructional conflicts have arisen, which sometimes entail elaborate constructions and which have significantly reduced the advantages of wall construction as compared to competing construction modes.

Vol. 3*, Sect. 12 Full-Leaf Systems*

▶ *Sect. 2.3 Double-Leaf Systems*

7.3.2 Eccentricities

Similar to skeletal supports, the structurally efficient dissipation of loads into the foundation requires load-bearing wall panels of each storey being placed upon each other and being eventually continued down to the foundation. **Eccentricities** in the masonry construction from offsets between superimposed walls, some of which then rest on floors, cause increased bending moments and shear forces in the floors. These are costly and not only disturb the load-bearing principle of masonry construction, but also the bracing concept, since a multi-storey wall loses its bracing racking stiffness with offsets of the wall plane (◘ Fig. 7.17), not to mention the additional strong bending moments acting on the floor.

▶ *Sect. 4.2.1.2.1 Extendibility*, in particular ▶ ◘ Fig. 4.36

Being obliged to arranging wall panels vertically one above the other naturally leads to **functional limitations** in the use of buildings, in particular because, as a matter of fact, the same layout of the floor plan must be adhered to across all floors. Even if complying with this rule, carefully stacked walls cannot be themselves situated in any arbitrary way, since they necessarily must support the floors—which is not true for non-load-bearing walls. From the mutual dependence of floor span, if applicable, span direction,

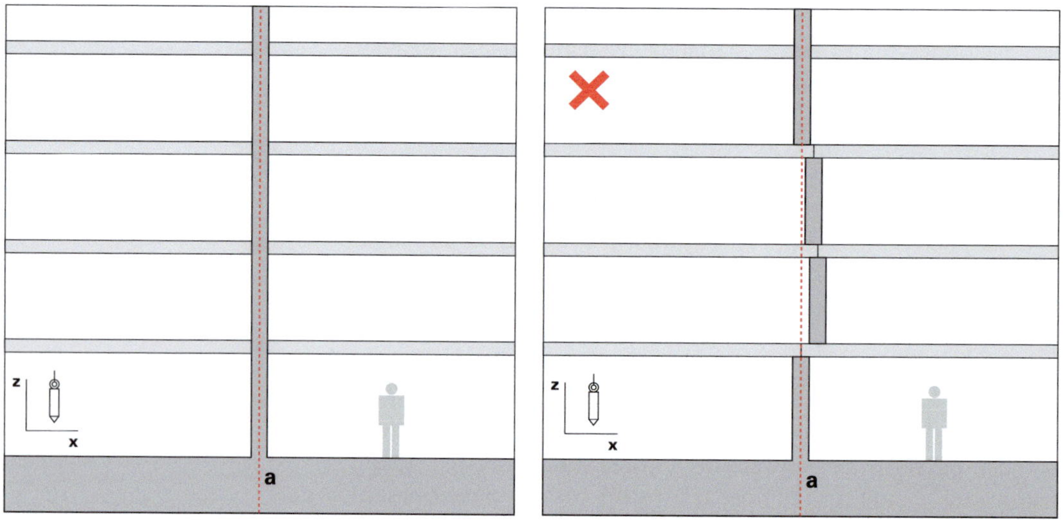

Fig. 7.17 (Left) Bracing shear-plate effect with a continuous wall. Forces in → **y** are discharged in the panel plane (axis **a**). (Right) **Offsets** in the shear-plate annul the bracing effect. Forces in → **y** can no longer be transferred. In addition, there are offset moments, i.e. additional stress, on the floors

and wall spacing in conventional wall construction, **maximum room widths** result, since these are usually also associated with **maximum floor-span widths**. For one-way spanning floors, this is not true for room *lengths*, which are in principle not limited from this point of view. In actuality, they are so nonetheless, but for other reasons: e.g. the need for transverse bracing walls shoring the contemplated load-bearing walls or from the requirement for reasonable, not too extreme room proportions for usage-related reasons.

► *Sect. 4.2.1.2.1 Extendibility*

7.3.3 Cell Formation

The bracing of wall structures is based on the principle of shear-plate formation, in such a way that wall panels are placed in two perpendicular directions. This principle is most consistently achieved with conventional masonry structures by combining the wall panels into closed rectangular **cells**. These create individual spaces or cellular rooms, which are enclosed on all sides. Herewith, a fundamental, use-related task of building construction, namely, the creation of space by enclosing it all around, is fulfilled in perfect congruence with static and constructional requirements, so to speak. In addition to the load-bearing function, which is ensured by the thickness and material strength, or stiffness, of the wall, other partial functions are also fulfilled, as noted, without additional measures: for example, sound and fire protection. The domain of masonry construction is therefore to be seen mainly in **housing construction**, or any build-

► *Sect. 4.2.1.2.4 Bracing*

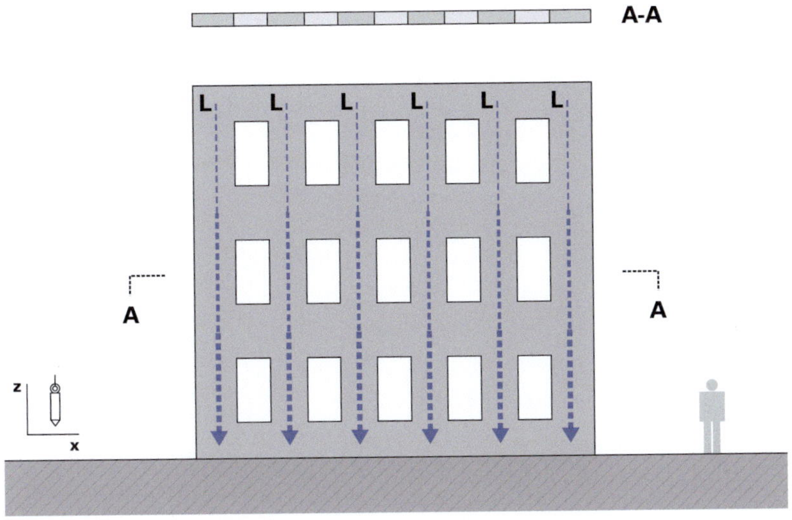

Fig. 7.18 System-compliant direct discharge of vertical loads **L** in the overlying wall pillars in classical masonry construction. Typical formatting of the openings in form of narrow upright rectangles

ing use which has a functionally caused multiplicity of separated rooms and room partitions and does not require short-term changes to the wall positions.

7.3.4 Flexibility

Classical masonry construction *cannot* meet demands for a high flexibility of use for system-specific reasons. For the reasons discussed above, wall positions are fundamentally unchanging. The production of wall breakthroughs in order to spatially connect adjacent space cells to each other is also severely restricted, in particular with regard to wide openings. The same principles apply as for window openings. These restrictions, however, do not apply in this form to reinforced concrete.

Cf. Figs. 7.12, 7.13, 7.14, 7.15, and 7.16; cf. also the next section and ▶ *Sect. 3.1.6.1 The Vertical, Planar Space-Enclosing Element*

7.3.5 Position and Shape of Wall Openings

Since wall openings always entail a weakening of the load-bearing wall structure for the reasons mentioned, it applies that:

EN 1996-1-1, 8.5

- The wider the openings and the narrower the wall pillars in between are, the greater the compressive forces within the remaining supporting cross-sections (Fig. 7.13) will be. Most unfavourable are opening patterns with extremely wide openings and very narrow wall piers in between. The height of the opening does not significantly affect the load-bearing capacity of the wall panel, as long as a suf-

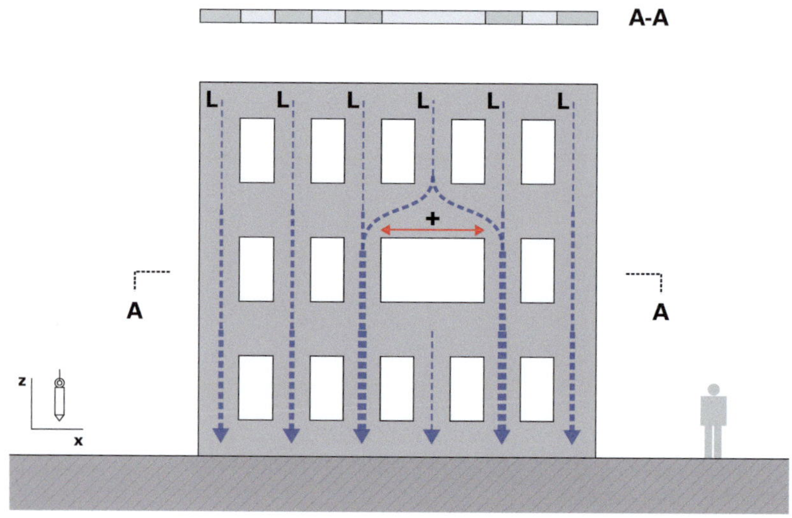

Fig. 7.19 Necessary diversion of forces by singular opening with large width. Tensile forces (+) are generated in the masonry

ficiently high transverse wall strip—usually a parapet—exists which secures the racking stiffness or at least the frame action of the surface element. These restrictions apply primarily to masonry construction, so that the distinctive, system-compliant perforated façade of masonry always has narrow, upright, openings with rectangular shape (Fig. 7.18).

- Disturbances or changes within a regular opening pattern from strictly vertically stacked, always equally wide opening formats necessarily result in **force deflections** (Fig. 7.19). These inevitably evoke **tensile forces** in the wall structure, against which masonry is very sensitive. As a rule, load concentrations in the affected wall piers also occur. Storey-wise offsets within an otherwise regular opening pattern (Fig. 7.20) also have comparable effects.
- Openings must not be too close to a building corner, since a sufficiently wide wall pillar is required there. This is necessary not only for transmitting vertical loads, but also for shoring the transverse wall perpendicularly abutting at the corner. This is called avoiding **free edges**. A typical feature of masonry construction is therefore a sufficient distance from edge openings to the building's corner.

Vol. 1, *Sect. 26.9.3.2 Bond—Lap Acting by Compression*, in particular, Fig. 26.145

As a result of these properties of load transfer in wall panels made of masonry, it can be stated in simplifying manner that the brickwork-style perforated façade is provided with a regular orthogonal grid of upright narrow openings standing not too close together (Figs. 7.21, 7.22, and 7.23).

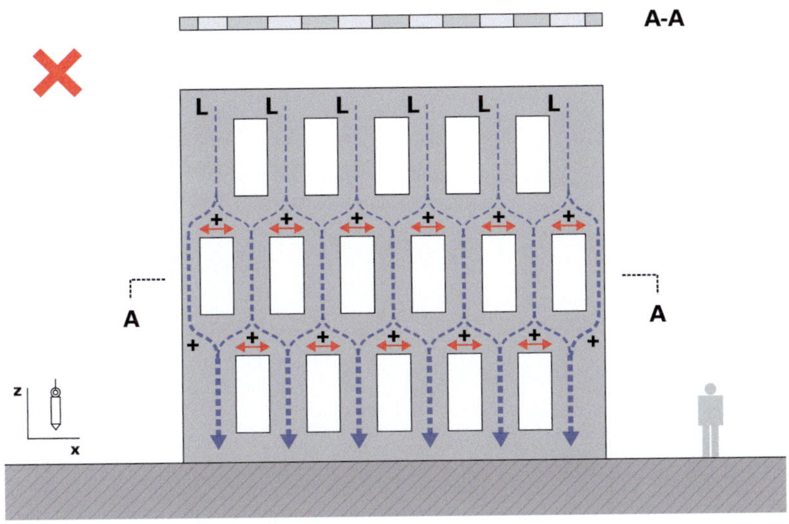

Fig. 7.20 Multiple diversion of forces by storey-wise offset of the openings. Tensile forces (+) are generated in the masonry. Non-masonry-compatible arrangement of openings

Fig. 7.21 Traditional perforated façade

7.3.6 Special Features of Arched Openings

In addition to the already discussed design rules of masonry-type perforated wall panels, further limitations are added in the case of arch-covered openings. They concern, in particular, historical designs, since the arch is used today for this purpose only in exceptional cases.

Fig. 7.22 Modern masonry building

Fig. 7.23 Chilehaus in Hamburg-Kontorhausviertel, 1922–24 (arch.: Fritz Höger)

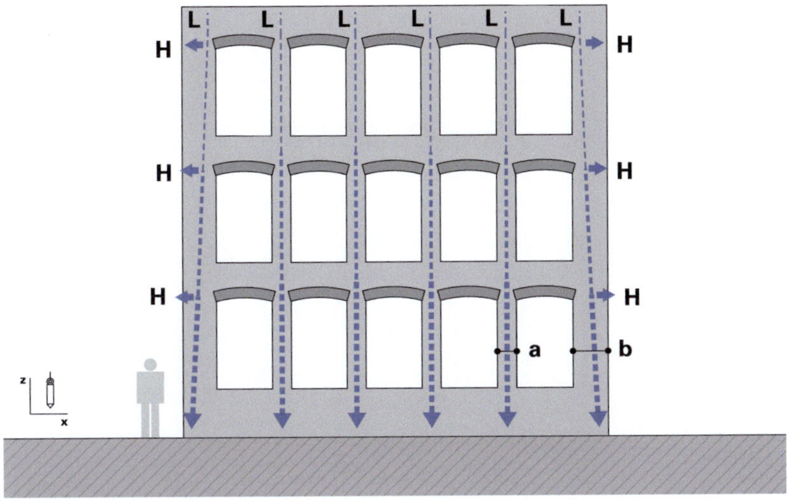

Fig. 7.24 Diversion of the load **L** in the outer wall pillars by the effect of the arch thrust **H** of the peripheral openings. A minimum pillar width **b** is required in order for the resultant to remain within the cross-section

- Arches create a **thrust** at the springing. This is neutralised, with the usual grid-like opening patterns, whenever two equal openings abut. The resultant force is then vertical and is transferred within the wall pillar between openings. This is not the case on the outside of the outermost openings, so that a sufficiently wide wall pillar is to be provided there (Fig. 7.24). It is to be dimensioned sufficient to not only stabilise the transverse wall abutting at right angles at the building corner (see above), but also to take up the arch thrust of the last opening.
- For a targeted force deflection within the wall structure, **relieving arches** can be incorporated (Fig. 7.25). These are bricked in a radial pattern and are enclosed within the regular brick bond of the wall. Through their effect, loads can be redirected over partial areas and introduced into selected wall piers, namely, to those where the relieving arches end.

7.4 Constructional Issues in Detail

Other constructional aspects are also decisive for masonry construction. They are causally linked to the marked difference in stiffness and stability of a wall panel, respectively, *within* its plane and *perpendicular* to it: A wall panel has a great stiffness within its plane but, transversely, it has to be stabilised by shoring components in most cases.

EN 1996-1-1, EN 1996-1-2, EN 1996-3; DIN 105-100, DIN 105-4, DIN 105-5, DIN 105-6; EN 771-1, EN 771-2, EN 771-3, EN 771-4, EN 771-5, EN 771-6

Sect. ▶ 7.2.1 Wall Construction

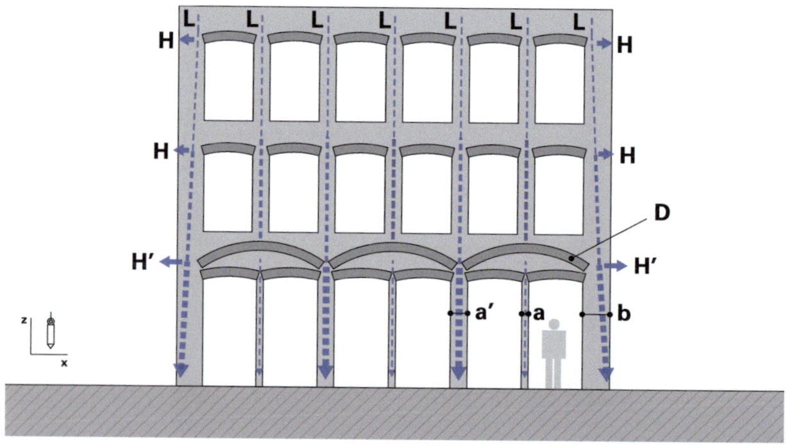

Fig. 7.25 Discharging arches **D** deflect the load **L** over two openings, respectively, in the lower floor and allow for thinner columns (**a**)

7.4.1 Materials

EN 771-1, EN 771-2, EN 771-3, EN 771-4, EN 771-5, EN 771-6; DIN V 20000-401, DIN V 20000-402, DIN V 20000-403, DIN V 20000-404

According to the standard, the following materials are permissible for masonry structures:

- **wall bricks** according to *EN 771-1* (in connection with *DIN 20000-401*);
- **lime-sand bricks** according to *EN 771-2* (in connection with *DIN 20000-402*);
- **concrete masonry blocks** with dense and porous aggregates according to *EN 771-3* (in connection with *DIN 20000-403*);
- **aerated concrete** blocks according to *EN 771-4* (in connection with *DIN 20000-404*);
- **concrete blocks** according to *EN 771-5* (in connection with *DIN 20000-403*);
- dimensionally accurate **natural stones** according to *EN 771-6* (including *EN 1996-1-1/NA* respective national annexes).

7.4.2 Stiffening and Stabilisation of Walls in Masonry Construction

Wall construction is based on the use of thin shear-plates with great stiffness within their plane. Transverse to their plane, however, they are sensitive to tilting, buckling, or bulging. They must therefore be **transversely reinforced** or **stabilised**. Various possibilities exist for stabilising wall structures in masonry against horizontal loads:

Cf. also the considerations in ▶ *Sect. 4.2.1.2.4 Bracing*

7.4.2.1 Heavyweight Wall

EN 1996-1-1/NA, respective national appendices

A tilt-resistant heavyweight wall with a resultant **R** of all forces from vertical and horizontal loads (**V**, **H**) which does not reach more than **t/6** to the lower boundary of the wall—

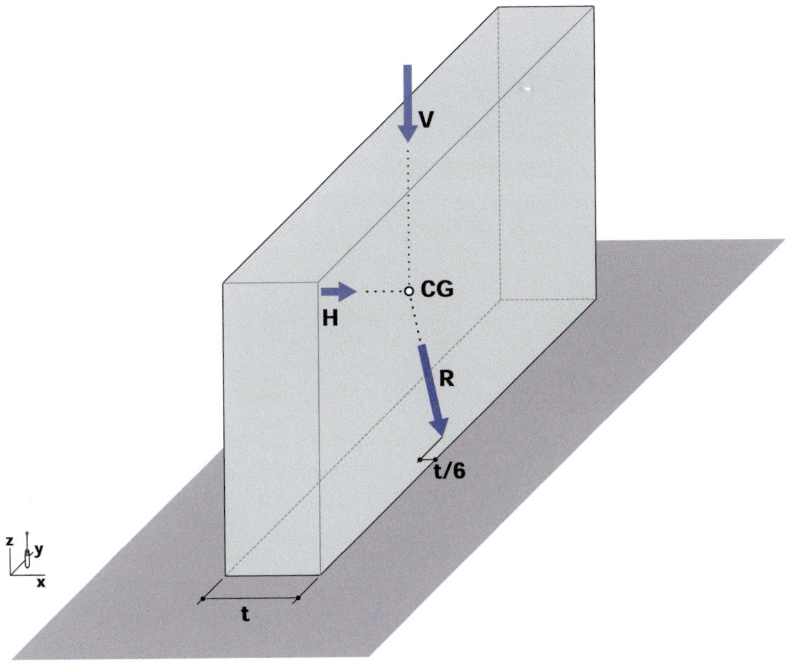

Fig. 7.26 Tilt-resistant **heavyweight wall**. As a result of the simultaneously acting loads **H** and **V**, the resultant **R** emerges from the parallelogram of forces. If the latter does not approach the wall edge in the bearing surface more than **t/6**, the wall remains stable due to its dead weight. It is clear that the enlargement of the load **V** has a favourable effect on the stability of the wall, since it forces the force **R** further into → –**x** within the core of the wall. **CG** centre of gravity; **t** wall thickness

it then lies within the **core surface**—is stable without further measures (Fig. 7.26). The resultant **R** is, so to speak, forced more strongly into the vertical line by the effect of the large dead-load component **V** than with a lighter wall. Because of the high consumption of material and the associated construction process, which is difficult due to heavy weights, this type of walls is only of secondary importance in modern building construction.

► *Sect. 6.2.5 Failure of the Ground*, in particular ► Fig. 6.21

7.4.2.2 Stiffening by Wall Pillars

The wall can also be stiffened by a local thickening with greater stiffness than the wall itself, i.e. with individual, spaced **wall pillars** or **wall pilasters**, usually on only one side of the wall (Figs. 7.3 centre and 7.27 top). The structural effect is comparable to a thickening of the wall, which is why the standard defines an **effective wall thickness** t_{ef} as a function of the pillar depth and spacing:

► *Sect. 4.2.1.2.4 Bracing*

EN 1996-1-1, 5.5.1.3

$$t_{ef} = \rho_t t$$

The value ρ_t is a coefficient, the **stiffness factor**, which can be taken from the table on Fig. 7.27, bottom.

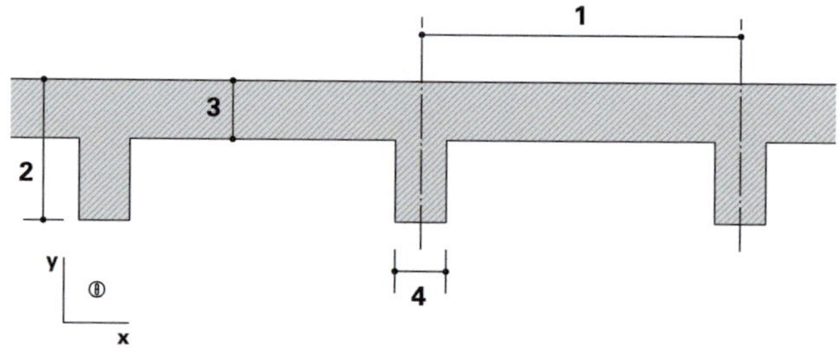

ratio of pillar distance (centre to centre) to pillar depth	ratio of the pillar depth to the actual thickness of the connected wall		
	1	2	3
6	1.0	1.4	2.0
10	1.0	1.2	1.4
20	1.0	1.0	1.0

Note: linear interpolation between these values is allowed.

Fig. 7.27 (Top) Schematic representation of the parameters used in the bottom chart. (Bottom) **Stiffness factor** ρ_t for the design of pier-reinforced walls according to *EN 1996-1-1*

7.4.2.3 Stiffening by Cross-Walls

The lateral shoring by means of **transverse** or **cross-walls** is the usual way in masonry construction to brace a wall or a load-bearing structure made from an array of walls. The walls can thereby be made substantially thinner and lighter than a free-standing heavyweight wall as described above. This principle of mutual support and formation of box-like wall structures has given its name to a particular masonry construction mode: the **box-construction mode**. It is considered to be the traditional and most original manifestation of masonry construction.

The box-construction mode is discussed in more detail below in ▶ *Sect. 7.5.1 Box Construction*

The masonry bond, which is composed of individual building blocks and is in principle compression-resistant but not tension-resistant, must be able to absorb in-plane shear forces for a stiffening effect. Under horizontal load, the following occurs in the wall (Fig. 7.28):

- **Horizontal shear** τ_h, which is absorbed by the shear stiffness of the masonry block itself and, moreover, by the tangential frictional connection in the horizontal joint, the bed joint. This can only be effective if there is sufficient transverse compression on the joint, i.e. sufficient vertical load (**q**).
- **Vertical shear** τ_v, which in its lifting component (Fig. 7.28 left side), i.e. on the side of the wall facing the external force, elicits transverse tensile stress on the bed joints and can cause cracks. The block's and mortar's tensile strength and the adhesive effect between mortar layer and stone are not sufficient for this. Instead, this force effect must be exceeded, and thus blocked by over-compression,

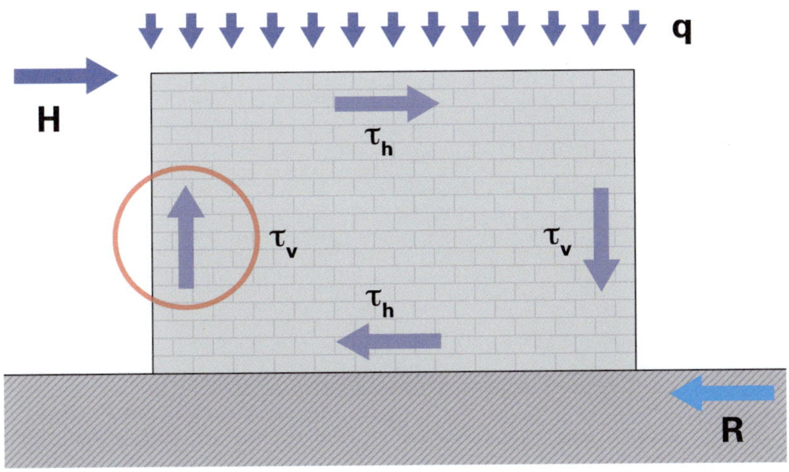

Fig. 7.28 Shear load on a bracing wall panel. The panel's shear forces τ_v resulting from the horizontal force **H** generate lift-off tensile forces (left). In the case of masonry, they must be exceeded by dead load **q** or by combined dead load and superimposed load

by a sufficient **dead load** or preferably by a **superimposed load** (e.g. floor loads). For this reason, it is recommended with this type of bracing to also load bracing walls—apart from load-bearing ones—with floors by changing the direction of the floor span, at least on some storeys. With the solid floors commonly used today, however, a two-way load transfer already ensures that loads are distributed to all walls, so that the distinction between load-bearing and bracing walls is no longer applicable.

Both factors, therefore, require sufficient **vertical load** in the wall to absorb the shear stress due to horizontal force.

In the lateral support, a **shear-resistant connection** of the load-bearing and bracing wall is essential. It is important to avoid a tilting or a shifting of the system at the contact joint (Fig. 7.29). This is achieved by means of toothing the walls within the masonry bond (Fig. 7.30 left) or by inserting band-iron strips into the butt joints of walls abutting without indentation (Fig. 7.30 right).

EN 1996-1-1, 5.5.3

Shear-stressed stiffening walls which are connected to the stiffened wall in a shear-resistant manner are additionally stiffened by a cooperating partial width of the transverse wall in the manner of a flange (Fig. 7.31).

If certain maximum distances between transverse supports are not exceeded, such a bracing of the wall is sufficient. If the distances are larger, the following additional measure is required.

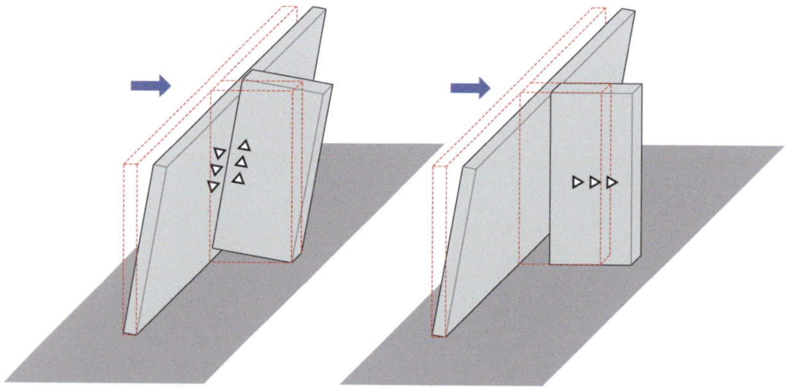

Fig. 7.29 **Longitudinal** and **transverse walls** which are not connected to one another cannot stabilise each other when a load is applied horizontally

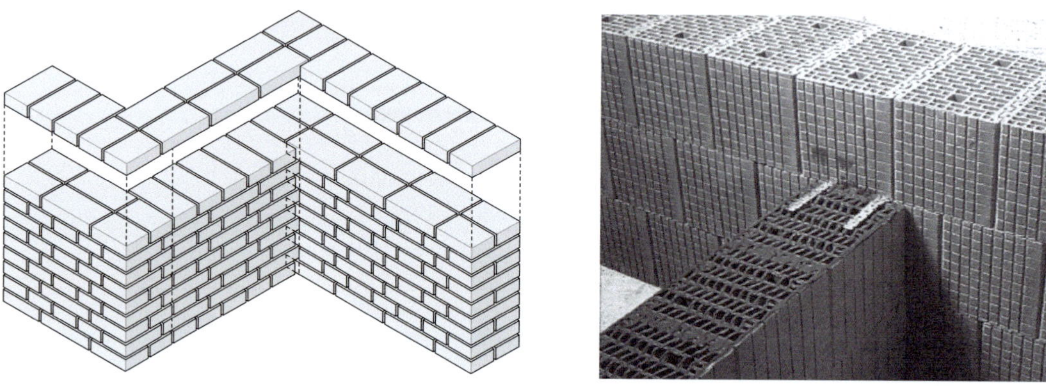

Fig. 7.30 (Left) **Toothing**: Connection between the longitudinal and transverse wall by form-fitting toothing in the bond. Both walls are bricked simultaneously. (Right) **Butt-joint** method: There is no interlocking of the wall bond. Instead, anchors are inserted into the bed joints

7.4.2.4 Stiffening by Ring Beams or Floor Shear-Plates

This option concerns holding the uppermost **free edge** of masonry walls. The possibility of stabilising masonry walls *solely* by means of transverse supporting walls, as described above, has the effect that the upper edge of a wall remains free at the top, i.e. is not held laterally there. This allows it to deform both under horizontal bending stress and to buckle or bulge laterally under vertical load (see Euler case **1**) (Fig. 7.32). This free edge appears not only in free-standing walls, but also in combination with floors which do not have shear stiffness—e.g. wooden-beam floors—or floors which do have it but are still unable to hold the wall, such as, for example, slabs on sliding wall supports introduced for accommodating temperature deformations.

***Vol. 1**, Sect. 26.8 Critical Mechanisms of Failure* and in particular *Sect. 26.9.3.2 Bond—Lap Acting by Compression*, Fig. 26.134 and Fig. 26.145

In order to keep the free edge of walls immovable, there are basically two possibilities:

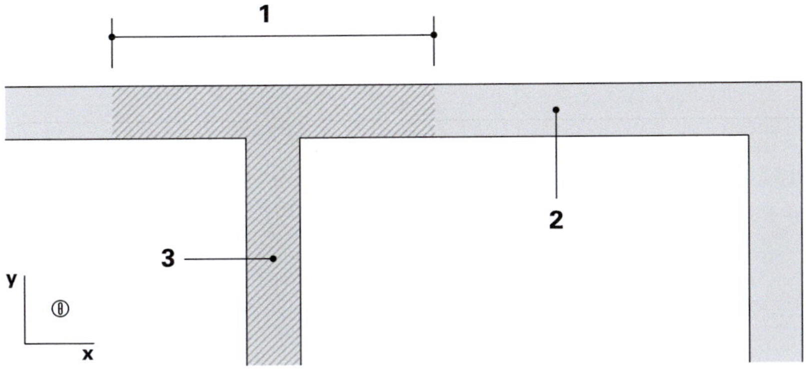

Fig. 7.31 **Cooperating width** of a shear-stressed stiffening wall **3** in the bracing wall **2** according to *EN 1996-1-1*. It reinforces the wall in the manner of a flange: (**1**) cooperating width, (**2**) cross-wall, and (**3**) bracing wall

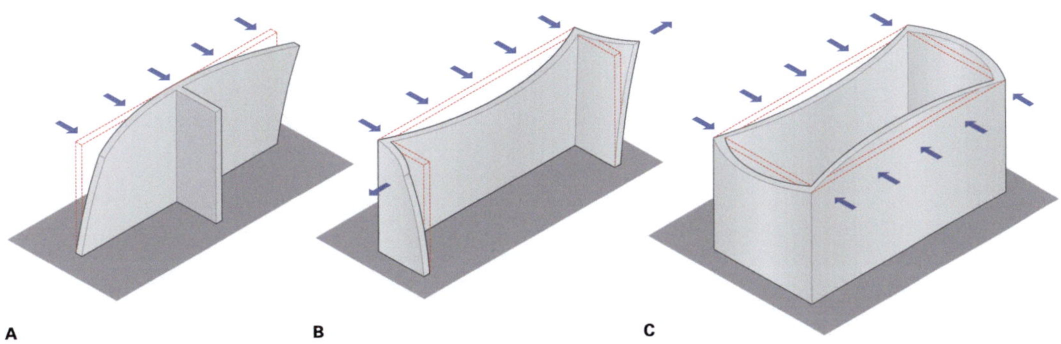

Fig. 7.32 Walls only braced vertically by transverse walls; deformation of the upper unretained edge in case of missing floor slab or ring beam (cf. also Fig. 7.5)

- A **ring beam** can be considered which stretches between two bracing walls—the transverse walls—like a flat bending beam, which transmits horizontal loads into the latter and prevents the deformation of the upper wall edge (Figs. 7.33, 7.34 left, and 7.35). Usually, this is a reinforced-concrete beam which is suitably reinforced—i.e. for bending within the horizontal plane. The maximum span of such a ring beam can be assumed as follows: *EN 1996-1-1*, 8.5.1.4

  $\mathbf{L} \sim\leq 30 \times \mathbf{t}\,\text{concrete}$

- Alternatively, the floor which is supported on the wall can be executed shear resistant, i.e. as a **floor shear-plate** with an associated **ring anchor** (Figs. 7.34 right, 7.36, and 7.37). This solution is used for the solid floors that are widely used today and is the standard case presently.
- Gable walls are force-locked to the roof truss, which in turn is stiffened by sheathing or wind bracing.

The transmission of horizontal forces from the free upper edge of the wall to the stiffening beam or the stiffening floor

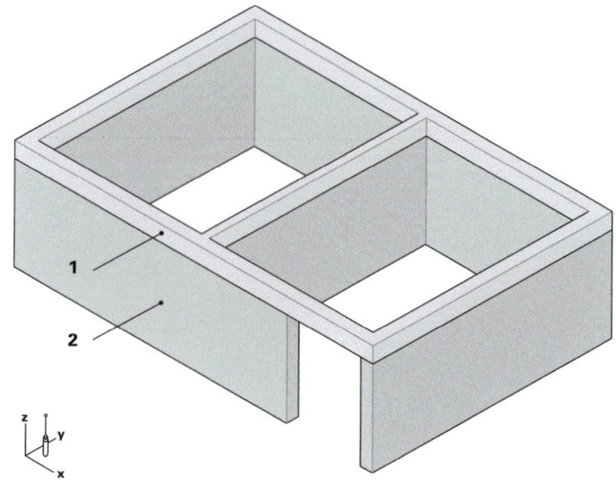

Fig. 7.33 Ring-beam arrangement on all walls to be stabilised. The ring beam is arranged circumferentially and stabilises all load-bearing walls due to its bending stiffness in the plane **xy**: (**1**) ring beam and (**2**) masonry

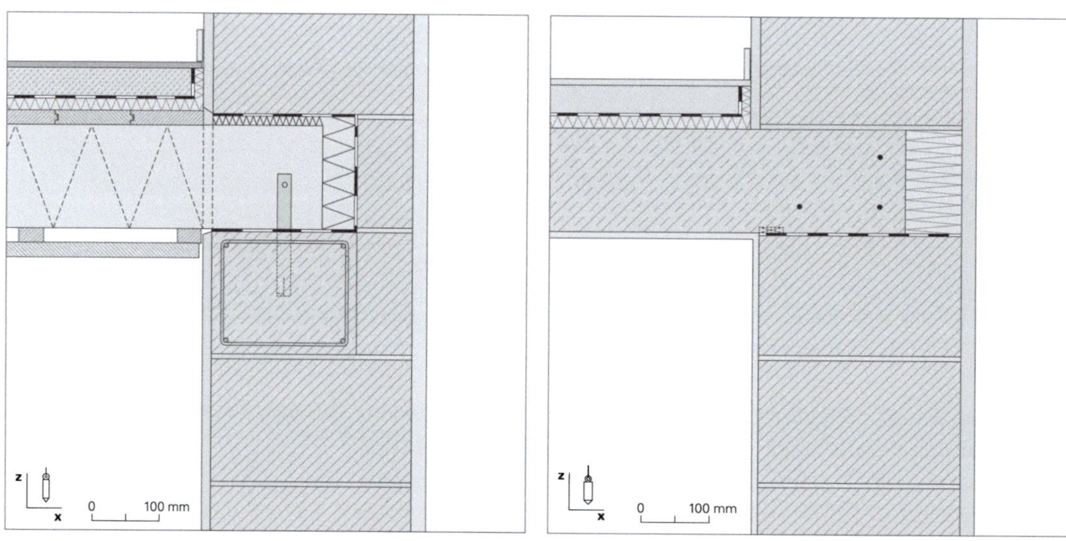

Fig. 7.34 (Left) Wooden-beam floor. The beams are placed on the ring beam and connected to it. The wall loads are not transferred through the beams (this would produce transverse compression), but by the wall piers remaining between the beams (see also ***Vol. 3**, Sect. 20.6.1.3.5 Wall Supports*). (Right) Ring-anchor arrangement in the edge area of a reinforced-concrete floor slab. At the same time, the reinforced-concrete floor, being a shear-plate, performs the task of the ring beam, i.e. the retention of the upper edge of the outer wall

can, in principle, take place via friction in the bearing surface between the two components (for which, in addition to the roughness of the contacting surfaces, an appropriate load is also required); or alternatively by means of suitable anchors with appropriate end fixings (Fig. 7.38).

Cf. also ***Vol. 4**, Sect. 8.3.5.1 Bracing by Beam Floors*

In principle, a non-displaceable upper wall edge is required for all wall constructions in order to maintain a wall surface which is held with a hinged connection at least on two sides, i.e. at the bottom and the top. In the presence of two addi-

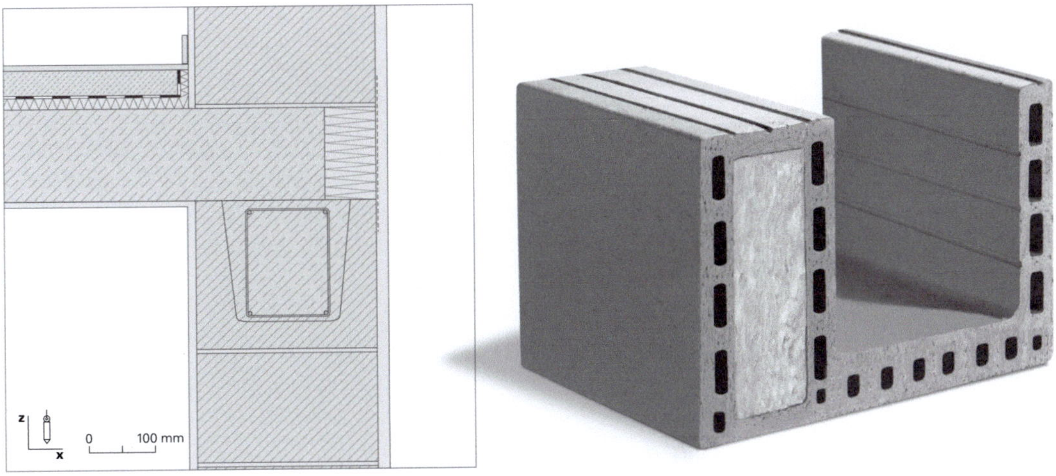

Fig. 7.35 (Left) Ring anchor in masonry U-shell, arranged under a reinforced-concrete floor slab. (Right) Trough-shaped brick shuttering from *Wienerberger Ziegelindustrie GmbH* for producing ring anchors and ring beams as part of the vertical-coring lightweight-brick programme

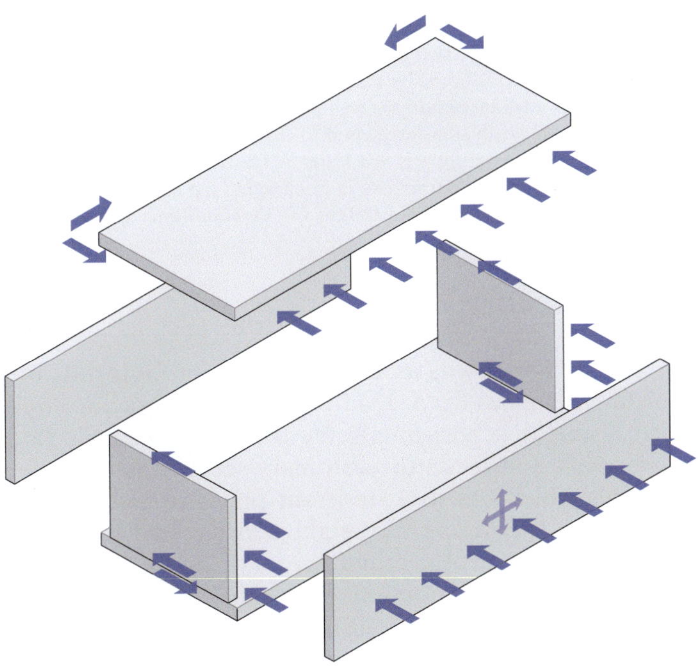

Fig. 7.36 Distribution of horizontal loads acting on the longitudinal wall into the floor slab/building ground, the lateral transverse walls, and the shear-resistant roof slab. The longitudinal wall is an example of a wall retained on four sides

tional bracing transverse walls, a wall surface with a four-sided shoring results. This means that, provided the free upper edge is braced by the floor shear-plate—in accordance with the above-mentioned second variant—a spatially closed **box** is formed overall. This variant, which is important for modern masonry construction, will be discussed in more detail below.

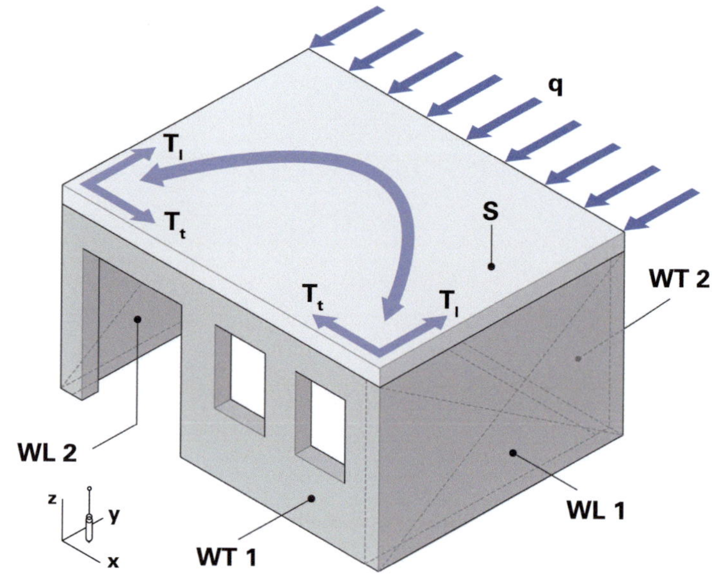

Fig. 7.37 Depiction of the compression vault and the resulting tensile stress at the roof or floor edge. The tie member (= ring anchor) is integrated in the roof or floor slab **S**. The component $\mathbf{T_t}$ oriented transversely to the force action (→ **x**) is borne by the ring anchor; the component $\mathbf{T_l}$ running along it (→ **y**) is introduced into the wall panels **WL 1** and **WL 2** oriented longitudinally (→ **y**), which transfer them into the foundation by virtue of their racking stiffness. The wall panels **WT 1** and **WT 2** oriented transversely to the force action (→ **x**) do not participate in discharging the load **q**. If the force is applied from the perpendicular direction (→ **x**), conditions are inverted accordingly

7.4.3 Generation of a Box

The box is composed, as considered in its footprint, of a quadrangle made from load-bearing and bracing walls, which is enclosed horizontally by the ceiling and the floor. A favourable and presently customary solution is using floor structures which are shear-resistant plates or slabs in solid construction, delivering its weight and its load linearly on the walls. This meets the requirements of masonry construction for even load distribution and a generally low load level. Due to the two-way load transfer of the reinforced-concrete floor slabs, all four walls of the box are evenly loaded, in contrast to the former wooden-beam floor, so that the usual distinction between load-bearing and bracing walls becomes obsolete. A two-way effect is generally possible and useful in the case of in-situ concrete slabs up to a side ratio of 1:1.5.

► *Sect. 4.3.1.1 Slab Spanning Two-Way, Linearly Supported*

7.4.3.1 Partial Tasks of a Ring Anchor

The walls of masonry structures affected by vertical loads are stressed within their planes and act as a shear-plate. They must be secured against buckling. In the case of multi-storey structures, the risk of buckling occurring over the

Fig. 7.38 Iron brackets connect the slender wall, which is at risk of buckling, to the stiffening floor structure at floor level

entire building height must be considered. Due to the small wall thickness, a virtual buckling length equal to the cumulative deflection over the entire height would not, however, be permissible. In order to reduce the buckling length to avoid buckling, an all-round tie member consequently has to be arranged at each storey around the total perimeter of the load-bearing walls—if necessary, also on the interior bracing walls—a so-called **annular** or **ring anchor**. Similar to the hoops of a wooden barrel, the ring anchor reduces the buckling length of a multi-storey outer wall to a single-storey height, thereby significantly reducing the risk of buckling. *EN 1996-1-1*, 8.5.1.4

A further partial task of the ring anchor is receiving the horizontal tensile forces in the edge zone of the floor slab subjected to in-plane shear stress (Figs. 7.36 and 7.37) which arise within it as a consequence of the emergence of a compression vault.

According to *EN 1996-1-1*, all external and internal walls which serve to transfer bracing forces must contain ring anchors when the following boundary conditions are present:

- buildings with more than two full storeys;
- buildings with lengths >18 m;
- walls with large openings;
- buildings with unfavourable ground conditions.

Ring anchors must be dimensioned for a tensile force of at least $\mathbf{F_k} = 30\,\mathrm{kN}$ or reinforced with at least two reinforcing bars with a minimum cross-section of $150\,\mathrm{mm}^2$. The reinforcement already present in a reinforced-concrete slab may be computed within certain limits.

7.5 Masonry Construction

7.5.1 Box Construction

The **box-construction** mode as described above is a consistent development of masonry construction, which, like hardly any other one, derives from the structural and constructional properties of the material used (◘ Figs. 7.39 and 7.40). From an historical point of view, it is the oldest of masonry constructions, apart from the heavyweight wall. However, the principle of interlacing, i.e. the principle of mutual bracing, is also valid for more recent masonry constructions such as the cross-wall or the detached-panel construction, or appears in different hybrid forms.

The box construction is a system of contiguous spaces closed on all sides, the connection of which is made to one another or to the outside only via local individual openings in the wall, such as windows or doors. The characteristic façade type is the perforated façade. The shapes of buildings designed according to this type of construction appear simple and cuboidal, this being a direct consequence of this stabilising system.

The box system optimally utilises the possibilities of masonry. All walls can be evenly loaded, especially when using reinforced-concrete floor slabs.

The following features are worth mentioning:

- Mutual bracing of the walls, thus being able to minimise wall thickness, which no longer plays a fundamental static role in this context—an important economic factor in box construction
- Restricted freedom of floorplan-layout, compulsory fixing of spatial configurations. In the pure form of this construction mode, only adding cell-like spaces is possible
- The spans of the floors were previously limited by the use of wooden-beam floors (approx. 4.5 m maximum span), today by adequate spans of solid slabs (approx. 5–7 m)
- Openings in the outer wall cannot be arbitrarily arranged, but only in a dimensionally limited manner suitable for load-bearing (◘ Figs. 7.18, 7.19, 7.20, 7.24, and 7.25).

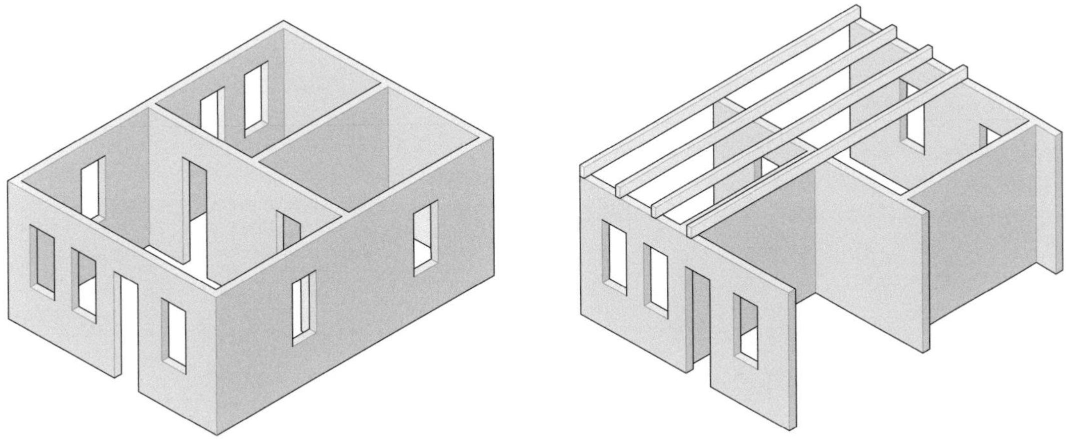

Fig. 7.39 (Left) **Box construction**. The arrangement of individual, mutually stabilising walls creates a box-like overall structure. (Right) **Longitudinal wall construction**. The construction method can be regarded as a special case of box construction and as a transitional form to cross-wall construction. The floor construction is placed on the supporting longitudinal walls. The transverse walls only serve to stiffen the longitudinal walls and to close off the room

The principle of the box can be seen most impressively in multi-storey residential buildings of fin de siècle, namely, masonry buildings with wooden-beam floors, which are found in all European cities being a produce of the construction boom of those years. The principle of the box was also used very consistently in the construction of traditional, handcrafted, residential dwellings of the 1920s, which can be understood as deliberate countermodels to classical modernism.[1]

7.5.2 Cross-Wall Construction

The bulkhead or cross-wall construction is understood to mean the linearly added arrangement of several parallel load-bearing walls, so that a number of spaces of similar proportions and the same daylighting conditions are produced. A characteristic feature of this construction mode is its economy as well as its aesthetically and technically simple basic form, which is based on catenation as a principle of organisation.[2]

The origin of this modern construction mode is the residential construction of the early 1920s. The building type was born out of the necessity to meet the requirements of mass housing with industrial manufacturing processes efficiently, quickly, and economically. The designers' wish for a uniform quality of living, the same orientation of the buildings, a uniform sun exposure and external view could be readily satisfied with the cross-wall construction mode, especially in densely

1 Schmitthenner, Paul (1984) *Das deutsche Wohnhaus*, DVA Stuttgart, p. 62.
2 Belz, Gösele, Jehnisch, Pohl, Reichert (1991) *Mauerwerks-Atlas*, Köln, p. 34.

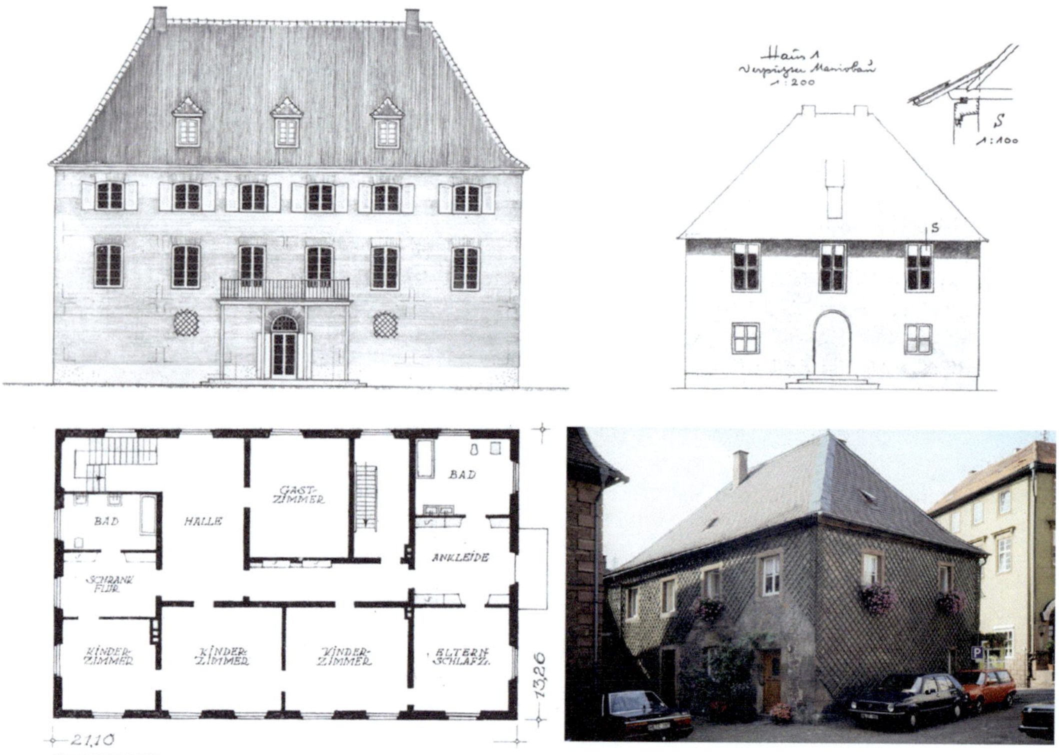

Fig. 7.40 (Top) Two classic masonry structures in box construction. The façade reflects the traditional design rules of masonry construction in a prototypical manner: perforated façade, strictly regular window arrangement, overlying window openings, wide wall pillars between the window openings, wide corner pillars; in the left example: flat brick lintel arches, the centrally arranged door opening is covered with a round arch to accommodate the load of the overlying wall pillar (House at the Hillside, P. Schmitthenner). A convincing example of consistency of shape, material, and construction. (Bottom left) Floor plan of the building shown on the top left with a clearly recognisable box structure. (Bottom right) Traditional building in box construction

built residential quarters. This construction consequentially led to organising buildings in **lines**, so it also referred to as **row construction.**

Structurally, the cross-wall construction mode is understood to be a system of load-bearing walls, each being oriented transversely to the longitudinal axis of the building, the bulkhead-like **cross-walls**, which is stabilised by means of additional bracing by **longitudinal walls** which are usually located in the region of the building's interior, for example, as a space enclosure of wet rooms (Fig. 7.41). A bonding with the bracing reinforced-concrete floor slab is necessary for this. In its bracing principle, the cross-wall construction shows parallels to the box construction, but differs significantly therefrom in that the outer walls are not executed as brick walls, primarily in favour of a better daylighting of the interior spaces. Instead, large glazed non-bearing façades are executed on both

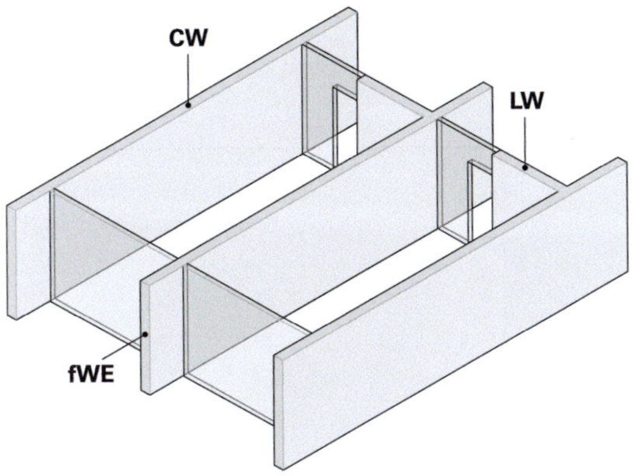

■ **Fig. 7.41** **Cross-wall construction** mode: equal room units with maximum opening width at the outer wall, **CW** cross-wall, **LW** longitudinal wall, **fWE** free wall edge

sides of the building, but also necessarily free wall ends remain, namely, those of the bulkheads or cross-walls that end there, something which generally poses structural problems in masonry construction (■ Fig. 7.42).

The use of this construction mode is therefore not limited to masonry, but forms a transition to concrete construction. The sequential configuration of the structure is very much in line with concrete construction with its increased use of industrial manufacturing methods.

The following features characterise the construction mode:

- narrowing/fixing of the width of a room/apartment by means of the **maximum span** of the **one-way** floor construction (i.e., spanning from cross-wall to cross-wall) (■ Fig. 7.43 centre);
- heavy, load-bearing interior walls/apartment partition walls with correspondingly good **sound** and **fire protection** (■ Fig. 7.42);
- outer walls without design restrictions, especially with regard to the size of daylighting openings, since these can be designed **non-load-bearing**. Cross-wall structures allow for executing fully glazed exterior walls (■ Fig. 7.43 top).

7.5.3 Detached-Shear-Plate Construction

The detached-shear-plate construction is characterised by a free arrangement of wall panels under a bracing floor slab (■ Fig. 7.44). The practical use of this type of construction has only been possible since the introduction of reinforced concrete. In contrast to the box construction with its completely enclosed spaces, spatially flowing connections and large-area openings to the outside are created here. Spatial areas are enclosed by load-bearing wall panels and non-

Fig. 7.42 Cross-wall building in masonry construction during construction. The otherwise free wall ends are reinforced with transverse, bonded masonry piers (made of white sand-lime bricks)

load-bearing envelope elements—e.g. glass façades.[3] In contrast to the box construction, in which individual openings are cut into the outer masonry wall, whole wall areas can be omitted here to create large-area openings. The option of creating flowing transitions from the inside to the outside is a special quality of this construction, something which is exemplified particularly well in the buildings of Mies van der Rohe,[4] Frank Lloyd Wright, and Richard Neutra.

The detached-shear-plate construction mode in masonry construction represents, similar to the cross-wall construction mode, a further transition to reinforced-concrete construction, since this building method would not be possible without at least partial use of reinforced concrete. With regard to the stress in the wall component, the detached-shear-plate construction is not a classical masonry construction in any case, since here almost necessarily load concentrations and heavy wall-edge compression must be expected, e.g. on the front sides of the walls or in other local areas. Classic masonry construction modes, on the other hand, operate according to the principle of systematic load distribution and avoid load concentrations consistently. Individual openings always are—as we have seen—disturbances in the load-bearing structure of a masonry building.

Sect. ▶ *7.3.1 Interaction of Load-Bearing and Enveloping Functions of the Wall*

The following features are typical of the detached-shear-plate construction:

3 Op. cit., p. 34.

4 David Spaeth (1986) *Mies van der Rohe*, Stuttgart, p. 27 ff.

■ **Fig. 7.43** (Top) View of terraced houses in Munich-Harlaching in cross-wall construction mode (arch.: Peter C. von Seidlein). (Centre) Floor plan and section: clearly recognisable cross-wall structure. The bracing transverse walls are limited to the inner area where stairs and wet spaces are arranged. The free ends of the transverse walls are held laterally by the floor slabs. For acoustic reasons, the cross-walls are decoupled between individual houses, i.e. a continuity effect of the reinforced-concrete floors cannot be used. (Bottom) Site plan

- There are no box-like spaces, but each room has one or more room openings or glazed wall areas (■ Fig. 7.45); this allows for freely designable spatial connections between individual rooms as well as an overall **free floorplan design**.
- Another characteristic feature is the **fluent spatial connection** of interior and exterior space, as is not possible with cross-walls or a box (■ Figs. 7.45, 7.46, and 7.47).
- Masonry without pier-like reinforcements made of reinforced concrete is only possible in exceptional cases. The load concentrations which are critical in masonry construction occur, in particular, at the walls' **free ends**, which are deliberately left visible here.

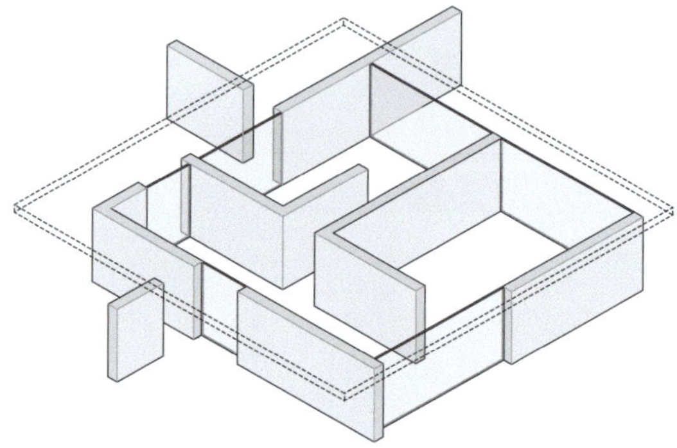

Fig. 7.44 The **detached-shear-plate** construction: free arrangement of wall panels which are not necessarily connected to each other and which are aligned in at least two main directions

Fig. 7.45 Barcelona Pavilion from 1929, historical shot of the interior with outfit by Lilly Reich (arch.: Ludwig Mies van der Rohe). Although this is not a detached-plate construction in the narrower sense, since the roof, which is not massive, is supported on pillars, this structure nevertheless embodies the formal canon of that type of construction in a prototypical manner

7.5.4 Combination of Construction Modes and Break-Up of Rules of Classical Construction Modes in Masonry Construction

Often, combinations of the three systems **box**, **cross-wall** and **detached-shear-plate** are used (Fig. 7.48). Pure designs are rather rare. In the case of such hybrid forms, the wall panel can appear as a detached element which reaches into the outer space; the cross-wall like a means for generating uniform

Fig. 7.46 Barcelona Pavilion, reconstruction of the building from 1986. Completion of the outdoor area with natural-stone wall panels

Fig. 7.47 Barcelona Pavilion, reconstruction. The wall panels and the cantilevering roof slab define interior and exterior spaces

spaces, and the box in the form of a cell can act as a stiffening, stabilising element.[5]

Contemporary masonry buildings are characterised by combinations of these construction modes originally conceived separately in their pure form (Figs. 7.49, 7.50, and 7.51). In some cases, this even entails incorporating elements of a skeletal structure, i.e. integrating individual

5 Belz, Gösele, Jenisch, Pohl, Reichert (1991) *Mauerwerksatlas*, Köln, p. 35.

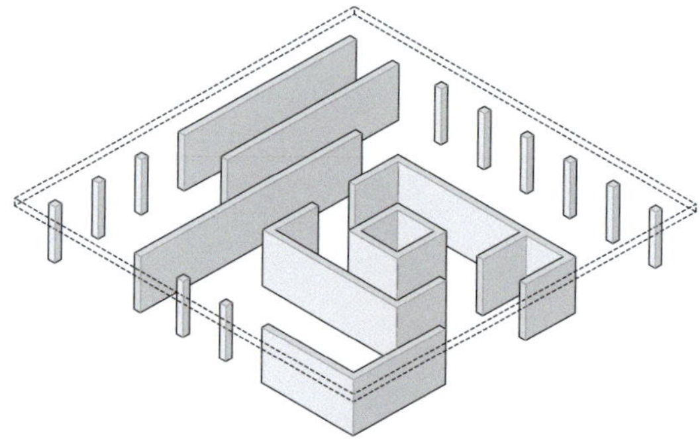

Fig. 7.48 Combination of box, cross-wall, wall panel, and column

Fig. 7.49 District Centre Stuttgart-West. An example of the contemporary use of brick material as veneer brickwork. In contrast to traditional masonry construction, horizontal window formats appear alternating with large-area glass façades

pillars or **columns** in the masonry walls for the purpose of dealing with load concentrations that were previously feared and considered uncontrollable in masonry construction. The combination of shear-plate-like exterior walls with skeletal load-bearing structures inside the building is particularly advantageous in terms of use, and is therefore often encountered in building practice. This structural solution, which can already be found in historical storage and industrial buildings, allows the advantages of both construction methods—wall and skeletal construction—to be combined: namely, the enclosure of space against the outside and, at the same time, the creation of large contiguous interiors or the free unhindered divisibility of interiors according to use-oriented needs. This factor plays a particularly important role in administrative and commercial buildings.

***Vol. 4**, Sect. 8.5 Skeletal Construction Modes*

Fig. 7.50 Administrative building in Stuttgart. The two-dimensional wall structure is largely perforated and becomes a partial skeletal structure with load concentrations in the wall pillars

Fig. 7.51 Example for the conversion of the wall into a skeletal structure using bricks as a veneer

Often, one is no longer dealing with masonry construction in the true sense of the word, but with structures behind the envelope of which skeletal buildings are concealed with reinforced-concrete structures featuring wide spans—as compared to traditional masonry construction. The horizontal stiffening of the wall panels at right angles to their plane is no longer mainly achieved by massive supporting transverse walls, as in classical box-type construction, but primarily by the solid floors which tie the structure horizontally to fixed points by virtue of their racking stiffness. Under these conditions, individual isolated wall panels, which only slightly restrict the free division of space, or complete cores, usually take over the role of fixed points.

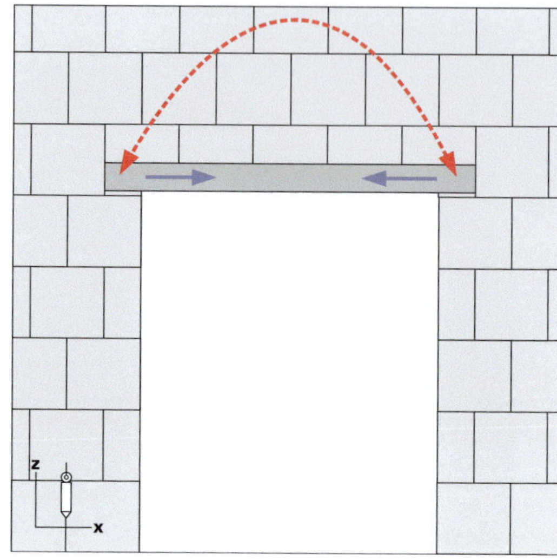

Fig. 7.52 **Compression-arch/tie-bar model** for flat brick lintels

These combinations are made possible primarily by the close connection of masonry with reinforced concrete, or by introducing composite structures, in particular with steel. The expansion of constructional options has thus deeply shaped and changed the appearance of modern masonry structures.

7.6 Openings in Masonry Walls

In masonry construction, openings for windows or doors are bridged with **lintels** or **arches** (Fig. 7.52).

EN 1996-1-17 NA respective national appendices, EN 1996-1-2, EN 1996-3/NA respective national appendices, *EN 845-2*

According to the standard, for dimensioning lintels or arches, only the weight of the wall section which is enclosed within an **equilateral triangle** above the lintel must be computed. Wall parts above that generate a vault-like support of their own, i.e. the lintel or arch construction carries only the load under the resulting imaginary vault (Figs. 7.53 and 7.54 left). The prerequisite is that there are no further openings above or adjacent to the opening, and the resultant vault thrust can be absorbed by the lateral wall surfaces or wall piers.

Floor loads which attack within the equilateral triangle must be calculated only for the sector within the equilateral triangle.

If **lintels** in masonry construction are executed as bending beams made of steel—e.g. introduced post hoc while renovating an old building—or reinforced concrete, the loads are transferred to the lateral supports by means of bending. At the lateral supports, i.e. the reveals of the opening, stress concentrations are generated within the masonry. Particularly in the case of refurbishment or conversion of masonry structures, steel beams are preferably used as lintel constructions,

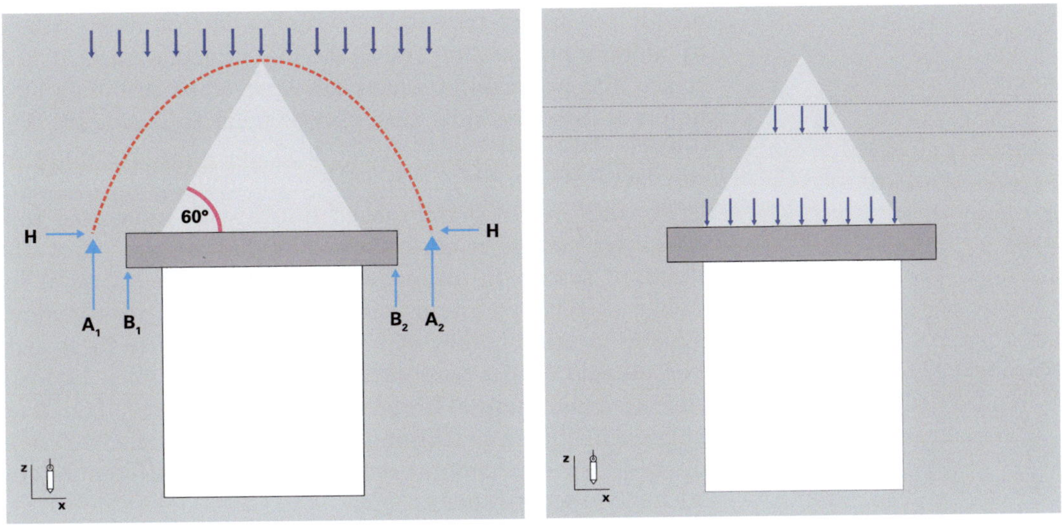

Fig. 7.53 (Left) Compression-arch model for dimensioning lintels in masonry construction. Since the wall portion above an opening is mostly self-supporting because it acts like a vault (forces **A**), the lintel over a wall opening must absorb only the loads **B** that are generated in the intake area of an equilateral triangle. (Right) Within the equilateral triangle, in addition to the dead loads of the wall, other acting loads—e.g. from a floor construction—must be taken into consideration depending on each case

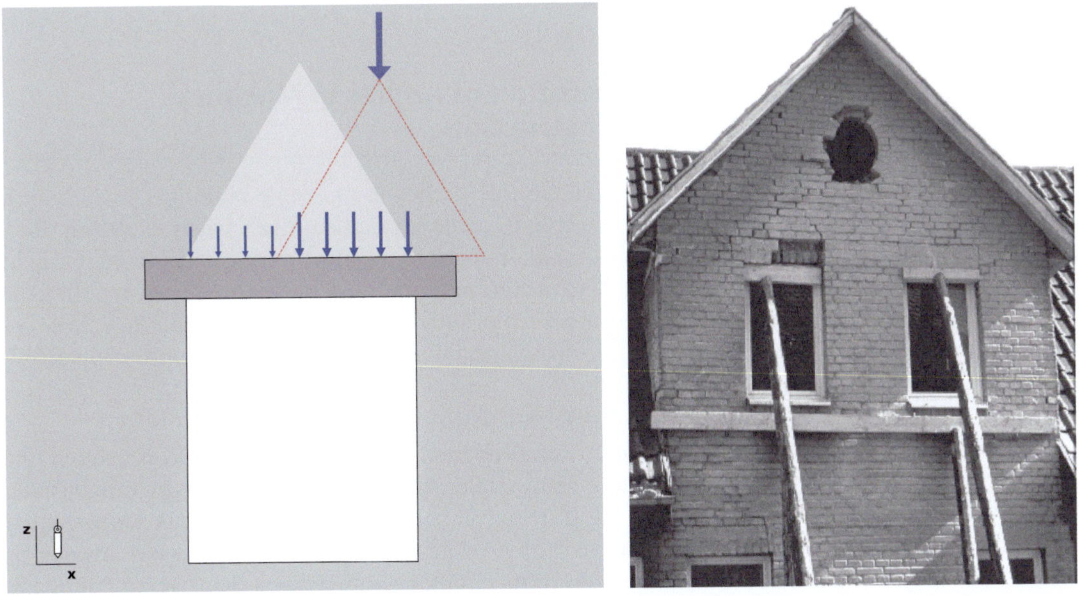

Fig. 7.54 (Left) Loads outside the triangular loading area may also be important for the design of a lintel, e.g. those which in their load spreading area intervene in the intake area of the lintel. (Right) A house damaged by a heavy earthquake. The bricked lintel over the left window was destroyed. A triangular wall section above the window, which previously was supported by the lintel, broke out. (The heavily damaged house had to be demolished later)

mainly because of the ease of assembly. The problems related to building physics—regarding thermal protection and, if necessary, fire protection—which can arise when installing steel lintels in masonry walls must also be taken into account. The diverging thermal behaviour of steel and masonry often results in cracks forming on the rendered surfaces.

In addition to prefabricated lintels with integrated roller-shutter boxes, **reinforced-concrete lintels** now form the standard system in masonry construction. They can be installed directly on site as prefabricated lintels or concreted directly in conjunction with the in-situ concrete floor slab. The outside of the reinforced-concrete lintels must be insulated to avoid thermal bridges. This inevitably results in a reduction in the cross-sectional width of the lintel. Also note the cracking of the outer rendering and the differential colouring of the lintel area.

***Vol. 3**, Sect. 12 Full-Leaf Systems* or *Sect. 18 Openings*

In contemporary brickwork construction, as a rule, **prefabricated lintels** are designed as so-called flat brick lintels (◘ Fig. 7.62). The structural depth of such a flat lintel can be small compared to simple bending beams, since almost only tensile forces have to be absorbed in the lintel itself. A **compression-arch/tensile member** model (◘ Fig. 7.52) can be assumed as the thrusts of a compression vault arising above the opening are short-circuited by means of the flat ceramic lintel, and the lateral supports, hence, do not have to accommodate a horizontal thrust.[6]

7.6.1 Execution of Arches in Masonry Construction

New construction of arches in masonry buildings plays only a subordinate role in today's practice. Even when executing fair-faced brick walls, arches are usually no longer executed. Only in the field of restoration or conversion of historical buildings is this elaborate form of lintel still being applied (◘ Fig. 7.55).

The following special arch shapes can be distinguished.

7.6.1.1 Semi-Circular Arch, Ogival Arch

The **round** or **semi-circular arch** (◘ Fig. 7.56 left) is shaped like a complete semi-circle. At the impost, it merges tangentially into the supporting vertical wall pillars. Its rise is always equal to the circle radius or half the span. Voussoirs can always be the same due to its circular geometry—a decisive advantage in terms of manufacturing, which has resulted in round arches being the most common arch shape historically.

---

6 Ohler, A (1988) *Richtlinien für die Ausführung von Flachstürzen. Mauerwerkskalender 1988*, p. 497–505, as well as Schmidt, U (2004) *Bemessung von Flachstürzen, Mauerwerkkalender 2004*, p. 275–309.

■ **Fig. 7.55** A rare example of new construction of a segmental arch over a centring. The picture was taken in 2007 near Mostar at the monastery of Blagaij in Bosnia-Herzegovina as part of a reconstruction project. The picture illustrates the centring works associated with the production of arches in masonry construction

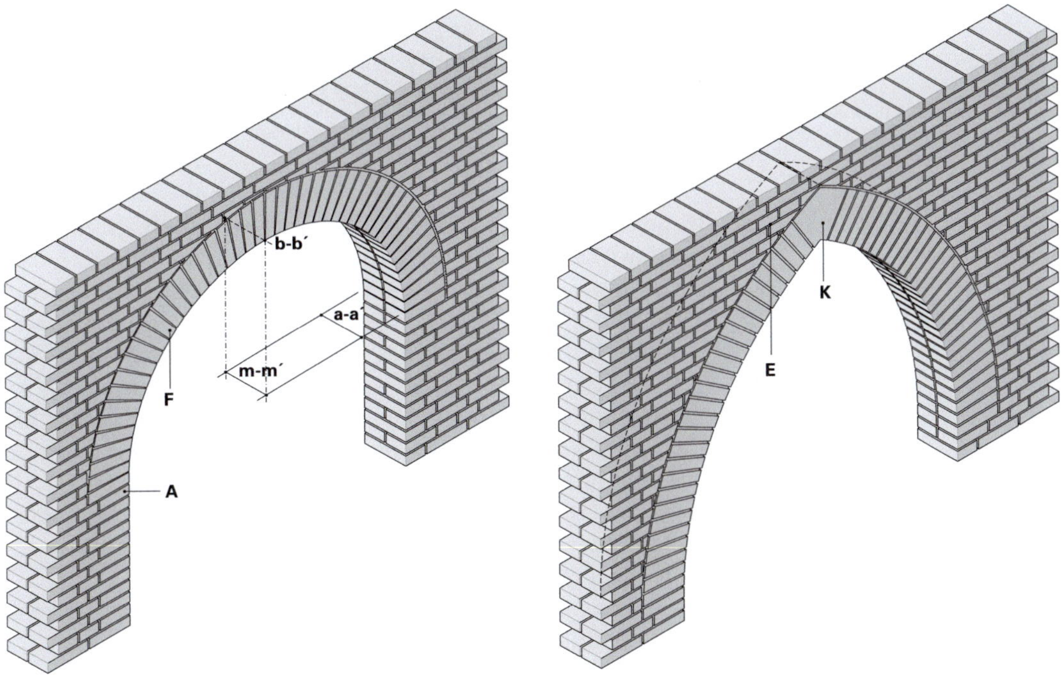

■ **Fig. 7.56** (Left) Classical **round** or **semi-circular arch**, as executed, e.g. in Roman antiquity or medieval Romanesque structures. (Right) Historic pointed or **ogival arch** as executed particularly in medieval Gothic structures

Key for ■ Figs. 7.56 and 7.57: **s** span, **m-m′** axis, **a-a′** impost line and arch depth, **a** and **a′** impost points, **b** and **b′** vertices, **m-b** camber, **A** abutment, **E** extrados, **R** reveal, **F** arch face, **K** keystone

The **pointed** or **ogival arch** (■ Fig. 7.56 right) consists of two circular arch segments with angles <90°, the centres of which lie on the connecting line of the two imposts, but can be arranged at variable distances and thus can result in different circle radii and different rises. This means that the arche's height can be set—in contrast to the semi-circular arch—almost at will, regardless of the span. As pointed arches consist of two circular segments, voussoirs again can always be the same—only the keystone has a different shape.

With the semi-circular arch and the pointed arch, the following points are to be considered:[7]

- execution from standard brick with wedge-shaped joint (0.5–2.0 cm joint width) or use of wedge-shaped voussoirs;
- joint lines are aligned pointing to the centre of the arch; in the pointed arch, either to the, respectively, opposite impost or other centre point situated on the impost height;
- the number of masonry units or voussoirs is in most cases odd; at the apex, there usually is no joint but a keystone; however, this is not a structural necessity, rather a convenience of erection, if any;
- circle centres are generally at impost height so that there is a tangential transition between the circular arch and the vertical reveal;
- maximum spans as guide values:
  - with 24 cm masonry: ≤200 cm
  - with 36.5 cm masonry: ≤200–550 cm
  - >36.5 cm masonry: >350–850 cm

7.6.1.2 Flat or Segmental Arch, Jack Arch

In addition to the classic round arch with semi-circular geometry (Fig. 7.56 left), there are also arches with flatter geometries or smaller segment angles (**flat** or **segmental arch**, Fig. 7.57 left) as well as arches that have completely straight horizontal outlines, so-called **jack arches** (Figs. 7.57 right and 7.59). These variants have the advantage of requiring considerably less height to arch-over an opening than the round arch. The jack arch has the smallest rise and does not differ in shape from a conventional beam lintel. If the stones of the jack arch are fanned out, a thrust line nevertheless emerges within the construction, so that there is a real arch effect (Figs. 7.58 and 7.59).

In the cases of the jack arch and of the flat or segmental arch, the following aspects must be considered:[8]

- Oblique abutments or skew-backs are required in form of masonry joints whose surfaces intersect with the arch centre. Particularly flat arch geometries generate correspondingly large horizontal thrusts at the skew-backs, so that care must be taken to ensure that they can be transferred to the flanking wall panels.

---

7 Belz, Gösele, Jenisch, Pohl, Reichert (1991) *Mauerwerksatlas*, Köln, p. 248.

8 Frick, Knöll, Neumann, Weinbrenner (1992) *Baukonstruktionslehre Teil 1*, Stuttgart, p. 124.

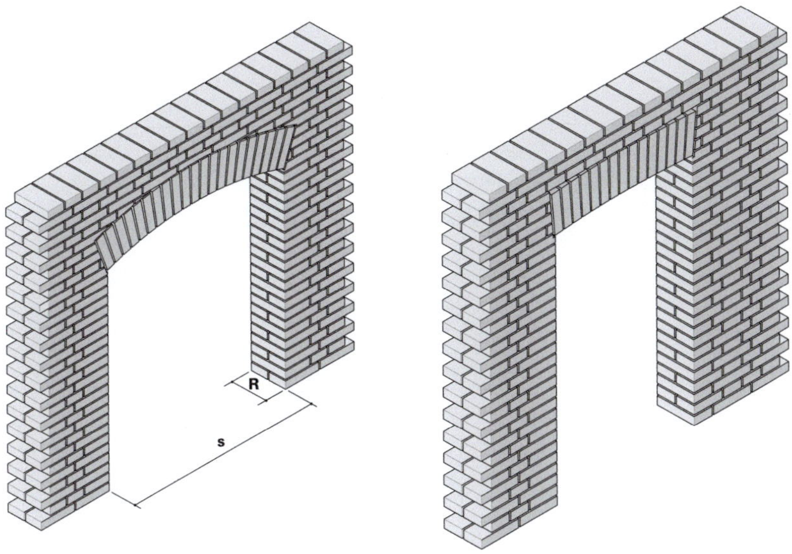

Fig. 7.57 (Left) **Segmental arch**. (Right) **Jack arch** with small span (for key, cf. Fig. 7.56)

- The extrados of the arch should be tangential to a regular bed joint of the masonry bond at its apex (as on Fig. 7.57 right), since the small-scale trimmed masonry units resulting in the apex area (as on Fig. 7.57 left) are usually difficult to cut and cannot otherwise be bricked workmanlike.
- The camber should be 1/50 of the width of the opening in the case of jack arches, 1/12 in the case of flat arches. Jack arches totally without cambers are also possible. For this purpose, the upright brick course, called soldier course, must gradually abandon the vertical line towards the reveals fan-shaped and the skew-back joints must be slightly inclined.
- Maximum spans as guide values:
  - with 24 cm masonry: ≤90 cm (jack arch)
  - with 24 cm masonry: ≤130 cm (segmental arch)
  - with 36.5 cm masonry: ≤130 cm (jack arch)
  - with 36.5 cm masonry: ≤160 cm (segmental arch)

7.6.2 Compensatory and Relieving Arches

In traditional masonry, so-called **compensatory arches** are often found. The use of compensatory arches can have various constructional reasons:

- Formation of **relieving arches** over window openings, the lintels of which cannot accommodate heavy bending, e.g. natural-stone beams (Figs. 7.25, 7.58, and 7.60).
- Installation of relieving arches into large continuous wall surfaces for targeted introduction of loads into wall areas or foundations able to take larger loads (Fig. 7.61).

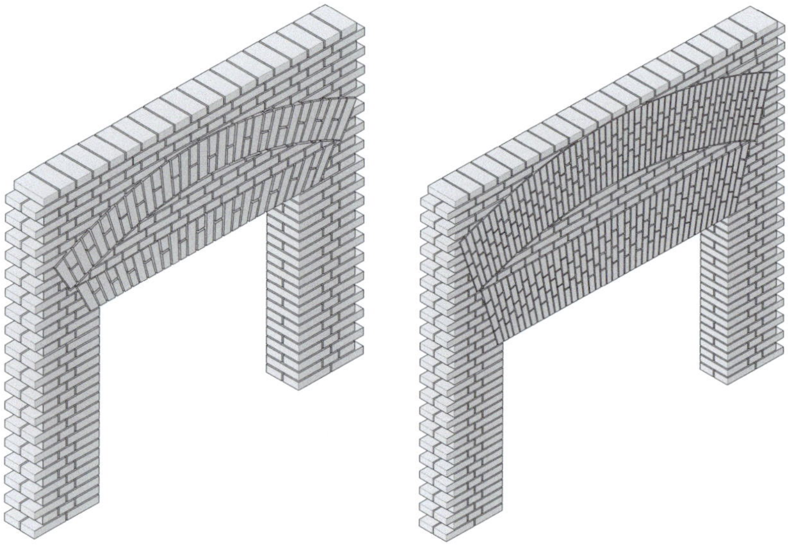

Fig. 7.58 (Left) Masonry opening with jack arch and overlying **compensating arch** for diverting vertical loads into the lateral wall areas. (Right) **Relieving arch** made of bricks for spans up to approx. 1.90 m. With larger spans, the straight arches which were built above the opening were suspended by means of iron bars from the relieving arches

Fig. 7.59 **Entablature** executed as a jack arch

Fig. 7.60 **Relieving arches** from brickwork on top of a jack arch from granite

- Creation of platforms for workers and construction gear during the construction phase. The openings are closed later; the relieving arch remains within the wall (see Fig. 7.61).[9]

7.6.3 Execution of Lintels

EN 845-2, EN 846-9, EN 846-11

Today, lintels over windows, doors, or comparable openings are executed as bending beams or tensile straps. Compared to older masonry structures, where—as described in this chapter—the openings have only small spans in the traditional variant of this type of construction, hybrid forms are nowadays often realised which allow for wide-spanning wall openings as compared to historic buildings. In masonry construction, this sometimes requires using reinforced-concrete piers to overcome the stress concentrations in the wall edges, which formerly used to be problematic (Fig. 7.62).

Bending beams which are used as lintels are often cast-in-situ concrete or precast-concrete lintels in conventional construction modes and have to be thermally insulated on the outside (Fig. 7.63). Since the introduction of aerated bricks, flat lintels made of brick U-leaves have been increasingly used. These prefabricated lintels are characterised by small overall height (Figs. 7.62 and 7.64). The brick U-leaves or brick flat lintels are reinforced and concreted and can be supplied as a semi-finished product. The structural behaviour of these lintels has been described in ▶ *Sect. 7.6* (Figs. 7.65 and 7.66).

9 Cf. the case of the Pantheon in Rome. The author Heene G (2008) assumes this type of function of the relief arches in the tambour (Fig. 7.61).

Fig. 7.61 Semi-circular **relieving arches** within the outer walls of the Pantheon in Rome. It is assumed that during construction large wheels were running for the operation of cranes and hoist in the wall sections under the small arches. The large relieving arches introduce vertical loads into the wall piers of the rotunda

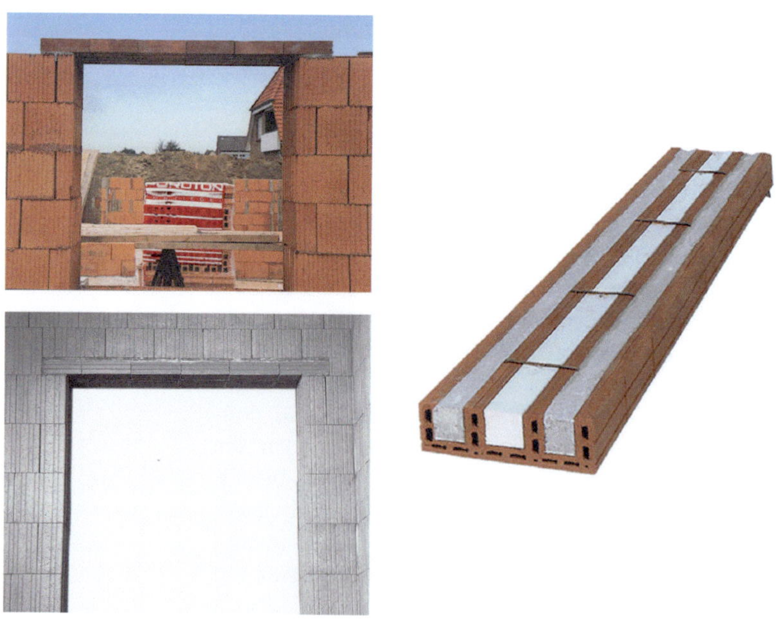

Fig. 7.62 Flat brick lintel of the company *Wienerberger* for 36.5 cm masonry. It consists of three chambers, the outer ones being reinforced and concreted, the inner chamber being filled with hard-foam insulation to reduce thermal bridges

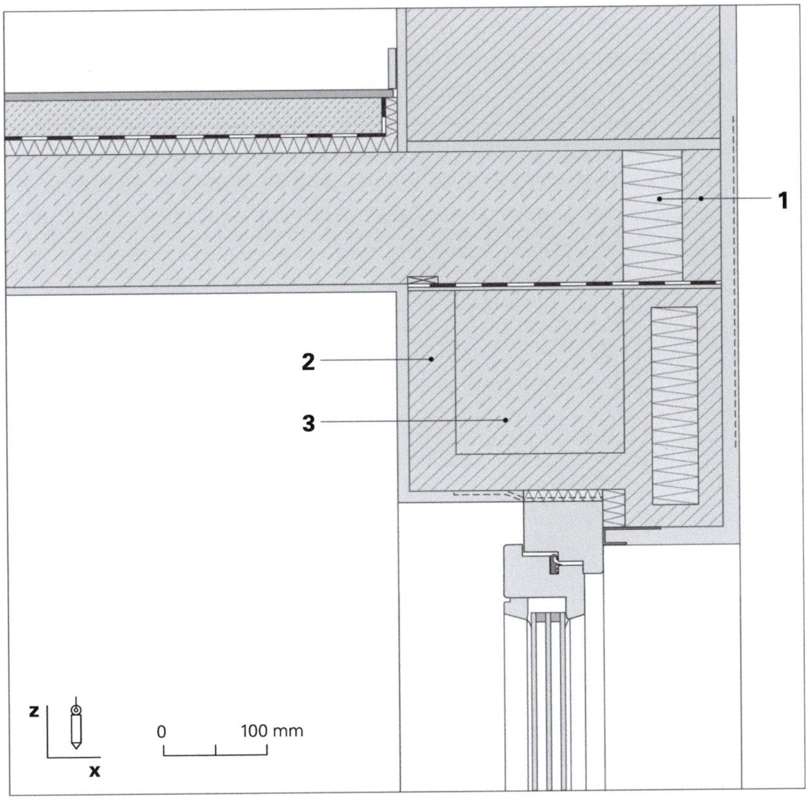

Fig. 7.63 Lintel construction by means of a trough-shaped shell made of aerated brick material with stop for single-leaf outer masonry 42.5 cm (standard detail Fa. *Wienerberger Ziegelindustrie GmbH*): (**1**) floor-edge brick shell with attached thermal insulation, (**2**) trough-shaped brick shell, and (**3**) reinforced-concrete beam

Particularly in residential construction, there are a large number of prefabricated lintels into which a roller shutter can be integrated (Figs. 7.67 and 7.68). The partially load-bearing construction of the roller-shutter box is often made of metal and is concealed by wood chipboard or hard-foam insulation, which is required as a carrier material for the rendering. The simplicity of this combined solution is now juxtaposed to the considerably increased requirements in terms of thermal protection. This is because roller-shutter boxes of this type often are extreme thermal weak points within an envelope because of the risk of thermal bridges originated by the offset of the insulation layer from the outside face of the wall to the inside face of the box (recognisable, for example, in Fig. 7.68), or because of an insulation layer that is too thin at some points (see, for example, the gap in the transition between the wall insulation and the box rear insulation in Fig. 7.67 left). In low-energy or passive houses today, this problem is often encountered with external roller-shutter boxes in order to altogether avoid thermal weak points within the insulation layer.

Often, lintels are simply avoided today as an independent structural element by designing wall openings room-high and topping the opening by the reinforced-concrete ceiling. The

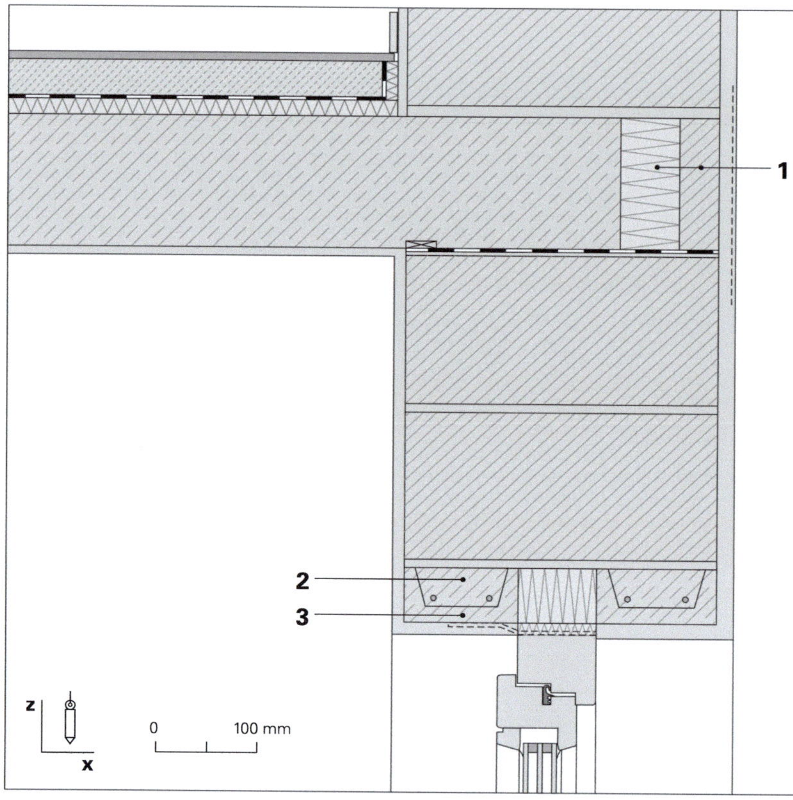

Fig. 7.64 Example of a lintel design according to the described compression-arch/tension-chord model through the use of brick flat lintels with single-leaf outer masonry of 42.5 cm thickness (standard detail Fa. *Wienerberger Ziegelindustrie GmbH*): (**1**) floor-edge brick shell with attached thermal insulation, (**2**) concrete filling with tensile reinforcement, and (**3**) trough-shaped brick shell

latter can be strengthened in the edge area (Figs. 7.65 and 7.66) by additional reinforcement for this purpose, especially in the case of larger spans.

7.7 Wall Construction with Artificial Masonry Units

Artificial masonry units are currently the building material used practically throughout modern masonry construction. Natural stones are only processed in very rare cases or in the renovation of historical buildings.

The design of masonry structures is based on the European standard *Eurocode 6*. The practice of *DIN 1053-2*, which has since been withdrawn, of allowing a simplified calculation procedure and a more precise calculation method for proving the stability of masonry structures has been continued by the currently applicable European standard. For most masonry structures, i.e. those with characteristics such as average building and storey heights, limited floor spans, etc., *EN*

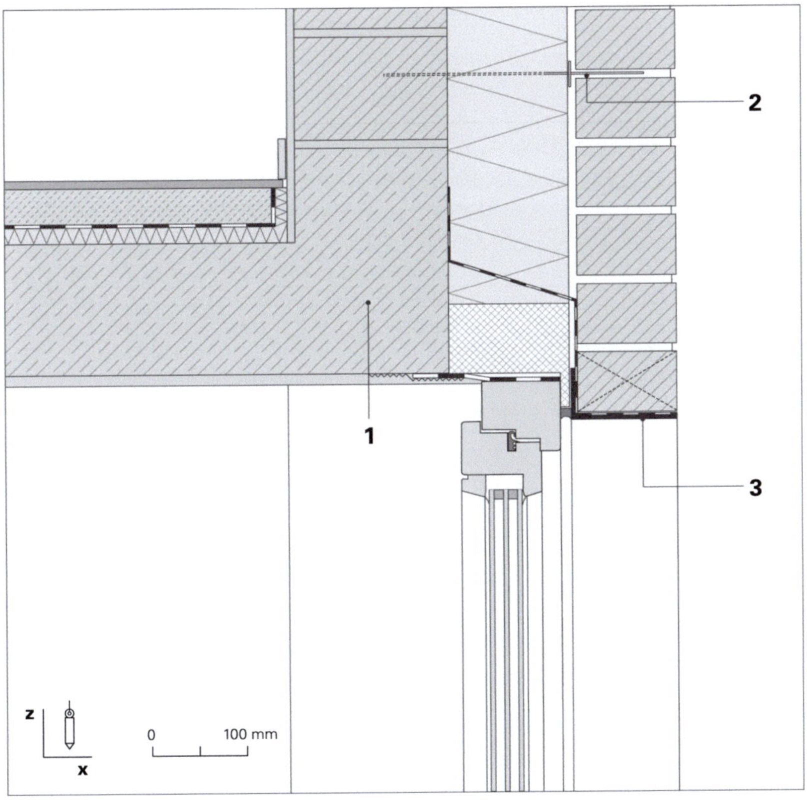

Fig. 7.65 Concrete lintel construction by means of a concrete suspender beam in combination with 11.5 cm veneer masonry and 17.5 cm interior masonry: (**1**) concrete suspender beam; (**2**) wall anchor (stainless steel); (**3**) steel angle profile acting as lintel for veneer leaf

1996-3 allows a simplified calculation method to be used as the basis for the stability analysis. This simplification results from the experience in handling the building material. Some basic requirements for the application of the method to vertical and wind-loaded walls are given (Fig. 7.69). Some are[10]

EN 1993-3, EN 1996-3/NA respective national appendices

- minimum wall thickness for interior and exterior walls;
- clear room height (as a rule ≤2.75 m);
- live load of the floor (as a rule ≤5.0 kN/m^2);
- building height ≤20 m;
- span of reinforced-concrete floors ≤6.0 m.

The boundary conditions stipulated by the standard are fulfilled by most masonry structures. The more accurate calculation method is employed accordingly when the above border conditions are surpassed or when more economical solutions are sought for. It takes into approximate consideration the portal-frame effect between floors and walls and can also be applied to individual components.

10 In the German National Appendix to the *DIN EN 1996-3/NA*, 4.2.1.1.

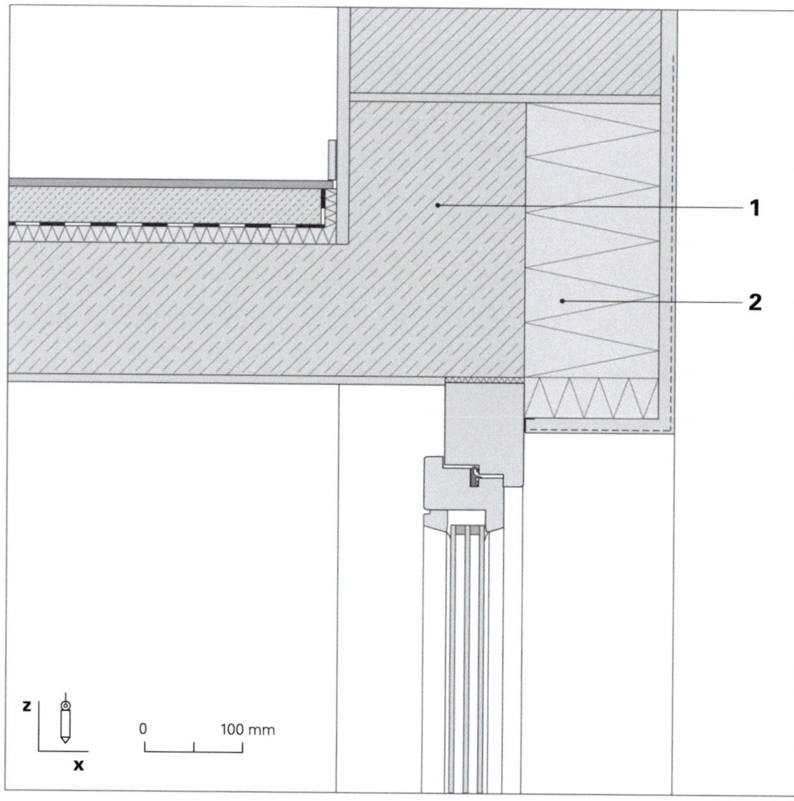

Fig. 7.66 Reinforced-concrete lintel design; concrete suspender beam with added insulation layer; 36.5 cm outer masonry wall: (**1**) concrete suspender beam, (**2**) thermal insulation

7.7.1 Exterior Basement Walls from Artificial Units Without Special Proof

Exterior basement walls today are mainly executed in concrete. They can, however, also be bricked (Fig. 7.70) if the horizontal forces acting on the outer wall of the basement are suppressed by the vertical forces within the structure.

EN 1996-3, 4.5, and *EN 1996-3/NA* respective national appendices

The standard allows a simplified calculation procedure in which a computational proof can be omitted as long as the following basic conditions are fulfilled (Fig. 7.71):

- clear height of the basement wall ≤2.60 m and wall thickness of the masonry ≥200 mm;
- basement ceilings can absorb the forces coming from the earth pressure as a shear-plate;
- the live load in the building's influence area does not exceed 5 kN/m^2; no single load >15 kN at a distance <1.5 m to the wall; the height of the earth fill must not exceed the height of the basement wall; the terrain surface shall not rise;

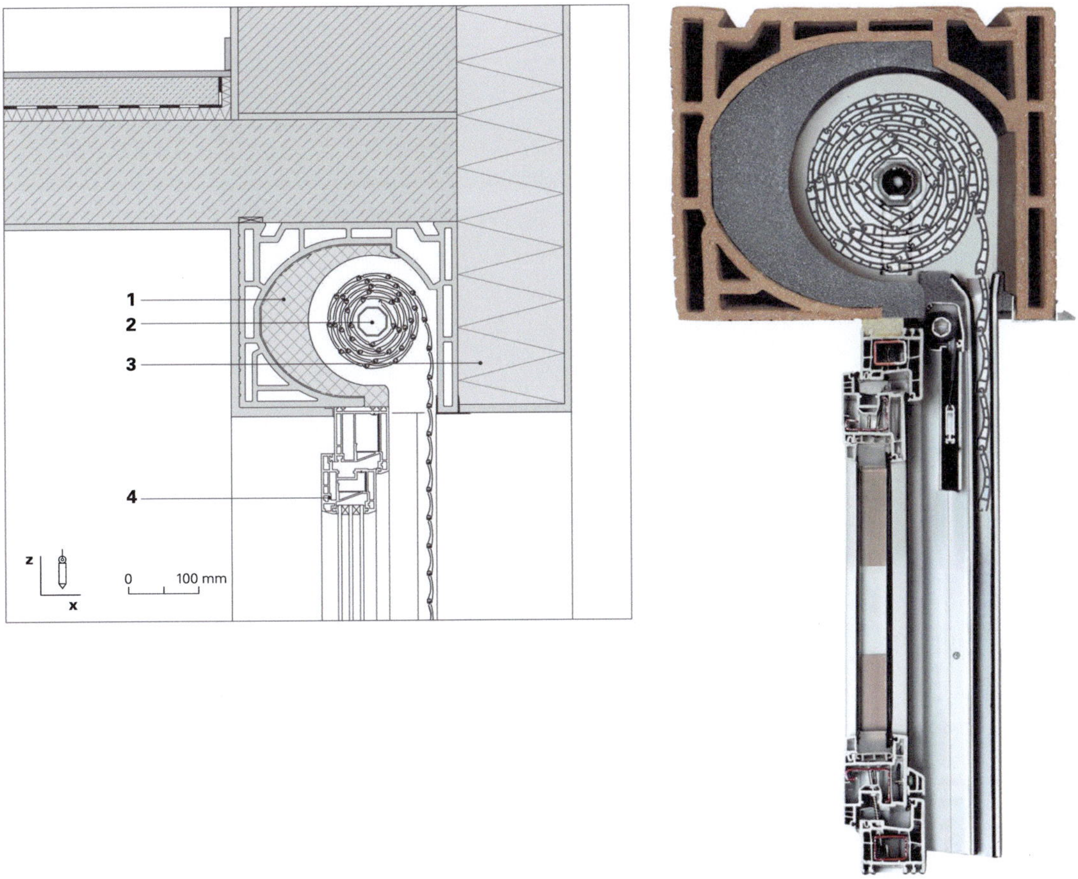

Fig. 7.67 (Left) Reinforced-concrete lintel in combination with a reinforced-concrete floor slab and a non-load-bearing roller-shutter box. (Right) Roller-shutter box made of aerated lightweight-brick material with integrated hard-foam insulation to create a continuous insulation plane for 36.5 cm masonry (manufacturer *Wienerberger Ziegelindustrie GmbH*): (**1**) thermal insulation, (**2**) roller-shutter box, (**3**) thermal-insulation composite system, and (**4**) window frame

- there must be no ground water permanently acting (pressing) on the basement wall;
- there is no sliding surface within the wall (e.g. due to cross-sectional waterproofing), or measures are taken to ensure the required shear strength.

7.7.2 Bracing Walls

In the context of the box construction introduced above, the principle, which is customary in masonry construction, has been described consisting in the stiffening of thin and buckling-endangered walls by means of mutual shoring, thereby, as a rule, effecting a fixation on several sides. The basic bracing principle consists in their nesting, that is, in the mutually transverse arrangement of individual walls which are connected to one another in a shear-resistant manner. In

EN 1996-1-1/NA respective national appendices

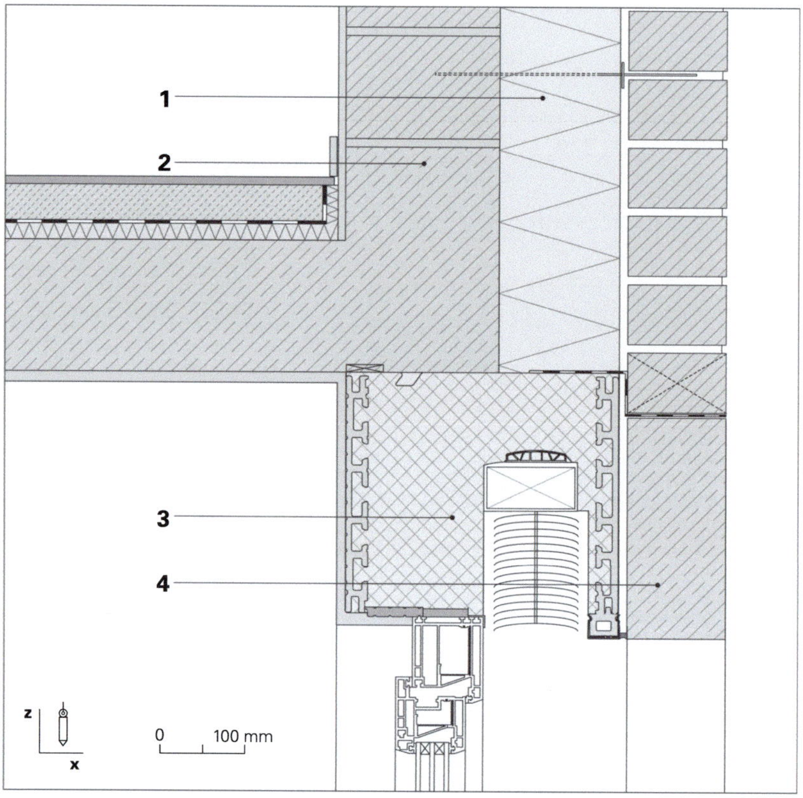

Fig. 7.68 Example of a lintel design with shutter box within a double-leaf masonry: (**1**) thermal insulation, (**2**) suspender beam, (**3**) non-load-bearing brick roller-shutter box, and (**4**) concrete beam supporting the veneer leaf

order for the mutual bracing to come into effect, bracing walls must have the following dimensions:[11]

- Bracing walls must have at least an effective length of 1/5 of the storey height (Figs. 7.72 and 7.73).
- A thickness of at least 1/3 of the thickness of the wall to be braced (at least 11.5 cm).
- They must be fixed in a non-displaceable manner and at a right angle to the braced wall. Both walls must be bricked at the same time (toothing). For this purpose, both wall materials must have the same deformational behaviour.

Alternatively, a bracing by reinforced-concrete piers or integrated steel profiles is also possible.

Sect. ▶ 7.5.1 Box Construction

As described in the section on box construction, the bracing by walls abutting at right angles alone does not suffice. The above-described problems of the upper free edge of a wall, multi-storey construction, and the occurrence of tensile forces in the edge zone of the floor construction require further supplementary or also overlapping bracing measures.

11 Mauerwerksbau aktuell, 1999, Beuth Verlag, Werner Verlag, p. I.91.

component	preconditions			
	wall thickness t mm	clear wall height h m	overlying ceiling span l_f m	usable load [a] q_k kN/m²
load-bearing interior walls	≥ 115 < 240	≤ 2.75	≤ 6.00	≤ 5
	≥ 240	---		
load-bearing exterior walls and double-leaf house partitions	≥ 115 [b] < 150 [b]	≤ 2.75	≤ 6.00	≤ 3
	≥ 150 [c] < 175 [c]			
	≥ 175 < 240			≤ 5
	≥ 240	≤ 12 t		

a Including supplement for non-load-bearing interior partition walls

b As a single-leaf exterior wall only for single-storey garages and comparable structures that are not intended for permanent residence of people.
As a load-bearing leaf for double-leaf exterior walls and for double-leaf house partition walls up to a maximum of two full floors plus converted attic; bracing transverse walls at a distance of ≤ 4.50 m or edge distance from an opening ≤ 2.0 m.

c For characteristic masonry compressive strength f_k < 1,8 N/mm², footnote b applies in addition.

Fig. 7.69 Conditions for the use of the simplified verification procedure according to *EN 1996-3/NA*

7.7.3 Non-Load-Bearing Interior Walls

A non-load-bearing interior wall is understood to be a wall component which fulfils exclusively space-divisive and shielding functions and does not assume any load-bearing function within the context of the primary structure. It is usually executed in masonry construction in 11.5-cm-thick brickwork—e.g. made of lime-sand bricks. Other non-fired masonry units are now increasingly being used in interior-wall construction, e.g. aerated concrete, gypsum plasterboards, etc. Non-bearing interior walls are now increasingly being executed as sandwich-type gypsum plasterboard studded walls. These wall types are particularly cost-effective and suitable for incorporating installation within the wall cavity.

__Vol. 3__, Sect. 62.2 Single-Leaf Partition Walls

__Vol. 3__, Sect. 62.4 Partition Walls in Ribbed Construction

The stabilisation of non-load-bearing inner walls made of masonry is effected, like when bracing load-bearing masonry structures, either by the mutual support of walls (Figs. 7.74 and 7.76), if possible, or by a connection of the walls to the overlying solid ceiling, which ensures that the inner walls do not tilt while at the same time allowing movements relative to the ceiling (Fig. 7.75 left). It is therefore important to decouple the non-bearing inner walls from the solid ceiling in

Fig. 7.70 Cellar erection in brick construction (manufacturer *Wienerberger Ziegelindustrie GmbH*)

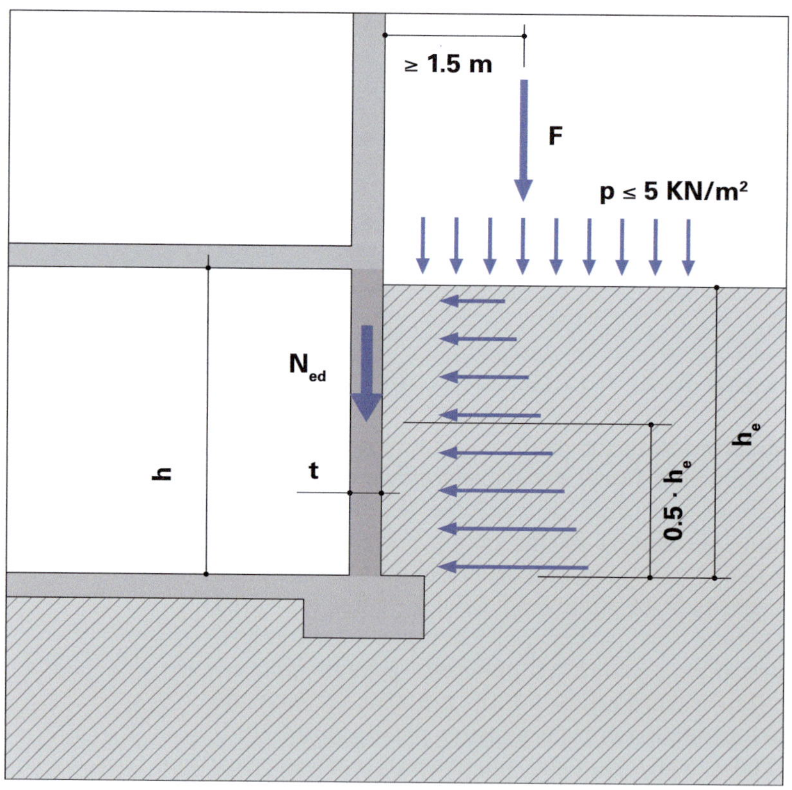

Fig. 7.71 Construction of bricked basement walls without calculational proof according to *EN 1996-3*. $\mathbf{N_{ed}}$ design value of the vertical load at half height of the backfill, **t** wall thickness, **h** clear height of the basement wall, $\mathbf{h_e}$ height of the backfill, **F** single load, **p** characteristic live load on the ground surface

the vertical direction at the upper connection by means of insulation strips and/or foils. The lower connection at the base of solid non-load-bearing inner walls is generally rigid (Fig. 7.75 centre and right). The wall, in this process, rests either on the solid floor slab (Fig. 7.75 right) or on a bonded screed

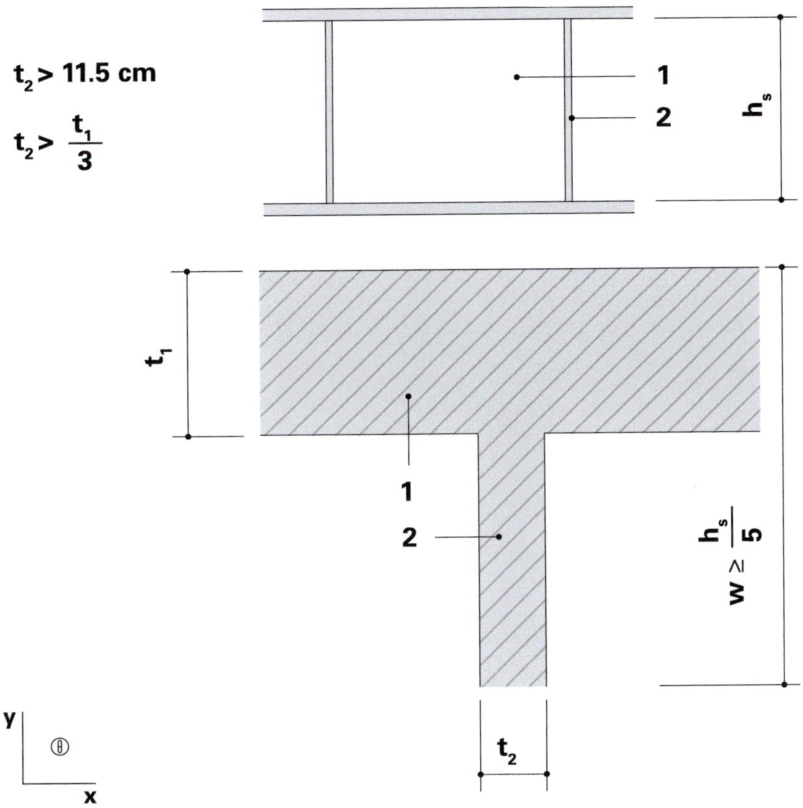

Fig. 7.72 Minimum dimensions of walls with bracing function: (**1**) load-bearing wall (wall to be braced) and (**2**) transverse wall (bracing wall)

via a separating layer (Fig. 7.75 centre). The long-term separation of the construction must ensure that interior walls are not loaded as a consequence of load changes, load deflections, or deformations over the period of use, which can eventually produce cracks in the walls.

According to *DIN 4103-1*, two different **mounting cases** are distinguished for non-load-bearing masonry walls:

DIN 4103-1, Sect. 4, p. 8, and Sect. 5.2, p. 9

- **Mounting case 1**: Residential, office, hotel, and social areas with a small number of people. Here, horizontal loads of 0.5 kN/m are set at a height of 90 cm above the wall foot.
- **Mounting case 2**: Assembly sites, exhibition, and commercial areas with large numbers of people. Here, horizontal loads of 0.5 kN/m are set at a height of 90 cm above the wall foot.

The following rules should be observed when constructing non-load-bearing partition walls made of masonry (Fig. 7.76):

- The bricking of non-load-bearing partitions should be started on the top floor. The floor deformation caused by the weight must not create additional loads in the partition walls underneath.

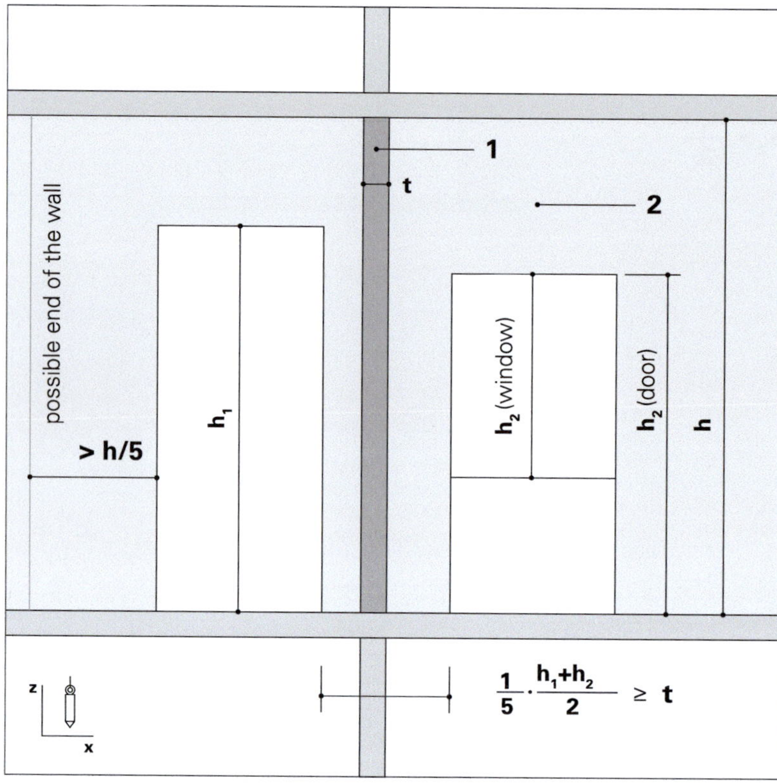

Fig. 7.73 Minimum length of walls with openings which have a bracing function, according to *EN 1996-1-1*. (**1**) Load-bearing wall (wall to be braced) and (**2**) transverse wall (bracing wall)

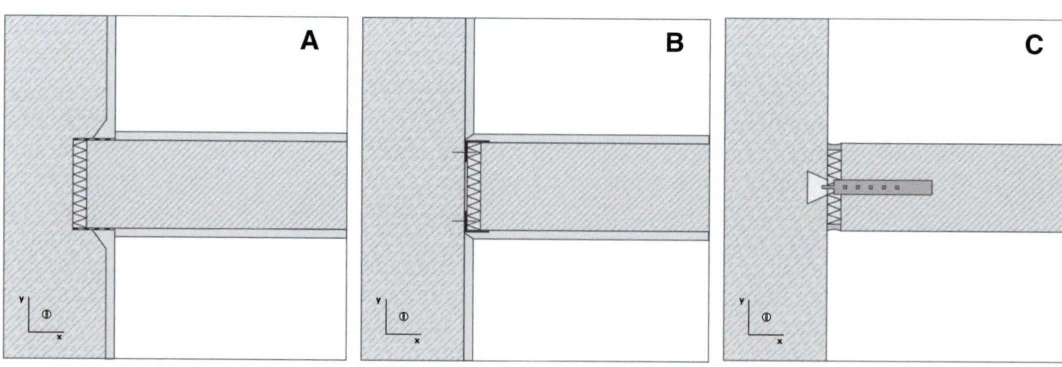

Fig. 7.74 (From left to right) Connections of non-load-bearing walls of 11.5 cm masonry to a load-bearing wall (horizontal sections): (**A**) sliding and form-locking connection via a groove. Only in this way can wall connections be produced with high-quality sound protection, (**B**) retrofitted wall connection and securing by means of angle, and (**C**) butted wall connection via an embedded connecting rail, e.g. when using fair-faced brickwork

- Up to a span of 7 m, the transfer of the load of a partition wall is possible via vault effect. The support of the horizontal thrust on the lateral connections must be ensured.
- The deflection of the floor should be limited to **l**/500.
- By observing the setting times of the concrete and by a subsequent aftertreatment (for example, to prevent rapid

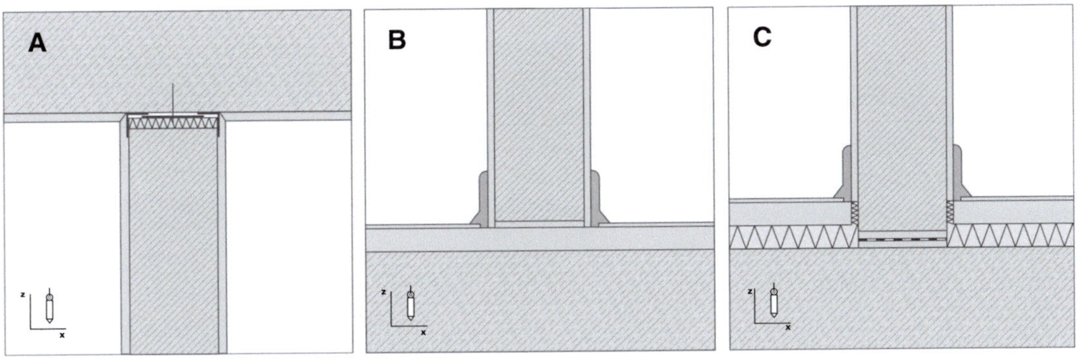

Fig. 7.75 (From left to right) Ceiling and floor connections of non-load-bearing bricked walls made of 11.5 cm masonry (vertical sections): (**A**) connection and anti-tilt protection of a free-standing inner wall to the reinforced-concrete floor slab; surfaces plastered subsequently including trowel cut at corners, (**B**) wall bricked upon a bonded screed, especially to meet the demands on usage flexibility, bonded screeds are often used today, and (**C**) sliding floor connection with separating layer and connection of floating screed

Fig. 7.76 Ceiling and wall connections via a novel form-fitting connection system from *Wienerberger Ziegelindustrie GmbH* (System ZIS): Easy installation of the connection system for sliding connections. The connection system absorbs deformations and simultaneously serves as a plastering stop. The system also serves to improve acoustics by reducing the flanking sound transmission as a result of complete decoupling

drying-out), floor deflection by creeping and shrinking can be reduced. The walls should be walled up as late as possible.

7.7.4 Non-Load-Bearing Exterior Walls

According to the standard, non-load-bearing, merely infilling exterior walls do not require a calculational proof if the following conditions are fulfilled:

EN 1996-3/NA respective national appendices

- the walls are held on four sides, e.g. by interlocking, offset, or anchors;
- the size of the infill areas $\mathbf{h_i} \cdot \mathbf{l_i}$ do not exceed the values given on Fig. 7.77, where $\mathbf{h_i}$ is the height and $\mathbf{l_i}$ the length of the infill area.
- the other conditions of Fig. 7.77 are fulfilled.

wall thickness t mm	Maximum permissible values [a,b] infill area m² at a height above ground level of			
	0 m to 8 m		8 m to 20 m [c]	
	h_i/l_i=1,0	$h_i/l_i \geq 1,0$ or $h_i/l_i \leq 1,0$	h_i/l_i = 1.0	$h_i/l_i \geq 2.0$ or $h_i/l_i \leq 0.5$
115 [c,d]	12	8	–	–
150 [d]	12	8	8	5
175	20	14	13	9
240	36	25	23	16
≥ 300	50	33	25	23

[a] For side ratios 0.5 < h_i/l_i < 1.0 and 1.0 < h_i/l_i < 2.0, the maximum permissible values of the infills may be interpolated in a straight line.

[b] The values given apply to masonry of at least stone compression strength class 4 with normal masonry mortar of at least group NM IIa and thin-bed mortar.

[c] In wind load zone 4 only permitted inland.

[d] When using bricks of strength classes ≥ 12, the values in this line may be increased by 1/3.

Fig. 7.77 Largest permissible values of the infill area of **non-load-bearing exterior walls** without calculational proof, according to *EN 1996-3/NA*

7.7.5 Double-Leaf Exterior Walls

► *Sect. 2.3 Double-Leaf Systems*

Double-leaf brickwork consists of an outer facing leaf and a back wall (Fig. 7.78). Both leaves are coupled to each other force-fittingly by means of wire anchors. Only the back-up masonry is mainly responsible for the load-bearing capacity and thus decisive for the structural dimensioning. The facing leaf is not capable of supporting itself and must be anchored back to the back wall. The space between the leaves can be executed alternatively as an air layer (not current today), as a thermal-insulation layer plus air layer, or as a core insulation layer without air layer. Due to the high current thermal-insulation standards, among other things, double-leaf masonry is predominantly executed today with core insulation.

Further issues of execution are dealt with in ***Vol. 3**, Sect. 12.4 Double-Leaf Systems*

The standard dimensions are given on Fig. 7.78. Any air layer must be at least 6 cm thick; it can also be reduced to 4 cm if the mortar is levelled on at least one side of the wall.

7.7.6 Wall Slots and Recesses (Integration of Installation)

To integrate wall-flush **installations**, slots and cut-outs are sawn or milled into the masonry, which has been usually walled shortly before. Larger recesses, e.g. for risers or shafts,

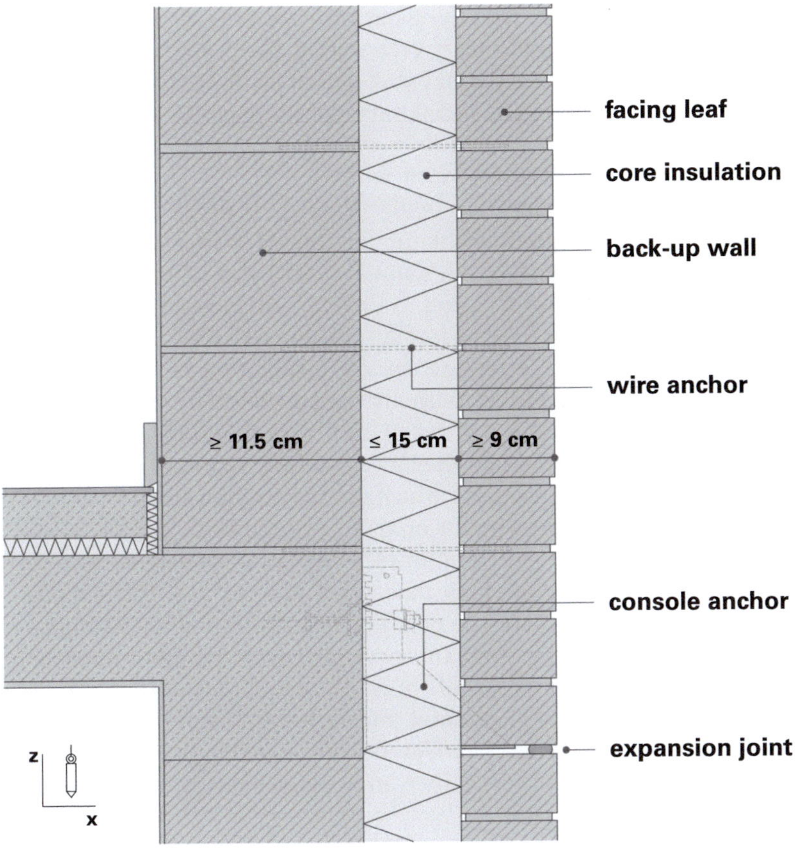

Fig. 7.78 Double-leaf masonry with core insulation: Main components and standard dimensions

Fig. 7.79 Exemplary integration of vertical risers within large-format moulded or U-shaped bricks. (Company: *Wienerberger Ziegelindustrie GmbH*)

are usually taken into account at an early stage in planning and already prepared in the process of walling—e.g. by using large-format purpose-made or U-stones (Fig. 7.79).

Slots could also be prepared in this way, but this is rather an exception in construction practice due to the general tendency to work with increasingly large block formats. This results in a really peculiar, actually unreasonable, partial dismantling of the just finished brick structure with a high percentage of manual work, but it is still the rule in masonry construction to this day. However, alternatives are also available (for example, added installation walls).

Post hoc electric installation in the wall area is thus only possible with increased effort by appropriate milling. New paths in installation and building-control technology may help that the laborious integration of installations in wall constructions, which is still common today, will in general become superfluous or at least be reduced.

7

EN 1996-1-1/NA respective national appendices

The standard specifies dimensions for slots and recesses which may be carried out without further proof, in particular for the post hoc insertion of slots (Figs. 7.80 and 7.81). For slots and recesses beyond these dimensions, appropriate static proofs must be carried out in each individual case.

Horizontal and **inclined** slots are particularly problematic, since they significantly weaken the load-bearing capacity of the vertically compression-bearing wall component.

<table>
<tr><th rowspan="3">wall thickness
mm</th><th colspan="2">later made slots and recesses [c]</th><th colspan="4">slots and recesses in the masonry bond created during construction of the masonry</th></tr>
<tr><th rowspan="2">maximum depth [a] $t_{ch,v}$
mm</th><th rowspan="2">maximum width [b] (single slot)
mm</th><th rowspan="2">remaining minimum wall thickness
mm</th><th rowspan="2">maximum width [b]
mm</th><th colspan="2">minimum distance of slots and recesses</th></tr>
<tr><th>from openings</th><th>among each other</th></tr>
<tr><td>115 to 149</td><td>10</td><td>100</td><td>–</td><td>–</td><td rowspan="7">≥ 2-fold slot width
or
≥ 240 mm</td><td rowspan="7">≥ slot width</td></tr>
<tr><td>150 to 174</td><td>20</td><td>100</td><td>–</td><td>–</td></tr>
<tr><td>175 to 199</td><td>30</td><td>100</td><td>115</td><td>260</td></tr>
<tr><td>200 to 239</td><td>30</td><td>125</td><td>115</td><td>300</td></tr>
<tr><td>240 to 299</td><td>30</td><td>150</td><td>115</td><td>385</td></tr>
<tr><td>300 to 364</td><td>30</td><td>200</td><td>175</td><td>385</td></tr>
<tr><td>≥365</td><td>30</td><td>200</td><td>240</td><td>385</td></tr>
</table>

[a] Slots reaching up to a maximum of 1m above the floor may be executed up to 80 mm deep and 120 mm wide with wall thicknesses ≥ 240 mm.

[b] The total width of slots according to column 3 and column 5 must not exceed the dimensions in column 5 per 2 m wall length. For wall lengths less than 2m, the values in column 5 must be reduced proportionally to the wall length.

[c] Spacing of slots and recesses of openings ≥ 115 mm.

Fig. 7.80 Permissible size $t_{ch,v}$ of vertical slots and recesses in masonry without verification, according to *EN 1996-1-1/NA*

wall thickness mm	maximum slot depth $t_{ch,h}$ [a] mm	
	unlimited length	length ≤ 1,250 mm [b]
115 to 149	–	–
150 to 174	–	0 [c]
175 to 239	0 [c]	25
240 to 299	15 [c]	25
300 to 364	20 [c]	30
≥ 365	20 [c]	30

a Horizontal and diagonal slots are only permitted in an area ≤ 0.4 m above or below the structural ceiling and on one side of the wall. They are not permissible for vertically perforated bricks.

b Minimum distance in longitudinal direction of openings ≥ 490 mm, from the next horizontal slot twice the slot length

c The depth may be increased by 10 mm if tools are used that allow the depth to be maintained exactly. When such tools are used, opposite slots of 10 mm depth each may also be made in walls ≥ 240 mm.

Fig. 7.81 Permissible size $\mathbf{t_{ch,v}}$ of horizontal and inclined slots and recesses in masonry without verification, according to *EN 1996-1-1/NA*

Fig. 7.82 Controlled manufacturing of openings and slots in masonry walls by drilling and milling. Uncontrolled chiselling must not be carried out. (Company: *Wienerberger Ziegelindustrie GmbH*)

In principle, the following regulations are to be considered:

- Slots must always be milled and may no longer be chiselled (Fig. 7.82).
- Recommended maximum slot depths for electrical cables are usually sufficient. Slots should not be oversized unnecessarily.
- Slots and recesses should not be carried out within load-bearing and highly loaded masonry zones—at support areas of columns, pillars, or similar.
- Horizontal slots should only be installed in masonry walls with thickness >11.5 cm. In general, horizontal and inclined slots must be arranged in the installation zones according to *DIN 18015-3*; but they are generally not permitted in horizontally perforated bricks. They must always be

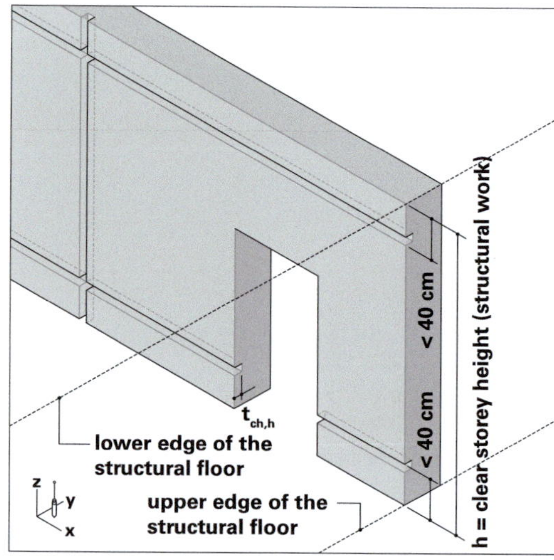

Fig. 7.83 Horizontal and oblique slotting of thin wall components may only take place in the area of 40 cm above the floor or below the ceiling

placed close under the ceiling or above the floor so that the buckling stiffness of the wall is not weakened (see Fig. 7.83).

- Out-of-wall installations—possibly also electrical installations—are always to be preferred if there is sufficient space.
- Recesses and slots are not permitted in chimney walls.Slots and recesses in walls not only impair the load-bearing capacity of the walls, but also their properties in terms of building physics.

Standards and Guidelines

DIN 105: Clay bricks

Part 4: 2019-01 Ceramic bricks

Part 5: 2013-06 Lightweight horizontally perforated clay masonry units and lightweight horizontally perforated day masonry panels

Part 6: 2013-06 High precision units

Part 41: 2019-01 Conformity assessment of ceramic bricks according to *DIN 105-4*

DIN 1045: Design of concrete structures

Part 100: 2017-09 Brick floors

DIN 1053: Masonry

Part 4: 2018-05 Prefabricated masonry compound units

DIN 4103: Internal non-load-bearing partitions

Part 1: 2015-06 Requirements and verification

DIN 4159: 2014-05 Floor bricks and plasterboards, statically active

DIN 18100: 1983-10 Doors; wall openings for doors with dimensions in accordance with *DIN 4172*

DIN 20000: Application of building products in structures

Part 403: 2019-11 Rules for the application of aggregate concrete masonry units (dense and lightweight aggregates) according to DIN EN 771

EN 845: Specification for ancillary components for masonry

Part 2: 2016-12 Lintels

EN 771: Specification for masonry units

Part 1: 2015-11 Clay masonry units

Part 2: 2015-11 Calcium silicate masonry units

Part 3: 2015-11 Aggregate concrete masonry units (dense and lightweight aggregates)

Part 4: 2015-11 Autoclaved aerated concrete masonry units

Part 5: 2015-11 Manufactured stone masonry units

Part 6: 2015-11 Natural stone masonry units

EN 1996: Eurocode 6: Design of masonry structures

Part 1-1: 2019-09 General rules for reinforced and unreinforced masonry structures, prEN

Part 1-2: 2022-10 General rules—structural fire design, prEN

Part 2: 2022-10 Design considerations, selection of materials and execution of masonry, prEN

Part 3: 2021-10 Simplified calculation methods for unreinforced masonry structures, prEN

Bibliography

Belz W, Gösele K, Hoffmann W, Jenisch R, Pohl R, Reichert H (1999) *Mauerwerk Atlas*. 5th rev. ed. Munich, Institut für internationale Architektur-dokumentation

Blum M (2005) *Kalksandstein: Planung, Konstruktion und Ausführung*. Düsseldorf, Bau und Technik

Deutsche Gesellschaft für Mauerwerks- und Wohnungsbau e.V. (DGfM), Zentralverband des Deutschen Baugewerbes (ZDB) (ed) (2017) *Merkblatt nichttragende innere Trennwände aus Mauerwerk*

Eifert H (2015) *Bauen in Stein: die Historie der mineralischen Baustoffe in Deutschland und Umgebung*. Düsseldorf, Bau und Technik

Glitza H (2004) *Grenzenloses Mauerwerk – Vom nationalen zum europäischen Mauerwerk, eine Bestandsaufnahme*. Andernach, KLB Klimaleichtblock GmbH

Gösele K, Schüle W (1989) *Schall, Wärme, Feuchte*. 9th rev. ed. Wiesbaden, Berlin; Bauverlag

Heene G (2008) *Baustelle Pantheon: Planung – Konstruktion – Logistik*. Erkrath, Verlag Bau + Technik

Hugues T, Greilich K, Peter C (2004): *Detail Praxis: Building with Large Clay Blocks and Panels*. Berlin, Basel, Boston; Birkhäuser-Verlag für Architektur

Jäger W (2016) *Mauerwerk-Kalender 2016, Baustoffe, Sanierung, Eurocode-Praxis*. 41st annual, Berlin, Ernst & Sohn

Kummer N (2017) *Masonry construction*. Basel, Birkhäuser

Pech A, Gangoly H, Holzer P, Maydl P (2015) *Ziegel im Hochbau: Theorie und Praxis*. Basel, Birkhäuser

Pfeifer G, Ramcke R, Achtziger J, Zilch K (2001) *Mauerwerk Atlas*. Basel, Boston, Berlin; Birkhäuser-Verlag für Architektur

Worch A (2013) *Mauerwerk im Bestand*. Munich, WTA-Publications

Timber Construction

J. L. Moro, *Building-Construction Design – From Principle to Detail*,
https://doi.org/10.1007/978-3-662-71006-7_8

Contents

8.1 History of Timber Construction

The tradition of timber construction extends far back to the origins of mankind's history. The high degree of wood-building knowledge which is reflected in the historical treatment of wood in construction is most remarkable and gives us an insight into the long-standing experience and building tradition in timber architecture. The knowledge associated to this building material can still be gleaned from the traditional wooden structures that have survived.

8.1.1 Early Timber Construction

Wood, being easily available in many world regions and easily workable in general, was used very early in building construction. Wooden buildings are therefore one of the oldest structures of mankind. The first wooden buildings were probably erected as simple dwellings for protection against nature or enemies (◘ Figs. 8.1 and 8.2). Because of the poor durability of this material, these structures have not been preserved, in contrast to numerous early stone buildings. At best, only very sparse remains of original wooden buildings can be found.

Viollet-le-Duc (1875) *Histoire de l'habitation humaine depuis les temps préhistoriques jusqu'à nos jours*, Paris

Wood, being the only biological building material, has to be protected—unlike buildings made of stone or other mineral materials—against rotting, mould formation, or pests, something which, since early days, used to be ensured during the life period of wooden buildings especially by constructional design measures.

Cf. also ***Vol. 4****, Sect. 8.4.2 Wall-Construction Modes in Timber* and ibid. *Sect. 8.6.1 Skeletal Construction Modes in Timber*

Used as a load-bearing element, wood allows the production of both compressive-stressed (vertical) posts and bending-stressed (horizontal) beams and thus enables, by a framework-like combination of both types of components, the creation of room enclosures and room coverings with a comparatively small constructional expenditure. To cover or over-span rooms with stone material, on the other hand, was always associated with considerably greater effort and technical knowledge. For this reason, although stone and masonry structures prevailed in many high cultures, wood nonetheless remained an indispensable basic building material even for stone buildings, at least for floors and roof structures. This applies even to those regions where it was always scarce and expensive, e.g. hot-arid regions.

Sect. ▶ *4.2.1.2 Flat Covering from Bar Skeins*

Particularly in the cold and woody areas of the northern or southern hemisphere, the wooden house remained the standard of profane construction until the time of the industrialisation. This was primarily due to the climatic conditions which, on the one hand, created the prerequisites for lush wood stocks and for which, on the other hand, wood was predestined thanks to its favourable thermal behaviour. The processing of wood, which is technically simpler than the one of

Fig. 8.1 A fallen tree—an extremely simple shelter made from wood

Fig. 8.2 Simple primitive hut from young trees and wattle, according to Viollet-le-Duc, *Histoire de l'Habitation humaine*; a hypothetical primal form of early wooden constructions

stone, also favoured the widespread use of this material, especially in everyday buildings.

In their basic structure and their constructional principle, early wooden constructions were comparatively heavily material-consuming. Early log and post-plank buildings (Fig. 8.3) are massive wall constructions made of solid

Fig. 8.3 Post-and-plank farmhouse in the Black Forest

wood. Only roofs were more or less light rib structures covered with shingles, straw, or reed.

In the further development of high cultures, wood turned into an ever more expensive raw material. Wood became scarce in Mediterranean countries, for example, due to its firing for producing iron and steel, for firing bricks, as well as for producing hydraulic binders, for which large amounts of wood were consumed. Shipbuilding also required considerable wood resources. In Northern Europe, extensive forest areas were lost by harvesting for the purpose of gaining agricultural land. Traditional half-timbered construction, which allows a more material-efficient use, can be understood as a response to these more restrictive economic conditions. The elaborate external representation of the wood, then already a high-priced commodity, was reserved for only a few representative buildings. The webs within frames were no longer closed with wooden planks, but with cheaper wattle-and-daub, clay, or field stones.

Starting from the late Middle Ages, especially in Central Europe, the social standing of timber construction changed considerably. Fire catastrophes further played an important role in the appraisal—or rather disdain—of the value of wood as a building material. Timber construction was tainted until not so long ago by negative connotations of cheapness, plainness, and poor durability, which did not make justice to the actual performance of wood. Some of the oldest structures still used today in Central Europe are wooden buildings. The house of Abbess Sophia from the year 1233 in Quedlinburg, Saxony-Anhalt, is the oldest wooden construction in Ger-

Fig. 8.4 (Left) Erecting a wall section sheathed on one side in platform-construction mode. (Right) Load-bearing structure of a half-timbered building from the 1920s, architect: Paul Schmitthenner, construction state. The building was completely rendered in a later construction phase

many, for example. The timber of the oak beams of the floors was felled in 1215, the one of the spruce beams in 1230.

8.1.2 American Timber Construction

► Sect. 8.3.4 Timber-Frame Construction

Wood construction resurged in North America with the new development of **wooden frame structures**, which derived from traditional European half-timbered construction (Fig. 8.4). This could not happen before the industrial production of machine-sawn, standardised wood squared sections as well as the mass-production of steel nails in high, consistent quality. With the invention of the **balloon frame** (formerly *Chicago Construction*) by George Washington Snow around 1830, the triumph of this extremely powerful construction began during the settlement of the American Far West. Although the standardisation of wood squared sections was already customary in ancient times, the consistent use of industrial semi-finished goods was a genuine qualitative step. Henceforth, industrially manufactured wooden houses made of wall elements completely prefabricated in New England were sent several thousand kilometres to the West of the USA. To this day, American wooden frame structures with their characteristic overlapping weather boardings are excellent examples of efficient timber construction.[1]

8.1.3 Industrialised Timber Construction

Impressed by world exhibitions and travels to the USA, European architects also tried to again promote wood construction in search for inexpensive solutions for residential construction. In this context, the work by Konrad Wachsmann and Walter Gropius (◘ Figs. 8.5, 8.6, and 8.7) in Europe and the USA from the period after the First World War should be mentioned. The experience of Bruno Taut in Japan and his studies of the Japanese residential house also deeply influenced modern European timber construction. The departure from traditional pure wall construction towards the development of modern skeletal construction, which can be traced back to East Asian models, has decisively advanced the possibilities of modern timber architecture.

► *Sect. 8.4 Skeletal Timber Construction*

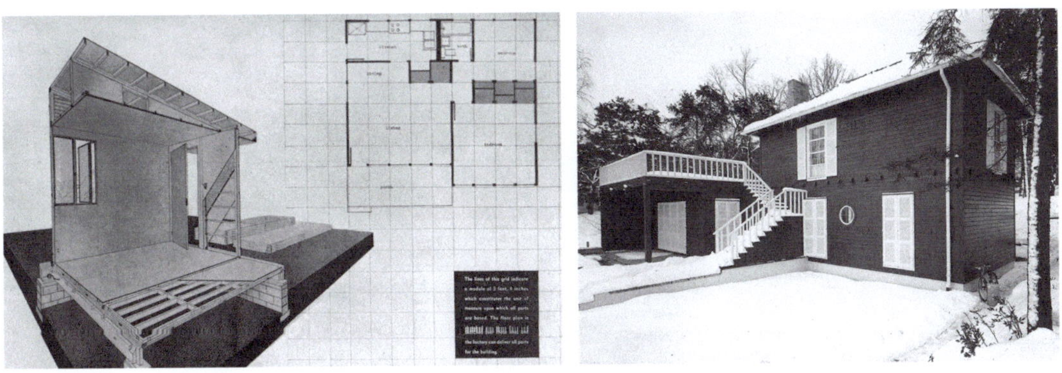

◘ **Fig. 8.5** Albert Einstein residence, Caputh near Potsdam, 1928. The famous contribution of Wachsmann to the revival of timber construction from the late 1920s (arch.: Konrad Wachsmann)

◘ **Fig. 8.6** Roof construction of a house in a housing compound in Merseburg, built in self-construction, 1922. Here, the Zollinger timber construction mode is used, which was developed in the 1920s

Fig. 8.7 Architecture exhibition *The Apartment*, Stuttgart 1927 (Weißenhof Estate), house no. 18, architect: Ludwig Hilberseimer. The building was executed in a new prototypical lightweight timber construction (*Feifel zigzag* construction)

In addition to material economy, the advantages of modern timber construction lie in its highly developed assembly technology, prefabrication, and standardisation. These advantages are fully exhausted by today's prefabricated housing manufacturers. Unlike other regions—e.g. Scandinavia—timber construction could not completely overcome its negative image in Central Europe until recently. The problem of chemical wood preservation transitorily pushed wood back in building industry. It was only in recent years[1] that a change in thinking has taken place among designers and builders, which has also led to corresponding changes in construction law. Particularly the small environmental footprint of wood has favoured the recent resurgence of timber construction, encompassing even technically sophisticated high-rise buildings and bridge constructions, in view of the challenges posed by competing building materials like concrete or steel in terms of sustainability.

1 Giedion S (1976) *Der Ballonrahmen und die Industrialisierung*. In: *Raum, Zeit, Architektur*, Zürich starting from p. 233.

■ **Fig. 8.8** Glued laminated timber cross-section composed of selected, and thus homogenised, individual lamellae, as is characteristic of modern timber construction with wood-based materials

8.1.4 Modern Timber Construction

The development of new **wood-based materials**, novel types of **connection**, and new **composite constructions**, in particular wood-concrete composite constructions, has opened up entirely new possibilities for timber buildings.

For wood-based materials, cf.: ***Vol. 1***, *Sects. 21.3–21.5*

Novel **wood-based materials** are a response to the discrepancy between the elusive, irregular natural structure of sawn timber, with the corresponding imponderables in its constructional use, on the one hand, and the requirements of conventional industrial production for uniformity and standardisation, on the other. To mitigate or overcome these uncertainties, wood is systematically cut into smaller pieces or shredded into particles and reassembled into load-bearing elements by various methods, especially gluing (■ Fig. 8.8). This process allows careful sorting (i. e. sorting-out of the defective parts); a removal of weak points (e. g. knotholes); a targeted manipulation of the grain orientation and thus a practically discretionary manipulation of the iso- or anisotropy of the end element; as well as, last but not least, the realisation of almost any, even complex curved component geometries, as long as the base parts are thin or elastic enough to be glued together to form corresponding shapes (■ Fig. 8.9).

Another development trend in modern timber construction, which is still in its infancy, endeavours to accept the given characteristics of naturally grown wood, in contrast to the wood-based materials technology described above, and to use them for constructional purposes. This includes the scanning of the found geometry or material structure of the basic material, its digital recording and inclusion in the further design and construction planning; the digital measurement of naturally grown bends and branches for planned use in curved building structures; and finally the monitoring of deformations, especially hygroscopic ones, during the manufacturing

Fig. 8.9 Curved glulam arch element composed of thin individual lamellae. The elasticity of the lamellae allows them to be shaped into this form before being glued in the press

8

process and the immediate (real-time) feedback of the data into the further production process (thus, for example, deformations of the base material after cutting can be taken into account in the subsequent milling of the joints).[2]

The far-reaching development of glued laminated timber construction in recent years has also produced composite, large-area, panel-like components made of solid wood, which make it possible to create solid wall and floor elements (Fig. 8.10): an absolute novelty in timber construction, which traditionally has always been a framework of linear solid timber components. Today, construction methods similar to shear-plate construction can therefore also be implemented in wood almost in their pure form. Furthermore, automated production systems allow both complex shapes to be realised and a high degree of precision to be achieved. Form-fit connection techniques, as were already common in historical carpentry, are therefore in some cases finding their way back into building practice today in the form of fully automatically produced plug connections without mechanical fasteners.

In addition to the newly developed wood-based materials, modern **joining technology** has also produced completely new timber construction methods, which are subsumed under the generic term of **engineered timber construction**. It represents a departure from the traditional, handcrafted, and labour-intensive techniques of **carpenter-made timber construction**. In modern timber construction, connections are no longer executed by form-fit assembly, but by industrially manufactured steel parts, often standardised industrial mass products such as bolts, dowels, or nails. It is precisely the most obvious weakness of the material, namely, the difficult force transmis-

2 Menges A, Schwinn T, Krieg O D (ed) (2017) *Advancing Wood Architecture*, p. 32, 120, 149.

Fig. 8.10 Mounting of a joistless, plate-like cross-laminated timber floor as it was only possible in concrete until recently

sion at the joints which is due not least to its pronounced anisotropy, which is so to speak being circumvented by the consistent use of steel joints. The result is a kind of **hybrid construction** in which highly stressed parts are not made of wood, but of the stronger and stiffer material steel (Fig. 8.11). (However, as noted, there has recently been a return to pure form-fit timber joints, which is a consequence of the availability of automated joinery systems.) An equally important drive towards engineered timber construction can be attributed to the availability of suitable **structural calculation methods** and **standards** introduced for this purpose (therefore the term 'engineered' timber construction).

Last but not least, the **sustainability** of wood, which cannot even begin to be achieved by competing materials, plays an important role today. The consumption of non-renewable energy to provide the technically processed construction timber is moderate to low as compared to other materials. The same applies to environmentally harmful emissions that would result from the processing and installation of wood. Not only is the impact of timber on the environment minimal over its entire service life in construction, but the material even has a very favourable balance of greenhouse gas emissions, as it has already bound carbon dioxide from the air, and thus removed it from the atmosphere, during its growth.[3] As a result, wood can be regarded as a carbon store

***Vol. 1**, Sect. 6 Ecology* as well as *Sect. 9.3 Comparative Consideration of the Life-Cycle Assessment Data of the Most Important Materials*

3 However, the true situation appears to be different from what the negative GWP-figures (Global-Warming Potential) usually found in the LCA databases suggest: Strictly speaking, the carbon storage from the atmosphere is attributable to the *tree* and not to the *timber*. The act of building with wood should therefore not benefit from any credit in terms of global-warming potential, as it does not—in itself—improve the CO_2

Fig. 8.11 **Engineered timber construction**: a hybrid construction method in which steel is used in highly stressed areas—e.g. for connections and tie rods (Scottish Parliament, Edinburgh; arch.: E. Miralles)

which temporarily—i.e. as long as it is installed and has not yet been thermally utilised or has decomposed—removes the harmful greenhouse gas carbon dioxide from the environment and thus makes a valuable contribution to slowing down climate change.

In addition, thanks to new wood-based material technology, wood types which until recently were considered unsuitable for construction purposes, e.g. due to a strong tendency to deform or lack of durability, as in the case of beech in respect to warping, are now being upgraded for construction purposes (e.g. as laminated veneer lumber, Fig. 8.12). This not only exploits their sometimes high strength, but also makes an important ecological contribution to the biological diversity of forests.[4]

As a consequence of these developments, wood is conquering more and more areas of the building industry, so that today even medium-sized high-rise buildings are being built

balance in any way. On the contrary: Around 50% of the wood logged becomes waste during processing. This is then—at least to a significant extent—thermally utilised, so that about half of the CO_2 already stored in the tree is instantly released again back into the atmosphere—a significant deterioration of the situation from the point of view of global warming as compared to the unharvested tree (which, by the way, could continue to bind CO_2 for some time) [information from Mr. D. Röver, *proHolz* Baden-Württemberg]. This suggests that the global-warming potential of construction timber should be set positively (not negatively, as in the LCA databases) at half of the stated (negative) values: i.e. around +390 kg CO_2-eq. for ordinary sawn construction timber (instead of −780 CO_2-eq.).

4 Menges et al. (2017), p. 158.

Fig. 8.12 Beech-veneer laminated lumber

from this material. In the training of architects and civil engineers, timber construction now again plays the role that the material deserves. The new cross-material thinking propagated in contemporary training of designers seeks to bring the respective possibilities of the materials into the planners' field of vision and has thus also given the building material wood an appropriate place in today's building scenery.

8.2 Timber Construction Modes

A fundamental categorisation of timber construction methods is based on the distinction between **wall** and **skeletal** construction modes. This differentiation essentially concerns the way in which vertical loads are transferred in vertical supporting elements: in the case of wall-construction methods, in flat, plate-like components; in the case of skeletal structures, in bar-shaped columns. Floor constructions, on the other hand, are disregarded in this subdivision, since in conventional building construction, they must always be two-dimensional and load-bearing due to obvious functional requirements.

Cf. also ► *Sect. 7.2 Categorisation of Construction Modes*

Curved structures such as **shells** are not differentiated into wall (or column) and ceiling or floor and are therefore assigned to a separate category in this context.

Cf. the comments in this respect in ► *Sect. 7.2.2 Skeletal Construction*

► *Sect. 8.6 Shell Construction Modes*

8.3 Timber Wall Construction

Within the category of wall construction, various construction methods can be identified, which differ mainly in the way the vertical wall elements are assembled from solid-wood bars. This is a typical task in timber construction, since the initial elements are always bar shaped due to the

Cf. also ***Vol. 4**, Sect. 8.4.2 Wall-Construction Modes in Timber*

Cf. also ▶ *Sect. 1.1.3 Surface Formation by Combining Individual Elements*

characteristics of the material and must be assembled to form a surface according to various addition principles.

Prehistoric wooden buildings mostly consisted of logs that were dug into the ground. This simplest way of building with wood, a kind of **palisade construction**, is associated with a serious disadvantage, namely, the unavoidable rotting of the wood in the damp soil. It was duly abandoned, in an important technical development step, in favour of timber construction methods in which the timber structure is always placed on an elevated solid base of mineral material to protect it against soil moisture. All timber construction methods existing today are based on this fundamental construction principle. At the same time, this innovation resulted in the removal of the wooden components from the ground and thus instantly raised the issue of bracing against horizontal forces, such as wind or earthquake forces, a question which did not arise with restrained dug-in timbers. Here, too, different timber wall-construction methods differ from one another, namely, in the way in which shear forces are conducted into the foundation within the wall panels.

The following **construction methods** or types of construction can be distinguished on the basis of these criteria:

- **Log construction**: Horizontal logs are layered vertically to form a wall element; at the corners and at inner wall connections, the wood stacks are connected to each other in a shear-resistant manner so that overall shear-resistant wall panels are created (◘ Fig. 8.13 left).
- **Post-plank construction**: Vertical loads are concentrated in uprights, which are set up at intervals. The gaps are filled with horizontal or vertical planks; due to the load concentration in the uprights, this can be considered a kind of transitional form to skeletal construction, especially if the

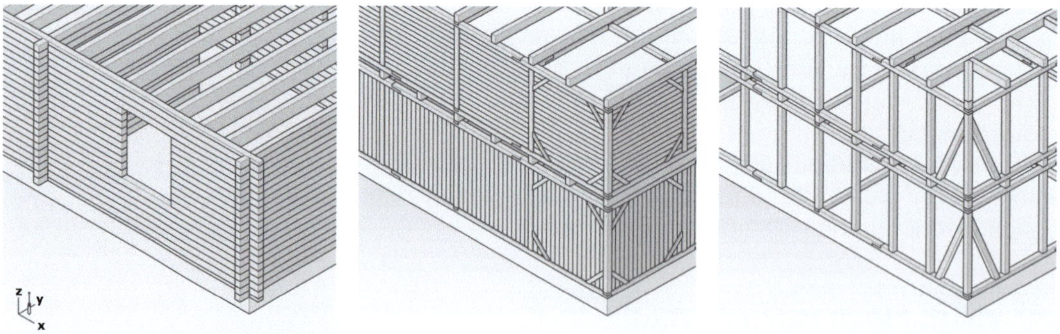

◘ **Fig. 8.13** (Left) **Log construction**: Stacked horizontal wooden parts bear vertical loads via transverse compression. Shear stiffness by connecting the wooden rods at the corners (right) and at partition wall connections (left). (Centre) **Post-plank construction**: Uprights support vertical loads along the grain; compartments are closed with standing (bottom) or lying (top) planks. Shear stiffening through diagonal braces mounted on the outside. (Right) **Half-timbered construction**: Posts support vertical loads along the grain; compartments are closed with different material and do not participate in load transfer. Shear stiffening by diagonal braces in wall plane

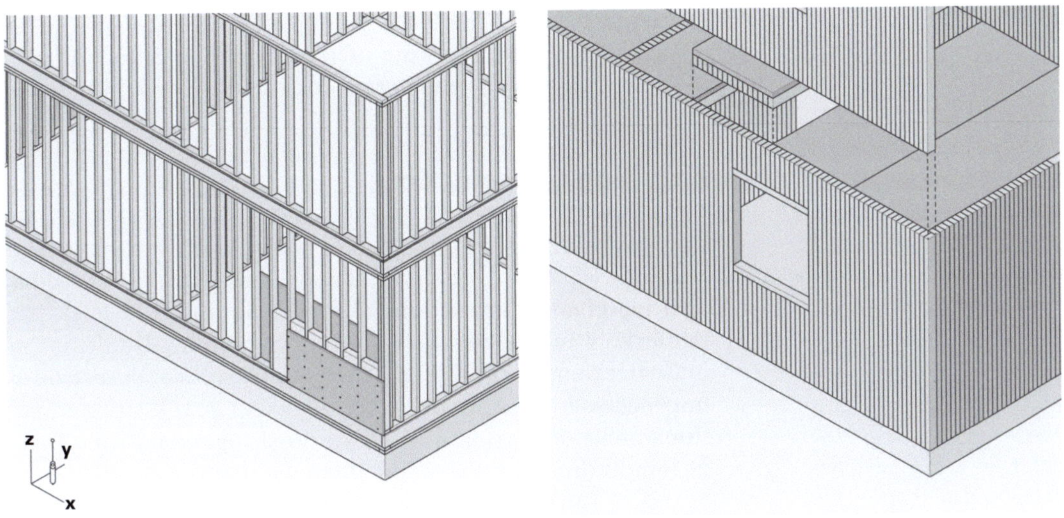

Fig. 8.14 (Left) **Timber-frame construction**: Slim, densely packed studs support vertical loads along the grain; compartments are closed with insulating material and panelling on both sides. Shear stiffening through shear-stiff sheathing. (Right) **Solid-wood construction**: Slim lamellae are glued together to form large-area wall and floor elements. Walls (here from standing plank stacks) conduct the vertical load along the grain; shear stiffness of the walls by gluing the slats or by additional elements (cross-bars, additional sheathing)

infilling planks are installed in a horizontal position and are therefore not involved in load transfer; racking stiffness in the wall plane was generally ensured by diagonal struts added on the outside (Fig. 8.13 centre).

- **Half-timbered construction**: Studs placed at intervals support the vertical loads. Horizontal bars complement the framework to form complete wall elements. The compartments remaining between the bars are filled with different material, for example mineral materials. The shear stiffness of the wall panel is ensured by diagonal braces, which are integrated in the bar framework (Fig. 8.13 right).
- **Timber-frame** or **timber-panel construction**: Here too, the wall panels consist of a framework of vertical, spaced studs and supplementary horizontal bars. In contrast to half-timbered construction, the racking stiffness of the wall element is not guaranteed by diagonal braces, but by shear-resistant sheathing made of wood-based materials (Fig. 8.14 left); a variant of this type of construction with a high degree of prefabrication is the **timber-panel construction**.
- Modern **solid-wood construction** methods: Bar-shaped solid-wood squared sections are assembled here into flat wall components by means of various joining techniques, primarily gluing, according to various addition patterns as well as in various types of layering in the wall plane. The result is largely solid wall cross-sections without cavities, or with only a small proportion of cavities (hence the name solid-wood construction). The racking stiffness of the wall

element results from the gluing of the solid-wood scantlings and/or the blocking of the element in different layers with crosswise oriented grain courses (◘ Fig. 8.14 right).

8.3.1 Log Construction

Vol. 3, *Sect. 12.2.1.5 Exterior Walls from Solid Wood*

Vol. 1, *Sect. 26.9.2 Element Composed of Abutting Bars Oriented Along y/z*

Log construction is one of the oldest timber construction methods (◘ Fig. 8.13 left). It is still used today, especially as a **log-plank construction** (◘ Figs. 8.16 and 8.17 left), particularly in regions with a long timber building tradition—as in Switzerland, Vorarlberg (Austria), or Russia.[5,6] Instead of unprocessed, possibly debarked logs, cut plank squared sections are processed and usually joined together with a form-fit longitudinal joint to make them wind and watertight.

As a further technical development of even older construction methods, in which wooden uprights were buried in the earth in the manner of palisades, the log construction is one of the oldest construction methods in which the wooden construction was consistently separated from the damp earth. The log construction consists of stacked horizontal logs or planks (or scantlings) (◘ Fig. 8.15), which are placed on a masonry base and thus removed from the soil moisture. At the corners and the junctions of partition walls, the wooden bars are connected to each other in a shear-resistant way. The log construction is thus composed of a structure of wall panels which is comparable with classical wall-construction methods, such as the box-construction method in masonry construction, and which also has a similar effect in its bracing principle.[7] The shear strength of the log walls stacked from individual rod-shaped elements, which would not exist without additional measures, is activated at the corners of the building—as well as at internal partition walls connecting in between—by interlocking wall planes abutting at right angles (◘ Figs. 8.16 and 8.17). This connection prevents the horizontal bars from sliding along each other. This results in a genuine wall shear-plate in the structural sense. At the same time, the structurally necessary shear-resistant connection between mutually reinforcing, rectangularly abutting longitudinal and transverse walls is created—a characteristic feature of box construction.

The log construction makes it possible to erect a building out of round logs, simply debarked, with a comparatively small amount of work, time, and tools—only an axe is necessary—as well as with the simplest form-fitting connections. It was not for nothing that this construction was the preferred one of the first settlers, for example, in North America and Siberia.

5 Natterer, Herzog, Volz (1991) *Holzbau Atlas Zwei*, Munich, p. 62.

6 Herzog, Natterer, Schweitzer, Volz, Winter (2003) *Holzbau Atlas*, Munich, from p. 222.

7 Wachsmann K (1930) *Holzhausbau – Technik und Gestaltung*, Berlin, p. 30.

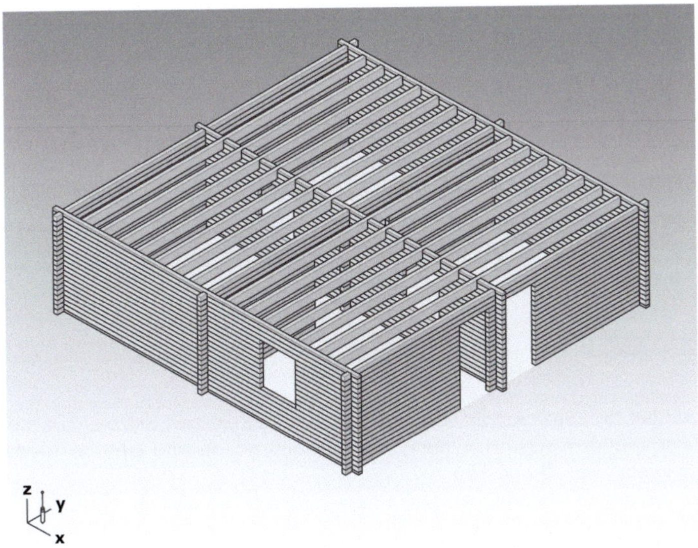

■ **Fig. 8.15** Structural axonometry: **wooden log construction** with typical form-fitting corner design which stabilises the mutually reinforcing wall panels as well as the shear connection between the adjacent planks. The height of the perforated windows is limited for reasons of stability, since the horizontal planking is interrupted by them

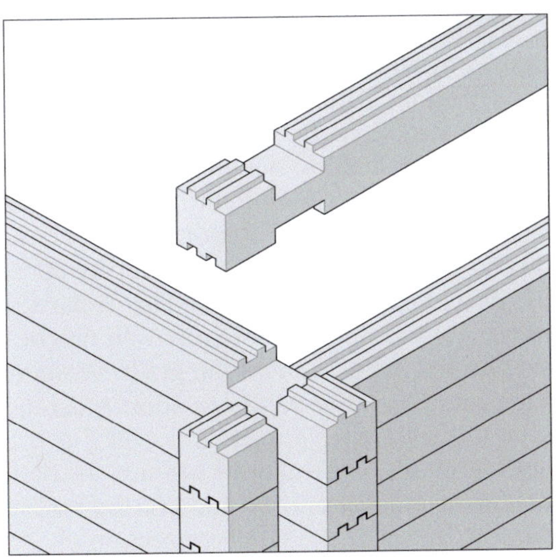

■ **Fig. 8.16** Corner detail of a modern wooden log construction from profiled wooden planks with sealing effect at the longitudinal joint

In early times, the horizontal joints between the roundwoods were closed for sealing purposes with organic materials or mud, e.g. moss and clay. The design of the form-fitting connections at the corner of the walls is characteristic of the structural behaviour of this type of construction. As a rule, a simple **lacing bond** is used here, that is, the round logs or planks are recessed up to a third of the cross-sectional top and bottom (■ Fig. 8.17 left) and stacked. The horizontal

Fig. 8.17 (Left) Corner formation in timber log construction: with notched solid timber cross-sections in a storage building in Finland. (Right) Example of a modern version of the timber log construction with grooved wooden planks for improved tightness of the plank joints, as on Fig. 8.16

joint between log-planks is today designed rather as a single or double tongue-and-groove joint. The joints can also be provided with additional sealing gaskets—e.g. for windproofing.

A relatively large material consumption is characteristic of this type of construction. Typically, softwood is used because it has an ideal rectilinear growth shape. In some log buildings, the base-plate ring, i.e. the lowest log ring, is made of oak, since this hardwood has a higher strength against transverse compression and, above all, a better durability against moisture penetration. Larch wood is the wood type most suitable for log construction. It becomes harder with increasing age. The wood is particularly rich in resin. The seepage of the resin under the influence of sunlight forms a natural coating against weathering and pest infestation.

The shrinkage of wood has to be especially taken into account with this construction method. This is particularly noticeable in the gradual reduction of the wall height, since the lying wood shrinks mainly across the grain, i.e. in vertical direction. Elements such as windows or doors must therefore be connected to the sides in a sliding manner (Fig. 8.18).

Modern log-plank constructions can now be prefabricated with great dimensional accuracy and can be assembled precisely on site by means of prepared connections.

A major structural drawback of this type of construction is the relatively rigid ground-plan configuration, typical of all wall constructions, which allows subsequent modifications only to a very limited extent.

8.3.2 Post-Plank Construction

Possibly in response to the difficulty of transverse shrinkage of the timber, which is typical of log construction, a further step in the development was taken with post-plank con-

Fig. 8.18 Window construction of a wooden log building in Finland. For thermal and constructional reasons, the size of the window formats in traditional log construction was always limited

struction, in which the main vertical loads are borne by individual, spaced-apart uprights, the collar and corner uprights (Figs. 8.3, 8.13 centre, and 8.19). The force was thus transmitted in the grain direction of the wood, so that larger settlements were excluded. The wall compartments thus created between the uprights are closed with planks, which are either laid horizontally and grooved in at the sides of the uprights, or alternatively aligned vertically and grooved in the sill and head plate. The usual lengths of head plates and (when laid horizontally) infill planks, i.e. 3–4 m, thus determine the distances between the uprights.

Lying planks (Fig. 8.13 centre, upper wall section) withdraw from the vertical load transfer over time due to shrinkage, whereby the ceiling and roof loads are completely absorbed by the head plate and transferred to the uprights by bending. This essentially corresponds to the load-bearing behaviour of a skeletal structure with non-load-bearing infill. This variant of the post-plank construction can therefore justifiably be described as a pure skeletal construction. This is because the ceiling here spans in both main spatial directions (once into the room; another time in the direction of the façade along the free-spanning head plate) with the usual spansizes—the indirect load transfer, a distinguishing feature of skeletal structures. However, this circumstance seems to be due more to a presumably unnoticed or overlooked material behaviour than to a deliberate technical developmental step. Upright planks (Fig. 8.13 centre, lower wall section), on the other hand, hardly shrink at all in the longitudinal direction of the grain, which is why they remain permanently involved in the transfer of vertical loads.

For the definition of skeletal structures, cf. Sec. 7.2.2 Skeletal Construction

Fig. 8.19 **Post-plank construction**: Wall frame consisting of corner post (left); collar post (right in the shade); head plate (upper bar); lying planks (bottom wall section); standing planks (upper wall section) as well as attached head angle braces (top) and foot angle braces (bottom)

This corresponds to the typical load-bearing behaviour of a wall construction within a two-dimensional wall element.

In contrast to the log construction method, this design does not provide any racking stiffness of the wall surfaces from the outset. Whereas in older post-plank constructions, a limited frame effect between stud and head plate or sleeper as well as a partial shear strength of the joint between the planks due to friction was relied upon, in the further development of this construction method, head and foot angle braces were introduced at the corners which provided the necessary shear stiffness of the wall surfaces.

8.3.3 Half-Timbered Construction

Cf. ***Vol. 4***, *Sect. 8.4.2.2 Post-Plank-Construction Mode*

The medieval half-timbered building method evolved from the early **post-plank wall** (Fig. 8.13 centre), in which the uprights transferred the vertical loads and the webs between them were closed with horizontal or vertical planks. The development step towards timber-frame construction can be understood as a rationalisation of the rather material-intensive log or post-plank construction methods. This was done by using a mixed construction of a supporting framework of wood and webs made of cheaper material, mostly clay or rubble stone. (The fact that half-timbered construction always had a close conceptual relationship to the masonry wall cannot be completely dismissed.) The English term *half-timbered* construction is revealing in this respect. Another characteristic feature of this construction method is the horizontal bracing of the structure by means of diagonal struts integrated in the webs.

Stud spacing was almost always significantly smaller than with the post-plank-construction method and therefore also much smaller than with skeletal constructions. Traditional half-timbered construction—unlike post-plank construction—can therefore only very rarely be clearly categorised as skeletal construction and is therefore almost always considered to be wall construction. It is true that in its original conception, load concentrations were consistently avoided with the help of wall-like timber frameworks made of comparatively small-format members; nevertheless, there are sometimes load-bearing concepts in half-timbered construction that at least come very close to skeletal construction in respect to their load concentration, especially inside the building, where larger connected rooms were often created.

Vol. 3, Sect. 14.2.1.3 Half-Timbered Walls

Cf. an introduction to the Alemannic wooden architecture by Hermann Phleps (1967); he also points here to the close kinship with stone buildings

See, for example, the exterior walls of the 'Old Building' in Geislingen (Germany), a skeleton-like half-timbered construction, or the interior supporting structure, which is clearly typical of skeletal construction; cf. ***Vol. 4**, Sect. 8.6.1.2 Segmented Column*

The traditional half-timbered construction mode produced various regional variants, e.g. the Franconian, Alemannic, French, or English (Figs. 8.20 and 8.21). They differ primarily in the distances between the uprights, the design of the braces and also in individual constructional details.[8]

Half-timbered structures consist of a storey-high load-bearing wall truss composed of uprights, sill, and top or head plate. As noted, the constructional characteristics are very diverse nationally and regionally and cannot be dealt with in detail in this context. As an example, the Alemannic half-timbered construction mode will be examined more closely here, which from a technical point of view reached a remarkable level of maturity and recognisable typification: In the Alemannic half-timbered construction, the posts often stand in a grid dimension of 1.25 m, which allows the integration of both a door and a window opening without the need to omit uprights (Fig. 8.21). Struts are either restricted to a single compartment between adjacent uprights, or laid over more than one compartment, mostly only two, in which case they must be half-jointed with the intersecting upright. Horizontal parapet or lintel transoms additionally stiffen the uprights at the sides.

The framework, therefore, consists of the following basic elements (Figs. 8.22 and 8.25):

- **Posts**, **uprights**, or **studs** as vertical elements of the rigid wall panels.
- **Base plate** and **header plate**: lower and upper horizontal end of the rigid wall trusses. The beams of the next higher floor structure rest on the header plate.

8 Pfeifer, Liebers, Reiners (1998) *Der neue Holzbau*, Munich, p. 58 ff.

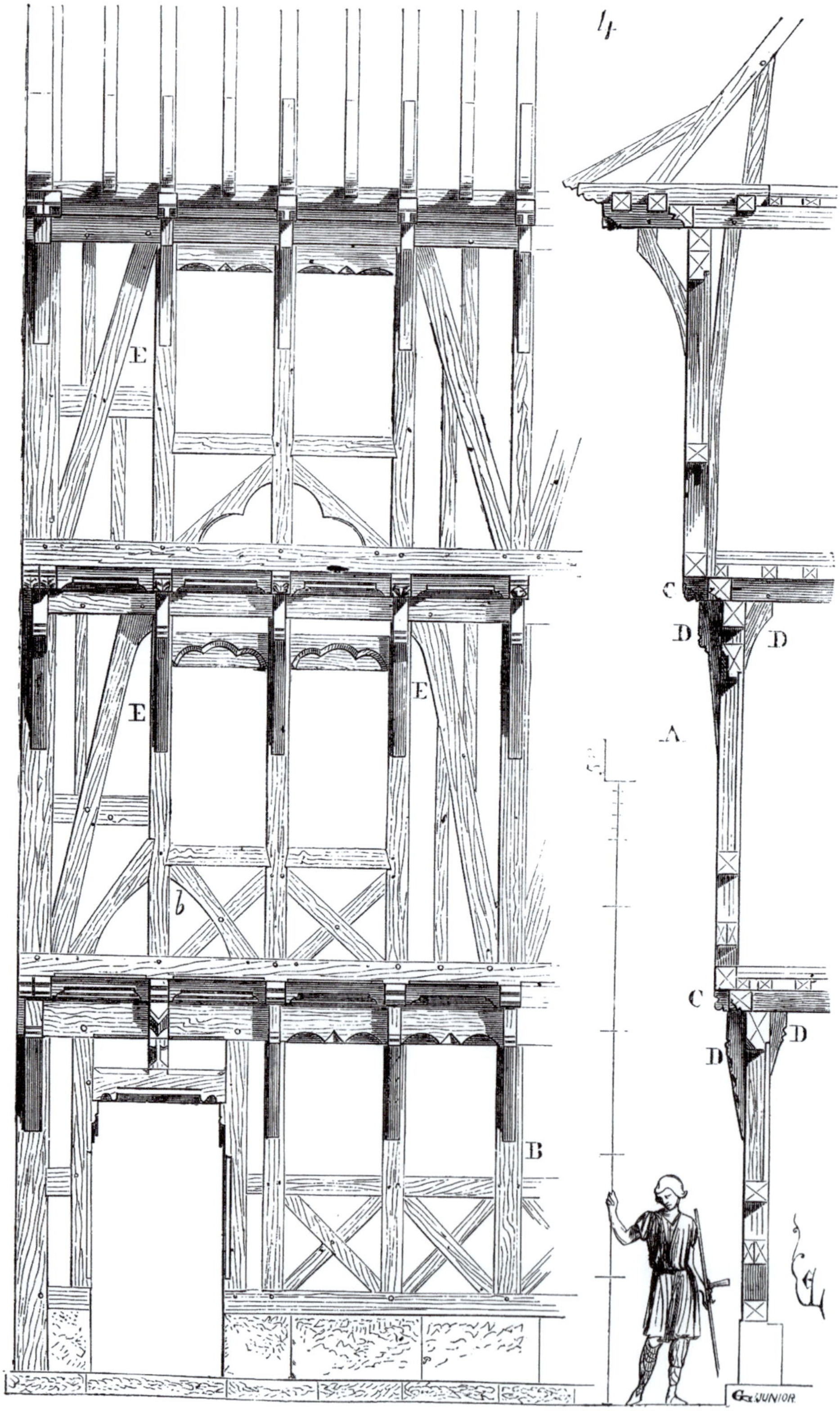

■ **Fig. 8.20** View and section of a French half-timbered building according to Viollet-le-Duc. All individual closed wall sections are braced by various strut designs

Fig. 8.21 Typical main construction of an **Alemannic half-timbered building**: Uprights set approximately in window width; storey-high wall element consisting of sleeper, upright and header plate; attached head and foot angle braces, connected to the collar upright (middle) in a shear-resistant manner by lap joints, secured with treenails; bracing always located within a single compartment, so that the straps do not have to cross an upright; parapet cross-bar under the window; in this case, the window is extended up to the header plate (lintel bar is omitted here); floor joists can be seen above lying on top of the header supporting the sleeper of the next upper wall

- **Floor beams** or **joists**: executed as solid-wood cross-sections; the spans are limited to approx. 5 m.
- **Diagonals** or **struts** for the bracing, i.e. for the discharge of the horizontal forces such as wind forces into the base plates or sleepers. The struts were shaped in very diverse forms in traditional half-timbered building and were often decorative (Figs. 8.19 and 8.20). They are also referred to as head or foot **angle braces** in traditional carpentry, depending on their location.
- **Lintel** and **parapet transoms**: Short horizontal intermediate timbers between uprights; they stiffen them laterally and provide a stop for door or window openings.

Traditional half-timbered construction, being an historical construction mode, is clearly marked by craftsmanship. Half-timbered structures were made of square or rectangular solid-wood sections. In the early Middle Ages, the squared timbers were hewn with the axe—the processing traces on the wood surfaces can still be recognised today. The squared sections were already standardised in the Middle Ages, which is

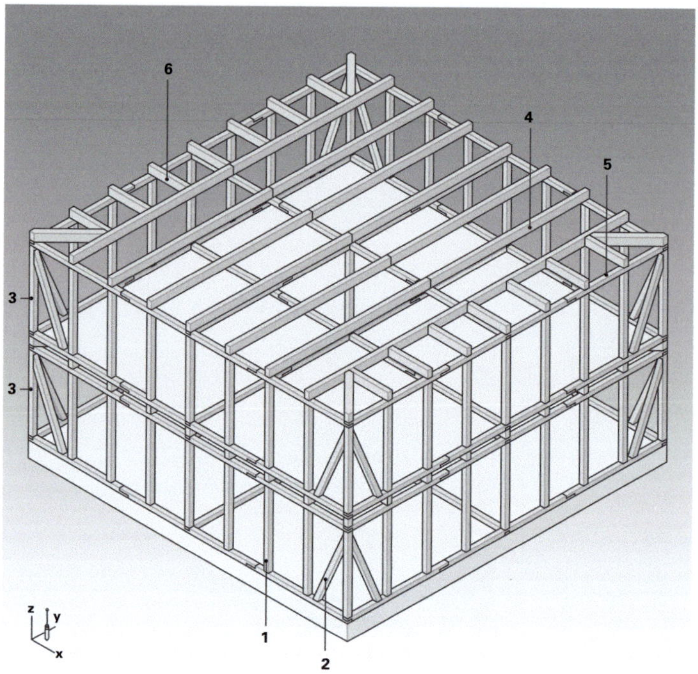

Fig. 8.22 Structural axonometry: Half-timbered construction with the typical wall structure of this construction mode, with posts **1** and diagonal struts **2**. The wall elements **3** are segmented in storey height. The floor structure **4** is executed as a one-way beam construction. Short beam sections **6**, so-called tail beams, are placed transversely to the main spanning direction at the walls **5** parallel to the beam-spanning direction. This creates a support for the sill plate of the upper wall section as well as an all-round uniform appearance of the beam heads on the façade. This solution is typical of historic half-timbered buildings

why we can speak of an early form of prefabrication.[9,10] The wooden structure was built in a manual way using standardised connections such as skew notch, tenon-and-mortise, and lap joint. The resulting weakening of the cross-sections, the inevitable transverse compression at the sleepers, and the shrinkage of solid wood are significant disadvantages of this construction mode (Fig. 8.23). Problems related to building physics arise on the outer walls, particularly in the interspaces of the trusses, i.e. the fields between the wooden components, which were closed by means of wood plaiting (wicker), wattle-and-daub, or field stones. Later, they were wallowed with stones or bricks and the whole wall was rendered (Figs. 8.24 and 8.25).

Today, half-timbered construction methods have practically no constructional significance. The typical truss stiffening by diagonal braces has today given way to the more efficient principle of bracing by sheathing, as practised in timber-frame and timber-panel construction.

9 Scheer, Muszala, Kolberg (1984) *Der Holzbau, Material-Konstruktion-Detail*, Leinfelden-Echterdingen, p. 86 ff.

10 Wachsmann K (1930), p. 14.

Fig. 8.23 Examples of carpentry-type, form-locking wooden connections in an historic half-timbered building

Fig. 8.24 Infill of the compartments of a half-timbered building from quarry stones and mortar

8.3.4 Timber-Frame Construction

Originally developed as a low-cost, material-saving, and easy-to-execute construction method for American settlers of the nineteenth century, timber-frame construction has still retained this prevailing position in housing construction, especially in North America and Scandinavia, which, after 170 years, provides evidence of the unusual technical and cost-related competitiveness of this design.[11,12]

Vol. 3, *Sect. 14.2.1.1 Wooden Frame Walls*

11 Wachsmann K (1930), p. 14.
12 Ruske W (1930), p. 26.

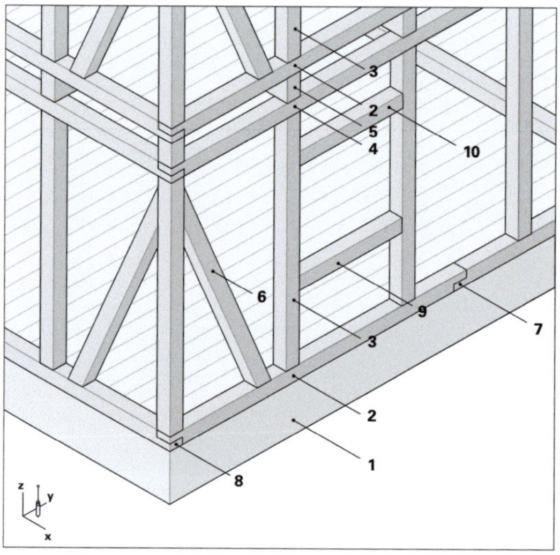

Fig. 8.25 Structural detail of a half-timbered building: The stacking or superimposition of load-bearing elements typical of historic half-timbered construction is clearly visible (sequence of sleepers, posts head plates, and joists). Furthermore, the form-locking connections—here mainly the horizontal lap joints—are pictured. The diagonal struts take over the building bracing and are, in their respective local design, defining for many half-timbered buildings (cf. Fig. 8.26): (**1**) stone plinth, (**2**) sole plate, (**3**) post, (**4**) head plate, (**5**) joist/tail beam, (**6**) strut, (**7**) halved joint, (**8**) corner halved joint, (**9**) parapet cross-bar, (**10**) lintel cross-bar

Fig. 8.26 Traditional half-timbered construction

In contrast to the predominantly handcrafted timber construction methods discussed so far, this is an early form of industrialised construction mode. A limited selection of

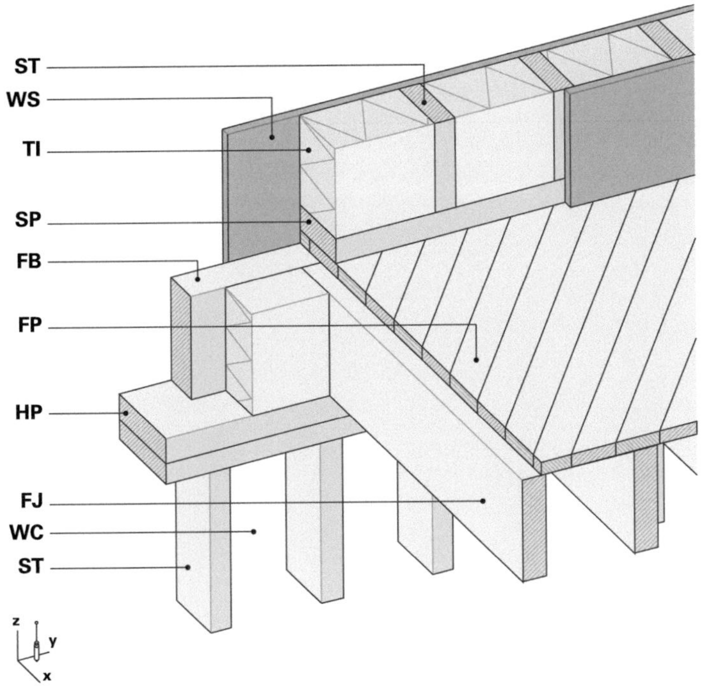

Fig. 8.27 Idealised basic construction of a platform-type wood frame construction with its essential components (no sealing layers and no external construction shown). **ST** stud, **WS** wall sheathing, **TI** thermal insulation, **SP** sill plate, **FB** front beam, **FS** floor planking, **HP** head plate, **FJ** floor joist, **WC** wall cavity, filled with thermal insulation

pre-manufactured, made-to-stock modular, graded wood squared sections was used taken directly from the catalogue (Fig. 8.27): 2/4 (two by four) in., i.e. approx. 5/10 cm, or multiples thereof for floor joists (2/8, 2/10, 2/12 in.), board cross-sections, later also wood-based plate material and industrially produced steel nails, which were originally used as the only fastener.[13] The extreme material economy of this type of construction is due to the combination of slim studs, which create the basic framework of the wall placed vertically at a distance of about 60 cm, and a two-sided sheathing: in the early phase, firstly from diagonally laid boards for the purpose of bracing. Alternatively, the bracing was carried out by means of diagonal struts at the building corners connected flush to the studs with lap joints. Later, industrially produced sheets made of wood-based materials were used for sheathing and creating shear-plates.

***Vol. 3**, Sect. 6.4.1 Nail Connections in Wood and Wood-Based Materials*

In this type of construction, a shear-plate-like outer wall component is produced in extreme lightweight construction, which nevertheless retains substantial advantages of massive walls. This is because wooden rib structures are characterised by some features of wall construction. The outer

13 Canada Mortgage and Housing Corporation (ed) (1980) *Canadian Wood-Frame House Construction*, Ottawa, p. 27 ff.

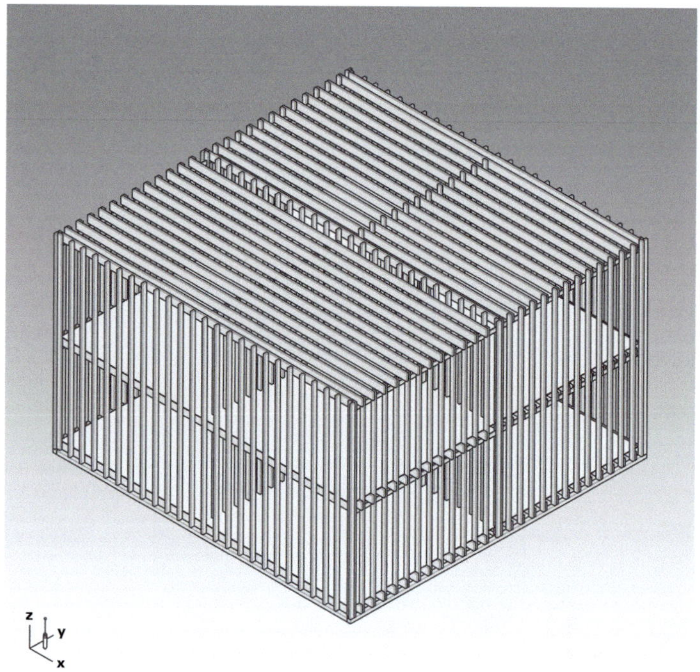

Fig. 8.28 Structural axonometry: Schematic representation of a **balloon frame**. The studs are continuous across two floors

wall component counteracts the action of the combined vertical and horizontal loads within its plane by means of its large shear stiffness—a direct consequence of the large lever arms of the reaction moment constituted by the wall length and height—and, like traditional half-timbered construction, consistently avoids load concentrations in the component. The bracing of the structure is achieved by mutual support of the individual perpendicularly oriented wall panels, similar to masonry box construction.

Sect. ▶ 7.2.1 Wall Construction

Two basic types of historic American timber-frame construction are differentiated:[14]

- The older design is the **balloon frame**. Here, the studs are continuous across two floors. For the support of the floor joists, planks with an upright cross-section are recessed into the ribs, thus providing a sleeper-like support surface. The floor joists resting on them, which cover spans of approx. 4–5 m commonplace in timber construction, discharge their load over the transoms to the studs, and are nailed to them to prevent tilting. Balloon-frame walls were the first to be delivered to the Far West in standardised, industrially prefabricated form (Figs. 8.28 and 8.29).
- The somewhat younger **platform frame** is similar in its structure and its storey-wise segmentation to traditional

14 Giedion S (1976), from p. 233.

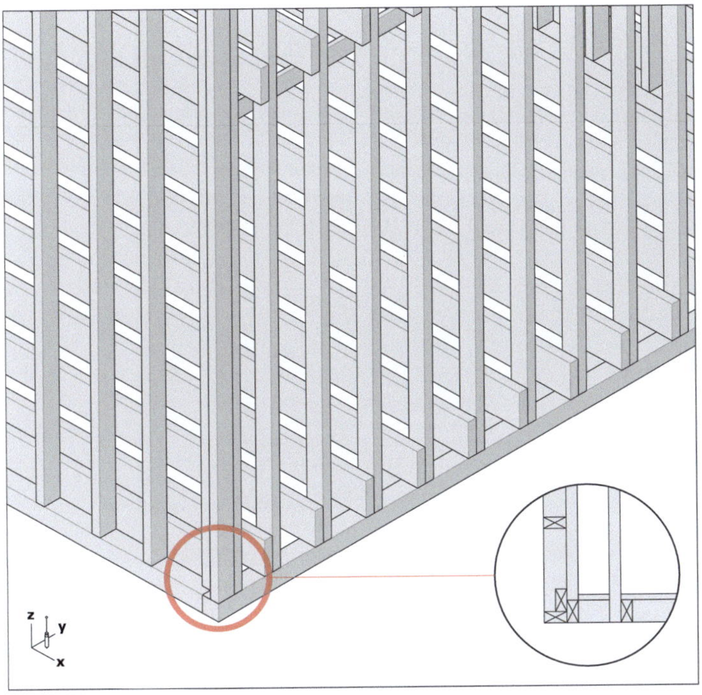

Fig. 8.29 Structural axonometry: **Balloon-frame** wood construction with representation of the rising studs in the wall area and the floor joists. In the detail section, note the design of the corner connection for the mutual holding and stiffening of the individual wall sections pre-assembled on the floor. The corner arrangement of the studs is necessary in order to be able to fasten the sheathing on the inside and outside at right angles

half-timbered building and is certainly inspired by it. In this construction, the wall elements are designed to be storey-high. The ribbed walls are equipped on top with an additional head plate. The floor joists are placed on top of it and secured against tilting, in contrast to the balloon frame, by means of an additional header joist. The floor joists are planked on the upper side and then the wall elements of the upper floor are placed on top (Figs. 8.30, 8.31, 8.32, and 8.33).

Since its inception, wood frame construction has always attracted interest among designers and builders because of its constructional simplicity and extreme material efficiency. Its inherently good thermal-insulation capacity, which can be achieved by filling the cavities between studs with insulation, is a major advantage of this construction method, not least because it is not achieved at the cost of an increased wall thickness, which is practically unavoidable in full-leaf constructions with an added thermal-insulation layer. This is because the load-bearing structure, i.e. the ribs, and the thermally insulating envelope, i.e. the thermal-insulation layer, are integrated in the same level in timber-frame construction. Later on, the thermal-insulation capacity was further en-

8

Fig. 8.30 **Platform** timber frame in the USA under construction. A large proportion of residential construction is still being built there today using this construction method

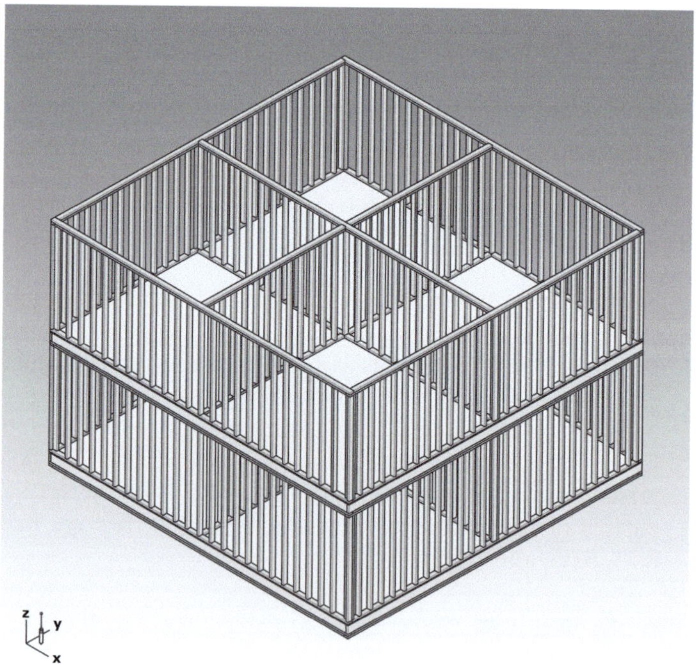

Fig. 8.31 Load-bearing structure of the **platform construction**. Schematic representation of a two-storey building with storey-high exterior wall components

hanced by the doubling of thermal-insulation material, following Scandinavian models. As a result, this lightweight construction can now meet the most stringent comfort and energy-saving requirements, up to the passive-house standard, in its still largely original form while maintaining a still tolera-

Fig. 8.32 Modern residential building in wooden-frame construction

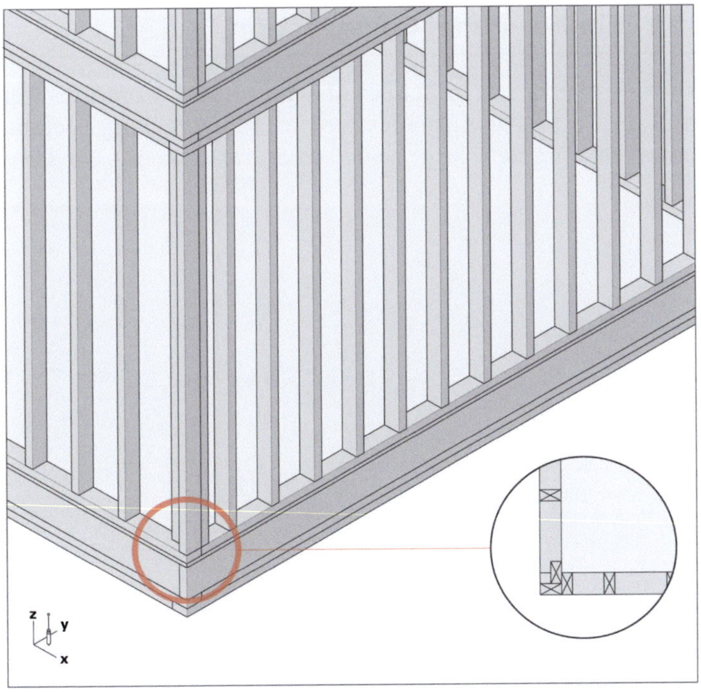

Fig. 8.33 Detailed view of the load-bearing structure of a **platform construction** with the upstanding studs in the wall area. In horizontal section, the design of the corner joint for connecting and bracing the individual wall elements pre-assembled on the floor. This corner arrangement of the studs is necessary in order to be able to fix the panelling to them at a corner on the inside and outside

8

ble wall thickness. This type of construction is now being used in up to five-storey high, or even higher, housing construction.

A modern modification of the above-discussed wooden frame constructions, in particular of the platform construction, is used in contemporary timber-frame buildings. With this variant, an attempt is made to adjust the advantages of American wood frame structures to European requirements. This primarily affects the thermal-insulation capacity of the outer walls, which needs to be increased by additional thermal-insulation layers, either on one or both sides of the wall, to achieve the required U-values. In contrast to the very limited catalogue of parts and the associated limited freedom of variation of the American wood frame structures, more varied grids and cross-sections can be applied here, which allows for a more versatile usage of this construction mode. In contemporary wood frame construction, joining is also primarily done with nails. The construction itself can be considered as uncomplicated and cost-effective (◘ Figs. 8.34, 8.35, and 8.36; cf. also ◘ Figs. 8.37, 8.38, and 8.39).

8.3.5 Timber-Panel Construction

The constructional principles of wood-frame structures were and are, so to speak, predestined to a far-reaching prefabrication of complete large-scale structural elements for floors

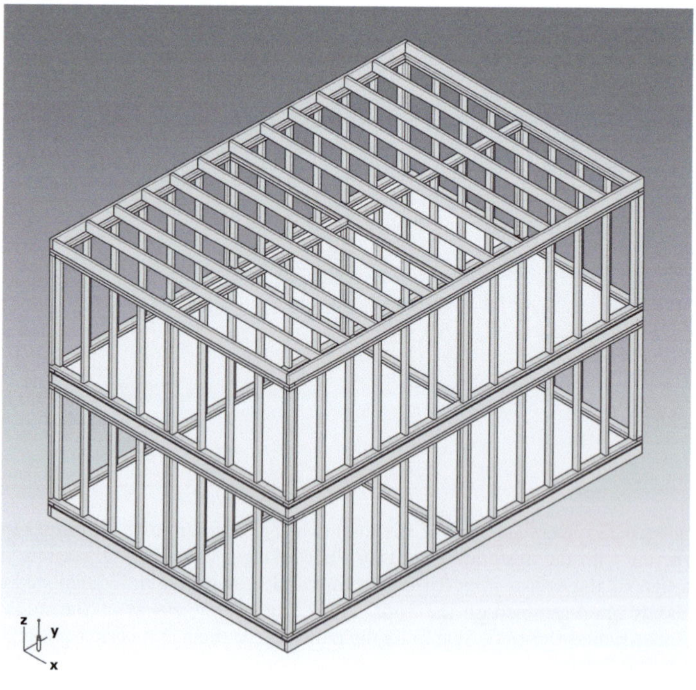

◘ **Fig. 8.34** Structural axonometry: contemporary wood frame construction. Schematic presentation of a two-storey building with floor-high exterior wall components

Fig. 8.35 Modern wood frame construction

or walls. The largely prefabricated variant of wooden frame construction is commonly referred to as wood-panel construction. The desire for cost efficiency and for keeping pervasive standards in building construction required transferring the manufacturing to the workshop or factory. Prefabricated composite elements made of studs and load-bearing and bracing sheathing also offer significant advantages in terms of rationalisation and industrialisation of construction processes. Some important milestones in this development were the following:

For general information on ribbed systems, cf. ***Vol. 3***, *Sect. 14.1 Generalities*

Vol. 3, *Sect. 14.2.1.2 Wooden Panel Walls*

- Development of standardised single-storey residential buildings for the company *Hirsch* by Walter Gropius in the year 1919. The wall elements were made of wooden frames with an aluminium inlay as well as an inner-side cladding made of asbestos-cement boards and an outer weather screen made of copper plates. The wall elements of the company *Kupfer* were already large sized.
- From 1943 to 1954, Konrad Wachsmann and Walter Gropius developed the *Packaged House System* in the USA, which provided the prototype for all other wood-panel buildings. In contrast to the above-mentioned *Kupfer* house, these were small-scale wooden panel elements which received a vertical wooden boarding on the outside. The small-sized wooden panels were joined by the so-called *connector* (Fig. 8.40), a four-piece steel fastening element which made possible a force and form-locking

8

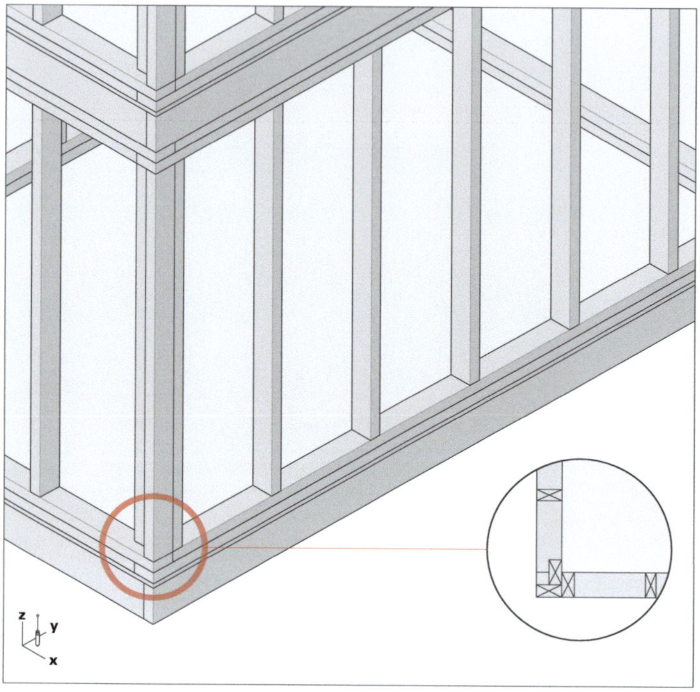

Fig. 8.36 Detailed view of the structure of a wooden frame construction showing the rising studs in the wall. In detail, the corner connection is designed to connect and reinforce the individual wall elements prefabricated in the factory. As a rule, these are delivered sheathed on the inside and are insulated on site. Subsequently, the outer weather screen is applied on site. The corner arrangement of the ribs is necessary in order to be able to fasten the sheathing on the inside and the outside at right angles

Fig. 8.37 Traditional American residential building in balloon-frame construction

■ **Fig. 8.38** Wood-frame structure under construction

■ **Fig. 8.39** Detail of the wall structure of the building on ■ Fig. 8.38 during construction. The diffusive (black) membrane is attached to the outside of the already installed ribbing and thermal insulation (both are therefore not visible here). The upright timbers are the substructure for the weatherboarding and provide ventilation if necessary

juncture of individual panels. This system allowed generating a variety of combinable and individual ground plans.

- Another important milestone in panel construction was the development of the *General Panel System* by Walter Gropius and Konrad Wachsmann (■ Figs. 8.41, 8.42, and 8.43).

Today, modern timber-frame construction appears primarily in the form of largely prefabricated timber-panel construction. Instead of assembling the ribbed elements on site,

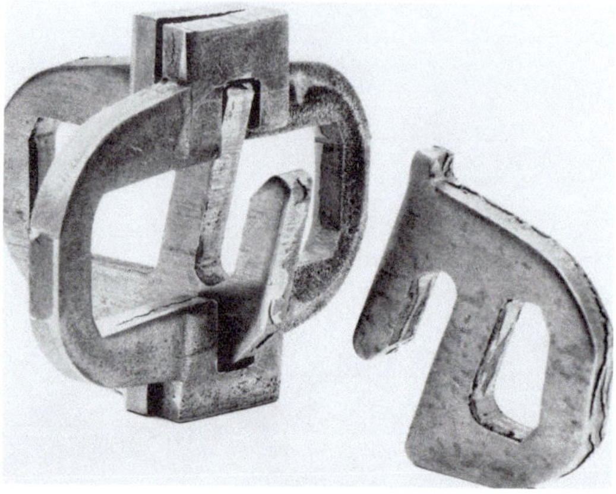

Fig. 8.40 The so-called *connector* by Walter Gropius and Konrad Wachsmann. A stamped metallic hook closure designed for the universal connection of modular wooden walls and panels

Fig. 8.41 Wood-panel construction: *General Panel System* by Konrad Wachsmann

large-format wooden panels up to the maximum transport sizes are now completely prefabricated in the interest of the fastest possible assembly.

Wood panels, until recently, were generally used for residential buildings with few storeys. Until not so long ago, only buildings with a maximum of two storeys were generally feasible. This maximum realisable building height was also specified by relatively strict fire-protection regulations. In recent years, however, multi-storey buildings with up to seven storeys have also been erected using this method of construction, an improvement for which the fire-protection legislation has been made more flexible and applied in a way that is clearly tailored to actual requirements and specific situations. With this building height, however, certain **constructional limits** of conventional timber-panel construction have been reached. On the one hand, this is due to the reasonable maximum dimensions of the solid-wood cross-sections of the ribs, which should not exceed 24 cm on the long side of the cross-section, although solutions with glued laminated timber or composite web or

Fig. 8.42 Panel construction: wooden panel elements of the *General Panel System* (K. Wachsmann)

Fig. 8.43 Wood-panel construction: assembly of small panels

box ribs are also possible, which allow even greater rib depth and stiffness. This height limit, on the other hand, is primarily determined by the maximum permissible transverse compression of the horizontal wooden members, namely, floor joists or floor plates, header plates, and sleepers, which form intrinsic parts of wooden panels. In particular, the horizontal wooden members of the wooden panel elements and floors of the lower storeys are subjected to especially heavy stress

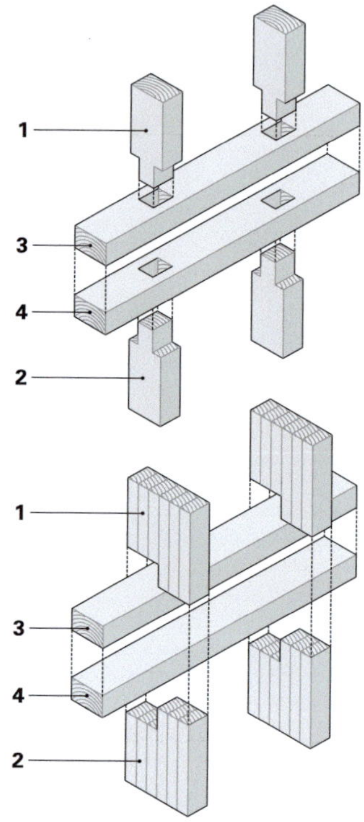

Fig. 8.44 Constructional solutions for the direct transmission of the vertical load circumventing sleeper and head plate: (**1**) upper stud, (**2**) lower stud, (**3**) sleeper of the upper wall element, (**4**) head plate of the lower wall element

transversal to the grain due to the cumulative vertical load, so that noticeable settlements can occur as a result. The height limit can be extended upwards even further if these elements are made of hardwood or if special constructional solutions are used (Fig. 8.44) which allow the force to be transmitted from stud to stud across the lying timbers without compressing them. The consistent change of the floor-span direction on different storeys and the resulting better load distribution also relieves the wall elements and makes greater building heights possible.

As long as the height limit of a maximum of five to seven storeys is not exceeded, panelled walls have an important advantage over solid-wood walls, namely, that they allow, as already mentioned and as will be discussed in more detail below, locating thermal insulation at the same level as the load-bearing elements, i.e. in the spaces between the studs. With solid-wood walls, however, the complete thermal-insulation layer must be added (usually on the outside). As a result, high insulation values can be achieved with wooden panel walls at relatively small wall thicknesses.

► *Sect. 8.3.6 Modern Solid-Wood Construction Modes*

A great advantage of this design lies in the extremely short assembly times and easy assembly. Transport and erection costs are relatively low and the buildings are cost-effective in comparison with other construction methods.

The usual dimensions of wood panels are.

- **Small-sized panels**: Width = 1.00–1.25 m. However, these small formats are no longer used today due to the resulting large number of erection joints and the comparatively high labour costs on the construction site.
- **Large-sized panels**: Length up to 10 m, reduction in the number of erection joints. This construction method consistently avoids joint-related weak points, especially with regard to wind protection and airtightness, which can quickly cancel out the advantages of this method. The dimensions of the panels are adapted to the maximum transport sizes that can be realised in each individual case (◘ Figs. 8.45 and 8.46).

8.3.5.1 Structural Behaviour of Timber Panels

Wooden panels are typical enveloping elements featuring the structure of a ribbed system. Their load-bearing behaviour is based on the interaction of the bar-shaped framework and the flat sheathing. According to a principle of division of labour, the ribs absorb both axial forces (compression, tension) and bending moments along their strong axis, i.e. in the direction of their long cross-sectional dimension transverse to the plane of the panel (→**x**). Provided that the connection between rib and sheathing is sufficiently shear resistant, the surface-forming plates help to reduce the bending stress, since they act like compression and tension chords at the two ends of the ribs to a certain extent (concurrent plate width) and thus significantly increase the section modulus of the plate cross-section. Against the danger of buckling under strong axial compressive forces, the rib is helped, in one di-

***Vol. 3**, Sect. 14 Ribbed Systems*

On the interaction of rib and plate, cf. ***Vol. 1**, Sect. 26.9.4 Element Composed of Ribs Oriented One-Way*

◘ **Fig. 8.45** Assembly of modern large-format wood panels in prefabricated construction

Fig. 8.46 Manufacture of large-format wood panels at the factory

rection, by its stiffness at right angles to the plane of the element (→**x**) and, in the other, by the sheathing oriented along the envelope plane at right angles to the rib (→**y**).

The sheathing primarily absorbs shear forces within the panel plane (**xy**). The diagonally oriented compressive component of this shear force, which endangers the thin plates on bulging, is blocked by the connection of the plates with the ribs, which have their strong axis in the possible buckling or bulging direction, i.e. perpendicular to the plane of the element (→**x**), and are correspondingly rigid.

Wooden panels thus develop both **bending-plate stiffness** out of their plane (→**x**) and **shear-plate**, i.e. **racking**, **stiffness** within their plane (**yz**). The former characteristic benefits them when used in floors as well as in outer walls exposed to wind; the latter, for load-bearing and, in particular, bracing walls or floor shear-plates.

8.3.5.2 Structure of Timber Panels

Modern timber-panel walls mostly consist, in their simpler version, of a framework of solid structural timber (structural solid wood) up to a rib cross-sectional dimension of 24 cm (Fig. 8.47). More resistant panels can be produced from glulam with even larger rib cross-sections; greater panel thickness to accommodate thicker thermal-insulation layers for highly insulated building envelopes can be achieved by using double-T-shaped web ribs, ladder ribs, or box ribs. For sheathing, wood-based panels are employed instead of diagonal boarding as used to be the case formerly; as inner panelling, OSB boards are preferably applied which are airtight and inhibit vapour diffusion, i.e. act as a vapour brake on their own. Multilayer boards or laminated veneer lumber are also suitable for this purpose.

Vol. 3, *Sect. 14.2.1.1 Wooden Frame Walls* and *Sect. 14.2.1.2 Wooden Panel Walls*

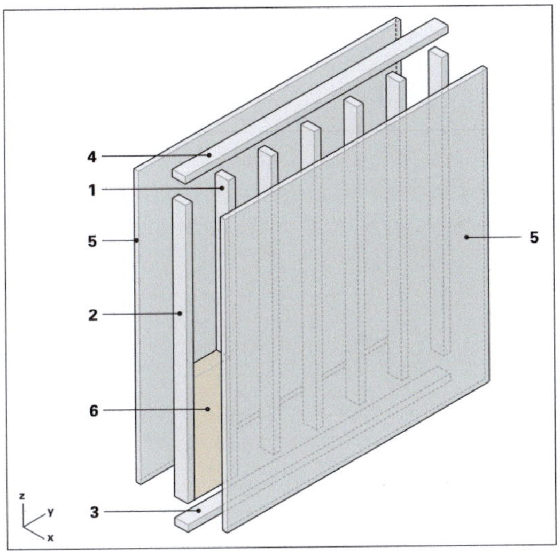

Fig. 8.47 Basic components and typical structure of a **wooden panel element**, applicable to walls and floors: (**1**) rib: stud (wall) or beam (ceiling), (**2**) edge rib, (**3**) sleeper (wall) or edge piece (ceiling), (**4**) head plate (wall) or edge piece (ceiling), (**5**) sheathing, (**6**) insulation filling

Wooden panelled floors, also known as **box floors**, are constructed analogously. Their basic structure, aligned along the ribs, makes them preferably one-way-spanning floor elements, which, however, can also have a continuous or cantilever effect. While structural solid timber is suitable for the ribbed structure in the case of smaller spans, higher and slimmer glued laminated timber or even veneered-laminated-timber cross-sections can be used for larger spans. Tilting of the slender ribs is prevented by the edge beams of the element; bending buckling due to lateral evasion of the compression chord of the rib, by the top sheathing.

***Vol. 3**, Sect. 20.6.1.4 Wood-Panel Floors*

A cooperating panel width of the sheathing is activated by a sufficiently shear-resistant connection between panel and rib, for example, by nail or screw gluing.

Improved transverse load distribution, e.g. for point support or a lateral cantilever effect, can be achieved by installing suitable cross-pieces (Fig. 8.48). It is possible to interrupt one of the ribs at the point of intersection, be it the main rib or the transverse rib, as—despite the gap in the rib—the continuous top and bottom sheathing act as a substitute for the interrupted rib's compression and tension chord at this point (only for a very limited length: the width of the penetrating rib). This is another example of the quasi-symbiotic interplay between ribbing and sheathing.

8.3.6 Modern Solid-Wood Construction Modes

A relatively new development in wood construction are two-dimensional **solid-wood wall** and **floor elements**, which are offered

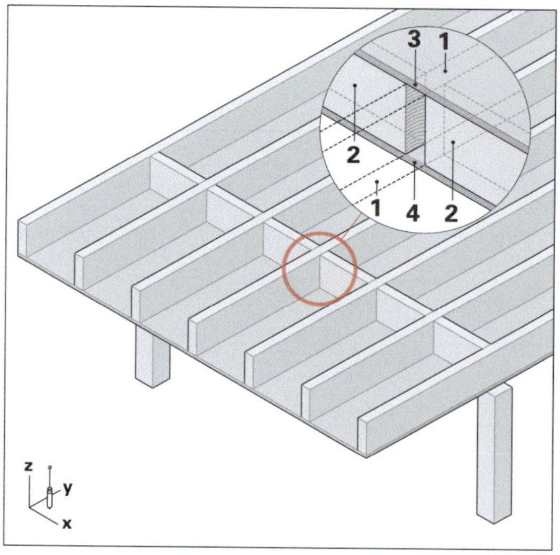

Fig. 8.48 Wooden panel floor on punctual support with transverse rib for two-way load transfer. One of the ribs (here transverse rib) can be interrupted as the two sheathing panels act as chords (detail): (**1**) longitudinal rib, continuous, (**2**) transverse rib, interrupted, (**3**) upper sheathing (upper chord), (**4**) lower sheathing (bottom chord)

8

Cf. also **Vol. 4**, *Sect. 8.4.2.4 Modern Wall Construction Modes in Timber*

Vol. 1, *Sect. 21 Wood Products*

by various manufacturers in open wooden building systems. Modern solid-wood construction methods are partly based on the availability of innovative wood-based materials, i.e. components from veneer plies or composite solid-wood cross-sections. In particular, the prefabrication and the preceding integrative design of the building's mechanical equipment plays an essential role in these building systems.[15] Compared to frame-like wooden panels, the mostly homogeneously plate-shaped solid wooden elements have considerably larger force-conducting cross-sections (hence the designation) and can also carry significantly larger loads. In terms of build-up, the construction of solid-wood elements is simpler and more cost-effective than that of frame elements, since the latter's close spatial intermeshing of supporting structure and finishing construction within the framework (infill with thermal insulation, vapour, and air barriers) (Fig. 8.47) disappears in the strict layer-by-layer construction of the solid-wood components—a not inconsiderable labour-cost factor.

8.3.6.1 Solid Wood in Constructional Application

The essential new contribution of modern solid-wood construction methods lies in the provision of load-bearing, homogeneous surface elements that can be processed into space-enclosing walls or floors. This is a novelty for timber construction, since otherwise only bar-shaped elements such as beams, uprights, or columns made of sawn timber were available, which subsequently were assembled to form func-

15 Natterer et al. (2003), p. 63.

Fig. 8.49 Wood-based material glued from individual board lamellae

tional enveloping surfaces. Wooden frame elements made of closely set ribs can be seen as a kind of transitional form between bar and solid-wood construction; but even when viewed as a framework, they became only really efficient to use since the availability of panel-shaped wood materials for panelling. In contrast to wooden panels, which are ribbed systems, most solid-wood elements belong to the category of full-leaf systems and are in many respects subject to the same constructional boundary conditions as, for example, solid constructions made of mineral materials.[16]

▶ *Sect. 2 Build-Up of Envelopes*

Due to the layered, composite structure of the solid-wood elements, various advantages can be made use of:

- Similar to other wood-based materials, more homogeneous components can be produced by cutting the log and appropriately **sorting** the basic components than with sawn timber, which generally has strong irregularities in its structure due to its natural growth properties. For example, it is technically much more efficient to sort out deficient smaller slats than whole tree trunks.
- The **regular structure** consisting of individual smaller elements (board lamellae, veneers) (Fig. 8.49) additionally enhances the homogeneity.
- The grain pattern of the veneers or board lamellae can be specifically adapted to the installation position or structural stress. For example, a uniform grain pattern can be achieved in laminated-wood or board-stack elements if

16 Although the term *solid* expresses this full-leaf characteristic, the technical term solid wood, which has been introduced in the meantime, still remains confusing, because, until now, solid materials have always been mineral.

Fig. 8.50 Isotropic, two-way spanning cross-laminated timber element (5-ply) for a floor plate; see also Fig. 8.10

the load acts in only one direction. This applies, for example, to vertically stressed walls or one-way spanning floors. The transverse load distribution of the elements can be improved with interspersed blocking layers, i.e. those running transverse to the main grain direction. This is the case, for example, with laminated veneer timber or cross-laminated timber elements. Finally, it is also possible to produce almost completely isotropic elements from alternately cross-oriented layers. These are suitable, for example, for two-way spanning floors (Fig. 8.50) or for simultaneously vertically and horizontally stressed bracing walls.

- Different layers can be produced specifically from stronger (or weaker) material to improve the load-bearing capacity (or for economic reasons). For example, upper- and lower-chord lamellae of a cross-section can be made of stronger wood to improve the bending stiffness.
- Due to the processing of thin lamellae, even curved components can be glued; in principle even doubly curved ones. The possible radii of curvature are only limited by the thickness of the lamellae; the direction of curvature (e.g. uniaxial or biaxial) by its cross-sectional shape: i.e. bending direction uniaxial for rectangular cross-sections (in the direction of the weak axis) and biaxial for very thin flat cross-sections or small square sections. However, this type of curved solid-wood elements is still an exception today.

E.g. so-called combined glued laminated timber according to *EN 14080*

Cf. Fig. 8.9

The use of two-dimensional components in solid-wood constructions results in a load-bearing structure of shear-plate-like and bending-plate-like elements, i.e. a wall construction that behaves statically similar to a wooden frame or panel construction or a solid box construction made of mineral material (masonry, concrete). It is thus characterised by consistent load distribution and mutual support of the wall elements, which are weak and unstable at right angles to their plane.

▶ *Sect. 7 Masonry Construction*

Compared to wooden panel elements made of solid-wood ribs, solid-wood elements can absorb considerably larger forces thanks to their significantly larger overall cross-section. With a correspondingly favourable grain orientation along the force axis, forces can be easily transmitted between abutting elements, e.g. in the case of stacked wall elements in multi-storey buildings. With wooden panels, on the other hand, force conduction is, as mentioned above, significantly impeded by the transverse compression of the horizontal bars of the element, i.e. the sleeper and the header plate. Beyond a certain load level, this results in settlements in the construction. These favourable structural properties have opened up new fields of application for modern solid-wood construction, especially for multi-storey wall construction.
However, the favourable fire-protection properties of the surface components, which minimise the risk of fire penetration thanks to their small number of joints, and slow down the rate of fire spread thanks to their solid cross-sections and favourable profile factor, have also favoured the use of solid-wood elements in multi-storey construction. Another advantage of these wooden products is their relatively good sound insulation, which is mainly due to their increased mass. Elaborate additional measures in the form of suspended ceilings, which are indispensable for fire protection and sound insulation with conventional ribbed or beam floors, are thus usually unnecessary.

The large formats of the solid-wood elements are also suitable for modern building procedures. Compared to the framework construction methods of conventional timber construction, they allow labour-cost savings and contribute to keeping the assembly time on the construction site short. The relatively small number of joints in these types of construction also makes it possible to ensure airtightness with relatively simple means and a comparatively low risk of leakage.

Finally, the composite structure of the solid-wood elements makes it possible to locally use wood types of greater strength in the element in a targeted manner. This mainly involves hardwoods, which are increasingly present in the market, also due to the already noticeable climate change which favours deciduous trees to the detriment of conifers, and will be more so in the future. The disadvantage of the stronger deformation tendency of some hardwoods (e.g. beech) compared to softwoods can be significantly reduced by using smaller lamellae.

Due to the plate characteristics of most solid-wood elements, however, the intertwining of ribs and thermal insulation in the same plane, which is typical for exterior wood-panel walls, is lost. In order to comply with the current high-insulation standards, an insulation layer and a weather skin must therefore normally be applied as a lining to the outside of exterior wall elements. The same questions of building physics and construction have to be solved for

Vol. 3, Sect. 12.3 Single-Leaf Systems with Functional Lining

this design as for other wall-construction methods, e. g. those made of mineral materials. However, the low thermal conductivity of the main material, i.e. wood, has a favourable effect in this respect.

Nonetheless, the predominant advantages of solid-wood construction methods have allowed timber construction to flourish again in recent years and have opened up fields of application that were previously denied to it (not only technically, but also by building law). Classic solid-wood frameworks, which were the standard in conventional timber construction until not so long ago, now practically only appear—if any, in form of skeletal constructions—in cases where high flexibility of use, large spans, or large building heights are required.

Cf. ▶ Sect. 8.4 Skeletal Timber Construction

8.3.6.2 Plate-Shaped Solid-Wood Elements

Solid-wood constructions basically use the following construction elements (Figs. 8.51, 8.52, 8.53, and 8.54).

Board-Stack Elements

Vol. 1, Sect. 21.4.2.4 Board-Stack Timber

Board-stack elements are laminated wood products in which board lamellae are joined together to form solid wall and floor components (Fig. 8.55). Pre-dried board squared sections are nailed, dowelled, and/or glued, oriented with their broad side at right angles to the component plane, to form a flat, load-bearing element. These are mainly used for floors and solid roof constructions, where the individual board lamellae are subjected to bending stress via their strong cross-sectional side. However, board-stack elements can also be used as vertically loaded wall elements with vertically running board lamellae, in which a force transmission takes place along the grain.

Fig. 8.51 U-shaped trough girder, filled with mineral thermal insulation (prod.: *Lignatur AG*)

Fig. 8.52 Wooden construction elements (cross-laminated timber): examples of boards glued together crosswise

Fig. 8.53 Wooden construction elements (cross-laminated timber): Example of a multi-storey residential building in shell construction phase before the additional insulation layer and the weather screen are applied

The boards or planks run over the complete element or are joined in longitudinal direction by finger-jointing. The lateral bond between the board lamellae endows the element with a certain racking stiffness, the extent of which depends strongly on the type of bond: An adhesive bonding is shear resistant; a dowelling or nailing is clearly more yielding. Board stacks can also be subsequently sheathed on their faces with panel-like wood materials, which give the element additional shear stiffness. Crossbar-like wooden rods at the ends of the element have a similar effect. The shear-resistant cross-connection of

8

Fig. 8.54 Installation of a floor made of box-shaped timber construction elements

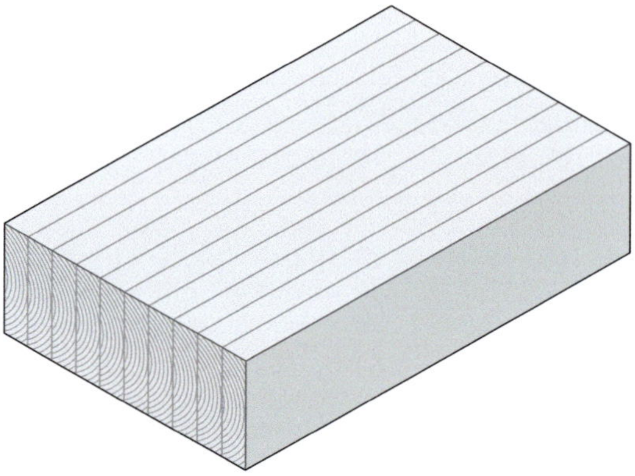

Fig. 8.55 Board-stack element

the lamellae allows a limited load cross-distribution in floors (see below); in the case of wall elements, it allows its being used as a bracing panel.

With this clearly anisotropic element (strong longitudinal direction along the grain, weak transverse to it) forces can practically only be received along the board lamellae. In the case of wall elements subjected to vertical loads, this means that boards must be installed upright; when used as a floor element under bending stress, the board stack acts as a one-way spanning plate with only limited transverse load distribution, which is considerably smaller than, for example, with cross-laminated timber (see below). This is due to the very poor transverse tensile strength of timber, which is decisive for the transverse load distribution in this variant featuring no crosswise oriented board layers.

A characteristic feature of board-stack elements is the maximum utilisation of the element's complete cross-section for force conduction along the grain. This benefits both vertically loaded wall elements, in which axial forces flow along the grain across the full wall cross-section, and one-way spanning floor plates, in which axial flexural tensile and flexural compressive forces act and which benefit from the maximum utilisation of the structural depth. This high load-bearing capacity of the element structure viewed in the direction of the grain comes, however, at the expense of the only slight load cross-distribution.

Cross-Laminated Timber Elements (X-Lam)

In the case of cross-laminated timber elements, 15–30-mm-thick board layers are glued crosswise to form wall and floor elements (◘ Fig. 8.56). Up to 80-mm-thick chipboard flat-pressed boards with building-authority approval can also be processed in the element. Prefabricated element sizes with transport dimensions of max. 5 m in height and max. 22 m in length enable the trimming of X-Lam in the factory, permit easy transport and, by means of screw connections, easy assembly on site. Wall openings and breakthroughs are prepared during the cutting-to-size process. The walls are either erected floor by floor on top of the storey floors, or the walls are continuous over the storeys and the floors are connected by means of a connecting construction (in steel) (◘ Fig. 8.64). This construction method is suitable for the construction of multi-storey buildings (up to high-rise buildings). As with most solid-wood construction methods, outer walls must be additionally insulated on the outside and be protected with a weather skin (◘ Figs. 8.57, 8.58, and 8.59).

Vol. 1, *Sect. 21.4.3 Cross-Laminated Timber (X-Lam)*; Cf. *EN 16351*

From a structural point of view, especially when used as floor panels subject to bending stress, the alignment of the two outer layers of boards must always be taken into account, which is always the same (since the number of layers is odd).

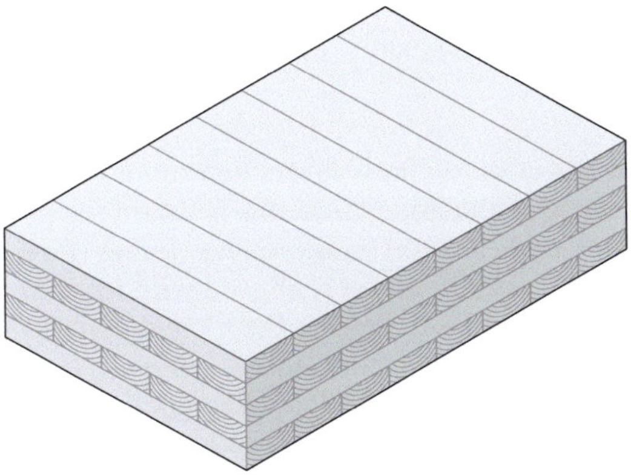

◘ **Fig. 8.56** Cross-laminated timber element

Fig. 8.57 Installation of solid cross-laminated timber elements made of boards bonded crosswise. The construction allows prefabrication of large wall and floor elements; maximum lengths of up to 20 m are possible

Fig. 8.58 Integration of technical building equipment at the factory is possible and is facilitated by the use of CAD/CAM

The grain direction of these layers determines the main spanning direction of the floor, since this is where the greatest flexural tensile and flexural compressive forces flow (extreme stresses) and where the wood is aligned with its grain direction along these forces. Thus, despite the almost complete isotropy of the overall element's build-up, a one-way (not two-way) spanning plate is created, but featuring good load cross-distribution. Thanks to this, lateral projections and point supports of the plate are also possible. Cross-laminated timber used as a floor shear-plate is well suited to absorb shear forces due to its extensive isotropy.

Fig. 8.59 Installation of a complete wall element made of cross-laminated timber including cut-out window openings

When used as a wall element, especially in comparison with board-stack and veneer-laminated-timber elements, it must be taken into account that the load-bearing cross-section of cross-laminated timber is only available to about half for the purpose of force absorption along the grain because of the cross-laminated structure. Cross-oriented board layers, being stressed across, not along, the grain, are unavailable for this purpose. This reduces its load capacity accordingly under these circumstances. However, the element again benefits from its isotropy when used as a bracing wall panel.

Laminated Veneer-Lumber Elements

Laminated veneer lumber, also referred to as micro-laminated timber, consists of several layers of softwood veneer glued together (Fig. 8.60). Recently, it has also been made of beech veneer, which has a significantly higher strength (Fig. 8.12). The grain of the layers basically runs in the same direction, so that a clearly anisotropic element is created. Similar to board stacks, veneer lumber offers a homogeneous load-bearing cross-section for force conduction along the grain, the structural possibilities of the element with regard to axial force conduction thus being practically optimally exploited. Laminated veneer-lumber elements are therefore able to absorb high loads. Unlike (non-glued) board-stack elements, laminated veneer lumber is characterised by good shear stiffness in its plane, which makes the element suitable for both bracing walls and floor shear-plates.

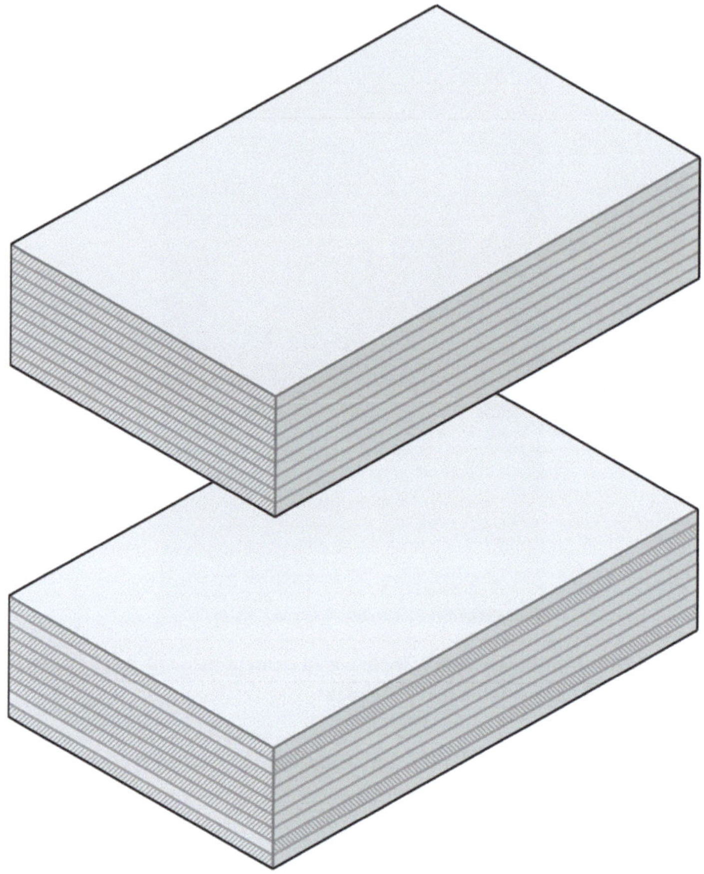

Fig. 8.60 Laminated veneer-lumber element

Due to their pronounced anisotropy, floor plates made of laminated veneer lumber are clearly characterised by one-way load transfer. Flexural compressive and flexural tensile forces are strictly directed along the grain. The transverse load distribution is only limited, which is why linear supports are required. However, this can be improved by interspersed blocking layers running transverse to the main grain direction. This in turn allows for lateral projections and point supports.

8.3.6.3 Timber Construction Elements

Vol. 3, *Sect. 20.6.1.5 Floors from Wooden Construction Elements*

There are various box- or trough-like wooden construction elements with cavities on the market, which, in contrast to the plate-like solid-wood elements just described, can rather be assigned to the category of ribbed systems. They are characterised by spaced ribs incorporated in the element, which, on the one hand, give it significantly increased bending-plate stiffness and, on the other hand, provide spaces for installation, ballasting, or insulation. With floor constructions, larger spans can be achieved, on the one hand, and, on the other hand, various building-physics characteristics of the

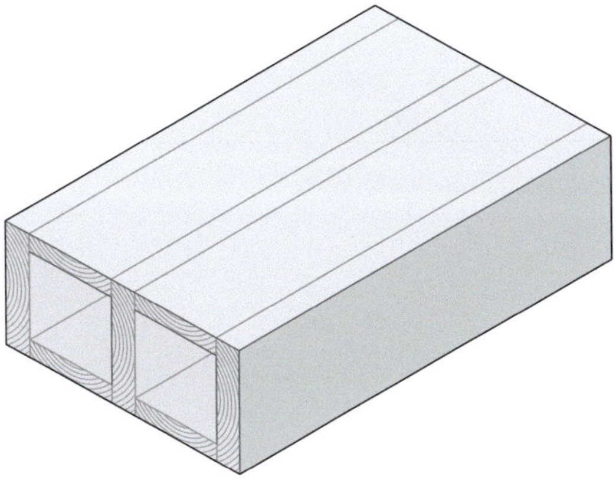

Fig. 8.61 Trough element

space-enclosing surface component can be favourably influenced. This applies in particular to sound insulation, which benefits from the flexible mass of a ballast and a damping of the cavities by fibre insulation material. Perforations on the underside of the ceiling can also improve the room acoustics.

Two variants of timber construction elements have currently established themselves on the market which are analysed in more detail below.

Trough Elements

The characteristic dimensions of the trough-like design of laterally coupled trough modules are as follows (Fig. 8.61): layer thickness of board lamellae: between 31 and 64 mm; element heights are between 120 and max. 480 mm; element widths between 20 and 100 cm. The individual box modules are coupled together via tongue-and-groove connections. Spruce wood is mostly used as a material. Different visual and surface qualities are possible.

*Cf. **Vol. 3**, Sect. 20.6.1.5 Floors from Wooden Construction Elements*

Due to the ribbed structure and the relatively large available construction depths, large spans of up to 14 m can be bridged with trough elements. The design of the ribs has a favourable effect on the load-bearing capacity, as their height is identical to the complete construction depth of the element, the surface-forming top and bottom boards being laterally abutted against them. In this way, the maximum structural depth can be used. Furthermore, the formation of continuous floor constructions is possible (e.g. two-span floors).

Depending on requirements, the cavities of the box elements can be filled with thermal insulation, absorbers, or grit. Furthermore, the integration of pipes, conduits, and channels in the cavity of the elements is conceivable. Open-top versions are available, which allow installation on site from the top (Fig. 8.51).

An adaptation to the respective fire-protection requirements is provided for (up to **F 90** or **REI 90**). For this purpose,

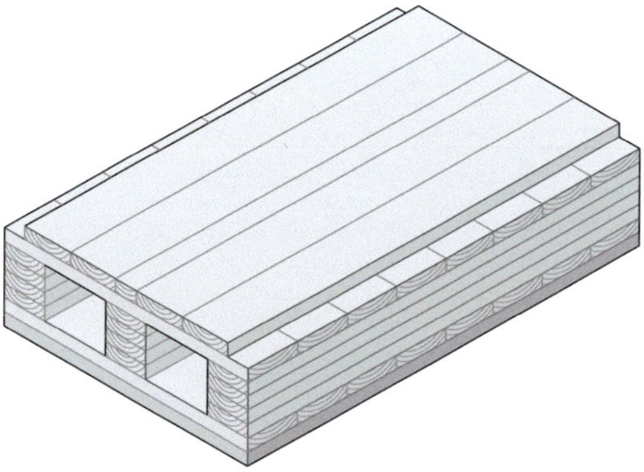

Fig. 8.62 Ribbed element from cross-laminated timber

the top and bottom cover boards can be made thicker to ensure the necessary additional burn-up thickness.

Box and **panel elements** are used both in the construction of flat floors and sloping roofs (Fig. 8.54).

Cross-Laminated-Timber Ribbed Elements

In contrast to the continuous solid timber construction methods and the board-stack construction method, the wall components of the cross-laminated-timber ribbed construction method (Fig. 8.62) each consist on the outside of one or more board layers running parallel to each other with the grain, which, in combination with rib-like board arrays arranged crosswise thereto and spaced at intervals, create a floor or wall component with cavities.

***Vol. 3**, Sect. 20.5.1.6 Floors from Wooden Construction Elements*; cf. also ibid. *Sect. 20.6.1.5 Floors from Wooden Construction Elements*

The floor or wall element, which is therefore not completely solid but interspersed with cavities, has a high rigidity at right angles to its plane. Mechanical ducts, mainly electrical installations, can be routed in these cavities. The standard length of an element is 2.50 or 3.00 m; overall lengths up to 18 m are possible by using finger-joints. This type of ribbed elements can be used for all parts of the primary structure, but can also be combined with elements of other solid-wood systems.

High noise-protection values can be achieved by filling the chambers with stone chippings or heavy sand. Normal sound-proofing requirements for housing floors can thus be easily ensured (Figs. 8.51 and 8.52).

8.3.6.4 Fit-Out

Surface elements such as walls or floors made of solid wood must, in most cases, be equipped with additional elements or layers in order to meet the requirements of building physics (Fig. 8.63). The following constructional subfunctions are affected:

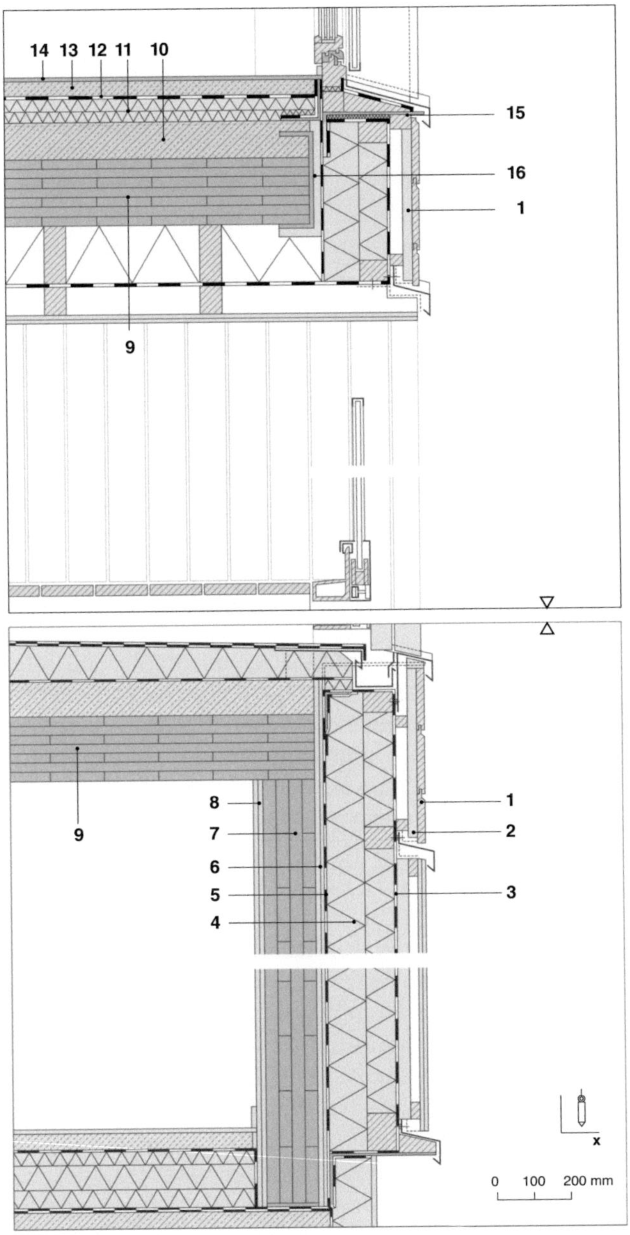

Fig. 8.63 (Top) Exemplary construction of a solid-wood floor (above balcony) (residential building IBA Hamburg, arch.: Adjaye Ass. London). (Bottom) Exemplary construction of a solid-wood exterior wall and balcony floor: (**1**) tongue-and-groove boarding, larch, 21 mm; (**2**) counter-battening/rear ventilation; (**3**) vapour-permeable sealing sheet; (**4**) thermal insulation; (**5**) airtight layer; (**6**) plasterboard 2 × 12.5 mm; (**7**) cross-laminated timber wall 120 mm; (**8**) plasterboard 2 × 12.5 mm; (**9**) cross-laminated timber ceiling 182 mm; (**10**) top-concrete layer in composite with cross-laminated timber; (**11**) impact-sound insulation and EPS layer as installation level; (**12**) separation layer; (**13**) cement screed 45 mm; (**14**) parquet 10 mm; (**15**) sheet-steel firewall 1.5 mm; (**16**) steel profile C 260/100 mm covered with plasterboard

- **Thermal insulation**: Exterior walls and roofs are usually equipped with additional thermal-insulation layers, located almost without exception on the outside. Despite the favourable thermal conductivity of wood, solid-wood elements in the statically necessary thickness are not suitable to guarantee the heat-transfer coefficients required by the law.
- **Wind protection**: Large-format solid-wood panel elements are characterised by a small number of joints. This makes it easier to guarantee the airtightness of the building envelope, a factor that is much more critical in the case of timber-frame constructions. As a rule, it is sufficient to professionally seal the few erection joints.
- **Fire protection**: This factor concerns in particular floor constructions. Unlike conventional wooden-beam floors, which usually have to be fitted with elaborate additional elements such as suspended ceilings for fire-protection purposes, solid-wood floors offer good fire protection by themselves. From below, they expose the smallest possible surface to the fire, namely, an unprofiled plane, and can be dimensioned with the necessary burn-up thickness. From above, either a floating screed or the concrete slab of a wood-concrete composite construction offers sufficient protection. In case of higher fire-protection requirements, where further thickening of the wooden construction does not seem reasonable, it can be clad with fire-protection boards (e.g. plasterboards).
- **Sound protection**: Due to their greater mass and the relatively small number of joints, solid-wood floors achieve better sound insulation than conventional wooden-beam floors from the outset, without any special additional measures. Nevertheless, the required sound-insulation values cannot usually be achieved without additional layers. Although the additional mass of the solid-wood cross-section is a favourable factor, this mass is bending-stiff, which significantly reduces its building-acoustic effect. Usual additional measures to improve the sound insulation of solid-wood floors are, in addition to floating screeds, laid stone chippings, bound or unbound, which contribute flexible mass. The concrete layer of wood-concrete composite floors also noticeably improves the sound insulation through its mass contribution. Suspended lightweight false ceilings improve both sound insulation and fire protection from below, although the sometimes desirable wooden soffit is lost.

8.3.6.5 The Wall/Floor Node

Special attention must be paid to the connection of a floor to a load-bearing wall when building with solid-wood components—similar to timber-panel construction, where even more attention is paid to it. This is due to a fundamental constructional conflict that arises at this node detail. It is causally connected with the pronounced anisotropy of wood, which shows clear weakness transverse to the grain direction. At the wall-floor junction, this results in the impossibility, beyond a certain number of storeys, mostly four or five, of applying vertical wall loads to horizontal timber cross-sections, as the transverse compressions become too great.

Applied to the constructional and static conditions of the wall-floor junction, it results that the floor construction, which consists of horizontal wooden parts as required by plan, cannot be integrated into the vertical outer wall. However, this would be desirable from a design point of view, as the lower section of the wall would provide an advantageous support for the floor. Typical traditional solutions for supporting wooden-beam floors on solid masonry walls, i.e. the local embedding of beams within the wall and the transmission of vertical loads within the wall sections remaining between the beams, cannot be implemented with plate-like continuous solid-wood floors. The linear engagement of the slab in an incision in the wall in turn leads to a significant weakening of the wall cross-section, which is particularly unfavourable in the lower floors due to the accumulation of the load affecting them.

Instead, the following design solutions are available for the wall/floor node in solid-wood construction:

- Lateral connection of the floor plate to the unweakened wall made from vertical board stacks or from cross-laminated timber by means of a kind of attached ledge or trimmer beam. This can either be a z-shaped steel angle (◘ Fig. 8.64**A**) or a wooden profile (◘ Fig. 8.64**B**), which either remains visible from below or otherwise engages in a notch practiced in the floor. In the latter case, the remaining plate cross-section must be able to transfer the vertical forces arising.
- Recess at the top edge of the wall panel, so that a support for the floor is created, but sufficient residual cross-section remains to carry the vertical loads along the grain without transverse compression of the slab (◘ Fig. 8.64**C**).
- Complete integration of the solid-wood floor plate into the wall cross-section so that the floor finds a suitable support on the lower wall section and the upper wall section is placed on the top edge of the floor plate. This solution is only suitable for small numbers of floors; otherwise, the transverse compression of the floor timber becomes too great. In higher buildings, the load can be no longer transmitted between the upper and lower wall elements by transverse compression of the floor plate, but can be instead delivered locally through concrete dowels which are cast into holes practiced in the floor edge before the up-

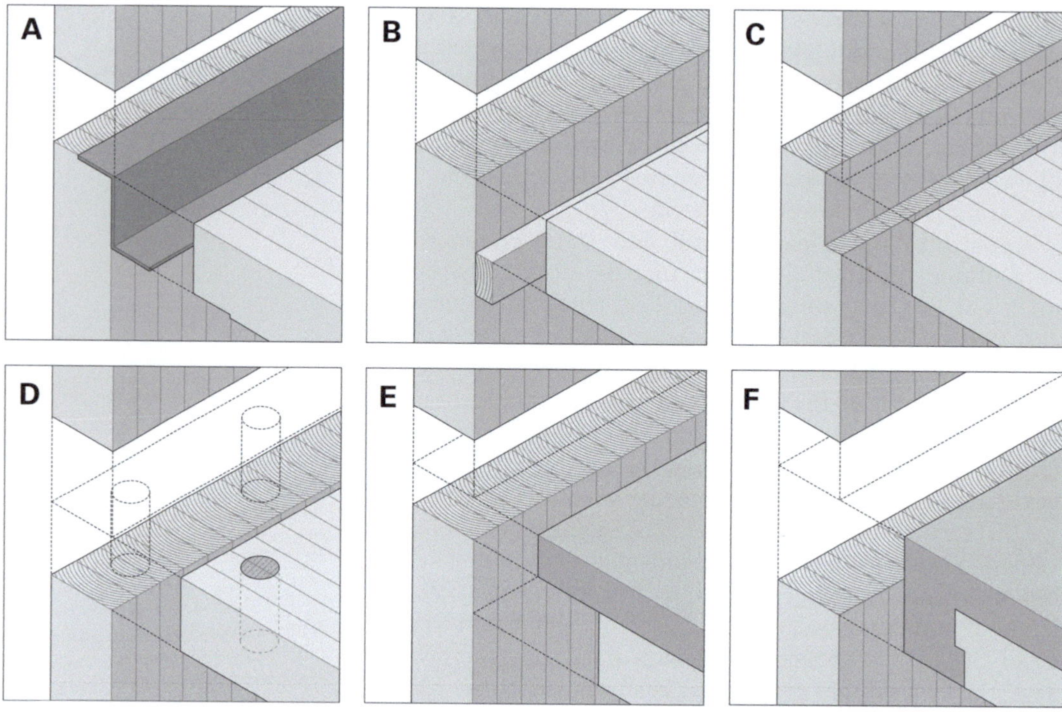

Fig. 8.64 Different constructional solutions for the bearing of a solid-wood floor on a solid-wood wall. **A** Bearing of the floor on z-shaped steel angle. Transmission of vertical loads in the wall without cross-sectional weakening along the grain. Suitable for high buildings. Angle must be protected from fire at the underside. No transverse compression of the floor. **B** Bearing of the floor on a wooden trimmer beam. Conditions comparable to those in **A**. The trimmer beam is visible from below and must be dimensioned for fire protection. No transverse compression of the floor. **C** Bearing of the floor on recess in the wall plate. Weakening of the load-bearing cross-section of the wall. Otherwise, transmission of vertical loads in the wall along the grain. Bearing invisible from below. Number of storeys limited. No transverse compression of the floor. **D** Bearing of the floor on the lower wall component through its complete incorporation into the wall plane. Transmission of the vertical load through the floor via local concrete dowels. Wall element here made of glued board-stack wood. **E** Edge mounting through protruding concrete plate. Good transmission of the vertical load in the wall through the compression-resistant concrete. Shear-force reinforcement in concrete required. Suitable for high buildings. **F** Edge mounting through concrete edge beam. Interlocking of the wooden-floor edge with the edge beam. Conditions as in **E**

per wall section is installed (Fig. 8.64**D**),[17] or by comparable steel elements. The racking stiffness of the wall elements, which is a prerequisite for this solution (therefore, only elements made of cross-laminated timber or glued, not dowelled or nailed, board stacks ought to be used), ensures that the vertical loads acting at the floor connection can be concentrated in the punctual dowels. This requires sufficient transverse load distribution.

- **Faceplate bearing** of a wood-concrete composite floor. In this case, either the concrete slab is extended at the edge of the slab so that it rests on the lower wall section, thus providing a support for the slab (Fig. 8.64**E**); or otherwise, the concrete slab is thickened at the slab edge to

17 Kaufmann H et al. (2017) *Atlas Mehrgeschossiger Holzbau*, p. 44.

form a kind of edge beam up to the complete slab thickness (◘ Fig. 8.64**F**). For a good reception of the shear forces between the concrete edge beam and the solid-wood plate of the floor, the latter can be notched at the edge for an interlocking. In both solutions, the vertical load is always transmitted between wall sections through a concrete cross-section so that transverse compression of the sensitive wood of the floor is avoided.

8.4 Skeletal Timber Construction

The timber construction modes studied so far are rather comparable to solid wall construction. As with these, the load transfer over the wall shear-plates and the bracing of the walls take place via their mutual support in generally orthogonal ground-plan arrangements. With their difficult-to-change wall structures, they are subject to similar usage restrictions as a classic masonry building. The resulting rooms are characterised by their cell-like shape.

A characteristic feature of skeletal construction is the way in which vertical loads are conducted, which, unlike in wall construction (◘ Fig. 8.65), are not distributed linearly over a wall cross-section, but concentrated at points in a column cross-section (◘ Figs. 8.66, 8.67, 8.68, and 8.69). Certain boundary conditions, which are causally derived from this basic structural solution, have a far-reaching influence on the structural variants typical of skeletal construction and will be briefly outlined in the following.

Cf. also ***Vol. 4***, *Sect. 8.6.1 Skeletal Construction Modes in Timber*

◘ **Fig. 8.65** Transverse compression of the sleeper timber through uprights in historical half-timbered construction. Due to the relatively low load level resulting from the narrow spacing between the uprights and the usually limited height of the buildings, the settlements are negligible in this construction method—actually more of a wall-construction method. However, the situation is different for pure skeletal construction, especially for greater building heights

Fig. 8.66 High-rise building in timber skeletal construction (*USB Brock Commons*, Vancouver; arch.: Acton Ostry Architects) (see also **Figs. 8.93 and 8.94**)

Fig. 8.67 East Asian timber-frame construction (Forbidden City, Beijing)

Fig. 8.68 Skeletal structure in timber pincer construction. State building and university construction department, Ulm (Germany)

Fig. 8.69 Skeletal construction in timber: girder on column. The girder is secured against tilting by means of the tong-shaped column head. Printing company building in Paderborn, Germany (arch.: P. C. v. Seidlein)

The first consequence that can be derived therefrom is the magnitude of the compressive force that is present in the column. It is necessarily several times greater than the compressive load existing in the wall. With a comparable load-infeed field of the floor, the load in skeletal construction is concentrated on the comparatively small cross-sectional area of the column and causes correspondingly high compressive stresses. For this reason, in multi-storey skeletal buildings, direct transmission of the force between column sections, preferably from end grain to end grain, is extremely important, even more so than in wall construction, where this question only becomes

critical beyond a certain number of storeys due to the comparatively lower load level. Under these conditions, interposed horizontal wooden parts, such as beams or sleepers, are subjected to a large local transverse compression which the wood cannot tolerate due to its material properties and which causes noticeable settlement over time (◘ Fig. 8.65). This aspect is examined in more detail below in connection with the design of the node of columns and beams.
Despite a relatively high load concentration in the columns, skeletal timber structures have now reached building heights that would have been unthinkable a few years ago. The highest current wooden skeletal construction is just under 100 m high (◘ Fig. 8.66). The good compressive strength of structural timber along the grain, which is comparable to that of normal concrete, makes these great heights possible. Newer developments such as laminated veneer lumber or glued laminated timber made of stronger hardwoods, such as beech, increase the compressive strength of the material to almost twice the value. At the same time, a significant advantage of skeletal structures over wall structures is also decisive here, namely, that the dead weight of the structure is proportionately much lower than with the latter. For this reason, the accumulation of dead weight in multi-storey buildings, which in wall construction sets the limits of what can be built at a relatively early stage, takes place much more slowly in skeletal construction, so that considerably greater building heights can be realised.

***Vol. 4**, Sect. 1.8 Scale Within the Context of the Historic Evolution of Building Forms*

8

Since columns, due to high load concentrations in skeletal construction, are preferably designed to be continuous over the storeys (although butted), a further consequence emerges, namely, the necessity to attach beams to the column either with their front side or with their side surface. This task was technically unsolvable in historical construction methods, which is why there were practically no beams butt-jointed or side-jointed to a column. However, modern timber construction has developed technical solutions for that kind of pure shear-force connections, made possible by special types of metal connectors according to the standard. Pincer constructions in which two twin beams are connected to both sides of a column, thus enabling both continuous columns and continuous beams, would not be feasible without these connecting means.

***Vol. 3**, Sect. 6.5.2 Connections with Connectors of Special Design*; cf. also *EN 912, EN 14545*

Punctual supports of solid-wood floor plates, as they are increasingly used in skeletal multi-storey construction today, are preferably executed by placing them on the lower column section. The vertical load arriving at the upper part of the column section must be passed through the floor plate by suitable steel parts. This prevents excessive transverse compression of the wood in the floor. Similar solutions with inserted steel parts can also be applied to wooden beams.

Another important characteristic of skeletal structures, which compares unfavourably with wall-construction methods, is the absence of shear-resistant vertical panels, i.e. walls, which play an important bracing role against horizontal loads in wall construction. In their pure form, skeletal structures are frameworks with hinged junctures that require additional bracing elements to ensure stability against horizontal loads. A bending-stiff connection between column and beam, i.e. the formation of **portal frames**, is generally difficult to realise in timber construction, which is why this option is a rare exception in timber skeletal construction. It is true, however, that in a certain modified form, namely, in the form of the angle brace, this solution does indeed occur occasionally. For reasons of usage, however, angle braces are generally avoided in modern skeletal construction as they obstruct the room profile. Instead, the most common type of bracing in modern skeletal timber construction is the **diagonal bracing**. The latter can be limited to a few bays, so that only minimal restrictions of use result. The ensuing load concentrations in the diagonals can also be absorbed by steel bars if necessary. Alternatively, individual column trusses can also be filled with shear-resistant wall elements, the task of which is to carry horizontal, not vertical loads. Modern solid-wood elements such as cross-laminated timber are suitable for this. In administrative buildings with few fixed points, these bracing wall panels can be bundled into a circulation core. The necessary fire protection can be guaranteed by additional burn-up thickness of the wall elements. Floor plates in skeletal timber construction can be designed today as shear-resistant panels. This is done either by a professionally executed sheathing of beam floors, usage of suitable solid-wood elements, or of wood-concrete composite constructions.

Cf. ▶ *Sect. 4.2.1.3 Bracing of Skeletal Structures*

An essential advantage of skeletal construction methods compared to wall-construction methods is, of course, the possibility of combining interiors into large, connected areas, which is a fundamental necessity for certain building uses, for example, office buildings, industrial buildings, or many public buildings. This is, so to speak, the positive flip side of the lack of shear-stiff surface elements, be it walls or diagonal braces. It is possible that the desire for larger connected interiors was, in any case, the driving motive in the evolutionary history of construction methods to develop the older wall-construction method into the more modern and, both in conceptual and constructional terms, more complex skeletal construction method. Medieval European storage buildings, which were actually mixed constructions of load-bearing outer walls and inner skeleton, are meaningful examples of early skeletal buildings. For pragmatic reasons, a skeletal construction was only implemented where it was really necessary, namely, in the interior. However, it should not be forgotten that even pitched roof constructions of ma-

***Vol. 4**, Sect. 8.5 Skeletal Construction Modes*

sonry buildings in Europe have always been skeletal constructions. Some of the medieval roof trusses preserved today show a high degree of complexity and technically demanding node solutions. Much earlier and much more complex and technically mature pure skeletal structures are found in East Asia (◘ Fig. 8.67).

Also typical of skeletal construction is the high degree of flexibility of use resulting from the small number of unchangeable fixed points, i.e. the columns and, if necessary, the circulation cores. Non-load-bearing vertical room closures can be modified as desired in skeletal construction. In contrast to wall construction, they do not have to be arranged strictly one above the other on different floors, so that floorplan configurations can be changed floor by floor.

In the case of skeletal construction, it is also possible to glaze large areas of outer wall surfaces and thus significantly improve the supply of daylight to the interior, or to make them highly thermally insulating while maintaining a relatively small wall thickness. This is made possible by the very limited cross-section of the load-bearing elements, namely, the columns. Careful consideration must be given in this context to the position of the exterior column in relation to the exterior wall. For reasons of building physics, the inner position of the column has proven to be the best choice for skeletal structures.

The following additional features characterise skeletal timber construction:

- great freedom in the design of the load-bearing structure, since this is decoupled from the enveloping function;
- wide range of modern engineering connections with fasteners from steel and steel castings;
- extensive possibilities of application to various building uses: residential buildings, hall buildings, industrial buildings, indoor baths, agricultural buildings, etc.

8.4.1 Main Structural Variants

In the following, the various types of load-bearing structures being currently used in skeletal timber construction are discussed in detail. They are decisive for the classification of particular skeletal construction methods. In this context, the positional relationship of the three main elements, **girder**, **floor joist**, and **column,** within the constructional node as well as their coordination within the overall structure provides the essential criterion for differentiation.

Although nowadays secondary beams are mostly avoided in floor constructions and, instead, large-area solid-wood panels are used, hierarchically tiered, stacked beam floors are nevertheless addressed below for reasons of completeness and for a better understanding of the constructional design. The structural conditions of secondary beam layers can be largely transferred to the ribbed elements of timber-panel construction, i.e. to box floors, and, in some cases, the constructional conditions as well.

8.4.1.1 Beam on Column

The variant of 'beam on column' (Fig. 8.70) is a simple, self-evident construction principle which is often used in single-storey wooden skeletal structures. In that case, there is no need to join the column at intermediate height, as is necessary with a continuous girder in case of there being more than one storey. The beam, or girder, can be mounted directly on the column, making the connection as simple as possible. The beam can be passed through, or above, the column node, that is, as a continuous beam, thus reducing its bending stress in the span. If necessary, Gerber hinges can be inserted at the points of contraflexure. In the main beam direction, cantilevers up to approximately 1.50 m can be produced with standard span sizes. A cantilever arm in the direction of the secondary beam, or floor joist, is also possible, since the deck joists are usually set up on the girders and, consequently, can in turn use the continuity or cantilever effect. In this case, it is advantageous to design the secondary beam located on the column axis as a twin beam and thus avoid a butted column connection. A semi-modular offset of the secondary beam position in relation to the column axis also yields the same result. Cantilevers, i.e. roof overhangs, both in the direction of the girders as well as the deck joists, are particularly advantageous as a constructional wood-preservation measure for the façade.

Cf. e.g. on Fig. 8.79

A secondary beam layer can also be laid at the same height as the girders in order to save construction depth (Figs. 8.71 and 8.72). To avoid a local coincidence of column joint and front-side secondary beam connection, the latter can be spatially separated by an offset (Fig. 8.72).

In multi-storey-type construction, the compression of the beam, inserted between column sections, transversely to its

Fig. 8.70 Beam on column

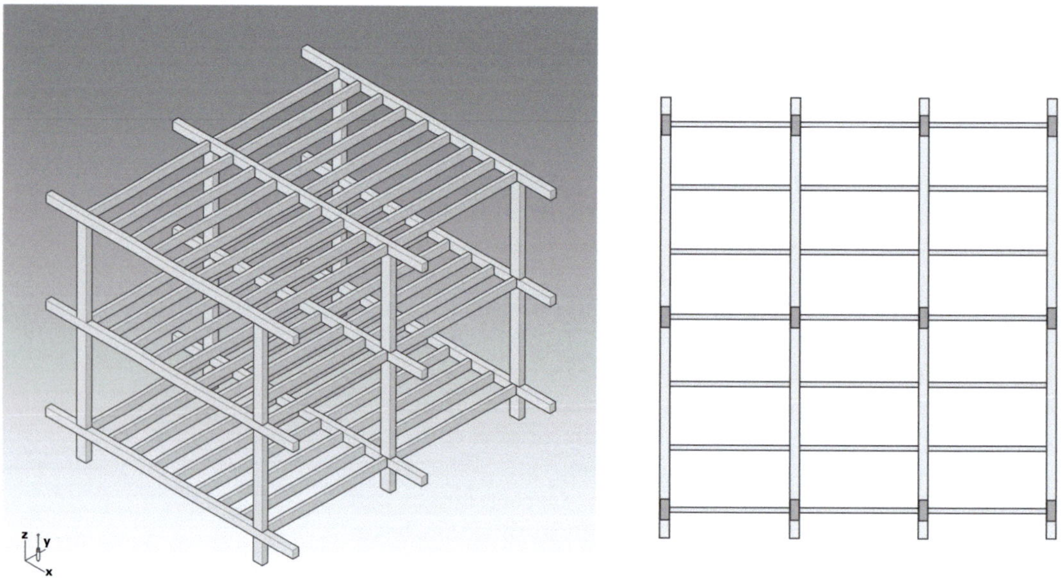

Fig. 8.71 Type of construction featuring a girder on a column: schematic representation of the load-bearing structure. Floor joists can alternatively also be carried out continuous by placing them on top of the girders

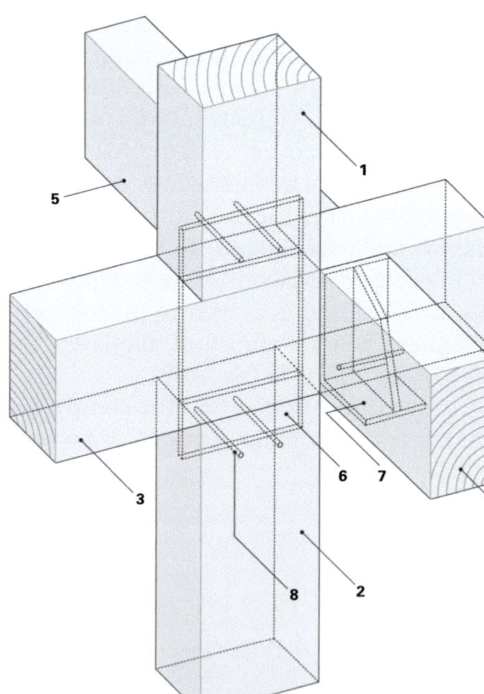

Fig. 8.72 Possible design of the main construction node (floor joists offset here to keep the column connection free): (**1**) upper column section; (**2**) lower column section; (**3**) girder, continuous; (**4**) floor joist, front; (**5**) floor joist, rear; (**6**) slotted steel plate; (**7**) slotted steel bracket; (**8**) rod dowel

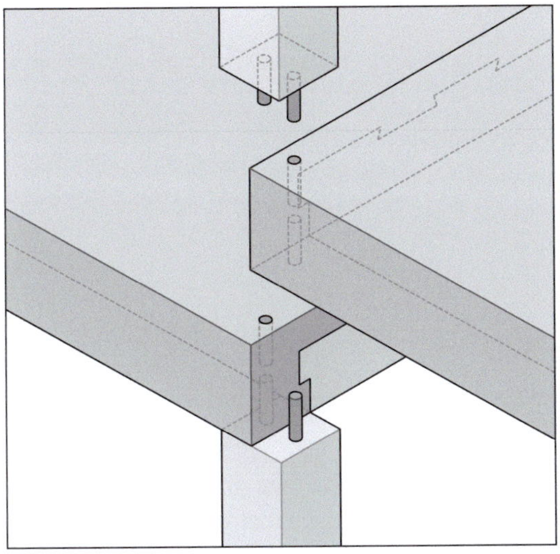

Fig. 8.73 A concrete edge beam of a wood-concrete composite floor, placed on top of the column, transfers vertical column loads through the floor plate without transverse compression of the wood

grain by the action of the abutting column sections must be prevented. Today, this is effected by using modern load-conducting fastening elements—for example, by slotted sheet-metal lugs (Figs. 8.71 and 8.72).

Cf. the discussion in ***Vol. 4***, *Sect. 8.6.1.1 Continuous Column* and *Sect. 8.6.1.2 Segmented Column*

A further example of this type of combination of girder and column is found in wood-concrete composite floors which are executed with a concrete edge beam (Fig. 8.73). The latter acts as a girder for the floor construction, can be easily placed on the lower section of the columns and, being made of concrete, is able to transfer even large concentrated vertical loads from the columns through the floor. This solution combines a simple beam support with high transverse compressive strength of the floor. It is used for high-rise skeletal structures.

Cf. edge beam on Fig. 8.64**F**

8.4.1.2 Two Beams End-Butting Against Column

In this variant, the column cross-section is continuous and the beams are connected at their ends on two opposite sides of the column (Fig. 8.74). This fully meets the requirement for unhindered transmission of the vertical load within the column along the grain. This solution allows for multi-storey buildings to be erected with heights beyond the threshold of high-rise buildings, i.e. with more than roughly 7 or 8 regular storeys. It is irrelevant whether the column effectively passes through the node without being butted or whether the end-grain surfaces of two sections of the column meet there with full compressive contact (and are secured against lateral sliding).

With this design solution, the shear force arriving at the end face of the beam must be introduced into the column cross-section at the contact surface. The following basic options are available:

Cf. the discussion in ***Vol. 4***, *Sect. 8.6.1.1 Continuous Column and Sect. 8.6.1.2 Segmented Column*

Fig. 8.74 Two beams butting against column

- **Recess** on the lateral surface of the column to receive a cleat (Fig. 8.75**A**) or a steel angle (Fig. 8.75**B**). This necessarily weakens the column cross-section. The load from the beam is introduced into the column at the bottom contact surface between the bracket or cleat and the end-grain wood, i.e. along the grain. It is possible to mount the beam simply by placing it on top of the cleat. In the case of slim beams, an additional anti-tilt device may be required.

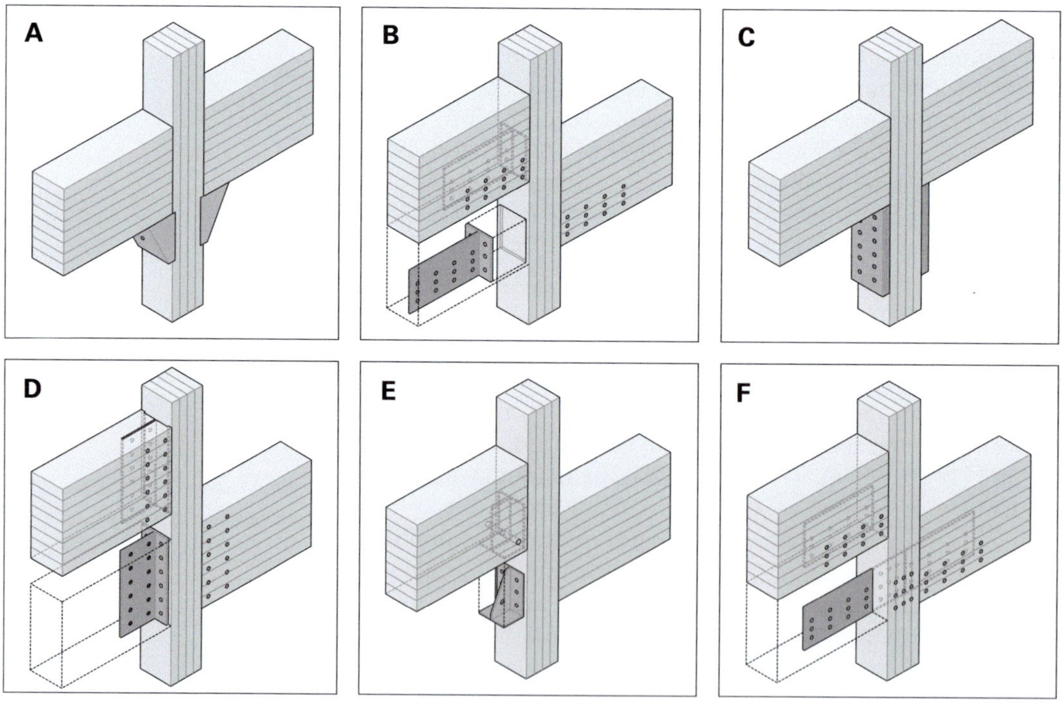

Fig. 8.75 Different constructional solutions for beams butting against a column. **A** Beam on hardwood cleat, recessed into the column side. **B** Beam attached to slotted plate; sheet metal embedded in the recess of the column at the front. **C** Beam on laterally fixed cleat made of derived wood material. **D** Beam attached to slotted plate; even distribution of the fasteners over the height of the beam. **E** Beam on steel brackets laterally connected to the column with centring plate for lateral slide prevention. **F** Beam attached to slotted sheet metal inserted through the column

Fig. 8.76 Connection of a beam to a column using a slotted plate, according to Fig. 8.75**B**, **D** or **F**

- Addition of **cleat-like wooden pieces** to the column side providing support for the beam (Fig. 8.75**C**). Although this solution leaves the column cross-section unweakened, it is roughly equivalent to the solution described above. Visually, the side pieces of wood are very visible, sometimes bothersome.
- Adding a **steel angle** to the side of the column to accommodate the beam (Fig. 8.75**D**, **E**). The column cross-section remains unweakened. The bearing surface should be in the lower area of the beam so that there is no transverse tension that could tear the beam fibres apart.
- Insertion of a steel strap through the column cross-section with double lateral lug connection for both abutting beams (Fig. 8.75**F**; cf. also Fig. 8.76). There is a slight weakening of the column cross-section due to the slitting. The shear-force transmission between lug and beam or column can be solved by means of rod dowels. Again, the rod-dowel arrangement should either be evenly distributed over the full height of the beam to prevent transverse tension within it or otherwise concentrated in the lower area.
- If the beam connection is only on one side and the loads are relatively small, a butt joint with **end-grain connectors** is also possible (only approved for glued laminated timber) (Fig. 8.77).

8.4.1.3 Twin Beams Attached Laterally to Column (Pincer Construction)

The pincer design is executed with double main carriers which grip the column on two sides like a pincer—hence its name (Fig. 8.78). These can therefore be carried out as continuous or cantilevering beams.

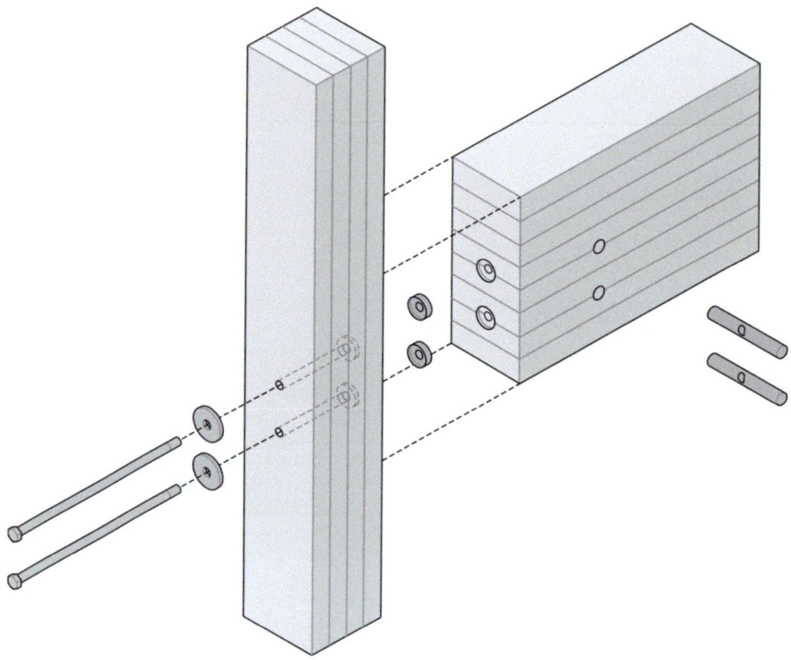

Fig. 8.77 Beam attached to the column on one side using end-grain-wood connectors

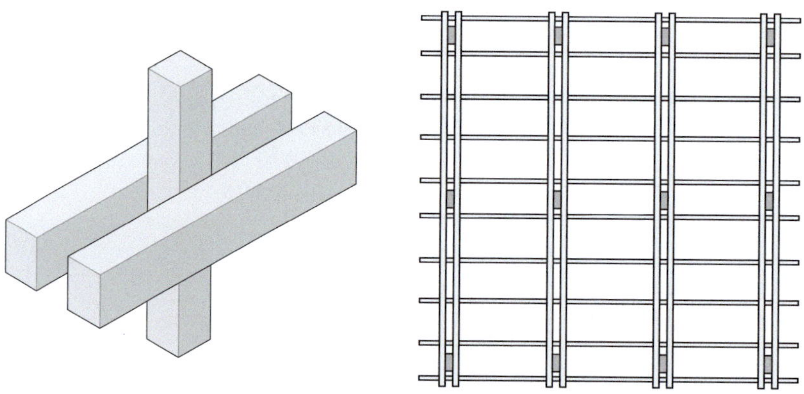

Fig. 8.78 (Left) Twin beams attached laterally to column. (Right) Generic beam layout (view from below)

The lateral connection of the main girder to the column is a shear-force connection, which is carried out with connectors of special design according to standard. As usual with this type of connection, it is secured against falling apart by means of inserted threaded bolts (Fig. 8.79). Inlet ring connectors are often used today, which are inserted into milled grooves. Milling is carried out automatically with CNC joinery machines.

In this connection, the vertical load is introduced into the column along the grain, while in the horizontal beam it is introduced transverse to it. For this reason, special attention must be paid to the load transmission into the beam: The con-

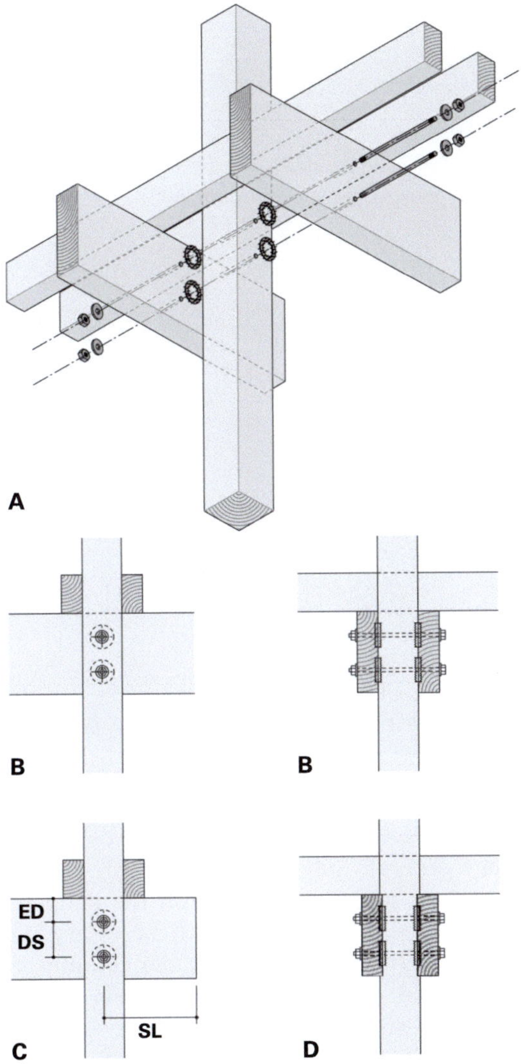

Fig. 8.79 **Pincer construction**: Possible execution of the main node. **A** Standard version, exploded drawing. **B** Standard version, two sections. **C** Standard version; **ED** edge distance, **DS** dowel spacing, **SL** shear length. **D** Special design with latching of the beams into the column

nector juncture creates transverse tension in the wood, which is why, on the one hand, a sufficient edge distance of the upper connector and sufficient connector spacing are required, as well as, on the other, a minimum shear length of the beam as measured from the connection to the front edge of the cantilevered beam (Fig. 8.79**C**).

Furthermore, slender twin beams reduce the eccentricity between their centre-of-gravity axis and that of the column. In principle, the beams can also be latched laterally into the support, but due to the differences in stiffness between the connector joint and the wood compressed crosswise to the grain, this measure hardly facilitates the transmission of forces

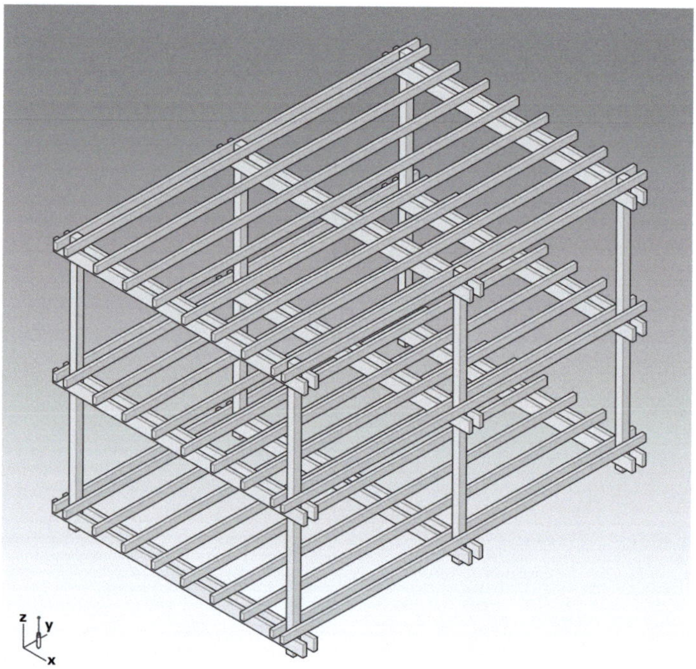

Fig. 8.80 Pincer construction in main and secondary beam hierarchies: schematic representation of the load-bearing structure

(Fig. 8.79**D**). Notches also weaken the column cross-section. They are generally only used as assembly aids.

The secondary beams are placed on the main beams and can be implemented continuous and, if necessary, with a cantilever arm, if they are offset against the column grid or are themselves doubled up at the column (Fig. 8.80). Cantilevers in both main directions are therefore possible. In contrast to the variant of beam on column, the column can also be produced in one piece or connected with a simple contact joint at its end surfaces so that the normal compressive load is transferred without interruption along the grain. Therefore, multiple floors can also be executed. This design variant is thus characterised by a consequent spatial disentanglement of the structural members which impinge on the main node, i.e. column, main, and secondary beams (Figs. 8.79, 8.80, and 8.81).

Special attention must be paid to the wood preservation of the end-grain surfaces of the cantilevering beams whenever exposed. These should, for example, be covered by head boards, or at least protected by overhangs.

8.4.1.4 Beam Through Twin Column

The elements involved in the node, namely, beams and columns, can also be spatially and structurally separated by doubling the column and placing a single beam between the two column halves (Figs. 8.82, 8.83, and 8.84). The shear-force

Fig. 8.81 Building in **pincer construction**: District Office Villingen-Schwenningen, Germany (arch.: Auer und Weber)

Fig. 8.82 Beam through twin column

connection that is created thereby is carried out in the same way as the pincer design discussed above using connectors of special design in accordance with the standard. The constructional conditions are comparable.

Although this solution also has the advantage of being able to transfer the vertical loads along the grain without in-

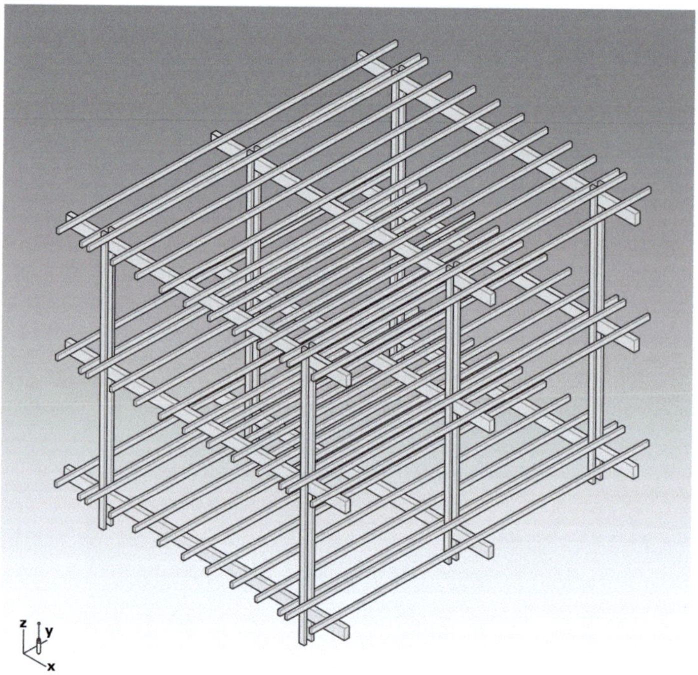

Fig. 8.83 Columns and secondary beams are twins: schematic representation of the structure

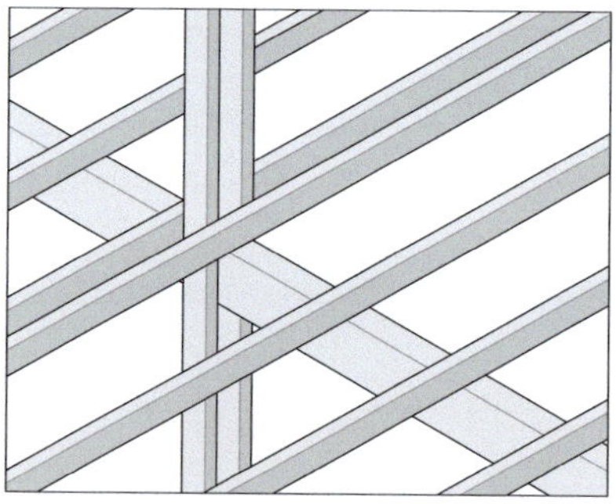

Fig. 8.84 Columns and secondary beams are twins: scheme of the main node

terruptions of the column, it has the disadvantage of a rather unfavourably designed column. If both column parts have the same total cross-section under compressive stress as a comparable one-piece column, the two resulting slender bars are exposed to a much greater risk of buckling than the latter. This risk is generally counteracted by the force-locking connection of both parts of the column, e.g. by inserting filler timbers between them at different points. In this way, a ladder-like column element is created, which necessarily takes up more space in the floorplan projection than a one-piece column—another disadvantage of this solution, not to mention the more prominent visual appearance of this column design.

Since the column of the skeletal system under consideration is assumed to be a hinged column, its cross-section should tend to have the same bending stiffness in both main directions (i.e.→**x** and →**y** on ■ Fig. 8.83). Since the strong side of the column cross-section (the connecting line between the two column parts, →**y**) has, from the outset, a significantly larger lever arm than the other direction of loading (the transverse width of each column part, →**x**), it is more advantageous to form the chords of the column as rectangular cross-sections with their longer side in the strong direction (→**x**) and with the narrow side oriented in the direction of the strong axis (→**y**). This also favours the shear-force connection between girder and column chords, the wider side of the column chords being thus available for that purpose.

If secondary beams are present, these are sensibly placed on top of the girders and either doubled in the column axis or moved past the column by means of a semi-modular grid offset.

8.4.1.5 Four Beams Butting with End Sides on Column

This type of timber construction consists of one-way spanning sets of beams which, unlike the previously discussed construction methods, are limited to a single support field between four columns and each change their spanning direction in a checkerboard-like pattern (■ Figs. 8.85 and 8.86). This explains why, in this case, not two but four main beams, or girders, are connected to the column from four directions.

The columns are continuous; all beams—both the girders, the so-called transoms, as well as the floor joists—have the same overall length, provided the column also has the same width as the beam. Since the columns are continuous, beams can be connected at any discretional height, that is, creating an intermediate floor is also possible.

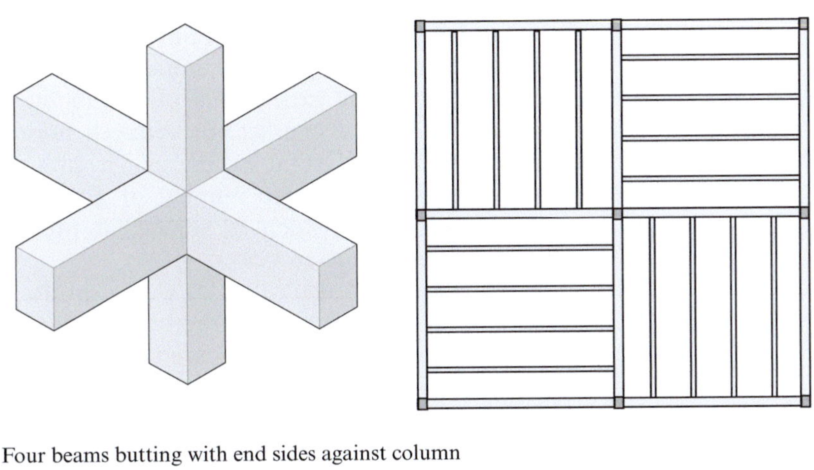

■ **Fig. 8.85** Four beams butting with end sides against column

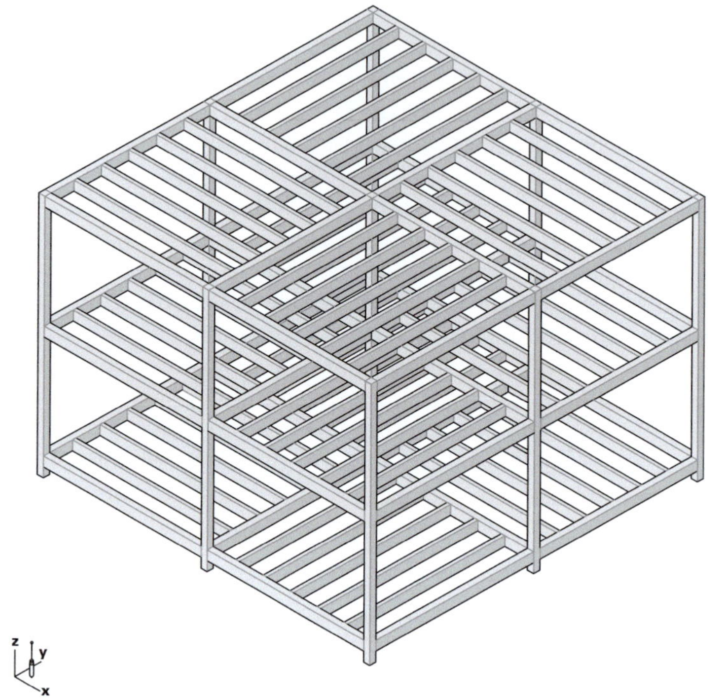

■ **Fig. 8.86** **Checkerboard construction**: schematic depiction of the structural principle

For the load-bearing principle, cf. also ***Vol. 3***, *Sect. 14.4.2 Roofs and Floors from Beam Grillages*

Since the beam direction changes in the respective neighbouring field, the load is introduced into each transom or girder from only one single set of joists, namely, from the one connecting to it from only one side—as on the opposite one there are no joists at all. In this way, the load discharging on a girder can be reduced to half as compared to a regular one-way beam floor in which joists always approach the girder from two sides. As a result, the latter can be assimilated in its overall structural depth to the floor joists—which have smaller load-infeed fields and consequently bear less load. Since, as a result of the changing spanning directions, there can be no continuity effect anyway and the girders and joists can be executed with the same structural depth, both beam hierarchies are expediently placed in the same plane. As a consequence, a significantly reduced floor thickness—compared to other skeletal structures—results. The price for this advantage is a doubled number of girders, as well as a node column/girder which is somewhat more difficult to solve, as will be discussed below.

Although each structural field, viewed in itself, is a directional floor construction, the structural system still counts as a non-directional structure in its entirety. According to the logic of the constructional principle, the structural fields are always square, that is to say having identical spans in both directions of span. In a way, this static and constructional principle was developed as a substitute solution to utilise the advantages of a non-directional load-bearing structure in timber construction, where two-way load-bearing surface elements were not available until recently due to the preconditions given by the particular material, but only bar-shaped construction elements.

The geometrical and modular arrangement of this type of construction results in the typical end-face connection of the beam to the column in both main directions, so that a total of four girders are connected to a column all around (◘ Figs. 8.87 and 8.88). The end-grain connection of the girder to the column flank under relatively restricted spatial conditions requires special junctures using modern engineering fasteners (see the example on ◘ Fig. 8.88).

Checkerboard constructions are often carried out in two-storey buildings with spans of up to 4 m. Roof overhangs can only be produced by means of additional measures, since projections are not possible in this system. Otherwise, this structural solution offers favourable conditions for coordination with the building envelope, since exactly the same geometrical conditions prevail in both main directions and no protruding—i.e. cantilevering—components exist.

8.4.1.6 Plate on Column

This variant involves a **punctual** plate support in which the floor plate is inserted between column sections (◘ Fig. 8.89). As a result, the floor is spanned **two-way**. This variant is relevant for multi-storey buildings, since the flat floor plate without beams, mostly made of solid wood, has an advantageous support on the lower section of the column, which does not

◘ **Fig. 8.87** **Checkerboard construction**: schematic depiction of the main node

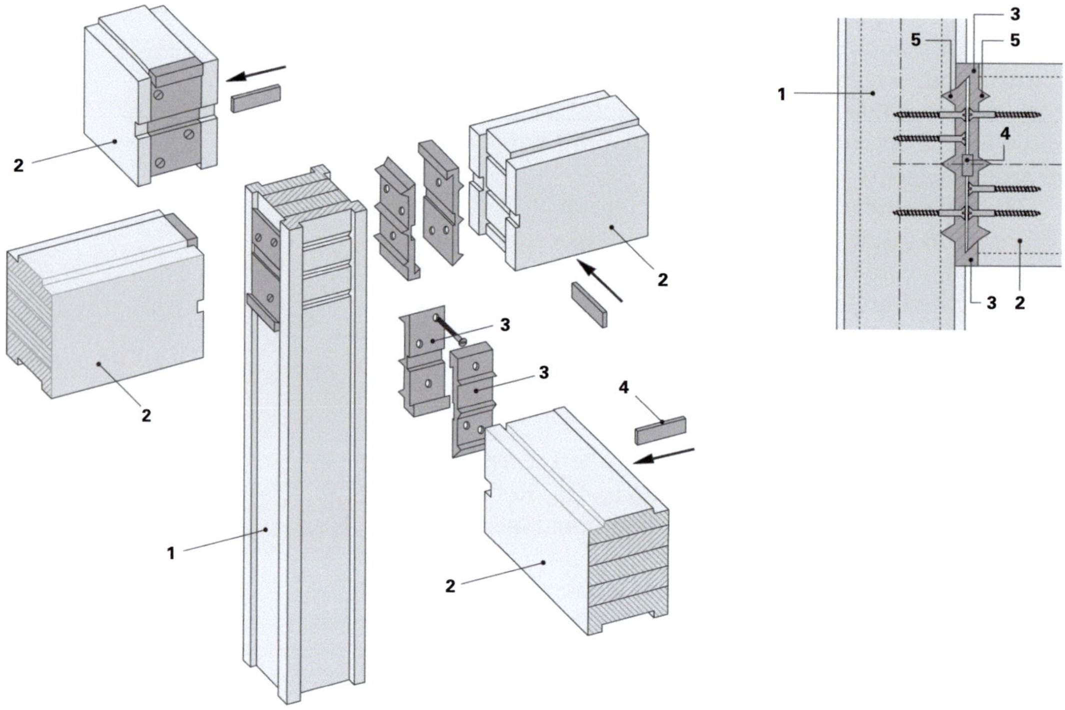

Fig. 8.88 Possible execution of the connection between column and girders (system *Moduli*). Transom connection only shown on one side in the sectional drawing: (**1**) Column. (**2**) Transom. (**3**) Profiled connection plate from aluminium for column and girder. Generates a wedging during assembly. (**4**) Tongue-like part from aluminium which is inserted into a groove of part **3**. The cohesion of the shear-resistant connection is ensured by the wedge effect of the connection plates **3**. (**5**) Form-locking toothing of the connecting plate **3** with the timber of the column and the transom

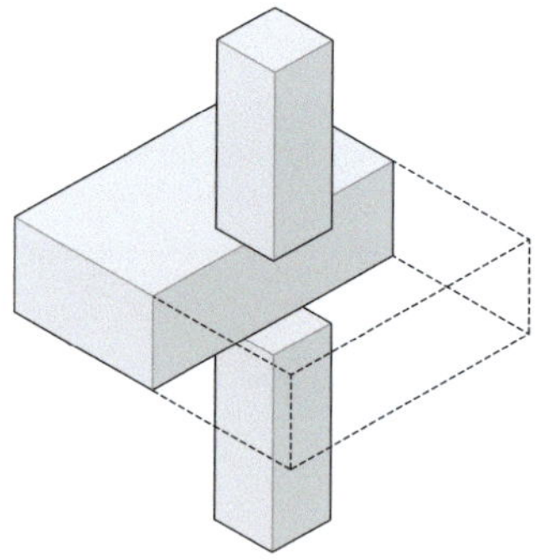

Fig. 8.89 Plate on column

protrude beyond the circumference of the column and is therefore invisible and can be protected against fire for the most part by means of the wooden construction itself. Particularly in multi-storey construction, for which this variant is especially suitable, high fire-protection requirements must anyway be expected from the outset. Possibly laterally exposed front edges of steel plates can be protected against fire either by recessing them in the wooden floor or by a subfloor construction on the top side and, if necessary, by a false ceiling on the underside.

However, it is not possible to transfer the vertical column load across the floor plate, i.e. across the necessarily lying grain direction of the floor timber, in the present case of a skeletal construction. This is due to the high force concentrations to be expected within the column, which are much greater than the more evenly distributed vertical loads in a wall element. The consequences would be strong transverse compression and significant settlement. A steel construction that provides both butting column sections with as full a support as possible and at the same time transmits the vertical loads within a cut-out in the plate, without compressing it transversely, is practically indispensable in these conditions (◘ Figs. 8.90, 8.93, and 8.94).

Only an **isotropic floor element** can be used for this type of point plate support due to the prevalent two-way load transfer. Cross-laminated timber is particularly suitable. Because of the maximum transport sizes, the element width for the short side of the plate is between 3 and 4m. This also

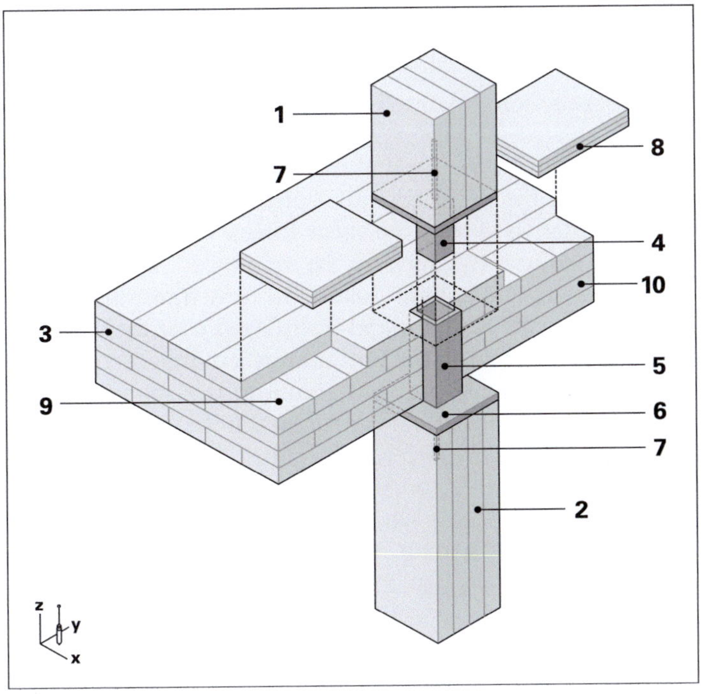

◘ **Fig. 8.90** Punctual floor-plate support on the lower column section with the aid of a steel part; the force transmission takes place via the steel tube and avoids transverse compression of the cross-laminated timber. A similar solution was implemented in the project on ◘ Figs. 8.93 and 8.94: **(1)** Upper column section, storey high; glulam or laminated veneer lumber. **(2)** Lower column section, storey high; glulam or laminated veneer lumber. **(3)** Ceiling plate; cross-laminated timber. **(4)** Steel bearing element, upper half, with telescopic plug. **(5)** Steel bearing element, lower half, with telescopic socket. **(6)** Bearing surface for the floor plate. **(7)** Centring pin and slide prevention. **(8)** Connecting board for shear-plate formation; plywood. **(9)** Recess to accommodate **8**. **(10)** Element butt joint predetermines the distance between columns in this direction (→**x**). In the other direction (→**y**), larger element dimensions can be realised, resulting in reasonable distances between columns in the range of 4–6m. In this direction, the element can be designed to run continuous across several fields. The resulting rectangular structural floor fields with two distinct spans also favour the actual load-bearing capacity of the cross-laminated panel element itself, which, due to its particular structure, always has a primary and a secondary span direction. The stronger primary direction of span, i.e. that of the two outermost board layers, is accordingly placed over the larger span.

► *Sect. 8.3.6.2.2 Cross-Laminated Timber Elements (X-Lam)*

This articulated connection of the plate to the column requires sufficient shear stiffness within the plate plane to connect the skeleton to fixed points, in this case preferably to access cores. For this purpose, the floor elements can be coupled at the butt joint, for example, with a plywood strip placed in a recess on the top side, so that they become shear-resistant (◘ Fig. 8.90, element **8**; cf. also ◘ Figs. 8.93 right and 8.94).

8.4.1.7 Plate Attached Laterally to Column

If the plate is to be connected laterally to the continuous column cross-section, a laterally protruding bracket-like support must be created (◘ Figs. 8.91 and 8.92). This in turn can be achieved by a box-like steel element, which has a cantilevered steel plate on the lower edge that provides the necessary support. In order to absorb the shear forces, the steel plate must either be sufficiently thick or be reinforced with stiffening plates. The vertical column load is in turn transmitted in the steel box through the floor and transferred over the entire surface to the end grain of the bottommost column section.

In contrast to the previous solution, the supporting steel plate remains exposed (and visible) on the underside and must be additionally protected against fire if necessary. It can be embedded in a recess in the floor panel and protected on the underside with a wooden board. However, a sufficient cross-section must remain on the notched floor slab so that

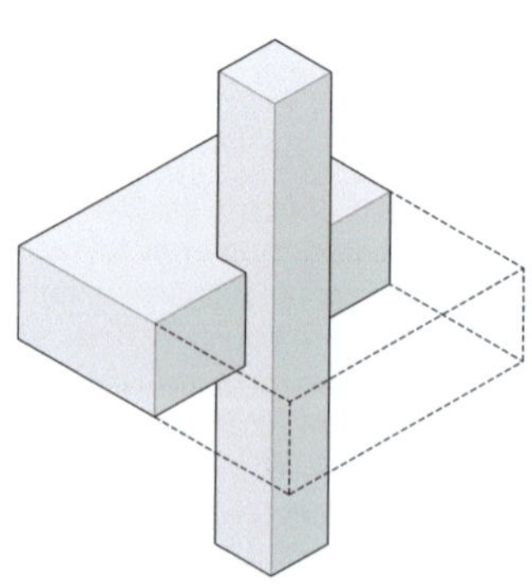

◘ **Fig. 8.91** Plate attached laterally to column

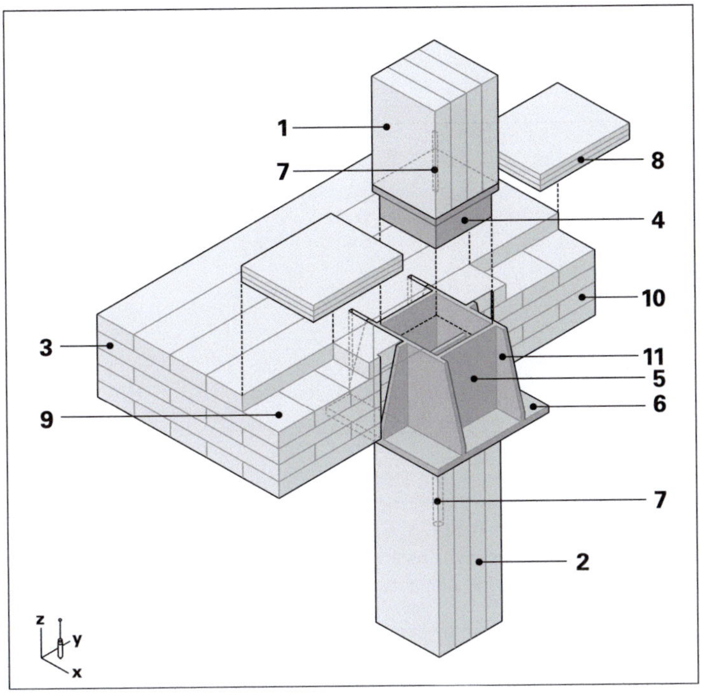

Fig. 8.92 Punctual plate support on a protruding bracket-like steel part **4/5**. The steel plate **6** providing the support is stiffened by plates. The force is transmitted in the steel part without transverse compression of the horizontal ceiling timber: (**1**) Upper column section, storey high; glulam or laminated veneer lumber. (**2**) Lower column section, storey high; glulam or laminated veneer lumber. (**3**) Floor plate, cross-laminated timber. (**4**) Steel bearing element, upper half, with telescopic plug. (**5**) Steel bearing element, lower half, with telescopic socket. (**6**) Bearing surface for the floor plate. (**7**) Centring pin and slide prevention. (**8**) Connecting board for shear-plate formation; plywood. (**9**) Recess to accommodate **8**. (**10**) Element butt joint. (**11**) stiffener

the relatively high shear forces emerging at the point support can be transferred and so that the lower layers of wood do not tear out due to transverse tension. Alternatively, a false ceiling can be suspended. In addition to enhancing the fire protection of the steel structure, the latter can also improve the sound insulation of the floor. This is recommended anyway for the dynamically relatively stiff cross-laminated timber floors (Figs. 8.93 and 8.94).

With regard to the element sizes, column spacings, and shear-plate formation, the same statements apply as for the previous variant.

8.4.2 Hall Construction

Hall construction in timber appears almost without exception in skeletal construction mode and will therefore be characterised here in this particular design variant by discussing its most important features and node solutions (Fig. 8.95).

Fig. 8.93 (Left) High-rise building in skeletal timber construction with two-way spanning, point-supported flat floors made of cross-laminated timber. (The horizontal bars visible on the top floor are assembly aids.) (*USB Brock Commons*, Vancouver; arch.: Acton Ostry Architects) (see also Fig. 8.66). (Right) Skeletal structure on Fig. 8.92, after column assembly. On the column heads, the steel tubes for receiving the next column section (analogous to element **5** on Fig. 8.90) as well as the pins for centring the floor elements are visible

Fig. 8.94 (Left) Skeletal load-bearing structure on Fig. 8.93, after floor installation. On the floor, e.g. in the column axes, the cover strips for creating the shear stiffness of the floor panels are visible (element **8** on Fig. 8.90). (Right) Skeletal structure on Fig. 8.93, column mounting

Load-bearing hall structures are predominantly directional structures with one-way load transfer, in which mutual penetrations of load-bearing elements of the same hierarchy (as are characteristic of non-directional load-bearing beam grillages) are systematically avoided. In accordance with the typical room layouts of directional systems, hall spaces are usually oblong rectangular formats with a much smaller span direction in which the main elements of the roof construction are spanned (Fig. 8.96; →**y**). At right angles to this, i.e. in the longitudinal direction of the hall (→**x**), a secondary structure spans between the main trusses made up of columns and beams, or portal frames. If the columns are not restrained, restraints being an exception in timber construction anyway, additional bracing elements must be provided, usually diagonal bracings, unless portal frames in one direction (→**y**) already perform this task.

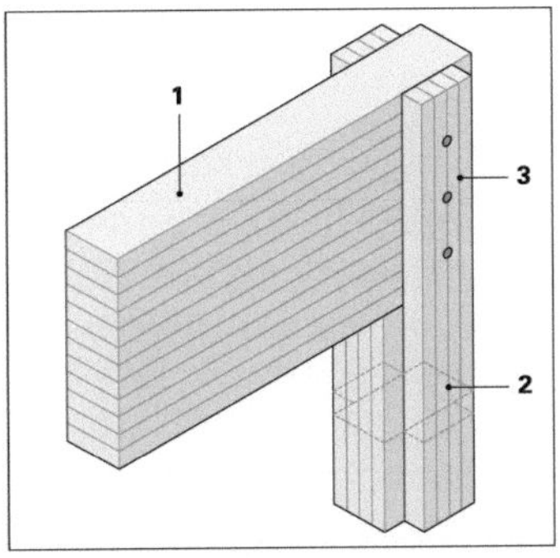

A Hall girder **1** on three-part cross column **2**. The column sections **3** running past the girder secure it against tipping sideways. All parts made of glued laminated timber.

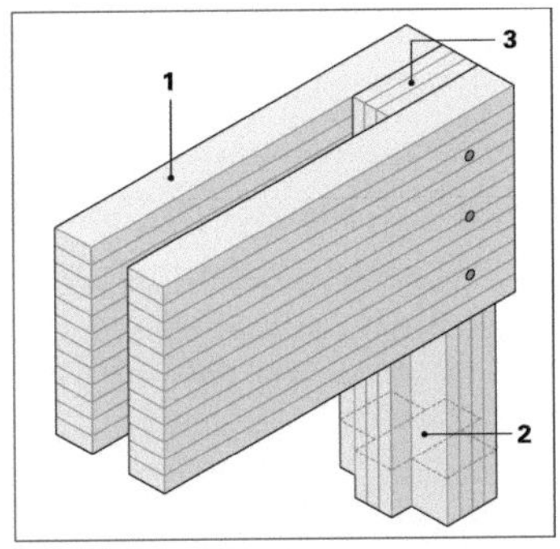

B Twin hall girders **1** on three-part cross column **2**. The central column section **3** passing between the girders secures them against lateral tipping. All parts made of glued laminated timber.

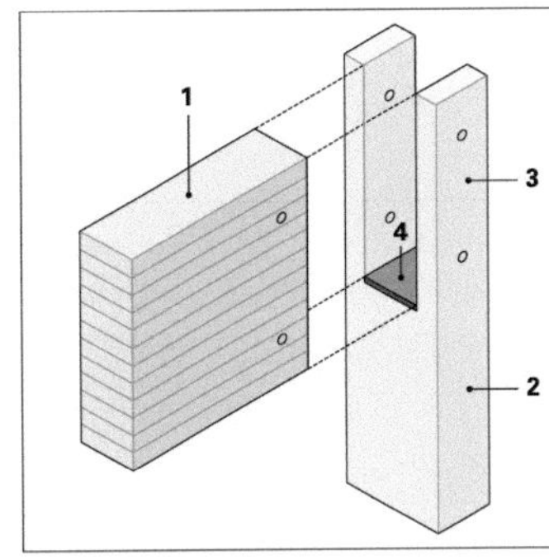

C Hall girder **1** on forked column **2** (precast concrete part). The fork **3** secures the girder against tipping sideways. It is mounted on an elastomeric bearing **4**.

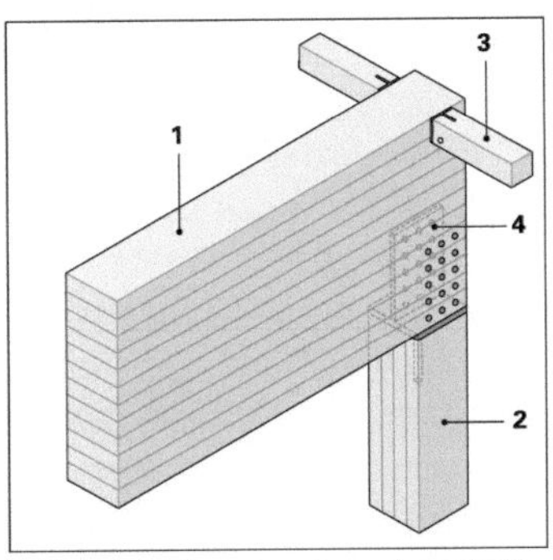

D Hall girder **1** on wooden column **2**. The crossbars **3** attached to the sides of the girder secure it against tipping sideways. These can be coupled to diagonal braces in individual bays. Support reinforcement and slide prevention by slotted plate **4**.

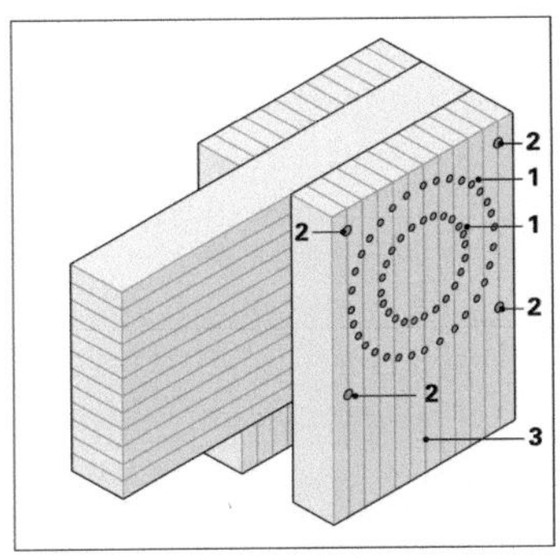

E Bending-stiff portal-frame corner, assembled on site with the aid of two rod-dowel circles **1** and secured against falling apart with four screw bolts **2**. Frame post **3** doubled.

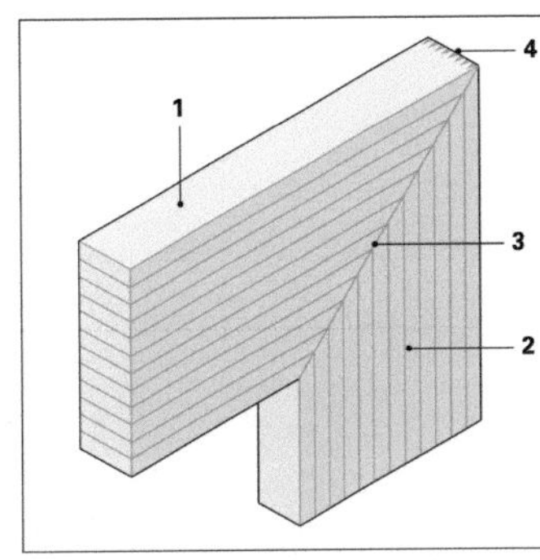

F Bending-stiff portal-frame corner with mitre cut **3** between frame beam **1** and post **2**, glued in the factory using a universal finger joint **4**.

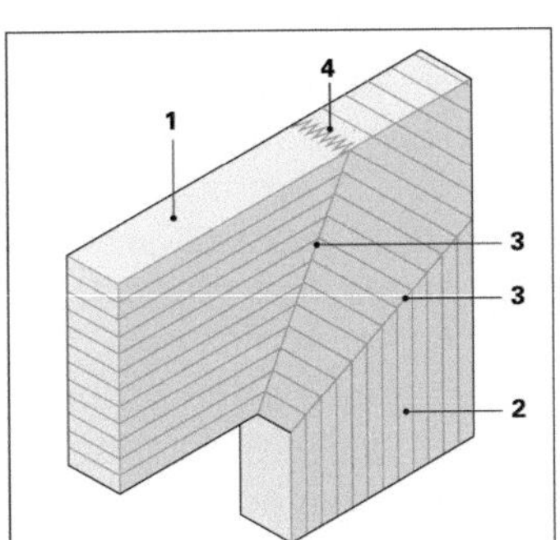

G Bending-stiff portal-frame corner with double mitre cut **3** between frame beam **1** and post **2**, glued in the factory using universal finger-joints **4**.

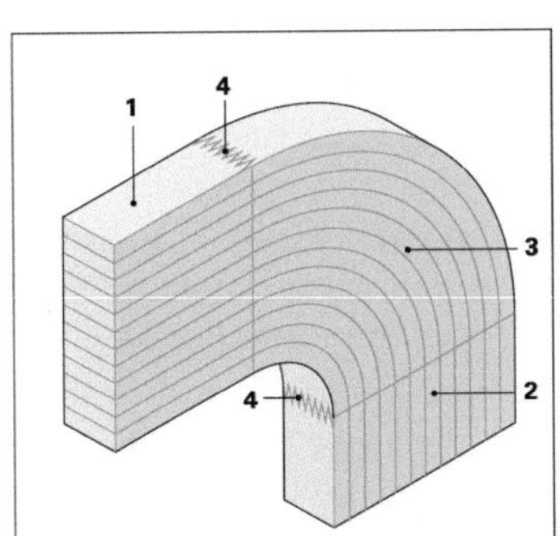

H Bending-stiff portal-frame corner with curved laminated transitional piece **3** between frame beam **1** and post **2**. Finger jointing **4** is not necessary if the element is laminated in one piece.

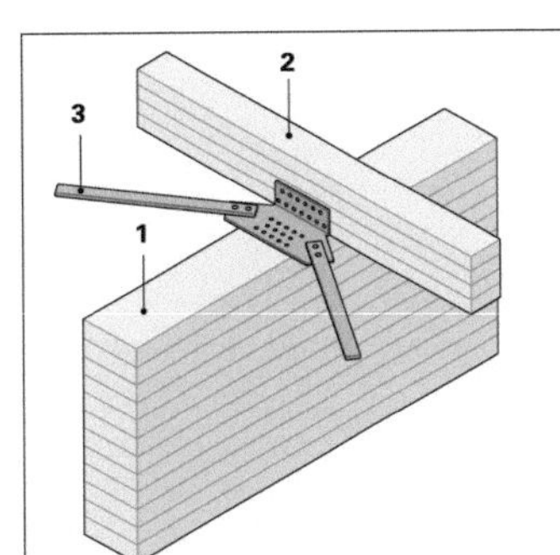

I Node of girder **1**, purlin **2** and wind bracing **3**. The latter is best placed on the level between girder and purlin.

Fig. 8.95 Different constructional node solutions for hall structures in timber

8

The main load-bearing structure of a typical hall in timber construction can basically be executed in two variants:

- The main framework consists of **hinged columns** and **girders**, connected to one another, respectively, with articulated joints (◘ Fig. 8.96 top): Both single and multi-field solutions are possible (◘ Fig. 8.97). The hinged connection between column and girder is relatively simple in terms of constructional design. In the case of the hall girders which are usually slender and high in cross-section and are generally limited to a maximum board-lamella width of 28 cm, care must be taken to ensure that they are secured against tipping. There are various constructional solutions for this (◘ Fig. 8.95**A**–**D**). Furthermore, it must be ensured that the slender girder does not tilt sideways due to lateral buckling of the top chord as a result of bending stress. This task is usually performed by the secondary structure (purlins or timber-panel elements) oriented at right angles to the girder (→**x**).

  For this issue, cf. ***Vol. 1***, *Sect. 21*, Note 6

  Because of the hinged connections, both at the base and the head of the column, the main girder is not stable in its plane (**yz**) from the outset. The same applies to the direction at right angles (**xz**). The usual bracing principle for wooden hall buildings under these conditions is the formation of a roof shear-plate by means of diagonal bracing to which all main trusses are connected with their girders, and the connection of the roof shear-plate to vertical fixed points in the two main directions of the ground plane (→**x**, →**y**). Since, for obvious reasons of usage, no fixed points are allowed within the hall space, these are arranged within the planes of outer walls, namely in this case both within the side walls (**xz**) and the end walls (**yz**).

  Usually, the roof shear-plate is formed by a circumferential ring of diagonal bracings. As a result, horizontal lattice trusses are created on each of the four façade sides, which transfer the wind loads via bending stress to the fixed points, i.e. to the vertical wind bracings arranged at their respective ends, usually at the hall corners, and aligned in the direction of force.

  Vertical wind bracings are preferably arranged in the outermost column fields of the hall, since from there the horizontal forces acting on the façades can be dissipated as quickly as possible into the building ground without first being conducted through the hall structure. However, arrangements in the middle of the façade are also possible, as these allow unobstructed thermal expansion of the construction in the direction of both ends.
- The main framework consists of a **portal frame** (◘ Fig. 8.96 bottom): Also in this case, single- or multi-field solutions are realisable. In the smaller main span direction (→**y**), the portal-frame-like main frames are already hori-

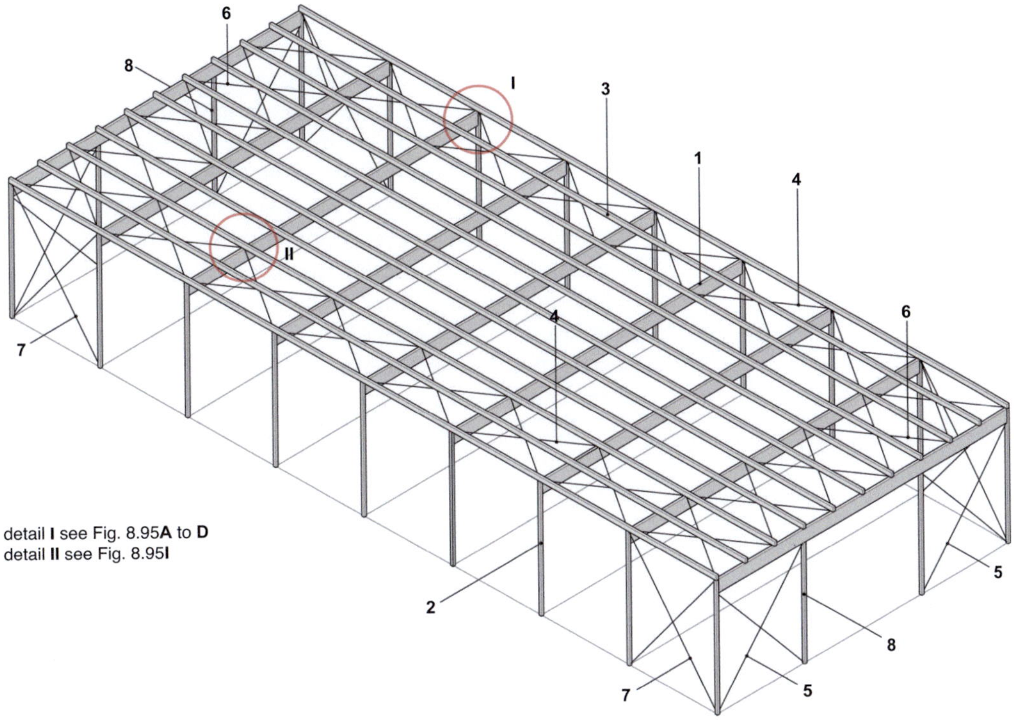

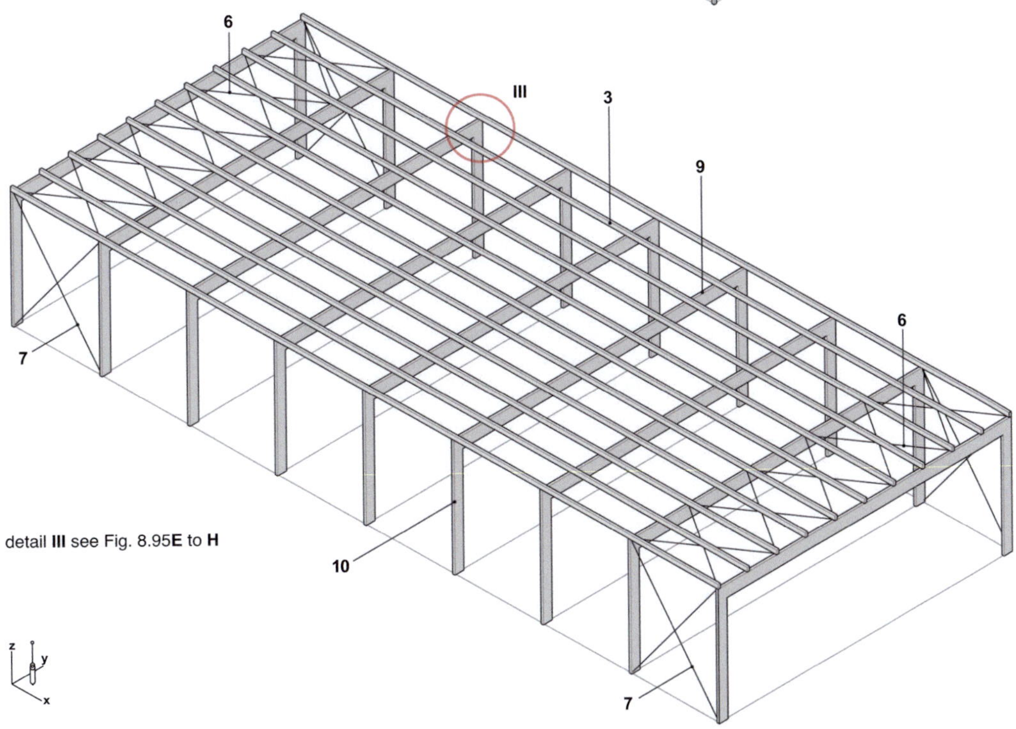

Fig. 8.96 (Top) Example of a conventional wooden hall structure with **hinged columns**. (Bottom) Example of a conventional wooden hall structure with **portal frames**: **(1)** girder (main beam) (glulam), **(2)** column (solid construction timber or glulam), **(3)** purlin (secondary beam) (solid construction timber or glulam), **(4)** horizontal cross-bracing to absorb the wind forces in →**y** and transfer to the bracings **5**, **(5)** vertical cross-bracing to absorb the wind forces in →**y**, **(6)** horizontal cross-bracing to absorb the wind forces in →**x** and transfer to the bracings **7**, **(7)** vertical cross-bracing to absorb the wind forces in →**x**, **(8)** posts for absorbing the vertical force component from the bracing **5**, **(9)** frame beam (glulam), **(10)** frame post (glulam)

zontally stiff within their plane (**yz**). Unlike the previous variant, this wind attack direction (→**y**) is therefore already secured and does not require any further bracing measures. This can prove to be an advantage, as the front façades of the hall (**yz**) can be kept completely open with this variant, for example, for wide entrance-gate openings.

Another advantage of this solution is the slightly smaller overall structural depth of the frame beam compared to the single-span girder of the previous solution, as the moments are distributed more favourably in the former. However, some of the moments are introduced into the frame post, so that its cross-section is much deeper (in the direction of the frame plane, →**y**) than that of the hinged column.

The assembly of the frame corner itself at the construction site is basically possible in timber construction (◘ Fig. 8.95**E**), but it is preferable (as with other construction methods) to prefabricate rigid corners in the factory. The main problem of a rigid connection with the anisotropic material wood lies in the difficulty of transferring flexural tensile and flexural compressive forces between the frame beam and the frame post without generating dangerous transverse compression or, worse still, transverse tension in the wood. This can hardly be solved in a reasonable way either with a continuous frame beam or with a continuous post. Only solutions with mitre cut (◘ Fig. 8.95**F**) or diagonal intermediate pieces (◘ Fig. 8.95**G**) or preferably curved transitions (◘ Fig. 8.95**H**) between frame beam and frame post create suitable geometric solutions to ensure a material-compatible force transmission along the grain, or at least with only slight angulation thereto. In timber construction, mitre cuts as well as joints between frame beams or posts and any connecting pieces are preferably executed as universal finger-joints (◘ Fig. 8.95**F–H**). Since these are factory connections, these solutions can only be realised if the frame is either transported prefabricated as a whole (in the case of the two-hinged frame), in one half (in the case of the three-hinged frame), or at least up to a point of contraflexure.

In the (still) non-braced longitudinal direction of the hall, i.e. at right angles to the portal frameworks (→**x**), which are at risk of tipping, (and only in this direction), the hall must be additionally braced. This is done in a similar way as in the variant previously discussed by diagonal bracing within the plane of the longitudinal walls (**xz**). Similar conditions apply as there.

The individual frame trusses are connected to the fixed points (in →**x**) by the secondary roof construction, this being made in timber construction either of secondary beams (purlins), timber-panel elements, or otherwise of separate coupling rods, which act as pure axially loaded compression and tension rods and do not take over bending moments.

Fig. 8.97 (Left) Trussed hall with hinged columns according to the structural diagram in Fig. 8.96, top. The horizontal bracing can be seen at the top and in the centre roof area. (Right) Connection of the truss of the hall on the left to the column according to the node solution in Fig. 8.95**A**. The two lateral parts of the three-part column cross-section are extended upwards to prevent the narrow truss from tipping over

Under these conditions, wind forces acting on the end façades stress the frames in their weak direction, i.e. transverse to their plane (→**x**), and must therefore be absorbed by horizontal wind bracings, usually localised in the end fields of the roof plane. It is advisable to place vertical and horizontal wind bracings in the same structural bays, preferably in the two end bays, so that, in total, something like continuous stiffening trestles is created.Mixed constructions are also realised from restrained precast-concrete columns with fork head to prevent tilting of the girders which are placed on them on hinged supports (Fig. 8.95**C**). In order to reduce the erection time, the columns can already be prefabricated with cast-on foundations. One advantage of this solution is that no additional bracing structures are required (cf. also Figs. 8.98, 8.99, 8.100, and 8.101).

8.5 Wood-Concrete Composite Construction

The constructional objective of the composite of wood and concrete is, similar to the reinforced-concrete composite, the beneficial combination of very different, at first sight even contradictory, materials in a way that both support each other in their constructional tasks.

The combination of wood and concrete is certainly something unusual for traditionally trained construction professionals, but it is not a completely new development. As early as 1922, Müller submitted a patent for a wood-concrete composite beam; in 1936, Datta carried out investigations on

Fig. 8.98 Fork bearing of a glulam roof truss on a concrete support with tilt protection according to the node solution on Fig. 8.95C

Fig. 8.99 Horizontal roof diagonals for hall bracing as shown on Fig. 8.96

bamboo-reinforced concrete components and, in 1943, Sperrle made a proposal for wood-reinforced concrete beams in hollow-core floors. Always in times of crisis, when steel was scarce and expensive as a tensile insert in concrete, this possibility was seriously considered. The main problem in the pioneering days of the wood-concrete composite was to develop suitable connecting means that ensured the shear bond between the two materials. Probably, the early developments failed because of this problem and therefore could not prevail.

***Vol. 1**, Sect. 6 Ecology*

Particularly against the background of resource-saving and sustainable construction, the environmentally friendly and renewable raw material wood has gained in importance again over the past 20 years. The combination with con-

Fig. 8.100 (Left) Bridge construction made of arched trusses (three-hinged arch). (Right) Apex joint of the arch construction on the left consisting of a hinge-pin connection

Fig. 8.101 (Left) Sports hall with slender glulam trusses on linear bearings. (Right) Hall with glulam girders on precast columns with forked bearings

crete in a composite construction allows wood to penetrate into areas of application that were previously denied to it. This is due, for example, to the improved fire-protection and sound-proofing properties resulting from the use of concrete, and to the increased stiffness of the components, which allows greater wall heights or, in particular, greater floor spans with smaller deformations.

8.5.1 Applications of Wood-Concrete Composite Construction

Wood-concrete-composite construction methods are used today for conversion and renovation measures, e.g. as wood-concrete-composite floors for the renovation of historical wooden floors, but also for the construction of new floor slabs in residential and administrative buildings. Another, relatively new field of application for this composite construction method is bridge construction.

Cf. further below, ▶ *Sect. 8.5.4 Wall Construction Methods*

The so-called **composite formwork construction** method allows the wood-concrete-composite application as a wall shear-plate-like component. These composite construction methods were frequently used in Switzerland, Italy, and Scandinavia. Furthermore, **wood-concrete shell block wall**-construction methods (incl. wood-fibre insulation blocks) ought to be mentioned here.

8.5.2 Fundamentals

Wood-concrete composite structures are structural elements in which a wooden substructure and a topmost concrete slab, usually made of a top layer of concrete poured on site, are joined together in a shear-resistant manner. Wood absorbs the flexural tensile forces with its good tensile strength; the comparatively thin concrete slab acts as a compression chord with its good compressive strength. Due to the elastic, i.e. yielding, bond, the wood also absorbs compressive stresses and the concrete correspondingly tensile stresses, which is why the latter must be reinforced. Nevertheless, the above-mentioned division of labour between the bonding partners essentially applies, which is why moment reversals such as in the case of continuous action and large cantilevers are problematic. The concrete layer can, however, transmit in-plane shear forces well, especially if it is cast as a homogeneous shear-plate on site.

The concrete, usually processed as a jointless slab, is also used to meet important building-physics requirements such as fire protection and sound insulation. In bridge construction, the constructional wood preservation is also an important factor.

For its part, the wood is characterised by its low dead weight combined with high strength. Another important advantage is the aesthetic quality of wood, as viewed from below on soffits of composite floors, which many people regard as a value in itself.

Wood-concrete composite structures usually allow for a high degree of prefabrication. With concrete applied on site, the wooden substructure acts as a lost formwork. Prefabricated systems with modular floor elements are also in use, in which components that have already been manufactured in composite combination are installed dry.

8.5.3 Wood-Concrete Composite Floors

Wood-concrete composite floors (◘ Fig. 8.102) consist of a wooden construction which is connected to a concrete slab on top it in a shear-resistant manner. A basic distinction is made between two types of construction for the floor:

Cf. ***Vol. 3***, *Sect. 20.5.1.4 Wood-Concrete Composite Floors*

Fig. 8.102 Wood-concrete composite floor element on the test bench

- **Solid-wood construction**: Formation of the bottommost tensile zone by means of a surface-forming panel-like cross-laminated timber, board-stack, or veneer-laminated-lumber plate.
- **Beam construction** (ribbed construction): The wooden construction consists of a classic, one-way spanning wooden-beam floor with surface-forming wooden boarding.

As a rule, top concrete is poured on site (thickness approx. 8–15 cm), or alternatively factory-made wood-concrete floor elements are used. To protect the wooden construction from the cement slurry during the pouring of the concrete, a separating layer is often inserted between the wood and the concrete. Floors with both one-way and two-way load transfer can be realised. The composite fasteners must be oriented according to the main stress direction (see, e.g. Fig. 8.103**B**, **C**); likewise, the grain of the wooden substructure. For one-way spanning floors, board-stack or veneered laminated timber elements are suitable; for two-way spanning ones, cross-laminated timber elements (Fig. 8.103**D**).

8.5.4 Wall-Construction Methods

The **composite formwork construction** should also be mentioned as a wall-construction method: The wooden formwork remains on the walls which are filled up with concrete. The construction method is used by two European manufacturers (*Kewo* and *Girhammar*). Advantages are the low construction costs and the high degree of prefabrication. All installation elements are inserted into the formwork at the factory. External walls receive additional thermal insulation. The walls fulfil a fire resistance up to **F 180**.

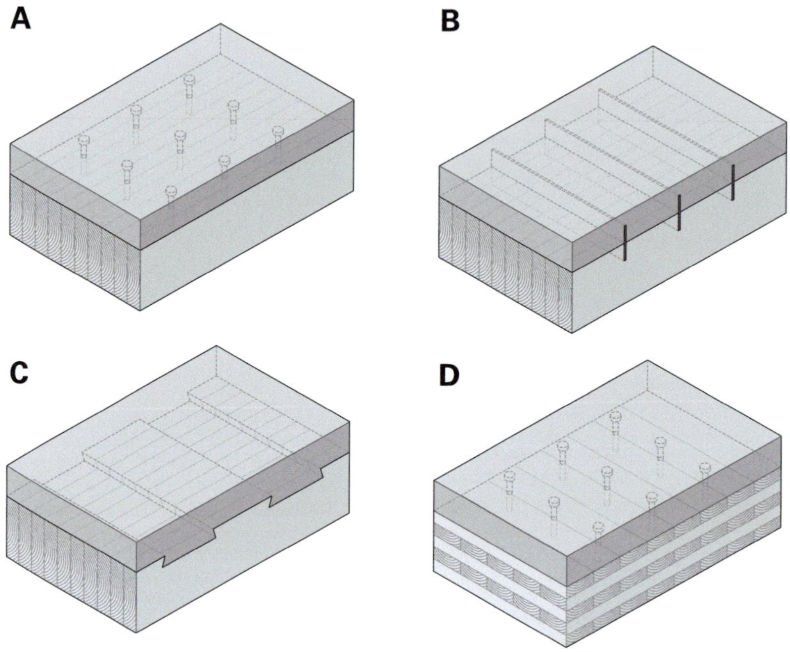

Fig. 8.103 **A** Wood-concrete composite slab from board stacks, spanning one-way; shear bond by pin-shaped composite fasteners. **B** Wood-concrete composite slab from board stacks, spanning one-way; shear bond through slotted flat steel strips. **C** Wood-concrete composite slab from board stacks, spanning one-way; shear bond through form fit (notches). **D** Wood-concrete composite floor made of cross-laminated timber, spanning two-way; shear bond through pin-shaped fasteners

8.5.5 Shear Bond

Cf. ***Vol. 3***, *Sect. 20.5.1.4 Wood-Concrete Composite Floors*

The fastener is of decisive importance with regard to load-bearing capacity and deformational behaviour. The following options of shear bond between wood and concrete can be distinguished today:

- mechanical connections with **pin-shaped connecting means** (Fig. 8.103**A**);
- mechanical connections with **special connectors** (flat steel strips in saw grooves, glued-in expanded metal strips) (Fig. 8.103**B**);
- connections by pure **form closure** without additional aids (notches) (Fig. 8.103**C**);
- connections by **adhesive bonding**; these are still under development.

The combination of the single variants is also possible and reasonable (e.g. notches and pin-shaped connections).

8.6 Shell Construction Modes

Cf. ► *Sect. 9.3.6 Lattice Shells*

Even if shell structures in wood, analogously to those in steel, do not represent a standard construction method in the sense of a design type for mass construction, they should nevertheless be addressed at this point, at least in overview, because of their constructional interest and because of the diverse material-related constructional solutions that are used in them. Shell structures in wood, especially in wood-based materials, are also of current interest because, within the framework of parametric design and the close coupling of planning and production, as has been implemented in an increasing number of examples in construction practice in recent years, they have manifested themselves in a whole number of experimental construction projects that show further very promising lines of development for modern timber construction.

8.6.1 Material and Load-Bearing Behaviour

As with curved arches, the difficulty of assembling double-curved shell structures from fundamentally rectilinearly growing rod-shaped wooden components also arises. Adapting to the curvature requires either the use of very slender and therefore flexible bars (a solution used for so-called slatted lattices) or otherwise working with short, straight bar sections that are connected at slight angles to form a continuous shell with the aid of numerous nodes.

In order to enable the two-way load transfer typical for shells, certain precautions in terms of constructional design must also be taken with the anisotropic material wood:

- On the one hand, a laminar shell can be produced by connecting layers of mutually butting wooden rods oriented in two main directions, in which biaxial membrane forces can always be transferred along the grain, respectively, in at least one of the layers. This results in a pure load-bearing surface structure[18] without discontinuities or, at best, only slightly protruding stiffening ribs. For the purpose of the classification of envelopes chosen in this book, these constructions are considered as **full-leaf systems**. Suitable base materials for this solution are largely isotropic components such as cross-laminated timber. These surface components are capable of absorbing both tangential normal forces and tangential shear forces. ▶ *Sect. 2.5 Ribbed Systems*
- On the other hand, it is possible to create—analogously to steel lattice shells—a framework of at least biaxially oriented straight or slightly curved wooden bars, the spaces between which are closed with a surface-forming secondary supporting structure. For the purpose of the classification of envelopes chosen in this book, these constructions

18 In the present context, in contrast to lattice shells, the term load-bearing surface structure is used somewhat more narrowly than in the technical literature, where the former are sometimes also referred to as surface structures. The transitions are fluent.

are considered to be **ribbed systems**. In these systems, rib bars are stressed by normal forces and have a grain pattern aligned along their system axis. They can therefore be made of solid wood, glued laminated timber, or layered board lamellae. The space-enclosing secondary load-bearing surface structure, which spans from rib to rib, can also be made of wood, either as boardings or wood-based materials. Often this two-dimensional sheathing also takes over the shear stiffening of the framework in tangential direction.

A decisive factor for the load-bearing behaviour of the shell is both the shape as well as the bearing of the structure in a manner suitable for the shell. Preference is often given to hanging surfaces, which are either determined experimentally on physical models or today mostly with the aid of digital models. These forms create the most favourable force conditions. Due to the changing load patterns to which the shell is exposed through various time-dependent effects such as wind or snow, these types of shells usually require a minimum bending stiffness, despite the prevailing membrane effect, in order to absorb the effect of the difference between the actual load pattern and the load pattern according to which the ideal membrane surface was determined. As a rule, this manifests itself in an increased cross-sectional depth of the rib.

8.6.2 Early Wooden Shells

As early as the sixteenth century, the French architect and theorist Philibert de l'Orme proposed shell-like wooden load-bearing structures consisting of composite curved ribs that were laterally stabilised by cross-bars (◘ Fig. 8.104). Although this did not yet result in a true two-way load transfer,

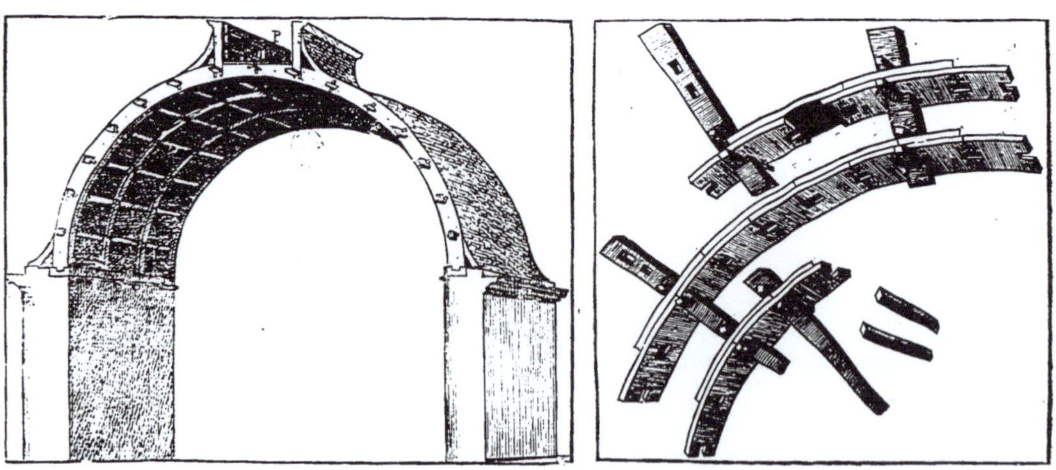

◘ **Fig. 8.104** Philibert de l'Orme: shell-like wooden construction from an assemblage of curved main ribs and stiffening cross-bars. The latter were inserted through the main ribs and wedge-locked on both sides

as is characteristic for a shell, because the force flowed mainly through the main ribs in the direction of the meridian, this early design solution nonetheless already showed shell-typical features such as the polygonal composition of curved ribs made from short, curvilinear individual segments as well as a grid structure that could be assembled on site.

At the beginning of the twentieth century, Friedrich Zollinger developed a genuine shell construction method of structurally equal-ranking, diagonally aligned sets of bars, composed of always the same, industrially prefabricated plank segments (◘ Fig. 8.105). The restriction of the shell shape to cylindrical surfaces, which resulted from the use of always the same rib bars, as well as the yielding bolt connection and shrinkage deformations of the solid-wood bars proved to be serious disadvantages of this design. Only modern timber construction techniques have overcome these difficulties and have recently produced new wide-span Zollinger shells (◘ Fig. 8.106).

Cf. also ***Vol. 3***, *Sect. 14.4.3.2 Manufacture of the Curved Shell Framework*

8.6.3 Recent Wooden Shell Structures

The earlier historical developmental solutions of newer shell structures are all lattice shells, such as the Zollinger design mentioned above (◘ Fig. 8.105), as two-dimensional wooden materials which could have absorbed the resulting forces were initially unavailable. This has changed fundamentally in recent years due to the development of modern wood-based materials, so that recently also pure load-bearing surface structures made of board material have been created. These are still small experimental pavilion constructions, but they can be regarded as harbingers of a rapid further development of this construction method.

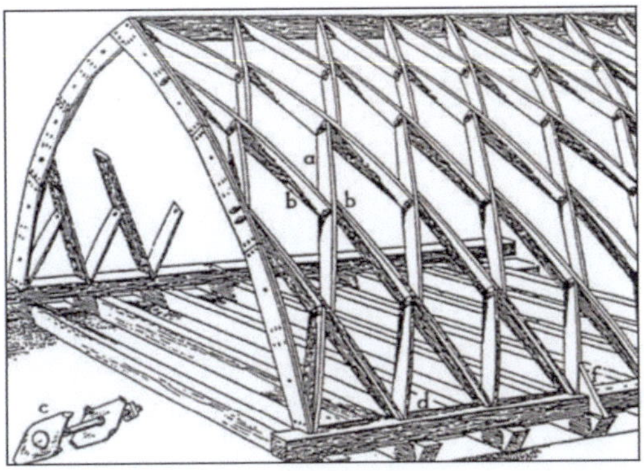

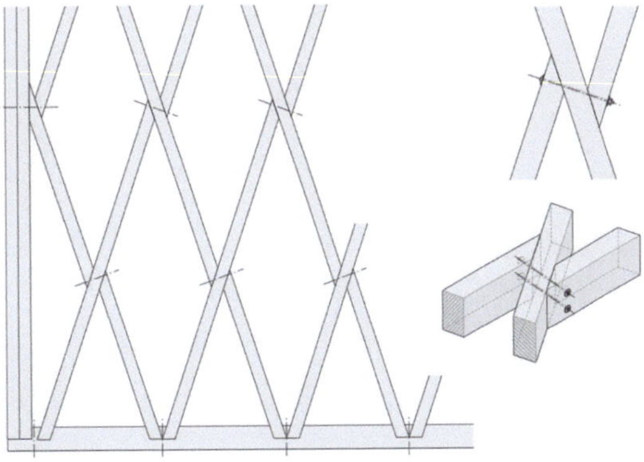

◘ Fig. 8.105 (Left) Zollinger construction: A genuine lattice shell is created by assembling short planks, cut curved at the top, which run over a maximum of two lozenge-shaped fields and are butted at the ends. (Right) Zollinger construction: The nodes, at which one piece of wood always passes through and two laterally abutting ones are interrupted, are resolved by slightly offsetting the latter in such a way that bolts can be inserted crosswise

From a design point of view, a fundamental challenge of shell construction in wood is, as noted, the creation of a two-way load-bearing structure from a rod-shaped, markedly anisotropic material like wood. In a short overview, different constructional approaches, in particular of node formation, which have produced different shell construction methods, will be discussed in the following.

8.6.3.1 Modern Zollinger Construction Modes

Characteristic for this construction method in its original form is the node consisting of a continuous wooden rod and two laterally abutting ones (◘ Fig. 8.105 right). The offset between both butted wooden members creates a slight eccentricity, but, in return, allows the three connecting partners to be easily coupled with a simple plugged-through bolt connection. In addition to the difficulties already mentioned above, this nodal construction does not allow good transmission of the axial force of the abutted bars from one side to the other, not only because of the geometric offset, but also because of the necessity to transmit this force via transverse compression of the passing bar. This is because the compressive force is transmitted via pressure contact between the lateral ribs and the continuous rib, not via the bolt connection. Through an improved, stiffer design of the nodes and an execution in glulam construction, this difficulty has been overcome in current shell constructions (◘ Figs. 8.106 and 8.107). The shear stiffening

◘ **Fig. 8.106** Modern Zollinger construction: Lattice shell of the Rostock Fair exhibition hall with 65 m span (Schlaich, Bergermann & P)

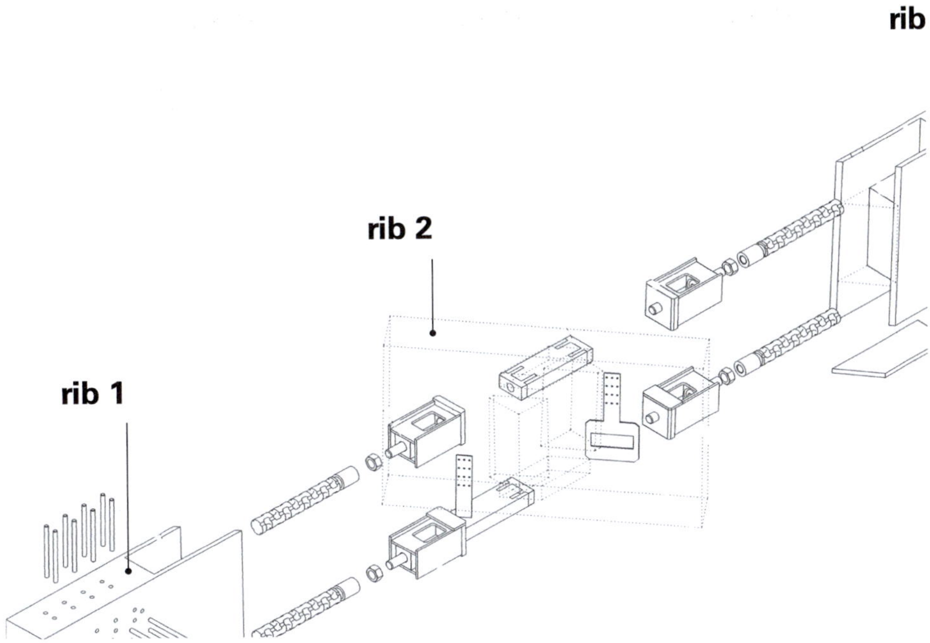

Fig. 8.107 Node detail of the lattice shell of the exhibition hall on Fig. 8.106. One rib (rib **2**) is continuous, the other (rib **1**) is butted. The normal force is transmitted in the node by means of steel pieces between the butted sections (rib **1**) along the grain without causing transverse compression of the crossing timber (rib **2**)

of the geometrically non-rigid lozenges of the lattice is provided by the top-side sheathing.

8.6.3.2 Slatted Lattice Shells

The shell is made of a regular quadrilateral-meshed slatted frame with articulated nodes. The thin wooden bars cross each other in two or more layers at the nodes and are therefore continuous. During erection, the elastic slender wooden parts bend in two directions and thus create the desired biaxial curvature without further ado. Thanks to their slenderness and elasticity, they can also withstand, apart from the bending, the associated twisting. The initially laid-out flat square-shaped meshes deform, in the process of jacking-up the shell into its design shape, into quadrangular lozenges, so that with the arbitrary, if necessary, locally different stepless angle change of this quadrangle, eventually practically any shape of the shell can be produced. Finally, the freely deformable lozenges are shear-stiffened with diagonal ropes to create a self-supporting shell. The fine-meshed latticework comes very close to a pure surface shell. The extremely thin rods can be bent or twisted into the shell curvature without difficulty. The surface can be covered with textiles, for example.

▶ *Sect. 4.3.2.5 Dome from Bars*, in particular ▶ Figs. 4.189, 4.190, and 4.191

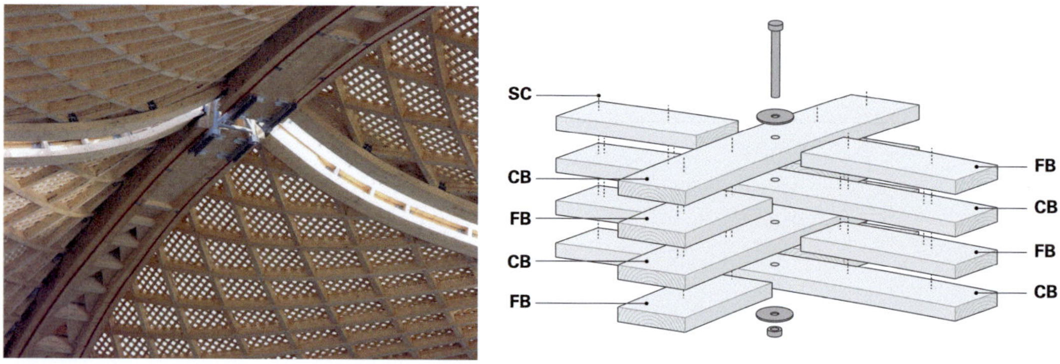

Fig. 8.108 (Left) **Board-rib construction**: EXPO Roof Hanover (arch.: Herzog & P; Ing.: Julius Natterer). (Right) **Board-rib construction**: node formation from layered flat board layers with alternately continuous and butted single boards: **CB** carrying board, **FB** filling board, **SC** screw connection between the board slats

8

8.6.3.3 Board-Rib Construction[19]

The underlying concept of this construction method is, analogously to the slatted frame construction, the transmission of the force in two axes through the node by stacking flat layers of boards in such a way that alternately one board always passes through the node (Fig. 8.108 right). The shallow board lamellae, which are flexible in the direction of their narrow side, adapt to the shell curvature without major bending edge stresses. Only the continuous supporting boards are available as force-conducting cross-section. The butted filling boards do not transmit any axial force, but are joined together by bolting to the supporting boards between the nodes to form a multilayer lattice bar with limited bending stiffness. A top layer of boards, plank formwork or battens laid in gaps closes the surface and stiffens the grid against shear (Fig. 8.108 left). The shell is assembled in layers on site on a falsework which determines the shell shape.[20] The shell geometry usually follows a hanging form.

8.6.3.4 Tensile-Stressed Shells

Wood is a material with approximately the same tensile and compressive strength. These material properties are therefore not only favourable for the most frequently encountered dome-like, i.e. compressive-stressed shell structures, but also for tensile-stressed structures. On the other hand, the transmission of tensile forces at connections between timber com-

---

19 We adopt the designation from the dissertation by Andreas Scholz (2004). He argues (correctly, in our view) that the term board-stack construction used by Julius Natterer is misleading, since this term is used today rather for board-stack floors or walls made of cross-sections connected at right angles to the component surface; cf. Natterer J et al. (2000) *Holzrippendächer in Brettstapelbauweise–Raumerlebnis durch filigrane Tragwerke*.

20 Natterer et al. (2000), p. 3.

Fig. 8.109 Hanging shell made of unjointed curved ribs from glulam (brine bath Bad Dürrheim, arch.: Geier & Geier; Ing.: Klaus Linkwitz)

ponents is always difficult. While this weakness of the material can be alleviated by layered lamella nodes, as described above, for example in board-rib constructions such as the Expo roof in Hanover, modern glued construction technology in particular has enabled convincing solutions for hanging shells: Glued laminated timber ribs can be produced without joints in almost any length and with practically any curvature, including torsion.

This solution was used, for example, for the hanging shell of the brine bath in Bad Dürrheim (Fig. 8.109). Together with the main ribs, which are subject to tensile stress, a transversely oriented secondary set of ring ribs and a diagonally running, space-enclosing and shear-strengthening boarding result in a flat, two-way acting overall load-bearing structure.[21]

8.6.3.5 Construction Modes with Steel Nodes

Numerous wooden shells, some of which have a very wide span, consist of a latticework of bars, based on various possibilities of geometrically subdividing a dome surface, which are connected with high-strength steel node elements to form a framework. As it is difficult to pass the force in the node on to the rib running through it, in order to transfer it from side to side, due to the sensitivity of wood to the necessarily resulting transverse compression and tension, all individual bars

Cf. also ▶ *Sect. 1.3.2.2 Sphere*

Cf. the complexity of the connection on Fig. 8.107, where such a force transmission takes place

21 Scholz A (2004) *Beiträge zur Berechnung von Flächentragwerken in Holz*, dissertation at TU Munich, p. 16.

8

Fig. 8.110 Triangulated lattice shell of the *Tacoma Dome* from the 1980s. Glulam lattice bars with steel nodes (SPS+ Architects)

are joined at the node in these construction methods. The ribs themselves are usually made of glulam. To absorb the tangential shear forces, top-side sheathing, planking, or boarding is again possible. With transparent or translucent covers, diagonal braces in the frame faces are also possible (Fig. 8.111 right). Alternatively, a triangulated latticework is also suitable to give the shell shear stiffness (Fig. 8.110).

A representative example of this shell construction, the *Konohama Dome* in Miyazaki City, Japan, is shown on Fig. 8.111.

8.6.3.6 Genuine Load-Bearing Surface Structures

In addition to the ribbed shell structures, as previously considered, a number of rib-less, genuine surface shell structures have also appeared in recent years, a form of structure that was previously confined to concrete shells. These are small experimental buildings, most of which have not yet reached real building size (Figs. 8.112, 8.113, and 8.117). As a result, it remains uncertain which of the implemented structural solutions will prevail in regular construction tasks in the future. Nevertheless, they show promising ways for a new construction method, and especially for a new computer-based methodology of design, construction, and execution.

Digital techniques are the focus of these construction methods. The form-finding and precise form definition, the structural analysis, the design development of the connections, the data sets for the executing robots as well as for the robotic assembly and the monitoring and timely feedback of sensor data

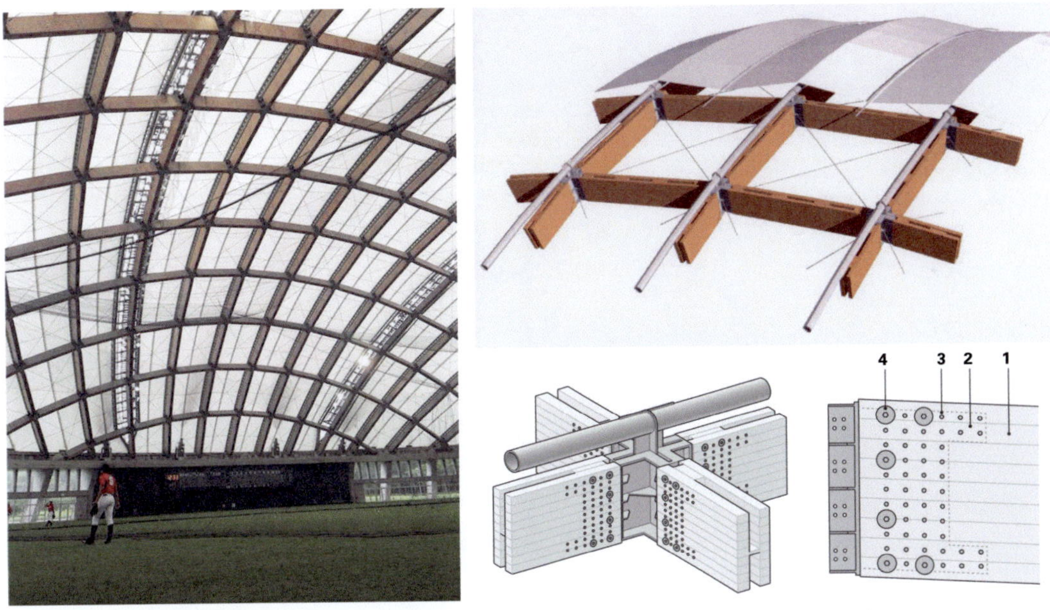

Fig. 8.111 (Left) *Konohama Dome*: Interior view; covering of translucent membrane material. The mesh size is 8×8m or smaller (planning: Structural Design Division Daiken Sekkei). (Right) *Konohama Dome*: Construction of the lattice shell. Lattice bars consist of slender twin cross-sections of glulam with filler woods. Slotted steel connecting lugs (see bottom right) are inserted into the steel cross-shaped nodes on site and fixed (bottom left). The transmission of force from wood to steel takes place via rod dowels. Shear reinforcement via diagonal bracing. The largest span of the shell is 200 m: (**1**) glulam lattice bar, twin profile 150×1200mm, (**2**) inlaid connection plate, (**3**) rod dowel, (**4**) screw bolt

Fig. 8.112 Surface shell composed of individual polygons (Pavilion of the State Garden Show Schwäbisch-Gmünd; ICD/ITKE, University of Stuttgart)

to manufacturing tolerances are understood as an integrated process in which the interdisciplinary interaction of architect, structural engineer, other specialist consultants, and executing companies work closely together from the earliest planning

Fig. 8.113 The polygonal wood-based panels made of plywood are form-fit connected by interlocking. The shell surface is created by faceting into a polyhedron surface. Three edges converging at one point do not lie in one plane and thus create the shell curvature

phase onwards. However, the application of this methodology is not limited to this particular construction method, but has every prospect of becoming established in the near future also in other areas of the construction industry in the long term.

The experiments carried out so far show different, partly very innovative methods of creating load-bearing surfaces from individual elements using various geometric principles of surface formation (Fig. 8.113).
The previously existing limitations to the definition of biaxially curved surfaces in particular, which not so long ago could only be defined either by simple geometric generation laws or by experimental methods on the physical model, have largely been overcome by the use of parametric CAD programmes. In analogy to physical models of form-finding, the designer manipulates various boundary conditions (parameters), which create a geometrically precisely defined surface via a mathematical algorithm (Fig. 8.114).

The different joining techniques that are used on the abutting edges of these facetted load-bearing structures made almost entirely of planar panel material are also remarkable (Figs. 8.113 and 8.114). The locally very different geometric

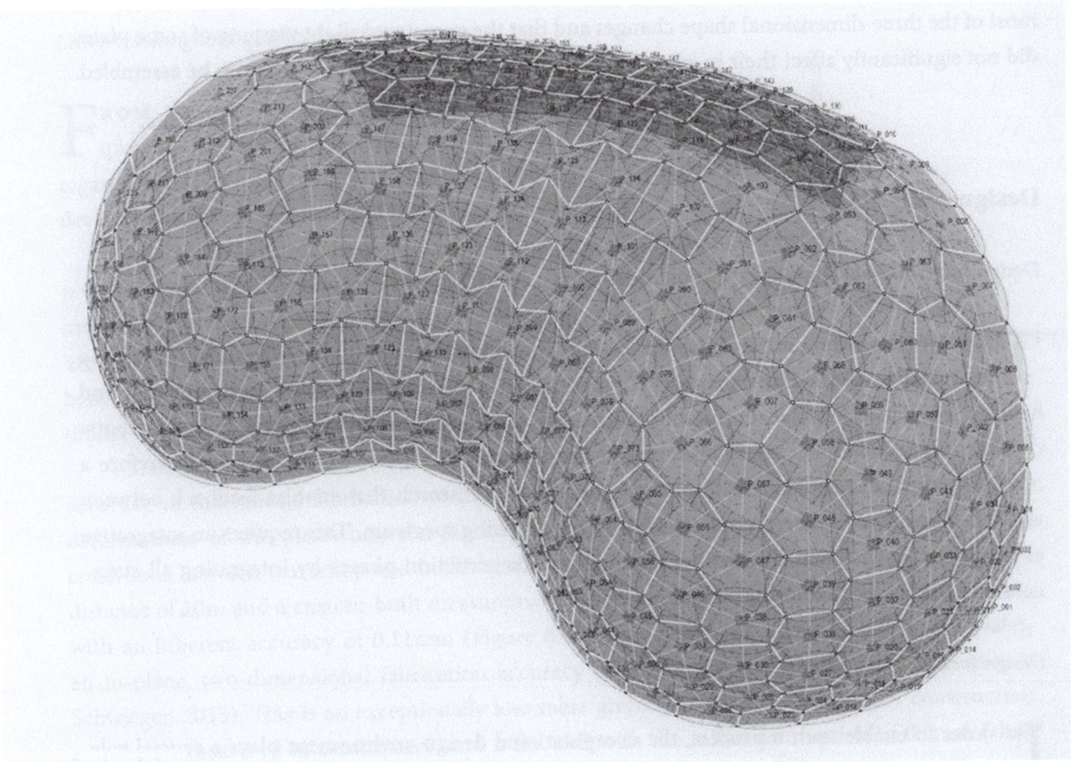

Fig. 8.114 The free form of the shell is facetted: Each dark point on the surface defines a tangent plane that intersects with the adjacent ones, thus defining the polygonal outline of the individual plate. The surfaces are planar; the edges are straight; the polygon shapes are individual, as are the dihedral angles (angle between two planes)

boundary conditions of these connections due to the complex geometry require an individualised production of the connections (Fig. 8.115). This is done using robots (Fig. 8.116).

It is an interesting phenomenon that features of the once highly developed handcraft techniques in timber construction, which were largely disappearing due to industrialisation, are now reappearing through this innovative use of digital technology for design and execution—albeit in a slightly modified, contemporary form. Geometrically demanding, extraordinarily precise form-fit connections, such as those traditionally used in East Asian timber construction, are now being digitally planned and executed again (Fig. 8.116). They provide a possibly future-oriented counter-model to the basic approach of contemporary conventional engineering timber construction, namely, to replace wood, the natural irregularities of which are difficult to control, with a much stronger and better controllable material, namely, steel, at those points where conditions become critical, i.e. above all at the joints. In a way, this means a return to the original, strictly material-oriented principles of artisan timber construction (cf. also Figs. 8.117, 8.118, and 8.119).

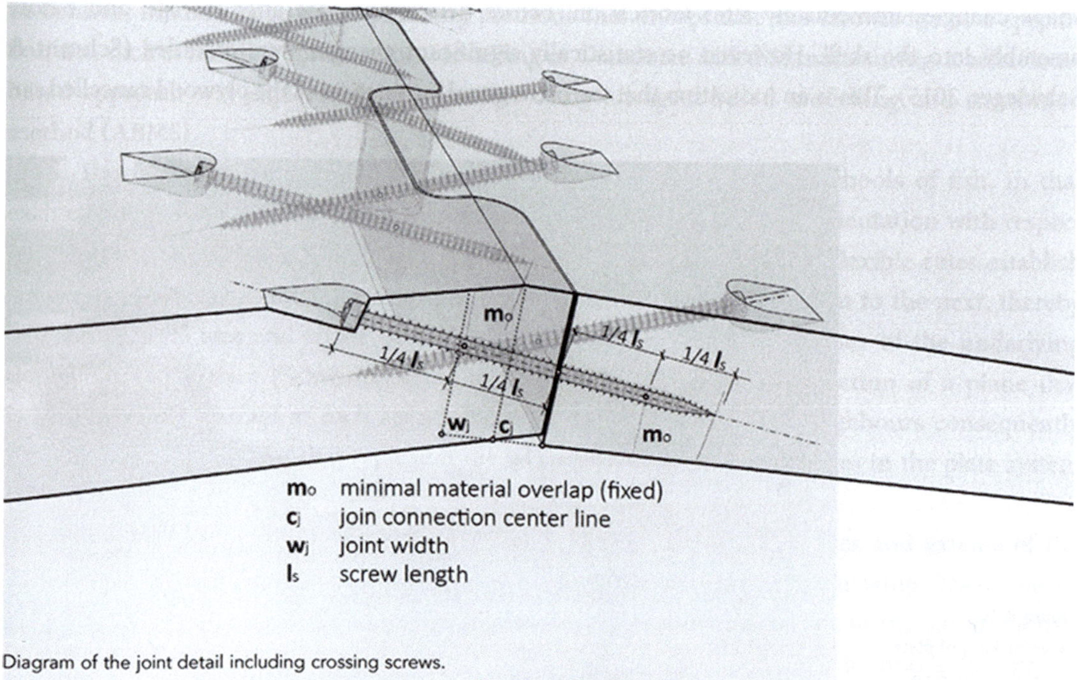

8

Fig. 8.115 Combination of form-fit connection by toothing (for tangential shear forces) and bolting (for right-angled shear forces); both the panel edges and the screw pockets are milled by the robot

Standards and Guidelines

DIN 436: 1990-05 Square washers for use in timber constructions

DIN 440: 2001-03 Washers—With square hole, especially for timber constructions

DIN 1052: Timber structures—Design of timber structures

Part 10: 2022-10 Additional provisions for fasteners and non-European regulated bonded products and types of construction

DIN 4103: Internal non-loadbearing partitions

Part 4: 1988-11 Partitions with timber framing

DIN 18203: Tolerances in building constructions

Part 3: 2008-08 Building components of wood and derived timber products

DIN 20000: Application of construction products in structures

Part 3: 2022-02 Glued laminated timber and glued solid timber according to *DIN EN 14080*

Part 5: 2022-11 Strength graded structural timber with rectangular cross-section

DIN 68364: 2003-05 Properties of wood species—Density, modulus of elasticity and strength

EN 1995: Eurocode 5—Design of timber structures

Part 1-1: 2023-10 General rules and rules for buildings, prEN

■ **Fig. 8.116** Robotic production of the finger-jointed plate edges. While comparable joints in traditional joinery (e.g. dovetail joints) were only possible in the direction of the grain (because otherwise the prongs would break off), the isotropy of the plywood material instead allows circumferential pronging

■ **Fig. 8.117** Mounting the facets of the wooden shell on ■ Fig. 8.112

Fig. 8.118 Production of the edge connection as shown on Fig. 8.115

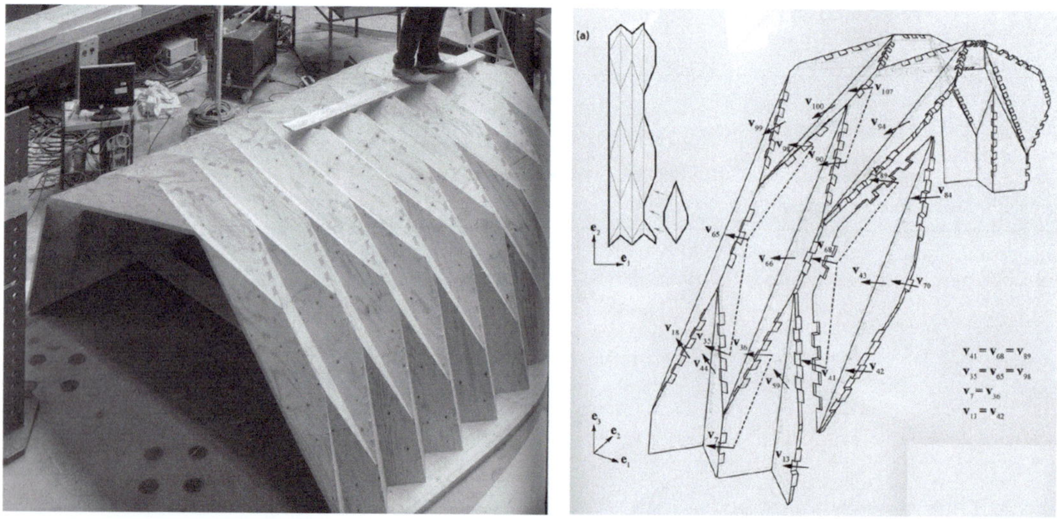

Fig. 8.119 (Left) Origami-like folded structure made of plywood boards (experimental building of the IBOIS Institute at EPFL). (Right) Interlocking of the edges between adjoining panels at the building on the left. Like at real form-fitting connections, there is only one single mounting direction, into which however each three edges are to merge (here each marked by $\mathbf{V}_i$). The tines must be cut accordingly

EN 336: 2013-12 Structural timber—Sizes, permitted deviations

EN 338: 2016-07 Structural timber—Strength classes

EN 380: 1993-10 Timber structures; test methods; general principles for static load testing

EN 594: 2011-09 Timber structures—Test methods—Racking strength and stiffness of timber-frame wall panels

EN 912: 2011-09 Timber fasteners—Specifications for connectors for timbers

EN 14080: 2013-09 Timber structures—Glued laminated timber and glued solid timber—Requirements

EN 15497: 2014-07 Structural finger-jointed solid timber—Performance requirements and minimum production requirements

EN 26891: 1991-07 Timber structures; joints made with mechanical fasteners; general principles for the determination of strength and deformation characteristics

EGH Holzbau Handbuch, Reihe 0, Teil 5, Folge 1: 2008-01 Holzhäuser—Werthaltigkeit und Lebensdauer

EGH Holzbau Handbuch, Reihe 1, Teil 1, Folge 5: 2006-09 Holzkonstruktionen in Mischbauweise

EGH Holzbau Handbuch, Reihe 1, Teil 1, Folge 7: 2015-02 Holzrahmenbau

EGH Holzbau Handbuch, Reihe 4, Teil 1, Folge 1: 2008-12 Holz als konstruktiver Baustoff

EGH Holzbau Handbuch, Reihe 4, Teil 2, Folge 1: 2013-02 Konstruktionsvollholz KVH, Duobalken, Triobalken—Technische Informationen—Überarbeitete Fassung auf der Grundlage der DIN EN 1991-1-1: 2010 (Eurocode 5-5-1)

EGH Holzbau Handbuch, Reihe 4, Teil 2, Folge 2: 2014-09 Herstellung und Eigenschaften von geklebten Vollholzprodukten

EGH Holzbau Handbuch, Reihe 4, Teil 6, Folge 1: 2012-07 Bauen mit Brettsperrholz—Tragende Massivholzelemente für Wand, Decke und Dach

Bibliography

Becker K, Rautenstrauch K (2012) *Ingenieurholzbau nach Eurocode 5 Konstruktion, Berechnung, Ausführung*. Berlin, Ernst & Sohn

Cheret P (ed) (2014) *Urbaner Holzbau: Handbuch und Planungshilfe; Chancen und Potenziale für die Stadt*. Berlin, DOM

de l'Orme P (1561) *Nouvelles inventions pour bien bastir et a petits fraiz, trouvées n'aguères par Philibert de l'Orme. A Paris, de l'imprimerie de Fédéric Morel*

de l'Orme P (1576) *L'Architecture de Philibert de L'Orme conseillier & aumosnier ordinaire du roy, & abbé de S. Serge lez Angiers. A Paris, chez Hierosme de Marnef, & Guillaume Cavellat*

de l'Orme P et al (1626) *Architecture de Philibert de l'Orme oeuvre entiere contenant onze livres, augmentée de deux; & autres figures non encores veuës, tant pour desseins qu'ornemens de maison, avec une belle invention pour bien bastir, & à petits fraiz tres-utile pour tous architectes, & maîstres iurez audit art, usans de la regle & compas. A Paris, chez Regnauld Chaudiere*

Giedion S (1998) *Raum, Zeit, Architektur*, 5th ed. Zurich, Munich, London

Hugues T, Steiger L, Weber J (2012) *Holzbau: Details, Produkte, Beispiele*. Munich; Detail, Institut für Internationale Architektur Dokumentation

Iimura Y, Kurita S, Ohtsuka T (2004) *Reticulated Timber Dome Structural System Using Glulam with a Low Specific Gravity and its Scalability*.

Jeska S et al (2015) *Neue Holzbautechnologien—Materialien, Konstruktionen, Bautechnik, Projekte*. Basel, Birkhäuser

Kaufmann H et al (2011) *Bauen mit Holz—Wege in die Zukunft*. Munich, Prestel

Kaufmann H, Krötsch S, Winter S (2017) *Atlas mehrgeschossiger Holzbau*. Munich, Detail Business Information GmbH

Kopff B (2018) *Holzschutz in der Praxis: Schnelleinstieg für Architekten und Bauingenieure*. Wiesbaden, Springer Vieweg

Kudla K (2017) *Kerven als Verbindungsmittel für Holz-Beton-Verbundstraßenbrücken*. Diss. at the University of Stuttgart

Lückmann R (2018) *Holzbau: Konstruktion, Bauphysik, Projekte*. Kissing, WEKA

Menges A, Schwinn T, Krieg OD (2016) *Advancing Wood Architecture: A Computational Approach*. Abindon, Taylor and Francis

Natterer J B, Müller A, Natterer J (2000) *Holzrippendächer in Brettstapelbauweise—Raumerlebnis durch filigrane Tragwerke*. Bautechnik 77(11), pp. 783–792.

Natterer J B, Herzog T, Schweitzer R, Volz M, Winter W (2003) *Holzbau Atlas*. 4th ed., newly ed. Basel, Birkhäuser

Pérouse de Montclos J M, De l'Orme P (2000) *Philibert De L'Orme architecte du Roi 1514-1570 Jean-Marie Pérouse de Montclos*. Paris, Mengès.

Pfeifer G, Liebers A, Reiners H (1998) *Der neue Holzbau—Aktuelle Architektur—Alle Holzbausysteme—Neue Technologien*, Munich

Phleps H (1967) *Alemannische Holzbaukunst*, Wiesbaden

Schmidt P et al (2012) *Holzbau nach EC 5*. Cologne, Werner bei Wolters Kluwer

Scholz A (2004) *Ein Beitrag zur Berechnung von Flächentragwerken aus Holz*. Munich, Technical University of Munich

Seike K (1970) *The Art of Japanese Joinery*, New York

Steiger L (2013) *Basics Holzbau*. Basel, Birkhäuser

Taut B (1997) *Das japanische Haus und sein Leben*, Berlin

Wachsmann K (1930) *Holzhausbau*, Berlin

Wachsmann K (1959) *Wendepunkt im Bauen*, Wiesbaden

Warth O (1900) *Die Konstruktionen in Holz*, Leipzig

Weinand Y (2017) *Neue Holztragwerke—architektonische Entwürfe und digitale Bemessung*.

8

Steel Construction

J. L. Moro, *Building-Construction Design – From Principle to Detail*,
https://doi.org/10.1007/978-3-662-71006-7_9

Contents

9.1 History of Iron and Steel Construction

An early flowering of the iron industry began in India around 2000 BCE. The Qutub column near Delhi consists of almost chemically pure pig iron. It was forged from individual iron blocks. It dates from the ninth century BCE; the weight of the column is 17,000 kg, its length is about 16 m. Up to now, there are no signs of corrosion.

***Vol. 1**, Sect. 16.1 Steps of Historical Development* as well as *Sect. 22.1 History of the Manufacture of Iron and Steel Products*

Iron was not used as a primary building material in preindustrial construction technology, but merely as a material for special applications, e.g. connecting elements in timber construction, or tie rods and clamps in masonry construction. The early manufacturing methods only allowed to produce iron parts with relatively brittle properties in limited dimensions. Iron remained a scarce, expensive, and coveted material for centuries because of its complex and difficult production.

9.1.1 Bridge Construction During the Industrial Revolution

This changed fundamentally with the introduction of new industrial manufacturing processes in the eighteenth century. The bridge in Coalbrookdale over the Severn from 1779 is the first important iron construction of the industrial age (◘ Fig. 9.1). Except for the piers, the complete load-bearing structure consists of iron. The span of the arched bridge is 100 ft (30 m). In Germany, the first steel bridge was built at Labbaan in Lower Silesia in 1796. The use of iron in Central Europe, compared with England, did not occur until later due to the delayed industrialisation.

New types of bridges soon followed, using the tough building material wrought iron:

- Early English and American chain bridges, such as the **Bangor Bridge** by Thomas Telford, with a span of 173 m.

  After the replacement of the chain by the rope, the span of bridge structures increased considerably within a very short time: In 1870, A. Roebling succeeded in reaching the 500 m limit with the **Brooklyn Bridge**.
- Another bold engineering achievement was Stephenson's first wide-span girder bridge, the **Britannia Bridge**, crossing the Menai Straits in Wales (◘ Fig. 9.2). The superstructure consists of a 9 m high box girder from steel plates and rolled profiles.
- In 1859, Isambard Kingdom Brunel built the **Saltash Bridge**, also known as the Royal Albert Bridge, in Plymouth as one of the first tubular truss structures (◘ Fig. 9.3).

By the middle of the nineteenth century, engineers had developed all essential constructional and structural concepts for

Fig. 9.1 The world's first iron bridge over the Severn at Coalbrookdale in Wales, the heart of the early British iron industry. Designed and built in 1779 by Thomas Pritchard and Abraham Darby III. The detail design of the bridge construction (right) is still characterised by the traditional understanding of form-fitting carpentry connections

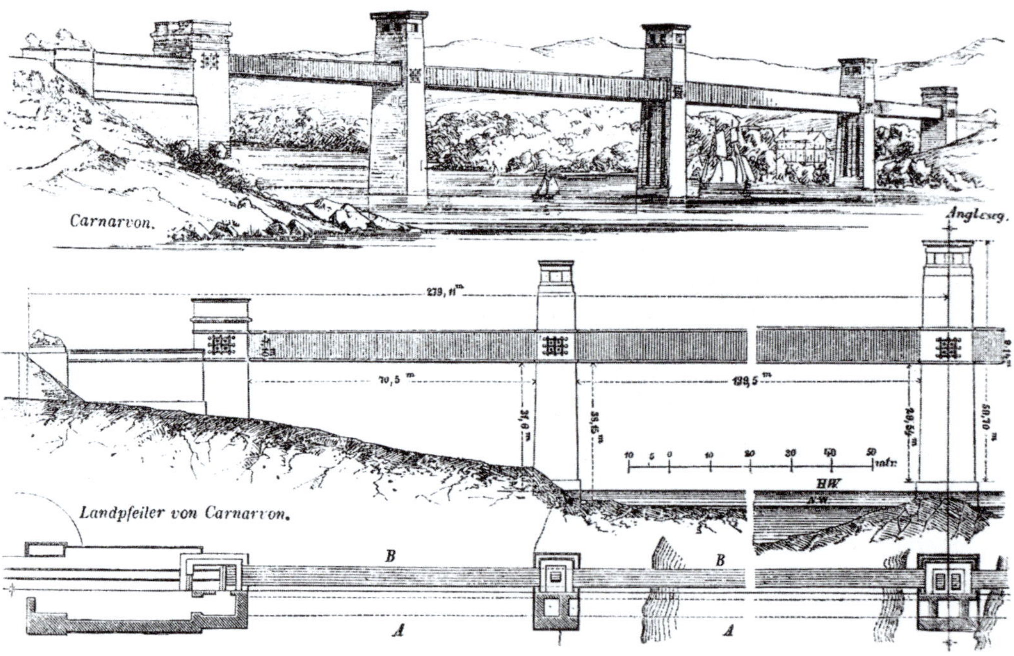

Fig. 9.2 Britannia Bridge by Robert Louis Stevenson over the Menai Straits with a span of 140 m. The bridge, together with the Conway Bridge, which was constructed almost at the same time, was the first hollow box constructed of sheet metal. Only 75 years later, a hollow box of this size was again created

bridge building which still determine the construction of steel bridges to this day.

9.1.2 First Buildings from Iron and Steel

It is probably not a complete coincidence that the concept for the construction of the pavilion of the World Exhibition in 1851 was developed by the gardener Joseph Paxton

Fig. 9.3 Royal-Albert-Bridge in Plymouth, designed and constructed by Isambard Kingdom Brunel for the Great Western Railway Company, 1859

Fig. 9.4 Crystal Palace in London, exhibition building designed by Joseph Paxton for the World Exhibition of 1851

(Fig. 9.4) following his early experiences with the construction of greenhouses made of steel and glass (Fig. 9.5). The **Crystal Palace** represents a key building of steel construction which has all the essential features of a modern skeletal construction and implements the inherent logic of this construction mode in a remarkably consistent form. The building is an

Fig. 9.5 Greenhouse in Kew Gardens during the construction. Erection of the palm house of the Royal Botanical Gardens in Kew; oldest example of a Victorian greenhouse (1841–1849)

early product of modern industrial prefabrication and industrial erection processes (cf. also Figs. 9.6, 9.7, and 9.8).

At the end of the eighteenth century, iron was also increasingly used for utility buildings such as storage buildings (Fig. 9.9). With this new material, a higher load-bearing capacity, larger spans, and slimmer dimensions compared to timber were achieved. Furthermore, the non-combustible iron offered advantages in terms of fire protection as compared to wood. The floor joists of these buildings were designed for the first time with a type of I-profiling. The floors themselves, on the other hand, were mostly bricked with flat vaults, for example in Germany, in form of the Prussian cap-vault.

In the years 1843 to 1850, Henri Labrouste built the **Bibliothèque Ste. Geneviève** (Fig. 9.10) in Paris, the first free-spanning iron construction which does not exert a side thrust on the surrounding classical masonry enclosure.

In 1871, Jules Saulnier erected the main building of **Menier chocolate factory** in Noisiel-sur-Marne, France (Fig. 9.11). The visible steel structure, which is filled with masonry, is braced by means of a lozenge-shaped mesh.

9.1.3 The Chicago School

For the historical development of modern steel construction, the revolutionary handling of the innovative material steel

Fig. 9.6 A total of three exhibition pavilions made of iron and glass, which were created during the Paris World Expositions in 1867, 1878, and 1898, were designated as Machine Hall or Galerie des Machines. The most famous of these halls was the one from 1898, a joint work of the Architect Charles Louis Ferdinand Dutert and the Engineer Victor Contamin. It was dismantled in 1910. The hall had a rectangular floor plan of 422.49 m × 114.38 m and was divided into a wide main nave and two narrow side naves. Their powerful portal frames, resting on 40 stone sockets, attracted a great deal of attention at the time. The trussed arches rose freely without intermediate columns up to the 46.67 m high vault crest. The Galerie des Machines surpassed all conceivable dimensions of a space free of columns (total surface area of the three naves: 48,324.9 m^2)

plays an important role in the urban development of Chicago as an economic centre after the American Civil War, although there had been widespread use of iron in building construction in the USA since the 1840s (Fig. 9.9). Chicago, until then a city with a building stock in typical American wood construction, completely burned down in 1871. Reconstruction was associated with an economic upswing of unprecedented proportions and required considerably densifying the inner-city areas. The first high-rise buildings which met this need were followed by the first **skyscrapers** of the history of architecture (Fig. 9.12). From the outset, an open floorplan concept was developed. The high-rise buildings, which were erected in steel-skeleton construction from an early stage, were designed as neutral as possible in terms of use. A prerequisite for the operation—and thus for the emergence—of these buildings was the invention of the elevator by Elisha Otis in 1857, but also the development of appropriate technical installations such as telephone, tube mail, central heating, and ventilation systems.

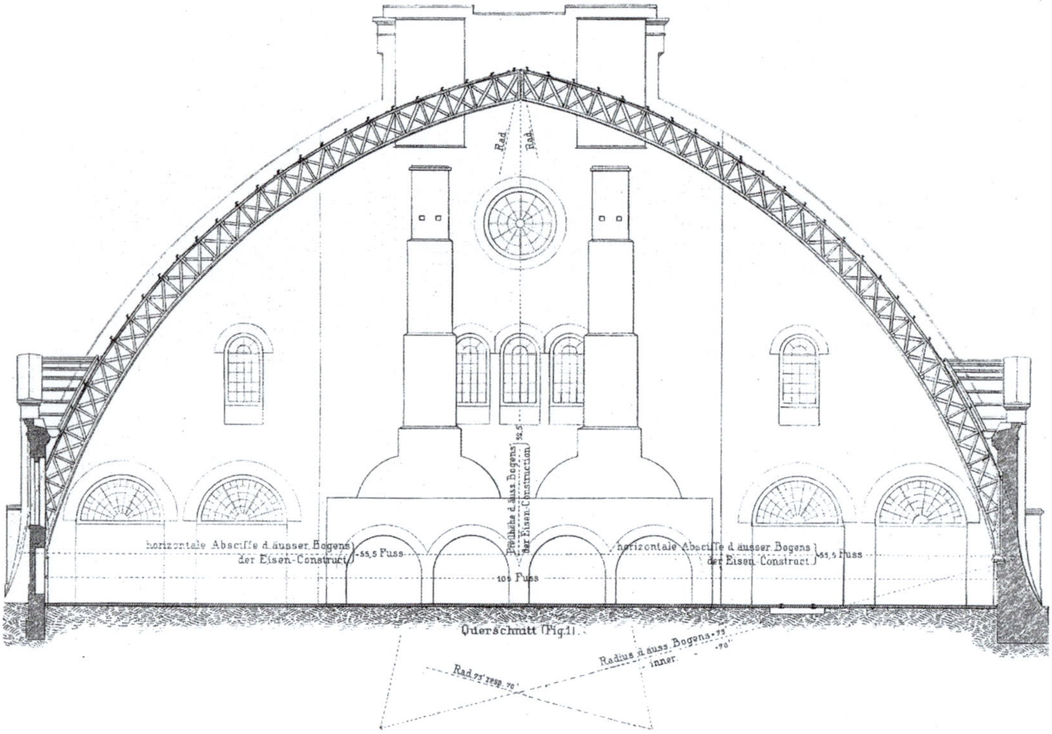

Fig. 9.7 Industrial hall in Berlin by Johann Wilhelm Schwedler, around 1863. For the first time, a statically determined triple-hinged arch was used here

In 1885, the Carnegie Steel Company rolled out the first mild-steel carrier, which replaced the forged-iron carriers. Mild steel possessed improved material properties, such as greater homogeneity. Standard steel sections also quickly replaced cast iron. The rivet connection became established as the standard fastener. The technology of high-rise buildings was already fully developed and should not experience any major innovations within the following 50 years. The development from skeletal construction to the cage had also already been completed.

9.1.4 Development of Steel Construction in the Twentieth Century

Steel construction was faster in gaining foothold in England, France, and Belgium than in Germany. It was only with the beginning of the twentieth century that steel structures in multi-storey and industrial construction were increasingly realised there (Fig. 9.13). These buildings, therefore, were all the more important for the subsequent development of modern architecture since 1910. Some key projects were the AEG

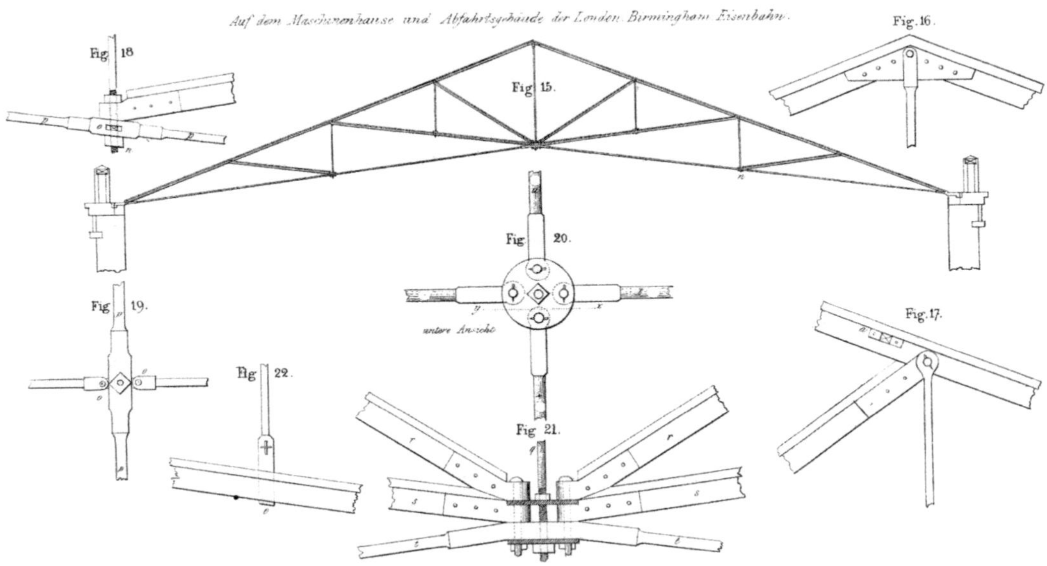

Fig. 9.8 English truss made of wrought iron, built as a roof construction of the London-Birmingham railway in London in 1835. Example of a filigree roof construction from the early days of industrialisation

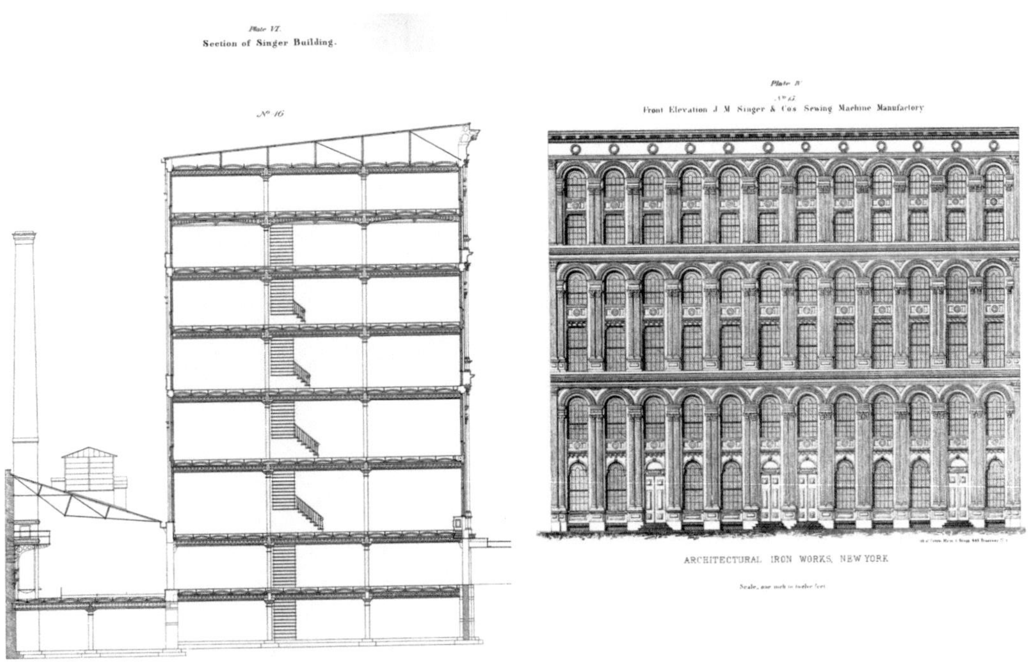

Fig. 9.9 J. M. Singer–sewing-machine factory in New York, approx. 1850. Elevation and section of the building, which was designed and erected as an early iron skeletal structure for flexible usage by the D. D. Badger Ironworks Company. The façade of the iron construction still shows the formal features of classical stone architecture, while the interior spaces in skeletal construction have an undoubtedly modern character. The exterior wall construction still partly consisted of masonry, as did the classic vaults of the floors. This construction method was further perfected over the years. Eventually, complete kits were offered. Many of these buildings have been preserved until to date and are under monument protection. The New York Ironworks emerged as a response to the city's economic development and fire-safety requirements long before the big fire in Chicago in 1871

9

Fig. 9.10 *Bibliothèque Ste. Geneviève*, Paris, 1845–51 (Arch.: Henri Labrouste)

Fig. 9.11 Main building of the chocolate factory Menier in Noisiel-sur-Marne, from 1871

Fig. 9.12 Frist skyscraper in Chicago, Home Insurance Building by William Le Baron Jenney, 1883–1885

turbine hall in Berlin by Peter Behrens (Fig. 9.14), the chemical factory in Luban near Posen by H. Poelzig, and the Fagus factory of Walter Gropius in Alfeld.

In particular, the further development of the skeletal construction and the curtain wall led to the modern steel skeletal construction with a steel-glass façade, as already anticipated by Mies van der Rohe in a high-rise project in 1920.

The real international breakthrough of steel construction was prepared by Mies van der Rohe from 1938 on at the IIT in Chicago, where Mies, as former head of the Bauhaus, had been appointed to directorship. After the Second World War, the real flourishing of steel construction set in, which developed very rapidly in high-rise building, administrative building, and also in residential construction developing an architectural appearance as an international style.

In the US, the buildings of Mies van der Rohe (Lakeshore Drive 1949/1950, Seagram Building 1954) (Fig. 9.15) and Skidmore, Owings, and Merrill (Alco-Building 1951, Lever Building 1951, John-Hancock Center 1968) (Fig. 9.16) are particularly worth mentioning. In housing construction in the USA, the innovative case-study programme must be considered in addition as a prototypical approach to new residential

■ **Fig. 9.13** Zeppelin hall in Germany, around 1910. Until the First World War, all the structural types that are still used today for steel construction were developed. The evolution to modern steel construction was already initiated in industrial construction

concepts. After the beginning of the reconstruction of Germany in the early 1950s—before that, steel and steel structures were too expensive there—new independent interpretations of steel skeletal construction emerged, especially by architects such as Sep Ruf and Egon Eiermann, who earned an international reputation with the Brussels International Exhibition Pavilion and used steel for their subsequent, ever path-setting office buildings (■ Fig. 9.17). This tradition of disciplined, sometimes ascetic, skeletal construction was later continued by other architects such as Fritz Haller (■ Fig. 9.18), Peter C. von Seidlein (■ Fig. 9.19), and Kurt Ackermann.

Steel construction found an innovative continuation in high-tech architecture since the 1960s, to a certain extent aiming at over-emphasising technological features. The ideas of industrialised building remained alive with this approach, albeit in a modified form. The ideas of prefabrication and modular construction were revisited in programmatic manner. The strict and sharp-edged orthogonality of classical modern steel construction in the Miesian tradition, which was a manifestation of the design discipline imposed on the designer by the very narrow range of standard steel profiles and which produced predominantly bending-stressed structures, gave way in high-tech architecture to a greater variety of structural concepts in which forces were often directed in a clearly vis-

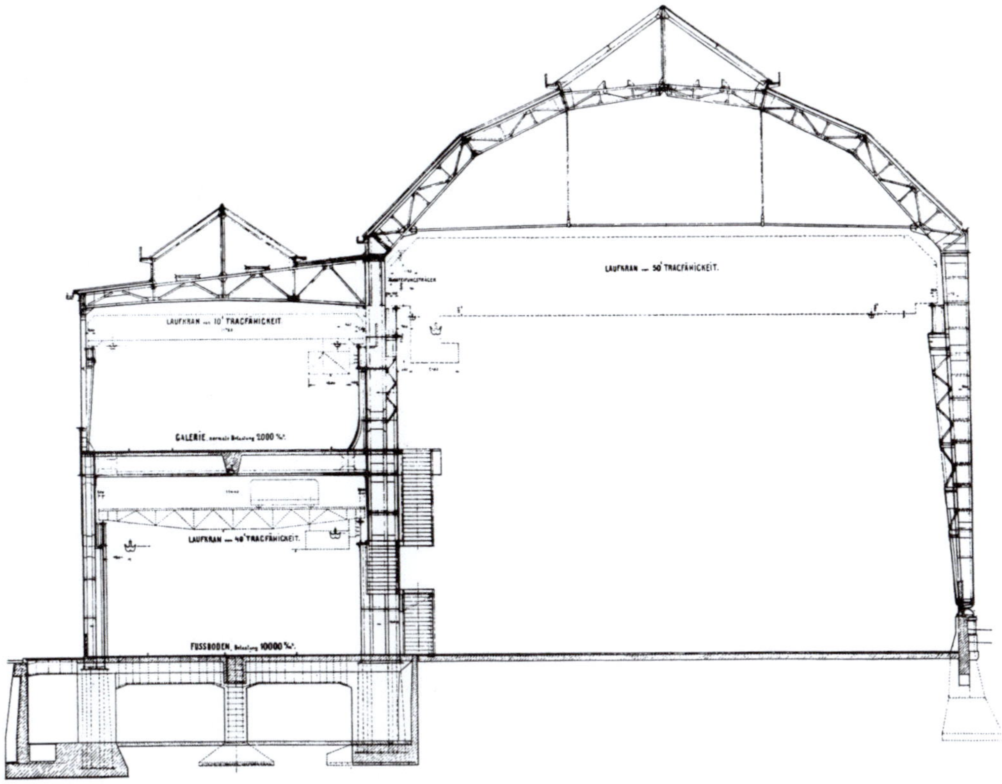

Fig. 9.14 *AEG*-production hall by Peter Behrens. Triple-hinged portal frame with crane runway, around 1910. Peter Behrens built a series of ground-breaking industrial halls for AEG Berlin. The most important was the AEG machinery hall

ible form in truss-like constructions through axially stressed compression and tension rods. Deliberately visible, emphatically technical connections underlined the highly technological character of these projects (Figs. 9.20 and 9.21). A clear influence on this steel architecture was exerted by the futuristic ideas of the *Archigram* Group and the space technology of NASA. This new type of steel construction made extensive use of profiles newly introduced at that time, in particular round, square, or rectangular tube profiles, and cables, which were characterised by their particular node solutions, both technically and in terms of aesthetics (Fig. 9.20, right). Steel castings were also often used for more complex node constructions (Fig. 9.20, left). Outwardly displayed, often emphatically elaborate load-bearing structures (Fig. 9.21), which sometimes deliberately—sometimes unnecessarily—solve highly complex requirements with high technical effort, are a trademark of this type of architecture, which sometimes leaves the viewer with the feeling that the construction task could as well have been solved with much simpler means (Fig. 9.21).

Fig. 9.15 *Seagram Building*, New York, 1954–1958 (arch.: Ludwig Mies van der Rohe)

Another characteristic feature of high-tech architecture was the endeavour to reduce secondary constructions of glass envelopes to the technically feasible material minimum, thus giving the building envelope the appearance of complete immateriality. The technical basis for this feature was the development of cable-tensioned, point-fixed glass constructions (Fig. 9.22).

Vol. 3, *Sect. 15 Point-Supported Glass Envelopes*

In recent years, there has been an increasing tendency to realise formally expressive design concepts with often curved,

Fig. 9.16 John Hancock Center, Chicago, 1970 (arch.: SOM)

geometrically partly very complex shapes. Some examples are projects by Santiago Calatrava (Fig. 9.23), Zaha Hadid, Frank Gehry (Fig. 9.24), Daniel Libeskind, Herzog de Meuron (Figs. 9.25 and 9.26), and others. The project of the Guggenheim Museum in Bilbao by Frank Gehry (Fig. 9.24), which with its worldwide renown has given very important impulses not only for this particular view of architectural creativity, but also for the urban development of its location, played a major role in this context. The media coverage of these extraordinarily spectacular projects supports their symbolic significance and causes formal factors to play an absolutely dominant role in the development of these designs. Another supporting factor that has favoured these architectural trends is modern digital planning tools and automated manufacturing facilities. Only these make it possible to master the at times extreme complexity of these projects both in design and in execution. Furthermore, the special properties of steel, in particular its high strength, and modern joining techniques make it possible to realise these sometimes very delicate and geometrically very complex designs.

With this development, traditional fundamental principles of design and construction, such as the classical notion of appropriateness to materials and the otherwise all-pervasive need for the greatest possible economy of construction or

Fig. 9.17 Headquarters of the company *Olivetti* in Frankfurt am Main by Egon Eiermann, 1968–1970. From the concrete base of the high-rise building, the main load-bearing structure made of steel arises. The bracing of the building takes place via the central core

Fig. 9.18 Example of the application of a steel construction system in housing (arch.: Fritz Haller)

appropriateness of means, have been effectively undermined. The design of classical skeletal construction of the Miesian School, inspired mainly by purely constructional principles, loses its significance with this type of projects. Conventional considerations of a reasonable relation between effort and result are not really applicable to this architecture, since it is not the apparent utilitarian value nor the structural and constructional efficiency that are in the foreground, but rather the

Fig. 9.19 Printing Company Süddeutscher Verlag, München-Steinhausen (arch.: Peter von Seidlein)

Fig. 9.20 (Left) The use of industrial tubular profiles, tension rods, and individually manufactured cast-steel parts for the nodes underlines the intentionally technological character of high-tech architecture (*Centre Pompidou*, Paris). (Right) Breakdown of the force action into an exposed array of tension and compression members, characteristic of high-tech architecture. The technical character of the connections is deliberately increased (*Centre Pompidou*, Paris)

spectacular building form, which promises to generate a corresponding added value through its visual expressiveness and suitable mediatic dissemination.

From a traditional engineering point of view, sometimes outright absurd concepts that turn out to be technically hardly feasible even with modern construction technology are (knowingly) put up with. Modern planning tools and steel construction technology make their realisation possible despite all ob-

■ **Fig. 9.21** *Hongkong and Shanghai Bank*, Hongkong, 1979–1986 (Arch.: Norman Foster)

stacles; upwardly open construction budgets of these usually very prominent, sometimes politically relevant, projects open the door to even the most bizarre designs. It seems doubtful, however, whether this design philosophy can provide a guideline for future developments in mass construction activity as it affects the vast majority of people in their daily environment. It seems more plausible that these spectacular and media-effective, sometimes extraordinarily expensive individual projects will remain rather capricious creations of wealthy societies and will not provide a really useful model for a sustainable, future-oriented use of steel as a building material.

Cf. also the considerations in ***Vol. 4***, *Sect. 5.9 The New Freedom of Forms—An Appeal*

9.2 Basic Aspects of Steel Construction

Vol. 1, *Sect. 16 Steel* and ibid. *Sect. 22 Steel Products*

The term *steel construction* is only 100 years old. It was not established until the 1920s. Previously, the term used was iron construction. Compared to the older forged or cast iron, modern steel is characterised by a relatively low carbon content. The material has ductility as well as high compressive and tensile strength. Mild steels offer high ductility with

Fig. 9.22 In high-tech architecture, point-fixed glass façades are intended to give the impression of complete immateriality of building envelopes (*Centro Reina Sofía*, Madrid; eng: RFR Paris)

moderate strength values; high-performance steels offer very high strengths with increased brittleness. Other types of steel can be optimised for special applications, such as weather-resistant steels and stainless steels.

9.2.1 Properties of Steel Structures

Modern steel construction is characterised by the use of standard, industrially manufactured semi-finished rod-shaped (Fig. 9.27, left) or flat-shaped basic elements (Fig. 9.27, right). The usage of semi-finished industrial products from steel for building construction almost invariably leads to skeletal structures, which are characterised in particular by the high load-bearing capacity and stiffness of their elements, and thus also by their resulting slenderness (Fig. 9.29). Thanks to this material-specific characteristics, steel can also be allocated particularly demanding building tasks, which up to the present time can only be achieved with the unique possibilities of this material. This applies to high-rise buildings as well as to wide-span bridges and tensile-stressed structures of extreme lightweight construction.

EN 1993-1-1, EN 1993-1-2

Fig. 9.23 BCE Place, Toronto (arch.: Santiago Calatrava)

9

The favourable properties of steel in multi-storey building have been significantly expanded by the development of modern steel-concrete composite construction. In particular, the inadequate fire-protection properties of steel were improved in combination with reinforced concrete.

EN 1090-2

Steel thus permits erecting load-bearing structures with large spanning widths as well as constructing structures with few fixed points, which, being applied to multi-storey construction, allow for a maximum degree of floorplan openness and flexibility. Creating perforated or lattice-shaped structural components with the possibility of simple installation routing (Fig. 9.28), such as perforated, castellated, or latticed girders, plays an important role in the construction of highly installed buildings or in issues of conversion of buildings (Fig. 9.29).

***Vol. 1**, Sect. 9.3 Comparative Consideration of the Life-Cycle Assessment Data of the Most Important Materials*

Less flattering are the LCA values of the material, for the production of which, at least at present, very large quantities of non-renewable energy have to be used and a severe environmental impact has to be accepted. This is, so to speak, the price that has to be paid today for making the extreme mechanical performance values of this material available.

Fig. 9.24 Guggenheim Museum Bilbao (arch.: Frank Gehry)

Fig. 9.25 Olympic Stadium Beijing. The formal design as a 'bird's nest' plays the all-dominant role in this project. Neither the constitutive construction principle of a nest (rods running past each other, connected by form fit and friction), nor considerations of lightweight construction by using a shell effect due to curvature have played the least role in this heavy frame construction (arch.: Herzog deMeuron, Ai Weiwei)

Fig. 9.26 (Left) Olympic Stadium Beijing. Extensive welding work was necessary on site under the most adverse conditions to create the numerous individualised nodes from box sections. Conventional principles of an efficient assembly process were of minor importance. (Right) Olympic Stadium Beijing. The biaxially curved geometry of the load-bearing structure results in both curvature and torsion of the box-shaped lattice bars and thus in warped surfaces

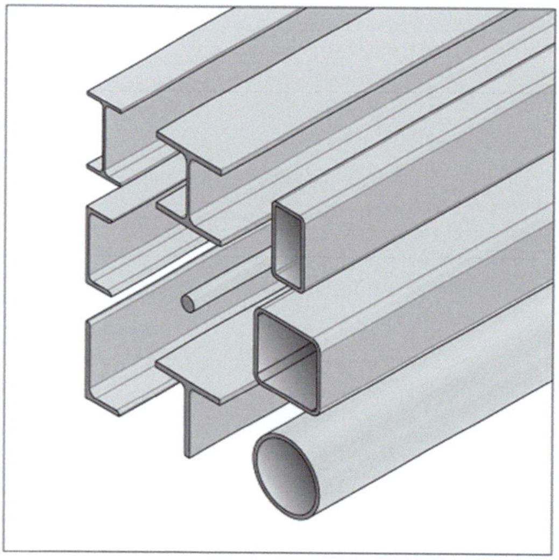

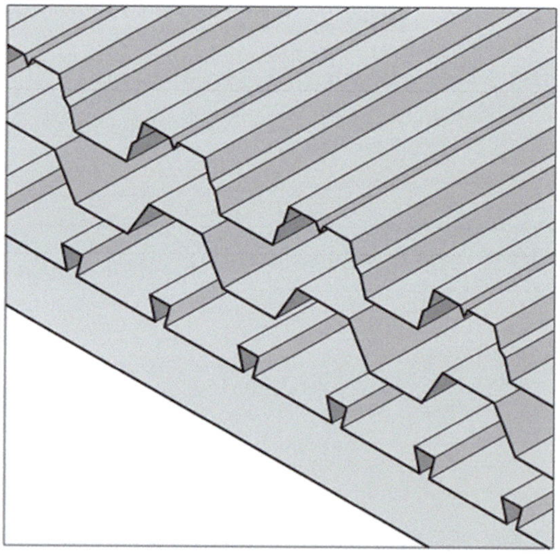

Fig. 9.27 (Left) Selection of standardised semi-finished profiles, which are available in iron and steel trade as hot-rolled products made by various manufacturers according to standardised dimensions and qualities. (Right) Hot-rolled steel strip (at the bottom) and cold-rolled trapezoidal sheets as examples of surface-forming semi-finished products from steel

9.2.2 Construction of Steel Structures

Vol. 1, Sect. 1.3.1.1 The Concept of Construction Type
EN 1090-2

As in the case of timber construction, steel construction is an assembled building method. The fabrication of the elements, components, or component groups is carried out at the factory. On site, they are assembled to form the finished building in a dry manner in short assembly times—that is, without delays produced by setting times or building moisture. Pure steel construction permits a largely weather-independent erection of the load-bearing structure. Due to the

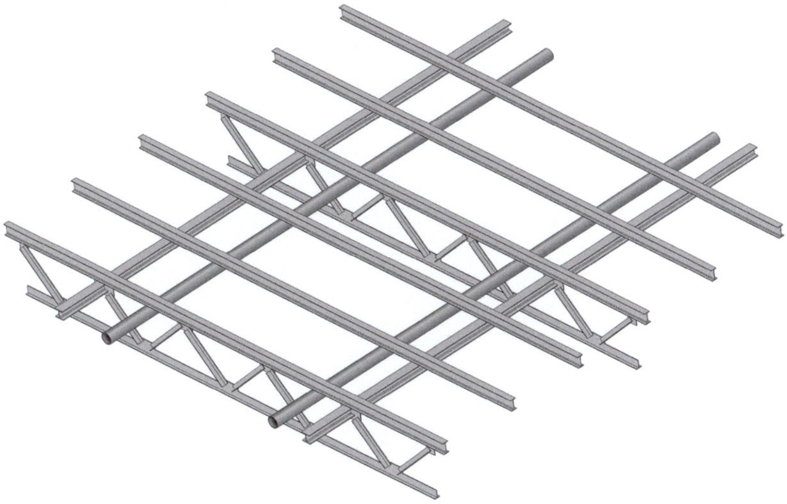

Fig. 9.28 Creation of pervious floor or roof structures which allow easy and dense mechanical installation

Fig. 9.29 The high strength and stiffness of steel allows extraordinary slenderness and delicacy of the structure

possibilities of assembled construction modes and the simplicity of the differential construction principle, which is well suited for modern steel construction, extremely short erection times are achieved in comparison with other construction methods (Fig. 9.31).

Fig. 9.30 Hinged bolt connection of a girder to a column, typical for steel construction. This type of differential connection can be made on site in a very short time

Fig. 9.31 Typical bolt mounting in steel construction

Building with steel allows extremely precise operation with tight tolerances as no other material suitable for primary structures permits in building construction. The accuracy of steel allows for a precise fitting of further fit-out elements subsequently in the construction process, especially of light outer walls.

Vol. 3, *Sect. 9.2 Welding of Steel Components*

The use of material-fitting welding technology in the factory, which produces non-detachable joints that get very close to the undisturbed material continuum, is a special feature in comparison with other materials and joining technologies. The welding of individual components or assembly groups takes place in the factory because of the easily controllable conditions, the final assembly on the construction site mostly by bolt connections (Figs. 9.30, 9.31, and

Fig. 9.32 Construction of the ice-skating rink by Ackermann & Partner (arch.) and Schlaich, Bergermann & Partner (eng.) at the Olympic District in Munich. The steel construction is carried out prefabricated with bolted connections

9.32). During assembly, quickly producible shear-force connections with shear/hole-bearing bolts are preferably used. Longer lifting times of steel parts for this purpose are unproblematic in steel assembly construction—in contrast to, for example, prefabricated reinforced-concrete parts—because of the comparative lightness of the components.

Vol. 3, *Sect. 6.2.5.1.3 Categories of Screw Joints in Steel Construction*; also ibid. *Sect. 5.4.1 Connections by Laying-On*

9.2.3 Usage-Related Aspects of Steel Structures

The following usage-related characteristics of steel buildings deserve mention:

- Due to the use of mainly rectilinear, rod-shaped components, and the high load-carrying capacity of steel parts, skeletal structures emerge in which the fixed points within floorplans can be reduced to a minimum, often solely to the columns.
- Due to load-bearing structures being rather easy to modify, there are numerous possibilities for supplementation and adjustment to changing usage—a factor which is particularly important in industrial and administrative building. Due to the joining and assembly techniques customary in steel construction, structures can be flexibly adapted to changing needs of usage.
- The easy dismantling of steel structures also offers significant advantages in recycling and disposal; in addition, unlike most competing materials, scrap steel is almost completely recyclable and can be reused in any number of recycling cycles without loss of quality. This noticeably improves the otherwise rather problematic environmental quality of steel.

Vol. 1, *Sect. 10.3 Recycling of Steel*

9.2.4 Fire Protection

When designing with steel, it must always be taken into account that the material must be adequately protected against fire and corrosion. In particular, fire protection requires, especially in the case of longer fire-resistance periods, bonding with concrete in a composite construction or coating or casing-like cladding of steel components, which greatly impairs the precise, filigree, and sharp-edged visual character of the material. The material-related principles for fire protection of steel are discussed elsewhere.

The material-related principles of fire protection for steel are discussed in ***Vol. 1**, Sect. 29.10.4 Components from Steel*

Especially when compared to its most successful competitor in building construction, namely, reinforced concrete, the material shows weaknesses in its resistance to fire, which have often resulted in building structures being constructed in concrete instead of steel. Catastrophic events such as the collapse of the towers of the World Trade Centre have contributed to the fact that concrete structures are often preferred to steel structures.

EN 1993-1-2

Requirements of fire protection of steel structures—steel structures must be protected from the effects of fire beyond two storeys—entail necessary measures such as:

- Paint coatings/insulating foaming coatings.
- Claddings.
- Sprinkler system/water-filled profiles.
- Composite constructions in combination with concrete.

9.2.5 Corrosion Protection

When steel structures are exposed to the weather, the material's greatest advantage, namely, its high strength, paradoxically turns out to be a disadvantage, since the usually extremely finely articulated structures are exposed to corrosion on a relatively large surface. This applies in particular to trusses at the level of the load-bearing elements; at the component level, it applies to open-section steels, a standard semi-finished product which is most frequently used when designing with steel. Closed tube profiles are more favourable in this respect.

***Vol. 1**, Sect. 30.2 Corrosion of Metallic Materials*

For the corrosion protection of steel structures, the engineer must take appropriate protective measures, such as

- Paint coatings (◘ Fig. 9.33).
- Synthetic coatings.
- Noble and base-metal coatings (especially zinc, chrome).
- Duplex systems: coating of previously hot-dip galvanised steel components with suitable coating materials (epoxy resin, acrylic resin, etc.).
- Use of protection-layer-forming steels/stainless steels.

■ **Fig. 9.33** Steel components provided with a corrosion-protection coating

9.2.6 Sustainability

There is no doubt that the life-cycle assessment of steel in its use as a building material is negative in terms of resource consumption and environmental impact compared with other materials suitable for primary load-bearing structures. Despite assertions to the contrary by the steel industry and certain advocates of its interests, this fact cannot be argued away and can be substantiated with verifiable figures. Even recycled steel (■ Fig. 9.34) has several times the environmental impact of competing materials. At the same time, however, once steel has been smelted, it is still available as recycled steel for a practically unlimited period of time, without any loss of quality, as any number of cycles of reuse are possible. Viewed over a longer period of time, this is at least a significant factor in terms of life-cycle inventory that improves the ecological quality of the material, at least to a limited extent, although it does not significantly reduce the associated harmful environmental effects.

Vol. 1, *Sect. 6.2.4 Life-Cycle-Assessment Indicators*; also ibid. *Sect. 9*, Fig. 9.7, and *Sect. 9.3 Comparative Consideration of the Life-Cycle-Assessment Data of the Most Important Materials*

On the concept of life-cycle inventory, cf. **Vol. 1**, *Sect. 6.2.4 Life-Cycle-Assessment Indicators > Life-Cycle Inventory*

These sustainability-related weaknesses of steel are, however, at least partially compensated for when the material is compared with other materials on the basis of functional equivalents, i.e. in terms of their respective performance in solving specific construction tasks. Here again, the extremely high strength and stiffness of steel play an important role, allowing extremely lightweight construction with the resulting material savings. It is also true that certain demanding to extreme construction tasks are simply not feasible without this material. This applies in particular to wide-span struc-

On the concept of functional equivalent, cf. **Vol. 1**, *Sect. 6.2.1 The System Under Consideration*; also ibid. *Sect. 9.3 Comparative Consideration of the Life-Cycle Assessment Data of the Most Important Materials*

Fig. 9.34 (Left) Residual sheet metal compacted into bales, prepared for recycling. (Right) Demolition of a steel hall. Recycling of the scrap steel

tures. It is this undeniable functional excellence of the material that is often argued in favour of steel, in compensation for its environmental weaknesses.

The great flexibility of use made possible by steel skeletal structures with few fixed points is also an important use-related, primarily economic factor that can significantly improve the ecological balance of steel structures, depending on their use during their later life cycle.

9

9.3 Constructional Design with Steel

9.3.1 Building with Standardised Profiles and Hinged Connections

EN 1993-1-11

***Vol. 1**, Sect. 22.4 Hot-Rolled Mild Steel Products*

An elementary form of building construction with steel is based on the use of standard rolled steel profiles as starting material. The relatively simple manufacture of components from standardised profiles at the same time entails a certain design restriction, since one can hardly deviate from a limited catalogue of profile shapes. In normal circumstances, rolled steel cannot be manufactured at all according to special requirements—unlike, for example, timber cross-sections. Nevertheless, numerous possibilities remain for producing special components adapted to particular requirements by welding together sheet metal to form composite profiles or by welding-on supplementary parts also made of sheet metal.

By means of standardised steel profiles, extremely material-efficient steel structures can be erected in a very simple manner which can easily bridge spans of 12–15 m or more in multi-storey construction. In this way, large, interconnected, largely fixed-point-free surfaces can be created with load-bearing structures which would hardly be possible with such slenderness in reinforced-concrete construction.

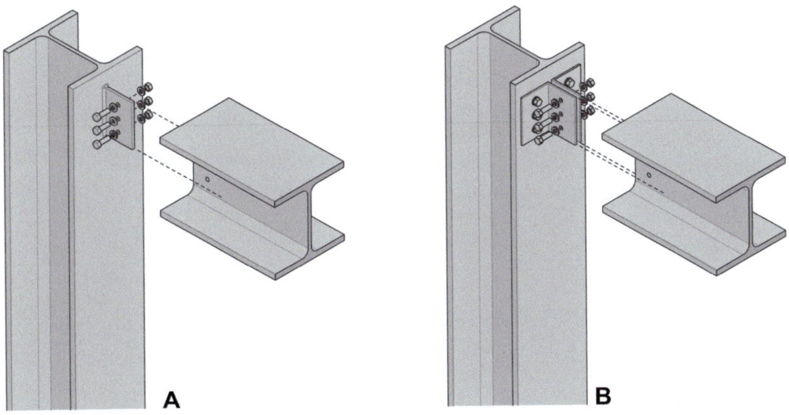

Fig. 9.35 Hinged beam connection to a steel column. Two options: **A** The carrier is connected via a single-shear connection by means of screws. The connection lug is already welded to the column profile at the factory. Eccentricities and bending stress in the fastener can be problematic. **B** In this example, the connection is made via a two-shear screw connection. The L-shaped connection lugs are also fastened to the column by means of screws. This is a completely detachable connection

A multi-storey building from steel can generally be reduced to a few elements: steel columns, floor beams, composite reinforced-concrete floor slab and bracing elements, such as core bracing.

Connections can be designed as hinged shear-force joints with lug-bolt connections, which ensure a fast assembly of the structure (Figs. 9.35 and 9.36). In contrast to other construction methods such as timber or prefabricated concrete construction, bending-stiff connections can also be realised on the construction site at a reasonable expense if the circumstances suggest it. Slip-resistant FG (friction-grip) joints, which in building construction—in contrast to welds—are also carried out on site, are comparable to welded joints in their load-bearing capacity and stiffness.

Vol. 3, *Sect. 6.2.5.1 Steel with Steel* and ibid. *Sect. 6.2.5.1.3 Categories of Screw Joints in Steel Construction*

Today, skeletal structures, for the purpose of utilising the large spans which can be realised, are designed in the main as well as in the secondary spanning direction with **wide column spacings**, which allows for a maximum of long-term floorplan flexibility (Fig. 9.37): The columns are arranged in both space directions with large spacings, structural grid modules being on the order of magnitude of >7.20 m. In this case, a hierarchical directional beam structure is used, where beams or joists are usually combined with floor slabs of concrete to form a composite construction.

Cf. further below, ▶ *Sect. 9.3.3.1 Floors in Composite Construction*

Due to the pronounced linear-rod character of the basic elements of steel construction (sectional steel), which is even more pronounced in this type of construction than, for example, in timber construction, non-directional, i.e. two-way spanning load-bearing structures, such as beam grillages, are rare. Here, as with other construction methods such as timber construction, the difficulty of executing the numerous girder nodes on the construction site plays a decisive role. Plate-shaped sup-

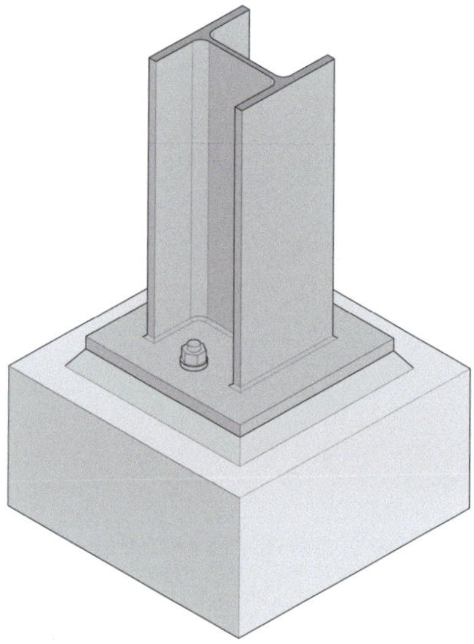

Fig. 9.36 Hinged connection of a steel column to a foundation using a connection plate. The connection is not bending-resistant, the screw connection serves only for the horizontal securing of the column profile

9

porting elements made of steel are used on the complexity level of the component only in the form of trapezoidal sheets. Solid plate-shaped sheets are not used in steel construction as a load-bearing component because of their heavy weight.[1]

9.3.1.1 Composite Construction

EN 1994-1-1, EN 1994-1-2

Steel-concrete composite construction is of great importance in steel building construction. The strengths of the respective composite partner, both in terms of load-bearing capacity and economy, are taken advantage of: with steel, the high tensile strength, both normal and under bending; with concrete, the good compressive strength at low cost. Consequently, areas subject to compressive stress (such as the top chords of single-span girders) are mainly made of concrete, while areas subject to tensile stress (such as bottom chords) are made of steel. In addition, the concrete contributes essential physical properties, namely, in particular fire- and sound-proofing related ones.

1 This is causally connected with the not particularly good ratio of load-bearing capacity to dead weight of steel, which—at least in regard to strengths customary in structural steel—is less favourable than, for example, that of wood (see ***Vol. 4***, *Sect. 1.6.2 The Influence of the Material*, as well as ibid. *Sect. 3.11.3.3 Ratio Between Strength and Bulk Density*). This factor is particularly important in the case of structurally inefficient load-bearing systems such as the concrete slab, where the dead weight consumes a good part of the load-bearing reserves and wood, in the form of solid-wood components, can play out its good ratio values. The situation is completely different with the much more efficient load-bearing system of folded thin sheets (trapezoidal sheet), where the relatively large dead weight of steel plays a much smaller role.

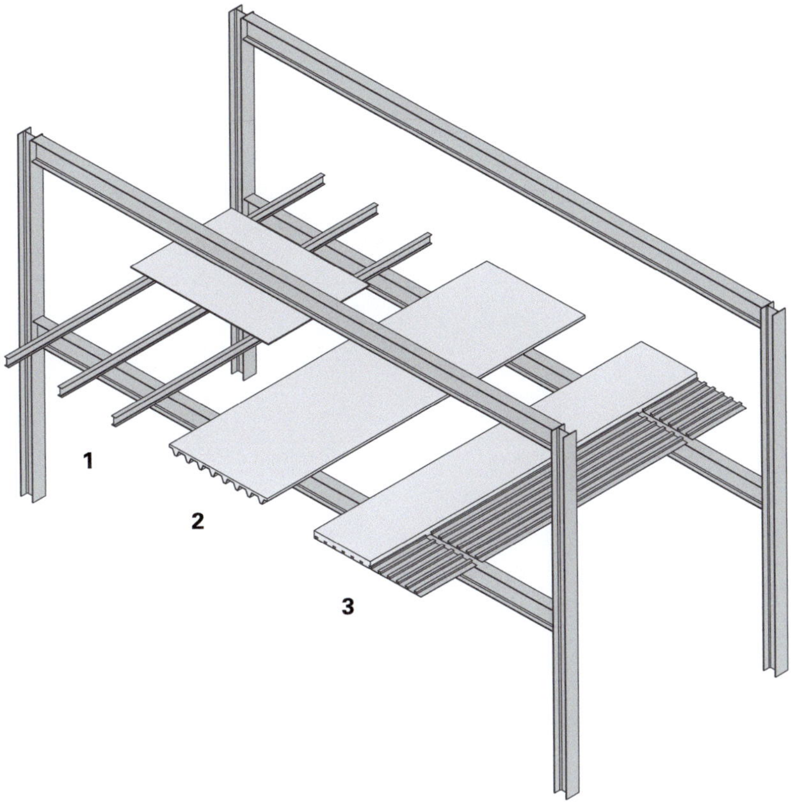

Fig. 9.37 Various execution types of floors in conventional steel building construction: (**1**) floor beam with plate placed on top. (**2**) beam-and-slab element from concrete. (**3**) steel-concrete composite floor: top concrete on trapezoidal-sheet metal

A prerequisite for the composite effect between the two materials is their tensile and shear-resistant connection at the composite joint. The most commonly used bonding agent for this purpose is the **headed stud anchor**. It is welded onto the steel part and then embedded in fresh concrete. At the same time, this composite construction method has the typical disadvantage of difficult material separation for recycling purposes; The steel and concrete of a composite component can practically only be separated by destroying them.

Vol. 1*, Sect. 10.8 Recycling and Environmentally-Friendly Design of Building Structures*

Composite structures can be produced both on site (as in the case of slabs, where a trapezoidal sheet serves as a lost formwork) (◘ Figs. 9.55 and 9.56) and prefabricated in the factory. In the latter case, the connections can also be made without major effort by grouting at the construction site (◘ Fig. 9.38). Grouting-free bonding is also possible with the aid of a non-slip bolt connection (dowel-less friction bond).[2]

Composite structures are characterised by high load-bearing capacity and stiffness and thus allow slim component dimensions. In turn, slender columns maximise the usable area

2 Petersen Ch (1994) *Stahlbau*, p. 788.

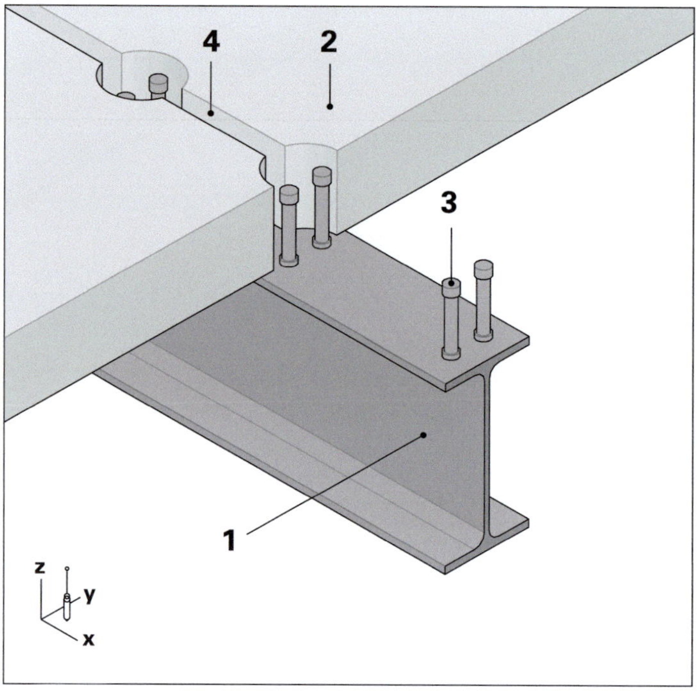

Fig. 9.38 Composite between sectional steel and precast floor produced on site by grouting. (**1**) Profile steel; (**2**) prefabricated floor element; (**3**) shear stud; (**4**) grouting joint

in the floor plans of steel skeletal buildings, which have few fixed points already from the outset. Composite structures also represent a realistic alternative in cases where high loads on concrete components result in increased reinforcement densities that cannot be sensibly achieved.

Cf. ▶ *Sect. 9.3.3.1 Floors in Composite Construction*

In addition to composite slabs, which will be dealt with below, combinations of concrete and rolled sections are used in particular for bar-shaped components. Hollow sections, such as tubes, can be filled with concrete (Fig. 9.39**D**); likewise, the chambers of I-sections (Fig. 9.39**B**, **E**, **F**); entire rolled sections can also be embedded in concrete (Fig. 9.39**A**, **C**).

9.3.2 Portal Frames and Bending-Stiff Connections

Portal frames are the result of the bending-stiff corner joint of the vertical frame components with the horizontal ones to form a self-supporting planar structural system. Therein, portal-frame structures are fundamentally different from frame structures with hinged connections between columns and beams. The profiles of the horizontal and vertical frame members can consequently not be selected independently

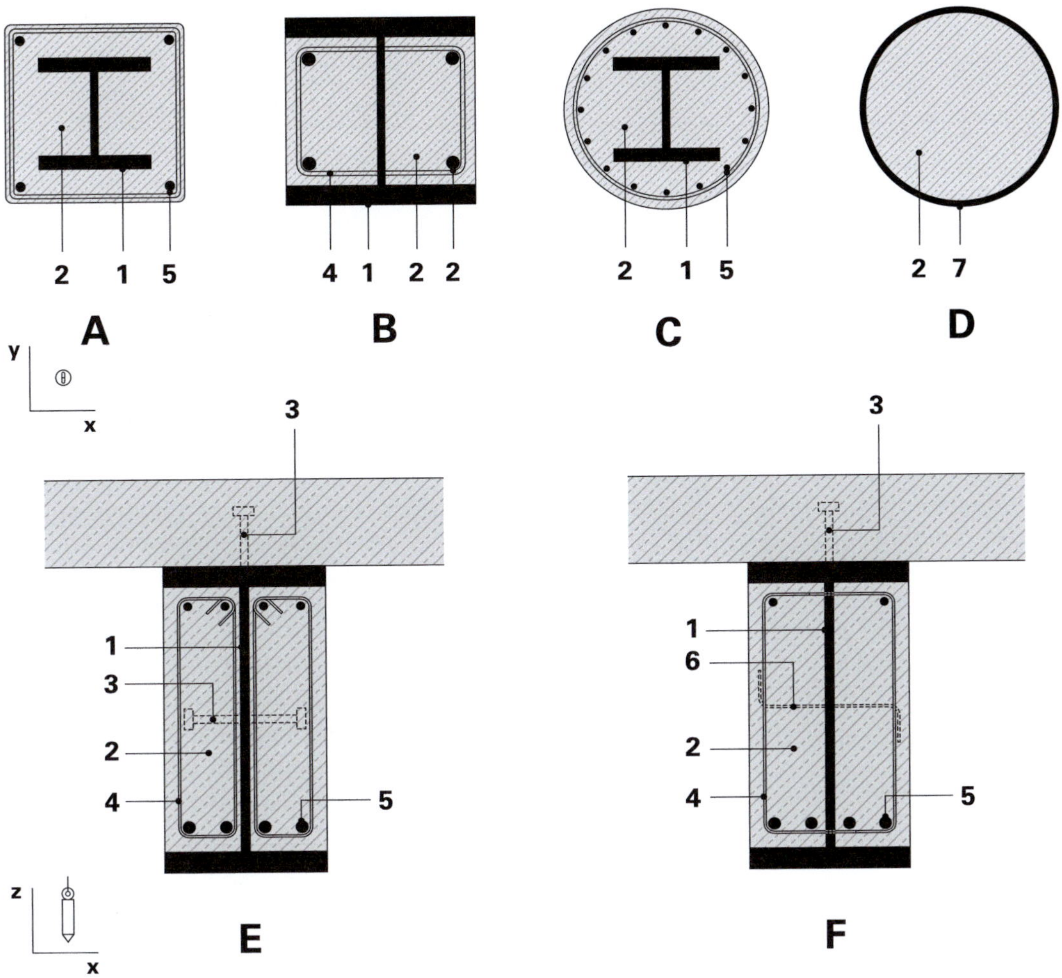

Fig. 9.39 Various designs of composite columns (top) and composite beams (bottom): **(1)** profile steel; **(2)** filling concrete; **(3)** shear stud; **(4)** reinforcement stirrup, closed; **(5)** additional reinforcement; **(6)** connection bracket; **(7)** steel tube

from each other, since the respective profile stiffnesses influence each other. Unlike hinged trusses made of hinged columns and girders, portal frames are stiff within their planes, a significant advantage especially for their frequent use in hall structures. Material-consuming bending stresses, which in hall cross-sections mainly affect the horizontal member and which result in larger structural depths, are mostly evenly distributed in the frame over the horizontal and vertical members. They can even be controlled in their distribution by choosing the degree of stiffness of both components. On the other hand, in the case of portal frames, in contrast to systems with hinged supports, systemic horizontal forces are always to be expected at the bearing, the discharge of which is associated with corresponding additional constructional complication.

In industrial and hall construction, portal-frame systems are comparatively frequent (Figs. 9.40, 9.41, and 9.42).

Fig. 9.40 Portal-frame corner bolted on site as a mounting connection. The constructional solution corresponds to the variant on Fig. 9.42

Vol. 3, *Sect. 9.2.8 Constructional Standard Solutions of Steel Construction*, in particular Figs. 9.12, 9.13, 9.14, and 9.15; and ***Vol. 4***, *Sect. 9.7.4 Transmission of Bending*

9

Necessary open hall clearances, undisturbed by any stiffening components, such as are necessary for use-related reasons, often suggest the use of portal-frame systems which are already stiff within the hall's cross-sectional plane. The moment distributions, which can be relatively favourable under suitable conditions, allow for relatively slender cross-sections. **Two-** and **three-hinged frame** systems are typical portal-frame structures. In industrial building, frame columns are often restrained not only within, but also transversely to the frame plane, which means that all further bracing elements are unnecessary.

On the other hand, multi-storey buildings, beyond a certain building height, are relatively seldom braced by means of bending-stiff frame corners because multi-storey frames (Figs. 9.45**E**, 9.52, 9.53, and 9.54) are rather pliable compared to other bracing solutions. They are regarded as Vierendeel systems and are used in particular when a high degree of flexibility in the use of the interior spaces, or a façade without interfering diagonal braces, is required.

In the case of high-rise buildings (Fig. 9.54), all beam connections, in addition to other bracing systems, are often made bending-stiff, with the aim of reducing horizontal deformations and vibrations or of coupling vertical bracing elements—for example twin cores.

9.3.2.1 Design of Portal-Frame Systems

Portal frames are trusses composed of beams and stanchions (Figs. 9.43 and 9.44). From the point of view of transport to the construction site, a portal frame is most advantageously divided into these individual components, i.e. into bar-shaped components that can be efficiently transported in compact packaging on the loading platforms of lorries. The consequence of this type of unitisation is that the assembly joint is located precisely at the point where the greatest stresses occur, i.e. in the rigid corner. As usual in steel construction, bolt connections are used for these junctures (Fig. 9.45**C**, **E**).

Fig. 9.41 Double-hinged portal frame in construction stage. AEG Berlin, around 1930

Fig. 9.42 Rounded portal-frame corner to support the covering membrane; radial compression stiffeners to reinforce the corner plate

Solutions with factory-welded frame corners (Fig. 9.45 **A**, **B**, **D**, **F**), in contrast, offer some advantages: being the most heavily stressed node, the corner joint is easier to manufacture in the workshop, so that the assembly joints can also be made more easily and quickly on site. This requires either working completely without assembly joints, as with a completely prefabricated frame (which, however, can only

9

Fig. 9.43 Manufacturing hall made of steel frames. The I-profiles of the frame stanchions and beams are corrugated to stiffen them against shear forces (arch.: Arlart)

Fig. 9.44 Portal frame corner with added haunch; the upper triple bolting receives two stiffeners on the stanchion side as back-up plates to prevent the weak stanchion flange from bending under tensile stress and to transfer the force to the stiffer web

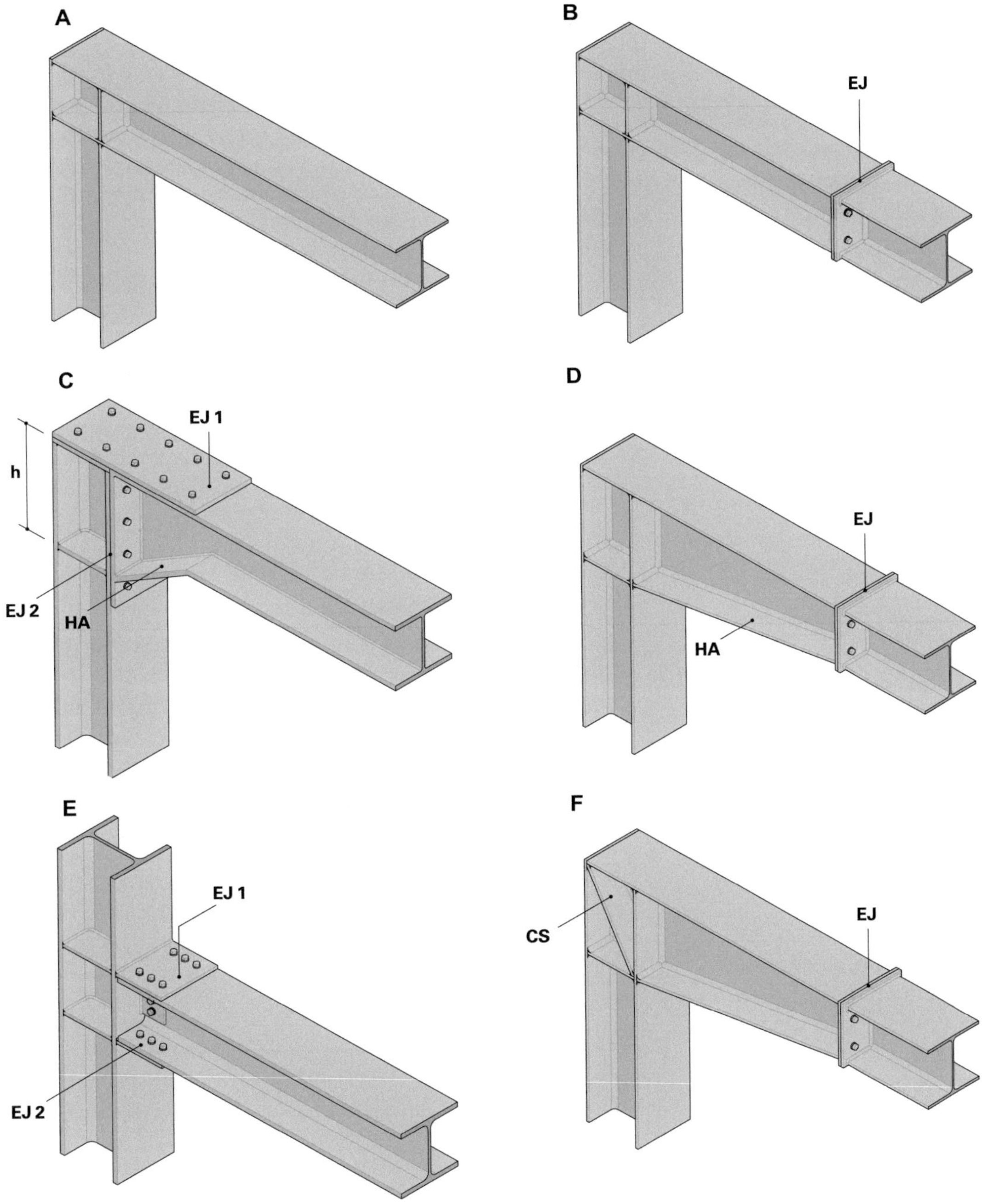

Fig. 9.45 Various bending-stiff corner connections of portal frames in steel. **A** Factory-welded portal-frame corner. Solution for completely prefabricated frames or three-hinged frames with joint at the apex. **B** Solution like **A**, but with erection joint **EJ** at the point of contraflexure of the frame beam (two-hinged or restrained frame). Simple end-plate joint with shear/hole-bearing screw connection. **C** Erection joint **EJ** in the rigid corner. Bending-induced tensile forces are transmitted via an attached plate and a screw connection subjected to shear (**EJ 1**); bending-induced compressive forces via compressive contact at the front plate (**EJ 2**); shear forces via shear-stressed screw connection (**EJ 2**). A haunch **HA** at the end of the beam increases the inner lever arm **h** of the connection and reduces the bending-induced tensile and compressive forces. **D** Solution like **B**, but with haunched frame-beam end **HA** for better absorption of the large bending moments, analogous to **B**. **E** Erection joint in the bending-resistant corner with continuous stanchion; typical solution for multi-storey portal frames. Bending-induced tensile and compressive forces are transmitted via shear-stressed screw connections at both flanges (**EJ 1** and **EJ 2**). **F** Solution like **D**, but with diagonal compression stiffener **CS** in the corner field for additional reinforcement

be transported as a whole when its dimensions are rather small), or moving them to a different location. As with the three-hinged portal frame, this can be the vertex in the middle of the element or otherwise a point of contraflexure in the beam of the two-hinged portal frame or restrained frame (◘ Fig. 9.45**B**, **D**, **F**). These bolted joints transmit the forces that occur, i.e. normal and shear forces and possibly small bending moments, via direct compressive contact or via simple shear/hole-bearing bolted connections.

Vol. 1, *Sect. 26.7.2.1 Two-Hinged Portal Frame Subjected to Linear Load*

In the case of double-hinged frames or restrained frames with mounting joints at the points of contraflexure of the frame beam, frame stanchions must be transported with short, rather unobtrusive cranks at the upper ends, which is usually problem-free. Complete frame halves of three-hinged portal frames are much bulkier and more difficult to transport. The decisive factor for transportability is the diagonal dimension of the frame section placed vertically on the loading area reverse V-shaped, i.e. its height in this position. The maximum height for conventional road transport is around 4 m.

Vol. 1, *Sect. 26.7.2.2 Three-Hinged Portal Frame Subjected to Linear Load*

9

The following portal-frame variants are common in conventional steel construction:

- **Restrained stanchions** (◘ Fig. 9.46): the simplest way of forming a stiff truss from columns and beams. In steel construction, the restraint is done by means of anchors and a base plate welded onto the steel stanchion (◘ Figs. 9.48 and 9.49). To prevent bending of the foot plate when strong moments occur, the plate is made either thick enough or otherwise additional stiffeners are incorporated (◘ Figs. 9.48 and 9.49). Alternatively, the restraining can also be produced by introducing the stanchion in a sleeve foundation and grouting the sleeve cavity (◘ Fig. 9.50). The single-span beam, which is thus placed on articulated supports, is exposed in this case to considerably

Vol. 3, *Sect. 5.4.1.3 Column Connections* and *Sect. 5.4.2 Connections with Hinge Pins*, in particular Figs. 5.34 and 5.35

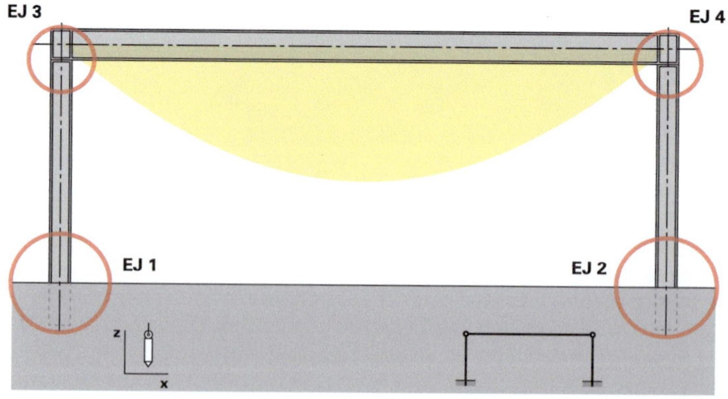

◘ **Fig. 9.46** Truss consisting of restrained columns and articulated single-span beam; no systemic bending moments on the columns due to vertical loads, only due to horizontal loads; large field moments in the middle of the span. **EJ** erection joints

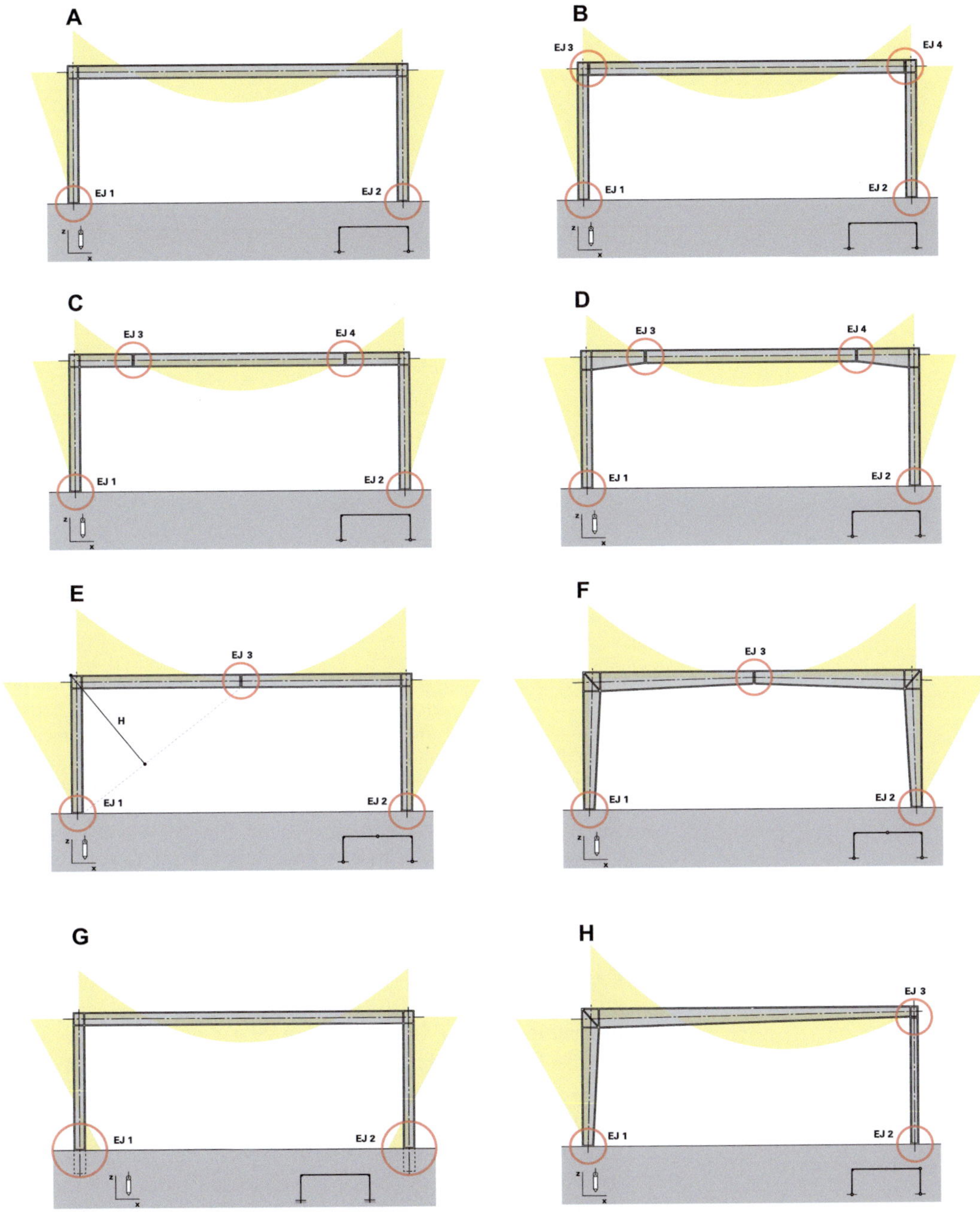

Fig. 9.47 Various execution types of portal frames in steel. **A** Double-hinged frame, completely prefabricated. Erection joints **EJ** at the base points. **B** Double-hinged frame. Erection joints **EJ** at the corners, greatest moments. Easy transport of bar-shaped frame beams and stanchions. **C** Double-hinged frame. Erection joints **EJ 3/4** at the points of contraflexure. Good transportability, only low moment stress on the two joints. **D** Double-hinged frame as on **C**, with haunches at the corners. **E** Triple-hinged frame. Erection joint **EJ 3** in the middle of the frame beam; transportability, no moment stress at the joint, but higher bending moments at the corners. Complete frame half can be transported if $\mathbf{H} \leq 4$ m. **F** Triple-hinged frame as on **E**, with haunch in the frame beam and diagonal compression stiffener in the corner field to absorb the larger bending moments. **G** Restrained frame. Execution of **EJ 2/3** at the construction site, by anchoring or insertion into sleeve foundation. Elevated moment parabola compared to the two-hinged frame: smaller field moment, larger corner moments; change of sign of the moments in the stanchion. **H** Asymmetrical frame with pin-ended column right; maximum moment size comparable with single-span beams, but frame, as a whole, stiff against horizontal force

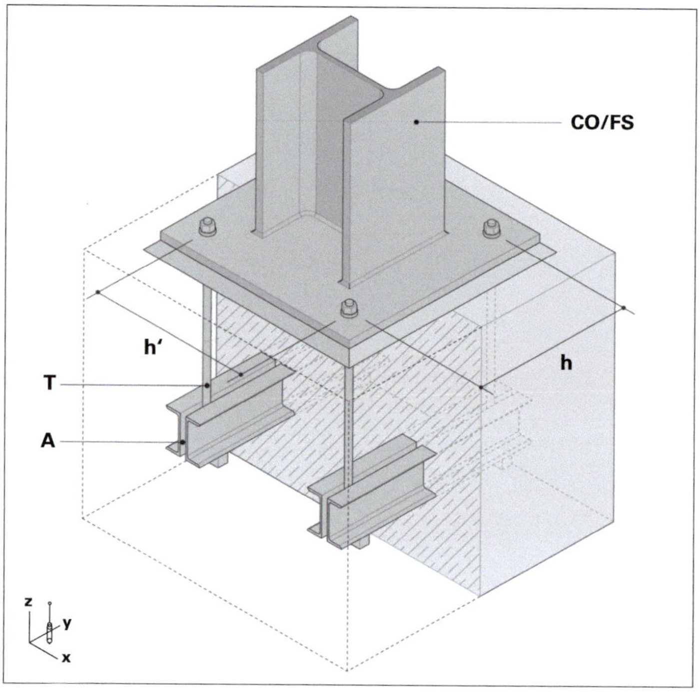

Fig. 9.48 Anchor bar **A** and T-head bolts **T** for a column **CO** or portal-frame stanchion **FS** restrained in the foundation. In this example, both the frame stanchion **FS** and the restraint are designed for bending moments in two main directions → **x** and → **y**: The stiffnesses of the wide-flange I profile are comparable in both directions; the lever arms **h** and **h′** of the anchor likewise. This design is suitable for making the frame stable even at right angles to its plane (i.e. in plane **yz**). The horizontal thrust of the frame is absorbed here by the shear strength of the tie rods. Alternatively, a thrust block as in Fig. 9.51 is also executable

9

greater bending moments than the beams of two-hinged frames or restrained frames, something which requires greater overall structural depths.

- **Restrained portal frame** (Fig. 9.47**G**): All connections, i.e. the footings of the stanchions (Figs. 9.48, 9.49, and 9.50) and the corner connections between horizontal and vertical members are made bending-stiff (Fig. 9.45). This is a portal-frame system typical of industrial construction when high loads, for example from a crane runway, have to be introduced into the frame and therefore high stiffness is required. Restrained frames are also added to form so-called storey ladders in multi-storey construction (Fig. 9.52, right).
- **Double-hinged portal frame** (Fig. 9.47**A**–**D**): In this case, the footings are hinged (Fig. 9.51), something which considerably simplifies the assembly on site compared to the restraining of the stanchions described above. Activating a—positive—field moment in the middle section of the horizontal member, as it occurs in the case of double-hinged frames, leads to a reduction of the

***Vol. 1**, Sect. 26.7.2.1 Two-Hinged Portal Frame Subjected to Linear Load*

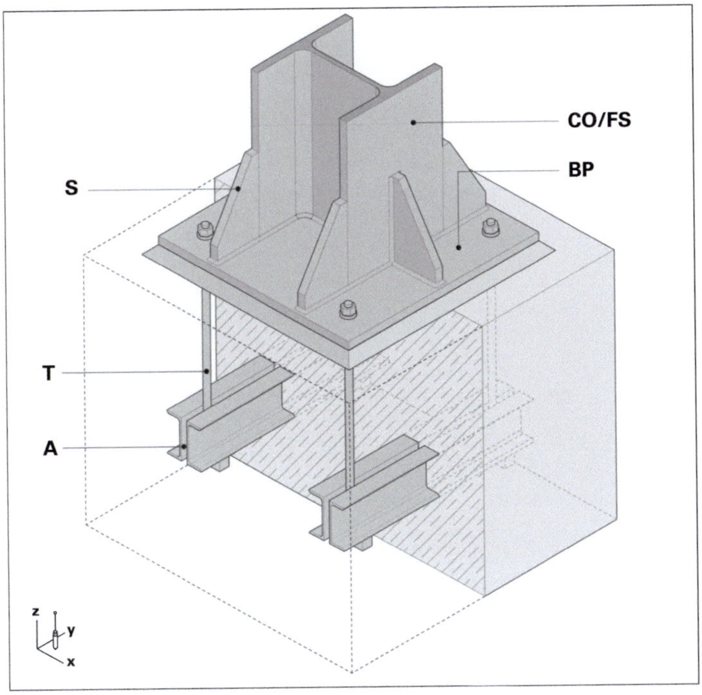

■ **Fig. 9.49** Stiffening of the base plate **BP** with stiffeners **S** in both main directions → **x** and → **y** to absorb larger bending moment. Alternatively, the base plate can be made without stiffeners, but thicker for greater bending stiffness. Type of restraint otherwise comparable with ■ Fig. 9.48. **CO** column; **FS** frame stanchion; **T** T-head bolt

corner moments compared to the three-hinged frame (see below) and, overall, to a more uniform moment distribution over the entire frame. This frame system can be used both in hall construction and in multi-storey building. The frame beams can also be executed inclined or kinked, for example to form a pitched roof for halls. As a rule, double-hinged frames require the production of bending-stiff erection joints. These can be located at the corners of the frame (■ Figs. 9.45**C** and 9.47**B**), an option which entails more complex assembly connections, but simplifies transport, since only rod-shaped components need to be transported (stanchions and beams). Alternatively, the mounting joints are often moved into the area of the points of contraflexure of the frame beam (■ Figs. 9.45**B**, **D**, **F** and 9.47**C**, **D**), where only small bending moments occur and the assembly connections are simplified accordingly. In this case, however, a somewhat bulkier stanchion with a crank must be transported.

- **Triple-hinged portal frame** (■ Fig. 9.47**E**, **F**): With the articulated point of contraflexure as erection joint in the centre of the frame beam and both hinges at the footings (■ Fig. 9.51), this type of portal frame can be regarded as a classic frame system for hall construction. The—neg-

***Vol. 1**, Sect. 26.7.2.2 Three-Hinged Portal Frame Subjected to Linear Load*

9

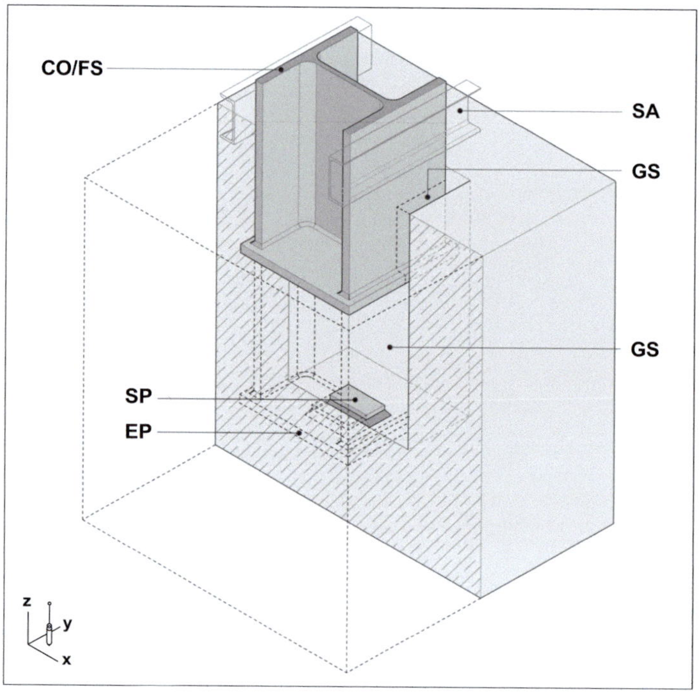

Fig. 9.50 Restraining of a column **CO** or frame stanchion **FS** in a sleeve foundation by concreting the profile in a grouting shaft **GS**, adjusting the frame stanchion by placing it on a levelled setting plate **SP** and aligning it using a setting angle **SA**; **EP** end position of the frame stanchion or column

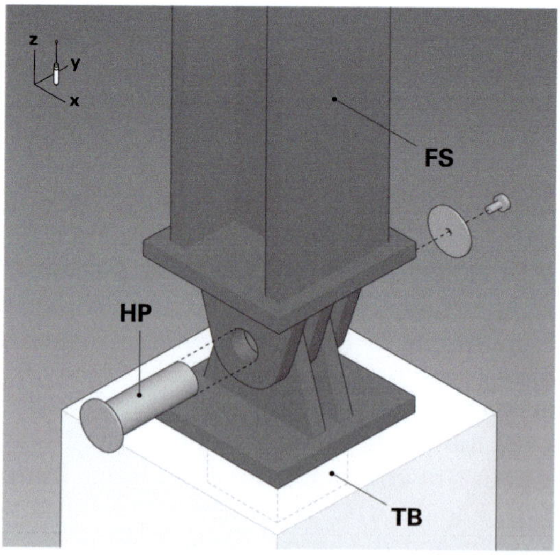

Fig. 9.51 Hinged connection of a frame stanchion **FS**, for example of a two- or three-hinged frame, to a foundation with hinge-pin connection **HP**. This allows large deformations of the frame stanchion and prevents bending moments. The horizontal thrusts are generally transferred into the foundation by means of a thrust block **TB** embedded in concrete. Plane of the portal frame: **xz**

ative—corner moments are larger than those of the double-hinged frame, which, of course, affects both the horizontal and vertical members. In their absolute size, they are comparable to the—positive—span moments of an articulated beam which covers the same span. Otherwise, there are no positive field moments. If the transport dimensions are not exceeded, the triple-hinged portal frame can be constructed from two prefabricated frame sections with purely articulated erection joints (◘ Fig. 9.47E).

- **Asymmetrical frame** (◘ Fig. 9.47**H**): Asymmetrical frame with frame beam connected bending-stiff to a stanchion on one side and supported on a pin-ended column on the other. The truss is, as a whole, stiff in its plane against horizontal loads.
- **Attached beam bays**: In the same plane, further articulated beam bays can be attached to a frame (◘ Fig. 9.52, right). The bracing of the entire structural system within the plane is ensured by the portal frame.
- **Portal-frame chains**: Further frames are connected to a portal frame, all connections being designed to be bending-stiff (see Vierendeel girder) (◘ Fig. 9.52, left).
- **Multi-storey portal frames**: As a replacement for a core as a bracing element, portal frames are added vertically in this solution (◘ Fig. 9.52). This form of bracing can be carried out completely in steel. As described above, additional articulated bays can be attached to the multi-storey frames (◘ Fig. 9.52 right). This form of building bracing is used primarily in the case of buildings with open usages or high degree of mechanical installation. In high-rise buildings, only medium heights can be realised with this construction due to its relatively high yield (◘ Figs. 9.53 and 9.54).

9.3.3 Design of Floor and Roof Structures in Steel

Depending on the use and intended building function, the floor or roof structures are designed differently in steel construction:

Vol. 3, Sect. 20.6.2 Floors in Ribbed Steel Construction

- **Multi-storey buildings**: Composite construction mode; composite floor construction; top concrete on lost formwork made of trapezoidal sheet metal; here, the high live loads as well as the physical requirements for the floor slabs (fire and noise protection) play a decisive role.
- **Hall buildings**: Pure trapezoidal-sheet roof construction; the relatively small live loads (snow, wind) and the low requirements in terms of building physics allow a light construction.

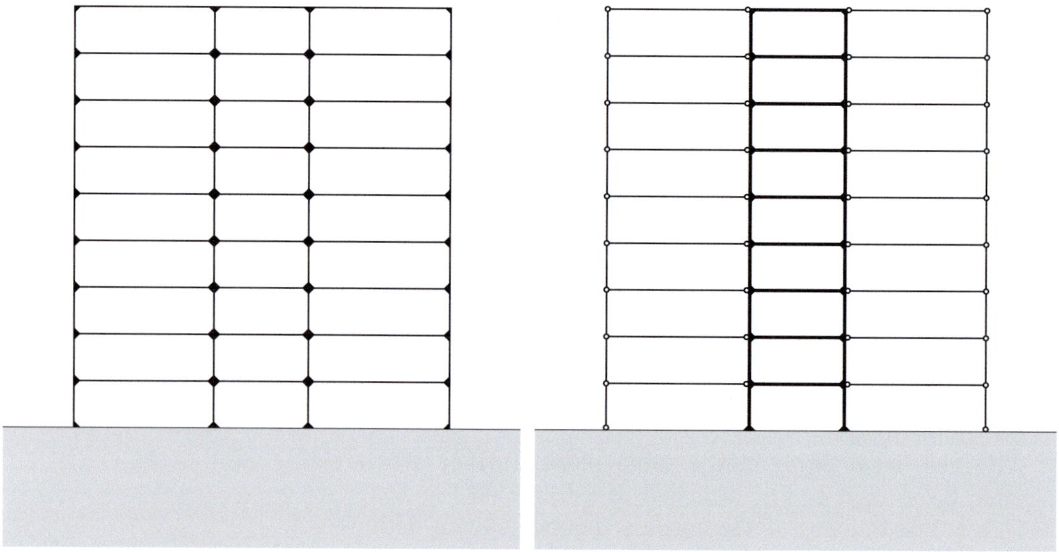

Fig. 9.52 (Left) Multi-storey portal frame, structural system. Three-span multi-storey frame. (Right) Multi-storey portal frame, structural system. Portal frame only between the inner columns

9

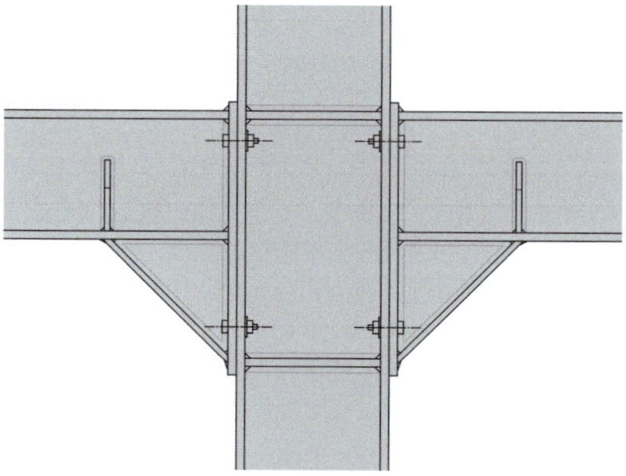

Fig. 9.53 Possible design of the node of a multi-storey portal frame

9.3.3.1 Floors in Composite Construction

Vol. 3, *Sect. 20.6.2.2 Steel-Concrete Composite Floors*

In the case of skeletal steel structures, floor constructions are generally also made of reinforced concrete. They are particularly favourable both in terms of load-bearing and building physics. Being in-plane shear-resistant plates, they contribute to the bracing of the building in the horizontal plane and, at the same time, ensure the necessary sound and fire protection.

The shear-resistant connection of the plate and the steel carrier serves the absorption of horizontal shear forces from bending. Flexural compressive stresses are absorbed by the concrete slab; flexural tensile stresses by the steel beam, which is why a correspondingly smaller profile can be selected for the steel beam than with a conventional girder floor. **Bonding ele-**

Fig. 9.54 Multi-storey portal frame (*JP Morgan Chase Building*, San Francisco)

ments are required to produce the shear-resistant bond between the steel beam and the concrete slab. In steel construction, this is almost exclusively the **shear stud**.

Composite floors (Figs. 9.55, 9.56, and 9.57) are now often made of *Holorib* trapezoidal sheets, which fulfil the function of a lost formwork during the concreting process. Under certain conditions, the sheet metal can also take over flexural tensile forces. For this purpose, it is necessary to secure the bond at the interface between concrete and sheet metal, either by the adhesive bond itself, by beads or studs embossed into the sheet metal, or by welded-on reinforcing-steel meshes. Finishing elements such as suspended ceilings, ventilation ducts, electrical cable trays, or similar can be attached to the dovetail-shaped grooves of the sheet metal in the ceiling cavity.

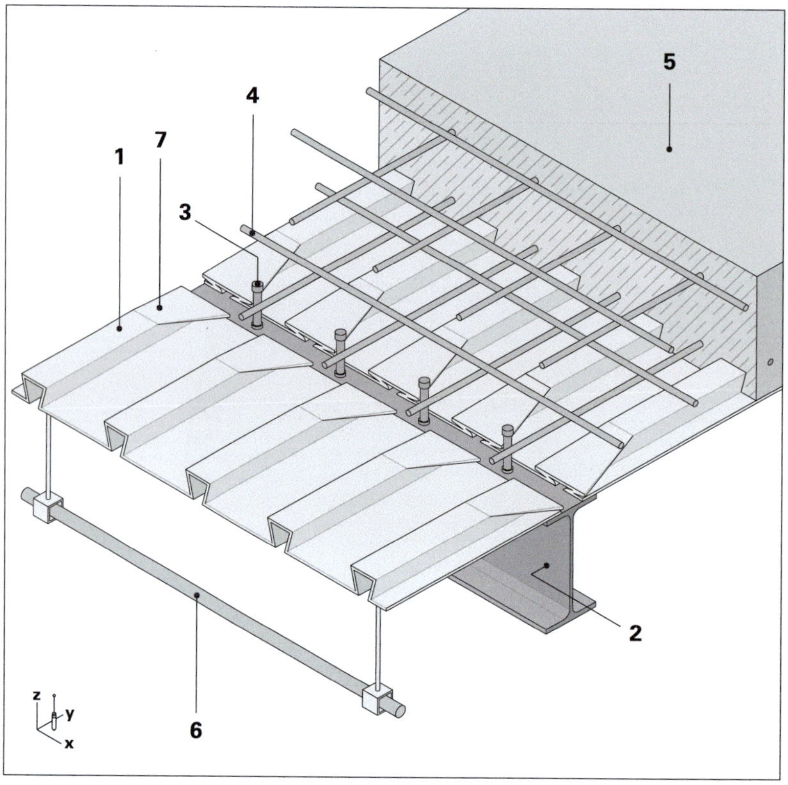

Fig. 9.55 Steel-concrete composite floor with depiction of the additional slab reinforcement and a possible suspension of interior fitting components from the swaging of the trapezoidal-sheet metal. (**1**) Trapezoidal sheet, (**2**) steel beam, (**3**) shear stud, (**4**) bar reinforcement, (**5**) top concrete, (**6**) suspended interior fitting part, (**7**) crimped trapezoidal-sheet end for final anchorage in the concrete (sheet deformation anchor) and for closure of the shuttering space

9

A major disadvantage of composite floor structures is the comparatively large overall height of these systems, which results from the combined depth of steel beams and concrete floors. Modern **slim-floor** constructions (Fig. 9.58) can be regarded as an attempt to counter the problem of the large construction depth of composite floors. With them, the joist is integrated flush with the construction depth of the floor slab on the top and bottom side. I-profiles or cap profiles with attached shear studs are concreted directly and flush within the reinforced-concrete slab, yielding a reduced construction height of approx. 20–25 cm total depth.

Cf. also ▶ *Sect. 10 Prefabricated Concrete Construction*, ▶ Fig. 10.37

In addition to the advantage of reducing the overall depth of the floor construction, this construction creates a flat ceiling without girders and without any obstruction to duct routing. This is a great advantage because, in practice, the planning coordination of installation and load-bearing structure rarely runs smoothly. Furthermore, mechanical equipment for buildings is subject to much shorter renewal cycles than the primary load-bearing structure. Because of the associated unpredictability of the future duct routing, the completely obsta-

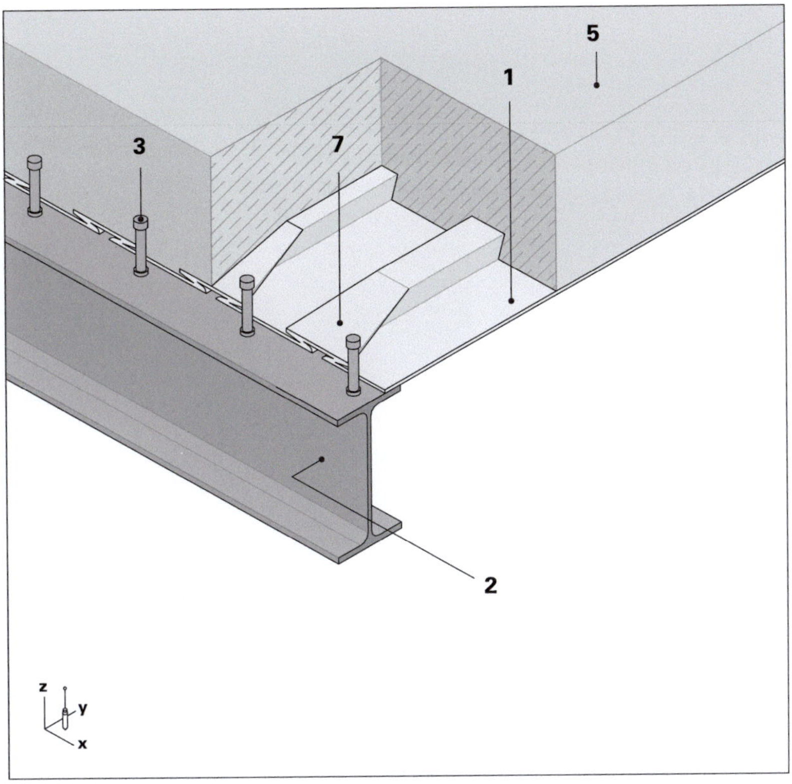

Fig. 9.56 Steel-concrete composite floor, schematic principle with main elements (same key as Fig. 9.55)

cle-free installation of this type of ceiling is particularly advantageous.

9.3.3.2 Trapezoidal-Sheet Floors and Roofs

In contrast to conventional multi-storey construction with its specific fire and sound-protection requirements for the floor construction, where the use of concrete, for example, as a top layer of concrete on a steel-concrete composite construction, is practically indispensable, floors in simple buildings with low requirements can be constructed as a simple trapezoidal-sheet-metal layer with a flat cover plate forming the floor surface (Fig. 9.59).

The roof plane can be designed in simplified form as a pure trapezoidal-sheet construction, particularly in hall buildings or, as a general rule, in lightweight construction (Fig. 9.60). The roof does not have to withstand heavy traffic loads, nor does it have to meet high requirements in terms of sound insulation or fire protection. Not only flat geometries, but also curved surfaces can be built in this way. From the point of view of building physics, these are ventilated or non-ventilated roof structures (see roofing by *KALZIP*). The trapezoidal sheet can, as the case may be, also take over the function of a vapour barrier, but must be suit-

Vol. 3, *Sect. 20.6.2.1 Trapezoidal-Sheet Floors*

DIN 18807-3

Cf. roof constructions with clamping-rib profiles, e.g. in ***Vol. 3***, *Sect. 14.3.2.2.2 Clamping-Rib Profiles*

9

Fig. 9.57 Steel-concrete composite floor during execution, before laying the sheets. The shear connectors arranged in a row are already welded onto the flange of the beam

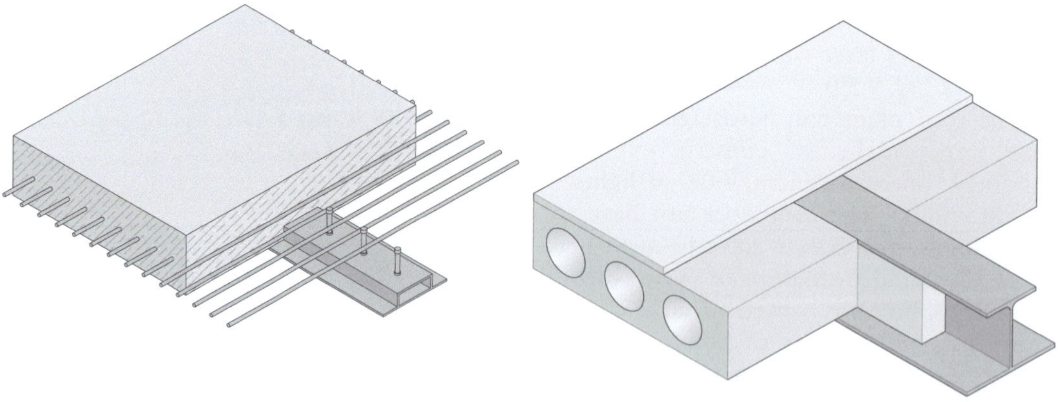

Fig. 9.58 (Left) Slim-floor structure, a composite construction with a flush girder, connected with shear studs. (Right) Slim-floor structure. Welded steel profile with hollow-core slabs, placed on the lower flange

Fig. 9.59 Trapezoidal-sheet metal as part of a floor construction

Fig. 9.60 Trapezoidal-sheet metal roof in an industrial hall. See also Figs. 9.68 and 9.72

***Vol. 3**, Sect. 14.3.3.2.1 Roofs from Steel Girders and Trapezoidal Sheets > Shear Fields*

ably sealed against vapour at the joints. Furthermore, the trapezoidal-sheet plane can also be assigned a bracing function by generating shear fields, but this has an effect on the dimensioning of the sheet metal and its connection to the load-bearing structure. Trapezoidal sheets can cover spans over 10 m at a structural depth of approx. 30 cm.

In the simplest version, the trapezoidal sheet is screwed directly onto the beams; the pleats are additionally filled with mineral thermal insulation. The vapour barrier is installed horizontally on the trapezoidal sheet and welded/glued homogeneously full surface. This layer is followed by a thermal insulation, which is protected against weathering by the roof waterproofing. This results in an exceptionally simple envelope construction which follows the principle of a non-ventilated roof (◘ Fig. 9.61) and, even with a simplified layer sequence, can fulfil all the essential envelope functions—except for the thermal-storage and sound-proofing functions.

9.3.4 Latticed Truss Constructions

Latticed truss systems are of great importance to skeletal steel construction. They offer a predestined application form for rod-shaped steel components under pure axial stress and allow maximum utilisation of the already per-se high strength values of the material due to the bending-free stressing of the cross-sections. Its approximately equal compressive and tensile strength has a positive effect on the use of steel in trussed structures, because this makes it possible to absorb the changing compressive and tensile normal stresses typical of lattice bars. In addition, the high strength of the material allows producing trusses from bars with a high degree of slenderness, which offers transport and assembly-related advantages apart from being an aesthetic factor.

The scope of application of truss girders ranges from small spans (approx. 10 m) to spans of more than 100 m. Some selected truss structures are described in more detail below.

9.3.4.1 Application of Trusses

Steel trusses can be found in the following forms, among others:

- Flexural beams for bearing vertical loads (◘ Figs. 9.62, 9.63, and 9.64).
- Horizontal trusses for creating shear-resistant floor or roof structures.
- Vertical trusses for bracing of buildings, i.e. as so-called shear-plate-replacement systems.

Full-webbed I-beams are efficient flexural beams within medium-scale limits, but cannot be used economically over very large spans because they become too heavy in these circum-

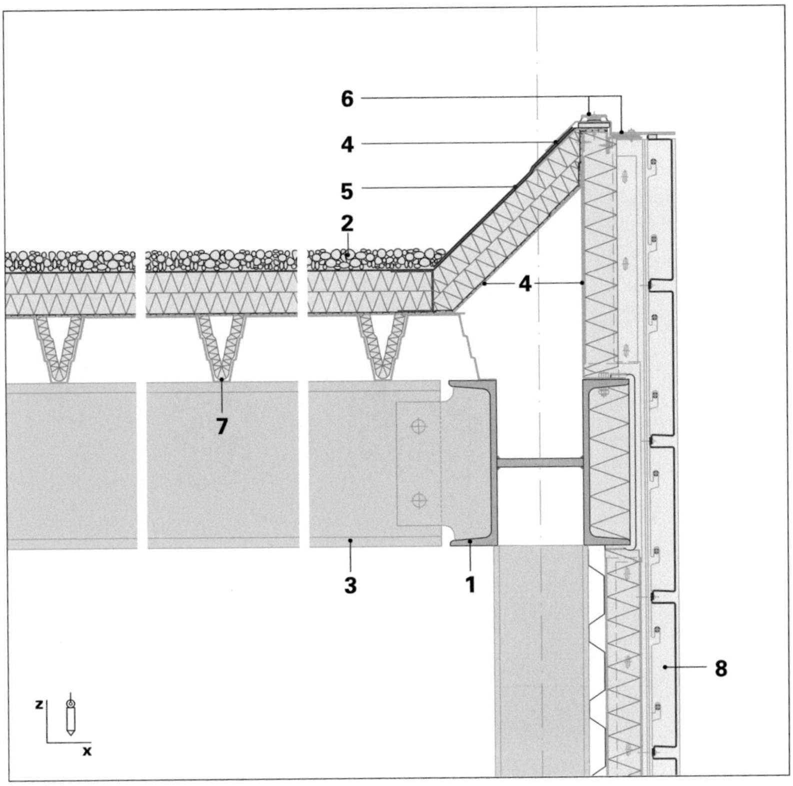

■ **Fig. 9.61** Newspaper printing company Süddeutscher Verlag, (arch.: Peter C. von Seidlein), roof detail (cf. also ■ Fig. 9.19)

(**1**) upper chord of truss girder

(**2**) roof construction:

gravel 16/32 mm, $h = 50$ mm,

waterproofing 1 layer PVC-sheet, 1.5 mm,

mineral-fibre insulation 2 × 50 mm,

vapour barrier, 1 layer PVC-film,

planar steel sheets, galvanised, 0.7 mm, as substrate for the roof layering

trapezoidal steel sheet, 160 mm, galvanised and colour-coated from below, webs perforated (acoustics), covered with a mineral-fibre mat, trickle-protected with lamination

(**3**) subcarrier

(**4**) folded steel sheet

(**5**) two sheet layers as UV protection

(**6**) aluminium clamping strip

(**7**) fastening of the trapezoidal sheets:

pop-rivet connection in web axis of subcarriers

(**8**) back-ventilated façade from folded aluminium sheets

Fig. 9.62 Centre Georges Pompidou, 1975–1977, view of the Western façade from Rue Saint Merri. Main truss girder with two-sided connection to the so-called Gerberette, which serves to reduce the field moments. Further truss formation to brace the building by storey-wise coupling of the beams to a form a building-high trussed framework (arch.: Piano and Rogers, eng.: Ove Arup)

stances.[3] A lattice girder allows for a substantial increase in material utilisation, since the full web is replaced by shear-resistant triangular trusses made of individual bars. These can then be designed and dimensioned according to nearly pure axial compressive and tensile forces. As a deliberate consequence of the structural system, the overall bending of the carrier is converted into normal forces acting within the truss bars, thereby transforming the beam into a lattice of interconnected rods joined with hinges, something which simplifies the constructional task.

Nowadays, in hall and multi-storey building, parallel trussed girders are used, i.e. their upper and lower chords run parallel to one another—in contrast to older triangular trusses (Figs. 9.65 and 9.66). Parallel girders have a construction depth of 1/10 to 1/12 of the structural span. It is a very economical form of load transfer. In the rods, as noted, only axial tensile and compressive forces occur provided the load is systematically introduced into the nodes.

3 Again, the above-mentioned unfavourable ratio between load capacity and dead weight plays the decisive role.

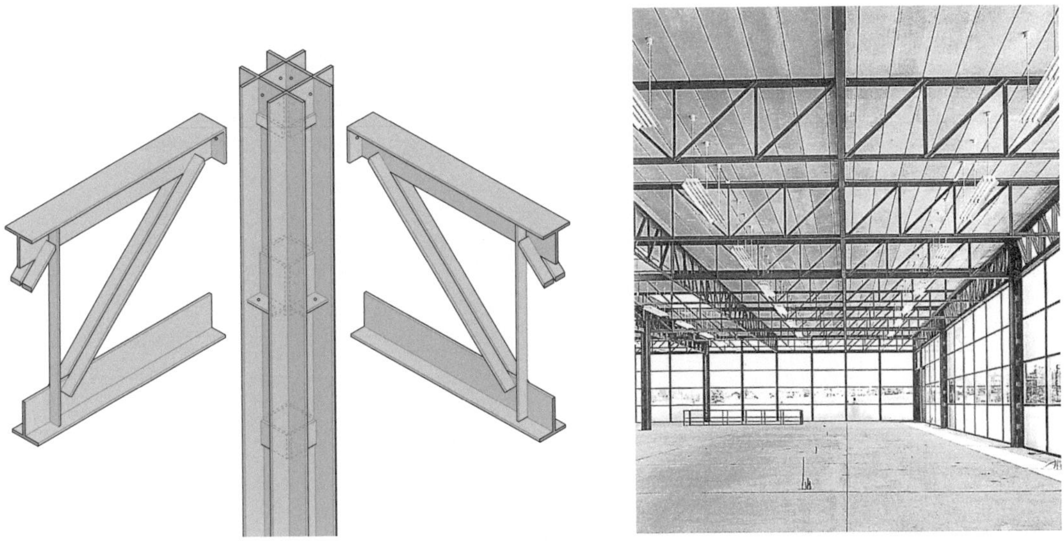

Fig. 9.63 Industrial steel construction system MAXI; non-directional open truss system with a span of 14.40 m (arch.: Fritz Haller). Interior view (right) and mounting of the connection to the column (left)

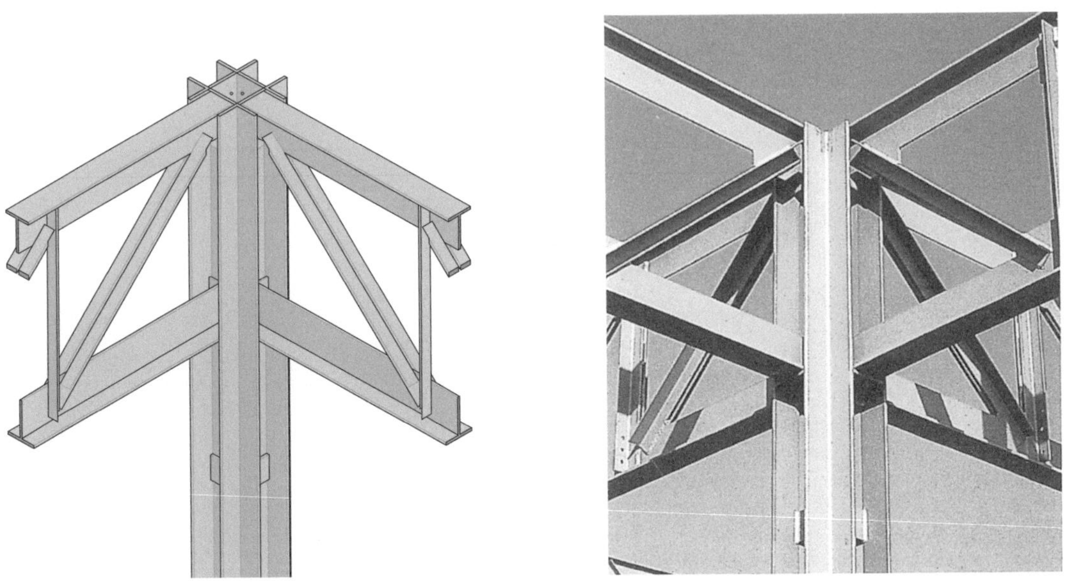

Fig. 9.64 Industrial steel construction system MAXI; axonometry and photo of the assembled construction

In this carrier form, the advantage of reduced material consumption is countered by the disadvantage of a comparatively high expenditure in manufacture, in particular when an ideally hinged design of individual nodes is intended. Conventional truss nodes in steel construction are, for this reason, rather approximations to this state, so that a certain limited bending stress in the rod is put up with in favour of a simplified joint design.

Fig. 9.65 Machinery hall of the Süddeutsche Print Company. The roof is composed of trussed girders with a span of 35 m (arch.: Peter C. von Seidlein). Cf. also Figs. 9.19 and 9.61

Fig. 9.66 *Patera* system in the London Docklands with an external structure made of coupled portal frames. The system can be installed by hand without any additional tools (arch.: Michael Hopkins)

Under outdoor weathering, the relatively large exposed surface of trusses and the numerous corners and edges of the construction—compared to full-webbed constructions—is a risk factor with regard to corrosion.

9.3.4.2 Rules for the Design of Trussed Systems

The following basic rules ought to be observed when designing trussed steel structures:

- Pervasive triangle formation, or triangulation, is an important design principle: For purely geometrical reasons, a triangle is rigid in itself within its plane. In contrast, quadrangles or polygons with more than four sides are not stiff without additional measures; they must be made shear-stiff by means of rigid nodes and bars sufficiently dimensioned for bending and shear force. This, in turn, significantly reduces the structural efficiency of the truss. Vierendeel beams, which are based on a rectangular geometry and hence lack triangulation, are only useful in practice if the absence of diagonal bars, and the ensuing loss of stiffness, offers in return reasonable advantages in other areas (e.g. usage or installations routing). However, compared to triangulated trusses, Vierendeel systems are always less material-efficient, heavier, and visually less delicate.
- Hinged node design: All rod connections are ideally hinged; otherwise, they are designed as an approximation to a hinged connection.
- Pure axial tensile and compressive forces flow within the bars: There is no bending stress by design.
- The load introduction into the truss must take place at the node points; there must be no transverse loading of individual bars, for instance, the ones of the top chord by the load of top-mounted floors or roofs, since these are otherwise subjected to bending and the advantage of the pure axial stress is lost thereby.
- Bars are always executed in a straight line, i.e. neither bent nor kinked. If this is not the case, the compressive and/or tensile stress in the bar generates bending moments by eccentricity, which flagrantly contravenes the basic design principle of the truss. For this reason, even overall curved trusses are never made of curved bars, but are assembled from straight elements to form a polygon which visually sufficiently approximates the curved shape (■ Fig. 9.67).
- Compression bars should be executed as short as possible, since these—in contrast to tensile bars—are subject to buckling. Already the correct geometric design of the truss, i.e. choosing a favourable angle of inclination of the diagonal rod which is neither too steep nor too shallow, is a major influencing parameter.

  In the case of a free choice between compressive and tensile diagonals, the latter are therefore always preferable.

Examples of designs of welded truss nodes can be found in ***Vol. 3***, *Sect. 9.2.8 Constructional Standard Solutions of Steel Construction*, in particular Figs. 9.17, 9.18, 9.19, 9.20, and 9.21

Fig. 9.67 Three-chord trussed arch of an ice-skating rink in Munich, 1984 (arch.: Kurt Ackermann, eng.: Schlaich and Bergermann)

- Because of the danger of buckling, to which compression bars are always exposed, using twin compression bars is also generally unfavourable. If the cross-sectional area required for the transmission of the compressive force is divided between two separate profiles, as it happens with the doubling of the bars, the result is necessarily slimmer bars than with a single bar and thus an increased risk of buckling. The profile factor of the cross-section worsens accordingly, which in turn has an unfavourable effect in terms of fire and corrosion protection.

  Naturally, increased buckling risk does not apply to pure tensile bars, where there is no danger of buckling in any case. For this reason, tensile and compressive members are sometimes made from different profiles (Fig. 9.71) which, respectively, better match their particular stresses.

Cf. ***Vol. 1***, *Sect. 29.10.4.1 Profile Factor* $A_{m,p}/V$

- Acute angles at the nodes should be avoided as a matter of principle. This is a geometrical and constructional connection problem. First and foremost, a mutual interference of adjacent connecting bars with extremely acute angles must be avoided.

Vol. 3, *Sect. 2.3.1.2 Geometrical Constraints*

- The system lines of the rods converging at the node points should preferably intersect at one point: This prevents misalignment moments which are a disturbance in the normal-force-loaded structure of the truss and cause local bending stresses.

Cf. marginal note above

- Trusses are largely prefabricated construction elements, which are only manufactured in sections to be assembled on the construction site if the overall element is beyond the reasonably transportable size. As a result, almost all nodes consist of factory connections. In practice, these are mostly welds. The choice of suitable rod profiles for chords and filling rods (see below) as well as the favourable geometric design of the connections plays a decisive role in the constructional design of trusses.

9.3.4.3 Profiles for Trussed Structures

Due to the typical normal force load on the bars, bar cross-sections with rather compact geometries are most suitable for trusses: for example, **I-sections** of the HE series, which have similar section moduli in both bending axes. This is particularly important for compression members which are subject to the same risk of buckling in all spatial directions due to their loading (cross-sectional *shapes* of pure tension members are irrelevant from the point of view of force transmission, only cross-sectional *areas* play a role in that case). Thanks to the perpendicular cross-section geometry of I-profiles, the connections at the nodes are simplified, provided the chords are also made of I-profiles. Diagonal bars can, for example, be inserted into the profile chamber of the chord bars and connected to the chord flanges. Gusset nodes can also be easily made with these cross-sections (◘ Fig. 9.68).

Rectangular tube or **box sections** offer similarly favourable conditions (◘ Fig. 9.69). In that case, butt welds are recommended at the nodes, where forces are transmitted directly from profile wall to profile wall, for example, between the filling rod and the chord rod. This is the case if both groups of members have the same profile width. Thanks to the perpendicular geometry of the cross-sections, the cuts are always straight and therefore easy to trim and weld. In comparison to I-profiles, closed tubular profiles expose considerably less

◘ **Fig. 9.68** Truss nodes made of gussets, bars made of rolled I-profiles. The roof bracings are also visible at the top

Fig. 9.69 Truss node from box sections. Chord and struts are designed with the same width so that a direct force transfer between side walls can take place at the connection. The corners are rounded to minimise notch stresses

surface to the weather and do not create any trough-like areas in which rainwater can accumulate. If the connections are welded airtight, an accumulation of dirt and moisture within the chambers of the profiles can be ruled out.

Round tubes offer ideal conditions for force transmission in latticed trusses. Their circular cross-sections are completely directionally neutral and therefore allow the best material utilisation under normal compressive force since it provides maximum bending stiffness in all conceivable buckling directions. It is somewhat more difficult to form the connection edges of round tube trusses than of rectangular ones. In principle, the tubes can be introduced into flat gusset plates in the area of the nodes which are welded into corresponding slots in the tube (Fig. 9.70, left). This simplifies the connection compared to pure butt joints, which for geometric reasons cause difficult, non-rectilinear trims. For this reason, round-pipe trusses with butt joints at the nodes were formerly generally avoided. Today, however, the production of butt joints has become much easier with round tubes thanks to automated fabrication systems: The tubes are first trimmed at their ends to the exact intersection geometry of the node by CNC machines and then butt-welded together in the factory (Fig. 9.70, right). In addition to their structural and constructional advantages, tubular-truss constructions also offer the weather only a small surface to attack, have a favourable profile factor with regard to fire protection and also create a clean and calm visual appearance without separate node elements.

Fig. 9.70 (Left) Truss node made of tubes with slotted and welded connecting plates. (Right) Truss nodes made of butt-jointed and welded tubes

Designs are also possible—as noted—in which different profiles are used for chord bars, compressive and tensile bars, depending on the load and geometric connection conditions in the node (Fig. 9.71). So-called R-beams, for example, combine tubular cross-sections for the chords and steel bars for the diagonals.

Frequently, tube profiles are used to form so-called **three-chord trusses** (Fig. 9.67). The triangular cross-section of the girder offers various advantages: for example, by its increased

Fig. 9.71 Cross-sectional design of the diagonal bars according to their stress: compression rods made of round tubes; tension rods made of finely structured bar elements, broken up into trusses (*Firth-of-Forth* Bridge, Edinburgh)

Fig. 9.72 Horizontal roof trusses in a hall

9

torsional stiffness; its tilt resistance when supported on both upper chords; the buckling stiffness of the double upper chord—provided it is compressive-stressed and the cross-sectional triangle points downwards with the tip, as is often the case with bending beams—or use-related aspects as in the case of incorporating top-light ribbons, etc.

9.3.4.4 Use of Horizontal Trusses to Form Shear-Resistant Roof Structures

Horizontal trusses act as horizontal beams destined to absorb horizontal loads and thus form part of the building's bracing. Such bracings are rather an exception in multi-storey buildings. The bracing is generally ensured in that case by means of the racking stiffness of the concrete floors. Only during the construction phase, diagonal bracings may occasionally appear in such buildings as temporary stiffening measures. The situation is different in hall construction. Heavy structures such as concrete slabs are avoided there in most cases. With the aid of horizontal diagonal bracings, so-called shear-plate-replacement systems are produced under these circumstances. Often, bracing is executed with pure tension rods in both diagonal directions, using simple L-profiles or steel bars (Fig. 9.72). These are then frequently located in the plane of the upper chord of the main carriers for an additional stabilisation (securing them against tilting/torsional buckling under heavy bending).

9.3.5 Spatial Trusses

The attempt to use truss systems in three dimensions (◘ Fig. 9.73) took place relatively early in the development of steel structures. Alexander Graham Bell (◘ Fig. 9.74) already experimented with spatially rigid tetrahedral structures for flying objects and transferred this extraordinarily firm basic structure to the construction industry in form of a framework of standardised tetrahedrons made of metal rods. He initially used this spatial framework construction—a so-called **spatial truss**—for temporary buildings (◘ Fig. 9.75).

The possibilities of prefabrication and lightweight construction in the application of spatial trusses were later taken up and further developed by Konrad Wachsmann and Buckminster Fuller. The two-way load transfer made possible by the spatial arrangement of the structure and the lightness of the design predestine these structures for wide-span constructions (see aircraft hangar by K. Wachsmann on ◘ Fig. 9.76). Based on the geodesic dome geometry, Fuller developed a rationalised, fully triangulated system with only a few different bar lengths which allowed the construction of spherical space grids (◘ Fig. 9.79).

◘ **Fig. 9.73** Regular spatial framework built on half octahedrons using the *Mero* system (University of Stuttgart, provisional lecture hall (arch.: F. Wagner))

Fig. 9.74 A. Graham-Bell, spatial truss structure made of tetrahedrons, used around 1900 for temporary buildings. Early example of modular structures in steel construction

For the concept of spatial tessellation, cf. ▶ *Sect. 3.1.6.1 The Vertical, Planar Space-Enclosing Element*

The volume enclosed by a space truss is subdivided into individual modular space units by means of the rod frames. It follows therefrom that only those body geometries which are liable of filling the three-dimensional space fully without gaps, i.e. which tesselate it in three dimensions, can be considered for these space modules. The geometrically regular polyhedra of the cuboid, tetrahedron, truncated octahedron, rhombic dodecahedron, and a few more belong to this category, which comprises only slightly more than 20 regular and semi-regular variants.

In its early stages of development, the basic spatial geometric elements of a spatial truss had to be designed as simple and recurrent as possible in order to keep the number of rods of different lengths and of node elements with different geometries to a minimum for serial production. Exemplary elementary variants that met these requirements consisted of (Fig. 9.77):

- A **Tetrahedron** from four equilateral triangles (Fig. 9.74): All bar lengths are equal. All node geometries (except the edge nodes) are identical. This variant has the advantage of being triangulated in all bar-frame planes, which significantly increases the stiffness of the overall structure. Or:
- A **half octahedron**, i.e. a four-sided pyramid, of four equilateral triangles and one square (Fig. 9.77). Here too, the

Fig. 9.75 Power mast for overland lines composed of three-dimensional trusses from simple hot-rolled L-profiles, formerly with rivet connections, today in a combination of welding and bolt connections

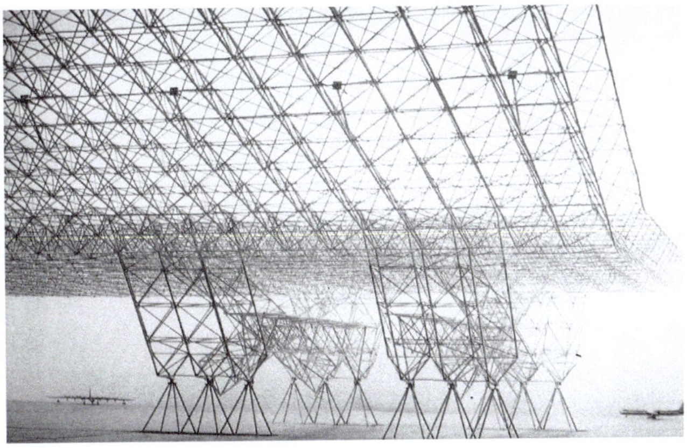

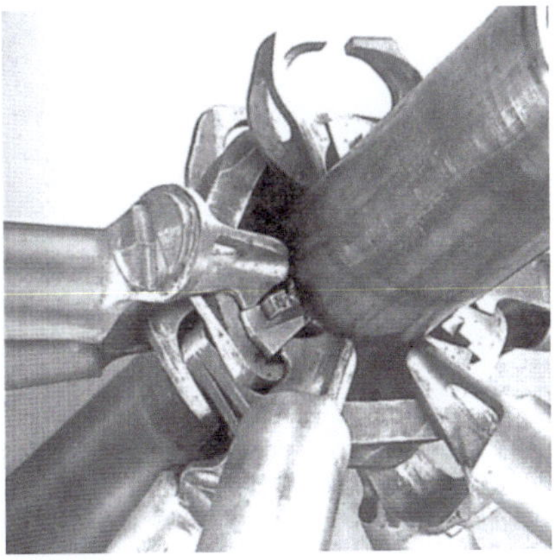

Fig. 9.76 (Left) Model view of an aircraft hangar: large-scale truss with supporting structure (arch.: Konrad Wachsmann). (Right) Standard node of the spatial truss on the left developed during the Second World War for the US Air Force. Upper-chord area with connection for roof panels or other covering elements

bar lengths and node geometries (again with the exception of the edge nodes) are the same, but the resulting squares at the top- and bottom-chord planes are not triangulated.

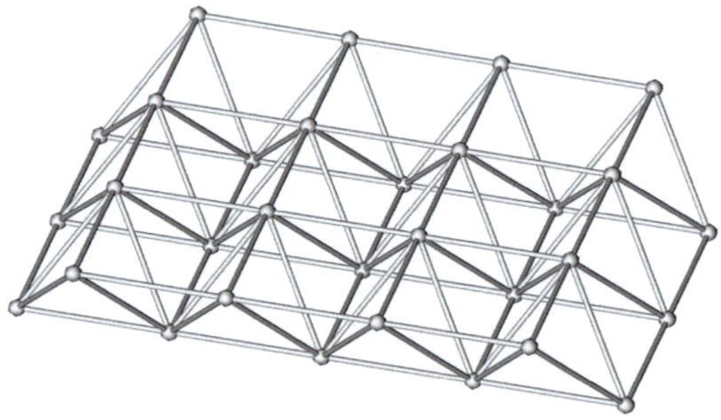

Fig. 9.77 Axonometry of a spatial truss, here *Mero*-System by Max Mengeringhausen

Overall, the result is a flat top and bottom surface, each corresponding, respectively, to the top-chord and the bottom-chord plane, and three-dimensionally, i.e. non-planarly aligned, diagonal bars.
The availability of automated, digitally controlled manufacturing facilities has partially removed the need for the strict geometric specifications to which the early spatial frameworks were subjected. It is no longer indispensable to always use the same bar lengths and the same node geometry, although a certain regularity helps to limit costs and complications during erection on site.

In recent years, various spatial frameworks have been created, the geometries of which are modelled on space-forming foam, cell, or crystal structures (Fig. 9.80, left). There is often the pretension to fill the space with polyhedral modules of the smallest possible outer surface, which in the end is supposed to result in the smallest possible total bar lengths. The truss geometries, which are thus not based on triangles but on quadrangles, pentagons, or hexagons, therefore lose the triangulation, so that the geometrically no longer rigid truss polygons must be stiffened by increased bending and shear stiffness of the truss rods. This can only be done at the cost of additional material consumption, a circumstance that makes these trusses heavier and ultimately leads the original claim of material economy by shortening the bar lengths to absurdity (Fig. 9.80, right).

9.3.5.1 Execution of Spatial Trusses

As noted, early spatial trusses could only be realised by using a strict construction system that allowed a standardisation of the truss. There were various approaches, such as the *Mobilar* structure by K. Wachsmann (Fig. 9.76) or the *Mero* system by M. Mengeringhausen (Figs. 9.77 and 9.78), which was later realised with great success and allows the articulated connection of 18 rods in given directions. The *Mero*-spaceframe has been used in many different forms. It is suitable for one- and two-way load transfer and can also be implemented in curved shapes (cf. also Fig. 9.79).

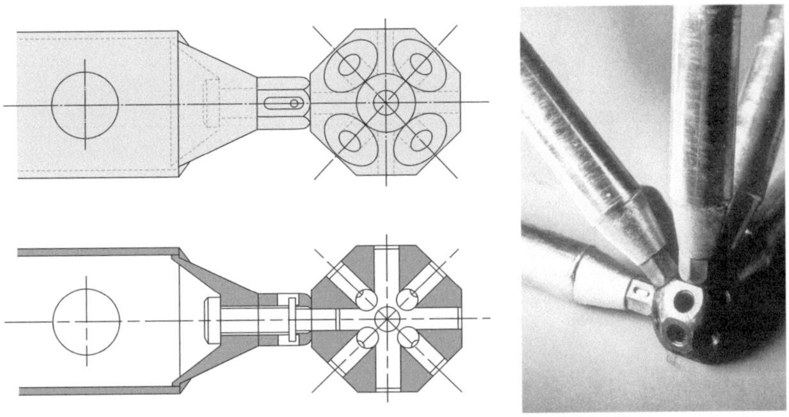

Fig. 9.78 *Mero* spatial-truss node: the world's most successful space-truss system with central node, with a maximum of 18 connection threads (Max Mengeringhausen)

Fig. 9.79 Pavilion of the USA at the World Exposition in Montreal, construction stage (arch.: Buckminster Fuller)

Today, modern CAD/CAM technology opens up much greater freedom in the geometric and constructional design of space trusses. In recent years, various complex shaped truss structures have been realised with numerous rods and nodes customised for local geometric conditions. This allowed, in addition to executing the above-mentioned special grid geometries (Fig. 9.80) inspired in natural forms, also realising non-regular curved surfaces. Even elementary grid geometries, such as continuous triangular meshes, require an individualisation of bar lengths and node geometries as soon as their overall geometry no longer corresponds to elementary surface types, such as spheres or translational surfaces. In the future, further significant innovations are expected in this area, which probably will significantly expand the range of use of space frames.

Fig. 9.80 (Left) Space framework of the Olympic Water Sports Centre Beijing. The grid geometry is based on a space-filling system of polyhedra (Weaire-Phelan structure) with the smallest possible surface (i.e. also theoretically the smallest possible total bar length) at a given truss density (arch.: PTW Architects; eng.: Ove Arup). (Right) Space framework of the project on the left. The missing triangulation requires heavier dimensioning of the lattice bars, which is quite clear from this picture. The rods were butt-welded to ball joints on site

9

9.3.6 Lattice Shells

► *Sect. 3.4.5.1 Shells* and ***Vol. 3**, Sect. 14.4.3 Coverings from Grid Shells*

► *Sect.* 3.4.4 *Surface Structures Under Membrane Forces*

Lattice or grid shells stand for a very recent shell construction method which was developed, in a certain way, as a replacement for concrete shell construction methods as were introduced and often realised in the middle of the last century. Complicated formwork with a high proportion of labour costs, as is almost inevitably associated with concrete shells, is circumvented by the latticed shell construction method. The shell surface is not generated by a two-dimensional solid-walled component. Due to the relatively high dead weight of steel, this design solution is practically impossible to find with this particular material.[4] Instead, the shell structure is created by an extremely light grid of short individual bars, which are coupled into a framework on the basis of various grid geometries. The grid is usually covered with glass at the end. The shell load-bearing behaviour means that there is hardly any significant bending stress transverse to the shell surface, but almost exclusively membrane forces tangentially thereto. This means that, in the bars, practically only axial compressive and tensile forces act while, tangentially to the shell surface, additional shear forces are involved. The latter must be absorbed either by frame action, i.e. bending stiffness of the bars and nodes, or alternatively by triangulation, i.e. triangular-mesh formation.

4 Similar to floors, and for the same reasons as for those, load-bearing pure surface structures for shells are found at best in trapezoidal-sheet-metal design. On the other hand, practically two-dimensional shells can be made of wood (see ► *Sect.* 8.6.3.6 *Genuine Load-Bearing Surface Structures*).

Fig. 9.81 Quadrangular-meshed frame shell without diagonal bracing. The bars, which are subjected to bending stress tangential to the shell surface, must be reinforced accordingly in their cross-sectional width, the dimension which is most easily perceived when looking from below (courtyard canopy in Dresden Palace)

The former is the case for non-rigid mesh geometries, i.e. for quadrilaterals or all polygons with more than four corners. Sufficiently rigid bars, however, require significantly larger cross-sections and are visually much more prominent than bars subjected to normal force—a fundamental contradiction to the idea of the extremely light bar network (Fig. 9.81).

Triangulation is therefore used more frequently: This is done either by arranging the bars in a triangular grid already from the outset, or alternatively by subsequent diagonal stiffening of a non-triangular grid, for example, a square mesh grid. For this purpose, continuous ropes are threaded through the rod nodes in both diagonal directions and subjected to pretension. In this way, more complicated node constructions with more than four abutting bars are avoided (Fig. 9.82).

Similar to planar trusses, the pure normal forces in the bars allow them to be designed extremely slender. This is additionally supported by the extremely high strength values of steel. Again, the danger of buckling of compression members suggests the choice of a compact cross-section. In order to make the bars as slim and visually as unobtrusive as possible, a factor that plays an important role with this type of visually almost immaterial mesh structures, even fully solid bars are often chosen. As these are already relatively short for other reasons—in particular, because the formats of the glass roofing need to be kept small from the outset—and thus also the buckling lengths are small, the rods can indeed be made extremely slim.

A factor most decisive for the structural design of the lattice shell is, besides the constructional formation of the lat-

Fig. 9.82 Quadrangular-meshed, diagonally rope-braced lattice shell. It is a translational surface that can be covered with planar glass panes (Bosch-Areal, Stuttgart; eng.: Schlaich, Bergermann & P)

► *Sect. 1.2 Shape Definition of Continuous Curved Layer Surfaces*

tice grid itself, also the definition of the overall shell geometry. Although, thanks to the possibilities of automated CNC manufacturing, practically any arbitrary free form can be built today—albeit, as the case may be, with a corresponding effort—when operating with completely irregular geometries this, however, requires a large number of individually manufactured customised parts, something which makes both production and assembly particularly complicated and expensive: This concerns both rod lengths and connection angles at the node elements. Free forms also force to triangulate the covering material, since it is usually planar, something which sometimes, like in the case of glass, is only reluctantly accepted because of the sensitive sharp corners of the panes and the more complicated connection geometry at the same (Fig. 9.83).

► *Sect. 1.2.3.3.3 Translational Surfaces* as well as ► *Sect. 1.2.3.3.4 Scaled Translational Surfaces*

For this reason, translational surfaces that can also be geometrically scaled (scaled or stretched translational surfaces) are often used, thus opening up a large number of geometric variants (Fig. 9.82). This category of surfaces offers the great advantage of allowing the use of planar quadrangular covering elements—albeit generally neither square nor rectangular ones—which is very much in line with the material characteristics of glass, as well as rationalising bar lengths.

The glass elements are usually laid flat on the rods and only fixed with discs at the corners (Fig. 9.84). The joints are filled with sealant, without pressure strips, which would interfere with water drainage and would accumulate dirt.

► *Sect. 4.3.2.5 Dome from Bars*

Without having to realise a multitude of different rod lengths and node geometries, free-form surfaces can also be constructed by rods that are coupled at two levels in hinged connections and run through the nodes unjointed. When shap-

Fig. 9.83 Triangular-meshed lattice shell with free form. The triangular facets allow adaptation to any shape. Both bar lengths and node geometries are usually locally customised. Tangential to the shell surface, shear stiffness is guaranteed

Fig. 9.84 Node of a lattice shell with disc-fixing of the glass panes

ing the shell, the originally square-shaped meshes assume—self-acting, so to speak—gradually different lozenge shapes and in this way create any desired surface geometry. These slatted lattice shells are also diagonally braced with ropes. The above-mentioned difficulties with the covering of the free form, however, remain unsolved with this variant as well.

9.3.7 Tensile-Stressed Structures

The quality and specificity of steel in building construction lies in the extensive possibilities that this tough material with its exceptional strength opens up to the planner and engineer. The ability of the material to absorb extremely large forces has allowed assigning tasks to steel construction which cannot be mastered with any other classic material.

▶ *Sect. 4.3.3 Tensile-Stressed Systems*

In the meantime, however, other materials are also being developed that compete with steel in the field of demanding building tasks, for example, high-strength concretes in high-rise construction or high-strength synthetic fibres in bridge construction.

The pronounced tensile strength of steel enabled the development of novel tensile-stressed constructions which are unique in the history of construction. At the same time, a completely new formal language emerged which is characterised by generic rope-type geometries and differs significantly from the planar or rectilinear forms of conventional bending-stiff types of load-bearing structures.

Purely tensile-stressed constructions benefit from the non-existence of buckling hazard. While the extremely slender steel members—which are so precisely because of their high strength and stiffness—are at high risk of buckling under compressive stress and therefore cannot really exploit their strength under these circumstances; the strength reserves of the material can be instead fully exploited under pure tensile stress. Tensile-stressed cross-sections are not predetermined by issues of cross-sectional geometry or stiffness, but can be designed freely. Only the absolute available cross-sectional area plays the decisive role.

Since tensile structures are extremely material-efficient load-bearing systems from a structural point of view, the relative weakness of steel, namely, the rather poor ratio of load-bearing capacity to dead weight, plays a negligible role in this case. Very large spans can be covered without the dead weight of the steel construction playing a significant role.

► *Sect. 4.3.3 Tensile-Stressed Systems* and ***Vol. 1**, Sect. 26.4.2 Movable Systems*

Since, in principle, no bending stiffness is required for pure tensile stress, most of the load-bearing structures mainly acting under tension are **movable structures** as defined elsewhere. Consequently, the special design features of these load-bearing systems must be taken into account, i.e. the fact that **forms of equilibrium** are involved which, when designing, can only be influenced indirectly but which are otherwise directly related to the prevailing balance of forces. The form is thus, so to speak, closely bound to the force.

***Vol. 1**, Sect. 22.9.4 Ropes, Bundles, and Cables*

The most important element of tensile-stressed steel structures, which for the reasons mentioned above work with non-bending-resistant, i.e. flexible, thin members, is the **rope** (■ Fig. 9.85).

In the succession of early tensile load-bearing structures—here names such as Roebling or Suchov can be mentioned—modern rope construction has produced unique buildings. Some highlights are addressed below:[5]

5 Although some of these are examples from structural engineering which do not belong to edifice building construction, they should nevertheless be mentioned here to illustrate the structural capabilities of steel.

Fig. 9.85 Suspension bridge in Stuttgart, 1992; mast with the connections of the main ropes (eng.: Schlaich, Bergermann and Partners)

9.3.7.1 Suspension Bridges

The *Akashi-Kaikyo* Bridge (jap. *akashi-kaikyo ohashi*) (Fig. 9.86), named after the road from Akashi, is a highway suspension bridge in Japan which connects Kobe on the main island Honshu and Matsuho on the island Awaji with 2 × 3 lanes. With a central span of 1990.8 m, it is currently the suspension bridge with the largest free span in the world. The bridge was inaugurated in 1998.

9.3.7.2 Cable-Stayed Bridges

Le pont de Normandie (the Bridge of Normandy) is the cable-stayed bridge with the largest span in Europe. It crosses the Seine Estuary and connects Le Havre (Upper Normandy) on the right bank to the north with Honfleur (Basse-Normandy) on the left bank to the south. The length of the steel structure is 2143.2 m, with a free span of 856 m between the pylons, which are 215 m high (Fig. 9.87; further example on Fig. 9.88).

Fig. 9.86 Suspension bridge: Akashi-Kaikyo Bridge, Kobe, Japan, 1998

Fig. 9.87 Cable-stayed bridge: Le Pont de Normandie, 1995 (eng.: Michel Virlogeux)

9.3.7.3 Rope-Net Constructions

EN 13411-1 to *EN 13411-8*

Modern lightweight rope-net constructions were developed to produce large contiguous roof surfaces (World Exposition in Montreal 1967, Olympic Canopy in Munich 1972, Fig. 9.89), at a time when the use of textile membranes in the re-

Fig. 9.88 Cable-stayed bridge: Ting-Kau Bridge, Hong Kong, 1998 (eng.: Schlaich Bergermann & P)

Fig. 9.89 (Left) Roofing of the main stand of the Olympic Stadium in Munich, completed in 1972 (arch.: Behnisch and Partner, Frei Otto; eng.: Leonhardt, Andrä & P). (Right) Mast and rope net of the Olympic Stadium in Munich. The panels were covered with acrylic panels

quired dimensions was not yet possible. They are, in this respect, load-bearing and simultaneously area-forming. Since they are forms of equilibrium subject to tensile stress, rope nets are always curved in opposite directions (anticlastic). This is necessary because in them, being movable load-bearing structures, two oppositely running pre-tensioned cable groups—a concave one (the suspension cable group) and a convex one (the pre-tensioning cable group)—keep each other in equilibrium and prevent otherwise unacceptable changes in shape.

▶ *Sect. 4.3.3 Tensile-Stressed Systems*

Rope nets are adapted to their shape by the exact cutting of the rope lengths in such a way that all ropes are placed under tensile force in their final state. This systematic prestressing is applied by presses via the edge cables and edge fastenings. The close-meshed network requires numerous node elements.

The roofing must reproduce the double-curved, geometrically non-elementary form of equilibrium in its curvature, and is very difficult to generate from planar glass panes when used as a covering material, even if, for example, the limited elastic deformability of laminated glass panes is taken advantage of. Therefore, only materials that can be cut to size, such as foils or membranes, or elastically deformable materials such as plastics (as on ◘ Fig. 9.89) or thin wooden parts can be considered for this task.

9.3.7.4 Rope Structures

A further special application of steel in lightweight construction are rope structures. In contrast to rope-net structures as discussed above, these structures do not have to be load-bearing and surface-forming at the same time. Often, the main load-bearing elements, which are set up at larger distances, are completed with a secondary construction (e.g. structural, i.e. self-supporting, membranes) to form a closed surface. The tensile forces in the load-bearing structure must either be short-circuited within the system or otherwise dissipated into the ground by means of suitable anchorages. Although the latter option offers greater design freedom, it is always associated with higher costs due to the resulting complex foundations; short-circuiting the forces within the system is much more efficient, but is only possible for special geometries. These are then called **self-contained** constructions.

Rope Trusses

Rope trusses are bending-stiff load-bearing elements made of ropes under tensile stress. They can be made either from two concave ropes coupled with tensile-stressed ropes in between, or alternatively from two convexly running ropes coupled with compressive-stressed bending-stiff bars in between (◘ Fig. 9.90). The ropes form the upper and lower chords and the tension ropes or rods form the cross-members. The chord ropes are thus forced into a polygonal shape by the transverse ropes or rods and are thereby able to absorb forces at right angles to the rope-truss axis. Depending on the direction in which the force is applied, only one of the two ropes, namely, the one which is concave against the direction of the force, carries loads. The rope chords can alternately absorb both flexural compressive and tensile forces, since they are always pre-tensioned and even in the compressive-stressed chord (convexly curved against the attack direction of the force) the compressive force is thus neutralised by merely reducing the pretension. The latter must be introduced into the bearings, i.e. into adjacent components.

Fig. 9.90 Rope trusses consisting of two polygon-shaped oppositely tensioned ropes with interposed compression rods. As façade mullions, they absorb horizontal wind forces here

The following geometric arrangements are possible with rope trusses:

- **Sequential arrangement**: simple parallel arrangement of the rope trusses (Fig. 9.90). The load-bearing behaviour corresponds to that of a conventional parallel group of bars. The main span is typically covered by the lightweight rope truss, the secondary span, for instance, by conventional bending beams.
- With a **radial arrangement** of the rope trusses, a self-contained system is created, which allows to short-circuit the prestressing forces. Well-known examples of radially arranged rope trusses are **spoked-wheel** systems, also called **radial ring-rope** systems, which were developed as grandstand and stadium roofing and have been built in numerous projects in recent years. Their most important features are briefly described below.

▶ *Sect. 4.3.1.6 Radial Skein of Bars, Annularly Supported*

Radial Ring-Rope Structures: Spoked Wheels

Cf. also ▶ *Sect. 4.3.3.2 Membrane and Rope Structure, Mechanically Tensioned, Linearly Supported*

Prestressing forces are always short-circuited in radial ring-rope structures, so that, in the end, the load-bearing structure of the roof only absorbs, in addition to vertical loads, horizontal forces from wind load, which are transferred into the ground as with a normal roof, for example, by bracings or cores. In this sense, they are regarded as self-contained load-bearing structures and are rigid within themselves, comparable to a spoked bycicle wheel, for example. The ring closure of the pre-tensioning forces allows a significant simplification of the design, but imposes restrictions on the designer in terms of shaping: Only roughly circular to oval, and with some difficulty almost rectangular ground plans can be realised, which, however, must not show any interruptions, since the forces must be short-circuited within the ring (◘ Fig. 9.91).

The prestressing force is introduced into one (or more than one) inner tension ring (◘ Fig. 9.92) and one or more outer compression rings (◘ Fig. 9.93, 1). The purpose of the combination of one ring with two rings is to spread the two radial belt ropes of the truss against each other, so that—analogous to rope nets—a concave bearing rope group (top) and a convex pretension rope group (bottom) are created. The curvature of the girder ropes, which—as with the rope trusses discussed above—enables them to take up (vertical) loads transverse to their axis, is created by vertical coupling ropes, which link the two together at intervals.

Numerous combinations of compression and tension rings (single and double) have already been implemented in various

9

◘ **Fig. 9.91** Inner rope ring of the stadium *Wanda Metropolitano*, Madrid. With the ring-cable system, the forces are short-circuited so that they do not have to be anchored externally. The prerequisite is the completeness of the ring, which cannot be interrupted (eng.: Schlaich, Bergermann & P)

Fig. 9.92 Connection of the two girder cables to the tension ring. The cables of the tension ring are continuous and are deflected in saddle-shaped clamps. The connecting piece is a cast steel part

projects (Fig. 9.93, 1, 2). Nested double radial ring-rope systems, in which one radial system is suspended within the other, have also been built (Fig. 9.93, 4).

In principle, it should be noted that the weakest member in the construction are the compression rings, which—unlike ropes—are exposed to the risk of buckling. Although this danger is partially eliminated thanks to the lateral support provided by the rope girders, the curvature of the compression ring, in particular, plays an important role: The greater the curvature (i.e. the smaller the radius of curvature), the smaller the normal forces in the compression ring are. From this, it follows that strong curvatures are basically advantageous. Perfectly circular ground-plan geometries (e.g. above bullrings) with always the same (greatest possible) curvature are ideal in this sense. With oval layouts, such as those found in stadia, the less curved areas of the compression ring on the long sides are the critical points. This difficulty is usually met by increasing the horizontal bending stiffness of the compression ring.

As shown by the example of the compression ring on Fig. 9.93, 1 spread out to form a horizontal truss

In principle, instead of short-circuiting the prestressing forces of the rope girders at the outer edge by means of one or two compression rings, they can also be directly anchored back into the ground locally, i.e. in each axis of the rope girders (Fig. 9.93, 3). Although it is possible and reasonable to combine this anchorage with the base of the associated column (a horizontal deflection strut is required for this) so that part of the uplifting force of the anchorage is over-compressed by the vertical load of the column, important advantages of the ring closure are lost. Nevertheless, this solution has sometimes been used because of the freedom of design it allows for.

In contrast to the rope nets discussed above, the rope-girder axes are laid at larger distances so that the covering

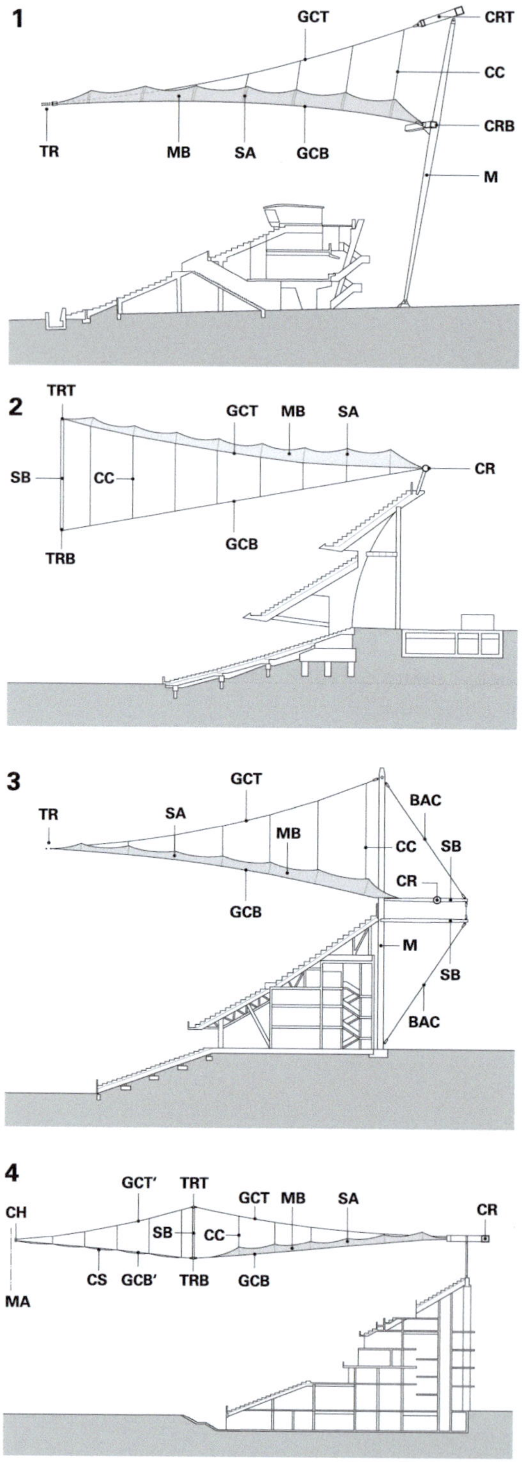

1 Neckar Stadium, Stuttgart.: Section of the ring cable construction; two compression rings, top and bottom respectively **CRT**, **CRB**, one tension ring **TR**.

CR compression ring
CRT compression ring top (2 chords from box profiles, lattice-stiffened)
CRB compression ring bottom (one box profile)
TR tension ring (cable group)
CH central hub
TRT tension ring top
TRB tension ring bottom
GCT girder cable top
GCB girder cable bottom
GCT' top girder cable of the subordinate ring cable system
GCB' bottom girder cable of the subordinate ring cable system
CC coupling cable
M mast
MB membrane
SA supporting arch
SB spreader bar (compression strut)
BAC back-anchoring cable
CS convertible sail
MA middle axis of stadium

2 Buki Jalil Stadium, Kuala Lumpur: Section of the ring cable construction; one compression ring **CR**, two tension rings **TRT**, **TRB**. The roof rests on the grandstand construction (eng.: Schlaich, Bergermann & P).

3 Gerry Weber Centre Court, Halle: Section of the ring cable construction; back anchoring of the upper girder cables in each girder axis into the ground; thus, no second (upper) compression ring is necessary on the outside, which would be visually very dominant. The guy rope **BAC** is angled via a horizontal strut **SB** and led to the base of the mast **M**, where the uplifting tensile forces are (at least partially) compensated by the vertical load of the mast (eng.: Schlaich, Bergermann & P).

4 Neues Waldstadion, Frankfurt/M: Section of the ring cable construction; double, nested ring cable system. From the two splayed tension rings **TRT**, **TRB** a second set of ring ropes **GCT'**, **GCB'** is extended to a central hub **CH**, thereby completely covering the stadium area. The inner, sail-like membrane cover **CS** is convertible (eng.: Schlaich, Bergermann & P).

Fig. 9.93 Four variants of radial rope structure; half cross-sections

Fig. 9.94 Mercedes-Benz Arena, Stuttgart: Covering with membranes; double curvature by means of supporting arches that span between cable trusses (eng.: Schlaich, Bergermann & P)

can be made with an extremely light membrane construction. Again, it must be made sure that the membrane has sufficient double curvature, e.g. by using additional support elements such as suspended columns, supporting arches, or similar if necessary (Fig. 9.94).

The extreme lightness and material efficiency of these steel structures illustrate in building construction, as do those of wide-span bridges in structural engineering, the special properties of steel as a building material as well as the enormous load-bearing reserves that can be activated by forms of equilibrium. The worldwide spreading of the ring-cable structures for stadia proves this fact in an impressive way.

Standards and Guidelines

EN 1090: Execution of steel structures and aluminium structures

Part 1: 2012-02 Requirements for conformity assessment of structural components

Part 2: 2024-09 Technical requirements for steel structures

Part 4: 2024-12 Technical requirements for cold-formed structural steel elements and cold-formed structures for roof, ceiling, floor and wall applications

Part 5: 2024-12 Technical requirements for cold-formed structural aluminium elements and cold-formed structures for roof, ceiling, floor and wall applications

EN 1993: Eurocode 3—Design of steel structures

Part 1-1: 2020-08 General rules and rules for buildings, prEN

Part 1-2: 2022-03 General rules-Structural fire design, prEN

Part 1-3: 2022-03 General rules-Supplementary rules for cold-formed members and sheeting, prEN

Part 1-4: 2023-03 Stainless steel structures, prEN

Part 1-5: 2022-03 Plated structural elements, prEN

Part 1-6: 2022-03 Strength and Stability of Shell Structures, prEN

Part 1-7: 2023-03 Plated structures subject to out-of-plane loading, prEN

Part 1-8: 2021-03 Design of joints, prEN

Part 1-9: 2023-03 Fatigue, prEN

Part 1-10: 2023-03 Material toughness and through-thickness properties, prEN

Part 1-11: 2024-02 Tension components, prEN

Part 1-12: 2010-12 Additional rules for the extension of EN 1993 up to steel grades S700

EN 1994: Eurocode 4: Design of composite steel and concrete structures

Part 1-1: 2024-03 General rules and rules for buildings, prEN

Part 1-2: 2024-04 Structural fire design

Part 1-2 Corrigendum 1: 2014-06

EN 13411: Terminations for steel wire ropes—Safety

Part 1: 2009-02 Thimbles for steel wire rope slings

Part 2: 2009-02 Splicing of eyes for wire rope slings

Part 3: 2023-03 Ferrules and ferrule-securing

Part 4: 2022-06 Metal and resin socketing

Part 5: 2009-02 U-bolt wire rope grips

Part 6: 2009-04 Asymmetric wedge socket

Part 7: 2022-01 Symmetric wedge socket

Part 8: 2011-12 Swage terminals and swaging

STLK LB 120: 2015-12 STLK–Standardleistungskatalog für den Straßen- und Brückenbau—Leistungsbereich 120: Ingenieurbauten aus Stahl

9

Bibliography

Ackermann K (1988) *Tragwerke in der konstruktiven Architektur.* Stuttgart

Baker, Godwin (1865) *Illustrations of Iron Architecture Made by The Architectural Iron Works of the City of New York.* New York

Beck W, Moeller E (2018) *Handbuch Stahl: Auswahl, Verarbeitung, Anwendung*. Munich, Hanser

Beratungsstelle für Stahlverwendung (1974) *Stahl und Form—Egon Eiermann,* 2nd ed., Munich

Beratungsstelle für Stahlverwendung (1985) *Stahl und Form—Zeitungsdruckerei Süddeutscher Verlag.* Munich

Blaser W (1991) *Mies van der Rohe,* 5th ed., Zurich

Boake, T M (2012) *Stahl verstehen—Entwerfen und Konstruieren mit Stahl.* Basel, Birkhäuser.

Boake T M (2014) *Diagrid Structures—Systems, Connections, Details*. Basel, Birkhäuser

Boake T M (2015) *Architecturally Exposed Structural Steel—Specifications, Connections, Details*. Basel, Birkhäuser

Bollinger K et al (2011) *Atlas Moderner Stahlbau: Material, Tragwerksentwurf, Nachhaltigkeit*. Munich, Institut für Internationale Architektur-Dokumentation

Dierks K, Schneider K J, Wormuth R (2002) *Baukonstruktion,* 5th ed., Düsseldorf

Eisele J et al (2016) *Bürogebäude in Stahl—Handbuch und Planungshilfe—nachhaltige Büro- und Verwaltungsgebäude in Stahl- und Stahlverbundbauweise*. Berlin, DOM publishers

Greiner S (1983) *Membrantragwerke aus dünnem Blech.* Düsseldorf, Werner

Hart F, Henn W, Sonntag H (1982) *Stahlbautlas Geschossbauten,* 2nd ed., Brussels

Gayle M und C (1998) *Cast-Iron Architecture in America.* London, New York

ICOMOS, Deutsches Nationalkomitee (1982) *Eisenarchitektur—Die Rolle des Eisens in der historischen Architektur der zweiten Hälfte des 19. Jahrhunderts.* Mainz, C. R. Vincentz-Verlag

Krahwinkel M, Kindmann R (2016) *Stahl- und Verbundkonstruktionen*. SpringerLink: Bücher. Wiesbaden, Springer Vieweg

Krausse J, Lichtenstein C (1999) *Your Private Sky—R. Buckminster, Design als Kunst einer Wissenschaft*. Zurich, Verlag Lars Müller

Lückmann R (2006) *Baudetail-Atlas Stahlbau.* Kissing, WEKA MEDIA

Mengeringhausen M (1983) *Komposition im Raum—Die Kunst individueller Baugestaltung mit Serienelementen.* Gütersloh, Bertelsmann Fachzeitschriften GmbH

Mengeringhausen M (1975) *Komposition im Raum, Vol. 1—Raumfachwerke aus Stäben und Knoten.* Wiesbaden and Berlin, Bauverlag GmbH

Minnert J, Wagenknecht G (2013) *Verbundbau-Praxis—Berechnung und Konstruktion nach Eurocode 4*. Berlin, Vienna, Zurich; Beuth Verlag GmbH

Neuburger A (1919) *Die Technik des Altertums,* Leipzig

Petersen C (2013) *Stahlbau. Grundlagen der Berechnung und baulichen Ausbildung von Stahlbauten*. 4th completely revised and updated edition. Wiesbaden, Springer Vieweg

Reichel A (2006) *Bauen mit Stahl—Details, Grundlagen, Beispiele. Detail Praxis*. Munich, Institut für Internationale Architektur-Dokumentation

Tirler W (ed) (2017) *Europäische Stahlsorten: Bezeichnungssystem und DIN-Vergleich*. Berlin, Vienna, Zurich; Beuth GmbH

Schlaich J et al (2004) *Leicht weit*. Munich, Prestel

Schöler R (1904) *Die Eisenkonstruktionen des Hochbaus,* 2nd ed., Leipzig

Schulitz H C, Sobek W, Habermann K (1999) *Stahlbauatlas*, Munich

Wachsmann K (1959) *Wendepunkt im Bauen.* Wiesbaden, Otto Krausskopf Verlag

Wietek B (2017) *Faserbeton: Im Bauwesen.* Wiesbaden, Springer Vieweg

Prefabricated Concrete Construction

J. L. Moro, *Building-Construction Design – From Principle to Detail*,
https://doi.org/10.1007/978-3-662-71006-7_10

Contents

10.1 History of Prefabricated Concrete Construction

In parallel to the technical development of modern concrete, attempts were made to use the advantages of factory prefabrication in concrete construction as early as the middle of the nineteenth century (◻ Figs. 10.1 and 10.2). Early prefabricated concrete building systems, such as those of W. H. Lascelles (1875) in Great Britain, were seeking to benefit from the specific advantages of the material, such as fire safety and durability, in sharp competition with the steel structures which were also being developed at the time. In France, F. Hennebique and A. Perret made significant contributions to the technical maturation and theoretical foundation of modern reinforced-concrete construction and experimented with combinations of prefabricated parts and pouring on site around the turn of the twentieth century. In particular, Le Corbusier, a pupil and former associate of Perret, developed an architectural formal language characteristic of concrete and laid the conceptual foundations for the remarkable spread of prefabricated-concrete buildings during the 1950s and 1960s (◻ Fig. 10.3). The pronounced, so to say, systemic modularity of this type of construction was conducive to the widespread building-system idea of that time. Various modular building systems were made from prefabricated reinforced concrete, a de-

Cf. also ***Vol. 4**, Sect. 8.4.3 Wall-Construction Modes in Reinforced Concrete* as well as ibid. *Sect. 8.6.3 Skeletal Construction Modes in Concrete*

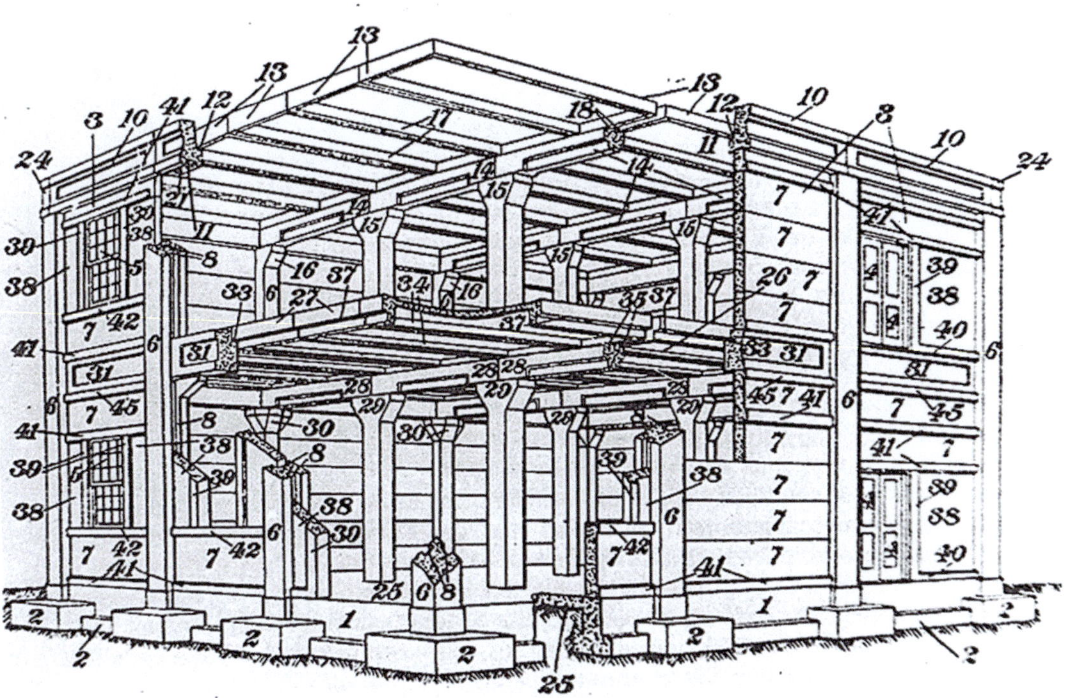

◻ **Fig. 10.1** Prefabricated concrete construction system by J. Colzeman from the year 1912

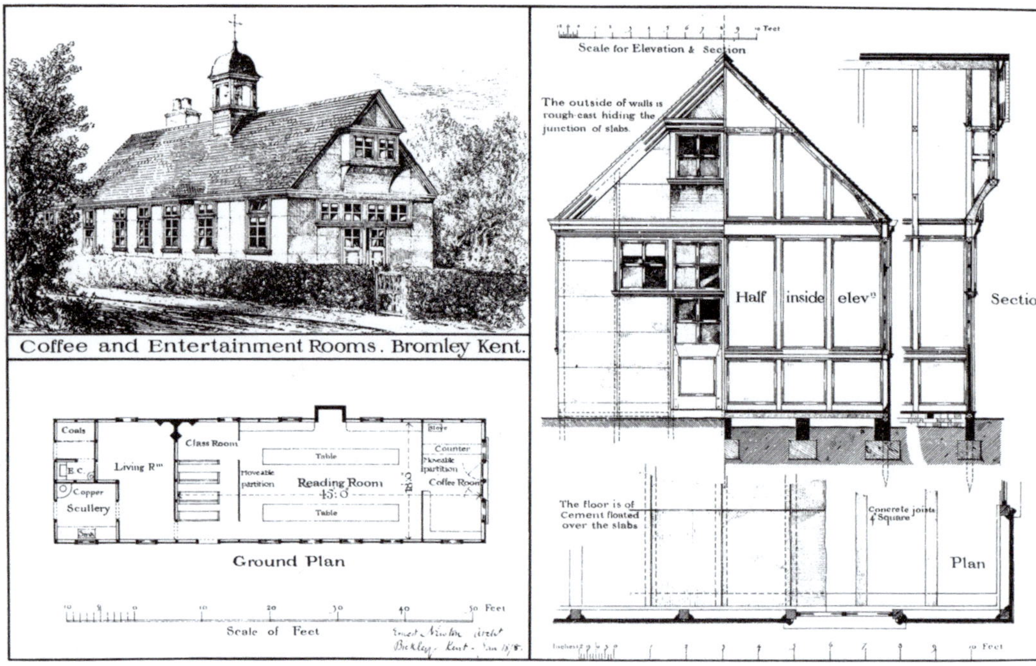

Fig. 10.2 Early prefabricated concrete system from W. H. Lascelles (1878)

Fig. 10.3 *Unité d'Habitation* by Le Corbusier: load-bearing structure from in-situ concrete; visible cladding elements on the façade in prefabricated construction

velopment that reached an apex particularly with the university building systems in Germany. At the same time, especially in socialist countries, large-scale housing was built following this construction mode.

The extensive spread of this construction method, however, made visible its many yet unsolved technical deficiencies, in particular the insufficient durability of the joint constructions. Spectacular building damages were the result, which together with the formal weakness and irritating monotony of most prefabricated buildings led to the construction method being thoroughly discredited. In contrast to this, the exemplary buildings of the Italian architect A. Mangiarotti have shown the actual formal and technical possibilities of modern prefabricated and prestressed concrete structures.

Today, prefabricated construction is still prevalent in those areas of building practice where its main actual strengths, such as rapid assembly and good fire protection, are decisive factors over competing construction methods, for instance in industrial building. On the other hand, construction with semi-prefabricated parts, which combines the advantages of prefabrication and casting on site, has experienced a real surge in the present construction practice.

► *Sect. 11.6.5 Semi-Prefabricated Parts*

10.2 Manufacture

The conceptual basis of the prefabricated concrete construction mode is similar to that of other **dry construction methods** with a high degree of prefabrication: Some operations of building production are transferred to the workshop, intending thereby to amortise, by appropriate planning, the pre-investments made in technical facilities for industrial manufacturing by taking advantage of recurring working operations and division of labour, thus achieving an added value in form of costs as low as possible and high processing quality.

***Vol. 1**, Sect. 3.3 Industrial Production*

Transport and **erection** play a decisive role, in contrast to in-situ concrete construction. The additional effort involved in factory prefabrication is compensated for by the multiple reuse of the formwork equipment.

10.2.1 Characteristics of Shop Manufacture in Concrete Construction

The production of prefabricated parts has the following essential characteristics:

- **Weather-independent** manufacturing under controllable, ideal working conditions.
- Possible **after-treatment** and **improvement** of concrete parts, so that the highest quality of execution is achieved.

- Use of high-quality manufacturing equipment requiring high investment.
- Manufacturing is possible in a **24-h cycle** so that short fabrication times and good utilisation of the production equipment are ensured.
- **Specialised reinforcement techniques** are available, such as pretensioned prestressing, which cannot be applied on the construction site.
- Due to conditions being controllable in the factory, a considerably better and more stringent **quality control** can be implemented than with concreting on site.
- Possible use of **specific concrete technologies** such as high-performance concrete or fibre concrete.
- Extremely **short construction times** thanks to short production periods and rapid assembly.

Vol. 1, Sect. 17.7.1 High-Yielding Concrete (HYC) and *Sect. 17.7.2 Fibre Concrete*

By using high-quality **formwork**, the following properties can be achieved:

- High **surface quality**.
- **Smaller tolerances**, i.e. greater **dimensional accuracy**. These can be adjusted specifically to the requirements of the respective construction project, but are a cost factor.

10.2.2 Consequences of Shop Manufacture

Furthermore, the following consequences result from factory production:

- Anticipation of **plastic deformation** (shrinkage) before final assembly.
- Possible disassembly for easy recycling or dismantling.
- High planning costs and long lead times for production. Careful planning of logistics required.
- Application of modern CAD-CAM techniques, which allow a much higher complexity of formwork geometries.
- **Hinged**, mostly **statically determinate** structural systems are executed with their specific advantages: They are comparatively less susceptible to deformations and settlements; they show a simple flow of force, etc. However, disadvantages are also to be put up with: no load relocations; portal-frame elements can only be realised to a limited extent, since bending-resistant erection joints are difficult to implement on site.
- Simplification of the assembly by **unifying sizes** and **weights**.
- Building structures must be thoroughly organised and rationalised according to **modular principles**, with aftereffects on other subsystems and trades.

***Vol. 1**, Sect. 10.2 Recycling of Concrete*

***Vol. 1**, Sect. 3.4.2 Application of New Digital Technologies of Planning and Manufacturing to the Building Sector*

10.2.3 Place of Manufacture

Manufacture takes place:

- In the **stationary plant**, this is the standard case.
- In exceptional cases, parts are produced on the construction site in a **field factory**, which eliminates the need for transport. This is only reasonable for large numbers of units or for large component dimensions.

10.3 Formwork Technology

Depending on the material and the design of the formwork, prefabricated parts with different **degrees of precision**, **surface qualities**, and **surface structures** can be produced. The service life and the cost of the formwork is also strongly dependent on the material and type of execution (table on ◘ Fig. 10.4).

Additional devices for the improvement of prefabricated parts are:

- **Vibrating tables**: They allow for better compaction of the concrete.
- **Vacuum compaction** equipment: Here, the formwork equipment is covered with an airtight tarpaulin and the air is pumped out. Due to the vacuum, an atmospheric pressure acts on the entire surface of the concrete, something which supports the compaction.
- **Heat treatment** devices with steam or oil, to accelerate hardening or to increase early strength.

10.3.1 Position of the Formwork

The position of the formwork is very important for the manufacture of prefabricated parts because the malleable fresh concrete must be introduced into the formwork from above, must penetrate into the cavity by gravity, and must spread evenly. From this, some important boundary conditions for formwork technology derive:

► *Sect. 11.4 Processing*

formwork type	execution	possible reusage	price
wood formwork	simple	35 x	15 €/m²
wood formwork	coated	50 x	25 €/m²
wood formwork	coated with metal sheet or plastic	80 x	70 €/m²
steel formwork	simple	100 x	75 €/m²
synthetic formwork	high-quality	150 x	100 €/m²
steel formwork	high-quality	500 x (with maintenance)	150 €/m²
formwork for tubbings	7 or 6 segments per ring	1,500 x (with maintenance)	100,000 €/set

◘ **Fig. 10.4** Various types of formwork equipment with details of the average number of possible uses. (Originally from: chart in p. 275 from: K. Zimmermann: *Konstruktionsentscheidungen* …, p. 25. Updated and complemented by data from the company *Züblin*, Stuttgart)

Fig. 10.5 Stocked prefabricated columns cast in upright position. Edge chamfers, cleats projecting on three sides, as well as embedded pipes for subsequent grouting can be seen

- **Narrow upright formwork** (Fig. 10.5) with a high ratio between height and width can be problematic since introducing and compacting concrete, which has to be carried out from the upper open side, is more difficult under these conditions. This may be relevant, for example, in the case of casting columns or walls, which are therefore preferably cast in a lying position. If upright parts are also installed upright, the rough screed side (see below) does not appear to be a problem in the normal case.
- **Lying formwork** (Fig. 10.6) offers more favourable conditions for casting than an upright one, since introducing and consolidating fresh concrete can be carried out without problems due to the small formwork depths—which are identical with the component thickness in this case. Furthermore, for practical purposes, a complete formwork surface is not required, since the top side of the component need not (and cannot) be formed. It is, however, to be taken into account again that the topmost **screed side** or **levelling side** has a rougher surface than the sides touching the formwork (Fig. 10.13). This can be seen with the naked eye. In the case of parts cast in a lying position which are later put upright during assembly—for example, columns or wall panels—this is possibly relevant and must therefore be taken into account during planning.
- In special cases, **inclined formwork** is sometimes used because it combines the advantages of both vertical and horizontal positions.
- The position of the formwork can also determine the **shaping** of a prefabricated part. The horizontal topmost, non-shuttered screed surface is always flat: Depending on the position of the shuttering, no protruding parts can be

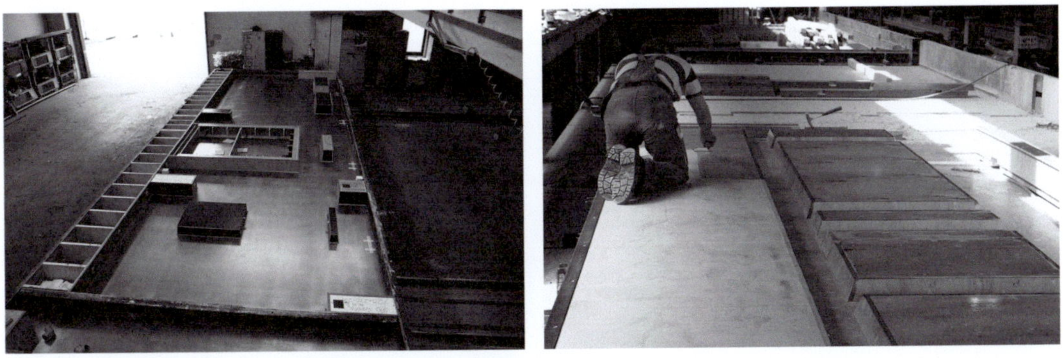

Fig. 10.6 (Left) Steel-sheet formwork for a wall panel, with inserts. (Right) Extraction formwork from wood during oiling. The bevelled edges which make the extraction process possible can be seen

concreted on this side—the levelling side. A good example of this are cleated columns (Fig. 10.5).

- The shuttering and casting position is not always identical with the final position of the finished part. This must be taken into consideration with regard to handling, transport, and assembly of the component.

10.3.2 Stripping Procedure

Depending on the stripping procedure of the formwork, the following distinction is made:

- **Extraction** formwork (matrices) (Figs. 10.6, right and 10.7).
- **Dismountable** formwork (Figs. 10.8, 10.9, 10.10, 10.11, and 10.12).

The nature of the stripping process has an effect on the **shaping** of the prefabricated **cross-sections. Some basic rules:**:

- In case using extraction formwork, the component surfaces must be inclined relative to the pull-out direction in the places where they are moved approximately parallel to the formwork surfaces when being pulled out of the formwork, so that they do not interfere by jamming with the demoulding process (Fig. 10.8**A**, **B**).
- The **edges** must be chamfered or rounded out, partially for the reason just mentioned, namely because they easily break off during stripping; partly because they can never be carried out cleanly or can be damaged during transport (Figs. 10.5 and 10.8**C**).
- **Penetrations** and **cut-outs** are possible, but are associated with additional effort (Fig. 10.13).
- **Inserts** and **compartmentalising** of a base mould are possible and allow casting different parts with the same formwork (Fig. 10.8**D**).

Fig. 10.7 Extraction formwork from steel for beam-and-slab elements

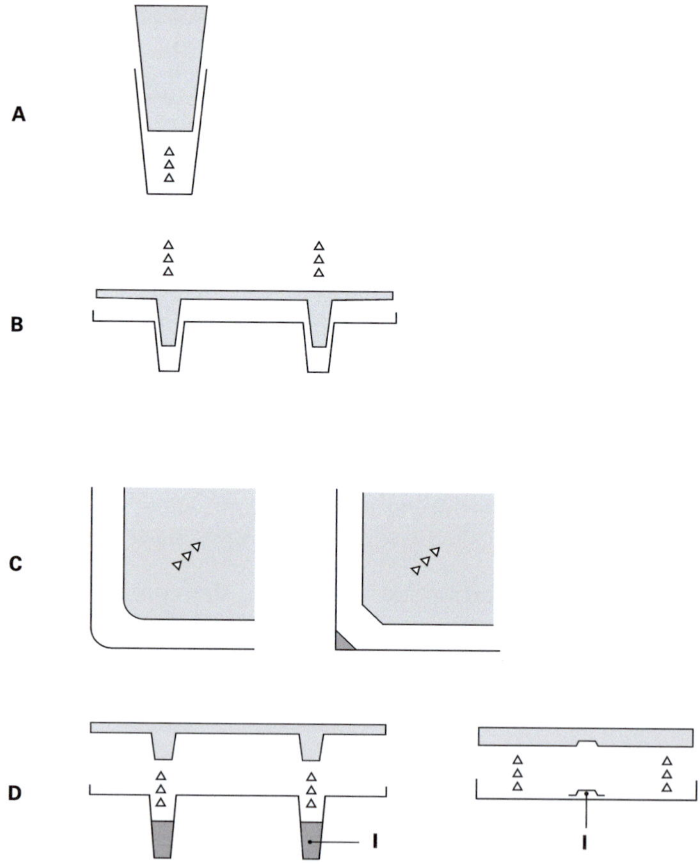

Fig. 10.8 Influences of the stripping process on the cross-sectional design of prefabricated parts: **a**, **b** bevelling of the lateral formwork surfaces for extraction formworks, **c** rounding off or chamfering the component edges to prevent tearing out, **d** various possibilities of modifying cross-sectional geometries by means of inserts **I** in the formwork

Fig. 10.9 Adjustable steel formwork for T-beams

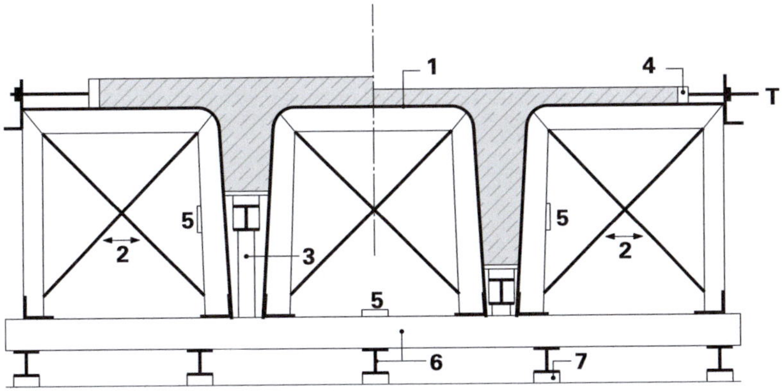

Fig. 10.10 Steel formwork for beam-and-slab elements with two possible cross-section variants (left-right), which can be produced with the same formwork device by changing various settings. (**1**) Steel-sheet forming surface, (**2**) displaceable side parts, (**3**) height-adjustable bottom, (**4**) adjustment up to **b** = 3.0 m, (**5**) vibrator, (**6**) beam grillage, (**7**) elastic bearings

- **Protuberances** of the component—e.g. support cleats (Fig. 10.14)—in two directions are easy to manufacture; in three directions, they are feasible in lying position; in four, only in upright position (Fig. 10.15).
- The **screed side** of the prefabricated part—as indicated—has a **poorer surface quality** than the formwork sides (Fig. 10.13) even with the most careful execution.

10.4 Reinforcement Technology

The following basically different reinforcement techniques are distinguished:

10

Fig. 10.11 Stacked beam-and-slab elements with a thin floor slab as a permanent formwork for in-situ top concrete application

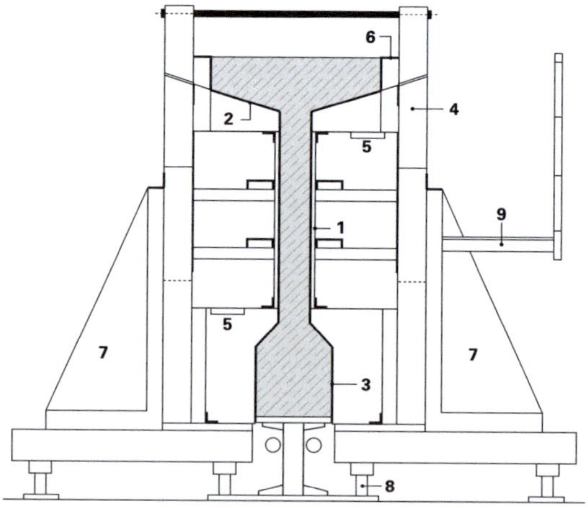

Fig. 10.12 Dismountable steel formwork for I-profiles. Height- and width-adjustable steel formwork; can be moved longitudinally for stressing-bed production; top-chord inclination by means of telescopic supports; formwork can be heated. **(1)** Plywood formwork, coated, **(2)** upper chord element from steel, **(3)** lower chord element from steel, **(4)** telescopic column, **(5)** vibrator, **(6)** adjustable upper chord, **(7)** lateral bracing, **(8)** castors and rail, **(9)** work dock

- **Unstressed** reinforcement.
- **Prestressed** reinforcement.
- **Fibre** reinforcement.

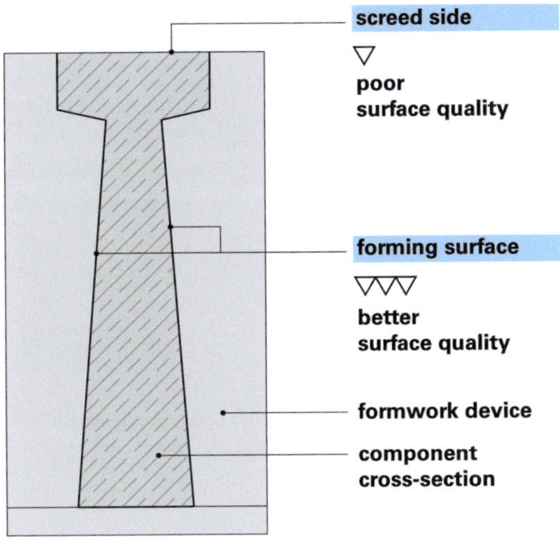

Fig. 10.13 The component surfaces adjacent to the forming surfaces always have a higher surface quality than the topmost screed side, from which the fresh concrete is filled in and subsequently simply smoothed

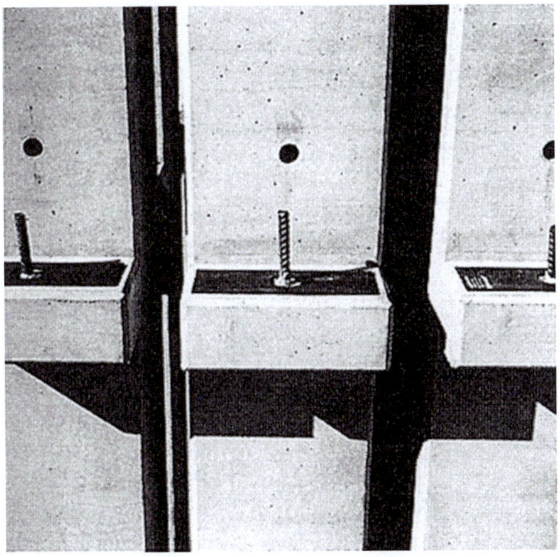

Fig. 10.14 Stocked cleat columns. The centring pin and the elastomeric bearing pad on the support surface of the cleat are recognisable

10.4.1 Unstressed Reinforcement

Steel reinforcements of different diameters and geometries are laid out, without prestress, in the formwork according to the expected tensile-stress trajectories within the component and are embedded within the concrete. This is the conventional type of reinforcement in reinforced-concrete construction.

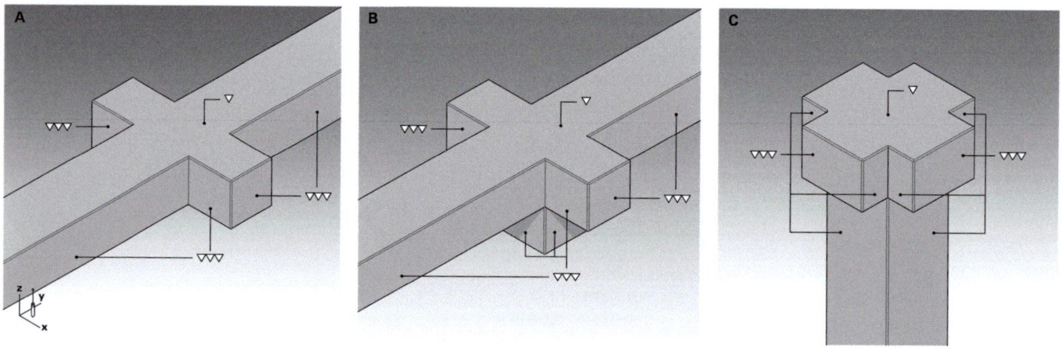

Fig. 10.15 Forming and screed surfaces of a console column depending on the concreting position: **a** horizontal concreting, two consoles: screed surface at one column flank, **b** horizontal concreting, three consoles: screed surface at one column flank, **c** vertical concreting: screed surface at the column head

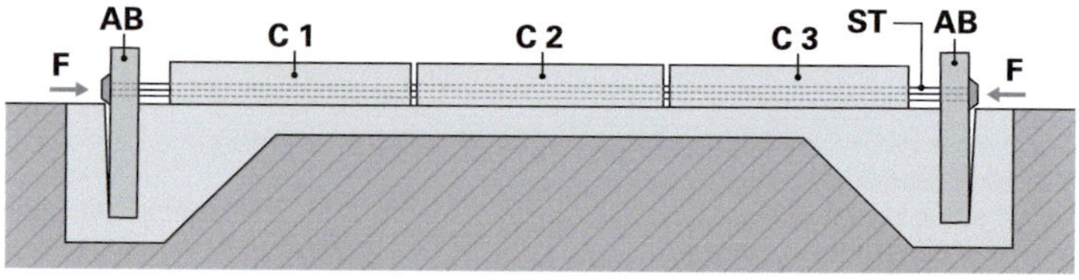

Fig. 10.16 Schematic representation of the stressing-bed production (with immediate bonding); here three components **C** 1–3 are simultaneously produced in-line. **F** prestressing force, **AB** anchor block, **C** component **1** to **3**, **ST** steel tendons

10

Vol. 1, *Sect. 17.2 Mechanical Properties* as well as in the present volume, ▶ *Sect. 11.5 Reinforcement Technology*

10.4.2 Prestressed Reinforcement

A deliberate **prestress** is applied to the reinforcement by various methods. Due to their anchoring in, or through the bond with, the concrete, the prestressing force of the steel elements is transferred to the concrete so that there is an equilibrium of forces within the component. The prestressing thus puts the concrete under compression by design, thereby improving its mechanical properties; for instance, cracks in the concrete are thereby over-compressed up to a certain load magnitude.

10.4.2.1 Method of Establishing the Prestressing

Two different prestressing techniques can be distinguished with regard to the way the prestressing is produced:

- **Pretensioned prestressing** or pretensioning with **immediate bonding** (Figs. 10.16, 10.17, and 10.22): The prestressing tendons are placed in the prestressing bed under prestress before concreting. The concrete is then poured. *Immediately* after the concrete has cured, a bond between the prestressing steel and the concrete is created along the entire length of the tensioning elements due to their complete encasement within the concrete body. The prestress-

■ **Fig. 10.17** Preparation of a **stressing bed**: The prestressed steel tendons are visible at the bottom; the standing bulkheads separate the individual bar sections, but the steel tendons are continuous. In a further operation, the lateral forming surfaces are mounted

■ **Fig. 10.18** Jacket tubes inserted in the lower chord area of a hollow-box girder used for **post-tensioning**

ing force of the tendons is released and the tension force is thus transferred to the concrete as a compressive force.

- **Posttensioning**: The prestressing force is only introduced into the tendons after the concrete component has hardened; the tendons initially do not have a firm bond with

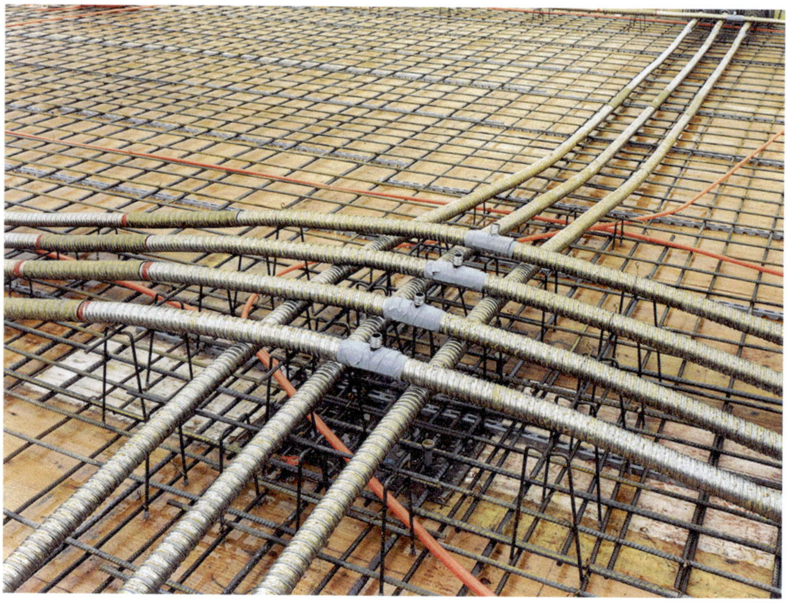

Fig. 10.19 Jacket tubes for prestressing a floor slab before concreting (internal prestressing, cf. Fig. 10.20). The picture shows the tendon trajectory over a column: correspondingly elevated following the negative support moments. The openings at the apex serve for the injection process

the concrete because they lie either loosely in cavities within the cross-section or completely outside of it. This can be done in two ways with regard to a composite effect:

- With **subsequent bonding**: The tendons are laid within sheaths (Figs. 10.18 and 10.19). The cavities between the tensioning elements and the sheathing walls are injected with suitable mortars, so that, after their curing, a bond with the concrete body is established over the entire length of the tensioning members. Through the bonding, the prestressing steel is better utilised for the load-bearing capacity of the component than when no bonding is present.
- **Without bonding**: The tendons are routed freely within empty sheathings, or even outside the cross-section. There is therefore no continuous bond with the concrete over the length of the tendons. The prestressing force is instead introduced into the concrete body solely at their end-anchoring. Cavities within sheathing tubes are injected almost without exception with anti-corrosive agents. Although the steel members are not as well-utilised as with bonding in this method, they can be located outside the cross-section, thereby significantly increasing the effective lever arm (external pretension, see below).

10.4.2.2 Mode of Tendon Guidance

Regarding the **conduction** or routing trajectories of prestressing tendons in relation to the component's cross-section, the following difference is made:

- **Internal prestressing** (◘ Fig. 10.20, top): The prestressing elements are guided within the cross-sectional area. This variant includes the prestressing with immediate bonding as well as that with a subsequent bonding or without a bonding.
- **External prestressing** (◘ Fig. 10.20, bottom): The prestressing elements are routed outside the cross-sectional area of the component. This can take place completely outside the cross-section or also in a chamber-like cavity, for example in the interior of a hollow box cross-section. A pretension without bonding can be carried out both internally—within sheathing tubes—as well as externally (◘ Fig. 10.21).

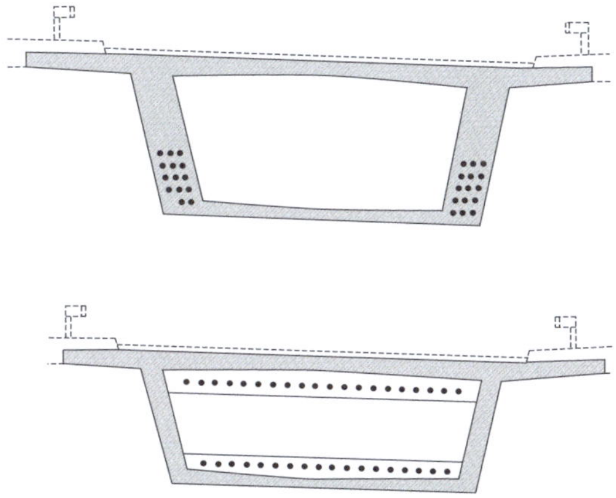

◘ **Fig. 10.20** (Top) Example of an **internal prestressing** in a hollow-box cross-section. (Bottom) Example of an **external prestressing** in a hollow-box cross-section

◘ **Fig. 10.21** (Left) Example of an **external post-tensioning**, arranged inside a hollow box cross-section, freely accessible. This picture shows the variant in ◘ Fig. 10.20, bottom. (Right) Tendons in jacket tubes during assembly of the example shown on the left

10.4.2.3 Pretensioned Prestressing

The prestressing tendons are stretched between two anchoring blocks in a straight line and set under tension (◻ Figs. 10.16, 10.17, and 10.22). Over this reinforcement, the component is poured within the corresponding form-

◻ **Fig. 10.22** (Top) Stressing bed: applying the hydraulic jacks. (Centre) Setting up the stressing bed. (Bottom) Extracting the finished beam from the formwork

work. After the concrete has hardened, the anchoring is released and the prestressing force is introduced into the component, usually in steps, by means of the friction between the steel and concrete.

10.4.2.4 Posttensioning

No special manufacturing equipment is required. Cavities are left within the prefabricated parts by means of incorporating sheathing tubes into which the **tensioning strands** of steel are subsequently threaded (Figs. 10.23 and 10.24). The tendons are then prestressed at both ends so that the concrete is compressed. In contrast to the pretensioned prestressing, with subsequent prestressing without bonding (see above), the prestressing force is not introduced into the concrete continuously but at both ends of the prestressing tendons.

10.4.3 Fibre Reinforcement

The reinforcement consists of fine metal fibres which are embedded in the concrete either randomly or more or less orderly. Fibre reinforcements are increasingly used in prefabricated construction, mostly in combination with conventional reinforcement techniques. The technical basics are discussed in another chapter.

Vol. 1*, Sect. 17.7.2 Fibre Concrete*

Fig. 10.23 (Left) Section of a post-tensioned beam (segmental construction method). One can clearly see the tendons, which are threaded through embedded sleeves. (Right) Segmental construction mode with prefabricated individual sections. This picture shows the scaffolding required for this type of assembly

Fig. 10.24 (Left) Roof girders of a hall assembled from individual sections in segmental construction (arch.: M. Fisac). (Right) End anchoring of the prestressed tendons posttensioned on site

10.5 Influences of Reinforcement Technology on Constructional Design

The choice of reinforcement technology also has an impact on the design and construction of the prefabricated part as well as of the overall structure. The following design rules can be derived from the application of different reinforcement techniques.

10.5.1 Unstressed Reinforcement

For parts with unstressed reinforcement, the following applies:

- As in the case of cast-in-place concrete, but in contrast to pretensioned prestressing, steel bars can be freely adapted to the stress trajectories within the component. Shifts from **positive** to **negative moments**—as occurred, for example, with a beam with cantilever arm, with a continuous beam, or a portal frame—can be absorbed by deviating the reinforcement from the lower to the upper chord region.
- On the construction site, components can be connected to one another in a **force-fitting manner** to form a **structurally homogeneous component**—for example by overlapping the reinforcement and grouting. This in principle permits creating bending-stiff monolithic structures: a significant advantage of this construction mode. However, it must be

borne in mind that this kind of construction work is always a hindrance to rapid assembly—a major advantage of prefabricated construction.

10.5.2 Pretensioned Prestressing

For prefabricated parts made by pretensioned prestressing, the following applies:

- Deflection of the rectilinearly running prestressing tendons is only possible to a limited extent. **Alternating moments** therefore cannot be dealt with. For this reason, beams with cantilever arms or continuous beams cannot be manufactured by pretensioned prestressing. The typical support of a beam produced by pretensioned prestressing is that of the **single-span beam**.
- The prestressing tendons of two adjacent parts cannot be connected to one another on the building site by simple grouting without special measures.
- Since the pretension force is applied after the concrete is hardened but prior to stripping, **transverse ribs** in the component—as in the case of coffered slabs, for example—are generally not possible. These would obstruct or even impede the stripping process because they would jam against the formwork walls as a result of introducing the prestressing into the component.
- Thanks to the very favourable structural behaviour of the prestressed concrete, extremely **slim cross-sections** can be achieved, comparable to those of steel components (Fig. 10.25).

Fig. 10.25 Project of a hall system by A. Mangiarotti. Note the slenderness of the pretensioned roof elements

- The unloaded **erection state** can become critical, since the upper chord made of concrete is subjected to tensile stress "as a consequence of the prestress in the lower steel chord. Therefore, an upper reinforcement against cracking may also be necessary. The cracks in the transport state are, however, often accepted, since in the final state the cracks are always over-compressed again anyway.

10.5.3 Posttensioning

For prefabricated parts with posttensioning, the following applies:

- In contrast to pretensioned prestressing, the prestressing members can be freely adapted to the course of the stress. Every load case can be handled—even beams with cantilever arm and continuous beams (Fig. 10.55).
- Connections capable of **bearing bending-moments** can be easily achieved on the construction site by simply simultaneously prestressing the parts or by coupling the prestressing elements (Figs. 10.57, 10.58, and 10.59).
- Large components which structurally behave as a whole in the final state can be divided into **small**, **easily transportable sections**, which are then prestressed together on site to generate the entire element (segmental construction) (Fig. 10.23).
- In general, extensive **auxiliary scaffolding** is required during assembly (Fig. 10.23, right).

10.6 General Principles for the Conceptual and Constructional Design of Prefabricated Concrete Components

In addition to the above-mentioned design and construction rules derived from reinforcement technology, it is also possible to infer certain necessities and regularities from the **transport** and, in particular, from the **erection** of prefabricated parts which significantly influence the design of prefabricated structures.

10.6.1 Transport

Some transport-related design principles of prefabricated parts are:

- **Rod-shaped** components such as trusses or narrow floor slabs can be transported in large lengths (Figs. 10.26 and 10.27). The greatest flexibility with regard to delivery logistics—a major factor in dry construction—prevails with **road transport**, but rail transport and water transport are also possible.

Fig. 10.26 Road transport of a prestressed roof girder

- **Two-dimensional** parts such as wall panels must be transported in road transport in upright position; this is possible up to a height **h** = 4 m, or a slightly larger one with a tilted position.
- The formwork and casting technology is subject to special boundary conditions which have already been mentioned above and which presuppose certain positions of the formwork, for example when casting columns. Often, therefore, there is a discrepancy in the case of prefabricated parts between the position **during casting** and the position in **installed state**: For this reason, special load cases are often to be taken into account which should not, however, govern the dimensioning.[1]

► *Sect. 10.3.1 Position of the Formwork*

10.6.2 Erection

Some erection-related design principles for prefabricated parts are:

- Prefabricated concrete parts should not be held too long by the hoist until they can be positioned. Prefabricated parts which, due to their **shape** and **support**, are immediately **tilt-proof** after being positioned and do not have to be supported additionally simplify the assembly considerably. **Form-fitting connections** which are effective solely by gravity, preferably self-centring ones, are therefore very suitable and are generally typical of prefabricated construction

***Vol. 3**, Sect. 5 Assembling Prefabricated Concrete Components*

1 Bindseil P (1991), *Stahlbetonfertigteile*, p. 39 f.

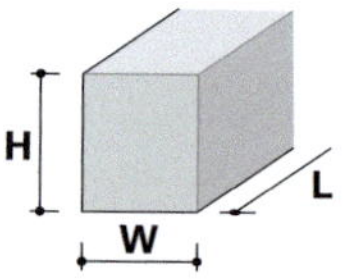

special permit	W	H	L
no (base profile)	< 250	< 400	< 1,800
yes, without police escort on federal and provincial roads	< 350	–	< 2,400
yes, without police escort on federal highways	< 450	–	< 2,800

Accompanyment by company-owned safety vehicle can be prescribed from case to case.

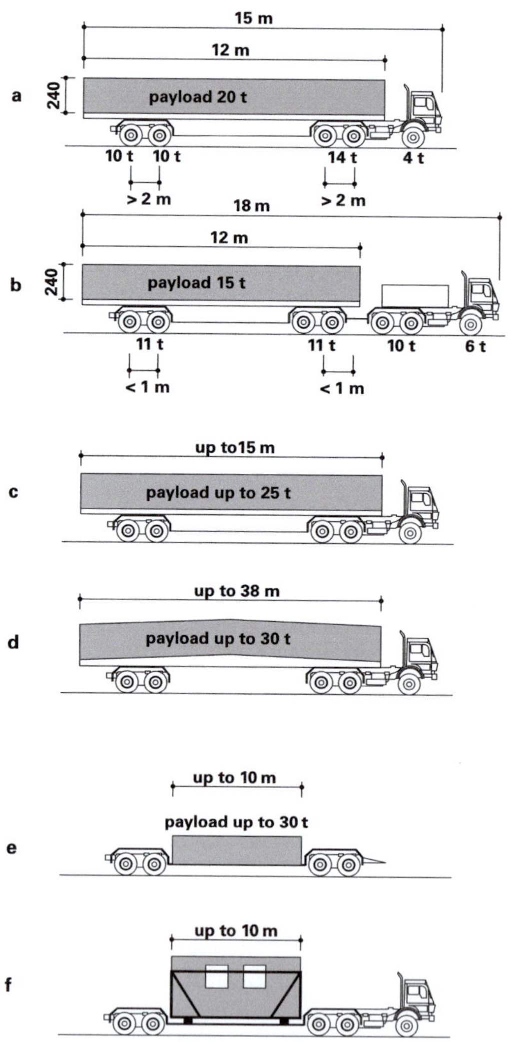

Fig. 10.27 Load dimensions and payloads for the transport of prefabricated concrete parts. (According to Koncz T (1976) *Bauen industrialisiert*, Wiesbaden)

(◘ Figs. 10.28, 10.29, 10.30, and 10.31). When the support conditions do not preclude tilting, it must be made sure that the centre of gravity of the component to be mounted is located **below the bearing plane** so that tilting is prevented already by its own weight.

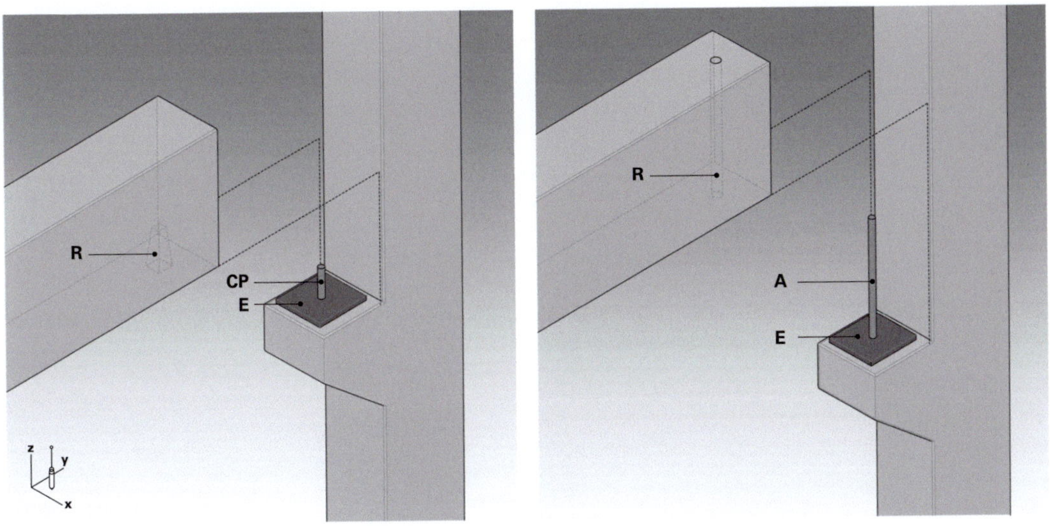

◘ **Fig. 10.28** (Left) Cleat support on elastomeric strip (**E**) with simple centring pin (**CP**). The pin recess (**R**) is filled with malleable material. (Right) Cleat support as on the left with continuous anchor rod (**A**). Additional securing against tilting of the carrier

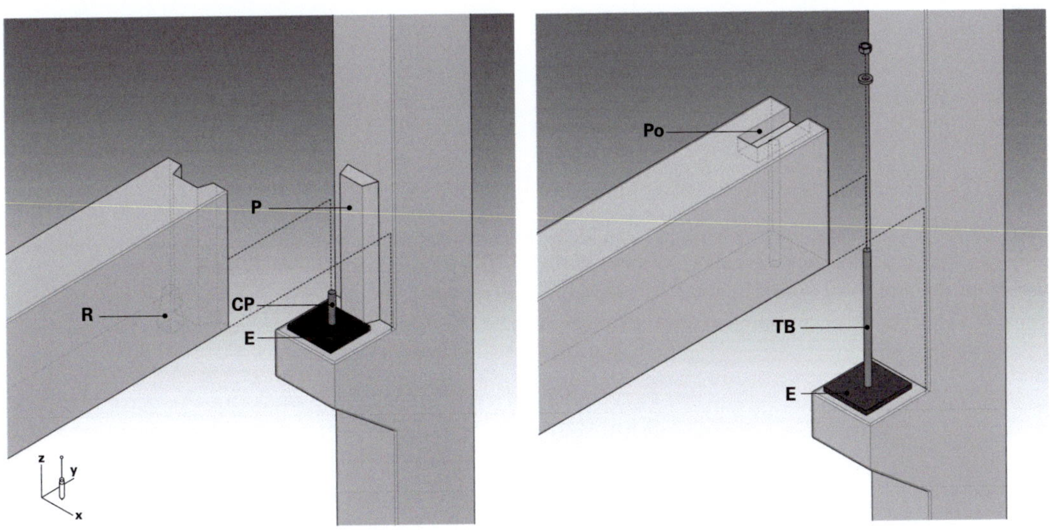

◘ **Fig. 10.29** Cleat support as on ◘ Fig. 10.28, left, with tilt protection by toothing at the carrier end. The trapezoidal cross-section of the protuberance (**P**) on the column facilitates centring. (Right) Cleat support with tilt protection by means of a threaded bar (**TB**). If necessary, the screw connection must be protected against fire and corrosion by potting the pocket (**Po**)

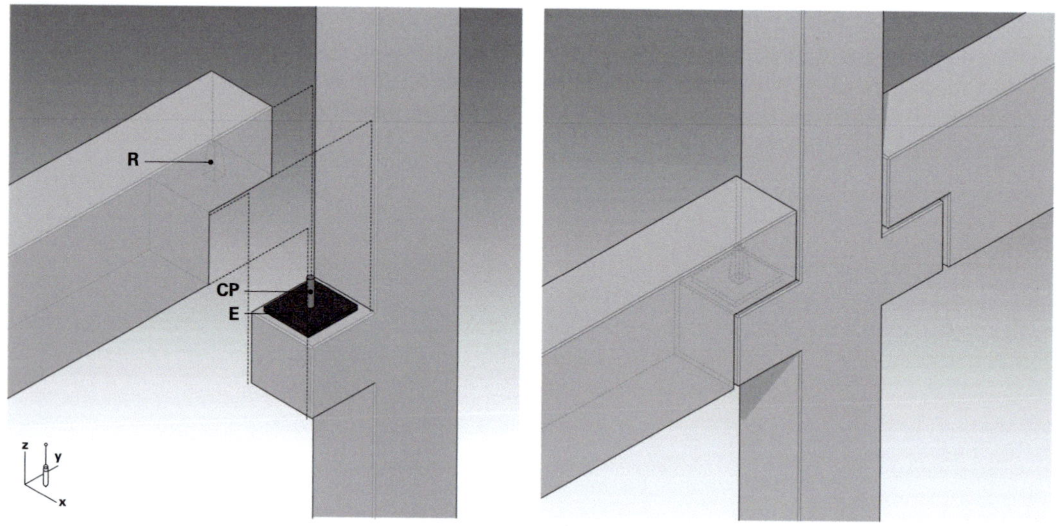

■ **Fig. 10.30** (Left) Cleat support with pin as on ■ Fig. 10.28, left, with notch on the beam so that the cleat does not protrude. (Right) Double cleat support with pin as on the left

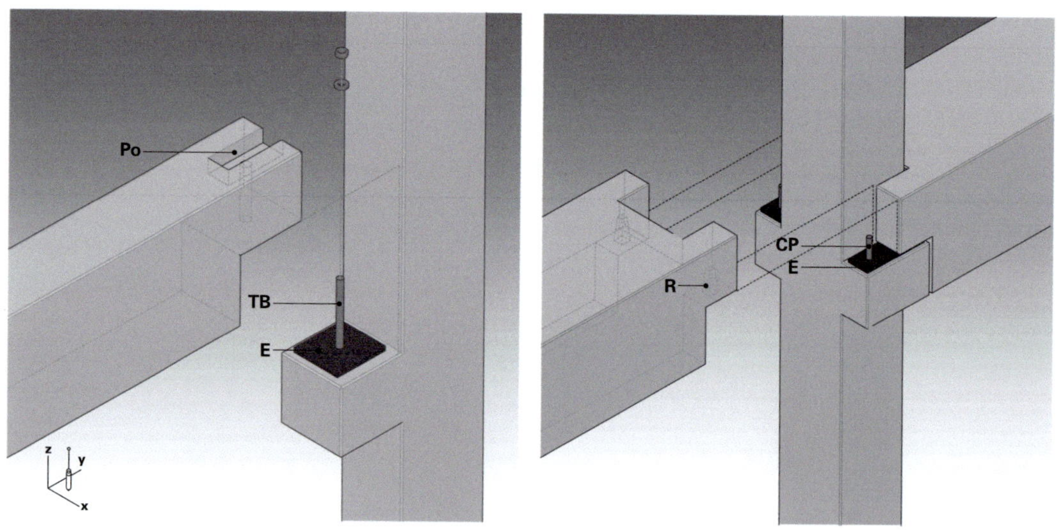

■ **Fig. 10.31** (Left) Cleat support with bolt lock as on ■ Fig. 10.29, right, with notch on the beam. (Right) Cleat support with pin analogous to the connection on ■ Fig. 10.28, left, with a trough-shaped twin girder. Shaping and support of the beam on a double cleat prevents tipping

- Erection is often simplified by merging the girder and the floor slab into one monolithic element as **beam-and-slab**. Structural advantages can also be achieved as the plate can act as a compression chord of the girder or of the rib.
- Components should be of a similar **order of magnitude** of **weight** for the purpose of good utilisation of the crane's capacity. Given the usual heavy weights prevalent, this aspect is more important for the weight ranges of the parts

to be assembled in prefabricated construction than in other, lighter construction modes.

- Erection **joints** should be executed **hinged**, whenever possible. Executing rigid connections on the construction site is certainly possible in—un-prestressed—prefabricated construction, but it is very labour-intensive and partially nullifies the main advantage of this construction method, namely rapid assembly. The possibility of creating **force-locking connections on site** without further effort, an advantage of the conventional in-situ concrete construction method, is largely lost in regular precast construction.

  In addition to grouting techniques, the production of bending-stiff connections on the construction site is potentially available—as indicated—only with posttensioning, so that fixed connections do not require any additional effort. However, in the last decades, some severe damage occurred to post-tensioned constructions which was due to hardly controllable corrosion processes affecting the prestressing tendons hidden within the sheathing tubes. It has therefore become increasingly customary today to install the prestressing structure in **freely accessible cavities** or **completely exposed**, e.g. as a trussing from beneath.
- A variety of **hybrid solutions** are possible ranging between pure dry mounting—by bolting, welding, setting on elastomer bearings, etc.—and more ample in-situ grouting work such as, for example, is possible with semi-prefabricated parts.
- **Continuous columns** extending over several storeys are often used in prefabricated construction: They can be easily installed with an auto-crane and the rest of the load-bearing structure can then be set up using the tower crane.[2] Continuous columns circumvent the need for a joint between column sections, which is always difficult to execute in structural and constructional terms (◘ Figs. 10.32, 10.42, right, and 10.43). As a result, a **cleat support**, typical of prefabricated construction, is created for twin girders—for example, in the case of trough elements—which laterally run past the column, so that cantilevers can also be implemented. Provisional support of the upright column parts is necessary; a full or partial restraint is sufficient for stabilising during erection.[3]

2 Bindseil P (1991), p. 41.
3 Ibid., p. 154.

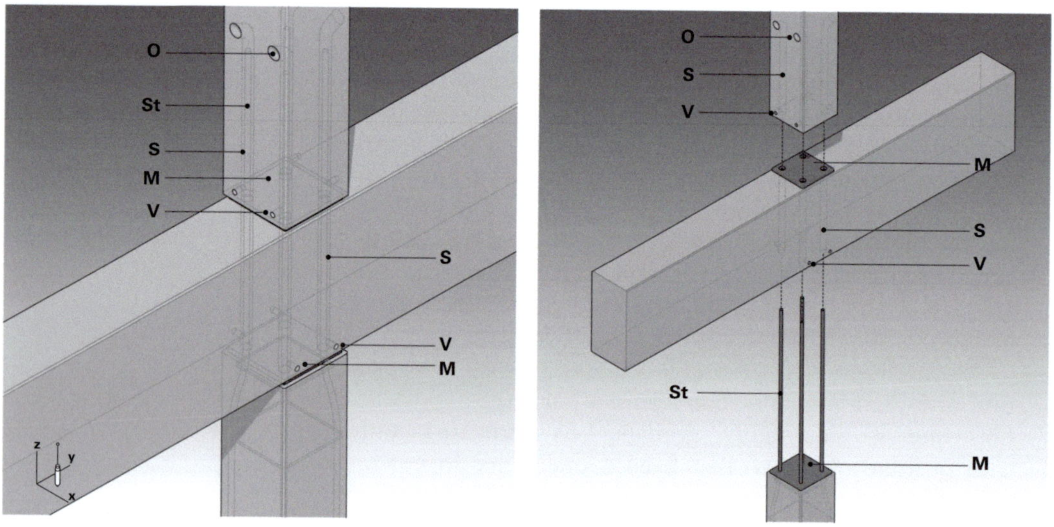

Fig. 10.32 (Left) Column joint with continuous beam. (Right) Explosion drawing of the node on the left. **M** mortar layer, **S** sleeve, **St** steel reinforcement, **O** filling opening for the grouting concrete, **V** vent for the grouting process

10.6.3 Standard Constructional Solutions

10

The following prefabricated construction modes and associated standard design solutions have become established in building construction (cf. also Figs. 10.44, 10.45, 10.46, 10.47, 10.48, 10.49, 10.50, 10.51, 10.52, 10.53, 10.54, and 10.55):

10.6.3.1 Wall Construction Modes

Vol. 3, Sect. 20.5.1.2 Prefabricated or Semi-Prefabricated-Concrete Floor Systems

Wall construction modes employing prefabricated elements have been extensively used in large-panel buildings in the sixties and seventies of the twentieth century. Today, for reasons already mentioned, they have lost their importance. Their functional principle in terms of structural behaviour and construction is comparable to other wall constructions such as masonry. In the field of prefabricated construction, **box** as well as **cross-wall systems** are realised. In contrast to masonry wall panels, reinforced-concrete panels are also able to absorb tensile stresses, so that, for example, a punctual—instead of a linear-only—support of the wall panel is possible. Similar to masonrywork—and unlike in the case of cast-in-situ concrete—it is also important to ensure that in-plane **shear forces** can be transferred at the joints of the surface elements, such as the intermediate floor connection (Figs. 10.33, 10.34, 10.35, 10.36, and 10.37) or the wall/floor node (Figs. 10.38 and 10.39), with adequate constructional solutions. Especially in the case of wall/floor nodes, constructional questions arise from the local concurrence of different materials such as the concrete of the prefabricated parts, the

Cf. exemplary joint design in precast wall construction in ***Vol. 3***, *Sect. 7.2.5.1 Linear Connections Between Surface Components*

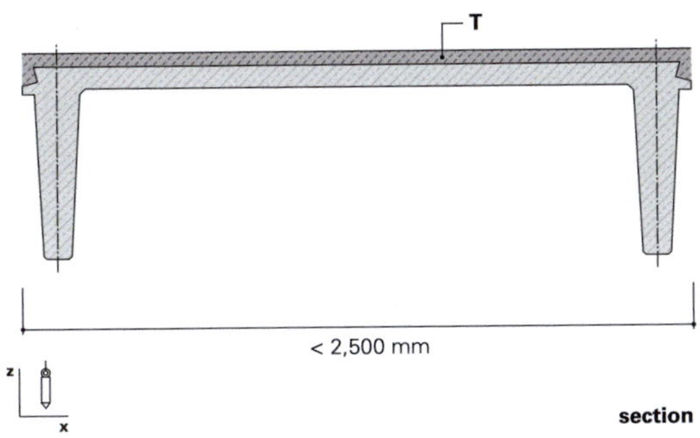

Fig. 10.33 Trough slab. **T** top concrete

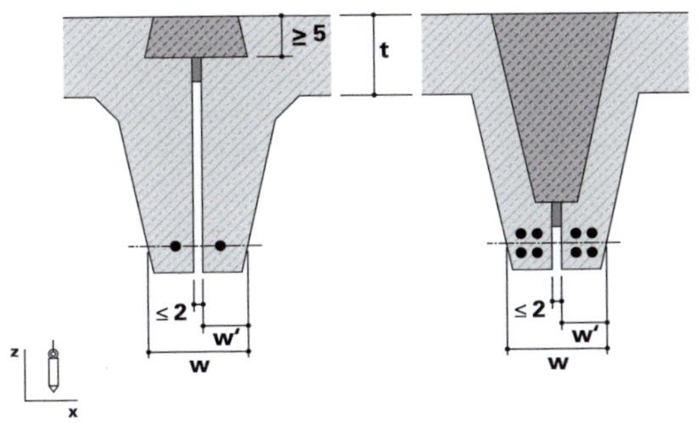

Fig. 10.34 Alternative joint designs of two adjoining trough slabs

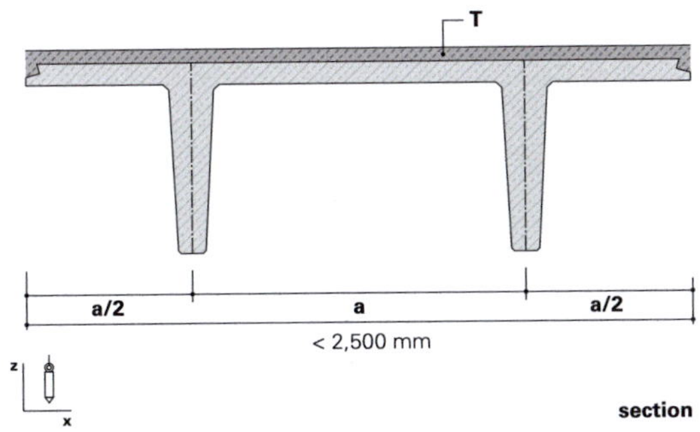

Fig. 10.35 Beam-and-slab element. **T** top concrete

grouting concrete, and the mortar for grouting gaps[4] as well as from the necessity to transfer vertical loads over the walls and in-plane shear forces within the floors (Figs. 10.38 and 10.39). Wall construction modes with increased shear stress as measured against normal forces and, if necessary, special connection solutions occur in the case of bracing walls in skeletal construction, for example in the form of core walls (cf. also Fig. 10.40, left).

► *Sect. 4.2.1.3 Bracing of Skeletal Structures*

4 Bindseil P (1991), p. 97.

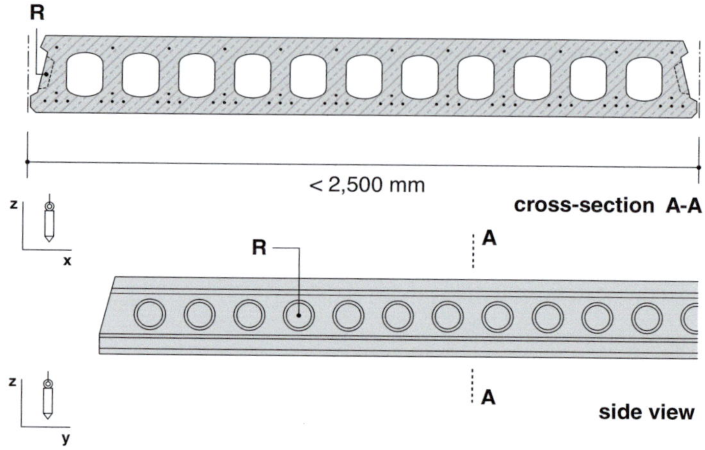

Fig. 10.36 Hollow-core slab; **R** recess for interlocking of adjoining elements with shear-resistance in → **y** after joint grouting

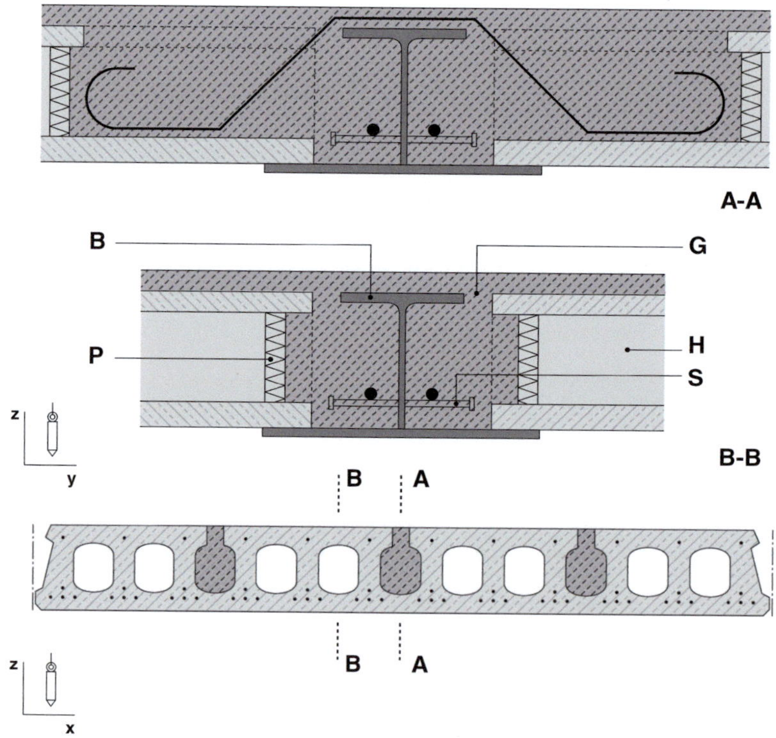

Fig. 10.37 Slim-Floor system consisting of hollow-core slabs (**H**) and flush composite steel beams (**B**), the lower flange of which serves as a support for the floor slabs during assembly. Individual chambers are grouted in sections (**G**, partition **P**) and reinforced (section **A-A**). Shear studs (**S**) welded to the beam web form a shear-resistant joint with the grouting in the longitudinal direction of the joint (→ **y**)

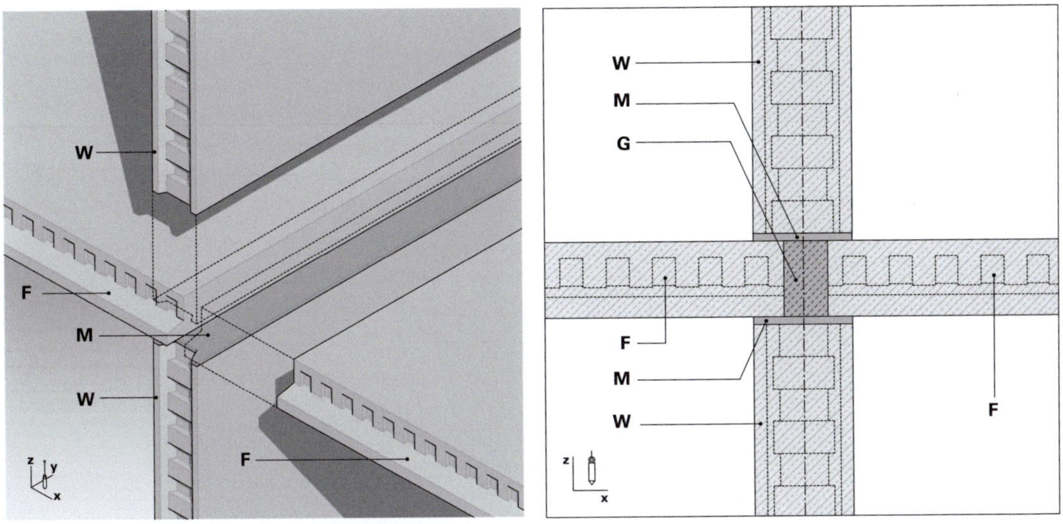

Fig. 10.38 (Left) Wall/floor node with solid floor slab: **W** wall component, **F** floor slab, **M** mortar layer. (Right) Sectional drawing of the node on the left. **G** grouting

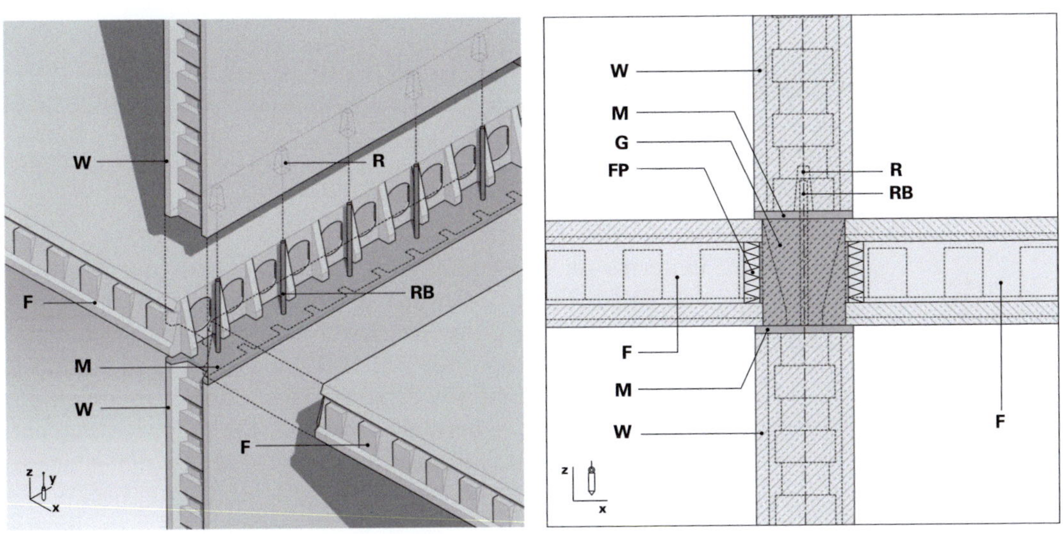

Fig. 10.39 (Left) Wall/floor node with hollow-core slabs: **W** wall component, **F** floor slab, **M** mortar layer, **R** recess, **RB** reinforcing bar. (Right) Sectional drawing of the node on the left. **G** grouting, **FP** foam-material partition

Sandwich façades, which were commonly used in the past in prefabricated wall constructions, consist of two concrete leaves with a thermal-insulation core layer. They are produced in the factory in two operations—layer-by-layer concreting of the facing and structural leaf including a hard-foam insulation layer between them (Fig. 10.40, right). Both leaves are connected to each other by means of retaining and load-bearing anchors and are mounted as a complete element. The comparatively large loads of these elements as well as the problems

***Vol. 3**, Sect. 12.4.1.2.2 Prefabricated Construction*

Fig. 10.40 (Left) Prefabricated wall components made of reinforced concrete. (Right) Intermediately stored sandwich elements made of reinforced concrete before assembly

in terms of building-physics associated with the non-back-ventilated front leaf, which with 8 cm is comparatively thin and corrosion-prone, have led to substituting simple load-bearing leaves with added thermal-insulation composite systems, or alternatively separate façade plates mounted at a distance from the thermal insulation, for sandwich elements. These can be ventilated and be executed with larger thickness (12–14 cm).[5] Their anchoring to the load-bearing back leaf is also carried out with retaining and support anchors. Restrictions of the achievable insulation-layer thickness always result from the relatively large weights of the facing leaves. Increased distances between the load-bearing and facing leaf result in correspondingly larger anchoring cross-sections with the associated thermal-bridge problems.

***Vol. 3**, Sect. 12.4.2.2.2 Walls with Suspended Facing Leaf*

► *Sect. 2.3 Double-Leaf Systems*, in particular Figs. ► 2.22, left, ► 2.24, and ► 2.32

10.6.3.2 Single-Storey Skeletal Constructions (Halls)

Today, single-storey skeletal structures are found particularly in industrial buildings. The possibility of reverting to read-

5 Pauser A (1998) *Beton im Hochbau—Handbuch für den konstruktiven V orentwurf*, p. 273 f.

ily available comprehensive construction systems as well as the short assembly times are specific advantages of this sort of construction, which are especially important in this type of building use. Particularly where increased fire protection is required, prefabricated concrete construction has an advantage over competing steel structures.

Large hall span widths can be bridged with girders fabricated by pretensioned prestressing, often with a saddle-shaped inclined upper chord, so that the overall height can be well adapted to the moment distribution in the single-span beam, which is typical of prefabricated construction. Columns are restrained within a sleeve foundation (◘ Figs. 10.52 and 10.53). Further bracing measures, such as for example forming a roof shear-plate, are mostly unnecessary; the foundation can thus also be installed in dry construction. Node variants for connecting the girder to the column are shown on ◘ Figs. 10.41, right and 10.42, left. As is typical for prefabricated parts, an anti-tipping device must also be provided for the girder.

For sleeve foundations, cf. ***Vol. 3****, Sect. 7.2.5.2 Grouting Joints for Column Restraints*

10.6.3.3 Multi-Storey Skeletal Constructions

Similar structural systems made of restrained columns, such as in single-storey hall construction (see above), can also be used with up to three-storey-high buildings. Typical features are multi-storey continuous columns cast in a lying position with support cleats for girders or floor elements. With this casting technique, they can also be executed with a load-bearing cross-section staggered storey-wise.

In the case of higher multi-storey buildings, other bracing concepts must be applied, preferably core or shear-plate bracings, or combinations thereof. In such cases, columns are hinged in the static sense (pin-ended columns), usually made of individual sections and thus vertically joined. In the case

Constructional solutions for this can be found on ◘ Figs. 10.42, right and 10.43

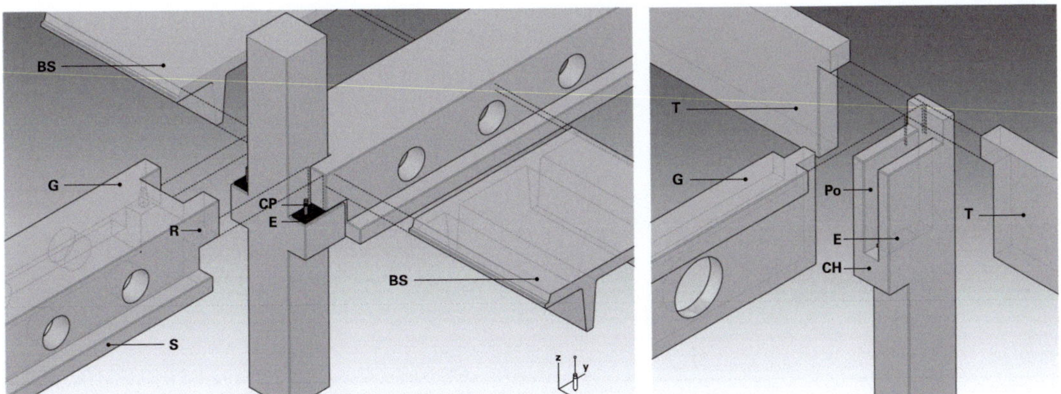

◘ **Fig. 10.41** (Left) Cleat support with pin similar to the connection on ◘ Fig. 10.31, right, with strip ledge (**S**) on the girder (**G**) and a beam-and-slab element (**BS**) supported thereon. (Right) Support of a hall girder (**G**) and two longitudinal transoms (**T**) placed on the column head (**CH**). Tilt protection of the girder by pocket (**Po**). The transoms are tilt-stabilised by a low centre of gravity

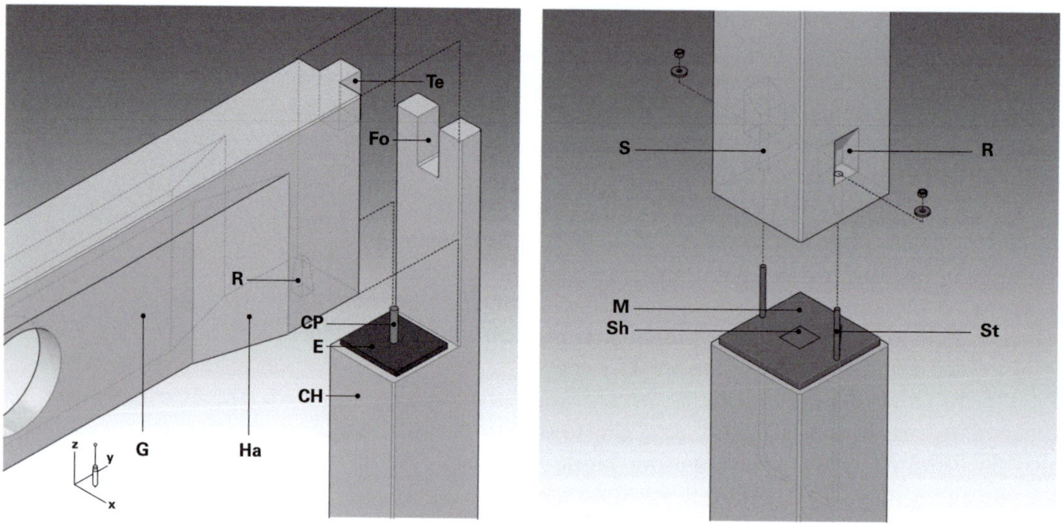

■ **Fig. 10.42** (Left) Support of a hall girder (**G**) with haunch (**Ha**) on a column head (**CH**). The tilt stabilisation is assumed by a tenon (**Te**) which engages in a fork (**Fo**). (Right) Column joint with bolt connection. **M** mortar layer, **S** sleeve, **Sh** shimming, **R** recess, **St** steel reinforcement

► *Sect. 10.6.2 Erection*

As, for example, in skeletal timber construction, cf. ► *Sect. 8.4 Skeletal Timber Construction*

of floors with girders—these can also be merged with the floor slab in prefabricated construction as already mentioned—a similar spatial coordination problem arises between columns and beams as with other construction modes. As in the case of timber pincer constructions, it is also possible in prefabricated construction to design girders as twin parts or as a trough element (■ Figs. 10.31, right and 10.41 left). A design variant of a penetration between the column and beams is shown on ■ Fig. 10.32.

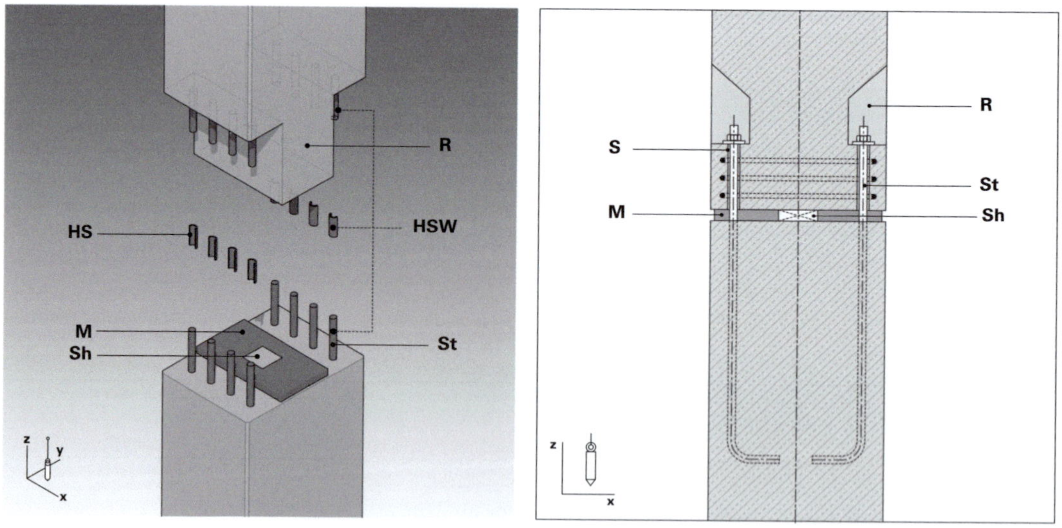

■ **Fig. 10.43** (Left) Column joint with half-sleeve welding of the reinforcing bars of both column sections. The recesses are then filled with concrete. **HS** half sleeve, **HSW** half-sleeve welding, **R** recess. (Right) Sectional drawing of the joint on ■ Fig. 10.42, right, above

Core and wall-panel bracing of skeletal structures require **floor shear-plates**. These are responsible for the force-locked connection of the skeleton to the fixed points. For this purpose, the strength of the joint between the floor elements against in-plane shear forces and the tensile strength of the panels' edges must be ensured by fitting suitable tie rods or sufficient reinforcement. More complex on-site grouting or welding work for the purpose of activating racking stiffness partly obliterates a substantial advantage of the prefabricated construction mode, namely the swift erection process. This is one of the reasons why more and more semi-prefabricated construction methods with an in-situ top concrete layer have gained widespread acceptance (◘ Figs. 10.44, 10.45, 10.46, 10.47, 10.48, 10.49, 10.50, 10.51, 10.52, and 10.53).

Constructional solutions for the design of floor panels can be found in ***Vol. 3****, Sect. 7.2.5.1.1 Floor Joints* as well as ***Vol. 3****, Sect. 20.5.1.2 Prefabricated or Semi-Prefabricated-Concrete Floor Systems*

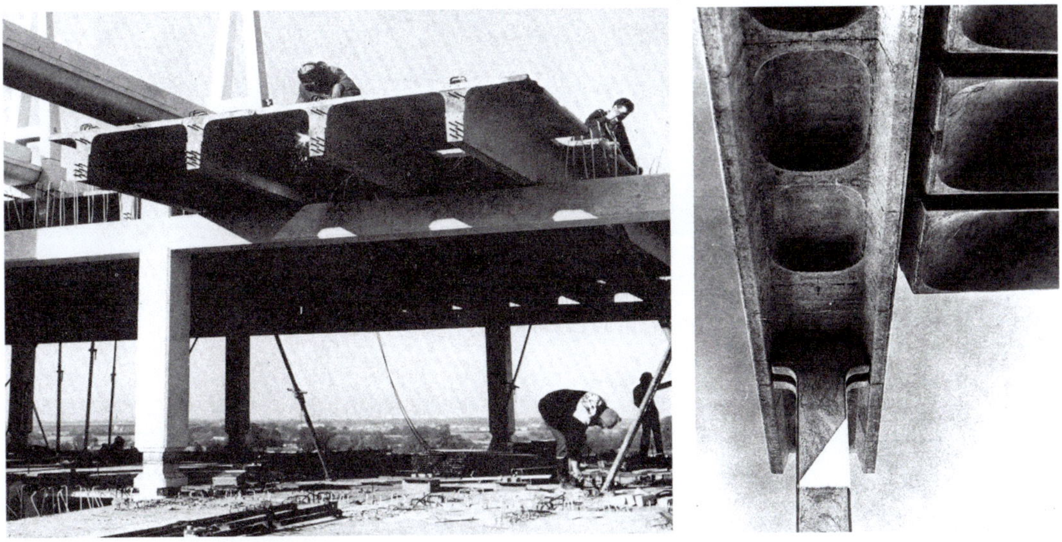

◘ Fig. 10.44 Two contiguous beam-and-slab elements being laid and joined together by welding reinforcement bars. (Right) Prefabricated concrete parts which are already mounted in a tilt-proof manner, simplify installation (architect: A. Mangiarotti)

◘ Fig. 10.45 Positioning of a trough-shaped girder

Fig. 10.46 Positioning of a roof element with autocrane

10.7 Influences on the Form

The aforementioned boundary conditions and their influence on the formal and constructional design of prefabricated-concrete structures make it clear that this type of construction is a way of dealing with concrete as a building material which differs significantly from that of in-situ concrete construction. The freedom that cast-on-site concrete offers to the designer and engineer in this respect is severely restricted with prefabricated construction by the extremely determinant manufacturing process. In this respect, prefabricated-concrete construction is more comparable to construction methods such as those in timber or steel. Nevertheless, this construction exemplifies, at least at the level of the single component produced in the shop, the possibilities of sculptural design typical of concrete, which are restricted only by the limits set by the formwork technique. The form-locking connections characteristic of prefabricated concrete clearly epitomise the chosen principle of force transmission and reveal their mode of action to the eye of the beholder much more clearly than do connections in other construction modes. The fact that this formal potential, which is impressively manifested in Angelo

Fig. 10.47 Skeletal structure typical of prefabricated construction with multi-storey cleat column and floor elements with moulded-on girders

Fig. 10.48 Prefabricated floor elements on trough girders. They are mounted on a ledge of the beam

Fig. 10.49 Prefabricated floor elements with moulded-on girders, supported by main girders with ledges

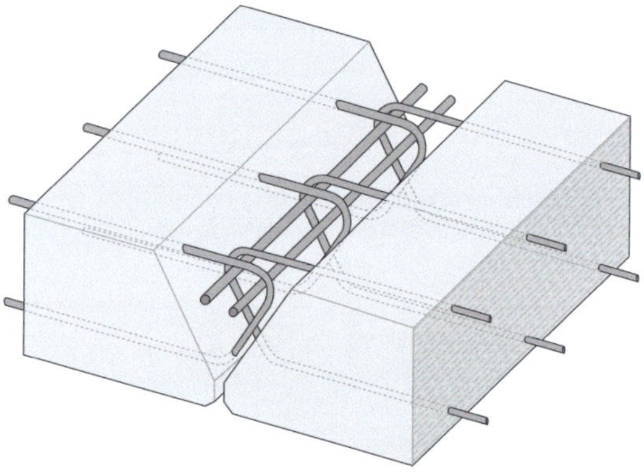

Fig. 10.50 Joint execution between two abutting floor elements. The reinforcement overlapping and the grouting create a racking stiffness of the floor as well as a vertically shear-resistant connection

Mangiarotti's work (Figs. 10.25, 10.54, and 10.55; cf. also Figs. 10.56, 10.57, 10.58, and 10.59), is not really—actually, not even nearly—exhausted in everyday building practice may be due to the fact that this construction mode, after a phase of failed heroic experiments, ultimately found refuge in the niche of industrial building of all things, a sector which, for whatever reason, is as a whole totally oblivious of, and—moreover—largely resistant to, good formal design.

On the other hand, the reason for this niche existence may also lie in the high requirements for systematic design discipline and overall planning accuracy which this construction mode exacts from the designer, the planner, and the structural engineer. The comprehensive and detailed previous plan-

Fig. 10.51 (Left) Restrained columns, provisionally secured with wooden wedges. (Right) The crane can release the element immediately after positioning it

Fig. 10.52 Laying of prefabricated sleeve foundations

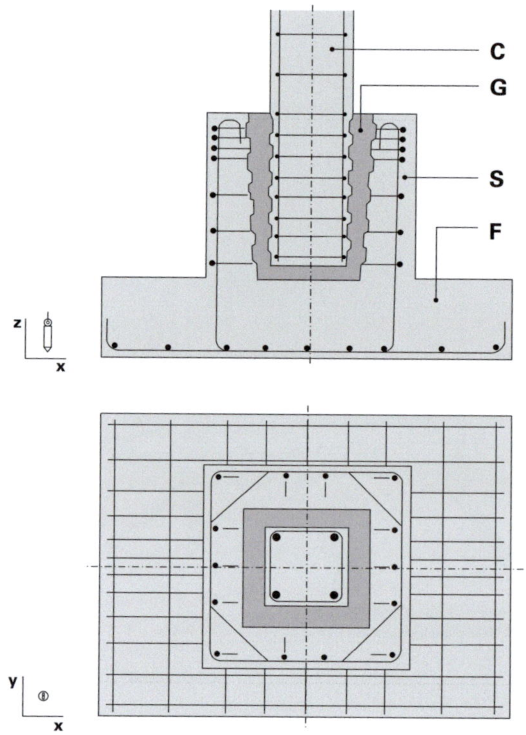

Fig. 10.53 Sleeve foundation for restrained prefabricated columns with representation of a usual reinforcement. The surfaces of the sleeve cavity and the column are profiled for a form-locking bond. **C** prefabricated column, **G** grouting, **S** sleeve, **F** foundation slab

10

Fig. 10.54 (Left) Strip-shaped prestressed roof elements resting on girders. Industrial building in Lissone, by Angelo Mangiarotti, 1964. (Right) Column head with mortise for the girder as shown on the left. Form-locking connection

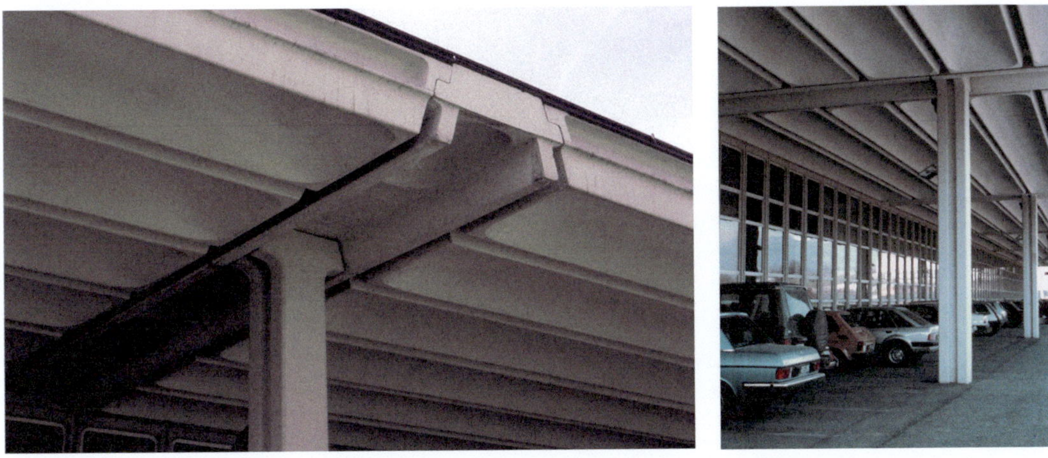

Fig. 10.55 (Left) Trough-shaped roof elements supported on the side of the girder. They are formed with the same height. The trough girder engages form-lockingly over the column head. Industrial building in Como, by Angelo Mangiarotti, 1969. (Right) Single-storey hall structure shown on the left with restrained columns

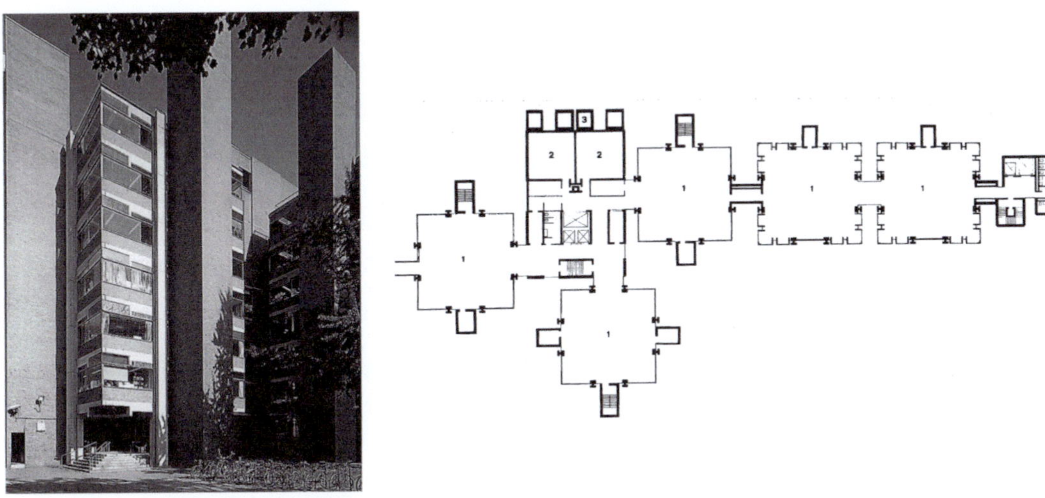

Fig. 10.56 Medical Research Center in Pennsylvania, arch.: Louis I. Kahn

ning of all relevant construction processes, which is indispensable in prefabricated construction, requires effort, relatively long lead times, and tends to stand in the way of short-term planning changes, such as unfortunately are commonplace in building construction.

As used on outer leaves of building envelopes, concrete has lost much of its significance as a result of developments of the last decades. This also applies to a prefabricated building in its variant as wall construction. The shortcomings in terms of building physics have already been mentioned above and structural problems difficult to resolve arise due to the conflict between the load-bearing and the thermal-insulation

***Vol. 3**, Sect. 12.4.1.2 from Reinforced Concrete*

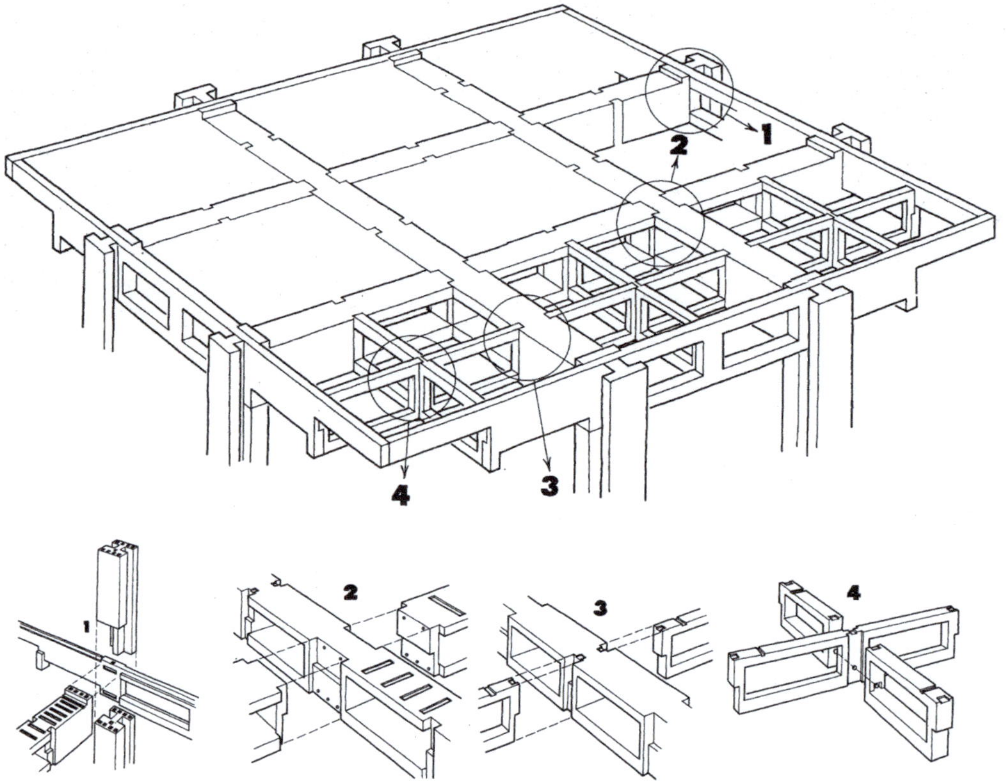

Fig. 10.57 (Top) Schematic depiction of the floor construction (Medical Research Center on Fig. 10.56). (Bottom) Explosion drawing of the nodes of the beam grillage shown on top

function. Furthermore, under direct weathering, concrete has not shown the durability that was promised in the pioneering era of concrete construction. The bad reputation of prefabricated wall construction in concrete, a damage thoroughly self-inflicted, has already been discussed above.

The overall appearance of most current prefabricated structures, particularly such as those found in industrial buildings, is characterised by the generally prevailing structural system of superimposed, both supporting and weighing elements joined to one another with hinged connections to form a whole, and is, so to speak, a modern variant of ancient post-and-lintel architecture, albeit an unloved and ugly one. Hardly any other type of construction mode being addressed in this book is, in actual practice, as far from its genuine formal potential as the prefabricated-concrete building. It is to be hoped that a renewed return to the basic idea of system-oriented architectural design, albeit under the sign of a much more differentiated approach than in the 1960s and 1970s, will help this type of construction experience a well-deserved renewal.

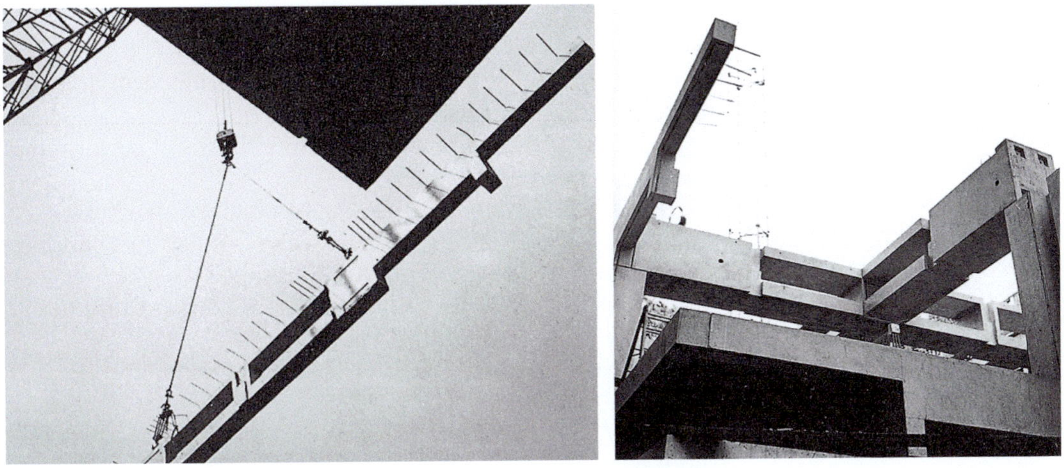

Fig. 10.58 (Left) A girder element being positioned (Medical Research Center on Fig. 10.56). (Right) The beam grillage being assembled

Fig. 10.59 (Left) A girder element being post-tensioned by means of hydraulic jacks (Medical Research Center on Fig. 10.56). (Right) Finished beam grillage from prefabricated parts joined together by post-tensioning

Standards and Guidelines

DIN 1045: Concrete, reinforced and prestressed concrete structures

Part 2: 2008-08 Concrete—specification, properties, production and conformity—application rules for *DIN EN 206-1*

Part 3: 2022-07 Execution of structures

Part 4: 2022-07 Precast concrete products—Common Rules

DIN 4172: 2015-09 Modular coordination in building construction

DIN 8580: 2019-11 Manufacturing processes—Terms and definitions, division

DIN 18540: 2014-09 Sealing of exterior wall joints in building using joint sealants

EN 206: 2021-06 Concrete—Specification, performance, production and conformity

EN 1992: Eurocode 2—Design of concrete structures

Part 1-1: 2021-10 General rules—Rules for buildings, bridges and civil engineering structures, prEN

Part 1-2: 2021-09 General rules—Structural fire design, prEN

Part 2: 2010-12 Concrete bridges—Design and detailing rules

EN 13224: 2012-01 Precast concrete products—Ribbed floor elements

EN 13225: 2013-06 Precast concrete products—Linear structural elements

EN 13369: 2022-04 Common rules for precast concrete products

EN 13747: 2010-08 Precast concrete products—Floor plates for floor systems

EN 14991: 2007-07 Precast concrete products—Foundation elements

EN 14992: 2012-09 Precast concrete products—Wall elements

EN 15258: 2009-05 Precast concrete products—Retaining wall elements

EN 15435: 2008-10 Precast concrete products—Normal weight and lightweight concrete shuttering blocks—Product properties and performance

ISO 4172: 2024-08 Technical product documentation (TPD)—Construction documentation—Drawings for the assembly of prefabricated structures

ISO 7437: 1992-06 Construction drawings; general rules for execution of production drawings for prefabricated structural components

Bibliography

Bindseil P (1991) *Stahlbetonfertigteile*. Düsseldorf, Werner

Bindseil P (2012) *Stahlbetonfertigteile nach Eurocode 2—Konstruktion, Berechnung, Ausführung*. Düsseldorf, Werner

Bona ED (1980) *Angelo Mangiarotti—Il processo del costruire*. Milan

Koncz T (1976) *Bauen industrialisiert*. Berlin, Wiesbaden; Bauverlag

Kordina K, Meyer-Ottens C, Richter E (1999) *Beton-Brandschutz-Handbuch*. 2nd rev. ed. Düsseldorf, Verl. Bau + Technik

Kind-Barkauskas F, Kauhsen B, Polónyi S, Brandt J (2009) *Stahlbeton Atlas: Entwerfen mit Stahlbeton im Hochbau.* Munich, Institut für Internationale Architektur-Dokumentation

Leonhardt F (2001) *Spannbeton für die Praxis*, reprint of the edition from 1955. Berlin, Ernst und Sohn

Pauser A (1998) *Beton im Hochbau—Handbuch für den konstruktiven Vorentwurf*. Düsseldorf

Rüsch H (1972) *Stahlbeton, Spannbeton—Werkstoffeigenschaften und Bemessungsverfahren*. Düsseldorf, Werner

Rationalisierungskuratorium der deutschen Wirtschaft (ed) (1972) *Transport von Fertigbauteilen*. Wiesbaden, Berlin

Stupré—Studienverein für das Bauen mit Betonfertigteilen, Niederlande (ed) (1978) *Kraftschlüssige Verbindungen im Fertigteilbau*. Düsseldorf

Zimmermann K (1973) *Konstruktionsentscheidungen bei der Planung mehrgeschossiger Skelettbauten aus Stahlbetonfertigteilen*. Wiesbaden, Berlin

Koncz T (1962) *Handbuch der Fertigteilbauweise: mit großformatigen Stahl- und Spannbetonelementen; Konstruktion, Berechnung und Bauausführung im Hoch- und Industriebau*. Berlin, Wiesbaden; Bauverlag

In-Situ Concrete Construction

J. L. Moro, *Building-Construction Design – From Principle to Detail*,
https://doi.org/10.1007/978-3-662-71006-7_11

Contents

11.1 History of Concrete Construction

11.1.1 Historical Precursors

The endeavours to develop an artificial stone which is freely-formable in the unset state go back far in history. The initial application of malleable mineral compounds was presumably intended to enhance the adhesion between, and to equalise dimensional imperfections of, layered natural stones or bricks. In the later development, already in late antiquity, however, the entire regular masonry bond was replaced by a mixture of aggregate and artificial petrified mass, so that stones, as a rule, were only randomly embedded within the concrete. Thus, the entire building structure was cast from one single piece, similar as with modern concrete.

Vol. 1, *Sect. 13.2 Stages of Technical Development of Masonrywork* as well as ibid. *Sect. 14.1 Stages of Historical Development*; cf. also ibid. *Sect. 17.1 Steps of Historical Development*

The first mortars, or plastically workable rock masses, consisted of aggregates and various binders, such as lime or lime compounds, which gave strength to the mixture. At first, one had to content oneself with air-hardening mortars, i.e., those which only harden by contact with the atmosphere. Thus, mortar layers lying deep within the cores of massive walls hardened late or not at all. The structural use of this type of mortar was therefore severely restricted. Complete structures could not be erected monolithically from these materials.

Cf. also ***Vol. 4***, *Sect. 8.4.3 Wall-Construction Modes in Reinforced Concrete* and ibid. *Sect. 8.6.3 Skeletal Construction Modes in Concrete*

Vol. 1, *Sect. 11.9.1.2 Manufactured Rocks*

Fundamental progress was made by the development of hydraulic binders. These are hardened by a chemical process—hydration—which does not depend on contact with air and can therefore also unfold in the interior of massive building structures. A binder acquires hydraulic properties by adding clay or volcanic earth.

Concrete technology, thus inaugurated, achieved a first highlight in Roman antiquity. The large-scale use of volcanic pozzolans with hydraulic properties and the labour force of thousands of slaves allowed the erection of numerous monolithic concrete structures made of a material with strength values similar to today's normal concrete (◘ Figs. 11.1 and 11.2).[1]

The knowledge of this technique was lost with the fall of the Western Roman Empire and was only redeveloped in the nineteenth century.

11.1.2 Evolutionary History of Modern Reinforced-Concrete Construction

Only after painstaking and protracted research and practical experimentation during the nineteenth century were modern hydraulic binders developed which had properties similar to those of the ancient ones. For marketing purposes,

Vol. 4, *Sect. 8.6.3 Skeletal Construction Modes in Concrete*

1 Lamprecht O (1993) *Opus caementitium: Bautechnik der Römer*.

Fig. 11.1 (Left) The Pantheon in Rome. The dome consists of *opus caementitium*. (Right) Wall core from Roman Concrete. The non-load-bearing facing brick leaves, effective only as lost formwork, are clearly recognisable (Taormina)

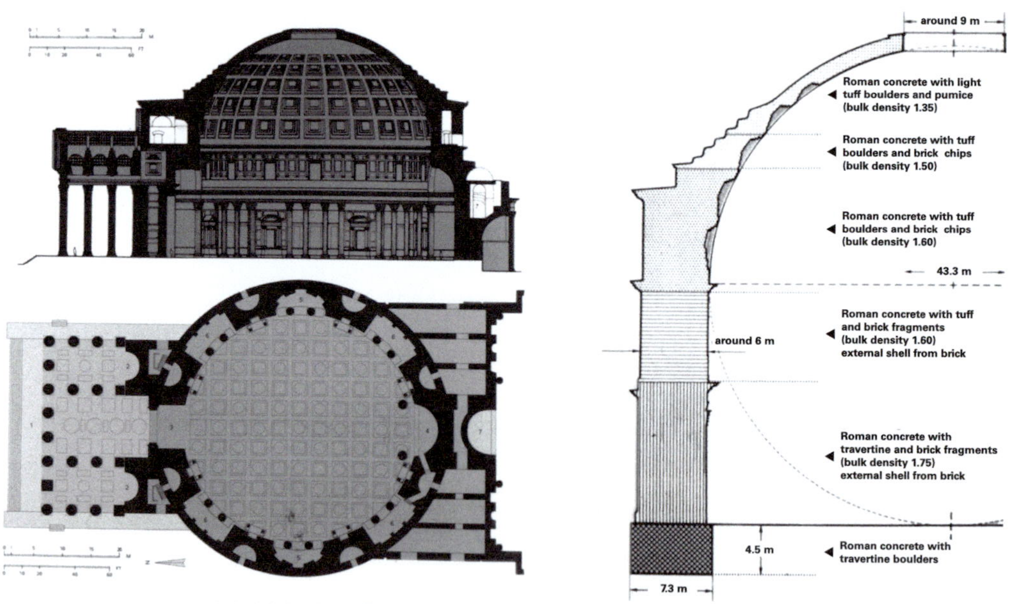

Fig. 11.2 Sections through the dome of the Pantheon

they were named Portland cements, in reference to the natural Portland stones which were widespread at the time and considered to be particularly strong and durable.

The excellent structural properties of the new material were only slowly recognised. In practical experiments, the gardener Monier developed the concrete reinforcement, and thus

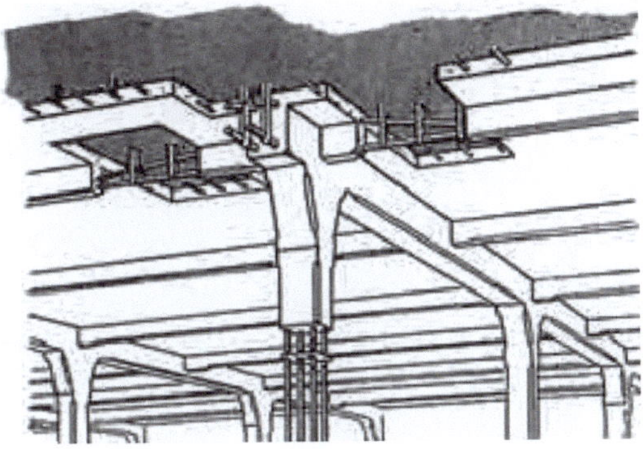

Fig. 11.3 Beam-and-slab floor of the System Hennebique, a patent from the year 1892

the composite material reinforced concrete, which he patented and then licensed—without any real insight into the structural significance and scope of his invention. However, licensees such as Wayss & Freitag subsequently made a decisive contribution to the technical development of the composite material.

F. Hennebique initiated a significant conceptual development in modern reinforced-concrete construction. He developed and patented an early beam-and-slab floor, a constructional principle tailored to the material's possibilities, which resulted from the fusion of the slab and beam to a monolithic component (Fig. 11.3) and featured the reinforced concrete-type of task division between a compressive chord made of concrete (slab) and tensile chord made of steel (beam). At the same time, the first basic principles were laid for a structural understanding of the three-dimensional load-bearing behaviour of the freely mouldable reinforced-concrete—as opposed to the two-dimensional one of bar-frameworks from iron and wood. The patented Hennebique system was licensed worldwide and used in numerous construction projects, primarily in industrial construction, where the new material was increasingly gaining ground in fierce competition with steel.

Auguste Perret[2] applied the material to conventional building construction and explored the new technical and aesthetic possibilities of concrete. In his apartment-house at *Rue Franklin* in Paris, he already early realised the skeletal-type separation of structure and envelope. Freyssinet developed prestressed concrete in the 1920s.

The rapid spread of reinforced concrete around the turn of the nineteenth and twentieth centuries revolutionised civil engineering and architecture, led to the development of entirely

2 Britton K (2001) *Auguste Perret*, Phaidon.

Fig. 11.4 Aircraft hangar in Orly during construction (1921–1923, E. Freyssinet). On-site concreted vault shell with folded cross-section (span 68 m). Visible in the foreground is the mobile load-bearing falsework (lattice truss), which supports the formwork surfaces above. In the background are the segments that have already been concreted; on the left are the previously concreted vault springings

new building types and to an unprecedented freedom of design and planning (Fig. 11.4). For the first time in almost 2,000 years, the span record of the Roman Pantheon was surpassed by the concrete structure of the *Jahrhunderthalle* in Breslau (arch.: Max Berg, Fig. 11.5), with a clear 65 m.

Reinforced concrete played an important role in the development of early Bauhaus Modernism and was considered the showpiece of the modern, industrially produced building material (Fig. 11.6, left).[3] It was particularly Le Corbusier, a pupil of Perret, who investigated the architectural possibilities of the material intensively in the first half of the twentieth century and explored its potentialities in architectural applications, also in aesthetic terms. Some design-related and conceptual innovations that resulted from the properties of the then new material can be outlined—even today—with his five points for a new architecture. His vision of the

3 Hilberseimer L (1928) *Beton als Gestalter: Bauten in Eisenbeton und ihre architektonische Gestaltung; ausgeführte Eisenbetonbauten.*

Fig. 11.5 *Jahrhunderthalle* (Century Hall) in Wrocław dating back to 1913. This steel-reinforced concrete structure exceeded the span of the Roman Pantheon for the first time in nearly 2,000 years

Fig. 11.6 Residential building at the Weißenhof compound in Stuttgart (arch.: Le Corbusier, 1927). (Right) Free design with reinforced concrete: curved shapes, continuous surface elements and monolithic structures (governmental building in Chandigarh, arch.: Le Corbusier)

Dom-Ino system seems particularly far-sighted from today's perspective: a skeletal construction with flat floors without beams, which shows great similarities to the concrete skeletons that are omnipresent today. In his later years, he also dealt with the plastic possibilities of this freely shapable material (Fig. 11.6, right).

For the *Dom-Ino System*, cf. ***Vol. 3****, Sect. 10 Fundamentals*, Fig. 10.30

11.1.3 Modern Concrete Technology

As a sequel, reinforced-concrete construction methods underwent significant technical developments, such as in shell construction, prestressed concrete construction or prefabricated construction, which have promoted the material to a hardly contestable dominance of modern building. Catastrophic events such as the attacks on the World Trade Center have further strengthened the already dominant position of this material, which is particularly resistant to fire and explosion, in recent years, especially in high-rise construction for safety reasons. The structural redundancy of monolithic concrete structures proves to be a particular advantage when the designer is faced with extreme situations that are difficult to predict. This trend is also driven forward by the development of modern self-compacting and high-strength concretes.

The nowadays extremely relevant disadvantage of conventional concrete construction from high labour costs resulting from formwork operations has been countered in recent years by semi-prefabricated construction, i.e. by systematic use of prefabricated lost formwork manufactured in automated plants while maintaining the monolithic load-bearing capacity of the overall structure.

The free formability of concrete is also well suited to the implementation of complex, double-curved construction forms, which have become more common in recent years with the advent of computer-aided design methods in other construction methods, mainly in steel and wood. The big stumbling block that has so far hindered the truly free formability of this castable material, namely the difficult realisation of free-formed formwork in wood or steel, is about to be overcome by modern CAD-CAM technology.

However, the partly rapidly deteriorating reputation of concrete in the public eye, which is increasingly being accused of being a climate killer due to its supposedly poor LCA values, should not go unmentioned. Nevertheless, the factual data refutes this accusation. It cannot be denied, however, that due to its enormous worldwide propagation and its dominance of the current construction industry, a great deal of concrete is used and its environmental footprint therefore appears to be *quantitatively* problematic.[4] The future will show to what extent it is possible and sensible to substitute concrete with other, more ecologically compatible materials.

11

4 Concrete is responsible for around 8% of global greenhouse gas emissions (The New York Times, 11.08.2020); steel causes around 7% and all car traffic around 6% [Columbia University, Center of Global Energy Policy, 10.2019]. In 2012, 10 billion cubic metres of concrete were used for construction, around twice as much as structural steel.

Fig. 11.7 (Left) Free design with reinforced concrete: a surface structure with shell-inspired design (arch.: Jørn Utzon). (Right) Free design with reinforced concrete: complex spatial and geometric compositions (arch.: F. Wotruba)

11.2 Reinforced Concrete as a Material

The building material reinforced concrete is a **composite** from steel and concrete and combines the characteristics of both materials in an extremely complex interplay. The compression-resistant, but very brittle concrete is strengthened by steel inserts to absorb tensile forces and is thus transformed into a ductile material. Furthermore, the two partners complement each other in a symbiotic interplay with regard to further structural properties.

On mechanical properties, deformational behaviour, and basic design aspects, cf. ***Vol. 1***, *Sect. 17 Reinforced Concrete*; on fire-protection behaviour, cf. ibid., *Sect. 29.5.1.3.2 Concrete/ Reinforced Concrete* as well as *Sect. 29.10.2 Components from Reinforced Concrete*; on corrosion behaviour, cf. ibid. *Sect. 30.3 Corrosion of Reinforced Concrete*

Processed on site, reinforced concrete offers the advantage, unique in the building industry, of being able to produce complete building structures in one single piece, without joints, in a process of **primary forming** on the building site. It is the only material suitable for primary structures which can be both highly prefabricated as well as alternatively be shaped on the building site in a process of primary moulding to form structural elements or entire structures. This extraordinarily wide range of production or processing has extensive planning-related and constructional implications and has therefore also produced distinctly different construction modes, namely **in-situ concrete** and **prefabricated** construction. These are clearly differentiated in building construction and are therefore dealt with separately in this book—despite sharing a common basic material.

11.3 Comparison with Other Construction Modes

In-situ concrete construction modes fundamentally differ from others in some important respects:

- The material on which the construction mode is based is **freely shapable** (▣ Fig. 11.7). In this respect, concrete differs from wood, which is produced by a natural rectilinear growth process which cannot be influenced by any technical means; it does so also from steel, which can only be formed within the narrow limits of a certain manufacturing process (e. g. rolling); as well as from brickwork, which likewise can be shaped and fired in only small formats and under tight technical constraints.
- **Areal elements** can be produced in one piece with dimensions suitable for building purposes (▣ Fig. 11.8). This is a decisive advantage of this construction mode, even compared to the related prefabricated construction. Other construction types must resort, in order to create surfaces, to coupling rod-shaped components, such as timber construction modes. Brickwork can certainly be processed as a planar shear-plate, albeit from individual building blocks (but always requires—in contrast to reinforced concrete—a linear support), but never as a plate which is subjected to bending stress. The latter is, however, ubiquitous as the solid floor slab in reinforced concrete in current construction practice. Bracing wall shear-plates in reinforced concrete are also able (in contrast to masonry) to absorb uplifting tensile forces without any significant contribution from large superimposed loads. Loads transverse to the wall plane, which would cause gaping bed joints in masonry, are absorbed by the bending stiffness of the cross-section in reinforced-concrete walls. Reinforced-concrete wall shear-plates can be supported at points and are therefore capable of covering large spans freely by virtue of their very large structural depths (acting as wall beams, see below). Even concentrated loads—practically impossible to deal with for masonry walls—can be transferred into

11

▣ **Fig. 11.8** In-situ concrete allows creating monolithic structures and two-dimensional components with sizes typical of building construction relatively thin reinforced-concrete walls with the aid of incorporated composite steel constructions.

- **Monolithic, jointless** structures can be erected if the concrete is poured on site (hence, in-situ concrete). This is also a clear advantage over other construction methods. Consequently, an integral construction principle can be realised, namely at the level of the building, that is to say, at the level of the overall construction, whereas other materials—at least under these conditions—allow only for the quasi-integral or differential construction principle. Under certain circumstances, the static redundancy inherent in the multiple statically indeterminate monolithic concrete structures offers significant advantages.

***Vol. 1**, Sect. 2.2.3.2 Derived from the Building Principle*

- A **two-way load transfer** can be realised without major additional effort. Both plate-shaped and grid-shaped components allow for two-way load-bearing behaviour. The force transmission in two directions, which in other construction methods, such as steel or timber construction, can only be achieved on site with considerable effort, is simpler in in-situ concrete construction, because junctions either do not exist at all (as with the slab) or are cast in one piece (as with the grid). The two-way load-bearing behaviour is made possible both by the isotropic concrete matrix and by the (non-jointed, only placed on top of one another) reinforcing bars which can be laid in two directions in superposed, not mutually intersecting layers (■ Figs. 11.9 and 11.10). In the current construction practice, however, these

■ **Fig. 11.9** Intersecting but not penetrating reinforcement in two directions situated in two separate layers enables, together with the isotropic concrete, a biaxial load transfer

Fig. 11.10 Cast-in-place girder grillage: a biaxially spanning grid-shaped supporting structure in which the rib penetrations are realised by joint casting

advantages are partially offset by the relatively high labour costs from the formwork and scaffolding work necessary on site.

- Wall shear-plates in reinforced concrete can be used as **wall beams** or **deep beams** thanks to their tensile strength. This allows the large structural depth of entire storey clearances to be used for load-bearing purposes. This happens, for example, on the lower floors of high-rise buildings, where column-free rooms are often desired for reasons of usage. In such cases, load-bearing wall panels can provide support for columns from the upper storeys. Vertical wall slabs can also be stacked at an angle to each other, i.e. in such a way that they are supported solely by point supports and thus bridge large spans (Figs. 11.11 and 11.12).

11

Cf. also ► *Sect. 11.8.2 Detached-Shear-Plate Structures*

Cf. for example Fig. ► 6.23 in ► *Sect. 6 Foundation*

***Vol. 4**, Sect. 8.4.3.4 Wall Beams*

11.4 Processing

This theoretically very large scope of design freedom which the material offers is, however, limited by some restrictions from processing:

- The **maximum sizes** of the sections that can be concreted in a single operation are limited. The concreting work is therefore subdivided into **concreting segments** which are each separated by so-called **working joints** (Fig. 11.13).
- The hypothetically completely freely-formable surface of concrete components is generally nevertheless limited in its shape by limitations from **formwork technology**. Formwork consists of a different material (wood, steel), which in turn is subject to specific processing rules. In particu-

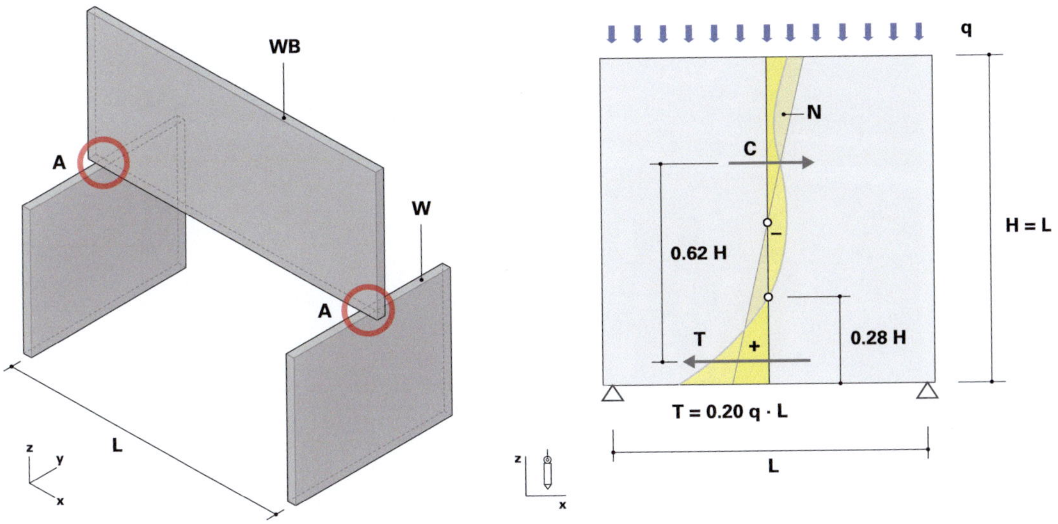

Fig. 11.11 (Left) Reinforced-concrete panels **WB** are able, functioning as wall beams, to bridge large spans **L**. They can be supported on further walls **W** oriented at right angles. Concentrations of force occur at the supports **A**. (Right) Distribution of compressive and tensile stresses in a point-supported panel with an approximately square shape under line load **q**. **C** Compressive force resultant; **T** tensile force resultant. For comparison: **N** regular triangular stress distribution in the cross-section of a comparable bending beam with normal slenderness (according to Navier)

lar, biaxially curved surfaces can only be realised with great effort (Figs. 11.14 and 11.15).

► *Sect. 1.3.2 Composing Biaxially Curved Surfaces*

- **Formwork** and **scaffolding work** on the construction site (Figs. 11.15 and 11.16), which is still carried out on a large scale on site in in-situ concrete construction despite extensive prefabrication of formwork elements—particularly in comparison with competing construction methods such as steel or timber construction, and even prefabricated construction—is currently a significant economic factor which, under certain conditions, places in-situ concrete construction at a serious disadvantage. This applies in particular to wide-span roofs, which are hardly ever executed in this way today. As a result, semi-prefabricated construction methods that do not require any shuttering work on the construction site are becoming increasingly widespread in concrete construction.

11.5 Reinforcement Technology

The concrete, which is sensitive to tensile stress because it is brittle, is made tensile-resistant by means of steel inserts placed in specific tensile zones (Fig. 11.17). These are completely embedded in a concrete matrix that surrounds them and are bonded to this matrix with a force fit. The concrete usually cracks in the tensile zones and thereby transfers the tensile force to the steel reinforcement. The enveloping

EN 1992-1-1, 8., 9.

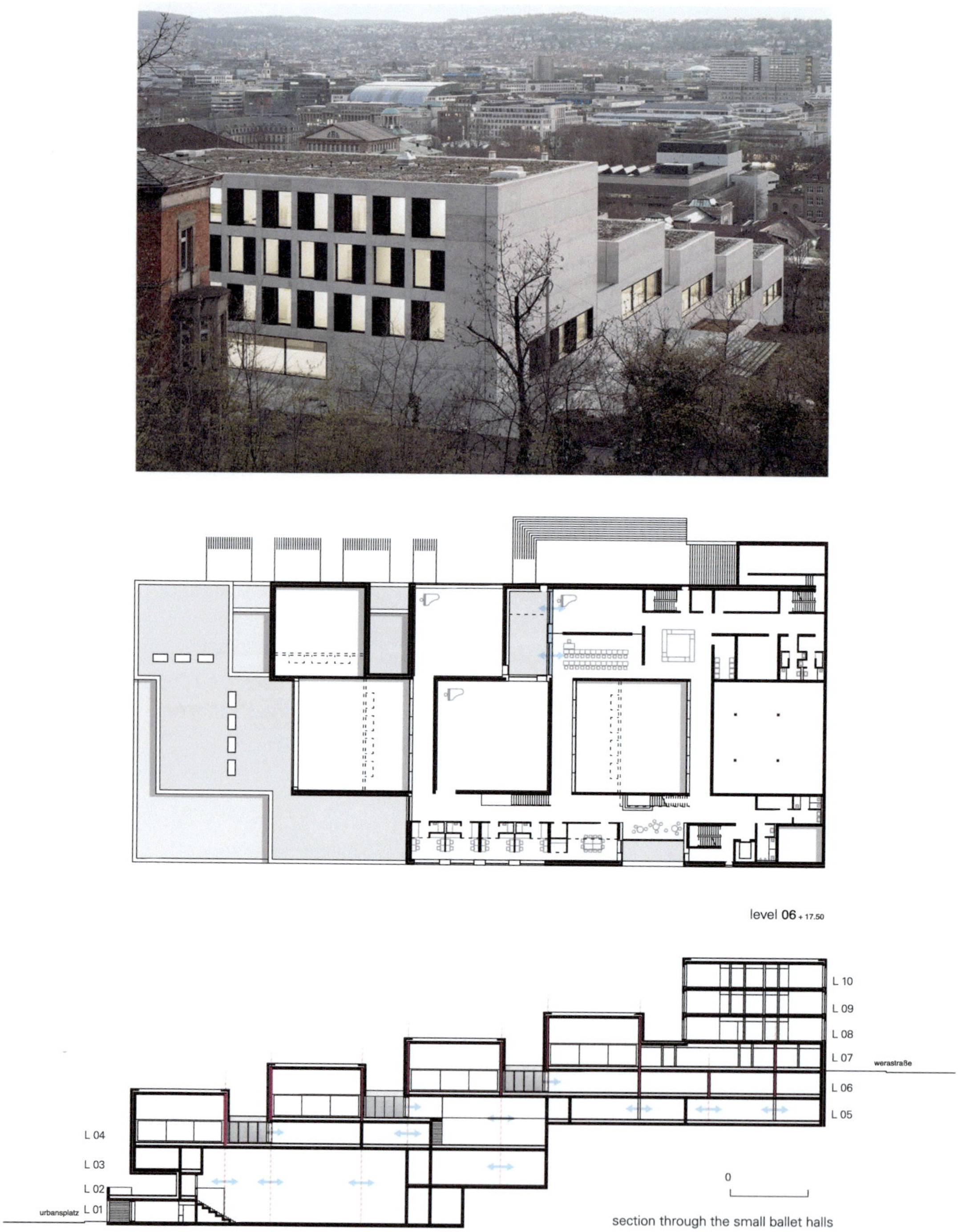

Fig. 11.12 (Top) John-Cranko Ballet School in Stuttgart: a building staggered along the slope, in which wall beams were used to create generous spatial connections (arch.: Burger Rudacs). (Bottom) Floor plan and section. Wall beams are highlighted in red; the spatial connections made possible by free spans are visualised by blue arrows

Fig. 11.13 Concreting joint (horizontal) between two concreting sections. The latter are also recognisable by their different colour

Fig. 11.14 Double-curved surface (hyperboloid of revolution) realised with the help of a board formwork, here aligned along the meridian lines (*La Zarzuela* racecourse, Madrid; eng.: Eduardo Torroja)

concrete matrix also performs an important protective function against corrosion of the reinforcing steel.

In contrast to other materials suitable for primary structures, such as wood or steel, the structural elements crucial to load-bearing, the reinforcing steels, are completely embedded within the volume of the structural component, which means that the elements are completely hidden from view. The assumption that a reinforcement has no effect on the shaping of a reinforced-concrete component is nevertheless wrong. For a proper execution, a sufficient bond between steel and concrete, as well as an adequate protection of the steel inserts, necessary spatial prerequisites are to be created within the formwork cavity so that the concrete matrix can reliably fulfil its tasks after pouring. As a rule, this presupposes certain

Fig. 11.15 Formwork for a chalice-shaped support of the Stuttgart 21 railway station: complex double-curved surface made of plywood panels milled into shape and then coated (see edge cut on both sides)

minimum dimensions in specific component areas and consequently clearly determines the design of the component.

Recent developments in concrete technology, however, have led to classical reinforcement rules losing part of their conclusiveness. For example, the use of self-compacting concrete allows significantly higher reinforcement densities than with normal concrete.

Vol. 1, Sect. 17.7.3 Self-Compacting Concrete (SCC)

Basic aspects of the mechanical behaviour and the reinforcement of reinforced-concrete parts are discussed in ***Volume 1*** *Chapter IV*. In the following, further questions concerning reinforcement technology will be dealt with. These mainly concern not only components made of in-situ concrete, but are also applicable to prefabricated parts.

Vol. 1, *Sect. 17.2 Mechanical Properties* as well as ibid. *Sect. 17.6 Constructional Consequences*

11.5.1 Arrangement

Reinforcement can be regarded as a substitute for the poor tensile strength of concrete, which is only about one tenth of its compressive strength. The basic requirement of the load-bearing capacity of a component is met by establishing a **minimum reinforcement**. This is additionally increased for the purpose of securing usability, i.e. to limit deformation.
Reinforcing steels are routed as accurately as feasible with reasonable effort along the main tensile stress trajectories within the component. This is referred to as **traction cover-**

Fig. 11.16 Laying the reinforcement of a wall panel in the formwork space before installing the second shuttering surface

age in the component (Figs. 11.18 and 11.19), so that bearing flexural tensile stresses as well as tensile stresses from shear force is secured at every point. For reasons of technical implementation, reinforcement cross-sections are not continuously adapted to the constantly changing stress magnitudes, but rather in successive tiered sections featuring respectively the same reinforcement density, so that a step-like enveloping distribution of tolerable tensile stresses in the component results. *EN 1992-1-1*

Determining the required reinforcement is usually carried out in the following steps, without taking into account any possible iteration step:

- Making of a **formwork drawing**, definition of the definitive geometry of the component.
- Possibly, applying an **FE-model** for calculating the stresses within the component.
- Determining the occurring tensile stresses.
- Defining of the necessary reinforcement cross-sections.
- Preparing a **reinforcement plan**.

Despite trajectories being mostly predominantly curved, it is not customary to skew or bend reinforcement bars in large radii. Instead, they are either laid in a straight line or folded to a polygon—that is, folded locally with small radii, otherwise laid out straight (see Fig. 11.17). Bend-roller diameters must not fall below minimum values in order to prevent the occurrence of bending cracks in the steel or the failure of the concrete in the area of the bar bend. Minimum values can be found in the code. *EN 1992-1-1*, 8.3, Table 8.1 N

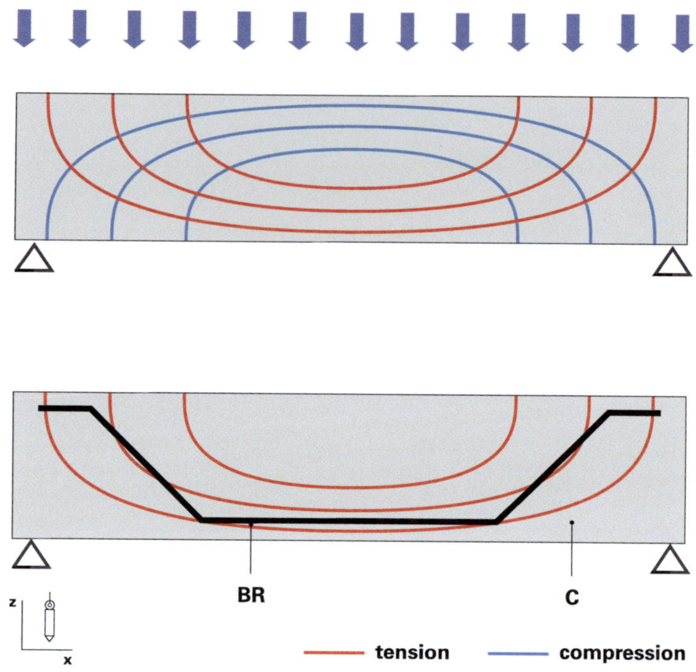

Fig. 11.17 Schematic representation of the course of the **main stress trajectories** in a beam under distributed load (top). The covering of the tensile stresses by the bar reinforcement **BR** can be seen below. **C** concrete

11

In order to avoid laborious bending works, which are associated with high labour costs, it is becoming more and more customary not to lay bars in an inclined position, but rather to reinforce perpendicularly, usually horizontally and vertically, with the aid of shear reinforcement. This applies, for example, to a shear-force reinforcement in a beam, which can be laid perpendicularly despite the force trajectory being diagonal (Figs. 11.19 and 11.20).

11.5.1.1 Types of Reinforcement

Depending on the assigned **function** in relation to the mechanical behaviour of the reinforced concrete, different types of reinforcement can be distinguished (Fig. 11.20):

- The **main reinforcement** is responsible for the main bearing action of the component. A distinction is generally made between **flexural tensile reinforcement** and **shear-force reinforcement**. The latter assumes an angle of between 45° and 90° with respect to the axis of gravity of the component and can consist of a combination of stirrups which encompass the longitudinal tensile reinforcement, and oblique bars or cages, ladders, etc. (Figs. 11.21 and 11.25).
- An enveloping fine-meshed so-called **shrinkage** or **surface** reinforcement made of meshes is tasked with preventing excessive cracks resulting from shrinkage in the concrete. This reinforcement is also referred to as **constructional reinforcement**.

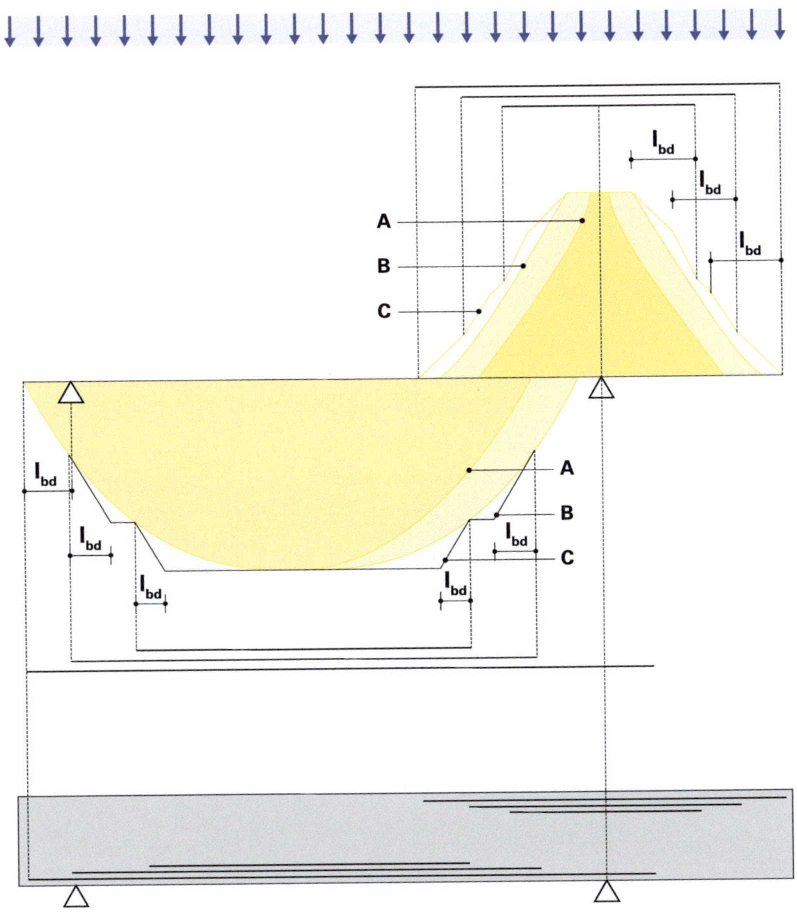

Fig. 11.18 Illustration of the **tensile-force coverage** in an exemplary single-span beam with cantilever under load **q** and the resulting staggering of the longitudinal reinforcement, taking into account inclined cracks and the load-bearing capacity of the reinforcement within the anchorage lengths $\mathbf{l_{bd}}$, according to *EN 1992-1-1*. The enveloping curve **A** covers the alternating load cases to be applied; **B** is the rated value of the tensile force due to shear-force reinforcement (surcharge); **C** is the tensile force actually absorbable by the reinforcement. The reduced load-bearing capacity of the bars within the anchoring length $\mathbf{l_{bd}}$ is taken into account in each case (inclined tensile force curve)

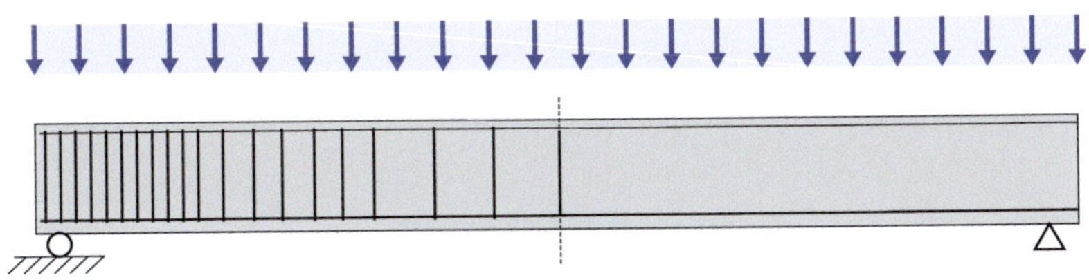

Fig. 11.19 **Shear-force** reinforcement on an exemplary single-span beam under linear load **q**

- In order to hold the main and constructional reinforcement in place during assembly and concreting, an additional **assembly** reinforcement may be required.
- **Through**-reinforcement across working joints.

With regard to supply by the manufacturer and on-site processing of reinforcing steel, a distinction can be made between: *DIN 488-1*

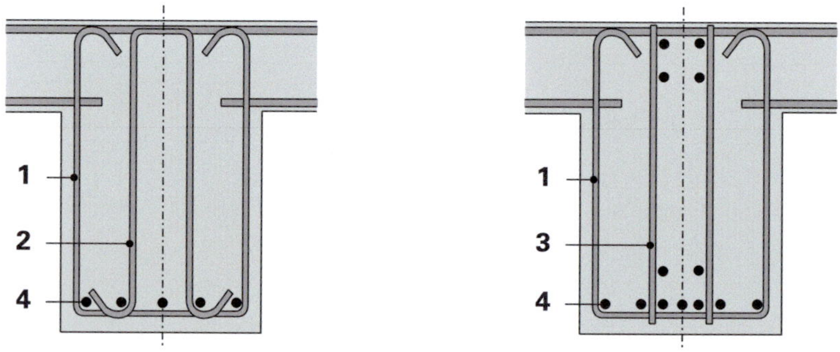

Fig. 11.20 Examples of **shear-force** reinforcement from combinations of stirrups and shear-force supplements. (**1**) Stirrup, (**2**) stirrup cage as supplement, (**3**) ladder-like shear-force supplement, (**4**) flexural tensile reinforcement

Fig. 11.21 Pre-assembled reinforcement cage for a girder with main and stirrup reinforcement

► *Sect. 11.5.2 Reinforcement Bars*

► *Sect. 11.5.3 Reinforcement Meshes*

► *Sect. 11.5.4 Lattice Girders*

- Reinforcing steel bars.
- Reinforcing steel in rings.
- Welded reinforcement meshes.
- Lattice girders.

11.5.1.2 Bar Spacing

Bar spacings must be sufficiently large so that the fresh concrete can be well introduced into the shuttering and subsequently sufficiently compacted. The relatively high viscosity of this material in plastic state must always be taken into account, which sets certain limits to the self-active compacting and filling of hollow spaces within the formwork cavity, i.e. to the self-compacting under the effect of gravitational force. According to the code, the clear distance—horizontal and vertical—between parallel individual bars, or layers of parallel bars, has to meet certain minimum values depending on the diameter of the bars and the diameter of the largest grain of the aggregate. Bar spacings are also essential to ensuring a sufficient bond between concrete matrix and reinforcing steel, since a complete wrapping of the rods is required as well as a minimum thickness of the matrix in order to transfer the force between concrete and steel via the bonding. *EN 1992-1-1*, 8.2

In addition, minimum bar spacings are also important for good compacting of the concrete by means of vibrating devices. The usual diameter of vibrating cylinders in the range of 6–8 cm requires, in practice, bar spacings equal or larger than 10 cm, so that the reinforcing bars are usually laid at distances between 10 and 15 cm.

The **bonding conditions** are the conditions which influence the bonding between steel and concrete. This relates in particular to the profiling of the reinforcing bars and also to the direction of the concreting in relation to the orientation of the bar: Depending on whether the bar is located at the bottom or at the top of the formwork chamber, or whether concreting takes place parallel or at right angles to the bar alignment (more favourable: parallel), the bonding conditions will be better or worse.

11.5.1.3 Anchoring of Longitudinal Bars

In order for the forces to be transmitted between the reinforcing steels and the concrete matrix, anchoring of the former in the concrete is necessary. In particular, the tensile forces resulting from the anchoring of the compression strut in the bar in the region above the support are absorbed (Fig. 11.22). At the same time, precautions must be taken to prevent chipping of the concrete in the anchoring area and longitudinal cracking in the concrete, which may require additional transverse reinforcement.

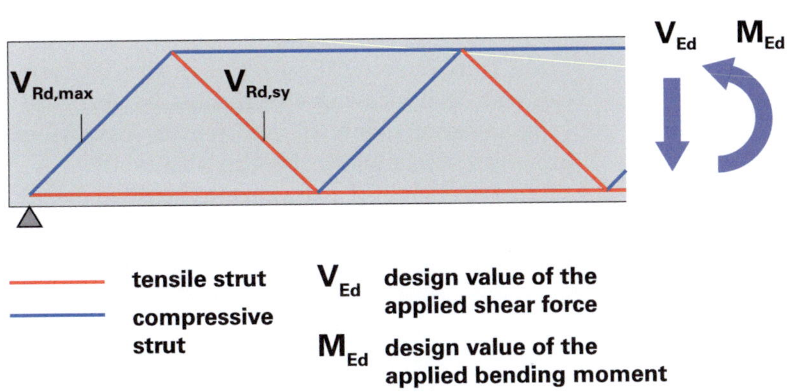

Fig. 11.22 Load-bearing behaviour of a beam in the support area with regard to the absorption of **shear forces**. Simplified representation using a truss model. The compressive-strut force $V_{Rd,max}$ is absorbed by the concrete, the tensile-strut force $V_{Rd,sy}$ by the shear-force reinforcement

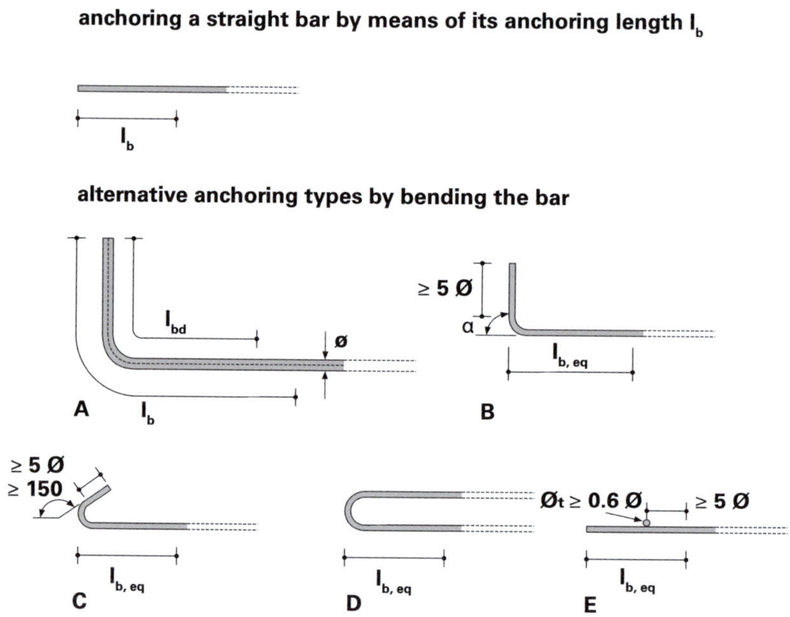

Fig. 11.23 Permissible **anchoring types** of reinforcement steel bars according to *EN 1992-1-1* with accountable anchorage length $\mathbf{l_a}$. **A** Basis value of the anchorage length $\mathbf{l_b}$, for all types of anchorage, measured along the centre line. **B** Replacement anchoring length $\mathbf{l_{b,eq}}$ for normal angle hooks. **C** Replacement anchoring length $\mathbf{l_{b,eq}}$ for normal hooks. **D** Replacement anchoring length $\mathbf{l_{b,eq}}$ for normal loops. **E** Replacement anchoring length $\mathbf{l_{b,eq}}$ for welded crossbar

11

Anchoring can be carried out with an effective or equivalent minimum anchorage length in the form of (Fig. 11.23):

- Straight bar ends.
- Hooks, angle hooks or loops.
- Straight bar ends with at least one welded bar in the anchoring area.
- Hooks, angle hooks or loops with at least one welded bar in the anchoring area before the start of curvature.

11.5.1.4 Transverse Reinforcement in the Anchoring Area

Local transverse tensile stresses in the anchoring area, which can lead to a cleaving effect, must be absorbed with a transverse reinforcement. If sufficient transverse compression is present in this region, this can also be dispensed with. In the case of rod-shaped components, such as beams or columns, a transverse reinforcement consists of stirrups which are laid approximately at intervals equal to the fivefold diameter of the rod of the anchored reinforcement (Fig. 11.24).

11.5.1.5 Anchoring of the Shear-Force Reinforcement

Analogously to longitudinal bars, stirrups and shear-force reinforcement must also be anchored in the concrete matrix. This can be done using hooks, angle hooks, or welded transverse reinforcement (Fig. 11.25).

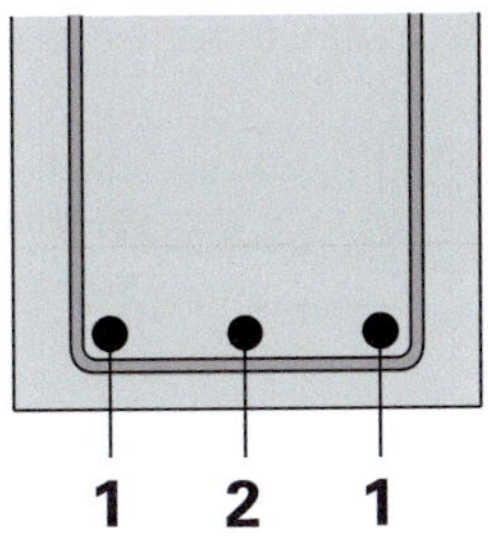

Fig. 11.24 Arrangement of the transverse reinforcement (stirrups) in the anchoring area of longitudinal bars in a beam according to *DIN 1045-1*. (**1**) Anchored reinforcement bars, (**2**) continuous reinforcement bar

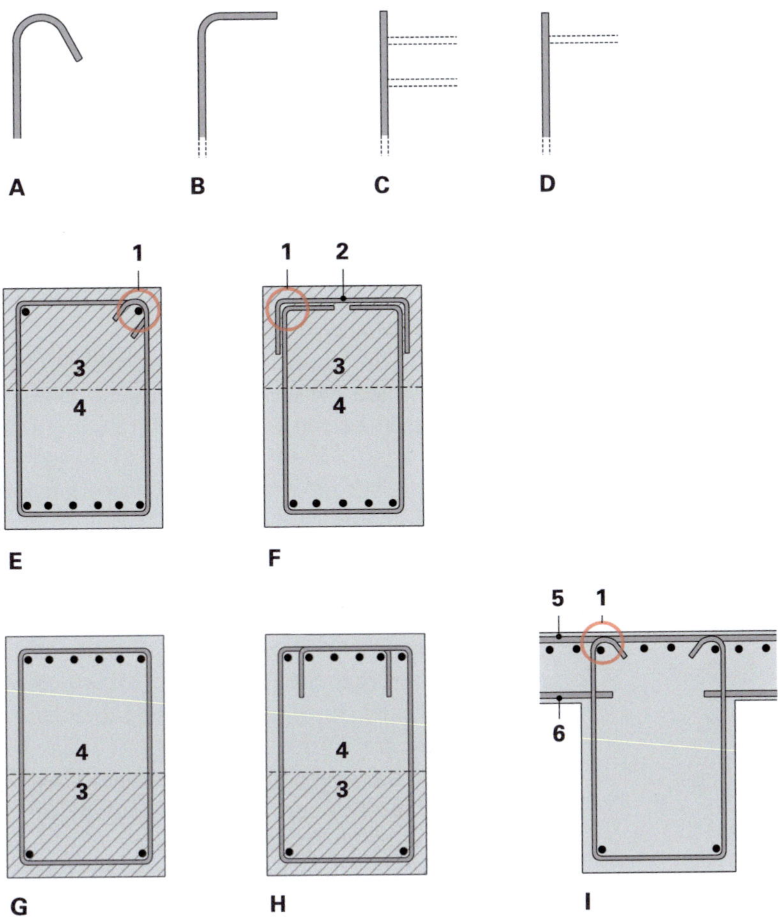

Fig. 11.25 Anchoring and closing of stirrups of a shear-force reinforcement in beam cross-sections according to *DIN 1045-1*. (**1**) Anchoring elements according to **A** and **B**, (**2**) cap stirrup, (**3**) concrete compressive zone, (**4**) concrete tensile zone, (**5**) upper transverse reinforcement, (**6**) lower reinforcement of the connecting slab. **A** Hook; **B** angle hook; **C** straight bar ends with two welded crossbars; **D** straight bar ends with one welded crossbar; **E**, **F** closing of the compressive zone; **G**, **H** closing in the tensile zone; **I** closing in the slab area in case of beam-and-slab construction

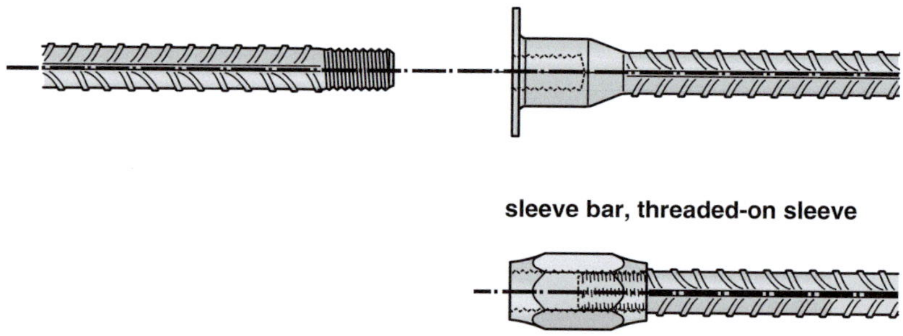

Fig. 11.26 Threaded-sleeve joint between abutting reinforcing bars

11.5.1.6 Reinforcement Joints

Reinforcement joints are necessary, either because the delivery lengths of the reinforcing steel are smaller than the component dimensions or because the concreting work has to be interrupted at a working joint. Force-fitting joints of reinforcing bars can be produced by:

- Mechanical connections such as coupling sleeves or special connectors (Fig. 11.26), or alternatively welded joints (Fig. 11.27): These are **direct joints**, i.e. those without involvement of the concrete.
- Overlap of the reinforcement bars within the concrete matrix: These are **indirect** or **overlapping joints**, i.e. joints effected with the co-operation of the concrete (Fig. 11.28).

Because of the simple execution, the overlap joint is the rule in construction practice. However, it presupposes sufficiently ample spatial conditions, which are sometimes not present—as is the case, for example, with grouting joints of prefabricated parts.

Neighbouring overlapping joints must always be offset against one another (Fig. 11.29). **Minimum overlap lengths** must be observed. In order to absorb the occurring transverse tensile forces, transverse reinforcement must be provided in the region of the overlap of longitudinal bars. Joints of reinforcing steel meshes can be carried out by interlacing in the same plane or by laying in two separate planes (Fig. 11.30). Additional transverse reinforcement in the joint area is not required for reinforcing steel meshes.

11.5.1.7 Grouping of Bars

EN 1992-1-1, 8.9

In the case of large reinforcement densities, **bar bundles** made of several connected, contacting bars are also used. The bonding conditions deteriorate significantly as compared to free individual bars. Details are regulated in the code.

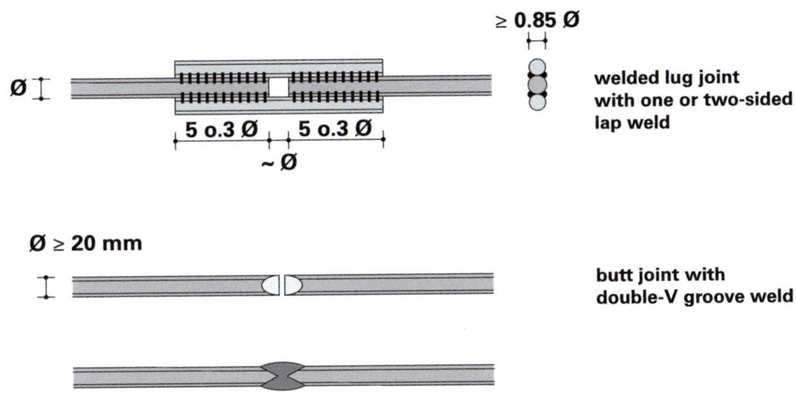

Fig. 11.27 Welded joint between abutting reinforcing bars

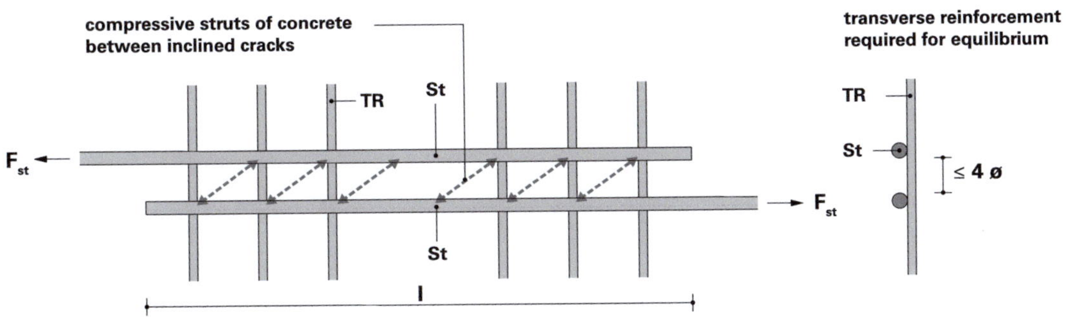

Fig. 11.28 Schematic representation of a lap joint of bar steels **St** with required transverse reinforcement **TR**

11.5.2 Reinforcement Bars

DIN 488-2; EN 10080

According to the definition of the code, reinforcement bars are a steel product with a circular or nearly circular cross-section suitable for the reinforcement of concrete. The following variants are distinguished (Fig. 11.31):

- **Ribbed reinforcing steel**: with at least two rows of **inclined ribs** spread evenly over the entire length. In addition, a longitudinal rib is provided: a continuous rib parallel to the axis of the bar, wire rod, or wire (Figs. 11.31, top and centre and 11.32).
- **Profiled reinforcing steel**: with defined indentations distributed uniformly over the entire length (Fig. 11.31 bottom).
- **Smooth** reinforcing steel.

Ribbed and profiled steel products are standardised according to surface geometry. The standard nominal diameters, cross-sectional areas, and unit masses can be found in the overview on Fig. 11.33.

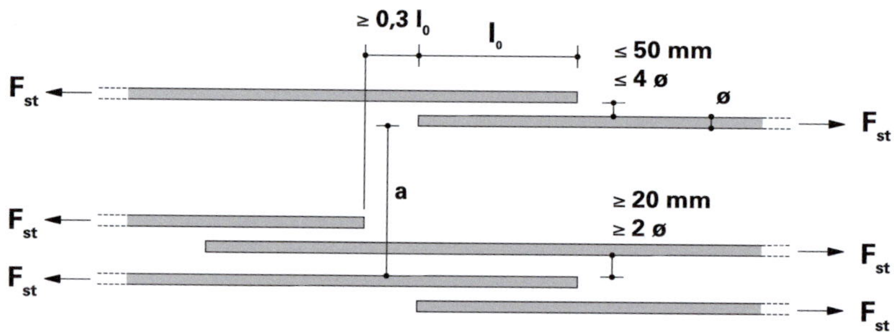

Fig. 11.29 Offset of lap joints of reinforcing bars according to *EN 1992-1-1*. The required overlap length $\mathbf{l_0}$ is regulated by the standard

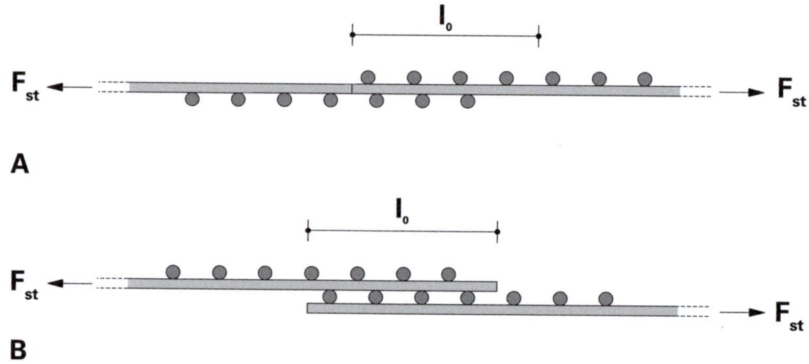

Fig. 11.30 Lap joints of steel reinforcement meshes according to *EN 1992-1-1*. **a** Interlocking of steel reinforcement meshes (longitudinal section). **b** Two-level joint of steel reinforcement meshes (longitudinal section)

11

11.5.3 Reinforcement Meshes

EN 10080, 3.17

According to the definition of the code, a welded mesh is an array of longitudinally and transversely extending bars, wire-rods, or wires of the same or different nominal diameter and length which are connected at all points of intersection by electrical resistance-welding executed in shop by means of automatic machines (Fig. 11.34). The bars can be single or double-welded, but double bars are only permitted in one position (Fig. 11.35).

Bars of reinforcing steel meshes are processed with three rows of sloping ribs, one row of ribs being counter-rotating (Fig. 11.36).

Mesh material is used when reinforcing flat parts. In this way, wage-intensive laying of bars over larger areas is rationalised.

Vol. 1, Sect. 22.9.1 Reinforcement Steel According to DIN 488

The predestined application of this type of reinforcement are two-way spanning slabs of reinforced concrete in which load transfer takes place in two equivalent directions. In this type of slabs, so-called uniform meshes are used. With them, the two bar skeins have the same diameter.

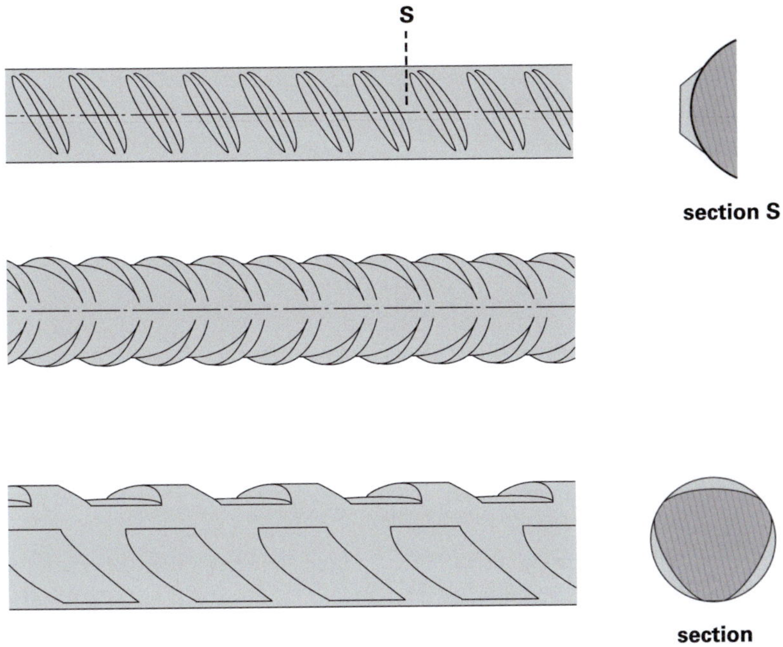

Fig. 11.31 (Top and centre) Rib geometry of a ribbed reinforcement bar. (Bottom) Profile geometry of a profiled reinforcement bar

Fig. 11.32 Ribbed reinforcement bars

One-way spanning slabs with a predominant direction of load transfer are reinforced with non-uniform meshes, in which the diameters of the bars can be differentiated according to the laying direction.

Regarding the **delivery form**, the following distinctions are made:

nominal diameter mm	nominal cross-sectional area [a] mm²	nominal mass [b] kg/m
6.0	28.3	0.222
8.0	50.3	0.395
10.0	78.5	0.617
12.0	113	0.888
14.0	154	1.21
16.0	201	1.58
20.0	314	2.47
25.0	491	3.85
28.0	616	4.83
32.0	804	6.31
40.0	1,257	9.88

[a] The nominal cross-sectional area is calculated as:

$$An = \frac{md^2}{4}$$

[b] Calculated with a density of 7.85 kg/dm³

Fig. 11.33 Standardised nominal values of reinforcement-bar steel according to *DIN 488-2*

Fig. 11.34 Laid-out reinforcement meshes for a floor slab

- **Stock meshes**: set-up by the manufacturer, available from stock.
- **List meshes**: set-up by the customer by name.
- **Drawing meshes**: set-up determined by drawing (Fig. 11.37).

11.5.4 Lattice Girders

A reinforcement lattice girder is a two- or three-dimensional metal structure consisting of an upper chord, one or more

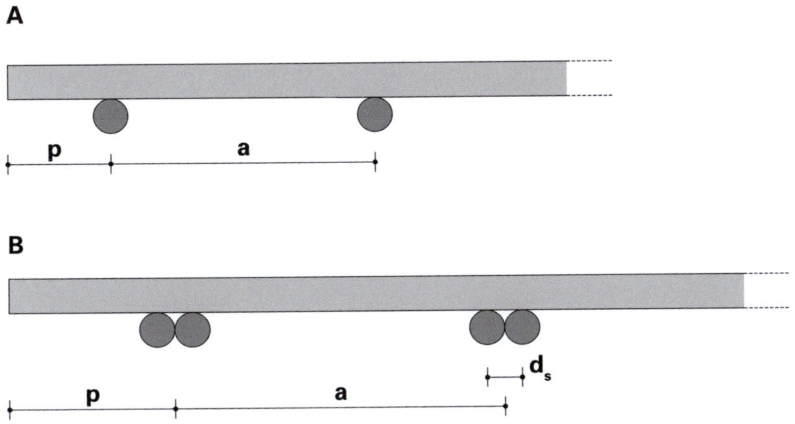

Fig. 11.35 Construction of steel reinforcement meshes with single and double bars. Definition of the bar spacings **a** and the projections **p**. **A** Distance **a** of the longitudinal or transverse bars and projections **p** for single bars. **B** Distance **a** of the longitudinal or transverse bars and projections **p** for double bars

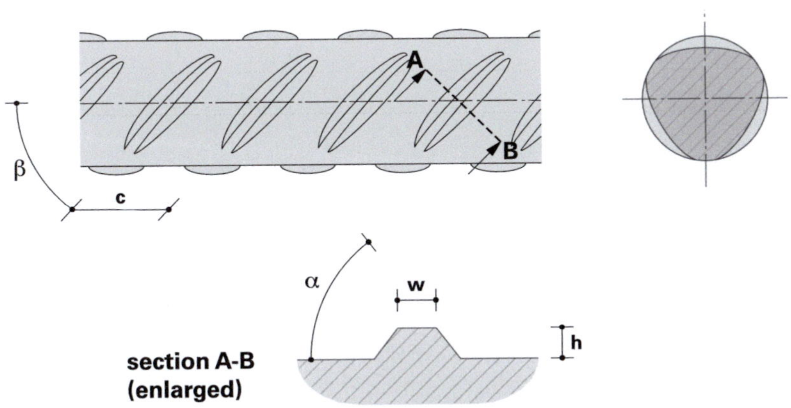

Fig. 11.36 Surface design of ribbed bars of steel reinforcement meshes according to *DIN 488-4*

lower chords, and continuous or interrupted diagonals connected to the chords by welding or a mechanical connection (Fig. 11.38). Lattice beams are used, for example, for unitised semi-prefabricated slabs or for beam-and-slab floors with intermediate components.

EN 10080, 3.18

11.5.5 Concrete Cover

A **minimum** concrete cover $\mathbf{c}_{min}$ is required to (Fig. 11.39):

EN 1992-1-1, 4.

- Protect the reinforcement against **corrosion**.
- Safely transfer the **bonding forces** between steel and concrete.
- Ensure sufficient **fire resistance**.

***Vol. 1**, Sect. 30.3 Corrosion of Reinforced Concrete*

***Vol. 1**, Sect. 29.10.2 Components from Reinforced Concrete*

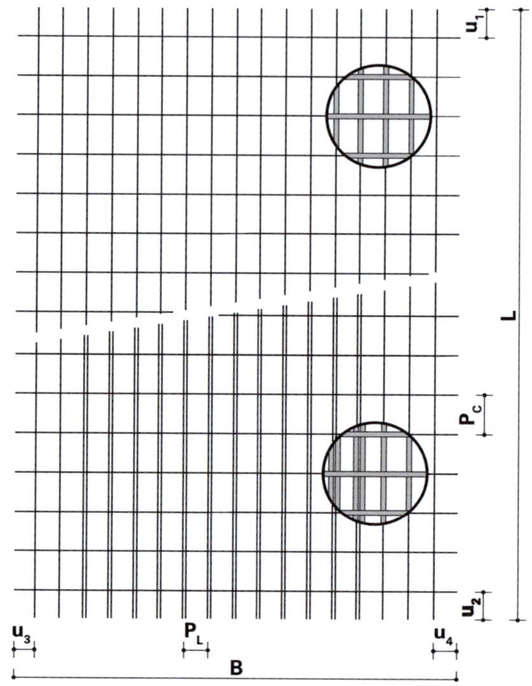

Fig. 11.37 Geometrical features of welded customised mats according to *EN 10080*. $\mathbf{P_L}$ spacing of the longitudinal wires, $\mathbf{P_C}$ spacing of the cross wires, **L** length of the longitudinal wires, **B** length of the cross wires, $\mathbf{u_1}$ projection of the longitudinal wires, $\mathbf{u_2}$ projection of the longitudinal wires, $\mathbf{u_3}$ projection of the cross wires, $\mathbf{u_4}$ projection of the cross wires

11

The given environmental conditions in the form of chemical and physical influences are recorded in the code by classification into different **exposure classes** (Fig. 11.40). Furthermore, different **requirements classes** (**S1** to **S6**, Fig. 11.41, top) are defined. The **concrete cover** is determined as a function of these two parameters, applying various correction factors, such as for example when using coated reinforcement or corrosion-resistant steel (Fig. 11.41, bottom).
Spacers made of plastic ensure that the **minimum cover** of the steel is ensured. These are clipped onto the outer bars when re-barring (Fig. 11.42).

11.5.6 Concreting Joints

Since the concreting work on a building can practically never be carried out in a single operation, that is to say this must happen in different **concreting sections**, it is necessary to introduce **working joints** into the construction (Figs. 11.43, 11.44, 11.45, and 11.46) at a suitable location. In this case, the reinforcing bars or meshes—the through-bars—protrude over the edge of the first section after curing and are

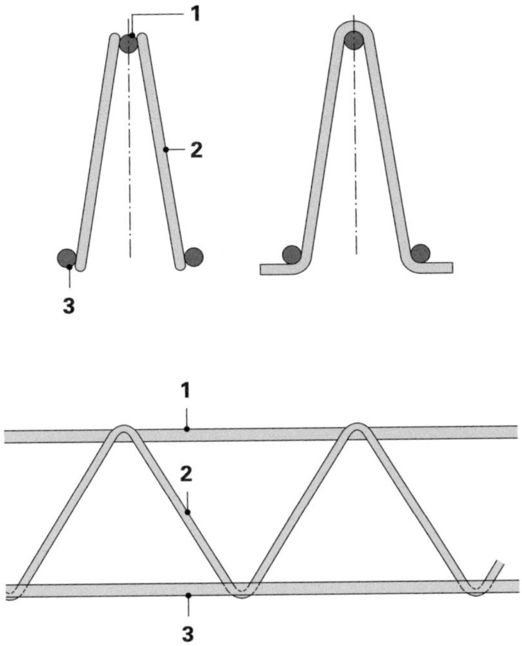

Fig. 11.38 **Lattice girder** according to *EN 10080*. Above, two cross-section variants. (**1**) Upper chord, (**2**) diagonal, (**3**) lower chord

embedded within the fresh concrete of the next concrete section. Tensile forces are transmitted at this joint by the overlapping or, as the case may be, mechanical connection of the protruding connecting bars with the bars of the newly concreted section. Compressive forces are transferred by contact between the joint surfaces which fit together in a form-fitting manner. This combined force transmission of tension via suitable reinforcement joints and compression via contact at the concrete joint creates a bond which comes very close to a fabric continuum. All in all, the construction can be regarded as **monolithic** even having working joints. The mechanical mode of operation of working joints is discussed in ***Volume 3**, Chapter II.*

***Vol. 3**, Sect. 7.2.1 Concreting Joints*

When determining the number and position of working joints, the **concreting position** also plays an important role: Whereas flat-lying components such as floor slabs can be concreted without difficulty even in larger sections, in particular standing formwork can only be filled with concrete in one operation up to a certain height. The proportion between component height and width plays a role here. Formwork cavities too high or too narrow make introducing and, in particular, compacting fresh concrete with a vibrating device more difficult. Uncontrolled air inclusions (blowholes) and associated structural deficiencies are the result.

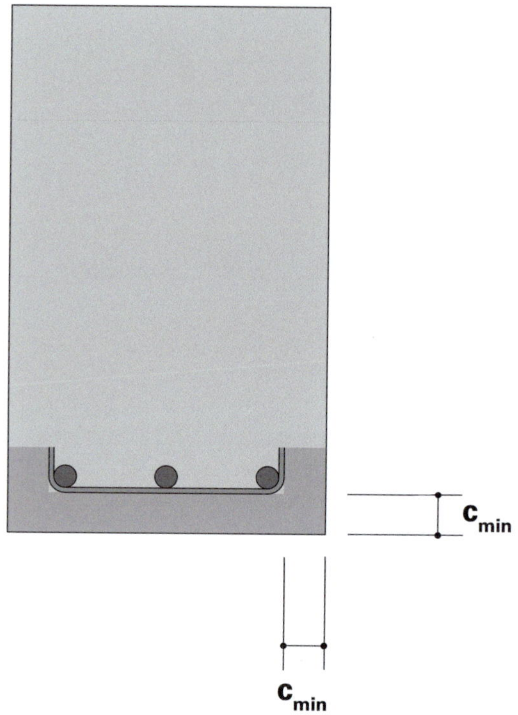

Fig. 11.39 The **concrete cover c_{min}** of the steel bars is decisive for their protection against fire and environmental influences (calculated from the stirrup reinforcement)

11

Working joints are always clearly visible on fair-faced concrete surfaces. For this reason—apart from issues of the construction process—they must be carefully planned and arranged, since they greatly influence the appearance of the exposed shuttered surface. Shadow gaps are often provided at working joints, since they can better visually conceal imperfections of execution.

11.6 Formwork Technology

A formwork or shuttering is a temporary negative image of the building to be constructed. The surface of the formwork permanently determines the visible surface quality of the cured concrete. Various formwork materials produce distinctly different concrete surfaces.

The formwork surfaces must be smooth and, if necessary, be treated with stripping oils in order to enable easy stripping (Fig. 11.47). This process takes place when the concrete has hardened to such an extent that it is stable without extensive support.

During casting and hardening, the formwork must be able to absorb the hydraulic pressure of the fresh concrete, which is

class	description of the environment	example of an assignment of exposure classes (informative)
	1 no corrosion or attack risk	
XO	for concrete without reinforcement or embedded metal: all exposure classes, except frost attack with and without de-icing agent, abrasion, or chemical attack for concrete with reinforcement or embedded metal: very dry	concrete in buildings with very low air humidity
	2 corrosion, caused by carbonation	
XC1	dry or constantly wet	concrete in buildings with low humidity; concrete constantly immersed in water
XC2	wet, rarely dry	surfaces which are wetted for a long time; often with foundations
XC3	moderate humidity	concrete in buildings with moderate or high humidity; rain-protected concrete outdoors
XC4	changing wet and dry	water-wetted surfaces which are not class XC2
	3 corrosion of reinforcement caused by chlorides, excluding seawater	
XD1	moderate humidity	concrete surfaces exposed to chloride-containing spray mist
XD2	wet, rarely dry	swimming pools; concrete which is exposed to chloride-containing industrial waste water
XD3	changing wet and dry	parts of bridges exposed to chloride-containing splash water, road pavements; parking decks
	4 corrosion of reinforcement caused by chlorides from seawater	
XS1	saline air, no direct contact with seawater	buildings close to the coast or on the coast
XS2	underwater	parts of marine structures
XS3	Tidal areas, spray water and spray mist areas	parts of marine structures
	5 concrete attack by frost with and without thawing agents	
XF1	moderate water saturation without thawing agents	vertical concrete surfaces exposed to rain and frost
XF2	moderate water saturation with thawing agents	vertical concrete surfaces of road constructions exposed to spray containing thawing agents
XF3	high water saturation without thawing agent	horizontal concrete surfaces exposed to rain and frost
XF4	high water saturation with thawing agents or seawater	road surfaces and bridge slabs exposed to de-icing agents; vertical concrete surfaces which are exposed to mist containing thawing agents and frost; splash water area of marine structures exposed to frost
	6 concrete attack by chemical attack of the environment	
XA1	chemically weakly attacking environment according to EN 206-1, Table 2	natural soils and groundwater
XA2	chemically moderately attacking environment and marine buildings according to EN 206-1, table 2	natural soils and groundwater
XA3	chemically strongly attacking environment according to EN 206-1, table 2	natural soils and groundwater

Fig. 11.40 **Exposure classes** of reinforced concrete according to *EN 1992-1-1*, in accordance with *EN 206-1*, for assessing the environmental exposure and the concrete cover derived therefrom

criterion	requirement class						
	exposure class according to table in Fig. 11.40						
	XO	XC1	XC2/XC3	XC4	XD1	XD2/XS1	XD3/XS2/XS3
utilisation period of 100 years	increase class by 2	increase class by 2	increase class by 2	increase class by 2	increase class by 2	increase class by 2	increase class by 2
compressive strength class[a,b]	≥C30/37 reduce class by 1	≥C30/37 reduce class by 1	≥C35/45 reduce class by 1	≥C40/50 reduce class by 1	≥C40/50 reduce class by 1	≥C40/50 reduce class by 1	≥C45/55 reduce class by 1
plate-shaped component (the position of the reinforcement is not affected by the construction work)	reduce class by 1	reduce class by 1	reduce class by 1	reduce class by 1	reduce class by 1	reduce class by 1	reduce class by 1
special quality control demonstrated	reduce class by 1	reduce class by 1	reduce class by 1	reduce class by 1	reduce class by 1	reduce class by 1	reduce class by 1

[a] It is assumed that the compressive strength class and the water/binder value may be associated to each other. A particular concrete composition (type of cement, water/binder value, filler) designed to produce low permeation may be considered.

[b] The required compressive strength classes may be reduced by one class if more than 4% air pores are produced by adding an air-entraining agent.

11

requirement class	durability requirement for $c_{min,dur}$ (mm)						
	exposure class according to table in Fig. 11.40						
	XO	XC1	XC2/XC3	XC4	XD1/XS1	XD2/XS2	XD3/XS3
S1	10	10	10	15	20	25	30
S2	10	10	15	20	25	30	35
S3	10	10	20	25	30	35	40
S4	10	15	25	30	35	40	45
S5	15	20	30	35	40	45	50
S6	20	25	35	40	45	50	55

Fig. 11.41 (Top) Recommended modification of the **requirements classes** of reinforced concrete according to *EN 1992-1-1*, for determining the concrete cover. (Bottom) **Minimum concrete cover $c_{min,dur}$** of reinforced concrete in the context of durability in accordance with *EN 1992-1-1* as a function of exposure and requirement class, for reinforcing steel according to *EN 10080*

about 2.5 times larger than that of water. For this purpose, a suitable **falsework** is required in addition to the shuttering skin.

Basically, a modern formwork consists of (Fig. 11.48):

- **Shuttering skin**.
- **Shuttering-skin support**, mostly falsework beams.
- **Transoms** or trestles.
- Other load-bearing elements.

Fig. 11.42 **Spacers** made of plastic are clamped on the reinforcing bars and ensure the minimum distance to the shuttering surface (= minimum concrete cover!)

Fig. 11.43 Finished concreting section: prepared working joint with starter bars for the next concreting section

In addition, as a rule, a **working platform** is installed at the upper edge of the standing formwork.

11.6.1 Formwork Skin

Older formwork consisted of a boarding as a shuttering skin and of squared timber as formwork beams. The resulting characteristic striped concrete surface is often observed on older exposed-concrete walls (Fig. 11.10).

Modern formwork for on-site concrete, however, consists of a shuttering skin from coated wood-based panels and of formwork beams in the form of standardised wooden girder beams (modern girder formwork). The joint arrangement visible on the concrete surface thus consists of

Fig. 11.44 Laid-out mesh reinforcement on strip-shaped spacers (dark strips) and starter bars for rising walls

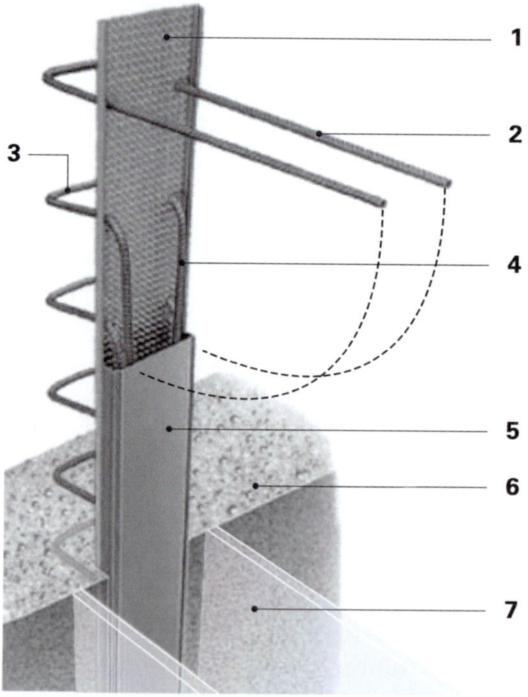

Fig. 11.45 Re-bend connection (system *Halfen*). (**1**) Steel cage profile with profiled back for shear transmission. (**2**) Connecting bar, bent back. (**3**) Anchoring loop in the concrete of the first concreting section. (**4**) Connecting bar, bent down. (**5**) U-shaped profile cover made of galvanised steel sheet for the preservation of the bent-down reinforcing bars from the fresh concrete of the first concreting section. Is removed before bending back. (**6**) First concreting section. (**7**) Formwork of the second concreting section

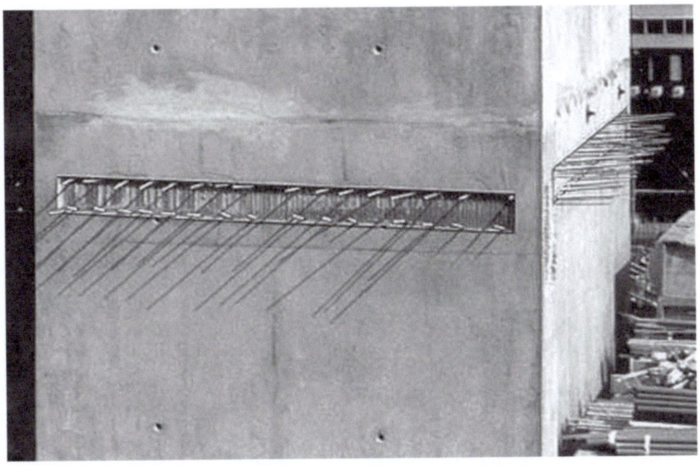

Fig. 11.46 Starter-bar reinforcement with re-bend connection as shown on Fig. 11.45 (system *Halfen*). Floor connection reinforcement at a building core

Fig. 11.47 Stripping of a wall

a coarse-meshed grid, which results from the used plate formats (Fig. 11.52).

As a principle, the shuttering skin should not be too absorbent. Otherwise, it would remove too much water from the concrete surface and thus hinder the hydration process. The result would be cracking.

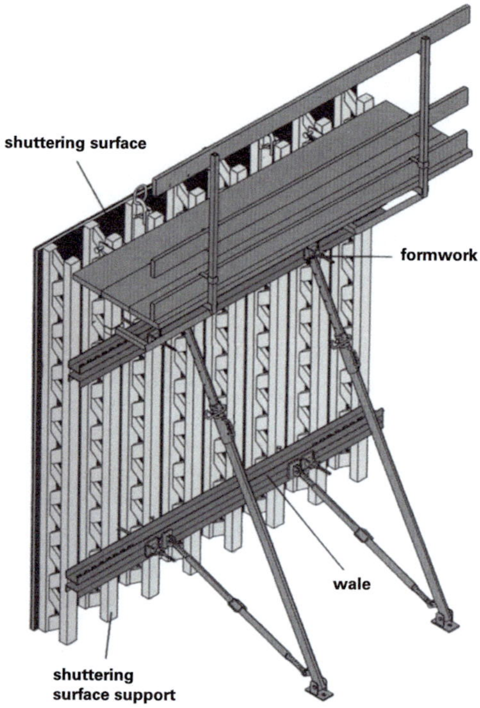

Fig. 11.48 Typical construction of a modern formwork composed of shuttering surface, wooden lattice girders, horizontal wale, formwork ties, guide struts, as well as a work platform at the top edge

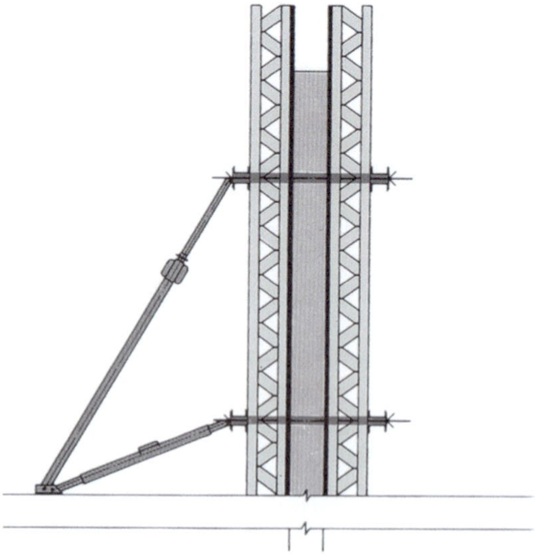

Fig. 11.49 Double-faced formwork

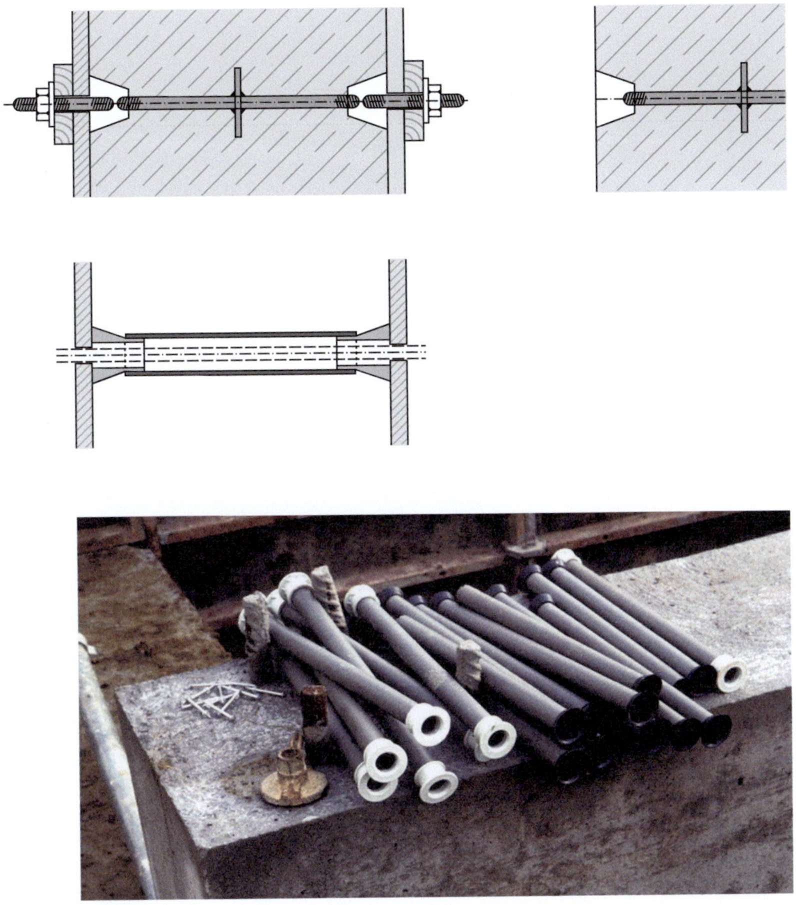

■ **Fig. 11.50** Formwork ties

11.6.2 Wall Formwork

If walls are to be shuttered on both sides—with double-faced formwork—the construction according to ■ Fig. 11.49 is used. The two formwork halves are coupled by means of penetrating **form ties**, so that the fresh concrete pressure can be better absorbed, so to speak by internal short-circuiting of the forces, without external supports (■ Fig. 11.50). These form ties are attached to the horizontal transoms. They are clearly visible on the finished concrete surface as punctiform depressions (■ Fig. 11.51). As in the case of defining the positions of joints between shuttering panels or of working joints, it is also recommended to plan the form ties' position early on, not only according to manufacturing requirements but also with regard to the visual appearance in the case of exposed concrete surfaces. Often, these elements—also again as working joints—are treated as an essential design feature and deliberately incorporated into the formal concept (■ Fig. 11.52).

Fig. 11.51 Close-up of a formwork tie on an exposed concrete surface

Fig. 11.52 Fair-faced concrete surface with carefully designed formwork panel and formwork-tie pattern

Fig. 11.53 Frame-type formwork

The complete formwork is also secured against external horizontal loads by means of telescopic props (Fig. 11.48). These also allow precise adjustment of the shuttering prior to concreting.

With the more modern **frame formwork**, shuttering skin, shuttering-skin support and steel transoms are combined into one single element (frame panel, Fig. 11.53). The peripheral bezel of the shuttering-skin edge considerably extends the life of these elements.

11.6.3 Special Wall Formwork

11.6.3.1 Climbing Formwork

In the case of high wall components which have to be concreted in several operations, it can be useful to work with climbing formwork. These are devices which, after curing of a particular working section, are dismantled, lifted, and reassembled on the upper edge of the finished wall component in a multi-phase procedure.

This is illustrated by the example of a modern climbing shuttering (Fig. 11.54, left). Here, after the concrete has hardened, the shuttering skin and the formwork beams can be moved about 75 cm away from the wall surface by means of a trolley. This allows the cleaning and preparation for the next concreting section.

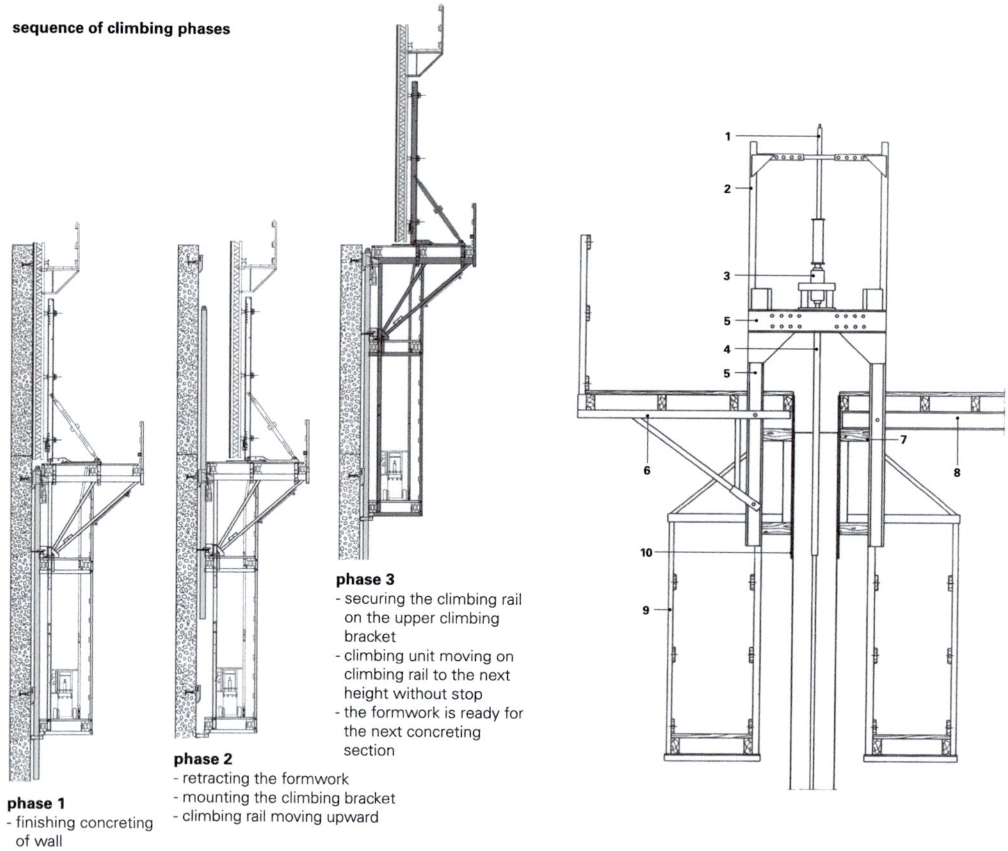

Fig. 11.54 (Left) Climbing formwork. (Right) Typical sliding formwork

The complete frame is then moved upwards and fixed in a crane lift. The shuttering form can be put back into position by means of the trolley, and the next work cycle can begin.

11.6.3.2 Slipforms

For larger heights, the use of a slipform can be advisable and economically justifiable. In this case, the form, which covers only a small section of approximately 1.20 m, is continuously moved vertically for 24 h a day.
The device essentially consists of (Fig. 11.54, right):

- The shuttering skin or form.
- A support of trestles in the form of an inverted U—this must take up the pressure of the fresh concrete.
- A climbing rod, which is supported in the core of the wall.
- As well as so-called lifters, which slowly move the whole device upwards using the climbing rod.

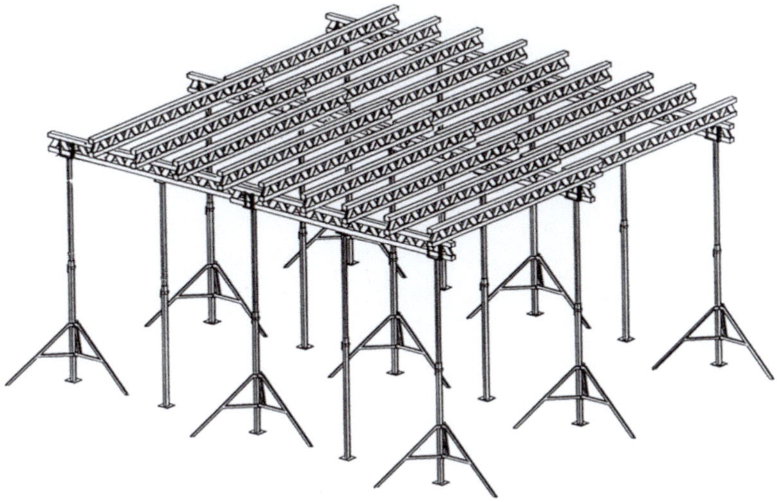

Fig. 11.55 Scaffolding of a floor formwork composed of cross beams, trusses and floor struts

11.6.4 Floor-Slab Formwork

Formerly, conventional floor formwork consisted mainly of board shuttering, beams made of squared wood, and round-wood props.

A modern floor formwork consists of (Figs. 11.55, 11.56, and 11.57:

- The **shuttering skin** or form.
- **Cross-beams** or transoms (subcarriers), mostly wooden lattice beams.
- **Trestles** (main carriers), mostly also wooden lattice beams.
- Adjustable steel **telescopic props** (ceiling columns).

Here too, as in the case of wall formwork, predominantly modularised shuttering systems are used.[5]

In addition, **panel formwork** is also used, in which the shuttering skin and transverse beams are combined into one single element (panel) (Fig. 11.58). In these systems, panels and main beams can be removed at an early curing stage, usually after 2 days, while the ceiling columns remain until the floor slab has sufficiently hardened.

Larger formwork units can also be designed as large-area **relocation units**, **frame tables**, or **portal tables**.

5 Rathfelder (1995) *Moderne Schalungstechnik*, p. 53ff.

Fig. 11.56 Example of a formwork structure as depicted on Fig. 11.55

11

Fig. 11.57 Floor formwork before concreting. The crossbeams are visible at the bottom; on top, the trusses made of wooden lattice beams and the telescopically adjustable floor rests

11.6.5 Semi-Prefabricated Parts

Efforts to reduce the cost of formwork, and in particular the associated labour costs, have led today to the widespread use of semi-prefabricated parts. The main goal of this technique is to replace the formwork gear by a **permanent formwork**. Par-

Fig. 11.58 Panel formwork for floors. The forming surface and the transverse beams are combined into one panel

ticularly in the case of concreting work at a high position, semi-prefabricated parts prove to be a favourable and efficient alternative to conventional formwork.

11.6.5.1 Unitised Floors

Unitised floors do not require formwork on the construction site. Because of the savings in comparatively expensive formwork, these systems are currently much more cost-effective than in-situ concrete. They consist of 4–6 cm thick prefabricated-concrete slabs as permanent formwork with integrated reinforcement and protruding lattice girders on which a concrete layer is applied. After hardening of the top concrete, the structure acts as a monolithic solid slab (Fig. 11.59).

***Vol. 3**, Sect. 20.5.1.2 Prefabricated or Semi-Prefabricated-Concrete Floor Systems*; cf. also *EN 13747-1, EN 13747-2, EN 13747-3*

The prefabricated panels are placed on the supports and initially span freely without propping (Fig. 11.60, top), whereby, in the construction state, the top-chord reinforcement of the lattice girders acts as a compression chord of the elements, the reinforcement contained in the prefabricated panel as the tension zone (Fig. 11.60, bottom). In addition, before concreting they must be provisionally supported with beams every 1.5–1.8 m to accommodate the load on the fresh concrete.

The precast slab can alternatively be executed with mesh or steel-bar reinforcement. The mesh reinforcement allows two-way spanning of the slab, for which the panel joints are covered with a cover reinforcement on site. The bar-reinforced plates can also have a two-way effect if a transverse

► *Sect. 11.8.4 Skeletal Structures with Core Bracing*

Fig. 11.59 Laying a unitised floor as a permanent formwork

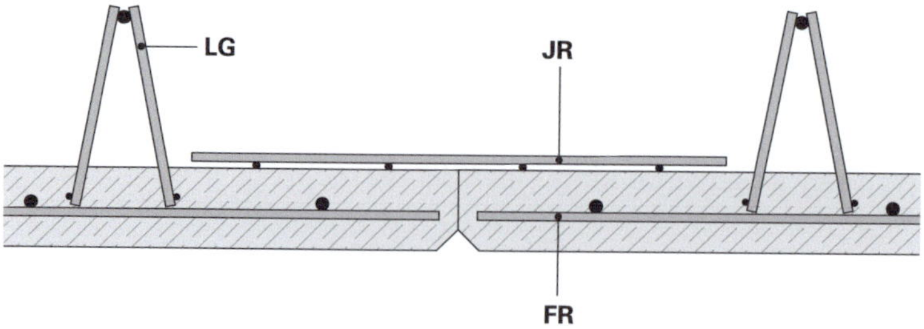

Fig. 11.60 (Top) Laying a unitised floor. On the right one can see the thin prefabricated concrete slab, i.e. the permanent formwork, as well as the protruding lattice girders, which are subsequently embedded in the concrete topping. Also visible is the connecting reinforcement protruding from the lost formwork, which is being threaded into the cage reinforcement of the girder beam by the fitters. (Bottom) Sectional drawing of the joint of two adjacent elements. Lattice girders **LG** provide the bond between the top concrete and the prefabricated panel. The field reinforcement **FR** of the floor slab is already integrated in the permanent formwork in the shop. **JR** joint reinforcement

reinforcement is applied at the building site prior to concreting. The structural depth usable in the transverse direction is thereby necessarily reduced, which leads to a 20–40% increase in reinforcement in the case of unitised slabs compared to a conventional solid slab. This increased material requirement is nevertheless made up for by the much more cost-effective processing.

Pin-supported flat floors can now also be implemented with unitised slabs.

11.6.5.2 Unitised Walls

Unitised walls are created in a process similar to unitised floors (Fig. 11.61). The wall surfaces consist of two thin prefabricated plates, which contain the necessary flexural tensile reinforcement. They are delivered as a sandwich-like element with a connecting lattice-girder reinforcement (Fig. 11.62), set up, temporarily shored (Fig. 11.61, right), and filled with concrete. If concreting takes place against an adjacent existing wall or formwork, one of the two leaves can be omitted in the semi-prefabricated element. *EN 14992*

In the factory, a pre-assembly of conduit lines for electric cables as well as steel door casings or simple windows such as cellar windows is possible. Common element heights are the normal storey height of about 3 m. In order to create a ceiling support, the inner leaf is frequently made shorter by the floor-slab thickness (Fig. 11.61).

Fig. 11.61 (Left) Erecting a unitised wall. (Right) Propped unitised wall before concreting

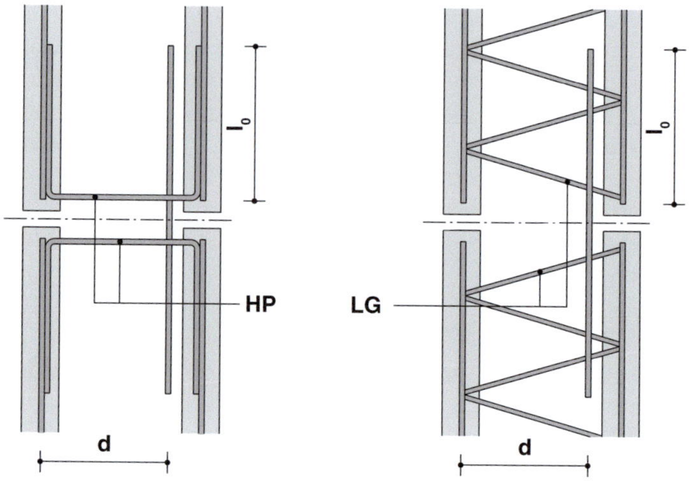

Fig. 11.62 Unitised wall: reinforcement connection by overlap at a connection joint and inner lever arm **d** in the joint area. Encircling of the reinforcement joint by means of the hairpin **HP** (left) or lattice girder **LG** (right); according to *EN 14992*

11.7 Design and Planning-Related Aspects

As a matter of principle, reinforced concrete allows the greatest freedom of design, conception, and planning of a building, in particular when the concrete is cast on site, as measured against other available building materials. This is due to the nature of the material, being formable and castable, a characteristic which—at least with the simplicity of processing typical of concrete construction and in a scale suitable for building works—no other material for primary structures has. The integral construction principle, which is made possible by this type of processing, does not even allow joint and connection problems to arise, as do especially in steel and timber constructions (Fig. 11.63). Due to the material continuum—or its nearly equivalent replacement through the working joint—stresses can be transferred very favourably in a distributed form either within monolithic surface components or in jointless nodes of bar frameworks. Force concentrations within the component can be avoided in this way, under comparable boundary conditions, better than applying differential constructional principles, such as in timber or steel construction.

Reinforced concrete has many advantages as a wall construction mode in the field of building construction. The simultaneous fulfilment of different partial tasks of load-bearing and building physics in one single monolithic, flat component is a great advantage of in-situ concrete construction. Compared to masonry construction, which is a pure wall construction method and, thanks to the masonry bond, offers at least partially the advantages of an integral construction principle, reinforced concrete can absorb significantly larger loads.

Fig. 11.63 Reinforced concrete allows the monolithic casting of statically multiply redundant structures with a minimum of joints (*ICROA*, Madrid; arch.: Fernando Higueras)

In addition to compression—the stress type that masonry is best and most economic at bearing—reinforced concrete can also be stressed by tension and, above all, in a combination of compression and tension, namely bending. In particular, the constructional problem of free edges that burdens wall panels made of masonry, which strongly constricts this type of construction in terms of design of the wall structure, is not so virulent in the case of reinforced concrete because, apart from its relatively high compressive strength, the material is also capable of taking bending, i.e. flexural tensile stresses. Consequently, free-standing arrangements of wall panels with laterally unsupported free edges can be perfectly realised in in-situ concrete. Even restraining walls—which is outright foreclosed in masonry construction—is possible thanks to the flexural tensile strength of the material.

As a material for floors, reinforced concrete shows a clear dominance over all other materials suitable for primary structures in current construction practice. Here too, as in the case of walls, the ability of the material to take on a variety of

Fig. 11.64 (Left) Formally sophisticated architecture made of exposed concrete, cast on site (German School Madrid, arch.: Grüntuch & P). (Right) Prefabricated construction (*Arena*-Hall, Stuttgart Vaihingen, arch.: HENN)

11

tasks such as force conduction as well as sound and fire protection plays a decisive role. However, the in-situ concrete slab, still predominant until recently, is increasingly being replaced by semi-prefabricated systems (unitised floors), which benefit from the generally high level of wage costs in this part of the world thanks to their significantly reduced formwork and scaffolding costs. The same applies to walls, which today are practically always designed as semi-prefabricated unitised walls.

In-situ concrete is further a predestined method for shell construction, since the properties and the processing of the material are very suitable for creating curved surface structures. However, shell structures are particularly affected by cost developments in the construction sector, since the high labour costs associated with the formwork and scaffolding required for elaborately shaped curved surface structures are now being mostly circumvented. Shells are therefore nowadays carried out in most cases as lattice shells in other materials or from flat wooden constructions.

► *Sect. 8.6 Shell Construction Modes* as well as ► *Sect. 9.3.6 Lattice Shells*

Overall, it can be observed that a combination of certain individual advantages of concreting respectively on the site and in the factory has already prevailed in construction practice, namely, in the form of semi-prefabricated construction modes. This has already been pointed out in ► Sect. 11.6. However, the pure construction methods of in-situ concrete and prefabricated construction are now largely restricted to certain niche applications: on the one hand, to formally ambitious free-form structures with representative aspirations in the case of in-situ concrete construction

► *Sect. 11.6.5 Semi-Prefabricated Parts*

(■ Fig. 11.64, left)—where high-quality fair-faced concrete plays an important role—as well as to industrial buildings in the case of prefabricated construction (■ Fig. 11.64 right).

In the following, some variants of in-situ concrete construction will be discussed in more detail in its application to building construction.

11.8 In-Situ Construction Modes

11.8.1 Cellular Structures

The consequential formation of the enclosing surfaces of a room, or of a room unit, as reinforced-concrete panels leads to a box construction or cellular construction mode. This is particularly useful in multi-storey housing construction (■ Fig. 11.65, left). The apartment units must in that case always be encapsulated all around for fire and noise-protection reasons. Solid concrete components offer ideal conditions for this.

As far as the structural mode of action of the cells is concerned, it is comparable to that of masonry wall construction (box construction), in which thin shear-plates stiffen each other in an orthogonal bond.

► *Sect. 7.5.1 Box Construction*

Nowadays, so-called tunnel formwork is used in rationalised building construction.

11.8.2 Detached-Shear-Plate Structures

Detached-shear-plate structures have already been discussed in the section dedicated to masonry construction. However, their predestined manifestation is rather in reinforced concrete (■ Fig. 11.65, right). Here, vertical wall and horizon-

► *Sect. 7.3 Fundamentals of Wall Construction*

■ **Fig. 11.65** (Left) Cellular construction (dormitory Stuttgart-Vaihingen, arch.: *Atelier 5*). (Right) Detached shear-plate construction with free wall ends (*Vitra*, arch.: Zaha Hadid)

tal floor shear-plates are combined in a free arrangement to form a rigid structure. The mutual shoring of wall plates perpendicular to, but not touching, each other in ground view is ensured by their coupling by means of a floor or roof slab. A prerequisite for this type of construction is—as noted—the possibility to keep edges of reinforced concrete slabs free (laterally unsupported).

Cf. also the comments on wall beams in ► *Sect. 11.3 Comparison with Other Construction Modes*

11.8.3 Portal-Frame Structures

Portal-frame structures benefit from the possibility of producing rigid corner connections between stanchions and beams in reinforced-concrete construction by monolithic casting (▫ Fig. 11.66, left). In addition to the favourable distribution of bending moments between the frame's vertical and horizontal members, which keeps the parts' dimensions small, this variant has the advantage of being stiff against horizontal loads in its plane without further measures. The relatively high complexity associated with the shuttering and, in particular, the reinforcing of rigid frame corners, where high reinforcement densities and unfavourable conditions are the rule, is one of the reasons why portal frames are now an exception in in-situ concrete. Once again, as in other construction methods, the share of labour costs on the construction site plays a decisive role.

► *Sect. 4.2.1.3 Bracing of Skeletal Structures > Portal-Frame Formation*

Portal frames can be stacked up to several floors (multistorey frames) and create structures braced from the beginning against horizontal loads.

▫ Fig. 11.66 (Left) Portal-frame structure (Satolas station, Lyon; arch.: S. Calatrava). (Right) Floor construction with point-supported flat slabs, which act as stiffening panes (Bollwerk, Stuttgart; arch.: Günter Behnisch)

11.8.4 Skeletal Structures with Core Bracing

Especially in administrative building, the construction method with core bracing and pin-supported flat floor slabs has now become established (◘ Fig. 11.66, right).

This construction mode takes advantage of:

- The presence of load-distributing **shear-resistant floor slabs** made of reinforced concrete, which are necessary for fire and sound-protection reasons in the first place. They are responsible for connecting the skeletal structure to the bracing core.
- The necessity to encapsulate the vertical circulation elements such as stairs, elevators, and installation shafts for fire protection purposes in multi-storey buildings by means of surrounding fire-resistant wall panels. These can be combined into a stiff solid **core**, which at the same time takes over the bracing of the building.
- The possibility to bundle further auxiliary rooms such as wet rooms, storage rooms, etc. within the core footprint.
- The **two-way load transfer** of solid floor slabs, which allows better material utilisation of the concrete and, consequently, a small floor-slab thickness. These non-directional skeletal structures usually have square-shaped structural bays, each having the same spans in both directions.
- The technical development of the **punching shear reinforcement**, which enables the solid slab to be placed on columns without the need for elaborate mushroom heads. The latter are connected with increased formwork costs and obstruct the conduct routing (see below).
- The **unobstructed duct routing** within the ceiling cavity, which is made possible by the girder-free flat ceiling; as well as:
- The possibility to connect partition walls to the solid slab without obstacles and without sound bridges: an important requirement of the free and flexible floorplan layout of modern administrative buildings.

Basically, this type of structure consists of a skeleton made up from flat floor slabs and columns which, from the outset, is not secured against horizontal loads, and a bracing core, to which the skeleton, by means of the floor shear-plates, transfers the horizontal loads. Combinations of cores and wall panels are also possible for bracing purposes. This can be particularly advantageous if restraints between opposite cores are to be prevented or the core itself has unfavourable ground-plan proportions.

► *Sect. 4.2.1.3 Bracing of Skeletal Structures* as well as ► *Sect. 5.2 Impact of Deformations on Building Construction*

These advantages with regard to requirements of administrative buildings is hardly provided by any other construction mode, which is why this variant of reinforced-concrete construction generally prevails today in this type of application.

Fig. 11.67 (Left) Biaxially curved shapes in exposed concrete (Mercedes-Museum, Stuttgart; arch.: Ben van Berkel). (Right) Shell structure (horse race track *La Zarzuela*, Madrid; eng.: Eduardo Torroja)

11

11.8.5 Free Structural Concepts

Highly sculptured, free building forms are also a predestined domain of reinforced concrete. Flat and rod-shaped components can be monolithically fused into an overall structure.

Even very complex curved shapes can be carried out relatively easily with in-situ concrete (Fig. 11.67, left). The limits for the free formability are only predetermined by the limitations derived from the formwork material.

11.8.6 Shell Structures

► *Sect. 3.4.5.1 Shells*

The most efficient utilisation of the specific material properties of reinforced concrete is found in shell structures (Fig. 11.67, right). These are one of the most powerful structural forms of construction in terms of the ratio between the thickness of the component and its span (**t/l**). In the following, some comparative values are provided:

- **Beam**: **t/l** = 1:10
- **Slab**: **t/l** = 1:30
- **Shell**: **t/l** = 1:500 or more

Areal processing and free shaping of the concrete allow taking advantage of the membrane effect in shells, in which only compressive and tensile stresses tangential to the surface of the component occur, but with no bending or shear stress transverse to it. The stiffness of the construction results primarily from its curvature, a basic requirement for the mode of action as a laminar structure.

► *Sect. 3.4.4.1 Membrane State*

■ Standards and Guidelines

DIN 488: Reinforcing steels

Part 1: 2009-08 Grades, properties, marking

Part 2: 2009-08 Reinforcing steel bars

Part 3: 2009-08 Reinforcing steel in coils, steel wire

Part 4: 2009-08 Welded fabric

DIN 1045: Concrete, reinforced and prestressed concrete structures

Part 2: 2008-08 Concrete—specification, properties, production and conformity—application rules for *DIN EN 206-1*

Part 3: 2022-07 Execution of structures

Part 4: 2022-07 Precast concrete products—Common Rules

EN 206: 2017-01 Concrete—Specification, performance, production and conformity

EN 1992: Eurocode 2—Design of concrete structures

Part 1-1: 2021-10 General rules—Rules for buildings, bridges and civil engineering structures, prEN

Part 1-2: 2021-09 General rules—Structural fire design, prEN

EN 10080: 2023-02 Steel for the reinforcement of concrete—Weldable reinforcing steel—General, prEN

EN 13670: 2011-03 Execution of concrete structures

EN 13747: 2010-08 Precast concrete products—Floor plates for floor systems

EN 14992: 2012-09 Precast concrete products—Wall elements

EN 15050: 2012-06 Precast concrete products—Bridge elements

Bibliography

Baar S, Ebeling K (2016) *Lohmeyers Stahlbetonbau: Bemessung—Konstruktion—Ausführung*. Wiesbaden, Springer Vieweg

Britton K (2001) *Auguste Perret*. London, Phaidon

Feix J, Walkner R (2012) *Lehrbuch Betonbau*. Innsbruck, Studia Universitätsverlag

Hanses K (2015) *Basics Betonbau*. Zurich, Birkhäuser

Hilberseimer L (1928) *Beton als Gestalter*. Stuttgart, Hoffmann

Lamprecht, H O (1993) *Opus caementitium: Bautechnik der Römer*. Düsseldorf, Beton-Verlag

Mettler D, Studer D (2018) *Made of Beton*. Zurich, Birkhäuser

Peck M (ed) (2013) *Moderner Betonbau Atlas—Konstruktion, Material, Nachhaltigkeit*. Munich, Institut für Internationale Architektur-Dokumentation

Rathfelder M (1995) *Moderne Schalungstechnik: Grundlagen, Systeme, Arbeitsweisen*. 2nd ed., Landsberg/Lech, Verl. Moderne Industrie, Die Bibliothek der Technik, Vol. 70

Wommelsdorf A (2012a) *Stahlbetonbau—Bemessung und Konstruktion Teil 1—Grundlagen—Biegebeanspruchte Bauteile*. Düsseldorf, Werner

Wommelsdorf A (2012b) *Stahlbetonbau—Bemessung und Konstruktion Teil 2—Stützen, Sondergebiete des Stahlbetonbaus*. Düsseldorf, Werner

Supplementary Information

J. L. Moro, *Building-Construction Design – From Principle to Detail*, https://doi.org/10.1007/978-3-662-71006-7

Index

A

B

C

D

E

F

G

H

I

J

K

L

M

Q

R

T

U

V

W

X

Y

MIX
Papier aus verantwortungsvollen Quellen
Paper from responsible sources
FSC® C105338

If you have any concerns about our products,
you can contact us on
ProductSafety@springernature.com

In case Publisher is established outside the EU,
the EU authorized representative is:
Springer Nature Customer Service Center GmbH
Europaplatz 3, 69115 Heidelberg, Germany

Printed by Libri Plureos GmbH
in Hamburg, Germany